W9-CKA-978

Handbook of Petroleum Processing

Handbook of Petroleum Processing

Edited by

DAVID S. J. "STAN" JONES†

retired chemical engineer (Fluor)
Calgary, Canada

and

PETER R. PUJADÓ

UOP LLC (retired)-Illinois, U.S.A.

Springer

A C.I.P. Catalogue record for this book is available from the Library of Congress.

ISBN-10 1-4020-2819-9 (HB)
ISBN-13 978-1-4020-2819-9 (HB)
ISBN-10 1-4020-2820-2 (e-book)
ISBN-13 978-1-4020-2820-5 (e-book)

Published by Springer,
P.O. Box 17, 3300 AA Dordrecht, The Netherlands.

www.springer.com

Contributing Editors:
L. C. James, Cambridge, Massachusetts, USA
G. A. Mansoori, University of Illinois at Chicago, USA

Printed on acid-free paper

Printed in the Netherlands.

Contents

Chapter 1

An introduction to crude oil and its processing

D.S.J. Jones

The wheel, without doubt, was man's greatest invention. However until the late 18th century and early 19th century the motivation and use of the wheel was limited either by muscle power, man or animal, or by energy naturally occurring from water flow and wind. The invention of the steam engine provided, for the first time, a motive power independent of muscle or the natural elements. This ignited the industrial revolution of the 19th century, with its feverish hunt for fossil fuels to generate the steam. It also initiated the development of the mass production of steel and other commodities.

Late in the 19th century came the invention of the internal combustion engine with its requirement for energy derived from crude oil. This, one can say, sparked the second industrial revolution, with the establishment of the industrial scene of today and its continuing development. The petroleum products from the crude oil used initially for the energy required by the internal combustion engine, have mushroomed to become the basis and source of some of our chemical, and pharmaceutical products.

The development of the crude oil refining industry and the internal combustion engine have influenced each other during the 20th century. Other factors have also contributed to accelerate the development of both. The major ones of these are the increasing awareness of environmental contamination, and the increasing demand for faster travel which led to the development of the aircraft industry with its need for higher quality petroleum fuels. The purpose of this introductory chapter is to describe and define some of the basic measures and parameters used in the petroleum refining industry. These set the stage for the detail examination of the industry as a whole and which are provided in subsequent chapters of this encyclopedia.

The composition and characteristics of crude oil

Crude oil is a mixture of literally hundreds of hydrocarbon compounds ranging in size from the smallest, methane, with only one carbon atom, to large compounds containing 300 and more carbon atoms. A major portion of these compounds are paraffins or isomers of paraffins. A typical example is butane shown below:

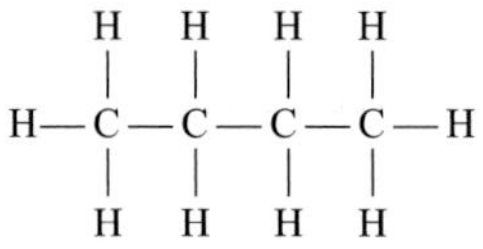

```
    H   H   H   H
    |   |   |   |
H — C — C — C — C — H     Normal butane (denoted as nC4)
    |   |   |   |
    H   H   H   H
```

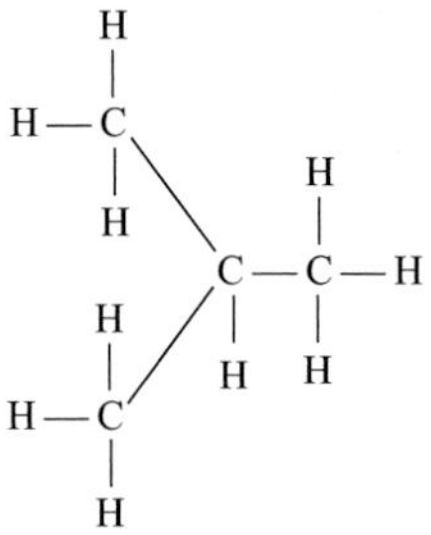

```
    H
    |
H — C
    |  \          H
    H   \         |
         C  —  C — H      Isobutane (denoted as iC4)
    H   /  |      |
    |  /   H      H
H — C
    |
    H
```

Most of the remaining hydrocarbon compounds are either cyclic paraffins called naphthenes or deeply dehydrogenated cyclic compounds as in the aromatic family of hydrocarbons. Examples of these are shown below:

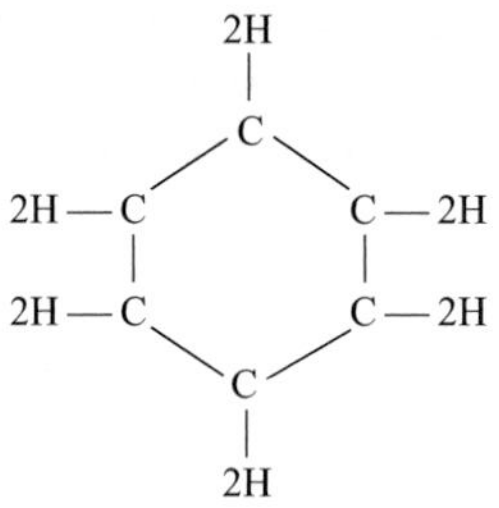

```
            2H
            |
            C
          /   \
2H — C           C — 2H
     |           |              Cyclohexane (Naphthene)
2H — C           C — 2H
          \   /
            C
            |
            2H
```

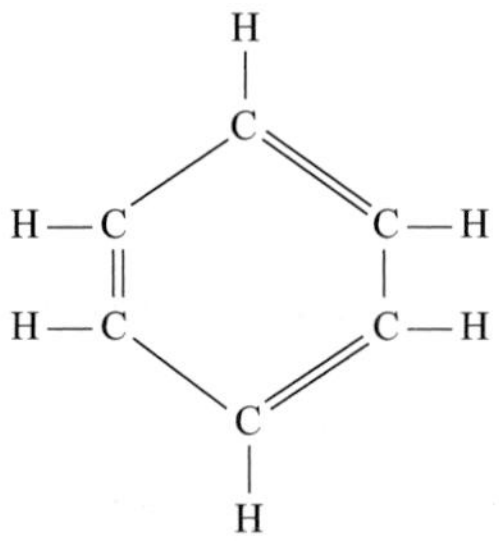

```
            H
            |
            C
          /   \\
 H — C           C — H
     ||          |              Benzene (Aromatic)
 H — C           C — H
          \   //
            C
            |
            H
```

Only the simplest of these homologues can be isolated to some degree of purity on a commercial scale. Generally, in refining processes, isolation of relatively pure products is restricted to those compounds lighter than C7's. The majority of hydrocarbon compounds present in crude oil have been isolated however, but under delicate laboratory conditions. In refining processes the products are identified by groups of these hydrocarbons boiling between selective temperature ranges. Thus, for example a naphtha product would be labeled as a 90°C to 140°C cut.

Not all compounds contained in crude oil are hydrocarbons. There are present also as impurities, small quantities of sulfur, nitrogen and metals. By far the most important and the most common of these impurities is sulfur. This is present in the form of hydrogen sulfide and organic compounds of sulfur. These organic compounds are present through the whole boiling range of the hydrocarbons in the crude. They are similar in structure to the hydrocarbon families themselves, but with the addition of one or more sulfur atoms. The simplest of these is ethyl mercaptan which has a molecular structure as follows:

```
     H   H
     |   |
H — C — C — SH        Ethyl Mercaptan
     |   |
     H   H
```

The higher carbon number ranges of these sulfur compounds are thiophenes which are found mostly in the heavy residuum range and disulfides found in the middle distillate range of the crude. The sulfur from these heavier sulfur products can only be removed by converting the sulfur to H_2S in a hydrotreating process operating under severe conditions of temperature and pressure and over a suitable catalyst. The lighter sulfur compounds are usually removed as mercaptans by extraction with caustic soda or other suitable proprietary solvents.

Organic chloride compounds are also present in crude oil. These are not removed as such but metallic protection is applied against corrosion by HCl in the primary distillation processes. This protection is in the form of monel lining in the sections of the process most vulnerable to chloride attack. Injection of ammonia is also applied to neutralize the HCl in these sections of the equipment.

The most common metal impurities found in crude oils are nickel, vanadium, and sodium. These are not very volatile and are found in the residuum or fuel oil products of the crude oil. These are not removed as metals from the crude and normally they are only a nuisance if they affect further processing of the oil or if they are a deterrent to the saleability of the fuel product. For example, the metals cause severe deterioration in catalyst life of most catalytic processes. In the quality of saleable fuel oil products high concentrations of nickel and vanadium are unacceptable in fuel oils used in the production of certain steels. The metals can be removed with the glutinous portion of the fuel oil product called asphaltenes. The most common process used to accomplish this is the extraction of the asphaltenes from the residue oils using propane as solvent.

Nitrogen, the remaining impurity is usually found as dissolved gas in the crude or as amines or other nitrogen compounds in the heavier fractions. It is a problem only with certain processes in naphtha product range (such as catalytic reforming). It is removed with the sulfur compounds in this range by hydrotreating the feed to these processes.

Although the major families or homologues of hydrocarbons found in all crude oils as described earlier are the paraffins, cyclic paraffins and aromatics, there is a fourth group. These are the unsaturated or olefinic hydrocarbons. They are not naturally present in any great quantity in most crude oils, but are often produced in significant quantities during the processing of the crude oil to refined products. This occurs in those processes which subject the oil to high temperature for a relatively long period of time. Under these conditions the saturated hydrocarbon molecules break down permanently losing one or more of the four atoms attached to the quadrivalent carbon. The resulting hydrocarbon molecule is unstable and readily combines with itself (forming double bond links) or with similar molecules to form polymers. An example of such an unsaturated compound is as follows:

```
     H   H
     |   |
H — C = C — H    Ethylene
```

Note the double bond in this compound linking the two carbon atoms.

Although all crude oils contain the composition described above, rarely are there two crude oils with the same characteristics. This is so because every crude oil from whatever geographical source contains different quantities of the various compounds that make up its composition. Crude oils produced in Nigeria for example would be high in cyclic paraffin content and have a relatively low specific gravity. Crude drilled in some of the fields in Venezuela on the other hand would have a very high gravity and a low content of material boiling below 350°C. The following table summarizes some of the crude oils from various locations (Table 1.1).

Worthy of note in the above table is the difference in the character of the various crudes that enables refiners to improve their operation by selecting the best crude or crudes that meet their product marketing requirements. For example, where a refining product slate demands a high quantity of ‘no lead’ gasoline and a modest outlet for fuel oils then a crude oil feed such as Hassi Messaoud would be a prime choice. Its selection provides a high naphtha yield with a high naphthene content as catalytic reforming feedstock. Fuel oil in this case also is less than 50% of the barrel. The Iranian light crude would also be a contender but for the undesirably high metal content of the fuel oil (Residuum).

In the case of a good middle of the road crude, Kuwait or the Arabian crude oils offer a reasonably balanced product slate with good middle distillate quality and yields.

Table 1.1. Characteristics of some crude oils from various world-wide locations

	Arabian light	Arabian heavy	Iranian light	Iranian heavy (Gach Saran)	Iraq (Kirkuk)	Kuwait	Algerian (Hassi Messaoud)	Libyan (Brega)	Nigerian (Bonny medium)	North Sea (Ekofisk)	South American (Bachequero)
% vol. boiling below 350°C	54.0	46.5	55.0	53.0	61.1	49.0	75.2	64.0	54.5	61.2	30.0
gravity, API	33.4	28.2	33.5	30.8	35.9	31.2	44.7	40.4	26.0	36.3	16.8
sulfur, wt%	1.8	2.84	1.4	1.6	1.95	2.5	0.13	0.21	0.23	0.21	2.4
PONA of heavy naphtha, vol%											
cut, °C	100–150	100–150	149–204	149–204	100–150	100–150	95–175	100–150	100–150	100–200	93–177
paraffins	69.5	70.3	54.0	50	69.0	67.9	56.5	53.0	27.5	56.5	27.6
olefins	–	–	–	–	265 ppm	–	–	20 ppm	1.5	–	–
naphthenes	18.2	21.4	30.0	35	21.0	22.1	32.9	39.3	57.0	29.5	58.5
aromatics	12.3	8.3	16.0	15	9.8	10.0	10.6	7.7	14.0	14.0	13.9
Metals in residuum											
residuum temp. °C	>565	>565	>538	>538	>370	>370	>350	>570	>535	>350	>350
vanadium, wt ppm	94	171	188	404	58	59	<5	24	7	1.95	437
nickel, wt ppm	22	53	70	138	<3	18	<5	32	52	5.04	75

The Bachequero pour point is 16°C.

For bitumen manufacture and lube oil manufacture the South American crude oils are formidable competitors. Both major crudes from this area, Bachequero, the heavier crude and Tia Juana, the lighter, are highly acidic (Naphthenic acids) which enhance bitumen and lube oil qualities. There is a problem with these crude oils however as naphthenic acid is very corrosive in atmospheric distillation columns, particularly in the middle distillate sections. Normal distillation units may require relining of sections of the tower with 410 stainless steel if extended processing of these crude oils is envisaged.

Refiners often mix selective crude oils to optimize a product slate that has been programmed for the refinery. This exercise requires careful examination of the various crude assays (data compilation) and modeling the refinery operation to set the crude oil mix and its operating parameters.

The crude oil assay

The crude oil assay is a compilation of laboratory and pilot plant data that define the properties of the specific crude oil. At a minimum the assay should contain a distillation curve for the crude and a specific gravity curve. Most assays however contain data on pour point (flowing criteria), sulfur content, viscosity, and many other properties. The assay is usually prepared by the company selling the crude oil, it is used extensively by refiners in their plant operation, development of product schedules, and examination of future processing ventures. Engineering companies use the assay data in preparing the process design of petroleum plants they are bidding on or, having been awarded the project, they are now building.

In order to utilize the crude oil assay it is necessary to understand the data it provides and the significance of some of the laboratory tests that are used in its compilation. Some of these are summarized below, and are further described and discussed in other chapters of the Handbook.

The true boiling point curve

This is a plot of the boiling points of almost pure components, contained in the crude oil or fractions of the crude oil. In earlier times this curve was produced in the laboratory using complex batch distillation apparatus of a hundred or more equilibrium stages and a very high reflux ratio. Nowadays this curve is produced by mass spectrometry techniques much quicker and more accurately than by batch distillation. A typical true boiling point curve (TBP) is shown in Figure 1.10.

The ASTM distillation curve

While the TBP curve is not produced on a routine basis the ASTM distillation curves are. Rarely however is an ASTM curve conducted on the whole crude. This type of distillation curve is used however on a routine basis for plant and product quality control. This test is carried out on crude oil fractions using a simple apparatus designed to boil the test liquid and to condense the vapors as they are produced. Vapor temperatures are noted as the distillation proceeds and are plotted against the distillate recovered. Because only one equilibrium stage is used and no reflux is returned, the separation of components is poor. Thus, the initial boiling point (IBP) for ASTM is higher than the corresponding TBP point and the final boiling point (FBP) of the ASTM is lower than that for the TBP curve. There is a correlation between the ASTM and the TBP curve, and this is dealt with later in this chapter.

API gravity

This is an expression of the density of an oil. Unless stated otherwise the API gravity refers to density at 60°F (15.6°C). Its relationship with specific gravity is given by the expression

$$\text{API}^\circ = \frac{141.5}{\text{sp.gr.}} - 131.5$$

Flash points

The flash point of an oil is the temperature at which the vapor above the oil will momentarily flash or explode. This temperature is determined by laboratory testing using an apparatus consisting of a closed cup containing the oil, heating and stirring equipment, and a special adjustable flame. The type of apparatus used for middle distillate and fuel oils is called the Pensky Marten (PM), while the apparatus used in the case of Kerosene and lighter distillates is called the Abel. Reference to these tests are given later in this Handbook, and full details of the tests methods and procedures are given in ASTM Standards Part 7, Petroleum products and Lubricants. There are many empirical methods for determining flash points from the ASTM distillation curve. One such correlation is given by the expression

$$\text{Flash point } ^\circ\text{F} = 0.77\ (\text{ASTM } 5\%\ ^\circ\text{F} - 150^\circ\text{F})$$

Octane numbers

Octane numbers are a measure of a gasoline's resistance to knock or detonation in a cylinder of a gasoline engine. The higher this resistance is the higher will be the efficiency of the fuel to produce work. A relationship exists between the antiknock characteristic of the gasoline (octane number) and the compression ratio of the engine in which it is to be used. The higher the octane rating of the fuel then the higher the compression ratio of engine in which it can be used.

By definition, an octane number is that percentage of isooctane in a blend of isooctane and normal heptane that exactly matches the knock behavior of the gasoline. Thus, a 90 octane gasoline matches the knock characteristic of a blend containing 90% isooctane and 10% *n*-heptane. The knock characteristics are determined in the laboratory using a standard single cylinder test engine equipped with a super sensitive knock meter. The reference fuel (isooctane blend) is run and compared with a second run using the gasoline sample. Details of this method are given in the ASTM standards, Part 7 Petroleum products and Lubricants.

Two octane numbers are usually determined. The first is the research octane number (ON res or RON) and the second is the motor octane number (ON mm or MON). The same basic equipment is used to determine both octane numbers, but the engine speed for the motor method is much higher than that used to determine the research number. The actual octane number obtained in a commercial vehicle would be somewhere between these two. The significance of these two octane numbers is to evaluate the sensitivity of the gasoline to the severity of operating conditions in the engine. The research octane number is usually higher than the motor number, the difference between them is termed the 'sensitivity of the gasoline.'

Viscosity

The viscosity of an oil is a measure of its resistance to internal flow and is an indication of its lubricating qualities. In the oil industry it is usual to quote viscosities either in centistokes (which is the unit for kinematic viscosity), seconds Saybolt universal, seconds Saybolt furol, or seconds Redwood. These units have been correlated and such correlations can be found in most data books. In the laboratory, test data on viscosities is usually determined at temperatures of 100°F, 130°F, or 210°F. In the case of fuel oils temperatures of 122°F and 210°F are used.

Cloud and pour points

Cloud and Pour Points are tests that indicate the relative coagulation of wax in the oil. They do not measure the actual wax content of the oil. In these tests, the oil is reduced in temperature under strict control using an ice bath initially and then a frozen brine bath, and finally a bath of dry ice (solid CO_2). The temperature at which the oil becomes hazy or cloudy is taken as its cloud point. The temperature at which the oil ceases to flow altogether is its pour point.

Sulfur content

This is self explanatory and is usually quoted as %wt for the total sulfur in the oil.

Assays change in the data they provide as the oils from the various fields change with age. Some of these changes may be quite significant and users usually request updated data for definitive work, such as process design or evaluation. The larger producers of the crude oil provide laboratory test services on an 'on going' basis for these users.

The next few sections of this chapter illustrate how the assay data and basic petroleum refining processes are used to develop a process configuration for an oil refining complex.

Other basic definitions and correlations

As described earlier the composition of crude oil and its fractions are not expressed in terms of pure components, but as 'cuts' expressed between a range of boiling points. These 'cuts' are further defined by splitting them into smaller sections and treating those sections as though they were pure components. As such, each of these components will have precise properties such as specific gravity, viscosity, mole weight, pour point, etc. These components are referred to as pseudo components and are defined in terms of their mid boiling point.

Before describing in detail the determination of pseudo components and their application in the prediction of the properties of crude oil fractions it is necessary to define some of the terms used in the crude oil analysis. These are as follows:

Cut point

A cut point is defined as that temperature on the whole crude TBP curve that represents the limits (upper and lower) of a fraction to be produced. Consider the curve shown in Figure 1.1 of a typical crude oil TBP curve.

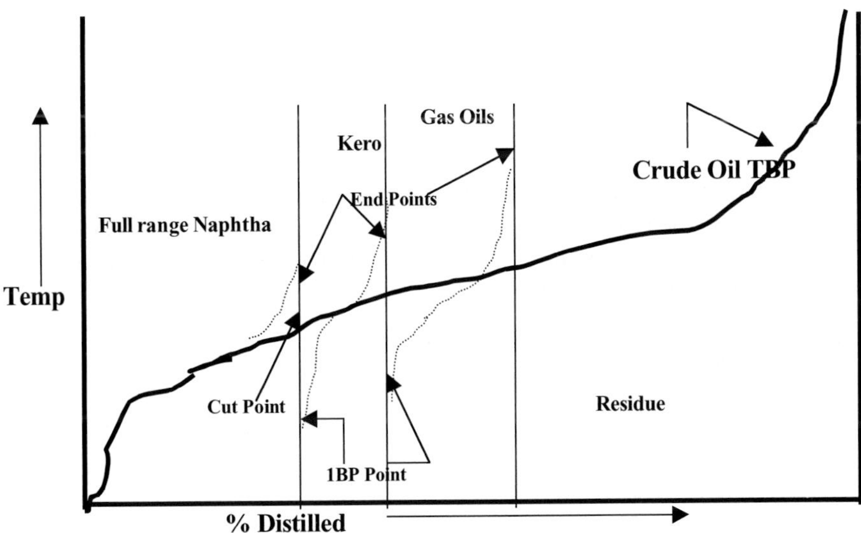

Figure 1.1. Cut points and end points.

A fraction with an upper cut point of 100°F produces a yield of 20% volume of the whole crude as that fraction. The next adjacent fraction has a lower cut point of 100°F and an upper one of 200°F this represents a yield of 30−20% = 10% volume on crude

End points
While the cut point is an ideal temperature used to define the yield of a fraction, the end points are the actual terminal temperatures of a fraction produced commercially. No process has the capability to separate perfectly the components of one fraction from adjacent ones. When two fractions are separated in a commercial process some of the lighter components remain in the adjacent lighter fraction. Likewise some of the heavier components in the fraction find their way into the adjacent heavier fraction. Thus, the actual IBP of the fraction will be lower than the initial cut point, and its FBP will be higher than the corresponding final cut point. This is also shown in Figure 1.1.

Mid boiling point components
In compiling the assay narrow boiling fractions are distilled from the crude, and are analyzed to determine their properties. These are then plotted against the mid boiling point of these fractions to produce a smooth correlation curve. To apply these curves for a particular calculation it is necessary to divide the TBP curve of the crude, or fractions of the crude, into mid boiling point components. To do this, consider Figure 1.2. For the first component take an arbitrary temperature point A. Draw a horizontal line through this from the 0% volume. Extend the line until the area between the line and the curve on both sides of the temperature point A are equal. The length of the horizontal line measures the yield of component A having a mid boiling point A °F. Repeat for the next adjacent component and continue until the whole curve is divided into these mid boiling point components.

Mid volume percentage point components
Sometimes the assay has been so constructed as to correlate the crude oil properties against components on a mid volume percentage basis. In using such data as this the TBP curve is divided into mid volume point components. This is easier than the mid boiling point concept and requires only that the curve be divided into a number of volumetric sections. The mid volume figure for each of these sections is merely the arithmetic mean of the volume range of each component.

Using these definitions the determination of the product properties can proceed using the distillation curves for the products, the pseudo component concept, and the assay data. This is given in the following items:

Predicting TBP and ASTM curves from assay data

The properties of products can be predicted by constructing mid boiling point components from a TBP curve and assigning the properties to each of these components.

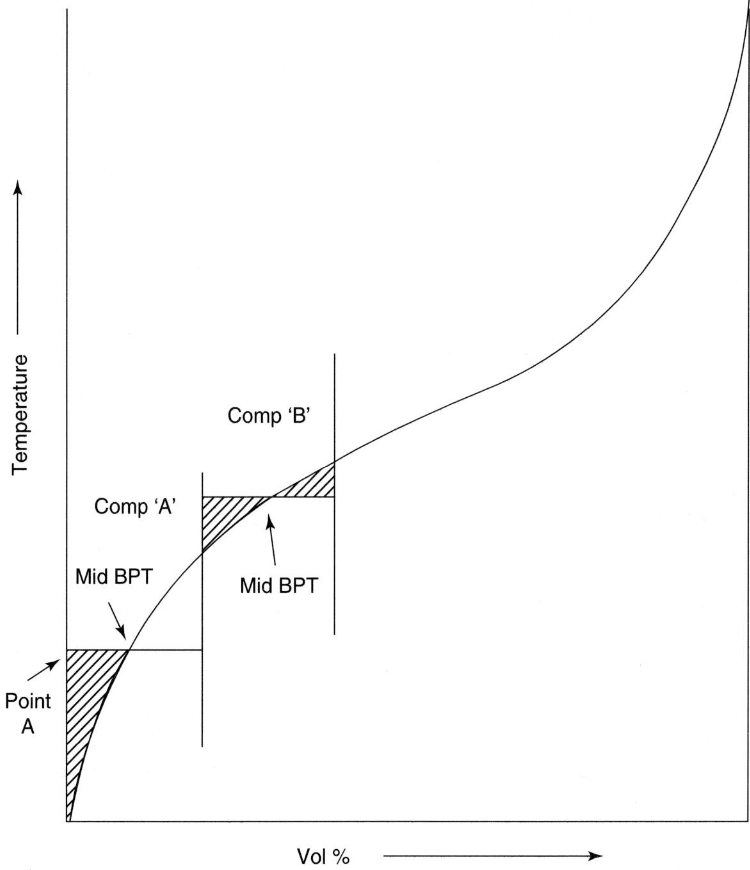

Figure 1.2. Example of mid boiling points.

These assigned properties are obtained either from the assay data, known components of similar boiling points, or established relationships such as gravity, molecular weights, and boiling points. However, before these mid boiling points (pseudo) components can be developed it is necessary to know the shape of the product TBP curve. The following is a method by which this can be achieved. Good, Connel et al. (1) accumulated data to relate the ASTM end point to a TBP cut point over the light and middle distillate range of crude. Their correlation curves are given in Figure 1.3, and are self explanatory. Thrift (2) derived a probable shape of ASTM data. The probability graph that he developed is given as Figure 1.4. The product ASTM curve from a well designed unit would be a straight line from 0 %vol to 100 %vol on this graph. Using these two graphs it is possible now to predict the ASTM distillation curve of a product knowing only its TBP cut range.

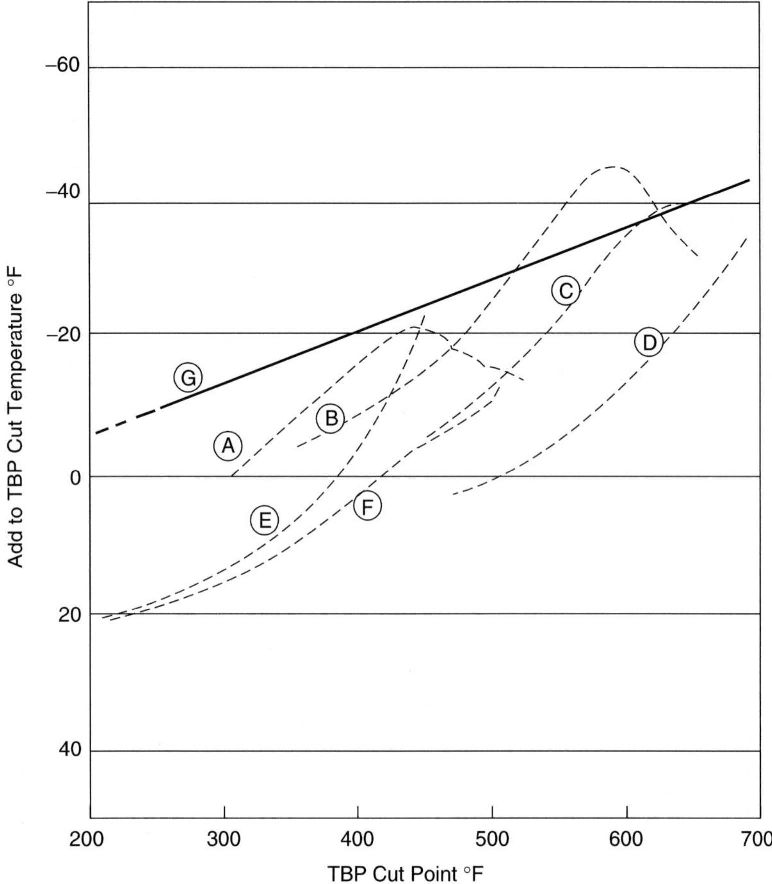

A End Points Vs TBP Cut Point for fractions starting at 200°F TBP or Lower
B End Points Vs TBP Cut Point for fractions starting at 300
C End Points Vs TBP Cut Point for fractions starting at 400
D End Points Vs TBP Cut Point for fractions starting at 500
E & F ASTM End Points Vs TBP Cut Point 300 ml STD col & 5 ft Packed Towers.
G 90% vol temp Vs 90% vol TBP cut (All Fractions).

Figure 1.3. Correlation between TBP and ASTM end points.

An example of this calculation is given below:

It is required to predict the ASTM distillation curve for Kerosene, cut between 387°F and 432°F cut points on Kuwait crude.

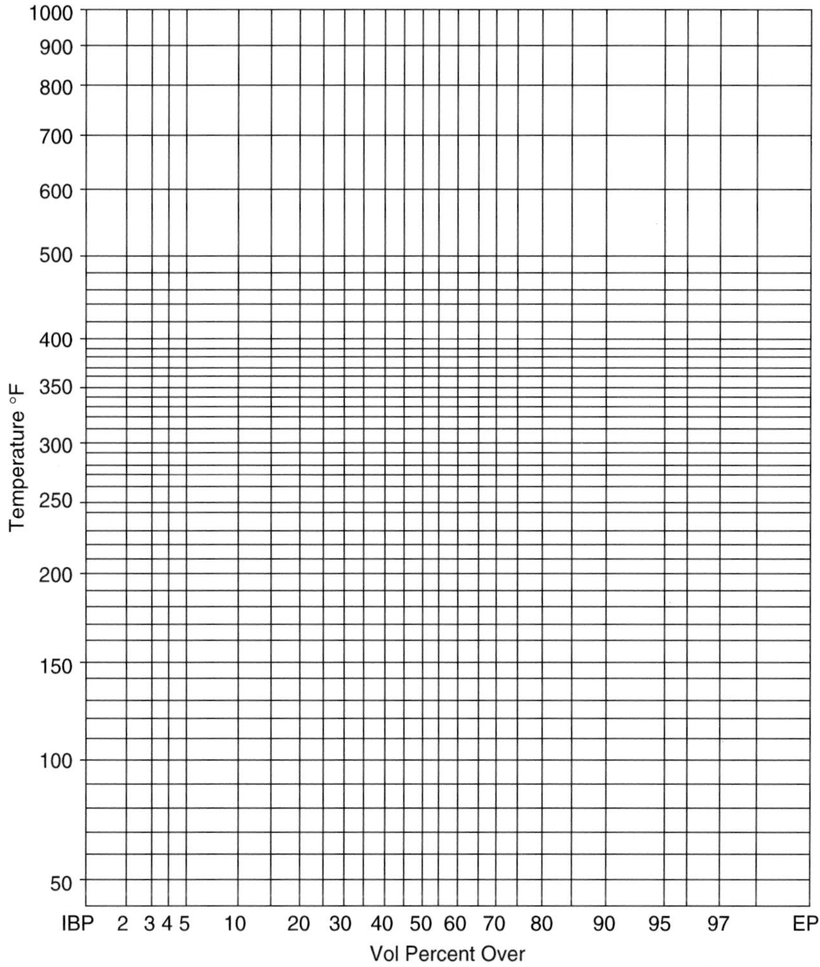

Figure 1.4. ASTM distillation probability curves.

Solution:

Yield on crude = 3.9% vol
Cut range = 27.3–31.2% vol on crude.
90%Vol of cut = 30.81 which is = 430°F
From Figure 1.3, curve B ASTM end point = 432 − 13°F = 419°F
From Figure 1.3, curve G ASTM 90% point = 430 − 24°F = 406°F

These two points are plotted in Figure 1.4 and a straight line drawn through them to define the probable ASTM distillation of the cut. This is plotted linearly in Figure 1.5

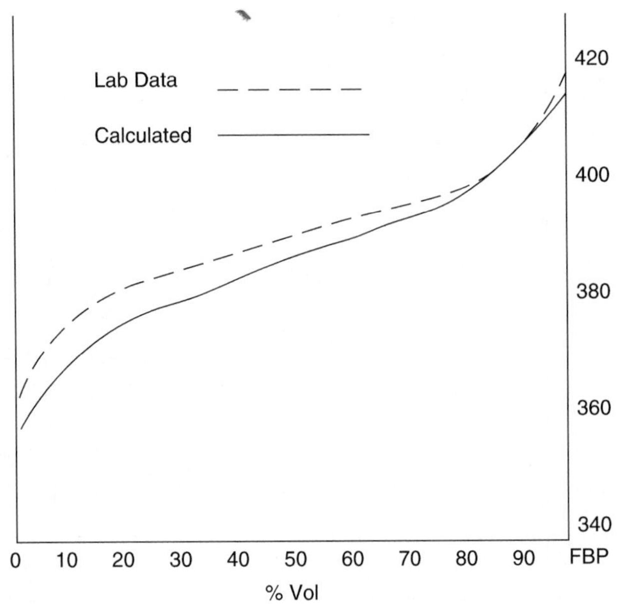

Figure 1.5. Comparison between calculated ASTM curve and lab data.

and can be seen to compare well with laboratory results of the actual product from a crude distillation unit.

Developing the TBP curve and the EFV curve from the ASTM distillation curve

Using a product ASTM distillation curve developed as shown above the TBP curve is developed as follows.

Converting the product ASTM distillation to TBP
Most crude distillation units take a full range naphtha cut as the overhead product. This cut contains all the light ends, ethane through pentanes, in the crude and of course the heavier naphtha cut. All the light ends are in solution, therefore it is not possible to prepare a meaningful ASTM distillation on this material directly. Two routes can be adopted in this case, the first is to take naphtha samples of the heavy naphtha and debutanized light naphtha from downstream units. Alternatively the sample can be subject to light end analysis in the lab such as using POD apparatus (Podbielniak) and carrying out an ASTM distillation on the stabilized sample. It is the second route that is chosen for this case.

There are two well-proven methods for this conversion. The first is by Edmister (3) and given in his book *Applied Thermodynamics* and the second by Maxwell (4) in his book *Data Book on Hydrocarbons*. The correlation curves from both these sources

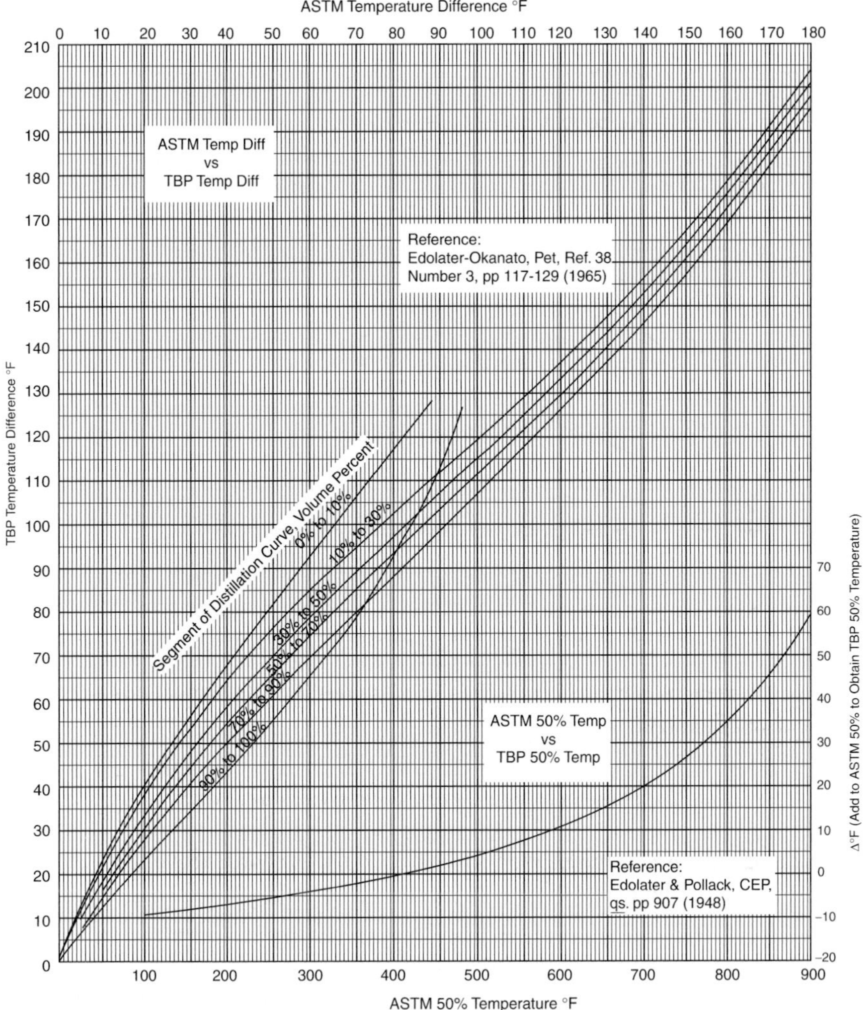

Figure 1.6. ASTM–TBP correlation—Edmister method.

are given as Figures 1.6 and 1.7. In this exercise Edmister's method and correlation will be used.

The ASTM distillation is tabulated as the temperature for IBP, 10%, 20% through to the FBP. IBP is the Initial Boiling Point (equivalent to 0% over) and the FBP is the Final Boiling Point (equivalent to 100% vol over). The multiples of 10% reflect the volume distilled and the temperature at which each increment is distilled. Using Figure 1.6 the 50% vol TBP point (in degrees Fahrenheit) is calculated from the 50% vol point of the ASTM distillation.

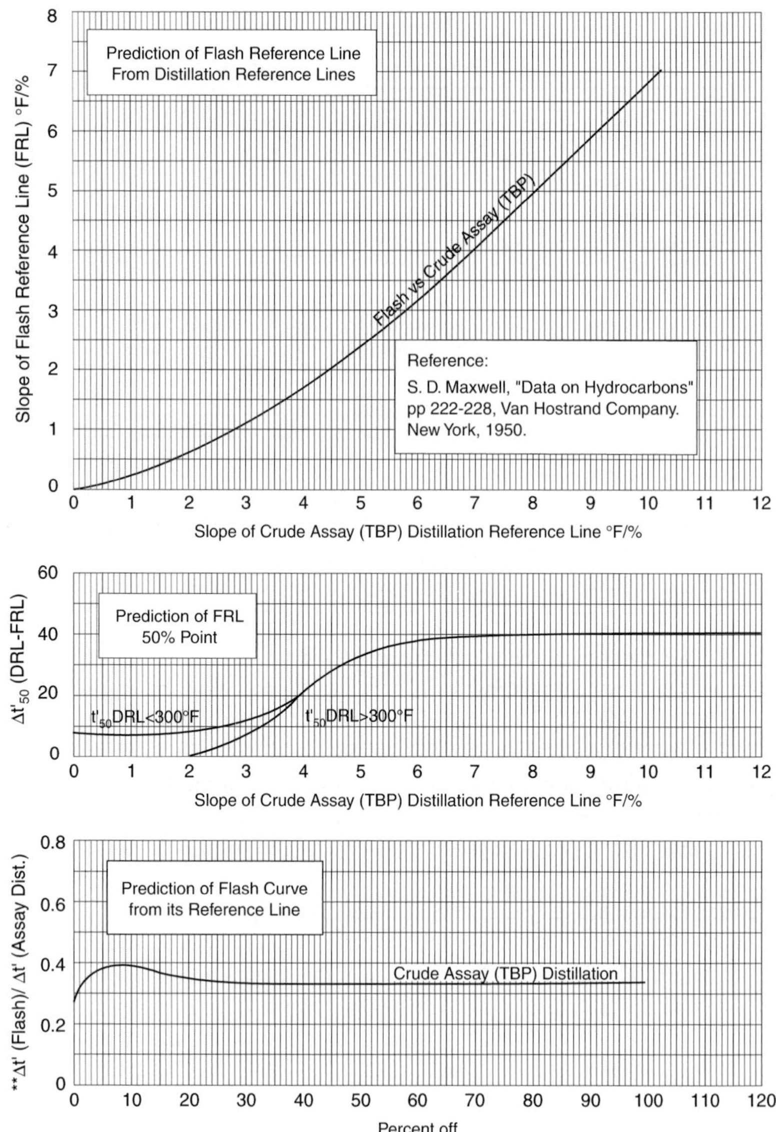

NOTE: * Flash and distillation reference lines (FRL and DRL) are straight lines through the 10% and 70% points. The temperature of the 50% points refer to these reference lines.

** Δt' is the departure of the actual flash and distillation curves from their respective reference lines. While the individual ($\Delta^{t'}$)'s may be either plus or minus, the ratio is always poistive.

Figure 1.7. EFV–TBP correlation—Maxwell method.

Table 1.2. Converting ASTM to TBP distillation

	ASTM (Lab Data)		TBP (from Figure 1.6)	
	°F	Δ°F	Δ°F	°F
IBP	424	29	61	361
10 %vol	453	31	52	423
30 %vol	484	18	52	475
50 %vol	502			507
70 %vol	504	2	31	538
90 %vol	536	32	41	579
FBP	570	34	40	619

Figure 1.6 is then used to determine the TBP temperature difference from the ASTM temperature difference for the 0–10% vol, 10–30% vol, 30–50% vol, 50–70% vol, 70–90% vol, and 90–100% vol. Moving from the established 50% vol TBP figure and using the temperature differences given by Figure 1.6 the TBP temperatures at 0, 10, 30, 50, 70, 90, and 100% vol are obtained (Table 1.2).

Developing the equilibrium flash vaporization curve

The Maxwell curves given as Figure 1.7 are used to develop the equilibrium flash vaporization curve (EFV) from the TBP. The EFV curve gives the temperature at which a required volume of distillate will be vaporized. This distillate vapor is always in equilibrium with its liquid residue. The development of the EFV curve is always at atmospheric pressure. Other temperature and pressure related conditions may be determined using the vapor pressure curves or constructing a phase diagram.

The TBP reference line (DRL) is first drawn by a straight line through the 10% vol point and the 70% vol point on the TBP curve. The slope of this line is determined as temperature difference per volume percent. This data are then used to determine the 50% volume temperature of a flash reference line (FRL). The curve in Figure 1.7 relating Δt_{50} (DRL–FRL) to DRL slope is used for this. Finally, the curve on Figure 1.7 relating the ratio of temperature differences between the FRL and flash curve (EFV) from that for the TBP to DRL is applied to each percent volume. From this the atmospheric EFV curve is drawn.

A sample calculation for the compilation of the EFV curve follows. Note the TBP curve is used to define product yields while the EFV curve is used to define temperature/pressure conditions in distillation. This example uses the TBP curve developed above as a starting point (Table 1.3).

The resulting TBP curves and EFV curves are shown in Figure 1.8.

Table 1.3. Converting TBP to EFV distillation

% Volume	Δt (TBP – DRL), °F	Δt (Flash – FRL) / Δt (TBP – DRL)	Δt (Flash – FRL), °F	Flash, °F
0	−46	0.2	−9.2	453
10	0	0.4	0	469
20	9	0.38	3.4	482
30	14	0.37	5.2	491
40	13	0.37	4.8	498
50	7	0.37	2.6	507
60	4	0.37	1.5	511
70	0	0.37	0	514
80	−2	0.37	−0.8	523
90	0	0.37	0	531
100	22	0.37	8.1	547

Predicting product qualities

The following paragraphs describe the prediction of product properties using pseudo components (mid boiling point) and assay data. A diesel cut with TBP cut points 432°F to 595°F on Kuwait crude (Figure 1.9) will be used to illustrate these calculations. The actual TBP of this cut is predicted using the method already described. The curve is then divided into about six pseudo mid boiling point components as described earlier and is shown as Figure 1.10.

Predicting the gravity of the product

Using the mid boiling point versus specific gravity curve from the assay given in the Appendix, the SG for each component is obtained. The weight factor for each component is then obtained by multiplying the volume percent of that component by the specific gravity. The sum of the weight factors divided by the 100% volume total is the specific gravity of the gas oil cut. This is shown in Table 1.4.

The prediction of product sulfur content

The prediction of sulfur content is similar to the method used for gravity. First the TBP curve for the product is determined and split into pseudo boiling point components. The weight factor is then determined for each component as before. Note that sulfur content is always quoted as a percent weight. Using the relationship of percent sulfur to mid boiling point given in the assay the sulfur content of each component is read off. This is multiplied by the weight factor for each component to give a sulfur factor. The sum of the total sulfur factors divided by the total weight factor gives the weight percent sulfur content of the fraction. For example, using the same gas oil cut as before its sulfur content is determined as shown in Table 1.5.

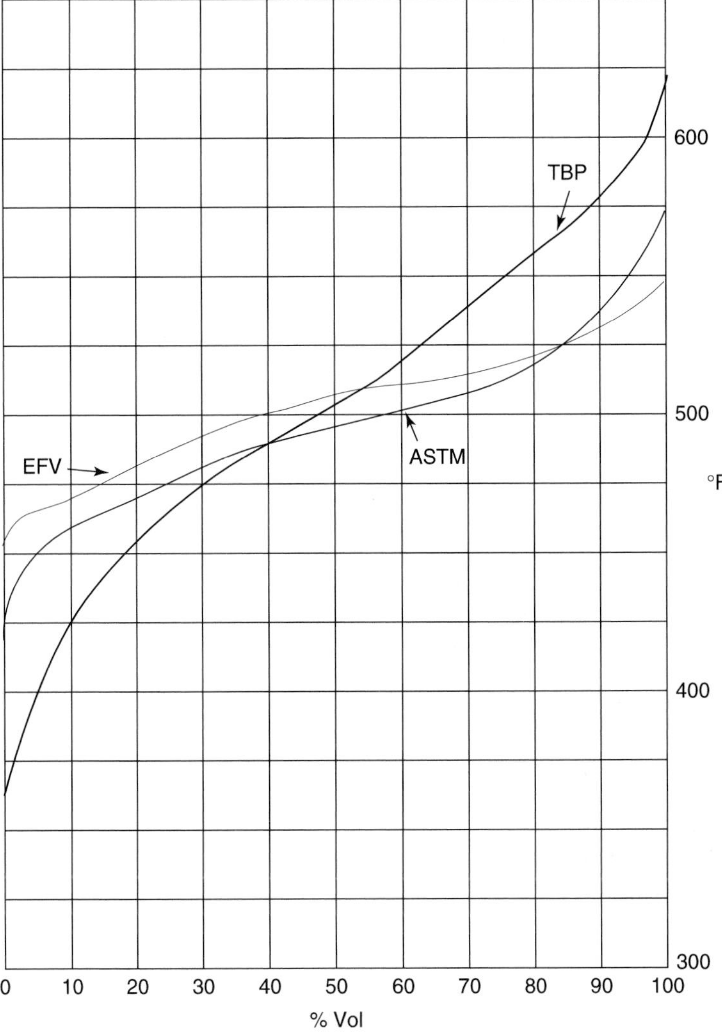

Figure 1.8. Types of distillation curves.

Viscosity prediction from the crude assay

Unlike sulfur content and gravity, viscosity cannot be arithmetically related directly to components. To determine the viscosity of a blend of two or more components, a blending index must be used. A graph of these indices is given in Maxwell "Data Book on Hydrocarbons," and part of this graph is reproduced as Figure 1.11. Using the

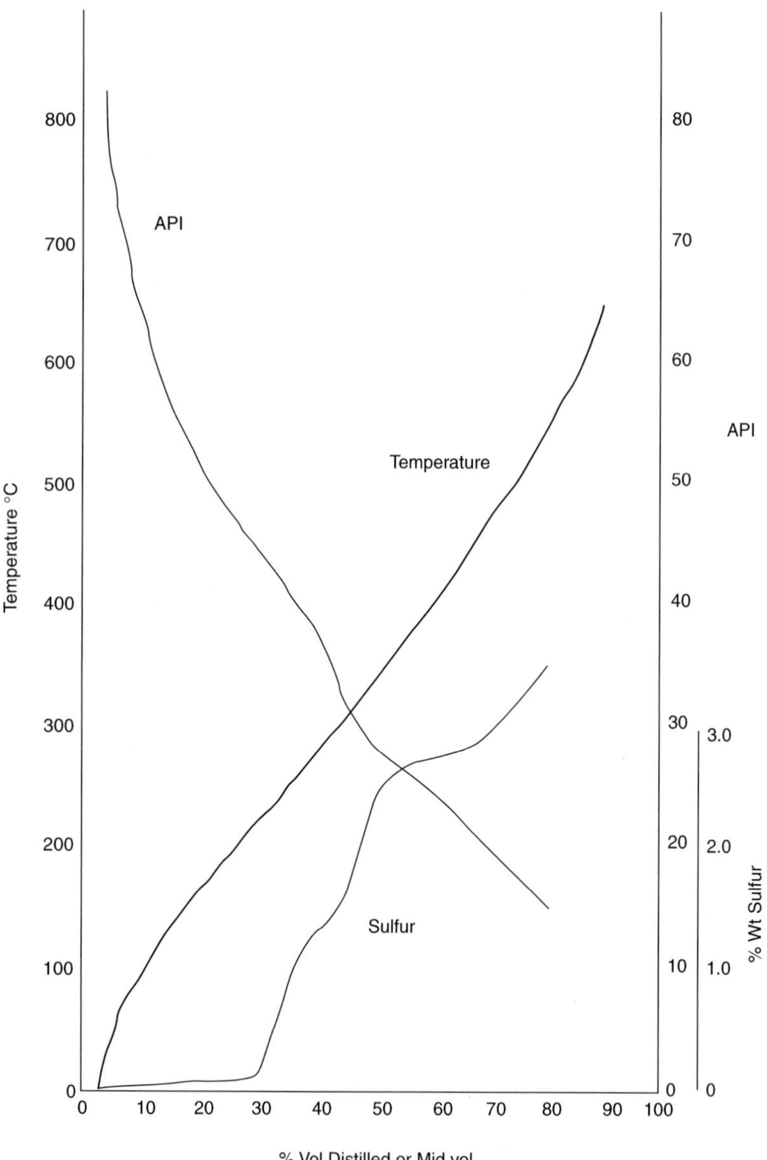

Figure 1.9. Typical crude assay curves (based on Kuwait crude).

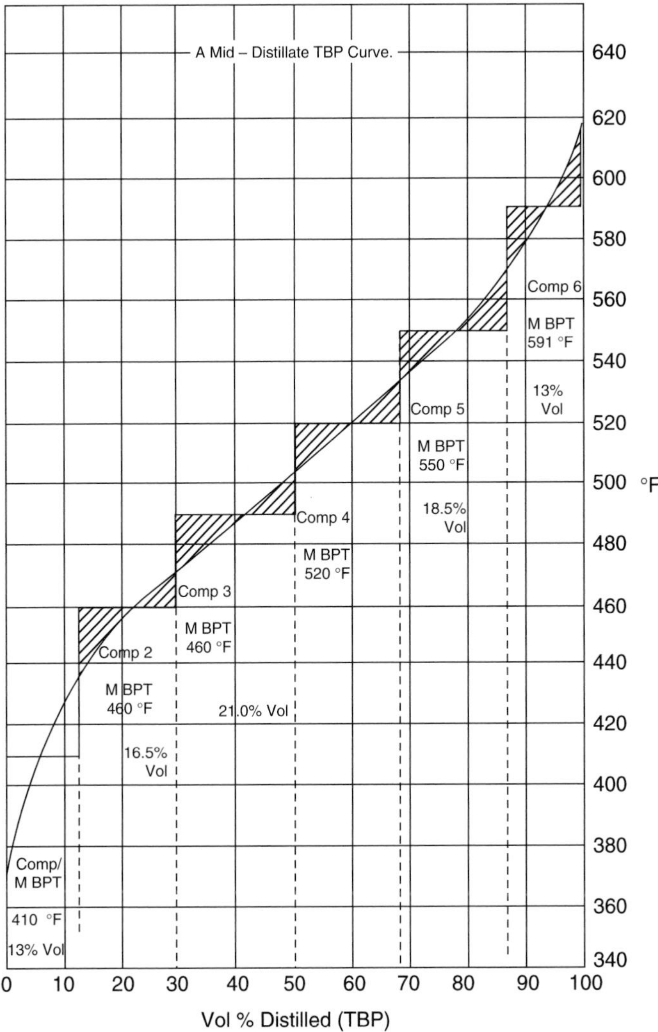

Figure 1.10. Typical pseudo component breakdown.

blending indices and having divided the TBP curve into components as before, the viscosity of the fraction can be predicted as shown in Table 1.6.

Cloud and pour points

In predicting these properties, it is not necessary to break down the product TBP as we have done for specific gravity, sulfur, etc. The accuracy of the tests and of blending

Table 1.4. Calculating the SG of a cut

Component	Volume %	Mid-BPt, °F	SG @ 60°F	Weight factor
1	13.0	410	0.793	10.3
2	16.5	460	0.801	13.2
3	21.0	489	0.836	17.6
4	18.0	520	0.844	15.2
5	18.5	550	0.846	15.7
6	13.0	592	0.850	11.1
Total	100.0			83.1

SG of cut $= \frac{83.1}{100} = 0.831$.

indices do not warrant this. These properties are therefore read off directly from the mid boiling point of the whole product. Considering the gas oil used in the previous example, its mid boiling point is about 510°F, from the crude assay its pour point is −5°F and cloud point is +4°F. Determining pour point for a blend of two or more products is rather more difficult. In this case blending indices are used for this purpose. A graph of these indices is given as Figure 1.12. It is self explanatory and its application is explained in Table 1.7.

Flash points

The flash point of a product is related to its ASTM distillation by the expression:

$$\text{flash point} = 0.77(\text{ASTM } 5\% \text{ in } ^\circ\text{F}–150^\circ\text{F})$$

Thus for the gas oil product in the above the example the flash point will be:

$$\text{flash point} = 0.77(420 - 150) = 208^\circ\text{F}.$$

Table 1.5. Calculating the sulfur content of a cut

Component	Weight factor	Mid BPt, °F	Sulfur, % wt	Sulfur factor
1	10.3	410	0.2	2.06
2	13.2	460	0.41	5.41
3	17.6	489	0.84	14.78
4	15.2	520	1.16	17.63
5	15.7	550	1.35	21.2
6	11.1	592	1.5	16.65
Total	83.1			77.73

Sulfur % weight $= \frac{77.7}{83.1} \times 100 = 0.935$ %wt. (actual plant data gave 0.931 %wt.)

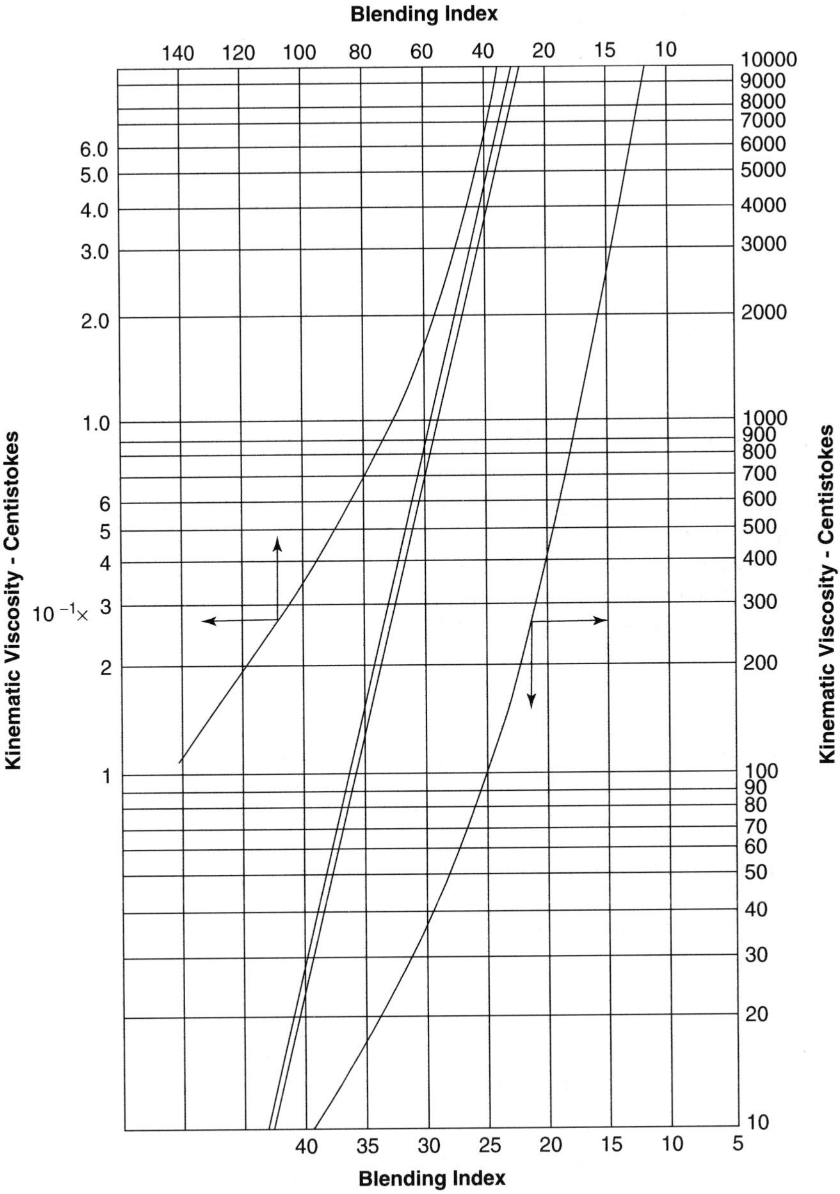

Figure 1.11. Viscosity blending index.

Table 1.6. Calculating the viscosity of a cut

Component	Volume %	Mid BPt °F	Viscosity Cs 100°F	Blending index	Viscosity factor
	(A)			(B)	(A × B)
1	13.0	410	1.49	63.5	825.5
2	16.5	460	2.0	58.0	957
3	21.0	489	2.4	55.0	1,155
4	18.0	520	2.9	52.5	945
5	18.5	550	3.7	49.0	906.5
6	13.0	592	4.8	46.0	598
Total	100.0				5,387.0

Overall viscosity index = $\frac{5{,}387}{100}$=53.87.
From Figure 1.8 an index of 53.87 = 2.65 Cs (actual plant test data was 2.7 Cs).

Blending products of different flash points

As with pour points and viscosity, the flash point of a blend of two or more components is determined by using a flash blending index. Figure 1.13 gives these indices. Again the indices are blended linearly as in the case of viscosity. Consider the following example:

> 2,000 BPSD of Kerosene with a flash point of 120°F is to be blended with 8,000 BPSD of fuel oil with a flash point of 250°F. Calculate the flash point of the blend (Table 1.8).

Predicting the mole weights of products

The prediction of molecular weights of product streams is more often required for the design of the processes that are going to produce those products. There are other more rigorous calculations that can and are used for definitive design and in building up computer simulation packages. The method presented here is a simple method by which the mole weight of a product stream can be determined from a laboratory ASTM distillation test. The result is sufficiently accurate for use in refinery configuration studies and the like.

A relationship exists between the mean average boiling point of a product (commonly designated as MEABP), the API gravity, and the molecular weight of petroleum fractions. This is shown as Figure 1.14.

Using a gas oil fraction as an example, the MEABP of the product is calculated from its ASTM distillation in degrees Fahrenheit given below:

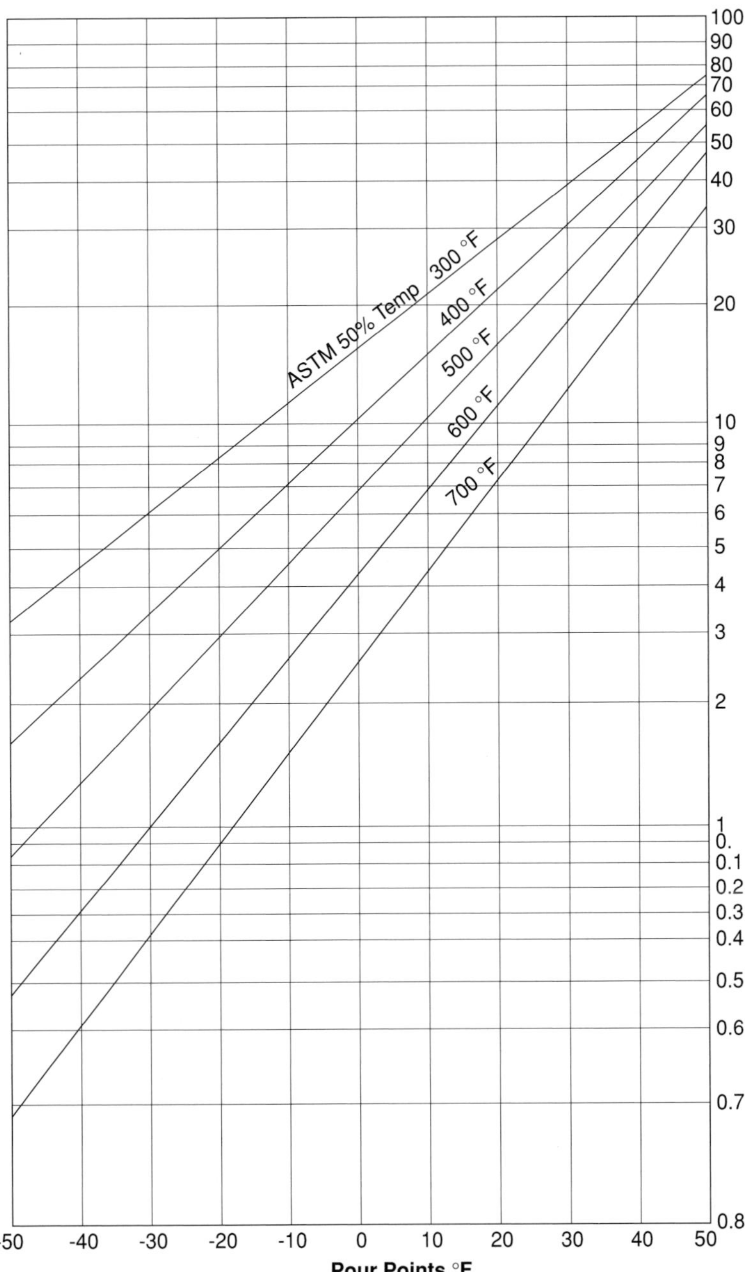

Figure 1.12. Pour point blending index.

% Vol	°F
0	406
10	447
30	469
50	487
70	507
90	538
100	578

The slope of the curve is calculated by $\dfrac{T°\text{F} @ 90\% - T°\text{F}@10\%}{80}$

$$= \frac{538 - 447}{80} = 1.14°\text{F}/\%$$

$$\text{Volume average boiling point} = \frac{T°\text{F} @ 10\% + (2 \times T°\text{F} @ 50\%) + T°\text{F} @ 90\%}{4}$$

$$= \frac{447 + 974 + 538}{4} = 490°\text{F}$$

From the upper series of curves given in Figure 1.14 the correction to the volumetric average boiling point (VABP) to obtain the Mean Average (MEABP) is –5°F. Thus, the MEABP is

$$490 + (-5) = 485°\text{F}$$

The °API of the stream from the calculation for gravity is 38.8. Using this figure and the MEABP in the lower series of curves in Figure 1.14 a molecular weight of 201 is read off.

Table 1.7. Calculating pour points of a cut

	Composition		ASTM dist		Pour point		
Components	BPSD	Fraction	50%°F	Factor	Pour point	Index	Factor
					°F		
Gas oil	2,000	0.33	500	85.8	−5	5.8	1.9
Waxy dist	4,000	0.67	700	249	30	12.7	8.5
Total	6,000	1.00		334	22		10.4

The pour point of the blend is read from Figure 1.9 where the ASTM 50% point is 334°F and the index is 10.4. In this case the pour point is 22°F.

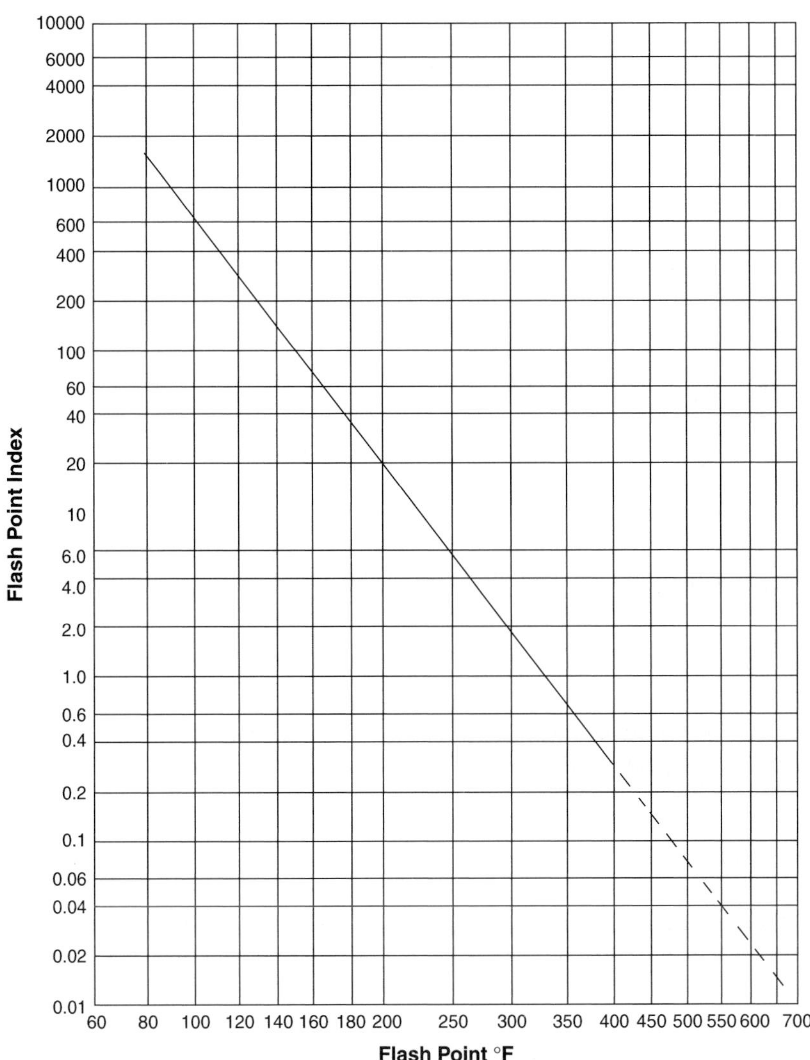

Figure 1.13. Flash point blending index.

Basic processes

This chapter provides an introduction to some of the most common of the processes included in fuel oriented and nonenergy oriented refineries. These processes are only discussed here in summary form. They are treated in more detail later in the book.

Table 1.8. Blending streams for flash points

Components	Volume	Fraction	Flash point, °F	Flash index	Factor
	BPSD	(A)		(B)	(A × B)
Kerosene	2,000	0.2	120	310	62.0
Fuel oil	8,000	0.8	250	5.5	4.4
Total	10,000	1.0			66.4

The flash point corresponding to an index of 66.4 (from Figure 1.10) is 166°F.

The processes common to most energy refineries

The atmospheric crude distillation unit

In refining the crude oil it is first broken up into those raw stocks that are the basis of the finished products. This break up of the crude is achieved by separating the oil into a series of boiling point fractions which meet the distillation requirements and some of the properties of the finished products. This is accomplished in the crude distillation units. Normally there are two units that accomplish this splitting up function: an atmospheric unit and a vacuum unit.

The crude oil first enters the Atmospheric unit where it is desalted (dissolved brine is removed by washing) and heated to a predetermined temperature. This is accomplished by heat exchange with hot products and finally by a direct fired heater. The hot and partially vaporized crude is 'Flashed' in a trayed distillation tower. Here, the vaporized portion of the crude oil feed moves up the tower and is selectively condensed by cooled reflux streams moving down the tower. These condensates are taken off at various parts of the tower according to their condensing temperature as distillate side streams. The light oils not condensed in the tower are taken off at the top of the tower to be condensed externally as the overhead product. The unvaporized portion of the crude oil feed leaves the bottom of the tower as the atmospheric residue.

The unit operates at a small positive pressure around 5–10 psig in the overhead drum, thus, its title of 'Atmospheric' crude unit. Typical product streams leaving the distillation tower are as follows:

Overhead distillate	Full range naphtha	Gas to 380°F cut point
1st side stream	Kerosene	380 to 480°F cut range
2nd side stream	Light gas oil	480 to 610°F cut range
3rd side stream	Heavy gas oil	610 to 690°F cut range
Residue	Fuel oil	+ 690°F cut point

Full details of this unit are given in Chapter 3.

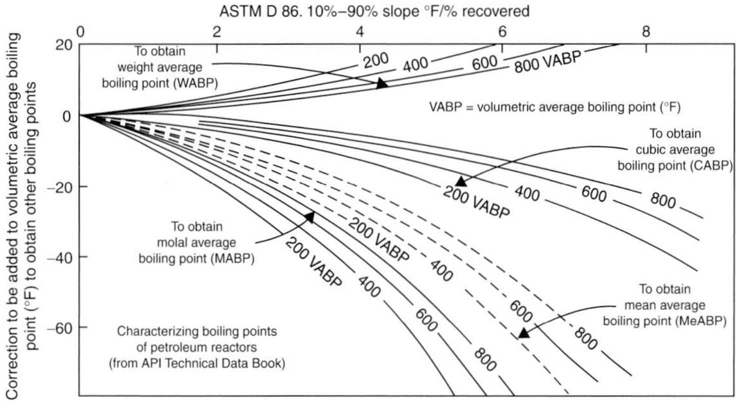

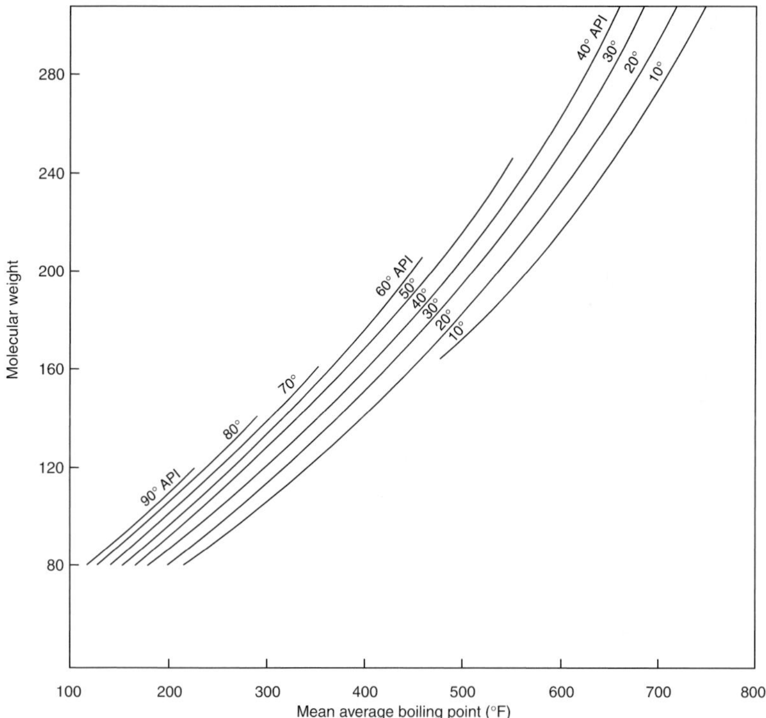

Figure 1.14. Correlation between boiling point, molecular weight, and gravity.

The crude vacuum distillation unit

Further break up of the crude is often required to meet the refinery's product slate. This is usually required to produce low cost feed to cracking units or to produce the basic stocks for lubricating oil production. To achieve this the residue from the atmospheric unit is distilled under sub atmospheric conditions in the crude vacuum distillation unit. This unit operates similar to the atmospheric unit in so much as the feed is heated by heat exchange with hot products and then in a fired heater before entering the distillation tower. In this case, however, the tower operates under reduced pressure (vacuum) conditions. These units operate at overhead pressures as low as 10 mmHg. Under these conditions the hot residue feed is partially vaporized on entering the tower. The hot vapors rise up the tower to be successively condensed by cooled internal reflux stream moving down the tower as was the case in the atmospheric distillation unit. The condensed distillate streams are taken off as side stream distillates. There is no overhead distillate stream in this case.

The high vacuum condition met with in these units is produced by a series of steam ejectors attached to the unit's overhead system. Typical product streams from this unit are as follows:

Top side stream	Light vacuum gas oil	690 to 750°F
2nd side stream	Heavy vacuum gas oil	750 to 985°F
Residue	Bitumen	+985°F

This unit is further described and discussed in Chapter 3.

The light end units

The full range naphtha distillate as the overhead product from the atmospheric crude unit is further split into the basic components of the refinery's volatile and light oil products. This is accomplished in the light end plant which usually contains four separate distillation units. These are:

- The de-butanizer
- The de-propanizer
- The de-ethanizer
- The naphtha splitter.

The most common routing of the full range naphtha from the atmospheric crude overhead is first to the de-butanizer unit. This feed stream is heated by heat exchange with hot products before entering the feed tray of the de-butanizer column. This is a distillation column containing between 30 and 40 trays. Separation of butanes and lighter gas from the naphtha occurs in this tower by fractionation. The butanes and lighter are taken off as an overhead distillate while the naphtha is removed as the column's bottom product. The overhead distillate is then heated again by heat exchange with hot streams and fed into a de-propanizer column. This column also has about 30–40 distillation trays and separates a butanes stream from the propane and lighter material stream by fractionation. The butanes leave as the column's bottom product to become the Butane LPG product after further 'sweetening' treatment (sulfur removal). The column's overhead distillate is fed to a de-ethanizer column after preheating. Here the propane is separated from the lighter materials and leaves the column as the bottom product. This stream becomes part of the refinery's propane LPG product after some further 'sweetening' treatment. There will be no overhead distillate product from this unit. The material lighter than propane leaves the overhead drum as a vapor containing mostly ethane, and is normally routed to the refinery's fuel gas system.

The de-butanized naphtha leaving the bottom of the de-butanizer is subsequently fractionated in the naphtha splitter to give a light naphtha stream as the overhead distillate and a heavy naphtha as the column's bottom product. The light naphtha is essentially C5's and nC6's, this stream is normally sent to the refinery's gasoline pool as blending stock. The heavy naphtha stream contains the cycloparaffin components and the higher paraffin isomers necessary in making good catalytic reformer feed. This stream therefore is sent to the catalytic reformer after it has been hydrotreated for sulfur and nitrogen removal.

The Light End units are further described and discussed in Chapter 4.

The catalytic reformer unit

The purpose of the catalytic reformer plant is to upgrade low octane naphtha to the high octane material suitable for blending into motor gasoline fuel. It achieves this by reforming some of the hydrocarbons in the feed to hydrocarbons of high octane value. Notably among those reactions is the conversion of cycloparaffin content of the feed to aromatics. This reaction also gives up hydrogen molecules which are subsequently used in the refinery's hydrotreating processes.

The feed from the bottom of the naphtha splitter is hydrotreated in the naphtha hydrotreater for the removal of sulfur and nitrogen. It leaves this unit to be preheated to the reforming reaction temperature by heat exchange with products and by a fired heater. The feed is mixed with a recycle hydrogen stream before entering the first of three reactors. The reforming reactions take place in these reactors and the reactor temperatures are sustained and controlled by intermediate fired heaters. The effluent leaves the last reactor to be cooled and partially condensed by heat exchange with cold feed and a condenser. This cooled effluent is routed to a flash drum from which a hydrogen rich stream is removed as a gas while the reformate is removed as a liquid stream and sent to a stabilizer column. The bottoms from this column is de butanized reformate and is routed to the gasoline pool for blending to meet motor gasoline specifications. Part of the gas leaving the flash drum is recycled to the reactors as the unit's recycle stream. The remaining gas is normally sent to the naphtha hydrotreater for use in that process.

Details of the catalytic reforming process are described and discussed further in Chapter 5.

The hydrotreating units (de-sulfurization)

Most streams from the crude distillation units contain sulfur and other impurities such as nitrogen, and metals in some form or other. By far the most common of these impurities is sulfur, and this is also the least tolerable of these impurities. Its presence certainly lowers the quality of the finished products and in the processing of the crude oil its presence invariably affects the performance of the refining processes. Hydrotreating the raw distillate streams removes a significant amount of the sulfur impurity by reacting the sulfur molecule with hydrogen to form hydrogen sulfide (H_2S) this is then removed as a gas.

Two types of de sulfurizing hydrotreaters are presented in this book. These are:

- Naphtha hydrotreating—Once through hydrogen
- Diesel hydrotreating—Recycle hydrogen

In naphtha hydro treating the naphtha from the naphtha splitter is mixed with the hydrogen rich gas from the catalytic reformer unit and preheated to about 700°F by heat exchange and a fired heater. On leaving the fired heater the stream enters a reactor containing a de-sulfurizing catalyst (usually a Co Mo on alumina base). The sulfur components of the feed combine with the hydrogen to form H_2S. The effluent from the reactors are cooled and partially condensed before being flashed in a separator drum. The gas phase from this drum is still high in hydrogen content and is usually routed to other down stream hydrogen user processes. This stream contains most of the H_2S produced in the reactors, the remainder leaves the flash drum with the de-sulfurized naphtha liquid to be removed in the hydrotreater's stabilizer column as a H_2S rich gas.

Diesel hydrotreating has very much the same process configuration as the naphtha unit. The main difference is that this unit will almost invariably have a rich hydrogen stream recycle. The recycle is provided by the flashed gas stream from the flash drum. This is returned to mix with the feed and a fresh hydrogen make up stream before entering the preheater system. The recycle gas stream in these units is often treated for the removal of H_2S before returning to the reactors.

A detailed discussion and description of these processes are given in Chapter 8.

The fluid catalytic cracking unit

This cracking process is among the oldest in the oil industry. Although developed in the mid 1920s it first came into prominence during the Second World War as a source of high octane fuel for aircraft. In the early fifties its prominence as the major source of octane was somewhat overshadowed by the development of the catalytic reforming process with its production of hydrogen as well as high octane material. The prominence of the fluid catalytic cracking unit (FCCU) was reestablished in the 1960s by two developments in the process. These were:

- The use of highly active and selective catalysts (Zeolites)
- The establishment of riser cracking techniques

These two developments enabled the process to produce higher yields of better quality distillates from lower quality feed stocks. At the same time catalyst inventory and consumption costs were significantly reduced.

The process consists of a reactor vessel and a regenerator vessel interconnected by transfer lines to enable the flow of finely divided catalyst powder between them. The oil feed (typically HVGO from the crude vacuum unit) is introduced to the very hot regenerated catalyst stream leaving the regenerator on route to the reactor. Cracking occurs in the riser inlet to the reactor due to the contact of the oil with the hot catalyst. The catalyst and oil are very dispersed in the riser so that contact between them is very high exposing a large portion of the oil to the hot catalyst. The cracking is completed in the catalyst fluid bed in the reactor vessel. The catalyst fluidity is maintained by steam injection at the bottom of the vessel. The cracked effluent leaves the top of the reactor vessel as a vapor to enter the recovery section of the plant. Here the distillate products of cracking are separated by fractionation and forwarded to storage or further treating. An oil slurry stream from this recovery plant is returned to the reactor as recycle.

The catalyst from the reactor is transferred to the regenerator on a continuous basis. In the regenerator the catalyst is contacted with an air stream which maintains the catalyst in a fluidized state. The hot carbon on the catalyst is burned off by contact with the air and converted into CO and CO_2. The reactions are highly exothermic rising the temperature of the catalyst stream to well over 1,000°F and thus providing the heat source for the oil cracking mechanism.

Products from this process are:

- Unsaturated and saturated LPG
- Light cracked naphtha
- Heavy cracked naphtha
- Cycle oil (mid distillate)
- slurry.

Details of this process together with typical yield data are given in Chapters 6 and 11.

The hydrocracking process

This process is fairly new to the industry becoming prominent in its use during the late 1960s. As the title suggests the process cracks the oil feed in the presence of hydrogen. It is a high pressure process operating normally around 2,000 psig. This makes the unit rather costly and because of this has diminished its prominence in the industry compared with the FCCU and thermal cracking. However, the process is very flexible. It can handle a wide spectrum of feeds including straight run gas oils, vacuum gas oils, thermal cracker gas oils, FCCU cycle oils and the like. The products it produces need very little down stream treating to meet finished product specifications. The naphtha stream it produces is particularly high in naphthenes making it a good catalytic reformer stock for gasoline or aromatic production.

The process consists of one or two reactors, a preheat system, recycle gas section, and a recovery section. The oil feed (typically a vacuum gas oil) is preheated by heat exchange with reactor effluent streams and by a fired heater. Make up and recycle hydrogen streams are introduced into the oil stream before entering the reactor(s). (Note in some configurations the gas streams are also preheated prior to joining the oil). The first section of the reactor is often packed with a de-sulfurizing catalyst to protect the more sensitive cracking catalyst further down in the reactor from injurious sulfur, nitrogen, and metal poisoning. Cracking occurs in the reactor(s) and the effluent leaves the reactor to be cooled and partially condensed by heat exchange. The stream enters the first of two flash drums. Here, the drum pressure is almost that of the reactor. A gas stream rich in hydrogen is flashed off and is recycled back to the reactors as recycle gas. The liquid phase from the flash drum is routed to a second separator which is maintained at a much lower pressure (around 150–100 psig). Because of this reduction in pressure a second gas stream is flashed off. This will have a much lower hydrogen content but will contain C3's and C4's. For this reason the stream is often routed to an absorber column for maximizing LPG recovery. The liquid phase leaves the bottom of the low-pressure absorber to enter the recovery side where products are separated by fractionation and sent to storage.

Further details of this process are given in Chapters 7 and 11.

Thermal cracking units

Thermal cracking processes are the true work horses of the oil refining industry. The processes are relatively cheap when compared with the fluid cracker and the hydrocracker but go a long way to achieving the heavy oil cracking objective of converting low quality material into more valuable oil products. The process family of Thermal Crackers has three members, which are:

- Thermal crackers
- Visbreakers
- Cokers.

The term Thermal Cracking is given to those processes that convert heavy oil (usually fuel oil or residues) into lighter product stock such as LPG, naphtha, and middle distillates by applying only heat to the feed over a prescribed element of time. The term Thermal Cracker when applied to a specific process usually refers to the processing of atmospheric residues (long residue) to give the lighter products. The term visbreaking refers to the processing of vacuum residues (short residues) to reduce the viscosity of the oil only and thus to meet the requirements of a more valuable fuel oil stock. Coking refers to the most severe process in the Thermal Cracking family. Either long or short residues can be feed to this process who's objective is to produce the lighter distillate products and oil coke only. The coker process is extinctive—that is it converts ALL the feed. In the other two processes there is usually some unconverted feed although the Thermal Cracker can be designed to be 'extinctive' by recycling the unconverted oil. The three Thermal cracking processes have the same basic process configuration. This consists of a cracking furnace, a 'soaking' vessel or coil, and a product recovery fractionator(s). The feed is first preheated by heat exchange with hot product streams before entering the cracking furnace or heater. The cracking furnace raises the temperature of the oil to its predetermined cracking temperature. This is always in excess of 920°F and by careful design of the heater coils the oil is retained in the furnace at a prescribed cracking temperature for a predetermined period of time (the residence time). In some cases an additional coil section is added to the heater to allow the oil to 'soak' at the fixed temperature for a longer period of time. In other cases the oil leaves the furnace to enter a drum which retains the oil at its cracking temperature for a little time. In the Coker process the oil leaves the furnace to enter one of a series of Coker drums in which the oil is retained for a longer period of time at its coking temperature for the production of coke.

The cracked oil is quenched by a cold heavy oil product stream on leaving the soaking section to a temperature below its cracking temperature. It then enters a fractionator where the distillate products are separated and taken off in a manner similar to the crude distillation unit. In the case of the cokers the coke is removed from the drums by high velocity water jets on a regular batch basis. The coking process summarized here refers to the more simple 'Delayed Coking' process. There are other coking processes which are more complicated such as the fluid coker and the proprietary Flexi coker.

Further details on Thermal Cracking are provided in Chapter 11. This chapter includes also the treating of residues by hydrocracking and fluid catalytic cracking.

Gas treating processes

The processes summarized above are the more common to be included in a fuel or energy refinery's configuration. In addition to these there will also be the gas treating processes and often sulfur recovery processes. These are described and discussed in Chapter 10.

Gas treating is always required to remove the H_2S impurity generated by hydrotreating or cracking from the refinery fuel gas or hydrogen recycle streams. The removal of H_2S for these purposes is accomplished by absorbing the hydrogen sulfide into an amine or similar solution that readily absorbs H_2S. Stripping the rich absorbent solution removes the H_2S from the system to be further reacted with air to produce elemental sulfur. This latter reaction takes place in specially designed sulfur plant.

The rich H_2S laden gases from all the refinery sources enters below the bottom tray (or packed bed) of the absorber tower. The lean H_2S free absorbent solution enters the tower above the top tray (or packed bed) to move down the tower counter current to the gas moving upwards. Mixing on the trays (or packed beds) allows the H_2S from the gas phase to be absorbed into the liquid solution phase. The H_2S free gas leaves the tower top to be routed to refinery fuel or other prescribed destination.

The rich absorbent solution leaves the bottom of the absorber to be heat exchanged with hot stripped absorbent solution before entering the feed tray of the Stripping column. The solution moving down the tower is stripped free of H_2S by a stripper vapor phase moving up the tower. This stripper phase is generated by conventional reboiling of the bottom tray solution. The hot stripped solution leaves the bottom of the tower to be cooled by heat exchange with the feed and then by an air or water cooler before entering the absorber tower. Conditions in the stripper column is maintained by partially condensing the rich H_2S overhead vapors. Then returning the distillate as reflux to the top tray of the rectifying section of the tower (which is the section of trays above the feed tray).

The vapor not condensed leaves the reflux drum to be routed to a sulfur plant. These vapors contain a high concentration of H_2S (usually in excess of 90% mol) and enter the specially designed fuel 'gun' of the sulfur plant heater. Here, about one third is mixed with an appropriate concentration of air and 'burned' in the plant's fire box to generate SO_2. The gases generated are combined with the remaining H_2S and passed over a catalyst bed where almost complete conversion to elemental sulfur occurs. This product, in molten form, enters a heated storage pit. The unconverted sulfurous vapors are further incinerated before venting to atmosphere from an acceptably elevated location.

Processes not so common to energy refineries

Octane enhancement processes

The octane enhancement processes detailed in this book are the alkylation process and the isomerization process. These processes are usually proprietary and are provided to refiners under license.

The alkylation process treated here is the HF process which utilizes hydrogen fluoride as the catalyst which is used to convert unsaturated C4's to high octane isobutanes. The unit 's recovery side is the aspect dealt with in some detail together with a descriptive item on the safe handling of hydrogen fluoride.

The isomerization process has a similar configuration to the catalytic reformer plant. This process uses hydrogen in its conversion of low octane hydrocarbons to the high octane isomers.

Both these processes are described and discussed in detail in Chapter 9.

Oxygenated gasolines

The concentration of vehicles on the roads in most of the cities in the modern world has increased dramatically over the last two decades. The emission of pollutants from these vehicles is causing a significant addition to the already critical problem of atmospheric pollution. The problem is now so acute that governments of most first world countries are seeking legislation to curb and minimize this pollution and most countries will see the implementation of 'Clean Air' acts in the 21st century.

Petroleum refining companies have been working diligently for many years to satisfy the requirements of 'Clean Air' legislation already in place. This began in the 1970s with the elimination of tetra ethyl lead from most gasoline requirements. Processes such as isomerization and polymerization of refinery streams were developed together with a surge in the use of the alkylation process. However, the further decrease of pollutants now requires a move away from the traditional gasoline octane enhancers such as the aromatics and the olefins.

Catalytic reforming produces gasoline streams to meet octane requirements mainly by converting cycloparaffin to light aromatics. Fluid catalytic cracking also produces gasoline blending stocks by cracking paraffins to light olefins and the products from these two processes still make up the bulk of a refinery's gasoline pool. Unfortunately the aromatics are 'Dirty Compounds' because they produce a sooty exhaust emission—unacceptable in meeting the 'Clean Air' requirements. Considerable work has been done with alcohol to take the place of aromatics and as octane enhancers and meeting other gasoline specs such as RVP. The work has had some success and some companies in North America use ethanol in the gasoline blend.

The greatest success in reducing aromatics in gasoline to date however has been in the production and blending of oxygenated compounds into the gasoline pool. A press release by a number of companies in 1,990 is quoted as follows:

- Adding oxygenates reduces the amount of exhaust emissions (hydrocarbons and carbon monoxide) and the benefits have been quantified.
- Changing the level of olefins in gasoline does not have much of an impact on vehicle exhaust emission.
- Reducing aromatics and/or boiling range of gasoline can either reduce or increase exhaust emissions depending upon vehicle type.

Oxygenates are ether compounds derived from their respective alcohol. There are three candidates of these ethers to meet the gasoline requirements. These are:

- Methyl tertiary butyl ether (MTBE)
- Ethyl tertiary butyl ether (ETBE)
- Tertiary amyl methyl ether (TAME).

Of these three candidates MTBE is the one that has been used more extensively to meet all the gasoline pool objectives. This compound has the blending quality of 109 octane, a RVP blending of 8–10 psi and a boiling point of 131°F.

ETBE octane blending properties seem to be slightly better than those of MTBE and so does its RVP blending qualities. The ethanol feed stream of course is not so readily available as methanol.

TAME has an average octane number of 104, and a RVP blending of 3–5 psi. Except for the lower octane value TAME has similar blending properties to ETBE. The ether in this case is formed by the reaction of 2-methyl-2-butene and 2-methyl-1-butene olefin feed and methanol. Commercial plants operate in the UK and parts of Europe producing TAME. The compound is used in Europe as a gasoline product and not as a gasoline blend stock. The front end of a cracked gasoline stream is used in this manufacture to provide the olefin.

Use of MTBE has been phased out in the U.S. because of ground water contamination but is still in use in Europe and elsewhere. However, emphasis on the use of renewable resources in Europe has prompted a gradual shift from MTBE to ETBE.

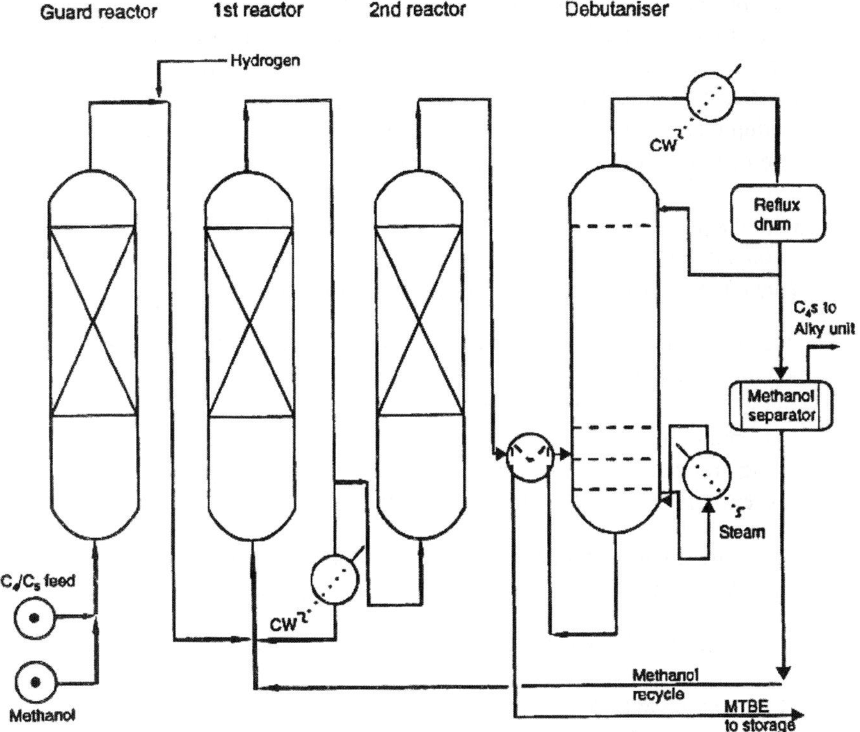

Figure 1.15. Typical flow diagram for the production of MTBE.

The production of the ethers

There are several licensed processes for the production of methyl tertiary butyl ether by the etherification of a C4/C5 olefin stream and methanol. These processes have very similar configurations and are flexible enough to be converted quite simply to the production of the other ethers. Figure 1.15 is a typical flow diagram for this process.

The olefin feed from a FCCU or a steam cracker is combined with a methanol stream to enter a guard reactor to remove impurities. A small hydrogen stream is added to the hydrocarbon from the guard reactor prior to entering the ether reactor. This reactor contains special resin catalyst and the reactor feed flows upward through this catalyst bed at moderate temperature and pressure and in the liquid phase. The reaction is exothermic and temperature control is maintained by externally cooling a recycle stream from the first of two reactor vessels.

The catalyst in this case performs three reactions simultaneously: Etherification of branched olefins, selective hydrogenation of the unwanted di-olefins, and hydro isomerization of olefin by a double bond switch. The reactor effluent leaves the top of the second reactor vessel to be heated in a feed heat exchanger with the de-butanizer bottoms product. The overheads from the de-butanizer is a C4 and methanol stream. The methanol stream is recycled to the first reactor while the C4s are returned to the FCCU light ends unit. The bottom product is C5+ enriched with MTBE (or TAME depending on the olefin feed used).

The production of Oxygenated gasolines is described and discussed in further detail in Chapter 9.

The non-energy refineries

In addition to the energy related refineries, which occupy most of this book, there are two major nonenergy producing refineries. These are, the lube oil refinery, and the petrochemical refineries. These are summarized below:

The lube oil refinery

The schematic flow diagram (Figure 1.16) shows a typical lube oil producing refinery configuration. Only about 8 or 9 base lube oil stocks are produced from refinery streams. The many hundreds of commercial grades of lubricating oils used in industry and transportation are blends of these base stocks with some small amounts of proprietary additives (mostly organic acid derivatives) included to meet their required specifications. There are also two quite important bye products to lube oil. These are bitumen and waxes. Most refineries include bitumen blending in their configuration, but only a few of the older refineries process the waxes. These are exported to manufacturers specializing in wax and grease production.

Lube oil production starts with the vacuum distillation of atmospheric residue. This feedstock is usually cut into three distillate streams each meeting a boiling range which gives streams with viscosity meeting the finished blending product specifications. The lighter stream is taken off as the top side stream and is further distilled again under vacuum to three light lube oil blending cuts. These are called spindle oils and when finished will form the basis of light lubes used for domestic purposes such as sewing machine, bicycle, and other home lubricant requirements. Some of the heavier spindle oils are also used as blend stocks for light motor oils. These spindle oils require very little treatment for finishing. Usually, a mild hydrotreating suffices to meet color requirement.

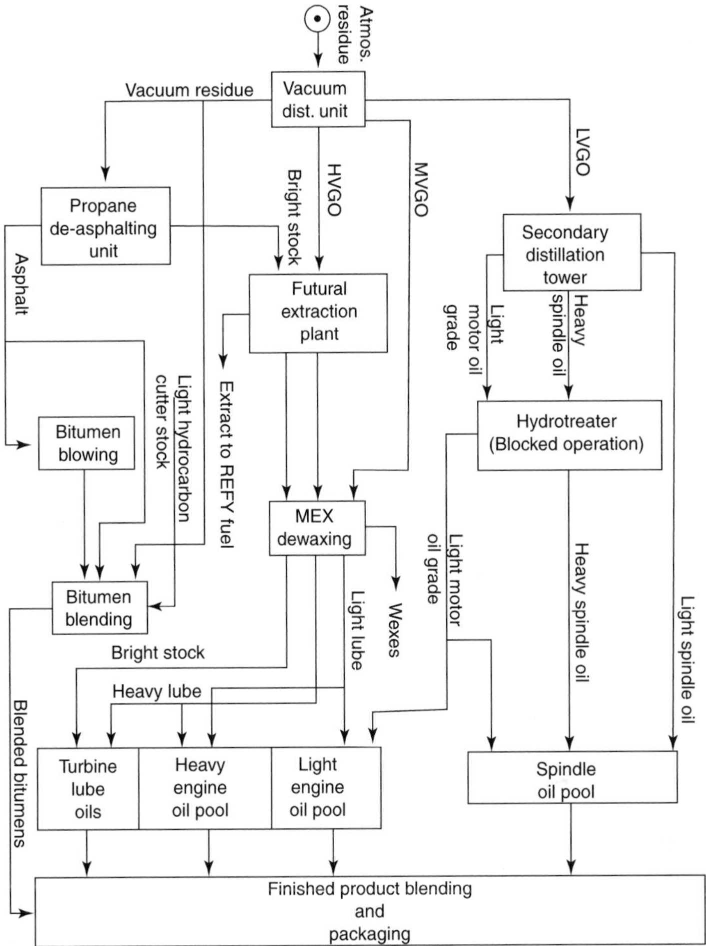

Figure 1.16. A lube oil refinery configuration.

The second distillate side stream is dewaxed and sent to the engine lube oil pool. It may also be blended with the heavier bottom side stream as heavy engine oil stock. The bottom side stream is one of the base blending stock for heavy engine oils and the turbine oil stocks. To meet color and other specifications these heavier oils must be treated for the removal of undesirable components (such as heavy aromatics and olefins) by solvent extraction. This is accomplished prior to the stream being dewaxed and routed to storage. The heavy vacuum residue from the vacuum tower is routed to a propane de-asphalting unit. Here, the very thick bituminous asphaltenes are removed by extraction with liquid propane. The raffinate from this extraction process is the heaviest lube oil blending stream commonly called Bright stock. This stream is also routed to the solvent extraction unit and the de waxing process before storage.

Solvent extraction is accomplished in a trayed column by contacting the oil feed and solvent counter currently in the tower. The lighter raffinate stream leaves the top of the tower to be stripped free of the solvent in an associated stripper column, before entering the de waxing unit. The extracted components leave the bottom of the tower also to be stripped free of the solvent in an associated stripper column. The extract in this case may be routed to the propane de asphalting unit or simply sent to the refinery fuel supply. The solvent in modern refineries is either Fufural, Phenol, or a proprietary solvent based on either of these chemicals. In earlier plants Oleum or liquid SO_2 was used for this purpose.

The oil streams routed to the de waxing plant are contacted and mixed with a crystallizing agent such as Methyl Ethyl Ketone (MEK) before entering a series of chiller tubes. Here the oil/MEK mix is reduced in temperature to a degree that the wax contained in the oil crystallizes out. The stream with the wax now in suspension enter a series of drum filters where the wax and oil are separated. Both streams are stripped free of the MEK in separate columns. The MEK is recycled while the de waxed oil is sent to storage and blending. The wax may be retained as a solid in a suitably furnished warehouse or re melted and stored in special tanks with inert gas cover.

The asphalt from the propane de-asphalting unit is stripped free of propane and any other light ends using inert gas as the stripping agent. It leaves the unit to proceed either directly to the bitumen pool or to be further treated by air blowing. The air blowing process increases the hardness of the bitumen where this is required to meet certain specifications. It is accomplished either as a batch process or on a continuous basis. The hot stripped asphalt from the de asphalting unit enters the air blower reactor under level control (if the process is continuous). Air is introduced via a small compressor to the bottom of the reactor vessel, and allowed to bubble up through the hot oil phase. The air removes some of the heavy entrained oils in the asphalt and reacts mildly to partially oxidize the asphaltenes. The hot oil vapors and the unused air leaves the top of the reactor to be burned in a suitably designed incinerator. The blown asphalt leaves the reactor as a side stream to bitumen storage or blending.

The production of lube oils usually takes place in a section of an energy refinery. The various grades of the oils are also produced in a blocked operation using storage facilities between the units. This is feasible as the amount of lube oils required to be produced are relatively small and normally do not justify separate treating facilities for each grade.

The lube oil refinery is further detailed in Chapter 12.

The petrochemical refinery

Feed stocks for the production of petrochemicals originate from refineries with processes similar to those described in this book producing fuels. Indeed there are only a few refineries world wide that cater only for petrochemical requirements. Most petrochemical feed stocks are produced by changing operating parameters of the normal fuel refinery processes. In catering for the petrochemical needs much of the refinery product streams are tailored as follows:

- *Aromatic streams*—High in benzene, toluene, xylenes
- *Olefin streams*—High in ethylene, propylene and C4s.

Producing the aromatic feed stock

The production of aromatics feed stocks originates with the catalytic reforming of a refinery stream of a heavy naphtha range (say 120°–420°F) and rich in naphthenes. A typical stream that meets this criteria would be a naphtha stream from a hydrocracker. Thus in meeting this petrochemical, needs a hydrocracker forming part of a fuel refinery configuration. This unit would be operated to maximize naphtha production. This would mean running the unit on a low space velocity with a higher oil recycle rate (that is most recovered product heavier than the naphtha would be recycled back to the reactors).

Another source of high naphthene feed to the cat reformer would be hydrotreated cat cracker naphtha. Of course the hydrotreating of unsaturates has a high demand on the refinery's hydrogen system, but this is balanced to some extent by the additional hydrogen produced in reforming the naphthenes. Should the refinery configuration include a thermal cracker and/or a steam cracker the hydrotreating of the naphtha cut from these units also yield high naphthene catalytic reformer feed stock.

Catalytic reforming of the high naphthene content naphtha produces aromatics but there is also present some unreacted paraffins and some naphthenes. The down stream petrochemical units that separate and purify the aromatic reformate are expensive both in capital and operating costs. The specification for the BTX (Benzene, Toluene, Xylene) feed therefore is very stringent and excludes non aromatic components as much as possible. Another process may therefore be included in the refinery configuration to 'clean up' this aromatic feed stream before leaving the refinery. This is an aromatic extraction plant. This is a licensed process using a solvent to separate the paraffins and aromatics by counter current extraction. The rich aromatic stream is then forwarded to the BTX plant where benzene, toluene, ethyl benzene, and *o*-xylene, are separated by fractionation while the para-xylene is usually separated by crystallization. The

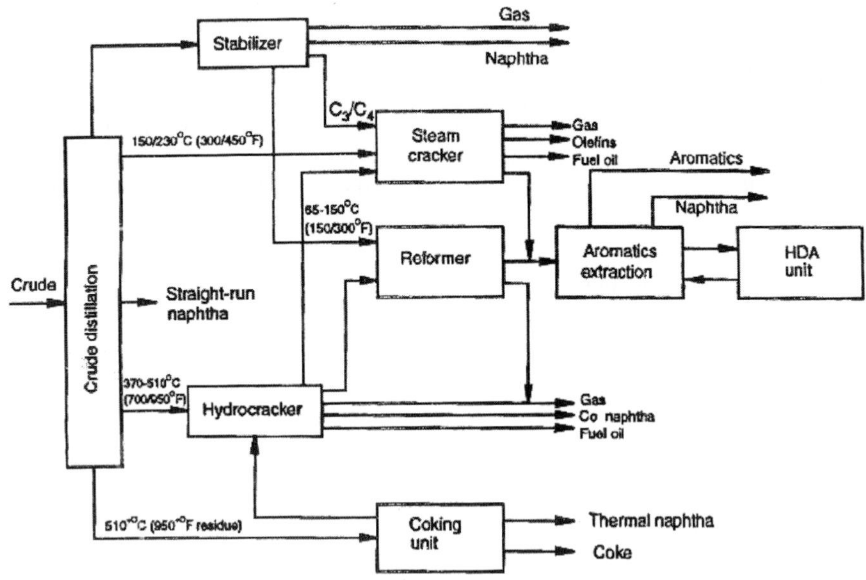

Figure 1.17. A petrochemical refinery configuration.

meta-xylene may also be recovered by super distillation but more often than not it is converted into *o*-xylene in an isomerization unit.

Producing the olefin feed stock

The source of olefins in a refinery configuration is from either the FCCU, a thermal cracker, or a steam cracker. The olefins produced as a gas are ethylene, propylene, and the C4's such as butylenes, butadiene etc. Liquid olefinic products from these units are normally hydrotreated to make reformer feedstock and thus the BTX feed. All products of course are treated for sulfur control and clean up before leaving the refinery as petrochemical feed stock. The specifications for these products are stringent and usually the 'clean up' plants are dedicated to the treatment of these products.

Olefins are used mostly in the production of polymers such as the vinyl polymers (vinyl chloride, vinyl acetate and the like), the poly ethylene products, and the polypropylene products. The heavier C4's are a major constituent in the production of synthetic rubbers. Figure 1.17 shows a configuration for a typical petrochemical refinery.

The petrochemical refinery is also further described and discussed in Chapter 12.

References

1. Good, Connel, et al., *Oil & Gas Journal* 30th Dec 1944. Penwell Publishing Company.
2. Thrift, *Oil & Gas Journal* 4th Sept 1961. Penwell Publishing Company.
3. Wayne C. Edmister, *Applied Thermodynamics.* By Gulf Publishing Company Houston, TX.
4. J. B. Maxwell, *Data Book on Hydrocarbons*. D. Van Nostrand Company, Ltd., London 1950, 9th Printing February 1968.

Chapter 2

Petroleum products and a refinery configuration

D.S.J. Jones

2.1 Introduction

This chapter defines the major products normally produced from the refining of crude oil. These products are the intermediary and finished products from energy refineries only. Chapter 1 of this book provided brief description of the products, this chapter expands on this with a more in-depth look at the products themselves. Their demand in the petroleum markets and also the environmental impact of the more prominent products are discussed in this chapter. Finally the chapter continues with an example of the development of a process configuration of the refinery to meet a particular product slate.

The first part of this chapter describes the basic fractions obtained from the atmospheric and vacuum crude distillation units. It continues with the description of products from the most common intermediate processes met with in many refineries to meet their various product requirements. This part of the chapter continues with an outline of the specifications for various finished petroleum products and discusses their salient points.

The second part of the chapter discusses the features of the motive fuels. It continues with the effect of environmental constraints and the development of changes in the content of these products.

The third part describes the development of a refinery's process configuration and discusses its purpose. The development of a refinery configuration is illustrated by an example. This example is compiled manually. In modern development practice however this would be accomplished by reducing the various product properties and their relationship to linear equations and solving these with high speed computers, a process called Linear Programming.

2.2 Petroleum products

Fractions from the atmospheric and vacuum distillation of crude oil

The most common fractions distilled from crude oil distillation processes are shown in Figure 2.1.

These cut lengths will vary slightly depending on the crude oil source and the finished product slate requirements. These are the basic components which after further processing and blending will make the finished products in composition and quantity required by the refinery. The cuts shown in Figure 2.1 are called 'straight run' products and these are described in the following paragraphs. The crude source, in this case is a medium Middle Eastern with a gravity of 33.9°API and a sulfur content of 1.9 %wt.

Atmospheric overhead distillate. This is not strictly a cut but consists of all the light material in the crude absorbed into the total overhead distillate from the crude tower. This distillate, and in most cases, together with similar distillates from other processes form the feed to the refinery's light end unit (see Chapter 4). It is in the light end unit that the straight run LPG, light naphtha, and the Heavy Naphtha are separated. However, the end point of the heavy naphtha is determined by the cut point of the overhead distillate and the fractionation between it and the first side stream product from the atmospheric unit.

Refinery gas and the LPGs. In many refineries most of the C_4's and lighter are removed from the atmospheric column overhead distillate in the first column of the light end unit. This is the unit's debutanizer column. Some refineries however chose to separate the light naphtha and lighter from the heavy naphtha first. There is no specific reason one can assume its really a question of the specific refinery's economic criteria. Having separated the C_4's and lighter as a distillate from the naphtha the distillate enters a depropanizer where C_3 and lighter are separated as a distillate from the C_4's. This distillate is further fractionated in a deethanizer column where the C_3 is removed as the bottom product. There is no distillate product from this unit but all the gas lighter than C_3 leaves the tower as a vapor usually routed to the refinery fuel gas system (see Chapter 10).

The C_4 portion of the overhead distillate is fractionated in the debutanizer so that it meets finished product specification with respect to its C_5 content. The fractionation in the depropanizer will be such that the C_3 content of the bottom product—C_4 LPG will meet the butane LPG specification with respect to RVP (Reid Vapor Pressure). The fractionation in this tower will also be such that the C_4 content in the overhead distillate will meet the propane LPG specification which leaves as the bottom product from the

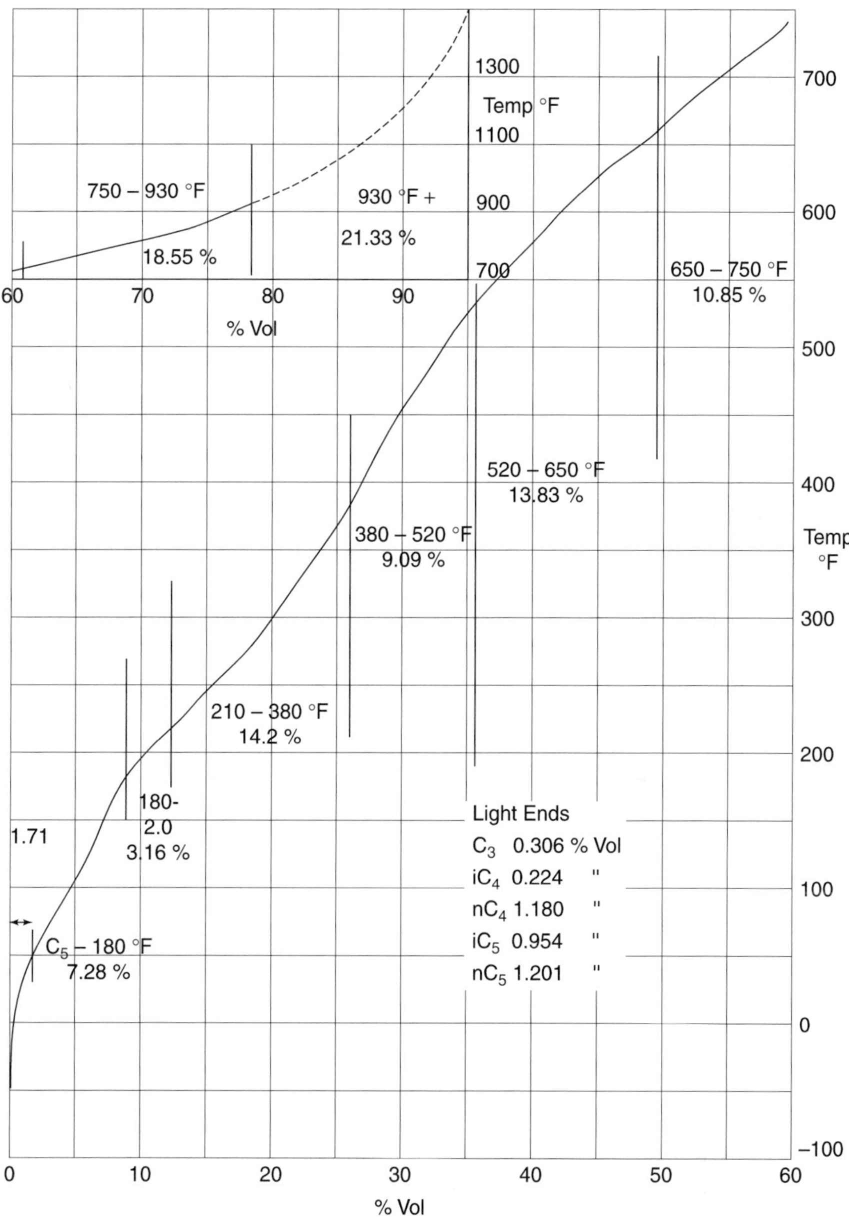

Figure 2.1. Sassan crude TBP curve and product split.

deethanizer with respect to its C_4 content. The fractionation in the de-ethanizer will be such as to ensure that the C_2 and lighter content of the propane LPG will be such as to meet that LPG's Reid Vapor Pressure specification.

The naphthas. There are usually two naphtha cuts produced from most crude. These are:

- Light naphtha (sometimes called light gasoline)
- Heavy naphtha.

Both these streams are the bottom product of the debutanizer unit. They are separated in a naphtha splitter fractionation tower. The light naphtha contains most of the crude's C_5's and much of the paraffin portion of the crude's C_6's. The purpose of making such a division is to produce a satisfactory heavy naphtha which will contain the heavier naphthenes and will be a suitable feed for a catalytic reformer (see Chapter 14).

The light naphtha has a TBP distillation range of C_5 to around 190°F. The heavy naphtha as the feed to the catalytic reformer and is a cut on crude of about 190°–360°F. This cut point of 360°F can vary depending on the severity operation of the catalytic reformer, the volatility specification of the finished gasoline which the reformate will be a major precursor, and the refineries production requirements. In this latter case for example the refinery's operating plan may call for maximizing kerosene in which case the atmospheric distillation unit would be operated to decrease the amount of overhead distillate in order to increase the kerosene (top side stream) fraction. Of course if the refinery plan is to maximize gasoline the atmospheric tower would be operated to increase the overhead distillate at the expense of the Kero fraction.

Straight run kerosene. This fraction is usually the first side stream of a conventional atmospheric distillation unit. It may be cut to meet a burning oil specification or become a component in Jet Fuel finished product. Its cut range is usually between 360°F and 480°F. Again this cut range may vary with the required heavy end of the naphtha and the front end of the Gas Oils. In most cases this Kero fraction must meet a flash point specification after it has been steam stripped, and its end cut point is usually set to meet a smoke point specification. Usually the sulfur content restriction is met by hydro-desulfurization, in some cases too the smoke point is reduced by other processes. These subsequent processes are aimed at removing or converting the aromatic content of the fraction. It can also be routed to the Gas oil pool as a precursor for a diesel finished product.

The atmospheric straight run gas oils. Usually there will be two gas oil side streams, a light gas oil side stream and below this take off a heavy gas oil side stream is withdrawn. Both these side—streams are steam stripped to meet their respective flash point specification (usually 150°F minimum). The lighter side stream (cut of about 480–610°F on crude) is the principle precursor for the automotive diesel grade finished product.

This side stream is desulfurized to meet the diesel sulfur specification in a hydro-treater (see Chapter 8). The lower gas oil stream is really a guard stream to correct the diesel distillation end point. This heavy gas oil may also be hydro-desulfurized and routed to either the fuel oil pool (as a precursor for marine diesel for example) or to a finished heating oil product from the gas oil pool.

The atmospheric residue. This is the bottom product from the atmospheric distillation of the crude oil. Most crude oils are distilled in the atmospheric crude oil tower to cut the atmospheric residue at a +650°F up to a +680°F cut point. Cutting the residue heavier than +680°F risks the possibility of cracking with heavier coke lay down and discoloring of the distillate products. Those atmospheric crude towers that do operate at higher cut points minimize the cracking by the recycle of cold quench into the bottom of the tower (below the bottom stripping tray) and minimizing the residue hold up time in the tower. The atmospheric residue may be routed to the fuel oil pool as the precursor to several grades of finished fuel oil products. The other options for this stream in a modern refinery are as follows:

- feed to a vacuum distillation unit. (This is the most common option.)
- feed to a thermal cracker (Visbreaker, or coking unit). See Chapter 11.
- feed to a deep oil fluid catalytic cracker. See Chapter 11.
- feed to a hydro-cracker or hydro-treater. See Chapter 11.

The vacuum distillation of atmospheric residue. In modern refinery practice the distillation of atmospheric residue is accomplished under high vacuum conditions in a specially designed tower whose internal equipment ensures a very lower pressure drop. Normally the vacuum conditions in the flash zone of the tower allows about the same percentage of distillate based on the tower feed to be cut in this tower as the distillate on whole crude in the atmospheric unit. Again the flash zone temperature in the vacuum unit is kept below 700°F. Usually there are two or three vacuum distillates from this tower. In a pure energy related refinery there will be two. The heavier of the two say to a cut range of 750–930°F will be the feed to a distillate hydro-cracker (see Chapter 7) or to a fluid catalytic cracker (see Chapter 6). In both these cases however a small heavier cut is taken off and returned to the tower bottom in order to correct the bottom distillate condradson carbon content to meet the specification required for either of the two downstream processes. This distillate product is usually titled HVGO (Heavy Vacuum Gas Oil).

For those refineries which produce lube oils as non energy products this bottom distillate may be split into two side streams in order to provide the flexibility required in the production of the lube oil blending stock specifications (see Chapter 12).

The light vacuum distillate is taken off as a top side stream and is usually routed to a hydro-desulfurizer to be sent either to the gas oil pool as heating oil stock or routed to the fuel oil pool as blending stock. This side stream is a cut range of 680–750°F and is usually labeled LVGO (Light Vacuum Gas Oil).

The vacuum residue. This is the bottom product from the vacuum distillation unit. Just as in the case of the atmospheric residue it has several options for its use in meeting the refinery's product slate. In the case of the energy refineries it can be upgraded to prime distillate products by a recycling thermal cracking process (see Chapter 11), coking, deep oil fluid catalytic cracking (Chapter 11) or hydro-cracking or indeed a combination of these processes.

In the case of the production of lube oils it can be processed to remove the heavy asphaltene portion of the stream. This deasphalted product is excellent lube oil blending stock, commonly called 'Bright Stock', which, when de waxed and subject to other lube oil processes becomes the precursor for a variety of lube oil products (see Chapter 12). The asphalt portion of the de asphalting process has a wide spectrum as precursor to the many grades of bitumen used in building, and road making (see Chapter 12). The vacuum residue itself also figures in the production of bitumen. Where non-energy products are significant in the refinery's product slate, the vacuum residue as produced is the key component in bitumen production.

Typical properties of some straight run product streams

The cuts and the stream properties are based on Arabian crude, and are shown here for the light, medium, and heavy Arabian crude oil (Table 2.1).

Product streams from intermediate and finishing processes

Most finished marketable petroleum products are blends of some straight run product streams with the products from further processing of some of these streams (see Part 2 of this chapter). This section deals with the intermediate or finishing product processes which provide the finished product precursors. Some of these more common intermediate products together with reference to their respective processes are described and discussed in the following paragraphs.

The LPG products. These two finished products, Propane LPG and Butane LPG are products from straight run sources combined with the products from intermediate processes such as the cracking and the naphtha reforming processes. Their finished specification except for some mild desulfurization treating are met by the distillation of the source streams and the central light end units (see Chapter 4). No further precursors are added, except for injection of odorizing chemicals for the detection of leaks and thus their safe storage. Finished LPGs are stored in spheres for propane and, usually, bullets for butane.

Table 2.1. Straight run product streams

	Light	Medium	Heavy
Crude, °API	38.8	30.7	28.2
Sulfur, %wt	1.1	2.51	2.84
Light naphtha			
Cut range, °F	68–212	68–212	68–212
Yield, %vol	10.5	9.4	7.9
Gravity, °API	77.4	78.4	80.1
Sulfur, %wt	0.056	0.007	0.0028
RVP, Psig	6.9	7.9	10.2
Paraffins, %vol	87.4	89.7	89.6
Naphthenes, %vol	10.7	8.8	9.5
Aromatics, %vol	1.9	1.5	0.9
RON clear	54.7	48.2	58.7
Heavy naphtha			
Cut range, °F	212–302	212–302	212–302
Yield, %vol	9.4	7.4	6.8
Gravity, °API	58.8	59.6	60.6
Sulfur, %wt	0.057	0.019	0.018
Paraffins, %vol	66.3	67.8	70.3
Naphthenes, %vol	20.0	20.8	21.4
Aromatics, %vol	13.7	11.4	8.3
Kerosene			
Cut range, °F	302–455	302–455	302–455
Yield, %vol	18.4	13.5	12.5
Gravity, °API	48.0	48.9	48.3
Sulfur, %wt	0.092	0.12	0.19
Paraffins, %vol	58.9	59.9	58.0
Naphthenes, %vol	20.5	21.9	23.7
Aromatics, %vol	20.6	18.2	18.3
Freeze point, °F	−67	−72	−84
Smoke point, mm	26	23	26
Luminometer No	57	55	60
Analine point, °F	133	139	138
Kin Cst @ −30°F	5.09	4.63	4.74
Kin Cst @ 100°F	1.13	1.09	1.12
Light gas oil			
Cut range, °F	455–650	455–650	455–650
Yield, %vol	21.1	17.4	16.4
Gravity, °API	37.3	37.2	35.8
Sulfur, %wt	0.81	1.09	1.38
Pour point, °F	10	0	5
Analine point, °F	166	156	156
Kin Cst @ 100°F	3.34	3.15	3.65
Kin Cst @ 210°F	1.32	1.22	1.4

(Cont.)

Table 2.1. (Cont.)

	Light	Medium	Heavy
Heavy gas oil			
Cut range, °F	650–1,049	650–1,049	650–1,049
Yield, %vol	30.6	30.5	26.3
Gravity, °API	24.8	22	21.8
Sulfur, %wt	1.79	2.87	2.88
Pour point, °F	100	75	90
Analine point, °F	195	172	172
Kin Cst @ 100°F	49.0	62.2	62.5
Kin Cst @ 210°F	6.65	7.25	7.05
Atmos residue			
Cut range, °F	+650	+650	+650
Yield, %vol	38.0	50.0	53.1
Gravity, °API	21.7	14.4	12.3
Sulfur, %wt	2.04	4.12	4.35
Pour point, °F	75	55	55
Con carb, %wt	4.5	10.0	13.2
Kin Cst @ 100°F	146	1,570	5,400
Kin Cst @ 210°F	12.4	54.0	106
Vacuum residue			
Cut range, °F	+1,049	+1,049	+1,049
Yield, %vol	7.4	19.5	26.8
Gravity, °API	11.5	3.8	4.0
Sulfur, %wt	3.0	5.85	5.6
Pour point, °F	80	120	120
Con carb, %wt	19	22.8	24.4
Kin Cst @ 210°F	392	19,335	13,400
Kin Cst @ 275°F	40.1	743	490
Vanadium, ppm	12	249	171
Nickel, ppm	7	55	53
Iron, ppm	36	79	28

Butane LPG is however used extensively as a finished gasoline product precursor to correct for volatility of the finished gasoline.

Gasoline precursors. Gasoline is a blend of straight run naphtha (usually light naphtha) and products of intermediary processes. These are usually suitably boiling range products from:

- Catalytic Reforming Process (see Chapter 5)
- Fluid Catalytic Cracking Process (see Chapter 6)

Modern refining configurations may also include processes which convert low quality products to gasoline precursors enhancing the gasoline's octane rating. These include the alkylation and Isomerization processes (see Chapter 9). Others which are less common are the oxygenated precursors (Chapter 1).

Catalytic reforming

This is a process which converts a relatively low octane straight run naphtha into a high octane liquid reformate which is a major precursor for finished gasoline. Of equal importance in upgrading naphtha this process also produces a high purity hydrogen stream. This makes the process a most important one in a modern day refinery configuration.

Briefly the straight run heavy naphtha is heated and is passed over beds of platinum catalyst contained in three separate reactor vessels. A recycle hydrogen stream accompanies the feed through these reactors and the reaction temperature is controlled by heaters between each reactor. The reaction of principle importance is the dehydrogenation of the naphthenes in the feed to their respective aromatic homologue. In so doing the excess hydrogen molecule is released as a hydrogen by product. Other reactions also occur which enhance the octane rating of the debutanized reformate liquid product. Among these are some isomerization and to some extent some hydrocracking. Details of this process are given in Chapter 5.

The properties of the debutanized reformate is given in the following table. The feed to the catalytic reformer in this case included the straight run naphtha (SRN) from the crude unit and the naphtha streams from gas oil hydro-treaters and a thermal cracker. This feedstream was also hydro-treated before entering the catalytic reformer (Table 2.2).

Table 2.2. Properties of a debutanized reformate

Feed	
Crude source	Sassan
Heavy naphtha blend	
Cut range, °F	212–375
Gravity, °API	56.0
Paraffins, %vol	59.0
Naphthenes, %vol	29.6
Aromatics, %vol	11.4
Octane number (Res) clear	34.5
Reformate	
Gravity, °API	47.0
Dist ASTM D86	
IBP, °F	124
10 %vol recovered, °F	174
30 %vol recovered, °F	214
50 %vol recovered, °F	255
70 %vol recovered, °F	273
90 %vol recovered, °F	314
FBP, °F	416
Octane number (Res) clear	91.0

Table 2.3. Light and heavy cracked naphtha stream properties (Feed is Sasson HVGO with gravity of 26.1° API)

	Light naphtha	Heavy naphtha
Gravity, °API	56.1	43.8
ASTM distillation D86		
IBP, °F	108	319
10 %vol recovered, °F	158	336
30 %vol recovered, °F	170	348
50 %vol recovered, °F	191	362
70 %vol recovered, °F	223	378
90 %vol recovered, °F	283	388
FBP, °F	330	448
Octane no. (Res) clear	94	89
Cut range, °F	C_5–300	300–420

Naphtha from a fluid catalytic cracking unit

The Naphtha from the FCCU is always a prime precursor for the gasoline blends. There are usually two naphtha streams which are the debutanized overhead distillate from the Cracker's main fractionator. This distillate is fractionated to produce a light naphtha and the heavy naphtha. As can be seen from Chapter 7 the Fluid Catalytic Cracker is a very flexible unit with respect to the feedstock that it can process. As a matter of fact the only real constraints on the process is a high metal and Condradson carbon content of the feed. And even this is removed with the improved catalysts now available and some modification to the conventional process (see Chapter 11 on deep oil cracking). The effect of the feedstock quality however is confined mostly to the severity of cracking and the resulting yields of the cracked products. The quality of the streams from the main fractionator can be and usually are constant. Table 2.3 shows the properties of the two naphtha streams from the catalytic cracker which in this case is operated on Sasson Heavy Vacuum Gas Oil at a 70% severity conversion.

Alkylation plant

The other most common precursor for the gasoline products is the liquid alkalate from the alkalation process. This process is described in some detail in Chapter 9, and its part in the gasoline blending pool is shown in Part 2 of this chapter (Table 2.4).

There are other processes that produce precursors for gasoline and diesel motive fuels, among these are the products from the Olefin Condenser process and other polymer processes. These are not so common however, but details of the olefin condensing process are given in Chapter 9.

Table 2.4. Alkylate

	Alkylate
Gravity, °API	49.5
ASTM distillation D66	
IBP, °F	211
10 %vol recovered, °F	237
30 %vol recovered, °F	266
50 %vol recovered, °F	289
70 %vol recovered, °F	296
90 %vol recovered, °F	304
FBP, °F	394
Octane no. (Res) clear	97

The middle distillate products
Finished products derived from the middle distillate boiling range include:

- Kerosene for illumination
- Kerosene as a precursor for Aviation Turbine Gasoline (ATG)
- Kerosene for Tractor Vaporizing Oil (TVO)
- Kerosene as a Precursor for Automotive Diesel
- Kerosene for Asphalt Cut Back
- Light Gas Oil for Diesel Fuel
- Heavy Gas Oil for Domestic Heating Oil
- All Middle Distillates for Fuel Oil Blending

The straight run kerosene products. Most straight run kerosenes are desulfurized and routed directly to the finished product blending pools. The exception is in the case of the aviation turbine gasoline (ATG) which will require the kerosene precursor to be treated for aromatics reduction or removal to meet the strict smoke point specifications associated with this finished product. In present day refineries this is accomplished by hydro-treating the kerosene using a nickel catalyst. In older refineries the aromatics contained in the cut were removed as an extract in a process using SO_2 as a solvent. In some cases also the specification for the tractor vaporizing oil (TVO) requires the kerosene, which is the major component of this finished product, to be de-aromatized also. Some details of the de-aromatization hydro-treating process are given in Chapter 8. The straight run kerosene product destined for fuel oil and asphalt blending are not usually desulfurized before routing to the respective blending pools.

Some general specifications for the various finished products containing predominately kerosene are given in Table 2.5.

Table 2.5. General specifications of kerosene finished products

Parameters	Reg Kero	ATG	TVO	
Flash point, °F	100	<66	100	D-56
Aromatics, %vol	–	20	–	D-1319
Temperature @ 20% Max		293°	–	D-86
Temperature @ 50% Max		374°	–	D-86
Temperature @ 90% Max		473°	540	D-86
Final boiling point	572°F	572°	–	D-86
Sulfur Max, %wt	0.04	0.04	0.3	D-1266
Smoke point, Min	–	25 mm	25 mm	D-1322
Freeze point, °C	–	−47	–	D-2386

Straight run light gas oil. This cut is always desulfurized and routed to the automotive diesel blending pool. It is the major component of this finished product. Its four major specifications are:

- Cetane number
- Pour point
- Sulfur content
- Flash point

Other important requirement is its volatility as defined by the ASTM distillation analysis. Except for the sulfur content these specification are met by the set points established in the atmospheric crude unit. Some kerosene may be blended to lighten the front end of the finished product if required, and some heavy gas oil or Light Vacuum Gas Oil (LVGO) may be added to maximize the finished product yield but all limited to the product's ASTM distillation and the other specifications.

Note: Where the refinery configuration contains a thermal cracker, the gas oil cut from this unit may, depending on the unit's fractionator cut points, be added to the light straight run gas oil as feed to the hydro-treater and its subsequent routing to the diesel blending.

Straight run heavy gas oil. This cut could be a heavy side stream from the atmospheric crude distillation tower or if there is a vacuum unit in the configuration it could be one or more of the gas oil distillate cuts from this tower. Most refineries that maximize motive fuels would almost certainly have a cracking unit of some kind in their configuration. More often than not this would be a fluid catalytic process taking as feed the heavy vacuum gas oil stream. The heavy gas oil from the atmospheric distillation unit would be hydro-treated before being routed to the heating oil blending pool or partly to heating oil and partly to fuel oil pool. If there is a catalytic cracker in the refinery units the cycle oil from this unit would be blended with the straight run heavy gas oil to be desulfurized and routed as described. This cycle oil as a cracked product will

Table 2.6. General specifications for gas oil finished products

Parameters	Auto diesel	Heating oil	ASTM tests
Color	2.0 Max	1–1/2	D-155
Specific gravity, °API	37.0–33.0	41.1–36.0	D-1298
Viscosity sus @ 100°F	32.0–43.9	37.5 Max	D-88
Flash point, °F	150 Min	150 Min	D-93
Pour point winter, °F	5.0 Max	5.0 Max	D-97
Pour point summer, °F	15.0 Max	15.0 Max	
Sulfur content, %wt	0.35 Max	0.5 Max	D-129
Diesel index	54 Min	57 Min	IP-21
Cetane number	45 Min	50 Min	D-613
Distillation			D-158
Recovered @ 446°F %vol	–	10 Min	
Recovered @ 464°F %vol	50 Max	50 Max	
Recovered @ 482°F %vol	–	40 Min	
Recovered @ 572°F %vol	–	70 Min	
Recovered @ 619°F %vol	–	80 Min	
Recovered @ 657°F %vol	50 Min	–	
Recovered @ 675°F %vol	90 Min	90 Max	
FBP, °F	725 Max	725 Max	

be rich in olefins and these would be saturated by the hydro-desulfurization process to meet heating oil specifications. Some traditional product specifications for the various gas oil finished products are given in Table 2.6. With the advent of low-sulfur diesel, these specifications are being modified considerably, with sulfur levels as low as 50 wt-ppm (0.005 wt%) and cetane numbers in excess of 50.

Products from the residues

The residues from the atmospheric and vacuum distillation of crude oil open a spectrum of saleable products which constitute a major source of revenue for many refineries. These are listed in order of their importance as follows:

- The fuel oils
- Petroleum coke
- The lube oil products
- The asphalt products

The fuel oil products. There are two common fuel oil products in use today. The first and perhaps the second most common product is the marine diesel. As the name suggests it is the fuel used by heavy marine diesel engines. It is the least viscous of the three grades and is usually a blend of a middle distillate (kerosene and/or gas oil) with atmospheric residue. It has a general specification as shown in Table 2.7.

Table 2.7. General specifications for fuel oil products

Parameters	Marine diesel	No 6 fuel oil	ASTM tests
Specific gravity, °API	39–40	11.4 Max	D1298
Viscosity redwood 1 @ 100°F, sec	30–40	2,400 Max	D445
Pour point summer, °F	25 Max	65 Max	D97
Winter, °F	10 Max	–	
Flash point, °F	150 Min	160 Min	D93
Calorific value (Gross) Btu/lb	19,200 Min	18,300 Min	D240
Sulfur content, %wt	1.0 Max	2.0 Max	D129
Diesel index	45 Min	–	IP21
Ash, %wt	0.01 Max	0.1 Max	D482
Sediment, %wt	0.01 Max	0.1 Max	D473
Water, %wt	0.05 Max	1.0 Max	D95

The most common fuel oil used in industry is the No 6 fuel oil which is either a suitably cut atmospheric residue or a vacuum residue blended with a distillate cut back. It may mean also that a cutback distillate has been desulfurized to meet the product sulfur specification of the fuel oil. A general specification for No 6 fuel is also shown in Table 2.7. Again, depending on use, sulfur specifications are being lowered in many instances.

Several other fuel oil grades are often produced by blending one or other of the distillation residues with suitable distillate streams. These however are customized for clients specific needs. For example a No 5 fuel oil grade was once produced in some quantity for the steel industry. This is a product less viscous and of a lower gravity than No 6 fuel oil. It is now seldom produced.

Petroleum coke. This is not found in crude oil but is formed by the thermal cracking of the residues from both atmospheric and vacuum distillation of the crude. The coke is really the thermal conversion of these heavy products which contain the resins and asphaltenes contained in the crude. The major purpose of the 'Coking' processes is to upgrade the residues by producing lighter marketable products in the gasoline and middle distillate range. The coke becomes the by-product of these thermal cracking mechanisms (see Chapter 12 for more details on these processes). The coke as produced in the refinery process is called *Green Coke* and this may be subclassified as *Sponge coke* from a delayed coking process and *Needle Coke* from a Fluid coking process. Approximately one half of the green coke (both Needle and Sponge) is further calcined to make the calcined coke product.

Uncalcined coke has a heating value of around 14,000 Btu/lb and is primarily used as a fuel. High sulfur uncalcined sponge coke is particularly popular in the cement industry since the sulfur reacts to form sulfates. Low sulfur low metals sponge coke may be calcined to remove volatiles and is used mostly as anodes in the aluminum

Table 2.8. Typical sponge coke specification

Parameter	Green coke	Calcined coke
Fixed carbon, %wt	86–92	99.5
Moisture, %wt	6–14	0.1
Volatile matter, %wt	8–14	0.5
Sulfur, %wt	<2.5	<2.5
Ash, %wt	0.25	0.4
Silicon, %wt	0.02	0.02
Nickel, %wt	0.02	0.03
Vanadium, %wt	0.02	0.03
Iron, %wt	0.01	0.02

industry while needle coke is calcined for anodes in the steel industry. Table 2.8 defines the general specification for green coke and calcined coke.

Lube oil production. The second most important products from the residue of crude oil is the lube oil products. Details of the lube oil production processes are given in Chapter 12. A brief description of this aspect of refinery products with a short history of its growth in the industrial countries. The development of the industrial revolution gave a big incentive for the production of high quality lubricants from the processing of crude oil. A further boost to these processes followed the rapid development of the automotive industry, particularly in the Western world. More and more sophisticated processes were developed for the purpose of this growing demand and an increasing quality requirement. This has culminated in the use of the hydro-processes used extensively in today's modern refineries.

Through the years the lubricant market has developed its own lexicon of terms. Some of these are:

A Lube Oil Base Stock—This refers to a base lube oil product from a lube oil process which meets all the requirement for blending to make a spectrum of lube oil finished product.

A Lube Oil Slate—is a set of lube oil base stocks that makes up a particular finished product. Usually these are between 3 and 5 such stocks.

Neutral Lubes—are lube oil streams obtained as side cuts from the vacuum distillation of the crude oil atmospheric residue.

Bright Stock Lubes—These are processed as de-asphalted oil from vacuum residues. (*Note*: de-asphalted oils are also used as feedstock to fluid catalytic crackers and more commonly to heavy oil hydro-crackes).

Paraffinic Lubes—Are all grades, both neutral and bright stock with a finished viscosity index of more than 75.

Naphthenic Lubes—Are all grades with finished viscosity index of less than 75.

Table 2.9. SAE viscosity specification for single-grade motor oil grades

Grade	Max viscosity (SUS) @ 0°F	Max viscosity (SUS) @ 210°F	Min viscosity (SUS) @ 210°F
5 W	6,000	–	–
10 W	12,000	–	–
20 W	48,000	–	–
20	–	58	45
30	–	70	58
40	–	86	70
50	–	110	85

The important properties of lube oils are:

Kinetic Viscosity (Centistokes)

- Kinetic viscosity (Centistokes)
- Color
- Pour point
- Flash point
- Volatility
- Oxidation stability
- Thermal stability

Although Kinematic Viscosity is measured in centistokes it is usually specified in Saybolt Seconds (SSU). Specifications for lube oils are established by the Society of Automotive Engineers who also perform research on a wide range of automotive topics. One very well known motor oil specification is the SAE viscosity. This is given in Table 2.9. Multi-grade oils (10–20, 10–30, 10–50, etc.) prevail in today's markets.

There are a great many specifications covering all grades and uses for the lube oil products. Many of these, particularly for the motor lube oils, are customized within certain basic maximum, and minimum constraints (such as viscosity), for the local climate, motor type, and legislative requirements.

The asphalt products. Asphalt is the heaviest boiling point product that can be produced in the processing of crude oil. Contrary to common beliefs asphalts can only be made from crude oils that contain asphalt. That is they contain the right amount of the carbon and the resin that allows the formation of asphalts. A full description of the types and the production of asphalt is given in Chapter 12. Briefly most grades of asphalt are produced from the deep cut vacuum residue of suitable crudes. The asphalt extract from the lube oil de-asphalting process may also be included with the residue for the production of certain asphalt grades. The finished product is often treated by air blowing to achieve the required hardness and ductility of the product. Asphalt becomes an important product because of its wide use in road and other building industries.

2.3 A discussion on the motive fuels of gasoline and diesel

Of all the petroleum products that are marketable, motive fuels are the ones that are most common to the general public. Because of this they are continually in the public focus for change with respect to their performance, their availability and cost, and their effect on the environment. This section of Chapter 2 continues with a closer look at the make up, manufacture and the environmental impact of these two motive fuels. It begins with the gasolines.

The parameters of gasoline

Parameters affecting engine performance
The following parameters are the major criteria in the production of finished gasoline with respect to its performance: These are:

- Octane number
- Thermal efficiency
- Volatility of the gasoline
- Engine deposits

Octane number. Among the most important parameters in the manufacture of gasoline is its resistance to 'Knocking'. This resistance is expressed as an 'Octane Number'. A definition of octane number and its measurement has already been given in Chapter 1.

Knocking limits the power that can be developed by the engine/fuel combination. Detonation is the spontaneous explosion of the residual fuel (after almost complete combustion) and the air in the combustion chamber as the normal mechanism of combustion nears its end. It's a very fast oxidation reaction which sets up its own flame front. The knocking occurs when this front and that of the normal combustion collide creating a pressure wave.

Another, and perhaps, a more destructive form of engine knock is 'Pre-ignition'. This is the spontaneous ignition of the fuel/air mixture before the ignition spark. This in turn is caused when the unburned gases are compressed in the cylinder and the resulting temperature reaches the auto ignition point before the ignition spark occurs. This pre-ignition usually causes major engine damage in just a few seconds or minutes.

Increasing humidity, cooler ambient air temperatures, and altitude reduce requirements for antiknock.

Thermal efficiency. As may be expected fuel economy is a factor in determining the thermal efficiency of the auto engine. The engines efficiency increases with the increase in compression ratio. However with the increase in compression ratio there is an increasing need for fuels that do not knock. Since around 1980s engine manufacturers have been incorporating knock sensors to the engine electronic management system to continuously adjust the ignition timing. This development allows the timing of the spark to occur in advance of the piston reaching the top of its travel. The maximum engine efficiency is produced with this timing control measured and adjusted by the computer control of fuel to air ratio. Further development and testing of individual engine design sets an optimum criteria for fuel octane number of the fuel and air ratio for lean mixture combustion.

Volatility of the gasoline. Another important property of gasoline is its volatility. The gasoline must be volatile enough to provide the engine capable of starting at the lowest temperature expected in its service. At too low a volatility the engine would have difficulty starting and would be prone to stalling in service. On the other hand too high a volatility would cause excessive vapor which in turn would cause vapor lock in pipes and pumps, etc.

Engine deposits. Engine deposits affect fuel efficiency and emissions. Deposits are of particular concern in the carburetor, fuel injectors, inlet valves, and combustion chamber. These deposit are essentially fine carbon granules which are formed by high inlet valve temperatures, airflow inconsistencies, and minor oil contamination.

Parameters affecting pollution by emission

Air pollution is an important factor in motive fuel design and manufacture. There are two main types of air pollutants that affect the manufacture of motive fuels. These are:

- Hazardous air pollutants (HAPs)
- Criteria air pollutants

Hazardous air pollutants

These are air pollutants that have been shown statistically to cause major health disorders such as cancer, neurological damage, respiratory irritation, and reproductive disorders. In general these pollutants are believed to have no threshold, and are harmful even in small doses or concentrations. The Environmental Protection Agency (EPA) has been regulating these pollutants since the 1977 Clean Air Act. There are 189 hazardous air pollutants among which Benzene, formaldehyde, and 1,3-butadiene which are formed by combustion of motive fuel such as gasoline.

Criteria air pollutants

These are pollutants that can injure health, harm the environment, and cause property damage. These pollutants do have a threshold level however, below which they are relatively harmless. The EPA labels these pollutants as *Criteria Air Pollutants* because the basis for setting standards are developed from health based criteria. There are six criteria of are pollutants that affect the design and manufacture of gasoline. These are:

- Lead
- Carbon monoxide (CO)
- Sulfur dioxide (SO_2)
- Nitrogen dioxide (NO_2)
- Particulate matter
- Ozone (O_3).

Lead. This emission from gasoline engines have been known to be toxic for several decades. It adversely affects kidneys, liver, and other organs, and leads to neurological impairment, learning deficits, and behavioral disorders. The reduction of lead programs conducted in the 1970s and 1980s have made leaded gasoline (TEL added) unavailable for on-road vehicles in the 1990s.

Carbon monoxide (CO). This is a colorless odorless and poisonous gas produced from the incomplete combustion of carbon in fuels. Elevated exposures are serious for those suffering from cardiovascular disease and leads to visual impairment, loss of manual dexterity, and the ability to perform complex tasks. At high enough levels it depresses cardiac activity and respiration, causes convulsions and ultimately death. Over half the carbon monoxide emissions are from motor vehicles.

Sulfur dioxide. This is formed during burning of sulfur containing fuels. Motor vehicles are a minor source of SO_2 due to the de-sulfurizing processes used in the production of motive fuel components. Short-term exposure to high concentration leads to reduced lung function, while longer exposures are associated with respiratory problems. Sulfur dioxide mixed with nitrogen oxide is a major contributor to the formation of *Acid Rain*. This in turn causes acidification of lakes, rivers, and waterways. It damages crops and trees, and accelerates the corrosion of building materials and coatings.

Nitrogen dioxide. Also known as nitrogen peroxide is a brown gas formed by high temperature combustion of fuels. It is one of a family of nitrogen oxides which are often treated as one species under the term NO_x. Nitrogen dioxide is a severe irritant to the lungs and can cause respiratory infections, pneumonia and bronchitis. Nitrogen dioxide coupled with volatile organic compound (VOC) and ozone are the main components of smog.

Particulate matter. Dust, dirt, smoke, and liquid droplets form the particulate matter in motor emission and pollutants. Some of the liquid droplets are formed in the atmosphere by the condensation of sulfates, nitrates, and hydrocarbons. Fine particles with diameter less than 2.5 μm are emitted by motor vehicles. Gasoline driven vehicles however do not emit significant levels of particulate matter compared with diesel driven engines. Particulate matter aggravates existing respiratory and cardiovascular disease, damages lung tissue and may cause cancer. Fine particles of 2.5 μm or less are considered more serious since they can penetrate easily into the respiratory tract and are retained longer. Particulate matter also contributes to haze.

Ozone. This photochemical oxidant is found in the atmosphere from ground level to 6 miles above the ground and again in the Stratosphere 6–30 miles above the earth. At ground level Ozone together with the Nitrogen oxides, the volatile organic compounds and sunlight are the major constituents of 'Smog'. Ozone is a strong oxidant and damages lung tissue, and reduces lung function.

More details of gas emission are given in Chapter 5.

Meeting the gasoline parameters

Prior to 1990 the performance and emission characteristics were mainly the responsibility of the engine manufacturers with some notable changes imposed on the gasoline manufactures. Notable among the restrictions imposed on the refineries in the gasoline manufacture was the restriction on the use of TEL (tetra ethyl lead) as an octane enhancer. The EPA (environmental protection agency) had, as early as 1970 enforced a program to reduce the addition of lead so that even before 1990 there was available only 'no lead' gasoline for road vehicles. The 'Clean Air Act' of 1990 however put the onus on the gasoline manufacturer for meeting the act's requirements. In the USA and indeed the North American Continent a program to introduce reformulated gasoline was launched. This RFG allows the performance of the gasoline to be retained whilst reducing the harmful emission caused by the traditional octane enhancers such as aromatics (from reformates) and olefines (from cracked naphtha).

Reducing aromatics in gasoline. The most stringent environmental restriction imposed on aromatics in gasoline is on benzene in gasoline. This component in benzene has now been reduced to levels below 1.0 %vol in the product. Refining processing philosophy has undergone some extensive changes to meet this single requirement. The following steps have needed to be taken in part or as a whole to meet this change:

- Reducing reformer severity. This also reduces the quantity of the heavier aromatics.
- Reducing the final boiling point of the gasoline. This essentially reduces the amount of the heavier aromatics in the gasoline.

- Increasing the quantity of isoparaffin. This is accomplished by saturating the benzene ring and isomerization of the naphthenes to isoparaffins.
- Reducing the aromatics in the cracked naphtha stock. Catalytic cracker naphtha is high in olefins in the front end but high in aromatics in the back end. The reduction of aromatics from this source is accomplished by lowering the cut's final boiling point.
- Aromatic extraction using an extraction process similar to that used in producing the petrochemical aromatic complex feed (e.g., the sulfolane process). Only about 20 of the largest refineries in North America has this extraction facility however.

Reducing the olefin in gasoline. Almost all the olefins and sulfur in the gasoline pool come from the Catalytic Cracker naphtha with a relatively small amount from thermal crackers. The following methods again in part or as a whole are used to reduce this olefin content:

- Alkylation and Etherification. The light olefins of C_5 can be fractionated out and following a simple sulfur removal is feed to an alkylation process to produce good high octane C_9 alkylate. Alternatively the light fraction from the cracked naphtha can be processed to the oxygenate TAME (Tertiary Amyl Methyl Ether).
- Isomerization of the C_6 and C_7 fractions. These two olefin components of cracked naphtha may be hydrogenated and isomerized to provide a good octane rating. The two olefin components are only small in quantity in gasoline they are however highly reactive and their removal and conversion is necessary to meet the present restrictions in gasoline manufacture.

Meeting the gasoline sulfur content. Almost all of the sulfur in gasoline comes from the catalytic cracker and other thermal crackers. The current phase 2 of the RFG program is expected to reduce sulfur in gasoline to around 30 ppm by weight; European specifications are being lowered to 10 wt-ppm. To meet this new criteria refiners are looking at the following approaches:

- Route the heavy end about 20 %vol of the catalytic naphtha into the middle distillate pool. This does not reduce the sulfur but it moves a large portion of it and the heavy aromatics into other parts of the product slate. Thus reducing the sulfur and aromatic content of the gasoline.
- Either hydrogenate or treat with a caustic type wash the lighter portion of the cracked naphtha. This is rich in olefins and sulfur. Treating by hydrogenation does reduce the octane value of the naphtha in that it saturates the olefins to lower octane paraffin or naphthene. Treating with sulfur extraction wash (e.g., UOP's Merox process) is the often preferred route.

Table 2.10. A gasoline recipe prior to 1990

Component	%vol	Aromatics %vol	Olefins %vol	ON res clear	ON ratio
Butane	3.75	0	0	92	3
LSR	12.5	10	2	69	8
Reformate	24.39	89	0	91	18
Lt crack naphthas	12.5		35	89	18
Hy crack naphthas	21.19	53		90	19
Alkylate	25.67	0	0	95	24
Total	100.00	31.0	12.0		90

SG @ 60°F 0.7749
RVP 7.0 psig

- The back end of cracked naphtha is high in sulfur, aromatics, and olefins. This may be hydrogenated using a catalyst selective in removing sulfur, and leaving the aromatics essentially as they were.
- Hydro-treating of the catalytic cracker gas oil feed is becoming quite common. This significantly reduces the sulfur content of all the cracked products from the catalytic cracker unit.

Manufacturing gasoline

The requirements of the clean air act of 1990 and additions to it since has changed refining requirements to meet this product's need quite significantly. Prior to this date much of the gasoline finished product recipe consisted of normal light naphtha, a reformate, usually some cracked naphtha, and possibly an alkalate some butane may be included if required to meet volatility. The Clean Air requirement and its subsequent additions forces a reduction of both the reformate and the cracked stock (see Tables 2.10 and 2.11).

Table 2.11. A gasoline recipe post 1996

Component	%vol	Aromatics %vol	Olefins %vol	ON res clear	ON ratio
Butane	1	0	0	92	0.9
LSR	15	0	0	69	10.0
Reformate	30	60	0	80	24.0
Cat naphthas	32	25	35	82	26.0
Alkylate	20	0	0	95	27.0
MTBE	2	0	0	110	2.1
Total	100	26	11.2		90.0

Table 2.12. Oxygenates commonly used in gasoline (subject to phase-out in the U.S.A.)

Name	Formula	RON	RVP psig	Oxygen %wt	Water solubility %*
Methyl tertiary butyl ether (MTBE)	$(CH_3)_3COCH_3$	110–112	8	18	4.3
Ethyl tertiary butyl ether (ETBE)	$(CH_3)_3COC_2H_5$	110–112	4	16	1.2
Tertiary amyl methyl ether (TAME)	$(CH_3)_2(C_2H_5)COCH_3$	103–105	4	16	1.2
Ethanol	C_2H_5OH	112–115	18	35	100

*Wt % soluble in water.

See also Part 3 of this chapter.

Using the oxygenates. Oxygenates were used originally simply as a additive to improve octane number. However, because of their oxygen content they are now added also ton reduce the carbon monoxide and hydrocarbon in the emission gases. There are a number of oxygenates now used in gasoline manufacture some of the more common are given in Table 2.12.

The EPA have established limits for the use of each oxygenates in gasoline blends. For example, MTBE may be blended up to 15 %vol subject to an overall limit of 2.7 %wt oxygen content. The role of MTBE and other oxygenates in the U.S. was discontinued in 2002 after it was discovered that underground storage tanks had not been upgraded to retain reformulated gasoline. As a consequence tanks were leaking gasoline that contained MTBE into the ground and drinking water systems.

Diesel fuel

The policies in place to control, or at least to endeavor to control, emissions from motive fuels has moved from gasoline engines to diesel engines. Similar to the gasoline fuels diesel has also changed in composition to meet these environmental constraints while still maintaining a required performance standard. These changes have imposed a major impact on the refining industry at a time when the demand for diesel is increasing. Again as in the case of the gasoline fuel reformulated diesel fuel becomes part of the modern diesel specification and composition. These are described and discussed in the following sections.

The diesel engine

Before dealing with the diesel fuel it may be advantageous to review briefly the principle of the diesel engine itself, diesel engines have historically been preferred in

many applications because of their simplicity of design, power, durability, and higher fuel efficiency. Diesel engines fit into three categories and these are:

- Low speed engines
- Medium speed engines
- High speed engines

Low speed engines. These operate at less than 300 revs per min (rpm), and are used for applications that require sustained heavy loads at constant engine speed. Examples are the main propulsion engines in marine vessels and engines used in electric power generation.

Medium speed engines. These operate at speeds between 300 and 1,000 rpm, and are used for applications with fairly high loads, and relatively constant speeds. Engines used in auxiliary power plants on marine vessels, and in smaller power generation plants are examples of this type of diesel engine.

High speed engines. Operate at speeds above 1000 rpm, and are designed for frequent and wide variations in load and speed. These are the type of engines used for road transport and Diesel locomotives.

Almost all diesel engines use a standard four stroke design which are, intake, compression, power, and exhaust. During the intakes stroke, air alone enters the cylinder, during the compression stroke the air is compressed by the upward movement of the piston. The final temperature and pressure reached in the cylinder is a function of the compression ratio, engine speed, and engine design. Pressures of 450 psig, and temperatures of 500°F are typical. Shortly before the end of the compression stroke, one or more jets of fuel are injected into the cylinder. The fuel pump introducing these jets is usually cam driven type and operates at pressures between 1,800 and 30,000 psig. The fuel does not ignite immediately. There is a delay period when the fuel droplets vaporize and reach their ignition temperature. The extent of this time delay is affected by the design of the engine, the temperatures of the inlet air and fuel, and the degree to which the injected fuel is atomized on entering the cylinder. This ignition delay must be kept as short as possible to avoid accumulation of fuel in the cylinder before ignition. Such a situation causes diesel knock when this large quantity of fuel detonates.

After ignition, fuel injection continues for a portion of the power stroke. This fuel burns almost instantaneously with the remaining air and combustion products in the cylinder. This more controlled burning period is referred to as diffusion combustion. Fuel injection is stopped part way through the power stroke. Finally the exhaust stroke purges the combustion products from the cylinder, and the cycle begins again.

Design improvements to the basic diesel engine

Fuel injection improvements. Most diesel engines built in the 1980s were equipped with indirect fuel injection system (IDI). Modern engines are equipped with the direct fuel injection (DI) which is designed for the geometry of the combustion chamber, and is significantly more efficient. (Note: This can well be compared with the direct injection concept in gasoline engines which was introduced around this time also).

Four valve cylinder heads. This promotes better movement of both combustion air and exhaust gases. This causes high turbulence of the fuel and air in the combustion chamber resulting in good distribution of the fuel and increased flame speed which reduces the ignition delay time.

Electronically controlled fuel injection timing. This allows for the injection of a small quantity of fuel before the main charge into combustion chamber. The result is a quieter combustion minimizing the amount of premixed combustion and maximizing the more controlled diffusion combustion.

Increasing the amount of air introduced into the cylinder. Using a Turbocharger, which is an air compressor driven by a turbine using exhaust gas energy, the mass of air introduced into the cylinder is increased. This enables the engine to burn more fuel thus increasing its power output. Using this concept the power output for a diesel engine can be increased by as much as 50% for the same engine displacement.

The parameters of diesel fuel

The following are the major parameters in meeting the diesel fuel specification with respect to engine performance or emission or both:

- The cetane number
- Aromatic content
- Density
- Sulfur content
- Distillation
- Viscosity
- Cloud and pour points
- Flash point

Most of these parameters have been defined elsewhere in this book. This section then deals with the effect these parameters have on the performance of the diesel engine or the emission of undesirable components from the engine or both. Where the definition of the parameter is not dealt with in detail elsewhere in the book, then definition is included here.

Cetane number. This is the result of an engine test that compares the ignition delay for a fuel. For this test two reference fuels are chosen. The first is normal cetane (C_{16}) and the second is an isomer of cetane which is heptamethylnonane. The normal cetane is arbitrarily given the cetane number of 100, while the isomer as the second reference fuel is assigned a cetane number of 15. The fuel being tested is run in a standard test engine. The cetane number is derived by comparing the ignition delay of the test diesel with a blend of the two reference fuels. The cetane number is then calculated using the equation:

$$\text{Cetane Number} = \%\ \text{normal cetane} + 0.15 \times \%\ \text{heptamethylnonane}.$$

Higher cetane numbers indicates that the fuel has a shorter ignition delay. The higher the cetane number also results in less CO and unburnt hydrocarbons in the engine emission gases. This has a greater effect in the older diesel engine. Modern engines are equipped with retarded ignition timing and increasing the cetane number has a smaller effect on these more modern engines.

Aromatics. The aromatic content of diesel fuel can be measured for single ring aromatics, multi-ring or poly-aromatic hydrocarbons (PAH). Some studies show that reducing the aromatics results in the reduction of all regulated emissions, but other studies have indicated that the reduction of emissions of unburned hydrocarbons, NO_x, and particulates can only be achieved by reducing multi-ring aromatics.

Density. As density is a measure of the mass per unit volume, diesel fuels of low density require a longer injection time to deliver the same mass of fuel into the cylinder. The longer the injection time the lower is the peak temperatures which, in turn, results in lower No_x formation. At high loads and engine speeds, the longer injection interval causes some incomplete combustion, resulting in a high emission of unburned hydrocarbons and CO. When the load is being increased however the lower density fuel results in less over-fueling, which actually decreases the emission of particulates, hydrocarbons, and CO.

Sulfur. The sulfur in diesel fuel is burned to SO_2, a portion of which is further oxidized to sulfates. This binds with water to form a portion of the particulate matter. Because only a small percent of the total sulfur in fuel is oxidized to sulfates the contribution of sulfates to the total particulates is quite small. However if an oxidation catalyst is used to reduce emission of hydrocarbons, CO, and particulate matter a significant amount of the SO_2 is converted to sulfates and consequently making a significant contribution to the particulates in the emission gases.

Distillation. The distillation range of diesel fuel has a significant influence on engine performance. This is especially so in medium and high speed engines. If the fuel is too volatile the engine loses power and efficiency because of vapor lock in the fuel system or poor droplet penetration into the cylinder. On the other hand if the volatility of the fuel is too low, the engine will lose power and efficiency as a result of poor atomization of the fuel. Both the front end and the back end of the distillation are important. If the 10 %vol point is too high, the engine will have difficulty starting. A low 50% point reduces particulate emissions and odor. Because heavier molecules are more difficult to burn, both soot and the soluble organic fraction (SOF) of the particulate emissions are increased if the 90% point is too high, the emission of unburned hydrocarbons will also increase.

Viscosity. Fuel viscosity has an important effect on the fuel pump and injector system. The shape of the fuel spray is affected by viscosity. If this is too high, the fuel will not be properly atomized into the cylinder, which will result in poor combustion, loss of power, and efficiency, with an increase in CO and hydrocarbon emission. Another effect of poor atomizing of the fuel will be to allow the fuel to impinge on the cylinder walls, and remove the lubricating film. This results in excessive wear and increase of hydrocarbon emissions.

If the viscosity of the fuel is too low, the injection spray is too soft and will not penetrate far enough into the cylinder. A loss of power and efficiency will occur due to this. Where the fuel system is also lubricated by the fuel, as is the case in some engine designs, increased wear to the system will result through low fuel viscosity.

Cloud and pour point. Cold weather performance of diesel fuel is a key consideration for users. Actual specifications for cold flow properties are based on expected temperature extremes and different test methods are used in different parts of the world. In the United States cloud point is used as an indicator of the cold flow properties of the fuel. Cloud point is the temperature at which wax begins to precipitate out of the fuel. The longer paraffin molecules in the fuel precipitate as a wax when the temperature falls below the cloud point. This wax clogs unheated fuel lines and filters. The more paraffinic a fuel the higher will be its cloud point. In some parts of the world, pour point is used as an indication of the lowest temperature at which a fuel can be pumped. Pour points are generally 4–5°C lower than the cloud point.

Flash point. The flash point of a fuel is the temperature that the fuel must be heated to produce an ignitable mixture of vapor and air above the surface of the liquid. This property is only important for safe handling and storage of the fuel. It has no effect on the performance of the fuel or its emission properties. If the flash point is too low, fire or explosion could occur when it is handled.

Meeting the diesel fuel parameters

There are various solutions for meeting the parameters of diesel fuel in the design of the fuel. These solutions range from fractionation, adding improvers, and more complex hydro-processing. Using modern developments in hydro-processing it is possible to convert low-grade blend stocks, such as FCCU or thermal cracker product streams to good diesel precursors. Among the more important parameters to be met both from the performance and the emission aspects of the fuel are discussed below.

Increasing cetane value. The simplest way to improve cetane number is to use an appropriate ignition improvement additive. These are mainly alkyl nitrates. The effectiveness of cetane improvement additives tend to be linear with addition rate, however the improvement can vary with different diesel blend stock. Paraffins which already have high cetane value respond best to the additives. Aromatics on the other hand, which have a low cetane number have a poorer response. About 500 wppm of standard alkyl nitrate will increase the cetane value by 2–5 numbers. Another way to increase cetane number is to modify the hydrocarbon type in the diesel blend. Figure 2.2 shows the relationship between molecular types and cetane number.

Straight chain paraffin molecules have the highest cetane numbers while multiple ring aromatics have the lowest cetane number improvement also occurs during the hydro-treating of the gas oil fractions used as blend components in the diesel product. During the hydro-treating to remove the organic sulfur and nitrogen species found in these untreated gas oil fractions, the molecules containing these items are opened and saturated to form paraffins and naphthenes. At the same time some of the

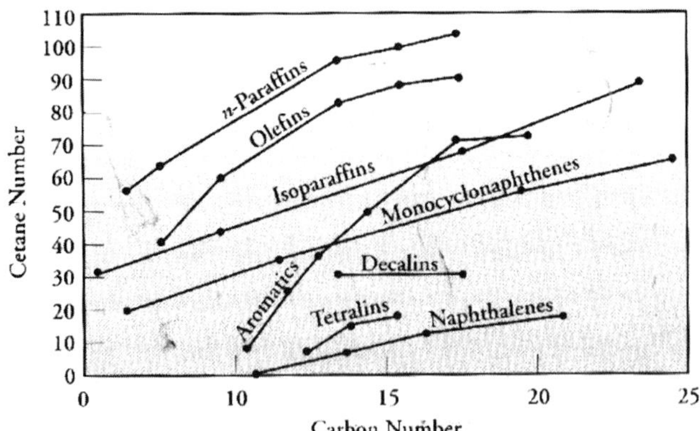

Figure 2.2. Cetane number of hydrocarbon types.

aromatic molecules contained in the gas oils are also opened and saturated increasing the paraffin/naphthene content of the hydro-treated product. As a result of this ring opening and saturation a cetane number improvement of between 3 and 5 numbers can be expected. Hydro-cracking of heavy gas oil fractions, and hydro-treating of thermal cracked gas oils and FCC gas oils significantly increase product cetane numbers.

Meeting sulfur content. Sulfur content of 0.05 %wt is now quite a common requirement for diesel fuel. In most cases this level of sulfur can only be met by extensive hydro-treating or hydro-cracking to remove the sulfur from the middle distillates that will be used to blend the diesel product. Some refineries select their crude slate with low sulfur crudes to reduce the hydro-treating demand. The impact on refiners to meet this parameter extends beyond the severity of hydro-treating that may exist. It impacts on the availability of hydrogen in the front end to the capacity for handling the sour gases (H_2S in particular) in the back end of the process. It is likely that the legislative sulfur content in diesel may be reduced still further in the near future. This makes the selection between crude slate demand for low sulfur crude and a modified refinery configuration and operation a critical one.

Reducing aromatic content. With the increasing demand for diesel fuel, particularly in North America and Europe, refiners will need to use streams normally routed to heating oils of fuel as components for the diesel product blend. These streams are of lower quality and in particular contain higher quantity of aromatics that will be acceptable as diesel blend stocks without severe treating to improve their blending characteristics. While high severity hydro-treating to remove sulfur and nitrogen helps in reducing aromatics in acceptable diesel blend stocks, it will not be sufficient to improve the quality of the heating oil and fuel oil blend stocks to meet the diesel pool requirement. Hydro treating using nickel molybdenum catalyst (de-aromatization process) is being considered for the purpose of upgrading these poorer blend stocks. This de-aromatization process consists of two stages. The first is processing over a conventional Co–Mo hydro-treating catalyst to remove sulfur and then to process over the Ni–Mo catalyst. The nickel catalyst is poisoned by sulfur, thus the two stages. The economics of using the de-aromatization process must be carefully evaluated because the metal content of the nickel catalyst represents a significant investment. The process however operates at a higher space velocity and lower pressure than the conventional Co–Mo hydro-treating process.

Improvement of cold flow properties. Problems with meeting the cold flow properties of diesel fuel are associated with the presence of straight chain paraffins. Although these have higher cetane numbers they pose cold flow limitations because they tend to precipitate as a wax at low temperatures. The simplest and probably the best solution to this cold flow property of these paraffins is the use of additives. These additives are however costly ranging from below $50 per thousand bbls for pour point reduction to as high as $250 per thousand bbls for cloud point reduction (these are year 2000 dollars). Another alternative is to reduce the diesel FBP and add kerosene. This action removes the heavier paraffinic molecules, which tend to precipitate at higher temperature. Reducing the cut point from 700°F to 640°F reduces the yield of diesel by 6 %vol on crude. Solvent dewaxing is another option. In this process the wax forming components are selectively removed from the diesel product. This is only considered however for stocks from high paraffin crudes. Selective hydrocracking of the straight chain paraffins to smaller paraffin molecules is an option. This however is accompanied by a reduction in yield and cetane number. Finally isomerization of the straight chain paraffins to branched paraffins minimizes cetane and yield loss is a viable option. However these isomerization catalysts contain noble metals and are therefore quite expensive and require the two stage process, desulfurization followed by the isomerization stage.

2.4 A refinery process configuration development

There are many steps necessary to complete a refining company's marketing strategy and production objectives. These steps may culminate in building new processing facilities or revamping existing ones. The course of action that will ultimately be adopted will, almost invariably, be decided on an economic basis reflecting the company's future profitability picture. Among the most important of these steps however will be the development of several refining scenarios and processing configurations.

A process configuration and its accompanying Block Flow Diagram is a blue print and a basis for economic study and decision making. The final and accepted configuration will also be the foundation for the design of new facilities or the revamping of existing ones. The block flow diagram shows the calculated results of process plant design capacity, the quantity of process streams to and from each process plant, the sequence of the plants to one another in the refining scheme, and finally the blending recipes (streams and their quantities that make up the finished product blends). Such a development requires the in-depth knowledge of the refining industry and relies, in no small measure, on the expertise of those companies that license processes and their technology.

The configuration example described here would be one of many that would be examined meeting the production and economic objectives of a particular refinery operation. In this example a refining company wishes to examine a new 'Grass Roots' configuration that will produce the following products from a medium Middle East crude, whose assay TBP curve is given as Figure 2.1.

The product specifications are given in the respective text, and the production limits are as follows:

Crude feed	50,000 BPCD
LPG	3,289 BPCD (Max)
Gasoline	22,170 BPCD (Max)
	20,390 BPCD (Min)
Kerosene	No target
Auto diesel	8,830 BPCD (Max)
	3,000 BPCD (Min)
Gas oil	8,553 BPCD (Max)
	2,630 BPCD (Min)
Marine diesel	3,000 BPCD (Max)
	2,300 BPCD (Min)
Heavy fuel	8,500 BPCD (Max)
	6,500 BPCD (Min)
Bunker fuel	No restrictions

A solution

For this configuration the following processes will be examined:

- Atmos crude dist unit (CDU)
- Vacuum dist unit (VDU)
- Light ends dist units
- Naphtha hydrotreater (Nap Hds)
- Catalytic reformer
- Lt gas oil hydrotreater (LGO Hds)
- Hy gas oil hydrotreater (HGO Hds)
- Fluid catalytic cracker (FCCU)
- Isom/alkylation plant
- Thermal cracker (TC)
- Gas treating and sulfur recovery processes

The stream quantities to each of these units will be developed as follows:

The atmospheric crude distillation unit. The crude is split to satisfy the quantities and distillation characteristics of the finished products required. For this example these are as follows:

Gas to C_5	1.71 %vol on crude	
C_5 to 210°F	10.44 %vol on crude	Light naphtha (LSR)
210 to 380°F	14.20 %vol on crude	Heavy naphtha (HSR)
380 to 520°F	9.09 %vol on crude	Kerosene
520 to 650°F	13.83 %vol on crude	Diesel
650°F+	50.73 %vol on crude	Residue (feed to crude vacuum unit)

Table 2.13. ASTM and TBP of the crude distillate cuts

	C_5 to 210°F		210 to 380°F		380 to 520°F		520 to 650°F	
	ASTM°F	TBP°F	ASTM°F	TBP°F	ASTM°F	TBP°F	ASTM°F	TBP°F
IBP	90	30	198	132	363	310	451	392
10 %vol	115	78	235	196	387	357	507	480
30	135	116	260	242	407	395	533	527
50	148	140	280	276	424	425	551	558
70	162	163	301	308	440	451	567	583
90	182	191	332	349	462	481	593	618
FBP	225	238	390	412	493	515	622	650

Then from the cut points corresponding to the stream's volume % on crude the following ASTM and TBP of the distillate streams leaving the unit are calculated. The ASTM (Starting at the C_5 + level) are developed using the 50% point of each cut and the 70% point converted to ASTM (using the Edmister correlation) and a straight line drawn through them to give the ASTM distillation. These are then converted to TBP (again using the Edmister correlation) and are shown in Table 2.13.

These curves are given in Figure 2.3.

From these curves the table below gives a component break down of the distillate streams (Table 2.14).

The properties of these distillate cuts are determined using the component mid vol% on crude multiplied by its volume on the respective cut or by use of indices as previously described.

The following table gives the component properties from the assay data (Table 2.15).

Using the methods described earlier in this chapter the products from the atmospheric crude unit including their salient properties are as follows (Table 2.16).

The vacuum crude unit, VDU. The feed to the vacuum distillation unit will be the atmospheric residue from the CDU. This stream will be treated in the same manner as the distillate streams shown above for the CDU. Thus the distillate streams and the vacuum residue stream including their salient properties are shown in Table 2.17.

The product Streams from this unit are:

LVGO (Light Vacuum Gas Oil)—650 to 750°F (49.27–60.12 %vol on crude)
HVGO (Heavy Vacuum Gas oil)—750 to 930°F (60.12–78.67 %vol on crude)
Vacuum Residue—+930°F (21.33 %vol on crude)

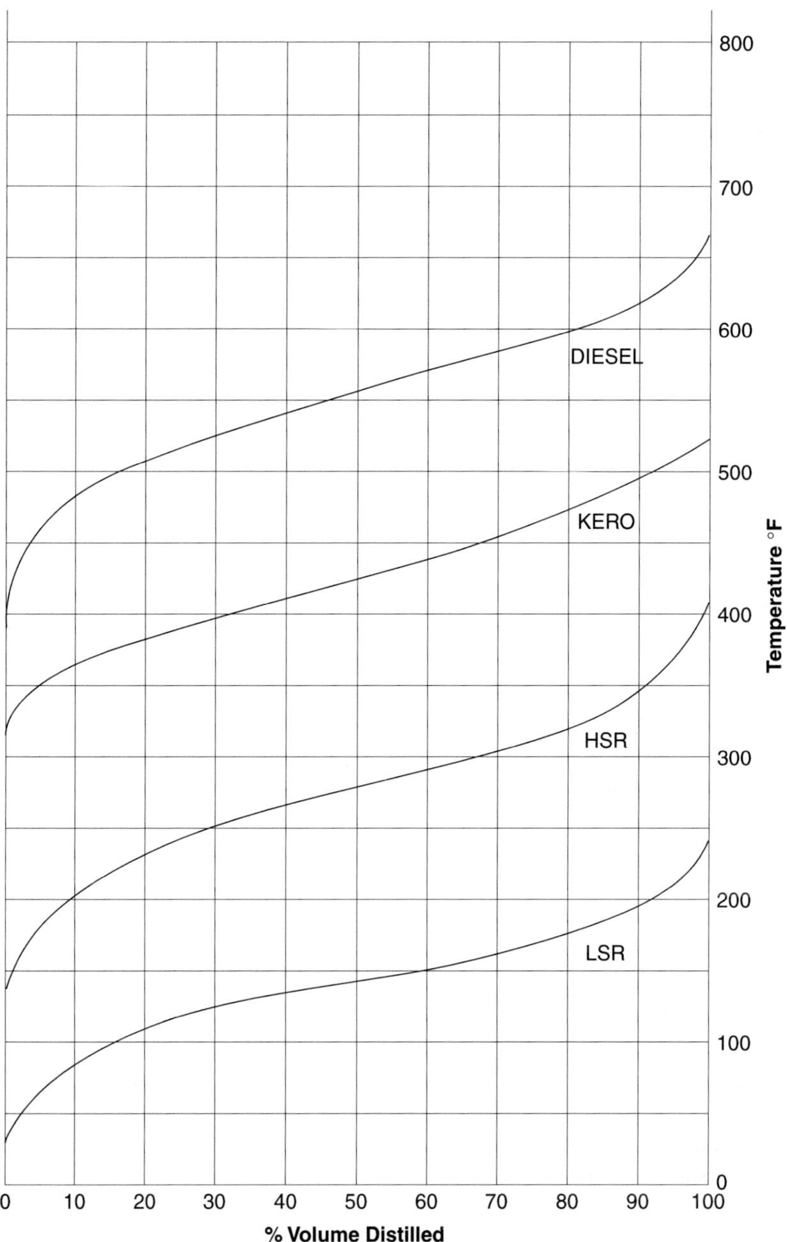

Figure 2.3. Distillate products TBP curves.

Table 2.14. Component split of the distillate cuts

Component	Range °F	LSR %vol	HSR %vol	Kero %vol	Diesel %vol
1	C_5	21.6			
2	C_5–125	15.4			
3	150	23.0	1.0		
4	175	20.0	4.0		
5	200	13.0	6.0		
6	230	7.0	14.0		
7	250		9.0		
8	275		16.0		
9	300		15.0		
10	325		15.0	2.0	
11	350		11.0	6.0	
12	412		9.0	29.0	1.0
13	450			30.0	4.0
14	480			23.0	5.0
15	515			10.0	14.0
16	550				19.0
17	575				20.0
18	600				19.0
19	650				18.0
Total		100.0	100.0	100.0	100.0

Table 2.15. Component properties

Component	Mid vol on crude	°API	SG @ 60°F	Sulfur %wt	Viscosity @ 100°F Cst	Viscosity @ 210°F Cst	Pour point °F
1	1.31	12	0.5811	0.04			
2	3.56	105	0.5983	0.05			
3	5.96	87.5	0.6461	0.05			
4	8.92	71.0	0.6988	0.06	0.27		
5	11.49	65.0	0.7201	0.06	0.33		
6	14.28	60.5	0.7370	0.07	0.45		
7	16.34	57.3	0.7495	0.07	0.48		
8	18.12	55.5	0.7567	0.08	0.52		
9	20.32	53.0	0.7669	0.09	0.70		
10	22.54	51.5	0.7732	0.10	0.78		
11	24.75	49.0	0.7839	0.12	0.90	0.60	−88.0
12	27.19	46.5	0.7949	0.14	1.25	0.68	−76.0
13	30.22	44.5	0.8040	0.20	1.82	0.78	−60.0
14	33.25	41.5	0.8178	0.26	1.95	0.90	−48.5
15	36.06	40.0	0.8251	0.40	2.23	1.13	−34.0
16	38.8	38.5	0.8324	0.62	2.85	1.25	−22.0
17	41.56	37.0	0.8398	0.89	4.00	1.52	−0.08
18	44.26	35.5	0.8473	1.16	5.20	1.70	2.0
19	46.85	34.5	0.8524	1.48	6.25	1.85	16.0

Table 2.16. CDU product streams

Product stream	BPCD	SG @ 60°F	mm lbs/CD	Sulfur %wt	Visc @ 100°F Cst	Visc @ 210°F Cst	Pour point °F
Gas + LSR	6,075	0.6512	1.383	0.053			<−100
HSR	7,100	0.7550	1.863	0.091	0.58	0.43	<−100
Kero	4,545	0.8048	1.279	0.21	1.60	0.80	−58
Diesel	6,915	0.8420	2.036	0.869	4.0	1.5	0
Atmos Residue	25,365	0.9485	8.405	3.92	>100	9.8	>100
Total	50,000	0.8565	14.966	2.35	8.00	2.20	−5.0

These streams totally or in part will be routed as follows:

- LVGO via the Gas Oil Hydrotreater to Gas Oil Blending pool
- HVGO via the Gas Oil Hydrotreater to FCCU
- Gas Oil Hydrotreater will be on block flow operation

Examination of the remaining units in this configuration now follows:

The thermal cracker. The following yields and properties have been developed from empirical correlation. These are presented and discussed in Chapter 12. Such correlation are acceptable for use in configuration development, particularly those that will be used in study work. For any definitive design work however it is essential that the data used be obtained from pilot plant runs using samples of the actual residue as feed (Table 2.18).

The following property curves were developed from pilot plant runs on a similar feed stock. These are based on total product %volume (Tables 2.19 and 2.20; Figures 2.4–2.6).

The gas to 380°F Stream will be routed as the overhead distillate from the Thermal Cracker Syn Crude fractionator to the Naphtha hydro treater and the saturated light end unit.

The gas oil stream will be routed to the straight run Diesel hydro treater.

Table 2.17. The VDU product streams

Product stream	BPCD	SG @ 60°F	mm lbs/CD	Sulfur %wt	Visc @ 100°F Cst	Visc @ 210°F Cst	Pour point °F
LVGO	5,425	0.8789	1.667	2.18	13.0	3.0	15
HVGO	9,275	0.9218	2.990	3.10	49.0	6.4	52
Vac Res	10,665	1.007	3.748	5.35	>100.0	28.5	>100
Total	25,365	0.9485	8.405	3.92	>100.0	9.8	>100

Table 2.18. Yield of thermal cracker product streams

Streams	%Vol on feed	%Vol on total products	Mid-vol % on total product
H_2S	0.18	0.17	–
Gas to C_6	7.81	7.40	3.87
Naphtha C_6 −380°F	9.51	9.02	12.08
Gas Oil 380–650°F	13.91	13.19	23.12
Cracked resid +650°F	74.08	70.22	71.49
Total	105.49	100.00	–

The cracked residue will be routed directly to the Fuel Oil Pool.

The HyGO hydrotreater on cat cracker (FCCU) feed operation. The Feed to the Heavy Gas Oil Hydrotreater shall be the HVGO stream from the Vacuum Distillation Unit (VDU) to meet the feed requirements of a FCCU. This hydro treater will operate on a blocked operation with the LVGO. That is it will operate so many days processing HVGO and the remainder of the STREAM year processing LVGO.

Details of the HVGO operation are as follows:

Feed to the hydrotreater = 9,275 BPCD
SG @ 60 = 0.9218
No. of sulfur/gal = 7.678
No. of sulfur /CD = 2,990,965
%wt sulfur = 3.10

Total sulfur in feed = 9,275 × 42 × 7.678 × 0.031
= 92,720 lbs per Calendar Day.

Table 2.19. Cracked product properties

Stream	BPCD	SG	Sulfur %wt	Vis @ 100°F	Vis @ 210°F
Gas to C_5	852	0.465	3.47*	–	–
Naphtha	1,014.2	0.724	1.58		
Gas Oil	1,483.5	0.825	2.55	5.4	0.2
Residue +650°F	7,900.6	1.046	6.19	110	12.5
Total	11,250.3	0.944	5.35		

*Includes 3,362 lbs/CD sulfur as H_2S.

Table 2.20. Sulfur balance over the thermal cracker

Streams	BPCD	lbs/Gal	mm lbs/CD	% Sulfur	lbs Sulfur/CD
Feed in	10,665	8.38	3.756	5.35	200,964
Streams out					
Gas to C_5	852	4.21	0.151	3.47	5,243.7
Naphtha	1,014.2	6.03	0.257	1.58	4,058.3
Gas oil	1,483.5	6.87	0.428	2.55	10,914.0
Cracked res	7,900.6	8.80	2.920	6.19	180,748.0
Total	11,250.3	7.94	3.756	5.35	200,964.0

85 %wt of the sulfur is to be removed. This amounts to 78,812 number of sulfur per or 2,462.9 moles/CD.

Hydrogen required for sulfur removal in the gas oil is

Hydrogen for sulfur removal only = 10914.0 × 0.85/32 = 289.9 mols/CD.
= 289.9 × 379
= 109,873 Scf (standard cuft @ 60°F and 14.7 psia).

Hydrogen is also required for nitrogen removal and to close ruptured hydrocarbon rings due to the sulfur removal. This reaction also produces light ends at the expense of the treated gas oil yield. The total hydrogen requirements are

Hydrogen = 933,439 Scf (this is from the Licensor's data).

There will also be a loss of hydrogen from the system (mostly due to the Flash drum operation and Purge (see Chapter 8 for more details). This amounts to 186,689 Scf (again from Licensor data).

Total Hydrogen Makeup required then = 933,439 + 933,439 + 186,689 = 2,053,567 Scf\CD.

Assuming the hydrogen makeup stream (usually from the naphtha hydrotreater) is 92% mole pure.

The make up gas stream required = $\frac{2{,}053{,}567}{0.92}$ = 2,232,000 Scf\CD or as usually expressed 240 Scf per Bbl of Feed.

The material balance over the unit shown in Tables 2.21 and 2.22.

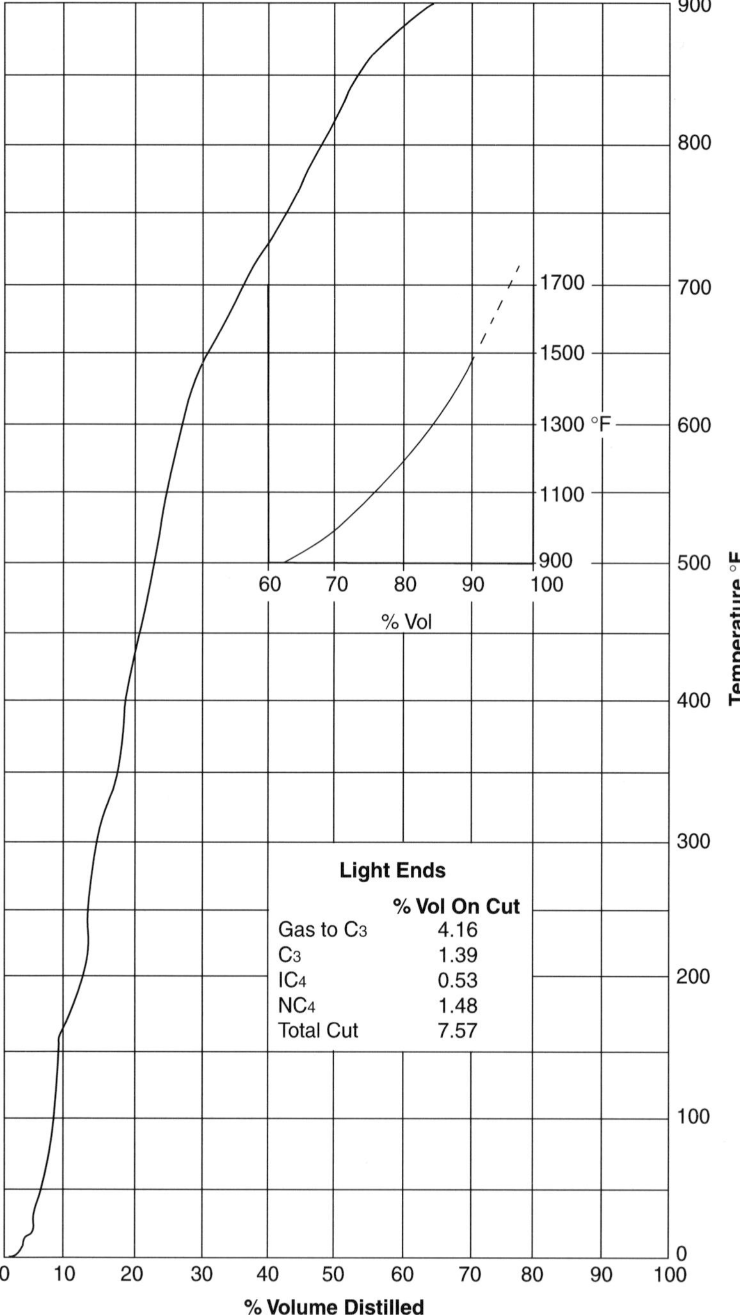

Figure 2.4. Thermal cracker effluent TBP curve.

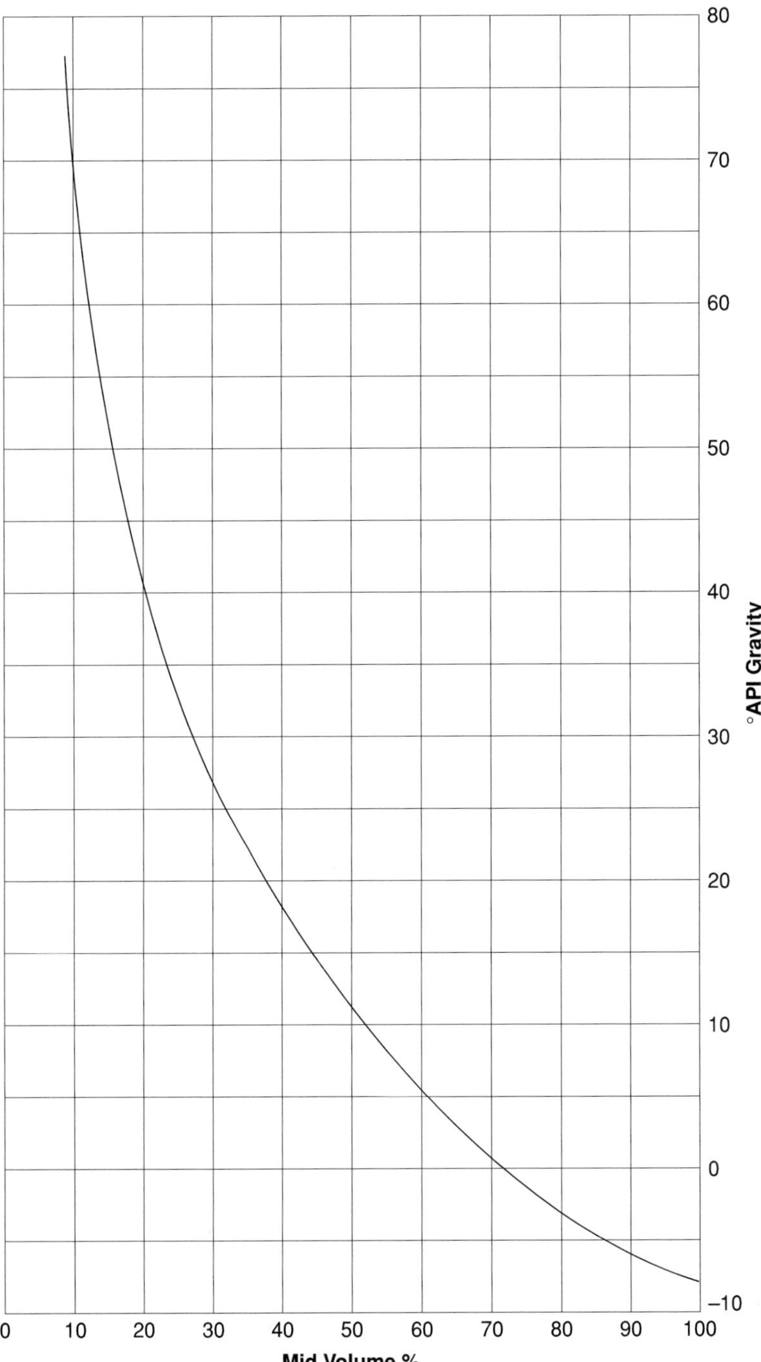

Figure 2.5. Thermal cracker effluent gravity curve.

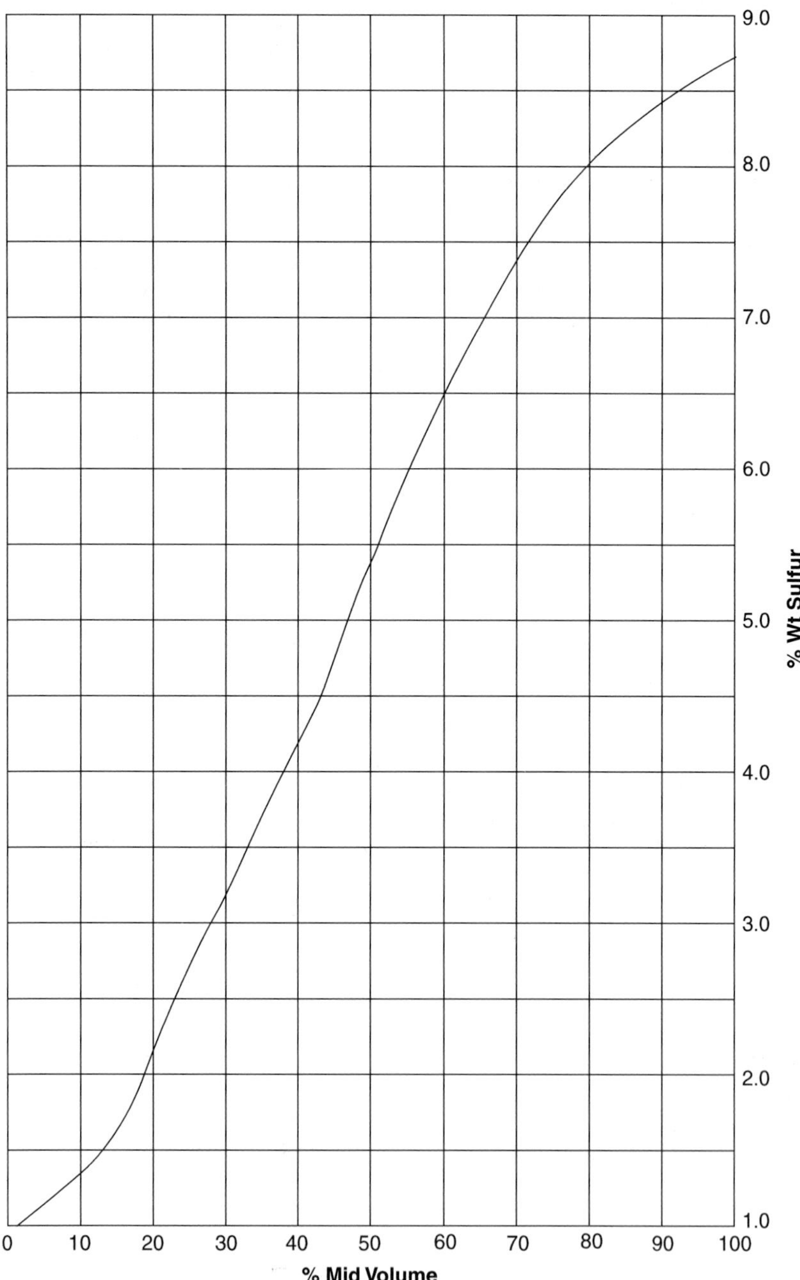

Figure 2.6. Thermal cracker effluent sulfur curve.

Table 2.21. HVGO desulf material balance

Stream	BPCD (Scf/CD)	SG60°F (MW)	lbs/CD	Routing
In				
Hydrogen MU	(2,232,000)	(11)	64,781	
HVGO	9,275	0.9218	2,990,965	
Total in	9,275 + MU		3,055,746	
Out				
H_2S (+LE)	(933,439)	(34)	23,739	Gas Treating
Naphtha	566	0.786	154,518	Nap Hydro treater
Gas oil	8,811	0.898	2,756,987	FCCU
H_2S vent	(162,777)	(28)	120,502	Fuel gas via GT
			3,055,746	

Hy gas oil hydrotreater on LVGO operation. Feed to Hydrotreater is as follows:

LVGO = 5,425 BPCD
SG @ 60°F = 0.8789
No. of sulfur/gal = 7.309
Sulfur %wt = 2.18

lbs per CD = 5,425 × 42 × 7.309 = 1,665,356

lbs/CD Sulfur in Feed = 1,665,356 × 0.0218 = 36,305 lbs
Sulfur removed at 85.0% = 30,859 lbs/CD
Sulfur in finished product = 5,446 lbs/CD

Table 2.22. Desulfurizer sulfur balance

Streams	mm lbs/CD	%wt Sulfur	lbs Sulfur/CD
Feed in			
HVGO	2.991	3.1	92,720
Mu Gas	0.065	Neg	nil
Total in	3.056		92,720
Out			
H_2S + Light ends	0.024	41.0*	9,840
Naphtha	0.155	0.08	125
Gas oil	2.756	0.5	13,908
Vent gas	0.121	57.0*	68,847
Total out	3.056		92,720

Note: Sulfur as H_2S.

Table 2.23. LVGO hydrotreater material and sulfur balance

Stream	BPCD (Scf/D)	lbs/Gal (MW)	mm lbs/CD	Sulfur %wt	Sulfur lbs/CD	Routing
Feed in						
LVGO	5,425	7.32	1.665	2.18	36,305	
MU gas	(925,666)	(11)	0.027	Neg	nil	
Total in			1.692		36,305	
Out						
Gas to C_5	(102,375)	(48)	0.013	84.4*	10,783	Gas Treating
Naphtha	239		0.063	0.08	51	Nap Hds
Gas oil	5,295		1.592	0.34	5,446	Gas oil Pool
Vent gas	(394,435)	(23)	0.024	84.1*	20,025	Gas Treating
Total out			1.692		36,305	

*Sulfur in the form of H_2S from purge and stripper overheads.

$$\text{Hydrogen make up required for sulfur removal} = \frac{30{,}859}{32} = 964 \text{ moles/CD}$$

$$\begin{aligned}
\text{Hydrogen make up required for cracking reactions} &= 1{,}079 \text{ moles/CD} \\
\text{Hydrogen losses} &= 204 \text{ moles/CD} \\
\text{Total hydrogen make up} &= 2{,}247 \text{ moles/CD} \\
&= 851{,}613 \text{ Scf/CD} \\
\text{Reformer gas as make up} &= 851{,}613 \div 0.92 \\
&= 925{,}666 \text{ Scf/CD}
\end{aligned}$$

Material balance for this unit is as follows (Table 2.23):

The FCCU process. From licensor data the following are the expected yields from the desulfurized feed:

Gas C_2 to C_3	8.1 %vol on feed.
C_3	2.3 %vol
C_{3-}	5.6 %vol
iC_4	2.8 %vol
nC_4	1.4 %vol
C_{4-}	7.6 %vol
C_5 + naphtha	59.0 %vol
Lt cycle oil	15.5 %vol
Slurry	14.0 %vol
Total	116.3 %vol on Feed.

Table 2.24. FCCU material and sulfur balance

Stream	BPCD	SG @ 60°F	mm lbs/CD	S %wt	S lbs/CD	Routing
Feed						
Des HVGO	8,811	0.898	2.757	0.50	13,908	FCCU
Products out						
Gas to C_3	714	–	0.13	2.19	2,850	Gas treating
C_3 to C_5	1,736	0.560	0.341	–	Neg	Alkylation
C_5 to 300°F	3,503	0.7856	0.929	–	Neg	Gasoline pool
300–420°F	1,695	0.8251	0.489	0.05	262	Gasoline pool
Lt cycle oil	1,366	0.9159	0.438	0.54	2,380	Gas oil blending
Slurry	1,234	0.990	0.428	2.0	8,416	Fuel oil
Total out	10,248		2.757	0.5	13,908	

In terms of BPCD:

$$\begin{aligned} \text{Gas to } C_3 &= 714 \\ \text{C3 to } C_4\text{'s} &= 1{,}736 \\ \text{Naphtha} &= 5{,}198 \\ \text{Lt Cycle} &= 1{,}366 \\ \text{Slurry} &= 1{,}234 \\ \text{Total} &= 10{,}248 \end{aligned}$$

Overall unit balance (Tables 2.24–2.26).

Lt gas oil hydrotreater. The LGO hydrotreater will operate on two blocked operations to desulfurize diesel feed and secondly to desulfurize kerosene.

Table 2.25. Diesel hydrotreater material and sulfur balance

Stream	BPCD (Scf/D)	lbs/gal (MW)	mm lbs/CD	%wt sulfur	lbs/CD sulfur	Routing
In						
Diesel	8,399	7.0	2.470	1.16	28,652	
MU Gas	(1,495,000)	(11)	0.043	Neg	nil	
Total in			2.513			
Out						
Gas to C_5	(42,037)	(55)	0.006	69.5	4,168	Gas treating
Naphtha	459	6.55	0.126	0.08	101	Nap Hds
Diesel prod	8,036	6.98	2.356	0.30	7,068	Diesel pool
Vent gas	(255,067)	(37)	0.025	69.4	17,315	Gas treating
Total out			2.513		28,652	

Table 2.26. Material balance over the reformer

Stream	BPCD (mm Scf/CD)	SG @ 60°F (MW)	mm lbs/CD	Routing
IN				
Hy Nap ex CDU	7,100	0.755	1.863	
Hy Nap ex TC	578	0.752	0.152	
Hy Nap ex Hds	940	0.793	0.261	
Total in	8,618	0.7543	2.276	
OUT				
Reformer gas	(7.33)	(11)	0.224	to Naphtha Hds
Light ends	837	0.521	0.152	to Sat Lt Ends
Reformate	6,852	0.7927	1.900	to Gasoline Pool
Total out	7,689		2.276	

(i) *Diesel operation.* The feed in this case will consist of Straight Run diesel from the CDU and the diesel cut from the thermal cracker. Thus:

	BPCD	SG @ 60°F	#/Gal	mm #/CD	S %wt	S/CD
Straight run	6,915	0.842	7.013	2.04	0.87	17,748
Cracked	1,484	0.825	6.87	0.43	2.55	10,967
Total	8,399	0.841	7.00	2.47	1.16	28,715

The licensor guarantees 75 %wt sulfur removal. Therefore:

Sulfur removed = 21,536 lbs/CD
Sulfur in product = 7,179 lbs/CD
Hydrogen consumption for sulfur removal = 157,664 Scf/D for SR Diesel
= 97,415 Scf/D for Cracked feed
= 673 moles/D
Hydrogen consumption for ring saturation = 384,474 Scf/D for Str Run (includes losses)
= 735,322 Scf/D for Cracked feed
= 2,954 moles/D
Total hydrogen make up = 1.375 mm Scf/D for Total Feed
Reformer gas make up = 1.495 mmScf/CD
= 3,945 moles/D

(ii) *Hydrotreater on Kero operation*

Kero feed ex CDU = 4,545 BPCD
SG @ 60 of Kero = 0.8048 then mm lbs/CD = 1.280
Sulfur content = 0.21 %wt then Sulfur = 2,685 lbs/CD

Licensor guarantees mid of run sulfur removal as 99 %wt.

Then lbs/CD sulfur removed = 2,631 lbs/CD. This generates 2,795 lbs/CD of H_2S.

There is no significant amount of naphtha made nor is there significant amount of hydrogen needed for saturation. Then make up hydrogen required is 82 moles/CD or $82 \div 0.92 = 89$ moles/CD Reformer gas. This amounts to 33,780 Scf/CD.

(iii) *Summary Kero operation*

IN	
Kero feed 4,545 BPCD	1.280 mm lbs/CD
Hydrogen 89 moles/CD	979lbs/CD of Reformer gas
Total	1,280,979 lbs/CD

OUT	
Kero Product 4,539 BPCD	1.277342 mm lbs/CD
Hydrogen sulfide 82 moles/CD	2,788 lbs/CD
Losses from stripper	813
Total	1,280,166 lbs/CD.

Naphtha consolidation and treating. Summary of naphtha and light ends streams to treating processes.

From	Stream	BPCD
CDU	Gas to C_5	855
	Light naphtha to 212°F	5,219.5
	Heavy naphtha to 380°F	7,100
Thermal cracker	Gas to C_5	852
	Light naphtha to 212°F	436.5
	Heavy naphtha to 380°F	577.7
The hydrotreaters	Gas to C_5	Mostly H_2S to gas treating
	Light naphtha to 212°F	323.5
	Heavy naphtha to 380°F	940.5

Unstabilized naphtha feed to naphtha hydrotreater

	Gas to C_5	1,707
	Light naphtha	5,981
	Heavy naphtha	8,618
	Total	16,306

Light naphtha to gasoline pool

Total	5,970

Heavy naphtha feed to cat reformer

Total	8,618

Catalytic reformer operation. The reformer will operate at a severity of 91 RON and at a reactor pressure of 350 psig.

Licensor data for these conditions are:

Yield of reformate = 79.5 %LV
Reformate make = 6,851.5 BPCD
Hydrogen made = 5.13 mmScf/CD Purity is 70 mol%
Then reformer gas is $5.13 \div 0.7 = 7.33$ mmScf/CD

Mol Wt of reformer gas is 12 as given by the Licensor. Then mm lbs/CD of the gas is 0.237.

C_5 = Reformate = 6,851 BPCD and at an SG of 6.604 lbs/gal lbs/CD of reformate = 1.9 mm lbs/CD.

Light ends leaving the reformer stripper will be in accordance with the following balance:

lbs per CD of fresh feed = 2.276 mm lbs
lbs per CD of reformate = 1.900 mm lbs
lbs per CD of reformer gas = 0.237 mm lbs
By difference Lt ends = 0.152 mm lbs/CD routed to Reformate Stabilizer.

Analysis of the Lt end stream gave:

	%wt	lbs/CD	BPCD
Gas to C_3	9.8	14,896	122
C_3	33.1	50,312	284
iC_4	22.0	33,440	170
nC_4	35.1	53,352	261
Total	100.0	152,000	837

The associate naphtha hydrotreater receives the fresh reformer feed and desulfurizes it using the reformer off gas on a once through basis. In this process the hydrotreater consumes 6 vol% of the hydrogen and absorbs light ends from the reformer gas to upgrade the off gas to 92 mol% hydrogen.

Hydrogen consumed in the hydrotreater = .308 mmScf/D
Hydrogen leaving in off gas = 5.13−0.308 = 4.822 mmScf/D
Vol of off gas from the hydrotreater = 4.822 ÷ 0.92 = 5.241 mmScf/D

By difference the volume of Lt End gas absorbed into the hydrotreater stabilizer feed will be 7.33−5.241 = 2.089 mmScf/CD or 5,526 moles/D. This absorbed gas will be slightly heavier than that for the gas absorbed in the reformate. This will have the following molar composition:

	%mol	moles/D	lbs/D	BPCD
Gas to C_3	8.2	453.1	6,797	56
C_3	37.3	2,061.2	90,693	514
iC_4	20.1	1,110.8	64,422	328
nC_4	34.4	1,900.9	110,255	540
Total	100.0	5,526	272,167	1,438

Saturated light ends consolidation. Although the LEs from the thermal cracker contains a significant amount of unsaturated hydrocarbons, it is usually blended with the CDU overhead distillate, hydrotreated and then split into its components. Thus the LEs are considered saturated for the study here.

The isomerization and alkylation processes. The fresh feed to the alkylation reactor is the C_3 and C_4 streams from the FCCU. The volumetric and mass composition of this stream is as follows:

Component	%vol	BPCD	#/Gal	lbs/CD
C_3	11.7	203	4.22	35,980
C_{3-}	28.4	493	4.34	89,864
iC_4	14.2	247	4.68	48,550
nC_4	7.1	123	4.86	25,107
C_{4-}	38.6	670	5.00	140,700
Total	100.0	1,736	4.67	340,201

From licensor data:

Alkylate from propylene = 1.76 × 493 = 868 BPCD
Alkylate from butylene = 1.77 × 670 = 186 BPCD

The volumetric composition of the alkylates is as follows:

From butylenes:

	%wt	#/Gal	Vol factor	%vol	%wt	Vol factor	%vol	
C_5	9.0	5.25	1.71	9.9	4.0	0.8	4.75	
C_6	4.0	5.53	0.72	4.16	2.0	0.4	2.37	
C_7	68.0	5.73	11.87	68.55	2.0	0.3	1.78	
C_8	10.0	5.89	1.69	9.76	80.0	13.58	80.65	
C_{9+}	9.0	6.8	1.32	7.63	12.0	1.76	10.45	
Total	100.0	5.78	17.31	100.0	100.0	16.84	100.0	#/Gal = 5.93

Total alkylate = 2,054 BPCD = 50,6101 lbs/CD

$$
\begin{aligned}
iC_4 \text{ Consumption} &= 1.34 \times \text{propylene} + 1.15 \times \text{butylene} \\
&= 1.31 \times 493 + 1.15 \times 670 = 1{,}416.3 \text{ BPCD} \\
&= 278{,}000 \text{ lbs/CD.}
\end{aligned}
$$

iC_4 available in FCCU Light Ends = 48,550 lbs/CD. Required from other source = 229,500 lbs/CD or 1,167 BPCD.

A butane isomerizer is included in the process to provide the necessary iso butane stream. In this process 50 %wt of the normal butane contained in a saturated butane stream is converted to iso butane.

Thus from Table 2.27 total butanes available from saturate LE is made up as follows:

Iso butane	670 BPCD	131,695 lbs/CD	= 29.2 %wt
Normal butane	1,559 BPCD	318,223 lbs/CD	= 70.8 %wt

Table 2.27. Material balance over naphtha Hds

Stream	BPCD (mm Scf/CD)	SG @ 60°F (MW)	mm lbs/CD	Routing
IN				
Gas to C_5	1,707	0.320	0.313	
Lt naphtha	5,970	0.661	1.379	
Hy naphtha	8,618	0.756	2.276	
Reformer gas	(7.33)	(11)	0.224	
Total in	16,295		4.192	
OUT				
Deb overheads	2,817	0.436	0.398	To Sats Lt End
Splitter feed	14,588	0.716	3.655	To Nap Splitter
Rich H_2 gas	(5.241)	(10)	0.139	To H_2 System
Total out	17,296		4.192	

Table 2.28. Saturated light ends summary

	BPCD				
Source	Gas to C_3	C_3	iC_4	nC_4	Total
Crude Distillation Unit	51	103	112	590	856
Thermal Cracker	468	156	60	168	852
Cat Reformer	122	284	170	261	837
Naphtha HDS	56	514	328	540	1,438
Total	697	1,057	670	1,559	3,983

50% of the normal butanes are isomerized giving an isomerate 29.2 + 35.4 lbs iso-butane per 100 lbs feed and 35.4 lbs normal butane (Table 2.29). Feed to the Isomerizer will be $229,500 \div (0.292 + 0.354) = 355,263$ lbs/CD. Or 528 BPCD iC_4 and 1,232 BPCD of nC_4.

Product blending and properties

(i) *Gasoline blending.* Both grades of gasoline will be blended to a 'no give away' on octane value. These grades and their blending specifications are:

	Premium grade	Regular grade
Octane research clear	90	82
Ried vapor pressure (Min)	7 psia	7 psia
Volatility ASTM		
50 %vol distilled (Min)	275°F	175°F
60 %vol distilled (Max)	300°F	210°F
95 %vol distilled (Max)	400°F	325°F

Table 2.29. Balance over isom/alky plants

Stream	lbs/CD	lbs/Gal	BPCD	Routing
In				
Unsats ex FCCU	340,201		1,736	
Butanes ex Sats	355,263		1,760	
Total In	695,464	4.74	3,496	
Out				
C_3 ex FCCU	35,980		203	LPG pool
nC_4 ex Sats	125,763		616	Gasoline/LPG
nC_4 ex FCCU	25,107		123	Gasoline/LPG
Alkylate	506,101		2,054	Gasoline
Total Out	692,951	5.51	2,996	

(ii) *Premium grade gasoline*. The following blend of gasoline precursor streams meet the Octane specifications and quantity for the Premium Grade Gasoline.

	BPCD	%vol	ON(RES) clear	Octane factor
Alkylate	2,054	26.7	97	0.26
Heavy cracked naphtha	1,695	22.0	89	0.20
Light cracked naphtha	1,000	13.0	94	0.12
Reformate	1,951	25.3	91	0.23
Light str run naphtha	1,000	13.0	69	0.09
Total	7,700	100.0	90	0.90

The remaining properties of this product are as follows:

(iii) *The distillation curves*. The TBP curves for each precursor making up the blend is divided into components of the same boiling range and combined to produce the composite TBP curve for the blend. Thus:

		LSR		Reformate		Lt cracked naptha		Hy cracked naptha		Alkylate		
Comp	Temp °F	%vol	BPCD	%vol	BPCD	%vol	BPCD	%vol	BPCD	%vol	BPCD	Total BPCD
1	45–100	20	200	6	117	5	50					367
2	125	15	150	4	78	5	50					278
3	150	15	150	7	137	19	190					477
4	175	30	300	8	156	14	140			2	41	637
5	200	10	100	8	156	15	150			8	164	570
6	250	10	100	17	331	19	190			2	41	662
7	175			20	390	8	80			30	616	1,086
8	300			12	234	5	50	5	85	45	925	1,294
9	350			12	234	10	100	42	712	13	267	1,313
10	400			3	59			43	728			787
11	475			3	59			10	170			229
Total		100	1,000	100	1,951	100	1,000	100	1,695	100	2,054	7,700

The TBP and the derived ASTM curves are given in Figure 2.7.

The above component breakdown will now be used to determine, the SG of the product and the quantity of C_4 that will be added to the blend to meet the RVP spec. The RVP

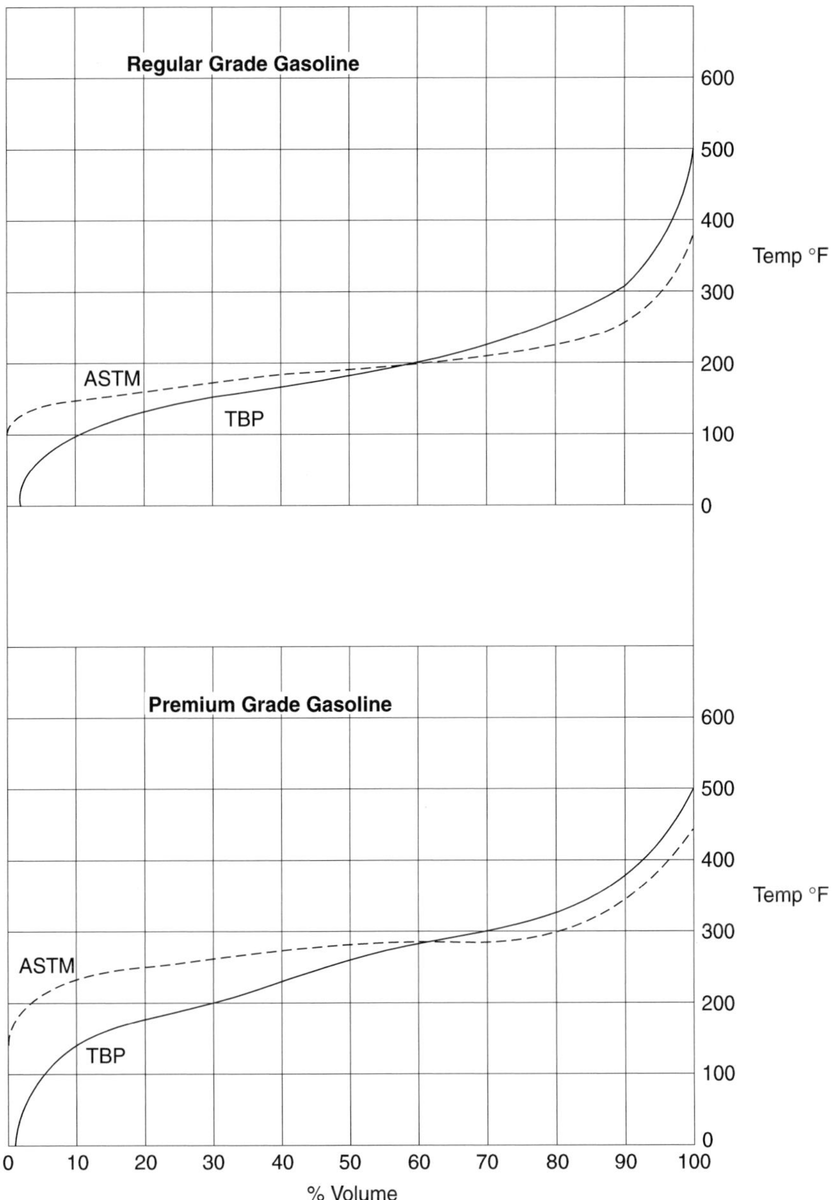

Figure 2.7. Premium and regular grade TBP and ASTM curves.

of the finished product is determined by a bubble point calculation in which the C_4 is an unknown x and the RVP is set at 7 psia and 100°F. Thus:

Comp	Mid Bpt	#/gal	Mol Wt	VP @ 100 psia	%vol	Wt factor	Mol factor
1	75	5.96	72	22	4.77	28.43	0.395
2	115	6.03	73	11	3.61	21.77	0.298
3	135	6.09	75	6.2	6.19	37.70	0.503
4	155	6.17	81	5.0	8.27	51.03	0.630
5	185	6.25	90	2.5	7.41	46.31	0.515
6	225	6.39	100	}	8.61	55.02	0.550
7	260	6.49	110	}	14.11	91.57	0.832
8	285	6.56	119	} 1.8	16.79	110.14	1.08
9	315	6.64	128	}	17.06	113.28	0.885
10	375	6.83	148	}	10.22	69.80	0.471
11	420	6.95	162	}	2.96	20.57	0.127
Total		6.45	103		100.00	645.62	6.287

Calculating for butane addition to make 7 psia RVP.

Comp	mid Bpt °F	VP @ 100°F psia	K = VP/7	moles liq	moles vap
C_4		60	8.57	x	$8.57x$
1	75	22	3.14	0.395	1.241
2	115	11	1.57	0.298	0.469
3	135	6.2	0.89	0.503	0.445
4	155	5.0	0.71	0.630	0.450
5	185	2.5	0.36	0.515	0.184
6	225	|	|	|	|
7	260	|	|	|	|
8	285	1.8	0.26	3.946	1.015
9	315	|	|	|	|
10	375	|	|	|	|
11	420	|	|	|	|
Total				$6.287 + x$	$3.804 + 8.57x$

$$
\begin{aligned}
6.287 + x &= 3.804 + 8.57x \\
x &= 0.328 \\
&= 0.328 \div (6.287 + 0.328) \\
&= 0.0495
\end{aligned}
$$

C_4 will be 4.95 mole% or 4.0 vol% of total product gasoline.

Then blended Premium Gasoline will have the following specification:

SG @ 60°F 0.7749
RVP 7.0 psia
50% ASTM vol distilled = 275°F
60% ASTM vol distilled = 290°F
95% ASTM vol distilled = 380°F
Octane number research clear = 90

These meet the required specifications for the product.

(iv) *Regular gasoline.* It is proposed only to include the component breakdown for the Regular Gasoline precursors here. The remainder of the blending recipe for the product will follow that for the Premium grade. Thus the component breakdown for regular Grade is as follows:

		LSR		Reformate		Lt crack naptha		
Comp	Temp °F	%vol	BPCD	%vol	BPCD	%vol	BPCD	Total BPCD
1	45–100	20	993.9	6	294	5	125.2	1,413.1
2	125	15	745.4	4	196	5	125.2	1,066.6
3	150	15	745.4	7	343	19	475.6	1,564.0
4	175	30	1,490.8	8	392	14	350.4	2,233.2
5	200	10	497	8	392	15	375.4	1,264.4
6	250	10	497	17	833	19	475.5	1,805.5
7	275			20	980	8	200.2	1,180.2
8	300			12	588	5	125.2	713.2
9	350			12	588	10	250.3	838.3
10	400			3	147			147
11	475			3	147			147
Total			4,969.5	100	4,900	100	2,503	12,372.5

Following the same calculation route as that for premium grade the TBP and ASTM distillation curves are also given in Figure 2.7. The calculation for RVP shows the blend requires an input of 0.36% vol C_4 to meet the RVP of 7 psia. Regular Grade Gasoline therefore has the following specification:

SG @ 60°F 0.7531
RVP 7 psia
50% ASTM vol distilled = 185°F
60% ASTM vol distilled = 200°F
95% ASTM vol distilled = 310°F
Octane number research clear = 82.

Total Butanes to Gasoline pool will be 310 BPCD to Premium grade and 45 BPCD to Regular grade totaling 355 BPCD. The Gasoline product make will be:

Premium Grade	8,010 BPCD
Regular Grade	12,417.5 BPCD

- *The LPG pool*
  The LPG make from the refinery will essentially be the net make from the Depropaniser, deethanizer, and the alkylation process, thus:

C_3 ex deethanizer	1,057 BPCD
C_3 ex Alky process	203 BPCD
Total C_3 LPG	1,260 BPCD

C_4 ex depropaniser (net)	469 BPCD
C_4 ex Alky process	739 BPCD
C_4 to gasolines	−355 BPCD
Total C_4 LPG	853 BPCD.

The respective processes will operate to meet LPG specification.

- *Fuel oil pool*

Three fuel oil products will be produced:

Marine diesel
Heavy fuel oil
High sulfur bunker oil

Marine diesel specification:

SG @ 60°F 0.830 Min
Viscosity Redwood @ 100°F 40 seconds Min
Pour point 45°F Max
Flash point 150°F Min
Sulfur %wt 1.0 Max

Heavy fuel oil specification:

SG @ 60°F 1.00 Max
Sulfur %wt 5.0 Max
Viscosity @ 122°F Cst 170 Max
Flash Point °F 160 Min
Pour Point °F 65 Max

High sulfur bunker fuel

SG @ 60°F = 1.05 Max
Viscosity Cst @ 122°F = 370 Max
Flash Point = 160°F Min
Sulfur %wt = 6.22 Max.

Blending to meet marine diesel spec.

This will be a blend of desulfurized LVGO and thermal cracker residue.
50% of available LVGO will go into this blend = 2,648 BPCD = 796,306 lbs/CD.
Calculating amount of TC Residue in blend to meet Sulfur Spec.

Let x lbs be the wt of Residue in blend. Then

$$\begin{aligned}
(796{,}306 \times 0.0035) + 0.0619x &= 0.01(796{,}306 + x) \\
x &= 99{,}730 \text{ lbs/CD} \\
\text{Total blend} = 796{,}306 + 99{,}730 &= 896{,}036 \text{ lbs} \\
\text{or} \quad 2{,}648 + 270 &= 2{,}918 \text{ BPCD SG @ } 60°\text{F} = 0.877.
\end{aligned}$$

This meets SG and sulfur spec. Checking for Pour Point as follows.

LVGO pour point = 15°F (no credit is taken for some reduction due to hydro-treating.)
Approx ASTM 50% point = 700°F
TC Residue pour point is 65°F
Approx ASTM 50% point = 900°F

	Composition		ASTM 50%		Pour point		
Components	BPCD	Fraction	50%	Factor	Pour point °F	Index	Blend
		A	B	A × B		C	A × C
LVGO	2,648	0.91	700	637	15	5	4.6
Tc Res	270	0.09	920	82.8	65	22*	2
Total	2,918	1.00		719.8			6.6

*The pour point corresponding to a 50% ASTM point of 720°F and an index of 6.6 is 22°F. Therefore blend meets spec for Marine Diesel.

Heavy fuel oil blending:

80% of thermal cracker residue will be routed to Heavy Fuel. Thus:

BPCD = 6,320
%wt Sulfur = 6.19
Viscosity @ 122°F = 75Cst
SG @ 60°F 1.046
Pour Point = 55°F

The product does not meet specification with respect to sulfur or specific gravity. The product will be blended with hydrotreated gas oil for sulfur.

Total sulfur in thermal cracker residue 6,320 × 42 × 8.8 × 0.0619 = 144,541 lbs.

Let x be lbs of hydrotreated LVGO in blend.

$$\begin{aligned} \text{Then } 144{,}541 + 0.0035x &= 0.05(2{,}335{,}072 + x) \\ x &= 597{,}578 \text{ lbs/CD of desulfurized LVGO.} \\ \text{Total fuel oil blend} &= 2{,}335{,}072 + 597{,}578 = 2{,}932{,}650 \text{ lbs/CD} \\ &= 6{,}317 + 1{,}990 \text{ BPCD} = 8{,}306 \text{ BPCD} \end{aligned}$$

SG @ 60°F of Fuel Oil Product = 0.998 which is within Spec.

Pour point of thermal cracker residue is 55°F and blended with LVGO will have a pour point <55°F and meets Fuel Oil spec of 65°F Max.

Blending for High Sulfur Bunker Fuel.

Uncommitted Thermal Cracked Residue amounts to 1,313.6 BPCD. This meets all the criteria for High Sulfur Bunker Fuel without further blending. Decanted FCCU Slurry will also be routed to this product. Then total product = 923,000 lbs/CD or 2,548 BPCD.

- *The gas oil pool*

Automotive Diesel specification:

SG @ 60°F, 0.83–0.86
Visc @ 100°F, 1.6–5.5 Cst
Flash PM 150°F Min
Pour point 15°F Max
Sulfur 0.35 %wt Max
ASTM dist:
50% recovered @ 464°F Min
657°F Max

90% recovered @	615°F Min
FBP	725°F Max
Diesel index	54 Min.

All the diesel stream and the remaining LVGO will be blended to meet this product spec. Some desulfurized kerosene may be added if required to make up gravity or distillation spec. The diesel index, pour point, and flash point are easily met. Viscosity spec will also be checked.

Specification check:

	BPCD	lbs/Gal	mm lbs/CD	Pour Point °F	%wt S
Gas oil	655	7.163	0.197	15	0.35
Diesel	8,101	6.98	2.374	0	0.31
Total	8,756	6.7	2.572	<15	0.313

Distillation:

ASTM distillation of blend is calculated to give the following:

% vol	Temp°F
0	460
10	515
30	540
50	561
90	615
FBP	685

This meets the spec required.

Viscosity check as follows:

	%vol	Visc @ 100°F Cst	Visc index	Factor (Index × %vol)
Gas oil	7.5	13.0	38	285
Diesel	92.5	4.0	48	4,440
Total	100.0			4,725

Index of blend $= 4{,}725 \div 100 = 47.25$ corresponding viscosity $= 4.8$ Cst which meets spec.

This completes the blend recipe calculation for automotive diesel product.

Gas oil—(Heating oil)

Specification:

SG @ 60°F 0.840–0.860
Flash pt °F 150 Min

Distillation ASTM:
10 %vol Rec @ 400°F Min
50 %vol Rec @ 475°F Max
90 %vol Rec @ 675°F Max
FBP 750°F Max
Sulfur 0.3% Max

Gas oil recipe:
It is proposed to route all the FCCU's Lt cycle oil make to this blend and sufficient Kero to meet the Sulfur Specification, thus:

Let x lbs/CD be the amount of Kero in the blend, then:

lbs/CD Sulfur in cycle oil $= 2,380$
lbs/CD sulfur in the Kero $= 0.00004x$

$$2{,}380 + 0.00004x = 0.003(438{,}000 + x)$$
$$x = 439{,}527 \text{ lbs/CD.}$$

Total blend:

Stream	mm lbs/CD	lbs/gal	BPCD
Lt cycle oil	0.438	7.63	1,366
Kerosene	0.440	6.70	1,563
	0.878	7.13	2,929

SG of blend = 0.856 which is with in spec

ASTM Distillation of blend is as follows:

10% rec @ 401°F
50% rec @ 473°F
90% rec @ 571°F
FBP 647°F

This meet Distillation spec.

$$\text{Checking flash point: Kero flash point} = 0.77(\text{ASTM } 5\% \text{ temp} - 150)$$
$$= 0.77(370 - 150) = 169.4°\text{F}$$

The Lt Cycle oil has an ASTM 5% point higher than 370°F therefore blend flash point will be > 150°F which makes it within spec.

Table 2.30. The calculated product slate and sulfur make

Item	Stream In	BPCD (mmScf/CD)	lbs/gal (MW)	Mm lbs/CD	%wt S	Sulfur lbs/CD
1	Crude Feed	50,000	7.13	14.973	2.35	351,701
2	Hydr Gas	(7.33)	(11)	0.224	neg	nil
3	Total In	50,000		15.197		351,701
	Out in LE					
4	Ex Nap Hds			0.272	4.65	11,273
5	Ex Kero Hds			0.004	65.78	2,631
6	Ex LGO Hds			0.031	69.30	21,483
7	Ex LVGO Hds			0.037	83.26	30,808
8	Ex HVGO Hds			0.144	54.64	78,687
9	Ex FCCU			0.13	2.19	2,850
10	Ex Reformer			0.152		nil
11	Total ex LE			0.770		147,732*
			* As H_2S			
	LE products					
12	Fuel gas	1,334		0.224		
13	LPG	2,113		0.398		
14	Sulfur			0.139	(or 62 Tons)	
	Liquid products					
15	Gasolines			5.440	neg	nil
16	Kero			0.837	0.004	36
17	Auto diesel			2.572	0.96	8,049
18	Gas oil			0.878	0.27	2,383
19	M diesel			0.896	1.00	8,960
20	Hy fuel			2.933	5.0	146,633
21	Bunker oil			0.923	4.17	38,469
15 to 21	Total Liq products			14.479		204,530
12 to 21	Overall Prod Slate.			15.240	2.32	352,262

Overall product slate balance and sulfur make (Table 2.30).

Deviation between In and Out = lbs/Cd = +0.2%
Sulfur/CD = +0.1%

Both are acceptable.

Unit capacities. The unit capacities required to meet this product slate will be based on a service factor for each unit. This service factor takes into consideration the downtime required for each type of process which has been based on statistical operating experience. Needless to say this will be different for different operating companies based on their experiences, their maintenance philosophy, and design features of these units that they wish to consider. The latter case for example will consider, their current pump sparing philosophy, the sophistication of their plant control system and the like. Those service factors given in the following table are taken from the author's file for the purpose of this exercise (Table 2.31). The block flow diagram showing this configuration is given as Figures 2.8 (Parts 1 and 2).

Table 2.31. Unit design capacities

UNIT	BPCD	Service factor %	Operating days per year	BPSD	Remarks
CDU	50,000	92	336	54,500	
VDU	25,365	92	336	27,570	
Nap Hds	17,583	90	330	19,540	Inc Gas ex Ref
Debutanizer	17,583	92	336	19,110	Inc O/Hds ex TC
Nap Splitter	14,589	92	336	15,860	
Depropanizer	39,60	92	336	4,310	
Deethanizer	1,774	92	336	1,930	
Cat Reformer	8,618	87	318	9,910	On 97 ON Sev
Thermal Cracker	10,665	87	318	12,260	
LGO Hds		90		13,980	
On Kero	4,545		119		
On LGO	8,036		211		
Hy Gas oil Hds		90		16,330	
on LVGO	5,425		122		
on HVGO	8,036		208		
FCCU	8,811	87	318	10,130	Inc Gas Conc
Isom/Alky	3,496	90	330	3,880	
Gas Treating	2,167	95	347	2,280	
Sulfur Recovery	60 (tons)	95	347	(70 tons)	

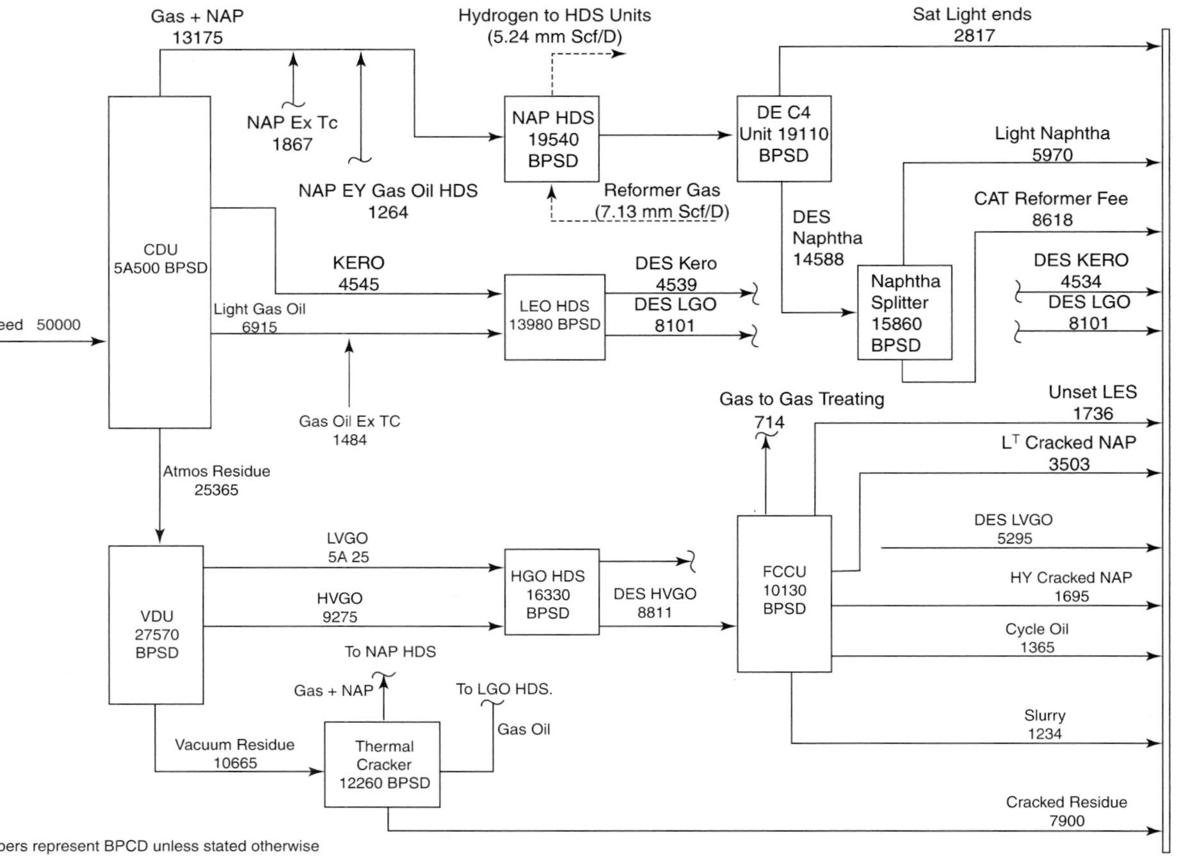

Figure 2.8. (2 Parts) The refinery configuration block flow diagram.

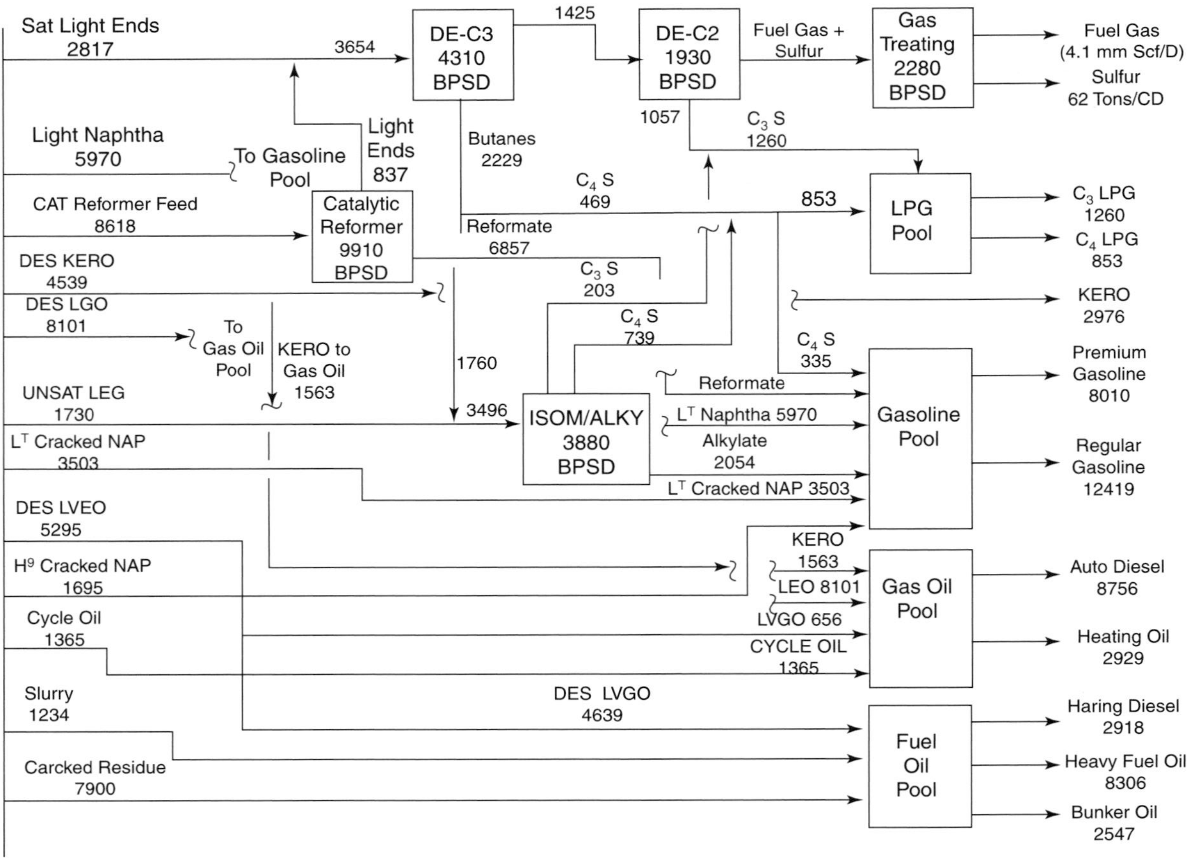

Figure 2.8. (Cont.)

Conclusion

This completes a very cursory look at the nature and refining of crude oil. The remainder of this first part of the encyclopedia continues with a more detailed examination of the various processes, the economic aspect of the refining industry, and the impact the Industry has on modern living. It also describes and discusses modern techniques used in the Industry and other allied industries (such as Engineering and Construction) to improve and optimize the design and operation of the refining processes.

Chapter 3

The atmospheric and vacuum crude distillation units

D.S.J. Jones

The distilling of petroleum products from crude oil to some extent or other has long been practiced. Certainly the ancient Egyptians, Greeks, and Romans had some form of extracting a flammable oil from, probably, weathered crude oil seepage. It wasn't though until the turn of the nineteenth and twentieth century that crude oil well drilling was first discovered and commercialized. Originally the crude oil was refined to produce essentially kerosene (lamp oil), and a form of gasoline known then as Benzine (as opposed to Benzene already being produced from coal) and the residue used as pitch for calkin and sealing. The lamp oil or kerosene was produced to provide a means of illumination, later a lighter cut known as naphtha was produced for the same purpose but used in special pressurized lamps.

The production of these early distillates was made by cascading the crude oil through successive stills each operating at successively higher temperatures. This is shown in the following diagram Figure 3.1.

The crude enters the first still to be heated to a temperature that vaporizes the light components. The residue from this still enters the second one and heated to a higher temperature to vaporize the Benzine fraction. The residue enters the third still and heated to remove the Kerosene fraction. The residue from this still is a very light fuel oil which may be further heated and partially vaporized to give a pitch of sorts as a residue and a distillate which could be used as fuel. This distillate would later become the Diesel or Gas Oil fraction and used in the developing diesel engine.

The vapor from each still passes through a small packed wash section before being condensed and collected in a condensate drum. A portion of the condensate is returned to the top of the wash section as the wash liquid, similar to the reflux stream in modern distillation towers. Usually steam was injected through the liquid phase of each still to facilitate vaporization and to strip out the light ends.

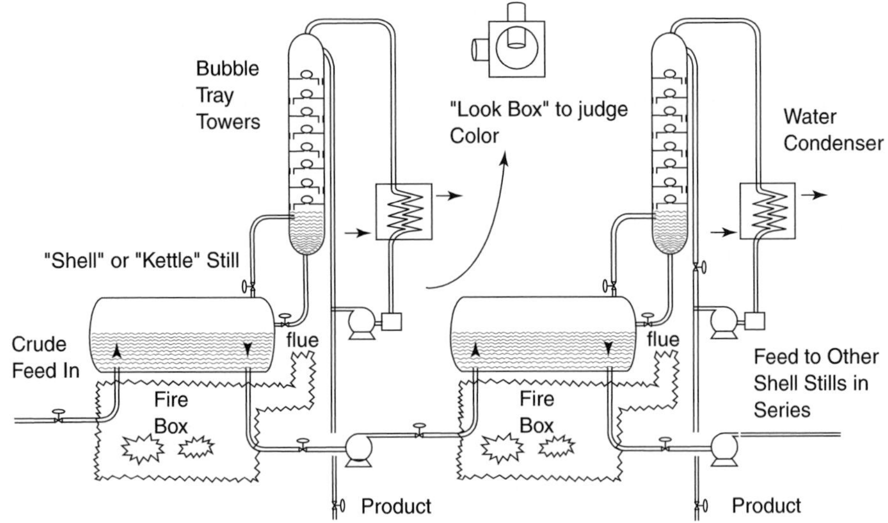

Figure 3.1. Early continuous pipestill schematic.

This type of crude distillation was superseded by the continuous fractionation tower used in modern petroleum refining. This type of crude 'Still' remained in operation well after the second World War in some oil refineries. The author recalls such a still in operation as late as 1956 in the refinery that he was employed at, although there were also in operation three other modern crude oil fractionating units. Indeed one major oil company even today call the crude distillation unit a 'Pipestill'. This chapter now continues with the description of a present day atmospheric and vacuum crude oil distillation units. It will be divided into two parts:

Part 1 The Atmospheric Crude Distillation Unit (CDU)

Part 2 The Vacuum Crude Distillation Unit (VDU)

Both parts will contain a process description with a schematic flow sheet, a discussion on the development of the material balance, a description and discussion on the design characteristics of the units, and finally a worked design example.

3.1 The atmospheric crude distillation unit

Process description

The first process encountered in any conventional Refinery is the Atmospheric Crude Distillation Unit. In this unit the crude oil is distilled to produce distillate streams which will be the basic streams for the refinery product slate. These streams will either be subject to further treating down stream or become feed stock for conversion units that may be in the Refinery Configuration. A schematic flow diagram of an atmospheric crude unit is shown in Figure 3.2.

Crude oil is pumped from storage to be heated by exchange against hot overhead and product side streams in the Crude Unit. At a preheat temperature of about 200–250°F water is injected into the crude to dissolve salt that is usually present. The mixture enters a desalter drum usually containing an electrostatic precipitator. The salt water contained in the crude is separated by means of this electrostatic precipitation. The water phase from the drum is sent to a sour water stripper to be cleaned before disposal to the oily water sewer.

It must be understood however that this ‘de-salting’ does not remove the organic chlorides which may be present in the feed. This will be discussed later when dealing with the tower’s overhead system.

The crude oil leaves the desalter drum and enters a surge drum. Some of the light ends and any entrained water are flashed off in this drum and routed directly to the distillation tower flash zone (they do not pass through to the heater). The crude distillation booster pump takes suction from this drum and delivers the desalted crude under flow control to the fired heater via the remaining heat exchange train.

On leaving heat exchanger train, the crude oil is heated in a fired heater to a temperature that will vaporize the distillate products in the crude tower. Some additional heat is added to the crude to vaporize about 5% more than required for the distillate streams. This is called over flash and is used to ensure good reflux streams in the tower. The heated crude enters the fractionation tower in a lower section called the flash zone.

The unvaporized portion of the crude leaves the bottom of the tower via a steam stripper section, while the distillate vapors move up the tower counter current to a cooler liquid reflux stream. Heat and mass transfer take place on the fractionating trays contained in this section of the tower above the flash zone. Distillate products are removed from selected trays (draw-off trays) in this sections of the tower. These streams are stream stripped and sent to storage. The full naphtha vapor is allowed to leave the top of the tower to be condensed and collected in the overhead drum. A portion of this stream is returned as reflux while the remainder is delivered to the light end processes for stabilizing and further distillation.

The side stream distillates shown in the diagram are:

- Heavy gas oil (has the highest Boiling Point)
- Light gas oil (will become Diesel)
- Kerosene (will become Jet Fuel)

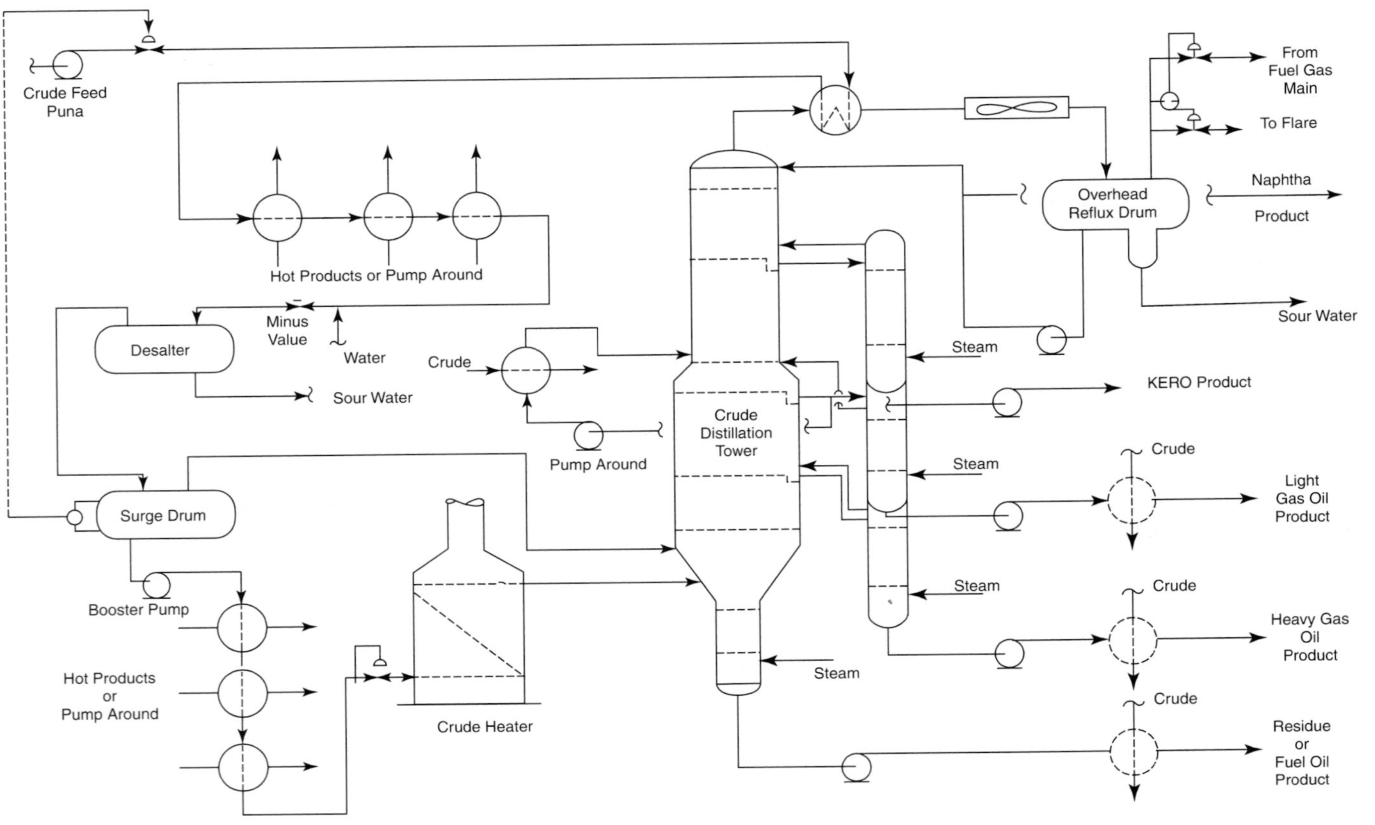

Figure 3.2. A typical atmospheric crude distillation unit.

A 'Pump around' section is included at the light gas oil draw off. This is simply an internal condenser which takes heat out of that section of the tower. This in turn ensures a continued reflux stream flow below that section. The product side streams are stripped free of entrained light ends in separate stripping towers. These towers also contain fractionation trays (usually four but sometimes as many as six) and the side stream drawn off the main tower enters the top tray of its respective stripper. Steam is injected below the bottom tray and moves up the tower to leave at the top, together with the light ends strip out, and is returned to the main fractionator at a point directly above the side stream draw-off tray. These side stream stripper towers are usually stacked one above the other in a single column in such a way as to allow free flow from the side stream draw-off tray to its stripper tower. On a few occasions, where the particular side stream specification requires it, the stripping may be effected by reboiling instead of using steam. One such requirement maybe in the Kero side stream if this stream is to be routed directly into jet fuel blending and therefore must be dry.

The residue (unvaporized portion of the crude) leaves the flash zone to flow over four stripping trays counter current to the flow of stripping steam. This stripping steam enters the tower below the bottom stripping tray. Its purpose primarily is to strip the residue free of entrained light ends. The fact that this steam enters the flash zone it also enhances the flashing of the crude in this zone by creating a reduced partial pressure for the liquid/vapor separation. This becomes an important factor in the design and operation of the atmospheric crude distillation unit. The stripped residue leaves the bottom of the unit to be routed either through the unit's heat exchanger system and the to product storage or hot to some down stream processing unit such as a vacuum distillation unit or a thermal cracker.

The development of the material balance for the atmospheric crude distillation unit

The knowledge of the material balance in any refining process is important both for ensuring its proper design and later for its proper operation. Because of the relative number of streams involved this is particularly so in the case of the atmospheric crude distillation unit. The operation of this unit also is critical to the performance of down stream units such as catalytic crackers and catalytic reformers. The material balance for any specific operation required of the unit, (for example, maximizing naphtha feed to the catalytic reformer), from a particular crude feed source also sets the performance parameters of the unit. This includes the amount of reflux to be generated, at which section this reflux is to be generated, distillate draw-off temperatures, tower overhead conditions, flash zone conditions and the like.

Whole crude TBP curve and assay data

The development of the unit's material balance begins with an 'in depth' examination and analysis of the crude oil feed's assay. The first step in doing this activity is to

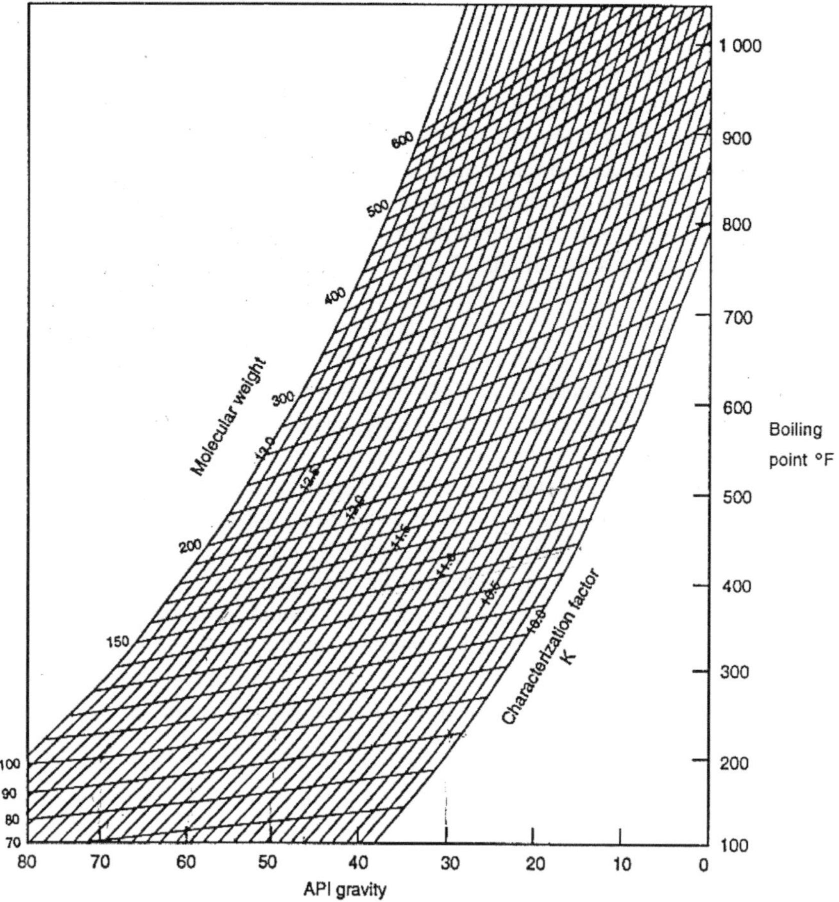

Figure 3.3. Boiling points V's molecular weight and gravity (°API).

break up the crude feed's TBP distillation curve into its pseudo components and to assign the product characteristics relative to each component's mid boiling point or mid volume point on crude.

The properties of each pseudo component are provided either by the assay data giving the gravity versus mid volume percent (or mid boiling point), and by the application of Figure 3.3. This chart gives the component's molecular weight based on boiling point (i.e., mid boiling point) and its gravity in °API. Every care must be made to establish these properties as accurately as possible. It will form the basis of any design process or definitive study work that may be carried out on this unit.

Developing the side stream product TBPs

The first step is to divide the crude feed's TBP curve into the product yield (in percent volume) and in terms of that product's temperature cut range. This starts with establishing the cut point of the residue. That is the temperature on the TBP curve at which it is intended to separate the total distillate from the residue. Then the distillate portion of the TBP is divided into its product cuts such as Gas Oils, Kerosene, and the total overhead Naphtha, or some other combination of products. The volume of these cuts and temperature ranges represent the yield of each product that will be produced as a percentage of the whole crude.

From the cut points of each of the products their ASTM curves are developed using the method described in Chapter 1. The initial boiling point (IBP) of each cut (except the overhead Naphtha) is fixed by the fractionation capability of the distillation unit. This term 'fractionation capability' for a crude oil distillation unit is measured as the difference in temperature between the 95% volume point of the lighter cut's ASTM distillation and the 5% volume point of the adjacent heavier cut. This difference may be positive (Gap) or negative (Overlap). A gap indicates good separation while the overlap indicates poor separation. The ability to separate the fractions efficiently decreases as the products become heavier. Thus, one can expect an ASTM gap between the overhead product and the first side stream to be around 25°F, while that between a third and fourth side stream to have an ASTM gap around −10°F (an overlap of 10°F).

The side stream TBP curves are now developed using these concepts and following these stepwise procedures.

Step 1. Establish the cuts and cut ranges on the crude TBP that represents the products that will be produced in this unit. For example—The overhead product will be a full range naphtha and will contain all the gas in the crude and the distillate to a cut point of 400°F. The first side stream will be a Kerosene cut beginning at 400 and ending at 500°F on the crude TBP curve. The next side stream will be a gas oil boiling between 500 and 650°F on the TBP curve. These will have yields as % volume on crude from the TBP curve. As an example typical cuts are given as Figure 3.4.

From this figure:

Naphtha (gas to 400°F) = 30 %vol on crude.
Kerosene (400–500°F) = 10.5 %vol on crude.
Gas oil (500–650°F) = 14.5 %vol on crude.
Residue (+650) = 45.0 %vol on crude.

Step 2. Predict the ASTM end point and the ASTM 90% point using the figures given in Chapter 1 of this Handbook for the full range Naphtha (i.e., the overhead product).

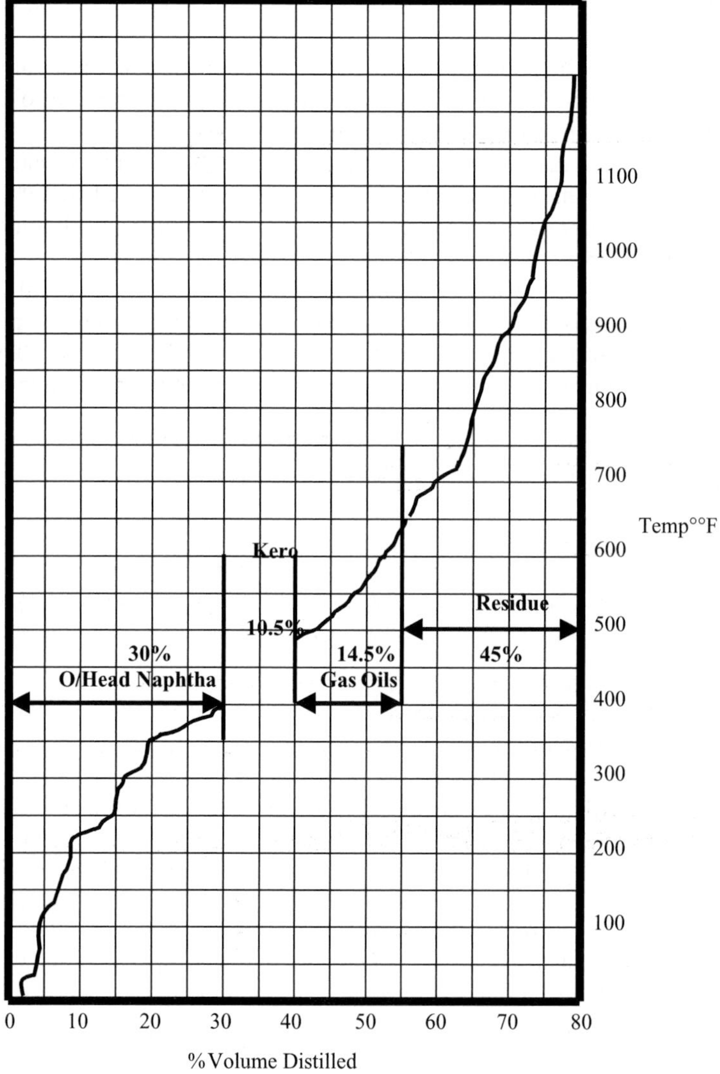

Figure 3.4. TBP and product cut points for a typical mid east crude.

Step 3. Plot these on the probability chart, also given in Chapter 1 and draw the straight line through the 30 %vol point. This will be sufficient to determine a meaningful ASTM curve from which a TBP curve can be produced.

Step 4. The ASTM curves for the remaining side stream products are developed using the curves in Chapter 1 for fixing their ASTM end points. The 5% points on both

ASTM curves are then fixed using a reasonable ASTM Gap or overlap between the products thus:

Step 5. The 5% ASTM point for the kerosene cut will be the 95% point of the full range naphtha plus 25°F gap. Similarly the 5% ASTM point for the gas oil will be the 95% point of the kerosene plus a 0°F gap.

Step 6. The ASTM curves for the side streams are drawn as straight lines on the probability chart between their respective end points and their predicted 5% points.

Step 7. Convert the developed ASTM curves to TBP curves using the Edmister correlations and as described in Chapter 1.

Step 8. Extend the front end of the full naphtha to include the gas portion and the light distillate below the 30 %vol point. Step this section off in mid boiling points to simulate real hydrocarbon components such as C5's, C6, C7, etc. This will become important later in establishing the reflux drum pressure and temperature.

An example of this method is given later in the worked example of an Atmospheric Crude Distillation Unit process design.

Developing the product volume, mass, mol balance

Using the component breakdown (pseudo components) and the product TBP curves, calculate each product volume rate, mass rate, and mole rates using the following steps:

Step 1. Establish the crude feed flow rate in terms of volume (usually BPSD), then calculate its mass flow (say in lbs/hour) and molal flow, using the crude feed breakdown table described in the previous section.

Step 2. Develop each products' specific gravity using its component composition and each component's specific gravity as given in the Crude Breakdown Table.

Step 3. Develop each product's mol weight similar to step 2 and again referencing the Crude Breakdown Table. There is relationship between gravity, boiling point, and mole weight. This is given in Figure 3.3.

Step 4. From the data developed in steps 2 and 3 calculate the quantity of each product in terms of BPSD, lbs/hr, moles/hr. The sum of each of these product quantities must equal the quantity of the crude oil feed calculated in Step 1.

This completes the description of the material balance development.

The design characteristics of an atmospheric crude distillation fractionating tower

In modern day refining the separation of the basic products from the crude feed is generally accomplished in a single atmospheric distillation tower. There are circumstances however that lead to the use of two towers to accomplish this. These circumstances usually occur where there is an abnormally high quantity of light components in the feed (such as crude spiked with naphtha) or where the unit capacity revamp calls for such a configuration. In these circumstances a light cut is flashed off and fractionated in a 'Primary' tower. The bottoms from this tower is then heated and vaporized to enter the flash zone of a 'Secondary' tower where it is fractionated to meet the conventional overhead, and side distillate products.

This section of this chapter will deal only with the single atmospheric crude distillation process however.

The tower flash zone

As described earlier the crude oil feed is heated to its required temperature for separating its distillate products from the unvaporized portion (i.e., the residue) by heat exchange with products being cooled or condensed and finally by a fired heater. This heated crude enters a section of the atmospheric fractionating tower at a point below the fractionating trays. This section of the tower is called The Flash Zone. The temperature and pressure conditions in this zone are such as to allow the correct amount of vapor to be flashed from the heated crude to provide the quantities of the distillate product streams and the internal reflux to control their quality.

These flash zone conditions of temperature and pressure are determined taking into consideration the shape of the crude oil's equilibrium flash vaporization (EFV) curve, the pressure (usually the partial pressure in the case of the atmospheric tower) existing in this section of the tower and the temperature on the EFV corresponding to the volume % cut of the total vapor to be flashed.

Setting the flash zone partial pressure

Because steam is used to strip the residue leaving the tower free of light ends this steam enters the flash zone from the top stripping tray. The steam influences the pressure condition at which the hydrocarbon vapors separate from the residue. This pressure condition is the partial pressure exerted by the steam. It is calculated by the equation as follows:

$$\text{Partial pressure of HC vapor} = \frac{\text{moles HC vapor}}{\text{moles HC vapor} + \text{moles steam}} \times \text{Total pressure}$$

The total pressure of the flash zone is determined by the pressure in the tower's overhead reflux drum plus the pressure drop over the overhead condenser(s), and the total distillate trays and internals above the flash zone. For an estimate of the flash zone partial pressure the following rules of thumb may be considered:

Overhead reflux drum pressure	5 psig
Pressure drop over the condensers	7 psi
Pressure drop over trays etc	10 psi (assumes 40 trays @ 0.25 psi per tray)
Assume ΔP over internals are negligible	
Then total pressure	22 psig.

Assume that the steam from the residue stripper will be 1.2 lbs/hot gallon of residue.

Setting the flash zone temperature

Calculate the EFV of the whole crude using the method described in Chapter 1 of this Handbook. This EFV curve is at atmospheric pressure. The temperature of the flash at atmospheric pressure is read off as the % volume vaporized to meet the amount of distillate products *and* the over flash required by the process. This over flash is usually fixed at between 3 and 5% volume on crude. Its purpose in the process is to provide that extra heat in order to generate sufficient reflux down flow over the trays to satisfy the prescribed degree of separation between the products.

This atmospheric flash temperature is now adjusted to the temperature at the previously calculated partial pressure existing in the flash zone. This may be done by reading the temperature at the partial pressure using the PVT curve shown as Figure 1.A.1 in the appendix to this chapter. This is the flash zone temperature.

Other features of the flash zone section

Most crude oils contain sulfur in some form or other and organic chlorides which are not removed in the pre-treating desalter. Corrosion from these impurities is particularly virulent in mixed phases (liquid/vapor) and at the elevated temperature experienced in the unit's flash zone, and made more so by the presence of steam. A cladding of 11/13 chrome is usually applied as a protective cover over the tower's carbon shell in this section. This cladding should be extended also to include the residue stripper and the section of the tower containing the first 4–6 wash trays above the flash zone. These trays and the residue stripping trays should also be of 11/13 chrome alloy.

Effective separation of the distillate vapor phase and the unvaporized residue phase is enhanced by the inclusion of a 'swirl' at the inlet of the flash zone. The crude from the fired heater is routed through this 'swirl' which is an inverted trough and extends around 2/3rds of the tower circumference. The swirling action caused by this forced flow of the mixed phases allows the lighter vapor phase to separate to a large extent from the heavier liquid phase. The final separation occurs on the top stripping tray.

The fractionator overhead system

The fractionator overhead equipment has four functions. These are:

- To condense the overhead vapors including the stripping steam
- To return and control the reflux condensate to the tower
- To collect and dispatch the overhead product
- To separate and dispose of the condensed steam

These functions are accomplished in several different ways. Some of these overhead configurations are described in the following paragraphs.

Configuration 1

This is probably the most common system and is shown in Figure 3.5.

The total overhead vapor leaves the top of the tower at its dew point. It first enters a shell and tube exchanger where it is partially condensed. The cooling medium is cold crude oil feed from storage. The partially condensed overhead leaves the shell side of this crude oil exchanger to be further and totally condensed by either a trim water cooler or, more usually an air condenser. The condensate from this final condenser enters an overhead condensate drum. This drum is designed to allow the complete separation of the condensed steam in the overheads from the hydrocarbon condensate. This condensed aqueous phase is collected in a 'boot' located below the main condensate drum and is pumped under level control to the refinery's sour water disposal system. A portion of the hydrocarbon phase is pumped to back to the fractionating tower to enter above the top tray. This stream is usually flow controlled which is reset by a tower top temperature control. The stream enters the tower through a spray system designed to ensure good distribution over the top tray.

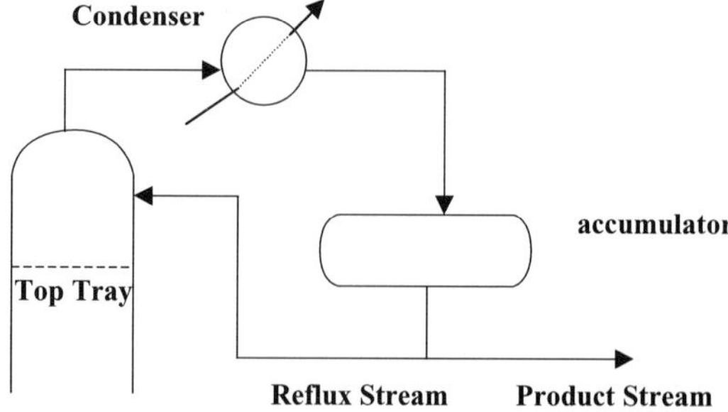

Figure 3.5. The most common overhead configuration.

The remaining hydrocarbon condensate leaves the drum under level control and is pumped to storage or a downstream process plant. The operating temperature and pressure of the condensate drum is vital to the operation of the fractionating tower. These conditions set the parameters for flash zone operation and intermediate product draw-off temperatures. These conditions are set and operated as follows.

The drum pressure. This is maintained by two pressure control valves operating on a split flow control. One of the valves is connected to the refinery flare while the other is connected to the unit's fuel gas supply. When the pressure in the drum exceeds the pressure set point of the controller, the first valve opens to the flare thus reducing the drum pressure. When the pressure falls below the set point then fuel gas is introduced by the second valve to correct this condition.

The drum temperature. The drum temperature is held at the bubble point of the condensate or slightly super cooled. This temperature is controlled either by fan pitch or by louvres on the final air condenser. Should the final condenser be a shell and tube water trim exchanger then a bypass of the condensate around the exchanger activated by a temperature/flow control is installed. Such an arrangement can also be used on an air condenser.

The tower overhead temperature. The tower overhead temperature in this configuration is the dew point of the overhead product vapor at the partial pressure of the hydrocarbon.

Note: The partial pressure for this case is the total moles of hydrocarbons (includes the Reflux moles) divide by the total moles of vapor (includes the steam present) times the total overhead pressure.

The tower overhead pressure. The overhead tower pressure in this case is the pressure of the overhead condensate drum plus the pressure drops of the heat exchangers, and the associated overhead condensate piping. This pressure drop may be taken as 7–9 psi.

Configuration 2

This second configuration is usually installed on units of high capacity at throughputs above 70,000 BPSD. Its purpose is to maintain the size of the overhead equipment to manageable dimensions. This includes drums and the heat exchangers associated with the system (Figure 3.6).

In this configuration the total overhead vapors are again condensed by two exchangers as described in Configuration 1. In this second case, however, the condensate from the first condenser is collected in a drum and returned to the tower as reflux. The vapor from this first drum includes the overhead product hydrocarbon and the uncondensed

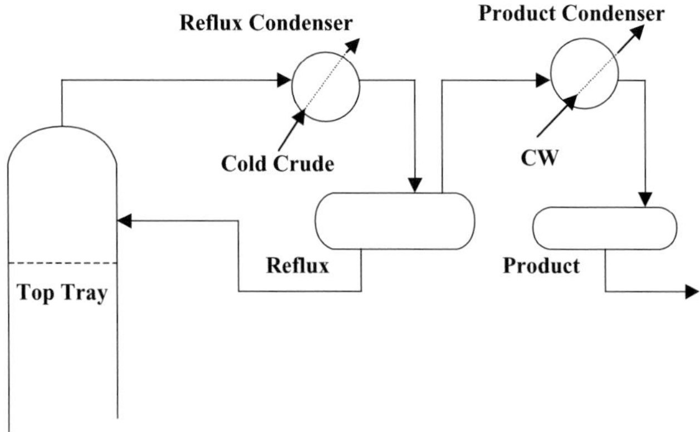

Figure 3.6. The 'Two Drum' configuration.

steam from the fractionator. This vapor is condensed in the second overhead condenser (again usually an air condenser) to be collected in a second condensate drum as the overhead distillate product and water. Disposal of the separated aqueous and hydrocarbon streams follow the same system as in Configuration number 1. In this configuration the reflux drum acts as an additional theoretical tray because the reflux liquid is in phase equilibrium with the product vapor.

The product drum pressure and temperature. This again is the controlling pressure for the fractionator as a whole. It is set at the pressure which allows the vapors to be condensed at a reasonable temperature at or below its bubble point. In moderate climates with average air temperatures of 60°F this temperature will be around 90–100°F with a pressure of 5 psig. Control of the drum pressure and temperature remains the same as that for configuration 1.

The reflux drum temperature. The conditions in this drum may be considered as those for the top tray of the tower in Configuration 1. The temperature therefore is the dew point of the product distillate at the partial pressure of the hydrocarbons vapors leaving the drum. This temperature is usually controlled by a vapor bypass over the exchanger on a flow reset by temperature control valve. Alternatively it may be controlled by the crude oil being bypassed on a similar flow/temperature reset basis. The partial pressure in this case is the moles of product vapor only divided by the total moles (including the steam) times the total absolute pressure. This total pressure is the pressure in the distillate product drum (say 5 psig) plus the pressure drop of the air condenser and piping (say about 3 psi).

The tower overhead temperature and pressure. The tower top conditions in the case of this second configuration is calculated as the dew point of the reflux stream entering

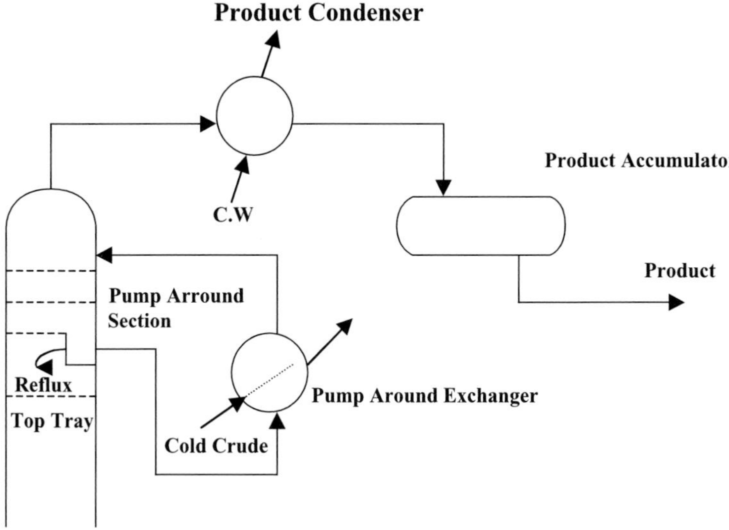

Figure 3.7. The tower top pumparound configuration.

the tower from the reflux drum. This temperature will be at the partial pressure of the total hydrocarbons leaving the tower.

Configuration 3

In this configuration the Tower itself is used to generate the internal reflux from the overhead product (Figure 3.7).

In this configuration the reflux is generated by a pump around system taking the liquid from the top section of the tower, and cooling it by heat exchange with cold crude oil. This cooled stream is then returned to the top tray of the tower. This is a conventional pump around system. About three trays are used for this pump around section in the tower. The tower top temperature remains as the dew point of the overhead product distillate at its partial pressure with steam. The temperature of the liquid leaving the tower as the pump around will be between 5 and 7°F higher than the tower top temperature. The internal pump around section may be assumed to be equivalent to one half an actual tray for mass transfer.

Discussion

All three of these configurations have their advantages and disadvantages. These are discussed briefly in the following paragraphs:

Configuration 1. As said earlier this is the most common configuration. It is relatively easy to control with really only two set points that one has to be concerned with. They are the drum pressure and the tower overhead temperature. There is good heat transfer between the condensing vapors and the crude oil. The overall heat transfer coefficient is high at around 70–80 Btu/hr/sqft/°F and advantage is taken in the LMTD of the latent heat of the vapors being transferred to the low temperature of the incoming crude. The biggest disadvantage with this system is that the crude to the first condenser is of course at a much higher pressure than the vapors condensing on the shell side. The rupture of a tube in the exchanger bundle or a leak in the tube plate will cause crude to contaminate the overhead product and 'dirty' the tower overhead trays. This can be quite a disaster if the downstream units (particularly the Cat Reformer) is fed directly from the condensate drum without some intermediate surge drum or tankage.

Configuration 2. This has the advantage of providing a very high LMTD in condensing with heat exchange against crude. This is due to the higher temperature now experienced in the vapors leaving the tower. As in the case of configuration 1, the overall heat transfer coefficient in this case is also high. There still remains though the problem of contamination but in this case downstream units have a measure of protection in that only the vapors from the reflux drum are condensed to make the product. The top trays of the tower however will be contaminated as before. The two drum system's biggest advantage however is more aligned to the installed cost of the unit. Moderately sized equipment are much cheaper than large items which may have to be field fabricated. Cost savings may in any case be achieved in the transportation costs of the drums and heat exchangers.

Configuration 3. The only major advantage in this configuration is that the liquid pumparound can be circulated at a higher pressure than the incoming crude. Thus, the problem associated with tube fracture or leakage contaminating the tower and product is eliminated. There is also an advantage in better control of the tower top temperature by the flow control of the pumparound itself. This too can be reset by temperature, but as an ex plant operator the author is not particularly in favor of temperature reset as a finite means of control. There is no other apparent advantage to this system. Indeed the inclusion of additional trays in the tower for the pumparound is quite expensive. The trays would need to be of monel and some monel lining of the shell will be required in this section as protection against chloride attack.

Some design considerations

Corrosion

The top section of the atmospheric distillation tower is very prone to corrosion from HCl. To combat this corrosion and to protect the carbon steel shell from damage monel is used in tray manufacture and as lining for the tower shell. This material is also used to line the first condenser shell interior and the tubes may also be fabricated in monel. Ammonia either in its anhydrous form or as a solution, is injected into the tower top and at specific intervals into the top section tray spaces. The purpose of the ammonia is to neutralize as far as possible the HCl. The injection of the ammonia is controlled by the Ph reading of the condensate water leaving the boot of the product condensate drum.

Corrosion is most virulent at the liquid interface on trays. Consequently in most modern fractionating towers monel lining of the shell is usually applied as strips on the top four or six trays. These strips installed at the interface of the liquid/vapor on these trays the strips will be about 4–6 ins wide. The injection of anhydrous ammonia in the case of configuration number 2 does cause problems in the formation of ammonium chloride crystals occurring in the 'dry' section of the condenser system. That is the section where the hydrocarbon is condensing but the steam is still in its vapor phase. The injection of aqueous ammonia solution instead of the anhydrous form solves this problem.

Sulfur compounds are also a problem with respect to corrosion particularly in the 'cold' liquid phase of the distillate product drum. The drum shell in this case is gunnite lined. This lining is applied to the bottom section of the main drum and extended to include the 'boot' in which the condensed water is concentrated. Gunnite is best described as a cement lining and is applied in the same way as refractory lining, by plastering it on to the shell.

The distillate drum

The design of the distillate drum includes space for the complete separation of the aqueous phase (condensed steam) from the hydrocarbon phase. It is important for the hydrocarbons to be as free of water as possible to protect the downstream units. The performance of these units such as hydro-treaters and, more particularly the catalytic reformer are badly effected by the presence of water in their feed streams.

The design of the drum to allow this separation is based on one of three laws. These laws are expressed by the following equations:

$$\text{Stoke's Law: } V = 8.3 \times 10^5 \times \frac{(d^2 \Delta S)}{\mu}$$

(used when Re number is < 2.0)

$$\text{Intermediate Law: } V = 1.04 \times 10^4 \times \frac{d^{1.14} \Delta S^{0.71}}{S^{0.29}c \times \mu^{0.43}}$$

(used when the Re number is 2–500)

$$\text{Newton's Law: } V = 2.05 \times 10^3 \times \{(d \Delta S) \div \mathrm{Sc}\}^{1/2}$$

(used when the Re number is > 500)

where

Re number $= \frac{10.7 \times dv\text{Sc}}{\mu}$

$V =$ Settling rate in ins per minute.

$d =$ Droplet diameter in inches.

$S =$ Droplet specific gravity.

$Sc =$ Continuous phase specific gravity.

$\Delta S =$ Specific gravity differential between the two phases.

$\mu =$ Viscosity of the continuous phase in Cps.

The following criteria may be used as a guide to estimating the droplet size:

Lighter phase	Heavy phase	Minimum droplet size
0.850 Sc and lighter	Water	0.008 inches
Heavier than 0.850	Water	0.005 inches

Piping

Vapor piping from the tower top should be sized for a minimum economic pressure drop. As a guide vapor rundown lines should be sized to meet an average pressure drop of 0.2 psi per 100 ft and should never exceed 0.5 psi/100 ft. The vapor (and mixed phase) piping and the drums in the tower overhead system should be so arrange as to allow free drainage towards the distillate drum (or in the case of configuration 2 to each successive drum). The vapor lines should never be pocketed.

Distillate drums should be located at least 15 ft above the center line of the respective distillate/reflux and the sour water pumps (pumps are normally located at grade). For safety this measurement should be taken from the bottom tangent line of the vessel. Piping to the pump should be sized for a 0.25 psi/100 ft pressure drop, while the discharge piping from the pump may be reduced to meet a pressure drop of 2.5–4.0 psi/100 ft.

The side streams and intermediate reflux sections

There are generally two or three side streams in the atmospheric crude distillation unit. There may be one more but a total of three side streams is the norm. The unit, that is going to be developed as an example in this work, has three side streams. These are:

- Kerosene
- Light gas oil
- Heavy gas oil.

With three side streams there will be usually two pumparound sections. The purpose of these pumparound sections is to create the proper internal reflux to that section of the tower below the location of the respective pumparound. The pumparound takes suction from a draw-off tray in the tower. Usually this tray is also a product draw off. The pump delivers the pumparound stream through a heat exchanger where it is cooled before returning it to a location in the tower about two to three trays above the draw-off tray. This cooling liquid flowing down the tower to the draw-off tray acts as an internal condenser. The hot vapors rising up the tower through the draw-off tray contain the side stream product, some internal overflow and the vapor phase of all the lighter products and their overflow. This particular side stream product and overflow is selectively condensed by the cold pumparound. It enters the draw-off tray where the product quantity and pumparound leave the tower, and the remaining condensate overflows to the tray below the draw off. This over flow material is the internal reflux stream and its quantity determines the fractionation between this lighter side stream product and the heavier side stream product drawn from the tower at some point below.

The side streams

Side streams are withdrawn from selective and specially designed trays in the main tower. Each stream is usually combined with the respective pump around as described earlier. When withdrawn from the tower the product is in equilibrium with the total vapor rising through the draw-off tray. Consequently the product liquid will contain entrained light ends which will affect the required product properties such as flash point and its ASTM distillation. To correct this the side stream product is steam stripped in a separate trayed stripping column. This arrangement is shown in Figure 3.8.

The side stream from the main tower is routed to a distributor over the top tray of the stripping column under level control. The stripping column contains four stripping trays. Steam is introduced below the bottom tray and flows up the tower counter current to the liquid product. The light ends contained in the stripper feed are removed by the steam and both the stripout and steam leave the top of the stripper to enter the main tower at the vapor space directly above the draw-off tray.

The stripped product leaves the well at the bottom of the stripper to be pumped under flow control to storage or a downstream unit's surge drum. The level in the well of the stripper activates the level control valve of the incoming stripper feed.

The side stream draw-off temperature. In theory the side stream product draw off from the main column will be in equilibrium with the vapor rising from the tray. In other words it will be at its bubble temperature at the partial pressure of the hydrocarbons on the tray. However, this liquid phase will contain some entrained light ends (Thus the need for steam stripping) which will affect the actual temperature of the draw off. J. W. Packie in his published work on crude oil distillation developed a curve which

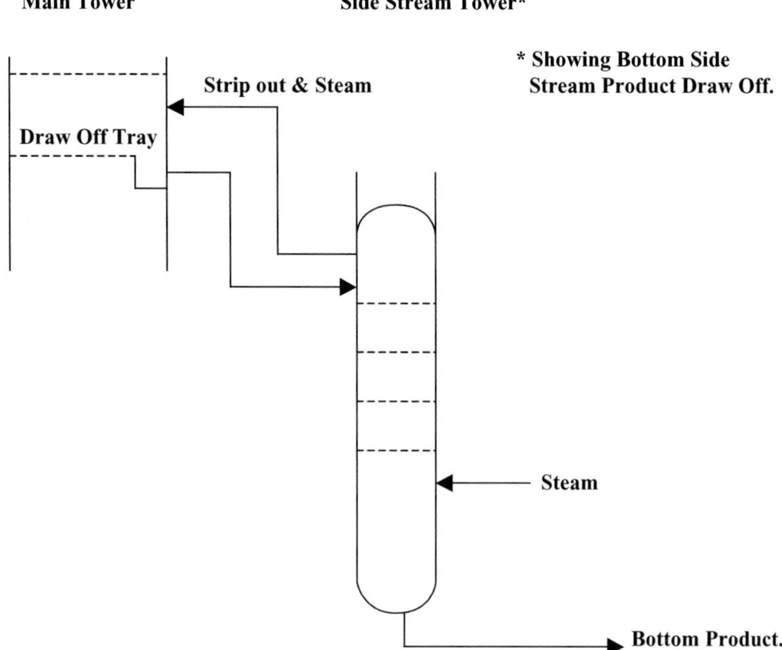

Figure 3.8. A sides tream stripper.

takes into consideration this difference between the theoretical draw-off temperature and the actual. This curve is given as Figure 3.9.

In arriving at the partial pressure in this calculation Packie assumes that all the vapor lighter than the draw-off is inert. With this as a parameter the draw-off temperature can be calculated using the following step wise process:

Step 1. Fix the amount of steam present in the vapor *passing through* the draw-off tray (*Note*: This does not include the steam returning from the particular draw-off stripper which enters the vapor space above the draw-off tray). The amount of stripping steam recommended is given in Figure 3.10.

Step 2. Fix the total pressure on the tray taking the estimated tower top pressure and each tray pressure drop above the draw-off tray (see the previous paragraphs on flash zone and overhead system). Estimate the position of the draw-off tray. As a guide the following rule of thumb can be used. With tray 1 as the top tray in the column:

Kero draw off	tray 12
Light gas oil	tray 22
Heavy gas oil	tray 32

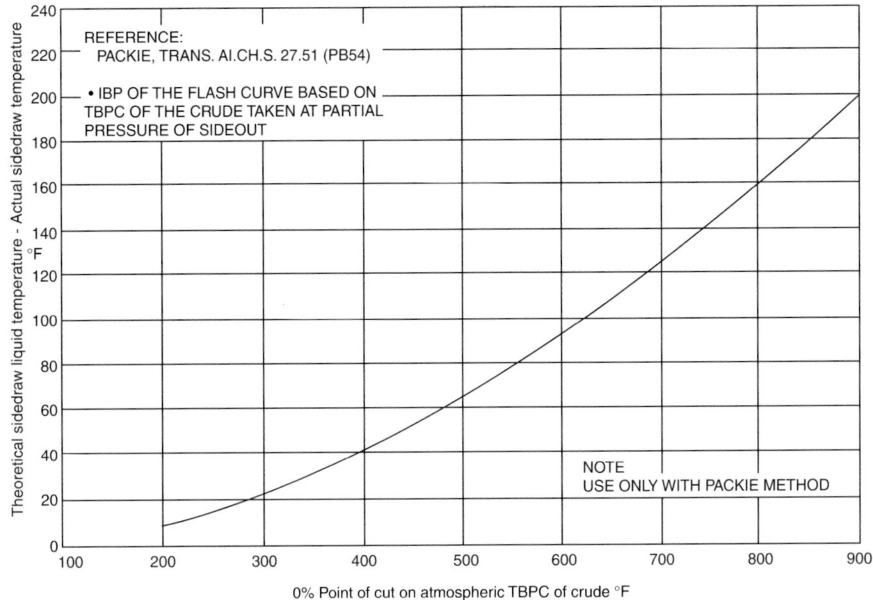

Figure 3.9. The difference between theoretical and actual draw-off temperatures.

There are usually 40 fractionating trays (i.e. Trays above the flash zone) and 4 bottom stripping trays in conventional distillation towers. These amounts will be checked later using the relationship of reflux to tray ratio for fractionation criteria.

Step 3. Calculate the FRL (Flash Reference Line) for the draw off TBP cut. In doing this use the TBP cut; not the TBP end points. This calculation has been described under the section on the Flash Zone. Establish its IBP at atmospheric pressure. This will be the 0 %vol temperature on the FRL.

Step 4. Predict the amount of overflow that will leave the draw-off tray as a liquid reflux to the tray below. Again this will be checked by the relationship of trays and reflux later. For this calculation the following rules of thumb for the ratio of moles overflow to moles product may be used:

Overflow from the HGO draw off 2.9
Overflow from the LGO draw off 1.2
Overflow from the Kero draw off 0.9 to 1.0

Step 5. Calculate the partial pressure on the draw-off tray as follows:

$$\text{Partial pressure} = \frac{(\text{total draw-off product vapor moles} + \text{total overflow moles})}{(\text{total hydrocarbon vapor moles} + \text{Steam passing through tray})} \times \text{total pressure}$$

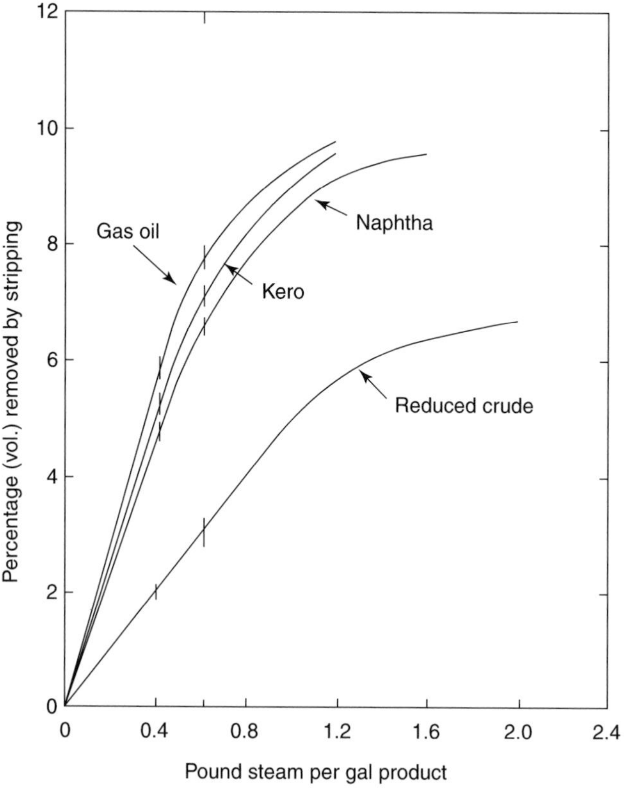

Figure 3.10. The amount of stripping steam.

Step 6. Using the vapor pressure curves (given in the appendix as Figure 1.A.1) Relate the IBP of the cut at atmospheric pressure (step 3) to the calculated partial pressure.

Step 7. The temperature arrived at in Step 6 is the theoretical temperature of a liquid in equilibrium with its associate vapor phase. In actual practice the fact that a vapor phase has been continually bubbling through the liquid on the tray, that liquid will contain entrained light ends from the vapor. The actual temperature will therefore be somewhat lower than the calculated, and this difference is provided by Figure 3.9. Establish the actual temperature of the side stream draw off. Note this will also be the temperature of the pumparound liquid also.

Side stream stripping. The side stream enters a steam stripper at the draw-off temperature. It is steam stripped and the amount of stripping steam and the corresponding strip out may be read off from Figure 3.10. The steam introduced below the bottom stripping tray is usually superheated at 50 psig. The temperature of the stripout and steam leaving the top of the stripper to enter the main tower can be taken as 5°F below the temperature of the stripper feed. The mol weight of the stripout is taken as the average of the vapors lighter than the draw off. A heat balance can now be carried out over the stripper with an unknown being the stripped product enthalpy and temperature. The product outlet temperature and enthalpy is calculated. By solving the equation:

Heat in = Heat out.

Side stream strippers for each product draw off are usually stacked in one single column. It becomes an exercise in layout to locate these stripper sections. It begins with establishing the height of the bottom stripper section which will ensure sufficient head to provide available NPSH for the product rundown pump. Thereafter each successive section is located to ensure free flow of the liquid feed into the respective stripper, and to minimize the length of the vapor return line to the main tower.

The pump arounds

The mechanism of the pumparounds and their purpose has already been described and discussed. It remains now to examine how to evaluate and detail quantitatively their size in terms of the duty they will be required to perform. This will be followed with some criteria that need to be observed in their design.

Total pumparound duties. This begins by establishing the Tower bottom temperature (Residue outlet temperature) and then conducting an overall Tower heat balance. To calculate the Tower bottom temperature consider the following diagram (Figure 3.11):

And then the following calculation steps:

Step 1. Establish the material entering the top stripping tray. This will be the total residue product which now includes the overflash as a liquid, and the stripout material (also as liquid). This is taken as being at the flash zone temperature.

Step 2. Calculate the total heat input to the bottom stripper. This will include the heat into the top stripping tray plus the heat in the stripping steam (enters below the bottom tray).

Step 3. Calculate the heat out of the stripper. This includes the bottom product at an unknown temperature and enthalpy, the stripout from the top stripping tray, and the steam from the top stripping tray. Assume the stripout and the steam will be at 5°F below the flash zone temperature and the steam will be at the flash zone partial pressure for the steam.

Step 4. From the equation

Heat in = Heat out

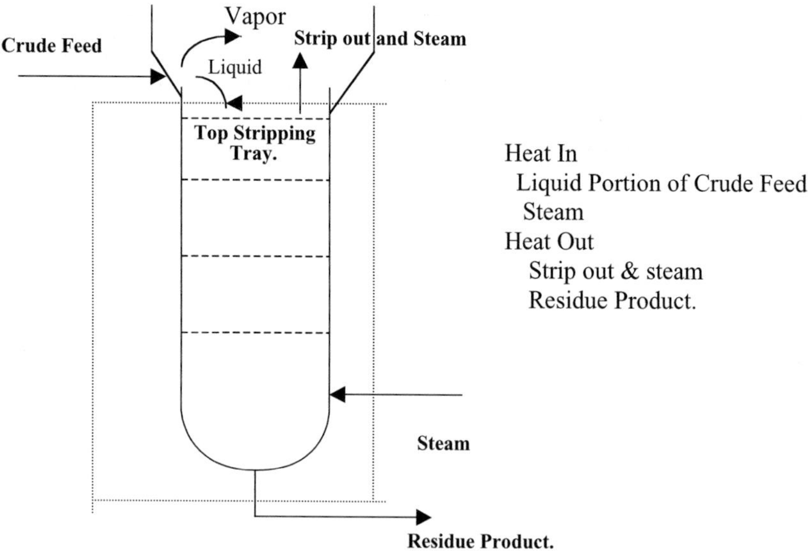

Figure 3.11. Heat balance diagram for below feed tray.

calculate the Enthalpy (in Btu/lb) and thus the temperature of the residue or bottom product leaving the tower.

Note: While the author uses enthalpy data from his own files, it is recommended that engineers use their own proven data from their own files, or data contained in Maxwell's *Hydrocarbon Data,* or those given in the *GPSA Engineering Data Book.*

With all the temperature and enthalpy criteria established an overall heat balance over the tower can proceed. In this balance the heat in with the crude will be that for the overflashed crude at the flash zone temperature. The other heat inputs are the steam streams to the various strippers. The heat out will consist of the overhead distillate product at the distillate drum temperature, the overhead condenser(s) duty, the condensed steam, all the side streams and residue as liquid and at their respective stripper bottom temperature, and finally the total of the pumparound duty.

As all the Enthalpy duties are known except that for the pumparound this can now be calculated from the expression

Heat in = Heat out.

Distributing the total pumparound to meet the required internal reflux

The degree of separation of one product from an adjacent one is determined by, the number of fractionating trays used for the separation and the liquid overflow (Reflux)

entering the section used for the separation. As has been said previously the degree of separation of products in the Atmospheric Crude Distillation tower is measured by the difference between the 95 %vol point of the lighter fraction and the 5 %vol point of the heavier fraction as measured by their ASTM distillation curve. Packie developed a relationship between the ASTM distillation difference, the number of trays, and the amount of overflow in terms of internal reflux ratio. This relationship is given by a series of curves in the appendix to this chapter (Figure 1.A.2). The use of these curves is illustrated by the following stepwise calculation procedure to arrive at the amount of overflow liquid.

Step 1. Establish the ASTM gap (or overlap) that is required. This is usually given in the design specification of the unit or by some constraint on one or other of the products. It is well to note here that ASTM gaps can usually be achieved between products lighter than light gas oil. Gaps can rarely be achieved within the tower for products heavier than the light gas oil cut. To trim the front or back ends of heavy products may require additional equipment such as vacuum flash equipment.

Step 2. Establish the number of trays separating the two adjacent products. This has usually been fixed by the design of the tower in operating units. For design purposes the following Fractionating number of trays can be used as a start point and guide:

	Tray number
Top tray	1
Kero draw off	12
LGO PA return	19
LGO draw off	22
HGO PA return	29
HGO draw off	32

As a rule of thumb the section of trays for the pumparound may be counted as one fractionating tray. Thus for example 10 + 1 trays can be counted for the separation of Kero and LGO.

Step 3. Evaluate the 50 %vol temperature difference between the TBP of the two adjacent cuts.

Step 4. Using the side stream to side stream family of curves in Figure 3.A.2 for max steam read off the factor corresponding to the ASTM difference required on the appropriate Δ50% TBP temperature curve.

Step 5. The factor read is the number of trays separating the two products' draw-off multiplied by the Reflux ratio. This Reflux ratio is defined as:

$$\frac{\text{Hot overflow in gallons per hour leaving the lightest product draw-off tray}}{\text{Total product gallons per hour entering the lightest product draw-off tray}}.$$

Equate to determine the overflow quantity, in hot gallons per hour. The units material balance is used to resolve the flows to lbs per hour.

It now only remains to utilize the overflow criteria in the overall tower heat balance to arrive at a duty for the respective pumparound. If there is only one other pumparound in the tower it can be determined by subtraction of the calculated pumparound from the total PA duty. Should there be more than two pumparounds involved then the calculation steps given above must be repeated for the second one, and the third obtained by subtraction.

On completion of setting the individual pumparound duties as described the items determined by subtraction must be checked to ensure that they do in fact generate enough overflow to meet the fractionation criteria. If they do not fulfill this then the overall pump: around duty must be increased. This can be achieved only by increasing over flash if the cut points are to remain the same. Should the check calculations show there is somewhat more internal reflux than required, providing it is not excessive, it should remain as is. If it is excessive the over flash can be reduced accordingly.

Checking the number of trays allocated to the pumparound

With the individual pumparound duties now established, the temperatures to and from the pumparound section trays are the next items to be fixed. Use the following steps to accomplish this:

Step 1. Draw the tower temperature profile from the flash zone to the tower top tray. Use the already calculated side stream draw-off temperatures for this.

Step 2. Assume the vapor temperature to a tray is the liquid temperature on the tray below it. This is a reasonable assumption and well within the accuracy required.

Step 3. Fix the pumparound inlet temperature. This will need to be a guess at this point, and may well change on completion of the heat exchange system calculations. From the pumparound duty calculate the flow of pumparound liquid in lbs per hour.

Step 4. Carry out a heat balance over the pumparound section of the tower. This will now include the following items:

Heat in

With the HC vapor from the tray below the section (includes the overflow).
With steam from the tray below the section.
With steam and strip out from the product stripper (if applicable).
With the pumparound return liquid.

Heat out
With lighter product vapors rising from the top PA tray.
Total steam from flash zone and side stream strippers below top PA tray.
Liquid pumparound leaving the tower.
Liquid product draw off plus strip out leaving to stripper (if applicable).
Liquid overflow from the PA draw-off tray.
The pumparound duty in Btu/hr.

In these items the only unknown is the enthalpy and therefore the temperature of the vapor leaving the top pumparound tray. Equate Heat in = Heat out to solve the unknown.

Step 5. With the duty of the pumparound section in Btu/hr and the temperatures in and out now known, the pumparound section can be treated as a heat exchanger. The area of heat transfer will be the total tray areas of the section, and the heat transfer coefficient in terms of Btu/hr/sqft/°F is given in Figure 3.12.

Step 6. The area is now calculated from the heat transfer equation;

$$Q = UA \quad \Delta t_{\mathrm{m}}$$

where

Q = Heat transferred in Btu/hr
U = The overall heat transfer coefficient in Btu/hr/sqft/°F
A = Heat transfer area in sqft.
Δt_{m} = Log mean temperature difference in °F.

Step 7. The number of trays required will be determined from the tray or tower diameter calculation provided by a future calculation to determine the tower dimensions.

Calculating the main tower dimensions

Having established the number of trays and the relative location of pumparound and side stream draw off, the overall dimensions of the tower can be calculated. Note: The number of trays allocated to the pumparound sections may be revised based on the tray or tower diameter calculation for those sections. This will not affect the tray loading calculations which will determine the tray diameters. The calculation to determine the tower diameter(s) is based on the vapor/liquid loading on each section of the tower. This loading is based on a series of heat balances to determine this traffic to and from the critical trays. These critical trays are:

- The top stripping tray of the residue stripping section (below the flash zone)
- The bottom side stream and pumparound draw-off tray
- The next pumparound draw-off (and the side stream if applicable) tray

Units are in BTU/h. °F. Ft²

Fluid being cooled	Fluid being heated	U_o
Exchangers		
Naphtha pumparound	Crude	70–80
Kero	Crude	70–75
Debutanizer bottom	Debutanizer feed	75
Gas oil (inc. BPA)	Crude	40–50
Reduced crude	Crude	20–30
Light end bottoms	Light end feed	70–75
Vacuum distillates	Crude	30–40
Bitumen	Crude	20
Cat oil slurry	FCCU feed	40
CDU overheads	Crude	80–90
Coolers		
Debutanizer bottoms	CT water	75
Light naphtha	CT water	80
Gas oils	CT water	40
DEA or MEA	CT water	110
Reduced crude	CT water	30
Vapour heat exchangers		
Reformer effluent	Naphtha feed	38
Reformer effluent	Recycle gas	38
Crude tower overheads	Crude oil feed	50
Condensers		
CDU overheads	CT water	65
Splitter overheads	CT water	85
Amine stripper overheads	CT water	100
Debutanizer overheads	CT water	90
Reformer effluent	Water	65
Air coolers		
Naphtha coolers		23
Debutanizer bottoms		30
Light end overhead		40

Reboilers (use heat flux in BTU/h/ft²)

	Hydrocarbon	Water
18 in. dia. bundles	20 000	30 000
30 in. bundles	17 500	26 500
>30 in. bundles	15 000	23 000

Figure 3.12. Overall heat transfer coefficients in oil refining.

- Any other lighter pumparound draw-off tray
- Any lighter side stream draw-off tray
- The top tray

The loading below the bottom pumparound and product draw-off gives the vapor loading for setting the *maximum* tower cross sectional area and diameter. This is fixed by the crude feed cut point and the degree of over flash.

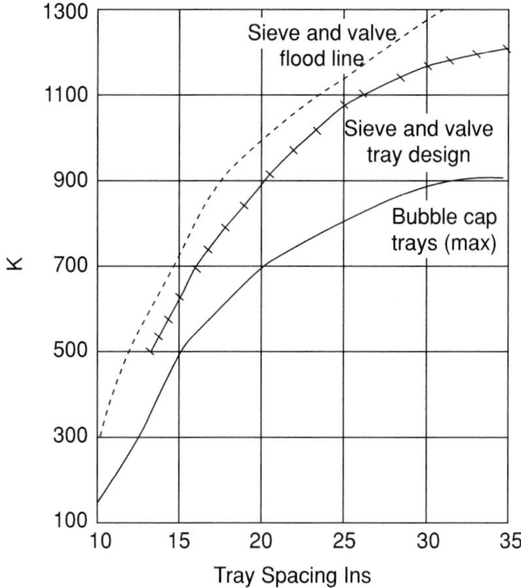

Figure 3.13. Values for K_f.

Calculating the tower vapor and liquid loading

These loadings at the critical trays are calculated from a series of heat balances starting at the bottom stripping section and continuing up the tower to the top fractionation tray. Consider the loading on the bottom side stream and pumparound draw-off tray. This commences with a heat balance from below this tray and includes the streams shown in the envelop in Figure 3.23.

The heat into and out of the envelope as shown is provided by the following streams:

Heat in: The crude vapor and liquid feed at the flash zone conditions.
The steam to the residue stripper.
The overflow liquid from the draw-off tray.
Heat out: The residue product.
The vapor of the total distillate products from the tray below the draw off.
The overflow vapor from the tray below.
The steam from the flash zone at the conditions of the tray below.

The temperature and quantities of these streams are known (from the material balance), except the quantity of overflow. Their temperatures are also known from the tower temperature profile or from previous calculations. To complete the calculation let the lbs of overflow be x and give all the streams their enthalpy value. Then solve the equation:

Heat in = Heat out

to find the value of x.

Repeat the heat balance calculation for the envelop ending below the next pumparound draw-off tray. In this case the vapor leaving the envelop will contain the overflow vapor from the pumparound tray, and all the distillate products except the heavier product drawn off from the pumparound tray below. It will also contain all the steam from the flash zone *and* from the side stream stripper of the heavier product. The remaining heat out of the envelope will be the stripped heavier product and the bottom pumparound duty in Btu/hr. The heat into the envelope will be as before except this time the unknown overflow quantity will be that flowing from the lighter pumparound draw-off tray. It will also include the steam from the heavier product side stream stripper.

The tower diameter. The tower diameter may now be calculated from the vapor flow to the tray and the total liquid load on the tray. The quantity of the vapor to the tray has already been established. The liquid load on the tray will be:

The product draw-off liquid.
The pumparound liquid.
The calculated over flow liquid.

These quantities must be in terms of weight and volume at the tray conditions of temperature and pressure.

The calculation procedure and data described below will give a good *estimate* of the tower diameter. For design purposes however the tray fabricator/designer's data and procedure should be used. Trays are usually proprietary items covered by patents, their performance therefore are subject to guarantees.

A calculation procedure to estimate the tower diameter is as follows:

Step 1. Summarize the liquid traffic through the tray.

Step 2. Select the type of tray that is to be considered for the design. (in the case of an existing unit refer to the fabricators drawings). Valve and sieve trays have reasonable similarity in their major characteristics. Bubble cap trays are seldom used these days. A table of the valve tray characteristics is given in the appendix as A.1.3.

Step 3. Compute the liquid loading on the tray being checked for size. This will include *all* the liquid entering the tray. For example on a side stream draw-off tray under a pumparound section it will include;

- The side stream feed to the stripper
- The pumparound liquid
- The liquid overflow from the tray

This loading should be in cubic feet per second (CFS).

Step 4. In the case of a new design, set the downcomer area in accordance with Table A.1.3. For an existing trays use fabricator's drawings. Calculate the linear velocity of the liquid in ft/sec. For good design this velocity should not exceed 0.6 ft/sec at the downcomer outlet. Tray spacing should be such that the liquid level in the downcomer should not exceed 50% of the tray spacing. To meet this criteria tray spaces on these critical liquid loading trays may be higher than the remaining tray spacing in the tower.

Step 5 . Summarize the vapor traffic to the critical tray under examination. This should be in lbs/hr and moles/hr. The vapor should be the total vapor as used to calculate the tray overflow.

Step 6. Calculate the flood vapor velocity Gf in lbs/hr sqft. For good tray design (or performance of an existing tray) the actual vapor velocity should not exceed 90% of this flood value. This vapor flood velocity is calculated using the following expression:

$$G_{\mathrm{f}} = K_{\mathrm{f}}\sqrt{\rho_{\mathrm{v}}} \times (\rho_{\mathrm{l}} - \rho_{\mathrm{v}})$$

where

G_{f} = Mass velocity in lbs/hr sqft of bubble area at flood.
ρ_{v} = Density of vapor at the tray conditions in lbs/cuft.
ρ_{l} = Density of liquid at the tray conditions in lbs/cuft.
K_{f} = constant based on tray spacing and given in Figure 3.13.

Step 7. Using the actual vapor load per sqft of bubble area as 90% of G_{f}, calculate the bubble area as G_{a} divided into the total vapor flow.

Step 8. Establish the following criteria using the characteristics given in Table A.1.3.

Where:

A_{s} = Total tray area in sqft.
A_{dc} = Inlet and outlet down comer areas in sqft.
A_{w} = Waste area (about 20% of A_{b}).
A_{b} = Bubble area.
$A_{\mathrm{s}} = A_{\mathrm{b}} + A_{\mathrm{w}} + A_{\mathrm{dc}}$

Use Figure 1.A.4 (relationship of chord height, area, and length) in the appendix as applicable.

Calculate the overall tower height

Using the calculated vapor and liquid calculations check and fix the number of trays used to meet the tower's fractionation requirements. Then check the tray spacing. For normal fractionation a 24 inch tray spacing is acceptable. For draw-off trays with pumparound the section should be checked for high downcomer filling (that is in excess of 50%). Trays in this section often require bigger tray spacing, usually 30 inch. Calculate the tower height allocated to the trays. Allow 6 ft between top tray and the top tangent line of the vessel to allow space for the reflux distributor and good liquid/vapor separation.

The space required for the flash zone (between the bottom fractionating tray and the top residue stripper tray) is usually 10–14 ft depending on throughput and the swirl size. The hot well below the bottom residue stripping tray should be sized to allow for the steam inlet distributor and a surge hold up of the residue product. This surge hold up is based on the company's operating policy, but as a guide:

- If the product is routed to storage 5 min hold up.
- If the product is to be fed into a fired heater of a downstream unit allow 15 min hold up.

The stripper column

This column is sized using the same procedure as used for the main column sizing. As mentioned earlier the strippers for the various side streams are usually stacked one above the other to form a single uniform diameter tower. The only diameter that may change is that for the hold up section in the bottom product stripper. This may be increased to adjust the height of the column to ensure the free flow of un stripped feed to each stripper from its respective draw-off tray in the main tower.

The number of trays for steam stripping is usually four. The tray spacing is also usually 24 inches. For the height of the tower begin with setting the height of the bottom stripper's bottom tangent line above grade. This should be at least 15 ft to allow a reasonable available NPSH for the product pump. From there allow a 5–15 min surge capacity at the calculated tower diameter and some 12 inches for the steam inlet distributor below its bottom stripping tray. Note the same comments apply to surge capacities in this tower as given for the residue stripper product.

The crude feed preheat exchanger system design

All crude distillation units pre heat the incoming crude oil feed by heat exchange with hot product and reflux streams. The preheated crude is then partially vaporised to satisfy the flash zone conditions by a fired heater. The degree of preheat is a question of an economic balance between the cost of the hardware and the savings in utilities due to the recovery of heat from the hot product and reflux streams. The exercise to arrive at this economic balance is one of critical analysis of the respective enthalpy and its temperature levels of the various streams to be considered for heat exchange. It requires also good cost data for the heat equipment included in the system. This equipment must include the cost of the heat exchangers (usually shell and tube), cost of the final product coolers, and the cost of the final crude feed fired heater.

More than one scheme is developed and their capital costs calculated from up to date equipment vandor data if possible. Next the utility requirements and cost for each system is developed using the company's unit utility costs. For an examination and comparison of these systems these cost data must be on the same bassis for each of the schemes. The following procedure is one of several that can be used for this economic analysis:

Step 1. Construct the enthalpy curve for the crude feed. This stream will remain in the liquid phase through the preheat train. The enthalpy curve therefore needs only to consider sensible heat. Start at say 60°F, calculate the enthalpy by multiplying the Btu/lb (from the enthalpy curve) by the total weight in lbs/hr, this gives the enthalpy point on the curve in Btu's/hr. Proceed with a further 10–15 other temperature points to make a smooth temperature V's enthalpy curve.

Step 2. Examine the temperature and weight/hr of the overhead, pumparound, and product streams. Select those that will be candidates for heat exchange against the crude feed. For example, the overhead stream from the main tower is a prime candidate for heat transfer against the cold crude because of its high heat content (latent heat) over a wide range of temperature. The kerosene product stream on the other hand is usually a poor candidate. It contains only sensible heat over a short temperature range and is also probably the smallest product stream by weight.

Step 3. Prepare enthalpy curves for the selected candidates. In the case of pump-arounds and the side stream products only the enthalpy (in Btu/hr) against say three or four temperatures need to be plotted.

Step 4. The overhead vapor enthalpy curve requires a more complex calculation. Here the enthalpy curve must be based on its condensation curve. This requires a calculation of the streams equilibrium (vapor and liquid) composition at the condensing range of temperatures and pressures. For this purpose assume a straight line pressure drop and temperature profiles between tower top and reflux drum. Select four to five pressures and their corresponding temperatures. For the purpose of this calculation it is assumed that no steam condenses in this segment of the system, therefore the equilibrium constants are taken at the partial pressure with steam. Using the overhead vapor composition (in moles/hr) from the sum of overhead product plus reflux vapor calculate its vapor/liquid composition at the selected temperatures and pressure. Apply the enthalpy value to both phases (not forgetting the steam) for the selected temperatures. Plot enthalpy in Btu/hr versus the selected temperature.

Step 5. Superimpose these product, overhead, pumparound enthalpy curves on the crude feed enthalpy curve. Start with the overhead curve and then with each other stream in their process sequence. Draw these enthalpy curves to provide a reasonable temperature approach to the crude feed curve. Figure 3.14 is an example of this concept.

Reasonable temperature approaches should not be less than 20°F for distillates and between 40 and 60°F for residues and heavy distillates.

Step 6. Several different schemes can now be developed using a ruler and set square for each of the product stream in different sequences. In the case of the exchange of heat against large volume streams such as the residue and perhaps the lower pumparound these streams can be split. They can also be shown to flow in series against the crude or in parallel.

Step 7. Size the heat exchanger equipment required for each of the schemes developed. This includes the sizing for additional equipment for each stream to meet its required end temperature. For example the final air coolers to meet product rundown temperature or trim coolers to meet pumparound return temperature. Sizing these items need not be precise as long as they are on the same basis. The simple equation for heat transfer

$$Q = UA\Delta t_{\mathrm{m}}$$

will suffice for this purpose. The value of U may be taken from Figure 3.12.

Step 8. Note the end temperature of the crude and using this enthalpy and the total enthalpy in the crude at the flash zone of the main tower determine the duty required by the fired heater in each case.

Step 9. Cost out the equipment required for each scheme. This includes all the heat exchangers (including trim and final coolers), and the fired heater. These can usually be obtained from equipment vendors on a unit size ($ per sqft for example) basis. Estimate the utility requirements for the equipment in each scheme.

Step 10. Tabulate the results starting with the highest capital cost scheme as a base case and calculate a simple incremental return on investment based on savings for the remaining schemes. The scheme with the highest positive ROI should be the system selected.

In almost all modern crude distillation units there are facilities to remove free salt. These desalting facilities are proprietary units which consist of fresh water injection into the crude feed and subsequent separation of the water, with the salt now in solution, from the oil. This separation takes place at a fixed temperature at the appropriate point in the crude feed preheat train. Demulsifying chemicals or (more usually) electrostatic precipitators or both are used to enhance this separation process. The temperature for this desalting process (usually around 220–250°F) must be catered for in the crude feed preheating.

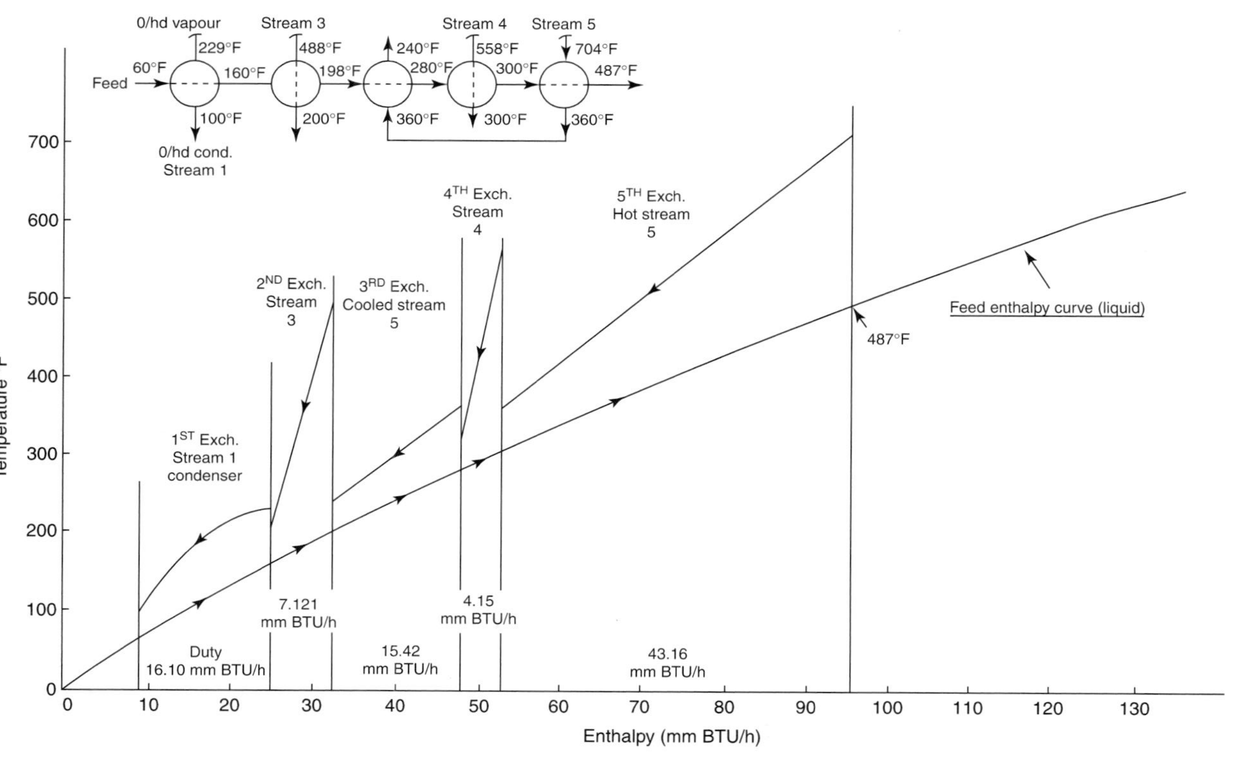

Figure 3.14. Calculation of the crude preheat train.

An example in the design of an atmospheric crude oil distillation tower

This example is confined to the design of the main fractionator and the associated stripper column. It is based on the processing of 30,000 BPSD of Kuwait crude providing the following products:

Overhead full range naphtha	Gas to 375°F
1st side stream kerosene	375–480°F
2nd side stream light gas oil	480–610°F
3rd side stream heavy gas oil	610–680°F

The distillation specifications shall be as follows:

- The ASTM end point of the Naphtha shall not exceed 400°F.
- The difference between the ASTM 95% point of the Naphtha and the 5% point of the Kero shall be at least 25°F.

The difference between the ASTM 95% point of the kerosene and the 5% point of the Light Gas Oil shall be −10°F (10° overlap). The ASTM end point of the Light Gas Oil shall not be greater than 620°F. The difference between the ASTM 95% point of the Light Gas Oil (LGO) and the 5% point of the Heavy Gas Oil (HGO) shall not exceed −35°F (35° overlap). The FBP of the TBP shall not exceed 710°F and a Condrason Carbon content of 8.0% maximum.

Developing the material balance

The crude TBP and sg curves are given in Figures 3.15 and 3.16, respectively. The TBP curve is divided into the distillate product cuts and also into narrow boiling point components.

The ASTM curves for the cuts are developed from the product cut points and probability paper. These are converted to TBP using the 'Edmister' method as shown in Table 3.1.

These values are plotted on Figure 3.17 and the narrow range components indicated for each of the product cuts. Note the front end of the naphtha is the front end of the whole crude so only the last segment of the cut is required in this calculation.

The cut characteristics are shown in Table 3.2.

Overflash   Cut range 690–725 (3 %vol on crude)

Sg @ 60 °F	0.891
Mol wt	295.

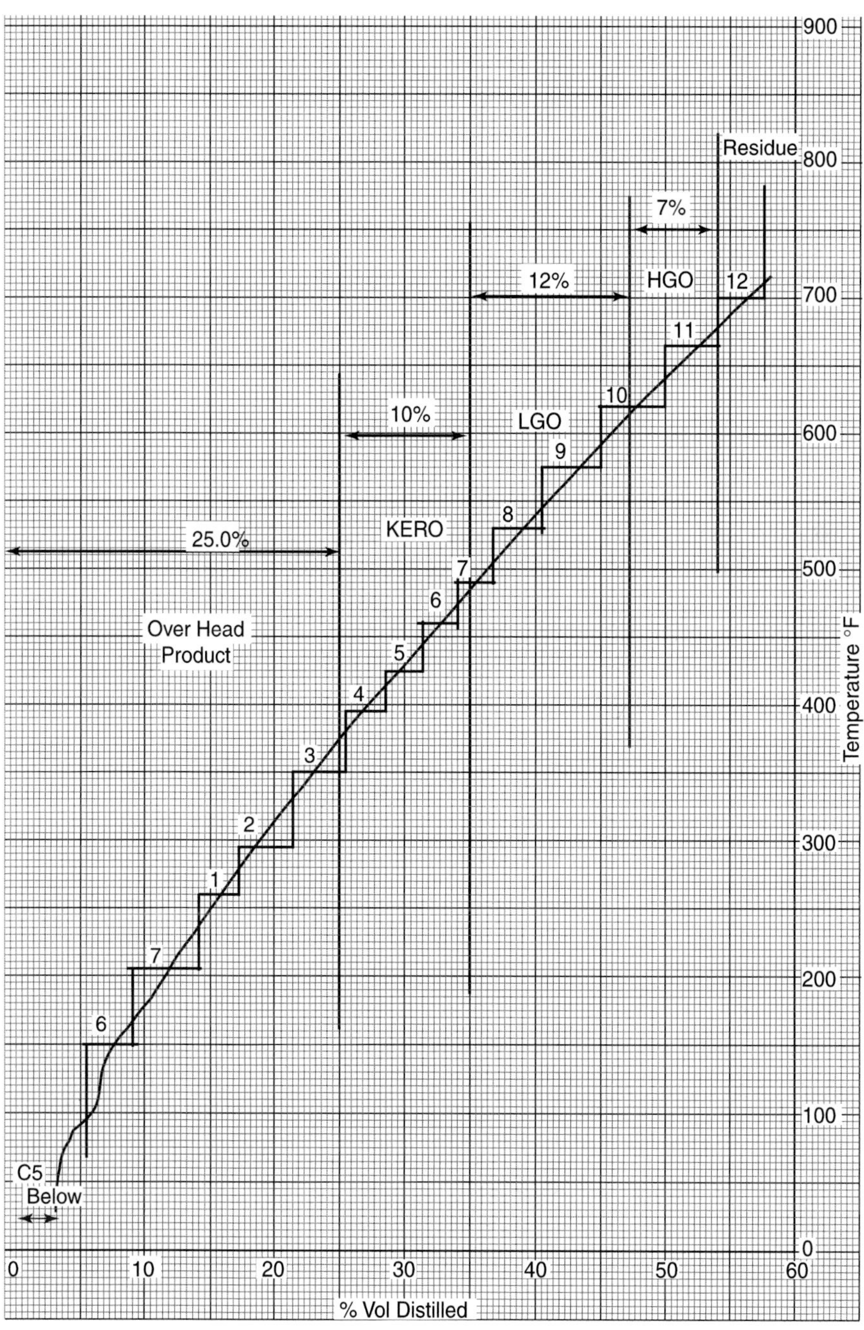

Figure 3.15. The TBP curve for the example calculation.

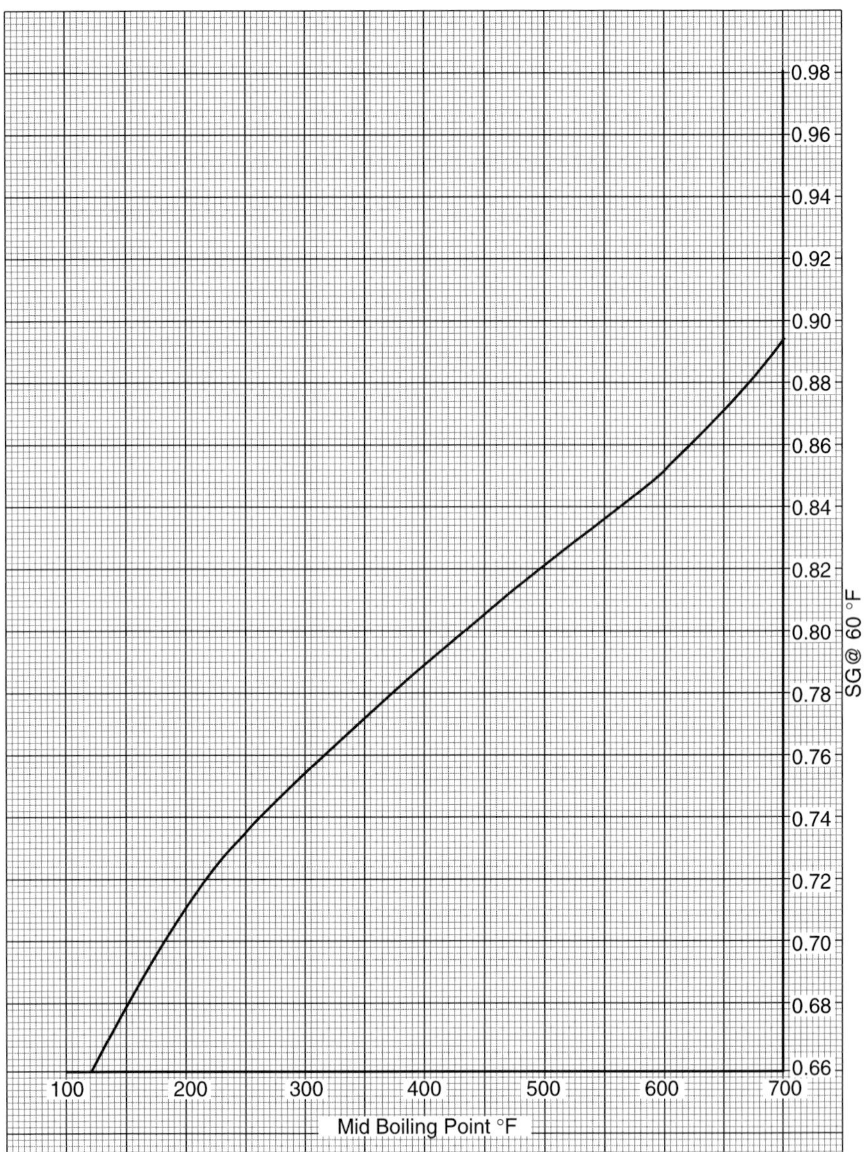

Figure 3.16. The SG curve for the example calculation.

Table 3.1.

	ASTM, °F	Δt ASTM, °F	Δt TBP, °F	TBP, °F
Naphtha (overhead)				
IBP		Not Required		
10%				
30%				
50%	300	10	14	305
70%	310	20	27	319
90%	330	28	32	346
FBP	358			378
Kero (1st side stream)				
IBP	345	31	57	287
10%		124	44	344
30%	400	19	32	388
50%	419	17	27	420
70%	436	24	33	447
90%	460	35	38	480
FBP	495			518
LGO (2nd side stream)				
IBP	495	25	48	457
10%	520	12	25	505
30%	532	11	20	530
50%	543	16	26	550
70%	559	13	19	576
90%	572	23	27	595
FBP	595			622
HGO (3rd side stream)				
IBP	595	20	40	568
10%	615	11	25	608
30%	626	7	14	633
50%	633	7	12	647
70%	640	18	25	659
90%	658	17	20	684
FBP	675			704

The complete material balance can now be written as shown in Tables 3.3.

Flash zone calculations

Total pressure at the flash zone
Estimate the overhead reflux drum pressure as 5 psig (This will be checked later by a bubble point calculation at 100°F).

Give the overhead air condenser a pressure drop of 5 psi. (This is a reasonable pressure drop for this equipment, and will be specified as such in the equipment data sheet to vendors.) Give the crude to overhead vapors heat exchanger a pressure drop of 2 psi (overhead vapors flow shell side).

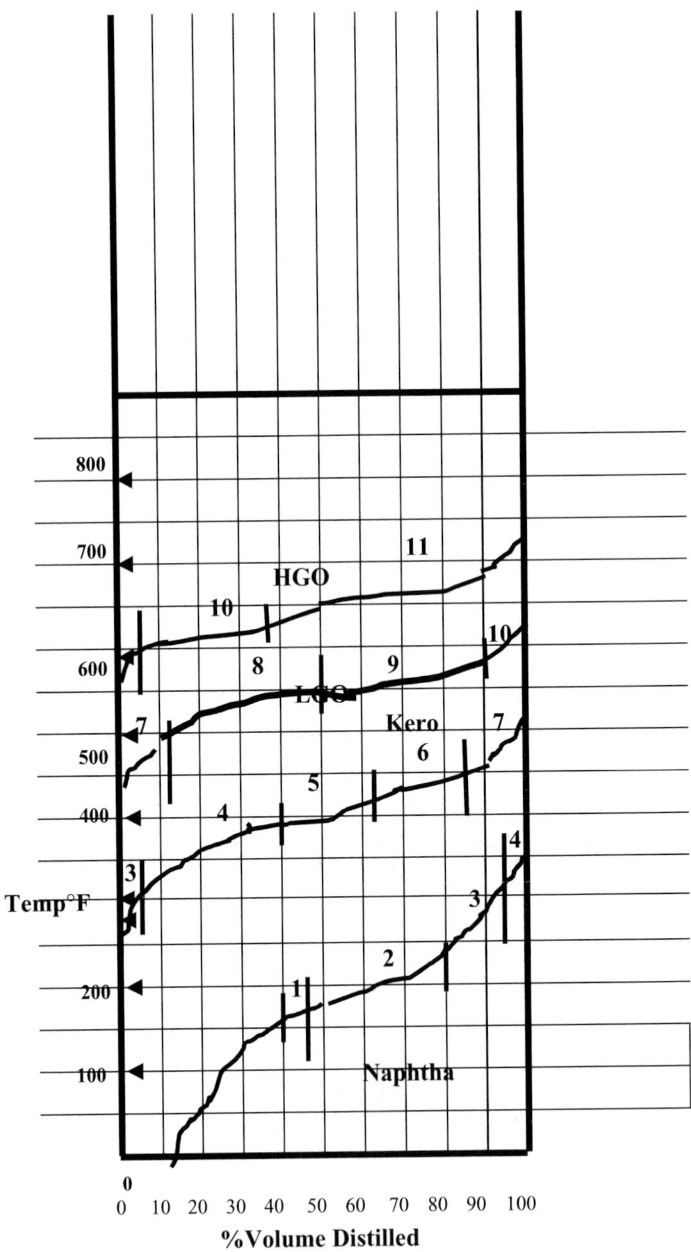

Figure 3.17. Product curves and narrow range components.

Table 3.2. Cut characteristics

Component	%vol on crude	SG @ 60°F	Wt factor	Mol wt	Mol factor
Overheads cut from gas to 375°F TBP cut point (25 vol% on crude)					
C2	0.11	0.374	0.411	30	0.014
C3	0.84	0.508	4.267	44	0.1
Ic4	0.40	0.563	2.252	58	0.039
nC4	1.53	0.584	8.935	58	0.154
C5's	3.02	0.629	18.996	72	0.264
C6	3.60	0.675	24.300	86	0.286
C7	4.50	0.721	32.445	100	0.324
Comp 1	3.50	0.743	26.005	114	0.228
2	3.77	0.765	28.841	126	0.229
3	2.41	0.776	18.702	136	0.138
4	1.32	0.788	10.402	152	0.068
TOTAL	25.0	0.702	175.556	95.2	1.884
Kerosene cut range 375–480°F (10 %vol on crude)					
2	4.0	0.765	30.6	126	0.24
3	14.0	0.776	108.64	136	0.80
4	22.0	0.788	173.36	152	1.14
.5	24.0	0.799	191.76	165	1.16
6	21.0	0.810	170.10	177	0.96
7	13.0	0.825	107.25	190	0.56
8	2.0	0.839	16.78	205	0.08
TOTAL	100.0	0.798	798.49	161.6	4.94
Light gas oil cut range 480–610°F (12 %vol on crude)					
6	2.0	0.810	16.2	177	0.09
7	9.0	0.825	74.25	190	0.39
8	38.5	0.839	323.02	205	1.58
9	40.5	0.848	343.44	238	1.51
10	10.0	0.860	86.00	250	0.34
TOTAL	100.0	0.843	842.91	215.6	3.91
Heavy gas oil cut range 610–680°F (7.0 %vol on crude)					
9	6.0	0.848	50.88	228	0.22
10	33.0	0.860	283.80	250	1.14
11	61.0	0.887	540.77	287	1.88
TOTAL	100.0	0.875	875.45	270.2	3.24

Neglect the pressure drop for fittings and piping—this will be quite small for a properly designed unit.

Then total tower top pressure is 12 psig.

Assume 40 valve trays in the section of the tower between flash zone and tower top. Let the pressure drop per tray be 0.25 psi. Then pressure drop in this section of the tower is 10 psi.

Table 3.3. The material balance

Stream	Cut range	%vol	Cum %vol	BPSD	GPH	SG	#/Gal	lbs/hr	%wt	Mol wt	Mol/hr
Whole crude	–	100.0	100.0	30,000	52,500	0.8685	7.23	379,575	100.0	225.3	1,684.8
O/head	IBP −360	25.0	25.0	7,500	13,125	0.702	5.84	76,650	20.2	95.2	805.1
Kero	−480	10.0	35.0	3,000	5,250	0.798	6.64	34,860	9.2	161.6	215.7
LGO	−610	12.0	47.0	3,600	6,300	0.843	7.02	44,226	11.7	215.6	205.1
HGO	−690	7.0	54.0	2,100	3,675	0.875	7.28	26,754	7.0	270.2	99.0
Resid	+690	46.0	100.0	13,800	24,150	0.957	8.16	197,085	51.9	547.6	359.9
The flash zone material balance											
O/flash	−725	3.0	57.0	900	1,575	0.891	7.4	11,655	3.0	295	39.5
Prod vapor	−690	54.0	54.0	16,200	28,350	0.773	6.43	182,490	48.1	137.7	1,324.9
Total vap	−725	57.0	57.0	17,100	29,925	0.780	6.49	194,145	51.1	142.3	1,364.4
Resid*	+725	43.0	43.0	12,900	22,575	0.988	8.22	185,430	48.9	578.7	320.4
Total		100.0	100.0	30,000	52,500	0.8685	7.23	379,575	100.0	225.3	1,684.8

*Does not include liquid overflow from bottom wash tray.

Total flash zone pressure is 12 psig + 10 psi = 22 psig. Call it 25 psig (40 psia) for design purposes.

Calculate the partial pressure of the hydrocarbon vapor at the flash zone
Take the quantity of stripping steam as 1.2 lbs/gal of residue (from Figure 3.10).

$$\text{The lbs/hr of stripping steam is } 1.2 \times 24{,}150 = 28{,}980 \text{ lbs/hr.}$$
$$= 1{,}610 \text{ moles/hr}$$

The partial pressure of the hydrocarbon vapor therefore is:

$$\frac{\text{Moles HC vapor}}{\text{Total moles vapor}} \times \text{Total pressure} = \frac{1{,}364.4}{1{,}364.4 + 1{,}610} \times 40 \text{ psia} = 18.35 \text{ psia.}$$

Calculate the EFV curve of whole crude at atmospheric pressure
From the crude TBP curve, the slope of the whole curve is 11.8°F/%vol (10–70 %vol on TBP temperatures divided by 60). From the Maxwell curves the slope of the flash reference line slope is 8.5°F/%vol. (See Chapter 1 of this Handbook.)

$$\Delta T_{50\%}\,(\text{DRL} - \text{FRL}) = 40^\circ\text{F}$$

$$T_{50\%}\,\text{DRL} = 667^\circ\text{F}$$

$$\text{Then } T_{50\%}\,\text{FRL} = 667 - 40 = 627^\circ\text{F}$$

Table 3.4. Flash curve at atmospheric pressure

	TBP				EFV		
%vol	Curve, °F	DRL	ΔT	Ratio	ΔT_1	FRL	Flash, °F
0	−127	75	−202	0.24	−48	200	152
10	190	190	0	0.4	0	285	285
30	420	430	−10	0.34	−3	450	447
50	645	667	−22	0.34	−8	627	619
70	900	900	0	0.34	0	795	795
90	1,235	1,140	95	0.34	32	915	947
100	2,192	1,250	942	o.34	320	1,040	1,360

Table 3.4 defines the flash curve at atmospheric pressure.

The flash curve calculated above is that at atmospheric pressure. To plot this at any other pressure take the 50 %vol temperature, and using the vapor pressure curves for hydrocarbons (see Figure 1.A.1 in the appendix) read off the temperature at the desired pressure. Draw the EFV curve through this new temperature parallel to the atmospheric curve. The flash zone temperature is the temperature at the % distilled on the partial pressure curve.

$$\text{Flash zone temperature} = 720°\text{F}$$

The EFV curve is shown in Figure 3.18.

Calculate total heat in the crude at the flash zone conditions
See Table 3.5 for the calculation of total heat in the crude at the flash zone conditions.

Please note: The enthalpy data used here are taken from the author's private files. It is recommended that the data given in Maxwell's *Hydrocarbon Data* or *The GPSA Engineering Data Book* be used for these heat balance calculations.

The tower heat balances

To calculate the temperature of the residue product leaving the tower
Consider the heat balance over the residue stripper as shown by the envelope in Figure 3.1.11. The product residue is 197,085 lbs/hr (from the material balance). The strip out vapor from the top stripping tray is 6% (from Figure 3.10) = 1,449 gals/hr. Assume the SG of the strip out is 7.5 lbs/gal (about the same as the overflash) and the mol wt is 305. Then the heat balance can be written as follows (see also Table 3.6):

$$\text{Heat in} = \text{Heat out}$$

$$197{,}085x + 45{,}404{,}000 = 119{,}916{,}000$$

$$x = 378 \text{ Btu/lb}$$

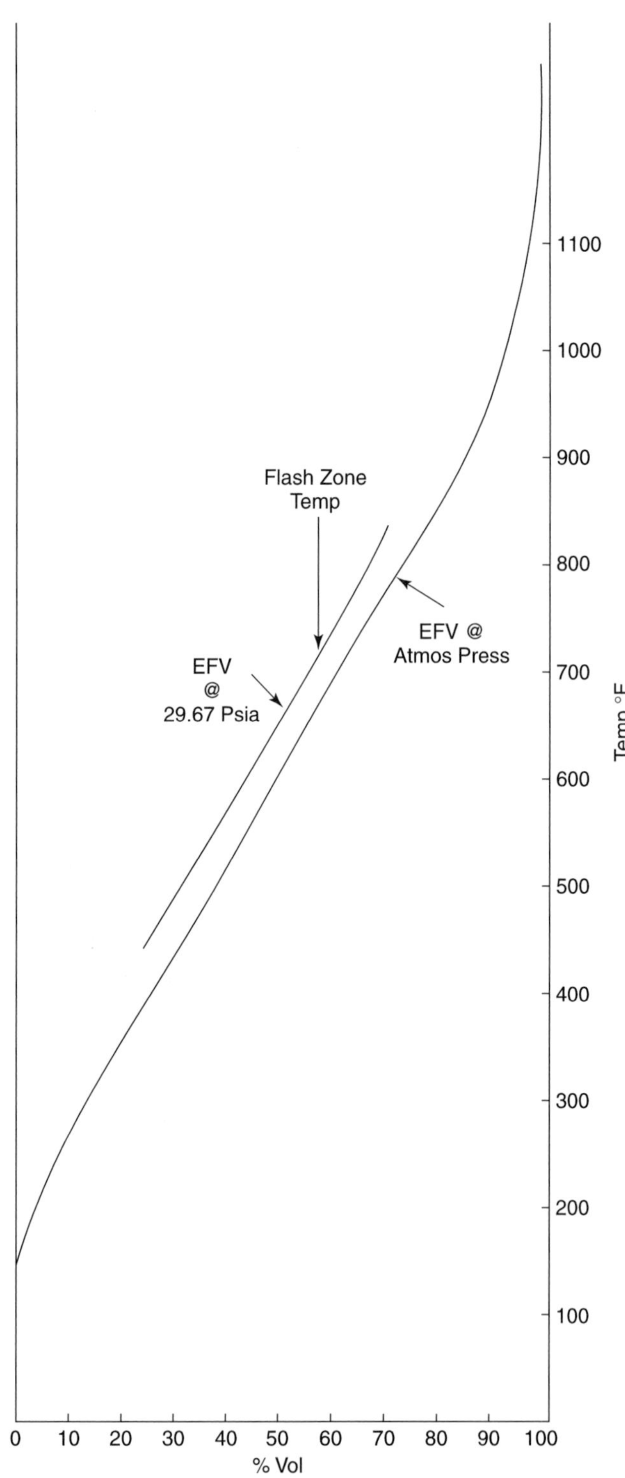

Figure 3.18. EFV curve for whole crude at flash zone conditions.

Table 3.5. Calculation of total heat in the crude at the flash zone conditions

Stream	V/L	°API	K	°F	Weight, lbs/hr	Btu/lb	mmBtu/hr
Crude vapor	V	5.0	11.8	720	194,145	528	102.509
Crude liquid	L	11.5	11.5	720	185,430	396	73.43
Total					379,575		175.939

From enthalpy tables this equates to 704°F.

To calculate the side stream draw-off temperatures
The steam rates to be used for side stream stripping will be:

All strippers will have three theoretical trays.
Both gas oil streams will use 0.5 lbs/gal respectively.
The Kero stripper will use 0.65 lbs/gal of steam.

Steam used is as follows:

Heavy gas oil = 3,675 × 0.5 = 1,838 lbs/hr
Light gas oil = 6,300 × 0.5 = 3,150 lbs/hr
Kero = 5,250 × 0.65 = 3,413 lbs/hr

- Tower pressure profile.
  There will be 40 fractionating trays in the main tower (trays above the flash zone) and four residue stripping trays (trays below the flash zone). These trays will be numbered from the bottom to the top. Thus, the bottom residue stripping tray will be tray 1. The tower top tray will be tray 45. The pressure profile is shown in Figure 3.1.19.

Table 3.6.

						Enthalpy	
Stream	V/L	°API	K	Temp, °F	Weight, lbs/hr	Btu/lb	mmBtu/hr
IN							
Residue	L	13	11.5	720	197,085	396	78.046
Stripout	L	25.5	11.5	720	10,868	410	4.456
Steam	V			450	28,980	1,290	37.384
Total in					236,923		119.916
OUT							
Residue	L	13	11.5	*t*	197,085	*x*	197,085*x*
Stripout	V	25.5	11.5	715	10,868	490	5.325
Steam	V			715	28,980	1,383*	40.079
Total out					236,933		45.404 + 197,085*x*

*At partial pressure of the flash zone = 13.53 psig.

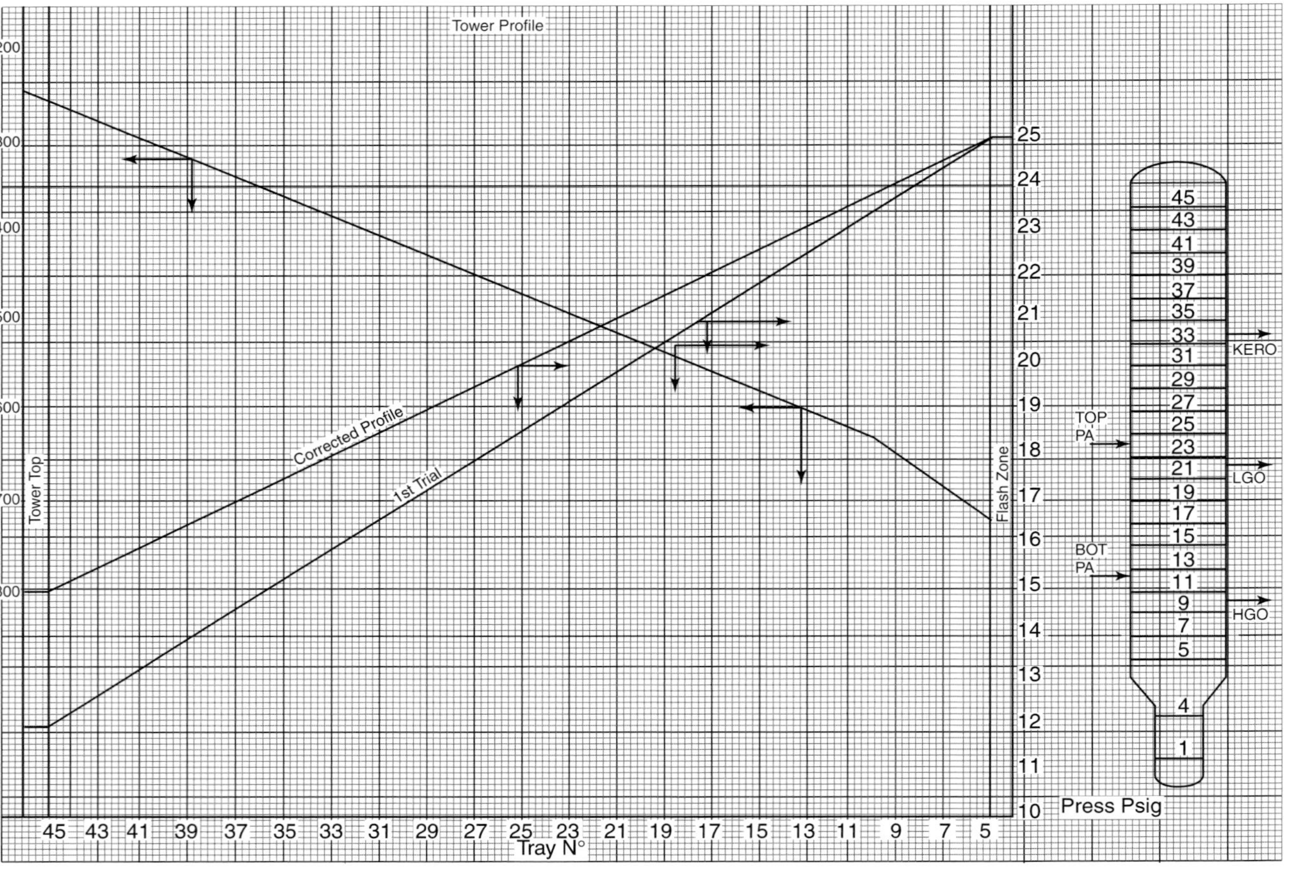

Figure 3.19. Tower pressure profile.

- Establish draw-off tray location
  Allow six wash trays above the flash zone to the HGO draw off. The HGO and bottom pumparound will be drawn from tray 10.
  Allow two trays for the pumparound return. Thus the pumparound will return on to tray 12.
  Allow 10 trays from tray 12 to the LGO and the top pumparound draw off. The draw-off tray will then be tray 22.
  Again allow two trays for the pumparound return. This will be tray 24.
  Allow 10 trays between top pumparound return and the Kero draw off. This locates the Kero draw-off at tray 34. It leaves 12 trays between the Kero draw off and the top tray (includes the draw-off tray and the top tray).
- Summary of tray locations and total pressure (1st trial).

	Tray no.	Pressure (psig.)
HGO draw off	10	23.4
BPA return	12	22.7
LGO draw off	22	19.5
TPA return	24	18.8
Kero draw off	34	15.5

- Calculate theoretical initial boiling points
  The FRL for each cut is developed from the TBP cut. Thus:

  HGO 50% TBP is 648°F. Slope of TBP is 0.6°F/%vol.
  Slope of FRL is 0.2°F/%. $\Delta t50$ (DRL – FRL) is 7°F.
  50% FRL is 641°F. From FRL curve IBP is 626°F.

  LGO This is developed in the same manner as HGO and the IBP of the FRL in this case is 527°F.

  Kero In the same way the Kero FRL has an IBP of 395°F.

  Now all these temperatures are at atmospheric pressure. It is now required to relate these temperatures to the partial pressure of the respective trays.
- To calculate the approximate partial pressures and draw-off temperatures

Assumptions:

1.0 All vapor lighter than the product cut is considered inert.

2.0 Internal reflux is assumed as follows:

To HGO tray 290 mol/hr.
To LGO tray 250 mol/hr
To Kero tray 200 mol/hr

$$\text{HGO tray partial pressure} = \frac{\text{Moles of HGO}}{\text{Total moles HC vapor + Steam}} \times \text{Total pressure}$$

$$= \frac{1,614.9}{3,224.9} \times 38.1 = 19.6 \text{ psia}$$

From vapor pressure curves theoretical draw-off temperature is 720°F.

From Figure 3.9:

Theoretical temperature − Actual temperature = 93°F
Actual draw-off temperature is 720 − 93 = 627°F

Draw-off temperatures for the LGO and Kero are calculated in the same way and are:

LGO = 493°F
Kero = 364°F.

To calculate the tower top temperature
Set the reflux drum temperature and pressure. In this case these will be set at 10 psig and 100°F. Taking the pressure drop across the exchangers and piping the tower top pressure will be 15 psig.

Fix the cold external reflux at 0.8 times the total moles overhead product;

The total moles HC in the overhead vapor is 1.8 × 805.1 = 1,449.18 moles/hr

Total moles steam in the overhead vapor is 2,076

Partial pressure of the hydrocarbons in the overhead vapor is:

$$\frac{1{,}449.19 \times 29.7}{3{,}525.18} = 12.2 \text{ psia}$$

The tower overhead temperature is the dew point of the hydrocarbons at the partial pressure and is shown in Table 3.7.

$K_2 = K_1 \times Sx = 0.135 \times 0.984 = 0.133 = 246°\text{F}$ which will be the tower top temperature.

To calculate side stream stripper bottom temperatures
These are calculated by heat balances over the respective side stream strippers. The following criteria are used in these calculations:

Table 3.7.

Comp	Mol frac	1st trial K	@ 250°F $X = y/K$	Mol wt	Weight factor	SG @ 60	Volume factor
C2	0.008	Neg	Nil				
C3	0.054	73.77	0.001	44	0.044	0.508	0.009
iC4	0.021	35.24	0.001	58	0.058	0.563	0.010
nC4	0.084	27.05	0.003	58	0.174	0.584	0.030
C5	0.143	11.80	0.012	72	0.864	0.629	0.137
C6	0.155	4.590	0.034	85	2.89	0.675	0.428
C7	0.175	2.213	0.079	100	7.90	0.721	1.100
Comp 1	0.124	1.066	0.116	114	13.224	0.743	1.780
Comp 2	0.124	0.557	0.223	126	28.098	0.765	3.673
Comp 3	0.075	0.311	0.241	136	32.776	0.776	4.224
Comp 4	0.037	0.135	0.274	152	41.648	0.788	5.285
Total	1.000		0.984	129.7	127.676	0.766	16.675

	HGO	LGO	KERO
Total strip out % vol	5	8	8
Mol wt of strip out	230	180	120
SG of strip out	0.865	0.820	0.750

Only the heat balance calculation of the HGO stripper is given below (see also Table 3.8). The other two stripper calculations will be similar in form (Figures 3.20–3.22).

$$\text{Solving for x: } 26{,}754x = 8{,}997{,}000$$

$$x = 336 \text{ Btu/lb}$$

From enthalpy tables = 615°F

Table 3.8.

Stream	V/L	°API	K	°F	lbs/hr	Btu/lb	mmBtu/hr
In							
Feed ex Hgo	L	30	11.5	627	26,754	347	9.284
Steam	V	–	–	450	1,838	1,290	2.371
Strip out	L	32	11.5	627	1,390	349	0.485
Total In					29,982		12.14
Out							
Hgo	L	30	11.5	t °F	26,754	x	$26{,}754x$
Steam	V			622	1,838	1,376*	2.529
Strip out	V	32	11.5	622	1,390	442	0.614
Total out					29,982		$3.143 + 26{,}754x$

*At partial pressure of 36.3 psia.

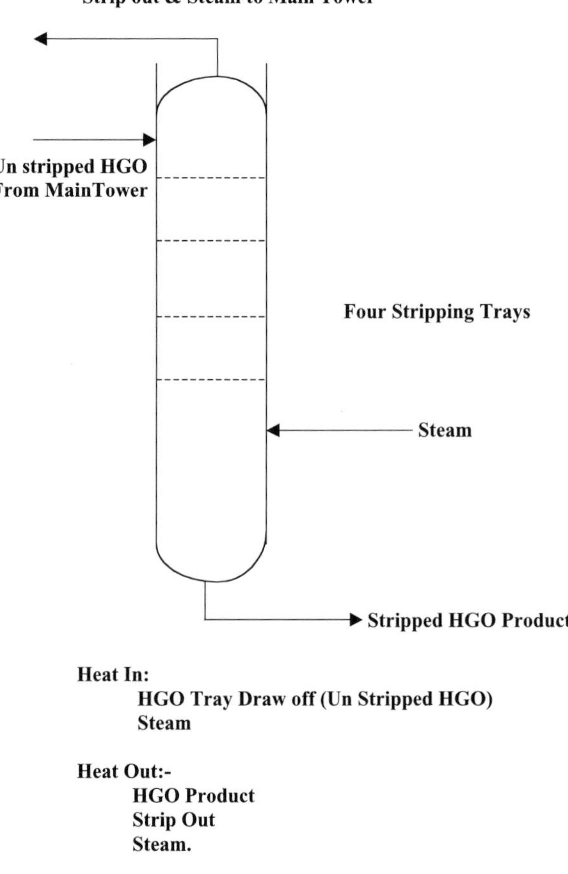

Figure 3.20. Heat balance diagram over HGO stripper.

Temperature of LGO leaving its stripper = 488°F.
Temperature of Kero leaving its stripper = 359°F.

Overall tower heat balance

Tower overhead heat balances

Balance included by envelop 1 determines the overhead condenser duty. See Table 3.1.9.

The overhead condenser is 71.360, therefore the heat removed by intermediate refluxes (pumparound) is 115.717 − 71.360 = 44.357 mmBtu/hr.

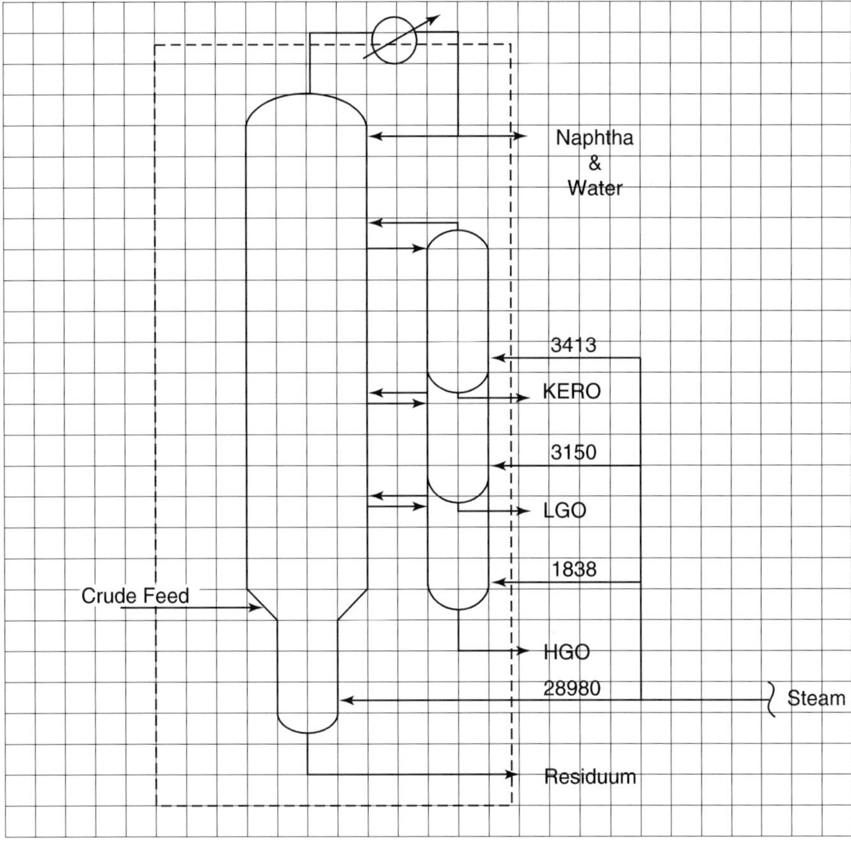

Figure 3.21. Overall tower heat balance diagram.

The heat balance included in envelope 2 determines the internal reflux from the top tray.

This is as follows:

Let x be the lbs/hr of overflow from the top tray. Then the heat balance is as follows (see also Table 3.10):

$$\text{Solving for } x = \frac{12{,}879{,}000}{121}$$

$$= 106{,}438 \text{ lbs/hr}$$

$$= 821 \text{ moles/hr or } 16{,}736 \text{ GPH.}$$

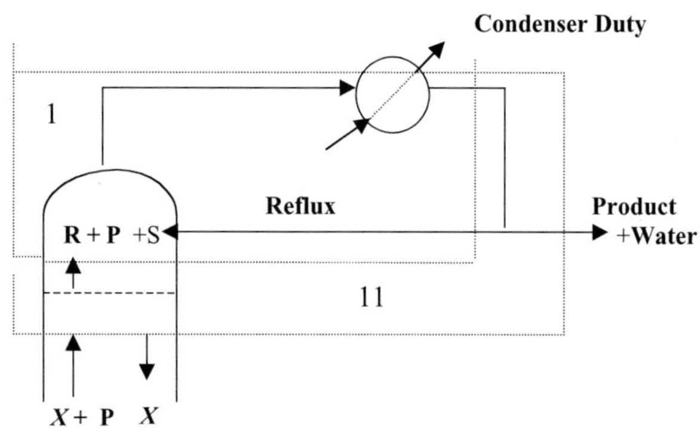

Envelope 1:-

Heat In

Reflux + Product Vapor

Steam

Heat Out

Reflux Liquid

Product Liquid

Water

Condenser Duty

Envelope 11:-

Heat In

Internal Reflux *X* Vapor

Product Vapor

Steam

Heat Out

Internal Reflux *X* Liquid

Product Liquid

Water

Condenser Duty.

Figure 3.22. Tower overhead heat balance diagram.

Check for fractionation: (using Figure 1.A.2 in the appendix)

Overflow @ 60°F = 16,736 GPH
Overflow @ 255° = 19,029 GPH
Prod vapor to top tray = 13,125 GPH
Reflux ratio = 1.45
Number of trays = 12
Factor is 12 × 1.45 = 17.4

50% TBP difference between Naphtha and Kero is 420 − 265 = 155°F.

ASTM Gap is 30°F which is within Spec.

Table 3.9a.

Stream	V/L	°API	K	°F	lbs/hr	Btu/lb	mmBtu/hr
IN							
Crude	V + L	–	11.7	720	379,575		175.939
Steam	V				37,381	1,290	48.221
Total in					416,956		224.160
OUT							
Residue	L	13	11.5	704	197,085	378	74.498
HGO	L	30	11.5	615	26,754	336	8.989
LGO	L	36.3	11.5	488	44,226	255	11.278
Kero	L	45.5	12	359	34,860	184	6.414
Naphtha	L	70.0	12	100	37,381	100	3.738
Refluxes							115.717
Total out					416,956		224.160

Table 3.9b.

Stream	V/L	°API	K	°F	lbs/hr	Btu/lb	mmBtu/hr
IN							
Naphtha	V	70	12	246	76,650	266	20.389
Reflux	V	70	12	246	61,230	266	16.311
Steam	V			246	37,381	1,197	44.745
Total in					175,351		81.445
OUT							
Naphtha	L	70	12	100	76,650	46	3.526
Reflux	L	70	12	100	61,320	46	2.821
Water	L			100	37,381	100	3.738
Cond duty							71.360
Total Out					175,351		81.445

Table 3.10.

Stream	V/L	°API	K	°F	lbs/hr	Btu/lb	mmBtu/hr
IN							
Naphtha	V	70	12	262	76,650	273	20.925
O/Flow	V	53	12	262	x	249	$249x$
Steam	V			262	37,381	1,199	44.82
Total In					$114,031 + x$		$65.745 + 249x$
OUT							
Naphtha	L	70	12	100	76,650	46	3.526
O/Flow	L	53	12	255	x	128	$128x$
Water	L			100	37,381	100	3.738
Cond duty							71.360
Total Out					$114.031 + x$		$78.624 + 128x$

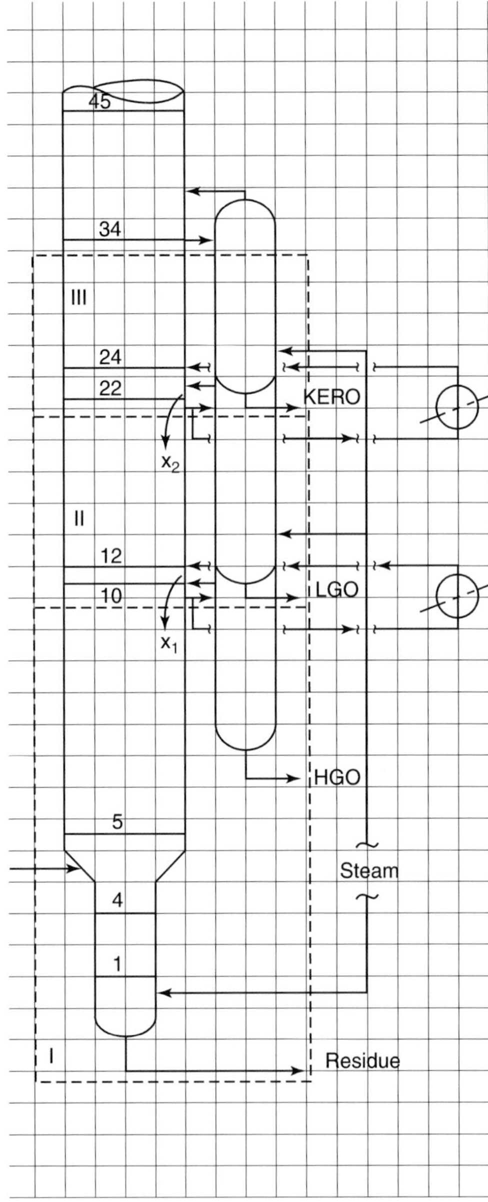

Figure 3.23. Remaining heat balance diagram for tower loading.

Calculating the remaining tower loadings
The tower loadings at the remaining critical sections of the tower are provided by heat balances over the envelopes given in the heat balance diagram (Figure 3.23). These critical sections are:

- Below bottom pumparound draw off (envelop 1)
- Below top pumparound draw off (envelop 2)
- Below Kero draw off (envelop 3)

Only the heat balance over envelop 3 is given in Table 3.11. The others follow in a similar fashion but the summaries of the calculated loads are given for all three sections (see Table 3.12).

$$\text{Solving for } x_3 = \frac{2{,}391{,}000}{111} = 21{,}540 \text{ lbs/hr.}$$

Checking fractionation:

Between HGO and LGO

$$\text{Reflux ratio} = \frac{22{,}809}{24{,}675} = 0.92$$

Number of trays = 11 (2 PA trays = 1 Frac tray).

Factor = 10.2

$\Delta 50\%$ TBP = 300°F

ASTM gap is +31°F which is well within the requirement for these two cuts.

Between LGO and Kero

Reflux ratio = 0.21
Number of trays = 11 (includes credit for PA trays.)
Factor = 2.3
$\Delta 50\%$TBP = 185°F
ASTM gap = −10°F

This is poor but as the kero cut is to be used as a blend stock it will be accepted . In the final design, however, provision will be made to increase the overhead reflux at the expense of the bottom pumparound to improve this separation.

Table 3.11. Heat balance over envelop 3

Stream	V/L	°API	K	°F	lbs/hr	Btu/lb	mmBtu/hr
ENV 3							
IN							
Heat in crude	V+L	–			379,575		175.939
Steam	V			450	33,968	1,290	43.819
Tray34 o/flow	L	46.3	12	367	$x3$	191	$191x3$
Total in					$413{,}543 + x3$		$219{,}758 + x3$
OUT							
O/flow to tray34	V	46.3	12	370	$x3$	302	$302x3$
Vapor to tray 34	V	62.5	12	370	111,510	320	35.683
Steam to tray 34	V			370	33,968	1,253*	42.562
LGO prod	L	36.3	11.5	488	44,226	255	11.278
HGO prod	L	30	11.5	615	26,754	336	8.989
Residue	L	13	11.5	704	197,085	378	74.498
Top PA							26.614
Bot PA							17.743
Total out					$413{,}543 + x3$		$217.367 + 302x3$

*Steam at the partial pressure of 18.3 psia.

Table 3.12. Summary of the main tower loading

	Liquid – From tray			Vapor – To Tray		
Draw-off tray	Lbs/hr	GPH	Hot GPH	Lbs/hr	Moles wt	Moles/hr
Tray 10 *		@627°F			@ 632°F	
Hydrocarbons	174,648	23,990	32,760	357,138	181	1,976.9
Steam	–	–	–	28,980	18	1,610.0
Total	174,648	23,990	32,760	386,118	107.6	3,586.9
Tray 22 *		@ 498°F		@500°F		
Hydrocarbons	122,412	17,487	22,809	278,148	154	1,802.9
Steam	–			30,818	18	1,027.3
Total	122,412	17,487	22,809	308,966	109	2,830.2
Tray 34		@367°F			@370°F	
Hydrocarbons	21,540	3,254	3,925	133,050	115	1,157.8
Steam				33,968	18	1,887.1
Total	21,540	3,254	3,925	167,018	55	3,044.9
Tray 45		@ 255°F			@ 262°F	
Hydrocarbons	106,438	16,735	19,029	183,089	112.6	1,626.1
Steam				37,381	18	2,076.7
Total	106,438	16,735	19,029	220,469	59.5	3,702.8

*Does not include pumparound liquid stream.

Tower diameter calculations

The calculation for tower vapor flood capacity is given by the expression

$$G_f = K_r \sqrt{\rho_v \times (\rho_l - \rho_v)}$$
G_f = Mass velocity of vapor at flood in lbs/hr · sqft of bubble area.
K_r = 1,100 for a 24″ tray spacing

The largest diameter will always be below the bottom pumparound in terms of the vapor load. For a quick estimate let the bubble area be 80% of the total tower diameter.

Below bottom pumparound loading

Assume bottom pumparound returns to tower at 300°F.

Total PA duty = 17,740,000 Btu/hr
Enthalpy at 627°F = 347 Btu/lb
Enthalpy at 300°F = 140 Btu/lb

The lbs/hr of the PA stream is $\frac{17,740,000}{207}$ = 85,715lbs/hr or 11,774 GPH
GPH of unstripped HGO product = 3,668
GPH of overflow = 23,990
Total liquid flow on the tray = 39,632 GPH @ 60°F

The loading data on this tray is as follows:

	Vapor	Liquid
Temperature °F	632	627
Pressure psig	23.8	–
Moles /hr	3,586.9	GPH hot 54,120
lbs/hr	386,118	lbs/hr 288,522
ACFS	302	2.0
ρ_v lbs/cuft	0.355	lbs/cuft 39.9

$$\text{Gf} = 1,100\sqrt{0.355 \times (39.9 - 0.355)}$$
$$= 4,121 \text{ lbs/hr} \cdot \text{sqft.}$$
$$\text{Ga} = 4,121 \times 0.8 = 3,297 \text{ lbs/hr sqft}$$

Tray area required to handle vapor = $\frac{386,118}{3,297}$ = 117 sqft

this may be considered as the bubble area.

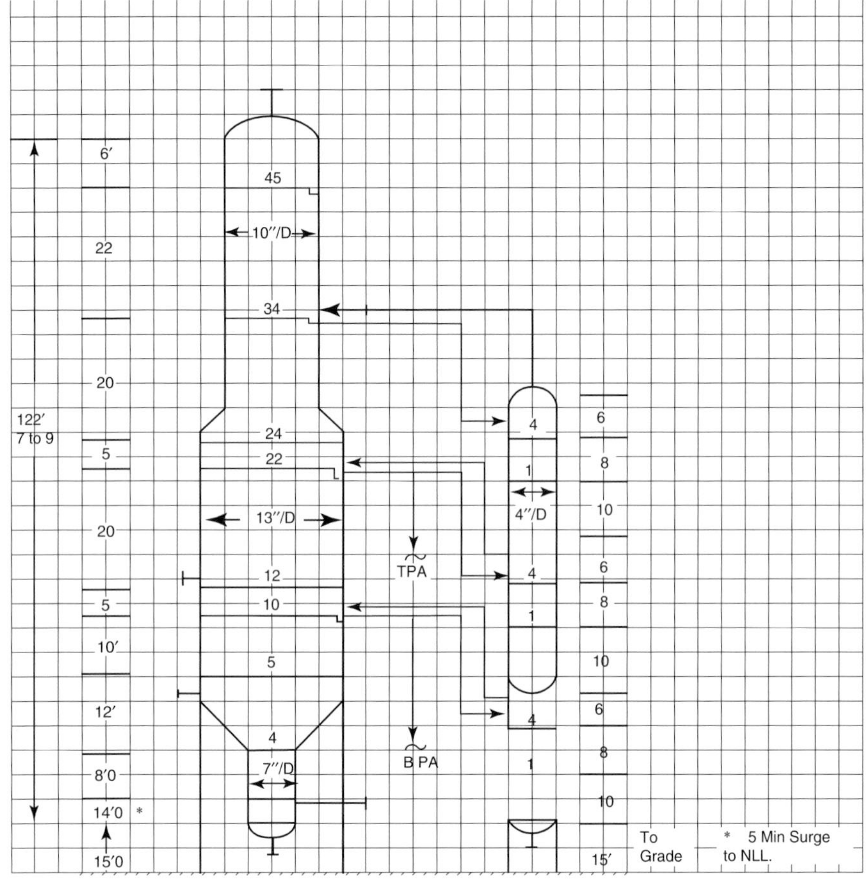

Figure 3.24. Main tower and side stream tower diagram.

Let d'comer area be such that liquid flow is 0.6 ft/sec. Area required is

$$\frac{2.0\text{ACFS}}{0.6} = 3 \text{ sqft}$$

For two (inlet + outlet) $A_{dc} = 6$ sqft.

Waste area is taken as 20% of $A_b = 117 \times .2 = 23.4$ sqft.

Then $A_s = 117 + 23.4 + 6 = 146$ sqft.

Tower diameter at this location is 13.6 ft.i/d.

Other sections of the tower

The other sections of the tower where there may be changes in diameter are below the top pumparound draw off, and of course the bottoms stripper top tray. The tower top tray should also be checked for loading.

The same calculation is followed for these other sections but it is not proposed to show them here. The results of the calculations though gave the following:

Top section of the main tower above tray 24 diameter is 10 ft i/d
Bottoms stripper section below tray 4 diameter will be 7 ft i/d.

A diagram of the main tower and the associated stripper tower is shown in Figure 3.1.24.

3.2 The vacuum crude distillation unit

As an introduction to this part of the chapter, it will be of interest to outline briefly an important development that occurred in this process during the early 1960s. Originally vacuum units followed closely on design to the atmospheric unit except of course it operated under a vacuum condition. The vacuum was obtained by a two or three stage steam ejectors and the internals of the tower were traditional trays, mostly bubble cap type. Under these conditions the vacuum obtained in the flash zone required the injection of steam to provide the required hydrocarbon partial pressure for adequate vaporization of the fuel oil feed. With the molecular weight of steam low at 18 the tower vapor traffic was extremely high in velocity requiring a large tower diameter to accommodate it.

The break-through to provide vacuum towers of much lower diameters came in the '60s with the use of high capacity steam injectors producing very low vacuum condition in the tower overhead. This coupled with the development of highly efficient expanded grid internals with very low pressure drop allowed the desired flash zone conditions to be met with out the injection of steam. This process became known as the 'Dry Vac' process and is the accepted process now for vacuum crude distillation. Such a process is describes below.

Process description

This process is often integrated with the Atmospheric Crude Distillation unit as far as heat transfer is concerned. Generally the atmospheric residue from the CDU is routed hot to the fired heater of the vacuum unit (Figure 3.25).

The atmospheric residue is further distilled to provide the heavy distillate streams used for producing lube oil or as feed to conversion units. This distillation however has to

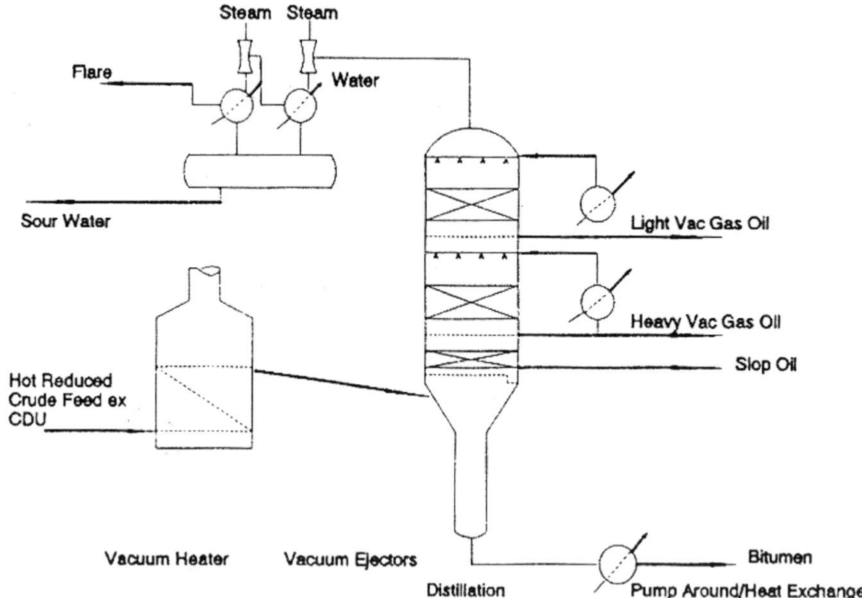

Figure 3.25. Vacuum distillation unit process schematic.

be conducted under sub atmospheric pressure conditions. The temperature required for vaporising the residue at atmospheric pressure would be too high and the crude would crack.

The process follows very much the same pattern as the atmospheric distillation. Should the cold feed be pumped from storage, it is heat exchanged against hot product and pumparound streams before being vaporised in the distillation unit heater. Normally though the feed is pumped hot directly from the CDU's residue stripper to the vacuum unit's heater. Thereafter the distillate vapours are condensed in the tower by heat and mass transfer with the cold reflux streams moving down the tower in the same way as the side streams in the Atmospheric unit. The products are taken off at the appropriate sections are cooled either by heat exchange with colder streams in the atmospheric unit, by air coolers or, in some cases as heating mediums to light end reboilers. They are then pumped to storage.

Neither the vacuum residue that leaves the bottom of the tower in this process nor the side-streams are steam stripped. The vacuum condition is produced by steam ejectors taking suction from the top of the tower. These ejectors remove inert and other vapour that may exist and pull a vacuum of about 5 mmHG absolute. The tower internals are usually expanded grid type which offer low pressure drop such that the flash zone pressure is about 25–30 mmHg absolute.

The vacuum crude distillation unit's flash zone

At atmospheric condition of pressure the flash temperature of normal atmospheric residue feed to achieve any meaningful degree of vaporisation would be extremely high (say in excess of 900°F). At these kind of temperatures the heavy residue will begin to break up or crack. This forms coke in the extreme and olefinic products which may not be desirable to the refiner. Effective vaporisation and fractionation can be achieved however at reduced pressures. Under this condition a reasonable flash temperature (say 650–750°F) can be easily obtained.

As described earlier vacuum distillation units handling reduced crude operate at 3–5 mmHg at the top of the tower and about 25–30 mmHg in the flash zone. No steam is used for stripping. The oil can still crack of course if the cut point desired is so high that excessively high flash temperature is required to meet it even at reduced pressures. The following graph is a guide to the critical cracking temperatures (Figure 3.26).

This graph shows a plot of a range of temperatures within which the oil will begin to crack. This is correlated to the Watson characteristic factor 'K'. Most residuum with a 700°F cut point for Middle East crudes have a 'K' factor of about 11.5. From the

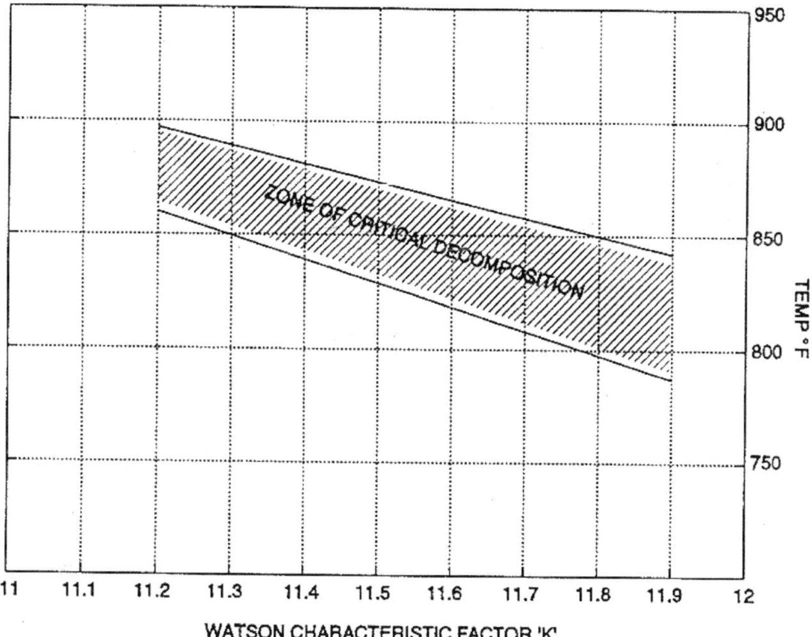

Figure 3.26. Critical cracking zone.

curve therefore it can be seen that these residuum would begin to crack at temperature between 830°F and 855°F. The degree of cracking at or above the 'Zone of critical decomposition' will be a function of temperature and the residence time of the oil at that temperature.

Significant cracking of the oil in a vacuum tower causes:

- High load to the ejectors (due to the formation of light ends)
- In lube oil production, de-colorising of the distillate streams
- In producing feed to hydro-treaters or hydro-crackers—high hydrogen consumption in these units due to the presence of unsaturates as the product of cracking

It is therefore very desirable to avoid these critical temperature in a vacuum unit.

The tower overhead ejector system

Most 'Dry Vac' Towers have a similar overhead ejector system, and as the design of this is critical to the units operation a calculation method to determine this design is described below as calculation steps:

The calculation procedure that is described here relates to the 'Dry' vacuum unit where no steam is used in the distillation process itself. This method can be used to determine the efficiency of the installed ejector set under test run conditions or indeed to specify the equipment to be purchased. The efficiency of the ejectors may be determined by the actual quantities of steam used to that calculated by this method.

The following data needs to be obtained to use this procedure:

- Quantity of inerts—either measured at the exhaust side of the last stage or established as a design criteria
- Tower top temperature and pressure required or observed
- Intermediate stage outlet temperatures and pressures of the process streams
- Total steam flow or steam flow to each stage ejector

A diagram of a typical ejector set is as shown in Fig. 3.27.

The calculation proceeds with the following:

Step 1. Determine the quantity of inerts entering the system from the tower. If this cannot be measured a rule of thumb is that total inerts is 0.5% to 1.0% by weight on feed. This is made up of air leaking into the system and some light ends. Again by rule of thumb light ends will be about 25% of total inerts.

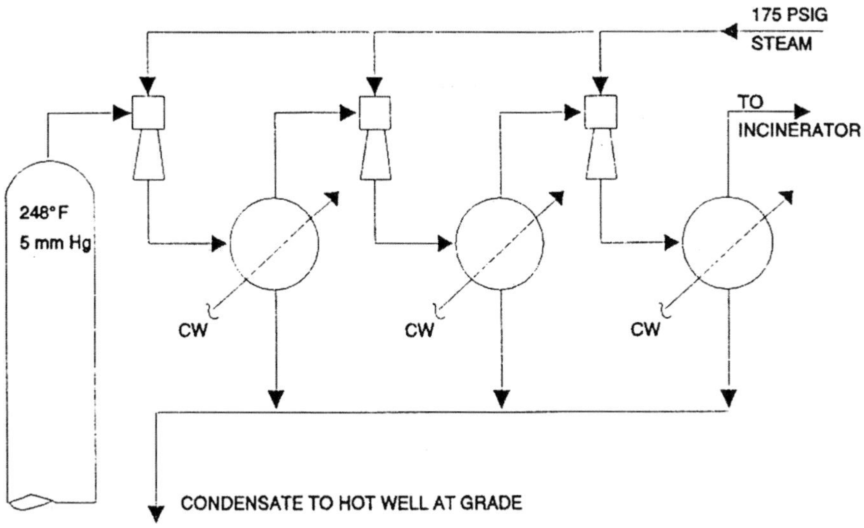

Figure 3.27. Typical ejector set.

Step 2. Calculate the 'Equivalent Dry Air Load' to the first ejector stage. Using the equation:

$$W_a = \frac{W_i}{Rm_i \times Rt_i} + \frac{W_s}{Rm_s \times Rt_s}$$

where

W_a = Equivalent air flow in lbs/hr.
Wi = Actual lbs/hr of o/head (includes air, and light ends).
Rm_i and Rt_i = Ratio factors for component i from ejector Figures 3.28 and 3.29
W_s = Weight flow of steam in lbs/hr.
Rm_s and Rt_s = Ratio factors for steam from Figures 3.30 and 3.31.

In this case W_s will be zero, as no steam will be used in the distillation.

Step 3. Calculate the steam consumption to the first ejector using tower top pressure as the suction pressure. The consumption is calculated by

$$W_{ms} = R_a M_p W_a$$

where

W_{ms} = Weight flow of motive steam in lbs/hr.
R_a = Ratio of lb motive steam/lb air equivalent using Figure 3.30
M_r = Steam usage multiplier from Figure 3.31
W_a = Air equivalent flow in lbs/hr.

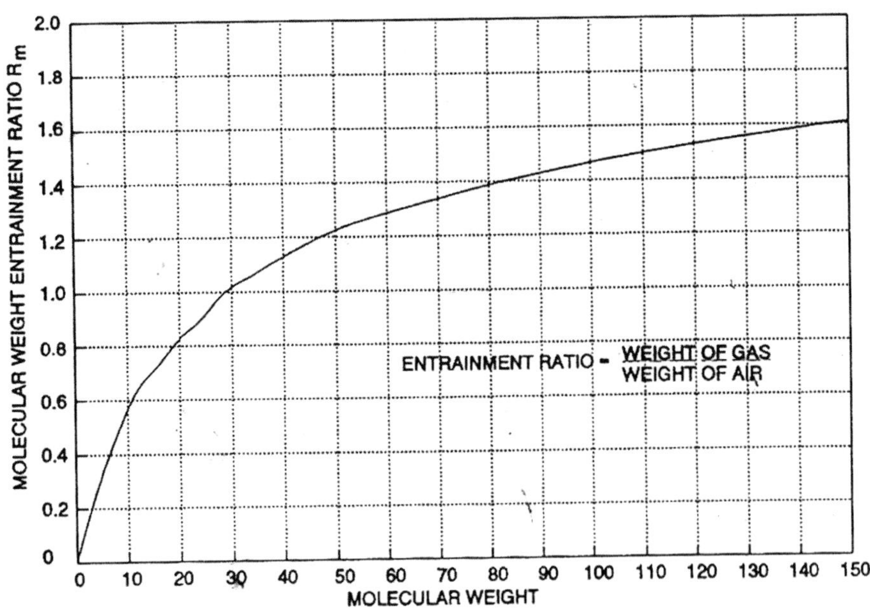

Figure 3.28. Ratio factors for steam.

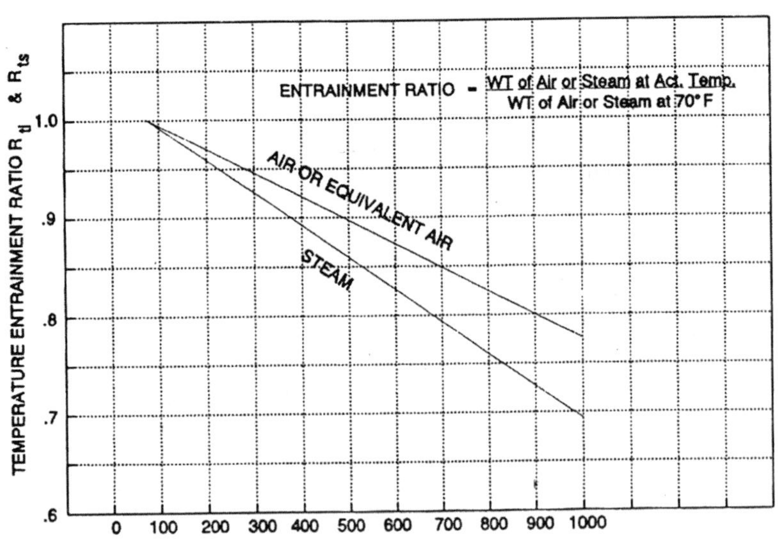

Figure 3.29. Temperature entrainment ratios Rt_i, Rt_s.

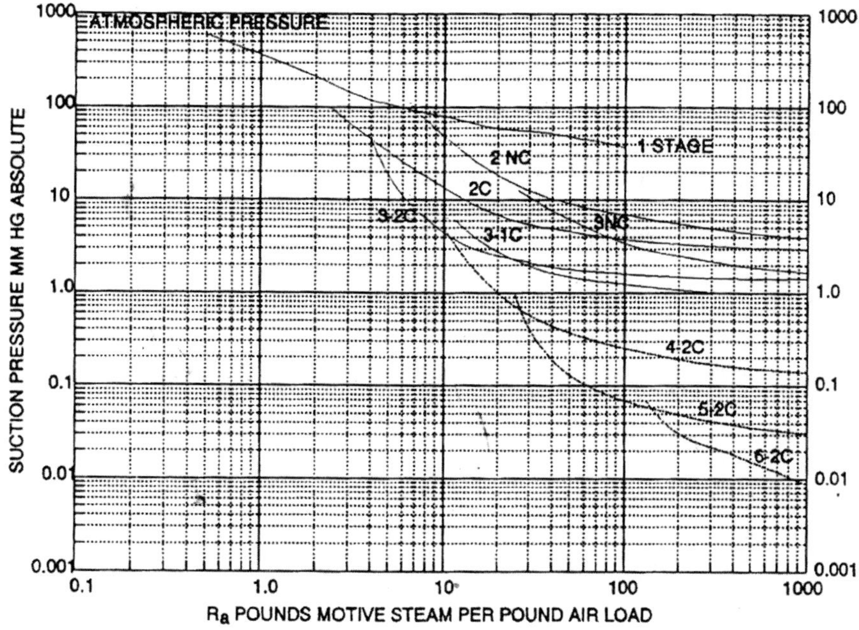

Figure 3.30. Suction pressure V's weight of motive steam.

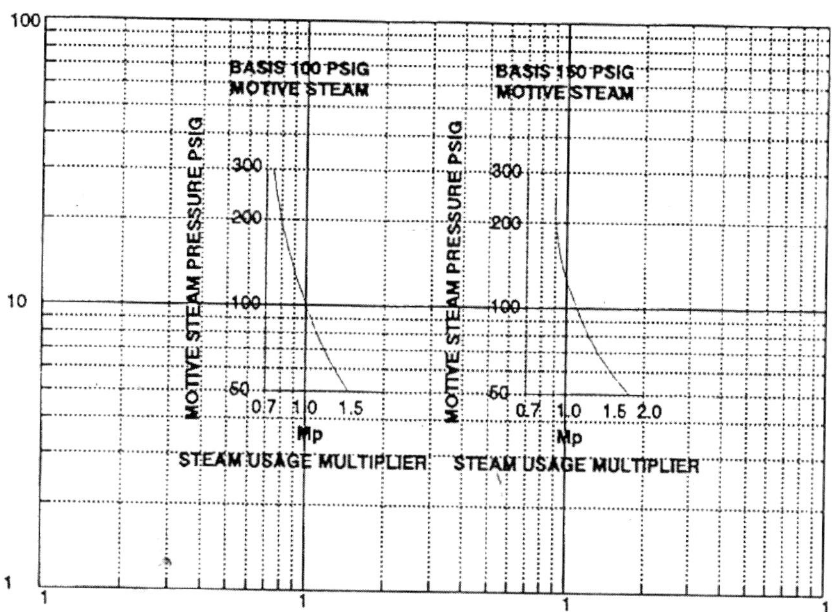

Figure 3.31. Motive steam usage correction factor.

Step 4. Calculate the partial pressure of steam at the condenser pressure. This pressure is the 1st/2nd stage intermediate pressure and is the suction pressure to the 2nd ejector. It is read on the plant or it may be assumed. In the calculation its assumed to be 50 mmHg.

Step 5. From the partial pressure of steam the condensing temperature in the condenser is read from steam tables. Assume 90% steam is condensed in this first condenser.

Step 6. (Optional) Calculate heat balance across the condenser and arrive at the condenser duty.

Step 7. Repeat step 2 for 'equivalent dry air load' to stage 2. Note there will now be steam present in this calculation.

Step 8. Repeat steps 3–5 for the second stage, making an assumption for the condenser pressure or reading off the actual pressure if the unit is an installed one. Assume also the amount of steam condensed in the second condenser—this will be high around 98%.

Step 9. Repeat steps 6 and 7 for the third stage condenser and ejector. This condenser will be about 1 or 2 psi above atmospheric pressure.

Step 10. Summarize the above results into a process specification for a required ejector set or if already installed compare the performance (e.g. calculated steam consumed versus actual).

Calculating flash zone conditions in a vacuum unit

Flash zone conditions are easier to calculate for a 'Dry' vacuum unit than for the atmospheric crude unit. Indeed the flash zone conditions can be measured in vacuum units with a greater degree of accuracy than in the case of the atmospheric column.

The procedure for predicting flash zone conditions in this case follows a similar route to that in the atmospheric unit case. The following steps describe this procedure.

Step 1. Develop the EFV from the TBP curve of the reduced crude. The same method that was used in item discussing the flash zone of the Atmospheric Crude Unit in this chapter will apply in this case also. Remember the EFV calculated is for atmospheric pressure.

Step 2. Develop the material balance for the vacuum unit. As the flash zone conditions are to be used in most calculations involving the vacuum tower it is best to develop the materials balance at this stage. To do this determine the distillate cuts required and by using the respective crude assay determine the specific gravity for each cut using mid boiling points. The mid boiling point for each distillate product only should be determined in this case. Use the method described in Chapter 1 of this Handbook to determine mol weight. Some of these will need to be extrapolated.

Step 3. Set the overflash. Now in this area of the TBP and EFV curves the slope of the curve is quite steep. That is there is a high temperature difference for each percent of volume increase. In vacuum units therefore a 1–2% over flash would be realistic to meet a realistic coil outlet temperature.

Step 4. Determine the new flash temperature to include the over flash from the EFV curve. This is the temperature at atmospheric pressure.

Step 5. Calculate the total pressure of the flash zone. Vacuum towers operate at or below 5 mmHg at the top. Pressure drop through the tower should not be more than 25 mmHg. A well designed off take trays and packing should be as follows:

Grid packing 6–7 mmHg per 10 foot of packed height.
Draw off (chimney trays) 2–3 mmHg/tray.

Step 6. There will be no partial pressure calculation of hydrocarbon vapor (as in the case of the atmospheric unit) as there is no steam in the flash zone of a 'Dry' vacuum tower. The total pressure calculated in Step 5 is the actual hydrocarbon flash pressure. Using the vapor pressure curves determine the flash zone temperature at the total flash zone pressure. This is the flash zone temperature that will now be used for all heat balances, etc.

Draw-off temperatures

Unlike the atmospheric crude distillation unit the temperature of the vacuum tower bottom (Bitumen) will be essentially the flash zone temperature. There will be a small difference, say 2–3°F, below actual flash zone temperature due to over flash returning from the wash trays. Very often the overflash amount is drawn off from below the wash section and either sent to fuel or blended into the bitumen stream external to the tower. In this case the unquenched bitumen leaving the tower will be at flash zone temperature. Again there is no steam present to influence this temperature.

Side stream draw-off temperatures are easier to calculate for a vacuum tower than was the case for the atmospheric tower. This is so because in a 'dry vacuum' column there is no steam to influence partial pressures and of course there is no side stream stripping.

A method similar to the 'Packie' method used for the atmospheric column is used for the vacuum column draw off. In this case however it is only necessary to determine the initial boiling point of the side stream EFV curve at the tower condition to arrive at the draw-off temperature.

Note: It is the IBP of the actual EFV curve in this case NOT the IBP of the flash reference line as in the case of the atmospheric unit. Also there will be no 'Packie' correction factors required in this case.

The calculation steps for this procedure are as follows:

Step 1. Draw the EFV curve from the sides tream TBP curve using the method described earlier in this chapter. Only the 0, 10, 30, and 50% vol section of the curve need to be developed.
Step 2. Set the total pressure at the draw-off tray. If this is not available as plant data then use the criteria for pressure drop given in the item dealing with the flash zone.
Step 3. Calculate the partial pressure of the side stream product at the draw-off tray. To do this consider all material lighter than the draw off side stream to be inert. Include in the inert estimates of air leakage and cracked hydrocarbon vapors as described earlier in this chapter in the item dealing with *Tower Overhead Ejector System*. The total hydrocarbon vapor will include the overflow from the draw-off tray. As a rule of thumb estimate overflow as:

Top side stream	0.8 times product
Mid side stream	1.0 times product
Bottom side stream	1.5 to 2.0 times product

Step 4. Using the vapor pressure curves relate the IBP temperature of the EFV to the partial pressure determined in Step 3. This is the draw-off temperature, and this will be the temperature for the respective side streams and pumparound draw off that will be used in the tower heat balances.

Determine pumparound and internal flows for vacuum towers

Now that the cut points and tower conditions of temperature and pressure are established the internal flow and pumparound duties can be calculated, although generally speaking fractionation requirements are not as strict in a vacuum crude unit as in the case for the atmospheric unit. Nevertheless proper wash streams are required in vacuum towers to protect distillates which nay become feed to cracking units, from entrained undesirable components such as metals. Test runs on vacuum units therefore should include the determination of reflux streams and, in turn, tower loading.

The following steps outline a calculation procedure to determine pumparound requirements and overflow (reflux) in the wash section of the tower.

Step 1. Set the overflow requirement for the LVGO draw-off tray using the rule of thumb given in the previous item. Alternatively if this can be measured on the plant use that data.
Step 2. From plant data or such data as can be developed from items on ejector system and draw-off temperatures, calculate the heat balance below the LVGO draw-off tray.
Step 3. In this heat balance the bottom pumparound duty will be the unknown. Equate heat in equals heat out to determine the duty of the pumparound required to produce the set overflow.

Step 4. This pumparound duty can be checked on the plant by multiplying the flow in the pumparound by the enthalpy difference over the exchangers.

Step 5. Carry out the overall heat balance over the tower. That is, calculate the difference between the total heat in with the feed and the total out with all the products. This difference gives the total heat to be removed by both pumparounds. Assuming there are two pumparounds (top and bottom). The duty of the bottom pumparound has already been calculated. Then the top pumparound duty will be the total heat to be removed minus the duty of the bottom pumparound.

Step 6. Usually the most critical flow in a vacuum unit is the wash oil flowing over the bottom wash trays or packing. This is the area where most undesirable entrainment can occur and this is the most vulnerable area for coking. Lack of wash oil enhances contamination of the bottom product and promotes coking in this area.

Step 7. Carry out a heat balance over the bottom wash section of the tower. The unknown in this case is the overflow liquid from the heavy vacuum gas oil. Equate the heat in with feed and overflow with the heat out with total product vapors, overflow vapor, and bitumen to solve for the unknown.

Note: The quantity of overflow in this case is independent of pumparound duties above it. It is dependant only on the amount of over flash.

Calculate tower loading in the packed section of vacuum towers

As discussed earlier, most modern 'dry' vacuum towers use low pressure drop grid or stacked packing. This packing enhances heat exchange in the tower and of course permits the tower to operate at very low pressures. Nevertheless this packing can become overloaded causing high pressure drop in the tower and poor all-around performance.

This item describes a general method of evaluating the grid performance in terms of its pressure drop. Please note this is a quick general method of estimating tower packing design or performance. Proprietary grid and packing manufacturers have their own correlations which they use in their design work. A more detailed examination of packed tower loading is presented in the author's published work titled '*Elements of Chemical Process Engineering*'*. The following are the calculation steps used for this quick general packed section evaluation.

Step 1. Determine the liquid and vapor flows across the section to be evaluated. As calculated in the previous section "determine pumparound and internal flows for vacuum towers."

Step 2. If the unit is existing and this calculation is to determine tower performance then use manufacturers drawings for tower details such as dimensions of the packed section.

Step 3. Calculate the liquid and vapor loads in terms of actual cubic feet per second for vapor and cubic feet per hour per square foot of tower for the liquid. All these will be at tray conditions of temperature and pressure.

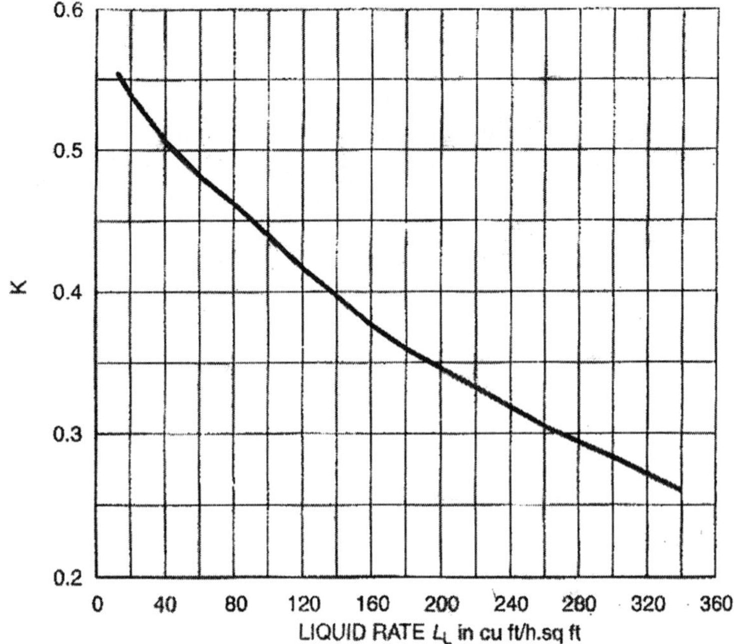

Figure 3.32. Capacity factor '*K*' *V*'s liquid rate.

Step 4. Using the liquid load as calculated in step 3, read off the value for '*K*' from Figure 3.32.

Calculate the linear velocity of the vapor at flood from the expression

$$K = V_f \sqrt{\rho_v / (\rho_l - \rho_v)}$$

where:

V_f = Vapor velocity at flood in ft/sec.
ρ_v = Density of vapor in lbs/cuft at section conditions of temperature and pressure
ρ_l = Density of liquid in lbs/cuft at section conditions.

Step 5. Calculate the actual vapor velocity required by multiplying the calculated velocity at flood by percent of flood permissible if this is to be a new design. The design cross sectional area of the tower is then the calculated vapor load in cuft/sec divided by the actual vapor velocity. If the unit is existing divide the vapor loading in cuft/sec by the cross sectional area of the tower to arrive at the actual vapor velocity in ft/sec. The existing unit operation as a percent of flood will be the actual velocity divided by the calculated flood velocity times 100.

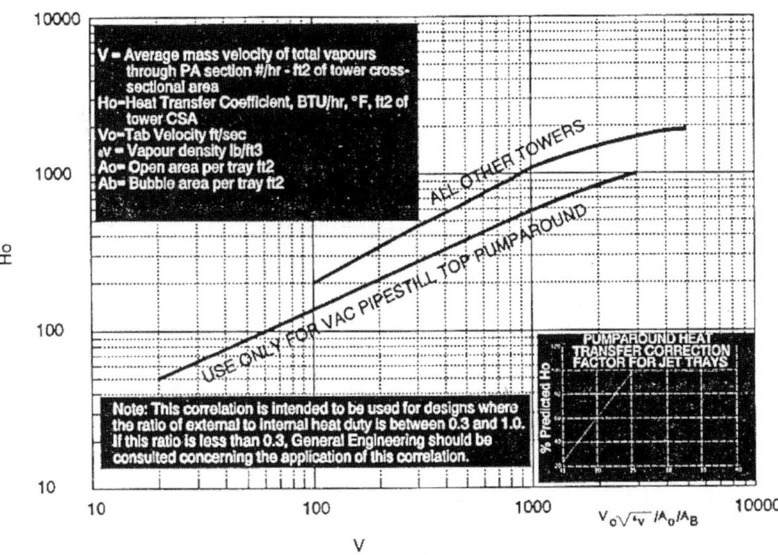

Figure 3.33. Transfer coefficient H_0 V's mass velocity for pumparound zones.

Step 6. Estimate the HETP (height equivalent to a theoretical tray) as being between 1.5 and 2.0 ft. Use the higher figure for vapor percent of flood 50–95% or higher and the lower value for vapor flood below 50%.

Step 7. Calculate the height of the packed section required for heat transfer using the quantity of heat to be transferred as being the pumparound duty in Btu/hr. Then treat the section above the pumparound draw off as a simple heat exchanger. Use the liquid flow to this section and the pumparound flow as calculated in the previous section (as part of the tower liquid load). The inlet temperature to the section can be taken as that for the pumparound liquid flow into the tower and the temperature out of the section as the pumparound draw-off temperature.

Step 8. Read off an overall heat transfer coefficient Ho from Figure 3.33.

Calculate the LMTD over the section from the temperatures used in the previous section on pumparound etc. Then calculate the total area of tower required for the heat transfer from the expression:

$$Q = AH_o\Delta t_m$$

where

Q = Heat duty in Btu/hr
A = Heat transfer area in sqft
H_o = Overall heat transfer coefficient in Btu/sqft·hr·°F
Δt_m = Log mean temperature difference in °F.

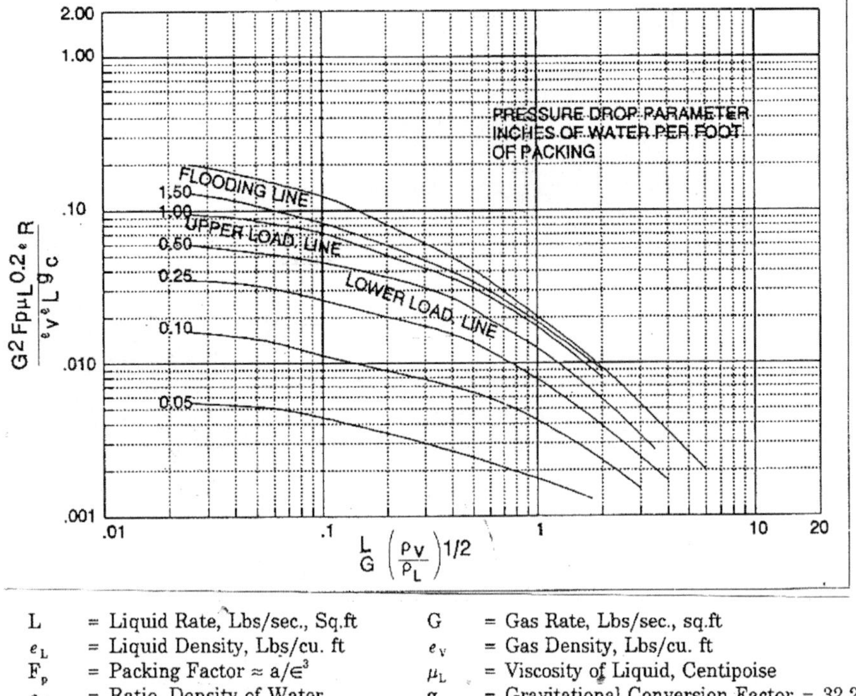

L	= Liquid Rate, Lbs/sec., Sq.ft		G	= Gas Rate, Lbs/sec., sq.ft
e_L	= Liquid Density, Lbs/cu. ft		e_V	= Gas Density, Lbs/cu. ft
F_p	= Packing Factor $\approx a/\epsilon^3$		μ_L	= Viscosity of Liquid, Centipoise
e_R	= Ratio, Density of Water / Density of Liquid		g_c	= Gravitational Conversion Factor = 32.2

Figure 3.34. Pressure drop through grid in inches of hot liquid per foot height.

Step 9. Calculate the theoretical number of trays required by dividing the total area calculated in step 8 by the design cross sectional area (step 5). Multiply these number of trays by the selected HETP to give the height of packing.

Step 10. Calculate the pressure drop through the grid using the actual vapor velocity in cuft/sec in the equation $K = V_a\sqrt{\rho_v(\rho_l - \rho_v)}$ to determine the constant K the read off the pressure drop in inches of hot liquid per foot height from Figure 3.34.

To express this pressure drop in mmHg multiply by the SG of the hot liquid and 1.865.

Appendix

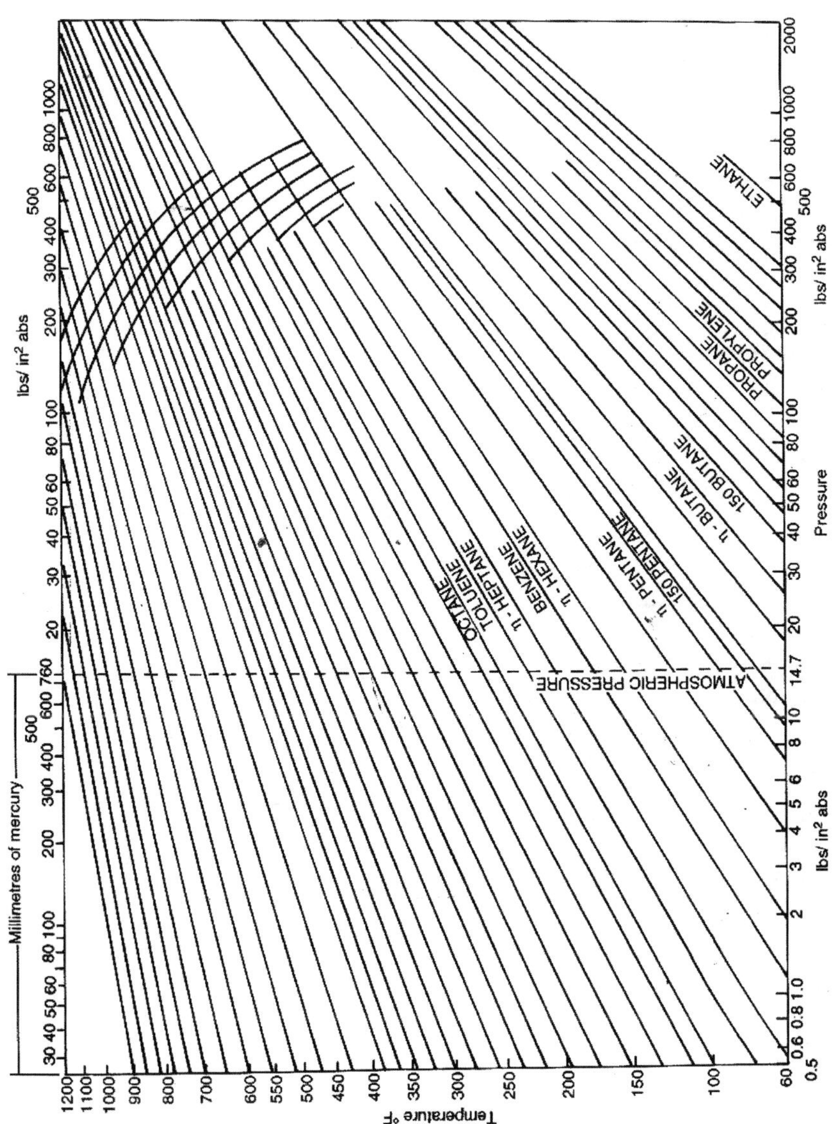

Figure 3.A.1. Pressure temperature curves (2 Pages).

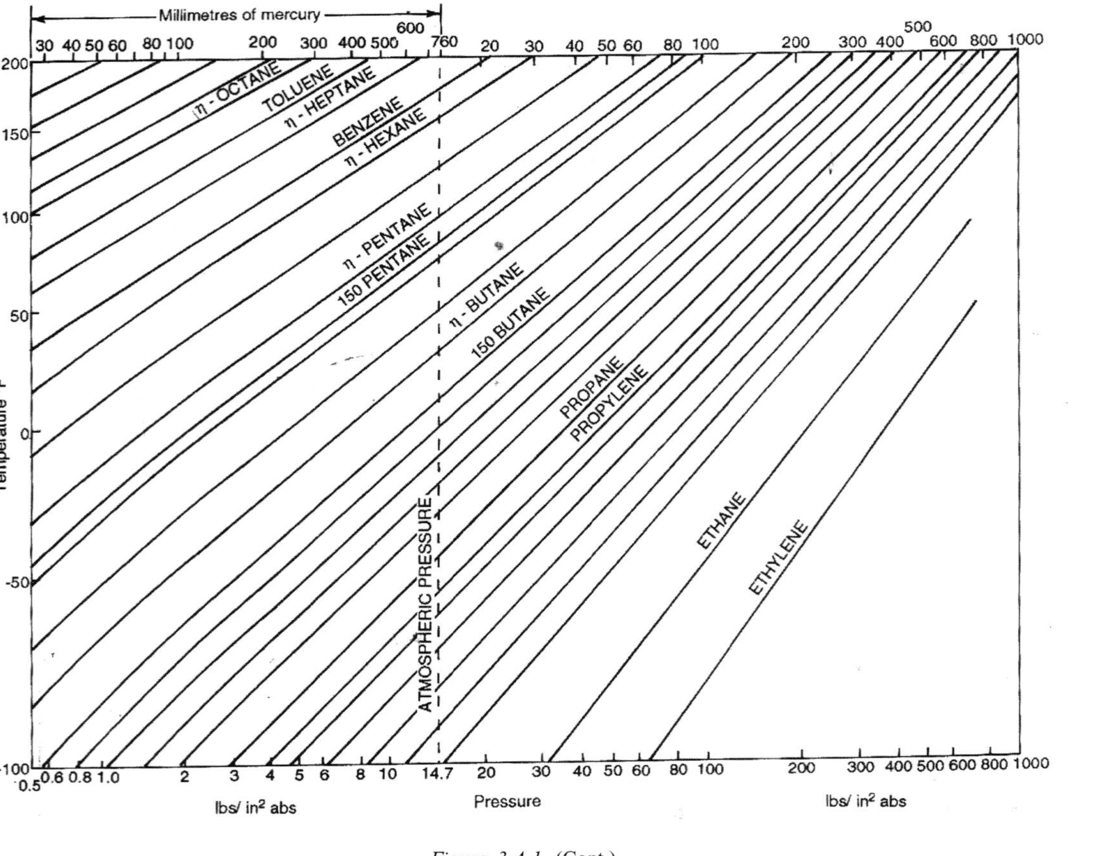

Figure 3.A.1. (Cont.)

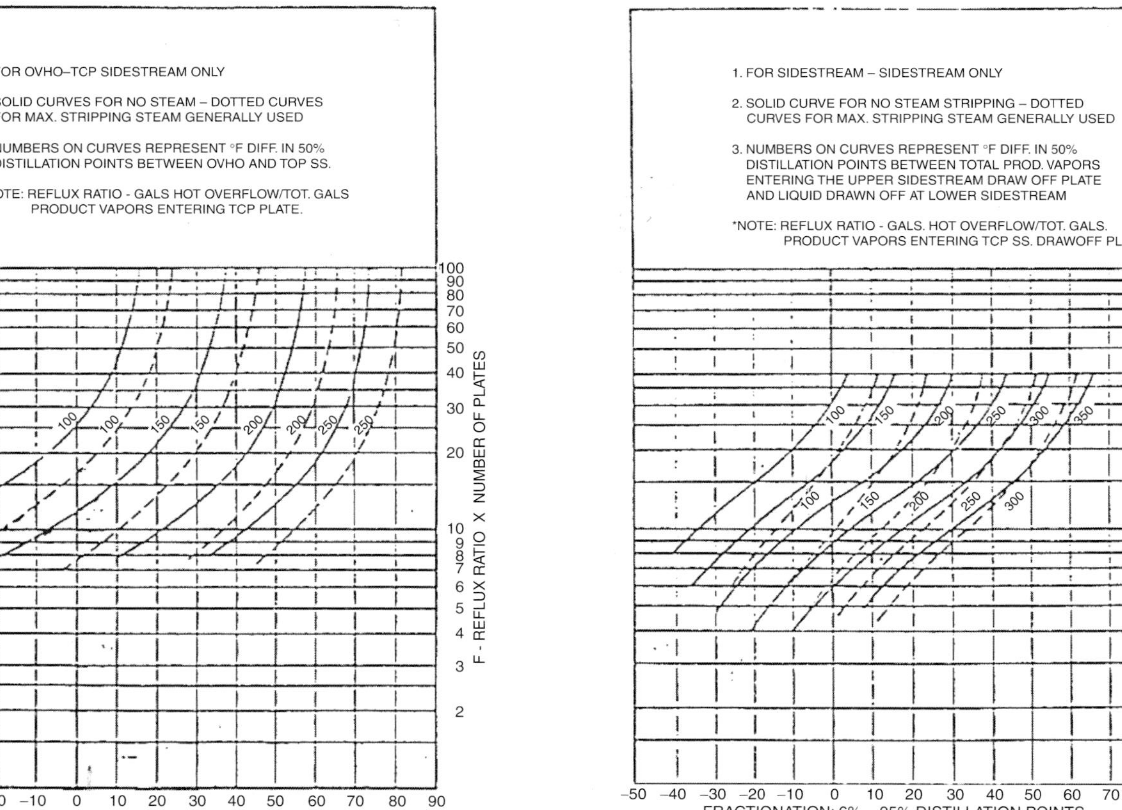

1. FOR OVHO–TCP SIDESTREAM ONLY
2. SOLID CURVES FOR NO STEAM – DOTTED CURVES FOR MAX. STRIPPING STEAM GENERALLY USED
3. NUMBERS ON CURVES REPRESENT °F DIFF. IN 50% DISTILLATION POINTS BETWEEN OVHO AND TOP SS.
*NOTE: REFLUX RATIO - GALS HOT OVERFLOW/TOT. GALS PRODUCT VAPORS ENTERING TCP PLATE.
F - REFLUX RATIO X NUMBER OF PLATES
FRACTIONATION: 6% – 95% DISTILLATION POINTS

Fractionation Curves for Overhead to Top Sidestream Products.

1. FOR SIDESTREAM – SIDESTREAM ONLY
2. SOLID CURVE FOR NO STEAM STRIPPING – DOTTED CURVES FOR MAX. STRIPPING STEAM GENERALLY USED
3. NUMBERS ON CURVES REPRESENT °F DIFF. IN 50% DISTILLATION POINTS BETWEEN TOTAL PROD. VAPORS ENTERING THE UPPER SIDESTREAM DRAW OFF PLATE AND LIQUID DRAWN OFF AT LOWER SIDESTREAM
*NOTE: REFLUX RATIO - GALS. HOT OVERFLOW/TOT. GALS. PRODUCT VAPORS ENTERING TCP SS. DRAWOFF PLATE.
F - REFLUX RATIO X NUMBER OF PLATES
FRACTIONATION: 6% – 95% DISTILLATION POINTS

Fractionation Curves for Sidestream to Sidestream Products.

Figure 3.A.2. ASTM gaps and overlaps.

FIGURE A 3.0
VALVE TRAY DESIGN PRINCIPLES

Design Feature	Suggested Value	Alternate Values	Comment
1. **Valve Size and Layout**			
a. Valve diameter	–		Valve diameter is fixed by the vendor
b. Percent Hole Area, A_0/A_B	12	8 to 15	Open area should be set by the designer. In general, the lower and open area, the higher the efficiency and flexibility, and the lower the capacity (due to increased pressure drop). At values of open area toward the upper end of the range (say 15%), the flexibility and efficiency are approaching sieve tray values. At the lower end of the range, capacity and downcomer filling becomes limited
c. Valve Pitch/diam. ratio	–		Valve pitch is normally triangular. However, this variable is usually fixed by the vendor
d. Valve distribution-		-	On trays with flow path length ≥ 5', and for liquid rates > 5000 GPH/ft. (diameter) on trays with flow path length < 5', provide 10% more valves on the inlet half of the tray than on the outlet half
e. Bubble Area, A_B	–		Bubble area should be maximised
f. Plate efficiency	–		Valve tray efficiency will be about equal to sieve tray efficiency provided there is not a blowing or flooding limitation
g. Valve blanking	–		This should not generally be necessary unless tower is being sized for future service at much higher rates. Blanking strips can then be used. Blank within bubble area, not around periphery to maintain best efficiency
2. **Tray Spacing, Inches**	-	12 to 36	Generally economic to use min. values given on p. III-E-2 which are set by maintenance requirements. Other considerations are downcomer filling and flexibility. Use of variable spacings to accommodate loading changes from section to section should be considered.
3. **Number of Liquid Passes**	11	to 2M	ultipassing improves liquid handling capacity at the expense of vapour capacity for a given diameter column and tray spacing. Cost is apparently no greater - at least, for tower diameters < 8 ft.
4. **Downcomers and Weirs**			
a. Allowable Downcomer inlet velocity, ft/sec of clear liq.		0.3 to 0.4	Lower value recommended for absorbers or other systems of known high frothiness
b. Type downcomer	Chord	Chord, Arc	Min. chord length should be 65% of tray diameter for good liquid distribution. Sloped downcomers can be used for high liquid rates - with maximum outlet velocity = 0.6 ft/sec. Arc downcomers may be used alternatively to give more bubble area (and higher capacity) but are somewhat more expensive. Min. width should be 6 in. for latter
c. Inboard Downcomer Width (Inlet and Outlet)		Min. 8 inches	Use of a 14-16" "jump baffle" suspended lengthwise in the centre of the inboard downcomer and extending the length of the downcomer is suggested to prevent possible bridging over by froth entering the downcomer from opposite sides. Elevation of base of jump baffle should be level with outlet weirs. Internal accessway must be provided to allow passage from one side to another during inspection
d. Outlet Weir Height	2"	1" to 4"	Weir height can be varied with liquid rate to give a total liquid head on the tray (h_c) in the range of 2.5" - 4" whenever possible. Lower values suggested for vacuum towers, higher ones for long residence time applications
e. Clearance under downcomer, in.	1.5"	1" min	Set clearance to give head loss of approximately 1 inch. Higher values can be used if necessary to assure sealing of downcomer
f. Downcomer Seal (Inlet or outlet weir height minus downcomer clearance)	Use outlet weir to give min. _" seal in plate liquid	Inlet weir or recessed inlet box	In most cases plate liquid level can be made high enough to seal the downcomer through use of outlet weir only. Inlet weirs add to downcomer build up; in some cases they may be desirable for 2-pass trays to ensure equal liquid distribution. Recessed inlets are more expensive but may be necessary in cases where an operating seal would require an excessively high outlet weir
g. Downcomer filling, % of tray spacing		40 to 50	Use the lower value for high pressure towers, absorbers, vacuum towers, known foaming systems, and also for tray spacings of 18" or lower

Figure 3.A.3. Valve tray design details.

WEIR LENGTH AND DOWNCOMER AREA

R*	L*	A*	R*	L*	A*	R*	L*	A*	R*	L*	A*	R*	L*	A*	R*	L*	A*
.070	.511	.0308	.120	.650	.0680	.170	.751	.113	.220	.828	.163	.280	.898	.230	.390	.977	.361
1	.514	.0315	1	.652	.0688	1	.753	.114	1	.829	.164	5	.903	.236	5	.979	.367
2	.517	.0321	2	.654	.0697	2	.755	.115	2	.831	.165						
3	.521	.0328	3	.657	.0705	3	.756	.116	3	.832	.166	.290	.908	.241	.400	.980	.374
4	.524	.0335	4	.659	.0714	4	.758	.117	4	.834	.167	5	.913	.247	5	.982	.380
5	.527	.0342	5	.661	.0722	5	.760	.117	5	.835	.169	.300	.917	.252	.410	.984	.386
6	.530	.0348	6	.663	.0731	6	.762	.118	6	.836	.170	5	.921	.258	5	.986	.392
7	.533	.0355	7	.665	.0739	7	.763	.119	7	.838	.171						
8	.536	.0362	8	.668	.0748	8	.765	.120	8	.839	.172	.310	.925	.264	.420	.987	.398
9	.539	.0368	9	.670	.0756	9	.766	.121	9	.841	.173	5	.930	.270	5	.989	.405
.080	.542	.0375	.130	.672	.0765	.180	.768	.122	.230	.842	.174	.320	.933	.276	.430	.991	.412
1	.545	.0382	1	.674	.0774	1	.770	.123	1	.843	.175	5	.937	.282	5	.993	.418
2	.548	.0389	2	.677	.0782	2	.772	.124	2	.845	.176						
3	.552	.0396	3	.679	.0791	3	.773	.125	3	.846	.177	.330	.941	.288	.440	.994	.424
4	.555	.0403	4	.682	.0799	4	.775	.126	4	.848	.178	5	.945	.294	5	.995	.430
5	.558	.0410	5	.684	.0808	5	.777	.127	5	.849	.179	.340	.948	.300	.450	.996	.437
6	.561	.0418	6	.686	.0817	6	.778	.128	6	.850	.180	5	.951	.306			
7	.564	.0425	7	.688	.0825	7	.780	.129	7	.851	.181				.460	.997	.450
8	.567	.0432	8	.691	.0834	8	.781	.130	8	.853	.182	.350	.955	.312			
9	.570	.0439	9	.693	.0842	9	.783	.131	9	.854	.183	5	.958	.318	.470	.998	.462
.090	.573	.0446	.140	.695	.0851	.190	.784	.132	.240	.855	.184	.360	.961	.324	.480	.998	.475
1	.576	.0454	1	.697	.0860	1	.786	.133	5	.860	.190	5	.964	.330			
2	.578	.0461	2	.699	.0869	2	.787	.134							.490	.999	.488
3	.581	.0469	3	.700	.0878	3	.789	.135	.250	.866	.196	.370	.967	.337			
4	.583	.0476	4	.702	.0887	4	.790	.136	5	.872	.202	5	.969	.343	.500	1.0	0.50
5	.586	.0484	5	.704	.0896	5	.792	.137	.260	.878	.207	.380	.971	.348			
6	.589	.0491	6	.706	.0905	6	.794	.138	5	.883	.213	5	.977	.354			
7	.592	.0499	7	.708	.0914	7	.795	.139									
8	.594	.0506	8	.710	.0923	8	.797	.140	.270	.888	.218						
9	.597	.0514	9	.712	.0932	9	.798	.141	5	.893	.224						
.100	.600	.0521	.150	.714	.0941	.200	.800	.142									
1	.603	.0529	1	.716	.0950	1	.802	.143									
2	.605	.0537	2	.718	.0959	2	.803	.144									
3	.608	.0545	3	.720	.0969	3	.805	.145									
4	.610	.0553	4	.722	.0978	4	.806	.146									
5	.613	.0561	5	.724	.0987	5	.808	.148									
6	.615	.0568	6	.726	.0996	6	.809	.149									
7	.618	.0576	7	.728	.1005	7	.810	.150									
8	.620	.0584	8	.729	.1015	8	.812	.151									
9	.623	.0592	9	.731	.102	9	.813	.152									
.110	.625	.0600	.160	.733	.103	.210	.814	.153									
1	.628	.0608	1	.735	.104	1	.816	.154									
2	.630	.0616	2	.737	.105	2	.817	.155									
3	.633	.0624	3	.738	.106	3	.819	.156									
4	.635	.0632	4	.740	.107	4	.820	.157									
5	.638	.0640	5	.742	.108	5	.822	.158									
6	.640	.0648	6	.744	.109	6	.823	.159									
7	.643	.0656	7	.746	.110	7	.824	.160									
8	.645	.0664	8	.747	.111	8	.826	.161									
9	.648	.0672	9	.749	.112	9	.827	.162									

* THIS TABLE RELATES THE DOWNCOMER AREA, THE WEIR LENGTH, AND THE HEIGHT OF THE CIRCULAR SEGMENT FORMED BY THE WEIR.

A_D

l_o

r

Diameter

$R = \dfrac{\text{* Downcomer Rise}}{\text{Diameter}} = \dfrac{r}{\text{Dia.}}$ $L = \dfrac{\text{* Weir Length}}{\text{Diameter}} = \dfrac{l_o}{\text{Dia.}}$ $A = \dfrac{\text{* Downcomer Area}}{\text{Tower Area}} = \dfrac{A_D}{A_S}$

Figure 3.A.4. Chord height, area, and lengths.

Chapter 4

The distillation of the 'Light Ends' from crude oil

D.S.J. Jones

The 'light ends' unit is the only process in a refinery configuration that is designed to separate 'almost' pure components from the crude oil. Its particular growth has resulted from the need of those components such as the butanes and propanes to satisfy a market of portable cooking fuel and industrial fuels. That these products can be suitably compressed and stored in small, easily handled containers at ambient temperatures provided the market popularity for these products, suitably titled Butane LPG and Propane LPG. The term LPG referring to Liquefied Petroleum Gas.

The introduction of the 'No Lead' in gasoline program during the late 1960s set the scene for the need of Octane sources additional to the Aromatics provided by high severity catalytic reforming. A source of such high-octane additives is found in some isomers of butane and pentane. This added to the need for light end processes which in many cases included the separation of iso butanes from the butane stream and also iso pentanes from the light naphtha stream.

A process description of a 'light ends' unit

The 'light ends' of crude oil is considered as those fractions in the crude that have a boiling point below cyclo-hexane. The 'light ends' distillation units however include the separation of the light naphtha cut, which is predominately pentanes and cyclo-pentanes, from heavy naphtha which contains the hexanes and heavier hydrocarbons necessary for the catalytic reformer feed. The feed to the 'light ends' distillation process is usually the full range naphtha distillate from the atmospheric crude distillation unit overhead condensate drum. In many cases the distillates from stabilizing cracker and reformer products are added to the crude unit overhead distillate to be included in the 'light end' unit feed.

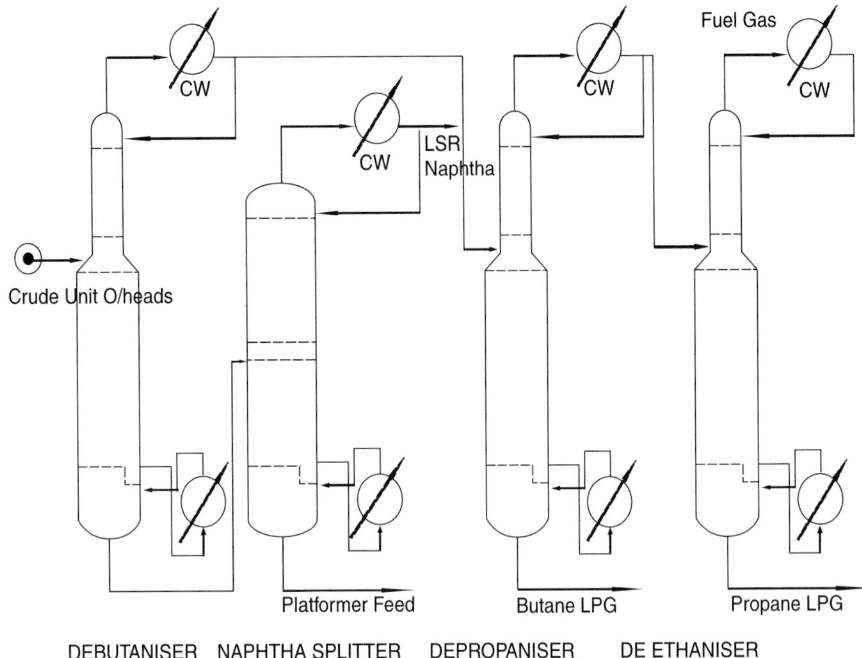

Figure 4.1. A typical light end unit configuration.

A typical process configuration for this unit is given in the flow diagram Figure 4.1.

In this configuration the total feed to the unit is debutanized in the first tower. The butanes and lighter hydrocarbons are totally condensed and collected in the column's overhead drum. Part of this condensate is returned to the tower top tray as reflux. The remainder is routed to a de-propanizer column. The bottom product from the de-butanizer is the full range naphtha product. This enters a naphtha splitter column where it is fractionated to give an overhead distillate of light naphtha and a bottom product of heavy naphtha.

The de-propanizer separates the debutanizer overhead distillate to give a propane fraction as an overhead distillate stream and the butane fraction (butane LPG) as the bottom product. The overhead distillate is fractionated in a de-ethanizer column to produce a rich propane stream (propane LPG) as the bottom product. The overheads from this column is predominately hydrocarbons lighter than propane. This stream is only partially condensed to provide reflux for the tower. The uncondensed vapor is normally routed to the refinery's fuel gas system.

The products from the 'light ends' unit are as follows:

Naphtha Splitter — Light Naphtha (overhead distillate)
Heavy Naphtha (bottom product–Reformer feed)
De-propanizer — Butane LPG (bottom product)
De-ethanizer — Propane LPG (bottom product)
Fuel Gas (overhead vapor).

Developing the material balance for light end units

In light end towers the material balance is developed as a molal balance. This type of balance is determined by the degree of separation of the feed molal components that enter the distillate fraction and those that leave with the bottom product.

Effective separation by fractionation in light end towers obey the same laws as those in the crude distillation units, that is: The degree of separation is the product of the number of trays (or stages) and the reflux (or overflow) in the column.

In the crude unit this separation was measured by the difference between the ASTM 95% point of the lighter fraction and the 5% ASTM point of the heavier fraction. This is the ASTM gap or overlap.

In light end towers the degree of separation is a little more precise. This is determined by the distribution of key components in the two fractions to be separated. Key components may be real components (such as C4's or C5's) or pseudo components defined by their mid boiling points. Normally key components are adjacent components by boiling point in the feed composition. Any two key components may be selected—a light key and a heavy key. By definition the light key has the lower boiling point. Both key components must, however, be present in the distillate and bottoms product of the column. If a side stream exists then these keys must also be present in the side stream product.

There are several correlation that describe the behavior of these key components in their distribution and relationship to one another. By far the more common of these correlation is the Fenske equation which relates the distribution of key components at minimum trays with infinite reflux. The equation is relatively simple and does not require iterative calculation techniques to solve it. The Fenske Equation is:

$$N_{\mathrm{m}+1} = \frac{\mathrm{Log}\,[((\mathrm{LT\ key}\,/\,\mathrm{HY\ key})_D \times (\mathrm{HY\ key}\,/\,\mathrm{LT\ key})_W)]}{\mathrm{Log}\,\frac{K_{\mathrm{LT}}\ \mathrm{key}}{K_{\mathrm{HY}}\ \mathrm{key}}}.$$

where

N_{m} = minimum number of theoretical strays at total reflux. The +1 is the reboiler which is counted as a theoretical tray

LT key = is the mole fraction of the selected light key
HY key = is the mole fraction of the selected heavy key
D = fractions in the distillate product
W = fractions in the bottom product
K_{LT} key = the equilibrium constant of the light key at mean system condition of temperature and pressure
K_{HY} key = the equilibrium constant of the heavy key again at mean system conditions

The ratio of the equilibrium constants is called the "relative volatilities" of the keys. Setting fractionation requirements for light end towers is usually done to meet a product specification. More often than not this specification is the vapor pressure of the heavier fraction on the tolerable amount of a heavy key allowed in the lighter fraction.

Sometimes, however, a specification for the separation may not be given. Under this circumstance some judgment must be made in determining the most reasonable separation that can be achieved with the equipment. This item of the manual addresses calculation techniques that satisfy either premise, and the procedure for these now follow.

Case 1: Setting separation requirements to meet a specification

In this case it is required to determine the amount of butane's that can be retained by a light naphtha cut to meet a RVP specification. The steps are as follows:

Step 1. Calculate the properties of the C_5+ naphtha from a component breakdown. These properties should give weight and mole rates per hour.

Step 2. Carry out a bubble point calculation of the C_5+ fraction and inserting butane's as the unknown quantity x.

Step 3. The equilibrium constantly used (K) in the calculation will be at the temperature and pressure conditions of the RVP (i.e., normally at 100°F which is the test temperature).

Step 4. Either the equilibrium constants given in the charts in the appendix may be used or the relationship vapor pressure divided by total systems pressure may be used.

Step 5. By definition the total moles liquid given is equal to the total moles vapor (calculated) in equilibrium at the bubble point. Thus equate and solve for x as the quantity of butane's tolerable to meet RVP.

Case 2: Setting fractionation where no specification is given

Step 1. Determine the composition of the feed in terms of real components, pseudo components or both. Calculate this in moles/hr and mole fractions.

Step 2. Select the key components, decide the minimum distribution of one or other key. For example in a debutanizer C_5's allowed into LPG must not be more than 2% of the total C_4's and lighter. This is to protect the Butane LPG "Weathering" Test Specification.

Step 3. Give values to the fraction of LT and HY keys in the distillate and bottoms. Use x as the unknown where appropriate.

Step 4. The Fenske equation will be used to calculate the distribution of keys. Determine the value of N_M by taking the actual number of trays and using an efficiency 70–75% to arrive at total theoretical trays. Divide this figure by 1.5 to arrive at the minimum theoretical trays N_M. Don't forget to add 1 for the reboiler to use in the equation.

Step 5. Estimate the mean tower conditions. This can be achieved by examining past plant logs, etc. Determine K values for the keys at this mean tower condition. Use published data for real components and the ratio of vapor pressure divided by total systems pressure for pseudo components.

Step 6. Solve for x in the Fenske correlation. This will be the split of the key components and the basis for the material balance.

Very often when setting up a design for a light end tower the actual number of trays are not known at the time when calculation number two is required. The following rule of thumb may be applied as a guide:

Tower	Number of actual trays
De-butanizer	30 to 35
De-propanizer	35 to 40
De-ethanizer	38 to 42
Naphtha Splitter	25 to 35

An example for illustrating Case 1 above is given in Chapter 1 of this Handbook where the amount of butane LPG allowed to meet Gasoline RVP is calculated. An example of Case 2 is given as follows:

The overhead distillate from an atmospheric crude distillation unit operating at 50,000 BPSD of Murban crude has the following composition (Table 4.1).

The key components for the de-butanizer will be nC4 and iC5 as Lt and Hy keys respectively. In the Fenske equation let x be the moles/hr of nC4 in the distillate. To satisfy the weathering test for butane LPG the maximum amount of C5's allowed in the de-butanizer distillate is 2.0 mol% of the total C4's and lighter. A reasonable amount of actual trays in a de-butanizer is 30. Allowing an efficiency of 70% the number of theoretical trays will be 21. It is reasonable to predict that the minimum

Table 4.1. Full range naphtha composition

Comp	BPSD	#/Gal	Gals/hr	lbs/hr	MW	moles/hr
C2	20	3.42	35	120	30	4
C	240	4.23	420	1,777	44	40.38
iC4	275	4.70	481	2,262	58	39.0
NC4	735	4.87	1,286	6,264	58	108.0
iC5	915	5.21	1,601	8,343	72	115.87
nC5	1,216	5.26	2,126	11,184	72	155.33
C6	2,170	5.56	3,798	21,114	84	251.36
C7	2,930	5.71	5,128	29,278	100	292.78
Mbpt 224	1,075	6.12	1,881	11,513	106	108.62
239	1,525	6.15	2,669	16,413	109	150.58
260	1,345	6.22	2,354	14,640	115	127.31
276	895	6.26	1,566	9,805	120	81.71
304	895	6.35	1,566	9,946	130	76.51
Total	14,235	5.73	24,911	142,659		1,551.45

theoretical trays will be the actual theoretical number divided by 1.5. Then N_m will be $21/1.5 = 14$ adding one for the reboiler gives $N_{m+1} = 15$.

From past data an average operating condition of temperature and pressure for a debutanizer are 210°F and 110 psig. At these conditions the equilibrium constants for both keys are read from curves given in the GPSA Engineering Data as:

$$\text{nC4} = 1.48$$
$$\text{iC5} = 0.94$$

Then the relative volatility $\phi = 1.48/0.94 = 1.57$.

Using the Fenske equation:

$$N_{m+1} = \frac{\text{Log}\left((\text{LT key / HY key})_D \times (\text{HY key / LT key})_W\right)}{\text{Log}\ \phi}$$

$$15 = \frac{\text{Log}((x/3.8 \times 112.1/108 - x))}{\text{Log}\ 1.57}$$

$$15 \times 0.196 = \text{Log}((112.1x)/(410.4 - 3.8x))$$

$$871 = 112.1x/(410.4 - 3.8x)$$

$$x = 104.48 \text{ moles/hr.}$$

Associated with nC4 in the bottom product will be an equilibrium amount of iC4. Although small this will have an effect on bottom product bubble point and therefore the tower bottom temperature. The amount of this iC4 in the bottom product can be calculated using a similar method as that for the split between nC4 and iC5, thus the keys for this calculation will be iC4 and nC4. Let x be the moles/hr of iC4 in the bottom product. Then:

	Feed	Dist	Bottoms
iC4	39	x	$39 - x$
nC4	108	104.48	3.52

Using the Fenske equation again;

$$15 = \frac{\log((x/104.48) \times (3.52/39 - x))}{\log(K\text{iC4}/K\text{nC4})}$$

$$= \frac{\log((x/104.48) \times (3.52/39 - x))}{\log 1.3}$$

$$1.709 = \log((x/104.48) \times (3.52/39 - x))$$

Then

$x = 38.99$ moles/hr. This gives zero iC4 in the bottom cut.

The material balance for the de-butanizer can now be written and is given in Table 4.2.

The material balances over the naphtha splitter, de-propanizer, and the de-ethanizer follow the same technique in identifying key components and utilizing the Fenske equation.

Table 4.2. De-butanizer material balance

	Feed			Dist			Bottoms		
Comp	BPSD	lbs/hr	mol/hr	BPSD	lbs/hr	mol/hr	BPSD	lbs/hr	mol/hr
C2	20	120	4	20	120	4.00			
C3	240	1,777	40.38	240	1,777	40.38			
iC4	275	2,262	39.00	275	2,262	39.00			
nC4	735	6,264	108.00	711	6,061	104.48	24	203	3.52
iC5	915	8,343	115.87	30	274	3.80	885	8,069	112.07
nC5	1,216	11,184	155.33				1,216	11,184	155.33
C6	2,170	21,114	251.36				2,170	21,114	251.36
C7	2,930	29,278	292.78				2,930	29,278	292.78
MBPt224	1,075	11,513	108.62				1,075	11,513	108.62
239	1,525	16,413	150.58				1,525	16,413	150.58
260	1,345	14,640	127.31				1,345	14,640	127.31
276	895	9,805	81.71				895	9,805	81.71
304	895	9,946	76.51				895	9,805	76.51
TOTAL	14,236	142,659	1,551.45	1,276	10,494	191.66	12,960	132,165	1,359.79

Calculating the operating conditions in light end towers

Light end units follow two calculation procedures for setting the conditions of temperature and pressure in the fractionating towers. The first procedure relates to towers in which the overhead product and reflux are totally condensed. The second procedure relates to those towers where the overhead product is not totally condensed.

Calculating the tower top pressure and temperature for totally condensed distillate product

This procedure commences with setting a realistic reflux drum temperature. This is fixed by the cooling medium temperature, such as ambient air temperature (for air coolers) or cooling water. As total condensation is required then the pressure of the reflux drum must be the bubble point of the distillate product (and reflux) at the selected drum temperature. Once the drum pressure has been calculated the tower top pressure can be determined by taking into account estimated or manufacturers specified pressure drops for equipment and piping between the drum and tower top. As a rough estimate condensers and/or heat exchangers in the system have between 3 and 5 psi pressure drop. Allow also about a 2 psi pressure drop for piping.

The tower top temperature is calculated as the dew point of the distillate product at the total overhead pressure. There are usually no steam or inert gases present in the light end tower overheads, so total pressure may be used. The following example uses the de-butanizer overhead from the material balance given in section "Developing the Material Balance for Light End Units" above.

The ambient air temperature for the site is 60°F, and the operating temperature for the reflux drum will be set at 100°F. The bubble pressure at this temperature is calculated and given in Table 4.3.

Table 4.3. De-butanizer reflux drum pressure. Distillate Bubble Point Calculation at 100°F (Trial 1)

Comp	Mol Frac x	K @ 125 psia	$Y = X \cdot K$
C2	0.0209	4.8	0.3906
C3	0.2107	1.48	0.4054
iC4	0.2035	0.68	0.0830
nC4	0.5451	0.5	0.1254
iC5	0.0198	0.23	0.0009
Total	1.0000		1.0053

Reflux Drum Pressure = 125 psia and 100°F.

Table 4.4. Tower top temperature

Comp	Mole Fract (y)	K @ 152°F	Mole Fract (x)
C2	0.021	6.2	0.003
C3	0.211	2.1	0.100
iC4	0.203	1.1	0.185
nC4	0.545	0.83	0.657
iC5	0.02	0.41	0.048
Total	1.000		0.993

Allowing 3 psi pressure drop over the overhead condenser and 2 psi for the associated overhead piping the tower top pressure becomes 130 psia (115 psig). The temperature of the tower top is the dew point of the distillate product at the top pressure. This is given in Table 4.4.

There is reasonable agreement that total y = total X therefore the dew point and therefore the tower top temperature at 115 psig is 152°F.

Calculating the overhead conditions for partially condensed distillate and product

There are two circumstances where the overhead stream from a light ends tower may not be totally condensed. The most common of these is in the case of the de-ethanizer. Here, usually, only sufficient overhead stream is condensed to provide the overhead reflux stream. The reason for this is that at a normal condensing temperature the pressure required at the reflux drum would be unacceptably high. It would be so high that the tower bottom pressure required would be higher than the bottom product's critical pressure and thus fractionation would not be possible. In this case the tower pressure is set again by the reflux drum pressure. Unlike the case of total condensation the reflux drum pressure for this partial condensation is found as the dew point at the condensing temperature of the distillate product. In fact the reflux drum becomes a theoretical fractionation stage. In certain cases refrigeration is used to totally condense the overhead stream at an acceptable pressure, but this selection would be as a result of a study of the operation's economics.

In certain processes where a high tower pressure (and consequently high temperature) may cause deterioration of one or more of the products from the tower, the tower pressure is reduced by condensing only a fraction of the overhead distillate product. For this purpose a vaporization curve for the distillate product is constructed at a selected reflux drum temperature. This curve is developed by calculating a series of equilibrium compositions of the distillate product at the reflux drum temperature but over a pressure range. A reflux drum pressure can then be selected from the curve that satisfies an acceptable tower pressure profile. A final equilibrium calculation is then made at this pressure to provide the component and quantity composition of the liquid and vapor streams leaving the reflux drum.

Example of a de-ethanizer overhead operating conditions is given below:

Consider the overhead vapor from the reflux drum, which is the product in this case, has the following molar composition:

Mole	Fraction
C_1	0.063
C_2	0.443
C_3	0.494

A reasonable temperature for operating the reflux drum is 100°F (based on local ambient conditions). The reflux drum pressure is calculated from as the dew point of the vapor at the reflux drum pressure, thus:

	Vapor mole fract y	K @100°F and 350 psia	Liquid mole fract x
C_1	0.063	7.0	0.009
C2	0.443	2.0	0.222
C3	0.494	0.64	0.772
Total	1.000		1.003

This was the second trial starting with the pressure at 450 psia. The total x value is close enough to y to allow the pressure to be 350 psia.

Now the Tower Top conditions will be at the dew point of the reflux plus the product leaving the tower. The pressure will be the reflux drum pressure plus say 7 psi for the condenser and piping pressure drop. In this case this pressure will be 357 psia. Set the reflux ratio (moles reflux/moles product) to be 2.0:1 where the moles/hr of product is 7.9. The composition of the reflux stream is the x value of the product vapor. The dew point calculation to establish the tower top temperature therefore will be as follows (Table 4.5).

Note this dew point calculation may change if subsequent calculations show that the reflux ratio required will be significantly different to the one assumed here.

It is not proposed to show the case of the partially condensed product here. This calculation would be similar to the de-ethanizer and in this case the reflux composition will be the liquid phase from the equilibrium flash of the product.

Table 4.5. Calculating the tower top temperature for a de-ethanizer

Component	mole/hr prod	mole/hr reflux	Total moles/hr	Mole fraction y	K @ 357 psia and 133°F	Mole fract liquid x
C1	0.5	0.14	0.64	0.027	0.42	0.064
C2	3.5	3.49	6.99	0.295	2.00	0.148
C3	3.9	12.17	16.07	0.678	0.86	0.788
Total	7.9	15.80	23.70	1.000		1.000

Calculating the tower bottom conditions for light end towers

The calculation to establish the tower bottom conditions for all light end towers are the same. Only the values for pressure and composition change. The temperature of the product leaving the bottom of the tower will be at its bubble point at the tower bottom pressure. Consider the de-butanizer column whose material balance is given in Table 4.2.

The number of actual trays (estimated) to accomplish the separation between the C4's and the C5's is 30. The pressure drop for fully loaded trays can be taken as being between 0.15 and 0.2 psi. Assume then a press drop of 0.17 per tray. The tower top pressure calculated earlier is 125 psia, then the bottoms pressure is $125 + (0.17 \times 30) = 130$ psia (Table 4.6).

Calculating the number of trays in light end towers

For definitive design work one of the many excellent simulation packages should be used. However most simulation packages require good quality input data. This often

Table 4.6. De-butanizer bottom temperature (bubble point)

Components	Moles/hr	Mol fract x	K @138 psia and 358°F	$Y = x/K$
nC4	3.5	0.003	3.7	0.010
iC5	112.1	0.082	2.4	0.198
nC5	155.33	0.114	2.0	0.228
C6	251.35	0.185	1.4	0.259
C7	292.78	0.215	0.77	0.166
Mpt 224°F	108.62	0.080	0.58	0.046
239°F	150.58	0.111	0.49	0.054
260°F	127.31	0.094	0.38	0.036
276°F	81.71	0.060	0.29	0.017
304°F	76.51	0.056	0.21	0.012
Total	1,359.79	1.000		1.026

The tower bottom condition is therefore 138 psia and 358°F.

means a fairly accurate estimate of the number of theoretical trays and the liquid/vapor traffic in the tower. An acceptably accurate 'short cut' method to arrive at the number of theoretical trays is given here. The estimate of the liquid/vapor traffic is discussed later in this chapter.

The short-cut method for predicting number of theoretical trays

Three calculation or relationships are used to determine the number of theoretical trays in this method. These are:

The Fenske calculation to determine the minimum number of trays at total reflux.
The Underwood calculation to determine the minimum reflux at infinite number of trays.

The Gilliland correlation which uses the result of the two calculations to give the theoretical number of trays.

The Fenske equation
This has been discussed earlier under the section dealing with the "Material Balance for Light End Towers." The equation is as follows:

$$N_{\mathrm{m+1}} = \mathrm{Log}\,[(\mathrm{LT\ key}\,/\,\mathrm{HY\ key})_D \times (\mathrm{HY\ key}\,/\,\mathrm{LT\ key})_W] \div \mathrm{Log}(K_{\mathrm{LT}}\mathrm{key}/K_{\mathrm{HY}})$$

where

N_{m} = minimum number of theoretical strays at total reflux. The +1 is the reboiler which is counted as a theoretical tray
LT key = is the mole fraction of the selected light key
HY key = is the mole fraction of the selected heavy key
D = fractions in the distillate product
W = fractions in the bottom product
K_{LT} key = the equilibrium constant of the light key at mean system condition of temperature and pressure
K_{HY} key = the equilibrium constant of the heavy key again at mean system conditions

The Underwood equation and calculation
The Underwood equation is more complex than the Fenske, and requires a trial and error calculation to solve it. The equation itself is in two parts: The first looks at the vapor volatilities (ratio of the K's) of each component to one of the keys, then by trial and error arriving at an expression for a factor B that forces the equation to zero. This first equation is written as follows:

$$\Sigma((\phi i) \cdot (\mathrm{x}iF) \div (xiF) - B)) = 0$$

where

ϕi = The relative volatility of component i.
xiF = The mole fraction of component i in the feed.
B = The factor that forces the expression to zero.

The second part of the equation is expressed as follows:

$$R_{(m+1)} = \Sigma((\phi i)(xiD)) \div ((xiD) - B))$$

where

R_m = Minimum reflux at infinite number of trays.
xiD = The mole fraction of i in the distillate.

The relationship between the Fenske equation and the Underwood is given by the Gilliland Correlation shown in Figure 4.2.

An example of the Underwood equation calculation
Consider the material balance of a de-butanizer developed in Table 4.2.

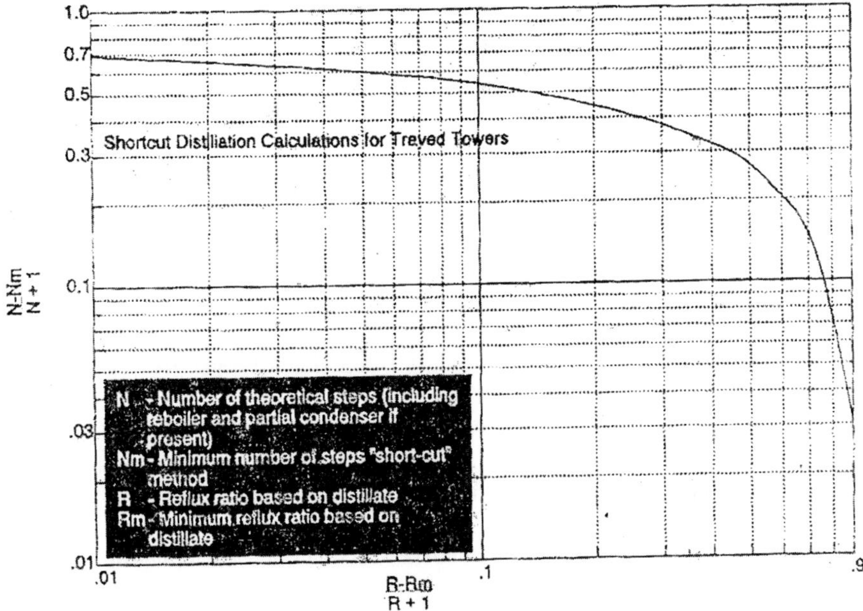

Figure 4.2. The Gilliland correlation for calculating theoretical trays.

Part 1 of the Underwood Equation is calculated as shown in the following table:

Comp	Mol fract x_f	K @ Ave* cond	Rel vol Φ	$x_f \times \Phi$	Trial 3(1) $B = 0.815$
C2	0.003	9.0	4.5	0.0135	0.0037
C3	0.026	4.2	2.1	0.0545	0.0424
iC4	0.025	2.4	1.2	0.0300	0.0776
nC4 (Key)	0.070	2.0	1.00	0.0700	0.3753
iC5	0.075	1.2	0.62	0.0450	−0.2108
nC5	0.100	1.0	0.52	0.0500	−0.1595
C6	0.162	0.49	0.245	0.0397	−0.0698
C7	0.188	0.25	0.125	0.0235	−0.0341
M Bpt 224°F	0.070	0.16	0.08	0.0056	−0.0076
239°F	0.097	0.13	0.065	0.0040	−0.0084
260°F	0.082	0.097	0.0485	0.0039	−0.0052
276°F	0.053	0.069	0.0345	0.0018	−0.0023
304°F	0.049	0.042	0.021	0.0010	−0.0013
Totals	1.000				0

*Ave conditions are 134 psia and 255°F.
$\Sigma((\phi i) \cdot (xiF) \div (xiF) - B)) = 0(1)$.

Part 2 of the Underwood equation is calculated as shown in the following table:

Comp	Mole fract x_D	Rel vol Φ	$(x_D)(\Phi)$	$(x_D)(\Phi)/(\Phi - B)$
C2	0.022	4.5	0.0945	0.0257
C3	0.217	2.1	0.4431	0.3444
iC4	0.207	1.2	0.2436	0.6303
nC4 (Key)	0.544	1.00	0.5450	2.9223
iC5	0.010	0.60	0.012	−0.0562
Total	1.000			3.8665

$$R_{m+1} = 3.8665 \quad R_m = 2.87 \quad \text{and} \quad R = 2.87 \times 1.5 = 4.3$$

Using the Gilliland Curve $\frac{(R-R_m)}{R+1} = 0.27$ and from the curve $\frac{(N-N_m)}{(N+1)} = 0.4$.

Then Number of Theoretical trays N will be $N - N_m = 0.4\ N + 0.4$.

N_m calculated from the Fenske equation (see section on 'Developing the Material Balance for Light End Units') is 14.

Then $N = 24$. Assume an average tray efficiency of 70% then total actual trays = 34.

Condenser and reboiler duties

Both the condenser and reboiler duties are the result of the light ends tower heat balance that meets the degree of separation required in the products. It may be said that the condenser duty and operation determines the amount of wash liquid flow in the tower to meet the degree of rectification required for the lighter product. The reboiler in turn generates the vapor flow in the tower to satisfy the degree of stripping required for the particular separation and the quality of the heavy product. At a constant feed temperature and pressure, changes to either the condenser duty or the reboiler duty will effect the duty of the other. In other words the tower must always be in heat balance.

Calculating the condenser duty

The duty of the overhead condenser is determined by a heat balance over the tower top (above the top tray) and reflux drum, with the condenser duty being the unknown quantity. Using the data already determined for the debutanizer in the previous sections of this the following is a calculation to determine the condenser duty for this unit.

Consider the following heat balance sketch (Figure 4.3).

In the following heat balance the liquid overflow from the top tray is based on the internal reflux ratio obtained from the Fenske, Underwood equations and the Gilliland correlation. This ratio is in moles and its molecular weight is derived from the molal

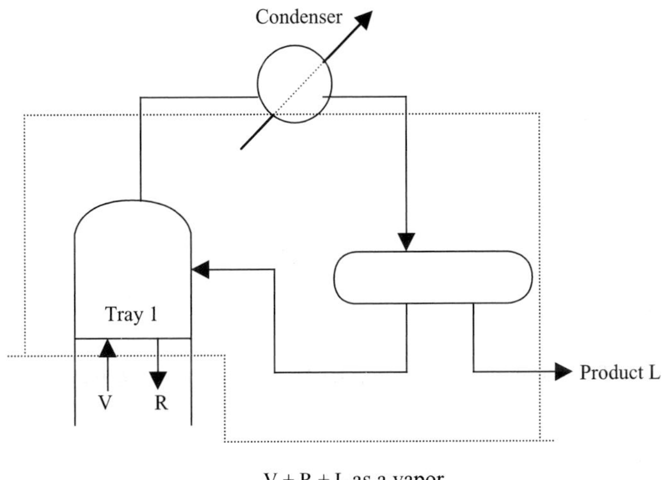

Figure 4.3. The overhead heat balance figure.

Table 4.7. Tower top heat balance

Stream	L/V	Mole wt	°F	lbs/hr	Btu/lb	mmBtu/hr
In						
Ref o/flow	V	56.8	160	46,803	305	14.275
Distillate	V	55	160	10,494	306	3.211
Total In				57,297		17.486
Out						
Dist Prod	L	55	100	10,494	150	1.574
Ref Liqid	L	56.8	155	46,803	190	8.893
Condenser				By Difference		7.019
Total Out				57.297		17.486

composition of the dew point calculation to establish the tower top temperature. It is the mole weight of the liquid in equilibrium with the overhead distillate vapor. Thus the total moles of the overflow liquid is $4.3 \times 191.66 = 824.14$ moles/hr. Its mole weight is 56.79. Therefore overflow liquid is 46,803 lbs/hr. The heat balance now follows (Table 4.7).

The top tray temperature is taken as 3 degrees above the tower top temperature. The vapor to the top tray is taken as about 5 degrees above the top tray.

Condenser duty is 7,100,000 Btu/hr.

Calculating the reboiler duty

It is important that the reboiler duty achieve a balance between generating an effective rate of vapor for stripping the bottom product while maintaining an economic vapor load on the stripping trays. As a rule of thumb this can be achieved with a stripping rate as a percent of bottom product of between 70 and 90 mole. The tower bottoms temperature has already been calculated as 358°F at 138 psia. The composition of the vapor in equilibrium at these conditions has also been calculated. Thus the moles/hr of bottom product is 1,359.79 then the moles strippout required will be $1{,}359.79 \times 0.9 = 1{,}223.8$ moles/hr. The molecular weight of strippout is 88 (from the equilibrium composition). The heat balance to determine the reboiler duty now follows. In this calculation the draw off tray to the reboiler is about 10°F lower than the tower bottom temperature that is 348°F (Table 4.8).

Reboiler duty is 13,403,000 Btu/hr.

Calculating the overall tower heat balance

Knowing the reboiler and condenser duties an overall heat balance over the tower can be calculated and will provide the pre-heat required in the feed and consequently

Table 4.8. Bottom tray heat balance

Stream	L or V	Temp °F	lbs/ hr	Btu/lb	mmBtu/hr
In					
Liquid ex tray 34	L	348	239,860	198	47.492
Reboiler			By Difference		13.403
Total in					
Out					
Bot product	L	358	132,165	200	26.433
Strip out	V	358	107,695	320	34.462
Total out			239,860		60.895

its temperature. A calculation to determine the bubble point of the feed at 134 psia (mean inlet pressure) was made. This temperature was 305°F. Should the calculated temperature be above this then an enthalpy curve would need to be developed for the feed to determine its actual enthalpy and temperature. The overall heat balance now follows (Table 4.9).

Enthalpy of the feed is $\frac{21{,}713{,}000}{142{,}659} = 152.2$ Btu/lb from the Enthalpy tables the temperature is found to be 276°F. The feed tray is based on a preliminary tower temperature profile and a 75% efficiency for rectifying trays with a 65% efficiency for the stripping trays (see Figure 4.4). Both the temperature profile and the tray efficiencies will be finalized by a suitable computer simulation package or a rigorous tray to tray calculation. Note: The calculation techniques and examples given here are good input to computer simulation packages. Wherever possible these packages should be used for final process design.

Tower loading and sizing

Light end towers all follow the following principles of cross sectional area sizing and indeed the tower height. These dimensions are inter-related by the height required

Table 4.9. Overall tower heat balance

Stream	V or L	Temp °F	lbs/ hr	Btu/lb	mmBtu/hr
In					
Feed	L	T	142,659	By Diff	21.713
Reboiler					13.403
Total in					35.116
Out					
Distillate	L	100	10,494	150	1.574
Bot prod	L	358	132,165	200	26.433
Condenser					7.109
Total out			142,659		35.116

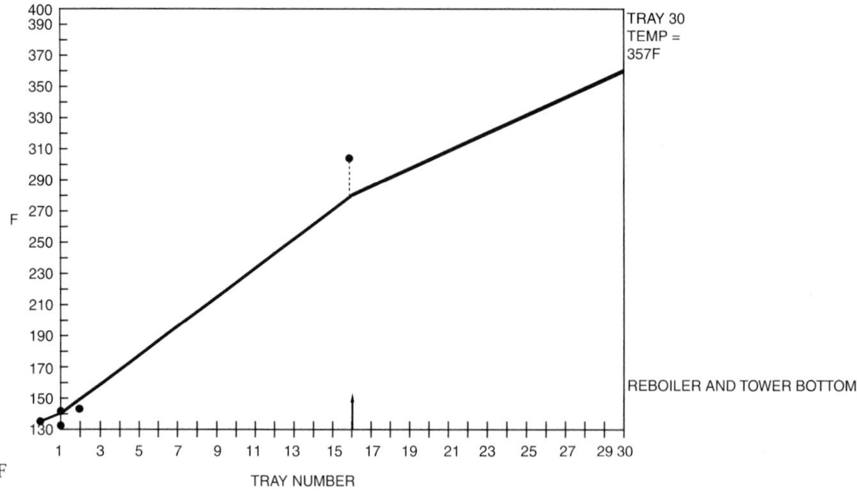

F

Figure 4.4. Tower temperature profile.

between trays to ensure proper separation of clear liquid from the frothy mixture of the tray inlet fluid. There are many procedures and co-relations to determine these dimensions. The following method is just one which will provide a good estimate for tower design. For definitive design however it is essential that tray manufacturers be consulted and their methodology be used. After all they, the manufacturers, will be required to guarantee the performance of the unit in terms of flooding capacity and tray efficiency. In all cases this has to be a significant consideration but in case of super fractionating units with over a hundred trays in many cases this has to be a primary consideration.

There are many light end tower computer simulation packages in the market, and most of these calculate vapor and liquid traffic in the tower on a tray to tray basis. The input for these programs however does require a fairly accurate estimate of these values for easy convergence and subsequent use of the program. The following procedure with the equations used can provide details of tower loading and tray criteria at critical trays in the tower. A linear co-relation between these points will then provide a reasonable loading profile over the tower sufficient for good computer input.

Tower loading and overall tower diameter

The diameter of a fractionating tower is usually based on the vapor loading on two critical trays. These are the tower top tray which will set the diameter for the rectifying section (i.e., the trays above the feed tray), and the tower bottom tray. This lower tray sets the tower diameter for the stripping section of the tower, (i.e., the feed tray and the trays below the feed tray). These two sections may have different diameters.

The vapor loading is based on the rate of vapor passing through the tray and the density of the vapor and the liquid on the tray through which the vapor bubbles. This relationship is given by the *Brown and Souder* equation. There are several forms of this equation which can be used with the appropriate physical tray constants. One of the forms used here is as follows:

$$G_f = K\sqrt{(\rho_v \times (\rho_l - \rho_v))}.$$

where

G_f = Mass of vapor per sq foot of tray at flood (lbs/hr · sqft)
K = A constant based on tray spacing at flood (see Chapter 3)
ρ_v = Density of vapor at tray conditions of temperature and pressure in lbs/cuft
ρ_l = Density of liquid at tray conditions in lbs/cuft

The area thus determined is the 'Bubble' area of the tray. Normally trays are designed at 85 to 90% of flood. Therefore for good design the Gf is divided by this percentage to give the actual or design area of the tray. The whole tray is made up of two other areas: That for the down comers, and a waste area which is allocated to calming the liquid leaving the bubble area before entering the down comer. The relationship of these areas to one another is given in Table 4.10.

Using the criteria in Table 4.10 and the value of the bubble area based on the vapor loading the total tray area and therefore the tower diameter can be determined. This relationship is summed up by the following expression:

$$A_s = A_B + A_{dc} + A_w$$

where

A_s = Total tray area
A_B = Bubble area
A_{dc} = Down comer area (inlet + outlet)
A_w = Waste or calming zone area (usually 15% of A_s)

Tray spacing

The tray spacing used in the initial determination of Flood Loading needs to be checked. If necessary the spacing and the calculation will be revised to meet the correct spacing criteria. Usually this first guess at tray spacing is taken as 24″. The following equations are then applied to determine whether this spacing is satisfactory. These equations calculate the pressure drop across the tray in terms of the clear liquid

Table 4.10. Valve and sieve tray characterization

Design feature	Suggested value	Alternate values	Comment
1. Valve size and layout			
(a) Valve diameter	—	—	Valve diameter is fixed by the vendor
(b) Percent hole area A_o/A_b	12	8–15	Open area should be set by the designer. In general, the lower the open area, the higher the efficiency and flexibility, and the lower the capacity (due to increased pressure drop). At values of open area toward the upper end of the range (say 15%), the flexibility and efficiency are approaching sieve tray values. At the lower end of the range, capacity and down comer filling becomes limiting
(c) Valve pitch diam ratio	—	—	Valve pitch is normally triangular. However, this variable is usually fixed by the vendor
(d) Valve distribution	—	—	On trays with flow path length $\geq 5'$, and for liquid rates > 5,000 GPH/ft. (diameter) on trays with flow path length $< 5'$, provide 10% more valves on the inlet half of the tray than on the outlet half
(e) Bubble area, A_b	—	—	Bubble area should be maximized
(f) Tray efficiency	—	—	Valve tray efficiency will be about equal to sieve tray efficiency provided there is not a blowing or flooding limitation
(g) Valve blanking	—	—	This should not generally be necessary unless tower is being sized for future service at much higher rates. Blanking strips can then be used. Blank within bubble area, not around periphery to maintain best efficiency
2. Tray spacing	—	12–36	Generally economic to use min. values which are which are usually set by maintenance requirements. Other considerations are down comer filling and flexibility. Use of variable spacing to accommodate loading changes from section to section should be considered.
3. Number of liquid passes	1	1–2	Multi passing improves liquid handling capacity at the expense of vapor capacity for a given diameter column and tray spacing. Cost is apparently no greater—at least, for tower diameters < 8 ft.

(Cont.)

Table 4.10. (Cont.)

Design feature	Suggested value	Alternate values	Comment
4. Down comers and Weirs			
(a) Allowable Down-comer inlet vel ft/sec of clear liq		0.3–0.4	Lower value recommended for absorbers or other systems of known high frothiness
(b) type of down-comer	Chord	Chord, Arc	Min. chord length should be 65% of tray diameter for good liquid distribution. Sloped down comers can be used for high liquid rates—with maximum outlet velocity = 0.6 ft/sec. Arc down comers may be used alternatively to give more bubble area (and higher capacity) but are somewhat more expensive. Min. width should be 6 in. for latter
(c) Inboard DC width (inlet and outlet)		Min. 8 inches	Use of a 14–16″ "jump baffle" suspended length wise in the center of the inboard down comer and extending the length of the down comer is suggested to prevent possible bridging over by froth entering the down comer from opposite sides. Elevation of base of jump baffle should be level with outlet weirs. Internal access way must be provided to allow passage from one side to another during inspection
(d) Outlet weir height	2″	1″ to 4″	Weir height can be varied with liquid rate to give a total liquid head on the tray (h_c) in the range of 2.5″–4″ whenever possible. Lower values suggested for vacuum towers, higher ones for long residence time applications
(e) Clearance under DC	1.5″	1″ min	Set clearance to give head loss of approximately 1 inch. Higher values can be used if necessary to assure sealing of down comer
(f) DC seal (inlet or outlet weir height minus DC clearance)	Use outlet weir to give min. ½″ seal in tray liquid	Inlet weir or recessed inlet box	In most cases tray liquid level can be made high enough to seal the down comer through use of outlet weir only. Inlet weirs add to down comer build up; in some cases they may be desirable for 2-pass trays to ensure equal liquid distribution. Recessed inlets are more expensive but may be necessary in cases where an operating seal would require an excessively high outlet weir
(g) DC filling (% of tray spacing)		40–50	Use the lower value for high pressure towers, absorbers, vacuum towers, known foaming systems, and also for tray spacing of 18″ or lower

hold up (or height) in the down comer. This height of liquid must be less than 50% of the tray space for most applications. In the case of a high foaming process this height must be less than 40% of the tray spacing. These pressure drop criteria as the determination of the tray hydraulics and their associated equations now follow:

Clear liquid height, h_{cl}

$$h_{cl} = 0.5 \times [V_L \div (N_p \times l_o)]^{2/3}$$

where

h_{cl} = Height of clear liquid on tray in inches of hot liquid
V_L = Liquid loading in gallons per minute of hot liquid
N_p = Number of passes on tray
l_o = length of outlet weir in inches

Effective dry tray pressure drop, h_{ed}

The effective dry tray ΔP shall be the greatest of the following two expressions:

(a) $\Delta P_{po} = 1.35 t_m \cdot \rho_m/\rho_l + K_1 \cdot (V_o^2) \cdot \rho_v/\rho_l$
(b) $\Delta P_{fo} = K_2 (V_o^2) \rho_v/\rho_l$

where

ΔP_{po} = Dry tray ΔP, valve partially open. In inches of hot liquid.
ΔP_{fo} = Dry tray ΔP, valve fully open. In inches of hot liquid.
t_m = Valve thickness in inches (see Table 4.11)
ρ_m = Valve metal density in lbs/cuft (see Table 4.12)
V_o = Vapor velocity through valves in ft/sec (= cuft/sec/A_o)

Constants K_1 and K_2 are given in Table 4.13.

In Calculating V_o assume ratio hole to bubble area (A_o/A_B) is 12%.

Table 4.11. Valve thickness in inches

Gage	t_m
20	0.037
18	0.050
16	0.060
14	0.074
12	0.104
10	0.134
8	0.250

Table 4.12. Valve metal densities (in lbs/cuft)

Metal	Density, ρ_m
C.S.	480
S.S.	510
Nickel	553
Monel	550
Titanium	283
Hastelloy	560
Aluminum	168
Copper	560
Lead	708

Total tray, $\Delta P \cdot ht$

Total Tray $\Delta P = h_{cl} + h_{ed}$

Head loss under downcomer, h_{ud}

$$h_{ud} = 0.06[\text{GPM} \div (cL_i - N_p)]^2$$

where

h_{ud} = Head loss under downcomer in inches of hot liquid
c = Constant = 1.5
L_i = Length of inlet weir in inches

Inlet head in inches of hot liquid, h_i

When there is an inlet weir use:

$$h_i = 0.06[\text{GPM} \div (N_p \cdot L_i)]^{2/3} + h_{wi}$$

where

h_i = Tray inlet head in inches of hot liquid
L_i = Length of inlet weir in inches
h_{wi} = Height of inlet weir in inches

Table 4.13. Values of K_1 and K_2

Type of unit	K_1	K_2			
Deck Thickness ins		0.074	0.104	0.134	0.25
Normal valves	0.2	1.05	0.92	0.82	0.58
Vacuum valves	0.1	0.50	0.39	0.38	–

where there is no inlet weir then $h_i = h_{cl}$.

Downcomer filling in inches of hot liquid

$$L_d = h_i + (h_t + h_{ud}) \cdot [\rho_l \div (\rho_l - \rho_V)] + 1.0$$

where

L_d = Downcomer filling in inches of hot liquid.

Example. The following is an example of tower sizing and hydraulic analysis. Tray loading data that is used for this example are those calculated for the debutanizer column earlier in this chapter. Thus, all data at tray conditions of temperature and pressure:

Vapor	Top Tray	Liquid	Top Tray
Moles/hr	1,012	lbs/hr	46,803
lbs/hr	57,297	GPM	187
ACFS	14.3	CFS	0.147
ρ_v lbs/cuft	1.11	ρ_L lbs/cuft	31.18
Vapor	Bottom Tray	Bottom Tray	
Moles/hr	1,223.9	lbs/hr	239,860
lbs/hr	107,695	GPM	768.4
ACFS	8.7	CFS	1.715
ρv lbs/cuft	3.44	ρ_L lbs/cuft	38.9

Bubble area on top tray at Flood:

$$G_f = K\sqrt{(\rho_v \times (\rho_l - \rho_v))}.$$

where

G_f = Load at Flood in lbs/hr sqft.
K = 1,110
ρ_v = 1.11
ρ_l = 33.37
G_f = 6,415.8 lbs/hr sqft

Tower diameter will be designed to 80% of Flood. Then $G_A = 5{,}132.6$ lbs/hr sqft
Bubble section A_B of tray at 80% flood will be $\frac{57{,}297}{5{,}132.6} = 11.16$ sqft.

Down comer area (inlet and outlet) A_{dc}:

Down comer velocity will be 0.4 ft/sec (see Table 4.10).

Then area of one downcomer will be $0.417 \div 0.4 = 1.04$ sqft.
Total downcomer area $A_{dc} = 2.08$ sqft.

Waste area of tray A_w will be 15% (see Table 4.10).

Total tray area $A_s = A_B + A_{dc} + A_w$

Then for the top section of the tower the diameter will be:

$$A_s - 0.15A_s = 2.08 + 11.6 \text{ sqft}$$
$$A_s = 15.58 \text{ sqft. diameter} = 4.45 \text{ ft. say } 4.5 \text{ ft.}$$

Similarly for the stripping side of the tower

$$G_f = 1{,}110\sqrt{3.44 \times (38.9 - 3.44)}$$
$$= 12{,}259 \text{ lbs/hr sqft. and at 80\% flood Ga} = 9{,}807 \text{ lbs/hr sqft}$$

Downcomer area $= 1.715/0.4 = 4.29$ sqft and $A_{dc} = 8.58$ sqft.

$A_s = 24$ sqft and diam is 5.53 ft say 6 ft.

Tower hydraulics and downcomer filling

Using the pressure drop equations defined earlier the percentage of downcomer filled by liquid is calculated. This calculation is based on the stripping section of the tower only. A similar one will be completed for the rectifying section. Thus:

Clear Liquid Height, h_{cl}

$$h_{cl} = 0.5 \times [V_L \div (N_p \times L_o)]^{2/3}$$

$V_L = 768.4$ GPM
$N_p = 2$ (Liquid loading is relatively high so the option of a 2 pass tray is used)
$L_o = 58.8$ inches. Use the correlation given in the appendix to this chapter.

Then $h_{cl} = 1.74$ inches of hot liquid.

Effective dry tray, ΔP

(a) $\Delta P_{po} = 1.35 t_m \cdot \rho_m/\rho_l + K_1 \cdot (V_o^2) \cdot \rho_v/\rho_l$
(b) $\Delta P_{fo} = K_2(V_o)\rho_v/\rho_l$

Use $A_o/A_B = 12\%$ giving A_o as 1.32 sqft.

Valve thickness is 0.05 inches, and metal density is 480 lbs/cuft.

$$K_1 = 0.2, \quad K_2 = 0.92$$
$$V_o = 6.6 \text{ ft/sec}$$
$$\Delta P_{po} = 1.603 \text{ inches of hot liquid}$$
$$\Delta P_{fo} = 3.54 \text{ inches of hot liquid.}$$

Total tray, $\Delta P \cdot h_t$

$$h_t = 1.74 + 3.54 = 5.28 \text{ inches of hot liquid}$$

Head loss under downcomer, h_{ud}

$$h_{ud} = 0.06[\text{GPM} \div (cL_i \cdot N_p)]^2$$
$$c = 3.2 \text{ (estimated)}$$
$$L_i = 58.8 \text{ inches.}$$
$$h_{ud} = 0.96 \text{ inches of hot liquid.}$$

Inlet weir head, h_i

There is no inlet weir therefore $h_i = h_{cl}$.

Total downcomer filling, L_D

$$L_D = 1.74 + (5.28 + 0.96)(38.9/35.46) + 1$$
$$= 9.54 \text{ inches of hot liquid.}$$
$$= 40\% \text{ of tray spacing, which is satisfactory.}$$

Checks for light end tower operation and performance

Most light end towers are very stable in their operation. That is, once they are lined out for an operating requirement normal unit control maintain their stability. When performance falls off it can be attributed to one of a few reasons. This item looks at some of these reasons and how they can be evaluated and checked. By performance in this case is meant the ability of the unit to make product quality at the prescribed throughput.

Cold feed

The condition of the feed entering the tower is very important to the tower operation. Ideally the feed should enter the tower at as close to a calculated feed tray temperature as possible. If the feed is well below its bubble point on entering the tower, several trays below the feed tray are taken up for heat transfer before effective mass transfer can begin. This could prevent the specified product separation occurring and tray efficiency in this section of the tower falls off dramatically. Feed condition can be checked by bubble point calculation and a flash calculation (Item 3.7).

Hot feed

This situation is probably the more serious regarding feed condition. If the feed enters at a temperature far above its bubble point its resulting enthalpy will be such as to reduce the reboiler duty. This will occur automatically as the tower must always be in heat balance. The tower controls will maintain the product quantity and split. However, if the reboiler duty is drastically reduced insufficient stripper vapors will be available for the stripping function. Poor separation will result.

As a rule of thumb the stripping vapor to the bottom tray must be at least 70% mole of the bottom product make. In super fractionation such as a de-isopentanizer this figure would be at least 80–100% of bottoms make.

Heat balances as shown in items and will quickly determine the stripping vapor status.

Ideal feed condition

Ideally the feed should enter the tower close to feed tray temperature. Usually then at the inlet pressure the feed will be in a mixed phase with the vapor portion very close in quantity to the distillate product. As the feed to these units are generally heated by the bottoms product heat exchange, the approach temperatures are always a consideration. To maintain good feed conditions however it is often beneficial to include a separate steam (or hot oil) feed pre-heater.

Entrainment

A common cause of poor plant performance at high throughout or high reflux rates is liquid entrainment or carry over from tray to tray. Very often in a high load and entrainment situation the problem is further agitated by increasing reflux to attempt separation improvement.

A well designed light end tower can operate up to about 120% of allowable flood before substantial carry over occurs. Loading above this figure would result in some degree of entrainment.

Downcomer backup and flooding

If tower loadings are increased well above allowable flood point there is a real danger that down comers become unable to cope with the liquid load. They would fill and the tower would be in a state of flood. This will be very apparent with very high abnormal pressure drop occurring across the tower. Separation by fractionation is not possible under these conditions. Heat input (and feed rate) to the tower must be reduced to bring the units back to a normal pressure drop.

Low tower loading

Most towers have been designed with at least a 50% turndown ratio for the trays. This means that the trays should operate satisfactorily at 50% of their loading. Nevertheless tray performance does fall off at these low loadings. At below this turndown ratio performance particularly in sieve trays is drastically reduced. This is almost certain to be due to "weeping" where liquid falls from tray to tray. If the low loads are to be for only a short time due to temporary reduced throughput tray loading can be increased by increasing reflux. If the low throughput is to continue for an extended period of time a tray blanking schedule should be considered to reduce the active tray area.

Operating close to critical conditions

De-ethanizer in particular operate close to critical pressure in the bottom of the tower. Careful attention should be paid to avoid any pressure surges in this unit. Feed to the unit and reflux streams should be on flow control.

No separation by fractionation can occur at pressures in excess of critical. Very often chilled water is used for overhead condensing to reduce reflux drum pressure but maintaining minimum C_3 loss in the case of de-ethanizers.

Chapter 5

Catalytic reforming

Peter R. Pujadó and Mark Moser*

Catalytic reforming is a process whereby light petroleum distillates (naphthas) are contacted with a platinum-containing catalyst at elevated temperatures and hydrogen pressures ranging from 345 to 3,450 kPa (50–500 psig) for the purpose of raising the octane number of the hydrocarbon feed stream. The low octane, paraffin-rich naphtha feed is converted to a high-octane liquid product that is rich in aromatic compounds. Hydrogen and other light hydrocarbons are also produced as reaction by-products. In addition to the use of reformate as a blending component of motor fuels, it is also a primary source of aromatics used in the petrochemical industry (1).

The need to upgrade naphthas was recognized early in the 20th century. Thermal processes were used first but catalytic processes introduced in the 1940s offered better yields and higher octanes. The first catalysts were based on supported molybdenum oxide, but were soon replaced by platinum catalysts. The first platinum-based reforming process, UOP's Platforming™ process, came on-stream in 1949. Since the first Platforming unit was commercialized, innovations and advances have been made continuously, including parameter optimization, catalyst formulation, equipment design, and maximization of reformate and hydrogen yields. The need to increase yields and octane led to lower pressure, higher severity operations. This also resulted in increased catalyst coking and faster deactivation rates.

The first catalytic reforming units were designed as semiregenerative (SR), or fixed-bed units, using Pt/alumina catalysts. Semiregenerative reforming units are periodically shut down for catalyst regeneration. This involves burning off coke and reconditioning the catalyst's active metals. To minimize catalyst deactivation, these units were operated at high pressures in the range of 2,760 to 3,450 kPa (400–500 psig). High hydrogen pressure decreases coking and deactivation rates.

Catalytic reforming processes were improved by introducing bimetallic catalysts. These catalysts allowed lower pressure, higher severity operation: ~1,380–2,070 kPa (200–300 psig), at 95–98 octane with typical cycle lengths of one year.

Cyclic reforming was developed to allow operation at increased severity. Cyclic reforming still employs fixed-bed reforming, but each reactor in a series of reactors can be removed from the process flow, regenerated, and put back into service without shutting down the unit and losing production. With cyclic reforming, reactor pressures are approximately 200 psig, producing reformates with octanes near 100.

Another solution to the catalyst deactivation problem was the commercialization of the Platforming process with continuous catalyst regeneration, or the CCR Platforming process, by UOP in 1971. The Institut Français du Pétrole announced the commercialization of a similar continuous regeneration reforming process a few years later. With CCR small amounts of catalyst are continuously removed from the last reactor, regenerated in a controlled environment, and transferred back to the first reactor. The CCR Platforming process has enabled the use of ultra low pressures at 345 kPa (50 psig) with product octane levels as high as 108. More than 95% of all new catalytic reformers are designed with continuous regeneration. In addition, many units that were originally built as SR reforming units have been revamped to continuously regenerable reforming units.

Figure 5.1 illustrates the evolution of catalytic reforming, in terms of both process yields and octane numbers.

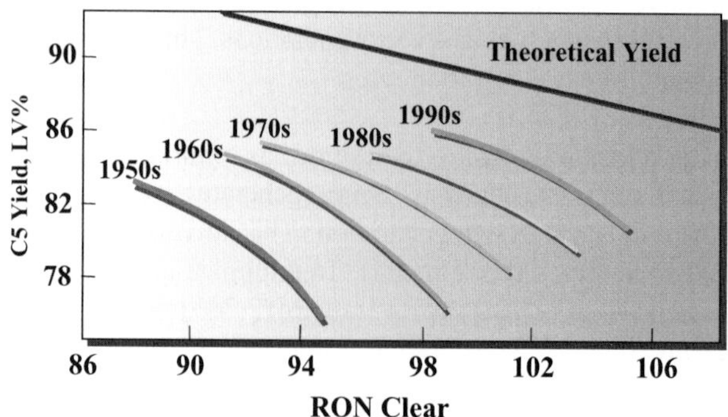

Figure 5.1. Increased yields and octane with Platforming advances (reprinted with permission from UOP LLC).

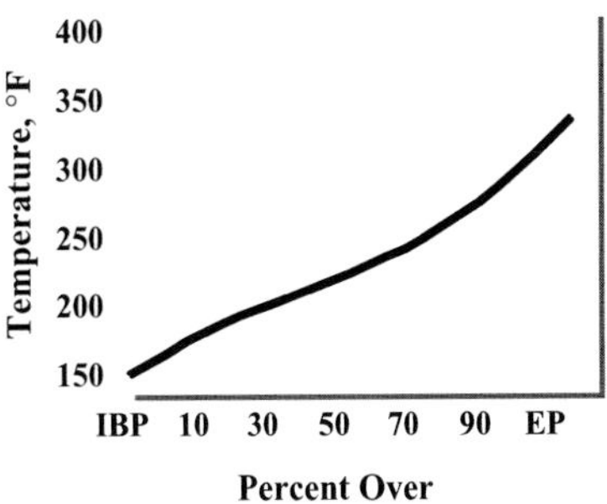

Figure 5.2. ASTM D-86 distillation curve for naphtha (1).

Feedstocks

Naphtha feedstocks to reformers typically contain paraffins, naphthenes, and aromatics with 6–12 carbon atoms. Most feed naphthas have to be hydrotreated to remove metals, olefins, sulfur, and nitrogen, prior to being fed to a reforming unit. A typical straight run naphtha from crude distillation may have a boiling range of 150–400°F (65–200°C).

In addition to naphthas from crude distillation, naphthas can be derived from a variety of other processes that crack heavier hydrocarbons to hydrocarbons in the naphtha range. Cracked feedstocks may be derived from catalytic cracking, hydrocracking, cokers, thermal cracking, as well as visbreaking, fluid catalytic cracking, and synthetic naphthas obtained, for example, from a Fischer–Tropsch process.

Light paraffinic naphthas are more difficult to reform than heavier naphthenic hydrocarbons. Distillation values for the initial boiling point, the mid-point at which 50% of the naphtha is distilled over, and the end point are often used to characterize a naphtha (Figure 5.2). If available, however, it is best to have a detailed component breakdown as provided by gas chromatographic analysis (Table 5.1).

Feed hydrotreating is used to reduce feedstock contaminants to acceptable levels (Figure 5.3). Common poisons for reforming catalysts that are found in naphtha are sulfur, nitrogen, and oxygen compounds (Figure 5.4). Removing these requires breaking of a carbon-sulfur, -nitrogen or -oxygen bond and formation of hydrogen sulfide, ammonia, or water, respectively. Hydrotreaters will also remove olefins and metal contaminants.

Some hydrotreaters are two-stage units. The first stage operates at low temperature for the hydrogenation of diolefins and acetylenes that could polymerize and plug the second, higher severity stage. The effluent from the first stage is cooled and fed to

Table 5.1. Composition of a typical naphtha

	Concentration (wt%)
Aromatics	
Benzene	1.45
Toluene	4.06
Ethylbenzene	0.52
p-Xylene	0.92
m-Xylene	2.75
o-Xylene	0.87
C9+ Aromatics	3.31
Total Aromatics	13.88
Total Olefins	0.11
Paraffins and Naphthenes	
Propane	0.79
Isobutane	1.28
n-Butane	3.43
Isopentane	5.62
n-Pentane	6.19
Cyclopentane	0.64
C6 Isoparaffins	6
n-Hexane	5.3
Methylcyclopentane	2.58
Cyclohexane	3.26
C7 Isoparaffins	4.55
n-Heptane	4.65
C7 Cyclopentanes	2.77
Methylcyclohexane	7.57
C8 Isoparaffins	4.24
n-Octane	3.43
C8 Cyclopentanes	1.52
C8 Cyclohexanes	5.23
C9 Naphthenes	3.63
C9 Paraffins	5.93
C10 Naphthenes	1.66
C10 Paraffins	3.41
C11 Naphthenes	1.04
C11 Paraffins	0.53
C12 P + N	0
> 200 P + N	0
Total Paraffins	55.35
Total Naphthenes	30.7

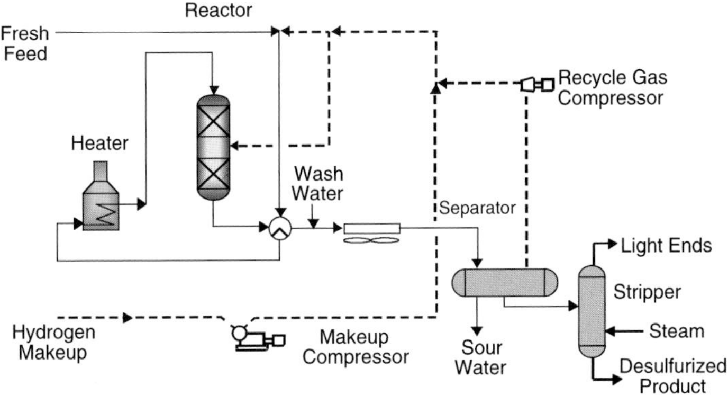

Figure 5.3. Naphtha hydrotreater flow scheme.

the second stage for the hydrogenation of olefins and the removal of sulfur and nitrogen compounds.

The reformate stream from a catalytic reforming unit is invariably used either as a high-octane gasoline blending component or as a source of aromatics—BTX (benzene, toluene, and xylenes), and C_9+ aromatics. Reforming for motor fuel applications still represents the majority of existing reforming capacity. Reformate specifications (octane, vapor pressure, end point, etc.) are set to provide an optimum blending product. The octane requirement is met through the production of high-octane aromatics, the isomerization of paraffins, and the removal of low octane components by cracking them to gaseous products. Feedstocks to these units are typically "full range" naphthas, consisting of hydrocarbons with 6–12 carbon atoms; however, the initial boiling point may be varied to limit the presence of benzene precursors.

Reforming units for the production of aromatics are often called BTX reformers. Naphthas for these units are specified to contain mostly naphthenes and paraffins of 6–8 carbons. The desired reaction is aromatization through dehydrogenation of the naphthenes, and cyclization and dehydrogenation of the paraffins to the analogous aromatic.

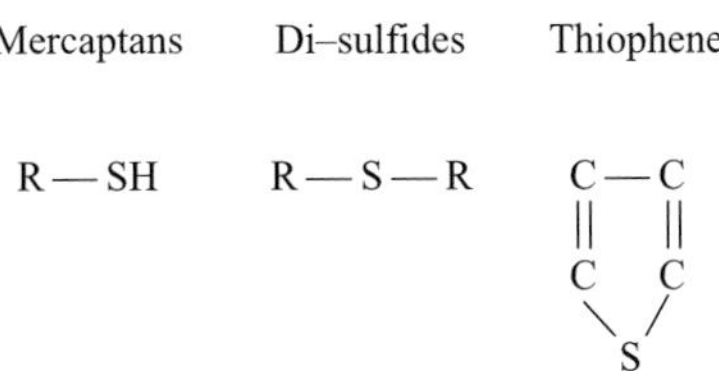

Figure 5.4. Sulfur types.

Table 5.2. Reformate composition

	mass%	liq-vol%
Aromatics		
Benzene	3.72	3.39
Toluene	13.97	12.93
Ethylbenzene	3.13	2.90
p-Xylene	3.39	3.14
m-Xylene	7.47	6.91
o-Xylene	4.83	4.47
C9+ Aromatics	36.05	33.30
Total aromatics	72.56	67.04
Total olefins	0.82	1.02
Paraffins and naphthenes		
Propane	0.00	0.00
Isobutane	0.14	0.20
n-Butane	0.94	1.32
Isopentane	2.52	3.29
n-Pentane	1.74	2.29
Cyclopentane	0.10	0.10
C6 Isoparaffins	3.91	4.77
n-Hexane	1.74	2.12
Methylcyclopentane	0.28	0.30
Cyclohexane	0.03	0.03
C7 Isoparaffins	7.70	9.02
n-Heptane	2.22	2.60
C7 Cyclopentanes	0.33	0.35
Methylcyclohexane	0.04	0.04
C8 Isoparaffins	2.86	3.24
n-Octane	0.62	0.70
C8 Cyclopentanes	0.14	0.14
C8 Cyclohexanes	0.06	0.06
C9 Naphthenes	0.04	0.04
C9 Paraffins	0.90	0.99
C10 Naphthenes	0.04	0.04
C10 Paraffins	0.24	0.26
C11 Naphthenes	0.00	0.00
C11 Paraffins	0.03	0.04
C12 P + N	0.00	0.00
Poly Naphthenes	0.00	0.00
> 200 P + N	0.00	0.00
Total Paraffins	25.56	30.84
Total Naphthenes	1.06	1.10

Reformate properties

Table 5.2 shows a typical reformate composition. For motor fuel applications, the octane number is the dominant parameter of product quality. A higher octane number reflects a lower tendency of the hydrocarbon to undergo a rapid, inefficient detonation in an internal combustion engine. This rapid detonation is heard as a knocking sound in the engine, so octane is often referred to as the antiknock quality of a gasoline. Motor fuel octanes are measured at low engine speeds (research octane number or RON) or at high engine speeds (motor octane number or MON). In the United States, the octane values posted on gasoline pumps are the arithmetic average of the MON and the RON. The acronym RONC, research octane number clear, is used to denote that there are no additives, such as lead, used to increase octane number. Table 5.3 provides a listing of the various octanes of pure hydrocarbons according to the American Petroleum Institute, API (2).

Octane numbers of a hydrocarbon or hydrocarbon mixture are determined by comparing its antiknock qualities with various blends of n-heptane (zero octane) and 2,2,4-trimethylpentane, or iso-octane (100 octane). Hydrocarbons may appear to have different octane numbers when blended with other hydrocarbons of a different composition—these are denoted as "blending octanes" and may be significantly different from the actual octane numbers of the individual hydrocarbon components (Table 5.4) (3).

Other property specifications of the reformate include volatility or vapor pressure, often given in terms of the Reid vapor pressure or RVP, end point, color, etc. (3) High-end point reformates, for example, may not combust well in an internal combustion engine.

Table 5.3. Examples of research and motor octanes of pure hydrocarbons

	RON	MON
Paraffins		
n-heptane	0	0
2-methylhexane	42.4	46.3
3-ethylpentane	65.0	69.3
2,4-dimethylpentane	83.1	83.8
Aromatics		
Toluene	120.1	103.2
Ethylbenzene	107.4	97.9
Isopropylbenzene	113.0	99.3
1-methyl-3-ethylbenzene	112.1	100.0
1,3,5-trimethylbenzene	>120	>120

Table 5.4. Octane and blending octane numbers by research method

	RON	Blending Octane
2,2-dimethyl butane	92.8	89
2-methyl-1-butene	102	146
Cyclopentane	101	141
1,4-dimethylbenzene	117	146

Reformulated gasolines, a requirement of the 1990 Clean Air Act, are the subject of much legislation. Specifications require a lower benzene content, lower volatility, and lower end point. Other specifications may pertain to the oxygenate content and other factors that affect the burning characteristics. The gasolines available to the consumer consist of a mixture of gasoline fractions from many refinery sources, including: straight run (unprocessed fraction), isomerate, alkylate, reformate, and FCC fractions, and, on occasion, polymer gasolines.

Reforming reactions

In BTX production, the objective is to transform paraffins and naphthenes into benzene, toluene, and xylenes with minimal cracking to light gases. The yield of desired product is the percentage of feed converted to these aromatics. In motor fuel applications, octane values of the feed may be raised via aromatization or through isomerization of the paraffins into higher octane branched species without sacrificing yield. Yield is typically defined as liquid product with five or more carbons.

Typical catalysts that consist of platinum supported on alumina (with or without other metals or modifiers) are bifunctional in that separate and distinct reactions occur on the platinum site and on the alumina. The platinum typically performs dehydrogenation and hydrogenolysis, while the acidic alumina isomerizes, cyclizes, and cracks.

The dehydrogenation of naphthenes to aromatics is probably the most important reaction. Feeds contain cyclopentanes and substituted cyclopentanes, as well as cyclohexanes and their homologues. Six carbon ring cyclohexanes, for example, can be directly dehydrogenated to produce aromatics and hydrogen.

$$\text{methylcyclohexane} \longrightarrow \text{toluene} + 3H_2$$

Dehydrogenation is typically catalyzed by the platinum function on the reforming catalyst.

Five member ring cyclopentanes must be hydroisomerized to give a cyclohexane intermediate prior to dehydrogenation to aromatics.

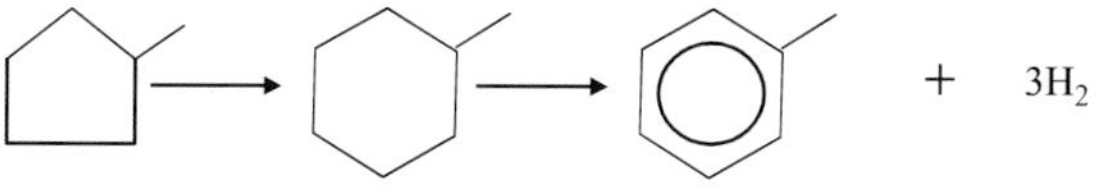

$+ \quad 3H_2$

Acid-catalyzed reactions together with the Pt-catalyzed dehydrogenation function are largely responsible for hydro-isomerization reactions that lead to the formation of aromatics.

Paraffin conversion is the most difficult step in reforming. For that reason, the ability to convert paraffins selectively is of paramount importance in reforming. Paraffins may be isomerized over the acidic function of the catalyst to provide higher octane branched paraffins.

H_3C CH_3 → CH_3 H_3C CH_3 H_3C

Another acid catalyzed paraffin reaction is cracking to lighter products, thus removing them from the liquid product. Octane is improved through the removal of low octane paraffinic species from the liquid product by their conversion to gaseous, lower molecular weight paraffins.

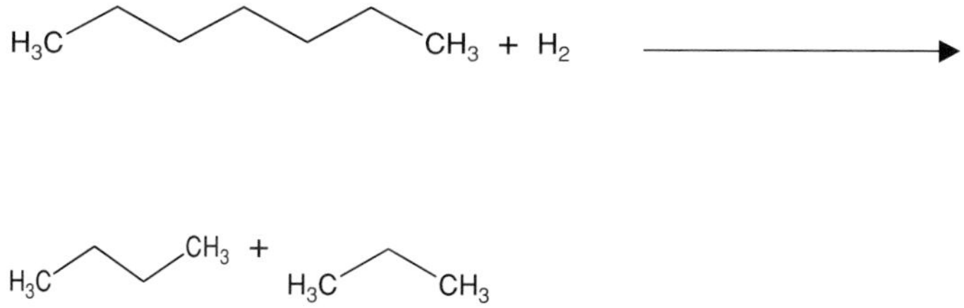

H_3C CH_3 + H_2 →

H_3C CH_3 + H_3C CH_3

Paraffins also undergo cyclization to cyclohexanes. This reaction is believed to proceed through an olefin intermediate, produced by Pt-catalyzed dehydrogenation (4). The cyclization of the olefin may be catalyzed by the alumina support.

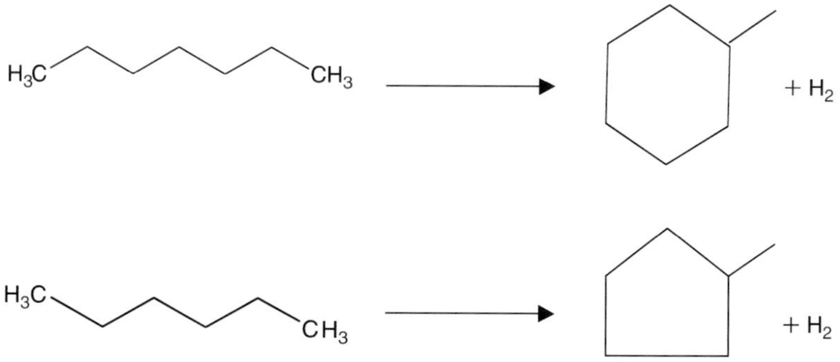

H_3C CH_3 → + H_2

H_3C CH_3 → + H_2

After cyclization, cyclohexane undergoes dehydrogenation to aromatics. Cyclopentanes undergo hydroisomerization to cyclohexane, followed by dehydrogenation to aromatics. Aromatics are stable species and relatively inert. Reactions of substituted aromatics involve isomerization, hydrodealkylation, disproportionation, and transalkylation.

Small amounts of olefins are formed that also undergo a number of isomerization, alkylation, and cracking reactions. In particular, they appear to play an important role as an intermediate in cyclization reactions.

The dehydrogenation of naphthenes and paraffins is rapid and equilibrium concentrations are established in the initial portions of a catalyst bed. Isomerization reactions are sufficiently fast that actual concentrations are near equilibrium. The observed reaction rate for dehydrocyclization is reduced by the low concentrations of the olefin intermediates that exist at equilibrium. Hydrogen partial pressure significantly affects olefin equilibrium concentrations and has a significant impact on aromatization and dehydrocyclization of paraffins. Lowering hydrogen partial pressures results in an increase in the rate of aromatization, a decrease in the rate of hydrocracking, and an increase in the rate of coke formation.

Table 5.5 provides thermodynamic data for typical compounds in reforming reactions at a reference temperature of 800°K. Thermodynamic data can be obtained from

Table 5.5. Thermodynamic data for reforming compounds at 800°K, ideal gas in kcal/mol

Reforming reactions are typically dehydrogenations of the form

$$A \leftrightarrow B + nH_2$$

with equilibrium expressed in the form

$$K_P = \frac{p_B (p_{H_2})^n}{p_A}$$

such that they are a strong function of the partial pressure of hydrogen.

	ΔH_f^o	ΔG_f^o
Typical C_6's		
n-hexane	−48.26	73.08
2-methylpentane	−49.68	72.74
3-methylpentane	−49.32	73.67
Cyclohexane	−37.19	75.94
Methyl cyclopentane	−33.73	71.92
Benzene	15.51	52.84
Typical C_7's		
n-heptane	−54.20	87.43
2-methylhexane	−55.91	87.23
3-methylhexane	−55.28	87.07
Methyl cyclohexane	−45.10	86.15
Toluene	6.65	61.98

standard sources (e.g., 2 and 5). Production of aromatics is favored by the reforming conditions. Current designs at low hydrogen partial pressures ensure full conversion to the equilibrium limits.

Catalysts

The platinum must be dispersed over the alumina surface such that the maximum number of active sites for dehydrogenation is available. Platinum cluster size dimensions are on the order of angstroms, or 10^{-10} meters. The interaction of the platinum with the alumina surface is such that the platinum clusters are relatively immobile and do not agglomerate during reforming. Sulfidation of the platinum is sometimes used to partially poison the platinum, or reduce its activity; this has the beneficial effect of reducing a major portion of the hydrogenolysis, or metal-catalyzed cracking reactions. Liquid product yields are improved and the light gas production, particularly methane, is reduced.

The alumina support is usually in the eta (η) or gamma (γ) phase, but most often gamma is used in reforming. Chloride is added to promote acidity. A simplified schematic diagram of the alumina functionality is given in Figure 5.5.

Catalysts that are used in reactors where the catalyst bed is not easily removed after deactivation must have long catalyst life cycles. A typical fixed bed catalyst life cycle may be a year or longer. Modifiers are added to reduce the effect of coke buildup and to lengthen the catalyst cycle length, either by hydrogenating the coke to a less graphitic species (6) or by cracking the coke precursors (7). Elements that are commonly added to the catalysts are rhenium and, to a lesser extent, iridium.

In moving bed units (8) the catalyst flows through the reactors and is regenerated continually in a sepaarte regeneration vessel that is part of the reactor–regenerator loop. Process conditions are much more severe, thus shortening catalyst life and requiring regeneration cycles of only a few days. In moving bed catalysts elements are added, such as tin and germanium, to increase liquid, aromatic, and hydrogen yields by reducing the activity of the platinum for hydrogenolysis or metal-catalyzed cracking reactions. These components also provide some stabilization of the catalyst relative to Pt alone.

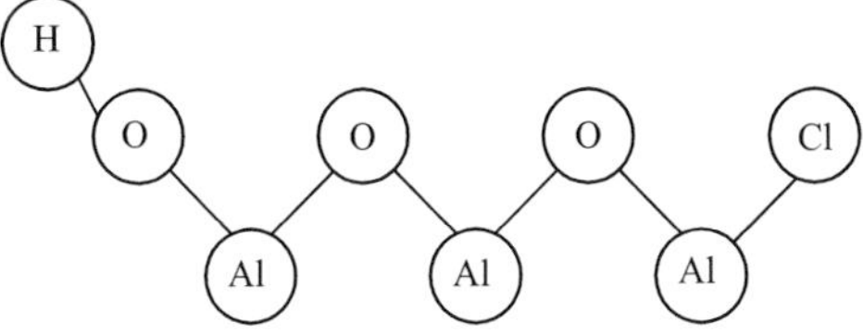

Figure 5.5. Alumina schematic.

Deactivation mechanisms for reforming catalysts include coking, poisoning, and agglomeration of the platinum. Under normal conditions coke accumulates on the catalyst. In a fixed bed or SR unit, this coke will deactivate the catalyst such that, in time, the temperature limit of the reforming unit will be reached, or the selectivity to desired products is too much reduced, or the octane of the liquid product is declining. When this occurs, the refiner will shut down the process unit and regenerate the catalyst to return it to its original state. In a moving bed unit, some of the catalyst is continually being regenerated outside the process and returned to the reactors. High selectivity and activity are maintained. New, degradation-resistant catalysts allow the refiner to operate continuous regeneration units for more than eight years before removing the catalyst. The actual chemistry and steps for regeneration for all process units are very similar. In the following discussion, catalyst regeneration for a SR reformer is described. For cyclic or continuous reformers, plant shutdown and start-up are unnecessary and the remaining steps are accomplished in equipment outside the process stream.

The objective of regeneration is to return the catalyst to its initial, fresh state. If the regeneration is successful, there is no difference between fresh and regenerated catalyst. To do this, the coke must be burnt off the catalyst, the platinum should be well dispersed and in a reduced state, and the acidity should be properly adjusted through chloride adsorption. These needs account for the steps in regeneration; carbon burn, chloride redispersion and metals reduction.

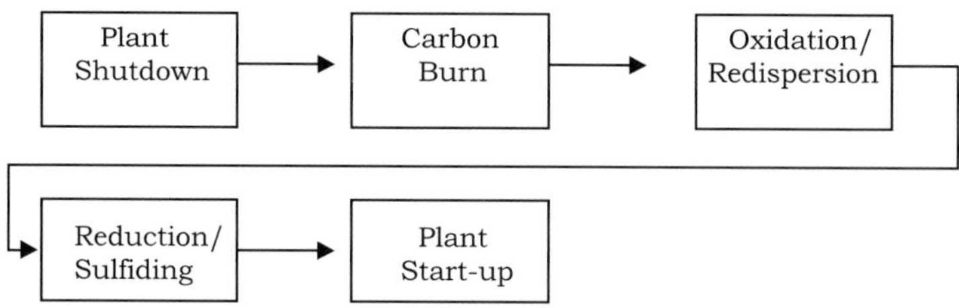

In order to conduct a regeneration, the heater temperatures and feed rates are reduced gradually. The circulation of recycle gas is continued to strip hydrocarbons from the catalyst, leaving only coke. If the coke is to be burnt in the unit, higher temperatures are maintained and the coke burning procedure is initiated. If the coked catalyst is to be removed from the unit, temperatures are lowered, to about 100–150°F before unloading. Since coked catalyst is often pyrophoric, nitrogen blanketing is often used to protect the catalyst from air contact and combustion.

The coke burning step must be carefully monitored. The combustion of coke to carbon dioxide and water is exothermic, and the oxygen concentration must be kept low to limit the reaction and temperature rise. Excessive temperature can cause agglomeration of the platinum or, in more extreme cases, can cause the alumina to change from the desired phase or crystal structure to a higher temperature phase. The water produced in the combustion also facilitates sintering of platinum. Due to the need to gradually burn coke, the carbon burn is usually the most time-consuming part of a regeneration.

Coke burning is usually done in the range of 400–500°C, and at oxygen concentrations initially in the 1–2 mol% range. Oxygen content and temperature are often increased during burning to ensure that all coke has been combusted by the end of the burn. Oxygen consumption is monitored to determine the total amount of coke combusted and the extent of the burn.

Since platinum can agglomerate even at relatively moderate exothermic conditions, the platinum must be redispersed after the carbon burn. The temperature is first increased to approximately 500°C, oxygen content to approximately 5–6 mol%, and chlorine or an organic chloride that breaks down to HCl and Cl_2 is injected into the air/nitrogen stream. Platinum oxychlorides or chlorides form that redisperse platinum over the alumina surface, ensuring that almost all the platinum is exposed for reaction. This also adds chloride to the catalyst to enhance its acidity.

Finally, the last step in the regeneration process is the reduction of the metals on the catalyst and sulfiding, if necessary. This is done in a dry hydrogen atmosphere. At the temperatures required for reduction, greater than 350°C, high moisture levels can lead to platinum agglomeration. Since water is formed in the reduction process as platinum oxide is reduced to platinum metal, water is drained from the unit during reduction. The reduction hydrogen is recirculated at as high a rate as possible in order to minimize moisture content.

Sulfiding is typically done by injection of H_2S or of an organic sulfide into the unit at the end of reduction. Sulfiding is continued until the specified sulfur level is reached or until sulfur is no longer adsorbed by the catalyst and is detected at the outlet of the catalyst bed.

Reactor performance

The major process variables that affect unit performance are reactor pressure, reactor temperature, space velocity, H_2/HC molar ratio, and catalyst type. The relationship between the variables and process performance is generally applicable to both SR and continuous regeneration modes of operation.

The reactor pressure declines across the various reaction stages. The change in reactor pressure across the unit, known as pressure drop, can be quite high for high-pressure reforming units, often 50–60 psig or more. The average reactor operating pressure is generally referred to as reactor pressure. For practical purposes, a close approximation is the last reactor inlet pressure. The reactor pressure affects reformer yields, reactor temperature requirements, and catalyst stability.

Practical operating constraints have led to a historical range of operating pressures from 345 to 4,830 kPa (50–700 psig). Decreasing the reactor pressure increases hydrogen and reformate yields, decreases the required temperature to achieve product quality, and shortens the catalyst cycle because it increases the catalyst coking rate. The high catalyst deactivation rate associated with lower operating pressure requires CCR.

The primary control for product quality in catalytic reforming is the temperature of the catalyst beds. Platforming catalysts are capable of operating over a wide range of temperatures. By adjusting the heater outlet temperatures, a refiner can change the octane of the reformate and the quantity of the aromatics produced.

The reactor temperature can be expressed as the weighted average inlet temperature (WAIT). The WAIT is the summation of the product of the fraction of catalyst in each reactor multiplied by the inlet temperature of the reactor. The weighted average bed temperature (WABT) is also used to describe catalyst temperature and is the temperature of the catalyst integrated along the catalyst bed. Temperatures in this chapter refer to the WAIT calculation. Typically, SR Platforming units have a WAIT range of 490–525°C (914–977°F). CCR Platforming units operate at a WAIT of 525–540°C (977–1,004°F). CCR Platforming units operate at even higher temperatures to produce a more aromatic-rich, high-octane product. The amount of naphtha processed over a given amount of catalyst over a set length of time is referred to as space velocity. Space velocity corresponds to the reciprocal of the residence time or time of contact between reactants and catalyst. When the hourly volume charge rate of liquid naphtha is divided by the volume of catalyst in the reactors, the resulting quotient, expressed in units of h^{-1}, is the liquid hourly space velocity (LHSV). Typical commercial LHSV range from 1 to 3 with the volumetric rates measured at standard conditions (60°F and 1 atm.abs.).

Alternatively, if the weight charge rate of naphtha is divided by the weight of catalyst, the resulting quotient, also expressed in units of h^{-1}, is the weight hourly space velocity (WHSV). Whether LHSV or WHSV is used is based on the customary way that feed rates are expressed at a given location. Where charge rates are normally expressed in barrels per stream day, LHSV is typically used. Where the rates are expressed in terms of metric tons per day, WHSV is preferred.

The combination of space velocity and reactor temperature is used to set the octane of the product. The greater the space velocity, the higher the temperature required to produce a given product octane. If refiners wish to increase the severity of a reformer operation, they can either increase reactor temperature or lower the space velocity by decreasing the reactor charge rate.

The hydrogen-to-hydrocarbon (H_2/HC) mol ratio is the ratio of mols of hydrogen in the recycle gas to mols of naphtha charged to the unit. The recycle gas is a mixture of hydrogen and light gases, typically 75–92 mol% hydrogen. The ratio of total recycle gas to hydrocarbon is sometimes called the gas-to-oil ratio. Recycle hydrogen is necessary to maintain catalyst-life stability by sweeping coke precursors from the catalyst metals. The exact mechanism is proposed to be hydrogenation and inhibition of polymerization. The rate of coke formation on the catalyst is a function of the hydrogen partial pressure present. Increasing the H_2/HC ratio increases the hydrogen partial pressure and removes coke precursors from the metal sites, thereby increasing stability with little effect on product quality or yields.

Except for units designed for continuous regeneration through the circulation of the catalyst between the reactors and the regenerator, catalytic reforming units normally will require a shutdown for regeneration every 6–12 months. This relatively long cycle can be obtained by operating under milder conditions of high partial pressure of hydrogen, lower reactor temperatures, and lower octane products. Continuous units operate under severe conditions to yield high octane, high aromatics production, at low hydrogen partial pressures, and higher reactor temperatures. Different catalysts are used depending on the application.

Semi regenerative reformers make use of catalysts the contain platinum or platinum modified by rhenium or, to a lesser extent, iridium. The support is most often gamma alumina, although there have been uses of eta alumina (4). Rhenium or iridium is used to enhance the life of the catalyst over that observed for Pt-only catalysts. All these catalysts are typically sulfided to minimize metal-catalyzed hydrogenolysis reactions that produce light gases and reduce gasoline yield. Additional components were used on catalysts commercialized in the 1990's. The use of two catalysts in a SR unit; one catalyst in the front reactors and another catalyst in the back reactors to provide maximum yield, activity, and stability was commercialized in 1994 (8).

There are two main shapes of catalysts, cylindrical and spherical. The cylindrical catalysts are usually extruded alumina. The spherical catalysts may be formed through a dropping method or by rolling wet, soft alumina dough. In some instances, factors such as the resistance to flow or flow distribution concerns may cause one form to be chosen over the other. The density of the catalysts may vary from approximately 0.5–0.8 g/cm^3. The variability in density allows the refiner to load more pounds of catalyst in a unit, should additional catalyst activity or stability be desired.

The process of moving catalyst from the reactors to the regenerator and back requires the use of spherical catalysts, rather than extrudate, to avoid catalyst dusting and breakage. Continuous regeneration units are operated at high severity and low pressures to produce the greatest amount of aromatics and hydrogen possible. The catalyst is circulated at a rate such that it corresponds to about one regeneration per week or even at a greater frequency if needed due to the rapid deactivation under these conditions.

Typical catalysts used in these units had a composition of platinum and tin on gamma alumina. The tin was used to reduce the hydrogenolysis activity of the platinum and to improve yields. The reduction in metal-catalyzed cracking is also considered to stabilize the catalyst relative to platinum only. Currently, new proprietary catalysts are used to increase yields, lower coke make, or allow higher throughput (9).

Process flow schemes

Fixed bed semiregenerative reforming

A typical SR Platforming flow diagram is presented in Figure 5.6. Feed to the unit is mixed with recycled hydrogen gas, raised to the reaction temperature first by a feed-effluent combined feed exchanger and then by a fired heater, and then charged to

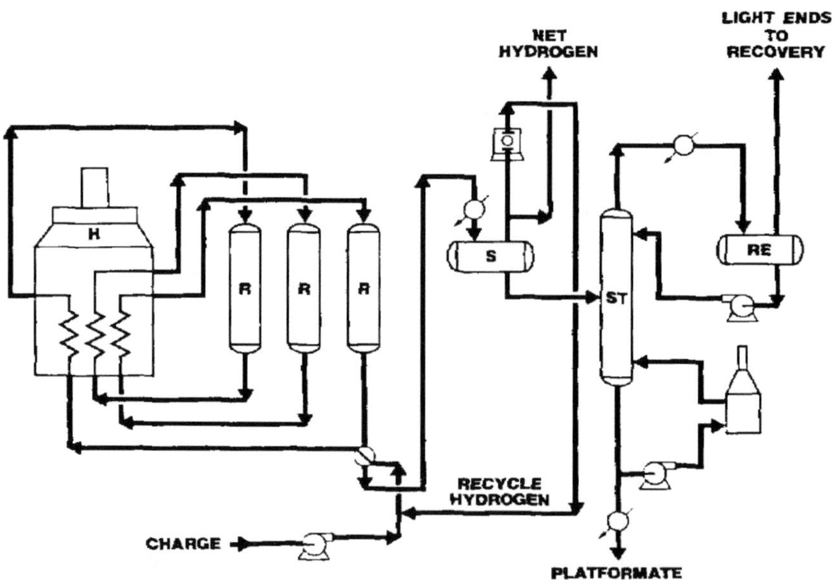

Figure 5.6. Semiregenerative reforming process (reprinted with permission from UOP LLC).

the reactor section. Because most of the reactions that occur in the Platforming process are endothermic, the reactor section is separated into several stages, or reactors. Interheaters are installed between these stages to maintain the desired temperature range across the catalyst in the reactor section. Effluent from the last reactor is cooled by the feed-effluent heat exchanger for maximum heat recovery. Air or water cooling provides additional cooling to near-ambient temperature. The effluent is then charged to the separation section, where the liquid and gas products are separated. A portion of the gas from the separator is compressed and recycled back to the reactor section. The net hydrogen produced is sent to hydrogen users in the refinery complex or for use as fuel. The separator liquid is pumped to a product stabilizer, where the more-volatile light hydrocarbons are fractionated from the high-octane liquid product.

Fixed bed cyclic reforming

Cyclic reforming is similar to SR reforming, but an additional reactor replaces one of the primary reactors while that reactor is being regenerated. The frequency with which a particular primary reactor is replaced and regenerated depends upon its rate of deactivation. Large diameter valves and piping are used to vary the process flow between reactors.

Platforming process with continuous catalyst regeneration

In parallel with bimetallic catalyst improvements and other process and regeneration advances, UOP began to develop the CCR Platforming™ process (Figure 5.7). In the CCR Platforming unit, partially aged catalyst in the reactors is continuously replaced with catalyst that has been freshly regenerated in an external regenerator (CCR section) to maintain a low average age for the reactor catalyst. Thus, the high selectivity and high activity characteristics associated with new catalyst can be maintained at significantly higher severities than with the SR Platforming process. For example, a SR Platforming unit operates at a severity that steadily builds coke up on the catalyst surface over the length of a cycle (6–18 months), at which point the unit is shut down and the catalyst regenerated. Throughout the cycle, yields decline. Instead, in a modern CCR Platforming unit, the catalyst is regenerated approximately every three to seven days and the yield does not decline.

The ability to continuously regenerate a controlled quantity of catalyst is the significant innovation of the CCR Platforming unit. The catalyst flows by gravity from the last reactor into a catalyst collector vessel. The catalyst is lifted by either nitrogen or hydrogen lifting gas to a catalyst hopper above the regeneration tower. Catalyst then flows to the regeneration tower, where the catalyst is reconditioned. Regenerated catalyst is returned to the top of the reactor stack by a transfer system similar to that used in the reactor-to-regenerator transfer. Thus, the reactors are continuously supplied with freshly regenerated catalyst, and product yields are maintained at fresh

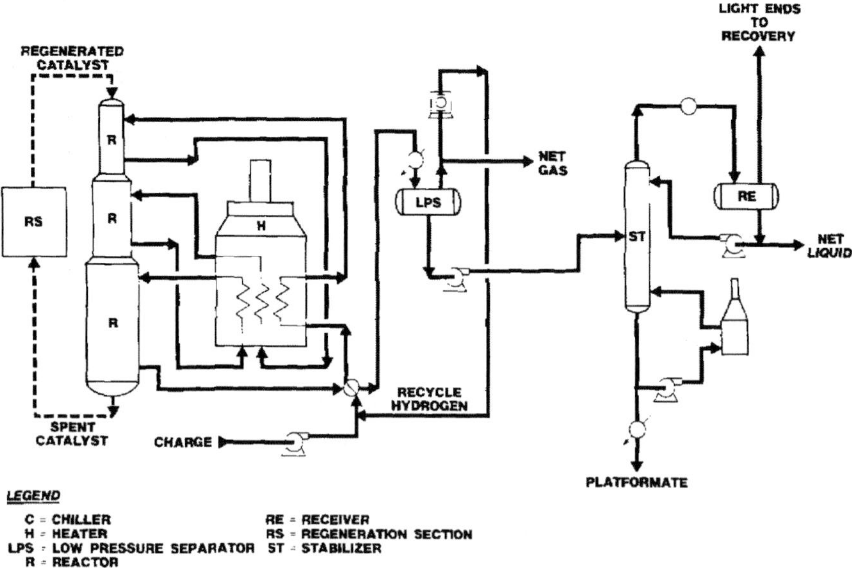

Figure 5.7. CCR Platforming process (reprinted with permission from UOP LLC).

catalyst levels. The regeneration and reactor sections of the unit are easily isolated to permit a shutdown of the regeneration system for normal inspection or maintenance without interrupting of the Platforming operation.

A few years after the introduction of the UOP CCR Platforming process, another continuously regenerable process design was offered by the Institut Français du Pétrole. Though similar to CCR Platforming, the continuous reforming units designed by the Institut Français du Pétrole differ most notably in that the reactors are located side-by-side and the catalyst transfer is effected through transfer piping between reactors.

Advantages of CCR Platforming

From both an economic and technical standpoint, the CCR Platforming process is superior to the SR and cyclic reforming processes. The CCR Platforming unit allows for low-pressure operation, leading to higher yields. At these conditions, the SR Platforming catalyst is completely deactivated after only a few days of operation. Both the hydrogen and C_5+ yields are maximized with the CCR Platforming process. Since the number of cyclic reformers is small relative to CCR Platforming process units and SR process units, the following comparison will focus on contrasting CCR Platforming units and SR units.

Table 5.6. Relative severities of CCR versus SR Platforming units

Operating mode	SR	CCR
Charge rate, barrels/day	20,000	20,000
LHSV, h^{-1}	Base	Base × 1.8
H_2/HC	Base	Base × 0.5
RONC	97	102
Reactor pressure, psig	Base	Base-50
Separator pressure, psig	Base	Base-145
Cycle life, months	12	Continuous

High yields and constant yields are important in the economics of reforming. As the catalyst is deactivated by coke deposition in the SR Platforming process, the yields begin to decline. With the CCR Platforming process, the reformate, aromatics, and hydrogen yields remain consistent and constant. This is particularly important for downstream users. The CCR section ensures proper redispersion of the metals and chloride balance to maintain fresh catalyst activity. CCR Platforming units have higher on-stream efficiency and are able to handle upset scenarios without long-term shutdown or significant decline in performance.

Table 5.6 shows the relative operating severities for the SR and CCR Platforming units. The CCR Platforming unit operates at higher severity and lower reactor catalyst inventory. In addition, the CCR unit runs continuously compared to 12-month SR Platforming cycle lengths.

Typical product yields for the SR and CCR Platforming units operating at the conditions presented in Table 5.6 are shown in Table 5.7. Many of the benefits of CCR Platforming are demonstrated in Table 5.7. More and higher-purity hydrogen is produced. The higher severity of the CCR Platforming unit results in similar liquid volume for the two units. However, the reformate produced by the CCR Platforming is more valuable than that produced by the SR Platforming unit. Taking into account both the higher octane value and the increased on-stream efficiency of the CCR Platforming unit, 80 million more octane-barrels, or 11.4 million more

Table 5.7. Yield comparison of CCR versus SR Platforming units

	SR	CCR	Delta
Hydrogen yield, SCF/bbl	1,085	1,709	+624
Hydrogen purity, mol%	80	92.6	+12.6
C_5+ yield, LV%	79.3	79.4	+.1
C_5+ yield, wt%	85.2	88.2	+3
Octane-barrel, 10^6 bbl/yr	513	583	+80

Table 5.8. Economic summary

Description	SR	CCR
Gross key product value, $MM/yr	120	141
Raw materials less by-products, $MM/yr	98	103
Consumables, MM$/y	0.3	0.75
Utilities, $MM/yr	2.8	6.2
Total fixed costs, $MM/yr	5.5	6.5
Capital charges, $MM/yr	3.5	5.2
Net cost of production, $MM/yr	110	122
Pretax profit, $MM/yr	10	20
Pretax ROI, %	30	41
Payout period, (gross) years	1.5	1.3

metric octane-tons, are produced per year with the CCR Platforming unit than with the SR Platforming unit. Octane-yield is defined as the product of the reformate yield, octane, and operating days.

A summary of the operating revenues and costs expected for the SR and CCR Platforming units in shown in Table 5.8. The nomenclature follows standard definitions. The economics of the CCR Platforming process are superior as a direct result of the differences in operating severity and flexibility of the two modes of operations. The CCR Platforming unit produces more valuable reformate at 102 RONC versus the SR Platforming reformate at 97 RONC. On-stream efficiency of the CCR Platforming unit is 8,640 hr per year compared to about 8,000 hr per year for the SR Platforming unit. Although the CCR Platforming utility costs are higher than those for the SR Platforming unit, these costs are offset by the increase in both product quantity and value as demonstrated by pretax profit and return on investment.

Catalysts and suppliers

For detailed updated lists of catalysts and suppliers consult the periodic reviews published by the Oil and Gas Journal.

The main catalyst suppliers are:

Axens/IFP Group Technologies
Criterion Catalyst Co.
Exxon Research & Engineering Co. (ERECO)
Indian Petrochemicals Corp., Ltd.
Instituto Mexicano del Petróleo (IMP)
UOP LLC

Some of the catalyst suppliers may restrict availability to process licensees only.

References

1. A. L. Huebner, "Tutorial: Fundamentals of Naphtha Reforming," AIChE Spring Meeting 1999, Houston, TX, 14–18 March 1999.
2. American Petroleum Institute Research Project 45, Sixteenth Annual Report, 1954.
3. E. L. Marshall and K. Owen, eds., *Motor Gasoline,* The Royal Society of Chemistry, London, 1995, p. 8.
4. G. A. Mills, H. Heinemann, T. H. Milliken, and A. G. Oblad, *Ind. Eng. Chem.*, 1953:**45**;134–137.
5. D. R. Stull, E. F. Westrum, and G. C. Sinke, *The Chemical Thermodynamics of Organic Compounds*. John Wiley & Sons, New York, 1969.
6. S. M. Augustine, G. N. Alameddin, and W. M. H. Sachtler, *J. Catal.*, 1989:**115**(1); 217–232. U.S. Pat. 4,469,812 (September 4, 1984) C. M. Sorrentino, R. J. Pellet, R. J. Bertolacini (to Standard Oil Company—Indiana).
7. J. A. Weiszmann, In Meyers, ed., *Handbook of Petroleum Refining Processes*, McGraw-Hill, New York, 1986, p. 31.
8. M. D. Moser, D. H. Wei, R. S. Haizmann, *CHEMTECH*, October, 1996, pp. 37–41.

Chapter 6

Fluid catalytic cracking

Warren Letzsch*

Crude oil comprises hundreds of molecules that boil over a wide temperature range. The lighter products can be separated directly by distillation into LPG, gasoline, naphtha, kerosene, and diesel fuels. Heavier products (BP $>$ 650°F/344°C) include vacuum gas oils and resids. Thermal and catalytic cracking processes in petroleum refining reduce the molecular weight of these heavier constituents and produce more valuable lighter products such as LPG, gasoline and diesel fuels.

Catalytic cracking was first commercialized in 1936 by Eugene Houdry. This fixed bed process was a major improvement over the thermal cracking processes it replaced due to the improved yield distribution and superior product properties. Multiple vessels were utilized that alternated between cracking, stripping, regeneration, and purge cycles. This configuration was quickly replaced by a moving bed reactor and a separate regenerator or kiln that first used a bucket lift to move the pelleted catalyst followed later by a pneumatic air lift system. The last of these units was built around 1960.

Standard Oil of New Jersey developed their own cracking process rather than pay the large royalty being asked at the time. They commercialized the fluid catalytic cracking (FCC) process in three years, starting in 1939 and culminating in 1942 with the start-up of PCLA#1 at their Baton Rouge, Louisiana refinery. The inherent superiority of the fluid process to transfer both heat and catalyst ultimately made it the catalytic cracking process of choice.

Many different designs of fluid catalytic crackers have been introduced over the years. Table 6.1 is a list of the various FCCU configurations and the approximate year of their commercial introduction.

Fluid catalytic cracking has evolved considerably over the more than 60 years since its inception. As seen in Figure 6.1 these changes have encompassed all aspects of

*Senior Refining Consultant, Stone & Webster Inc—A Shaw Group Company.

Table 6.1. Evolution of fluid catalytic crackers

Commercial fluid catalytic crackers	
1942	Model I upflow
1943	Model II downflow
1945	Sinclair design
1947	Model II side by side
1951	Kellogg orthoflow A
1952	Exxon Model IV
1953	Kellogg orthoflow B
1955	Shell two stage reactor
1956	UOP straight riser (SBS)
1958	Exxon riser cracker
1961	HOC cracker (Phillips)
1962	Kellogg orthoflow C
1967	Texaco design
1971	Gulf FCC process
1972	Exxon flexicracker
1972	Amoco ultracracking
1973	UOP high efficiency design
1973	Kellogg orthoflow F
1981	Total petroleum resid cracker
1982	Ashland/UOP RCC unit
1985	IFP R2R
1990	Kellogg/Mobil HOC
1991	Deep catalytic cracking (RIPP and S&W)
1993	Exxon flexicracker III
1996	MSCC UOP/Coastal
2002	Catalytic pyrolysis process

the process as it has adapted to meet ever-changing demands and to accommodate new technologies.

The initial units (1940s) were tall, had dense bed reactors and were made of carbon steel. Dilute phase catalyst coolers and regenerator steam coils were employed to limit the regenerator temperatures. Recycle rates of 100–150% were needed to achieve the desired conversions. Later designs (late 1940s to early 1950s) were undertaken to reduce the height of the crackers, make them more compact and cater to the many small refiners around the United States.

In the late 1950s side by side designs with straight feed risers were introduced to improve gasoline selectivity. Residue cracking in fluid cracking units was first practiced in the early 1960s. This unit was designed for 100% atmospheric bottoms and had large amounts of heat removal.

Catalytic cracking was truly revolutionized in the early 1960s with the advent of zeolite containing fluid cracking catalysts. Catalyst activities were raised by an order of

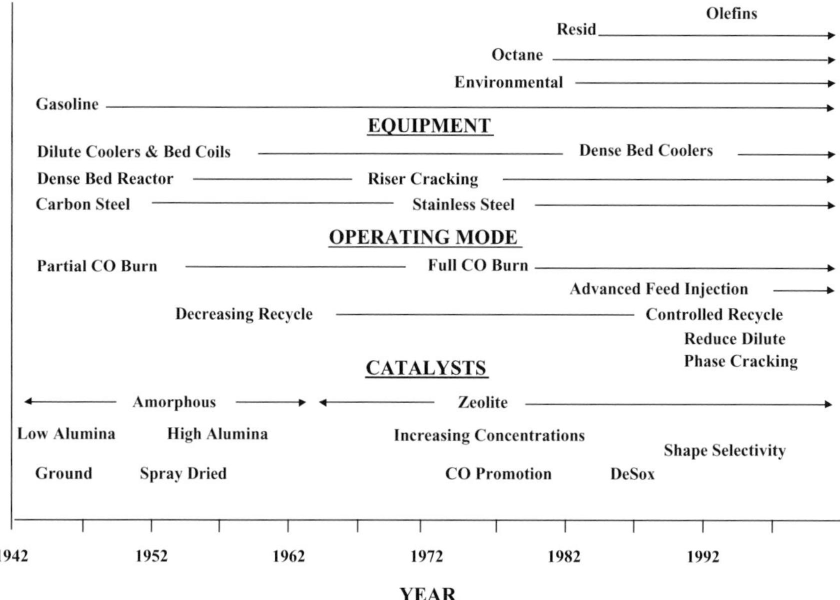

Figure 6.1. Fluid catalytic cracking development.

magnitude and units needed to be redesigned to take full advantage of the new catalyst technology. These design changes included the elimination of reactor dense beds and the use of the feed riser as the sole conversion vessel. Recycle was greatly reduced and replaced with more fresh feed. It was found that coke left on the regenerated catalyst impaired the catalyst's activity and selectivity and the average carbons on regenerated catalyst were reduced from 0.3–0.5 wt% to 0.1 wt% or less. Complete CO combustion in the regenerator was introduced to provide the needed regenerator burning conditions which necessitated higher regenerator temperatures. Alloy internals replaced the carbon steel and chrome-moly hardware in the regenerator and the regenerated catalyst standpipe. Catalyst inventories were minimized due to the favorable coke burning kinetics and modifications in the regenerator design (1970s).

In the early 1980s several new resid cracking designs were introduced. The spent catalyst regeneration was staged in these designs and dense bed catalyst coolers were optional. These coolers have also been placed with single stage regenerators for residual catalytic cracking units. The growth of this segment of the catalytic cracking process is shown in Figure 6.2. Over two million barrels per day of on purpose resid crackers have now been licensed and another million barrels of FCC capacity processes some resid along with their normal gas oils. Typical yields from resid cracking along with the feedstock properties are given in Table 6.2.

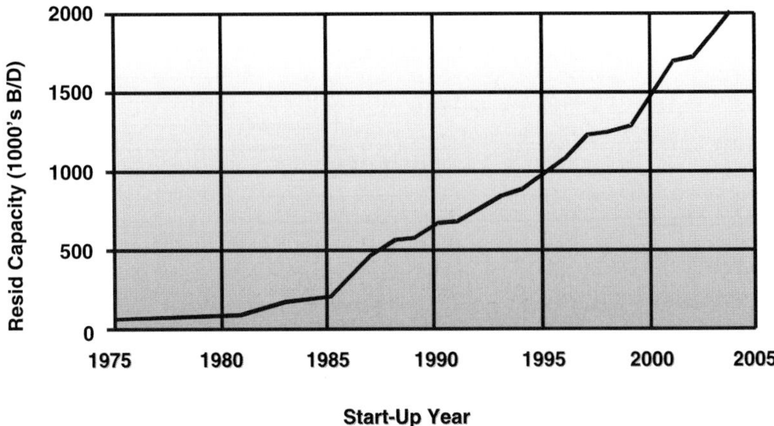

Figure 6.2. Growth in resid cracking worldwide.

While the olefins in gasoline have become a concern in the gasoline pool, the light olefins for petrochemicals are a valuable product, often exceeding the revenue obtained for transportation fuels. As a consequence, distinctive processes for making much larger amounts of propylene than a normal FCC unit have been developed. The Deep Catalytic Cracking Process (1991) was the first commercial scale process that was designed to maximize propylene. Specially formulated catalysts, more severe process conditions and equipment made to handle the unique product distribution are all components of this technology. Table 6.3 compares the yields from the DCC technology with those obtained in a normal FCCU. Commercial yields have verified the laboratory data.

Table 6.2. Typical resid cracking yields

	Unit A	Unit B
Feed properties		
API gravity	0.922	0.963
Conradson carbon	3.8	8.0
Ni + V	4	7.5
	Wt%	Wt%
Conversion	70.0	70.0
Dry gas	2.1	3.4
LPG	15.3	14.4
Gasoline	46.2	43.1
LCO	16.3	16.0
Decant oil	13.7	14.0
Coke	6.5	9.1

Table 6.3. Deep oil cracking versus FCCU

Process	FCC	DCC type 1	DCC type 2
Yields, wt%			
Dry gas	3.5	11.9	4.0
LPG	17.6	42.2	34.5
C_5+ Gasoline	55.1	27.2	41.6
LCO	10.2	6.6	9.8
DO	9.3	6.1	5.8
Coke	4.3	6.0	4.3
Ethylene	1.1	6.1	1.6
Propylene	4.9	21.0	14.3
Butylenes	8.1	14.3	14.7
Feedstock: Chinese Waxy VGO			

Even more severe cracking conditions are used in the Catalytic Pyrolysis Process (2002), where the desire is to produce all petrochemical products, i.e. ethylene, propylene, butenes, and aromatics. This process is really a substitute for a steam cracking furnace in an ethylene plant. It allows the operator to use cheaper feedstocks and vary the ratio of ethylene to propylene over a wider range than is possible with only thermal cracking.

A diagram of the reactor-regenerator of a modern fluid cracking unit is shown in Figure 6.3. Hot regenerated catalyst contacts the oil near the base of the reactor riser. Virtually all of the cracking takes place in the feed riser which connects to a

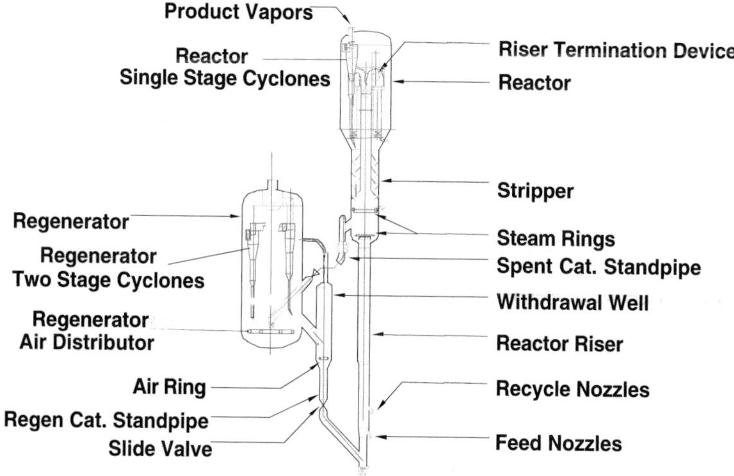

Figure 6.3. Reactor-regenerator of modern FCCU.

catalyst/vapor separator. The hot catalyst is discharged into the catalyst stripper while the vapor is routed through secondary cyclones to remove any remaining catalyst and then to the main fractionator and gas plant for separation of the products.

The spent catalyst enters a multi-stage stripper where the absorbed hydrocarbons are displaced with steam and leave with the product from the overhead of the reactor. This prevents unwanted hydrocarbons from entering the regenerator, consuming air and possibly causing excessive catalyst deactivation. Staging is accomplished by the use of baffles or packing and is similar in concept to a multi-tray distillation tower.

Spent-stripped catalyst enters the regenerator where the coke is burned off the catalyst to restore its activity. The heat generated in the combustion process can supply all of the needed heat for the process. If excess heat is produced it can be removed by external catalyst coolers. The regenerated catalyst flows back to the base of the riser where the cycle is completed. Typical catalytic cracking units undergo 100–400 such cycles a day.

There have been many advances in the fluid catalytic cracking process. The high capacities of a unit (typically a third of the crude oil a fuels refinery runs goes to a standard gas oil FCC unit), and its positive influence on overall refinery economics has made it a prime target for innovation. A list of most of the major innovations in the process is given in Table 6.4. These include catalyst, equipment and process changes that have occurred on a continual basis over the 60 years that FCCUs have been operating. By being able to both improve its performance and evolve its functions, the catalytic cracking process has remained a staple in the modern refinery.

Fluidization

The basic circulating, fluid bed reactor system that is widely used today in many other applications got its start in fluid catalytic cracking. It was the development of the standpipe that made the process possible. The fine powder acts as a fluid when contacted with a gas and it is the standpipe that allows the catalyst to be circulated from a vessel at lower pressure to one of higher pressure, thus completing the circulating catalyst loop.

In Figure 6.4, the head gain from the fluidized catalyst is given as a function of the bed density. The latter depends on the catalyst properties, the amount of aeration being used and the rate at which catalyst is being added to the system.

The change in density of a bed of cracking catalyst with gas velocity is shown in Figure 6.5. At very low velocities the bed is stationary but soon reaches a point where

Table 6.4. Major advances in fluid catalytic cracking

1942	First FCC unit on stream (Exxon)
1947	Stacked configuration (UOP)
	Compact with small inventory
1948	Spray dried catalyst (improved fluidization)
1952	Synthetic high alumina
1955	Reactor riser cracking (shell)
1959	Semi-synthetic catalyst (addition of clay)
1960	Improved metallurgy
	(Higher regenerator temperatures)
1961	Heavy oil cracking (Phillips-Kellogg)
1964	Zeolitic catalysts introduced (Mobil)
1972	Complete CO combustion process (Amoco)
1974	Combustion promoters (Mobil)
1975	Metals passivation (Phillips)
1981	Two independent regenerators for unlimited
	Regeneration temperature (Total)
1982	High performance feed injectors
1982	Dense bed catalyst coolers (Ashland/UOP)
1987	Vapor quench (Amoco)
1987	Mix temperature control (Total)
1988	Close-coupled cyclones (Mobil)
1991	Deep catalytic cracking (RIPP/Sinopec)
1996	Enhanced stripping designs
2002	Catalytic pyrolysis (RIPP/Sinopec)

the upward force of the gas balances the weight of the catalyst (minimum fluidization velocity) and the catalyst becomes suspended in the gas. More gas expands the fluidized bed to a point where the next bit of gas enters and forms bubbles. This is the minimum bubbling velocity and the two phases are referred to as the emulsion and bubble phases, respectively. Once bubbles are formed all of the added gas

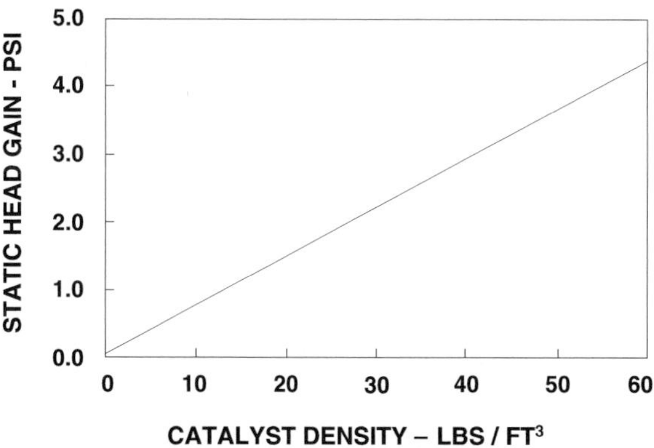

Figure 6.4. Static head gain per 10′ of standpipe versus catalyst density.

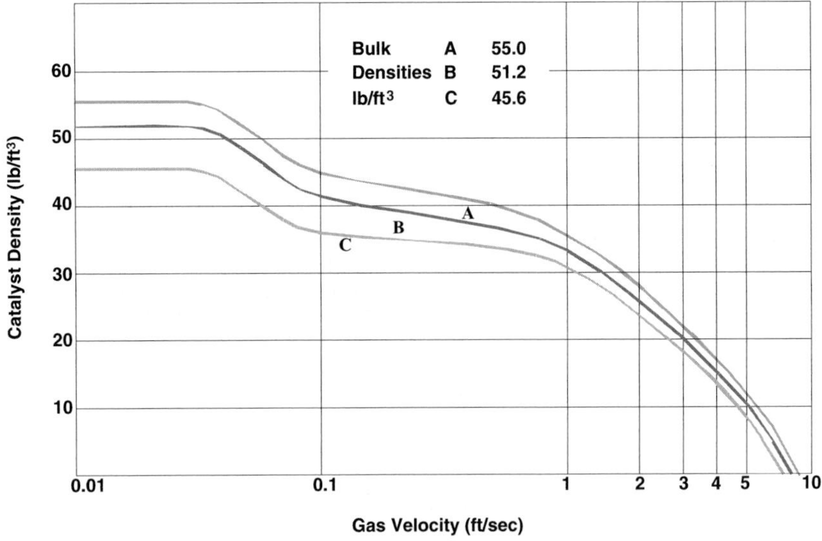

Figure 6.5. Fluidization curves FCC catalysts.

contributes to this phase. At even moderate velocities the catalyst is entrained with the gas and must be captured and returned to the bed. This allows operation of the catalyst beds at much higher superficial velocities than would otherwise be possible. Typical velocities in the reactor-regenerator system are given in Table 6.5.

Bubbles play a key part in the operation of a fluid bed system. They are the 'engine' that stirs the bed of catalyst making the high heat and mass transfer rates possible. In the standpipes bubbles can have a profound effect on the smoothness of catalyst flow and care must be exercised to avoid over-aerating as well as under-aerating the catalyst. A formula for adding the proper amount of air to a standpipe, Q, was presented by Zenz and is shown below:

$$Q = 2000\left(\frac{P_{\mathrm{B}}}{P_{\mathrm{t}}}\left[\frac{1}{\rho_{\mathrm{f}}} - \frac{1}{\rho_{\mathrm{p}}}\right] - \left[\frac{1}{\rho_{\mathrm{t}}} - \frac{1}{\rho_{\mathrm{p}}}\right]\right) \tag{1}$$

Table 6.5. Superficial velocities for FCC (ft/sec)

Minimum fluidization	0.01
Minimum bubbling	0.10
Bubbling bed	0.3–2
Turbulent bed	2–4
Fast-fluidized bed	4–8
Pneumatic conveying	12

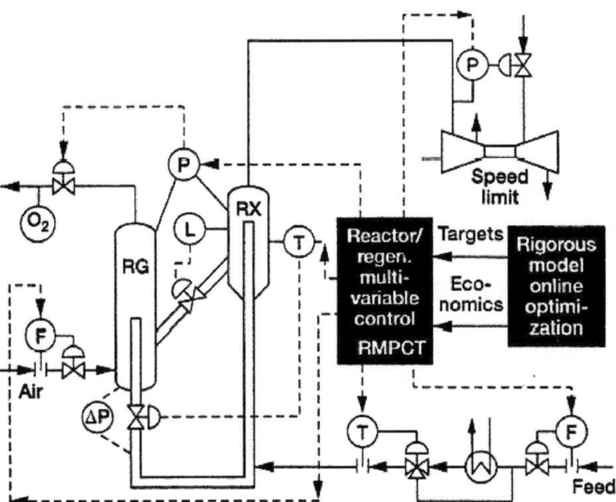

Figure 6.6. Advanced FCC process control scheme. Reprinted with permission from HYDROCARBON PROCESSING, by Gulf Publishing Co., Copyright 2001; all rights reserved.

Typically 70% of the above calculated aeration rate is about optimum and it should be added at even intervals, around the standpipe, about six to eight feet apart. Fine-tuning is done once the unit is in operation and a single gauge pressure survey can be taken to aid in the operation of the system. It is critical that the catalyst be properly fluidized entering the standpipe since refluidizing catalyst is very difficult. Over-aeration is a more common problem than under aeration in commercial operations.

Process control

While there are a number of ways to configure the controls of the cat cracker, Figure 6.6 shows the typical scheme used in the industry for a slide valve controlled unit.

The reactor pressure is controlled by the inlet pressure to the wet gas compressor but the actual pressure at the top of the reactor in the dilute phase is this pressure plus the pressure drop taken through the main fractionator overhead system, the main fractionator and the reactor overhead system (cyclones, plenum, and overhead vapor line). At steady state this pressure should be constant. The regenerator pressure is controlled by the flue gas slide valve and is maintained to give a constant differential pressure between the reactor and regenerator. This is adjusted to provide optimum catalyst circulation, adequate pressure drops across the control valves and to balance the wet gas compressor and air blower.

The reactor temperature is controlled by the regenerated catalyst slide valve. More hot catalyst is circulated to raise the reactor temperature or accommodate increased feedrates. The feed going to the unit is preheated by exchange with the hot products. While heat exchange is adequate for feed temperatures ranging up to about 500°F, a fired heater is included for higher temperatures.

The regenerator temperature is a function of the coke on the catalyst entering the regenerator, its composition and the mode of coke burning. Full combustion or the complete burning of the carbon to CO_2 is most common due to the necessity of limiting the CO concentration in the flue gas to 500 ppm or less to meet environmental regulations. Partial combustion resulting in CO/CO_2 ratios ranging up to one are practiced in units that do not have the needed metallurgy in the regenerator to operate at high temperatures or wish to maximize regenerator coke burning capacity or limit the heat produced in the regeneration process. Such units are equipped with CO incinerators or boilers where the CO is converted to CO_2 before being discharged to the atmosphere. Typical regenerator temperatures in these two operating modes are 1150–1250°F (620–675°C) and 1275–1350°F (690–732°C), respectively for partial and complete combustion.

The spent catalyst slide valve controls the catalyst level in the stripper or reactor bed if one is being used. No direct control of the regenerator catalyst bed level is made and its level floats depending on the catalyst losses and the catalyst withdraw and addition policies. Most units are capable of retaining most of the fresh catalyst added daily and must periodically withdraw equilibrium catalyst to keep the regenerator level from getting too high. The regenerator is used as the 'floating' vessel because it is larger and it is desirable to withdraw regenerated rather than spent catalyst.

With the advent of computers and advanced process controllers, many refinements to the basic scheme are possible. The biggest benefits come from operating closer to several limits at one time. Better analysis of the feedstocks could allow feed-forward control in the future. While these control systems can improve refining profitability from 20 to 40 cents/barrel processed, they require more instrumentation, which must be maintained to achieve the stated benefits.

Reaction chemistry and mechanisms

Many studies have been performed elucidating the difference between thermal and catalytic cracking. Two separate reaction mechanisms are attributed to the methods of cracking, i.e., thermal cracking goes through free radicals and catalytic cracking via carbenium ions. The latter are generally associated with the bronsted acid sites on the catalyst. In Table 6.6, the major differences between the two mechanisms are shown. Thermal cracking is minimized as much as possible in current FCC units by the use of advanced equipment such as radial feed injectors, riser termination devices and post riser quench. Catalyst selection is also critical.

Table 6.6. Characteristics of cracking mechanism

Thermal—Free radical	Catalytic—Carbenium ion
• C_1–C_2 Principle products	• C_3^+ Principle products
• Little skeletal isomerization	• Much skeletal isomerization
• Cracking at beta position with little preference for free radical type	• Aromatics dealkylate at ring if chain is at least 3 carbons long
• Alpha olefins primary product	• cracking occurs at tertiary > secondary > primary carbeniums ions
	• Products contain an olefin
	• Largest molecules crack fastest unless steric hindrance controls

There are many reactions that occur during the cracking process. These are listed below as primary or secondary reactions (Table 6.7). Most of the secondary reactions are undesirable and are controlled through reactor and catalyst design. Numerous cracking reactions occur with the large feed molecules before the desired products are achieved. Typical feeds to a catalytic cracking unit contain molecules boiling above the diesel end point (650–700°F or 343–371°C) and may boil as high as 1500°F. The consecutive reactions that occur are shown in equation (2):

$$\underset{\text{Gas Oil or Resid}}{\text{A}} \longrightarrow \underset{\text{Diesel}}{\text{B}} \longrightarrow \underset{\text{Gasoline}}{\text{C}} \longrightarrow \underset{\text{LPG + Coke}}{\text{D}} \qquad (2)$$

Table 6.7A. Primary cracking reactions for hydrocarbon types

Reactant		Products
Paraffin		Paraffin + Olefin
$C_n H_{2n+2}$	$\longrightarrow$	$C_pH_{2p+2} + C_mH_{2m}$
Naphthene (cyclic paraffin)		Olefin + Olefin
C_nH_{2n}	$\longrightarrow$	$C_pH_{2p} + C_mH_m$
Alkylaromatic		Aromatic (base) + Olefin
ArC_nH_{2n+1}	$\longrightarrow$	$ArH + C_nH_{2n}$
Olefin		Olefin + Olefin
C_nH_{2n}	$\longrightarrow$	$C_pH_{2p} + C_mH_{2m}$
Aromatic	$\longrightarrow$	No reaction

Table 6.7B. Secondary reactions of olefins

Naphthene + Olefin	$\longrightarrow$	Aromatic + Paraffin
Hydrogen + Olefin	$\longrightarrow$	Paraffin
Normal Olefin	$\longrightarrow$	Iso-Olefin
Olefin + Olefin	$\longrightarrow$	Larger Olefin
Aromatic + Olefin $\longrightarrow$	Alkylaromatic	$\longrightarrow$ Cyclization ⌉
Multi-ring aromatic (Coke)	$\longleftarrow$ Dehydrogenation	$\longleftarrow$ ⌋

While this is an over simplification of what occurs, it does give an overall view to the cracking process. From the above equation it is clear that separate processes or applications can center around either product or feed differences.

As such, there are three basic catalytic cracking applications, today. These are:

	Application	Feedstock	Products
1.	Gas oil cracking	Vacuum gas oils	Motor gasoline LCO and LPG
2.	Resid cracking	Atmospheric resid VGO + Vacuum resid	Motor gasoline LCO and LPG
3.	Cracking for petrochemicals	Vacuum gas oils and added resids.	Light olefins—C_2's, C_3's, and C_4's Plus aromatics

The driving forces behind these applications are the need for gasoline, the lack of demand for bottom of the barrel and the increase in demand for light olefins and aromatics. The technologies for each of these process configurations are reviewed in turn.

Gas oil cracking technology features

Reaction technologies

All of the reaction systems offered today consist of a feed injection system, reactor riser, riser termination device, and sometimes vapor quench technology. Each licensor approaches the design with different equipment and configurations.

The feed injection system is probably the most important part of the reaction system since it provides the initial contacting between the oil and catalyst. A good system must vaporize the feed quickly, quench the hot catalyst as fast as possible and provide plug flow of the hydrocarbons and catalyst. Other features are important but are beyond this limited description.

The hot-regenerated catalyst generally comes to the feed riser either through a wye section or a J-bend depending on the proximity of the regenerated catalyst standpipe and the feed riser (see Figure 6.7). Feed nozzles are usually located from 10 to 30 ft downstream of the base of the riser. Since raising the nozzles increases the back pressure on the regenerated catalyst slide valve, lift gas may be incorporated to control the pressure balance and minimize the back-mixing of the catalyst. The lift gas is normally light gas from the product recovery section and adds to the load in the main fractionator and wet gas compressor. Steam is sometimes substituted either whole or in part to reduce the processed gas but can cause some catalyst deactivation if the regenerator temperature gets too high.

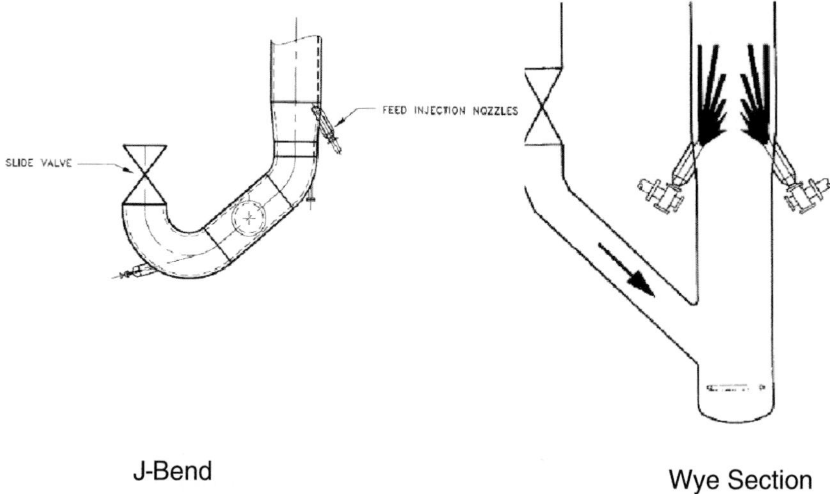

Figure 6.7. Regenerated catalyst/feed contacting configurations.

The feed injection systems (Figure 6.8) employed by the various licensors all use different nozzle designs. Oil pressure and dispersion steam rates vary by both licensor and application. Typical oil pressure drops and dispersion steam rates are 30–150 psi and 1–7 wt% and depend on the type of feedstock processed, i.e. vacuum gas oils or residual feeds (BP > 1050°C (565°C)).

The feed injection systems being offered by the licensors are shown in Figure 6.8.

These are all two fluid nozzles that mix an oil feed and steam and distribute them across the riser cross-sectional area. Differences exist in the methods of contacting the steam and oil, the amount of steam and pressures used and in the tip design. Refiners are concerned about both performance and reliability when choosing a system.

Reactors are vertical pipes that are generally straight and are about 100 ft long. The diameters at each end are controlled to give adequate lift to the catalyst. Velocities of 20–30 ft/sec at the base and 50–65 ft/sec at the vapor outlet are typical with vapor residence times of about 2 sec based on the vapor outlet velocity. Bends in the riser pipe do not change the vapor residence time but increases the catalyst residence time (or slip) and usually results in poorer oil/catalyst contacting.

A termination or separation system is used at the top of the riser. These have been simple Tee's, rough cut cyclones or specialized inertial separators. Most recently the riser termination devices have been close-coupled cyclones, linear disengagers, riser separator strippers, or vortex separators. These are pictured in Figure 6.9 and all can be or are directly connected to the secondary cyclones.

Stone & Webster
Dual Slot Impact

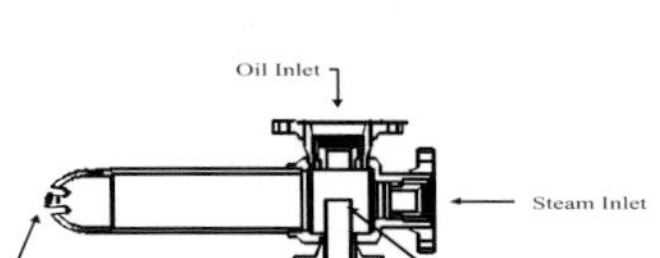

Kellogg ATOMAX II

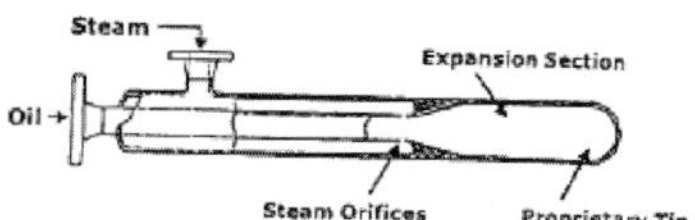

Lummas MicroJet

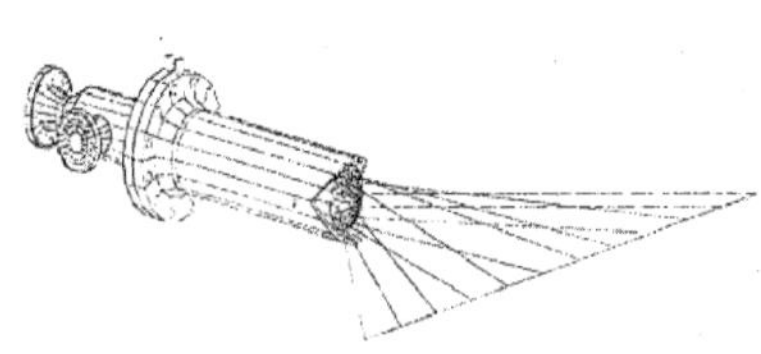

UOP OPTIMIX

Figure 6.8. Commercial feed injection systems.

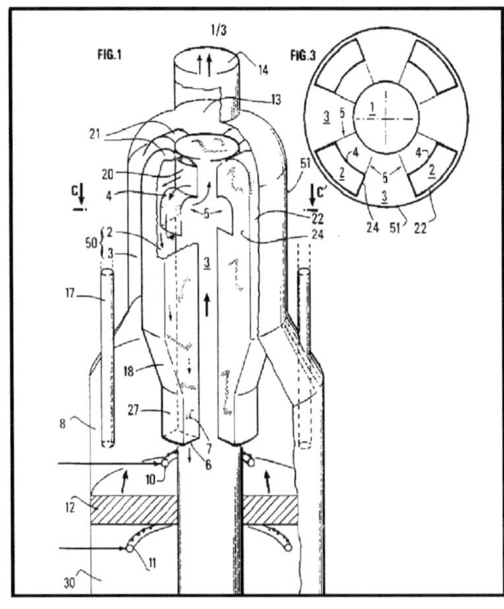

RISER SEPARATOR STRIPPER

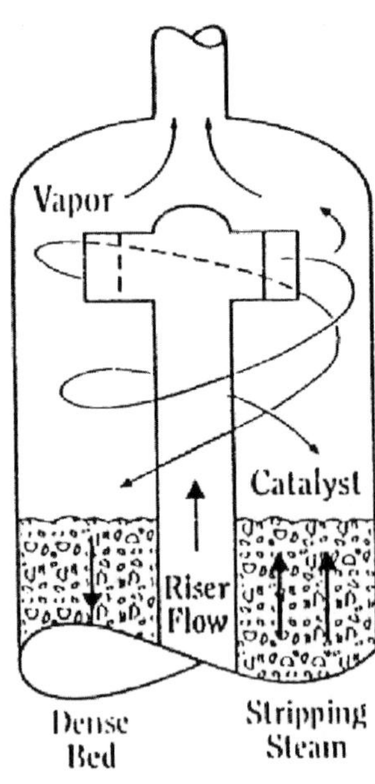

VORTEX SEPARATOR

Figure 6.9. Riser separators.

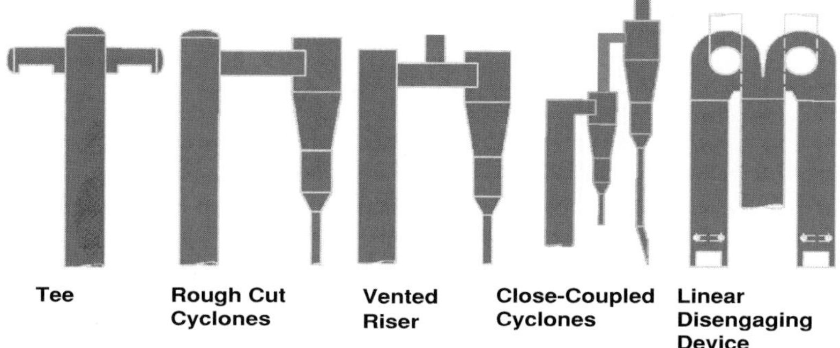

Figure 6.9. (Cont.)

The dilute phase of the reactor-stripper is a place where secondary thermal reactions can occur and the extent of these reactions can be estimated from equation (3).

$$\text{Delta Dry Gas} = Ke^{E/RT}t = (8.3 \times 10^{12})\left(e^{-93{,}000/(1.987^{*}T(R))}\right)t \qquad (3)$$

A plot of this equation (Figure 6.10) shows that high reactor temperatures and long dilute phase residence times lead to high dry gas yields. At low reactor temperatures,

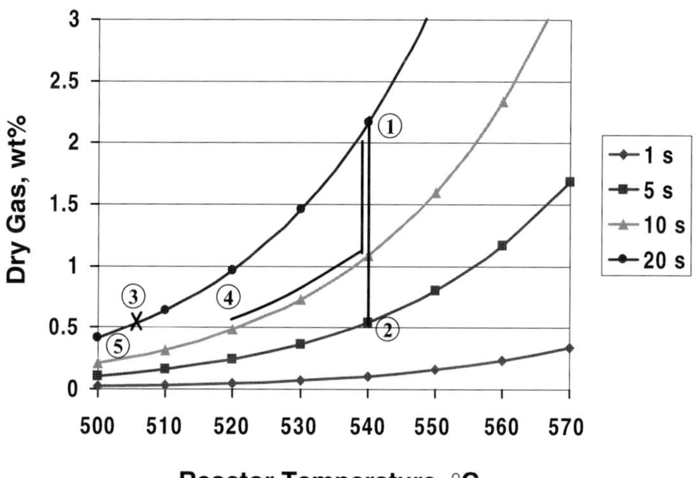

Figure 6.10. Post riser dry gas production.

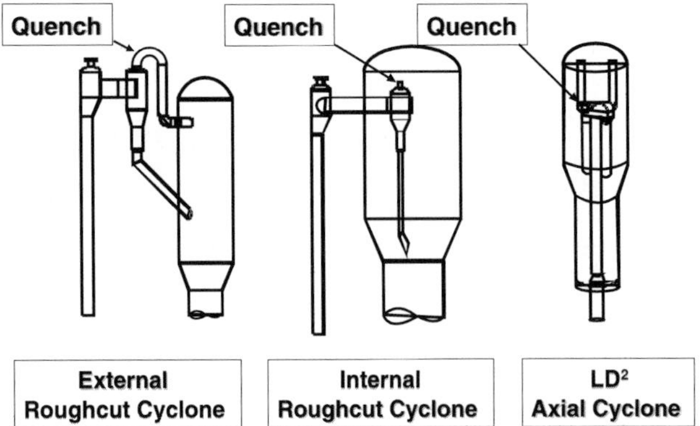

Figure 6.11. Reaction systems with post riser quench.

below 505°C (941°F), the impact of the dilute phase is small and the use of advanced termination technology is difficult to justify especially if there is a chance of downtime due to equipment malfunction. The value of K can be determined for an individual unit by measuring the amount of dry gas produced at two different temperatures and measuring the vapor residence time with tracers. A number of factors can add to the dry gas make such as catalyst type, feed properties, the amount of dilute phase catalyst and carbon on regenerated catalyst.

Another approach to reducing the post gas dry gas make is to reduce the temperature at the end of the riser. This has been done at the end of the riser before the separator and just downstream of the catalyst/vapor separator at the end of the riser. Quenching in the riser will reduce gas make and may be needed if there is a metallurgical limit in the reactor-stripper vessel. However, the amount of quench is high due to the catalyst that is cooled and the stripper must operate at a lower temperature. The coke make and air requirements are increased since this is a recycle stream from a heat balance perspective.

Downstream quenching only cools the vapor and does not add to the coke yield. It can be applied to any unit that has a rapid/effective catalyst/vapor separator as shown in Figure 6.11. The quench material is taken from the main fractionator and from a heat balance perspective acts the same as a pump around on the main column. It can be adjusted to accommodate changes in the reactor temperature and discontinued at low temperatures. Benefits from three different units are listed in Table 6.8. As equation (3) implies and Figure 6.10 illustrates, the effects of time and temperature are equivalent. However, the vapor quench also reduces the cracking that occurs in the secondary cyclones, plenum changer and overhead vapor line. Care must be exercised when using quench to avoid coke formation. These precautions include the choice of the quench medium, operating above minimum operating temperatures and automatic shut-offs and purges.

Table 6.8. Impact of vapor quench on FCC yields

Unit	A	B	C
Temperature (°C)			
Riser outlet	513	549	532
After quench	484	519	494
Yield shifts (wt%)			
Dry gas	−0.23	−0.80	−0.66
Gasoline	+0.43	+1.80	+2.89
LPG	–	–	−1.58
LCO	–	–	+0.25
DO	–	–	−0.86

Stripping technology

After the cracking reactions are completed, the spent catalyst needs to be stripped of the hydrocarbons that would accompany it to the regenerator. This is done in a staged-fluidized bed where steam enters from the bottom and pushes the hydrocarbons in the gas phase out of the top of the bed. Design parameters for new units are given in Figure 6.12, which is a common disk and donut design.

The baffles improve contacting between the steam and catalyst and increases the number of contacting stages. As shown in Figure 6.13, seven stages of stripping is sufficient to remove at least 95% of the hydrocarbons. Each of the design parameters are important to the proper operation of the stripper. A minimum amount of steam is necessary to displace the hydrocarbons in the emulsion and bubble phases of the fluid bed. The flux rate determines the catalyst velocity through the bed. If the downward velocity of the catalyst gets too high, it will sweep hydrocarbons and steam with it and adversely affect the stripper performance. The residence time is a function of the stripper's catalyst inventory and the catalyst circulation rate and relates to the number of stages and their efficiency that the stripper can obtain.

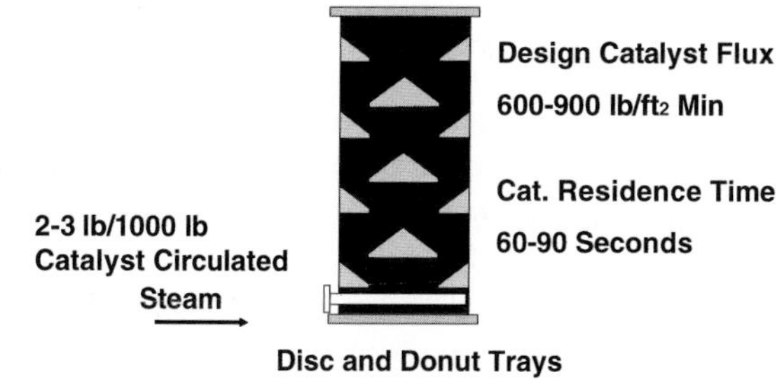

Figure 6.12. Disk and donut stripper.

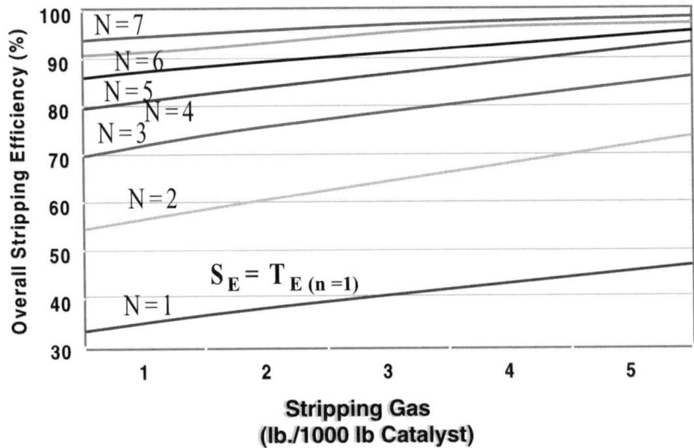

Figure 6.13. Disk and donut fray efficiency/stage efficiency.

While the disk and donut design shown has proven to be both reliable and effective, there are other variations on this design. Holes can be placed in the baffles to improve contacting, the skirts can be lengthened to provide a larger gas ΔP, vent tubes have been used to allow the gas from the bottom of the baffles to pass to the next stage and rods or shed decks can be substituted as the contacting devices. The use of structured packing has been reported recently with excellent results. Lower steam usage, better contacting, and utilization of almost the entire cross-sectional area of the stripper are benefits claimed with the new design. Horizontal trays with small holes configured as distillation trays have been tested in the lab and will also be implemented in the field.

Regeneration technology

The object of the regenerator is to remove the coke that builds up on the catalyst in the reactor without damaging the catalyst. Many studies have been made on the burning rates of coke in a fluidized bed of cracking catalyst. Equation (4) describes the major regeneration operating variables

$$\frac{dC}{dt} = K \times C_i \times \mathrm{L_m O_2} \times e^{A/RT} \tag{4}$$

The contacting between the oxygen and catalyst is improved significantly as the air rate or superficial velocity is increased in the regenerator. As the velocity increases the bed goes through three stages. A bubbling bed occurs at low superficial velocities (up to about 1.5 ft/sec or 46 cm/sec). Here relatively distinct bubbles are formed and pass through the bed. A turbulent bed (1.5–4.0 ft/sec or 46–122 cm/sec) exists at higher superficial velocities in which an emulsion is formed and the diffusion rate of oxygen is significantly increased. At higher velocities a fast-fluidized bed (4–8 ft/sec or 122–244 cm/sec) exists in which turbulence is maximized. A return line from the recovered catalyst to the combustor is required to provide enough residence time and a

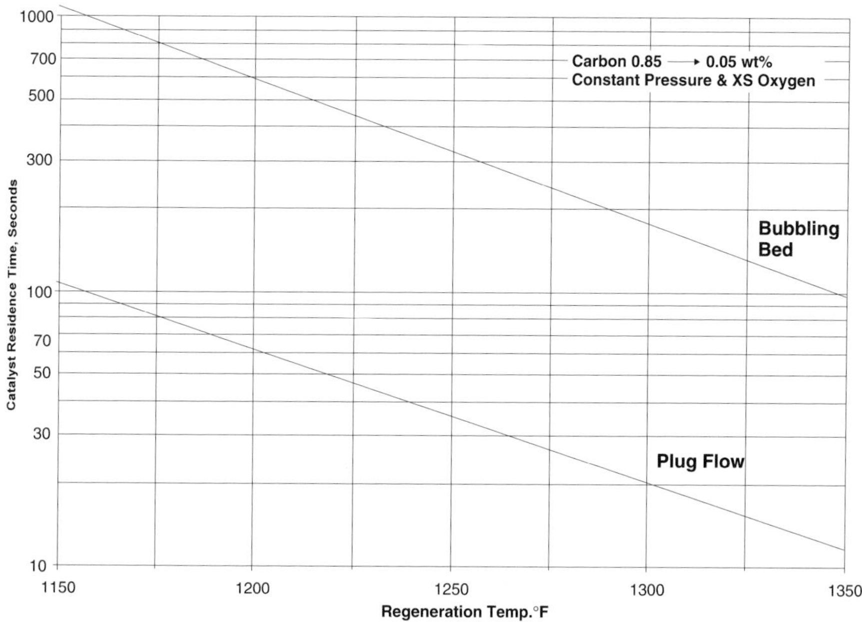

Figure 6.14. Comparison plug flow and Backmix regenerators (ideal).

sufficient mix temperature for the coke to completely burn. The difference in burning rates between these idealized regenerator designs is shown in Figure 6.14.

A few observations can be made from Figure 6.14 and equation (4). At high temperatures (>1,300°F or 704°C) the burning rate is very high. Oxygen availability limits the burning rate though the small size of the fluid cracking catalyst eliminates or minimizes diffusion as a reaction barrier at normal regenerator temperatures. The plug flow curve moves up, as catalyst is recycled in the combustor since the effect is to increase the residence time. Regenerators with larger inventories will reduce the carbon satisfactorily even if there is a partial malfunction of the air distributor while smaller inventories allow faster change outs of the catalyst inventory. Over the years, the regenerator temperatures have increased due to better metallurgy and the need to burn all of the carbon off the catalyst to restore the zeolite catalysts' activity and product selectivity.

At 1,100°F (593°C) the regenerator inventories had to be large to provide enough time to burn the coke. Full CO combustion raised these temperatures to 1,300–1,350°F (704–732°C). CO promoters are also frequently employed to assist the carbon burn and prevent the afterburning of CO in the dilute phase or downstream hardware where serious equipment damage can occur. The differences in catalyst inventory are illustrated in Table 6.9 for a 50,000 B/D unit that circulates between 30 and 40 tons/min of catalyst.

Table 6.9. Regenerator parameters (50,000 B/D)

Regenerator type	Residence time (Min)	Temperature (°F)	Regenerator inventory (tons)
Bubbling bed	5–20	1,100	300–800
Turbulent bed	3–5	1,250–1,350	200
Fast-fluidized bed	1–3	1,275–1,350	120

Shorter contact times and smaller catalyst inventories limit operable regeneration conditions and need higher internal catalyst recycle rates for increased throughput or coke burn. Since catalyst is frequently added on a pound per barrel basis each of these units would use about 5 tons/day of fresh catalyst. Equation (5) relates the catalyst activity to S, the daily fractional replacement rate or age of the catalyst:

$$A = \frac{A_O S}{K_D + S}. \tag{5}$$

This equation implies the smaller inventory would give the highest equilibrium or unit activity. However, K is also a function of the contacting between the spent catalyst and air, the mix temperature, the catalyst type and activity and the number of cycles the catalyst makes through the system. This latter fact implies there is an optimum unit inventory for a given processing capacity.

Commercial regenerator designs are shown in Figure 6.15. These utilize either turbulent beds or fast-fluidized beds. Cocurrent or countercurrent contacting of the catalyst and air is practiced and care is taken to prevent short-circuiting of the catalyst from the regenerator inlet to the outlet to ensure an even, low carbon distribution on the regenerated catalyst.

Resid catalytic cracking

Processing heavier feeds poses challenges to the normal FCC design due to the higher coke laydown on the catalyst during the cracking reactions. The coke layed down in the cracking process has been shown to come from four main sources as shown in Table 6.10.

The catalytic coke comes from the secondary cracking reactions and are caused by polymerization and condensation of hydrocarbons. Strippable coke are the hydrocarbons that are entrained with the spent catalyst that enters the regenerator. Heavy metals that lay down on the catalyst surface promote dehydrogenation and lead to extra coke and hydrogen. Nickel, vanadium, and Iron are the main contaminates though occasionally copper, zinc, and lead have been known to cause problems. Feed coke has been associated with the carbon residue in the feed as measured in the Conradson Carbon Test (ASTM). This has been also referred to as additive coke.

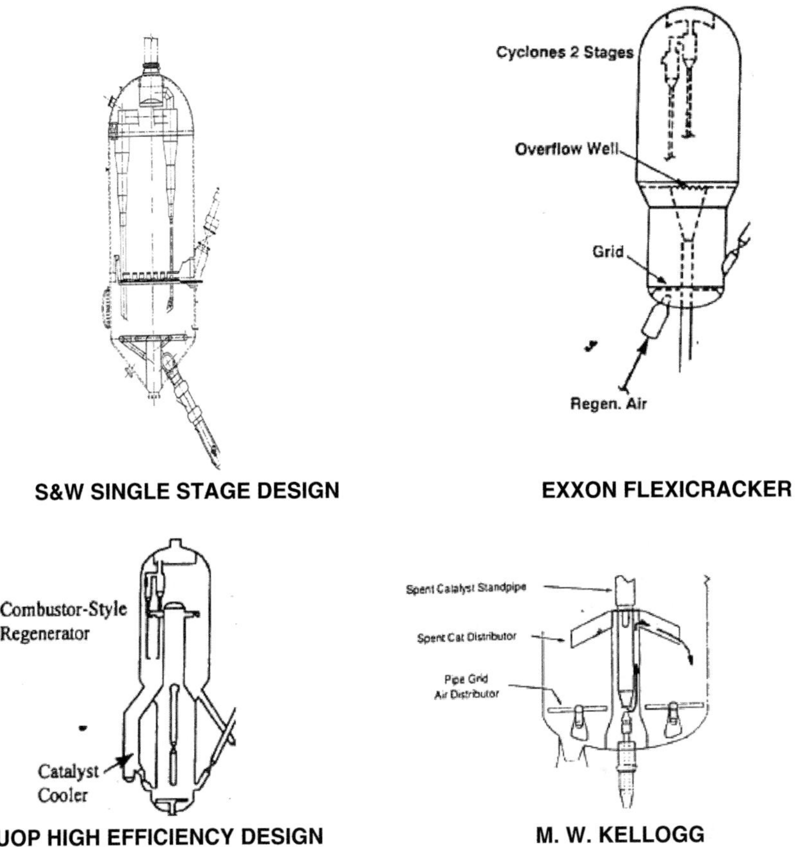

Figure 6.15. Commercial FCC regenerator designs.

As Table 6.10 shows the sources of coke shift dramatically when resid is in the feed. The percentages given in Table 6.10 are not fixed and shift as the composition of the feed and operating parameters change. The total coke make is different in each case. There are other factors that lead to coke formation that are included in the four categories. Basic nitrogen is known to cause coke since these molecules are strongly adsorbed on the acid sites in the reactor and are burned off in the regenerator.

Table 6.10. Sources of coke production

Feedstock	Gas Oil	Residue
Coke categories		
Catalytic	65	45
Strippable	25	5
Contaminant	5	20
Feed Coke	5	20

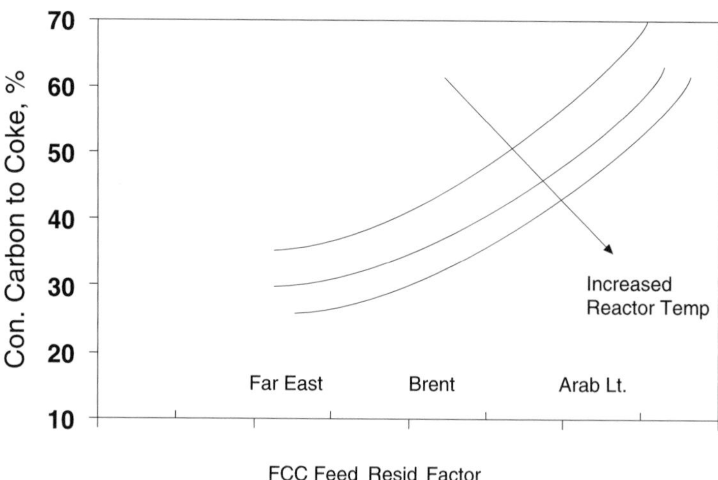

Figure 6.16. Percentage of conradson carbon going to coke. Reprinted with permission from HYDROCARBON PROCESSING, by Gulf Publishing Co., Copyright 1987; all rights reserved.

Some of the very heavy hydrocarbons in resid may not be vaporized and be laid down on the catalyst surface where they eventually coke. Figure 6.16 shows that the percentage of Conradson Carbon that goes to coke is a function of the feedstock and reactor temperature.

Both desorption of basic compounds and feedstock vaporization would be improved by raising the reactor temperature (and consequently the stripper temperature) so the relationship shown in Figure 6.16 is directionally correct. The carbon laydown or delta coke can be represented by equation (6). The first term in is the Voohries relationship for carbon laydown for gas oil feedstocks while the second term reflects the feed coke contribution. A, B, and C are constants that depend on operating conditions, feed properties and catalyst tested.

$$\text{Delta Coke} = A\left(\frac{\text{Catalyst Res. time}}{\text{in the reactor}}\right)^{B} + \frac{C * \text{Conradson Carbon}}{\text{Catalyst/oil ratio}}. \qquad (6)$$

The overall heat balance equation for any cracking unit is:

$$\text{Wt\% Coke} = \text{Delta Coke} \times \text{Catalyst/Oil ratio}. \qquad (7)$$

The dependent variable is the catalyst/oil ratio, which needs to be high enough to give the desired conversion. Weight percent coke is strictly a function of the operating variables (i.e. fresh feed and recycle rates, feed temperature, reactor temperature, steam rates, heat of cracking, air rates, and carbon burning mode). Since higher delta

Table 6.11. Methods of increasing coke make for resid processing

Modification	Objective	Consequence
Partial burn	Lower regen temp. Burn more coke	CO emissions rise, carbon on catalyst increases
Mix temp. control	Cool regen Aid vaporization feed	Higher coke
Water injection	Cool regen	More coke CAT deactivation?
Oxygen enrichment	Increase coke burn Maintain superficial Air velocity	Higher regen temperature
Auxiliary air	Burn more coke	Inc. regen superficial velocity
Catalyst cooler	Reduce regen Temperature	Higher coke More air
Second regenerator	Increase throughput and resid	Cost

cokes are caused by heavier feeds, the resulting cat/oil ratio becomes too low for medium to high conversion levels. The coke make must be increased to raise the catalyst/oil ratio and this can be done by any of the means shown in Table 6.11. For the heaviest feeds a catalyst cooler will be required.

In Figures 6.17–6.21 the commercially offered resid FCC units are pictured.

Much of the reactor-stripper design is the same for resid crackers as it is for gas oil designs. However, there are feed injectors designed specifically to process residual feeds that require more dispersion steam than the normal gas oil models.

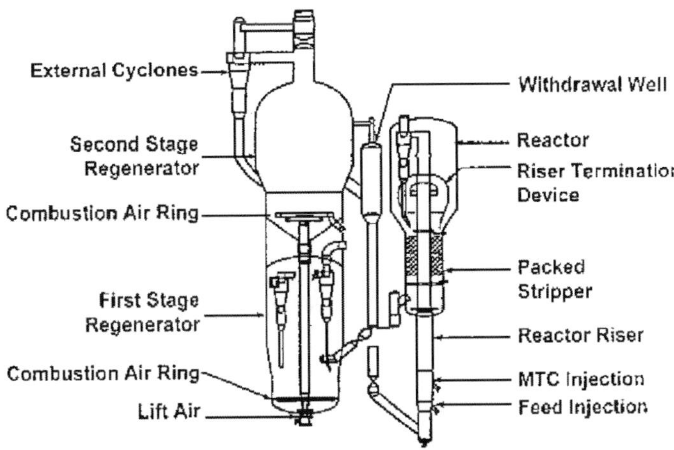

Figure 6.17. S&W/Axens R2R.

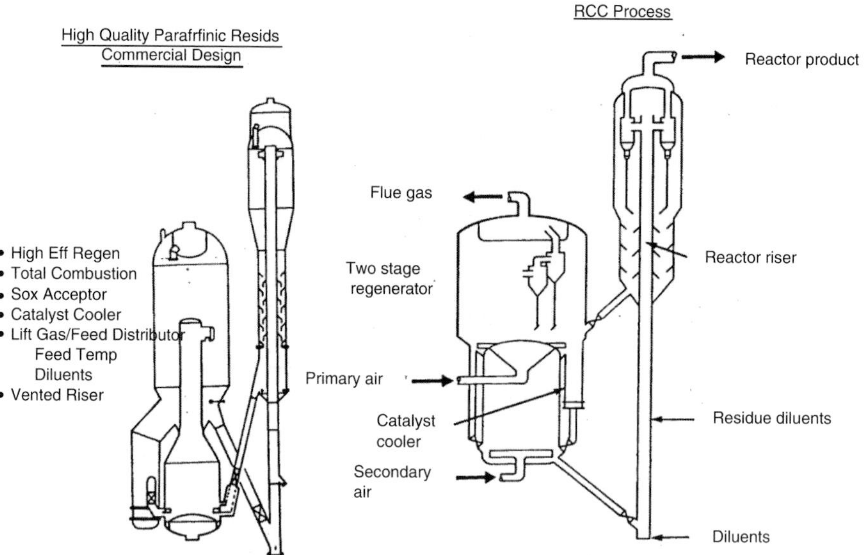

Figure 6.18. UOP resid designs.

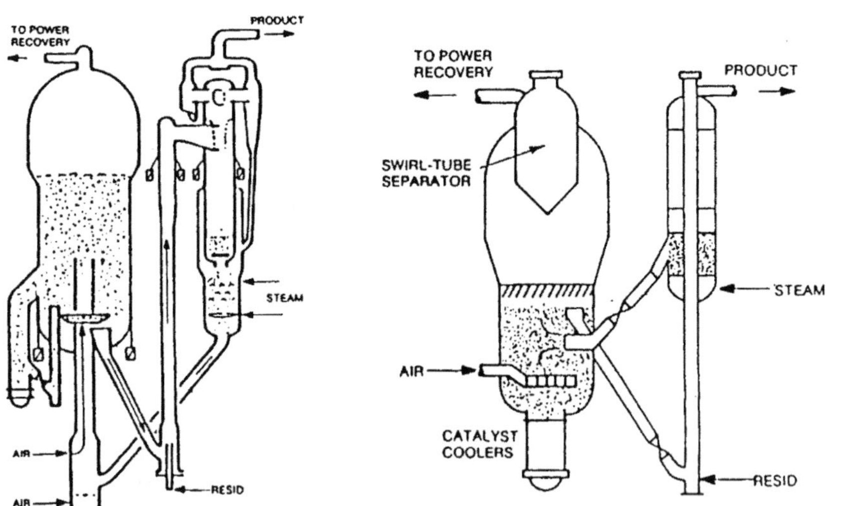

Figure 6.19. Shell resid cracker.

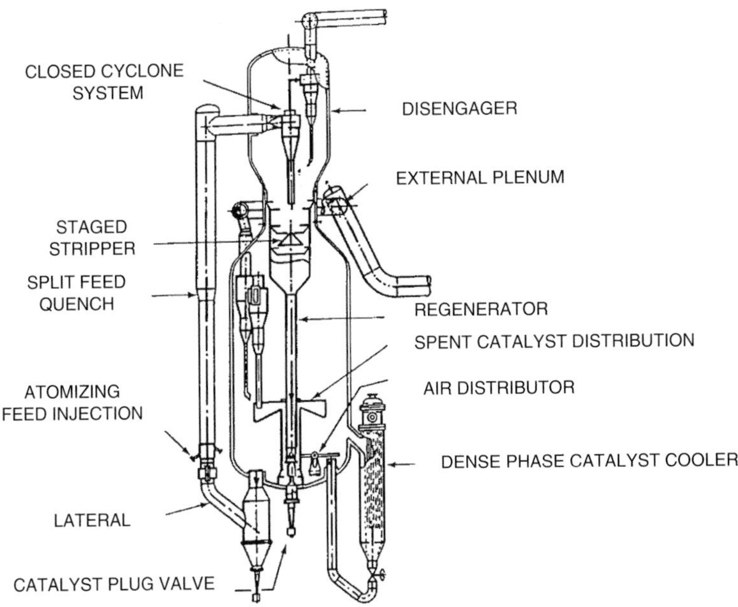

Figure 6.20. Kellogg resid cracker.

On the regenerator side, the two approaches are to use a single stage regenerator and add a catalyst cooler or split the regenerator into two stages and make the catalyst coolers optional. The two stage designs offered differ in the sequence of catalyst flow and how the air is introduced and utilized.

The Ashland/UOP design has the first regenerator on top of the second. Spent catalyst flows into regenerator one, is partially regenerated and flows to the second regenerator where the carbon burn is completed. Air is introduced into both regenerators but the flue gas from number two passes up through arms into regenerator one. Entrained catalyst is carried with the flue gas. All of the regeneration air goes through the top regenerator and a CO boiler to reduce the CO content of the flue gas to permitted levels.

Stone & Webster/Axens have reversed the regenerators so that the spent catalyst enters the bottom regenerator and the partially regenerated catalyst is vertically conveyed with lift air to regenerator two where the coke burn is finished. Air goes to both regenerators and the catalyst lift line. The flue gas from regenerator one contains CO since the regenerator is run with no excess oxygen. A CO incinerator or CO boiler is used to produce an acceptable flue gas composition. Lift gas and the air to regenerator two are burned and exit the second regenerator and is recombined with the flue gas from regenerator one. Several flue gas combination schemes have been used.

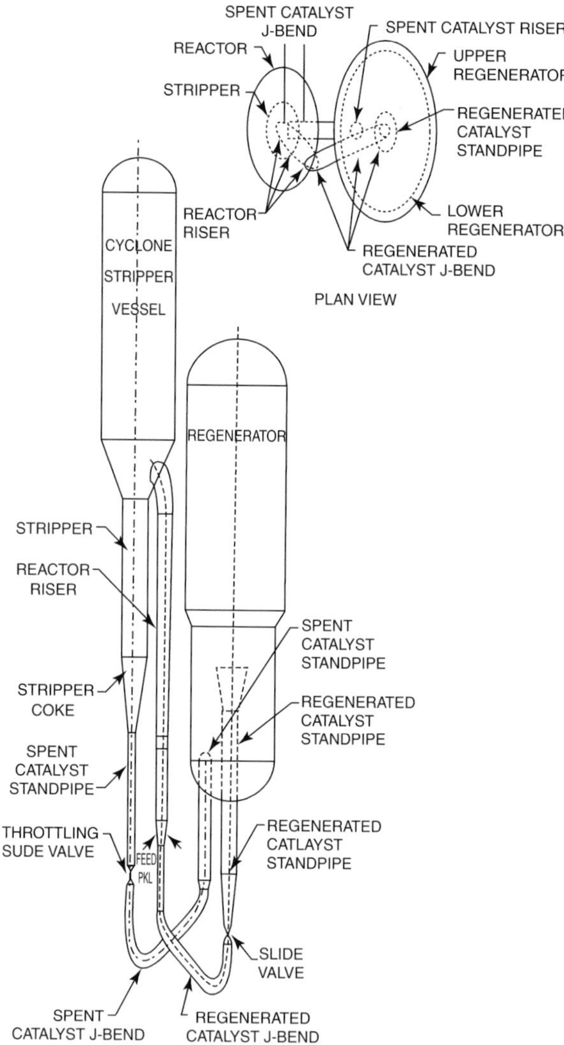

Figure 6.21. Exxon flexicracking IIIR unit.

The reason for splitting the regeneration is to produce CO rather than CO_2 so that the heat removal can be eliminated or reduced. Depending on the capacity of the FCC Unit, the heat removal by this technique can reach 100 million BTUs per hour. Another benefit is that the hydrogen in the coke burns faster than the carbon as shown in Figure 6.22. This hydrogen is the chief source of moisture in the regeneration process that has been shown to deactivate the catalyst. The carbon burn is typically adjusted so that 45–75% is accomplished in the first stage.

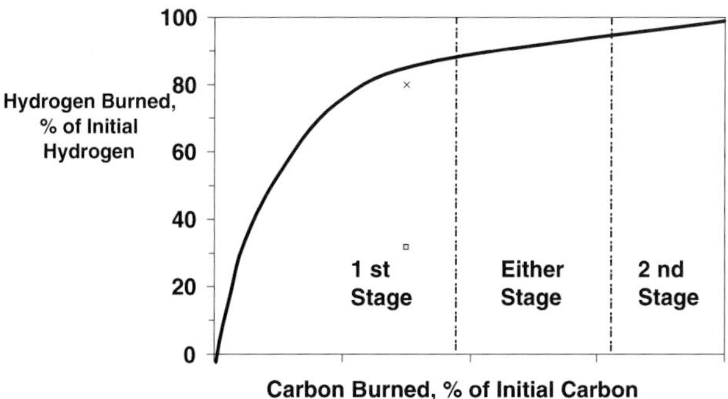

Figure 6.22. Carbon and hydrogen burning rates of coke.

If even more heat removal is required one or more catalyst coolers can be added to the top regenerator. The two basic designs used are shown in Figure 6.23.

Both are dense bed catalyst coolers. Dilute phase coolers were used in the past but frequent leaks made them too unreliable to use for commercial applications. One of the designs features a shell and tube exchanger while the other has tube clusters that have isolation valves. The latter is more expensive but allows isolation of the leaking cluster without shutting down the catalyst cooler. If steam leaks into the regenerator unabated, excessive catalyst deactivation occurs. The advantage of a catalyst cooler

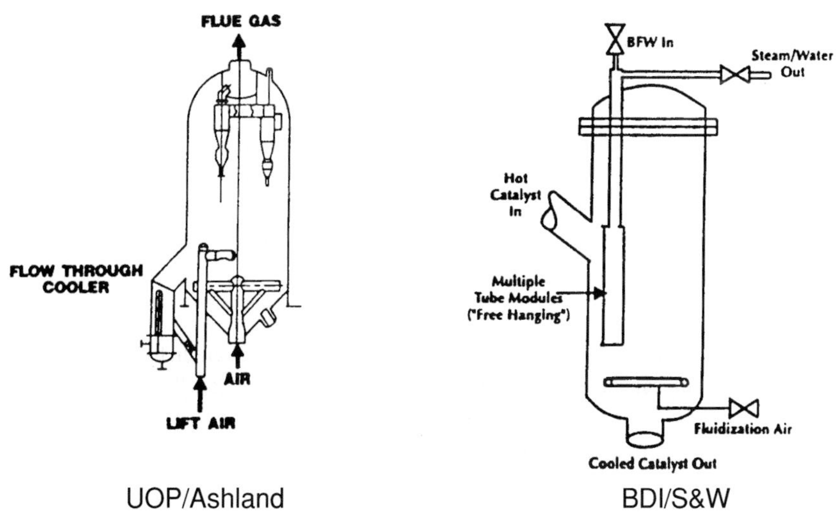

Figure 6.23. Commercial catalyst coolers.

is that its duty can be varied over virtually its entire heat load range. This allows the refiner to adjust the coke make to the feedstock and desired reaction severity.

Many early FCC units had steam coils but these must be run at full load to prevent a mechanical failure. All of these systems require a high water to steam ratio to ensure vaporization does not occur in the regenerator coils or tubes, which leads to hot spots and subsequent holes. Older heat removal systems were designed with coils and trim coolers, but this concept has been rejected in modern resid crackers.

The demand for residual fuels has steadily declined and it is unlikely that it will ever return. Consequently, residual catalytic cracking is a high growth area as shown in Figure 6.2.

Heavy fuel oil will not be sold without hydroprocessing in the future. Concern over acid rain and the severe desulfurization required for transportation fuels will focus attention to all the other fuels (off-road diesel, bunkers, etc.) and make desulfurization mandatory. Refiners will find feedstock preparation, while costly, will greatly improve overall yields and install hydroprocessing where the crude warrants. For high metals laden crude or those deficient in hydrogen, coking will be the preferred bottoms processing route followed by hydroprocessing of the coker gas oils or the FCC products.

Fluid cracking catalysts

The FCC process has been shaped and reshaped to accommodate the advances made in fluid cracking catalysts. Early catalysts were relatively inactive and amorphous in nature and required a lot of recycle of the uncracked feed to achieve the desired conversions. Carbon on regenerated catalyst was usually around 0.3–0.6 wt% and had little effect on unit performance. In the early 1960s zeolite containing catalysts were introduced that were much more active and selective than previous catalysts but required the removal of residual coke for optimum commercial performance. This allowed the refiner to substitute fresh feed for the large amounts of recycle being used and resulted in greatly expanded capacity and gasoline yields.

The preferred FCC zeolite is a crystalline silica-alumina compound that has the sodium removed. The Type Y or Ultrastable Y zeolite commonly employed has a faujasite structure and as produced formula as shown in Figure 6.24. The important properties of these zeolites that make them suitable for use in fluid cracking catalysts are:

- High stability ($>1{,}600^\circ$F) to heat and steam
- Three-dimensional structure
- High activity (acidity)
- Large pores (7.5 Å)

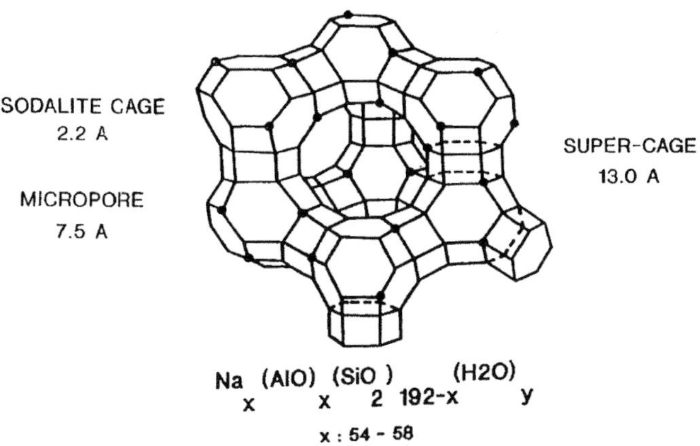

Figure 6.24. Fautasite zeolite structure.

Without all these characteristics the zeolite will not standup to the high temperatures in the regenerator or let in large molecules so that the interior of the crystal can be utilized. Rapid coking is mitigated by the three-dimensional structure. The acidity or activity of the zeolite is associated with the hydroxyl groups attached to the aluminum atoms in the crystal structure.

The type of zeolites contained in fluid cracking catalysts are variations of the basic faujasite or Type Y zeolite structure. These are made and characterized as illustrated in equation (8). The products are referred to as Ultrastable Y's (US–Y), Hydrogen Y's (H-Y), Calcined Rare Earth Y (CREY) or Rare Earth Ultrastable Y (RE–US–Y). Each of these has been used as the primary cracking component in commercial cracking catalysts. Variations of the structures and compositions of the above products are made by new methods of zeolite syntheses and secondary treatment. These include controlling the amount of alumina in the crystal structure and occluded in the zeolites pores, substituting other cat ions for alumina in the zeolite framework and using other cat ions for ion exchange to replace the sodium.

$$\begin{array}{lccccccc} & \text{Na-Y} & \xrightarrow{\text{Exc}} & \begin{array}{c}\text{NaHY}\\ \text{NaREY}\end{array} & \xrightarrow[\text{Exc.}]{\text{Cal}} & \begin{array}{c}\text{HY}\\ \text{CREY}\end{array} & \xrightarrow[\text{Stm}]{\text{Cal}} & \begin{array}{c}\text{US-Y}\\ \text{RE-US-Y}\end{array} \\ \text{Na}_2\text{O} & 13.0 & & 3.0 & & <1 & & <1 \end{array} \qquad (8)$$

where

Exc = Exchange (NH_3 or RE)
Cal = Calcine
Stm = Steam

Table 6.12. Effect of rare earth exchange of Y zeolite on activity and selectivity

Rare earth on zeolite	0	4	8	16
Activity (C/(100-C))	2.55	2.8	3.0	4.0
Conversion	71.8	73.7	75	80
Coke	2.55	3.5	4.8	6.1
C_3=/TC_3	0.81	0.75	0.72	0.65
C_4=/TC_4	0.39	0.32	0.29	0.23

In Table 6.12 the effect of rare earth exchange of the Type Y zeolite used in FCC catalysts is shown. Rare earths stabilize the zeolite and results in higher activity and more hydrogen transfer. Coke selectivity declines along with the olefinicity of the LPG and catalytic gasoline.

In commercial use, the zeolites undergo dealumination due to contacting with steam at elevated temperatures such as those encountered in the regenerator. This is shown in Figure 6.25 where the dealumination is measured by x-ray defraction to obtain the corresponding unit cell size of the zeolite crystal.

The commercial performance of the cracking catalyst depends on a number of factors, but the equilibrium unit cell size of the catalyst is a principle variable. As illustrated

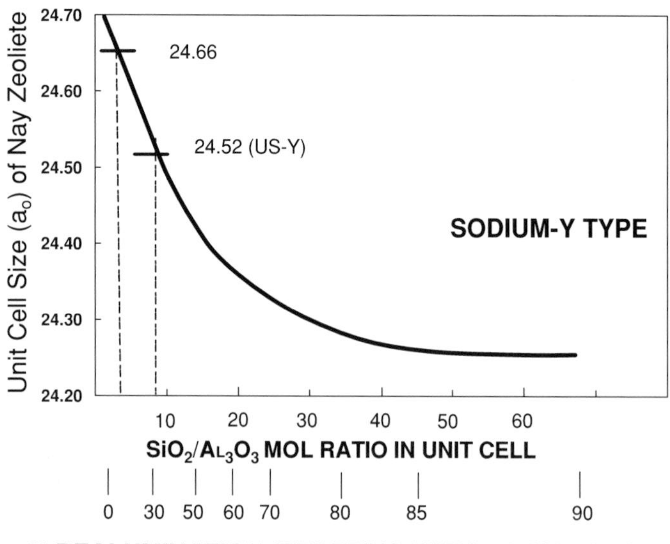

Figure 6.25. Unit cell size versus SiO_2/Al_2O_3 ratio for NaY zeolite.

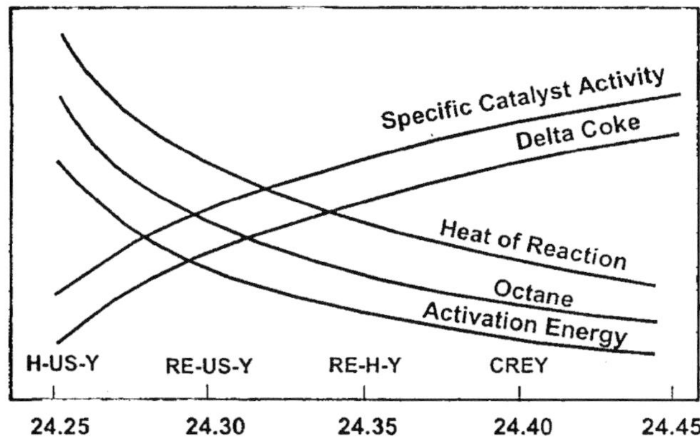

Figure 6.26. Effect of unit cell size on catalytic properties.

in Figure 6.26 many of the important properties of the catalyst are determined by this number. The zeolite types tend to equilibrate around the levels shown but it should be understood that both hydrogen and rare earth can be used for exchange on the same zeolite to give to a mixed result.

As the aluminum atoms are removed from the zeolite structure the activity goes down and much more zeolite needs to be used to give an equivalent conversion. Lower unit cell sizes increases C_3 and C_4 olefinicity, gasoline octane, and reduces coke formation.

The catalyst's activity is measured by a standard laboratory (MAT) test. The base catalyst without zeolite can have an activity as low as four MAT though numbers ranging from 20 to 45 MAT are more typical. Commercial FCC operations have activities that range from 58 to 77 MAT with 62 to 72 being the most common range.

The matrix is the rest of the catalyst and contains clay, additives and/or a binder that holds all the components together. Important properties such as the catalyst's attrition characteristics, density, CO burning rate, coke selectivity, bottoms cracking and dry gas make are a direct function of the matrix composition.

Clay is used as filler and provides some pore structure. Additives such as alumina and silica-alumina are used to increase the matrix cracking activity and crack large molecules. Binders can consist of silica, alumina or silica-alumina, and can also enhance activity. In the new catalysts, more than one additive can be incorporated into the catalyst and small amounts of secondary compounds such as titanium or phosphorous can modify the catalyst's performance.

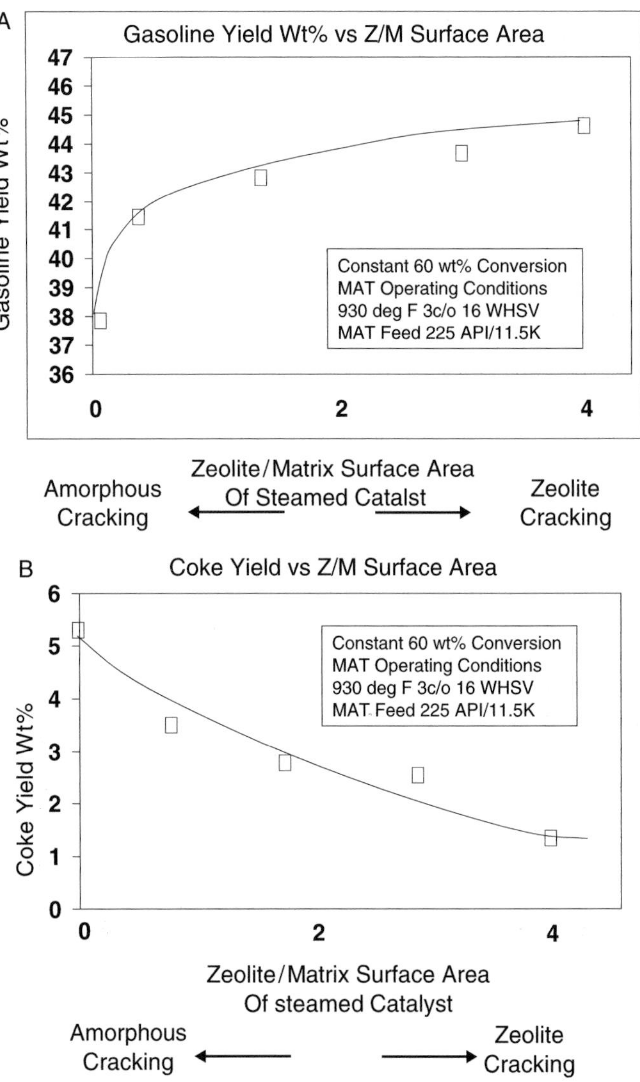

Figure 6.27. (A–C) Effect Zeolite/Matrix SA ratio on gasoline, coke, and 640+ bottoms yield.

The ratio of zeolite to matrix is used to vary the yields from an FCC and must be optimized for each matrix and zeolite system. Each feedstock with different properties such as molecular weight and aromaticity require a unique catalyst formulation. Figures 6.27A–C show how the ratio of zeolite to matrix affects yields for a 22.5° API (11.5 K) feed.

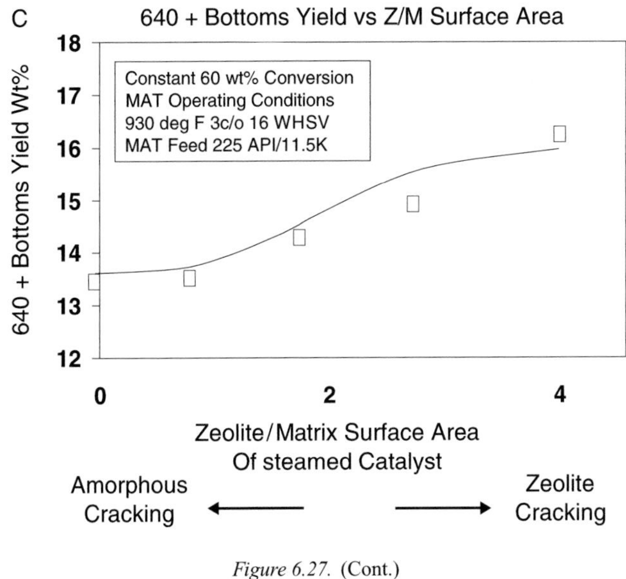

Figure 6.27. (Cont.)

In addition to FCC catalysts, there is now a long list of FCC additives to meet specific processing needs. These include products for:

- SO_x removal
- NO_x removal
- Octane enhancement
- Metals passivation
- Bottoms cracking
- Fluidization aids
- Olefin generation

A compilation of all the commercially available products is given in Appendix A.

Cracking for light olefins and aromatics

The demand for propylene as a petrochemical feedstock is outpacing the need for ethylene. As a result, the traditional source of propylene, i.e. as byproduct of the steam cracker, is not sufficient for future propylene projections.

Catalytic cracking is the other major source of the propylene with propane dehydrogenation making up the balance. Worldwide market shares (2001) are 68:28:4, though recently in the United States, catalytic cracking has overtaken steam cracking as the largest single source of propylene used for petrochemicals.

By changing the catalyst formulation, operating parameters, and equipment design, the product distributions shown in Table 6.3 are obtained. In the Deep Catalytic Cracking Process, the yields of light olefins (C_2–C_4) and BTX aromatics are much higher than in standard gas oil FCC units and can result in significantly improved processing economics. As shown in equation (2), this mode encourages the deliberate overcracking of gasoline.

Many international petrochemical producers are challenged with the need to obtain secure, low cost naphtha supplies. Integrating petrochemical operations with refineries can ameliorate this situation. To become and remain competitive it is essential to have access to low cost feedstocks and to build vertically integrated, strategically located world scale refinery and petrochemical complexes. Such integration adds value and provides raw materials, both keys to gaining a competitive edge in global markets. In the case of increased propylene demand, heavier, less expensive feedstocks, i.e. gas oils can be used in a continuous catalytic process run at more favorable operating conditions.

The product qualities of the naphtha from the Deep Catalytic Cracking Process are compared to a conventional FCC unit and a steam cracker in Table 6.13. The higher aromatics concentration in the naphtha is mostly a result of the concentration of existing aromatics due to the cracking of the paraffins and cycloparaffins rather than through dehydrogeneration of naphthenes or cyclization of paraffins. Xylenes are the major aromatic compounds produced catalytically while benzene is the primary thermal product. Diolefins are low in both of the catalytic processes since these result from the largely absent thermal reactions.

Table 6.13. Comparison of DCC, FCC, and SC naphthas

	DCC	FCC	SC
Components			
Paraffins	14.3	28.6	3.5
Olefins	32.4	35.3	13.3
Naphthenes	5.0	9.8	4.1
Aromatics	48.3	26.3	79.1
Aromatics breakdown			
Benzene	1.9	0.6	37.1
Toluene	9.4	2.4	18.9
C_8	15.6	6.7	13.5
C_9	12.1	12.5	5.4
$C_{10}+$	9.3	4.1	4.2
Total	48.3	26.3	79.1

The normal operating conditions of the cracking process are contrasted below:

Typical operating conditions

	FCC	DCC	SC
Reactor Temp. (°F)	950–1,020	980–1,100	1,400–1,600
Residence Time (Sec)	1–10	1–10	0.1–0.2
Pressure, ATM	1–2	1–2	1
Catalyst/oil	5–10	8–15	–
Steam (% fd)	0–5	10–30	30–80

A flow diagram of the DCC process is shown in Figure 6.28. The catalyst flow and oil vapors and products follow the same path as the gas oil FCC unit described earlier. Considerably more steam is added and the reactor can be run with a bed level to increase the hydrocarbon residence time and facilitate the overcracking reactions. Other options include recycling naphtha to the same feed riser or the installation of a separate smaller riser to run at higher temperature.

Recracking of naphtha can optionally include the C_4 olefins. Further, the recycled naphtha can be full range or cut to maximize the crackable molecules. The result of the recycle is to increase propylene and the concentration of BTX in the naphtha. The product slates before and after naphtha recycle are shown in Table 6.14 from the Jinan DCC unit.

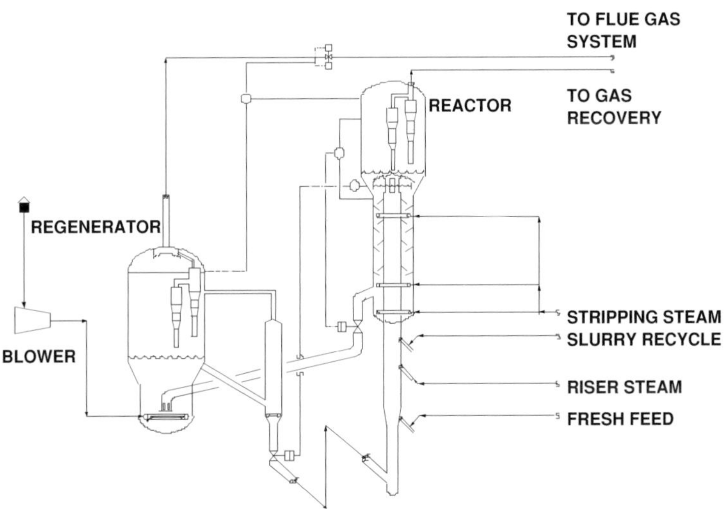

Figure 6.28. DCC process.

Table 6.14. DCC operation with naphtha recycle

Naphtha recycle	No	Yes
Products (wt%)		
Dry gas	6.8	7.5
LPG	39.2	43.9
Naphtha	27.8	20.9
LCO	16.8	18.2
Coke	8.9	9.0
Loss	0.5	0.5
Olefin yield		
Propylene	17.4	21.0
Butylene	12.4	13.2

A pilot plant test showed that when DCC naphtha with a 150°C end point was recycled the aromatics concentration increased dramatically. The total aromatics went from 42 to 81 wt% of the naphtha with the BTX content being 4.4, 28.9, and 40.4 wt%, respectively.

Recent advances in the technology have been directed to expanding the feedstock types that can be processed. Successfully run feeds now include:

- VGO
- Hydrotreated VGO
- Deasphalted Oil
- Dewaxing wax
- Coker gas oil
- Atmospheric resid

Catalyst formulations have had to be altered to match the changing feedstock slate. The DCC catalysts are designed with the following characteristics: high matrix activity for primary cracking of the heavy hydrocarbons and higher metals tolerance; a large quantity of a modified mesopore zeolite with a pentasil structure for cracking the primary product (gasoline), isomerization activity for the light olefins and minimization of hydrogen transfer. Ten different catalysts have been used commercially in the seven operating DCC installations.

Representative yields from various commercial operations are tabulated in Table 6.15. Paraffinic feedstocks give high propylene yields and the isobutylene concentration in the C_4 stream is near the thermodynamic limit. Atmospheric resid makes an excellent feed as long as it is paraffinic. The results from commercial runs on Daqing atmospheric resid are given in Table 6.16 along with the feedstock analysis. The combination of an ARDS unit in front of a DCC unit would allow a refinery to process

Table 6.15. DCC light olefin yields

Refinery	Daqing	Anqing	TPI	Jinan	Jinan
Operating mode	DCC + I	DCC-I	DCC-I	DCC-I	DCC-II
Feedstock	Paraffinic VGO + AR	Naphthenic VGO	Arabian VGO + DAO + Wax	Intermediate VGO + DAO	Base
Reaction temp.	545	550	565	564	530
Olefins					
Ethylene	3.7	3.5	5.3	5.3	1.8
Propylene	23.0	18.6	18.5	19.2	14.4
Butylene	17.3	13.8	13.3	13.2	11.4

a wide variety of crudes and still produce high quantities of petrochemical base stocks.

Other technologies designed to increase the propylene yield from heavy feeds have been announced. Two such processes are the PetroFCC and Maxofin. Neither of the processes has been commercialized to date but have been discussed in various papers.

The PetroFCC shown in Figure 6.29 has a recycle stream of spent catalyst from the stripper to the base of the feed riser. The principle behind this design, as Table 6.17 shows, is that spent catalyst is less active but somewhat more selective than the clean regenerated catalyst at constant conversion.

This makes recycling spent catalyst an option which increases the effective catalyst/oil ratio and reduces the initial regenerated catalyst activity.

The PetroFCC utilizes many of the same technologies that are used in the standard UOP FCC units. This includes both the optimix nozzles and vortex separator system.

Table 6.16. DCC yields for atmospheric resid

Daqing ATB		Yields (wt%)	
Density	0.9012	Dry Gas	11.7
CCR (wt%)	4.7	$C_3 + C_4$	48.3
Hydrogen (wt%)	12.84	C_5 + Naphtha	18.9
Saturates	55.5	LCO	12.1
Aromatics	28.0	Coke	8.0
Resins	15.7	Loss	1.0
Asphaltenes	0.8	Olefins	
Nickel (ppm)	6	Ethylene	6.85
		Propylene	24.83
		Butylene	15.27

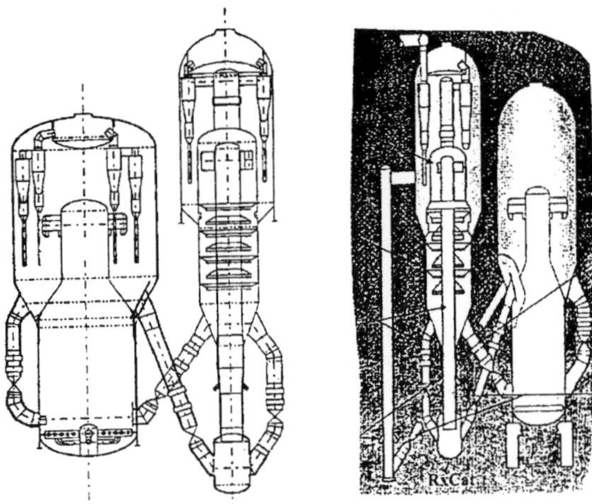

Figure 6.29. PetroFCC process.

A second riser for recracking the naphtha is provided that operates at higher temperature.

To enhance light olefin production, low hydrocarbon partial pressures (10–30 psia), elevated temperatures (1,000–1,150°F) and 10–25 wt% of shape selective (pentasil zeolites) additives are employed. Maximum olefin yields require high conversion since it is the gasoline produced in the cracking process that ultimately is made into propylene and butylene.

Cited yields for the PetroFCC are (Table 6.18).

While the feedstock properties and operating conditions are not known, the feedstock had to be very paraffinic to yield such a high gasoline yield in the regular FCC processing mode.

Table 6.17. Selectivity of spent catalyst

	Regenerated	Spent
Initial carbon	0.0	0.91
MAT conversion	64.5	55.0
Yields @ 70 wt% Conversion		
C_2 Minus	2.0	2.1
Gasoline	49.3	50.0
Coke	2.9	1.6

Table 6.18. Pilot yields–PetroFCC (wt%) versus FCC

	PetroFCC	FCC
Propylene	22	4.7
Butylene	14	~6
Ethylene	6	~1
Gasoline	28	53.5

Another process designed for high olefin yields is the Maxofin FCC. This unit is shown in Figure 6.30 and has a second riser for propylene maximization. The process uses the Orthoflow FCC Hardware, e.g. atomax nozzles and close-connected cyclones and the stacked reactor-regenerator configuration. A special ZSM-5 additive is used along with the FCC catalyst.

Pilot yields from the Maxofin process are shown in Table 6.19. The data are shown for a paraffinic gas oil operating in three different modes.

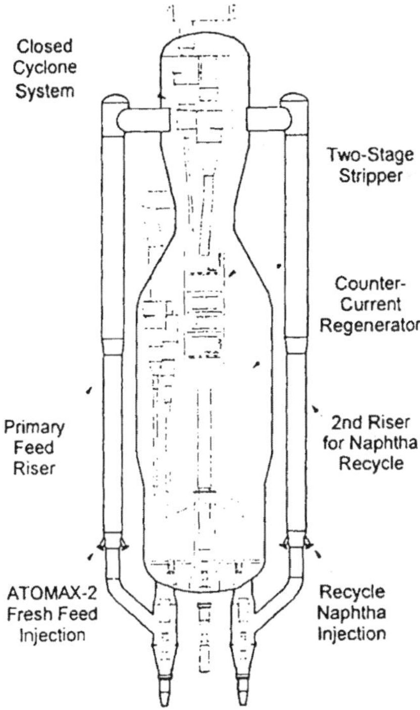

Figure 6.30. Maxofin process.

Table 6.19. Yields from Maxofin unit

	Max C_3 =	Intermediate	Max fuels
Recycle	Yes	No	No
ZSM-5	Yes	Yes	No
Riser temp., °C	538/593	538	538
Yields (wt%)			
C_2 Minus	7.6	2.3	2.2
Ethylene	4.3	2.0	0.9
Propylene	18.4	14.4	6.2
Butylene	12.9	12.3	7.3
Gasoline	18.8	35.5	49.8
Coke	8.3	6.4	5.9
Conversion	86.4	87.7	85.4

As the need for light olefins and LPG increases, these processes will assume even more attention.

Catalytic cracking is well positioned as we enter a new decade of petroleum refining. The continued demand for gasoline, the need to reduce the residual portion of the crude barrel and an ever increasing thirst for petrochemical base stocks will insure a premier position for the fluid catalytic cracking unit in the refining and petrochemical industry.

Nomenclature

A = activation Energy for coke burning
A, A_O = equilibrium and initial catalyst inventories
BP = boiling point
C, C_i = carbon on catalyst and initial carbon on catalyst (wt%)
CO = carbon monoxide
DO = decant oil
E = activation energy for thermal cracking
e = natural exponent
K = rate constant
K_D = deactivation constant
LCO = light cycle oil
L_mO_2 = log mean oxygen concentration
LPG = liquefied petroleum gas
P_B, P_T = pressure at bottom and top of standpipe (psi)
ΔP = pressure difference
Q = aeration requirements for standpipe (scfm)
R = gas constant
S = daily fractional catalyst addition rate (ton/ton)

T = temperature
t = time (seconds)
ρ_{mf}, ρ_t = catalyst densities at minimum fluidization and top of standpipe (lb/ft^3)
ρ_p = skeletal density of catalyst (lb/ft^3)

References

1. A. A. Avidan, M. Edwards, and H. Owen, Innovative improvements highlight FCC's past and future, Oil and Gas Journal, 1/8/90, 33–58.
2. M. F. Raterman, US Patent 5,409,872.
3. A. G. Shaffer, Jr., and C. L. Hemler, Seven years of operation prove RCC capability, *Oil and Gas Journal*, 1/28/90, 62–70.
4. R. Miller, FCC's role in refinery-petrochemical integration, *Grace Refining Technology Conference*, Singapore, 9/18-20/02.
5. A. Fu, D. Hunt, J. A. Bonilla, and A. Batachari, Deep catalytic cracking plant produces propylene in Thailand, *Oil and Gas Journal*, 1/12/98, 49–53.
6. S. Benton, Advanced Technology for increasing LPG and propylene production, *2nd Bottom of the Barrel Technology Conference (BBTC 2002)*, 10/8–9/02, Istanbul, Turkey.
7. I. B. Cetinkaya, UOP PetroFCC process, *Grace Refining Seminary*, Singapore, 9/18-20/02.
8. A. S. Krishna, C. R. Hsieh, A. R. English, T. A. Pecoraro, and C. W. Cuehler, Additives improve FCC process, *Hydrocarbon Processing*, November 91, 59–66.
9. W. Shi, C. Xie, Y. Huo, and X. Zhong, DCC family technology for producing light olefins from heavy oils, *Chinese Petroleum Processing and Petrochemical Technology*, No. 2, 6/01, 15–21.
10. R. D'Aquino, Refiners get cracking on petrochemicals, *Chemical Engineering*, May 99, 30–33.
11. J. L. Mauleon and J. C. Courcelle, FCC Heat balance critical for heavy fuels, *Oil and Gas Journal*, 10/21/85, 64–70.
12. T. E. Johnson, Improve regenerator heat removal, *Hydrocarbon Processing*, 11/91, 55–57.
13. L. C. Yen, R. E. Wrench, and A. S. Ong, Reaction kinetic correlation equation predicts fluid catalytic cracking coke yield, *Oil and Gas Journal*, 1/11/88.
14. L. Chapin, W. Letzsch, and D. Dharia, Deep catalytic cracking for petrochemical and refining application, *Proceedings Petrotech—95*, India.
15. Z. T. Li, W. Y. Shi, R. N. Pan, and F. K. Jiang, DCC flexibility for isoolefins production, *American Chemical Society 206th National Meeting*, Chicago, 8/22-27/93.
16. Y. Gao, C. Xie, and Z. Li, DCC Update and its commercial experiences, *5th Stone and Webster/Axens FCC Forum*, May 2002.
17. R. E. Ritter, J. C. Creighton, D. S. Chin, T. G. Roberie, and C. C. Wear, Catalytic octane from the FCC, Grace Catalagram, No. 74, 1986.
18. *Advanced Control and Information Systems 2001*, Hydrocarbon Processing, 102.
19. R. J. Glendinning, H. L. MCQuiston, and T. Y. Chan, Implement new advances in FCC process technology, *Fuel Reformulation*, 3/4-95, 45–53.
20. J. S. Magee and M. M. Mitchell, Jr., *Fluid Catalytic Cracking Science and Technology*, Elsevier, Amsterdam, 1993.
21. J. W. Wilson, *Fluid Catalytic Cracking Technology and Operation*, Pennwell, Tulsa, 1997.

22. P. B. Venuto and E. T. Habib, Jr., *Fluid Catalytic Cracking with Zeolite Catalysts*, Marcel Dekker Boca Raton, FL, 1979.
23. C. Leckenbach, A. C. Worley, A. D. Reichle, and E. M. Gladrow, Cracking-catalytic, *Encyclopedia of Chemical Processing and Design*, Marcel Dekker, New York, Vol 13, 1–133.
24. R. A. Meyers, *Handbook of Petroleum Refining Processes*, 2nd ed., Chapters 3 thru 7, McGraw-Hill, New York.
25. J. A. Rabo, Zeolite chemistry and catalysis, *ACS Monograph 171*, 1979, Chapter 11, J. S. Magee and J. J. Blazek, Preparation and Performance of Zeolite Cracking Catalysts
26. D. Decroocq, *Catalytic Cracking of Heavy Petroleum Fractions*, Editions Technip, 1984, IFP.
27. W. S. Letzsch, Fluid Catalytic Cracking in the New Millennium, *NPRA Annual Meeting*, Paper AM-99-15.
28. E. J. Demmel and H. Owen, US Patent 3,791,962.
29. S. V. Anderson, Improved FCCU feed and catalyst contact, *Petroleum Technology Quarterly*, Spring 1999, 55–59.
30. J. W. Wilson, First FCC Licenses Forum, May 1994 Paper, KOA, Osaka Revamp.
31. J. Haruch, US Patent 5,673,859.
32. C. Y. Sabottke, European Patent 0444860A1.
33. B. W. Hewrick, J. P. Koebel, and I. B. Cetinkaya, Improved catalyst stripping from cold flow modeling, *Petroleum Technology Quarterly*, Autumn 2001, 87–95.
34. American Chemical Society, *The Fluid Bed Reactor*, November 3, 1998.
35. L. R. Anderson, H. S. Kim, T. G. Park, H. J. Ryu, and S. J. Jung, Operations adjustments can better catalyst cooler operations, *Oil and Gas Journal*, 4/19/99, 53–56.
36. J. R. Murphy, Evolutionary design changes mark FCC process, *Oil and Gas Journal*, 5/18/92, 49–58.
37. K. V. Krikorian and J. C. Brice, FCC's effect on refinery yields, *Hydrocarbon Processing*, September 1987, 63–66.
38. S. J. McCarthy, M. F. Raterman, C. G. Smalley, J. F. Sodomin, and R. B. Miller, FCC technology upgrades: A commercial example, *NPRA Annual Meeting*, Paper AM-97-10.
39. M. J. P. C. Nieskens, F. H. H. Khouw, M. J. H. Barley, and K. H. W. Roebschlaeger, Shell's resid FCC technology reflects evolutionary development, *Oil and Gas Journal*, 6/11/90, 37–44.
40. R. J. Glendinning, H. L. McQuiston, and T. Y. Chan, Paper entitles New Developments in FCC Process Technology.
41. D. L. Johnson, FCC Catalyst Stripper, International Patent WO96/04353.
42. R. E. Wrench and P. C. Glasgow, FCC Hardware options for the modern Cat cracker, *AIChE National Meeting*, Los Angeles, CA, November 17–22, 1991, Paper 125C.
43. P. E. Glasgow and A. A. Murcia, Process and mechanical design considerations for FCC regeneration air distributors, *Katalistiks 5th FCC Symposium*, Vienna, Austria, May 1984.
44. A. K. Rhodes, Number of catalyst formulations stable in a tough market, *Oil and Gas Journal*, 10/6/97, 41–72.
45. J. Stell, Worldwide catalyst report, *Oil and Gas Journal*, 10/8/01, 56–76.
46. P. K. Niccum, et. al., Maxofin: A novel FCC process for maximizing light olefins using a new generation of ZSM-5 additive, *NPRA Annual Meeting*, Paper AM-98-18.

47. C. A. Cabrera, Recent innovations UOP RCC/FCC technology, Paper Katalistiks Technology Seminar, New Orleans, 12/15/88.
48. R. J. Campagna and A. S. Krishna, Advances in resid cracking technology, *Katalisticks 5th Annual FCC Symposium*, 5/23-24/84.
49. C. L. Hemler, D. A. Lomas, and D. G. Tajbi, FCCU reflects technological response to resid cracking, *Oil and Gas Journal*, 5/28/84, 79–86.
50. W. L. Pierce, D. F. Ryan, R. P. Souther, and T. G. Kaufmann, Innovations in flexicracking, *API Div. of Refining 37th Midyear Meeting*, New York, 5/10/72.
51. W. Letzsch, J.-L. Mauleon, G. Jones, and R. Dean, Advanced residual fluid catalytic cracking, *Katalistiks 4th Annual FCC Symposium*, Amsterdam, 5/83.
52. W. S. Letzsch and P. A. Minton, FCC revamps, *Hydrocarbon Engineering*, 1/2000, 32–35.
53. J. W. Wilson, Modernizing older FCCU's, *NPRA Annual Meeting*, Paper AM-00-09.
54. M. G. Bienstock, D. C. Draemel, P. K. Ladiwig, R. D. Patel, and P. H. Maher, A history of FCC process improvement through technology development and application, *AIChE Spring Meeting*, Houston, 3/28-4/1/93.
55. R. Dean, J.-L. Mauleon, and W. Letzsch, Resid Puts FCC Process in new perspective, *Oil and Gas Journal*, 10/4/82.
56. R. Dean, J.-L. Mauleon, and W. Letzsch, Total introduces new FCC process, *Oil and Gas Journal*, 10/11/82.
57. O. J. Zandona, W. P. Hettinger, Jr., and L. E. Busch, Reduced crude processing with Ashland's RCC process, *API 47th Midyear Refining Meeting*, New York, 5/13/82.
58. Additives play important role in FCC development, *Oil and Gas Journal*, 9/23/91, 50–52.
59. M. W. Schnaith and D. A. Kauff, Resid FCC Regenerators: Technology options and experience, *NPRA Annual Meeting*, Paper AM-97-13.
60. L. L. Upson and H. V. D. Zwan, Promoted combustion improves FCCU flexibility, *Oil and Gas Journal*, 11/23/87, 65–70.
61. W. S. Letzsch, D. J. Dharia, W. H. Wallendorf, and J. L. Ross, FCC modifications and their impact on yields and economics, *NPRA Annual Meeting*, Paper AM-96-44.
62. A. A. Avidan, F. J. Krambeck, H. Owen, and P. H. Schipper, FCC closed-cyclone system eliminates post riser cracking, *Oil and Gas Journal*, 3/26/90, 56–62.
63. M. W. Schnaith, A. T. Gilbert, D. A. Lomas, and D. N. Myers, Advances in FCC reactor technology, *NPRA Annual Meeting*, Paper AM-95-36.
64. B. Dahlstrom, K. Ham, M. Becker, T. Hum, L. Lacijan, and T. Lorsbach, FCC reactor revamp project execution and benefits, *NPRA Annual Meeting*, Paper AM-96-28.
65. A. M. Squires, Circulating fluidized bed technology, *The Study of Fluid Catalytic Cracking: The First Circulating Fluid Bed.*
66. A. V. Sapre, P. H. Schipper, and F. P. Petrocelli, Design methods for FCC feed atomization, *AIChE Symposium Series*, No. 291, Vol 88, 103–109.
67. L. E. Chapin, W. S. Letzsch, and T. E. Swaty, Petrochemical options from deep catalytic cracking and the FCCU, *Harts Fuel Technology and Management*, 4/98, 30–33.
68. T. E. Johnson and R. K. Miller, New developments in resid FCC technology, Paper at the Institute for International Research, 5/9-10/94, Singapore.
69. F. H. H. Khouw, M. J. P. C. Nieskens, M. J. H. Borley, and K. H. W. Roebschlaeger, The shell residue fluid catalytic cracking process commercial experience and future developments, *NPRA Annual Meeting*, Paper AM-90-42.

70. W. S. Letzsch, D. Dharia, and J. L. Ross, The future of catalytic cracking, *NPRA Annual Meeting*, Paper AM-97-65.
71. P. R. Satbhai, J. M. H. Dirkx, R. J. Higgins, and P. D. L. Mercera, Best practices in shell FCC units, *Akzo Nobel Catalyst Seminar*, 10/98, Mumbai, India.
72. J. L. Mauleon and J. B. Sigaud, Mix temperature control enhances FCC flexibility in use of wider range of feeds, *Oil and Gas Journal*, 2/23/87, 52–55.
73. C. R. Marcilly and R. R. Bonifay, Catalytic cracking of resid feedstocks, *The Arabian Journal for Science and Technology*, Vol 21, No. 48, 12/96.
74. J. R. Murphy, Designs for heat removal in HOC operations, *Petroleum Refining Conference*, JPI, Tokyo, 10/86.
75. I. B. Cetinkaya, Plug flow vented riser, US Patent 5,449,497.
76. R. Miller, Y.-L. Yang, E. Gbordzoe, D. L. Johnson, and T. Mallo, New Developments in FCC feed Injection and stripping technologies, *NPRA Annual Meeting*, Paper AM-00-08.
77. Exxon Research and Engineering Company, Flexicracking IIIR State-of-the-Art Cat Cracking Commercial Brochure.
78. C. C. Wear and R. W. Mott, FCC catalysts can be designed and selected for optimum performance, *Oil and Gas Journal*, 7/25/88, 71–79.
79. E. G. Wollaston, W. J. Haflin, W. D. Ford, and G. J. D'Souza, FCC model valuable operating tool, *Oil and Gas Journal*, 9/22/75, 87–94.
80. I. A. Vasalos, E. R. Strong, C. K. R. Hsieh, and G. J. D'Souza, New cracking process controls FCCU SOX, *Oil and Gas Journal*, 6/27/77, 141–148.
81. Complete combustion of CO in cracking process, *Chemical Engineering*, 11/24/75.
82. P. H. Barnes, Tutorial: Basic process principles of residue cat-cracking, *AIChE, 1998 Spring National Meeting*, New Orleans, 3/8-12/98.

Appendix 6.1. Commercially available FCC catalysts and additives

Note: All catalysts are micro spheres with densities from 0.70 to 0.90 gm/cc Matrix binders are silica or aluminum sols or silica-alumina. Clays and other aluminas are added.

Feedstocks		Products	
GO	Gas oils	MDG	Minimum dry gas
R	Resids	MLPG	Maximum LPG
HN	High nitrogen	MG	Maximum gasoline
HNR	High nickel resid	MLCO	Maximum LCO
HR	Heavy resid	MBC	Maximum bottoms cracking
A	All	MO	Maximum octane
		MC	Minimum coke
		MLY	Maximum liquid yields
		MOB	Maximum octane—barrels
		MIO	Maximum Iso-Olefins

Akzo-Nobel

Applications

Catalyst name	Feedstocks	Products	Product features
Access	HN, R	MLY, MBC, MDG	Metals tol., high stability
Advance	GO, R	MG, MDG, MC	Metals tol., high act with octane
Aztec	ALL	MLCO, MDG	Middle distillate maximization
Centurion	R	MC	Max metals tol. high stability
Centurion—max	HNR	MC	Max Ni tol., low ΔC high stability
Cobra	ALL	MOB, MDG, MBC	Metals tol—max. prod. value in OB mode
Conquest	ALL, R, HN	MG, MBC, MDG	Metals tol. max conv + stability
Conquest HD	ALL, R, HN	MG, MBC, MDG	Metals tol., reduce cat losses, opacity
Eclipse	ALL, R, HN	MIO, MO, MBC, MDG	Maximum isomerization
Foc	HR	MOB	Max throughput with resids
Horizon	HN, R	MOB, MBC	High conv with resids
Vision	HR	MOB, MO	Increase motor octane

Catalysts and Chemicals Industries Co. Ltd.

Applications

Catalyst name	Feedstocks	Products	Product features
HMR	GO, R	MOB, MBC	High liq. yield with minimum dry gas and coke
BLC	GO, R	MOB, MBC	High liq. yield with minimum dry gas and coke
STW	R	MOB, MBC, MDG, MC	Metals tol. and high stability
DCT	GO, R	MOB, MBC, MDG	Metals tol. and high stability
ACZ	HR	MOB, MBC	Metals tol., high stability

Davison, W. R. Grace & CO.

Applications

Catalyst name	Feedstocks	Products	Product features
Atlas	A	MBC	High stability and activity
Aurora	A	MBC	High stability and activity
GDS	A	MBC	Selective bottoms cracking
Orion	A	MDG, MC	Low coke, high activity
Polaris	GO, R	MBC	Maximum bottoms conv.
Spectra	A	MDG, MC	High activity, sel. botoms conversion
XPD	GO	MO, MBC	Max. olefins, high activity
Vanguard	GO, R	MDG, MC	High activity

Engelhard Corp.

Applications

Catalyst name	Feedstocks	Products	Product features
Maxol	A	MOB, MO	Maximum C_4 olefins
Millenium	GO, R	MOB, MDG, MC	Metals tol. low coke and dry gas
Naphtha max	GO, R	MG, MBC, MC	Selective matrix
Naphtha max—LSG	GO	MG, MBC	Reduce gasolines
Petromax	GO, R	MOB, MBC	High zeolite/matrix ratio
Precision	A	MOB	High activity
Reduxion	A	MO, MBC, MC	Matrix acidities controlled
Syntec	GO	MC	Low coke
Ultrium	R	MOB	Metals tol., low coke
Vektor	R	MOB, MBC	Max conv

Instituto Mexicano del Petróleo

Applications

Catalyst name	Feedstocks	Products	Product features
IMP-FCC-05	GO	MOB	Gasoline selectivity
IMP-FCC-06	GO	MOB	
IMP-FCC-10	GO	MOB, MLPG	High octane, olefins
IMP-FCC-11	GO, R	MOB	
IMP-FCC-12	GO, R	MOB	Metals tol.
IMP-FCC-14	GO	MG	Gasoline selectivity
IMP-FCC-16	GO, R	MG	Gasoline selectivity
IMP-FCC-51	GO, R	MOB	Metals tol.
IMP-FCC-53	R	MG	
IMP-FCC-54	GO, R	MOB	
IMP-FCC-55	GO	MO	Olefin selectivity

Flue gas SO_x reduction

Akzo	KDSOX	
CCIC	PLUS 1	
Davison	DESOX	
Englehard	SoxCAT	Mg based
IMP	IMP-RESOX-01	
Intercat	Soxgetter	Hydrotalcite Tech.

Flue gas NO_x reduction

Akzo	Kdnox
Grace	Denox

Gasoline sulfur reduction		
Akzo	Resolve 700, 750, 800	
Davison	D-Prism	
	RFG	
	Surea	
Intercat	LGS-150	
Metals passivation		
IMP	IMP-ADF-01	
Ondeo-nalco	MVP	
	Nickel passivation plus	Antimony
Fluidization aid		
Akzo	Smoothflow	
Fcc additives		
Bottoms cracking		Comments
Akzo	BCMT-100	
	BCMT-200	V trap
	BCMT-500	Fe, V tolerance
Intercat	BCA-105	
Octane (Pentasil containing products)		
Akzo	K1000, 2000	
CCIC	OCTUP-3	
Davison	Olefins plus	Moderate sieve content
	Olefins extra	High sieve content
	Olefins max	Highest sieve content
	Olefins ultra	
	Butmax	Max. butylene
Engelhard	Z-2000	
	Z-2000 super	
	Z-2000 super pro	IMP-10-02
	IMP	
	IMP-10-03	
	IMP-10-04	With rare earth
Intercat	Isocat	Moderate LPG
	LGS-150	Reduce gaso. sulfur
	Octamax	Low LPG
	Pentacat	High LPG
	Pentacat plus	Max propylene
	Pentacat-HP	Max propylene
	SoCat-HP	High LPG

	Z-Cat-HP	Max propylene
	ZMS-B-HP	Moderate LPG
	ZMX-C-HP	Low LPG
CO promoters		
Azko	KOC	Pt containing
	KNO_x-DOWN	Non Pt < NOX
Amber	CCA-1	Pt–Pd High Act
	CCA-1000	High act
	CCA-350	Low act
	CCA-500	Medium act
	CCA-8	Pt–Pd
	CCA-850	Pt
CCIC	SP-IOS-60	Pt
Davison	CP-3, CP-5, CP-A	Varying activities
	XNox	
Engelhard	Procat plus	Pt
	Procat 500, 700, 900	Pt varying activities
IMP	IMP-PC-500	Pt
Intercat	COP 375, 550, 850	Pt varying activities
UOP	Unicat C1-3	Pt

Chapter 7

Distillate hydrocracking

Adrian Gruia (retired)*

Hydrocracking is a versatile catalytic refining process that upgrades petroleum feedstocks by adding hydrogen, removing impurities, and cracking to a desired boiling range. Hydrocracking requires the conversion of a variety of types of molecules and is characterized by the fact that the products are of significantly lower molecular weight than the feed. Hydrocracking feeds can range from heavy vacuum gas oils and coker gas oils to atmospheric gas oils. Products usually range from heavy diesel to light naphtha. Hydrocrackers are designed for and run at a variety of conditions depending on many factors such as type of feed, desired cycle length, expected product slate but in general they will operate at the following range of conditions: liquid hourly space velocity (LHSV)—0.5 to 2.0, H_2 circulation—5,000 to 10,000 SCFB (850–1,700 Nm^3/m^3), H_{2PP}—1,500 to 2,000 psia (103–138 bars), and SOR temperatures ranging between 675°F and 725°F (357–385°C). Hydrocracking is particularly well suited to generating products that meet or exceed all of the present tough environmental regulations.

Brief history

While the first commercial installation of a unit employing the type of technology in use today was started up in Chevron's Richmond CA refinery in 1960, hydrocracking is one of the oldest hydrocarbon conversion processes. Hydrocracking technology for coal conversion was developed in Germany as early as 1915 designed to secure a supply of liquid fuels derived from domestic deposits of coal. The first plant for hydrogenation of brown coal was put on stream in Leuna Germany in 1927, applying what may be considered the first commercial hydrocracking process. Conversion of coal to liquid fuels was a catalytic process operating at high pressures, 3000–10,000 psig (207–690 bar) and high temperatures, 700–1000°F (371–538°C). Other efforts were undertaken subsequently to develop hydrocracking technology designed

*UOP LLC.

to convert heavy gas oils to lighter fuels. The emergent availability of Middle Eastern crude after World War II removed the incentive to convert coal to liquid fuels, so continuing the development of hydrocracking technology became less important.

In the mid-1950s the automobile industry started the manufacture of high-performance cars with high-compression ratio engines that required high-octane gasoline. Thus catalytic cracking expanded rapidly and generated, in addition to gasoline, large quantities of refractory cycle stock that was material that was difficult to convert to gasoline and lighter products. This need to convert refractory stock to quality gasoline was filled by hydrocracking. Furthermore, the switch of railroads from steam to diesel engines after World War II and the introduction of commercial jet aircraft in the late 1950s increased the demand for diesel fuel and jet fuel. The flexibility of the newly developed hydrocracking processes made possible the production of such fuels from heavier feedstocks. The early hydrocrackers used amorphous silica alumina catalysts. The rapid growth of hydrocracking in the 1960s was accompanied by the development of new, zeolite based hydrocracking catalysts. They showed a significant improvement in certain performance characteristics as compared with amorphous catalysts: higher activity, better ammonia tolerance, and higher gasoline selectivity. While hydrocracking was used in the United States primarily in the production of high-octane gasoline, it grew in other parts of the world, starting in the 1970s primarily for the production of middle distillates. The amorphous catalysts remained the catalysts of choice for this application, though some 'flexible' catalysts were developed that made it possible to maximize the yield of different products by using the same catalyst but changing the operating conditions. As of the beginning of 2001, there were more than 150 hydrocrackers operating in the world with a total capacity in excess of 3,800,000 B/D (500,000 MT/D).

Flow schemes

Various licensors have slightly different names for their hydrocracker flow schemes, but in general, they can be grouped into major two categories: single stage and two stage. Table 7.1 shows the general evolution of flows schemes, generally driven by improvements in catalysts.

Single stage once-through hydrocracking

Figure 7.1 shows a schematic of a single stage, once-through hydrocracking unit, which is the simplest configuration for a hydrocracker. It is a variation of the single stage hydrocracking with recycle configuration (described in "Two Stage Recycle Hydrocracking"). The feed mixes with hydrogen and goes to the reactor. The effluent goes to fractionation, with the unconverted material being taken out of the unit as unconverted material. This type of unit is the lowest cost hydrocracking unit, can process

Table 7.1. Evolution of hydrocracker flow schemes

Date	Process scheme	Reason
Early 1960s	Separate hydrotreating	Low activity amorphous catalysts
Mid 1960s	Two-stage hydrocracking	Advent of zeolitic HC catalysts More economical scheme Hydrocracking in 1st stage makes a smaller 2nd stage
Late 1960s	Single stage	Efficient design, capacity limitation of 35,000 BPSD fresh feed for a single train
1970s and 1980s	Once-through, partial conversion	Upgraded unconverted oil for FCC or ethylene plant feed, lube oil

heavy, high boiling feed stocks and produces high value unconverted material which becomes feed stock for FCC units, ethylene plants or lube oil units. In general, the conversion of the feed stock to products is 60–70 vol%, but can range as high as 90 vol%.

Single stage with recycle hydrocracking

The most widely found hydrocracking unit is the single stage with recycle in which the unconverted feed is sent back to the reactor section for further conversion.

Figure 7.2 depicts this type of unit. It is the most cost-effective design for 100% (or near 100%) conversion and is especially used to maximize diesel product.

A more detailed flow diagram of the reactor section in a single stage hydrocracking unit (than those shown in either Figures 7.1 and 7.2) is shown in Figure 7.3. The fresh is passed downward through the catalyst in presence of hydrogen, after being preheated to reaction temperature by passing it through heat exchangers and a heater. The effluent from the reactors goes through a series of separators where hydrogen is recovered and, together with make up hydrogen, is recycled to the reactors. The liquid product is sent to fractionation where the final products are separated from unconverted oil. In once-through units, the unconverted oil is sent out of the unit,

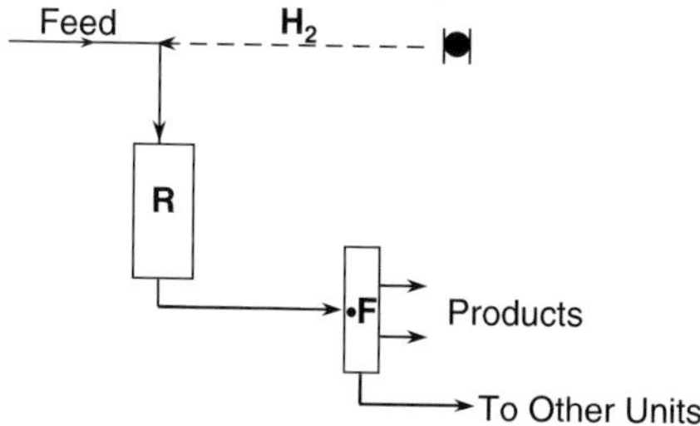

Figure 7.1. Single stage once-through hydrocracking unit.

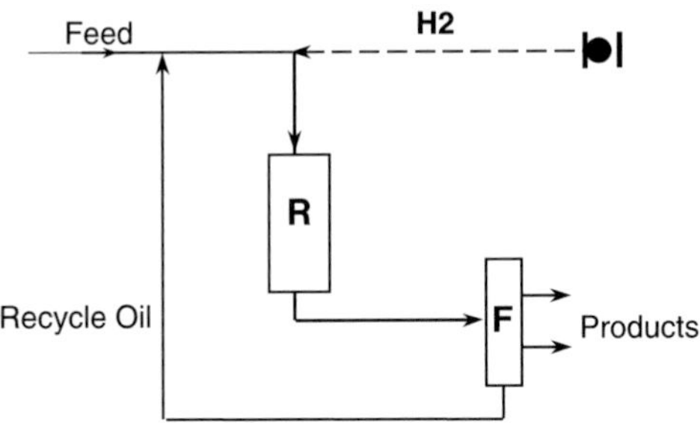

Figure 7.2. Single stage hydrocracking unit with recycle.

as previously described. In units designed to operate with recycle, the unconverted oil combines with the fresh feed, as shown in Figure 7.3. As described in the next section ("Chemistry"), the reaction section fulfills two functions: pre-treating and cracking. This is shown in Figure 7.3 as separate reactors, though both functions can be achieved in a single reactor when using amorphous catalyst. When using a pre-treat and cracking catalyst configuration, the first catalyst (a hydrotreating catalyst) converts organic sulfur and nitrogen from hetero compounds in the feedstock to hydrogen sulfide and ammonia, respectively. The deleterious effect of H_2S and NH_3 on hydrocracking catalysts is considerably less than that of the corresponding organic hetero compounds. The hydrotreating catalyst also facilitates the hydrogenation of aromatics. In the single stage, two-reactor configuration, the products from the first

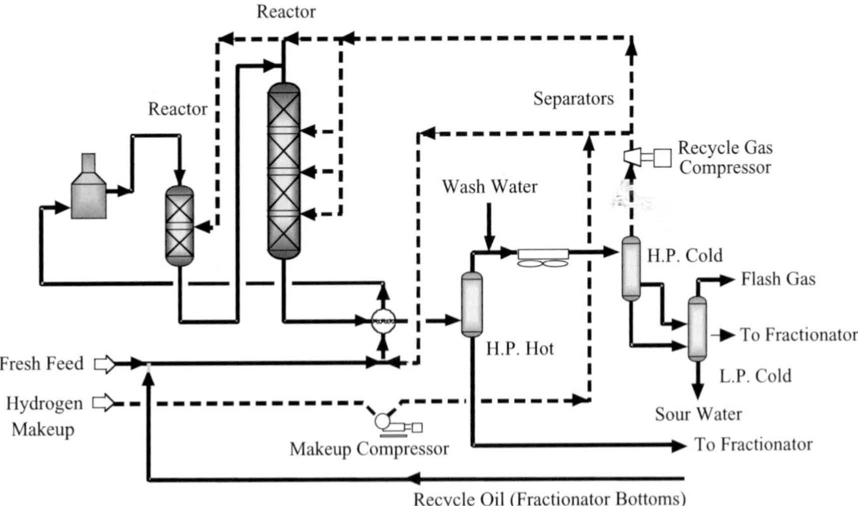

Figure 7.3. Typical flow diagram of reactor section of single stage hydrocracker (or first stage of a two stage hydrocracker).

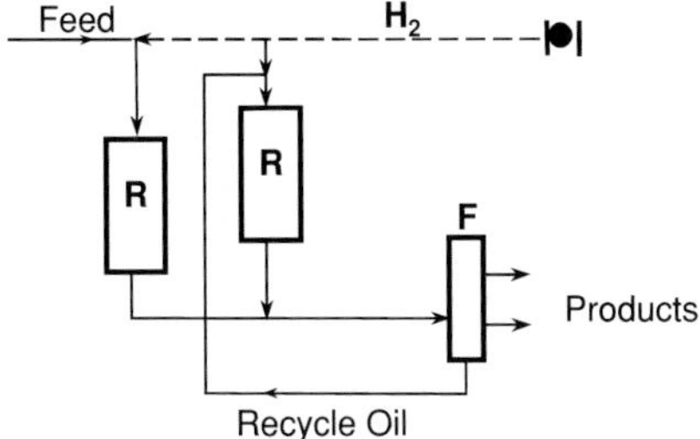

Figure 7.4. Two stage hydrocracking.

reactor are passed over a hydrocracking catalyst in the second reactor where most of the hydrocracking takes place. The conversion occurs in the presence of NH_3, H_2S, and small amounts of unconverted amounts of hetero compounds. The hydrotreating catalyst in the first reactor is designed to convert the hetero compounds in the feed stock. Typically, such catalysts comprise sulfided molybdenum and nickel on alumina support. The reactor operates at temperatures varying from 570°F to 800°F (300–425°C) and hydrogen pressures between 1,250 and 2,500 psig (85–170 bar). Under these conditions, in addition to heteroatom elimination, significant hydrogenation occurs, and some cracking also takes place. The cracking reactor operates at the same hydrogen pressures but at temperatures varying from 570 to as high as 840°F (300–450°C) for amorphous hydrocracking catalysts and up to 440°C (825°F) for zeolite containing catalysts.

Two stage recycle hydrocracking

The two stage hydrocracking process configuration is also widely used, especially for large throughput units. In two stage units, the hydrotreating and some cracking takes place in the first stage. The effluent from the first stage is separated and fractionated, with the unconverted oil going to the second stage. The unconverted oil from the second stage reaction section goes back to the common fractionator. A simplified schematic of a two stage hydrocracker is shown in Figure 7.4. The catalysts in the first stage are the same types as those used in the single stage configuration. The catalyst in the second stage is operating in near absence of ammonia, and depending on the particular design, in the absence or presence of hydrogen sulfide. The near absence of NH_3 and H_2S allows the use of either noble metal or base metal sulfide hydrocracking catalysts.

Separate hydrotreat two stage hydrocracking

A variation of the typical two stage hydrocracking with common hydrogen circulation loop is the separate hydrotreat hydrocracking shown in Figure 7.5 in which each stage

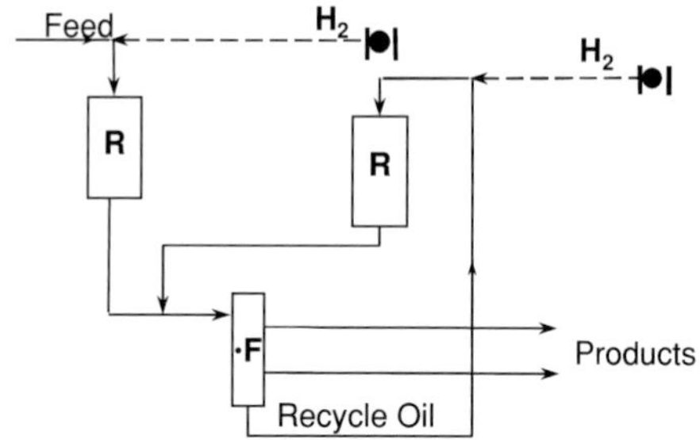

Figure 7.5. Separate hydrotreat two stage hydrocracking.

has a separate hydrogen circulation loop, allowing for operation of the second stage in the near absence of hydrogen sulfide (and ammonia).

Chemistry

Hydrocracking converts the heavy feed stock to lower molecular weight products, removes sulfur and nitrogen and saturates olefins and aromatics. The organic sulfur is transformed into H_2S, the nitrogen is transformed into NH_3 and the oxygen compounds (not always present) are transformed into H_2O. The reactions in hydrocracking can be classified in two categories: desirable and undesirable. Desirable are the treating, saturation, and cracking reactions. Undesirable reactions are contaminant poisoning as well as coking of the catalyst. There are two types of reactions taking place in hydrocracking units: treating (also called pre-treating) and cracking (also called hydrocracking). The cracking reactions require bi-functional catalyst, which possess a dual function of cracking, and hydrogenation.

Treating reactions

The treating reactions that will take place (if the respective contaminants are present) are the following: sulfur removal, nitrogen removal, organo-metallic compound removal, olefin saturation, oxygen removal, and halides removal. The first three types of compounds are always present though in varying amounts depending on the source of feed stock. The others are not always present. In general, the treating reactions proceed in the following descending order of ease: (organo) metals removal, olefin saturation, sulfur removal, nitrogen removal, oxygen removal, and halide removal. Some aromatic saturation also occurs in the pre-treating section. Hydrogen is consumed in all treating reactions. In general, the desulfurization reaction consumes 100–150 SCFB/wt% change (17–25 Nm^3/m^3/wt% change) and the denitrogenation reaction consumes 200–350 SCFB/wt% change (34–59 Nm^3/m^3/wt% change). Typically, the heat released in pre-treating is about 0.02°F/SCFB H_2 consumed (0.002°C/Nm^3/m^3 H_2).

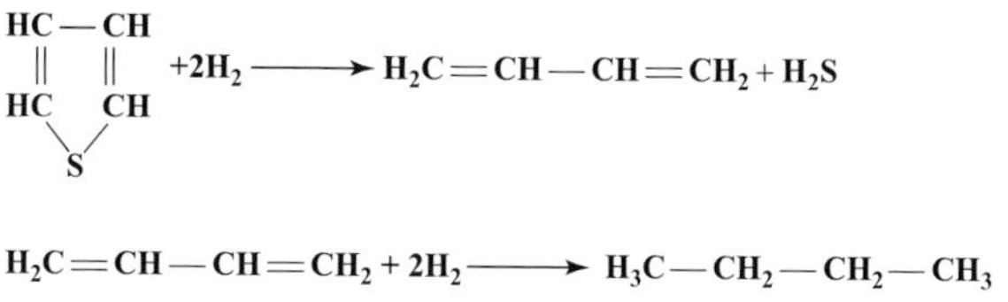

$$\text{HC}{=}\text{CH}{-}\text{CH}{=}\text{CH (thiophene ring, S)} + 2\text{H}_2 \longrightarrow \text{H}_2\text{C}{=}\text{CH}{-}\text{CH}{=}\text{CH}_2 + \text{H}_2\text{S}$$

$$\text{H}_2\text{C}{=}\text{CH}{-}\text{CH}{=}\text{CH}_2 + 2\text{H}_2 \longrightarrow \text{H}_3\text{C}{-}\text{CH}_2{-}\text{CH}_2{-}\text{CH}_3$$

Figure 7.6. Postulated mechanism for hydrodesulfurization.

The postulated mechanism for the desulfurization reaction is shown in Figure 7.6: first, the sulfur is removed followed by the saturation of the intermediate olefin compound. In the example below the thiophene is converted to butene as an intermediate which is then saturated into butane.

Listed below in Figure 7.7 are several desulfurization reactions arranged in increasing order of difficulty.

The denitrogenation reaction proceeds through a different path. In the postulated mechanism for hydrodenitrogenation the aromatic hydrogenation occurs first, followed by hydrogenolysis and, finally denitrogenation. This is shown in Figure 7.8.

Figure 7.9 shows a few typical examples of denitrogenation reactions.

Cracking reactions

Hydrocracking reactions proceed through a bifunctional mechanism. A bifunctional mechanism is one that requires two distinct types of catalytic sites to catalyze separate steps in the reaction sequence. These two functions are the acid function, which provide for the cracking and isomerization and the metal function, which provide for the olefin formation and hydrogenation. The cracking reaction requires heat while

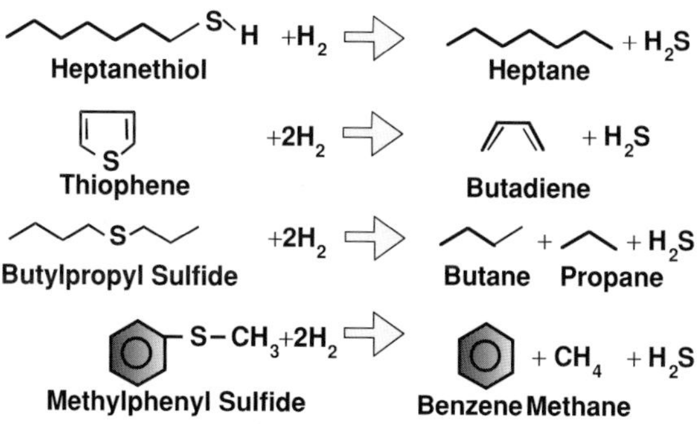

Figure 7.7. Typical desulfurization reactions.

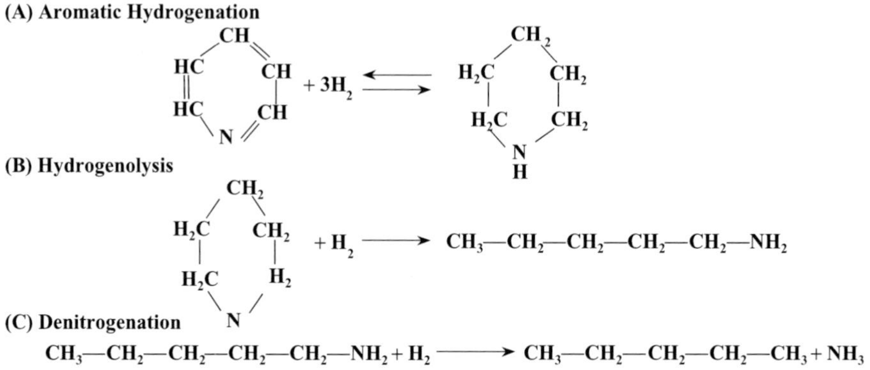

(A) Aromatic Hydrogenation

Pyridine (HC=CH–CH=CH–N=CH ring) $+ 3H_2 \rightleftarrows$ piperidine (H_2C–CH_2–CH_2–CH_2–H_2C–NH ring)

(B) Hydrogenolysis

Piperidine (H_2C–CH_2–CH_2–H_2–H_2C–N ring) $+ H_2 \longrightarrow CH_3—CH_2—CH_2—CH_2—CH_2—NH_2$

(C) Denitrogenation

$CH_3—CH_2—CH_2—CH_2—CH_2—NH_2 + H_2 \longrightarrow CH_3—CH_2—CH_2—CH_2—CH_3 + NH_3$

Figure 7.8. Postulated mechanism for hydrodenitrogenation.

the hydrogenation reaction generates heat. Overall, there is heat release in hydrocracking, and just like in treating, it is a function of the hydrogen consumption (the higher the consumption, the greater the exotherm). Generally, the hydrogen consumption in hydrocracking (including the pre-treating section) is 1200–2400 SCFB/wt% change (200–420 Nm^3/m^3/wt% change) resulting in a typical heat release of 50–100 Btu/SCF H_2 (2.1–4.2 kcal/m^3 H_2) which translates into a temperature increase of about 0.065°F/SCF H_2 consumed (0.006°C/Nm^3/m^3 H_2). This includes the heat release generated in the treating section.

In general, the hydrocracking reaction starts with the generation of an olefin or cycleolefin on a metal site on the catalyst. Next, an acid site adds a proton to the olefin or cycloolefin to produce a carbenium ion. The carbenium ion cracks to a smaller

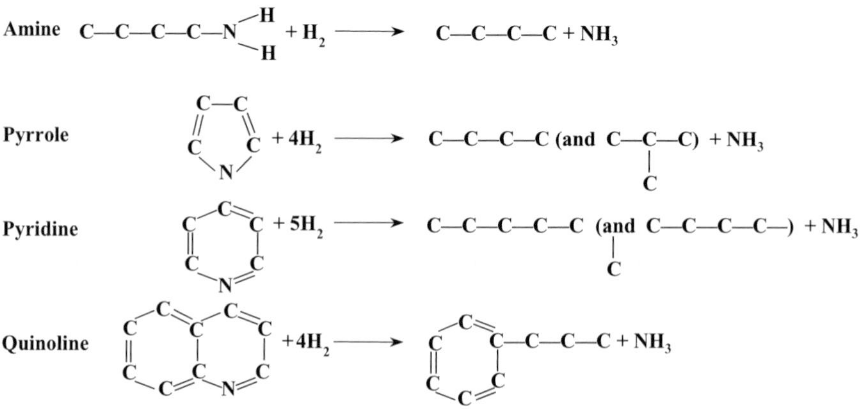

Amine C—C—C—C—N(H)(H) $+ H_2 \longrightarrow$ C—C—C—C $+ NH_3$

Pyrrole (C=C–C=C–N ring) $+ 4H_2 \longrightarrow$ C—C—C—C (and C—C(—C)—C) $+ NH_3$

Pyridine (C=C–C=C–N=C ring) $+ 5H_2 \longrightarrow$ C—C—C—C—C (and C—C(—C)—C—C—) $+ NH_3$

Quinoline (fused benzene–pyridine rings) $+ 4H_2 \longrightarrow$ (benzene ring)—C—C—C $+ NH_3$

Figure 7.9. Typical denitrogenation reactions.

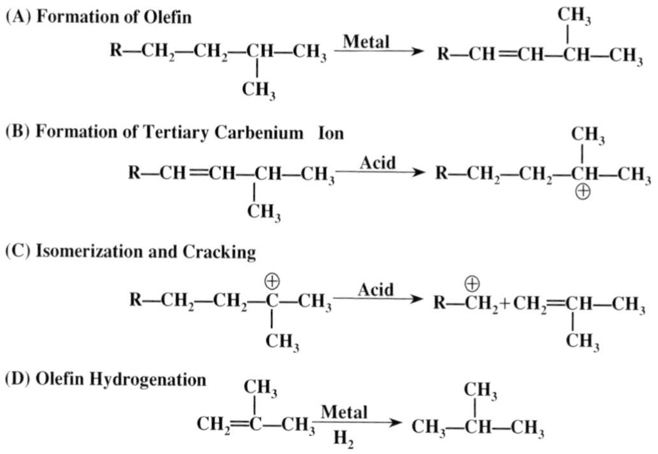

(A) Formation of Olefin

$$R\text{—}CH_2\text{—}CH_2\text{—}\underset{\displaystyle CH_3}{\underset{|}{CH}}\text{—}CH_3 \xrightarrow{\text{Metal}} R\text{—}CH{=}CH\text{—}\overset{\displaystyle CH_3}{\overset{|}{CH}}\text{—}CH_3$$

(B) Formation of Tertiary Carbenium Ion

$$R\text{—}CH{=}CH\text{—}\underset{\displaystyle CH_3}{\underset{|}{CH}}\text{—}CH_3 \xrightarrow{\text{Acid}} R\text{—}CH_2\text{—}CH_2\text{—}\overset{\displaystyle CH_3}{\overset{|}{\underset{\oplus}{CH}}}\text{—}CH_3$$

(C) Isomerization and Cracking

$$R\text{—}CH_2\text{—}CH_2\text{—}\overset{\oplus}{\underset{\displaystyle CH_3}{\underset{|}{C}}}\text{—}CH_3 \xrightarrow{\text{Acid}} R\text{—}\overset{\oplus}{C}H_2 + CH_2{=}\underset{\displaystyle CH_3}{\underset{|}{CH}}\text{—}CH_3$$

(D) Olefin Hydrogenation

$$CH_2{=}\overset{\displaystyle CH_3}{\overset{|}{C}}\text{—}CH_3 \xrightarrow[H_2]{\text{Metal}} CH_3\text{—}\overset{\displaystyle CH_3}{\overset{|}{CH}}\text{—}CH_3$$

Figure 7.10. Postulated hydrocracking mechanism of n-paraffins.

carbenium ion and a smaller olefin. These products are the primary hydrocracking products. These primary products can react further to produce still smaller secondary hydrocracking products. The reaction sequence can be terminated at primary products by abstracting a proton from the carbenium ion to form an olefin at an acid site and by saturating the olefin at a metal site. Figure 7.10 illustrates the specific steps involved in the hydrocracking of paraffins. The reaction begins with the generation of an olefin and the conversion of the olefin to a carbenium ion. The carbenium ion typically isomerizes to form a more stable tertiary carbenium ion. Next, the cracking reaction occurs at a bond that is β to the carbenium ion charge. The β position is the second bond from the ionic charge. Carbenium ions can react with olefins to transfer charge from one fragment to the other. In this way, charge can be transferred from a smaller hydrocarbon fragment to a larger fragment that can better accommodate the charge. Finally, olefin hydrogenation completes the mechanism. The hydrocracking mechanism is selective for cracking of higher carbon number paraffins. This selectivity is due in part to a more favorable equilibrium for the formation of higher carbon number olefins. In addition, large paraffins adsorb more strongly. The carbenium ion intermediate causes extensive isomerization of the products, especially to α-methyl isomers, because tertiary carbenium ions are more stable. Finally, the production of C_1 to C_3 is low because the production of these light gases involves the unfavorable formation of primary and secondary carbenium ions. Other molecular species such as alkyl naphthenes, alkyl aromatics, and so on react via similar mechanisms e.g., via the carbenium ion mechanism.

In summary, hydrocracking occurs as the result of a bifunctional mechanism that involves olefin dehydrogenation–hydrogenation reactions on a metal site, carbenium

Table 7.2. Thermodynamics of major reactions in hydrocracking

Reaction	Equilibrium	Heat of reaction
Olefin formation	Unfavorable but not limiting	Endothermic
Aromatic saturation	Unfavorable at high temperature	Exothermic
Cracking	Favorable	Endothermic
HDS	Favorable	Exothermic
HDN	Favorable	Exothermic

ion formation on an acid site, and isomerization, and cracking of the carbenium ion. The hydrocracking reactions tend to favor conversion of large molecules because the equilibrium for olefin formation is more favorable for large molecules and because the relative strength of adsorption is greater for large molecules. In hydrocracking, the products are highly isomerized, C_1 and C_3 formation is low, and single rings are relatively stable.

In addition to treating and hydrocracking several other important reactions take place in hydrocrackers. These are aromatic saturation, polynuclear aromatics (PNA) formation and coke formation. Some aromatic saturation occurs in the treating section and some in the cracking section. Aromatic saturation is the only reaction in hydrocracking which is equilibrium limited at the higher temperatures reached by hydrocrackers toward the end of the catalyst cycle life. Because of this equilibrium limitation, complete aromatic saturation is not possible toward the end of the catalyst cycle when reactor temperature has to be increased to make up for the loss in catalyst activity resulting from coke formation and deposition. Table 7.2 shows the thermodynamics of the major reactions taking place in a hydrocracker. Of course, the principles of thermodynamics provide the means to determine which reactions are possible. In general, the thermodynamic equilibrium for hydrocracking is favorable. Cracking reactions, desulfurization and denitrogenation are favored at the typical hydrocracker operating conditions. The initial step in the hydrocracking of paraffins or naphthenes is the generation of an olefin or cycloolefin. This step is unfavorable under the high hydrogen partial pressure used in hydrocracking. The dehydrogenation of the smaller alkanes is most unfavorable. Nevertheless, the concentration of olefins and cycloolefins is sufficiently high, and the conversion of these intermediates to carbenium ions is sufficiently fast so that the overall hydrocracking rate is not limited by the equilibrium olefin levels (Table 7.2).

Polynuclear aromatics (PNA), sometimes called polycyclic aromatics (PCA), or polyaromatic hydrocarbons (PAH) are compounds containing at least two benzene rings in the molecule. Normally, the feed to a hydrocracker can contain PNA with up to seven benzene rings in the molecule. PNA formation is an important, though undesirable, reaction that occurs in hydrocrackers. Figure 7.11 shows the competing pathways for conversion of multiring aromatics. One pathway starts with metal-catalyzed ring

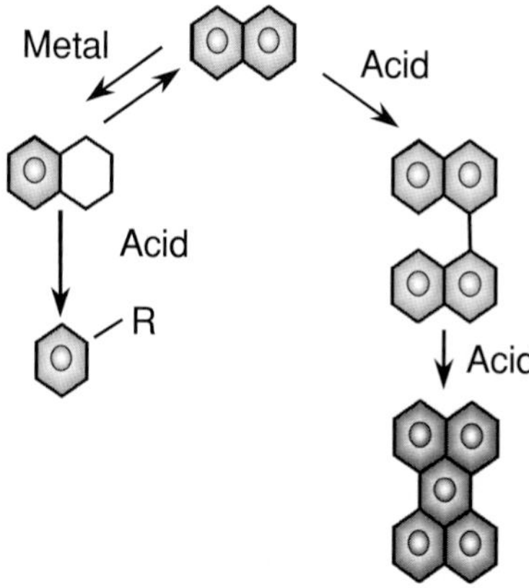

Figure 7.11. Possible pathways for multiring aromatics.

saturation and continues with acid-catalyzed cracking reactions. The other pathway begins with an acid-catalyzed condensation reaction to form a large aromatic-ring compound. This molecule may undergo subsequent condensation reactions to form a large PNA.

The consequence of operating hydrocracking units with recycle of the unconverted feed is creation of PNA with more than seven benzene rings in the molecule. These are called heavy polynuclear aromatics (HPNA) whose formation is shown in Figure 7.12. The HPNA produced on the catalyst may exit the reactor and cause downstream fouling; or they may deposit on the catalyst and form coke, which deactivates the catalyst. Their presence results in plugging of equipment. For mitigation a stream of 5 to as much as 10% of unconverted material might have to be taken out of the hydrocracker, resulting in much lower than expected conversion of the feed.

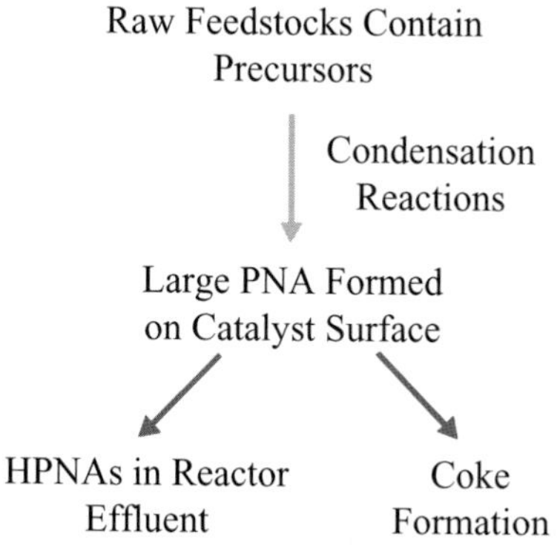

Figure 7.12. HPNA formation.

Catalysts

Hydrocracking catalysts are dual function catalysts. For the cracking reaction to occur (as well as some of the other reactions taking place in hydrocracking, such as hydroisomerization and dehydrocyclization), both metallic sites and acidic sites must be present on the catalyst surface. Hydrocracking catalysts have a cracking function and hydrogenation function. The cracking function is provided by an acidic support, whereas the hydrogenation function is provided by metals.

The acidic support consists of amorphous oxides (e.g., silica-alumina, a crystalline zeolite (mostly modified Y zeolite)) plus binder (e.g., alumina), or a mixture of crystalline zeolite and amorphous oxides. Cracking and isomerization reactions take place on the acidic support. The metals providing the hydrogenation function can be noble metals (palladium, platinum), or non-noble (also called "base") metal sulfides from group VIA (molybdenum, tungsten) and group VIIIA (cobalt, nickel). These metals catalyze the hydrogenation of the feedstock, making it more reactive for cracking and heteroatom removal, as well as reducing the coking rate. They also initiate the cracking by forming a reactive olefin intermediate via dehydrogenation.

The ratio between the catalyst's cracking function and hydrogenation function can be adjusted in order to optimize activity and selectivity. Activity and selectivity are but two of the four key performance criteria by which hydrocracking catalysts are measured:

- Initial activity, which is measured by the temperature required to obtain desired product at the start of the run.
- Stability, which is measured by the rate of increase of temperature required to maintain conversion.
- Product selectivity, which is a measure of the ability of a catalyst to produce the desired product slate.
- Product quality, which is a measure of the ability of the process to produce products with the desired use specifications, such as pour point, smoke point, or cetane number.

For a hydrocracking catalyst to be effective, it is important that there be a rapid molecular transfer between the acid sites and hydrogenation sites in order to avoid undesirable secondary reactions. Rapid molecular transfer can be achieved by having the hydrogenation sites located in the proximity of the cracking (acid) sites.

Acid function of the catalyst

A solid oxide support material supplies the acid function of the hydrocracking catalyst. Amorphous silica-alumina provides the cracking function of amorphous catalysts and serves as support for the hydrogenation metals. Amorphous silica-alumina catalysts

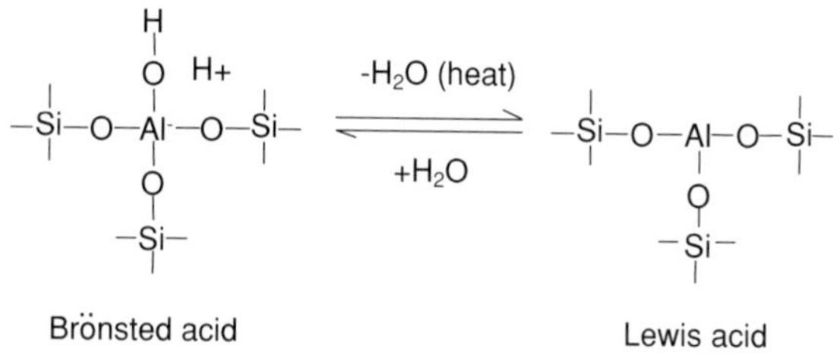

Figure 7.13. Silica-alumina acid sites.

are commonly used in distillate producing hydrocracking catalysts. Amorphous silica-alumina also plays a catalytic role in low-zeolite hydrocracking catalysts. In high-zeolite hydrocracking catalysts it acts primarily a support for metals and as binder. Zeolites, particularly Y zeolite, are commonly used in high activity distillate catalysts and in naphtha catalysts. Other acidic support components such as acid-treated clays, pillared clays, layered silicates, acid metal phosphates, and other solid acids have been tried in the past, however, present day hydrocracking catalysts do not contain any of these materials.

Amorphous mixed metal oxide supports are acidic because of the difference in charge between adjacent cations in the oxide structure. The advantages of amorphous silica-alumina for hydrocracking are that it has large pores, which permit access of bulky feed stock molecules to the acidic sites, and moderate activity, which makes the metal-acid balance needed for distillate selectivity easier to obtain. Figure 7.13 is an illustration of silica-alumina acid sites. The substitution of an Al_3^+ cation for a Si_4^+ cation leaves a net negative charge on the framework that can be balanced by an acidic proton. The removal of water from this Bronsted acid site creates a Lewis acid site. A Bronsted acid site on a catalyst is an acid site where the acidic entity is an ionizable hydrogen. A Lewis acid site on a catalyst is an acid site where the acidic entity is a positive ion such as Al_3^+ rather than an ionizable hydrogen. Although plausible hydrocracking mechanisms can be written for both Bronsted or Lewis sites, Bronsted acidity is believed to be more desirable because Lewis acid sites may catalyze coke formation.

Zeolites are crystalline aluminosilicates composed of Al_2O_3 and SiO_2 tetrahedral units that form a negatively charged microporous framework structure enclosing cavities occupied by large ions and water molecules, both of which have considerable freedom of movement, permitting ion-exchange, and reversible dehydration. Mobile cations, which are not part of the framework but are part of the zeolites, are readily exchanged. If the mobile cations are exchanged with NH_4^+, followed by calcination to remove NH_3, a Bronsted acid site is formed. The zeolite used in hydrocracking, Y zeolite, is synthetic. It has a structure nearly identical to the naturally found zeolite faujasite. The Y zeolite has both a relatively large free aperture, which controls access of reactants to acid sites, and a three-dimensional pore structure, which allows diffusion of the reactants in and products out with minimal interference. Both Bronsted and Lewis acids are possible in zeolites. The number of acid sites and the strength of the acid sites may be varied. These sites are highly uniform, but each zeolite may have more than one type of site. The following factors influence the number and strength of acid sites in zeolites: the types of cations occupying the ion exchange sites, thermal treatments that the sample has received, and the ratio of silica to alumina of the framework. For example, Y zeolite can be dealuminated by a variety of methods, including thermal and hydrothermal treatments. Dealumination decreases the total number of acid sites because each proton is associated with a framework aluminum. However, dealumination also increases the strength of the acid sites to a certain point. As a result, the total acidity of the zeolite, which is a product of the number of sites and strength per site, peaks at an intermediate extent of dealumination. Clearly, the acid site concentration and strength of zeolites affect the final hydrocracking catalyst properties. The principal advantage of zeolites for hydrocracking is their high acidity.

Metal function of the catalyst

A metal, a metal oxide, a metal sulfide, or a combination of these compounds may supply the metal function of the catalyst. The key requirement for the metal function is that it must activate hydrogen and catalyze dehydrogenation and hydrogenation reactions. In addition, metal-catalyzed hydrogenolysis (carbon–carbon breaking) is undesirable because the distribution of the hydrogenolysis products is less desirable relative to hydrocracking.

The most commonly used metal function for hydrocracking catalysts is a combination of Group VIA (Mo, W) and Group VIIIA (Co, Ni) metal sulfides. The major advantage of this combination of metal sulfides is that it is sulfur tolerant; however, it has only moderate activity compared to Pd or Pt. The combination of Group VIA and Group VIIIA metal sulfides has been extensively characterized because of its importance to hydrocracking. Although Group VIIIA metal sulfides have some hydrogenation activity, these sulfides alone are much less active than the Group VIA metal sulfides and are considered to be promoters. The Group VIIIA metal promoter interacts synergistically with the Group VIA metal sulfide to produce a substantial increase in activity.

Because the Group VIA and Group VIIIA metals are most conveniently prepared as oxides, a sulfiding step is necessary. That will be discussed in section "Catalyst Loading and Activation".

Catalyst manufacturing

Hydrocracking catalysts can be manufactured by a variety of methods. The method chosen usually represents a balance between manufacturing cost and the degree to which the desired chemical and physical properties are achieved. Although there is a relationship between catalyst formulation, preparation procedure, and catalyst properties, the details of that relationship are not always well understood due to the complex nature of the catalyst systems. The chemical composition of the catalyst plays a decisive role in its performance; the physical and mechanical properties also play a major role.

The preparation of hydrocracking catalysts involves several steps:

- Precipitation
- Filtration (decantation, centrifugation)
- Washing
- Drying
- Forming
- Calcination
- Impregnation

Other steps, such as kneading or mulling, grinding, and sieving, may also be required. Depending on the preparation method used, some of these steps may be eliminated, whereas other steps may be added. For example, kneading or comulling of the wet solid precursors is used in some processes instead of precipitation. When the metal precursors are precipitated or comulled together with the support precursors, the impregnation step can be eliminated. Described below are the steps that are an integral part of any hydrocracking catalyst manufacturing process.

Precipitation

Precipitation involves the mixing of solutions or suspension of materials, resulting in the formation of a precipitate, which may be crystalline or amorphous. Mulling or kneading of wet solid materials usually leads to the formation of a paste that is subsequently formed and dried. The mulled or kneaded product is submitted to thermal treatment in order to obtain a more intimate contact between components and better homogeneity by thermal diffusion and solid-state reactions. Precipitation or mulling is often used to prepare the support for the catalyst and the metal component is subsequently added by impregnation.

The support determines the mechanical properties of the catalyst, such as attrition resistance, hardness, and crushing strength. High surface area and proper pore size distribution is generally required. The pore size distribution and other physical properties of a catalyst support prepared by precipitation are also affected by the precipitation and the aging conditions of the precipitate as well as by subsequent drying, forming, and calcining.

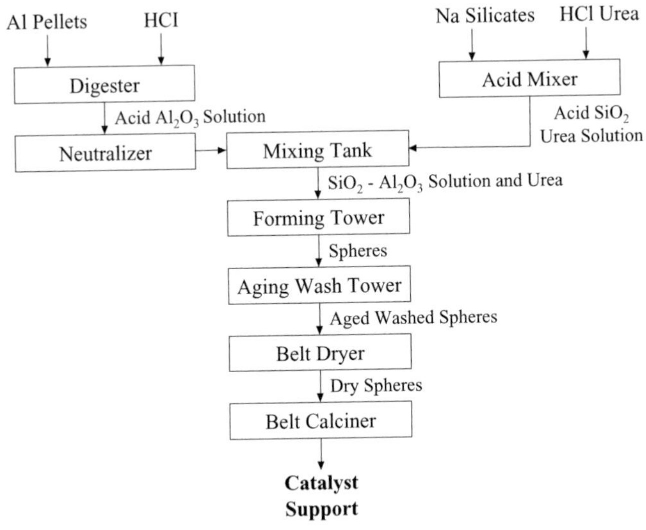

Figure 7.14. Spherical catalyst support manufacturing.

Forming

The final shape and size of catalyst particles is determined in the forming step. Catalysts and catalyst supports are formed into several possible shapes such as spheres, cylindrical extrudates, or shaped forms such as trilobes or quadrilobes. Spherical catalyst support catalyst is obtained by 'oil dropping' whereby precipitation occurs upon the pouring of a liquid into a second immiscible liquid. Spherical bead catalyst are obtained by this process which is shown in Figure 7.14.

Generally, because of cost considerations, the majority of catalysts are currently formed in shapes other than spheres. Only amorphous silica-alumina catalysts are formed as spheres.

Extrudates are obtained by extruding a thick paste through a die with perforations. Peptizing agents are usually included in the paste. The spaghetti-like extrudate is usually dried and then broken into short pieces. The typical length to diameter ratio of the extrudates varies between 2 and 4. The extrudate is then dried and/or calcined. The water content of the paste submitted to extrusion is critical because it determines the density, pore size distribution, and mechanical strength of the catalyst. The water content of the paste is usually kept close to the minimum at which extrusion is still possible. Figure 7.15 shows a typical extrudate support manufacturing.

The form of extrudates may vary. The simplest form is cylindrical, but other forms such as trilobes, twisted trilobes, or quadrilobes, are also found commercially. Catalysts

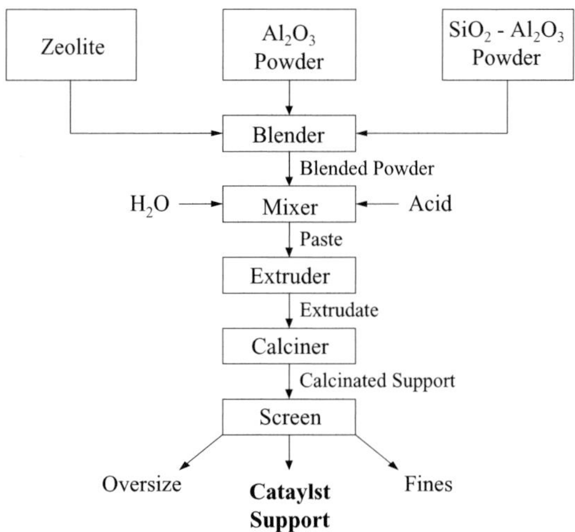

Figure 7.15. Extrudate catalyst support manufacturing.

with multilobal cross-sections have a higher surface-to-volume ratio than simple cylindrical extrudates. When used in a fixed bed, these shaped catalyst particles help reduce diffusional resistance, create a more open bed, and reduce pressure drop. Figure 7.16 depicts several shapes of commercial catalysts used in hydrocracking.

Drying and calcining

Thermal treatment is generally applied before or after impregnation of the formed catalyst. For catalysts prepared by precipitation or comulling of all the components

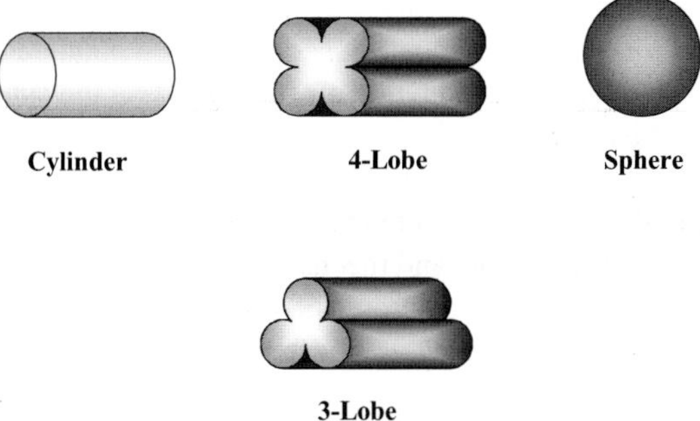

Figure 7.16. Commercial catalyst shapes.

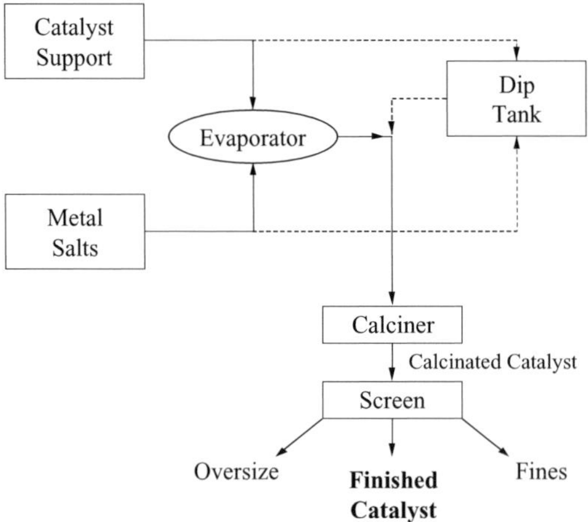

Figure 7.17. Example of catalyst finishing (impregnation).

(including the metal components), only drying may be required prior to forming, with subsequent calcination of the formed product. Thermal treatment of the catalyst or support eliminates water and other volatile matter. The drying and calcination conditions are of critical importance in determining the physical as well as catalytic properties of the product. Surface area, pore size distribution, stability, attrition resistance, crushing strength, as well as the catalytic activity are affected by the drying and calcination conditions.

Impregnation

Impregnation is used to incorporate a metal component into a preformed catalyst support. Several impregnation methods may be used for catalyst preparation: (a) impregnation by immersion (dipping), (b) impregnation by incipient wetness, and (c) diffusional impregnation. In the first method, which is the most commonly used, the calcined support is immersed in an excess of solution containing the metal compound. The solution fills the pores and is also adsorbed on the support surface. The excess volume is drained off. Impregnation to incipient wetness is carried out by tumbling or spraying the activated support with a volume of solution having the proper concentration of metal compound, and equal to or slightly less than the pore volume of the support. The impregnated support is dried and calcined. Because metal oxides are formed in the process, the calcination step is also called oxidation. In diffusional impregnation the support is saturated with water or with acid solution, and immersed into the acqueous solution containing the metal compound. That compound subsequently diffuses into the pores of the support through the aqueous phase. Figure 7.17 shows an example of catalyst finishing (impregnation).

Catalyst loading and activation

Catalyst loading

There are two methods of catalyst loading sock loading and dense loading. Sock loading is done by pouring catalyst into a hopper mounted on top of the reactor and then allowing it to flow through a canvas sock into the reactor. Dense loading or dense bed packing is done with the help of a mechanical device. The dense loading method was introduced in the mid-1970s. Catalyst loaded by sock loading will have a higher void fraction than catalyst that was dense loaded. Dense bed packing and the resulting higher-pressure drop provides a more even distribution of liquid in a trickle flow reactor which is the flow regime for most hydrocracker applications. If diffusion limitations are negligible, dense loading is desirable in order to maximize the reaction rate per unit reactor volume. This is often the case in hydrocracking reactors. The other advantage of dense loading is that it orients the catalyst particles in a horizontal and uniform manner. This improves the vapor/liquid distribution through the catalyst beds. Catalyst particle orientation is important especially for cylindrically shaped extruded catalyst in vapor/liquid reactant systems. When the catalyst particles are oriented in a horizontal position in the catalyst bed, liquid maldistribution or channeling is almost completely eliminated. This maldistribution tends to occur when the catalyst loading is done by the sock loading method, which generally causes the extrudates to be oriented in a downward slant toward the reactor walls increasing bed voids and creating liquid maldistribution. Of all the factors influencing catalyst utilization, catalyst loading has generally proven to be the most important factor. Except for the hydrocrackers that have reactor pressure drop limitations mainly due to operation at higher than design throughputs, the great majority of units worldwide are dense loaded.

Catalyst activation

Hydrocracking catalysts have to be activated in order to be catalytically active. Several names are used for that purpose, such as sulfiding, presulfiding, presulfurizing in addition to activation. The metals on the greatest majority of catalysts are in an oxide form at the completion of the manufacturing process. The noble metal catalysts are activated by hydrogen reduction of the finished catalyst, in which the metal is also in an oxide form. Calcination in air prior to reduction is necessary to avoid metal sintering. The presence of water vapors is generally avoided, also to prevent metal sintering. By using an excess of hydrogen, the water formed during reduction can be swept away. The activation of noble metal catalysts by hydrogen reduction occurs at 570–750°F (300–400°C).

The non-noble (base metal) catalysts are activated by transforming the catalytically inactive metal oxides into active metal sulfides (thus the name sulfiding, etc.). This is accomplished mainly in-situ though some refiners have started to do the activation outside the unit (ex-situ). It is likely more and more refiners will opt to receive the catalyst at the refinery site in pre-sulfided state to accelerate the start up of the unit. In-situ sulfiding can be accomplished either in vapor or liquid phase. In vapor phase sulfiding, the activation of the catalyst is accomplished by injecting a chemical which easily decomposes to H_2S, such as di-methyl-di-sulphide (DMDS) or di-methyl-sulfide (DMS); use of H_2S was fairly common until a few years ago, but now it is only rarely used because of environmental and safety concerns. Liquid phase sulfiding can be accomplished with or without spiked feedstocks. In the latter case, the feedstock is generally a gas oil type material that contains sulfur compounds in ranges from a few thousand to twenty thousand ppm. The H_2S necessary for the activation of the catalyst is generated by the decomposition of the sulfur compounds. This method is in very little use today, but it was 'state-of-the-art' in the 1960s and early 1970s. The preferred sulfiding procedure in the industry is liquid phase with a spiking agent (generally DMDS or DMS). It results in important savings of time when compared to either vapor phase or liquid phase without spiking agents. Another advantage of liquid phase over gas phase sulfiding is that by having all the catalyst particles wet from the very beginning there is very little chance of catalyst bed channeling which can occur if the catalyst particles are allowed to dry out. The in-situ sulfiding occurs at temperatures between 450 and 600°F (230–315°C) regardless of the method used. Some catalyst manufacturers recommend the sulfiding be conducted at full operating pressure while others prefer it be done at pressures lower than the normal operating pressure. Ammonia injection is practiced during the sulfiding of high activity (high zeolite content) catalysts to prevent premature catalyst deactivation.

In the case of ex-situ presulfurization of catalyst, sulfur compounds are loaded onto the catalyst. The activation occurs when the catalyst, which has been loaded in the reactor, is heated up in the presence of hydrogen. The activation can be conducted either in vapor or liquid phase. Generally, activation of ex-situ presulfurized catalyst is accomplished faster than if the sulfiding is done in-situ, however there is the additional expense due to the need for the ex-situ presulfurization step.

The economics vary from refiner to refiner, however ex-situ presulfurization is rarely used for hydrocracking catalysts.

Catalyst deactivation and regeneration

Catalyst deactivation is the gradual loss of the catalyst's ability to convert the feed into useful products. Catalyst activity is a measure of the relative rate of feedstock conversion. In practical terms it is the temperature required to obtain a fixed conversion. As the run progresses, the catalyst loses activity. Catalyst will lose activity in several ways described below.

Coke deposition

Coke deposition is a byproduct of the cracking reactions. The laydown of coke on a catalyst is a time–temperature phenomenon in that the longer the exposure and/or the higher the temperature the catalyst is subjected to, the more severe the deactivating effect. It begins with adsorption of high-molecular weight, low hydrogen/carbon ratio ring compounds; it proceeds with further loss of hydrogen content, and ends with varying degrees of hardness of coke. This coke can cover active sites and/or prevent access to these sites by physical blockage of the entrance to the pores leading to the sites. Coke is not a permanent poison. Catalyst, which has been deactivated by coke deposition, can be, relatively easily, restored to near original condition by regeneration.

Reversible poisoning

Catalyst poisoning is primarily the result of strong chemisorption of impurities on active sites. This type of poisoning is reversible—that is, when the deactivating agent is removed, the deactivating effect is gradually reversed. In some cases, raising the catalyst temperature can compensate for the deactivating effect. But, raising temperatures will, however, increase the rate of coke deposition. One example of a reversible poison is carbon monoxide, which can impair the hydrogenation reactions by preferential adsorption on active sites. Another example is H_2S, which in moderate to high concentrations can reduce the desulfurization, rate constant. In this case, the removal of H_2S from the recycle gas system solves the problem.

Agglomeration of the hydrogenation component

Another reversible form of catalyst deactivation is the agglomeration of the hydrogenation component of the catalyst. It can be caused by poor catalyst activation conditions in which a combination of high water partial pressure and high temperature may exist for a prolonged period. Regeneration can restore the catalyst to near original condition.

Metals deposition

Deposition of metals is not reversible, even with catalyst regeneration. The metals may come into the system via additives, such as silicon compounds used in coke drums to reduce foaming, or feedstock contaminants such as Pb, Fe, As, P, Na, Ca, Mg, or as organo-metallic compounds in the feed primarily containing Ni and V. The deposition of Ni and V takes place at the pore entrances or near the outer surface of the catalyst, creating a ‘rind’ layer—effectively choking off access to the interior part of the catalyst, where most of the surface area resides. Metals deposition can damage the acid sites, the metal sites, or both.

Catalyst support sintering

This is another reason for loss of catalyst activity and it also is irreversible. This is also a result of high temperatures and particularly in connection with high water partial pressures. In this case the catalyst support material can lose surface area from a collapse of pores, or from an increase in the diameter of pores, with the pore volume remaining approximately constant.

Catalyst regeneration

A coked catalyst is usually regenerated by combustion in a stream of diluted oxygen or air, although steam or steam–air mixtures have also been used in the past. Upon combustion, coke is converted to CO_2 and H_2O. In the absence of excess oxygen, CO may also form. Except for the noble metal catalysts, hydrocracking catalysts contain sulfur, as the metals are in a sulfide form. In the regeneration process, the sulfur will be emitted as SO_2. In general, sulfur oxide emission starts at lower temperature than CO_2 emission. Regeneration of commercial catalysts can be done in-situ or ex-situ. The majority of commercial catalysts regeneration is performed ex-situ because of environmental considerations as well as because it results in a better performing catalyst. There are several companies that perform ex-situ regeneration by using different equipment for burning off the coke. One company uses a continuous rotolouver, which is a cylindrical drum rotating slowly on a horizontal axis and enclosing a series of overlapping louvers. The spent catalyst passes slowly through the rotolouver, where it encounters a countercurrent of hot air. Another company uses a porous moving belt as a regenerator. The catalyst is moved with the stainless steel belt through a stationary tunnel furnace vessel where the regeneration takes place. A third company regenerator uses ebullated bed technology to perform the catalyst regeneration. Regardless of the process, the spent catalyst is submitted to de-oiling prior to regeneration. This is to eliminate as much hydrocarbon as possible as well as to remove as much sulfur as possible to prevent formation of sulfates which could deposit on the catalyst and not be removed during regeneration. Sulfates are deleterious to catalyst performance.

Design and operation of hydrocracking reactors

Design and construction of hydrocracking reactors

Hydrocracking reactors are downflow, fixed-bed catalytic reactors, generally operating in trickle flow regime. Because hydrocracking occurs at high pressure and relatively high temperature and in the presence of hydrogen and hydrogen sulfide the reactors are vessels with thick wall constructed from special materials. The reactors are usually cylindrical vessels fabricated from $2^1/_4$ Cr–1 Mo or 3Cr–1 Mo material with stabilized

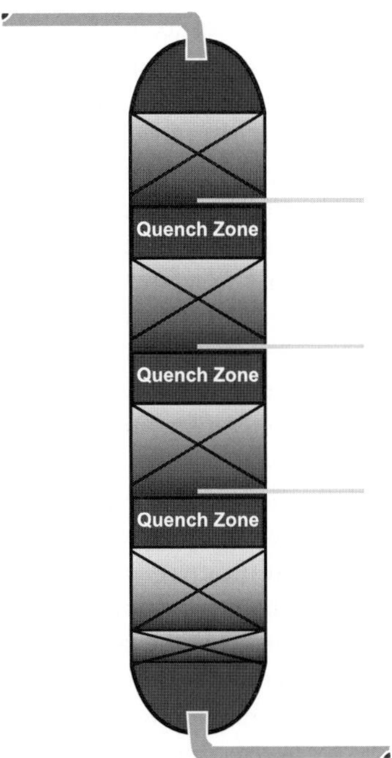

Figure 7.18. Typical hydrocracking reactor.

austenitic stainless steel weld overlay or liner, for added corrosion protection. More specialized materials, in which a small amount of vanadium is added to the $2^1/_4$ Cr–1 Mo or 3 Cr–1 Mo reactor base metal to increase its strength characteristics, started being used by some fabricators in the last few years. A typical drawing of a hydrocracking reactor is shown in Figure 7.18.

The size of the hydrocracking reactors varies widely depending on the design conditions and is dependent on the desired mass velocity and acceptable pressure drops. Commercially, reactors with inside diameters of up to 16 ft (4.9 m) have been fabricated. Depending on the design pressure and inside diameter, the thickness of the reactor walls can be as much as 1 ft (30 cm). Since heat release is a common feature for all hydrocrackers, reactor temperature control has to be exercised. As shown schematically in Figure 7.18, a hydrocracking reactor will contain several separate catalyst beds. The number of catalytic beds in a reactor and their respective lengths are determined from the design temperature rise profile. The maximum acceptable temperature rise per bed defines the length of the catalyst bed. The acceptable temperature maximum, in turn, depends on the operating mode of the hydrocracker. For example, operations designed to maximize naphtha have a different maximum from those designed the production of middle distillate. A typical reactor operated to maximize naphtha yields will have as many as five or six beds. A typical reactor operated to produce middle distillate will have three or four beds. Commercial catalyst beds can reach lengths up to 20 ft (6 m). A typical pre-treating reactor will have two or three beds if the feed is straight run material, and up to five beds if the feed contains appreciable amounts of cracked material. Cold hydrogen gas, introduced in the quench zones, is used for reactor temperature control. The quench zones separating successive catalyst beds have the following functions: (a) to cool the partially reacted fluids with hydrogen quench gas; (b) to assure a uniform temperature distribution the fluids entering the next catalyst bed; and (c) to mix efficiently and disperse evenly the fluids over the top of the next catalyst bed. Since hydrocracking is an exothermic process, the fluids exiting one catalyst bed have to be cooled prior to entering the next catalyst bed, in order to avoid overheating and to provide a safe and stable operation. This is accomplished by thorough mixing with cool hydrogen. Furthermore, the temperature distribution in the cooled fluids entering the next catalyst bed has to be uniform in order to minimize the radial temperature gradients in successive catalyst beds. Unbalanced temperatures in a catalyst bed may result in different reaction rates in the same bed. This can lead to different deactivation rates of the catalyst, and, in worse cases, to temperature excursions. In addition to a uniform temperature distribution, it is also important to achieve a good mass flow distribution. The effective vapor/liquid mixing and uniform distribution of fluids over the top of the catalyst bed, accomplished in the quench zone, reestablishes an even mass flow distribution through the bed. There is a multitude of companies providing vapor/liquid distribution devices, from process licensors, to catalyst manufacturers and, engineering contractors. Most distribution devices perform well, provided they are properly installed. Another important parameter is liquid flux (lbs/hr/sqft of cross-sectional area). While gas mass flux has practically no influence on liquid distribution, liquid mass flux is determinant in avoiding mal distribution in the catalyst bed. Operation at a liquid mass flux of more than 2,000 lbs/hr/sqft is recommended; operation at liquid fluxes lower than 1,500 lbs/hr/sqft is discouraged. Furthermore, it should be noticed that if the liquid mass flux is below the recommended limit, increasing the gas mass flux will have very little effect, if any, on the liquid distribution (e.g., it will not improve it).

Hydrocracking reactor operation

During operation, the hydrocracking catalyst gradually loses some of its activity. In order to maintain the conversion of feedstock to products constant, the average bed temperature is gradually increased. The temperature increase in many cases is very small, less than 2°F/month (1°C/month). When the average bed temperature reaches a value close to the design maximum, the catalyst has to be replaced or reactivated. Because the required temperature increase per unit time is relatively small, the reactor can be operated with the same catalyst for several years before replacement of the deactivated catalyst becomes necessary. Similar changes take place in the pre-treating reactor.

Kinetics is the study of the rates of reaction. The rates of reaction determine the key properties of a hydrocracking catalyst: initial activity, selectivity, stability, and product quality. The temperature required to obtain the desired product at the start of the run measures the initial activity. In general, the catalyst activity is a measure of the relative rate of feedstock conversion. In hydrocracking, activity is defined as the temperature required to obtain fixed conversion under certain process conditions. Hydrocracking conversion is usually defined in terms of change of endpoint:

$$\%\,\text{Conversion} = \left(\left(\text{EP}^{+}_{\text{feed}} - \text{EP}^{+}_{\text{product}}\right)/\text{EP}^{+}_{\text{feed}}\right) \times 100$$

where EP^{+} indicates the fraction of material in the feed or product boiling above the desired endpoint.

Catalyst selectivity is a measure of the rate of formation of a desired product relative to the rate of conversion of the feed (or formation of other products). Hydrocracking selectivity is expressed as the yield of desired product at a specific conversion. Yield is determined by the rate of formation of the desired product relative to the feed rate. At 100% conversion, catalyst yield equals catalyst selectivity. Hydrocracking selectivity is affected by operating conditions. In general, more severe operating conditions cause higher selectivity to secondary products.

Catalyst stability is a measure of change of reaction rate over time. Hydrocrackers are typically operated in the constant conversion mode, with temperature adjustments made to maintain the desired conversion. Hydrocracking activity stability is defined as the temperature change required to maintain constant conversion. Changes in product yield over time on-stream occur when using zeolitic catalysts. Hydrocracking yield stability is defined as the yield change with time at constant conversion and is usually expressed as a function of temperature change.

The product quality is a measure of the ability of the process to yield products with the desired use specification such as pour point, smoke point or octane. Table 7.3 shows some of the important product quality measurements and the chemical basis for these measurements.

Table 7.3. Chemical basis for product quality

Quality measurement	Chemical basis
High smoke point	Low concentration of aromatics
Low pour point	Low concentration of *n*-paraffins
Low freeze point	Low concentration of *n*-paraffins
Low cloud point	Low concentration of *n*-paraffins
Low CFPP	Low concentration of *n*-paraffins
High octane	High ratio of *i*/*n*-paraffins
	High concentration of aromatics
	High concentration of naphthenes

Hydrocracking process variables

The proper operation of the unit will depend on the careful selection and control of the processing conditions. By careful monitoring of these process variables the unit can operate to its full potential.

Catalyst temperature

The amount of conversion which takes place in the reactors is going to be determined by several variables: the type of feedstock, the amount of time the feed is in the presence of catalyst, the partial pressure of hydrogen in the catalyst bed, and, most importantly, the temperature of the catalyst and reactants. The obvious generalization about temperature is that the higher the temperature, the faster the rate of reaction and therefore, the higher the conversion. Since hydrocracking is exothermic, overall, the temperature increases as the feed and recycle gas proceed through the catalyst beds. It is very important that the temperature increase (ΔT) be controlled carefully at all times. It is possible to generate more heat from the reactions than the flowing streams can remove from the reactors. If this happens, the temperature may increase very rapidly. This condition is called a temperature excursion or a temperature runaway. A temperature runaway is a very serious situation since extremely high temperatures can be generated within a short period of time. These high temperatures can cause damage to the catalyst and/or to the reactors. To avoid these situations, temperature guidelines have to be observed. These guidelines are dependent on the type of feedstock, and the type of catalyst, and vary from catalyst supplier to catalyst supplier, but by and large, limit the temperature rise of catalyst beds loaded with noble metal catalyst to about 30°F (17°C). The temperature rise of catalyst beds loaded with high activity base metal catalysts (for naphtha production) is limited to about 40°F (22°C) and those loaded with low zeolite content catalyst (for middle distillate production) the temperature rise is limited to 50°F (28°C). Finally, maximum bed temperature rises of about 75°F (42°C) are recommended for amorphous catalysts. The same maximum bed temperature rise is also recommended for most pre-treating reactors. To properly monitor the reactions as the reactants pass through the catalyst bed, it is not sufficient to just measure the temperature of the flowing stream at the inlet and outlet of each bed and/or the reactor. It is necessary to observe the temperature at the inlet, outlet, and radially throughout the catalyst bed. A temperature profile plot is a useful tool for evaluating performance of catalyst, effectiveness of quench, and reactor flow patterns. A temperature profile can be constructed by plotting the catalyst temperature versus distance into the catalyst bed (or more accurately vs. weight percent of catalyst). The hydrocracking reactor should be operated with equal catalyst peak temperatures. In this manner the total catalyst volume is utilized during the entire cycle. The weight average bed temperature (WABT) is typically used to compare the catalyst activity. Figure 7.19 gives a general description of how the WABT is calculated for a reactor.

W EIGHT
A VERAGE
B ED
T EMPERATURE

Attribute A Weight Fraction Of The Catalyst Bed To Each TI

For Example

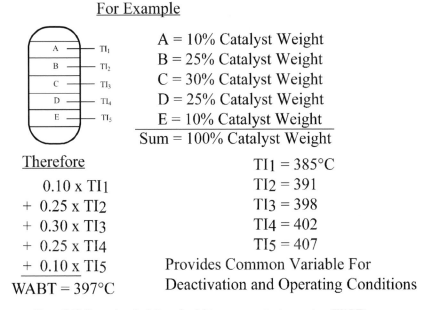

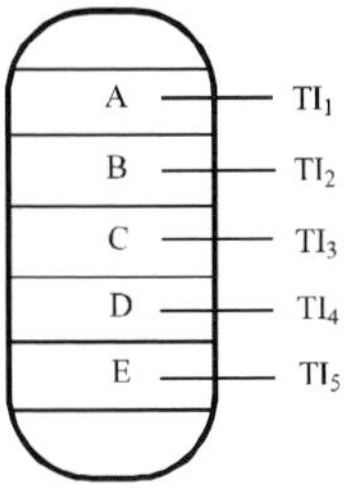

A = 10% Catalyst Weight
B = 25% Catalyst Weight
C = 30% Catalyst Weight
D = 25% Catalyst Weight
E = 10% Catalyst Weight
Sum = 100% Catalyst Weight

Therefore

0.10 x TI1
+ 0.25 x TI2
+ 0.30 x TI3
+ 0.25 x TI4
+ 0.10 x TI5
WABT = 397°C

TI1 = 385°C
TI2 = 391
TI3 = 398
TI4 = 402
TI5 = 407

Provides Common Variable For Deactivation and Operating Conditions

Figure 7.19. Example calculation of weight average reactor temperature (WABT).

The rate of increase of the reactor WABT to maintain both hydrotreating and hydrocracking functions, in order to obtain the desire conversion level and product quality, is referred to as the deactivation rate. It is one of the key variables used to monitor the performance of the catalyst systems. The deactivation rate can be expressed in °F per barrel of feed processed per pound of catalyst (°C per m^3 of feed per kilogram of catalyst) or more simply stated as °F per day (°C per day). The decrease in catalyst activity for hydrotreating catalyst will show up in a decrease in its ability to maintain a constant nitrogen level in the hydrotreating catalyst effluent. For hydrocracking catalyst, a decrease in catalyst activity will generally show up in its ability to maintain a constant conversion to the desired product slate. To hold the same conversion level to the desired product slate the reactor WABT is gradually increased.

Conversion

The term "conversion" is usually defined as:

$$\text{Conversion, vol\%} = (\text{Fresh Feed} - ((\text{Fractionator Bottoms})/\text{Fresh Feed})) \times 100$$

where:

FF = Fresh feed rate, BPD or m^3/hr
Frac Bottoms = Net fractionator bottoms product to storage, BPD or m^3/hr

Conversion is useful as a measure of the severity of the operation. It requires higher severity (meaning higher catalyst temperature) to go to higher conversion levels and higher severity to reduce the endpoint of the product at a constant conversion. Conversion is normally controlled by catalyst temperature.

Fresh feed quality

The quality of the raw oil charged to a hydrocracker will affect the temperature required in the catalyst bed to reach the desired conversion, the amount of hydrogen consumed in the process, the length of time before the catalyst is deactivated, and the quality of products. The effect of the feedstock quality on the performance of the unit is important and should be well understood, especially with regard to contaminants that can greatly reduce the life of the catalyst.

Sulfur and nitrogen compounds

In general, increasing the amount of organic nitrogen and sulfur compounds contained in the feed results in an increase in severity of the operation. The sulfur content of the feed for a normal vacuum gas oil charge stock can vary up to as high as 2.5–3.0 wt%. The higher sulfur levels will cause a corresponding increase in the H_2S content of the recycle gas that will normally have little or no effect on catalyst activity.

The organic nitrogen compounds are converted to ammonia which, if allowed to build up in the recycle gas, competes with the hydrocarbon for the active catalyst sights. This results in a lower apparent activity of the catalyst as the ammonia concentration increases. Because of this, feedstocks with high organic nitrogen contents are more difficult to process and require higher catalyst temperatures.

Hydrogen content
The amount of unsaturated compounds (such as olefins and aromatics) contained in the feed will have an effect on the heat released during the reaction and on the total hydrogen consumption on the unit. In general, for a given boiling range feedstock, a reduction in API gravity (increase in specific gravity) indicates an increase in the amount of unsaturated compounds and, therefore, higher heats of reaction and higher hydrogen consumption. Large amounts of unsaturated hydrocarbons can also cause a heat balance problem if the unit has not been designed to process that type of feed.

Boiling range
The typical charge stock to a Hydrocracker is a 700°F+ (370°C+) boiling range HVGO. Increasing the boiling range usually makes the feed more difficult to process which means higher catalyst temperatures and shorter catalyst life. This is especially true if the feed quality is allowed to decrease significantly due to entrainment of catalyst poisons in the feed. Higher endpoint feeds also usually have higher sulfur and nitrogen contents, which again make it more difficult to process.

Cracked feed components
Cracked feedstocks derived from catalytic cracking or thermal cracking can also be processed in a Hydrocracker. These cracked components tend to have higher contaminants such as sulfur, nitrogen, and particulates. They are also more refractory, with high aromatics content and PNA precursors. These compounds make cracked stocks harder to process to produce quality products.

Permanent catalyst poisons
Organo-metallic compounds contained in the feed will be decomposed and the metals will be retained on the catalyst, thus decreasing its activity. Since metals are normally not removable by oxidative regeneration, once metals have poisoned a catalyst, its activity cannot be restored. Therefore, metals content of the feedstock is a critical variable that must be carefully controlled. The particular metals which usually exist in vacuum gas oil type feeds are naturally occurring nickel, vanadium, and arsenic as well as some metals which are introduced by upstream processing or contamination such as lead, sodium, silicon and phosphorous. Iron naphthenates are soluble in oil and will be a poison for the catalyst. Iron sulfide as corrosion product is normally not considered a poison for the catalyst and is usually omitted when referring to total metals.

The tolerance of the catalyst to metals is difficult to quantify and is somewhat dependent upon the type of catalyst being employed and the severity of the operation, i.e., the higher the severity, the lower will be the metals' tolerance since any impairment of activity will affect the ability to make the desired conversion. It is recommended to keep the total metals in the feedstock as low as possible and certainly not higher than 2 wt-ppm.

Fresh feed rate (LHSV)

The amount of catalyst loaded into the reactors is based upon the quantity and quality of design feedstock and the desired conversion level. The variable that is normally used to relate the amount of catalyst to the amount of feed is termed liquid hourly space velocity (LHSV). LHSV is the ratio of volumetric feed rate per hour to the catalyst volume. Hydrocrackers are normally designed for a LHSV that depends on the severity of the operation. Increasing the fresh feed rate with a constant catalyst volume increases the LHSV and a corresponding increase in catalyst temperature will be required to maintain a constant conversion. The increased catalyst temperature will lead to a faster rate of coke formation and, therefore, reduce the catalyst life. If the LHSV is run significantly higher than the design of the unit, the rate of catalyst deactivation may become unacceptable. LHSV can be defined as:

$$\text{LHSV hr}^{-1} = \frac{\text{Total Feed to Reactor Inlet, m}^3/\text{hr}}{\text{Total Catalyst Volume, m}^3}$$

Liquid recycle

Most hydrocrackers are designed to recycle unconverted feed from the product fractionator bottoms back to the reactors. This stream is normally material distilled above the heaviest fractionator side cut product. For a distillate producing hydrocracker, the recycle stream is normally a 600–700°F (315–370°C) heavy diesel plus material.

The liquid recycle rate is normally adjusted as a ratio with fresh feed. This variable is called combined feed ratio (CFR), and is defined as follows:

$$\text{CFR} = \frac{\text{Fresh Feed Rate} + \text{Liquid Recycle Rate}}{\text{Fresh Feed Rate}}$$

It can be seen that if the unit has no liquid recycle from the fractionator back to the reactors, the CFR is 1.0 and the unit is said to operate once-through, i.e., the fresh feed goes through the catalyst bed only once. If the amount of liquid recycle is equal to fresh feed, the CFR will be 2.0.

An important function of liquid recycle is to reduce the severity of the operation. Considering conversion per pass that is defined as follows:

$$\text{Conversion per Pass} = \left(\frac{\text{Feed Rate} - \text{Frac Bottoms Rate to Storage}}{\text{Feed Rate} + \text{Liquid Recycle Rate}}\right) \times 100$$

It can be seen that if a unit were operating once-through (CFR = 1.0), and 100% of the feed were converted into products boiling below, i.e. 700°F (370°C), the conversion per pass is 100% since the feed only makes one pass through the catalyst. At the other extreme, if a unit is designed at a CFR of 2.0 and 100% of the feed converted into products, the conversion per pass is only 50%. In this way, it can be seen that as the CFR increases, the conversion per pass decreases. It is also seen that the catalyst temperature requirement is reduced as the CFR is increased (at a constant fresh feed conversion level). Therefore, reducing the CFR below the design value can lead to higher catalyst temperatures and shorter catalyst cycle life. Increasing the CFR above design can be helpful when operating at low fresh feed rates since it does not allow the total mass flow through the catalyst bed to reach such a low value that poor distribution patterns are established.

Hydrogen partial pressure

The reactor section operating pressure is controlled by the pressure that is maintained at the high-pressure separator. This pressure, multiplied by the hydrogen purity of the recycle gas, determines the partial pressure of hydrogen at the separator. The hydrogen partial pressure required for the operation of the unit is chosen based on the type of feedstock to be processed and the amount of conversion desired.

The function of hydrogen is to promote the saturation of olefins and aromatics and saturate the cracked hydrocarbons. It is also necessary to prevent excessive condensation reactions from forming coke. For this reason, running the unit for extended periods of time at lower than design partial pressure of hydrogen will result in increased catalyst deactivation rate and shorter time between regeneration.

Hydrogen partial pressure has an impact on the saturation of aromatics. A decrease in system pressure or recycle gas purity has a sharp effect on the product aromatic content. This will be especially true for kerosene aromatic content, which will in turn affect the kerosene product smoke point.

A reduction in operating pressure below its design will have a negative effect on the activity of the catalyst and will accelerate catalyst deactivation due to increased coke formation.

Operating at higher than design pressure may not be possible. There will be a practical equipment limitation on most units that will not allow significantly higher pressure than design, such as the pressure rating of the heaters, exchangers, and vessels. The major control variable for hydrogen partial pressure is the recycle gas purity that should be monitored closely to assure it is always maintained above the minimum value. The hydrogen purity can be improved by increasing the hydrogen purity of the makeup hydrogen, venting gas off the high-pressure separator, or reducing the temperature at the high-pressure separator.

Recycle gas rate

In addition to maintaining a prescribed partial pressure of hydrogen in the reactor section, it is equally important to maintain the physical contact of the hydrogen with the catalyst and hydrocarbon so that the hydrogen is available at the sites where the reaction is taking place. This is accomplished by circulating the recycle gas throughout the reactor circuit continuously with the recycle gas compressor. The amount of gas that must be recycled is a design variable again set by the design severity of the operation. The standard measure of the amount of gas required is the ratio of the gas being recycled to the rate fresh feed being charged to the catalyst.

As with hydrogen partial pressure, the recycle gas/feed ratio should be maintained at the design ratio. The actual calculation for the gas-to-oil ratio, can be defined as:

$$\text{Gas-to-Oil Ratio} = \frac{\text{Total Circulating Gas to Reactor, Nm}^3\text{/hr}}{\text{Total Feed to Reactor Inlet, m}^3\text{/hr}}$$

$$= \text{Nm}^3\text{/m}^3 \text{ Feed}$$

As with hydrogen partial pressure, any reduction of the gas-to-oil ratio below the design minimum will have adverse effects on the catalyst life. During normal operations and through out the cycle length, there will be a gradual increase in the reactor section pressure drop. As the pressure drop increases, there will be a tendency for the gas-to-oil ratio to decrease. When the pressure drop through the system increases to the point where the minimum gas-to-oil ratio cannot be kept, either the unit throughput will have to be decreased to bring the gas-to-oil ratio back above the minimum, or the unit shutdown for catalyst regeneration or replacement.

Gas-to-oil ratio recommendations vary between licensors and/or catalyst vendors but in general the minimums recommended are as follows: (a) 4,000 SCFB (675 nm^3/m^3) for amorphous catalyst systems and 5,500 SCFB (925 nm^3/m^3) for zeolitic catalyst systems.

Makeup hydrogen

The quality of the hydrogen-rich gas from the hydrogen plant is an important variable in the performance of Hydrocrackers since it can affect the hydrogen partial pressure and recycle gas/feed ratio and thereby influence the catalyst stability (deactivation rate). The following guidelines should be used in operating the hydrogen plant to produce acceptable feed gas to a hydrocracker.

Hydrogen purity

The purity of hydrogen in the makeup gas to a Hydrocracker will have a major influence on the hydrogen partial pressure and recycle gas/feed ratio. Therefore, the minimum purity on the makeup gas should be set to provide the minimum recycle gas purity allowed. If the hydrogen plant is unable for some reason to produce minimum hydrogen purity product, it may be possible to purge sufficient recycle gas off the high-pressure separator to maintain the recycle gas purity requirements.

Nitrogen and methane content

The total of the nitrogen and methane contained in the makeup gas is only harmful as a diluent, i.e., it will reduce the hydrogen partial pressure and as long as the minimum hydrogen purity is maintained, it will not affect the unit. However, it should be noted that excessive quantities of molecular nitrogen entering a hydrocracker in the makeup gas stream can cause a buildup of nitrogen in the recycle gas since the nitrogen is non-condensible. If this is the case, the nitrogen will have to be removed from the reactor circuit by a small, continuous purge of recycle gas off the high-pressure separator.

$CO + CO_2$ content

The normal specification for CO plus CO_2 in the makeup gas stream to a Hydrocracker is in low two-digit mol-ppm maximum. Larger quantities can have a harmful effect on catalyst activity. CO is considered the worst impurity due to the fact that it has a limited solubility in both hydrocarbon and water and will, therefore, build up in the recycle gas. CO_2, on the other hand, is much more soluble and is readily removed from the system in the high-pressure separator liquids.

Both CO and CO_2 have similar effects on the Hydrocracking catalyst; they are converted on the active sites of the catalyst in the presence of hydrogen to methane and water. This methanation of CO and CO_2 competes with the normal hydrocarbon reactants for the catalyst. Therefore, if $CO + CO_2$ is allowed to build up, higher catalyst temperatures will be required. In an extreme case where a large quantity of CO or CO_2 would be introduced to the Hydrocracker in a short period of time, it is theoretically possible that a temperature excursion would result since the methanation reaction is highly exothermic.

It is recommended practice that if the $CO + CO_2$ content exceeds the maximum design limit, the catalyst temperature should not be increased to compensate for a resulting decrease in conversion. Catalyst temperature should be maintained at the same level or reduced until the problem causing the high $CO + CO_2$ is eliminated. In this way the catalyst will not be harmed by increased deactivation at a higher temperature and it will also eliminate the possibility of a temperature runaway due to methanation.

Hydrocracker licensors and catalyst manufacturers

Licensors

Hydrocracking licensing started in 1960. Chevron, UOP, Unocal, Shell and Exxon were active from the beginning. Since that time, some 250 hydrocrackers have been licensed worldwide. As of the beginning of 2001, 154 hydrocrackers were in operation. Through the years, the licensing 'landscape' has changed. Currently, the active licensors are Chevron, EMAK (ExxonMobil-Akzo Nobel-Kellogg), IFP and UOP.

Catalyst suppliers

Catalysts used in hydrocrackers are pre-treating catalysts and cracking catalysts. Following is a list of the current major suppliers of pre-treating catalysts: Advanced Refining Technology (in conjunction with Chevron), Akzo Nobel, Criterion, Haldor Topsoe, Axens/Procatalyse (in connection with IFP) and, UOP. The major cracking catalyst suppliers are: Akzo Nobel, Chevron, Criterion and Zeolyst, Axens/Procatalyse (in connection with IFP) and UOP.

Chapter 8

Hydrotreating

Adrian Gruia (retired)*

Hydrotreating or catalytic hydrogen treating removes objectionable materials from petroleum fractions by selectively reacting these materials with hydrogen in a reactor at relatively high temperatures at moderate pressures. These objectionable materials include, but are not solely limited to, sulfur, nitrogen, olefins, and aromatics. The lighter materials such as naphtha are generally treated for subsequent processing in catalytic reforming units, and the heavier distillates, ranging from jet fuel to heavy vacuum gas oils, are treated to meet strict product quality specifications or for use as feedstocks elsewhere in the refinery. Hydrotreating is also used for upgrading the quality of atmospheric resids by reducing their sulfur and organo-metallics level. Many of the product quality specification are driven by environmental regulations that are becoming more stringent every year. Hydrotreaters are designed for and run at a variety of conditions depending on many factors such as type of feed, desired cycle length, expected quality of the products but in general they will operate at the following range of conditions: LHSV—0.2 to 8.0, H_2 circulation—300 to 4,000 SCFB (50–675 Nm^3/m^3), H_{2PP}—200–2,000 psia (14–138 bars) and SOR temperatures ranging between 550 and 700°F (290–370°C), with the lower limits representing minimum operating conditions for naphtha hydrotreating and the higher values showing operating conditions used for hydrotreating atmospheric resids. Until about 1980, hydrotreating was a licensed technology being offered by a fairly large number of companies. In the past 25 years, hydrotreating catalysts have become commodities and the process has been offered without licensing fees.

The common objectives and applications of hydrotreating are listed below:

- Naphtha (catalytic reformer feed pretreatment)—to remove sulfur, nitrogen, and metals that otherwise would poison downstream noble metal reforming catalysts
- Kerosene and diesel—to remove sulfur and to saturate olefins and some of the

*UOP LLC.

aromatics, resulting in improved properties of the streams (kerosene smoke point, diesel cetane number or diesel index) as well as storage stability

- Lube oil—to improve the viscosity index, color, and stability as well as storage stability
- FCC feed—to improve FCC yields, reduce catalyst usage and stack emissions
- Resids—to provide low sulfur fuel oils to effect conversion and/or pretreatment for further conversion downstream.

Brief history

Hydrotreating has its origin in the hydrogenation work done by Sabatier and Senderens, who in 1897 published their discovery that unsaturated hydrocarbons could be hydrogenated in the vapor phase over a nickel catalyst. In 1904, Ipatieff extended the range of feasible hydrogenation reactions by the introduction of elevated hydrogen pressures. At the time, the progress of the automobile industry was expected to entail a considerable increase in the consumption of gasoline. This led to the experimental work by Bergius, started in 1910 in Hanover, Germany who sought to produce gasoline by cracking heavy oils and oil residues as well as converting coal to liquid fuels. He realized that to remedy the inferior quality of the unsaturated gasoline so produced, the hydrogen removed mostly in the form of methane during the cracking operation has to be replaced by addition of new hydrogen. Thus, formation of coke was avoided and the gasoline produced was of a rather saturated character. Bergius also noted that the sulfur contained in the oils was eliminated for the most part as H_2S. Ferric oxide was used in the Bergius process to remove the sulfur. Actually, the ferric oxide and sulfides formed in the process acted as catalysts, though the activity was very poor. The first plant for hydrogenation of brown coal was put on stream in Leuna Germany in 1927. The past large scale industrial development of hydrogenation in Europe, particularly in Germany, was due entirely to military considerations. Germany used hydrogenation extensively during World War II to produce gasoline: 3.5 million tons were produced in 1944. The first commercial hydrorefining installation in the United States was at the Standard Oil Company of Louisiana in Baton Rouge in the 1930s. WWII plants were developed by Humble Oil and Refining Company and Shell Development Company, though there was considerably less dependence on hydrogenation as a source of gasoline. Even though hydrogenation has been of interest to the petroleum industry for many years, little commercial use of hydrogen-consuming processes has been made because of the lack of low-cost hydrogen. That changed in the early 1950s with the advent of catalytic reforming which made available by-product hydrogen. That brought up an extensive and increased interest in processes that will utilize this hydrogen to upgrade petroleum stocks. As a result of the enormous growth of hydrotreating, as of the beginning of 2001, there were more than 1,600 hydrotreaters operating in the world with a total capacity in excess of 39,000,000 B/D (4,800,000 MT/D).

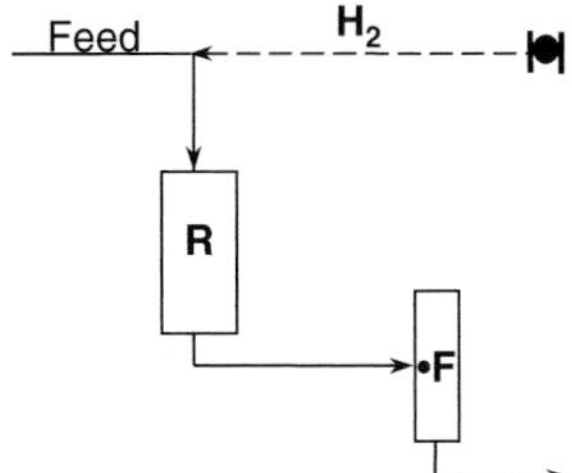

Figure 8.1. Schematic flow diagram of a hydrotreater.

Flow schemes

Although the 'hydrotreating process' has several different applications (e.g. desulfurization, olefin saturation, denitrogenation, etc) and is used for a variety of petroleum fractions from naphtha all the way to atmospheric residue, practically all units have the same flow scheme. It consists of a higher-pressure reactor section and a lower pressure fractionation section. This is shown very schematically in Figure 8.1 and is described below in general terms.

Reactor section

The reactor section consists of the following major pieces of equipment: feed/effluent exchangers, reactor charge heater, reactor(s), reactor effluent condenser, products separator, recycle gas compressor, and make-up gas compressors. Additionally, several other pieces of equipment are found in some hydrotreating units: fresh feed filters, reactor effluent hot separator, recycle gas scrubber. Figure 8.2 shows a schematic flow diagram of a reactor section including all the equipment, which is listed above.

Feed filters
It is preferable to route the feed directly from an upstream unit without going through intermediate storage. When storage facilities are used, feed filters are (should be) used. The purpose of the filters is to retain the particulate matter (mostly corrosion products) picked by the feed while in storage. The feed filters are either automatic backwash filters operating on a pressure drop setting or manual cartridge (disposable) filters.

Feed/effluent exchangers
In the most commonly used heat recovery scheme, the reactor effluent in a series of feed/effluent exchangers preheats the reactor charge before entering the reactor charge heater. This recovers as much heat as possible from the heat of reaction. Liquid feed may be preheated separately with reactor effluent exchange before combining with the recycle gas depending on the heat integration scheme.

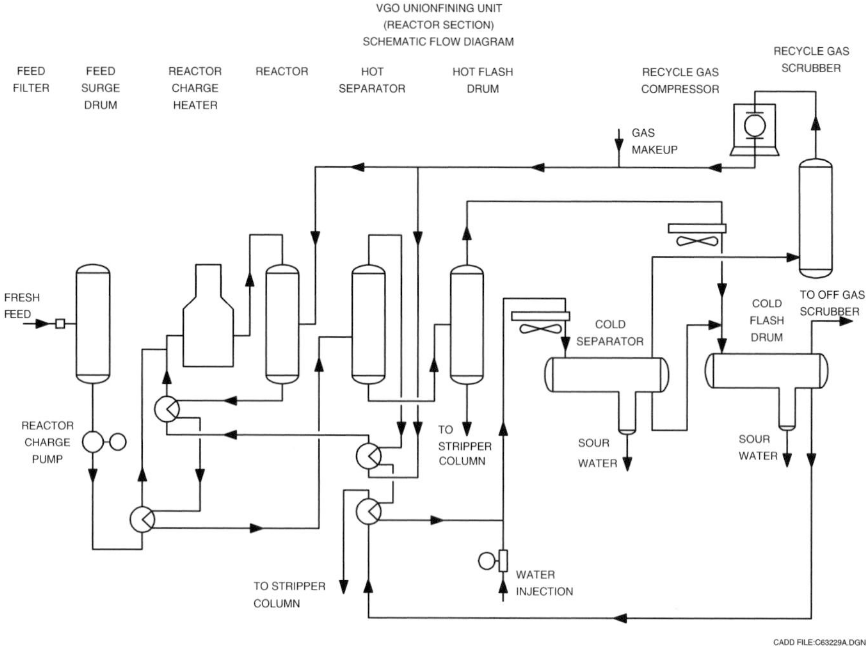

Figure 8.2. Schematic flow diagram of a reactor section.

Reactor charge heater

In most units, the combined feed and recycle gas is heated together to desired reactor inlet temperature in a combined charge heater. In units processing heavy feed, especially the atmospheric residue units, the liquid feed is preheated separately with reactor effluent exchange and only the recycle gas is heated in the heater upstream of the reactor.

Make-up hydrogen system

Make-up H_2 is typically obtained from hydrogen manufacturing plants and/or naphtha reforming units. Depending on the pressure of the hydrotreating unit, the make-up hydrogen might have to be compressed before introduction into the unit. Reciprocating compressors are used for this service. The make-up gas is introduced into recycle gas system.

Recycle hydrogen system

After the separation of gas and liquid phases in the separator, the gas goes to the recycle gas compressor. In some cases, the recycle gas will be sent first to an amine scrubber to remove most of the H_2S. Most often, the recycle gas compressor is a separate centrifugal machine, but it could also be a part of the make-up gas compressors, as additional cylinders. The recycle gas compressor is designed to pump a large volume of gas at a relatively low compression ratio.

Recycle gas scrubbing

The recycle gas stream will contain H_2S. The H_2S reduces the hydrogen partial pressure and thus suppresses the catalyst activity. This effect is more pronounced with high sulfur containing feed stream and for the same feedstock, the heavier the cut, the higher the sulfur content is. Recycle gas scrubbing is typically included in the design if the H_2S recycle gas content is expected to be in excess of 3 vol%.

Reactor(s)

Once the feed and recycle gas have been heated to the desired temperature, the reactants enter the top of the reactor. As the reactants flow downward through the catalyst bed, various exothermic reactions (which will be explained later) occur and the temperature increases. Multiple catalyst beds (and possibly additional reactor may be required depending on the heat of reaction, unit capacity and/or type of hydrotreating unit (its intended goal). Specific reactor designs will depend upon several variables. Reactor diameter is typically set by the cross-sectional liquid flux. As the unit capacity increases, the reactor diameter increases to the point where two parallel trains would be considered. Reactor height is a function of the amount of catalyst and number of beds required. Depending on the expected heat of reaction, cold recycle gas is brought into the reactor at the interbed quench points in order to cool the reactants and thus control the reaction rate. Good distribution of reactants at the reactor inlet and at the top of each subsequent catalyst bed is essential for optimum catalyst performance. There are many companies that design proprietary internals. These internals are reactor inlet diffuser, top liquid distribution tray, quench section which includes quench inlet assembly, quench and reactants mixing device and redistribution tray, as well as the reactor outlet device. As indicated, not all reactors contain all the internals described above.

Reactor effluent water wash

Most of the cooling of the effluent is accomplished in the feed/effluent exchangers. Final cooling of the reactor effluent is obtained in air fin coolers and/or water trim coolers. Water is injected into the stream before it enters the coolers in order to prevent the deposition of salts that can corrode and foul the coolers. The sulfur and nitrogen contained in the feed are converted to hydrogen sulfide and ammonia in the reactor. These two reaction products combine to form ammonium salts that can solidify and precipitate as the reactor effluent is cooled. Likewise, ammonium chloride may be formed if there is any chloride in the system. The purpose of the water is to keep the H_2S and NH_3 in solution and not allow them to precipitate. Various companies have slightly different guidelines for the quality of the water injection, but in general use of boiler feed water is preferred.

Vapor/liquid separation

The exact method of separating vapor and liquid will vary depending on the optimum heat integration scheme. Up to four separate vessels may be used to disengage and individually remove vapor, water and hydrocarbon liquid. A hot separator is sometimes installed after the feed/effluent exchangers to collect the heavier hydrocarbon material from the reactor effluent and send it to fractionation via the hot flash drum. The overhead vapor from the hot separator continues through an air cooler into a cold separator. The two-separator system offers an improved heat integration scheme (all this is shown in Figure 8.2).

Hydrogen purification

Increasing the recycle gas hydrogen purity will decrease the catalyst deactivation rate. Depending upon the feedstock and type of unit, additional measures may be taken to increase the hydrogen purity. These measures may include hydrogen enrichment and/or membrane separation.

Fractionation section

A schematic flow diagram of a typical fractionation section is shown in Figure 8.3.

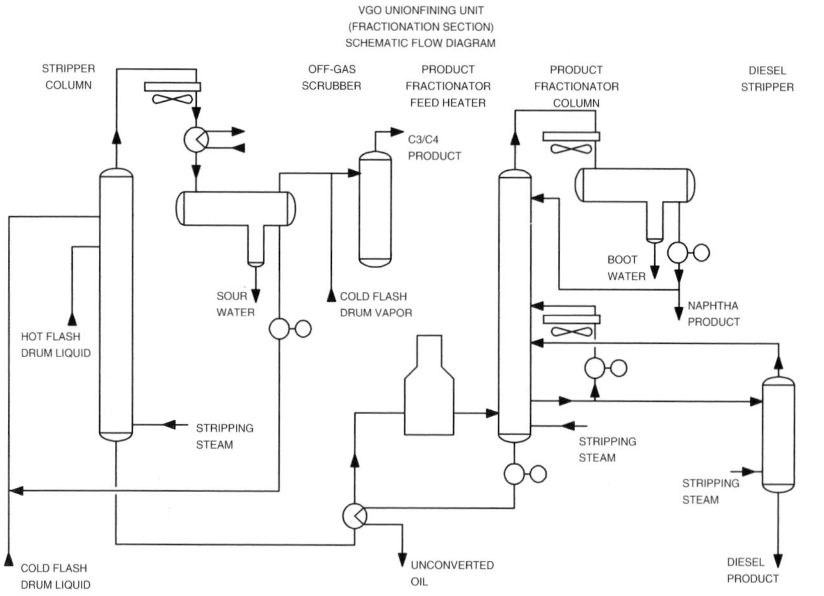

Figure 8.3. Schematic flow diagram of a fractionation section.

The function of the fractionation section is to separate the reactor effluent into the desired products. This can accomplished with either a one or a two-column fractionation scheme, depending on the type of hydrotreating unit.

In the two columns scheme the flash drum liquids combine and go to a stripper column. Steam and/or a fired heater is used to strip naphtha (if desired) and lighter material overhead. The stripper bottoms go to a fractionator where it is further separated into naphtha (if desired) and heavier products. The fractionator feed is typically preheated with fractionator bottoms and a fired heater before entering the column. Stripping steam is used to drive lighter material up the column and various product strippers are used to pull sidecut products to the desired specifications.

Chemistry

The following chemical steps and/or reactions occur during the hydrotreating process (depending on the impurities present):

- Sulfur removal, also referred to as desulfurization or hydro-desulfurization (HDS) in which the organic sulfur compounds are converted to hydrogen sulfide
- Nitrogen removal, also referred to as denitrogenation or hydro-denitrogenation (HDN) in which the organic nitrogen compounds are converted to ammonia
- (organo-metallic) metals removal, also referred to as hydro-demetallation or hydro-demetallization, in which the organo-metals are converted to the respective metal sulfides
- oxygen removal, in which the organic oxygen compounds are converted to water
- olefin saturation, in which organic compounds containing double bonds are converted to their saturated homologues
- aromatic saturation, also referred to as hydro-dearomatization, in which some of the aromatic compounds are converted to naphthenes
- and halides removal, in which the organic halides are converted to hydrogen halides

The first three types of compounds are always present though in varying amounts depending on the source of feed stock. For example, naphtha will contain extremely low amounts of organo-metallic compounds while atmospheric residues will contain levels in % range. Some crudes contain much more sulfur in all the fractions when compared with other crudes (for example, most middle eastern crudes contain much more sulfur than some crudes from Indonesia or North Africa). The same is true for nitrogen levels, as well. The other impurities are not always present. In general, the hydrotreating reactions proceed in the following descending order of ease: (organo-metallic) metals removal, olefin saturation, sulfur removal, nitrogen removal, oxygen removal, and halide removal. Some aromatic saturation also occurs. The chemistry of residue hydrotreating is essentially that of contaminant removal and selective hydrogenation which includes both olefin and aromatic saturation. Hydrogen is consumed in all the reactions. The contaminant removal in residue hydrotreating involves controlled breaking of the hydrocarbon molecule at the point where the sulfur, nitrogen or oxygen atom is joined to carbon atoms. Some cracking occurs in residue hydrotreating, but it normally is less than 20 vol% of the fresh feed charge.

In general, the desulfurization reaction consumes 100–150 SCFB/wt% change (17–25 Nm^3/m^3/wt% change) and the denitrogenation reaction consumes 200–350 SCFB/wt% change (34–59 Nm^3/m^3/wt% change). Typically, the heat released in hydrotreating is about 0.02°F/SCFB H_2 consumed (0.002°C/$Nm^3/m^3 H_2$).

In general, the 'main messages' concerning hydrotreating reaction rates, heats of reaction and hydrogen consumption are:

- Desulfurization and olefin saturation are the most rapid reactions
- Olefin saturation liberates the most heat per unit of hydrogen consumed
- Denitrogenation and aromatic saturation are the most difficult reactions
- Hydrogen consumption and heat of reaction are related

Sulfur removal

Sulfur removal occurs via the conversion to H_2S of the organic sulfur compounds present in the feedstock. This conversion is sometimes referred to as desulfurization or hydro-desulfurization (HDS). Sulfur is found throughout the boiling range of petroleum fractions in the form of many hundreds of different organic sulfur compounds which, in the naphtha to atmospheric residue range, can all be classified as belonging to one of the following six sulfur types: mercaptans, sulfides, di-sulfides, thiophenes, benzo-thiophenes, and di-benzo-thiophenes. Typical reactions for each kind of sulfur compound are shown below.

Mercaptans

$$R\text{-}SH + H_2 \rightarrow R\text{-}H + H_2S$$

Sulfides

$$R1\text{-}S\text{-}R2 + 2H_2 \rightarrow R1\text{-}H + R2\text{-}H + H_2S$$

Di-sulfides

$$R1\text{-}S\text{-}S\text{-}R2 + 3H_2 \rightarrow R1\text{-}H + R2\text{-}H + 2H_2S$$

Thiophenes

$$R\text{-}C_4H_3S + 4H_2 \longrightarrow R - CH(CH_3)\text{-}CH_2\text{-}CH_3 + H_2S$$

Benzo-thiophenes

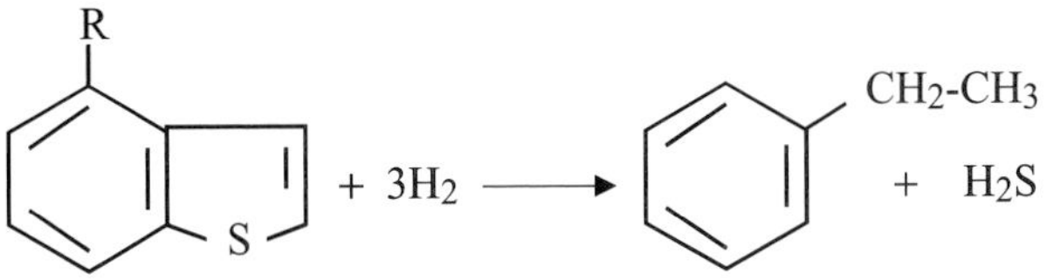

R

$$+ 3H_2 \longrightarrow$$ $CH_2\text{-}CH_3$ $+ H_2S$

Di-benzo-thiophenes

R R

$$+ 2H_2 \longrightarrow$$ $+ H_2S$

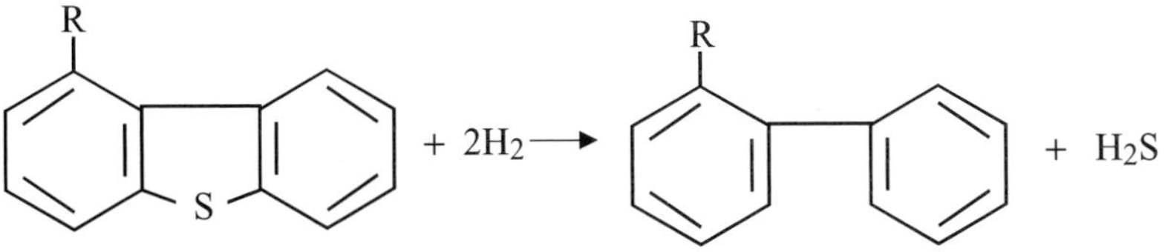

Most of the reactions are straightforward with the exception of the desulfurization of aromatic sulfur species. This reaction is more complex because it must start with ring opening and sulfur removal followed by saturation of the resulting olefin. The postulated mechanism for the desulfurization reaction is shown below: first, the sulfur is removed followed by the saturation of the intermediate olefin compound. In the example below the thiophene is converted to butene as an intermediate which is then saturated into butane.

Desulfurization mechanism

(A) Sulfur removal

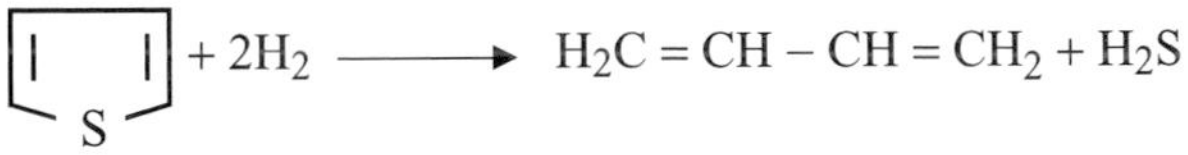

$$+ 2H_2 \longrightarrow H_2C = CH - CH = CH_2 + H_2S$$

S

(B) Olefin saturation

$$H_2C = CH\text{-}CH = CH_2 + 2H_2 \rightarrow H_3C\text{-}CH_2\text{-}CH_2\text{-}CH_3$$

Shown below is a ranking of the six sulfur types ranked on the basis of ease of removal:

Easiest to remove → Hardest to remove

Mercaptans → Sulfides → Disulfides → Thiophenes → Benzo-thiophenes → Dibenzo-thiophenes

The relative ease of removing sulfur from a particular hydrocarbon fraction depends greatly on the sulfur types present. In naphtha fractions, much of the sulfur is present as mercaptans and sulfides which makes for relatively easy sulfur removal. In gas oil fractions, the majority of the sulfur is present as benzo-thiophenes and di-benzo-thiophenes. Hence the sulfur is much more difficult to remove from gas oils than from naphtha fractions. And the more difficult sulfur species are found in the heavier fractions, which means that heavy gas oils are more difficult to treat than light gas oils. As an example of the relative degree of desulfurization: if the degree of difficulty of converting the di-ethylsulfide were 1, thiophene is 5 times more difficult to convert, benzo-thiophene is 15 times more difficult to convert, and di-benzo-thiophene is 20 times more difficult to convert.

Nitrogen removal

Nitrogen is mostly found in the heaviest end of petroleum fractions in five- and six-membered aromatic ring structures. Both the molecular complexity and quantity of nitrogen containing molecules increases with increasing boiling range, making them more difficult to remove. The denitrogenation reaction proceeds through a different path from that of desulfurization. While in desulfurization the sulfur is removed first and the olefin created as an intermediate is saturated, in denitrogenation, the aromatic is saturated first and then the nitrogen is removed. This is shown below.

Denitrogenation mechanism

(A) Aromatic hydrogenation

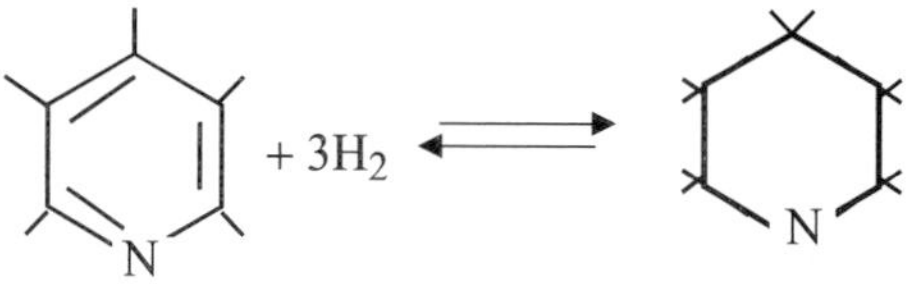

N + $3H_2$ ⇌ N

(B) Hydrogenolysis

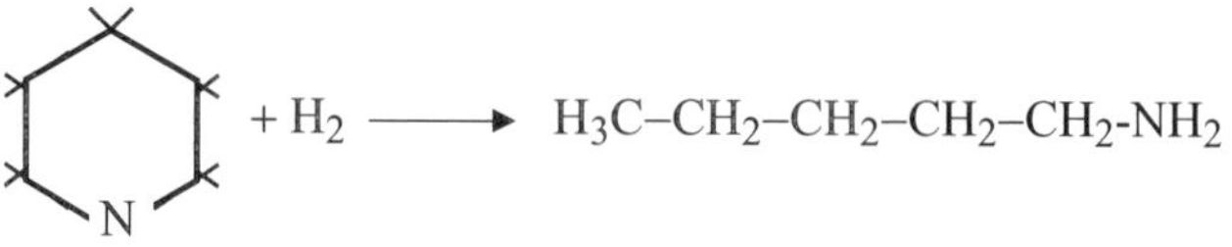

N + H_2 ⟶ $H_3C–CH_2–CH_2–CH_2–CH_2\text{-}NH_2$

(C) Denitrogenation

$$H_3C\text{-}CH_2\text{-}CH_2\text{-}CH_2\text{-}CH_2\text{-}NH_2 + H_2 \rightarrow H_3C\text{-}CH_2\text{-}CH_2\text{-}CH_2\text{-}CH_3 + NH_3$$

Some typical examples of denitrogenation reactions are shown below.

(A) Amine

$$H_3C\text{-}CH_2\text{-}CH_2\text{-}CH_2\text{-}NH_2 + H_2 \rightarrow H_3C\text{-}CH_2\text{-}CH_2\text{-}CH_3 + NH_3$$

(B) Pyrrole

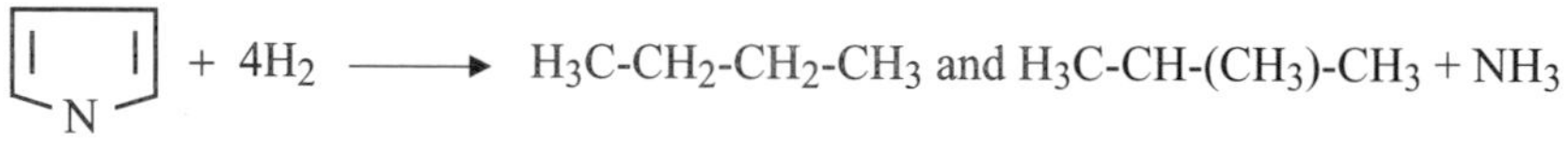

N + $4H_2$ ⟶ $H_3C\text{-}CH_2\text{-}CH_2\text{-}CH_3$ and $H_3C\text{-}CH\text{-}(CH_3)\text{-}CH_3 + NH_3$

(C) Pyridine

$$C_5H_5N + 5H_2 \rightarrow H_3C\text{-}CH_2\text{-}CH_2\text{-}CH_2\text{-}CH_3 \text{ and } H_3C\text{-}CH\text{-}(CH_3)\text{-}CH_2\text{-}CH_3 + NH_3$$

(D) Quinoline

$$C_9H_7N + 4H_2 \rightarrow C_6H_5\text{-}CH_2\text{-}CH_2\text{-}CH_3 + NH_3$$

Nitrogen is more difficult to remove and consumes more hydrogen than sulfur removal because the reaction mechanism involves aromatic ring saturation prior to nitrogen removal. In desulfurization, the sulfur is less often associated with aromatic rings and when it is, the sulfur can be removed without ring saturation. Hydrogenation of associated aromatic ring structures is very dependent on hydrogen partial pressure and is the rate limiting reaction step in nitrogen removal. Nitrogen removal is therefore dependent on H_2 partial pressure.

Oxygen removal

Most petroleum crudes contain low levels of oxygen. The oxygen-containing compounds are converted, by hydrogenation, to the corresponding hydrocarbon and water. The lower molecular weight compounds are easily hydrogenated, however, the higher molecular weight compounds—e.g. furans—can be difficult to remove. Shown below are typical examples of de-oxygenation.

Phenols

$$C_6H_4(\text{-OH})(\text{-R}) + H_2 \rightarrow C_6H_5\text{-R} + H_2O$$

Oxygenates

$$(CH_3)_3\text{-C-O-}CH_3 + 2H_2 \rightarrow (CH_3)_3\text{-CH} + H_2O + CH_4$$

Naphthenic Acids

$$R-C(=O)-OH + 3H_2 \rightarrow R\text{-}CH_3 + 2H_2O$$

Olefin saturation

Olefins are not found in petroleum, but are formed when processed in thermal or catalytic units. In general, fractions containing olefins are unstable and thus must be protected from contact with oxygen prior to hydrotreating to prevent the formation of polymer gums. That is especially true for feedstocks derived from thermal cracking operations such as coking and ethylene manufacturing. Typical olefin saturation reactions are shown below.

Hexene

$$C_6H_{12} + H_2 \rightarrow C_6H_{14}$$

Cyclohexene

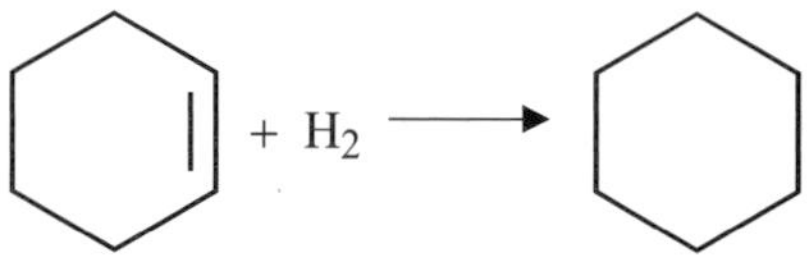

$$+ H_2 \longrightarrow$$

Olefin saturation reactions are very rapid and highly exothermic. While the denitrogenation reaction shows a heat of reaction of 1 Btu/lb of feed for each 100 ft^3 of H_2 consumed, and the desulfurization reaction generates 1 Btu/lb of feed for each 10 ft^3 H_2 consumed, the olefin saturation generates 1 Btu/lb of feed for each 2 ft^3 of H_2 consumed. If proper care is not exercised during operations, it can result in mechanical problems such as excessive coking that can lead to pressure drop build up and/or poor liquid flow distribution through the catalyst bed(s). Diolefins are readily hydrogenated to olefins at low temperatures (<400°F, or less than about 200°C).

Aromatic saturation

Saturation of aromatics is desirable for improvement of the properties of petroleum products e.g. smoke point, diesel index, etc. The aromatics found in the naphtha to gas oil boiling range are present as one, two, and three ring aromatics—often referred to as mono, di, and tri aromatics. Typical reactions are shown below

One ring—Toluene

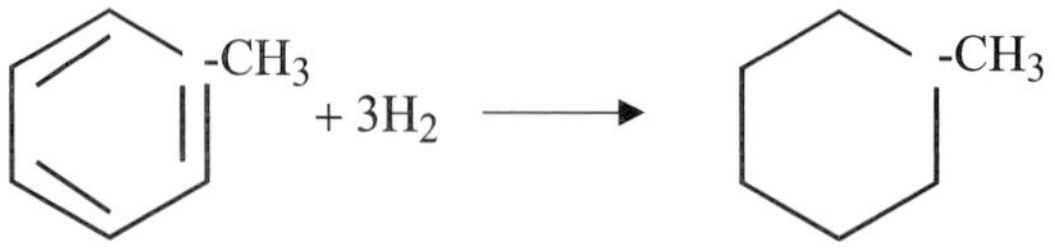

$$-CH_3 + 3H_2 \longrightarrow -CH_3$$

Two ring—Naphthalene

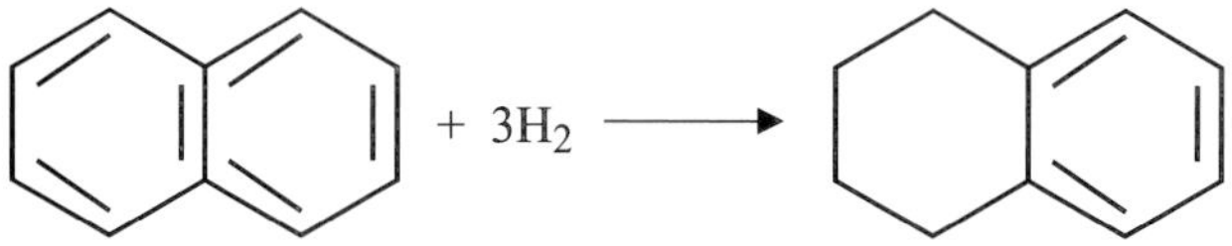

$$\text{Naphthalene} + 3H_2 \longrightarrow \text{Tetralin}$$

Three ring—Phenanthrene

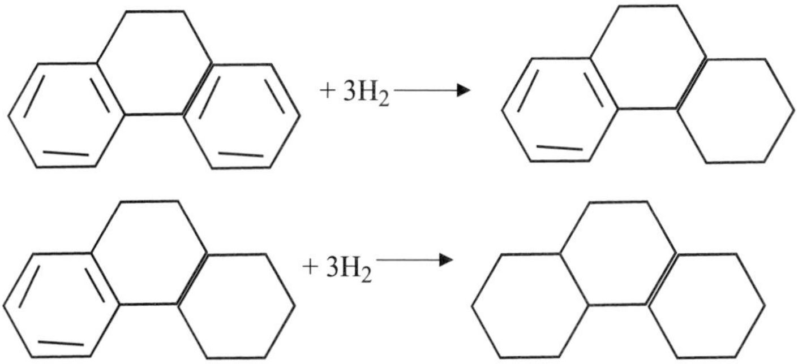

$$\text{Phenanthrene} + 3H_2 \longrightarrow \text{partially saturated product}$$

$$+ 3H_2 \longrightarrow$$

The reactions shown above provide the mechanism by which poly saturates aromatics. That occurs via a stepwise mechanism; i.e. from 3-ring to 2-ring to 1-ring; and the end products are naphthane rings. Ring opening does not occur in hydrotreating (it does in hydrocracking) because there is very little hydrocracking function within a standard hydrotreating catalyst. The aromatic saturation reaction is strongly favored by high hydrogen partial pressure. Unlike all the other hydrotreating reactions, the amount of conversion of aromatics becomes equilibrium limited at higher operating temperatures within the commercial operating range. This is because the reverse reaction of naphthene dehydrogenation becomes favored when temperature is increased. The optimum temperature for maximum aromatic saturation depends on LHSV, hydrogen partial pressure and catalyst type, but typically lies in the range 320–350°C.

Mono aromatic rings are much more difficult to saturate than the di and tri aromatic rings because the saturation of the last aromatic ring requires the most energy. This means that as aromatic saturation proceeds, there is little progress in total aromatics reduction until most, if not all, of the di and tri aromatics have been saturated. The complete saturation of aromatics requires significantly more severe processing conditions than those used in 'normal' hydrotreating.

Metals removal

Most metallic impurities occur in naphthas and middle distillates at ppm or even ppb levels. They are present as organo-metallic compounds. In naphtha hydrotreating, the most commonly occurring metals are arsenic from certain crude sources, mercury from certain condensates and silicon from anti-foam agents used in visbreakers and cokers. These compounds decompose in the hydrotreater and the metal is deposited on the catalyst in the form of metal sulfide as shown below.

$$\text{R-Me} + H_2S \rightarrow \text{R-}H_2 + \text{MeS}$$

Once deposited, these metals contribute to catalyst deactivation and unlike coke are not removed by regeneration. Gas oil streams can contain traces of nickel and vanadium in the heavier feedstock fractions. These too are deposited on the catalyst and contribute to deactivation. Atmospheric residua can contain metals, almost exclusive Ni and V, in the three-digit ppm range. Demetallation of that type of feedstock is an important goal of processing and special demetallation catalyst is employed for that purpose. Demetallation occurs before desulfurization and any conversion of the feedstock takes place.

Halide removal

Organic halides, such as chlorides or bromides, can be present in petroleum fractions at trace levels. Under hydrotreating conditions, organic halides are largely converted to the corresponding hydrocarbon and hydrogen halide. The typical reaction is shown below.

$$C_6H_5\text{-}CH_2\text{-}CH_2\text{-}CH_2\text{-}Cl + 4H_2 \longrightarrow C_6H_5\text{-}CH_2\text{-}CH_2\text{-}CH_3 + HCl$$

Catalysts

Hydrotreating catalysts are high surface area materials consisting of an active component and a promoter, which are uniformly dispersed on a support. The catalyst support is normally gamma alumina (γ-Al_2O_3), sometimes with small amounts of silica or phosphorous added, which is prepared in such a way so as to give a high surface area and an appropriate pore structure. The active component is normally molybdenum sulfide, although tungsten containing catalysts are also used (though seldom, and that generally for special applications such as lube oil processing). For molybdenum catalysts both cobalt (CoMo) and nickel (NiMo) are used as promoters. The promoter has the effect of substantially increasing (approximately 100-fold) the activity of the active metal sulfide. The acidity of the support (which is provided by the silica and/or phosphorous) can be increased to boost the catalyst activity for (hydro)cracking and isomerization reactions. The commercially available catalysts have varying amounts of promoters and active components, depending on the desired applications, but in

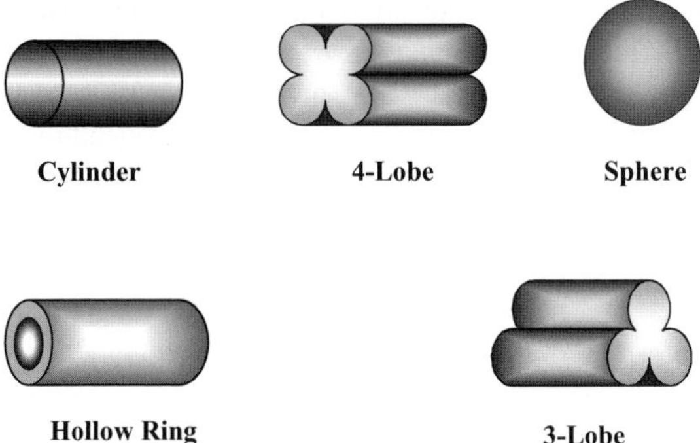

Figure 8.4. Hydrotreating catalysts shapes.

general they can contain up to about 25 wt% promoter and 25 wt% active component. Hydrotreating catalysts come in different sizes and shapes and vary depending on the manufacturer (Figure 8.4):

- Cylindrical 1/32″–1/4″
- Trilobe 1/20″–1/10″
- Quadrilobe 1/20″–1/10″
- Spheres 1/16″–1/4″
- Hollow rings Up to 1/4″

The size and shape of the catalyst pieces is a compromise between the wish to minimize pore diffusion effects in the catalyst particles (requiring small sizes) and pressure drop across the reactor (requiring large particle sizes). The physical characteristics of catalysts also vary from manufacturer to manufacturer and the intended use of the catalyst, but in general are as follows:

- High surface area 150 m^2/g or more
- Pore volume 0.6–1.0 ml/g
- Average pore radius 30–100 Angstrom
- Compacted bulk density 35–55 lbd/ft^3
- Crushing strength 4–20 lbs/in^2
- Average length (except spheres) 1/8–3/8 in

Cobalt–Moly Catalysts

By and large, CoMo catalysts have been designed primarily for desulfurization, but some denitrogenation and demetallation is also achieved. These catalysts can treat feedstocks of widely varying properties. CoMo catalysts have the lowest hydrogenation activity, therefore they have the lowest hydrogen consumption for a given sulfur removal. They also have the lowest sensitivity of H_2 consumption to changes in operating pressure. In general, CoMo catalysts have the highest desulfurization performance at the lower operating pressures (<600 psig, or <∼ 40 barg). These catalysts also have the lowest denitrogenation performance due to low hydrogenation activity. Because CoMo catalysts exhibit the highest sulfur removal per unit of hydrogen consumed, they are best suited for desulfurization at lower pressures and when hydrogen is in short supply.

Nickel–Moly catalysts

NiMo catalysts have been designed for desulfurization, but particularly for hydrogenation and denitrogenation. Metal removal can also be achieved. These catalysts can treat feedstocks of widely varying properties. NiMo catalysts have higher denitrogenation activities than CoMo and are therefore used for cracked stocks or other applications where denitrogenation and/or saturation is as important as desulfurization. The higher hydrogenation power of NiMo catalysts allows them to be used as a topping layer to saturate olefins and other gum precursors to mitigate catalyst bed fouling leading to pressure drop accumulation and poor liquid flow distribution through the catalyst bed. The performance of NiMo catalysts is very good at high pressures. NiMo catalysts show a greater response in denitrogenation and desulfurization performance to changes in H_2 partial pressure than CoMo. High-pressure operations, such as FCC and hydrocracking feed pretreatment, therefore favor the use of NiMo catalysts. NiMo catalyst use is also favored for reforming units pretreating as the modern reforming catalysts are very sensitive to the nitrogen content of the feedstock.

Other catalysts

Other catalysts used in hydrotreating are NiW and NiCoMo. NiW catalysts have applications in treating feeds where higher hydrogenation activity is required than is available from either NiMo or CoMo. In general, their desulfurization activity is poor at the pressure levels used in hydrotreating—they perform very well however, at the high pressures used in hydrocracking. NiW in sulfided form exhibits hydrocracking activity surpassing that of both CoMo and NiMo. Increasing the activity of the support material with promoters or zeolite can further enhance the hydrocracking activity. NiW can be made selective for saturating one of the double bonds in diolefins in light feeds, which may be desirable in some hydrotreating operations. NiCoMo catalysts attempt to combine the benefits of CoMo and NiMo, however, they are rarely used.

Measuring catalyst performance

Catalyst performance is measured by several criteria shown below, which are more or less self-explanatory:

- Initial activity, which is measured by the temperature required to obtain desired product at the start of the run. During the cycle, the catalyst activity can be calculated as shown below:

$$D_s = D_S^0 e^{At}$$

where

D_s = desulfurization activity
D_S^0 = initial desulfurization activity
A = deactivation rate, °F/bpp
t = catalyst life, bpp

- Stability, which is measured by the rate of temperature increase required to maintain product quality
- Product quality, which is a measure of the ability of the catalyst to produce products with the desired use specifications, such as pour point, smoke point, or cetane number

Catalyst manufacturing

Hydrotreating catalysts contain metals dispersed on a support. That support is γ-alumina which is arrived at by synthesis. Several raw materials can be used to produce the γ-alumina:

- Gibbsite (α-alumina trihydrate)
- Bayerite (β-alumina trihydrate)
- Boehmite (α-alumina monohydrate)

Hydrotreating catalysts can be manufactured by several methods:

- Impregnation
- Co-mulling
- Hot soaking

Impregnation

When catalyst is manufactured by the impregnation the support is first made, followed by loading of the support with metals, by wet impregnation. The support can be manufactured either in spherical shape or by extrusion.

Figure 8.5 shows the preparation of spherical support by the oil drop method. Figure 8.6 shows support preparation by extrusion. The support, either spherical or extruded is then finished by wet impregnation as shown in Figure 8.7. Figure 8.8 depicts hydrotreating catalyst manufactured by co-mulling. Figure 8.9 shows the schematic of catalyst manufacturing by hot soaking.

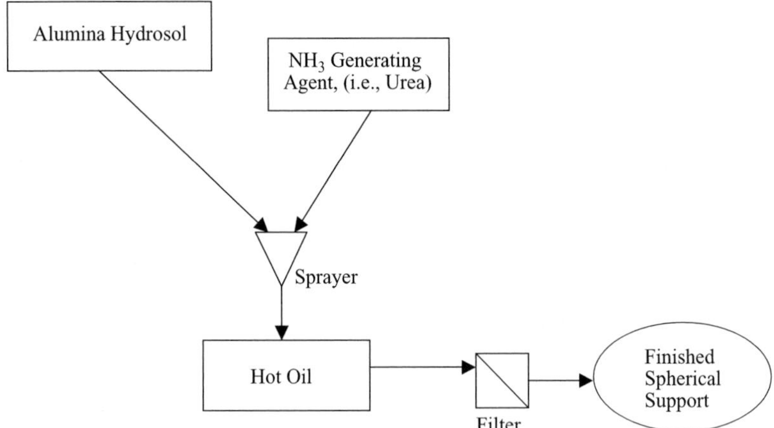

Figure 8.5. Oil dropping (spherical support preparation).

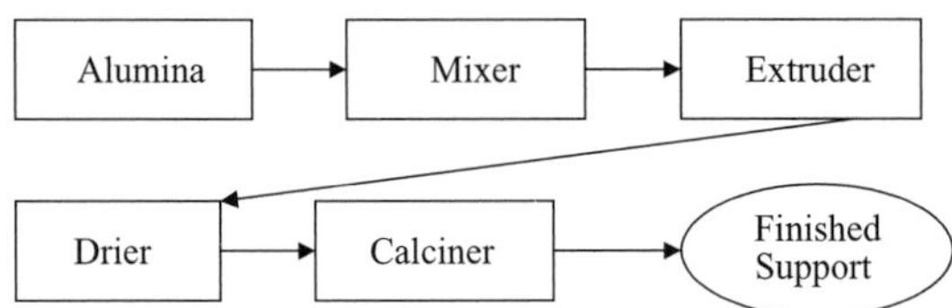

Figure 8.6. Support preparation by extrusion.

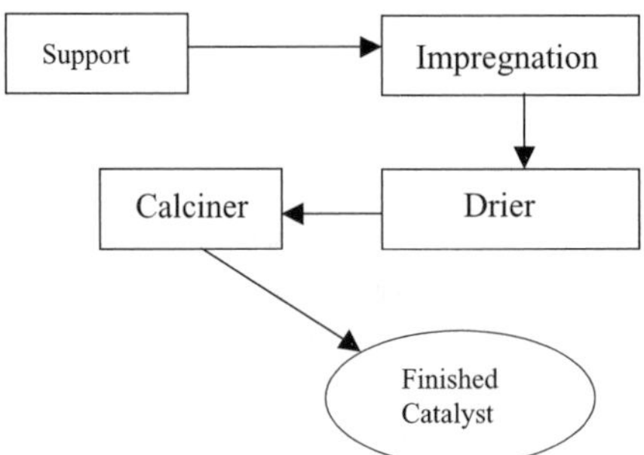

Figure 8.7. Wet impregnation.

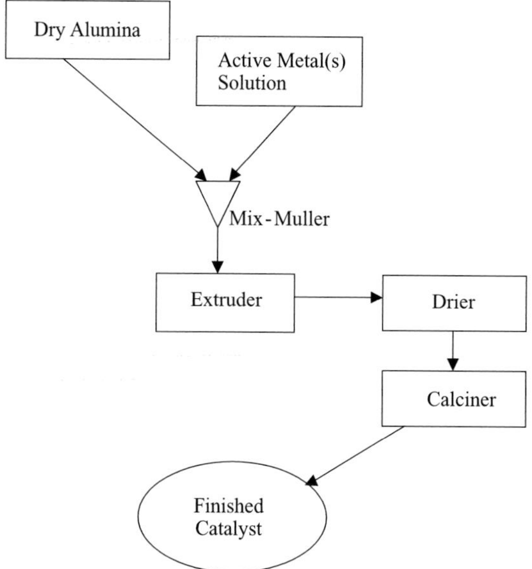

Figure 8.8. Co-mulling.

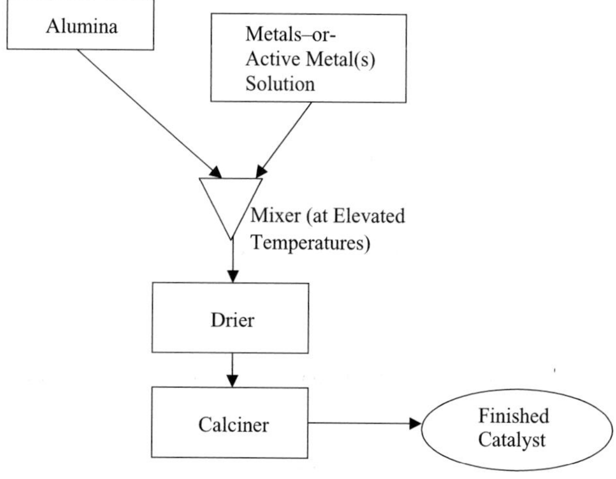

Figure 8.9. Hot soaking.

During catalyst preparation, there are several variables that have an influence on the finished product. They are:

- Mixing intensity (influences the pore size)
- Peptization
- Calcination (time, temperature, concentration)
- Additives to mixing
- Metals application
- Solution preparation (contacting, time, order, drying)
- Handling and screening.

Catalyst loading and activation

Catalyst loading

There are two methods of catalyst loading: sock loading and dense loading. Pouring catalyst into a hopper mounted on top of the reactor and then allowing it to flow through a canvas sock into the reactor is sock loading. Dense loading or dense bed packing is done with the help of a mechanical device. The dense loading method was introduced in the mid 1970s. Catalyst loaded by sock loading will have a higher void fraction than catalyst that was dense loaded. Dense bed packing and the resulting higher pressure drop provides a more even distribution of liquid in a trickle flow reactor which is the flow regime for most hydrotreating applications. If diffusion limitations are negligible, dense loading is desirable in order to maximize the reaction rate per unit reactor volume. This is often the case in hydrotreating reactors. The other advantage of dense loading is that it orients the catalyst particles in a horizontal and uniform manner. This improves the vapor/liquid distribution through the catalyst beds. Catalyst particle orientation is important especially for shaped extruded catalyst in vapor/liquid reactant systems. When the catalyst particles are oriented in a horizontal position in the catalyst bed, liquid maldistribution or channeling is eliminated. This maldistribution tends to occur when the catalyst loading is done by the sock loading method, which generally causes the extrudates to be oriented in a downward slant toward the reactor walls increasing bed voids and creating liquid maldistribution. Of all the factors influencing catalyst utilization, catalyst loading has generally proven to be the most important factor. Another advantage of dense loading is that it allows loading more catalyst in the reactor because of the reduced void fraction in the catalyst bed. As much as 15–20% more catalyst can be loaded when dense loading, compared with sock loading. Thus, the catalyst life can be extended or else the unit can be operated at more severe conditions (lower product sulfur level, increased feed rate) than if the catalyst had been sock loaded. Except for the hydrotreaters that have reactor pressure drop limitations mainly due to operation at higher than design throughputs, most hydrotreaters are dense loaded. Ex-situ presulfurized catalysts (see catalyst activation) are self-heating materials. Thus, they should be loaded in an inert atmosphere though some loading contractors do load them under air atmosphere.

Catalyst activation

Hydrotreating catalysts have to be activated in order to be catalytically active. The activation of the catalyst (going from the oxidic to the sulfide state) is commonly called sulfiding, though several other names are used to describe the same thing. Other names that are used to describe catalyst activation are presulfiding or presulfurizing. The metals on the catalysts are in an oxide form at the completion of the manufacturing process. The catalysts are activated by transforming the catalytically inactive metal oxides into active metal sulfides (thus the name sulfiding). This is accomplished mainly in-situ though more and more refiners have started to use catalyst which had the sulfiding compound loaded onto the catalyst outside the unit (ex-situ presulfurization). It is likely more and more refiners will opt to receive the catalyst at the refinery site in presulfided state to accelerate the start up of the unit and because it is more environmentally friendly (eliminates the sometimes unpleasant odor evolved when the sulfiding compound is introduced into the unit).

In-situ sulfiding can be accomplished either in vapor or liquid phase. In vapor phase sulfiding, the activation of the catalyst is accomplished by injecting a chemical which decomposes easily to H_2S, such as di-methyl-di-sulphide (DMDS) or di-methyl-sulfide (DMS); use of H_2S was fairly common until a few years ago, but now it is only rarely used because of environmental and safety concerns. Liquid phase sulfiding can be accomplished with or without spiked feedstocks. In the latter case, the feedstock is generally a gas oil type material that contains sulfur compounds in ranges from a few thousand to twenty thousand ppm. The H_2S necessary for the activation of the catalyst is generated by the decomposition of the sulfur compounds. This method is in very little use today, but it was 'state of the art' in the 1960s and early 1970s. The preferred sulfiding procedure in the industry is liquid phase with a spiking agent (generally DMDS). It results in savings of time when compared to either vapor phase or liquid phase without spiking agents. In addition to the time savings, liquid phase sulfiding is desirable because the liquid phase provides a heat sink for the exothermic sulfiding reactions which helps prevent high catalyst temperatures and temperature excursions which could otherwise result in metals reduction. Another advantage of liquid phase over gas phase sulfiding is that by having all the catalyst particles wet from the very beginning there is very little chance of catalyst bed channeling which can occur if the catalyst particles are allowed to dry out. The in-situ sulfiding occurs at temperatures between 450 and 600°F (230–315°C) regardless of the method used. Some catalyst manufacturers recommend the sulfiding be conducted at full operating pressure while others prefer it be done at pressures lower than the normal operating pressure.

In the case of ex-situ presulfurization of catalyst, sulfur compounds are loaded onto the catalyst. The activation occurs when the catalyst, which has been loaded in the reactor, is heated up in the presence of hydrogen. The activation can be conducted either in vapor or liquid phase. Generally, activation of ex-situ presulfurized catalyst is accomplished faster than if the sulfiding is done in-situ, however, there is the additional expense due to the need for the ex-situ presulfurization step.

Catalyst deactivation and regeneration

Catalyst deactivation is the gradual loss of the catalyst's ability to produce the desired specification product unless reactor temperatures are increased (or feed rate is decreased). The catalyst activity determination is shown under "catalysts." It can be seen that as the run progresses, the catalyst loses activity. Catalyst will lose activity in several ways described below.

Coke deposition

Coke is the term used to describe the formation of hydrogen deficient carbonaceous materials, most particularly on the catalyst surface. Coke is generally formed by thermal condensation, catalytic dehydrogenation and polymerization reactions. A schematic of this is shown below.

$$\text{Hydrocarbons} \underset{(2)}{\overset{(1)}{\rightleftarrows}} \text{Coke precursors} \xrightarrow{(3)} \text{Coke}$$

where

(1) Dehydrogenation
(2) Hydrogenation
(3) Condenstaion/polymerization

The coke level rapidly rises to an equilibrium level during the early part of a catalyst cycle. This initial coke is often referred to as 'soft' coke. During the rest of the cycle, the total amount of coke remains almost constant, however further structural changes occur to produce what is often referred to as 'hard' coke. Thus, the observed catalyst deactivation during a cycle is primarily the result of structural changes to the coke rather than an actual marked increase in the total amount of coke. Short term recovery of catalyst activity has been observed on a number of occasions after a period of hot hydrogen stripping. This fits with the expectation that soft coke should be able to be partially stripped or washed from the catalyst.

As can be seen from the reaction schematic, the route to coke precursors formation is dehydrogenation. Hydrogen deficient feedstocks (i.e., cracked stocks) therefore result in faster coke deactivation. High temperatures favor faster coke deactivation because the laydown of coke on a catalyst is a time-temperature phenomenon in that the longer the exposure and/or the higher the temperature the catalyst is subjected to, the more severe the deactivating effect. Coke is not a permanent poison. Catalyst, which has been deactivated by coke deposition, can be, relatively easily, restored to close to original condition by regeneration. Low hydrogen partial pressures also favor coke formation. In general, the heavier feedstock will produce higher levels of coke on the catalyst. In general, the maximum coke laydown is about 20 wt%.

Metals deposition

Deposition of metals is not reversible, even with catalyst regeneration. The metals may come into the system via additives, such as silicon compounds used in coke drums to reduce foaming, or feedstock contaminants such as Pb, Fe, As, P, Na, Ca, Mg, or as organo-metallic compounds in the feed primarily containing Ni and V. The deposition of Ni and V takes place at the pore entrances or near the outer surface of the catalyst, creating a 'rind' layer—effectively choking off access to the interior part of the catalyst, where most of the surface area resides.

Catalyst support sintering

This is another reason for loss of catalyst activity and it also is irreversible. This is also a result of high temperatures and particularly in connection with high water partial pressures. In this case the catalyst support material can lose surface area from a collapse of pores, or from an increase in the diameter of pores, with the pore volume remaining approximately constant.

Catalyst regeneration

The activity decline due to coke laydown can be recovered by burning the coke off in a controlled atmosphere. The regeneration can be accomplished in either of three ways: in-situ with steam/air, in-situ with nitrogen/air or ex-situ. The majority of commercial catalysts regeneration, at least in the industrialized world is performed ex-situ, by specialized contractors, because of environmental considerations as well as because it results in a better performing catalyst. Upon combustion, coke is converted to CO_2 and H_2O. In the absence of excess oxygen, CO may also form. Hydrotreating catalysts contain sulfur, as the metals are in a sulfide form. In the regeneration process, the metal sulfides are converted into the corresponding metal oxides and the sulfur will be emitted as SO_2. In general, sulfur oxide emission starts at lower temperature than CO_2 emission. There are several companies that perform ex-situ regeneration by using different equipment for burning off the coke. One company uses a continuous rotolouver, which is a cylindrical drum rotating slowly on a horizontal axis and enclosing a series of overlapping louvers. The spent catalyst passes slowly through the rotolouver, where it encounters a countercurrent of hot air. Another company uses a porous moving belt as a regenerator. The catalyst is moved with the stainless steel belt through a stationary tunnel furnace vessel where the regeneration takes place. A third company regenerator uses ebullated bed technology to perform the catalyst regeneration. Regardless of the process, the spent catalyst is submitted to de-oiling prior to regeneration. This is to eliminate as much hydrocarbon as possible as well as to remove as much sulfur as possible to prevent formation of sulfates which could deposit on the catalyst and not be removed during regeneration. Sulfates are deleterious to catalyst performance. While the in-situ regeneration results in about 90% catalyst activity recovery, ex-situ catalyst regeneration results in 95–97% catalyst activity recovery.

Design and operation of hydrotreating reactors

Design and construction of hydrotreating reactors

Hydrotreating reactors are downflow, fixed-bed catalytic reactors, generally operating in trickle flow regime. Because hydrotreating occurs at moderately high pressure and relatively high temperature and in the presence of hydrogen and hydrogen sulfide, the reactors are vessels with relatively thick wall. The reactors are usually cylindrical vessels and while those use for naphtha hydrotreating as well as many of the older vintage reactors are made from lower alloys, most of those designed in the last 10 years, are typically constructed of $1^1/_4$ Cr–$^1/_2$ Mo or $2^1/_4$ Cr–1 Mo base metal with a lining of stabilized austenitic stainless steel for added corrosion protection. This choice of alloys gives the high strength of the base metal and the excellent corrosion resistance of the inner lining. There are several items concerning the selection of materials that must be taken into consideration during the operation of the unit. Concerning the use of austenitic stainless steels in Hydrotreating units, the possibility exists for corrosion cracking to occur if the proper procedures are not followed. Corrosion cracking in a hydrotreating unit can occur through chloride attack or polythionic acid attack. Chloride attack can be prevented by minimizing the amount of chloride in the process material that will come in contact with the austenitic stainless steel during normal operations. In addition, during startup and shutdown operations precautions should be taken to limit the chloride content in any flushing, purging, or neutralizing agents used in the system.

Polythionic acids occur as the result of the action of water and oxygen on the iron sulfide scale that forms on all items made of austenitic stainless steel. Once formed these acids can attack the austenitic steel and cause intergranular corrosion and cracking. To prevent polythionic acid attack, it is necessary to maintain the temperature above the dew point of water in those areas containing stainless steel. Under normal operating conditions, the system is essentially free of oxygen. However, when the system is depressurized and the equipment is opened to air, it becomes necessary to maintain a nitrogen purge to prevent air from entering. In cases where adequate temperatures or purges cannot be maintained, a protective neutralizing environment should be established. Generally, a 5% soda solution is used to neutralize the austenitic stainless steel.

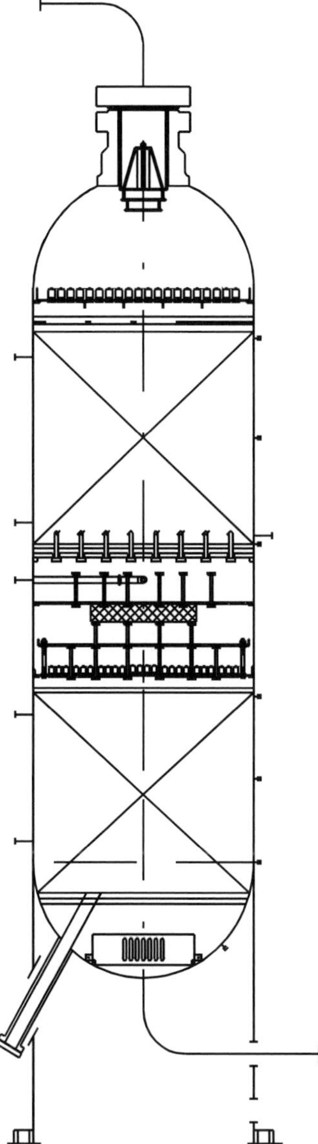

Figure 8.10. Two bed hydrotreating reactor with interbed quench.

Figure 8.10 shows a hydrotreating reactor with two beds of catalyst and one interbed quench zone is pictured, but the number of beds can vary for different designs. As already indicated, most naphtha hydrotreaters only have one catalyst bed. Many reactors processing cracked feedstocks will have several beds to facilitate temperature control by cooling with hydrogen quench between the catalyst beds. For example, a reactor design could require three catalyst beds and two interbed quench zones.

The reactor vessel is designed to allow maximum utilization of catalyst. Creating equal flow distribution, providing maximum liquid/vapor mixing, and providing multiple beds with quench zones for efficient catalyst usage achieve this. The internals of the reactor found in a reactor are the following (though not all reactors necessarily have all of them).

- Inlet diffuser
- Top vapor/liquid distribution tray
- Quench section (present only when there are multiple catalyst beds)
- Catalyst support grid (present only when there are multiple catalyst beds)
- Outlet collector

The size of hydrotreating reactors varies widely depending on the design conditions and is dependent on the desired mass velocity and acceptable pressure drops. Since heat release is a common feature for all hydrotreaters, reactor temperature control has to be exercised. Generally, the maximum allowable ΔT is 75°F (42°C). If that temperature is not expected to be exceeded, the reactor will be mono bed and temperature control will be exercised by changing the reactor inlet temperature. If the maximum reactor ΔT is expected to exceed 75°F (42°C), a multiple bed reactor should be installed with cold hydrogen quench inserted in the quench section for temperature control.

Hydrotreater reactor operation

During operation, the hydrotreating catalyst gradually loses some of its activity. In order to maintain the desired quality of the products at the design feed rate, the average bed temperature is gradually increased. The temperature increase in many cases is very small, less than 2°F/month (1°C/month). When the average bed temperature reaches a value close to the design maximum, the catalyst has to be replaced or reactivated. Because the required temperature increase per unit time is relatively small, the reactor can be operated with the same catalyst for several years before regeneration or replacement of the deactivated catalyst becomes necessary. Quite often, catalyst regeneration or replacement is dictated by a high reactor pressure drop, due to catalyst fouling.

Kinetics is the study of the rates of reaction. The rates of reaction determine the key properties of a catalyst. In hydrotreating, the key properties are initial activity, stability, and product quality. The temperature required to obtain the desired product at the start of the run measures the initial activity. Catalyst stability is a measure of change of reaction rate over time. The product quality is a measure of the ability of the process to yield products with the desired use specification such as pour point, smoke

Table 8.1. Chemical basis for product quality

Quality measurement	Chemical basis
High smoke point	Low concentration of aromatics
Low pour point	Low concentration of *n*-paraffins
Low freeze point	Low concentration of *n*-paraffins
Low cloud point	Low concentration of *n*-paraffins
Low CFPP	Low concentration of *n*-paraffins

point, or octane. Table 8.1 shows some of the important product quality measurements and the chemical basis for these measurements.

Hydrotreating process variables

The proper operation of the unit will depend on the careful selection and control of the processing conditions. By careful monitoring of these process variables the unit can operate to its full potential.

Reactor temperature

Reactor temperature should be minimized while maintaining desired product quality. Increasing reactor temperature will accelerate the rate of coke formation and reduce the length of the operating cycle. The required temperature is dependent upon feed rate and quality. The reactor inlet temperature is most easily and commonly controlled by the operator to adjust for obtaining the desired product quality. The reactor outlet temperature is a function of the feed quality and cannot be easily varied except by changing the reactor inlet temperature. The inlet temperature must always be controlled at the minimum required to achieve the desired product properties. Temperatures above this minimum will only lead to higher rates of coke formation and reduced processing periods. The weight average bed temperature (WABT) is typically used to compare the relative catalyst activity. The WABT can be calculated as shown in Figure 8.11.

If the reactor only has inlet and outlet thermometry (as is the case in perhaps as many as 2/3 of hydrotreaters), the WABT represents the average of inlet and outlet temperatures. The rate of increase in this temperature is referred to as the deactivation rate expressed as °F per barrel of feed per pound of catalyst (°C per m^3 of feed per kilogram of catalyst), or simply as °F per day (°C per day). During the course of an operating cycle, the temperature required to obtain the desired product quality will increase as a result of catalyst deactivation. Increasing reactor temperatures up to a limit of about 800°F (428°C) maximum bed temperature can compensate for the gradual loss in catalyst activity. In general, above this level, coke formation becomes very rapid and little improvement in performance is obtained.

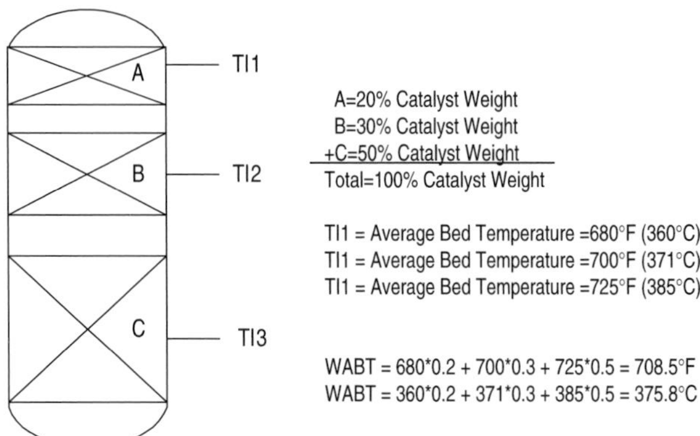

Figure 8.11. WABT.

The design temperature of the reactor(s) will determine the maximum allowable operating value. The temperature rise across the reactor(s) must be monitored continuously in order to assure that the design limitation of the unit is not exceeded. This can be especially important when changing feedstocks since olefin saturation results in considerably higher heats of reaction. Units are typically designed for a maximum reactor bed temperature rise $<60°F$ (33°C).

Feed quality and rate

The amount of catalyst loaded into the reactors as well as other design parameters are based on the quantity and quality of feedstock the unit is designed to process. While minor changes in feed type and charge rate can be tolerated, wide variations should be avoided since they will tend to reduce the useful life of the catalyst. An increase in the charge rate will require higher reactor temperature to achieve a constant desulfurization (or denitrogenation) as well as higher recycle gas rate to maintain a constant ratio of H_2 to hydrocarbon. The increased reactor temperatures will lead to a faster rate of coke formation that will reduce the cycle length. A reduced feed rate may lead to bad flow distribution through the catalyst, such that higher temperatures will be required to obtain good product quality. Its distillation range and API gravity best indicate the type of feed being processed. An increase in the end point of the feed will make sulfur and nitrogen removal more difficult, thus requiring higher reactor temperatures which, in turn, accelerate coke formation. Coke deposition is also accelerated by the fact that heavier feed contains more of the precursors that favor coke formation. In addition to the above, high boiling fractions also contain increased quantities of metals which lead not only to higher reactor pressure drop, but to rapid catalyst deactivation as well. A reduction in the API gravity of the feed for the same boiling range is an indication of higher unsaturates content. This type of feed will result in increased hydrogen consumption and higher temperature rise

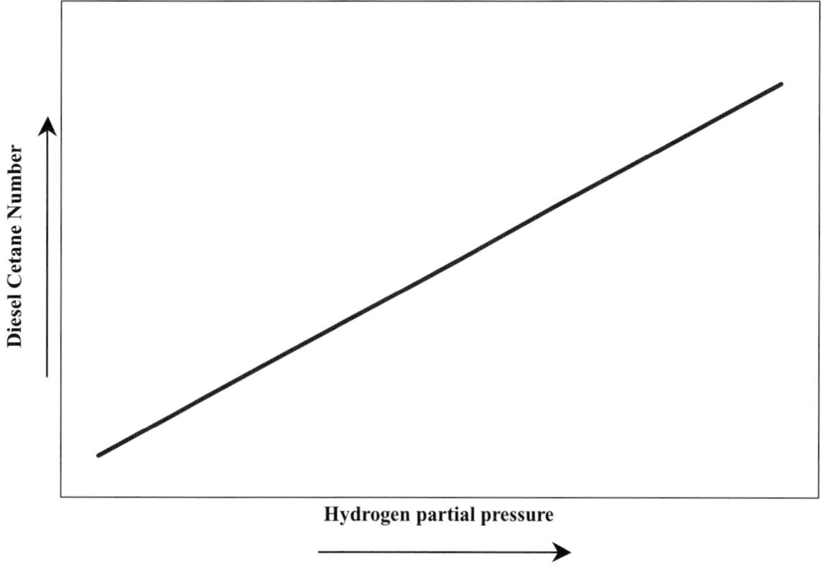

Figure 8.12. Effect of pressure on aromatics saturation.

across the catalyst bed. It also contains more of the materials that easily condense to form coke in the reactor and associated equipment.

Hydrogen partial pressure

The hydrogen partial pressure is calculated by multiplying the H_2 purity of the recycle gas times the pressure of the product separator. The hydrogen partial pressure required for the operation of a unit is chosen based on the degree of sulfur (or nitrogen, or aromatic saturation etc.) removal that must be achieved and is generally an economic optimum that balances capital cost and operating costs against catalyst life. Hydrogen partial pressure is also a critical design parameter for achieving the desired degree of feed saturation. Figures 8.12 and 8.13 illustrate the effect of hydrogen partial pressure on the quality of the products. A reduction of the operating pressure below the design level will have a negative effect on the activity of the catalyst and will accelerate catalyst deactivation due to coke formation.

Gas-to-oil ratio

This is an important variable for the satisfactory performance of a hydrotreater. If the unit is operated at lower than design ratios more rapid catalyst deactivation will result. The circulating gas also provides the heat sink for the removal of the heat of reaction.

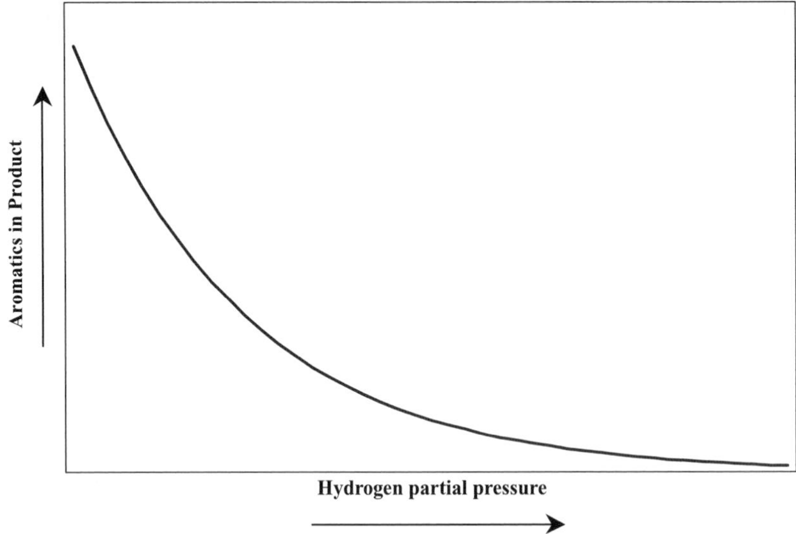

Figure 8.13. Effect of pressure on distillate quality.

The gas-to-oil is calculated as follows:

$$\text{Gas to oil ratio [SCFB]} = \frac{\text{Total gas to reactors (SCFH or Nm}^3\text{/h)}}{\text{Raw oil charge (BPH or m}^3\text{/h)}}$$

Though various hydrotreating unit designers and catalyst manufacturers use different values, it is generally accepted that the minimum gas to oil ratio should be at least 4 times the amount of hydrogen consumption.

Liquid hourly space velocity

The design quantity of catalyst per unit of feed will depend upon feedstock properties, operating conditions, and product quality required. The liquid hourly space velocity (LHSV) is defined as follows:

$$\text{LHSV [1/hr]} = \frac{\text{Charge rate (ft}^3\text{/hr or m}^3\text{/h)}}{\text{Volume catalyst (ft}^3\text{ or m}^3\text{)}}$$

A simplified kinetic expression based on sulfur and/or nitrogen removal determines the initial liquid hourly space velocity for most feedstocks and processing objectives. This initial value may be modified due to other considerations such as unit size, extended catalyst cycle life, abnormal levels of feed metals and requirements of other processing units in the refinery flow scheme. A unit design is based on operation to achieve optimum performance. One criterion is liquid mass flux across the catalyst bed. At reduced throughput, unit operation may become difficult due to hydraulic considerations. Also, liquid distribution in the reactor may become unequal as preferential flow paths are established. For these reasons, the unit should not be operated below the minimum turndown capacity for extended periods. Unit turndown will vary for each design and is typically 50–70% of design capacity. Operation at too high of space velocity (compared to original design) is not advisable because of increased catalyst deactivation rates as well as increased system pressure drop.

Recycle gas purity

The effective completion of the hydrogenation reactions occurring over the catalyst requires that a certain quantity of hydrogen be present at a minimum partial pressure. As noted previously, both the quantity (gas-to-oil ratio) and partial pressure are dependent upon the hydrogen content, i.e., purity, of the recycle gas. Practical considerations, such as the cost of compression, catalyst life, etc., limit the purity of the recycle gas to a minimum value usually in the range of 70–80 mol%. Lower hydrogen purities are detrimental to the performance of the unit since higher temperatures must be used to achieve the desired product quality. The purity of the recycle gas is determined by the following factors:

- The purity of the makeup gas
- The amounts of light hydrocarbons and H_2S that are allowed to accumulate in the recycle gas.

In most instances, the makeup gas H_2 purity cannot be easily manipulated since it is fixed by the operation of the Reformer or the Hydrogen manufacturing plant. The light hydrocarbons present in the recycle gas enter the system with the makeup gas in addition to those being formed in the reactor, and must be vented from the high-pressure separator to prevent their accumulation in the recycle gas. The amount of hydrogen required is determined by:

(a) Chemical hydrogen consumption—The hydrogen consumed during the hydrotreating reactions
(b) Solution losses —The hydrogen that is removed from the reactor circuit dissolved in the liquid hydrocarbon leaving the high-pressure separator
(c) Mechanical losses—The hydrogen lost through the makeup and recycle gas compressors' packing vents and seals. This value may be roughly estimated at 3–5% of the combined chemical consumption plus solution losses
(d) Venting losses—The hydrogen lost in the purge stream from the high-pressure separator to maintain recycle gas purity

The H_2S formed in the reactors can reach equilibrium values as high as 5 mol% in the recycle gas. This concentration of H_2S has a depressing effect on the activity of the catalyst. Therefore, in many cases it is desirable to remove the H_2S from the recycle gas. The removal of H_2S is performed in a scrubber where the recycle gas in contacted with an amine (generally MEA or DEA) solution. In this manner, the H_2S content of the recycle gas can be reduced to the parts per million range.

Another method to increase hydrogen purity is membrane separation. This system removes the hydrogen from the vent gas and recycles it back with the makeup hydrogen.

Catalyst contaminants

Temporary
Process variables influence catalyst life by affecting the rate of carbon deposition on the catalyst. There is a moderate accumulation of carbon on the catalyst during the initial days of operation. The rate of increase will be reduced to very low levels under normal processing conditions. A carbon level of 5 wt% may be tolerated without a significant decrease in desulfurization. However, denitrogenation activity would be reduced.

The sulfur and nitrogen found in the feed could be considered contaminants to the extent that they produce hydrogen sulfide and ammonia which can react to form ammonium bisulfide. The water injected into the reactor effluent dissolves the ammonium bisulfide and prevents exchanger fouling. Organic nitrogen in the feed, if present in amounts higher than expected, will require higher reactor temperatures for processing, and will lead to a reduction in catalyst life.

If the water injection should be stopped for any period of time, the H_2S and NH_3 may accumulate in the recycle gas and result in a sudden loss in catalyst activity. The activity will return to normal once wash water is reestablished. Catalyst bed temperatures should not be increased to compensate for the temporary activity loss.

Small amounts of molecular nitrogen, CO and CO_2 that enter the system with the makeup gas are not harmful to the catalyst, but must be vented to prevent accumulation in the recycle gas. Excessive amounts of CO and CO_2 may have an adverse effect on catalyst performance, as they may be methanated by the catalyst taking up active sites and liberating heat. This will raise the outlet temperature and reduce the apparent catalyst activity. The unit should never be pressured up with high $CO + CO_2$ containing makeup gas as a temperature runaway may result.

Permanent
Permanent loss of catalyst activity is usually caused by the gradual accumulation of inorganic species picked up from the feed, makeup hydrogen or effluent wash water. Examples include arsenic, lead, calcium, sodium, silicon, and phosphorus. Low

concentrations of these elements (and other alkaline metals) can cause deactivation over time as they are deposited on the catalyst.

Organic metal compounds are decomposed and typically deposit in the upper section of the catalyst bed as a metal sulfide. The graded catalyst bed, if used, may contain demetallization catalysts that have a high metals retention capacity. Some of these catalysts may retain as much as 100 wt% of the fresh catalyst weight as metals from the feed. These demetallization catalysts typically have a lower activity for desulfurization and denitrogenation.

Hydrotreating catalysts exhibit a moderate tolerance for metals such as arsenic and lead. Total metal content of 2–3 wt% of the hydrotreating catalyst have been observed. However, product analysis frequency should be increased to monitor breakthrough when calculations show the metals level on the hydrotreating catalyst exceeds 0.5 wt%. Metals cannot be removed by catalyst regeneration. Catalyst replacement should be considered when metal loading of 1–2 wt% is exceeded on the hydrotreating catalyst.

Apparent catalyst deactivation may be caused by the accumulation of deposits on top of the catalyst bed. Solid material, such as corrosion products and coke, will lead to rapid fouling of the catalyst bed if allowed to enter the reactor. This problem is remedied by skimming a portion of the catalyst, screening, and reloading. Feed filtering is quite effective in removing solid material, and as such results in longer operating cycles due to the lower rate of reactor pressure drop build-up. The use of feed filtering will depend on the type of feed processed and its source.

Hydrotreaters licensors and catalyst manufacturers

Licensors

Until the 1970s all the types of hydrotreating were offered as licensed processes by several major licensors. Since that time, more and more of the types of hydrotreating have been offered without a license and the catalyst used in the process was sold as a commodity. Today, Chevron, IFP and UOP offer only residue hydrotreating as a licensed process. A few specialized types of hydrotreating such as some aromatic saturation units are also offered as licensed units, however, the vast majority of new hydrotreaters are sold without a license. Engineering contractors do most of the designs.

Catalyst suppliers

Through the years there have been many manufacturers of hydrotreating catalysts. Other manufacturers have absorbed some, some have changed names, and some have discontinued hydrotreating catalyst manufacturing. Some suppliers simply sell catalyst they have re-branded with their name. Following is a list of the current major suppliers of hydrotreating catalysts: Advanced Refining Technologies LP, Akzo Nobel Catalysts BV, Axens/Procatalyse SA, Catalysts and Chemicals Industries Co. Ltd., Chevron Research and Technology Co., Criterion Catalysts Co. Ltd., Exxon Research and Engineering Co., Haldor Topsoe AS, Orient Catalyst Co. Ltd., Sud Chemie Inc., and UOP.

Chapter 9

Gasoline components

9.1 Motor fuel alkylation

James F. Himes and Robert L. Mehlberg*

9.1.1 Introduction

Motor fuel alkylation in the petroleum refining industry refers to the acid catalyzed conversion of C_3-C_5 olefins with isobutane into highly branched C_5-C_{12} isoparaffins collectively called alkylate, a valuable gasoline blending component. A major constituent of alkylate is 2,2,4-trimethyl pentane which is defined as 100 on the octane scale.

Alkylation reactions are catalyzed by liquid and solid acids, including H_2SO_4, $AlCl_3$-HCl, HF, HF-BF_3, H_2SO_4-HSO_3F (Fluorosulfuric acid), Trifluoromethane sulfonic acid chlorided Pt alumina, BF_3 on alumina, zeolites, and ion exchange resins. However, the catalysts and associated processes commercialized during WWII for aviation gasoline, HF alkylation, and sulfuric acid alkylation, are the focus of this section as these remain the primary commercial motor fuel alkylation processes. A solid catalyst alkylation process (UOP Alkylene™) has been developed and is being offered to the industry.

History

In 1932–6, alkylation was independently discovered by UOP,[1] Shell, the Anglo Iranian Oil Company (AIOC), and Texaco whose first publications issue in that order. Herman Pines told the story of UOP's discovery of the alkylation of ethylene by pentanes in 1932.[2] At that time leading universities taught that isoparaffins were inert except at high temperatures and pressures. After finding anomalies in an olefin assay based upon H_2SO_4 extraction Pines and his mentor V. I. Ipatieff hypothesized that paraffins may not be inert to acids. Despite ridicule, they tested that hypothesis by bubbling ethylene into chilled pentanes over $AlCl_3$. All the ethylene was converted into saturated

*UOP LLC.

hydrocarbons. Over the next few years they tested $AlCl_3$-HCl, H_2SO_4, HF, and HF-BF_3 as alkylation catalysts.

Alkylate was found to have excellent aviation gasoline properties. It was the highest octane fuel component then known, with high motor octane and excellent lead response. All of the properties derive from the highly-branched paraffins that form its composition.

Humble Oil built the first commercial H_2SO_4 alkylation unit in 1938 at Baytown, Texas.[3] Alkylation for aviation gasoline grew rapidly with the Allies war effort. In 1939, six petroleum companies formed a consortium to pool their alkylation technology and develop both sulfuric acid and HF acid processes for 100 octane aviation fuel. The first commercial HF alkylation unit started up in 1942. During the war 60 alkylation units were built for the Allies' war effort. Half were built with sulfuric acid as the catalyst and half with HF.

Following World War II, most alkylation operations were discontinued although a few refiners continued to use the process for aviation and premium automobile gasolines.

In the mid-1950s, use of higher performance automotive engines required the refining industry to both increase gasoline production and quality. The development of catalytic reforming, such as UOP Platforming™, provided refiners with an important refining tool for production of high octane gasolines. However, the motor fuel produced in such operations, called reformate, is highly aromatic with a higher sensitivity (the spread between research and motor octane) and a lower lead response than alkylate. Many refiners expanded their alkylation operations and began to broaden the range of olefin feeds to both existing and new alkylation units to include propylene and occasionally even some pentenes along with the butenes.

With the phase-out of leaded gasolines and the advent of environmental gasolines the lead response of alkylate is no longer valued, but the importance of alkylate and its production have both grown because of its other properties. Its high unleaded motor octane, low volatility, low-sulfur, and nearly zero olefins and aromatics make alkylate critical to the production of quality environmental gasolines. Alkylate can reach 60% of low-sulfur reformulated premium.

Licensors of motor fuel HF Alkylation processes are UOP LLC and Phillips. Licensors of H_2SO_4 alkylation processes are Exxon Mobil and Stratco Engineering.

Process chemistry

The reactions taking place in the alkylation reactor are many and relatively complex. First the olefin reacts with the acid to form an ester; then the ester is alkylated by a t-butyl carbenium ion chain mechanism.[4] The principal overall reactions are described below.

Primary alkylation reactions

Most of the alkylate product is made by primary alkylation reactions. In these reactions one mole of olefin reacts with one mole of isobutane to form an isoparaffin exactly 4 carbon numbers heavier. Example primary alkylation reactions, showing only the carbon framework and one or two of the principal product isomers, are shown for each of the principal feed olefins in equations (1.1)–(1.5).

ALKYLATION OF BUTENE-2

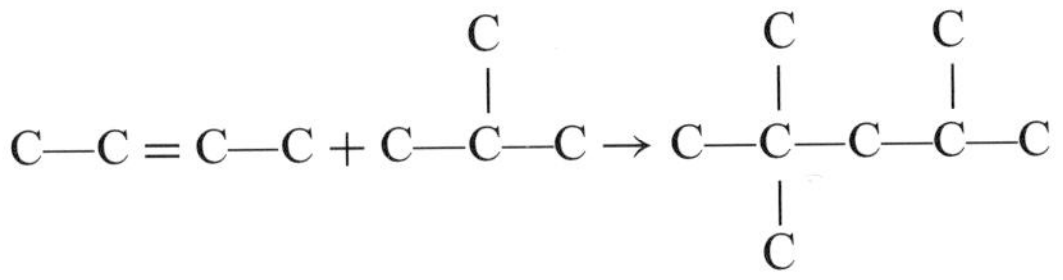

```
                    C          C     C
                    |          |     |
C—C = C—C + C—C—C → C—C—C—C—C
                                     |
                               C
```

cis-Butene-2 + Isobutane → 2,2,4-Trimethylpentane (100 RON) (1.1)

ALKYLATION OF BUTENE-1

```
                    C          C     C
                    |          |     |
C = C—C—C + C—C—C → C—C—C—C—C—C
```

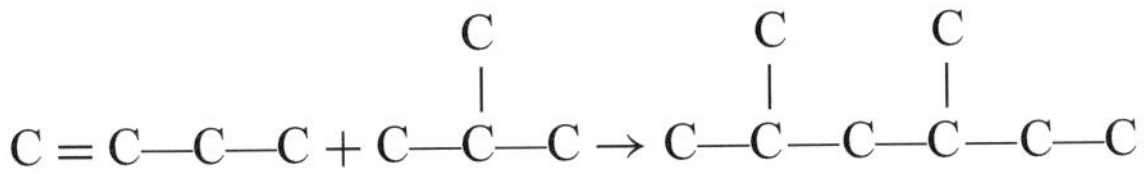

Butene-1 + Isobutane → 2,3-Dimethylhexane (71.3 RON) (1.2)

```
                  C          C  C  C
                  |          |  |  |
C = C—C—C + C — C—C → C—C—C—C—C
```

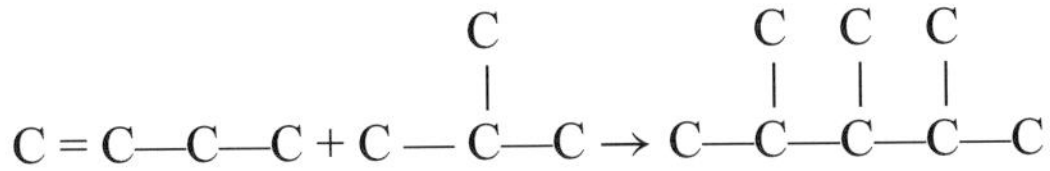

Butene-1 + Isobutane → 2,3,4-Trimethylpentane (102.7 RON) (1.3)

ALKYLATION OF ISOBUTYLENE (I-BUTENE)

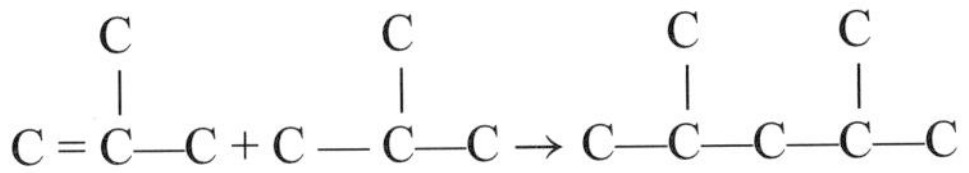

```
     C           C           C     C
     |           |           |     |
C = C—C + C — C—C → C—C—C—C—C
```

Isobutylene + Isobutane → 2,2,4-Trimethylpentane (100 RON) (1.4)

ALKYLATION OF PROPYLENE

```
                  C            C  C
                  |            |  |
C—C = C + C—C—C → C—C—C—C—C
```

Propylene + Isobutane → 2,3-Dimethylpentane (1.5)

(91.1 RON)

ALKYLATION OF 2-METHYL-BUTENE-2

```
              C           C           C
              |           |           |
C—C = C—C + C—C—C → C—C—C—C—C—C
```

2-methylbutene-2 + Isobutane → 2,2,4-Trimethylhexane (1.6)

(92 RON)

Complex alkylation reactions

The balance of the alkylate and small amounts of undesirable byproduct conjunct polymer are formed by more complex reactions of the ionic intermediates. Some of these reactions are given by equations (2.1)–(2.5). For brevity only a few key steps and no structures are shown.

HYDRIDE TRANSFER (PROPYLENE EXAMPLE)

$C_3H_6 + H^+ \rightarrow C_3H_7^+$ (Formation of propenium ion)

$C_3H_7^+ + iC_4H_{10} \rightarrow iC_4H_9{}^+ + C_3H_8$ (Hydride transfer from isobutane to propenium ion)

$iC_4H_9^+ \rightarrow iC_4H_8 + H^+$

$iC_4H_8 + iC_4H_{10} \rightarrow iC_4H_{10}$

Overall

$C_3H_6 + 2iC_4H_{10} \rightarrow C_3H_8 + C_8H_{18}$

Propylene + 2 Isobutane → Propane + 2,2,4-Trimethylpentane (2.1)

POLYMERIZATION (ISOBUTYLENE EXAMPLE)

$iC_4H_8 + iC_4H_9^+ \rightarrow iC_8H_{17}^+$

$iC_8H_{17}^+ + iC_4H_8 \rightarrow iC_{12}H_{25}^+$

$iC_{12}H_{25}^+ + iC_4H_{10} \rightarrow iC_{12}H_{26} + iC_4H_9^+$

Overall

$2iC_4H_8 + iC_4H_{10} \rightarrow iC_{12}H_{26}$

2 isobutylene + Isobutane → Pentamethylheptane (2.2)

CRACKING (CRACKING OF HEAVY ENDS FROM ISOBUTYLENE)

$$2iC_4H_8 + iC_4H_9^+ \rightarrow iC_{12}H_{25}^+$$
$$iC_{12}H_{25}^+ \rightarrow iC_5H_{11}^+ + iC_7H_{14}$$
$$iC_5H_{11}^+ + iC_4H_{10} \rightarrow iC_5H_{12} \text{ (isopentane)} + iC_4H_9^+ \text{ (}t\text{-butyl ion)}$$
$$iC_4H_9^+ + iC_7H_{14} \rightarrow iC_{11}H_{23}^+$$
$$iC_{11}H_{23}^+ + iC_4H_{10} \rightarrow iC_{11}H_{24} + iC_4H_9^+ \text{ (}t\text{-butyl ion)}$$

Overall

$$2iC_4H_8 + 2iC_4H_{10} \rightarrow iC_5H_{12} + iC_{11}H_{24}$$

2 isobutylene + 2 isobutane → isopentane + isoundecane (2.3)

COMPETITION (ISOPENTANE ALKYLATION)

$$iC_5H_{12} + C_4H_8 \rightarrow iC_9H_{20}$$

Isopentane + butene-2 → 2,2,5-Trimethlyhexane (2.4)

CONJUNCT POLYMERIZATION

$$17iC_4H_8 + H^+ \rightarrow\rightarrow\rightarrow C_{20}H_{31}^+ + 4C_{12}H_{26}$$

isobutylene → conjunct polymer ion (in acid phase) + Heavy alkylate (2.5)

Isomerization

In addition to the listed reactions the acids catalyze isomerization of olefins or their esters before alkylation as well as isomerization of the alkylate products. Most of the product isomerizations are methyl shifts of ionic intermediates.

- Reaction selectivites—The selectivities of these reactions depend upon the feedstock, catalyst, and reaction conditions.
- Isobutylene and isopentenes are very reactive and are prone to polymerization to heavy alkylate (reaction (2.2)) and to conjunct polymer (reaction (2.5)), especially when catalyzed by H_2SO_4.
- Propylene is less reactive and tends to make more conjunct polymer, especially for the H_2SO_4 catalyst.
- Butene-1 makes higher yields of low-octane dimethyl hexanes (reaction (1.2)) than butene-2 in HF, but there is little difference in these olefins in H_2SO_4. To maximize octane many refiners have added selective hydrogenation units (such as UOP-Huls SHP™) upstream of the alkylation unit to saturate diolefins reducing the amount of conjunct polymers as well as isomerizing most of the Butene-1 to Butene-2.
- HF enhances hydride transfer reactions (2.1) much more than H_2SO_4. These reactions increase alkylate yield, RVP, and octane. As a result, HF catalyzes higher yields of isopentane from pentenes than H_2SO_4; it also produces low amounts of n-paraffins from n-olefins while H_2SO_4 produces none. The additional yield of hydride transfer reactions naturally comes with increased isobutane consumption.

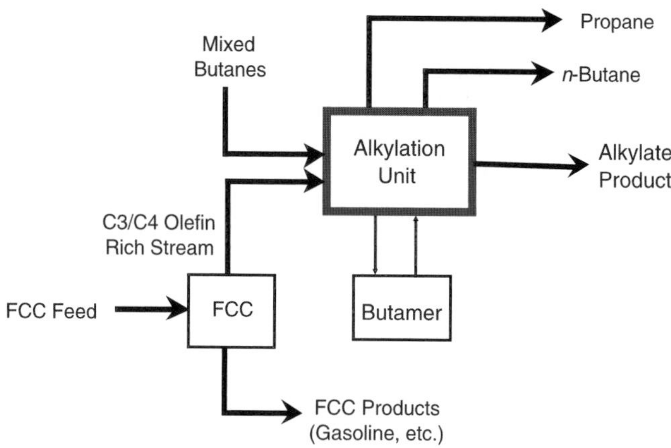

Figure 9.1.1. Alkylation unit location in refinery.

- H_2SO_4 produces more heavy ends (reaction 2.2) and more conjunct polymer (reaction 2.5) and less isopentane than HF, especially from iso-olefins and propylene. For all olefins it produces lower yields, higher endpoints, and consumes less isobutane.
- Increasing reactor temperature increases cracking and conjunct polymer formation at the expense of primary alkylation lowering product octane and increasing RVP and acid consumption.
- Increasing isobutane concentration increases primary alkylation and hydride transfer reactions while reducing polymer formation, cracking, and competition which increase yield and octane and reduce acid consumption.

HF alkylation process flow description

Figure 9.1.1 illustrates how an alkylation unit fits into the refinery in a typical FCC-Alkylation arrangement. The location of the butane isomerization unit (Butamer™) is also indicated as many refiners have this arrangement.

Figure 9.1.2 is a simplified process flow diagram for the UOP Propylene-Butene HF Alkylation process.

Feed pretreatment
In the UOP HF alkylation process olefin-rich feeds from the FCC gas plant are typically deethanized, Merox treated to remove H_2S and mercaptans and dried. Some refiners have also added MTBE or Selective Hydrogenation (SHP) plants upstream the alky.

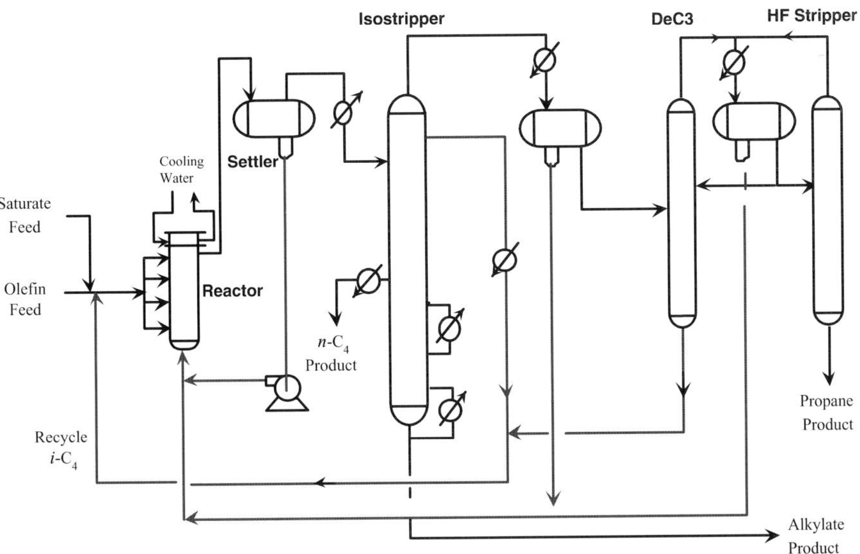

Figure 9.1.2. UOP HF alkylation.

Reaction

After pretreatment, the olefin feeds are combined with a large excess of recycle isobutane to provide an 6–14 isobutane:olefin molar ratio and injected into the circulating HF acid catalyst at the shell side inlet to the water-cooled reactor. Cooling water flows through the reactor tubes to remove the highly exothermic heat of reaction and to maintain reaction conditions at 80–100°F. The alkylation reaction is very fast with 100% olefin conversion. The excess isobutane, alkylate product, non-reactive hydrocarbons (propane, n-butane) in the feeds and the acid catalyst pass on to the settler vessel. The dense acid phase separates from the hydrocarbons rapidly by gravity and is then pumped back to the reactor. The hydrocarbons containing dissolved HF flow off the top of the settler to the isostripper.

Fractionation

It consists of an isostripper, a depropanizer, and an HF Stripper. The isostripper is a large tower with two sidedraws with the primary function recycling isobutane to support the high isobutane:olefin molar ratio of the reactor. The tower typically has two reboilers; the upper reboiler typically maximizes the use of relatively low cost low or medium pressure steam and the lower reboiler uses a heating medium that can give 400–450°F process temperatures. Alkylate is drawn off the bottom of the tower, cooled in exchangers, and sent to product storage. The next product draw up the tower is the n-butane sidedraw and above that is the isobutane recycle draw. On most UOP units the isobutane draw is located below the feed tray to minimize HF in the isobutane recycle.

The isostripper overhead vapor is a propane-enriched isobutane stream and HF which is condensed and separated into settling drum. The HF phase is pumped back to the reactor section. The HF saturated hydrocarbon phase is charged to the depropanizer.

The depropanizer and its associated HF stripper remove propane from the isobutane recycle. The depropanizer bottoms is returned to the reactors as part of the recycle isobutane. The depropanizer overhead containing the propane product and HF are condensed and separated in the overhead receiver. The acid phase is returned to the reactor section and the acid-saturated propane is stripped free of acid in the HF Stripper column. The HF stripper bottoms is an acid-free propane product which is treated with hot alumina to remove organic fluorides, cooled and treated with KOH pellets to remove traces of HF and water.

Acid regeneration
A key advantage of HF alkylation over sulfuric is the ability to recover the acid from the byproduct polymer, water, and other contaminants. In the UOP HF Alkylation Process, a small stream of circulating acid is stripped with superheated isobutane in a small monel tower called the Acid Regenerator. The regenerator overhead is HF and isobutane that are recycled to the reactor; the regenerator bottoms is polymer and the HF–water azeotrope which are neutralized with aqueous KOH. The neutralized polymer has good fuel value. The amount of polymer produced is generally only 1 to 2 barrels per 1,000 barrels of alkylate product.

KOH regeneration
The UOP process minimizes chemical costs by regenerating the KOH used to treat products and all waste and storm water. This KOH regeneration is accomplished using lime. As the lime is mixed into the KOH, regeneration of the KOH takes place by the following reaction:

$$2KF + Ca(OH)_2 \rightarrow 2KOH + CaF_2$$

The calcium fluoride forms a precipitate and can be easily separated from the regenerated KOH.

Process variables in HF alkylation
Key variables are reactor temperature, isobutane to olefin molar ratio, acid strength, and acid to hydrocarbon volume ratio.

Reaction temperature is one of the more important process variables as it has a significant influence on the octane number of the product. Almost all HF alkylation reactors are operated below 100°F. At higher temperatures a decrease in alkylate octane number will occur. Above 120°F polymerization and cracking side reactions become excessive reducing alkylate quality. In many cases, acid regeneration capacity of the HF alkylation unit would not be able to maintain proper control of the acid strength at temperatures above 110°F. Extremely low reaction temperatures may cause incomplete alkylation. Thus reaction temperatures below 80°F are typically not used.

Isobutane to olefin molar ratio is generally the most important variable that the refiner has the most control over within limitations of isostripper fractionation loadings. As the isobutane to olefin molar ratio is increased, octane increases; thus the flow of isobutane recycle is usually kept at a practical maximum at all times, up to the capacity of the isostripper. The reasons to reduce the recycle ratio are for the conservation of fractionation energy or for a reduction in the hydride transfer reaction. The higher energy consumption, the greater consumption of isobutane (due to the hydride transfer) and the increased production of propane (due to the hydride transfer) must be justified economically against higher product quality. Another practical limitation of isobutane circulation is possible entrainment of acid from the reactor section acid settler to the isostripper because of inadequate settling time.

Acid strength is usually kept between 85 and 95 mass% HF. Maintenance of this strength level results from a balance between the performance of the unit feed treating systems for sulfur and water removal and acid regeneration operation. In some cases oxygenate removal systems or diene removal systems are also used on the feed where there are known to be high oxygenates (such as downstream of an MTBE unit, or high diolefins (from severe FCC conditions). The action of the acid on reactions is a complex phenomenon and is dependent on the type as well as the amount of diluents. The fresh acid is supplied by acid manufacturers at 99.0^{+} wt% HF. This purity is too high for optimum performance of the HF alkylation process. As the water content of the circulating acid increases, carbon steel that is not attacked by anhydrous HF, becomes less resistant to acid attack.

The acid to hydrocarbon volume ratio used in the reactor of the UOP process is generally around 1:1. At some point below 0.8–1, excess polymerization occurs. In the most extreme cases, alkylate production could stop.

Two other variable related topics are reaction time and pressure. As the reaction time decreases, the combined fluorides leaving the reactor section will increase. However, any reduction in time is limited by the settling capacity and there is generally little effect within the permissable operating ranges of a particular unit. Excessive velocities in the settler will cause free (above saturation level) acid to be carried over into the isostripper. This carryover may result in corrosion of the upper trays of the isostripper and discolored alkylate as heavier contaminants in the acid that is carried over drop into the alkylate product. Pressure is not really a process variable as long as it is kept high enough that all of the hydrocarbon and acid in the reactor section remain in the liquid state.

HF feed contaminants

As with many refining processes, the control of contaminants coming into the unit with the normal feedstocks is critical to the long and dependable operation of the HF alkylation unit. Above the recommended maximum levels of feed stock contaminants, acid consumption, acid regeneration requirements, and in some cases unit corrosion and product quality are all measurably affected. Generally all contaminants are kept as low as possible within the capabilities of the feed treating systems. The major feed contaminants normally found in alkylation feeds are water or oxygenates, sulfur compounds, nitrogen compounds, non-condensibles, and diolefin.

HF alkylation maintenance
Because HF is highly corrosive to most materials, careful control and maintenance of equipment metallurgy and condition is required. Carbon steel is the primary material used for vessels and piping and it can be used only because of a corrosion barrier layer of iron fluoride that forms on any carbon steel surface exposed to HF. The iron fluoride layer is tenacious and serves as a barrier against further carbon steel corrosion as long as the deposit remains undisturbed. Under certain conditions, such as when wet acid is in the unit, this iron fluoride scale can soften and break off leading to fouling and corrosion issues. In severe cases this can lead to considerable unscheduled feed outages as well as clear safety issues. Most refiners aggressively monitor their equipment's remaining corrosion allowance and use regularly scheduled valve and flange replacement to head off any problems. For maintenance during an FCC turnaround, the normal time most alkylation turnarounds are maintained, many refiners choose to dissolve all the iron fluoride scale by using a chemical cleaning company.

There are other areas within an alkylation unit where conditions are too severe for carbon steel and monel is the primary metallurgy for such areas as monel has good resistance even for high water content HF. Temperature is also an important variable for corrosion rate of both carbon steel and at higher temperatures for monel. Another use of monel is for moving parts where use of carbon steel would cause cementing of parts together a iron fluoride scale formed on each steel surface.

Sulfuric acid alkylation

Today there are two processes for H_2SO_4 alkylation the Cascade process licensed by Exxon Mobil and MW Kellogg and the Stratco effluent refrigerated process.

Figure 9.1.3 shows the cascade alkylation process.

Feed pretreatment
It usually consists of deethanizing and Merox-treating of FCC olefins. Some refiners have added selective hydrogenation units (SHP) to saturate dienes and reduce acid consumption. Feeds are generally not dried.

Reaction
The FCC olefins are chilled and coalesced to remove water and injected through sparge rings to 3–6 agitated reaction zones in a large horizontal reactor/settler vessel. Recycle isobutane from the deisobutanizer and the refrigeration system and recycle acid from the settler are fed to a pre-flash zone and "cascade" from one zone through

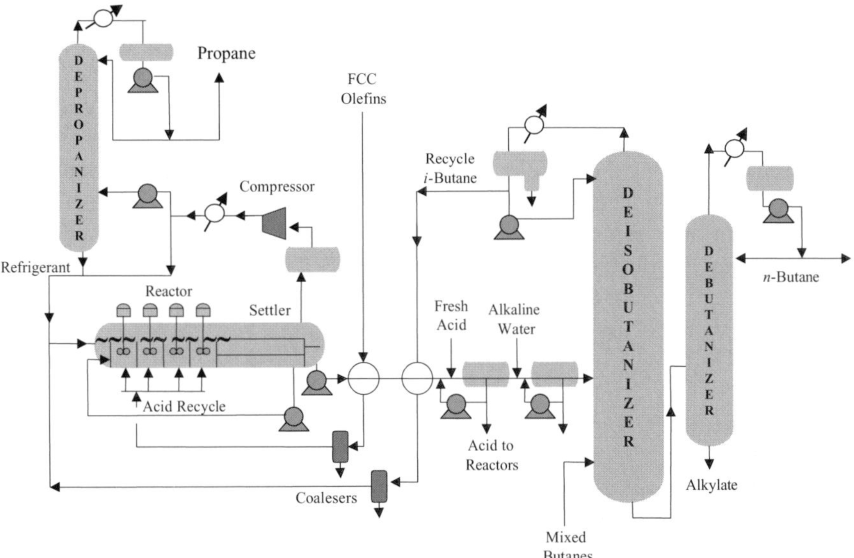

Figure 9.1.3. Cascade auto-refrigerated alkylation process.

specially designed weirs from which the process name derives. Typical isobutane olefin ratios are 8–12 for the process. The first zone in the cascade reactor has the lowest operating temperature and the highest isobutane concentration and produces the highest octane alkylate. As additional olefin is injected in subsequent zones the temperature increases and isobutane concentration decreases and successively lower octanes are produced. Because isobutane and H_2SO_4 are highly immiscible, each zone requires a mixer with high power inputs to produce a tight emulsion. After the final reaction zone the emulsion is allowed to settle. The settler acid phase is pumped back to the lead zone and the hydrocarbon phase effluent is pumped to effluent treating.

Refrigeration

The heat of reaction is removed by "auto-refrigeration" at reaction temperatures of 35–65°F. While refrigeration is often viewed as costly, in this process it conserves the heat of reaction to distill 4–5 moles of isobutane recycle per mole of olefin alkylated and concentrates propane. Isobutane and propane vaporized from the reactor are compressed, and condensed with cooling water and recycled as "refrigerant" to the reactors. A fraction of the refrigerant is charged to the depropanizer to remove propane contained in the feeds from the unit.

Effluent treating

The hydrocarbon effluent containing alkylate and excess isobutane is warmed by chilling recycle isobutane and feed and treated to remove traces of entrained acid and ester reaction intermediates. Treating systems include washing with fresh acid and aqueous caustic (as shown). Caustic and water washes, bauxite, and KOH pellets have also been used.

Fractionation
After effluent treating the balance of reactor isobutane requirement is distilled from the Alkylate and n-butane deisobutanizer tower (DIB). Most refiners charge saturated butanes from other units to the DIB for isobutane/n-butane splitting. N-butane is distilled from the Alkylate for control of product RVP in a debutanizer, and in some cases an n-butane vapor draw from the DIB. Finally, in a few units, aviation alkylate is produced by removing heavy ends in a Rerun column.

Stratco effluent refrigerated alkylation process

In the Stratco process (Figure 9.1.4.), the principal differences from the Cascade process are in the reactor and refrigeration design and that the reaction is carried out without vaporization.

Reaction
Treated feeds and recycle isobutane are first chilled and coalesced to remove water and charged to several Stratco contactors. Feed and isobutane from the DIB and

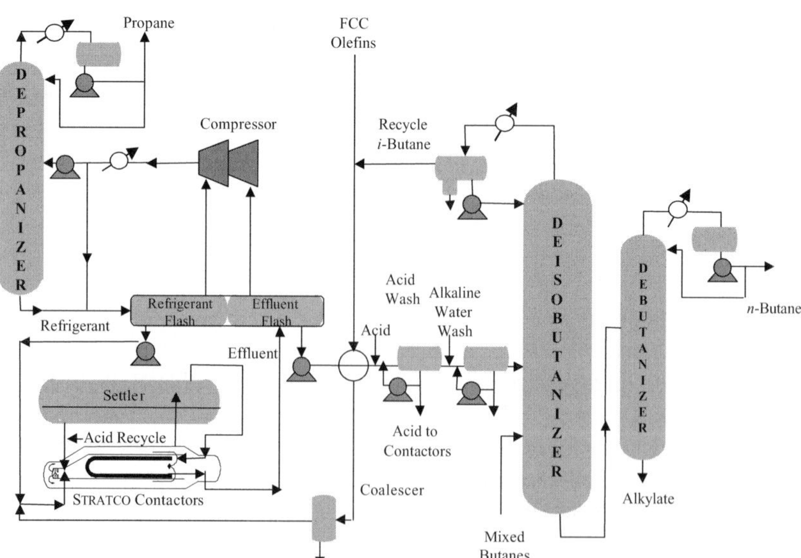

Figure 9.1.4. Stratco effluent refrigerated alkylation process.

refrigeration and recycle acid from the settler are emulsified together by the high power impeller of the Stratco Contactor. After reaction and chilling the emulsion passes to the settler located above the contactors for acid separation. The acid phase is recycled by gravity to the contactor impeller. And the hydrocarbon phase (effluent) routed in the tube-side of the contactor heat exchanger.

Refrigeration
The heat of reaction is removed by chilling the emulsion in shell side of the contactor heat exchanger by partially vaporizing settler effluent on the tube side. Refrigerant vapor is separated from the effluent liquid in a flash drum, compressed, and condensed. A portion of the condensed refrigerant is routed to a depropanizer. The balance of the refrigeration and depropanizer bottoms are flash cooled and returned to the reactors.

Effluent treating and fractionation
These steps are essentially the same as in the Cascade process.

Process variable in H_2SO_4 alkylation

Temperature
Maximum octanes for butenes and pentenes are obtained at about 35–40°F and for propylene at about 50°F. Typically, octane decreases at about 0.04–0.06 octanes per °F at higher temperatures.

Isobutane concentration
Reaction selectivity and octane increase with isobutane concentration. Typical responses are about 0.1 octane per 1 vol% isobutane in effluent.

Acid strength
Octane is maximized at about 93 wt% acidity. It declines at about 0.2–0.3 RON per 1% acidity decline below that. The economics of acid cost usually require lower final spending strengths (as low as 88 wt%), so multi-reactor alkylation units typically feed fresh acid to several reactors in parallel operating at 93% then reuse the 93% as makeup to other reactors in series.

Space velocity
Octane decreases with the charge rate of olefin. Alkylation reactors are typically designed at 0.15–0.30. LHSV defined as the volume of olefin charged per hour per volume of acid. Operations at much higher throughputs are possible but low octanes and high acid consumption generally result.

Acid fraction
Octane increases with acid fraction and peaks at about 60 vol% acid in emulsion.

Power per unit volume
Octane depends upon the mass transfer rate of isobutane from the hydrocarbon phase into the acid phase. According to Sprow[5] reaction selectivity and octane increase with interfacial area which depends upon power per unit volume to the 0.25 power.

Sulfuric acid consumption and regeneration

Sulfuric acid alkylation uses about 100 times as much acid as HF alkylation because H_2SO_4 cannot be stripped from the conjunct polymer, water, and contaminants. Sulfuric acid is typically spent at 88–91 wt% H_2SO_4, below which it is too weak to carry out alkylation. The spent acid is not a waste. It is regenerated by burning the organics which reduces the H_2SO_4 to SO_2 and water vapor. The SO_2 is then dried and oxidized to SO_3 and scrubbed with water to produce 98.5–99.5% H_2SO_4.

For sulfuric alkylation, aside from the direct hazards of contacting the acid, there are transportation safety issues around the need to ship and regenerate 20–500 tons of H_2SO_4 daily. Several refiners have built dedicated acid plants to regenerate spent acid from the alkylation unit or pipelines to and from an acid manufacturer.

Acid consumption is generally expressed as the pounds of acid diluted from 99% to 90% by conjunct polymer, water, and other contaminants per gallon or barrel of alkylate. Consumption can be as low as 16 lbs per barrel alkylate for diene-free, MTBE raffinates run at ideal process conditions. Acid consumption can exceed 40 lb/bbl of alkylate when running high levels of isobutylene, pentenes, and propylene at high temperatures.

H_2SO_4 contaminants

Dienes
Butadiene and pentadienes are a major contaminants affecting acid consumption. Dienes and can account for 10–30% of the acid consumed. Dienes can dilute 5–20 lbs of 99% acid down to 90% per lb of diene. Some refiners use diolefin selective hydrogenation units (such as UOP-Huls SHP™) to eliminate this source of acid consumption.

Water and oxygenates
Water can also account for 10–20% of acid consumption. Wet feeds, wet isobutane recycle, and high reactor temperatures are the main sources. Oxygenates including acetone from the FCC, methanol and dimethyl ether from MTBE and TAME units also consume acid. Water and oxygenates dilute 10 lbs of 99% acid to 90 per lb of contaminant.

Sulfur and nitrogen compounds
Dilute about 20 lbs of 99% acid to 90% per lb of sulfur or nitrogen.

C6+
Dilutes 2 lbs of 99% acid to 90% per lb of contaminant. This can be a significant contributor to acid consumption when alkylating pentenes.

Ethylene
Dilutes 20 lbs of 99% acid to 90% per lb of contaminant by forming stable sulfates that stay in the acid.

Ethane and lighter
While inert, light ends increase compressor discharge pressure for total condensation. These indirectly increase acid consumption when limited by horsepower by increasing reactor temperature.

Table 9.1.1. Alkylation Product Properties/Yields

HF Alkylation			
Volume Yields per vol olefin	FCC Propene/Butene	FCC Butene only	Butene only with Butene Isomerization
Isobutane Consumed	1.28	1.15	1.15
C5 plus Alkylate Produced	1.78	1.77	1.77
C5 plus Alkylate Properties			
Specific Gravity	0.693	0.697	0.697
RON-0	93.3	95.5	96.5
MON-0	91.7	93.5	94.5
RVP (PSI)	2.8	2.7	2.7

H_2SO_4 Alkylation		
Volume Yields per vol olefin	FCC Propene/Butene	FCC Butene only
Isobutane Consumed	1.2	1.12
C5 plus Alkylate Produced	1.72	1.72
C5 plus Alkylate Properties		
Specific Gravity	0.693	0.697
RON-0	92	96
MON-0	90.4	94
RVP (PSI)	2.8	2.7

Alkylate properties

Table 9.1.1 illustrates typical product properties for alkylate from HF and H_2SO_4 alkylation units for propylene-butylene and butylene only units. Alkylate is composed of primarily C_7's and C_8 paraffins with a relatively narrow boiling range. Motor fuel alkylate has desirable T50 and T90 distillation properties in addition to the long recognized low endpoint. RON is in the mid to low 90s and MON is 1–2 units lower. Its high MON is an important property of alkylates. The reid vapor pressure of debutanized alkylate can be as low as 5–6 PSIA except when alkylating pentenes. Its low RVP and very low sulfur and the absence of olefins and aromatics make alkylate a valuable component of reformulated gasolines by reducing Evaporative, NO_x and toxics emissions. These properties (or rather lack thereof) cause alkylate to trade at a premium over its already high octane value.

Almost all recently designed new alkylation units have been butylenes only or butylene with some portion of the pentenes. In most cases for butylene only units, the feedstock is pretreated in a selective hydrogenation unit to both remove the diolefins which increase acid passivation tar precursors and for HF alkylation units to isomerize most of the low octane producing 1-butene to the more desirable 2-butene isomers. Such a pretreating unit can raise the octane of HF alkylate product by one 1 Road ON as can also be seen in the third column of Table 9.1.1.

Recent developments

Motor fuel alkylation has recently received a boost from the planned phase-out of MTBE in gasoline blending formulations. Without MTBE, the best conventional components for today's reformulated gasolines are high octane paraffinic components like alkylate.

While environmental legislation has increased demand for alkylate, its catalysts H_2SO_4 and HF came under intense environmental scrutiny in recent years. While both acids are corrosive and toxic, HF is especially hazardous since it has the potential to form toxic aerosols, which can travel significant distances downwind of an accidental release. This scrutiny challenged the petroleum industry to enhance safety and even to seek new alkylation catalysts.

- UOP-Texaco and Mobil-Phillips developed cosolvents called Alkad™ and Revap™, respectively, to reduce HF aerosoling.
- A consortium of refiners and HF manufacturers lead by Amoco developed HF dispersion models and water spray mitigation with extensive, large-scale testing to design of release control systems. Together, cosolvents and water spray mitigation can reduce downwind HF concentrations by over 95% should a release occur.

- Through the American Petroleum Institute, the industry developed a comprehensive document for safe operation of HF alkylation units entitled 'Recommended Practice for Safe Operation of Hydrofluoric Acid Alkylation Units', API Recommended Practice 751, first published in June, 1992 and reissued in February, 1999.
- Finally the efforts to find new catalysts for alkylation have yielded solid catalyst processes such as UOP's Alkylene™ process that are ready for commercialization.

Conclusions

Motor fuel alkylation using liquid hydrofluoric or sulfuric acids is one of the oldest catalytic processes used in refining and petrochemical operations. The liquid acid processes remain important despite concerns with safety and environmental properties of the liquid acids. Solid catalyst alkylation using processes such as Alkylene™ (UOP LLC) are emerging technology.

References

1. V. M. Ipatieff and A. V. Grosse, *J Am. Chem. Soc.*, 1935 37 1616.
2. Pines, Herman, *ChemTech*, March 1982 150.
3. Hooper, John H. D. *Chemistry and Industry*, 1986 20, 683.
4. Jernigan, E. C., Gwyn, J. E. and Claridge, E. L. *Chem. Eng. Prog.* 1965 61(11) 94.
5. Sprow, F. B., *Ind. Eng. Chem. Process Des. Dev.*, 1969 8(2), 254.
6. V. N. Ipatieff and R. E. Schaad, Mixed polymerization of butenes by solid phosphoric acid catalyst. *Ind Eng Chem* (May 1938) 30 596–599.
7. A. Chauvel and G. Lefebvre, *Petrochemical Processes*. Vol 1, *Synthesis-Gas Derivatives and Major Hydrocarbons*, ed. Technip (1989) 183–187.

9.2 Catalytic olefin condensation

Peter R. Pujadó and Dennis J. Ward (retired)*

Introduction

Catalytic olefin condensation refers in general to the reaction of one molecule of an olefin with one or more molecules of the same olefin or of other olefins to yield heavier olefinic compounds. The term "condensation" reflects the fact that liquid products are obtained from gaseous olefins.

Olefin condensation is in some ways similar to motor fuel alkylation, except that in alkylation we react an olefin with an isoparaffin to yield a paraffinic compound. In olefin condensation, on the other hand, the products that are made are olefinic, but may be later hydrogenated in order to meet the requirements and specifications of the end products.

Olefin condensation can be carried out in many ways and over a diversity of catalysts. The common characteristic is that olefin condensation takes place over acid catalysts. These catalysts may be solid phosphoric acid (SPA), liquid acids, organometallic catalysts, zeolites, sulfonic acid resins, etc.

Although olefin condensation finds broad application in the petroleum refinery environment, it is also extensively used in petrochemical applications (often combined with refinery operations) as, for instance, in the production of heptenes, nonenes, dodecenes, etc. often used for the production of alkylaromatic derivatives. In fact, some of the same catalysts (e.g., SPA) can also be used for the alkylation of olefins with aromatics as in the production of cumene (isopropyl-benzene), ethylbenzene, etc. This review will concentrate on refinery and olefin oligomer applications and will not discuss aromatic alkylation applications.

*UOP LLC.

History

The discovery of acid catalysis, and in particular the development of solid phosphoric acid (SPA) catalysts, is attributed to the pioneering work of Vladimir Ipatieff at what was then called Universal Oil Products and is today known as UOP. The first commercial unit for the production of octenes by the dimerization of butenes came on-stream in 1935. By introducing this new ingredient in the formulation of gasolines, it greatly contributed to the popularization of the automobile that had hitherto been constrained by the limited availability of straight-run gasolines from the fractionation of petroleum crudes. In addition, the octenes obtained by butene dimerization had a higher octane number than naturally occurring straight-run gasolines. Several hundred units were soon built to satisfy the growing demand for gasoline. Commercially, the dimerization and oligomerization of olefins over SPA catalysts is known as the Catalytic Condensation process. This is still the dominant process for the dimerization and oligomerization of olefins. It is only recently that more work has been done on the development of acid zeolite catalysts and, in some cases, of sulfonic acid resin catalysts.

The other type of catalysts most often found in refinery applications for the production of gasoline components are organometallic catalysts developed by the Institute Français du Pétrole (IFP) and used, for example, in the dimerization of propylene to isohexenes (Dimersol G process), or in the oligomerization of light olefins from cracked gases (Dimersol E), or in the production of heptenes and octenes from propylene and/or butenes (Dimersol X).

Because the production of olefin dimers and oligomers almost invariably produces a blend of isomers and often also a lighter and a heavier fraction, the production of these olefins for petrochemical applications cannot be easily segregated from their application in gasoline and diesel fuel uses. However, it is usually the case that catalysts can be tailored within certain limits so as to yield more branched compounds (higher octane) for gasoline applications and less branched products for petrochemical applications. Because the products from such processes always have some branching, the quality of the products in the diesel range is somewhat poor (low cetane numbers) and their use in diesel applications is not that extensive.

Catalytic condensation process

The catalytic condensation process was developed in the early 1930s as a means to convert light gases produced by thermal cracking and thermal reforming into useful products. In the early years of the 20th century, most operations in the petroleum refining industry were limited to the distillation of crude oils into gasoline boiling range materials (straight-run gasolines), kerosenes, and lamp fuels. The yields in the gasoline fraction were low (sometimes as low as 5%, depending on the crude) and did not meet the requirements of the growing automobile industry.

One of the early ways to increase gasoline yields was by thermal cracking. The dominant thermal cracking technology was the Dubbs Process. In fact, out of the development of the Dubbs Process emerged the Universal Oil Products company (now UOP). A by-product from the Dubbs Process was a light gas stream that contained saturated and olefinic components that ranged from methane to butenes. Although there was no immediate use for those gases, refiners knew that by circulating these gases through a strong mineral acid it was possible to condense the olefins into heavier liquid products, and actually this was the principle applied in the Orsat apparatus used to measure the olefin content of those gases. Use of strong mineral acids (e.g., sulfuric acid) poses significant corrosion problems and presents considerable operational difficulties. Ideally, it was desired to operate with a single carbon steel vessel in which the strong acid could be contained. R. E. Schaad and V. N. Ipatieff discovered that phosphoric acid mixed with kieselguhr (an abundant naturally occurring siliceous diatomite mineral) solidified upon heating. When the solid mass was cooled, broken up, and placed inside a reactor, the propylene and/or butenes condensed into a liquid in the gasoline boiling range. This discovery was scaled up and eventually commercialized (1).

Shortly after the introduction of the Catalytic Condensation process came the catalytic cracking process, initially as a fixed process and later as the fluidized-bed catalytic cracking process, or FCC. These catalytic cracking processes, and in particular FCC, yielded even more olefins than thermal cracking, principally propylene and butenes, which at that time had very little commercial value. Thus, further application of the Catalytic Condensation process ensued.

The early operations were, by today's standards, primitive. To manufacture the catalyst, phosphoric acid and kieselguhr were mixed together, formed into a cake that was heated and then cooled, and broken up with a sledge hammer. The pieces were screened into the proper size range and loaded into the reactor. The reactor operated at very low pressures and produced such amounts of carbon that the catalyst had to be burnt off every other week, and be completely replaced every four to five weeks. The technology evolved by first improving the catalyst both in terms of manufacture (e.g., by producing an extrudate) and of formulation and also by optimizing the operating conditions.

Early units operated at about 10 atmospheres, since that was the pressure level of many liquefied petroleum gas (LPG) recovery systems. However, it was soon discovered that, if the reactor was operated at 30 or more atmospheres, catalyst lives could be extended by a factor of four to six times. Thus, began the evolution of higher pressure units. Some units that employed tubular reactors at 60 atmospheres were constructed and were found to be even more successful.

SPA catalyst was also tried for other reactions. In the late 1930s, it was discovered that SPA would catalyze the reaction of olefins with aromatic compounds to yield mono-olefinic alkyl aromatics. This fortunate discovery was widely used to make cumene (isopropyl-benzene) for aviation fuel used in World War II. Cumene has a very high performance number necessary for the aviation gasoline used in internal combustion engines in aircraft. After the war, use of cumene in aviation fuel declined, but cumene found an even more important outlet as feedstock for the production of phenol and acetone—to this date, this petrochemical use of cumene continues to dominate this industry, even though in recent years there has been a significant shift from SPA to acid zeolite catalysts for this aromatic alkylation. During World War II there was also a high demand for ethylbenzene and styrene. Some units were built to produce ethylbenzene by reacting ethylene with benzene over SPA catalyst. The reaction was slow and fairly inefficient, so SPA never gained relevance for this application; most ethylbenzene units at the time made use of aluminum chloride catalysts. Today, practically all existing capacity makes use of acid zeolite catalysts.

World War II also led to increased demand and production of polyvinylchloride (PVC). PVC is an excellent plastic material but lacks the flexibility required for many applications. It was discovered, however, that octyl phthalates and the esters of higher alcohols could be used to "plasticize" PVC; that is, to make it more flexible and supple. Octyl to decyl alcohols could be obtained by the "oxo" reaction (hydroformylation reaction) of olefins with carbon monoxide and hydrogen. The corresponding olefins, with one carbon less than the desired alcohols, could be recovered as by-products from the gasoline operation simply by fractionating the desired olefins out of the olefinic gasoline blend obtained by the Catalytic Condensation process. To this date, this is still the way that these olefins, from heptenes to nonenes, are produced, except that now these olefins are the main products from this operation, and the remaining lighter and heavier fractions are blended back into the gasoline pool.

A further application that derived from World War II was the advent of synthetic detergents to overcome the shortage of soap caused by the diversion of glycerin supplies to the manufacture of explosives. Though not readily biodegradable, one of the earliest and still one of the most active detergent ingredients is dodecylbenzene sulfonate. This was obtained by first alkylating dodecene with benzene according to technology developed independently by UOP and Standard Oil of California (now Chevron). Dodecene, or propylene tetramer, was readily produced by the oligomerization of propylene over SPA catalyst.

Still, in the apex of its success, the Catalytic Condensation process was used mostly for the production of gasoline fractions. During World War II, an important component in the aviation fuels used in fighter airplanes of that time was hydrocodimer, described more fully below. This together with cumene, also derived from phosphoric acid catalyst operations, was credited with being a significant factor in Britain's success in the Battle of Britain. The typical refinery product was known as polymer gasoline or "poly gasoline" even though in reality it consisted mostly of fairly simple dimers and trimers. At one time there were as many as 300–400 Catalytic Condensation units producing gasoline alone. Today, production of polymer gasoline is less significant, but the Catalytic Condensation process is still the preferred route to olefin dimers and oligomers and, despite the inroads made by zeolitic catalysts, is still widely used for the production of cumene. Based on the current demand for SPA catalyst, it may be estimated that about 40% is for gasoline, 40% for higher olefins (plasticizer olefins), and the balance 20% for cumene applications.

SPA catalyst may contain as much as 65 wt% phosphoric acid, but not all is active. Only the so-called "free phosphoric acid" provides catalytic activity, and this is typically around 16–20 wt% of the catalyst as manufactured. The activity of SPA is dictated by its equilibrium with water. Anhydrous phosphoric acid, P_2O_5, is not catalytic for it lacks the hydroxyl groups that are necessary for activity. Too much water, on the other hand, may cause the catalyst to get "soupy" and lose mechanical strength or even dissolve and come out of the unit. The successful operation of a Catalytic Condensation unit with SPA catalyst requires a close control on the water balance going into the reactor, both in order to obtain the desired level of activity and also in order to achieve long catalyst lives. Use of SPA catalyst in aromatic alkylation units may present additional constraints owing to the partial solubility of the phosphoric acid in aromatic streams; a small purge stream of phosphoric acid is usually maintained in these units.

Catalytic condensation process for gasoline production

Figure 9.2.1 illustrates a simplified process flow diagram for gasoline production using a tubular reactor. Figure 9.2.2 illustrates a similar diagram using an adiabatic plug flow reactor. Either type can be used, except that smaller size catalyst particles are normally used with a tubular reactor so as to minimize wall effects.

The tubular reactor system has the catalyst in the tubes and condensate or boiler feed water is used to remove the exothermic heat of reaction and to raise steam on the shell side. The reactants are circulated through the catalyst and the effluent is cooled and passed to a stabilizer column. An LPG (propane–butane) recycle may be maintained to control the concentration of olefins in the feed and the heat release in the system. Usually, the olefin concentration in the feed to the reactor is maintained at not more than about 50%. The gasoline produced is removed via the bottoms of the stabilizer.

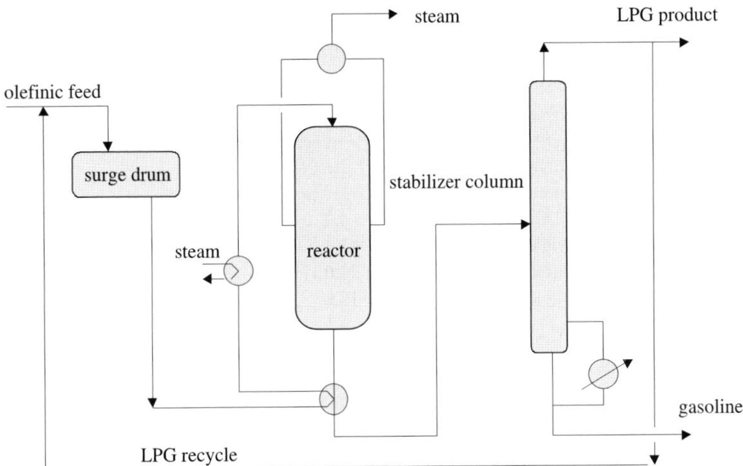

Figure 9.2.1. Catalytic condensation with tubular reactor.

Adiabatic plug flow reactors do not have provisions for steam generation within the reactor. Instead, an LPG interbed quench is used to control the temperature rise within the reactor with maybe as many as five catalyst beds being arranged within the vessel. The product is sent to a flash drum in which partial condensation occurs, and the liquid is sent to a stabilizer column. The bottom from this column is the final gasoline product. With this type of reactor, the concentration of olefins in the feed is usually limited to about 30%. From an energy point of view, it is less efficient than a tubular

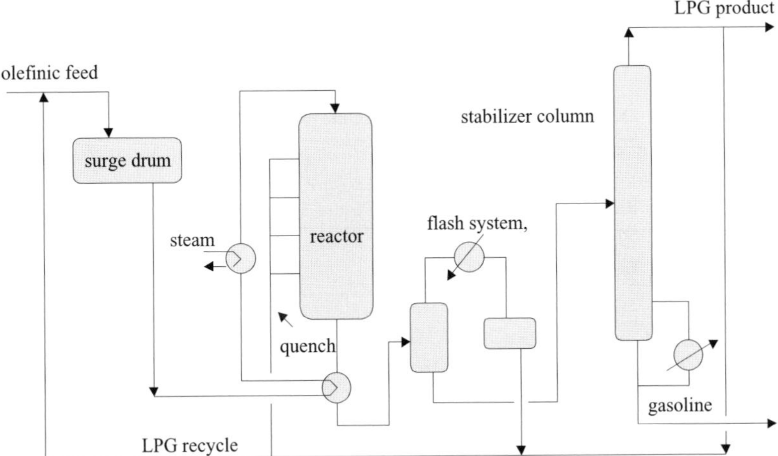

Figure 9.2.2. Catalytic condensation with plug flow reactor and quench.

Table 9.2.1. Oligomerization reactions

Propylene/butene feed:	FCC–C_3/C_4 fraction
Operation:	without polymer recycle
Wt% yield based on feed olefins	
Unreacted propylene	1.3 (97.3% conv.)
Unreacted butenes	3.7 (93.1% conv.)
C_6 olefins	4.4
C_7 olefins	41.3
C_8 olefins	24.0
C_9 olefins	15.8
C_{10}+ olefins	9.5

reactor, but it is more economical to build. For practical purposes, the products made by either system are almost identical.

Table 9.2.1 illustrates a typical product composition distribution based on a mixed C_3-C_4 feed from an FCC unit expressed as weight percent of the olefins in the charge stock. Feed olefin conversion is about 97% for propylene and 93% for butenes, but this may vary somewhat depending on the state of the catalyst, the feed composition, and the operating conditions. Table 9.2.2 provides typical distillation properties of depentanized gasoline; actual composition will vary depending on what vapor pressure adjustments are desired to adapt the product to the gasoline pool requirements. This means that the end product may also contain some pentenes and butenes if the higher vapor pressure is acceptable for blending into the gasoline pool.

Table 9.2.2 also serves to illustrate three different typical operations of these units: (1) polymer gasoline made from propylene only with no attempt made to produce higher olefins beyond the gasoline range; (2) polymer gasoline from a mixed C_3–C_4 feed; and (3) polymer gasoline from a unit operating with butenes only. The specific

Table 9.2.2. Typical polymer gasoline properties

	C6+ basis		
Polymer type:	C3 poly	C3/C4 poly	C4 poly
Specific gravity	0.739	0.738	0.738
Engler distillation (°C)			
IBP	57	93	62
10	114	101	103
30	132	109	113
50	141	119	119
70	151	137	127
90	191	168	181
EP	218	211	216

gravities of all three products are virtually identical. Although these data are from actual commercial operations, some differences arise from unit to unit. Also, it may be appreciated that the mid-point (50%) and the end point (100%) for a propylene feed are higher than for mixed feeds or butene feeds. In actual practice, these end points can be adjusted in order to meet product specifications. Often too the amount of polymer gasoline blended into the pool is small, so that it does not significantly alter the overall properties of the pool; only seldom is the product rerun in order to meet more stringent specifications. Interestingly, the product from a mixed C_3-C_4 feed, although having a higher molecular weight, has a lighter boiling range than that of the propylene product. The butene product in Table 9.2.2 is typical of a codimer operation. This type of operation was rare in the past, but because the octane number from this mode of operating is exceptionally high, it is currently generating considerable interest as a replacement for the production of methyl-*tert*-butyl ether (MTBE) in existing MTBE units, and will be discussed later in more detail.

Hydrogenated versus nonhydrogenated polymer gasolines from the catalytic condensation process

Polymer gasoline as obtained is almost totally olefinic. Polymer olefins, however, can be easily hydrogenated to yield a very interesting range of products. Most polymer gasolines blended in the gasoline pools have traditionally been olefinic, without any attempt of hydrogenation. When the Catalytic Condensation process was first developed, polymer gasoline was far better than the natural straight-run materials then available. Gasoline characteristics have changed dramatically since the 1930s, and while polymer gasoline used to be the best material in the leaded gasoline pool because of its excellent lead susceptibility, it is now one of the poorest in today's unleaded gasolines. This perception, however, can be altered significantly if, instead of considering untreated polymer gasoline, one looks at the properties of the hydrogenated products.

Olefinic products have a high Research (R) octane number (RON), but a low Motor (M) octane number (MON). As a result, the average octane index, $(R + M)/2$, of unsaturated polymer gasolines is low. The difference (RON–MON) is called the octane sensitivity; saturated or paraffinic products usually do not have a high octane sensitivity and in general, but not always, have much higher octane indices.

Hydrogenation is frequently performed at conditions that lead to the skeletal isomerization of the product. Polymer gasolines from SPA catalysts are highly branched, much more than predicted from thermodynamic equilibrium. Therefore, skeletal isomerization of these products would tend to move in the direction to equilibrium and, thus, to less branched products and to lower octane numbers. Thus, in general, hydrogenation catalysts that avoid isomerization should be selected.

Table 9.2.3. Octane numbers of hydrogenated polygasolines

	Polymer type		
	C3 poly	C3/C4 poly	C4 poly (codimer)
Olefinic polygasoline			
RON	93	97	99
MON	82	83	84
Hydrogenated polygasoline			
RON	49	70	99
MON	59	76	94

Table 9.2.3 illustrates typical octane numbers of clear olefinic and hydrogenated products. These values were measured on actual commercial samples and correspond to the trends that can be observed with these products. It is clear that hydrogenation does not enhance the octane numbers in all cases, so polymer gasolines from propylene feeds are best used without hydrogenation. It is interesting to note though that the MON of the hydrogenated material is much higher than the RON—an octane sensitivity of minus 10 is quite unusual in gasoline components of any kind.

Polymer gasoline from propylene–butene mixtures has higher RON typically in the 96–97 range, and a MON of about 83. In the past, when tetra ethyl lead (TEL) was added, both of these numbers increased by about 3, so leaded polymer gasoline from propylene–butylene had excellent characteristics with (R + M)/2 of about 93. When this gasoline is hydrogenated, both the RON and MON drop significantly, again rendering this product unattractive for motor fuel usage, but here too we encounter a reverse octane sensitivity of minus six. We may add that, when used in its leaded form, the RON and MON of the hydrogenated product increased to 86 and 90, respectively, thus making it an acceptable blending component for the older leaded pools.

Of more interest is perhaps the operation with butene feed components. The RON and MON of the product, sometimes called "codimer," are up to 99 and 84, respectively. Contrary to the negative experience with propylene and propylene–butene feedstocks, the hydrogenated product from an all-butene feed operation, sometimes called "hydrocodimer," has increased RON and MON of 99 and 94, respectively, which is excellent for gasoline blending in clear pools. This response was even more striking in the past when lead was added. With the addition of 0.5 g/L TEL, the RON and MON increased to 110–112 and to about 103, respectively. However, even in the absence of lead, both the olefinic and the saturated products have excellent blending characteristics.

Figure 9.2.3 illustrates an approximate linear correlation between the butene content in the feed and the RON of the hydrogenated product. The more butene there is in

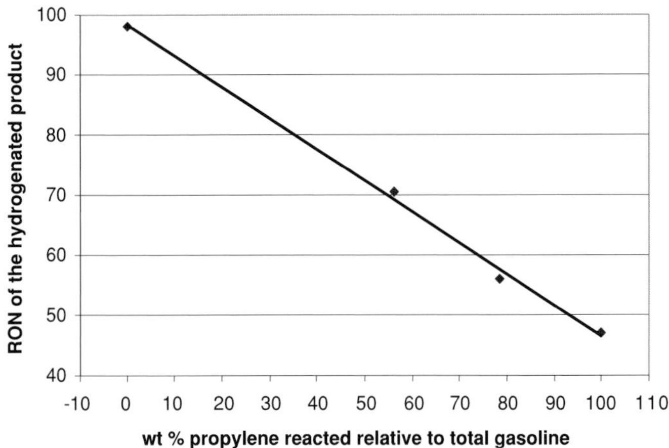

Figure 9.2.3. Effect of propylene content on RON of the hydrogenated polygasoline.

the feed to the catalytic olefin condensation unit, the higher will be the RON of the hydrogenated product and the better the product will be as a paraffinic component of the gasoline pool. Because of the absence of olefins and aromatics, this product meets or exceeds all current environmental regulations.

It is seldom the case that the various components that integrate the composition of a gasoline pool are used straight without blending. Normally, a motor fuel gasoline is a complex blend of hydrocarbons generated from a number of sources, for example: straight-run gasoline (mostly paraffinic with some aromatics, depending on the cut), alkylate (paraffinic), isomerate (paraffinic), reformate (aromatic), FCC gasoline (aromatic), polymer gasoline (olefinic or paraffinic), and sometimes also hydrocracked naphtha (mostly paraffinic). The octane numbers of these various components do not blend as the octane numbers of the pure fractions; a "blending octane" has to be used instead. Thus, for example, if two gasoline materials both with an octane number of 90 are blended together, there is no assurance that the mixture will have on octane of 90; the octane of the blend may be the same, lower, or higher, depending on the synergism among the various components as parametrized by the blending octane.

Table 9.2.4 illustrates the blending octanes of various polymer gasoline fractions when blended with different base stocks. The reference base stock was a mixture of approximately 60% FCC gasoline and 40% reformate, having a clear octane of 93 RON and 80.8 MON. Blend stocks consisting of the various polymer gasolines were added at 10 or 20 vol% levels and the blending octanes were back calculated from the octanes measured in the resulting mixtures. It must be noted, however, that small errors in the determination of the octane values can lead to significant fluctuations in the estimation of the blending octanes.

Table 9.2.4. Blending octanes of polygasolines

	Polymer type		
	C3 poly	C3/C4 poly	C4 poly
Measured clear octanes			
RON	93	96	98
MON	83	83	83
Blending octanes			
10 vol% blend			
RON	94	103	126
MON	83	95–99	112.5
20 vol% blend			
RON		103	
MON		~92	

Note: Base stock was a mixture of about 60% FCC gasoline and 40% reformate with MON/RON of about 93/81.

Based on these studies, and for these particular base stocks, it was determined that propylene-derived polymer gasoline had blending octanes of 94 RON and 83 MON; these values are almost identical to the clear values measured on the blend stock so that there is no synergism between the propylene condensation product and the base stock. The situation with the propylene–butenes polymer gasoline is quite different. Here, the blend stock had RON and MON of 96 and 83, respectively, but it blended with the base stock at the 10% level as though the octane numbers of the blend stock were 103 RON and 95 MON, a clear display of synergism with the base stock. Slightly different but still very high blending values are obtained when blended at the 20 vol% level. Naturally, the blending values decline progressively when the blend stock is added in higher and higher proportions, until they reach the clear measured values on the blend stock at a 100% blend stock proportion.

The impact of synergism is even higher for the butene-derived polymer gasoline. Here, instead of the measured octanes, 98 RON and 83 MON, we calculate blending octanes of 126 and 99, respectively, a synergism of 28 numbers for RON and 17 numbers for MON.

The above observations are for the olefinic catalytic condensation products. Similar observations can be made with the hydrogenated products. In general, however, the paraffinic product does not exhibit significant synergism with this type of base stock.

A corollary from these observations is that propylene polymer gasoline is fairly poor for most motor fuel applications. Propylene–butene feeds can make reasonably good gasolines, and the gasoline products from the Catalytic Condensation process based on butenes can be best characterized as superb.

Selective and nonselective gasoline production with the catalytic condensation process

Most Catalytic Condensation operations with SPA catalysts are conducted in a nonselective mode in which the main objective is to maximize conversion and the yield of useful products. Nonselective units tend to be operated at higher temperatures, typically between 160°C inlet and 200°C outlet temperatures for propylene or propylene–butene feeds. The objective is to maximize the yield in the gasoline fraction at a target octane number with little regard for the occurrence of side reactions that may modify the skeletal structure of the isomers.

For butenes, however, there can be significant differences between selective and nonselective operations. Nonselective operations at higher temperatures typically will produce some C_6's and C_7's, a lot of C_8's, C_9's, C_{10}'s, and C_{11}'s, and significant amounts of C_{12}'s. Selective operations are carried out at lower temperatures and usually at a lower space velocity. The butenes react selectively with each other to form virtually nothing but the product dimers, C_8, and trimers, C_{12}. Most of the dimers are the so-called "codimers," or selective condensation products of one molecule of isobutene with one molecule of *n*-butene. The resulting codimers are trimethylpentenes. Typical properties of the codimers and of their hydrogenated counterparts (the "hydrocodimers") are illustrated in Table 9.2.5 (note that this material is slightly different from those shown in Table 9.2.3 or Table 9.2.4, always as a function of the precise feed composition and the range of operating conditions). The total codimer product comprises the C_8 and C_{12} fractions, and has a very flat boiling range even when expressed in terms of an Engler or ASTM distillation, which is a less precise fractionation than a TBP distillation. The total product had a 99 RON and a 82 MON. The hydrogenated total

Table 9.2.5. Properties of codimer products

	Total product		Dimer product	
	Olefinic	Hydrogenated	Olefinic	Hydrogenated
Specific gravity	0.744	0.728	0.737	0.717
Engler distillation (°C)				
IBP	104	104	102	102
10	109	109	107	107
30	113	112	109	109
50	117	117	110	109
70	124	124	111	110
90	174	170	112	111
EP	210	213	134	132
Octane numbers				
RON	99	98	96.8	99
MON	82	93	82.4	93

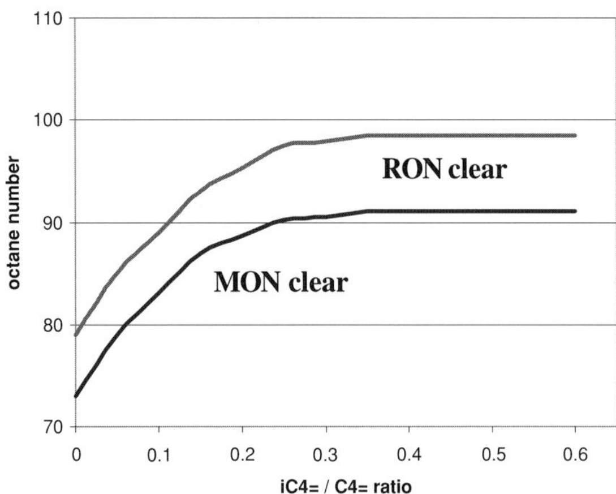

Figure 9.2.4. Octane numbers of 90% hydrogenated dimmer versus iC_4=/nC_4 = ratio in feed.

product, including C_{12}'s, had 98 RON and 93 MON. As expected, the specific gravity of the paraffinic product is slightly lower than the gravity of the olefinic material, but the boiling range remains virtually unchanged. If one considers the C_8's only, after fractionating out the C_{12}'s, we obtain about 97 RON and 82 MON for the olefinic product, and about 99 and 93, respectively, for the saturated material. Not shown in here but worth noting is that the lead response of the paraffinic product is always much higher than for the olefinic material.

The C_{12} materials thus made are quite unusual. They are relatively high boiling, and thus hardly qualify as favored gasoline fractions, but the hydrogenated C_{12}'s have RONs in the 95–97 range, which is indicative of a highly branched structure. Many different isomers are made and a precise characterization of the product is difficult. The operation can be conducted in the "dimer" mode with *n*-butenes only (1-butene or 2-butene) reacting among themselves, but the end product is not as attractive as the "codimer" that is obtained by reacting isobutene with *n*-butenes, or essentially by reacting isobutene with 2-butene since 1-butene rapidly isomerizes to 2-butene as the reaction proceeds. Figure 9.2.4 plots the octane number of the hydrogenated dimer product (in this case after only 90% hydrogenation), versus the weight ratio of isobutene reacted relative to the total weight of butenes reacted. The reciprocal of this value is sometimes called the "reaction ratio." It was soon learned during World War II that a low reaction ratio is desirable in order to obtain a high octane product.

The reason that Figure 9.2.4 is based on only 90% hydrogenation is that, as shown in Figure 9.2.5, a maximum RON is obtained when butylene dimers are hydrogenated to the corresponding paraffins. This slight optimum is a function of the relative synergism

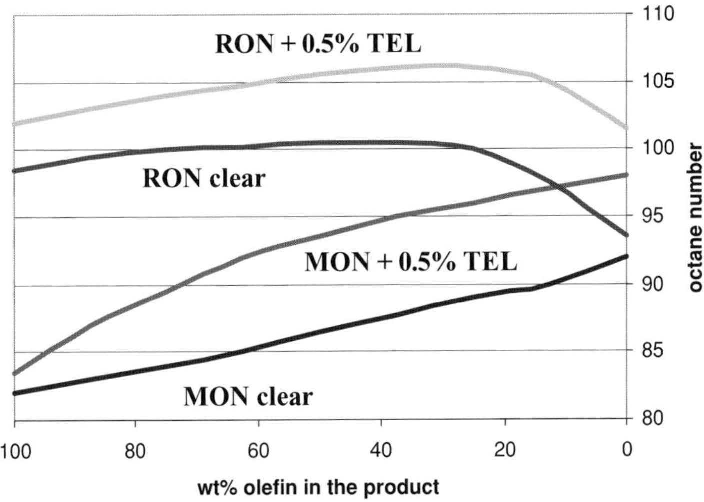

Figure 9.2.5. Heptene yields effect of olefin composition.

between olefins and paraffins and, although no longer used, is especially significant for the leaded material. Thus, for leaded material, partial hydrogenation of about 85–90% leads to a maximum octane value. For clear, unleaded material, a maximum octane value is obtained at around 60–70% hydrogenation. Because it is difficult to control the degree of hydrogenation, it is easier to simply bypass some of the material around the hydrogenation reactor.

Catalytic condensation process as a source of diesel fuels

The catalytic condensation process can also produce fuels in the diesel boiling range, although because of their relatively high degree of branching, they all have fairly poor cetane characteristics, even after hydrogenation.

Table 9.2.6 illustrates a typical range of products in the diesel range obtained by the Catalytic Condensation process with SPA catalyst followed by hydrogenation. The worst product is the heavy polymer from a C_3–C_4 gasoline operation. The other three materials are products derived from propylene feeds: propylene tetramer (dodecene), heavy polymer from a tetramer operation, and the entire feed to the tetramer rerun distillation column.

A commercial operation for the production of synthetic diesel fuel from propylene was built in South Africa. The commercial results agree reasonably well with the numbers in Table 9.2.6.

Table 9.2.6. Comparison of hydrogenated polymer products in the diesel range

	Propylene tetramer	Heavy propylene polymer	Tetramer plus heavies	Heavy polymer from C_3/C_4 gasoline
API gravity	54.2	44.4	51.8	49.3
Engler distillation (°C)				
10	184	262	188	154
30	187	266	192	206
50	189	271	198	215
70	193	279	208	229
90	198	300	238	271
EP	213	329	277	304
Cetane numbers	32.8	37.0	36.9	28.7

It is possible to improve the quality of the diesel range products by operating in the selective mode at lower temperatures and lower per-pass conversions with a higher recycle rate of unconverted material. This approach, however, has not been commercialized.

Petrochemical operations

Petrochemical applications of catalytic olefin condensation center on the production of heptenes, nonenes, and dodecene (or propylene tetramer). Though initially recovered by fractionating olefins out of the olefinic product from the Catalytic Condensation process for the production of polymer gasoline, modern units are specifically designed and operated to maximize the yield of the desired olefin fractions. The by-products are still blended into the gasoline pool.

Heptenes and nonenes are the main products used for the production of octyl and decyl alcohols via the "oxo" or hydroformylation process. Both these alcohols find end use as esters in plasticizer applications, principally for PVC. Dodecene at one time was a major product for the manufacture of dodecylbenzene sulfonates as active detergent ingredients; however, their relatively poor biodegradability led to the discontinuance of this operation. Dodecenes are still used in the production of tridecyl alcohols via the oxo process. Some nonenes are also used in the production of nonyl phenols. Because this chapter is confined to olefin condensation, mostly for motor fuel applications, only heptenes and nonenes will be discussed briefly.

Heptenes

Table 9.2.7 illustrates a wide range of specifications for heptenes. Though fairly common to all producers, the values can vary depending on the desired end-use application.

Table 9.2.7. Typical heptene specifications

Parameter	Specification	Typical values
Specific gravity	0.705 min 0.715 max	0.711
Boiling range (°C)		
IBP	85 min	85
50%	88 min 90.5 max	89
95%	95 max	93
Bromine number	150 min	154
Sulfur	0.02 wt%	–
Peroxide number	40 mg/L	–
Color	+20 Saybolt min	–
Ratio of dimethyl-pentenes to methyl-hexenes	2/1 to 3/1	2.6

Thus, for example, the boiling range and the desired ratio of dimethylpentenes to methylhexenes may determine the operating conditions in a catalytic condensation unit and the net recovery and yield of the useful fraction. The product heptenes consist of a fairly complex mixture of heptene isomers. Table 9.2.8 shows a typical composition and approximate boiling points for individual major isomer components; many

Table 9.2.8. Major isomers in the heptene fraction (> 0.1 wt% approx.)

No.	Component	Wt%	Boiling point (°C)
1	4,4-Dimethyl-1-pentene	0.8	72.5
2	4,4-Dimethyl-*trans*-2-pentene	4.4	76.8
3	3,3-Dimethyl-1-pentene	0.2	77.6
4	4,4-Dimethyl-*cis*-2-pentene	0.1	80.4
5	3,4-Dimethyl-1-pentene	0.2	81.1
6	2,4-Dimethyl-1-pentene	6.2	81.7
7	2,4-Dimethyl-2-pentene	16.5	83.4
8	2,3-Dimethyl-1-pentene	7.8	84.3
9	5-Methyl-*trans*-2-hexene	1.0	86.1
10	2-Methyl-*trans*-3-hexene	0.2	86.1
11	3,4-Dimethyl-*trans*-2-pentene	16.3	87.2
12	3,4-Dimethyl-*cis*-2-pentene	4.4	87.2
13	3-Methyl-1-ethyl-1-butene	3.1	88.9
14	3-Methyl-*trans*-3-hexene	0.5	93.6
15	3-Methyl-*trans*-2-hexene	1.5	93.9
16	3-Methyl-*cis*-2-hexene	2.9	93.9
17	3-methyl-*cis*-3-hexene	2.2	95.3
18	2-methyl-2-hexene	2.2	95.4
19	3-Ethyl-2-pentene	1.2	96.0
20	2,3-Dimethyl-2-pentene	27.5	97.4
21	*Trans*-2-heptene	0.6	97.9
22	*Cis*-2-heptene	0.1	98.5

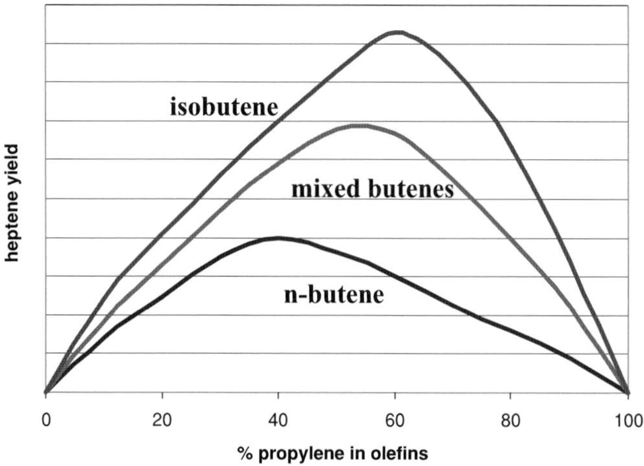

Figure 9.2.6. Hydrogenation of butane dimer.

other isomers can be present at lower concentrations. As a rule, dimethylpentenes are produced mainly from isobutene, while methylhexenes tend to originate from *n*-butenes.

The relative yield of heptenes is illustrated in Figure 9.2.6 as a function of the olefinic feed composition; as expected, higher yields of heptenes are obtained when the feed mixture consists of equal proportions of propylene and butenes. Isobutene tends to produce significantly higher yields of heptenes than *n*-butene. However, because isobutene produces more dimethylpentenes, the final yield can only be calculated on the end product after fractionation that meets the required specifications.

Nonenes

Typical nonene specifications are shown in Table 9.2.9. Actually, there are several different specifications that could be called "typical," a main characteristic being the breadth or narrowness of the cut.

A minimum olefin concentration of 95% or higher is usually required. An olefin concentration of 95% can be easily obtained for nonene units that process only propylene in the feed. However, in units that may also process olefinic recycle streams, as for example from a tetramer operation, there may be a reduction in the olefin content of the product owing to hydrogen transfer mechanisms in which hydrogen is transferred from the heavier species to the lighter olefins. Units with a high gasoline recycle may experience increased paraffin concentrations in the nonene fraction as high as 30–40%, which render the nonenes unsuitable as oxo plant feed. Most nonene units

Table 9.2.9. Typical nonene specifications

	Specification		Typical values	
Parameter	Narrow range	Wide range	Narrow range	Wide range
Specific gravity	0.738 min	0.740 min		
	0.745 max			0.745
Boiling range (°C)				
IBP	134 min	134 min	134	134
50%	–	–	135	138
EP	143 max	155 max	142	150
Bromine number	125 min	–	127	
Sulfur	0.005 wt% max		–	–
C_8 olefins, wt%	0.5 max		–	
C_9 olefins, wt%	90.0 min		97.3	
C_{10} olefins, wt%	3.5 max		2.7	
Olefin content (wt%)		95 min		

today are disassociated from tetramer production and therefore this problem is seldom encountered.

Dimersol™ process

While the Catalytic Condensation process described above makes use of a supported phosphoric acid catalyst, the Institut Français du Pétrole (IFP) developed and commercialized a different approach based on Ziegler-Natta catalyzed oligomerization using organometallic catalysts. Because of the nature of the catalyst, this process can be tailored to specific highly selective applications that have been identified by various names (2):

Dimerization of ethylene to 1-butene

This version is known as Alphabutol™ and is intended for the selective production of 1-butene for use as a comonomer in the production of linear low-density polyethylene (LLDPE). Though not in the motor fuel range for refinery applications, its mention is included here for the sake of completeness and continuity with the other Dimersol process schemes. The reaction takes place under mild conditions at about 50–60°C, under sufficient pressure to maintain the reaction medium in the liquid phase, and in the presence of a homogeneous titanium-based catalyst. Under these conditions, the *n*-butene equilibrium favors an almost total conversion of 1-butene to 2-butene; therefore, it is critical that the catalyst have a total lack of isomerization activity. Some C_6+ oligomers are also formed which are used to maintain the catalyst in solution.

Molar yields in excess of 93% can be achieved in this operation. This type of operation is best suited for locations where LLDPE is being produced and in which there is no other local source of 1-butene. Alphabutol units have been reportedly built in Saudi Arabia and Thailand.

Dimerization of ethylene to n-butenes

Whereas the IFP Alphabutol process is selective for the production of 1-butene, the IFP Dimersol E process produces *n*-butenes and heavier oligomers. The Dimersol E process also operates at about 50°C in the liquid phase with a Ziegler-type catalyst that can be a nickel derivative activated by an organometallic reducing agent. Whereas the Alphabutol process avoids isomerization, the Dimersol E features both dimerization and isomerization, and it produces a blend of *n*-butenes and heavier oligomers, depending on the feed composition, the actual type of catalyst, and the operating conditions. For polymer-grade ethylene, the Dimersol E process yields 30–70% *n*-butenes with a once-through conversion of 90–100%, with the rest being in the C_6+ gasoline fraction. Gasoline production can be further enhanced by adding propylene to the feed. It is worth noting though that in the Dimersol E process the ratio of 1-butene to 2-butene in the product is about 1/1, still well above the equilibrium ratio at 50°C.

Dimerization of propylene and butenes, separately or combined

The Dimersol process (Figure 9.2.7) uses an organometallic catalyst consisting of a nickel salt in combination with an alkylaluminum compound. Various modalities are available:

- Dimersol G—for the dimerization of propylene to isohexenes used in gasoline blending
- Dimersol E—for the oligomerization of ethylene and propylene olefins contained in FCC gases and already described above for the nonselective dimerization and oligomerization of ethylene
- Dimersol X—for the selective dimerization of propylene and butenes for the production of heptenes or for the dimerization of *n*-butenes for the production of octenes. These octenes are fairly linear, have a poor octane number, and are unsuited for gasoline applications. Heptenes and octenes obtained from the Dimersol X process are feedstocks for the production of plasticizer alcohols.

In all these Dimersol processes, the olefinic feedstock and the liquid catalyst are introduced into the reactor in a fashion compatible with the heat removal requirements in the reaction system. The level of conversion and the distribution of dimers and oligomers depend on the number of reactors used in series, the residence time, and the concentration of catalyst. The effluent from the reactor is circulated through a catalyst neutralization and elimination section and a water wash before fractionation for the desired cuts.

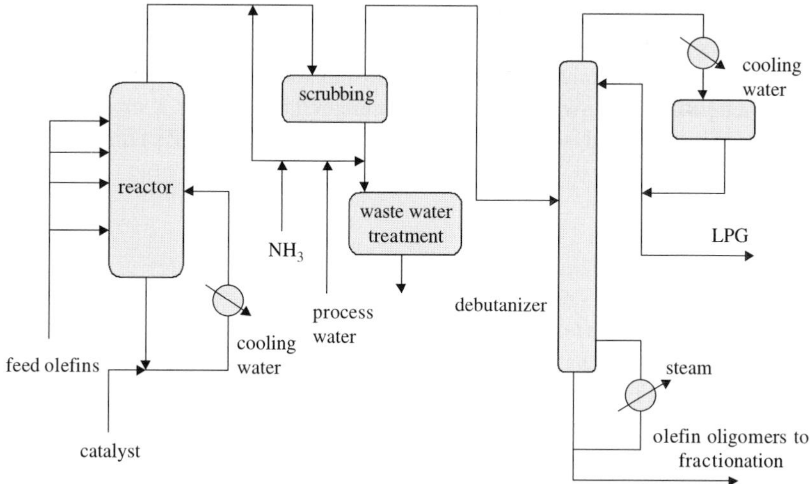

Figure 9.2.7. Dimersol process (IFP).

Other dimerization or oligomerization processes

No other dimerization or oligomerization processes are used commercially for the catalytic condensation of light olefins into gasoline range products, but Chevron in Richmond, California, and NPRC in Japan, use or have used liquid phosphoric acid to produce propylene tetramers. There are also a few highly selective processes that are used for the production of specific dimers in petrochemical applications. Some of the most notable are:

- Sulfuric acid dimerization of isobutene for the production of trimethylpentenes, in particular a mixture of 2,2,4-trimethyl-1-pentene and 2,2,4-trimethyl-2-pentene. These are excellent gasoline blending components (once hydrogenated, 2,2,4-trimethylpentane is the reference 100 octane hydrocarbon for both RON and MON), but the technology is fairly messy and not used commercially. Liquid acid catalysis with either sulfuric acid or hydrofluoric acid is extensively used in refining applications for the alkylation of isobutane with *n*-butenes or other olefins.
- Selective dimerization of propylene to 2-methyl-1-pentene over triethylaluminum in a process developed by Scientific Design and Goodyear. This dimer can be used as an intermediate for the synthesis of isoprene.
- Selective dimerization of propylene to 4-methyl-1-pentene over organopotassium compounds or, in general, strong alkaline catalysts, in the liquid phase. 4-Methyl-1-pentene is a monomer used in the preparation of transparent plastics with a high melting point used in microwable cookware, labware, and medical products, and manufactured commercially by Mitsui Plastics under the name "TPX" and by Phillips 66 under the name "Crystalor."

- Hüls AG's Octol process for the selective dimerization of *n*-butenes to highly linear octenes suitable for the manufacture of isononyl alcohol plasticizers over a solid heterogenous catalyst. The product may contain up to 20% *n*-octene and over 50% methyl-heptenes, with a balance of dimethylhexenes and a small amount of trimethylpentenes, and is quite similar to that obtained with IFP's Dimersol X process.

Recent developments

Catalytic condensation of olefins for the production of gasoline fractions and higher olefins is an old art that has recently received a boost from the planned phase-out of MTBE in gasoline blending formulations.

MTBE (methyl-*tert*-butyl ether) is a high octane blending component used extensively throughout the world since about 1975 and more widely since the mid-1980s. In addition to providing a significant increase in the octane of the gasoline relative to that of the base stock, it also has a beneficial impact in the reduction of hydrocarbon emissions. MTBE, however, is slightly soluble in water (about 0.55 wt%) and has a foul odor. Leakage of gasoline from underground tanks into a potable water aquifer normally has little impact because of the lack of solubility of the hydrocarbon components, but becomes very noticeable when MTBE is present at even a low ppb level. This has led to the planned discontinuance in the use of MTBE in certain regions, principally in the United States.

MTBE is made by the reaction of isobutene with methanol over a sulfonic acid resin in the liquid phase and at temperatures in the 40–80°C range in a process that was independently pioneered by Snamprogetti SpA and by Hüls AG. Mostly, adiabatic plug flow reactors are used, but isothermal tubular reactors have also been used; later units use reaction combined with distillation in what is a quasi-isothermal operation. MTBE initially was obtained by reacting the isobutene contained in the C_4 fraction from an FCC unit or from a steam cracker after butadiene extraction—the other C_4 components are not reactive under these conditions and only MTBE is formed. However, with the advent of legislation that mandated the use of oxygenates in gasoline for environmental reasons, use of MTBE skyrocketed and it became necessary to build large on-purpose isobutane dehdyrogenation units to meet the demand for isobutene. These MTBE units often had capacities for about 600,000 to 900,000 metric tons per year, of which about 65% corresponds to the weight of the required isobutene.

The planned phase-out of MTBE presents a costly dilemma for the existing MTBE producers, principally for those who rely on isobutane dehydrogenation units. They can either shut down the units or they can convert them into something else. An obvious solution is available in two forms:

- The old codimer/hydrocodimer experience can be revisited either by dimerizing isobutene or by reacting isobutene with *n*-butenes over a SPA catalyst to yield the corresponding isooctenes. However, while the older dimerization facilities operated in the gas phase and had relatively poor selectivity to isooctenes, the newer versions now being implemented operate in the liquid phase and have much higher isooctene selectivity. Also, the older hydrocodimer facilities required adding a costly hydrogen plant just to supply the hydrogen for the hydrogenation of the olefins. In isobutane-based MTBE complexes this is not necessary because hydrogen is available as a by-product from the isobutane dehydrogenation step.
- Similar experience can be derived from Bayer AG's use of sulfonic acid resins for the dimerization of isobutene to the corresponding isooctenes in the liquid phase at 100°C as done commercially at a plant in Dormagen operated by Erdölchemie. That experience showed that sulfonic acid resins can be used in hydrocarbon environments and can achieve over 99% isobutene conversions and about a 75 mol% selectivity to dimers and 25 mol% selectivity to trimers. Under appropriate modifications, use of sulfonic acid resins can be adapted to the selective production of gasoline components.

Technologies for the dimerization of isobutene with isobutene, *n*-butenes, or other olefin components are now being offered commercially by Snamprogetti SpA, KBR (Kellogg—Brown & Root), UOP LLC, and others. Because UOP offers both SPA catalyst and sulfonic acid resin catalyst, the following discussion will center on UOP's Indirect Alkylation (InAlk) process technology. It may be further added that isobutane dehyrogenation units operate equally well with *n*-butane or with mixtures of isobutane and *n*-butane in the feed; the corresponding olefins are obtained in these situations, typically with a 1-butene/2-butene ratio that corresponds to the high temperature dehydrogenation equilibrium.

Catalytic olefin condensation with the InAlk process

Barring the use of MTBE and of the former hydrocodimer products, the best components for today's environmentally friendly gasoline formulations are high-octane paraffinic components like alkylate and isomerate. From the viewpoint of octane number and vapor pressure, alkylate is the preferred blending material.

Alkylate is made by reacting isobutane with *n*-butenes or other olefins over concentrated sulfuric acid or hydrofluoric acid in the liquid phase, or over novel solid heterogenous acid catalysts now being introduced.

The Indirect Alkylation or InAlk process obtains a product similar to motor fuel alkylate by reacting isobutene with isobutene, n-butenes, or other olefins, and also by dimerization reactions of n-butenes. Idealized products after hydrogenation are:

2,2,4-trimethylpentane	RON = 100	MON = 100
2,2,3-trimethylpentane	RON = 109.6	MON = 99.9
3,4-dimethylhexane	RON = 76.3	MON = 81.7

Either SPA catalysts or sulfonic acid resin catalysts can be used for the condensation. As usual, either noble or non-noble metal catalysts can be used for the hydrogenation.

Resin-catalyzed condensation

Sulfonic acid resin catalysts are attractive because they are the ones used for the synthesis of MTBE. Thus, ideally, one could make isooctenes instead of MTBE with virtually no plant modification regardless of whether the original MTBE unit uses a conventional adiabatic plug flow (packed bed) reactor or a reactor that combines reaction with distillation as in CD Tech's "Catalytic Distillation" process or in the Hüls/UOP "Ethermax" process. In the case of the Ethermax process, it has been established that the same process unit can be used for the production of isooctenes without significant modifications except perhaps for the addition of an additional front end packed-bed reactor for added conversion and, obviously, the hydrogenation reactor, and possibly a product stripper for the removal of light ends.

Overall isobutene conversions in excess of 95% are achievable, but the n-butene conversion is much lower, in the order of 10%, so that an n-butene recycle should be maintained if maximum conversion is desired. Selectivity to the desired diisobutenes or isobutene/n-butene codimers is in the 85–90% range, while production of trimers (C_{12} =) is about 10–15%, and the production of heavier compounds (tetramers, etc.) is negligible. A typical trimethylpentane (TMP) product from an FCC C_4 feed at about 10% n-butene per-pass conversion can be characterized as follows:

C_8 (TMP)	88%
C_{12}	12%
C_{16}	traces

Blending characteristics

RON	97.5–98.0
MON	95.5–96.0
RVP	~2 psi (~14 kPa)
T_{50}/T_{90}	227/255°F (108/124°C)

If not hydrogenated, the same product characteristics are obtained except that the RON increases to about 102.0–102.5 while the MON decreases to about 85.5–86.0, as corresponds to the higher octane sensitivity of olefinic materials that was discussed earlier.

With a high concentration isobutene feed as might be available from an isobutane dehydrogenation unit, the product characteristics are:

C_8	(TMP) 88%
C_{12}	12%
C_{16}	traces

Blending characteristics

RON	99–101
MON	97–100
RVP	~2 psi (~14 kPa)

If not hydrogenated, the olefinic product shows similar characteristics but with RON in the 111–114 range and MON of about 92–95.

SPA-catalyzed condensation

Although analogous in concept to the earlier work with SPA catalysts, significantly enhanced results can be obtained by fine tuning the operation for gasoline production, in particular by increasing the isooctene selectivity at lower temperatures and operation in the liquid phase. While traditional Catalytic Condensation units with SPA catalyst had isooctene selectivity of typically about 55%, in the InAlk process, operating in the liquid phase at lower temperatures, the selectivity can be increased to about 90%, with the rest being C_{12}. This material can be put directly into the gasoline pool, usually after hydrogenation.

Again, an MTBE unit can be converted to an InAlk unit with SPA catalyst with only minor modifications, of which the most significant is perhaps the addition of another front end reactor in parallel.

A typical trimethylpentane (TMP) product from an FCC C_4 feed at about 15% *n*-butene per-pass conversion can be characterized as follows:

C_8 (TMP)	95%
C_{12}	5%

Blending characteristics

RON	100.8
MON	95.7
RVP	~2 psi (~14 kPa)
T_{50}/T_{90}	232/275°F (111/135°C)

With SPA catalyst it is possible to increase the n-butene conversion to about 30% by operating at slightly higher temperatures. Virtually the same results are obtained except for about a one point decrease in both RON and MON.

If the feed has a high isobutene content, the product distribution after hydrogenation is:

C_7 (TMP)	3%
C_8	87%
C_9	10%

Blending characteristics

RON	101.5
MON	97.5
RVP	~1.4 psi (~10 kPa)

The relative operation between the Catalytic Condensation process and the InAlk process can be differentiated as follows:

	Traditional catalytic condensation	InAlk olefin condensation
Phase	vapor	liquid
Temperature	base	lower
Pressure	base	lower
Product recycle	no	yes
Space velocity	base	higher
iC_4= conversion	base	base
nC_4= conversion	base	lower

The effect of temperature on isooctene selectivity can be best appreciated by the following simple comparison using traditional catalytic condensation without recycle:

	190°C average bed temperature	150°C average bed temperature
C_7-	8.5%	3.5%
Isooctenes	56.0%	71.1%
C_9+	35.5%	25.4%

Operation at the lower temperature but with added recycle improves the selectivity to about 2.2% C_7-, 84.1% isooctenes, and 13.7% C_9+. Further improvements can be achieved by operation at even lower temperatures or at other recycle rates.

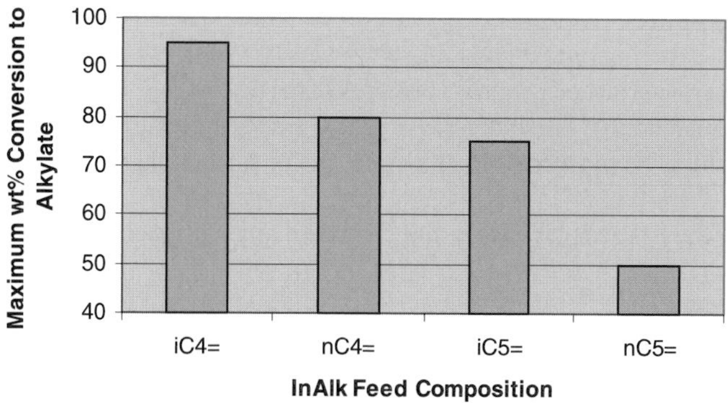

Figure 9.2.8. Comparison of InAlk conversion for C_4 and C_5 streams.

It may be further noted that by operation at these lower temperatures, the life of the SPA catalyst can be greatly lengthened relative to that observed in traditional Catalytic Condensation operations. The increased catalyst life is the net effect of having lower carbon deposition on the catalyst and of requiring lower SPA hydration rates. Under these conditions, the catalyst life is so much longer that the impact of feed impurities may become the issue in determining the overall catalyst life.

Lastly, we can point out that, because it does not have the temperature limitations inherent to a sulfonic acid resin, SPA catalysts can operate over a significantly broader range than just propylene or butenes. Figures 9.2.8 and 9.2.9, for example, illustrate the comparative conversions and product RON values for isobutene, *n*-butenes, isopentene, *n*-pentenes, and mixtures thereof.

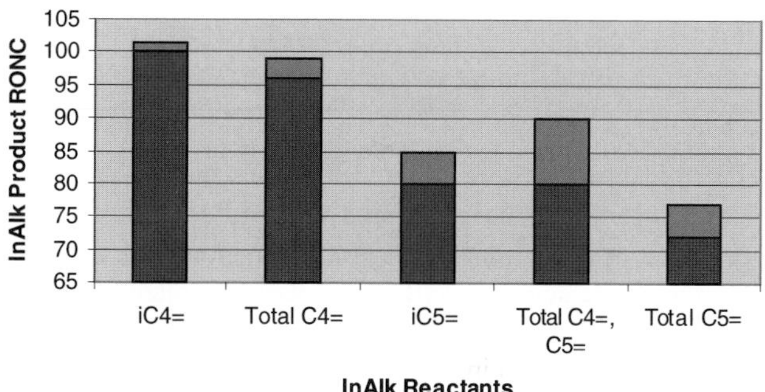

Figure 9.2.9. Comparison of InAlk product RON for C_4 and C_5 streams.

Catalyst suppliers

The information provided herein is correct as of the time of this writing but is subject to change without notice. The addresses provided are for the central offices; the catalysts themselves may be sourced from other geographical locations.

Dimersol catalysts are proprietary and can be obtained through IFP

Institut Français du Pétrole (IFP)
1 et 4, avenue de Bois-Préau
92852 Rueil-Malmaison Cedex
France

IFP North America, Inc.
100 Overlook Center, Suite 400
Princeton, NJ 08540
USA

There are two recognized suppliers of SPA catalyst

UOP LLC
25 East Algonquin Road
P. O. Box 5017
Des Plaines, IL 60017-5017
USA

Süd-Chemie AG
Lenbachplatz 6
80333 München
Germany

Süd-Chemie United Catalysts Inc.
P. O. Box 32370
Louisville, KY 40232
USA

Sulfonic acid resins come in many forms. Preferred catalysts include, but are not limited to, highly crosslinked macroreticular resins such as those available from Rohm and Haas (Amberlyst), Dow Chemical (Dowex), Bayer AG, Purolite, and others. Proprietary formulations may be desirable though for specific applications in this area, and the various suppliers should be consulted accordingly.

Conclusions

Catalytic olefin condensation is one of the oldest catalytic processes used in refining and petrochemical operations and pioneered the used of acid catalysts. Many catalysts have been used over the years that have been examined, albeit briefly, in this chapter. Solid phosphoric acid catalysts (SPA) were for many years the work horse of the industry, and still command a very respectable market share. Zeolitic catalysts are slowly making some inroads in this area, but much more slowly than originally expected in view of their initial success in the alkylation of aromatics. More surprisingly, this old technology is currently enjoying a bit of a revival by the introduction of highly selective, low temperature, liquid phase operations in which both sulfonic acid resins and SPA are finding new niche opportunities. If catalysts can be found with higher selectivity to more linear oligomers in the C_{12} to C_{18} range, additional markets may develop in the production of high-quality synthetic diesels and jet fuels.

References

1. V. N. Ipatieff and R. E. Schaad, Mixed polymerization of butenes by solid phosphoric acid catalyst. *Ind. Eng. Chem.* 30 (May 1938) 596–599.
2. A. Chauvel and G. Lefebvre, *Petrochemical Processes*. Vol. 1, *Synthesis-Gas Derivatives and Major Hydrocarbons*, ed. Technip (1989) 183–187.

9.3 Isomerization technologies for the upgrading of light naphtha and refinery light ends

Peter R. Pujadó*

Upgrading light hydrocarbon (C_4–C_6) streams in refineries and gas processing plants has increased in importance as new regulations affecting gasoline composition are enacted in many regions of the world. These regulations include lead phasedown, benzene minimization, and oxygen content requirements. Light-naphtha isomerization plays a key role in meeting octane demands in the gasoline pool that result from both lead phasedown and increasing market share for premium gasoline grades. By isomerization we mean the skeletal isomerization of a paraffin to a more branched paraffin with the same carbon number; however, because of its nature, isomerization may also affect some of the naphthene components and will also hydrogenate aromatics. In fact, isomerization is the most-economic means available for reducing the benzene content in gasoline.

Several isomerization technologies are available from various licensors. UOP's Penex process for the isomerization of C_5–C_6 paraffins and Butamer process for the isomerization of *n*-butane to isobutane are some of the most widely used and will form the basis for this chapter. Developments in isomerization are usually based on the introduction of new, more active or more stable catalysts, some of which will be discussed in more detail below. Similar comments could be made for other equivalent processes or catalysts available from other sources.

Introduction

The skeletal isomerization of alkanes is one of many important industrial applications of acid-function-promoted catalysis. Other examples of the wide commercial application of acid-catalyzed industrial hydrocarbon reactions are the alkylation of paraffins and aromatics with olefins, transalkylation and disproportionation of aromatics, metathesis of olefins, oligomerization of olefins, etherification and hydration

*UOP LLC

of olefins, and hydrocracking. Although these various applications have some similarities, they are all promoted by an acidic catalyst and often by a metal function. The specific catalytic requirements for achieving a highly economic result have led to the proliferation of specialized catalytic materials.

This paper focuses on light paraffin isomerization. The primary commercial use of the branched isomers of C_4, C_5, and C_6 paraffins is in the production of clean-burning, high-performance transportation fuels. The elimination of tetraethyl lead over the last 30 years as a means of improving the antiknock properties of gasoline and more recent regulations restricting motor fuel composition have led refiners to select alternative means of producing high-quality gasoline. As a result of benzene concentration restrictions, end-point and olefin content limitations, and potential limitations on total aromatics concentration, the choices of high-quality gasoline blending components available in the typical refinery are limited. Isomerate, the gasoline blending component from light paraffin isomerization, is an ideal choice. Another equally valuable blending components is alkylate resulting primarily from the acid-catalyzed reaction of isobutene with an aliphatic olefin. Both isomerization and alkylation yield highly branched, high-octane paraffinic blending components that by themselves can satisfy the strictest environmental requirements. Often, *n*-butane isomerization is one of the sources for the isobutane requirements in alkylation.

Because of the heightened demand for isomerate, refiners continue to look for increasingly effective and economic means of producing this valuable blending component. Over the years, UOP has developed and continues to develop new catalyst systems in order to improve process economics and operability. This paper discusses two examples, the LPI-100 and I-80 catalysts in terms of their fundamental improvement over existing catalysts in various processing options and, most important, the increased value available to refiners who use these new high-performance isomerization catalysts.

Process chemistry of paraffin isomerization

The octane rating of the components used in the manufacture of the various commercial gasoline grades is indicative of the antiknock quality of a given fuel or component. The inherent octane values of different hydrocarbons have led to a variety of processing strategies to produce high-octane components for the production of high-performance motor fuel. In the case of C_5 and C_6 paraffins, the most highly branched isomers have the highest-octane values. In the case of butanes, octane is somewhat irrelevant because the majority of isobutane is consumed in the production of motor fuel alkylate and oxygenates. Octane values for C_5 and C_6 paraffin isomers are shown in Table 9.3.1 (1). The two empirical octane measurement methods, research (ASTM Method 2699), and motor (ASTM Method 2700), measure antiknock characteristics

Table 9.3.1. Hydrocarbon octane values

	Research octane by ASTM 2699	Motor octane by ASTM 2700	$(R + M)/2$
n-Butane	93.8	89.6	91.7
i-Butane	100.4	97.6	99.0
n-Pentane	61.7	62.6	62.2
i-Pentane	92.3	90.3	91.3
n-Hexane	24.8	26.0	25.4
2-Methylpentane	73.4	73.5	73.4
3-Methylpentane	74.5	74.3	74.4
2,2-Dimethylbutane	91.8	93.4	92.6
2,3-Dimethylbutane	101.0	94.3	97.6

at two severity levels. Often, the average of these two values, or $(R + M)/2$, is used to express the overall engine performance of a gasoline component or blend. Table 3.7.1 shows that isopentane has an octane value nearly 30 numbers higher than n-pentane (1).

Similarly, the hexane isomer, 2,2-dimethylbutane (2,2-DMB) has an octane about 67 numbers higher than n-hexane. Clearly, highly branched isoparaffins are the desired isomers for motor fuel production.

Thermodynamic equilibria for the branched paraffin isomers are generally favored by low temperatures. Figures 9.3.1, 9.3.2, and 9.3.3 illustrate this trend (2). The most active catalyst, when all other variables are equal, is capable of producing the highest-octane products.

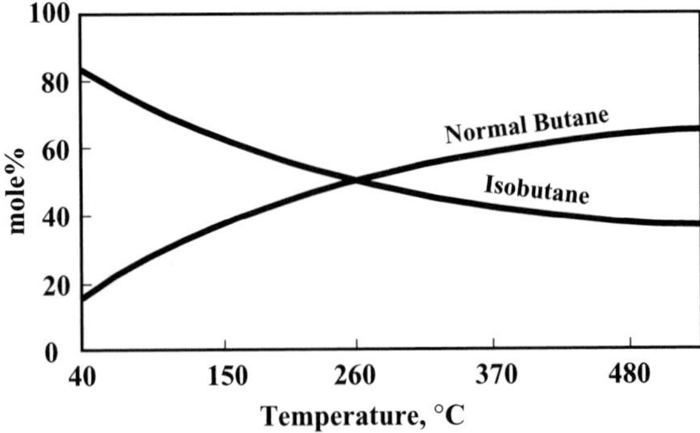

Figure 9.3.1. Butane equilibrium.

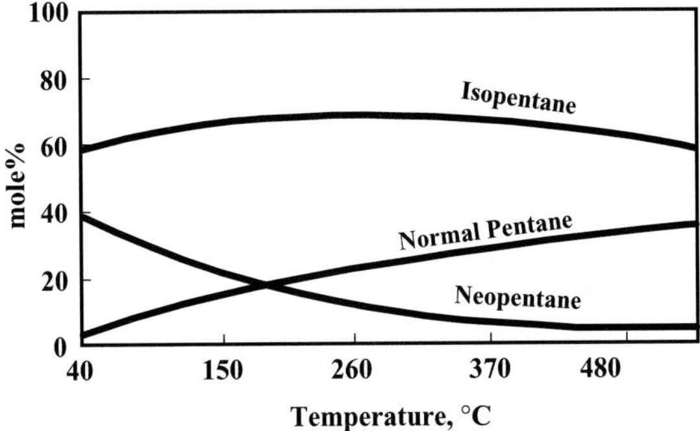

Figure 9.3.2. Pentane equilibrium.

Primary reaction pathways

Paraffin isomerization is most effectively catalyzed by a dual-function catalyst containing a noble metal and an acid function. The reaction is believed to proceed through an olefin intermediate, which is formed by paraffin dehydrogenation on the metal site (reaction (1)):

$$CH_3\text{-}CH_2\text{-}CH_2\text{-}CH_3 \overset{Pt}{\leftrightarrow} CH_3\text{-}CH_2\text{-}CH = CH_2 + H_2. \tag{1}$$

Although the equilibrium conversion of the paraffin in reaction (1) is low at paraffin isomerization conditions, sufficient olefins are present to be converted to a carbonium ion by the strong acid site (reaction (2)):

$$CH_3\text{-}CH_2\text{-}CH = CH_2 + [H^+]\,[A^-] \rightarrow CH_3\text{-}CH_2\text{-}\overset{+}{C}H\text{-}CH_3 + A^-. \tag{2}$$

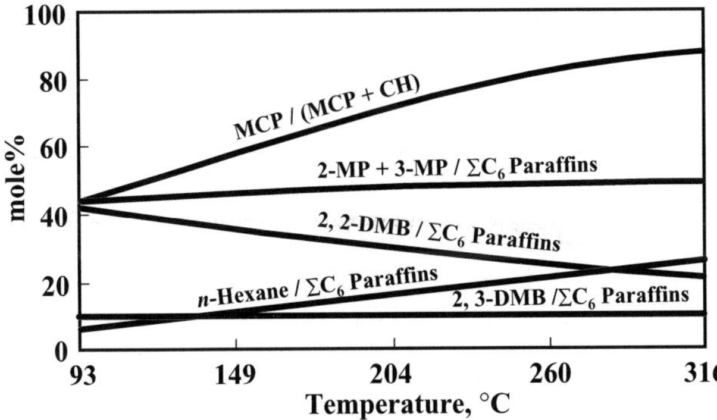

Figure 9.3.3. C_6 fraction equilibrium.

The formation of the carbonium ion removes product olefins from reaction (1) and allows the equilibrium in reaction (1) to proceed. The carbonium ion in reaction (2) undergoes a skeletal isomerization, probably through a cycloalkyl intermediate as shown in reaction (3):

$$CH_3\text{-}CH_2\text{-}\overset{+}{C}H\text{-}CH_3 \rightarrow \underset{\text{cyclopropyl cation}}{CH_2\text{—}\overset{+}{C}(CH_2)\text{-}CH_3} \rightarrow CH_3\text{-}\underset{+}{\overset{CH_3}{\overset{|}{C}}}\text{-}CH_3. \tag{3}$$

Reaction (3) proceeds with difficulty because it requires the formation of a primary carbonium ion at some point in the reaction. Nevertheless, the strong acidity of isomerization catalysts provides enough driving force for the reaction to proceed at high rates. The isoparaffinic carbonium ion is then converted to an olefin through loss of a proton to the catalyst site (reaction (4)):

$$\underset{+}{\overset{CH_3}{\overset{|}{CH_3\text{-}CH\text{-}CH_3}}} + A^- \rightarrow \overset{CH_3}{\overset{|}{CH_3\text{-}C}} = CH_2 + [H^+][A^-]. \tag{4}$$

In the last step, the isoolefin intermediate is hydrogenated rapidly back to the analogous isoparaffin (reaction (5)):

$$\overset{CH_3}{\overset{|}{CH_3\text{-}C}} = CH_2 + H_2 \xrightarrow{pt} \overset{CH_3}{\overset{|}{CH_3\text{-}CH\text{-}CH_3}}. \tag{5}$$

In addition to these primary reaction pathways, some evidence indicates the existence of a bimolecular reaction mechanism, in which olefinic intermediates dimerize, internal carbon atoms are protonated, skeletal isomerization occurs, and the dimer undergoes beta scission that results in the product isoparaffin. In addition to the C^{13} labeling experiments that support this mechanism, a relatively small amount of hydrocarbons containing carbon numbers higher than the feed are always found in the reaction products. The bimolecular mechanism has a minor impact on commercial isomerization processing.

Isomerization catalysts

Operation at the lowest temperature results in the formation of the highest-octane product (Table 9.3.1 and Figures 9.3.1, 9.3.2, and 9.3.3). Catalyst activity and the lowest operating temperature are important to achieve an economic operation. In the previous discussion of the primary reaction pathway, the primary functionality of a paraffin isomerization catalyst is to protonate a secondary carbon atom. All known paraffin isomerization catalysts have a combination of strong Lewis and Brönsted acid sites, which result in varying levels of protonation activity.

In addition to a strong acid function, isomerization catalysts must also be capable of hydrogenolysis, which not only assists in the protonation step, but also serves to saturate olefin intermediates and aromatic hydrocarbons and to assist in the ring opening of cycloparaffins. This function also gives activity stability to isomerization catalysts, thereby improving the process economics.

An aluminum chloride catalyst for alkane isomerization was first developed in the 1930s (3). The original application was for the conversion of n-butane to isobutane, which was, and still is, reacted with C_3, C_4, and C_5 olefins to produce motor fuel alkylate. The first application of this high-octane product was in the production of high-octane aviation gasoline. Subsequent developments of this predominantly Lewis-acid catalyst resulted in the current alumina-supported, bifunctional catalyst. UOP's I-8 catalyst is one commercial example of this catalyst system, which has seen wide commercial application since 1981.

Chlorided alumina, the highest-activity paraffin isomerization catalyst available, increases the octane of a typical light-naphtha stream from about 70 to as high as 85 RONC in a once-through paraffin isomerization unit. Higher product octanes, up to 93 RONC, can be obtained by recycling low-octane hydrocarbons. The C_5^+ yield from chlorided alumina catalysts is the highest from any commercial catalyst because of high catalyst selectivity and low operating temperature. Because chlorided alumina systems are not economically regenerable, eventual reloading of the isomerization catalyst must be considered. Nevertheless, the chlorided alumina system is often the most-economic choice because of its inherent high activity. In addition, only chlorided alumina catalysts have enough activity to economically isomerize butanes.

As a result of ongoing intensive research and development in paraffin isomerization technology, UOP's I-80 catalyst is one of the highest-activity chlorided alumina catalyst currently available. The I-80 catalyst, is significantly more active than the I-8 catalyst, and is based on a unique formulation and manufacturing technique. By simply reloading the I-80 catalyst in existing reactors, a gain of 0.5–1.0 RON can be realized compared with the product RON when I-8 catalyst is used.

Zeolitic isomerization catalysts, such as UOP's HS-10™ or I-7™ catalysts, operate at higher temperatures than chlorided alumina catalysts. The maximum product octane that can be achieved is limited by the unfavorable equilibrium at these conditions. Yields are also lower as a result of the higher operating temperature and the less-selective characteristics of zeolitic catalysts. A typical octane upgrade for a once-through zeolitic isomerization unit is from 70 to about 79 RONC. Higher product octanes (86–88) can be obtained in a recycle operation, such as a TIP™ unit.

The most-attractive benefit of zeolitic isomerization catalysts is that they are not permanently deactivated by water or other oxygenates and are fully regenerable.

Consequently, zeolitic catalysts have often been used when revamping other process units, such as hydrotreaters or reformers to isomerization service. Sulfur can be present in the feedstock with catalysts such as the HS-10 catalyst, but performance is always affected to some degree. Sulfur suppresses the platinum function as it does in any platinum-containing catalyst. Although isomerization activity is maintained, a net C_5^+ product yield loss results. Hydrotreating is not an absolute requirement with zeolitic catalysts, but it is necessary to get the optimum performance from the catalyst.

Sulfated metal oxide catalysts, which have been described as solid super acids, exhibit high activity for paraffin isomerization reactions. These metal oxides form the basis of the new generation of isomerization catalysts that have been actively discussed in the scientific literature in recent years. These catalysts are most commonly tin oxide (SnO_2), zirconium oxide (ZrO_2), titanium oxide (TiO_2), or ferric oxide (Fe_2O_3) that has been sulfated by reaction with sulfuric acid or ammonium sulfate. Sulfated alumina is not an active catalyst for hydrocarbon reactions.

A typical metal oxide catalyst, UOP's LPI-100 catalyst (4), has a considerably higher activity than that of traditional zeolitic catalysts, and this activity advantage is equivalent to about 80°C lower reaction temperature. The lower reaction temperature allows for a product with a significantly higher product octane, about 82 RONC for a typical feed or three numbers higher than that of a zeolitic catalyst. Similar to zeolitic catalysts, the LPI-100 catalyst is not permanently deactivated by water or oxygenates in the feedstock. These catalysts are also fully regenerable using a simple oxidation procedure that is comparable to the one used for zeolitic catalysts. The high activity of the LPI-100 catalyst makes it an ideal choice for revamping existing zeolitic isomerization units for higher capacity and higher octane isomerate or for new units where the full performance advantage of chlorided alumina catalysts is not required or where the refiner is concerned about contaminants in the feedstock.

The isomerization performance of the previously discussed I-8, I-80, HS-10, and LPI-100 catalysts is compared in Figure 9.3.4, which plots the i-C_5 to total C_5 product ratio for all the catalysts as a function of reactor temperature. This ranking clearly illustrates the striking activity advantage for the LPI-100 catalyst over that of the zeolitic HS-10 catalyst, an advantage that translates directly to octane improvement. Likewise, this ranking shows the performance benefit of the I-80 catalyst over the predecessor I-8 catalyst. The development and benefits of the I-80 and LPI-100 catalysts, are discussed in the following sections.

I-80 catalyst development and applications

Chlorided alumina represents the industry standard for isomerization catalysts. UOP introduced the second-generation, high-activity I-8 chlorided alumina catalyst in

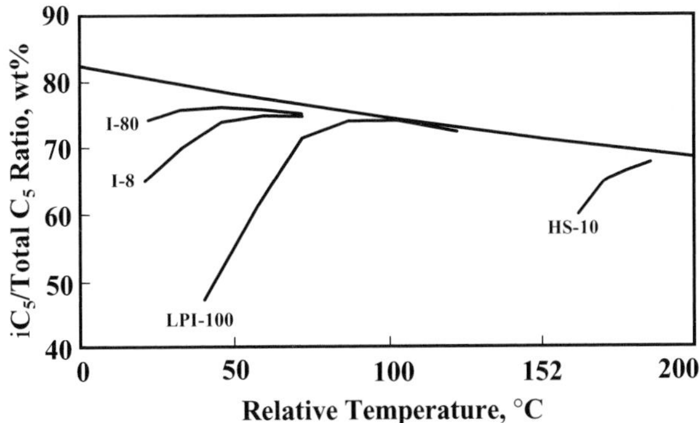

Figure 9.3.4. Light paraffin isomerization catalyst performance.

1981. The I-8 catalyst has been successfully used in more than 50 operating Butamer™ butane isomerization units and 90 operating Penex™ light-naphtha isomerization units. The flow schemes range from simple once-through hydrocarbon processes to complex recycle operations to achieve high product octane that have been described in detail elsewhere (5). In 1998, UOP introduced the I-80 catalyst, a new generation, high-activity chlorided alumina catalyst. The I-80 catalyst has stability and selectivity identical to that of the I-8 catalyst.

Higher catalyst activity can be used in a number of ways in existing or new isomerization units. The best application depends on site-specific factors and whether the unit operates in a once-through or recycle mode. Typical means to exploit higher activity are:

- Higher throughput at constant octane (Case A)
- Reduced catalyst volume at constant throughput and octane (Case B)
- Higher octane at the same throughput (Case C)
- Longer catalyst life (Case D)

Case A operates at higher throughput but maintains a constant product octane. The performance of the I-8 and I-80 catalysts is compared in Figure 9.3.5 as a plot of product octane against relative reactor temperature. The newer I-80 catalyst has a substantial 30% activity advantage over the previous catalyst. This advantage is equivalent to about a 12°C reduction in reactor temperature. A 30% activity increase corresponds directly to an increase in space velocity, or throughput.

If the I-80 catalyst is reloaded in place of the existing I-8 catalyst, the charge rate to the reactor can be increased by up to 30% with no decrease in product octane. This throughput advantage is a major benefit in equipment cost savings if a higher-capacity revamp is being considered because additional reactor volume is not required. Another

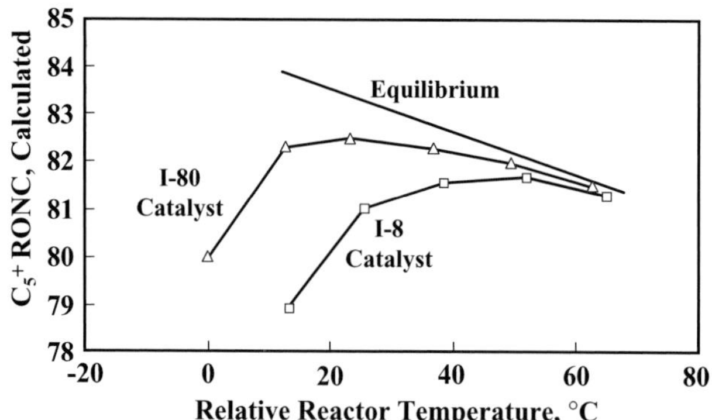

Figure 9.3.5. Chlorided alumina catalyst performance.

benefit of a higher-activity catalyst is the increased flexibility to respond to seasonal fluctuations in product demand.

The ability to operate at higher space velocities with the I-80 catalyst can be used in existing units by reducing the volume of catalyst in the reactor of existing units by up to 30% and using a smaller overall reactor size in new units (Case B). These cost savings are obvious. The refiner also realizes a savings in platinum inventory.

As mentioned previously, isomerization reactions are equilibrium limited and the extent of the reaction improves at lower temperatures. The 12°C activity advantage of the I-80 catalyst can be exploited by operating at a lower temperature to achieve higher product octanes (Case C). This advantage is illustrated in Figure 9.3.6 as a plot of product octane against C_5^+ yield. Achieving a 0.5–1.0 RON higher product

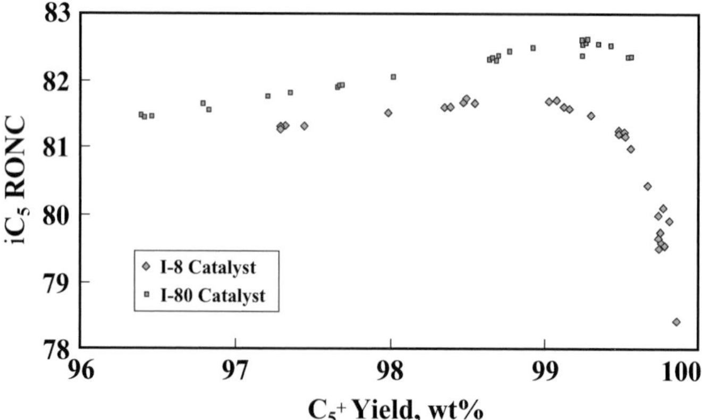

Figure 9.3.6. Chlorided alumina catalyst yields.

quality at similar yields is possible with the I-80 catalyst in once-through units. This improvement in quality is equivalent to an additional revenue of about $600,000 U.S. per year at an octane value of $0.25 per octane-barrel. Thus, the I-80 catalyst is an obvious choice when catalyst replacement is required. The benefit for recycle units is lower because the savings are related to the cost reduction associated with maintaining higher conversion per pass.

The higher activity of the I-80 catalyst is achieved by increasing the number of catalytically active sites on the catalyst surface. Because the deactivation of chlorided alumina catalysts is mostly due to ingress of moisture or oxygenates, this higher acid-site density can translate into longer catalyst life at a constant catalyst loading (Case D). If all other factors are equal, catalyst life may be extended by about 30%. Increased catalyst life has the obvious benefit of reducing operating costs. However, the benefit is relatively small in well-run commercial units that typically achieve a catalyst life in excess of 5 years.

LPI-100 catalyst development and applications

The development of the LPI-100 catalyst represents the first successful application of sulfated metal oxide in industrial catalysis. This new catalyst is the culmination of a joint development effort between Cosmo Research Institute (CRI) and Mitsubishi Heavy Industries (MHI) in Japan and UOP.

The basic formulation for this sulfated metal oxide catalyst was developed by CRI and MHI researchers in the late 1980s. Several years of intensive development were required to translate the CRI–MHI laboratory formulation into a commercially viable extruded catalyst that met all the original performance targets. The commercial product is now the UOP LPI-100 catalyst.

The first commercial loading of the LPI-100 catalyst was completed at the Flying J Refinery in North Salt Lake City, Utah. The catalyst was loaded in a zeolitic isomerization reactor that had previously been operated with UOP's I-7 catalyst. The loading was completed and the reactor placed into operation in December 1996. Performance data for the LPI-100 catalyst in the Flying J unit are compared to pilot plant predictions in Figure 9.3.7 as a plot of the i-C_5 to total C_5 product ratio against reactor temperature. The performance at Flying J was in line with UOP's expectations. Since the initial operation at Flying J, the LPI-100 catalyst has been used in another once-through zeolitic isomerization unit in Canada and in other units in Canada and the Middle East.

Contaminant-sensitivity studies with the LPI-100 catalyst have shown responses comparable to those experienced with conventional zeolitic catalysts. Water and oxygenates at typical concentrations are not detrimental. Sulfur suppresses activity, as

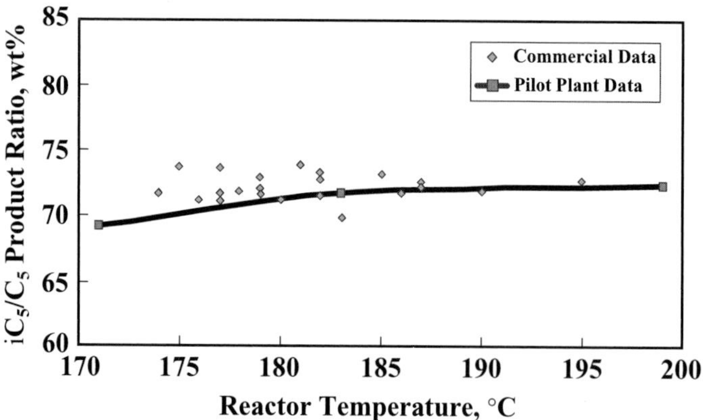

Figure 9.3.7. LPI-100 performance at Flying J refinery.

expected for any catalyst containing noble metals. The suppression effect of sulfur is reversed by subsequent processing with clean feedstocks. The effects of other common contaminants are similar to those experienced with conventional zeolitic catalysts.

In pilot plant evaluations, the stability of the LPI-100 catalyst is comparable that of the HS-10 zeolitic catalyst. This stability translates to a commercial process cycle of at least 18 months before regeneration is required. Multiple-cycle regeneration testing was completed and showed that the catalyst was fully regenerable through more than three regenerations. Regeneration consists of a simple carbon-burn step at conditions comparable to those used for regenerating zeolitic catalysts.

New isomerization process technologies

The introduction of new catalyst systems, such as the LPI-100 catalyst, allows the development of new process technologies that fully use the characteristics of the catalyst. The process technology developed for the LPI-100 catalyst is the Par-Isom™ process. A simplified flow scheme for the Par-Isom process operating in a once-through configuration is shown in Figure 9.3.8. For simplicity, a once-through hydrocarbon flow scheme is discussed in this chapter. Recycle flow schemes previously developed for naphtha isomerization use either molecular sieve separation, such as in the TIP process, or fractionation, such as a deisohexanizer column.

The process flow scheme shown in Figure 9.3.8 is similar to that used for conventional once-through Penex and zeolitic isomerization units. The fresh C_5–C_6 feed is combined with makeup and recycle hydrogen and directed to the charge heat exchanger, where the reactants are heated to reaction temperature. A fired heater is not required in the Par-Isom process, because of the much lower reaction temperature needed with

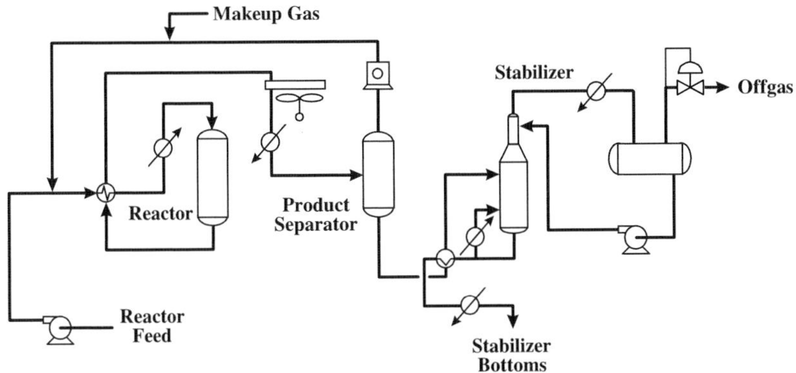

Figure 9.3.8. Par-Isom process flow scheme.

the LPI-100 catalyst than with zeolitic catalysts. Hot oil or high-pressure steam can be used as the heat source in this exchanger. The heated combined feed is then sent to the isomerization reactor.

Either one or two reactors can be used in series, depending on the specific application. Two reactors obtain the best performance from the process but at a higher capital and additional catalyst inventory costs. In a two-reactor system, the first reactor is operated at higher temperature (200–220°C) to improve the reaction rate, and the second reactor is operated at a lower temperature to take advantage of the more-favorable equilibrium distribution of higher octane isomers. One reactor is generally specified for the Par-Isom process to reduce capital and operating cost of the unit, but a small debit in performance can result.

The reactor effluent is cooled and sent to a product separator, where the recycle hydrogen is separated from the other products and returned to the reactor section. The liquid product is sent to a stabilizer column, where the light ends and dissolved hydrogen are removed. The stabilized isomerate product can be sent directly to gasoline blending. In recycle flow schemes, either molecular sieve or fractionation options are used to separate the lower-octane isomers for recycle to the reactor. The selection of the separation scheme depends on the feed composition, availability of utilities, and the product octane desired.

The higher activity of the LPI-100 catalyst makes it an ideal candidate for revamps of existing zeolitic isomerization units to achieve higher capacity or for revamps of idle process units, such as reformers or hydrotreaters to isomerization service. The integration of a Par-Isom reactor into an existing semiregeneratative reforming unit can be a particularly attractive revamp opportunity for economically adding isomerization capacity to a refinery. A simple flow scheme illustrating the integrated

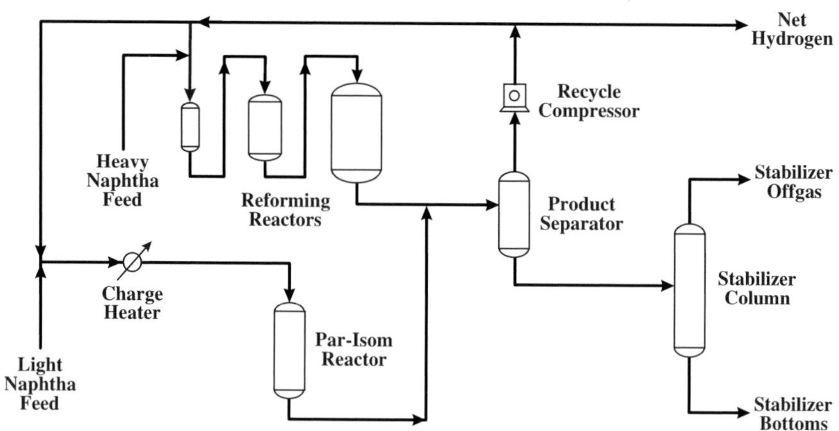

Figure 9.3.9. Par-Isom and reformer revamp flow scheme.

units is shown in Figure 9.3.9. In this configuration, a new Par-Isom reactor and charge heater are added to operate in parallel with the existing reforming unit reactors.

The hydrogen required for the Par-Isom reactor is taken from the reforming unit recycle compressor, combined with the light-naphtha charge, and sent to the Par-Isom reactor. The Par-Isom reactor effluent is returned to the reforming unit. The integrated units share a common product condenser, product separator, and stabilizer. The isomerization unit at the Flying J Refinery has operated successfully in this mode since 1986 and is currently operating as a Par-Isom unit.

Isomerization process economics

The economics of a process are the most-important consideration in selecting a technology for a given application. The Par-Isom process offers an economic alternative to conventional isomerization processes. Two case studies illustrate the benefits of this technology: one for reloading existing once-through zeolitic units (Case study A) and the other for integration of a Par-Isom reactor into a reforming unit to add isomerization capacity (Case study B).

Case study A

The revamp of an existing once-through zeolitic isomerization unit is an excellent application for the Par-Isom process because both increased product quality and throughput can be realized. This increase is shown in Table 9.3.2 as a comparison of the base case (continued operation with current zeolitic catalyst) and the unit after revamp to the Par-Isom process. This example was selected to represent a typical

Table 9.3.2. Par-Isom revamp opportunities for once-through isomerization units

	Base case	Revamp
Catalyst	Zeolitic	LPI-100
Capacity, bpd	5,000	6,250
KMTA	190	238
Product RONC	79.5	81.7
MONC	77.1	79.8
Yield, LV-%	98.0	98.0
Incremental		
Catalyst cost, $MM	Base	1.33
Net income, $MM/yr	Base	1.28
Shutdown case		
Payback, years	Base	1.7
IRR, %	Base	58
Incremental case		
Payback, years	Base	1.0
IRR, %	Base	96

zeolitic process unit. The Par-Isom process allowed throughput to be increased by 25% along with an increased product octane of 2 RONC. An alternative would be to realize a 3 RON upgrade by keeping the capacity constant. However, the economics of this option are generally not as attractive as increasing throughput. Octane was valued at $0.25 per octane-barrel, and no substantial changes to existing equipment were considered in this case.

Two scenarios were considered: a shutdown case and an incremental case. In the shutdown case, the current zeolitic catalyst was assumed to be operating satisfactorily, and so the revamp would have to justify replacement of the existing catalyst. The payback for this case was 1.7 years, which is equivalent to an internal rate of return (IRR) of 58%; typically a viable alternative for most refiners. In the incremental case, the zeolitic catalyst was assumed to be deactivated and new catalyst required for continued operation of the unit. Revamp to the Par-Isom process is an attractive option in this situation. The payback was 1 year, or an IRR of 96% on the incremental investment required for the revamp.

Case study B

New isomerization capacity is being considered in many areas of the world, such as South America, Russia, and the Far East, as lead is being phased out of the gasoline pool or regulations on benzene or aromatics content are enacted. Often the refineries involved are hydroskimming-type refineries that have limited conversion capacity and one or more semi-regenerative reforming units to upgrade the heavy naphtha and provide a source of hydrogen. These refineries are ideal candidates for adding isomerization capacity by integrating a Par-Isom unit with the existing reforming unit (Figure 9.3.9).

The revamp in this example involved reloading a 600 kMTA (15,200 bpd) reforming unit with UOP's R-56™ catalyst in place of the existing unit's less-stable reforming catalyst so that excess recycle hydrogen was available for the Par-Isom reactor. The Par-Isom reactor section is installed to operate in parallel with the reforming unit reactors. The addition of the Par-Isom unit involved installing a charge heater using high-pressure steam, a light-naphtha feed pump, the Par-Isom reactor, and the necessary controls. The balance of the existing equipment, including the product condenser, product separator, and stabilizer, was adequate for the integrated unit. The estimated erected cost (EEC) for the Par-Isom reactor system, including an allowance for royalties, was \$2.3 million U.S. (4th Quarter 1997, U.S. Gulf Coast basis). The catalyst cost was \$2.1 million U.S., excluding the cost of the platinum, and the total investment cost was \$4.4 million U.S.

Adding isomerization capacity is a low-cost means of increasing gasoline pool octane (Table 9.3.3). The base case, without isomerization, produces a gasoline product consisting of 494 kMTA (11,750 bpd) of 96 RONC reformate and 300 kMTA (8,780 bpd) of 66 RONC light naphtha. The combined streams produce a gasoline product at

Table 9.3.3. Integration of a Par-Isom unit with a semiregenerative reforming unit

	Heavy naphtha	Reforming unit	Light naphtha	Par-Isom unit
Catalyst		R-56		LPI-100
Product, bpd	15,200	11,750	8,780	8,325
kMTA	600	494	300	287
Product RONC	–	96	66	79
MONC	–	85	65	78
Blended gasoline		without Par-Isom unit		with Par-Isom unit
RONC	–	84.6	–	89.8
Bpd	–	20,530	–	20,075
kMTA	–	794	–	781
Δ RON-bpd	–	–	–	65,897
Revamp				
Catalyst cost, \$MM	–	–	–	2.10
Equipment cost, \$MM	–	–	–	2.30
Total cost, \$MM	–	–	–	4.40
Net income, \$MM/yr	–	–	–	5.50
Revamp economics				
Payback, years	–	–	–	0.8
IRR, %	–	–	–	125

84.6 RONC and would require the addition of lead to make regular-grade gasoline. In the revamp case, the light naphtha is upgraded in the Par-Isom reactor to a 79 RONC. Because of the nature of the feedstocks, achieving this high isomerate octane with conventional zeolitic catalysts is not possible. Some yield loss occurs in the Par-Isom reactor, and so 287 kMTA (8,325 bpd) of isomerate is available for blending with the reformate. Adding the Par-Isom reactor provides sufficient octane to produce a total gasoline product at 781 kMTA (20,075 bpd) with a 89.8 RONC. Regular-grade gasoline can be produced in the integrated reforming and Par-Isom units without the addition of lead.

The economics involved in the revamp are also shown in Table 9.3.3. The required investment of $5.5 million U.S. includes catalyst, equipment, and royalty costs. The Par-Isom unit provides an additional 65,897 octane-barrels to the gasoline pool, or $5.5 million U.S. if the octane is valued at $0.25 U.S. per octane-barrel. The simple payback for the revamp is less than a year, and the internal rate of return is attractive at 125%. Integrating the Par-Isom reactor with an existing reforming unit is clearly an attractive solution to the challenges posed by lead phaseout or benzene reduction.

Other applications

The Par-Isom process can be considered for other applications, particularly for new units. In general, chlorided alumina catalysts, such as the I-80 catalyst used in the Penex process provide the most cost-effective isomerization technology for grassroots refineries or for installing new units without revamping of existing equipment. The additional octane provided by these catalysts results in the most-value-added and most-attractive economics.

In some cases, the Par-Isom technology should be considered for new units. Since the Par-Isom process is the lowest capital cost option for once-through isomerization units, it is the process of choice when capital cost is a major concern. Another situation is when the primary processing objective is to minimize the benzene content in the gasoline pool, and sufficient octane to meet the gasoline product blending requirements already exists. Because the additional octane achieved by a chlorided alumina catalyst may not have any value in these situations, the lower-cost Par-Isom process is the most economically attractive option.

Conclusions

Light-naphtha isomerization processes have become increasingly important in helping refiners improve octane and meet the production demands for reformulated or low-benzene-content gasoline. The discussion in this chapter has been based on the relative performance of UOP's catalysts. Similar conclusions could be drawn if using equivalent catalyst from other suppliers.

The I-80 catalyst is one of the highest-activity chlorided alumina catalysts available today. This catalyst provides a means to improve the profitability of Penex and Butamer units by improving octane or allowing for increased capacity or both. The 0.5–1 RONC additional octane that can be provided by the I-80 catalyst makes this catalyst an obvious choice for reloading existing units or for new installations that require high octane.

The commercialization of the LPI-100 catalyst and the Par-Isom process provides opportunities for upgrading existing zeolitic isomerization units or the low-cost addition of isomerization capacity in refineries by revamping other process units or integrating with semiregenerative reforming units. The Par-Isom process provides a low-cost, economically attractive route for refiners who must meet demands for higher octane or reformulated gasoline. In most cases, the revamp can be accomplished with minimal changes to existing equipment. The high activity of the LPI-100 catalyst can result in up to 3 RON higher isomerate octane than can be achieved with zeolitic catalysts as well as substantially higher throughput in revamp situations.

References

1. *Physical Constants of Hydrocarbons C_3-C_{10}*, ASTM Committee D-2 on Petroleum Products and Lubricants and API Research Project 44 on Hydrocarbons and Related Compounds, American Society of Testing Materials, 1971.
2. D. R. Stull, E. F. Westrom, Jr., and G. C. Sinke, *The Chemical Thermodynamics of Organic Compounds*, Robert E. Krieger Publishing Co., Malabar, FL, 1987.
3. H. Pines, *The Chemistry of Catalytic Hydrocarbon Conversion*, Academic Press, New York, 1981, pp. 12–18.
4. C. D. Gosling, R. R. Rosin, P. Bullen, T. Shimizu, and T. Imai, Revamp Opportunities for Isomerization Units, *Petroleum Technol. Q*. Winter 1997–1998 55–59.
5. P. J. Kuchar, and others, Paraffin Isomerization Innovations, *Fuel Proc. Technol.* 35 (1993) 183–200.

Bibliography

N. A. Cusher and others, Isomerization for future gasoline requirements, *Proceedings of the NPRA Annual Meeting*, Mar. 25–27, 1990.

R. A. Meyers, *Handbook of Petroleum Refining Processes*, 2nd ed, McGraw-Hill, New York, 1996.

A. S. Zarchy and others, Impact of desulfurization on the performance of zeolite isomerization catalysts, *Proceedings of the AIChE Annual Meeting*, Nov. 11–16, 1990.

C. L. Moy and others, Benzene reduction for reformulated gasoline, *Proceedings of the AIChE Spring National Meeting*, Mar. 29–Apr. 2, 1992.

Chapter 10

Refinery gas treating processes

D.S.J. Jones

Introduction

Refinery gas sweetening is the process used to remove the so called 'Acid Gasses' which are hydrogen sulfide and carbon dioxide from the refinery gas streams. These acid gas removal processes used in the refinery are required either to purify a gas stream for further use in a process or for environmental reasons. Clean Air legislation now being practiced through most industrial countries require the removal of these acid gases to very low concentrations in all gaseous effluent to the atmosphere. Hydrogen Sulfide combines with the atmosphere to form very dilute sulfuric acid and carbon dioxide to form carbonic acid both of which are considered injurious to personal health. These compounds also cause excessive corrosion to metals and metallic objects.

The following sections of this chapter describe the processes, their chemistry and a method of calculating the design of the more common processes.

The process development and description

The use of chemically 'basic' liquids to react with the acidic gases was developed in 1930. The chemical used initially was Tri ethanol amine (TEA). However, as Mono ethanol amine (MEA) became commercially more available it became the preferred liquid reactant due to its high acid gas absorption on a unit basis. The molecular weight of MEA is 61 while that for TEA is 149. As both react mole for mole with the acid gases 2.44 lbs of TEA must be circulated to achieve the same absorptive capacity as 1 lb of MEA. TEA also suffers degradation at temperatures above its boiling point and as this is its regeneration temperature MEA has replaced TEA in this service. Since 1955 numerous alternative processes to MEA have been developed. These have fewer corrosion problems and are to a large extent more energy efficient. Inhibitor systems have however been developed which have eliminated much of the MEA corrosion

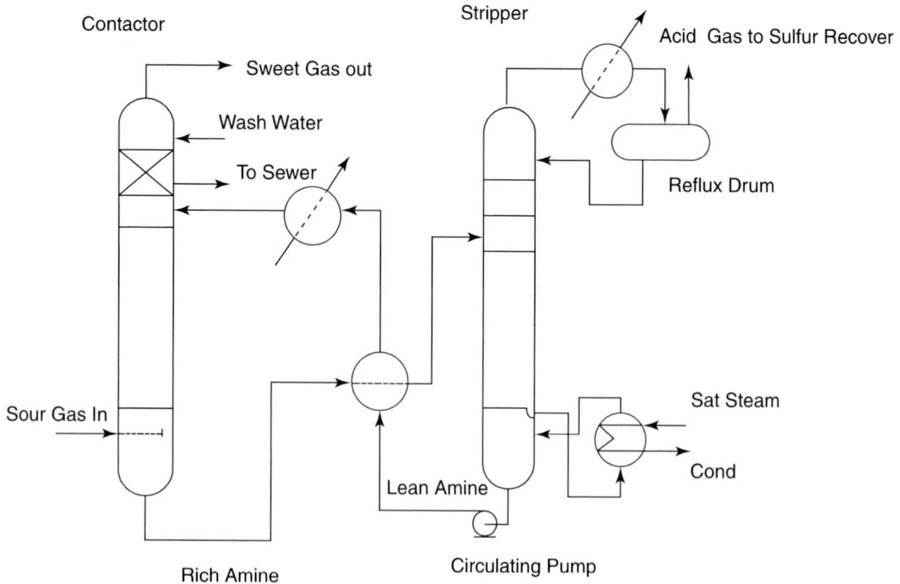

Figure 10.1. Schematic drawing of a typical amine treating unit.

problems. Some of these newer processes also are designed to selectively remove the H_2S leaving the CO_2 to remain in the gas stream.

The process flow and description of the more common processes are essentially the same. This process flow is shown in Figure 10.1.

For product gas streams which must meet lower than 1 grain per scf of H_2S, MEA must be used. This amine, however, is degraded by certain sulfide compounds found in gas from thermal crackers. The most common compound that degrades MEA is carbonyl sulfide (COS). MEA can, however, be regenerated by batch or continuous vaporization and disposal of the sludge formed by the degeneration.

Referring to the above flow sheet, sour gas (rich in H_2S) enters the bottom of the trayed absorber (or contactor). Lean amine is introduced at the top tray of the absorber section to move down the column. Contact between the gas and amine liquid on the trays results in the H_2S in the gas being absorbed into the amine. The sweet gas is water washed to remove any entrained amine before leaving the top of the contactor.

Rich amine leaves the bottom of the contactor to enter a surge drum. If the contactor pressure is high enough a flash stream of H_2S can be routed from the drum to a trayed stripper. The liquid from the drum is preheated before entering a stripping column on the top stripping tray. This stripper is reboiled with 50 psig saturated steam. Saturated 50 psig steam is used because higher temperatures cause amines to break down. The H_2S is stripped off and leaves the reflux drum usually to a sulfur production plant. Sulfur is produced in this plant by burning H_2S with a controlled air stream.

The lean amine leaves the stripper bottom and is cooled. The cooled stream is routed to the contactor.

Common processes

There are several liquid solvents in commercial use for the removal of H_2S and CO_2 from refinery gases. Among the more common are the amines. These include:

MEA Mono ethanol amine
DEA Diethanol amine
DGA Diglycol amine

In addition to these amine base solvents there are also the hot potassium carbonate process, sulfinol and ADIP. These latter two processes are marketed by the Shell company and are quite common in world wide usage.

It is not the intent here to detail all of these processes with their various properties. The three amine solvents will however be described and discussed and some mention will be made of the other processes.

Mono ethanol amine

This is the most common acid gas absorption process. Normally 15–20 wt% MEA in water is circulated down through a trayed absorber to provide intimate contact with the sour gas. The rich solution is routed to a steam stripping column where it is heated to about 250°F at 10 psig to strip out the acid gases. The lean MEA solution is then returned to the absorber.

MEA is the most basic (and thus reactive) of the ethanol amines. MEA will completely sweeten sour gases removing nearly all acid gases if desired. The process is well proven in refinery operations.

Like all of the amine solvents used for acid gas removal MEA depends upon its amino nitrogen group to react with the acidic CO_2 and H_2S in performing its absorption. The particular amines are selected with a hydroxyl group which increases their molecular weight and lowers their vapor pressures yielding minimum solvent losses to the gas stream. MEA is considered a chemically stable compound. If there are no other chemicals present it will not suffer degradation or decomposition at temperatures up to its normal boiling point.

The process reactions are given below:

$$HOCH_2CH_2NH_2 = RNH_2$$

$$2\ RNH_2 + H_2S \underset{\text{High temp}}{\overset{\text{Low temp}}{\rightleftharpoons}} (RNH_3)_2S$$

$$(RNH_3)_2S + H_2S \underset{\text{High temp}}{\overset{\text{Low temp}}{\rightleftharpoons}} 2RNH_3HS$$

Some of the degradation products formed in these systems are highly corrosive. They are usually removed by filtration or reclaimer operations. Filtration will remove corrosive by products such as iron sulfide. Reclaiming is designed to remove heat stable salts formed by the irreversible reaction of MEA with COS, CS_2 (carbonyl sulfide and carbon disulfide). The reclaimer operates on a side stream of 1–3% of the total MEA circulation. It is operated as a stream stripping kettle to boil water and MEA overhead while retaining the higher boiling point heat stable salts. When the kettle liquids become saturated at a constant boiling point with the degradation products it is shut in and dumped to the drain. Union Carbide has developed a well proven corrosion inhibitor system for MEA that allows solution strengths of 28–32%. The inhibitor requires payment of a royalty. The inhibitor chemicals are both expensive and hazardous for personnel to handle. The system does reduce corrosion problems to nearly zero and allows much higher system capacity for the same size equipment. In addition to the chemical degradation mentioned above, MEA oxidizes when exposed to air. Storage and surge tanks must be provided with inert blanket gases such as N_2 or sweet natural gas to avoid this degradation.

Amine systems foam rather easily resulting in excessive amine carryover in the contactor. Foaming can be caused by solids such as carbon or iron sulfide; condensed hydrocarbon liquids from the gas stream; degradation products; almost any foreign material introduced to the system such as valve grease, excess corrosion inhibitor, etc. Some of these items such as iron sulfide or carbon particles are removed by cartridge filters. Hydrocarbon liquids are usually removed by the use of a carbon bed filter on a lean amine sidestream (about 10% of total flow). Corrosion byproducts are removed by reclaiming as noted above.

High skin temperatures on the reboiler/reclaimer tubes promote amine degradation. Steam or hot oil used for the reboiler should be limited to a maximum of 285°F (140°C) to avoid excessive temperatures. The reclaimer should not see hot oil or steam above 415°F (213°C).

MEA is nonselective in absorbing acid gases. It will absorb H_2S faster than CO_2 but the difference is not significant enough to allow its use to separate them. With the lowest molecular weight of the common amines, it has a greater carrying capacity for acid gases on a unit weight or volume basis. This generally means less pure amine circulation to remove a given amount of acid gases.

Because the solvent is in solution with water, the gas with which it comes in intimate contact will leave the contactor at its water saturation point. If dehydration is necessary it must be done after the MEA system.

MEA has a vapor pressure 30 times that of DEA (300 times that of TEA and DGA) at the same temperature. This causes MEA losses of 1 to 3 lbs/mmscf compared to 1/4 to 1/2 lb/mmscf for the other amine systems using the same system design parameters. Entrainment and leakage losses prevent low vapor pressure amines from attaining the predicted losses.

Diethanol amine

DEA does not degrade when contacted with CS_2, COS, and mercaptans as does MEA. However for product gas streams which must meet lower than 1 grain per scf of H_2S, MEA must be used. Because of this, DEA has been developed as a preferred solvent when these chemicals are present in the stream to be treated.

The reaction with acid gas for any of the amines is a mole to mole reaction. As shown in Table 10.1 the molecular weight of DEA is 1.7 times that of MEA. Even after correcting for density it requires 1.6 lb of DEA to react with the same amount of acid gas as 1 lb of MEA.

DEA is a weaker base (less reactive) than MEA. This has allowed DEA to be circulated at about twice the solution strength of MEA without corrosion problems. DEA systems are commonly operated at strengths up to 30 wt% in water and it is not unusual to see them as high as 35 wt%. This results in the DEA solution circulation rate usually being a little less than MEA for the same system design parameters.

The process reactions are shown below.

$$HOCH_2CH_2NHCH_2CH_2OH = R_2NH = DEA$$

$$2R_2NH + H_2S \underset{\text{High temp}}{\overset{\text{Low temp}}{\rightleftharpoons}} (R_2NH_2)_2S$$

$$(R_2NH_2)_2S + H_2S \underset{\text{High temp}}{\overset{\text{Low temp}}{\rightleftharpoons}} 2R_2NH_2HS$$

Because the system has much fewer corrosion problems and removes acid gases to nearly pipeline specifications it has been installed as the predominant system in recent years.

Table 10.1. Comparison of H_2S and CO_2 solvents

Formula	MEA	DEA	DGA	DIPA	SULFINOL	SULFOLANE
Molecular wt.	61.1	105.1	105.14	133.19		120.17
Boiling point (°F)	338.5	515.1	405.5	479.7		545
Boiling range, 5–95% (°F)	336.7–341.06	232–336.7	205–230	–		–
Freezing point (°F)	50.5	77.2	9.5	107.6		81.7
Sp. Gr., 77°F	1.0113	1.0881 (86°F)	1.0572	–		1.256 (86°F)
140°F	0.9844	1.0693	1.022	0.981 (129°F)		1.235
Pounds per gallon, 77°F	8.45	9.09 (86°F)	8.82	8.3 (86°F)		10.46 (86°F)
Abs. visc., cps., 77°F	18.95	351.9 (86°F)	40	870 (86°F)		12.1 (86°F)
140°F	5.03	53.85	6.8	86 (129°F)		4.9
Flash point (°F)	200	295	260	255		350
Fire point (°F)	205	330	285	275		380
Sp. ht. Btu/lb—(°F)	0.663	0.605	0.571	0.815		0.35
Critical-temp. (°F)	646.3	827.8	765.6	–		982.4
Critical–press. atm.	44.1	32.3	37.22	–		52.2
Ht. of vaporiz. (Btu/lb)	357.94	267.00	219.14	202.72		225.7
Ht. of reaction—CO_2 Btu/lb (Approx.)	825	620	850		580	
Ht. of reaction—H_2S, Btu/lb (Approx.)	650	550	674		500	

MEA = $HOC_2H_4NH_2$ DIPA = $(HOC_3H_6)_2NH$
DEA = $(HOC_2H_4)_2NH$ SULFOLANE = $(CH_2)_4SO_2$
DGA = $HOCH_2CH_2OCH_2C_2H_4NH_2$

Diglycol amine

This process has been developed by the Fluor Company. It originally began as a combination of 15% MEA, 80% triethylene glycol, 5% water. The system would both sweeten and dehydrate (to the same level as 95% TEG) the gas in a single step. The high vapor release during regeneration (both water vapors and acid gases) causes severe erosion/corrosion problems in the amine/amine exchanger and in the regeneration column. This system has generally been abandoned.

The present system uses 2-(2-amono ethoxy) ethanol at a recommended solution strength of 60 wt% in water. DGA has almost the same molecular weight as DEA and reacts mole for mole with acid gases. DGA seams to tie up acid gases more effectively so that the higher concentration of acid gas per gallon of solution does not cause corrosion problems as experienced with the usual amine systems.

The system reactions are given below.

$$HOCH_2CH_2OCH_2CH_2NH_2 = RNH_2 = DGA$$

$$2RNH_2 + H_2S \underset{\text{High temp}}{\overset{\text{Low temp}}{\rightleftharpoons}} (RNH_3)_2S$$

$$(RNH_3)_2S + H_2S \underset{\text{High temp}}{\overset{\text{Low temp}}{\rightleftharpoons}} 2RNH_3HS$$

DGA does react with COS and mercaptans similarly to MEA but forms N, N1, *bis* (hydroxy, ethoxy ethyl) urea, BHEEU. BHEEU can only be detected using an infra-red test rather than chromatography. Normal operating levels of 2–4% BHEEU are carried in the DGA without corrosion problems. BHEEU is removed by the use of a reclaimer identical to that for an MEA system but operated at 385°F (196°C). Materials of construction are the same as those for MEA systems.

There has been a concern that DGA might be a good solvent for unsaturated hydro-carbons. A survey of the DGA users indicates that many of the systems are operated on gas containing concentrations of C_5+ above 2% without any indication of hydro-carbon loading of the system.

Those systems near their hydrocarbon dew point are usually installed with a flash tank on the rich amine from the absorber. The flash tank is operated at a reduced pressure just high enough to get into the plant fuel gas system. It reduces the vapor load on the regenerator column. (A similar system is recommended on MEA systems operating near the hydrocarbon dew point.)

DGA allows H_2S removal to less than 1/4 grain per 100 scf and removes CO_2 to levels of about 200 ppm using normal absorber design parameters.

Other gas treating processes

Hot potassium carbonate (Benfield)

The basic process concept has been known since the early 1900s. It was not an economical, practically demonstrated process until the mid 1950s. Since that time the process has been used for bulk removal of acid gases where residual CO_2 content was not needed in the ppm range.

The process is very similar to the amine processes. High temperatures favor high solubility of potassium carbonate (PC) in water leading to high concentrations of PC. High PC concentrations mean higher carrying capacity of acid gases in the system.

The system is ideal for streams having CO_2 partial pressures of 30–90 psi (205–620 kPa). It has a high affinity for H_2S so that pipeline specification for H_2S can easily be reached at about 4 ppm. A stream with little or no CO_2 is not suited for the process due to making regeneration of the lean PC extremely difficult.

This process is usually found as part of the hydrogen plant in those refineries that need to produce hydrogen.

Sulfinol

This is a proprietary system developed by Shell Oil Company. The process uses a solution containing both a chemically reactive component, Di-isopropanolamine (DIPA) and a physical solvent, Tetra hydro thiophene 1-1 dioxide (sulfolane).

Sulfolane is a very active solvent for H_2S, COS and the mercaptans. CO_2 is also soluble in it, but not nearly as much as the S compounds are. Because of this, sulfinol systems are most economically attractive (compared to amine systems) for H_2S/CO_2 ratios greater than 1:1. If the bulk of the acid gas can be dissolved in the sulfolane the system is much cheaper to operate than amine. The acid gases are picked up and released with very little heat increase or heat required. The solubility of acid gases is much higher in sulfinol than for the amines. Sulfinol loading are limited to 4–6 scf acid gas/gal solvent versus 2.5 scf acid gas/gal amine solution. In addition the heat capacity of sulfolane is about half that of the amines, further reducing the regeneration heat required.

Other, sweetening liquid processes such as Vetracoke, Stretford, and Rectisol have found high usage in the coal gasification and natural gas industries. They have not reached the prominence of the amines or the PC processes in oil refining. Table 10.1 summarizes a comparison of the common solvents.

Calculating the amine circulation rate

The circulation rate for amine solvents is important to ensure effective treatment of the sour gas. It is important also because it is a major contribution to the operating cost of the plant. These costs are incurred by pumping cost, steam to reboiler and air cooler/condenser costs. This item provides a calculation method to establish this circulating rate. It is based on some fixed parameters which have been accepted as the optimum. These steps are:

Step 1. In DEA and MEA treating processes a ratio of the amine to H_2S is about 3:1 mole.

Step 2. Obtain gas feed rate and its composition (from plant and lab tests). Determine its mole weight and the volume percent H_2S.

Step 3. Fix the weight percent of amine in the recirculating amine solution. This will be between 15% and 20% weight of DEA or MEA in water.

Step 4. Calculate the H_2S in moles/hr that is in the feed. This is done by resolving volume flow of gas to moles and using lab data to provide moles/hr of H_2S.

Step 5. Calculate the amount of H_2S to be left in the lean gas. This is usually 10 grains/100 SCF with DEA as absorbent and 3 grains/100 set with MEA as absorbent (1 grain = 0.0648 g or 0.0022857 ounces).

Step 6. Calculate H_2S absorbed. This is the difference between Step 4 and Step 5.

Step 7. Using the ratio 3:1 fixed in Step 1 calculate the rate of DEA (or MEA) that will be required.

Step 8. Mole weight of DEA is 105.1 and the mole weight of MEA is 61.1. Calculate weight of the amine using percent weight of amine in solution calculate weight per unit time of solution.

Step 9. Using the data from step 2 calculate the solution's gallons per hour or per minute.

Calculating the number of theoretical trays in an amine contactor

There are several accepted methods to calculate theoretical trays in amine contactors. Among these are the McCabe Thiele—Graphical Method and the calculation method described by the following steps. This calculation method is considered by many to be the sounder and more accurate of the methods available. The following steps describes this calculation procedure:

Step 1. The equation to determine the number of theoretical trays is:

$$N = \frac{(\text{Log } 1/\text{q}\,(A - 1))}{(\text{Log } A)} - (1)$$

where

N = number of theoretical trays
q = mole H_2S in lean gas/mole H_2S in feed gas
A = the absorption factor $L/V \cdot K$

Step 2. Calculate H_2S in lean gas in moles/hr and moles H_2S in the rich gas. Divide H_2S in lean gas by that in the feed. This is q.

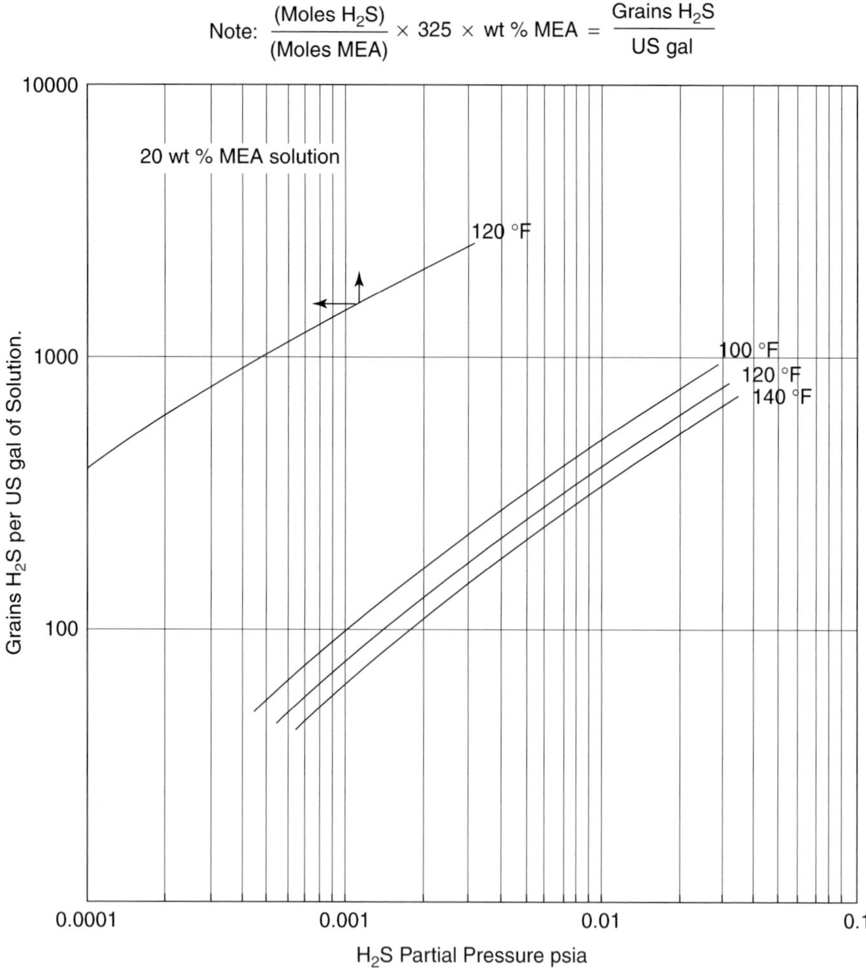

Figure 10.2. Partial pressure of H_2S in amine solutions.

Step 3. Calculate the amine circulation rate in moles/hr amine, lbs/hr amine and lbs/hr solution (about 20 vol%). Resolve to gals/hr of solution.

Step 4. Using the grains/hr of H_2S leaving in the richer amine (that is grains in feed gas less grain in product gas), calculate grains per gallon absorbed by amine solution. Add to that value the H_2S residual in the lean amine. The sum of these is the grains of H_2S per gallon of amine, and is the value used in Figure 10.2 to determine the partial pressure of H_2S in MEA.

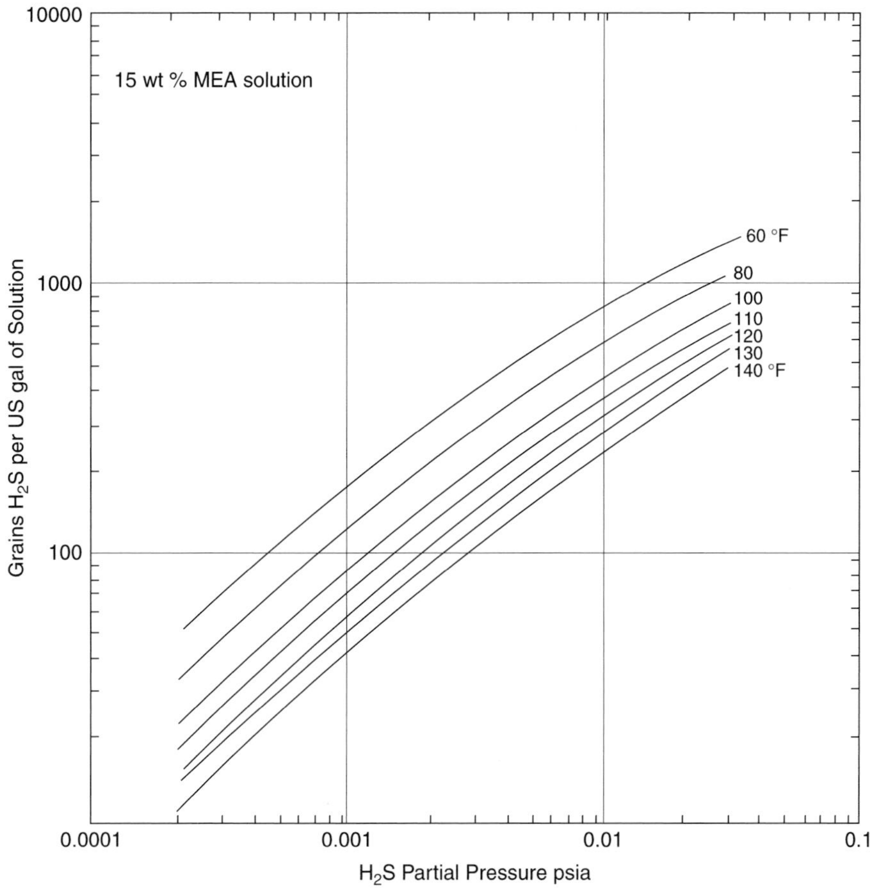

Figure 10.2. (Cont.)

Step 5. The absorption factor is obtained from the equation:

$$A = \frac{a(1 + Rr)(1 - q)}{pp/P}$$

where

A = the absorption factor
a = mole fraction of H_2S in gas feed
R = moles MEA/Moles H_2S absorbed
r = residual H_2S in lean MEA solution (mole H_2S per mole MEA)
pp = partial pressure of H_2S in rich solution PSIA
P = systems pressure (tower pressure) PSIA

Step 6. Solve for A. Then solve equation given in Step 1 for N. Divide N by efficiency factor (usually about 15% for absorbers) to determine actual number of trays.

An example calculation is given in Appendix 10.1 of this chapter.

Calculating absorber tray size and design

This procedure is similar to those described in other sections of this Handbook. Emphasis is made here, however, to the foaming tendencies of MEA and DEA. Most of these units contain facilities to inject anti-foaming chemicals. Under normal operation these chemicals subdue foaming to a large extent particularly in the contactors. The addition of a filter in the rich amine line also helps combat severe foaming.

It is believed that foaming is enhanced by the absorption of large quantities of hydrocarbons into the amine stream. This can occur quite easily if the temperature or pressure change forces the hydrocarbon gas into a condition below its dew point. The presence of large quantities of impurities such as COS (Carbonyl Sulfide) or the tarry derivative of COS may also induce foaming. The addition of the filter and reclaimer usually solves this problem to a large extent.

Loading of amine towers nevertheless is critical due to the foaming nature. Consequently the towers are designed with more latitude than most other towers. In evaluating its performance also the acceptable level of flooding is lower by 20–30% than that for normal towers.

The calculation to evaluate the towers flooding follows the same steps as those described in Chapter 4. In this case, however, the calculated value for V_L is multiplied by a system factor between 20% and 30%. If the unit is being used in service where high concentration of impurities is present a figure of 30% should be used.

Calculating the heat transfer area for the lean/rich amine exchanger

The amine in the contactor picks up the heat of reaction which occurs with the absorption of H_2S. On leaving the amine contactor or absorber the rich amine is heat exchanged against hot lean amine leaving the bottom of the stripper. The performance of this exchanger is critical to the process as a whole. Usually the rich amine receives no other heat before it enters the stripper. As in light end towers the feed condition is vital to the proper operation of the tower. The purpose of this item is to provide a calculation procedure to evaluate the heat transfer coefficient of this exchanger. The following steps gives the procedure:

Step 1. Using lab data for operational units or the design specification for those units to be designed, determine the quantity of H_2S absorbed in the contactor.

Table 10.2. The heat balance

Stream	V or L	lbs/hr	°F	Btu/lb	mmBtu/hr
IN					
Rich product gas	V	34,722	100	350	12.153
Lean amine soln	L	194,747	105	98	19.085
Heat of reaction				650	4.696*
Total In					35.928
OUT					
Lean Product gas	V	34,722	105	355	12.326
Rich amine soln	L	194,747	By	Diff	23.602
Total out					35.928

*7,225 lbs of H_2S absorbed/hr.

Step 2. From the data given in Table 10.1 calculate the total heat of reaction in Btu/hr.

Step 3. Set the temperature of the lean gas leaving the absorber to be the same as the amine entering.

Step 4. Carry out a heat balance over the contactor with the rich amine temperature being the unknown. Equate and solve for this.

Step 5. From plant data (or the Design Spec) apply the temperatures and flow in and out of the exchanger to calculate its heat duty in Btu/hr.

Step 6. Calculate the LMTD (Log Mean Temperature Difference) and from Table 10.2 establish the overall heat transfer coefficient U in Btu/hr ft^2 °F.

Step 7. Using the energy equation

$$Q = UA\Delta T_M$$

where

Q = the duty in Btu/hr
A = the area in ft^2
ΔT_M = the log mean temperature difference in °F
U = the overall heat transfer coefficient in Btu/hr ft^2 °F.

Calculate the value for A in sqft.

The stripper design and performance

This evaluation is based on reconciling the steam used for stripping the rich amine. The quantity of steam is a major operating cost in this type of plant and therefore deserves attention. An acceptable level of steam usage is about 0.8–1.1 lbs steam/gallon of circulating solution for MEA and 1.1–1.3 lbs steam/gallon in the case of DEA. This item describes a procedure to calculate this steam rate.

Step 1. From plant data design specification ascertain the feed rate of rich amine to the stripper, and its composition in terms of H_2S, water, MEA (or DEA), and hydrocarbons.

Step 2. Calculate recycle rate of lean amine leaving the bottom of the tower. Obtain from the lab data or design specification its composition in terms of residual H_2S, water, MEA (or DEA) and hydrocarbon (if any).

Step 3. Develop the material balance over the tower.

Step 4. Using the data in Step 3 set the external tower top reflux ratio as 2:1, and calculate the heat balance over the tower top to find the condenser duty.

Step 5. Calculate the heat in with the rich amine feed before the preheat exchangers (see previous item).

Step 6. Using plant data for lean amine temperature in and out of the preheat exchangers calculate its duty. Add the duty to the enthalpy from Step 5 to give feed enthalpy into the tower.

Step 7. Calculate the overall heat balance over the stripper to find the reboiler duty. Remember to add in heat of dissociation which is equal to the heat of reaction in the contactor (See previous item).

Step 8. Saturated 50 psig steam is usually used as the heating medium. Calculate the amount of steam from its enthalpy data (steam tables) and the reboiler duty.

Step 9. If the steam usage is excessive check overhead stream and the lean amine concentration. It is possible that a high volume of water is being evaporated. If this is the case reduce the reboiler duty to maintain the amine concentration and H_2S concentration.

Removing degradation impurities from MEA

Although MEA (mono ethanol amine) is the most efficient absorbent in the amine family. It has one major shortcoming. It is readily degraded by certain sulfur compounds such as carbonyl sulfide (COS) and by carbon disulphides. Both these compounds are found in significant quantities in gases from cracking processes such as thermal crackers and catalytic crackers.

It is difficult if not impossible under normal refinery conditions to remove these sulfides from the gas. What can be done and is the normal practice is to remove the product of degradation and return the 'clean' amine to the system. Two items of equipment are added to the process to achieve this. These are a filter and a reclaimer. The filter is a normal leaf type filter contained in two filter casings. These are piped up in parallel with one on stream and the other shut down for cleaning and as spare.

Reclaimers are really kettle type reboilers. It takes as feed a portion of the lean amine leaving the stripper. This stream is vaporized and the vapor returned to the stripping tower. The residue or sludge is the product of degradation and is dumped to waste. Reclaimers can be designed to operate continuously or on a batch basis. Steam is used as the heating medium. As in the case of the stripper reboilers the heating steam temperature to the reclaimer is also carefully controlled. Thus the steam medium is saturated 50 psig steam. It must be remembered that the duty to condense the vaporized amine from the reclaimer must be added to the stripper overhead condenser duty. It must also be included in any tower loading exercise that may follow.

Appendix 10.1 The process design of an amine gas treating unit

The following is an extract from a design specification and defines the parameters of this example:

1.0 Unit required is a gas treating unit for the removal of H_2S from a Hydrotreater recycle gas stream. The feed to the hydrotreater consists of gas oil from straight run source and streams from a thermal cracker and a catalytic cracker.
The recycle gas will therefore contain some COS.

2.0 The feed gas shall have the following properties:
- The mole weight of the gas is 10.5
- The H_2S content of feed gas is 4,048 Grains/100 scf.
- The gas rate is 30 mmscf/day
- The pressure of the gas at the outlet of the contactor is 320 psig.

3.0 Product gas shall have a H_2S content of no greater than 0.1 grain/100 scf.

4.0 The amine solvent to be used shall be mono ethanol amine.
- Amine ratio to H_2S shall be 3.0 moles amine to 1.0 mole H_2S.
- The amine solution shall be 20% by weight in water.
- Residual H_2S in the Lean Amine Solution shall be no greater than 0.09 mole per mole of MEA.
- Protection against degradation of the amine shall be included.

The contactor design

Calculating the amine solution circulation rate

$$\begin{aligned}\text{Feed gas rate} &= 30 \text{ mmscf/day}\\ &= 3{,}306.9 \text{ moles/hr}\\ &= 34{,}722 \text{ lbs/hr}\end{aligned}$$

$$\begin{aligned}H_2S \text{ in feed} &= \frac{4048 \times 0.0022857}{16} = 0.578 \text{ lbs/100 scf}\\ &= 5{,}100 \text{ moles/day} = 1.928 \text{ mmscf/day}\\ &= 212.5 \text{ moles/hr}\end{aligned}$$

$$H_2S \text{ in product gas} = 0.1 \text{ grain/100 scf}$$
$$= 0.126 \text{ moles/day}$$
$$= 0.00521 \text{ moles/hr}$$

$$H_2S \text{ absorbed in amine} = 212.5 - 0.00521$$
$$= 212.49 \text{ moles/hr}$$

amine ratio is 3.0 moles amine per mole H_2S

Moles amine circulating = 637.47 moles/hr. = 38,949 lbs/hr (mol wt. MEA = 61.1) 20% wt solution = 194,747 lbs /hr made up of 155,798 lbs water and 38,949 amine.

lbs/gal MEA = 8.45

lbs/gal water = 8.328

MEA gals/hr = 4,609

Water gals/hr = 18,708

Amine solution circulation rate = 23,317 gals/hr.

Calculating the number of trays and the overall dimensions of the contactor

The number of theoretical stages required in the contactor will be calculated using the equation:

$$N = \frac{(\text{Log } 1/q\ (A - 1))}{(\text{Log } A)} - 1 \qquad (1)$$

where

N = number of theoretical trays
q = mole H_2S in lean gas/mole H_2S in feed gas
A = the absorption factor L/V·K

$$\text{`}q\text{' in this case} = \frac{\text{moles } H_2S \text{ in lean gas}}{\text{moles } H_2S \text{ in feed gas}}$$
$$= \frac{0.00521}{212.5}$$
$$= 2.45 \times 10^{-5}$$

$$\begin{aligned}
\text{Total moles of acid gas absorbed} &= 212.49 \text{ moles/hr} \\
\text{Acid gas residual in the lean amine} &= 57.37 \text{ moles/hr} \\
\text{Total acid gas in grains/hr} &= \frac{269.87 \times 34 \times 16}{0.0023} \\
&= 63.83 \times 10^6 \text{ grains/hr} \\
\text{Total MEA solution} &= 23{,}317 \text{ gals/hr}
\end{aligned}$$

Then grains acid gas per gal of amine solution = 2,738 grains H_2S/gal MEA.

From Figure 10.2 for 20% MEA solution H2S partial pressure is 0.33 psia.

Using the following equation the absorption factor A is calculated as follows:

$$A = \frac{a(1 + Rr)(1 - q)}{pp/P}$$

$$\text{'}a\text{' is the mole fraction of } H_2S \text{ in feed gas} = \frac{212.5}{3{,}306.9} = 0.0643$$

'R' = moles MEA per mole acid gas absorbed = 3.0
'r' = 0.09 moles H_2S per mole lean MEA.

Then

$$A = \frac{0.0643 \times \{1 + (3.0 \times 0.09)\} \times \{1 - (2.45 \times 10^{-5})\}}{0.33 \div 335} = 82.1$$

Therefore

$$N = \frac{\log\{(1 \div 2.45 \times 10^{-5}) \times (82.1 - 1)\}}{\log 82.1} = \frac{\log(3.283 \times 10^6)}{\log 82.1} = \frac{6.5}{1.91} = 3.4 \text{ theoretical trays.}$$

Set tray efficiency at 15% then actual number of trays = 23

MEA has a tendency to foam therefore set tray spacing at 30 inches.

Then trayed section will have a height of 22 × 1.5 ft = 33 ft.

Calculating the contactor diameter
Use foaming factor of 60%.

Feed gas to the contactor is 30 mmscf/day

Temperature of gas is 100°F and its pressure is 335 psia (these are average conditions).

$$\text{Then actual cubic feet per second (ACFS)} = \frac{30 \times 10^6 \times 14.7 \times 580}{24 \times 3600 \times 520 \times 335}$$
$$= 16.41 \text{ cfs}$$

Feed gas in lbs/hr is 34,722 = 9.645 lbs/sec

$$\rho_v = \frac{9.645}{16.41} = 0.588\text{lbs/cuft}$$

lbs/hr of MEA solution is 194,747
Gals/hr is 23,317
And cubic ft /hr is 3,117

Then lbs/cuft = 62.48 and at 120°F (MEA inlet temperature) = 62.1 lbs/cuft.

Loading at flood: $K_f\sqrt{\{\rho_v \times (\rho_l - \rho_v)\}}$

K_f is 1,280 from Figure 10.3 and inserting the 60% foam factor $K_f = 768$.

Loading = 768 × $\sqrt{36.17}$ = 4,619 lbs/hr·sqft.

Let design load be 70% of flood = 3,233 lbs/hr·sqft

$$\text{Cross sectional area of tray (and tower I/D)} = \frac{34{,}722}{3233}$$
$$= 10.74 \text{ sqft.}$$

Internal diameter of tower (calculated) = 3.7 ft

Call it 4 ft = 12.6 sqft cross sectional area.

The actual tray design will be done by others (tray manufacturer) to protect the guarantee requirements.

Calculate the amine hold up in the bottom of the tower
The liquid hold up will have be 1 min to NLL.

$$\text{The volume of amine in 10 min} = \frac{3{,}117}{60} = 51.95 \text{ cuft}$$

$$\text{Then NLL will be } \frac{51.95}{12.6} = 4.1 \text{ ft say 4 ft}$$

Then HLL will be set at 8 ft and LLL at 4 ft above Tan.

Add a further 4 ft from HLL to bottom tray.

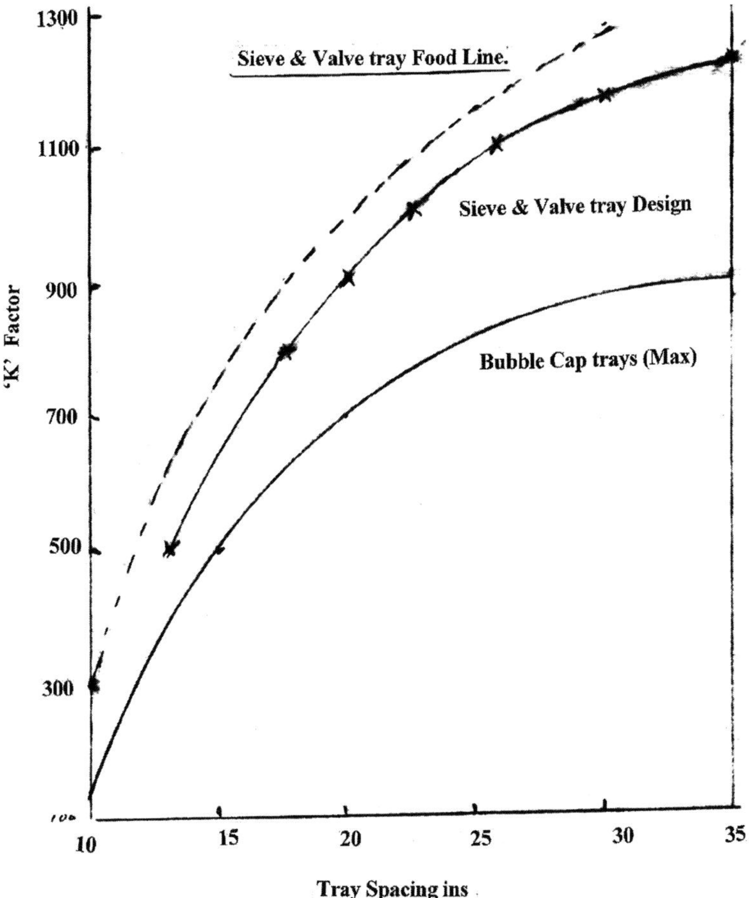

Figure 10.3. The Brown and Souder flood constant.

The overall dimensions of the contactor

The height of the contactor is as follows:

From Bottom Tan line

To bottom tray = 12 ft
To top absorbing Tray = 45 ft
To wash water draw off = 48 ft (bottom of chimney tray)
To top of wash section = 54 ft
To top tan = 58 ft

Overall dimensions for the contactor is 4 ft i.d × 58 ft Tan-Tan.

Calculating the heat balance over the contactor

The rich gas flow will enter at 100°F and 320 psig and will have an enthalpy of 350 Btu/lb. Its flow rate is 34,722 lbs/hr.

The heat of reaction is calculated at 650 Btu/lb of H_2S absorbed.

The lean amine solution enters at a temperature of 105°F which will be set by heat transfer. The lean product gas will leave at about the same temperature.

The rich amine leaving the contactor is determined by difference

$$\text{Temperature of rich amine out} = \frac{23.602 \times 10^6}{(38{,}949 \times 0.663) + 155{,}798}$$
$$= 129.95 \text{ say } 130^\circ\text{F}$$

The heat exchanger design

The hot lean amine stream to the contactor will be cooled from the stripper bottom temperature first by heat exchange against the rich amine leaving the contactor. It will then be trim cooled to the contactor inlet temperature by either water or air. The following is the calculation to determine the size of the lean/rich amine exchanger.

The lean amine from the stripper will be cooled to 175°F in the heat exchange with the rich amine leaving the contactor. The duty of this exchanger is:

$$\{38{,}949 \times 0.663 \times (249 - 175)\} + (155{,}798 \times 74) = 13.44 \text{ mmBtu/hr}$$

The figure 240°F is a value for the stripper bottom temperature estimated by a quick bubble point calculation of the lean amine at the stripper bottom conditions of temperature and pressure. This will be checked later.

Temperature of the rich amine feed to the stripper (T) is as follows:

$$(23.602 + 13.44) = (38{,}949 \times 0.663 \times T) + 155{,}798T$$
$$T = 204^\circ\text{F}$$

The overall heat transfer coefficient for the amine exchanger can be taken as 100 Btu/hr sqft °F (this will be checked by the exchanger manufacturer). The exchanger size is:

$$\begin{aligned} \text{LMTD}\ \ 249 &\longrightarrow 175 \\ 204 &\longleftarrow 130 = 45^\circ\text{F} \end{aligned}$$

Table 10.3.

Com	MW	lbs/Gal	Feed			O/heads			Lean amine		
			Mole/hr	lbs/hr	GPH	Mole/hr	lbs/hr	GPH	Mols/hr	lbs/hr	GPH
H_2S	34	6.55	270	9,180	1,401	212.5	7,225	1,103	57.5	1,955	298
H_2O	18	8.33	8,655	155,790	18,702	8.43	151	18	8,646.6	155,639	18,684
HC	72	5.25	1.8	130	25	1.8	130	25	Nil	Nil	Nil
MEA	61	8.45	637.5	38,888	4,602	Nil	Nil	Nil	637.5	38,888	4,602
Total			9,564.3	203,988	24,730	222.73	7,506	1,146	9,341.6	196,482	23,484

$$\text{Area} = \frac{13{,}440{,}000}{45 \times 100}$$
$$= 2{,}987 \text{ sq ft.}$$

The stripper design

Total moles of acid gas in feed = moles absorbed = 212.49 moles/hr
Residual acid gas = 57.37 moles/hr
= 269.86 moles/hr
Moles amine = 637.47 moles/hr
(This assumes no losses)
Moles water = 8,655 moles/hr
Moles hydrocarbon dissolved = 1.8 moles/hr as C5

The material balance

The material balance over the stripper is given in Table 10.3.

The following calculation establishes the composition of the overhead product and the composition of the liquid reflux stream thus:

Let x be the moles per hour of water in the overhead product. The H_2S content is established by the total in the feed less the residual H_2S in the bottom product—the lean amine. It is assumed that all the hydrocarbon will leave with the overhead vapor product. The value of x is found by the following Dew point calculation of the o/head product at the reflux drum conditions of temperature and pressure (Table 10.4).

The reflux drum conditions were set at 23 psia pressure and 100°F and the dew point calculation at these conditions gave a reflux stream composition of:

Table 10.4.

Comp	Moles/hr	K100	$x = y/k$
H_2S	212.5	14.75	14.40678
H_2O	x	0.041	$x/.041$
C5's	1.8	0.67	2.686567
	214.3		17.09335
	$214.3 + x$		$17.09 + x/0.41$
	197.21		$x/.041 + x$
		$x = 8.43$	

			Moles/hr	lbs/hr
H_2S	0.065 mole fraction	=	29.5	1,003
H_2O	0.923	=	419.6	7,553
HC	0.012	=	5.5	396
		=	454.6	8,952

Moles reflux is 2 × product vapor.

Tower top conditions and condenser duty

The tower top pressure shall be the reflux drum pressure plus say 3 psi pressure drop over the condenser and about 0.7 psi for piping etc. Then tower top pressure will be 26.7 psia. The total overhead vapor will be product plus reflux thus:

	Product (Moles/hr)	Reflux (Moles/hr)	Total	mole Fraction
H_2S	212.5	29.5	242	0.357
H_2O	8.43	419.6	428.03	0.632
HC	1.8	5.5	7.3	0.011
	222.73	454.6	677.33	1.000

The dew point calculation gave the tower top temperature of 218°F.

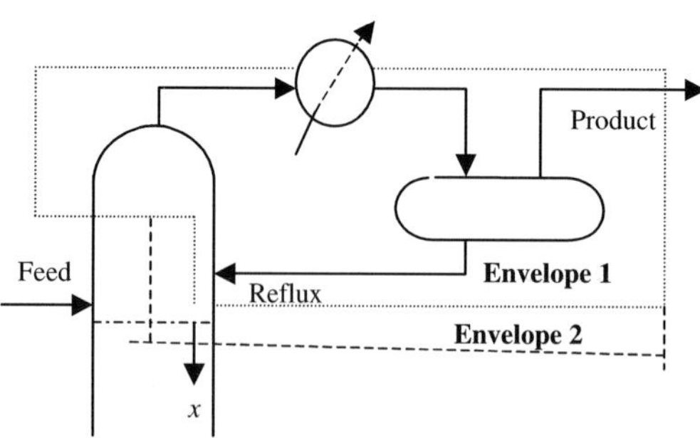

The condenser duty is calculated from the following heat balance over the tower top. The (Refer to envelop 1) heat balance is shown in Table 10.5.

Table 10.5.

Stream	V or L	°F	lbs/hr	Btu/lb	mmBtu/hr
IN					
Total O/head	V	218	16,458	615	10.122
Total in					10.122
OUT					
Prod	V	100	7,506	155.7	1.169
Reflux	L	100	8,952	92.5	0.828
Condenser	By	Diff			8.125
Total out			16458		10.122

Condenser duty to strip vapors from the feed is 8.125 mmBtu/hr. To this will be added the vapor from the reclaimer that has to be condensed. This will be done later.

To calculate the internal reflux from the top tray

Knowing the condenser duty, the internal reflux x lbs/hr can be calculated from the heat balance over the tower top as shown in envelope 2 of diagram 1 (Table 10.6).

Solving for x:

$$9.294 + 222x = 0.923 + 1162x$$
$$x = 8{,}905 \text{ lbs/hr}$$

Mole weight of reflux = 18.5 (from the dew point calculation)

Moles/hr reflux = 481

The moles of vapor from the reclaimer will be added to this figure when calculating the vapor loading over the top tray. Two trays above the feed tray will be provided as wash trays.

Table 10.6.

Stream	Vor L	°F	lbs/hr	Btu/lb	mmBtu/hr
IN					
Int ref	V	222	x	1,162	1,162x
Product	V	222	7506	123	0.923
Total in			7,506 + x		0.923 + 1,162x
OUT					
Prod	V	100	7,506	155.7	1.169
Reflux	L	222	x	222	222x
Condenser					8.125
Total out			7,506 + x		9.294 + 222x

Table 10.7.

Temperature			249°F	
Comp	Moles/hr	Mole frac	K@ 34 psia	$y = xk$
H_2S	57.5	.0062	45.4	0.2815
Water	8646.6	.9256	0.79	0.7312
MEA	637.5	.0682	0.09	0.0057
Total	9,341.6	1.0000		1.0184

The stripper bottom conditions and reboiler duty

The pressure at the bottom of the tower is fixed at 34 psia. This allows a pressure drop of about 0.35 psi per tray which is estimated as a total of 20 trays. The tower bottom temperature is calculated by a bubble point calculation of the bottom product at this pressure of 34 psia (see Table 10.7).

Enthalpy of bottom product = 230 Btu/lb as liquid.

Calculating the reboiler duty

This is determined from the overall tower heat balance as shown in Table 10.8.

Vapor/liquid on bottom tray

The bottom tray will have a temperature of 240°F. (See calculation diagram 2 and Table 10.9). In the following heat balance let the lbs/hr of the stripout vapors to the tray be x,

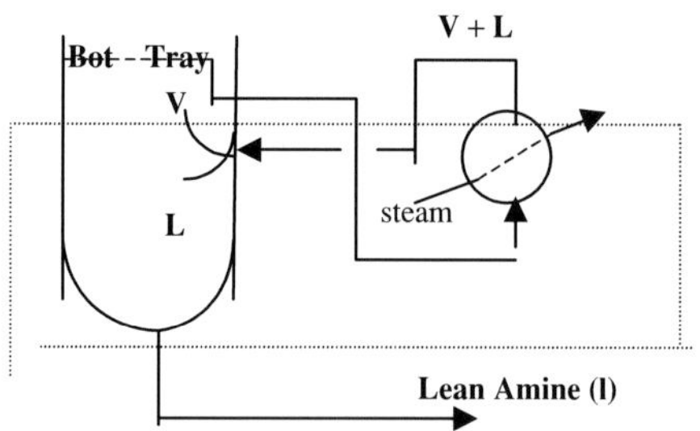

$$x = 25{,}748 \text{ lbs/hr} \qquad \text{Mole wt} = 23.08$$
$$\text{Moles/hr} = 1{,}115.6$$
$$\text{V/L at bottom tray} = 1{,}115.6/10{,}457$$
$$= 0.107$$

To calculate the number of theoretical trays in the stripper

The Kremser equation which is shown graphically by Figure 10.4 will be used for this calculation.

Table 10.8.

Stream	V or L	°F	lbs/hr	Btu/lb	mmBtu/hr
IN					
Feed	V + L	204	203,988	181.7	37.06
Reboiler			By Diff		22.049
Total in			203,988		59.109
OUT					
O/head prod	V	100	7,506	155.7	1.169
Bottoms	L	249	196,482	230	45.179
Ht reaction					4.696
Condenser					8.125
Total out			203,988		59.109

The V/L factor calculated above will be used for this equation. A Tower average K value for each component in the feed will also be used. The equation is shown by Table 10.10.

Three theoretical trays will achieve the stripping required. Stripping trays have poor efficiency between 12% and 18%. Use 15% in this case, then actual trays will be $3/.15 = 20$ actual trays.

The anomaly for the amount of HC stripped in the above calculation stems from the assumption that the HC is pentane. It is probably a heavier hydrocarbon.

Calculating the reclaimer duty and size

A slip stream of 2 wt% of lean amine solution will be routed through the reclaimer. It will be vaporized to leave a sludge stream of 2% of the reclaimer feed. The operation will be continuous and the vapor will be routed back to the tower entering below the bottom tray. The material balance over the reclaimer is shown in Table 10.11.

Table 10.9.

Stream	V or L	°F	lbs/hr	Btu/lb	mmBtu/hr
IN					
Bot prod	L	240	196,482	224	44.012
Stripout	L	240	x	224	$224x$
Reboiler					22.049
Total in			$196{,}482 + x$		$66.061 + 224x$
OUT					
Strip out	V	249	x	1,035	$1{,}035x$
Bottoms	L	249	196,482	230	45.179
Total out			$196{,}482 + x$		$45.179 + 1{,}035x$

A typical heat flux over a kettle reboiler is 26,500 Btu/hr sqft therefore area for heat transfer = 80 sqft.

The vapor generated by the reclaimer will be added to the vapor load at the bottom tray. Likewise this amount as liquid will be added to the trays liquid load. The condenser

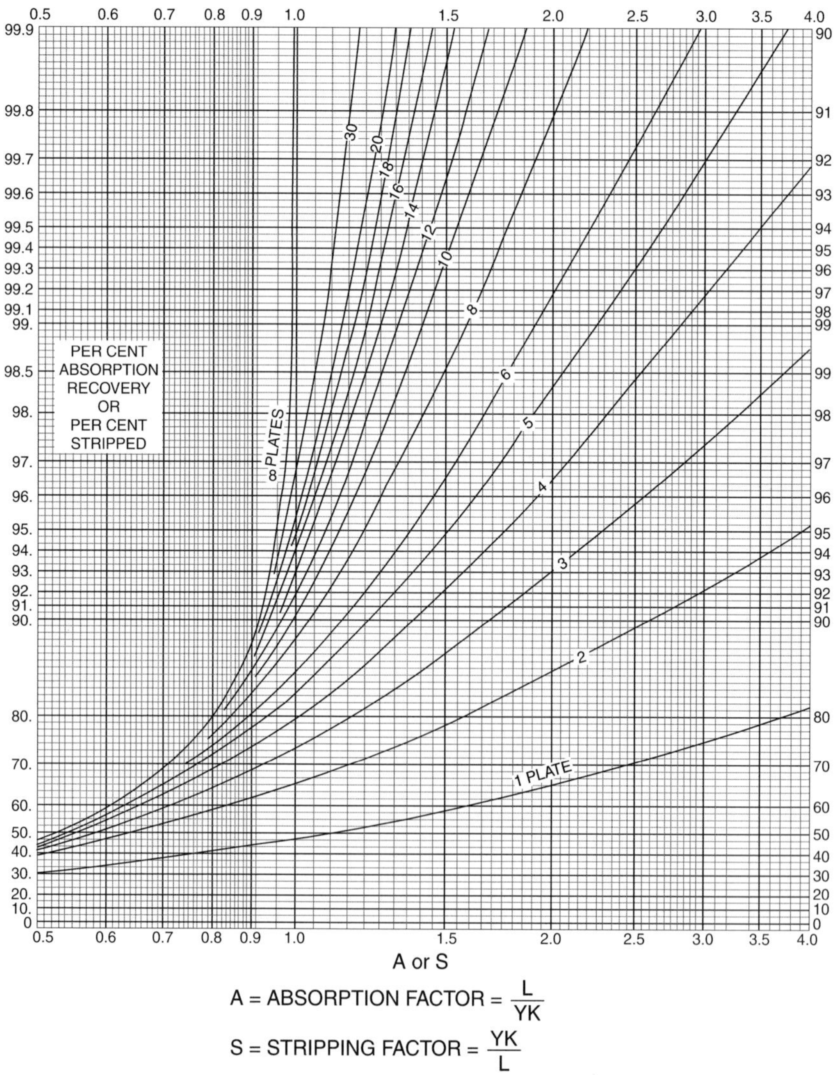

Figure 10.4. The Kremser equation and correlation.

Table 10.10.

Component	Moles/hr	Ave K	VK/l	Strip out %	Moles/ hr
H_2S	270	44	4.7	100	270
H_2O	8,655	0.72	0.08	< 0.5	8.0
HC	1.8	3.5	0.37	50	0.9
MEA	637.6	0.1	0.01	0	0
Total	8,564.3				278.9
Notes			Use 3 trays	From Figure 10.4	

duty will also be increased to accommodate the reclaimer duty. Thus:

$$\begin{aligned}
\text{Total condenser duty} &= 8.125 + 2.130 = 10.255\ \text{mmBtu/hr} \\
\text{Vapor load to bottom tray} &= \text{Stripout} + \text{reclaimed vapor} \\
&= 25{,}748\ \text{lbs/hr} + 2851 \\
&= 28{,}599\ \text{lbs/hr} \\
\text{Moles/hr} &= 1{,}115.6 + 135.8 \\
&= 1{,}251.4
\end{aligned}$$

Stripper tower dimensions

As in the case of the contactor the cross sectional area of the stripper will be calculated using the Brown and Souder method as follows:

Cross sectional area of the tower will be based on the loadings over the bottom tray thus:

$$\begin{aligned}
\text{Total vapor to tray} &= 1{,}251.4\ \text{moles /hr} \\
\text{In actual cubic feet/sec} &= \frac{1251.4 \times 378 \times 700 \times 14.7}{520 \times 34 \times 3{,}600} \\
&= 76.47\ \text{ACFS} \\
\rho_v = 7.85/76.47 &= 0.104\ \text{lbs/cuft}
\end{aligned}$$

Table 10.11.

Stream	V or L	°F	lbs/hr	Btu/lb	mmBtu/hr
IN					
Feed	L	249	3,930	230	0.935
Reclaimer			By diff		2.130
Total in			3,930		3.065
OUT					
Rec vapor	V	260	2,851	1070	3.051
Sludge	L	260	79	182	0.014
Total out			3,930		3.065

Total liquid from the bottom tray = Stripout + Product + reclaimer liq
= 25,748 + 196,482 + 2,851
= 225,081 lbs/hr

Lbs/gal of liquid at 240°F = 7.87
Gals per hour = 28,600 and cubic ft/hr is 3,823
ρ_l = 58.87 lbs/cuft.

Loading at flood = $K_f\sqrt{\{\rho_v \times (\rho_l - \rho_v)\}}$
From Figure 3.8.3 K_f = 1,280 and at 60% = 768

Then loading = 768 × 2.47
= 1,899 lbs/hr sqft

let loading be 80% flood = 1,519

$$\text{Cross sectional area} = \frac{28{,}599}{1519}$$

= 18.8 sqft and diameter
= 4.9 ft call it 5 ft or 19.6 sqft

Tower will contain 20 valve type trays made up of 18 stripping trays and 2 rectifying trays. As in the case of the contactor trays they will be spaced at 30 inches. Trayed section will therefore be 19 × 30 inches = 570 inches or 47 ft 6 inches from bottom to top tray.

The bottom of the tower (from bottom tray to bottom tan will be sized to cater for a 2 min hold up of liquid to the HLL. Thus:

Total liquid from bottom reboiler = 196,482 + 2,851
= 199,333 lbs/hr

lbs/gal at 60°F = 8.35 at 249 = 7.89 lbs/gal
Then gals/hr of total bottoms = 199,333/7.891
= 25,264 gals/hr
= 421 gpm
Or 56.3 cubic ft/min

Volume resident over three min = 168.9 say 169 cubic ft
Height of HLL = 8.6 allow 4 ft from HLL to bottom tray.

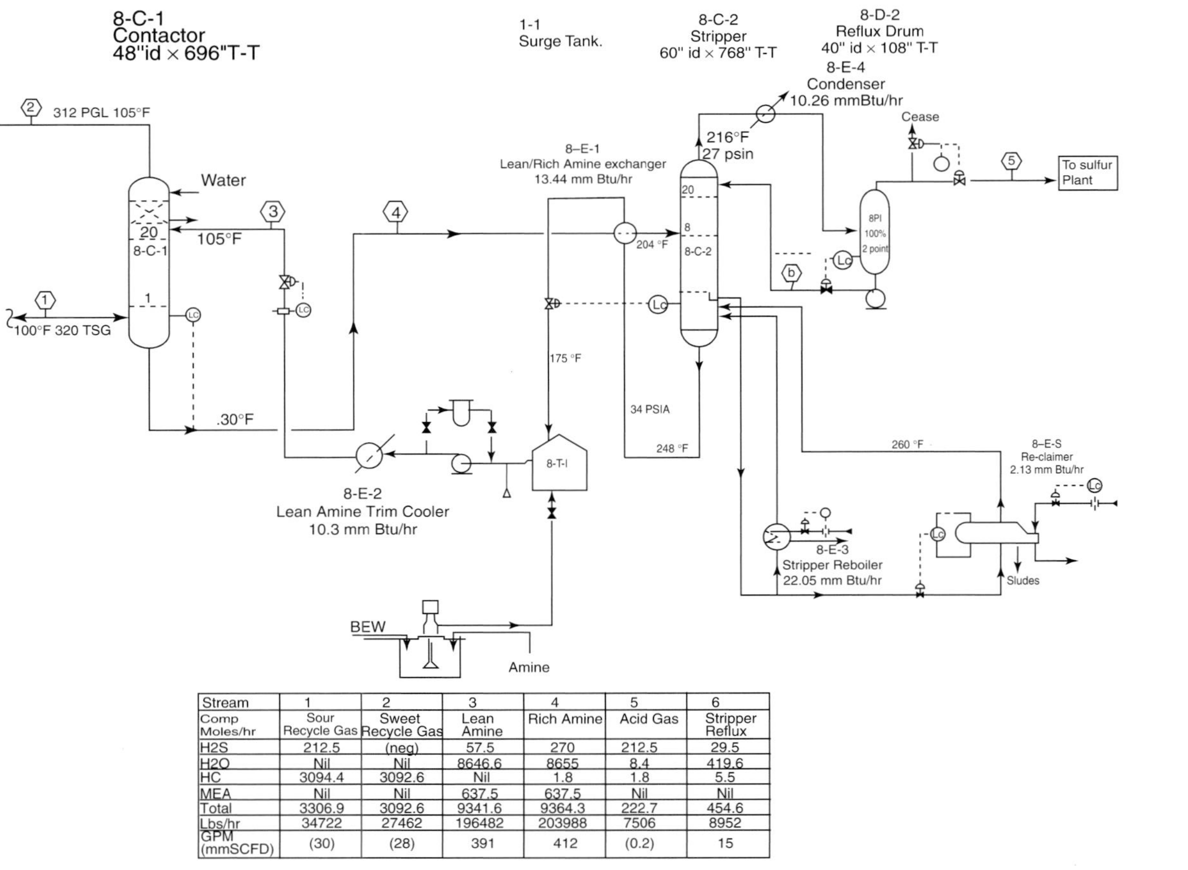

Stream	1	2	3	4	5	6
Comp Moles/hr	Sour Recycle Gas	Sweet Recycle Gas	Lean Amine	Rich Amine	Acid Gas	Stripper Reflux
H2S	212.5	(neg)	57.5	270	212.5	29.5
H2O	Nil	Nil	8646.6	8655	8.4	419.6
HC	3094.4	3092.6	Nil	1.8	1.8	5.5
MEA	Nil	Nil	637.5	637.5	Nil	Nil
Total	3306.9	3092.6	9341.6	9364.3	222.7	454.6
Lbs/hr	34722	27462	196482	203988	7506	8952
GPM (mmSCFD)	(30)	(28)	391	412	(0.2)	15

Figure 10.5. The preliminary process flow sheet for a MEA treating plant.

Summary of Stripper height (all dimensions from bottom tan)

To bottom tray	12 ft 6 ins
Trayed section	47 ft 6 ins
Top tray to top tan	4 ft
Total height	64 ft

There will be a 15 ft skirt to give the total height above grade of 79 ft.

This completes the process design of the major items of equipment. These and the smaller items of equipment are shown in the following preliminary process flow sheet Figure 10.5.

Chapter 11

Upgrading the 'Bottom of the Barrel'

D.S.J. Jones

The highest yield of straight run product as a percentage on crude is the residue from the atmospheric distillation of the crude feed. In most Middle East crudes this ranges from around 40 vol% for the lighter crude oils to 50% and higher in case of the heavier oils. It can be, and often is, the major factor in crude oil slate selection and in the operation of the refinery itself. In most cases a large portion of the residue can be blended off to meet the fuel oil product in the refinery's production plan. In other cases, perhaps equally as common, a major portion of the atmospheric residue is further distilled, under vacuum, to distillates which can be processed to meet gasoline and middle distillate product slates or lube oil blending stocks. The residue from this vacuum distillation, now considerably smaller in volume, is routed to fuel oil or to a bitumen product pool.

There are cases however where the quantities of both atmospheric and vacuum residues are high enough to limit the refinery's throughput or limit its production of the more valuable products and thus limit the refinery's profitability. In such cases, the conversion of these residues becomes attractive, and, in some cases absolutely imperative. This latter case refers to those refineries that have no or a very small fuel oil market. The upgrading of these residues is accomplished by the indirect processing of the atmospheric residue, processing the distillates, by catalytic cracking or hydrocracking and then thermally cracking the vacuum residue. Most of the processes involved with the conversion of the vacuum distillates are described in other chapters of this Handbook dealing with hydrocracking and fluid catalytic cracking.

Direct upgrading of the atmospheric residue by the thermal cracking processes has long been the 'work horse' of the industry in processing the 'Bottom of the Barrel'. However, some catalytic processes have been developed which demonstrate a more efficient and effective method than the thermal cracking routes. This chapter deals with the direct processing of crude oil residuum, and is divided into the following parts:

- Thermal cracking
- 'Deep oil' fluid catalytic cracking
- Residuum hydrocracking and desulfurization

The thermal cracking processes

Thermal cracking processes refer to those that convert the residuum feed (whether atmospheric or vacuum residues) into higher grade products such as naphtha and middle distillates, by heat at high temperature alone. That is, no catalyst or chemicals are used in the conversion. The processes themselves are:

- Visbreaking
- Thermal cracking
- Coking

Certain confusion exists in the definition of visbreaking and thermal cracking. Differentiation is based on the type of feedstock, severity of cracking or the final result. Strictly speaking, the term visbreaking should refer strictly to the viscosity reduction of heavy stock as the process's main objective.

Applications of the thermal cracker processes

In a simple refinery, without vacuum distillation facilities, the residue from the crude oil atmospheric distillation typically boiling above 650°F or 700°F constitutes the bulk of the heavy fuel oil produced. In cases where incremental production of light and middle distillates at the expense of fuel oil is desired, one stage thermal cracking of the residue is an easy and cost effective solution. The residue is cracked in a specially designed heater, the effluent from the heater is quenched and routed to a fractionator, sometimes with a pre-flash and the products of cracking such as light gases, naphtha, gas oil, and residue are separated in the conventional manner. Some 20% of the residue feed can be converted into lighter products, mostly gas oil, by this process. Figure 11.1 shows a typical one stage thermal cracker.

For increased gas oil production, a somewhat more complicated scheme can be applied. The residue feed in this case is first cracked in the heater and the effluent flashed, the hot vapors from the flash drum are routed to a fractionating tower where a heavy gas oil is recovered as the bottom product. This is, in turn, cracked in a second heater and under more severe conditions to yield additional quantities of light distillate products and gas oil. Typically the first, residue, heater is operated at 15–20 psig and a coil outlet temperature of 900°F. The second, gas oil, heater operates at around 250–300 psig and a coil outlet temperature of 930°F. This process is a two stage thermal cracker and is shown as Figure 11.2.

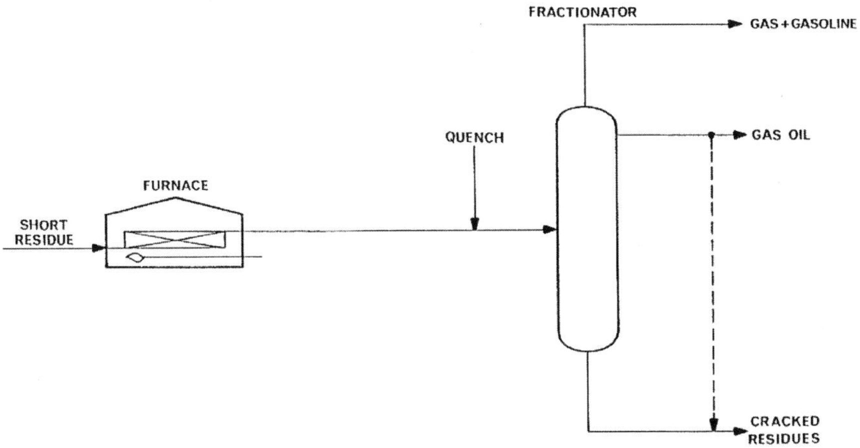

Figure 11.1. One stage thermal cracker.

The visbreaker process

The visbreaker process configuration is very similar to the single stage thermal cracker as shown in Figure 11.1. More often than not though an additional piece of equipment is added immediately after the heater. This is a simple soaking drum which prolongs the time the heater effluent remains at the cracking temperature without being subjected to further heat input and temperature. The objective here is to maintain good fuel oil stability while still converting sufficient of the feed to gas oil and thus lowering the residue viscosity to fuel oil specification. By providing a soaker drum suitable conversion is obtained by residence time at moderate temperature and pressure conditions

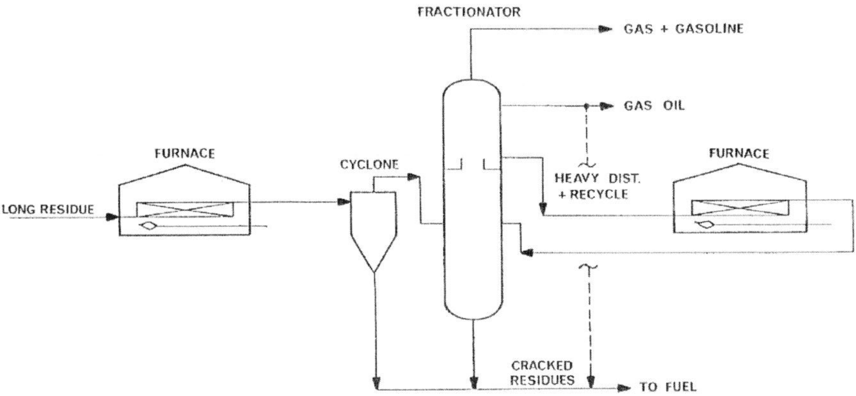

Figure 11.2. Two stage thermal cracker.

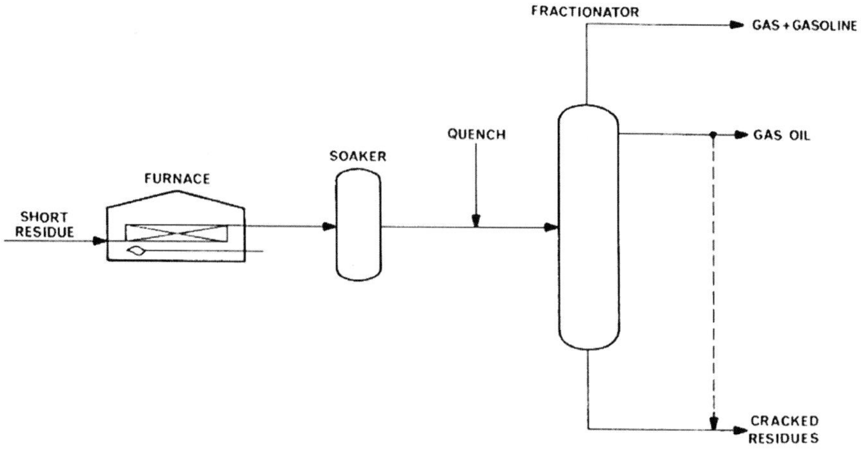

Figure 11.3. A typical visbreaker.

without impairing the resulting fuel oil stability. Visbreaker heater outlet temperature can be as low as 830°F to meet viscosity specification. This process configuration is shown as Figure 11.3.

The coking processes

Coking is the most severe form of thermal cracking, in which all the residue feed is converted into light ends, naphtha, middle distillate, and a solid material, coke. No residual fuel oil is left. This process is particularly useful in handling very heavy crudes such as Bachequero or the Canadian 'Heavy Oil' crudes. A suitable market for the coke must be found, and coke quality requirements may impose further processing such as calcining with, of course, a resulting additional cost.

Two major types of coking processes are in operation today, these are the delayed coking process and the fluid coking process. The intricacies, delicate operation and cost of the conventional fluid process makes it an unlikely contender when simple and cheap ways to convert residues into more valuable products are desired. A fairly recent development of the fluid coking process, however, allows for the conversion of the coke into low Btu gas. Only the delayed coking process and a proprietary process for Fluid coking will be discussed here.

Delayed coking process
The delayed coking process is illustrated by Figure 11.4.

The fuel oil or heavy oil feed is routed to a cracking furnace similar to other thermal cracking processes. The effluent from the furnace is sent to one of a set of several

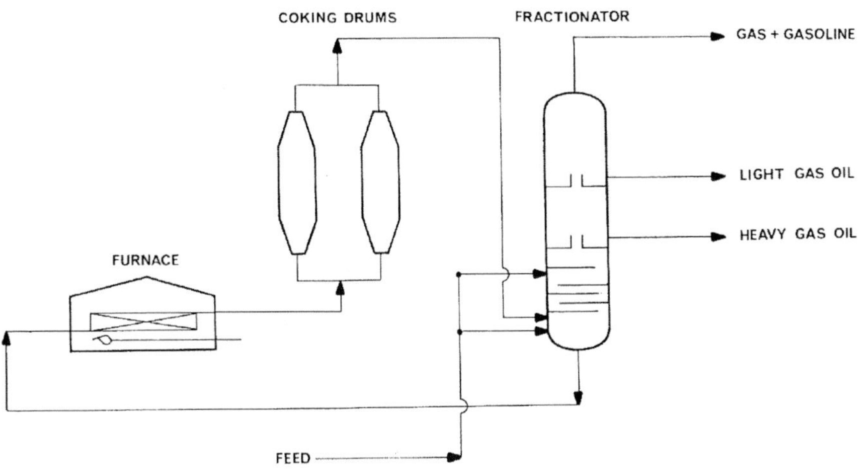

Figure 11.4. A delayed coker.

coking drums without quenching. The effluent is normally at a coil outlet temperature of around 920°F and at a pressure of 30–50 psig and the coking drum is filled in about 24 hr. The vapors leaving the top of the drum whilst filling are routed to the fractionator, where they are fractionated to the distillate products. The very long residence period of the effluent in the coking drums results in the complete destruction of the heavy fractions, and a solid residue rich in carbon is left in the drum as coke.

When the drum is full of coke it is cooled and the top and bottom of the drum are opened. Special high pressure water jets are used to cut and remove the coke through the drum's bottom opening.

The following table gives a rough comparison of yields from the various thermal cracking processes. The two stage thermal cracker as shown in Figure 11.2 is considered in Table 11.1.

Table 11.1. Comparison of yields from thermal cracking processes, wt% on feed

	Thermal cracker	Visbreaker	Delayed coker
Feedstock	Atmos residue	Vacuum residue	Atmos residue
Butane and lighter	4.0	2.5	7.5
C5–330°F naphtha	7.0	4.5	15.5
330–660°F gas oil	26.0	13.0	59.0
Cracked residue	63.0	80.0	18.0 As coke

Flexi-coking (fluid-coking)

Flexi-coking is a thermal conversion process licensed by Exxon Research and Engineering Company. The process itself is an extension of the traditional fluid coking process. The extension allows for the gasification of the major portion of the coke make to produce a low Btu gas. The gasification step follows closely to the concept used in the coal gasification processes.

Figure 11.5 below is a simplified flow diagram of the flexi-coking process.
Heavy residuum feed is introduced into the reactor vessel where it is thermally cracked. The heat for cracking is supplied by a fluidized bed of hot coke transferred to the reactor from the heater vessel. The vapor products of the reaction leave the reactor zone to enter the scrubber section. Fine coke and some of the heavy oil particles are removed from the cracked products in the scrubber zone and returned to mix with the fresh feed entering the reactor. The reactor products subsequently leave the scrubber and are routed to a conventional fractionating facility. Steam is introduced to the bottom of the reactor to maintain a fluid bed of coke and to strip the excess coke leaving the reactor free from entrained oil.

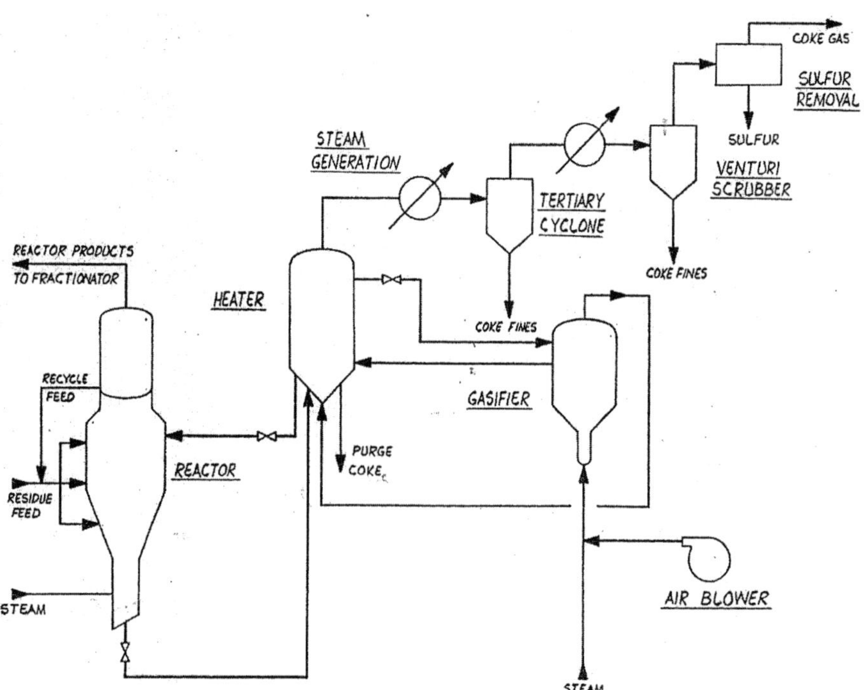

Figure 11.5. The flexi coker process.

Table 11.2. Typical operating conditions and yields from flexi-coking

Feed	
Cut range, °F TBP	+ 1,050
°API	3.0
Sulfur, wt%	6.0
Nitrogen, Wppm	4,800
Condradson carbon, wt%	27.7
Metals, Wppm	269
Yields based on fresh feed	
H_2S, wt%	1.45
C_4s and lighter, wt%	10.29
C_5 to 370°F, LV%	15.0
370–650°F, LV%	16.7
650–975°F, LV%	28.4
Purge coke, wt%	0.69
Coker gas, Scf/Bbl	10.79 (LHV = 127 Btu/Scf)

The coke leaving the reactor enters the heater vessel, where sufficient coke is converted into CO/CO_2 in the presence of air. This conversion of the coke provides the heat for cracking which is subsequently transmitted to the reactor by a hot coke stream. The net coke make leaves the heater and enters the gasifier vessel. Air and steam are introduced into the gasifier to react with the coke producing a low Btu gas consisting predominately of hydrogen, CO, CO_2, and nitrogen. This gas together with some excess air is transferred to the heater, and leaves this vessel to be suitably cleaned and cooled.

Flexi-coking is an extinctive process. By continuous recycle of heavy oil stream all the feed is converted into distillate fractions, refinery gas, and low Btu gas. There is a very small coke purge stream which amounts to about 0.4–0.8 wt% of fresh feed. When suitably hydrotreated the fractionated streams from the flexi-coker provide good quality products. Hydrotreated coker naphtha provides an excellent high-naphthene feed to the catalytic reformer. Typical yields from a flexi-coker based on Arabian residue are shown in Table 11.2.

This process is quite flexible as to the quality of the feedstock, most of the metals are removed with the coke purge. Condradson carbon content of the feed does not affect the yield as may be expected. Where an economic use can be found for the low Btu gas, the process offers an attractive route for upgrading low quality fuel oil.

The principles of and correlations associated with thermal cracking processes

The soaking volume concept

The design of a thermal cracker is keyed to the configuration and temperature profile across the heater and soaking drum or soaking coil. The degree of cracking is dependant on this temperature profile and the residence time of the oil under these conditions.

The thermal cracking reaction is accepted as being of the first order thus it complies with the equation:

$$\text{In (conv)} = (\text{A})e^{-E/RT \times t}$$

where

A, E, R = constants
t = reaction time
T = reaction temperature

The thermal cracking reaction occurs in the heater along a curve of increasing temperature. In this concept of design the cracking progression is expressed by a soaking volume factor (SVF) which is defined by the following equation:

$$F = 1/D \int_0^V \frac{P_s}{P_o} \cdot \frac{K_t}{K_0} dv$$

where

V = coil volume
D = feed rate
K_t = reaction rate constant at any given temperature
$K_o P_o$ = standard reference value for K_t and P_t
P_t = pressure at a given point in coil

The standard reference temperature for thermal cracking is taken as 800°F. A curve giving value of K_t/K_o at a typical thermal cracking heater is given in Figure 11.6. The curve was produced from experimental results using a normal accepted pressure drop profile across the heater. For simplicity the curve is related to temperature versus the factor K_t/K 800°F.

The soaking volume factor

The soaking volume factor (SVF) is related to product yields and the degree of conversion. Definition of these items are:

- *The degree of conversion*
  The relationship of the soaking volume factor to the degree of conversion is given by Figure 11.7. These curves were the result of experimental data from the laboratory cracking of many feedstocks. The family of curves given in Figure 11.7 demonstrates the comparative ease of cracking the large molecular structure of short residue to that of increasing wax distillate content. Conversion is measured by the result of gas and gasoline (to an end point of 257°F) produced.

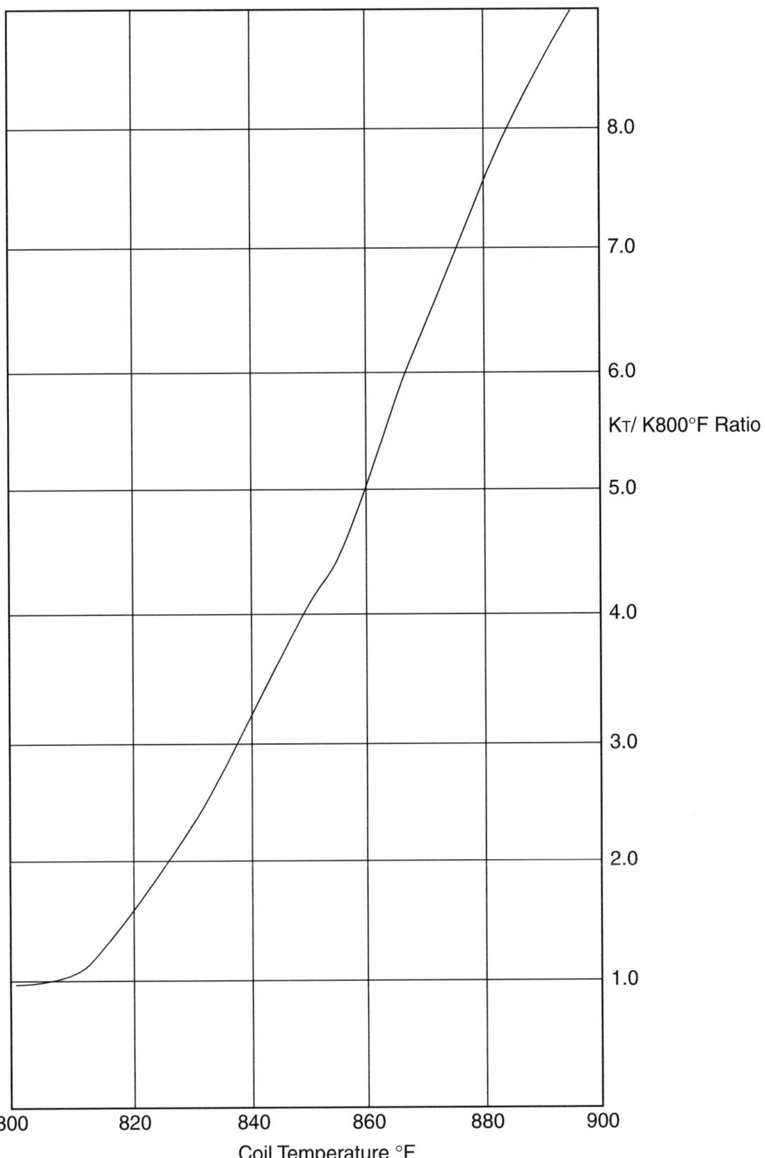

Figure 11.6. Temperature versus the $K_T/\ K_{800°F}$ ratio.

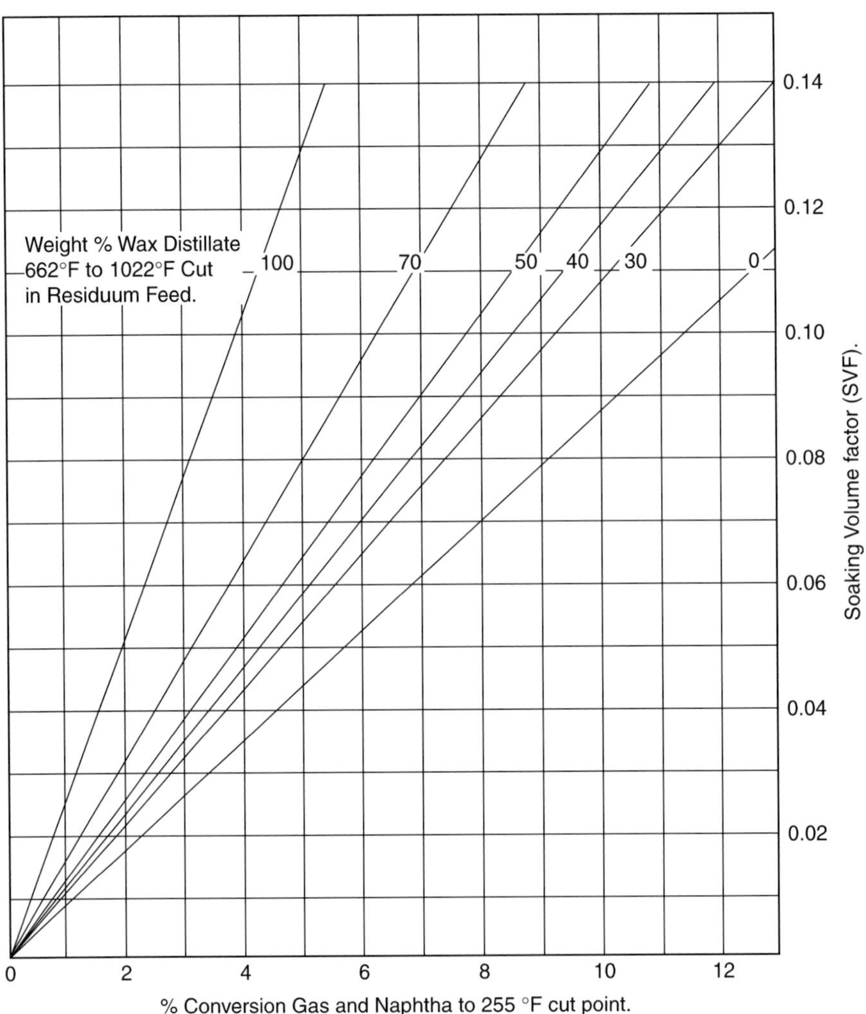

Figure 11.7. Soaking volume factor (SVF) versus wt% conversion.

- *Product yields*
  A family of curves given below as Figures 11.8 and 11.9 shows the relationship of the soaking volume factor (SVF) to the yield of products of thermal cracking. Figure 11.8 shows the yields when cracking wax distillate 662–1,022°F portion of the feed. Figure 11.9 shows the yields when cracking the bitumen portion of the feed. Both series of curves relate to the soaking column factor.

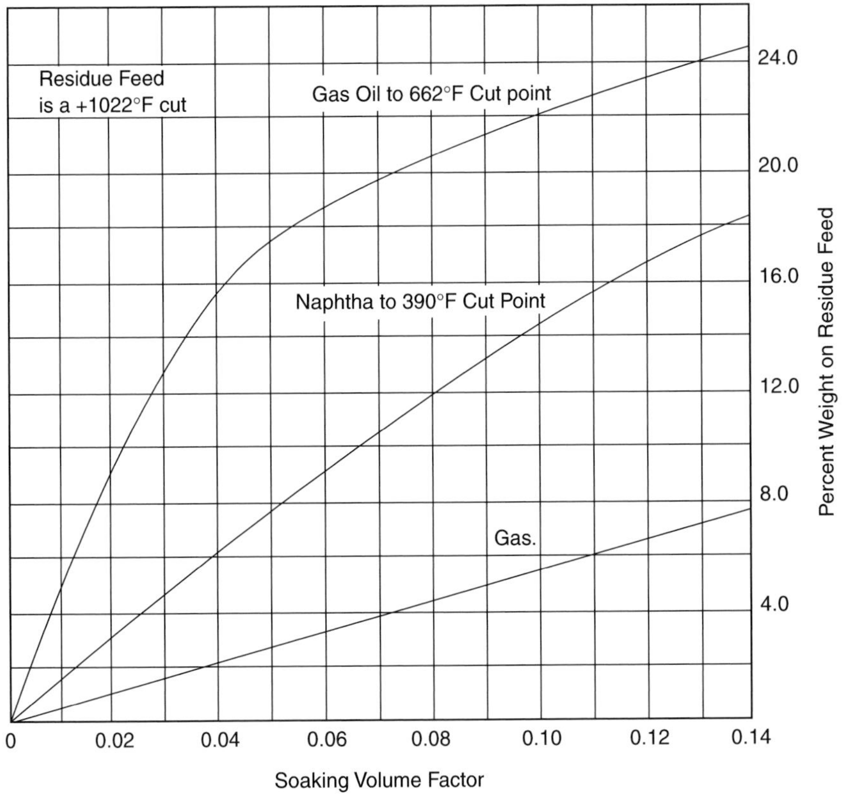

Figure 11.8. Conversion yields from wax distillate versus SVF.

- *The zone of critical decomposition*
  Experimental data shows that cracking and stability of the cracked product varies with the characteristics of the crude source material and boiling range. Figure 11.10 demonstrates this criteria as a relationship with the Watson characterization factor '*K*' of the feed stock.

The shaded area of the figure is the range in which the major portion of cracking occurs. Above this area the cracked residue becomes unstable and precipitates sediment when stored. It is undesirable to operate above this zone in most cases where a cracked residue is required.

Discussion of the concept

This is not a new concept for it has certainly been accepted as a basis for design since the late 1950s. To date there are at least six visbreakers and/or thermal crackers to the author's knowledge that were designed using these parameters and are still in commercial use. In evaluating this concept, it is interesting to note the following test run data compared with that calculated using SVF concept (Table 11.3).

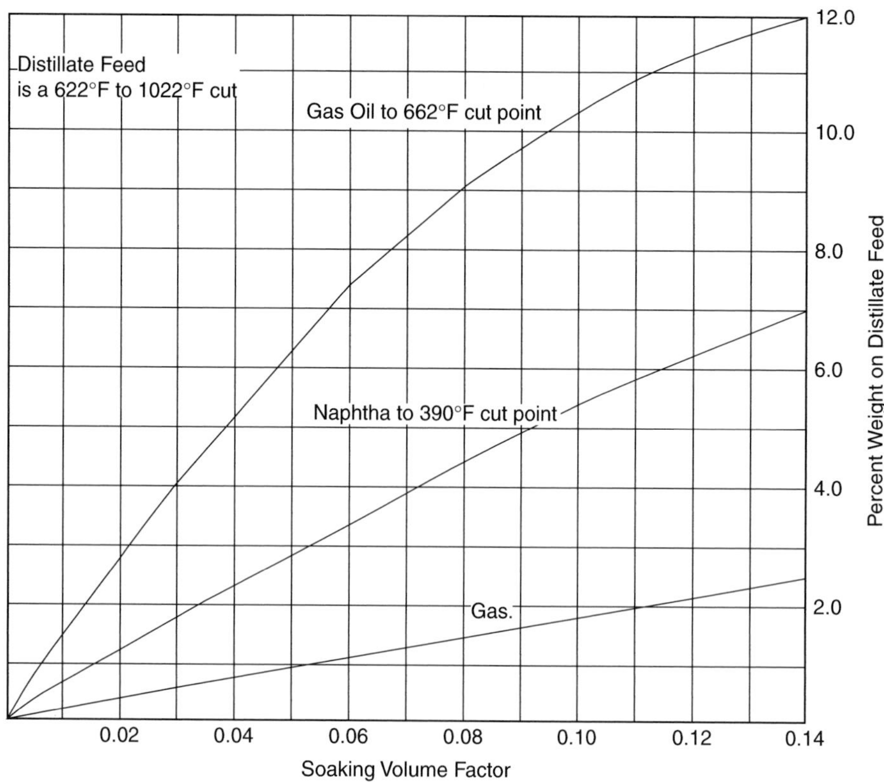

Figure 11.9. Conversion yields from residuum versus SVF.

It can be seen that the data calculated compare well with those actually observed under test conditions. One interesting point in the test run data is the high coil outlet temperature that was used. As this is slightly above the critical decomposition zone for this type of crude it would lend one to suspect that the cracked residue would be unstable. There was no recorded evidence of this however.

An example calculation using the SVF concept is given as Appendix 11.1 of this chapter.

'Deep oil' fluid catalytic cracking

Up to the late 1980s feedstocks to FCCU were limited by characteristics such as high Conradson carbon and metals. This excluded the processing of the "bottom of

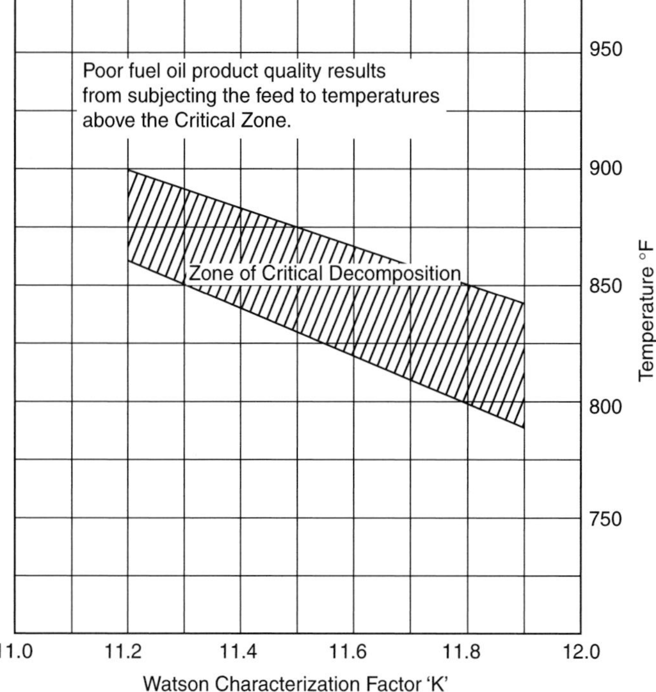

Figure 11.10. Zone of critical decomposition.

the barrel" residues. Indeed, even the processing of vacuum gas oil feeds was limited to

Conradson carbon < 10 wt%
hydrogen content > 11.2 wt%
metals Ni + V < 50 ppm

During the late 1980s significant research and development breakthroughs produced a catalytic process that can handle residuum feed.

Feed stocks heavier than vacuum gas oil when in conventional FCCU tend to increase the production of coke and this in turn deactivates the catalyst. This is mainly the result of:

- A high portion of the feed that does not vaporize. The unvaporized portion quickly cokes on the catalyst choking its active area.

Table 11.3. Comparison of test run data calculated using SVF method

	Test run	Calculated
Feed stock	Kuwait residue + 350°C	
Conversion wt%	11.07	10.5
Feed rate	11 500	12 000
Heater inlet press PSIG	512	Not calculated
Outlet press PSIG	252	Not calculated
Conversion inlet °F	548	590
Conversion outlet °F	651	655
Heater outlet °F	880	860
Soaking drum outlet press PSIG	242	250
Inlet temperature °F	870	850
Outlet temperature °F	849	830
Yields		
Gas wt%	4.4	4.58
Total distillate wt%*	33.59	32.43
Residue wt%	62.01	62.99

*This is total distillate C_5 to 350°C EP.

- The presence of high concentrations of polar molecules such as poly cyclic aromatics and nitrogen compounds. These are absorbed into the catalyst's active area causing instant (but temporary) deactivation.
- Heavy metals that poison the catalyst and affect the selectivity of the cracking process.
- High concentration of poly-naphthenes that dealkylate slowly.

In FCCU that process conventional feedstocks cracking temperature is controlled by the circulation of hot regenerated catalyst. With the heavier feedstocks, with an increase in Conradson carbon, there will be a larger coke formation. This in turn produces a high regenerated catalyst temperature and heat load. To maintain heat balance therefore catalyst circulation is reduced leading to poor or unsatisfactory performance. Catalyst cooling or feed cooling is used to overcome this high catalyst heat load and to maintain proper circulation.

The extended boiling range of the feed, as in the case of residues, tends to cause an uneven cracking severity. The lighter molecules in the feed are instantly vaporized on contact with the hot catalyst, and cracking occurs. In the case of the heavier molecules vaporization is not achieved so easily. This contributes to a higher coke deposition with a higher rate of catalyst deactivation. Ideally, the whole feed should be instantly vaporized so that uniform cracking mechanism can commence. The *mix temperature* (which is defined as the theoretical equilibrium temperature between the uncracked vaporized feed and the regenerated catalyst) should be close to the feed dew point temperature. In conventional units this is about 20–30°C above the riser outlet temperature. This can be approximated by the expression

$$T_M = T_R + 0.1\Delta H_C$$

where

T_M = the mix temperature
T_R = riser outlet temperature °C
ΔH_C = heat of cracking in kJ/kg

This mix temperature is also slightly dependent on the catalyst temperature.

Cracking severity is affected by poly cyclic aromatics and nitrogen. This is so because these compounds tend to be absorbed into the catalyst. Rising the mix temperature by increasing the riser temperature reverses the absorption process. Unfortunately, a higher riser temperature leads to undesirable thermal cracking and production of dry gas.

The processing of residual feedstocks therefore requires special techniques to overcome:

- Feed vaporization
- High concentration of polar molecules
- Presence of metals

Some of the techniques developed to meet heavy oil cracking processing are as follows:

- Two stage regeneration
- Riser mixer design and mix temperature control (MTC) (for rapid vaporization)
- New riser lift technology minimizing the use of steam
- Regenerated catalyst temperature control (catalyst cooling)
- Catalyst selection for:
  - ➢ Good conversion and yield pattern
  - ➢ Metal resistance
  - ➢ Thermal and hydro-thermal resistance
  - ➢ High gasoline RON

These are discussed in the following items.

Conventional fluid catalytic crackers can be revamped to incorporate the features necessary for heavy oil (residual) cracking. World wide, there were approximately some 23 FCC residue units in operation (to the beginning of 1992), having a total processing capacity of 760,000 BD.

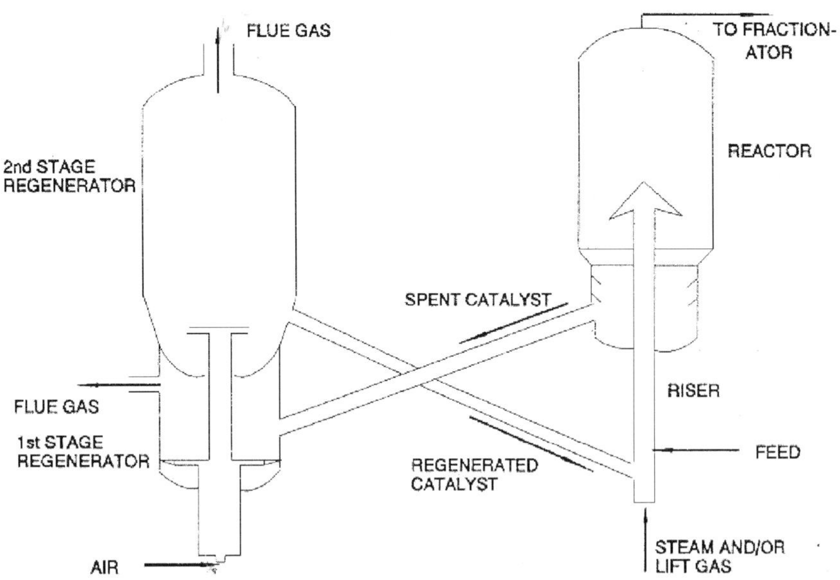

Figure 11.11. Two stage catalyst regeneration.

Two stage regeneration and regenerated catalyst temperature control

An important issue in the case of deep oil (residue) cracking is the handling of the high coke lay down and the protection of the catalyst. One technique that limits the severe conditions in regeneration of the spent catalyst is the two stage regeneration. Figure 11.11 shows the layout of such a regenerator.

The spent catalyst from the reactor is delivered to the first regeneration. Here the catalyst undergoes a mild oxidation with a limited amount of air. Temperatures in this regeneration remain fairly low around 700–750°C range. From this first regeneration the catalyst is pneumatically conveyed to a second regenerator. Here excess air is used to complete the carbon burn off and temperatures up to 900°C are experienced. The regenerated catalyst leaves this second regeneration to return to the reactor via the riser.

The technology that applies to the two stage regeneration process is innovative in that it achieves the burning off of the high coke without impairing the catalyst activity. In the first stage the conditions encourage the combustion of most of the hydrogen associated with the coke. A significant amount of the carbon is also burned off under mild condition. These conditions inhibit catalyst deactivation.

All the residual coke is burned off in the second stage with excess air and in a dry atmosphere. All the steam associated with hydrogen combustion and carry over

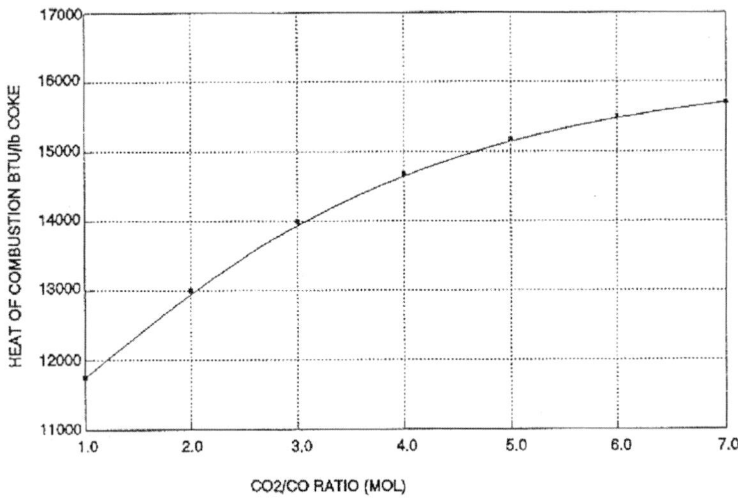

Figure 11.12. Heat of combustion versus CO_2/CO ratio.

from the reactor has been dispensed within the first stage. The second regenerator is refractory lined and there is no temperature constraint. The catalyst is allowed to come to equilibrium. Even at high regenerator temperatures under these conditions lower catalyst deactivation is experienced. The two stage regeneration technique leads to a better catalyst regeneration as well as a lower catalyst consumption. Typically the clean catalyst contains less than 0.05 wt% of carbon. This is achieved with an overall lower heat of combustion. See Figures 11.12 and 11.13.

Since the unit remains in heat balance coke production stays essentially the same. The circulation rate of catalyst adjusts itself to any changes in coke deposition on the catalyst according to the expressions:

$$\text{coke make} = \text{delta coke} + \text{c/o}$$

and

$$\text{regenerator temperature} = \text{riser temperature} + C \times \text{delta coke}$$

where

delta coke = difference between the weight fraction of coke on the catalyst before and after regeneration.
C = unit constant (typically 180–230)
c/o = catalyst to oil ratio

In this regard a small circulation of extremely hot catalyst may not be as effective as a large circulation of cooler catalyst. It has been found that there is a specific catalyst temperature range that is desirable for a given feed and catalyst system. A

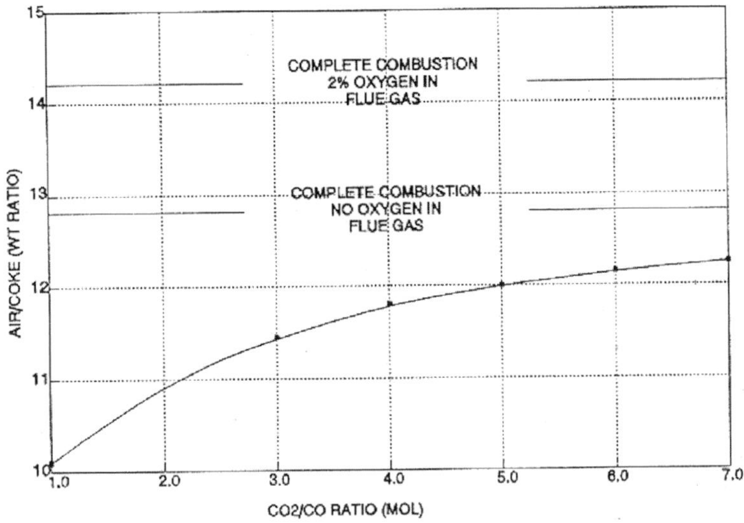

Figure 11.13. Combustion air requirement versus CO_2/CO ratio in flue gas.

unique dense phase catalyst cooling system provides a technique through which the best temperature and heat balance relationship can be maintained.

Consider the enthalpy requirements for a FCC reactor given in Table 11.4.

It can be seen from this table that 69% of the enthalpy contained in the heat input to the reactor is required just to heat and vaporize the feed. The remainder is essentially available for conversion. To improve conversion it would be very desirable to allow more of the heat available to be used for conversion. The only variable that can be changed to achieve this requirement is the feed inlet enthalpy. That is through preheating the feed. Doing this, however, immediately reduces the catalyst circulation rate to maintain heat balance. This of course has an adverse effect on conversion. The

Table 11.4. Enthalpy requirements for fluid cracking reactor

	Per pound of feed	
	Btu	%
Feed heating/vaporizing	530.0	69.00
Stripping steam enthalpy	5.0	0.65
Feed steam for dispersion	12.7	1.65
Feed water for heat balance	18.4	2.40
Heat of reaction	200.0	26.04
Heat loss	2.0	0.26
Total	768.1	100.00

Table 11.5. Effect of catalyst cooling on reactor yields

Feedstock		
°API	24.5	
Conradson carbon	1.6	
Yields:	Without catalyst cooling	With catalyst cooling
H_2S wt%	0.1	0.19
C_2- wt%	3.4	2.00
C_3 LV%	9.9	10.34
C_4 LV%	13.9	14.51
C_5+ (430:EP) LV%	58.2	60.87
LCO (650:EP) LV%	17.1	15.54
CLO LV%	8.6	8.10
Coke wt%	5.9	6.07
Conversion LV%	74.3	76.36

preheating of the feed can, however, be compensated for by cooling the catalyst. Thus, the catalyst circulation rate can be retained and in many cases can be increased. Indeed, by careful manipulation of the heat balance the net increase in catalyst circulation rate can be as high as 1 unit cat/oil ratio. The higher equilibrium catalyst activity possible at the lower regeneration temperature also improves the unit yield pattern. This is demonstrated in Table 11.5.

In summary, catalyst cooling will:

- slightly increase unit coke
- give a higher plant catalyst activity
- be able to handle more contaminated feeds
- improve conversion and unit yield
- provide better operating flexibility

In resid cracking commercial experience indicates that operations at regenerated catalyst temperatures above 1,350°F result in poor yields with high gas production. Where certain operations require high regen temperatures the installation of a catalyst cooler will have a substantial economic incentive. This will be due to improved yields and catalyst consumption.

There are two types of catalyst coolers available. These are:

- the back mix type
- flow through type

These are shown in Figure 11.14. Both coolers are installed into the dense phase section of the regenerator.

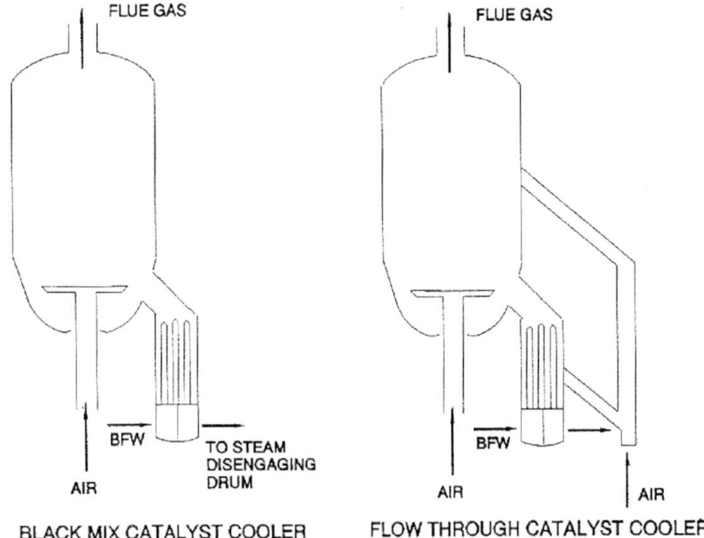

Figure 11.14. Typical catalyst coolers.

The back mix cooler. Boiler feed water flows tube side in both cooler types. The catalyst in the back mix cooler circulates around the tube bundle on the shell side. The heat transfer takes place in a dense low velocity region so erosion is minimized. The back mix cooler can remove approximately 50 million Btus/hr.

The flow through cooler. As the name suggests the catalyst flows once through on the shell side of this cooler. Again erosion is minimized by low velocity operation in the dense phase. This type of cooler is more efficient than the backmix. This unit can achieve heat removal as high as 100 million Btus per hr.

Mix temperature control and lift gas technology

The equilibrium temperature between the oil feed and the regenerated catalyst must be reached in the shortest possible time. This is required in order to ensure the rapid and homogeneous vaporization of the feed. To ensure this it is necessary to design and install a proper feed injection system. This system should ensure that any catalyst back mixing is eliminated. It should also ensure that all the vaporized feed components are subject to the same cracking severity.

Efficient mixing of the feed finely atomized in small droplets is achieved by contact with a pre-accelerated dilute suspension of the regenerated catalyst. Under these conditions feed vaporization takes place almost instantaneously. This configuration is shown in the diagram below:

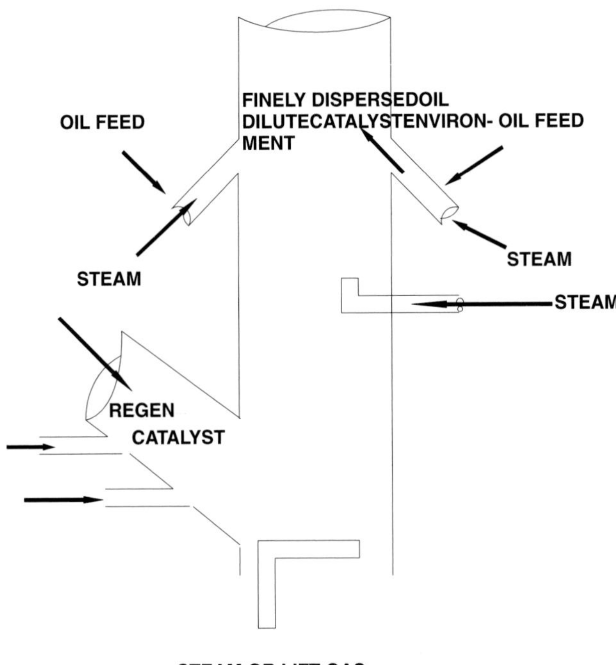

The regenerated catalyst stream from the regenerator is accelerated by steam or lift gas injection to move up the riser. The oil feed is introduced atomized by steam into the catalyst environment. The main motive steam into the riser is introduced below the feed inlet point. Good mixing occurs in this section with maximum contact between oil, catalyst and the steam.

In residue cracking the proper selection of catalyst enables even the most bulky molecules to reach the active catalyst zone. Catalysts such as zeolites have a high silica to alumina ratio which cracks the heavy molecules into sizes that can enter the active zone.

Efficient mixing of the catalyst and feed together with the catalyst selection ensures:

- Rapid vaporization of the oil
- Uniform cracking severity of the oil

Another problem that is met within residue cracking is the possibility of the heavier portion of the oil being below its dew point. To ensure this is overcome the "mix temperature" (see previous item) must be set above the dew point of the feed. As stated earlier the presence of poly cyclic aromatics also affects cracking severity. Increasing the mix temperature to raise the riser temperature reverses the effect of poly cyclic aromatics. In so doing, however, thermal cracking occurs which is undesirable. To solve this problem it is necessary to be able to control riser temperature independently of mix temperature.

Mix temperature control (MTC) is achieved by injecting a suitable heavy cycle oil stream into the riser above the oil feed injection point. This essentially separates the riser into two reaction zones. The first is between the feed injection and the cycle oil inlet. This zone is characterized by a high mix temperature, a high catalyst to oil ratio, and a very short contact time.

The second zone above the cycle oil inlet operates under more conventional catalytic cracking conditions. The riser temperature is maintained independently by the introduction of the regenerated catalyst. Thus increase in cycle oil leads to decrease in riser temperature which introduces more catalyst, this finally increases the mix temperature, and the catalyst to oil ratio and decreases the regenerator temperature.

The lift gas technology

As described earlier it is highly desirable to achieve good catalyst/oil mixing as early and as quickly as possible. The method described to achieve this requires the preacceleration and dilution of the catalyst stream. Traditionally steam is the medium used to maintain catalyst bed fluidity and movement in the riser. Steam, however, has deleterious effect on the very hot catalyst that is met in residue cracking processes. Steam under these conditions causes hydro-thermal deactivation of the catalyst.

Much work has been done in reducing the use of steam in contact with the hot catalyst. Some of the results of the work showed that if the partial pressure of steam is kept low, the hydro-thermal effects are greatly reduced, in the case of relatively metal free catalyst. A more important result of the work showed that light hydrocarbons imparted favorable conditioning effects to the freshly regenerated catalyst. This was even pronounced in catalysts that were heavily contaminated with metals.

Light hydrocarbon gases have been introduced in several heavy oil crackers since 1985. They have operated either with lift gas alone or mixed with steam. The limitations to the use of lift gas rests in the ability of downstream units to handle the additional gas. Table 11.6 compares the effect of lift gas in residual operation with the use of steam.

As can be seen the use of lift gas as an alternative to steam gives:

- Lower hydrogen production
- Lower hydrogen/methane ratio
- Increase in liquid yield

Table 11.6. Comparison of the effect of 'lift gas' vs. steam
Feed: Atmospheric residue 4.3 wt% con carb

Product distribution:	Lift gas	Steam
C_2- wt%	3.2	4.0
C_3/C_4 LV%	11.4/15.1	11.6/15.4
C_5- gasoline LV%	56.9	55.0
LCO + slurry LV%	23.9	24.4
Total C_3 + LV%	107.3	106.4
Coke wt%	8.6	8.5
H_2 SCFB	70	89
H_2/C_1 mol	0.74	0.85
Catalyst	Same	
SA, m_2/g*	91	90
Ni + v wt ppm	7,100	7,300

*SA—surface area of equilibrium FCC catalyst, m^2/g.

Residuum hydrocracking

Residuum hydrocracking is often combined with a thermal cracker to upgrade the 'bottom of the barrel' and provide high quality light and middle distillates. Figure 11.15 shows such a configuration.

Bitumen feed from the crude vacuum distillation unit enters the hydrocracker section of the plant to be preheated by hot flash vapors in shell and tube exchangers and finally in a fired heater. A recycle and make up hydrogen stream is similarly heated by exchange with hot flash vapors. The hydrogen stream is mixed with the hot bitumen stream before entering the hydrocracker heater. The feed streams are risen to the reactor temperature in the heater and leave to enter the top of the reactor vessel. The feed streams flow downwards through the catalyst beds contained in the reactor. Additional cold hydrogen is injected at various sections of the reactor to provide temperature control as the hydrocracking process is exothermal.

The reactor effluent leaves the reactor to enter a hot flash drum. Here the heavy bituminous portion of the effluent leaves from the bottom of the drum while the lighter oil and gas phase leaves as a vapor from the top of the drum. This vapor is subsequently cooled by heat exchange with the feed and further cooled and partially condensed by an air cooler. This cooled stream then enters a cold separator operating at a pressure only slightly lower than that of the reactor. A rich hydrogen gas stream is removed from this drum to be amine treated and returned as recycle gas to the process. The distillate liquid leaves from the bottom of the separator to join a vapor stream from the hot flash surge drum (thermal cracker feed surge drum). Both these

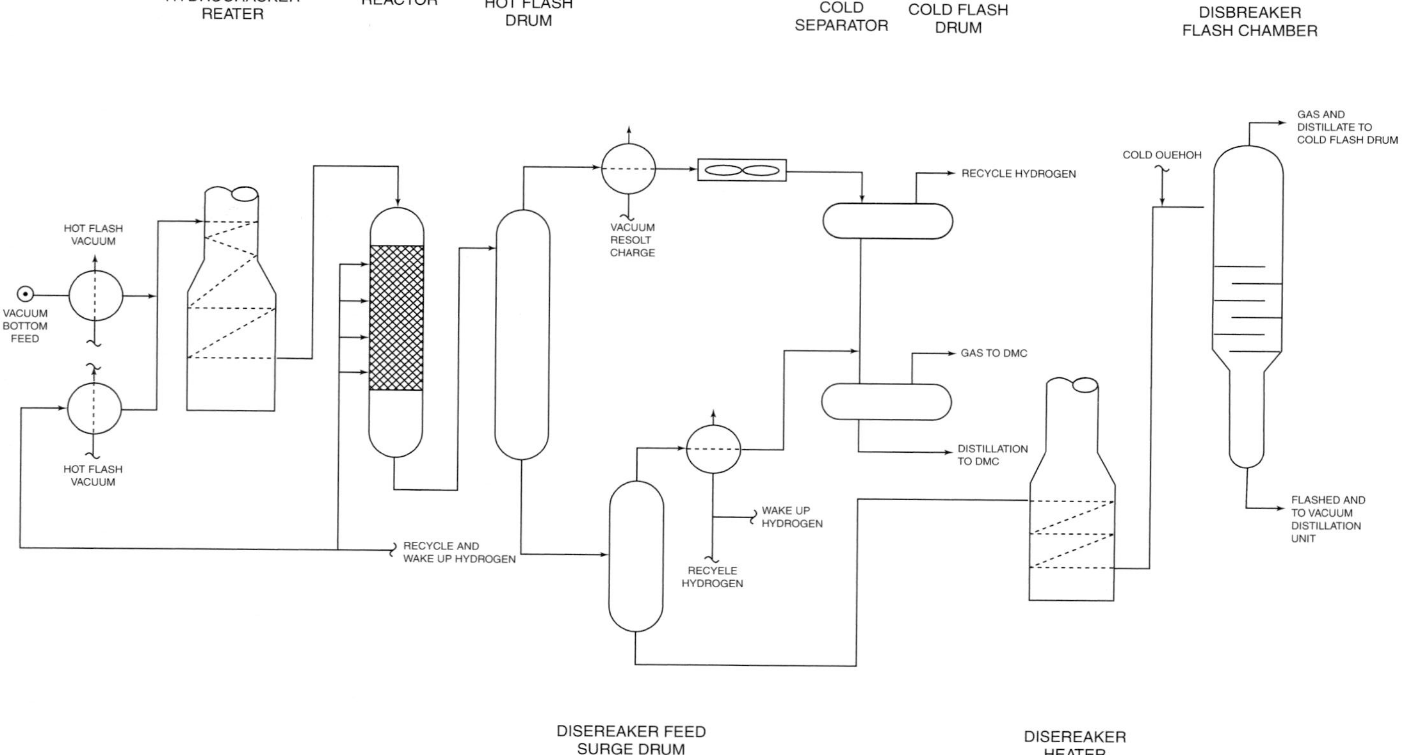

Figure 11.15. Typical residuum hydrocracking configuration.

streams enter the cold flash drum which operates at a much lower pressure than the upstream equipment. A gas stream is removed from the drum to be routed to an absorber unit. The liquid distillate from the drum is routed to the debutanizer in a light ends recovery unit.

The thermal cracking section of the unit takes as feed the heavy bituminous liquid from the hot flash drum. This enters the cracker heater via a surge drum. This heater has two parallel coils. The oil feed enters these coils to be thermally cracked to form some lighter products. The stream leaving the heater is quenched before entering a flash chamber. This vessel contains some baffled trays and a light gas and oil vapor stream leaves overhead. This stream is subsequently cooled and the distillate formed routed to the cold flash drum. The bottoms from the flash chamber is fed to the thermal cracker vacuum distillation unit where vacuum gas oil is removed as feed to a FCCU or to a lube oil refining facilities.

Hydrocracking yields, and product properties

The following data illustrate the yield and operating conditions for a fixed bed residue hydrocracker. Both the start of run (SOR) and end of run data are shown. These were recorded during a test run on a Middle East vacuum residue feed. This hydrocracker contained a guard reactor which essentially converted most of the nitrogen and sulfur content of the feed to ammonia and hydrogen sulfide, respectively (Table 11.7).

The hot flash liquid is subsequently further cracked in a thermal cracker or can be vacuum distilled to obtain vacuum distillates to be fed to a FCCU or a distillate hydrocracker. A typical TBP curve for this liquid is given below as Figure 11.16.

Effect of heavy metals on the catalyst

Metals such as Vanadium, Nickel, and Sodium seriously reduce the life of the catalyst in most residue fixed bed reactors. This fact makes the process less competitive to other residue upgrading ones. The addition of a front-end guard reactor does help to prolong the catalyst life by removing some of these metal contaminants. The catalyst in this guard reactor does become poisoned also, but this catalyst usually Cobalt Molybdenum, is considerably less expensive than that used for hydrocracking.

The most effective solution to date however is to extract the very heavy ends of the residue. This is the asphalt portion and almost all the metals are contained in these asphaltene molecules. The extraction of these asphaltenes is accomplished by the counter current flow of propane as a liquid. The heavy asphalt is then routed to refinery fuel or blended into the asphalt pool for marketing.

Table 11.7. Conditions and yield data on hydrocracking M.E. residue

	SOR	EOR
Reactor conditions		
Pressure, psia	3,000	3,000
Inlet temperature, °F	700	750
Outlet temperature, °F	750	800
Space velocity, V/V/hr	0.5	0.4
Feed		
Vacuum residue	+1,025°F TBP cut point	
BPSD	13,000	13,000
lbs/hr	190,266	190,266
Make up hydrogen		
Scf/Bbl	1,290	1,875
lbs/hr	4,010	5,606
Products		
Ammonia		
lbs/hr	209	203
M wt	17	
H_2S		
lbs/hr	6,964	6,680
M wt	34	
Net separator gas		
lbs/hr	Nil	2,405
M wt		4.3
Cold flash gas		
lbs/hr	3,772	5,569
M wt	9.6	
Cold flash liquid		
lbs/hr	22,586	39,886
lBPSD	1,852	3,321
Hot flash liquid		
lbs/hr	160,744	141,129
BPSD	11,672	10,280

Conclusion

Catalytic cracking of residues described above is a poor competitor to the other direct processes for residue upgrading. The process itself is costly to build and has relatively high operating and maintenance costs. The products however are good quality, and require very little downstream treatment. The naphtha produced is (as in the case of distillate hydrocracking) high in naphthene content ideal for gasoline production or as feed to an aromatic production process. The middle distillates also are of high quality with the kerosene fraction having very low smoke point and the gas oil fraction having a good cetane number. In the process illustration given as Figure 11.15 the light distillate liquid from the cold flash drum and the associated vapor are routed to a distillate hydrocracker recovery side for further processing and fractionation.

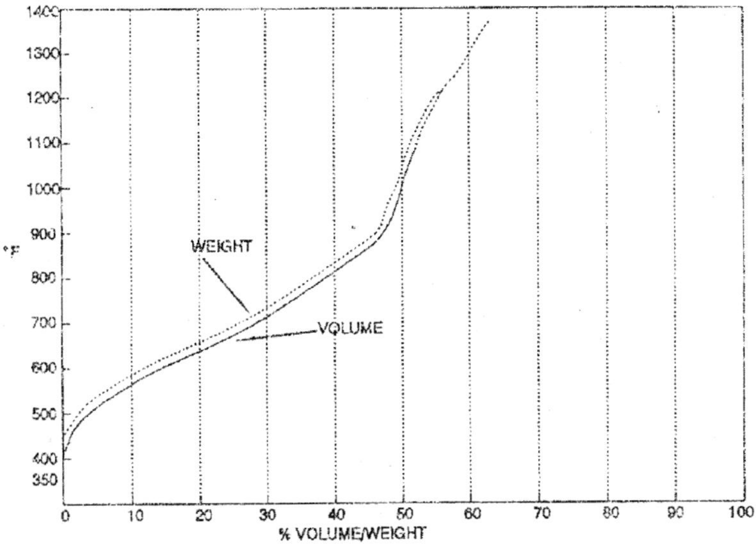

Figure 11.16. A typical TBP curve for 'hot flash' liquid.

These recovery units consist of a main fractionation tower, and alight ends unit which includes an absorber column.

Appendix 11.1: Sizing a thermal cracker heater/reactor

In this example it is required to define a thermal cracker in terms of coil volume and temperature profile in processing an atmospheric residue from Sasson Crude. The 25,500 BPSD of the +600°F residue is preheated to 500°F by heat exchange with the cracker's products and reflux stream before entering the convection side of the cracker's heater/ reactor. It is required to produce a conversion (based on gas to naphtha of 260°F cut point) of 9 wt% and the heater will be fitted with 4″ schedule 80 tubes throughout. It will be designed to have three sections which are:

- The convection section with a heat flux of 12,000 Btu/hr sqft.
- The radiant heater section with a heat flux of 15,000 Btu/hr sqft.
- The soaker section with a heat flux of 10,000 Btu/hr sqft.

The heater section and the soaker section are divided by a fire wall. A predicted temperature profile is given in the diagram below as Figure 11.17.

The salient temperature and pressure conditions are as follows:

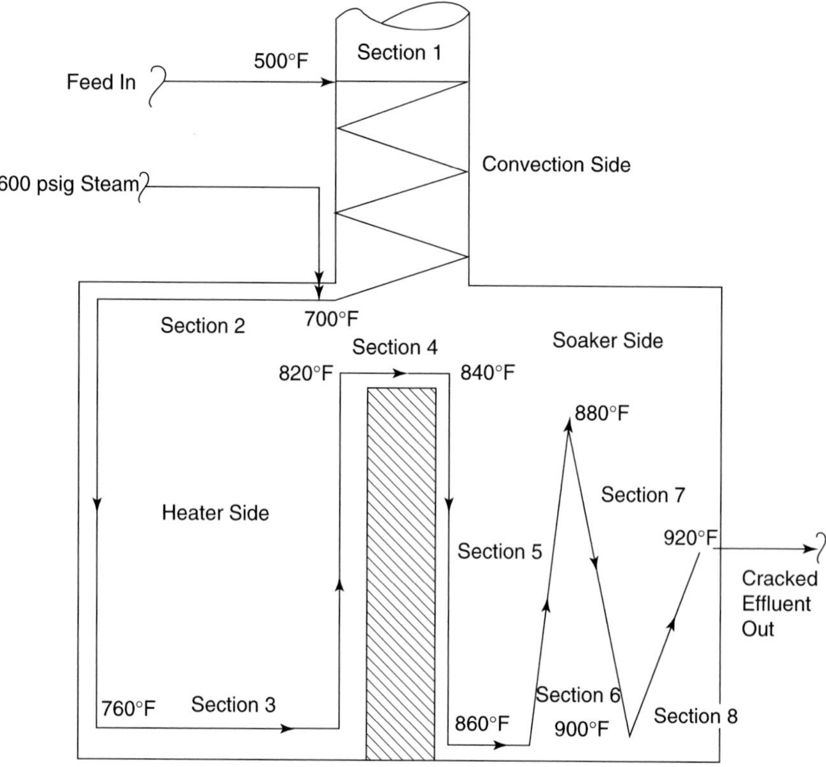

Figure 11.17. Temperature profile across the thermal cracker heater.

Convection inlet = 500°F at 330 psig pressure
Heater inlet = 700°F at 320 psig
Soaker inlet = 820°F at 290 psig
Soaker outlet = 920°F at 250 psig

A 600 psig steam is introduced into the heater inlet coil. This will be at 10 wt% of the residue feed. Details of the residue feed are as follows (a TBP curve of the feed is given in Figure 11.18):

Feed gravity = 18°API (7.88 lbs/gal).
Feed rate = 25,500 BPSD = 351,645 lb/h.
Volume of waxy distillate in the feed to 1,022°F cut point = 68 vol%
= 17,340 BPSD

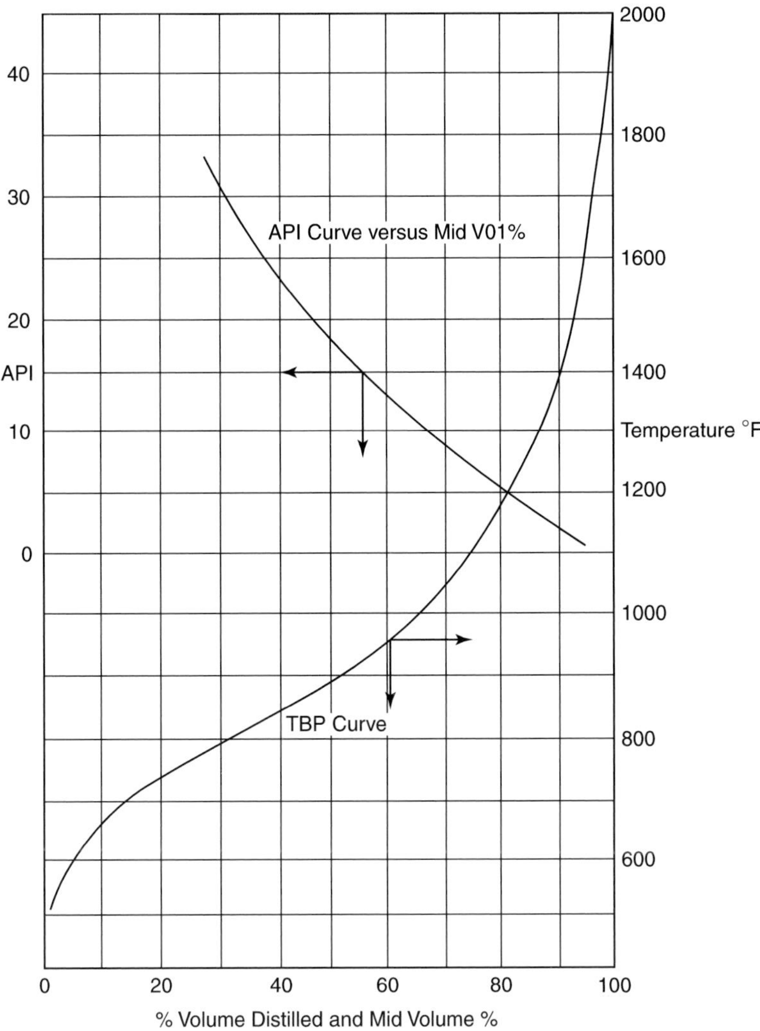

Figure 11.18. The TBP curve for the sassan crude residue feed.

Gravity of the distillate = 26°API (7.48 lbs/gal)
Weight of distillate = 17,340 × 7.48 × 1.75 = 226,981 lbs/hr
Percent weight distillate on feed = 65%

Referring to Figure 11.7 in this chapter the soaking volume factor (SVF) corresponding to a 9% conversion with a distillate content of 65% is 0.135.

The heater coil is divided into the following sections:

		Temp in, °F	Temp out, °F
Section 1	Convection side	500	700
2	Heater side	700	760
3	"	760	820
4	Soaker side	820	840
5	"	840	860
6	"	860	880
7	"	880	900
8	"	900	920

Heat balance over the convection side is as follows:

1st Section
Pressure = 365 psia

	V/L	°F	°API	lbs/hr	Btu/lb	mmBtu/hr
IN						
Feed	L	500	15	351,645	250	87.911
Heater duty				By diff		47.824
Total In						135.735
OUT						
Feed	L	700	15	351,645	386	135.735
Total out						135.735

$$\text{Sqft of coil} = \frac{47{,}824{,}000}{12{,}000} = 3{,}985.3$$

2nd Section
Pressure = 335 psia
600 psig steam is introduced into this section of the furnace.

	V/L	°F	°API	lbs/hr	Btu/lb	mmBtu/hr
IN						
Feed	L	700	15	351,645	386	135.735
Steam	V	700		35,165	1,383	48.633
Heater duty				By diff		16.652
Total in						201.020
OUT						
Steam	V			35,165	1,417	49.829
Feed	L	760	15	351,645	432	151.191
Total out						201.020

3rd Section
Pressure = 315 psia

	V/L	°F	°API	lbs/hr	Btu/lb	mmBtu/hr
IN						
Feed	L	760	15	351,645	386	151.191
Steam	V	760		35,165	1,417	49.829
Heater duty				By diff		22.584
Total in						223.604
OUT						
Liq feed	L	820	20	344,612	482	165.103
Vap	V	820	73	7,033	425	2.989
Steam	V	820		35,165	1,440	50.638
Ht of crack					547	3.874
Total out				386,810		223.604

Cracking begins at a temperature of 800°F and the oil feed and steam enters the soaking section at 820°. The purpose of the soaking section is to provide a space for the cracking function to occur at a moderate increase in temperature. To calculate the required coil volume the first step is to assign the degree of cracking that occurs at the end of each section. This is a trial and error process and provides an SVF value to each section which in turn is used to calculate the amount of cracked products leaving each section. Using these amounts the heat balances for each section of the soaker coil are calculated. Thus:

Final trial

Percent crack at section outlet

	% Conversion	SVF (from Figure 11.7)
Section 4	3.0	0.045
Section 5	4.0	0.059
Section 6	6.0	0.090
Section 7	8.0	0.120
Section 8	10.0	0.150

Material compositions (Figures 11.8 and 11.9).

Section 4 temp 840°F

	Gas		Naphtha		Gas Oil		Residue	
	wt%	lbs/hr	wt%	lbs/hr	wt%	lbs/hr	wt%	lbs/hr
Distillate	0.9	2,043	2.5	5,675	5.9	13,392	–	205,871
Residue	2.2	2,743	6.6	8,228	14.9	18,575	–	95,118
Total		4,786		13,903		31,967		300,989

Section 5 temp 860°F

	Gas		Naphtha		Gas Oil		Residue	
	%wt	lbs/hr	wt%	lbs/hr	wt%	lbs/hr	wt%	lbs/hr
Distillate	1.1	2,497	3.3	7,490	7.3	16,570	–	200,424
Residue	5.9	3,740	8.6	10,721	17.2	21,442		88,761
Total		6,237		18,211		38,012		289,185

Section 6 temp 880°F

	Gas		Naphtha		Gas Oil		Residue	
	%wt	lbs/hr	%wt	lbs/hr	%wt	lbs/hr	wt%	lbs/hr
Distillate	1.7	3,859	5.0	11,349	9.9	22,471	–	189,302
Residue	4.4	5,785	23.2	16,456	21.2	26,429		75,994
Total		9,644		27,805		48,900		265,296

Section 7 temp 900°F

	Gas		Naphtha		Gas Oil		Residue	
	wt%	lbs/hr	wt%	lbs/hr	wt%	lbs/hr	wt%	lbs/hr
Distillate	2.1	4,767	6.4	14,527	11.5	26,103	–	181,584
Residue	5.8	7,231	16.8	20,944	23.6	29,421	–	67,068
Total		11,998		35,471		55,524		248,652

Section 8.0 temp 920°F

	Gas		Naphtha		Gas Oil		Residue	
	wt%	lbs/hr	wt%	lbs/hr	wt%	lbs/hr	wt%	lbs/hr
Distillate	2.6	5,902	7.3	16,570	12.5	28,373	–	176,136
Residue	7.2	8,976	19.0	23,686	25.4	31,665	–	60,337
Total		14,878		40,256		60,038		2,36473

Heat balances over the soaker section

The heat balances for Sections 4, 5, 6, 7, and 8 can now be developed to establish the heat surface area for each of these coil sections. Only the balance for Section 4 is shown here in detail. The remaining coil sections are given in the summary table that follows.

The heat balances for each section of the soaker coil can now be written. Only the first section heat balance is shown here:

Section 4

	V/L	°F	°API	lbs/hr	Btu/lb	mmBtu/hr
IN						
From sec 3		820		3,386,810		223.604
Heater duty				By diff		12.275
Total in						235.879
OUT						
Gas	V	840	73	4,786	622	2.978
Naphtha	V	840	65	13,903	616	8.564
Gas oil	V	840	38	31,967	598	19.116
Res	L	840	17	300,989	476	143.271
Steam	V	840		35,165	1,471	51.728
Ht of crack					547*	10.223
Total out				386,810		235.879

*Heat of cracking is 547 Btu per lb of the converted gas & naphtha.

$$\text{Heater coil duty} = 12{,}274{,}500$$

heat flux is 10,000 Btu/sqft

$$\text{Sqft of coil} = \frac{12{,}274{,}500}{10{,}000} = 1{,}227.5$$

A summary of coil section exit temperature, surface areas and coil volumes is given in the following table:

The volume data in the table below is based on coils constructed using 4″ schedule 80 steel pipe. The ratio of area to volume is 0.11 cuft/sqft

Coil section	Exit temp °F	Duty mmBtu/hr	Sqft of coil	Volume of coil cuft	Cumulative volume cuft	K_T/K_{800}
1	700	47.824	3,985	438.4	438.4	–
2	760	48.633	1,110	122.1	560.5	–
3	820	22.584	1,506	165.7	726.2	1.55
4	840	12.275	1,227	135.0	861.2	3.02
5	860	5.682	568	62.5	923.7	5.0
6	880	21.739	2,174	239.1	1,162.8	7.3
7	900	13.855	1,386	152.5	1,315.3	9.0
8	920	11.634	1,163	127.9	1,443.2	10.2
Total		184.226		1,443.2		

Temperature vs. volume of coil is plotted over each coil section, and is given as Figure 11.19 below. The plot of coil volume above 800°F V's the K_T/K_{800} ratio is also plotted

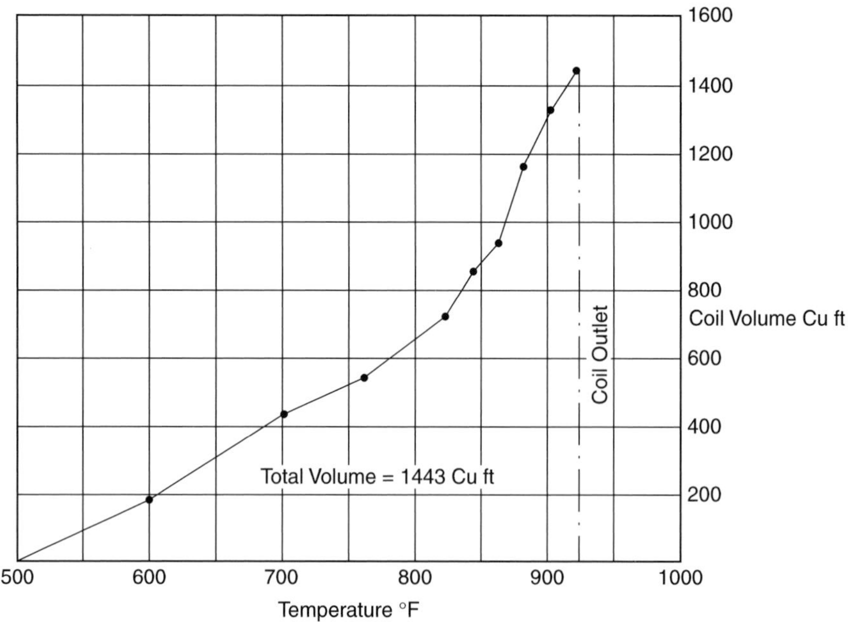

Figure 11.19. Coil temperature versus coil volume.

as Figure 11.20. The area under the curve developed in Figure 11.20 is calculated and then divided by the throughput in terms of BPSD gives the SVF for the conversion. Thus:

$$\text{Area under the curve of Figure 11.20} = \int_{800}^{920} K_T/K_{800} \times \text{Coil vol}$$

$$= 3{,}867$$

$$\text{allow 10\% for the steam} = 3{,}480$$

$$\text{then calculated SVF} = \frac{3{,}480}{25{,}500}$$

$$= 0.136$$

which compares well with the estimate for a 9% conversion originally used.

The duty specification for the heater can now be developed with the coil profile and other data to meet the required conversion. The final material composition can also be used now to develop the syncrude composition for the design of the recovery side

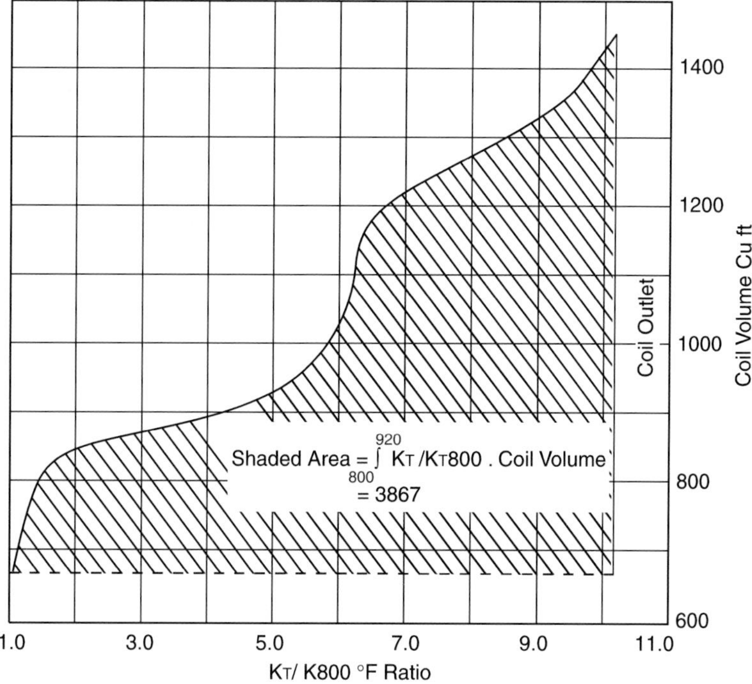

Figure 11.20. Coil volume above 800°F versus the $K_T/K_{800°F}$ ratio.

which will probably consist of a main fractionator with possibly a vacuum distillation unit for the cracked residue. The main fractionator usually produces a 'Wild' full range naphtha which is routed to the naphtha product stream leaving the Atmospheric Crude Unit. A full range gas oil would be blended with the straight run atmospheric light gas oil to be hydrotreated and routed to the diesel pool.

Chapter 12

The non-energy refineries

D.S.J. Jones

Introduction

Non-energy refineries refer to those plants that produce non-energy products from crude oil. These fall into two categories, which are:

The Lube Oil Refinery
The Petrochemical Refinery

Both of these types of refineries have already been mentioned in Chapter 1 of this Handbook. This chapter then provides some further details on these two important refining aspects of crude oil.

The lube oil refinery

The lube oil refinery is usually included as part of a conventional energy refinery. That being the case this lube oil section of the process will take its feed from the atmospheric distillation tower in the form of an atmospheric residue stream. Conventional lube refining utilizes solvent extraction processes to separate undesirable materials from the natural desirable ones present in the crude oil. However many of the world's crude oils do not have enough of these desirable molecules in them to justify the cost of elaborate extraction processes, and therefore lube oil refining leans towards processing the 'desirable' lube oil crude oils. An example of desirable lube oils are those from Venezuela (i.e., Tia Juana, and the heavy Bachequero crude). This restriction on lube oil production often leads to large refineries dedicating a separate atmospheric distillation unit to provide the necessary lube oil feed.

This problem of lube oil selective crude oils has diminished somewhat with the development of heavy hydrotreating and hydrocracking for lube oil treating. These processes virtually eliminate the need for solvent extraction and de-waxing of the lube oil stock. Hydrotreating of light vacuum distillates to make lube oil specification has been in commercial use since the mid-1960s, but with the advancement of this technology the heavier lube oil stock can also be treated now in a similar manner. Lube base oil production by severe hydrotreating, hydrocracking, and hydro dewaxing, also improves color and stability but each crude source of the lube oil stock requires different processing conditions in these hydro processes. This in turn leads to different yields of the various cuts to meet their specified viscosity.

Definitions and specifications

Lube oil refining has developed its own series of definitions quite separate to those of the conventional energy refineries. Some of the major definitions are as follows:

- *Lube base stock*
  This refers to a lube product that meets all lube specifications and is suitable for blending to meet performance specifications.
- *Lube slate*
  A lube slate is the set of lube base stocks produced by the refinery. This is usually a set of 4–6 base stocks.
- *Neutral lubes*
  These are obtained from straight run cuts of the vacuum tower.
- *Bright stock lubes*
  These are processed from the raffinate of the vacuum residue de-asphalting unit.
- *Virgin base lube oils*
  Are those base stocks obtained by processing straight run distillates from crude.
- *Re-refined base stocks*
  Are those processed from used or recycled lube oils.
- *Finished lube oil*
  This is a lube oil that has been blended to meet a final specification.
- *Paraffinic lube oils*
  These are all grades of lube oils, from both neutral and bright stocks, that have a finished viscosity index (VI) greater than 75.
- *Naphthenic lube oils*
  These are lubricants with a VI less than 75.

There are two important specification definitions for lube oils. These are:

- The Quality Specification
- The Performance Specification

The quality specification

Quality specification includes those for kinematic viscosity, viscosity index, pour point, color, flash point. These specifications are determined by standard tests from

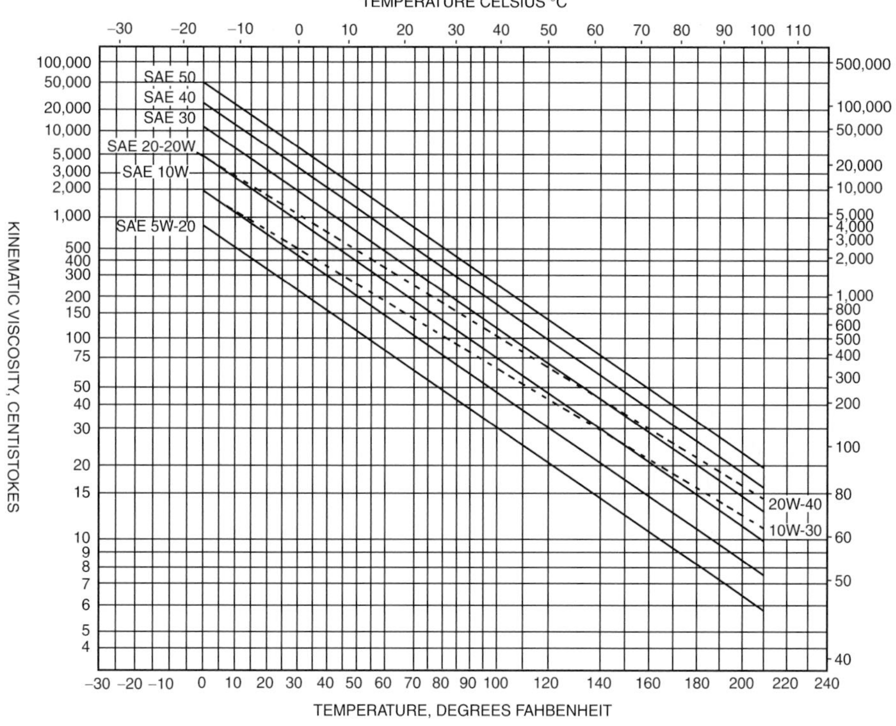

Figure 12.1. Viscosity versus temperature of crankcase oil.

organizations such as ASTM, SAE, or API. The specifications generally detail the characteristics considered important to the application of the specific lube oil. For example one very well known motor oil specification is SAE viscosity, established by the Society of Automotive Engineers, an organization that also carries out research, and publishes specification on a wide range of automotive topics. Figure 12.1 gives a range of Viscosities versus Temperature for a family of motor crankcase oils.

Although the kinematic viscosity is measured in centistokes, specifications are labeled in Saybolt seconds (SSU).

Performance specifications

Finished lube oil products are usually certified or qualified against performance specification requirements established by equipment and vehicle manufacturers, government bodies, and industrial organizations such as SAE. These qualifications usually require extensive performance testing using specialized apparatus and engines (including road tests). These tests refer only to those base oils produced in a specific manner and from specific crudes. Significantly altering the process scheme or the crude source could violate the intent and validity of the performance qualification certificate.

The performance of the lube base oils in service cannot be measured until after the production sequences are long completed. Manufacturing lube base oils therefore requires good correlations between measured properties at each stage of production and the final product performance quality. Developing these operating guidelines and correlations for plant units requires pilot plant and full plant test runs for each crude oil feedstock and finished products.

Lube oil properties

The important properties that are recognized in lube oil production are:

- Viscosity
- Viscosity Index
- Pour Point
- Oxidation Stability
- Flash Point
- Thermal Stability
- Volatility
- Color.

These are discussed in the following paragraphs.

Viscosity. This is the single most important characteristic of any lube oil. At its service temperature it is the balance to prevent wear and to enhance fuel economy. Most bearings operate with a hydrodynamic wedge shaped film of lubricant between the bearing surfaces. This mechanism minimizes wear of the surfaces under the operating conditions of the machinery. To accomplish this the lube oil must have adequate viscosity for the operating load, bearing speed, and the bearing surface smoothness to allow the hydrodynamic head to form and remain between the surfaces. Too high a viscosity will cause extra friction and heat, while too low a viscosity will allow the surfaces to touch. A bearing may operate over a wide temperature range, as in an engine or automatic transmission, and the lube oil has to have adequate viscosity over the entire temperature range to be effective.

Viscosity index. This is a measure of the amount the viscosity of a lube oil will change with temperature. A high viscosity index denotes less change of viscosity with temperature. High viscosity index is necessary for lube oils that are to be used for services with large temperature operating range.

Pour point. This property is important for any application in cold climates. Too high a pour point will result in the oil being gelatinous or even solid at ambient temperatures.

Oxidation stability. This is also an important requirement for a lube oil that is to operate under severe high-temperature conditions. If the oil has a tendency to oxidize the product of oxidation is often hard granules (similar to coke) which will damage bearing surfaces and block up the lube oil system filters and ports.

Flash point. The flash point of a lube oil must be high enough to be safe under any of the temperature condition that the lube oil will be required to operate.

Thermal stability. This property requirement is similar to that for oxidation stability. In the case of thermal stability the lube oil must be able to operate within its specified temperature range without the danger of the oil cracking or breaking down and giving rise to harmful polymers, acids, or other compounds.

Volatility. This property is important for ensuring minimizing any vaporization of the oil and thus maintaining the lowest oil consumption.

Color. Unusually dark coloring in a lube oil is indicative of the presence of olefins. These are undesirable in lube oils as they are thermally unstable. Lube oils are often filtered through a clay bed (usually bauxite) to improve color. Hydroprocessing always results in lube stocks with good color.

A description of major processes in lube oil refining

It is not proposed to describe or discuss the hydro units now used extensively in modern lube oil refining. These units have a similar process configuration as those described elsewhere in this book, only the conditions applicable to their specific application will be different. The major processes described here are:

- The propane deasphalting process
- An aromatic extraction—Furfural
- A de-waxing process—MEK

The deasphalting process

Deasphalting the heavy end of crude, that is the vacuum residue, is not entirely reserved for the production of lube oils. In energy refineries it is used to remove the asphaltene portion of the residue to prepare a suitable feedstock for catalytic conversion units. In most of these conversion units the performance of the catalyst is greatly impaired by the presence of heavy metals and the high Conradson carbon content of the residue feed. Most of the metals are entrained in the asphaltene compounds which makes up most of the asphalt portion of the residue. These asphaltenes are also high in Conradson carbon content, so the removal of these serves to eliminate both the heavy metal content and the high Conradson carbon.

In the case of lube oil production the light liquid phase resulting from the extraction of the asphalt makes excellent lube base oil. This deasphalted oil is termed 'Bright Stock' and can now be further refined in the same way as neutral base stock which are vacuum distillates to meet the specifications for blend stocks. The process configuration is shown as Figure 12.2.

The configuration shown here is that for a propane deasphalting unit. Vacuum residue feed enters the unit directly from the vacuum distillation tower and is heated to an inlet temperature of around 140°F before entering the top section of a trayed extraction tower at a pressure of 485 psig. Liquid propane is introduced into the bottom of the tower below the bottom tray.

The propane solvent moves up the tower counter current to the precipitated asphalt. The extracted asphalt is removed from the bottom of the tower and is routed through a fired heater to enter the top tray of a baffle-trayed separator. Some of the propane entrained in the asphalt phase is removed from the overhead of the flash tower. The asphalt phase leaves the bottom of the flash tower, and is reduced in pressure before entering the asphalt stripper tower above the top row of baffles. Steam is introduced to the bottom of the tower, and the remaining entrained propane is stripped out of the asphalt. The stripped asphalt leaves the bottom of the tower, is cooled to battery conditions and routed to storage or the asphalt blending plant. The propane stripout enters the suction side of a propane compressor and on discharge routed to the propane accumulator.

The deasphalted oil phase from the extraction tower overhead enters a high-pressure oil vaporizer which removes most of the entrained propane. The propane stream leaves the vaporizer to be cooled and enter the propane accumulator. On leaving the vaporizer, the oil phase enters the top row of baffles in a low-pressure stripper tower. Steam is introduced to the bottom trays of the tower and moves up the tower counter current to the oil phase. The residual propane is stripped out of the oil phase and leaves the top of the stripper tower with the steam. This overhead vapor stream joins the stream from the low-pressure asphalt stripper at the suction of the propane compressor.

The compressed propane is cooled and drained free of the water from the stripping steam in the accumulator. The dry propane is then recycled to enter the extraction tower.

Typical data on deasphalting Aramco crude residuum for lube oil with a propane treat of 600% volume on feed are as follows (Table 12.1).

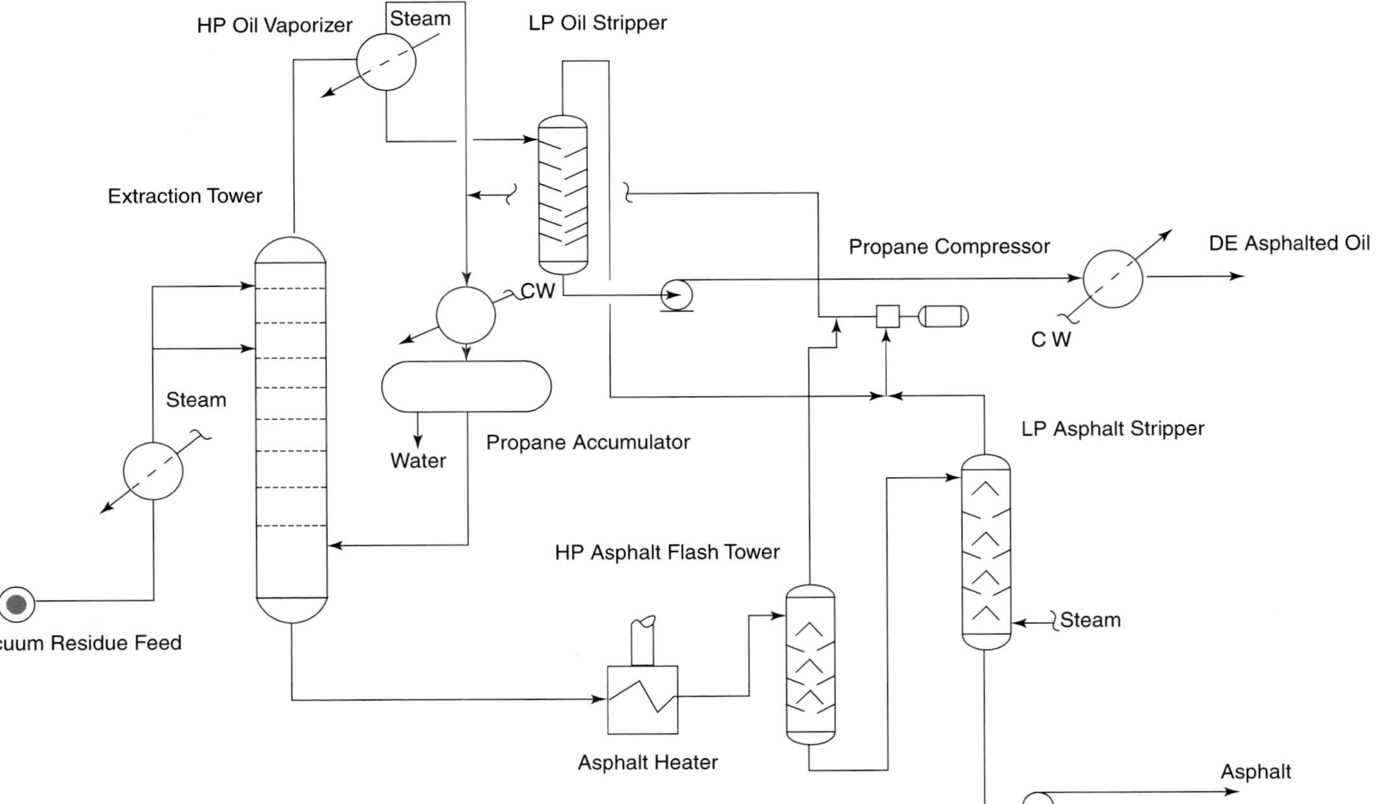

Figure 12.2. Propane de-asphalting process.

Table 12.1. Product yield and properties—propane deasphalting

	Feed	Deasphalted oil	Extract
Yield vol%	100	52	48
Gravity °API	11	21	1.5
Viscosity, SSU @ 210°F	1,150	175	–
Viscosity, SSU @ 275°F	–	–	250
Con carb wt%	10	2.7	–
Color TR dilute	–	9	–
Softening pt, R&B, °F	–	–	135

The salient operating conditions for this unit are as follows:

Treater tower:

Top temperature, 140°F
Bottom temperature, 105°F
Tower pressure, 485 psig

Deasphalted oil vaporizer, 315°F
Deasphalted oil stripper, 310°F
Vaporizer pressure, 220 psig
Stripper pressure, 5 psig

Asphalt flash tower, 525°F
Asphalt stripper, 515°F
Flash tower pressure, 220 psig
Stripper pressure, 5 psig

The lube oil extraction process—furfural extraction

The furfural extraction process described here is designed to produce lube oil base oils (before dewaxing) having high viscosity indices and other desirable lube oil qualities, such as color and stability. The furfural process is one of a few means of removing aromatic and naphthenic compounds from both neutral feed stocks and the bright oil feed stocks (de asphalted oils) by extracting them using furfural as solvent. The other most common processes to attain this is extraction using liquid SO_2 as solvent and a process using phenol as solvent. The furfural process configuration is shown below as Figure 12.3.

Either straight run distillate feed or Bright Stock oils (or a blend of both) enter an trayed extraction tower below the bottom tray. A dry furfural stream enters above the top tray to move counter current to the oil feed. The undesirable compounds are

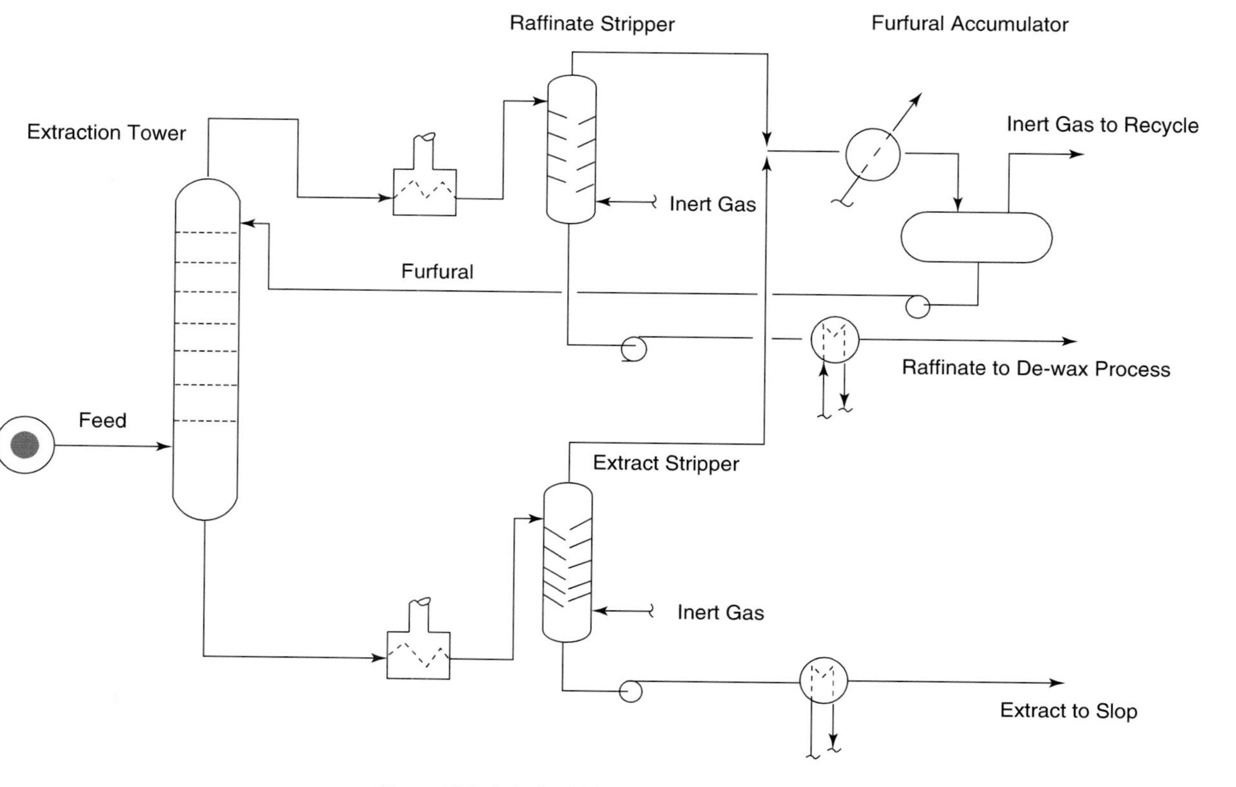

Figure 12.3. A furfural lube oil extraction process.

Table 12.2. Product yields and properties—furfural extraction

Untreated oil	Feed A	Feed B
Viscosity index	70	89
Gravity °API	25	28
Viscosity, SSU @ 210°F	55	44
Treated oil		
Yield vol %	69	55
Viscosity index (de-waxed)	102	113

removed from the lube oil feed to leave the bottom of the tower as a rich furfural stream. This stream is heated in a fired heater before entering an extract stripper. The furfural is stripped off and leaves the top of the stripper to be cooled and enter the furfural accumulator vessel. Stripping of the furfural is achieved using an inert gas stripping stream introduced below the bottom row of trays in the stripper. Inert gas is used as the stripping medium because furfural is highly susceptible to polymerizing on exposure to steam or air. The extract leaves the bottom of the tower to be cooled and routed to refinery fuel or the fuel oil pool.

The raffinate oil essentially free of aromatic and naphthene compounds leaves the top of the extraction tower to be heated in a fired heater before entering the raffinate stripper. The entrained furfural is again stripped out using an inert gas as the stripping medium leaving the raffinate to be routed from the bottom of the stripper to a dewaxing plant. The stripped furfural stream leaves the top of the stripper to be cooled and collected in the furfural accumulator.

The entrained inert gas from the stripping is flashed off the furfural collected in the accumulator to be either vented to the atmosphere or returned to the refinery's inert gas system.

Some data on this process is given in Table 12.2. These data are based on two typical feedstocks A and B which cover the usual range of feedstock parameters.

The Extraction Tower operates at between 150 and 200 psig and at a temperature Top/Bottom of between 190/140°F and 195/135°F.

The dewaxing process—MEK dewaxing

There are several dewaxing processes in use today. Among them are the propane dewaxer, Edeleanu's SO_2/benzol combination of extraction and dewaxing, the MIBK (methyl iso butyl ketone), and the more common MEK (metyl ethyl ketone) process. These processes, except the Edeleanu process, have a similar process configuration. The one based on MEK is shown in Figure 12.4.

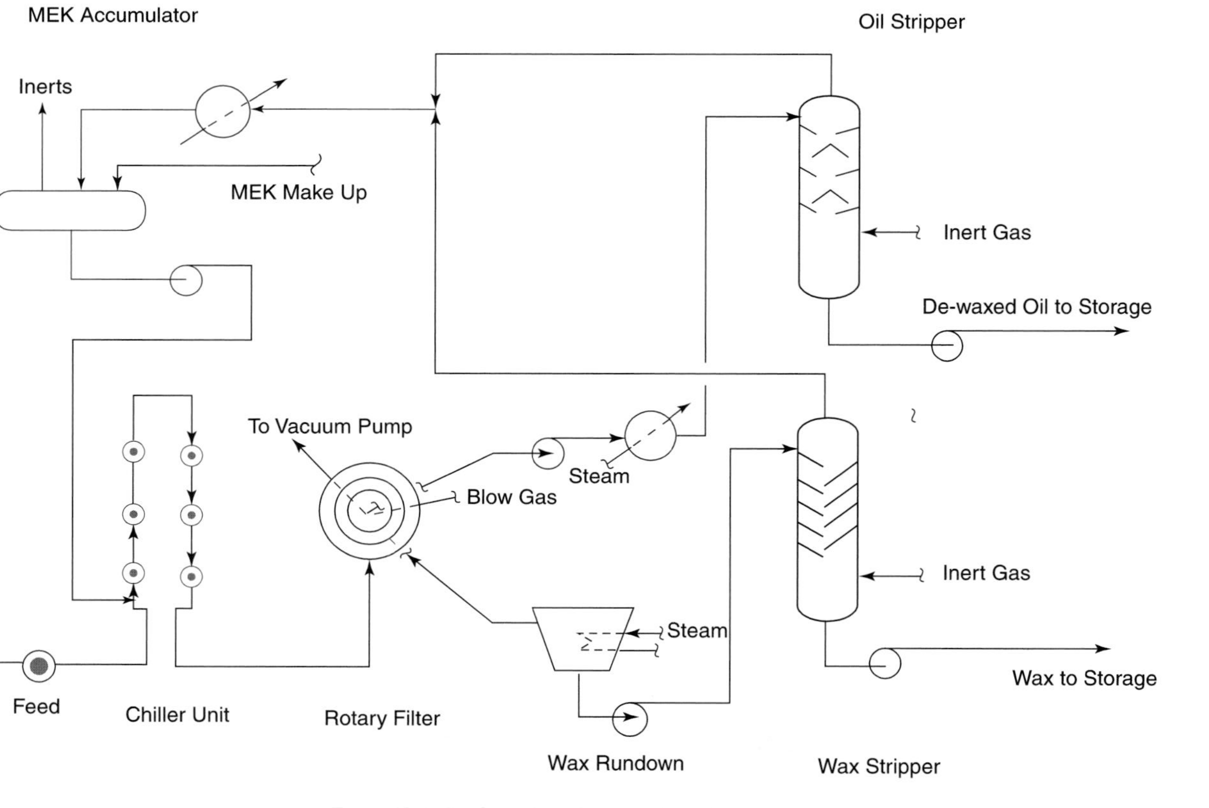

Figure 12.4. Configuration of a MEK dewaxing process.

Either neutral base oil or bright stock from the extraction unit is mixed with MEK before entering a tubular chilling unit. During the chilling of the mixture the wax content in the oil forms crystals in the presence of the MEK. The chiller tubes are furnished with internal chain scrapers which collect the oily slurry. This stream is then introduced to the bottom trough of a rotary filter. The outer drum of the filter is covered by a filter cloth. The drum in turn is divided into two or sometimes three chambers. The first is under vacuum and draws the oily slurry in the trough through the filter cloth. The oil phase free of wax crystals flows though the cloth to the inner section of the drum. The wax is retained firmly on the cloth. In some processes a stream of the dewaxed oil is returned as a wash to the wax cake. As the drum rotates it enters the second phase of its operation. Here a positive pressure in the form of an inert gas stream is applied to the rear of the wax cake. This lifts the cake off the filter cloth sufficient for a disengaging scraper to remove the cake and deliver it to a heated trough where the wax is melted. The molten wax is pumped from the trough to the top row of baffles in a wax stripper tower. Here the wax is stripped free of MEK using a stream of inert gas passing upwards through several rows of baffle trays counter current to the wax stream. The stripped wax is pumped from the bottom of the stripper to storage and further treating.

The dewaxed oil phase from the filter is pumped from the inner chamber of the filter to a steam heater and then to the top row of disc and donut trays in the oil stripper tower. Inert gas flows upwards through several rows of trays counter current to the oil flowing downward. The oil is stripped free of MEK and leaves the bottom of the tower to storage and use as blend stock.

The MEK from both strippers are collected, cooled, and routed to an accumulator drum. Here the absorbed inert gas and the blow gas are flashed off to be recycled or vented. The MEK is returned to the system by a pump to join the waxy oil feed at the chillers (Table 12.3).

12.2 Asphalt production

Almost invariably associated with the Lube Oil Refinery are facilities for the production of Asphalt. It is a popular belief, even among engineers that asphalt can be made from the heavy end of any crude oil. Unfortunately this belief is far from being correct. The asphalt product must have a series of properties which meet the specifications for the different grades. Since many of these properties are interrelated, it is difficult to meet all the specifications at the same time. The penetration/softening point relationship is probably the most important property to be considered from this stand point.

The ideal crude for asphalt production should be heavy, have a high, good quality asphalt yield, and, of course, should be inexpensive. The oxidizing properties of the

Table 12.3. Product yields and properties—MEK dewaxing process

	Feed A	Feed B	Feed C
Feed stock			
Viscosity, SSU @210°F	41	58	126
Pour point, °F	95	110	125
Flash point, °F	400	450	540
Products			
De-waxed oil, vol%	70	77	80
Pour point, °F	15	15	10
Cloud point, °F	20	18	12
Slack wax			
Yield, vol%	30	23	20
Oil content, wt%	25	20	15

asphalt must also be considered. Some crudes will produce fine paving stock, but will not be good feed for air blowing, when the specification of the product calls for an 'oxidized' asphalt.

Types of asphalt products

There are two major categories of asphalt products. These are:

- Paving and liquid asphalt
- Roofing asphalt

The paving and liquid asphalt

The paving and liquid asphalt products are again sub divided into five paving grades of paving and three liquid grades. The paving grades will have a penetration specification of 300 or less @ 77°F and 100 g weight, while the softer liquid grades will have a penetration of 300 and higher. The five paving grades fall into the following categories based on their penetration:

40–50, 60–70, 85–100, 120–150, 200–300.

The liquid asphalt grades are typified as, RC (Rapid Curing), MC (Medium Curing), and SC (Slow Curing) each of these grades will also have 4 viscosity grades as given below:

70 grade—RC/MC/SC	70 to 140 cS @ 140°F.
250 grade—RC/MC/SC	250 to 500 cS.
800 grade—RC/MC/SC	800 to 1,600 cS.
3,000 grade—RC/MC/SC	3,000 to 6,000 cS.

Rapid curing cutbacks are penetration asphalt and naphtha blends having a viscosity range from 75 SSF @ 77°F to 600 SSF @ 180° F. The naphtha content may be as high as 75% vol. Cutbacks are hot sprayed onto existing roads as a binding medium for new wearing surfaces.

Medium curing cutbacks are penetration grade asphalt and kerosene blends with four grades having the same viscosities as the RC cutbacks. These are used in road building in the same way as the RC cutbacks.

Slow curing cutbacks are penetration asphalt and gas oil blends normally produced directly from the crude oil atmospheric or vacuum distillation. The volume produced is small and they are used mainly as a gravel dust layer or mixed with aggregates for cold patching of asphalt surfaces.

Roofing asphalt

The second largest asphalt use is that for roofing. Most of these asphalts are produced by air blowing. There are three major roofing grades and may be classified by penetration and softening points. These are:

	Pen @ 77°F	Softening point °F
Type 1	25–50	140–150
Type 2	20–30	166–175
Type 3	15–25	190–205

Type 1 grade is used on 'Dead Flat' roof while the other two are used on intermediate and steep slope roofs respectively.

Asphalt oxidizing mechanism

Asphalt is basically a colloidal dispersion of asphaltenes in oil with resins as the stabilizing agent. The quantities of these can vary widely with the type of crude.

Asphaltenes can be hydrogenated to resins, resins to oils. Resins can be oxidized to asphaltenes, oil to resins. Thus:

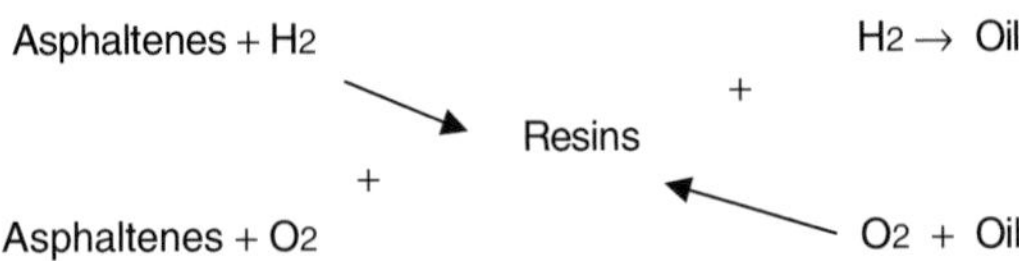

Oxidation is really a misnomer, as air blown asphalt has essentially the same oxygen content as the charge. Air blowing asphalt increases the asphaltene content, hardens it, decreases penetration, increases softening, and reduces ductility. Basically air blowing is a polymerization process following the route below:

1. Addition of O_2 to form unstable compounds from which H_2O is eliminated leaving unsaturated compounds which polymerize.
2. Addition of O_2 to form carboxyl derivatives from which CO_2 is eliminated followed again by polymerization.
3. Formation and elimination of volatile oxidation products other than H_2O and CO_2 followed again by polymerization.

It is worth noting, that the best crudes for asphalt air blowing are those with high percentage of asphaltene fraction and low in paraffin hydrocarbons, and waxes. Resins can be oxidized to asphaltenes relatively easily, so crudes rich in resin are good raw materials for asphalt manufacture. Oils can also be oxidized to asphaltenes, but they must be oxidized to resins first, which requires a more severe operation. Cracked residua simply do not make good asphalt.

Asphalt yields

In general heavier crudes contain a higher content of asphalt or asphaltic material and this is demonstrated in Figure 12.5.

The actual yield of these materials depends mainly on the composition of the crude itself in terms of paraffinic, olefinic, asphaltenic, and naphthenic fractions it contains. Good reliable crude oil assays provide such data for specific crude oils. However if such data is absent then a reasonably close estimate of the asphalt yield can be obtained using a method by S. Patel (*Oil & Gas Journal* Feb 1964).

This predicting method correlates Asphalt yield vs. 'Asphalt Factor' a parameter obtained by Patel's formula, which is:

$$F = \frac{(A)(CC)^n}{(B/C)(K - 10.4)}$$

where:

F = asphalt factor
A = the slope between 400°F and the vapor pressure at which 100 penetration starts to boil (see Figure 12.6).
CC = Conradson carbon of 750°F plus residue.
B = Slope of high boiling material. (see Figure 12.6).
C = Slope of lower boiling material (see Figure 12.6).
K = Characterization factor at 750°F.
n = Empirical exponent, normally 1.0.

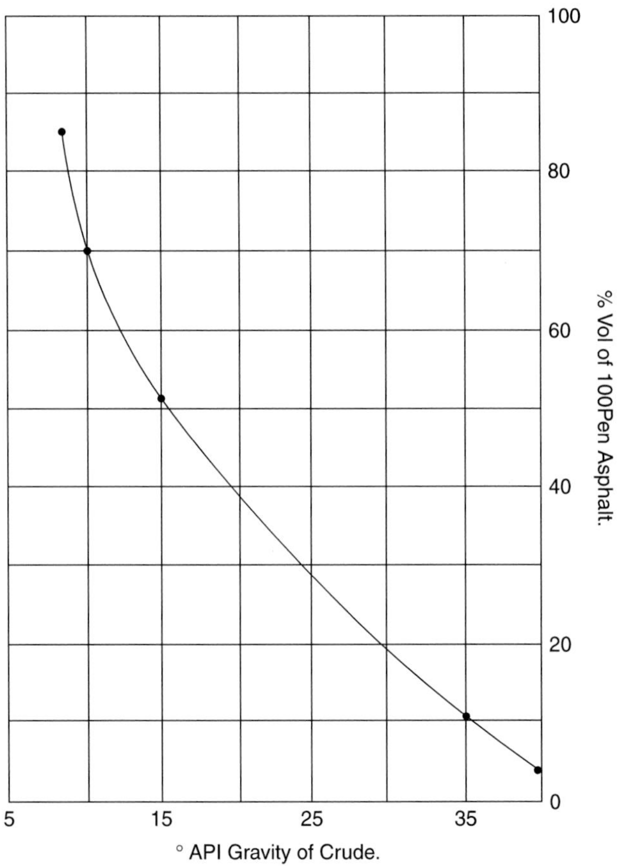

Figure 12.5. Asphalt in average crudes (approximate).

Figure 12.7 shows the relationship between the '*F*' factor and the approximate yield of 100 pen asphalt.

The calculation is by trial and error, and this is given in the example for the design of an asphalt air blowing process in Appendix 12.1 of this chapter.

Asphalt blowing process

Blowing air into a vessel containing asphalt from vacuum distillation or the de-asphalting process will change its penetration and softening properties. Among the variables that affect the manufacture of air blown asphalt are:

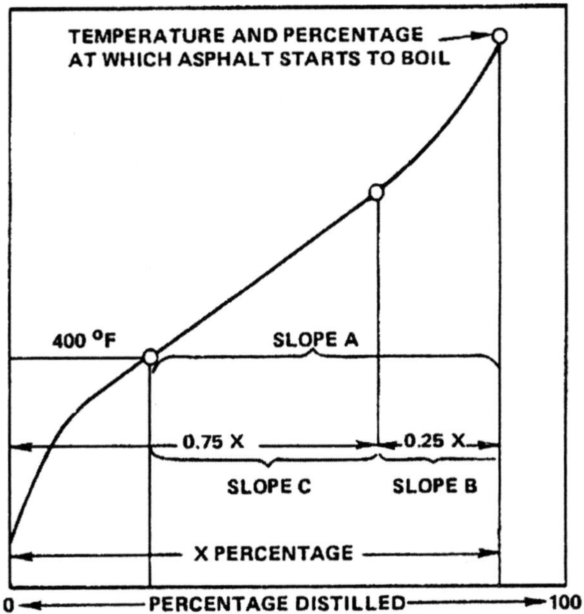

Figure 12.6. How to obtain crude oil slopes for use in the Patel equation.

- The rate of the air injected
- The temperature of the asphalt
- Retention time of the asphalt
- The system pressure

These variables and their effect on the process are discussed below.

The rate of air injected

This is the most important variable in the process. If the temperature, pressure, and residence time are kept constant in the oxidizer, the softening point of the asphalt (the ring & ball) can be remarkably increased by increasing the air rate. This increase is larger at low air rates than higher values A point is reached where the increase in air rate has little effect on the softening point and this is shown in Figure 12.8. The penetration properties of the asphalt are also affected by the air rate and this is shown in Figure 12.9.

For continuous operation normal air rates will be between 0.3 SCFM/BPSD of oil to 1.0 SCFM/BPSD. For Batch processes this rate will be between 1.5 SCFM/Bbl to 9.5 SCFM/Bbl this will depend of course on the length of each batch and the number of cycles to be processed each day. The air supply must not contain free water but need

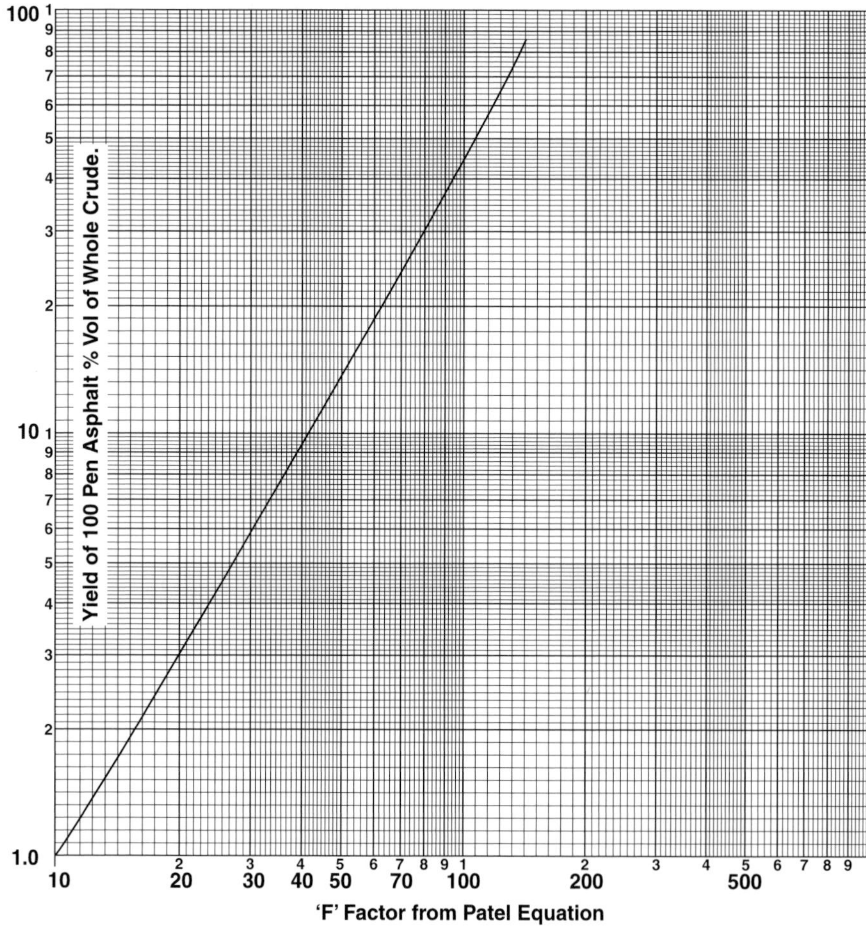

Figure 12.7. Correlation of '*F*' factor versus yield of bitumen.

not be dried. If the air is delivered by a reciprocating compressor then a well designed knock out pot is sufficient to remove the free water.

The reaction temperature

If the other variables are kept constant increase in the temperature of the oxidizer increases the softening point of the asphalt. The effect of temperature increase on the softening point is larger at the lower temperature levels than when the temperature is already high. A point is reached when the increase in temperature has little effect on the softening point. Figure 12.10 shows this effect. The ratio of softening point and penetration increases by decreasing the temperature. To increase the softening point without affecting the penetration it is necessary to reduce the reaction temperature, blow air for a longer period, or increase the residence time of the asphalt. Figure 12.11 shows this behavior.

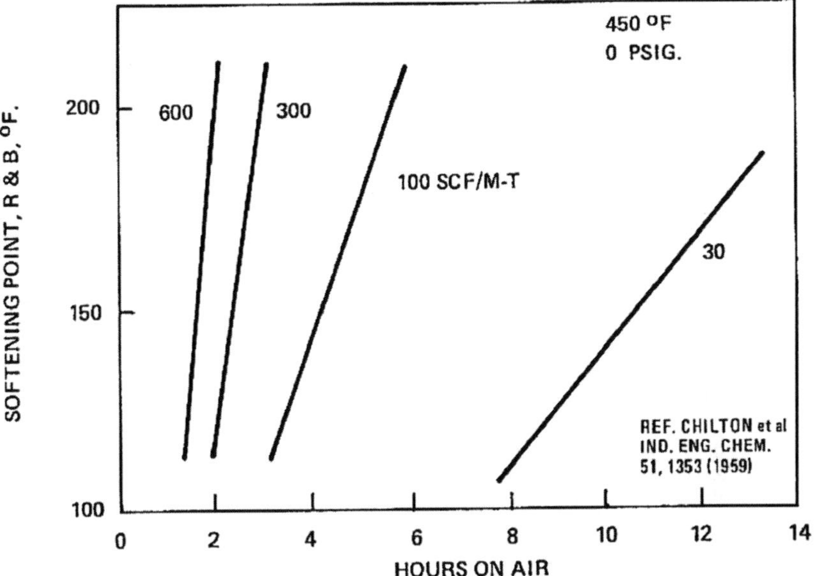

Figure 12.8. Effect of air rates on softening point.

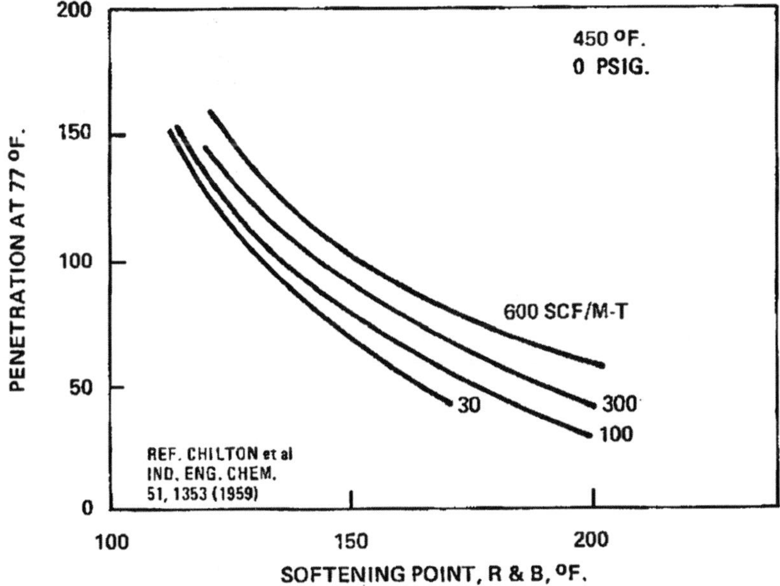

Figure 12.9. Effect of air rates on softening point/penetration ratio.

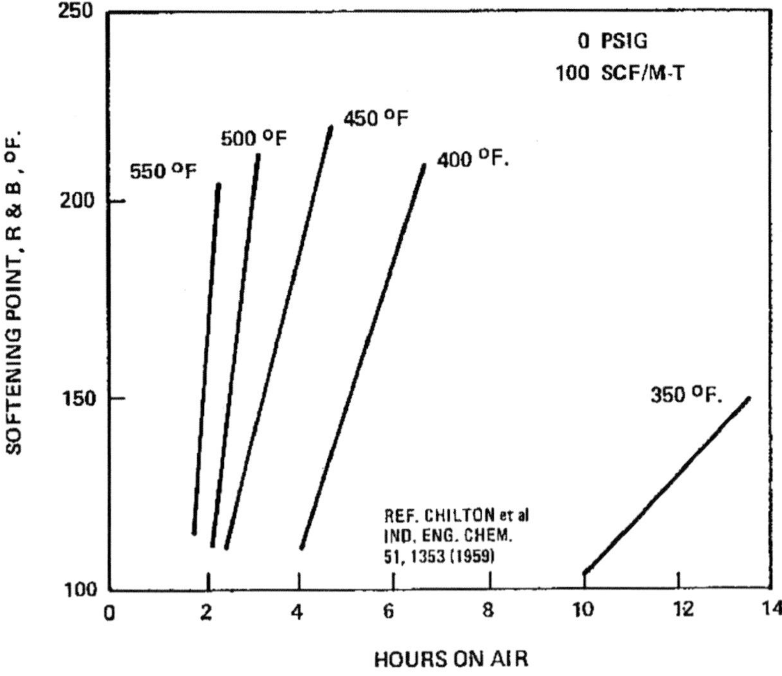

Figure 12.10. Effect of temperature on softening point (or hardness).

There are certain limitations to the temperatures at which the oxidizer can be designed or operated. The design of the oxidizer should not be higher than 50°F below the flash point of the feedstock or any of its components. It must however be recognized that effective oxidizing reaction does not occur at temperatures below 420°F. Generally speaking batch processes operate at a temperature between 430°F and 475°F while continuous processes operate at between 500°F and 550°F. The asphalt oxidizing reaction is exothermic and as such must be controlled by some form of cooling. The most common means of cooling in continuous processes is to recycle a portion of the reacting feed through a cooler and returning it to the reactor mixed with the feed. Another method is to install cooling coils into the reactor itself. This method is used mostly for batch processes. Steam is injected above the oil level in the reactor. The main purpose of this steam is to act as snuffing steam to avoid explosion in the air/oil atmosphere existing in this zone.

Retention and contact time

If all other variables are kept constant, the softening point can be remarkably increased by increasing the retention time of the asphalt in contact with air in the oxidizer. In the case of continuous processes this is achieved by varying the throughput or by

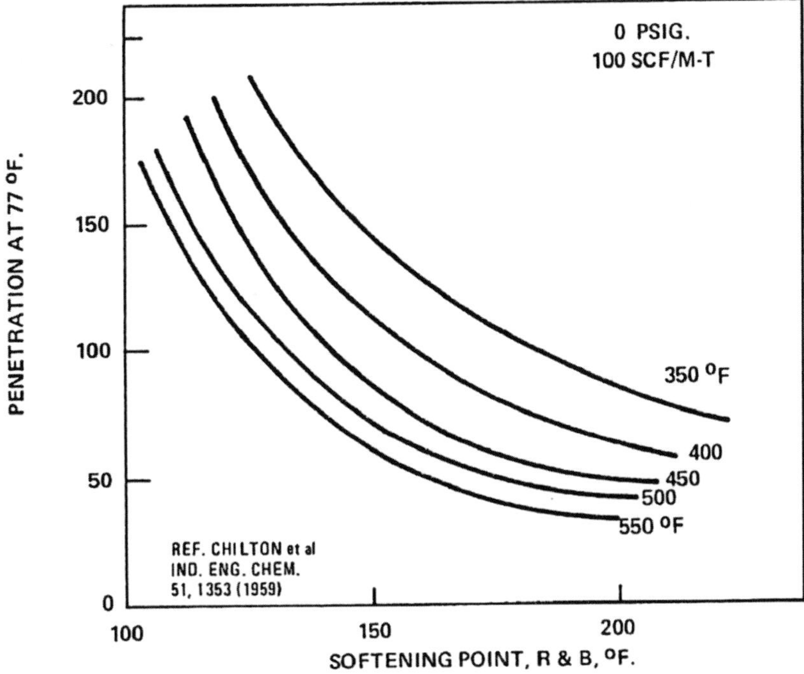

Figure 12.11. Effect of temperature on softening point/penetration ratio.

increasing the volume of asphalt (depth) in the oxidizer. In the case of the batch process this is achieved by varying the batch residence time. Figures 12.12–12.15 show the effect of the changing retention time.

The retention time or residence time required for oxidation is different in every case depending on how much change in softening point or penetration is required. A good rule of thumb is to calculate retention time assuming it takes one hour to increase 12°F in the softening point of the asphalt. For example to increase the softening point of 100 °F asphalt to a 220°F product will require a residence time of $\frac{220-100}{12} = 10$ hr.

A second rough rule for batch processes is that it takes approximately 4–8 hr of blowing utilizing 15 SCFM of air per ton of asphalt. That is if 25 SCFM of air is used the residence time would be around 2.4–4.8 hr since the effect of residence time is proportional to the oxidation rate. A continuous process usually requires less time, about one half that of a batch process.

Increasing the contact time or the depth of the asphalt level in the oxidizer vessel will increase the softening point of the asphalt if all other variables remain

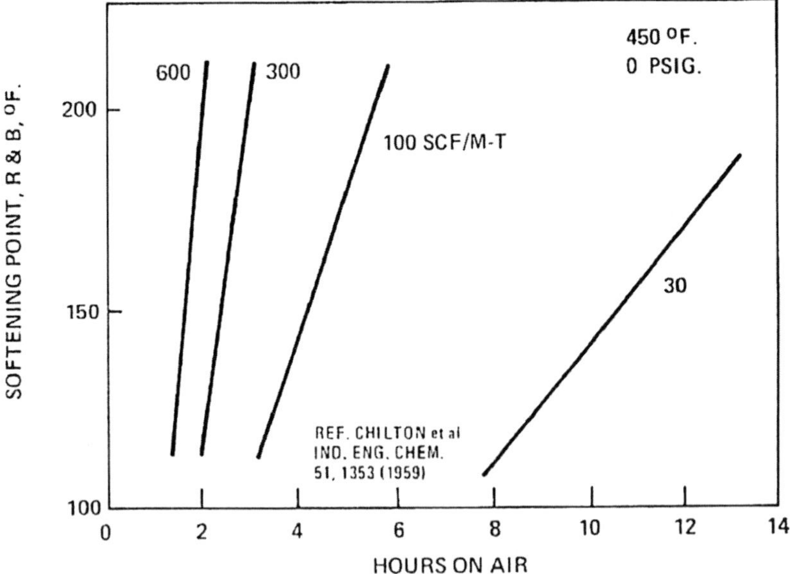

Figure 12.12. Effect of retention time on softening point (or hardness) at different air rates.

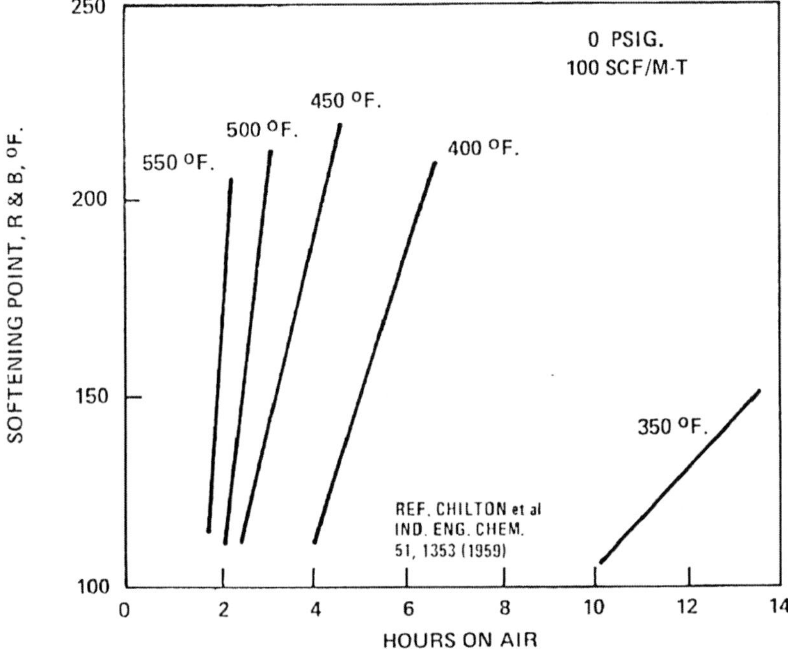

Figure 12.13. Effect of retention time on softening point (or hardness) at different temperatures.

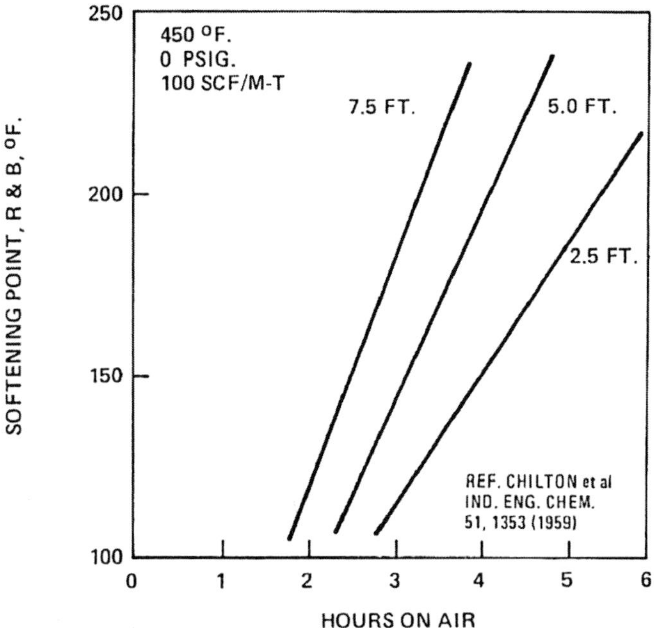

Figure 12.14. Effect of retention time on softening point (or hardness) at different liquid depth in oxidizer.

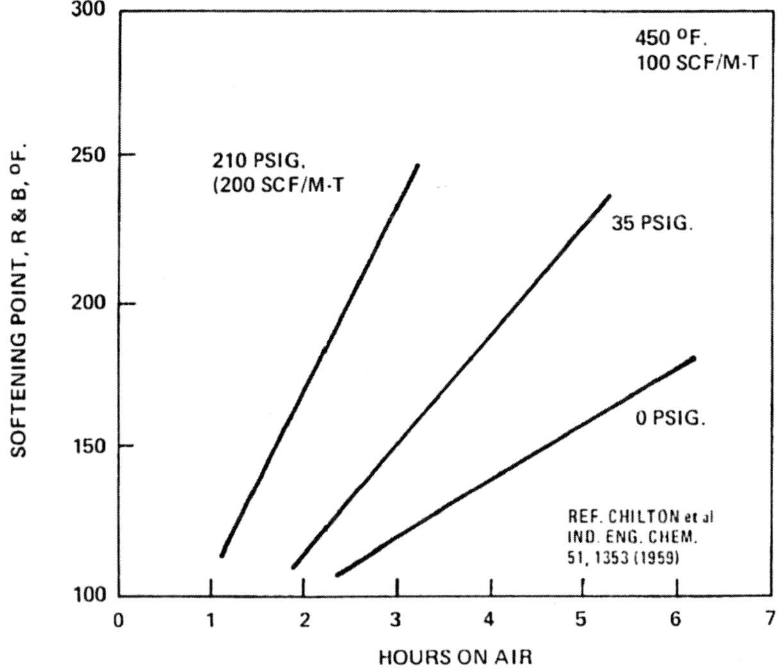

Figure 12.15. Effect of retention time on softening point (or hardness) at different oxidizer pressures.

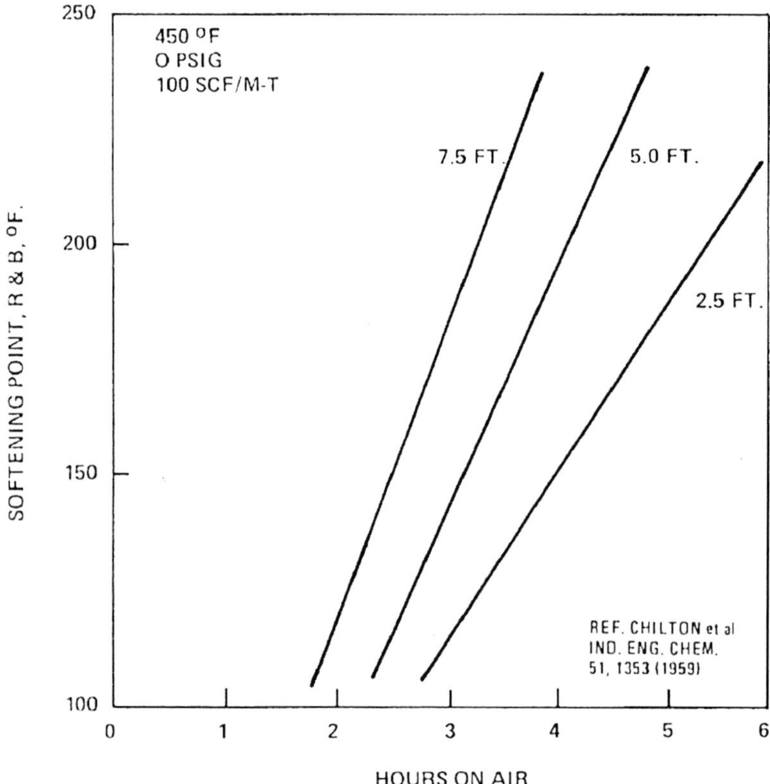

Figure 12.16. Effect of oil level in oxidizer on softening point (or hardness).

constant. In this case the softening point is changed without altering the softening point/penetration ratio. The effect of this variable is the same as increasing the residence time of the asphalt. Figure 12.16 shows this effect.

Although the height of the liquid and the residence time of the air are closely related they don't exactly have the same effect on the oxidizing process. The higher the liquid level the more efficient is the oxidizing process. This is so because the same amount of air is better utilized by remaining in contact with the oil for a longer period. Since the air is not completely used by the time it leaves the oxidizer, asphalt can be oxidized faster with the same air rate in a tall small diameter oxidizer than one of a larger diameter but shorter, even though the residence time will be the same in both cases.

Normally a height to diameter ratios of between 3.5/1 and 5/1 are used in the design of continuous process oxidizers and a somewhat lower 2.5/1 for batch processes. The liquid level in the oxidizer is usually no more than 2/3 of the total height of the vessel.

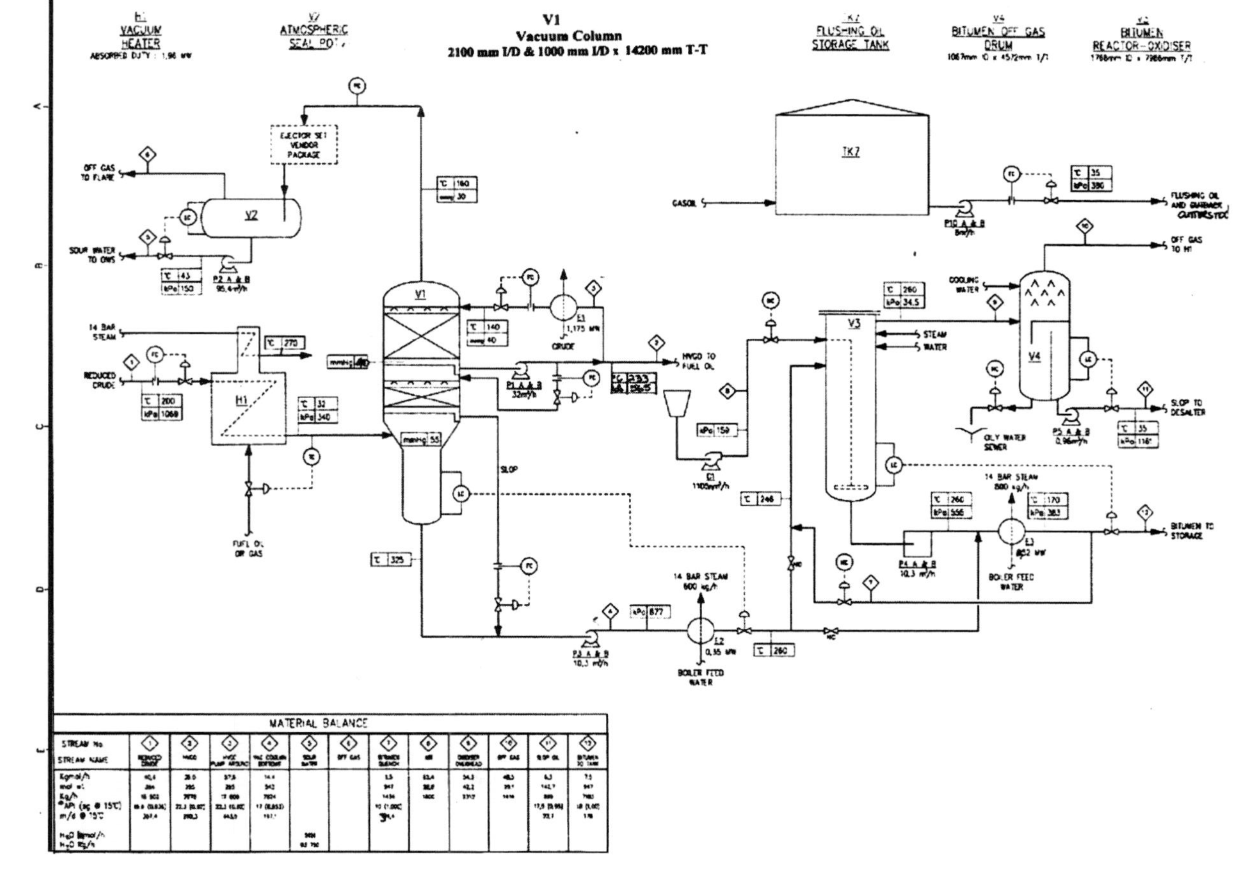

Figure 12.17. A process flow diagram of a typical bitumen manufacturing plant.

At least 10 ft must remain free of liquid above the high liquid level. This is to allow for liquid disengaging and foaming.

The system pressure

With all other variables constant, increasing the pressure of the oxidizer will increase the asphalt softening point. The effect of increasing pressure however is not as great as the effect of increasing any of the other variables. Perhaps the more significant change in increasing the oxidizer pressure is in the ratio of softening point to penetration of the asphalt. Usually it is not economical to operate at high pressure since the gains on hardening rates are small relative to the effect of the other variables. For practical purposes, pressure is not considered a design variable, and most oxidizers operate at near atmospheric pressure. A positive pressure of 5 psig is the more usual.

A flowsheet of a typical asphalt blowing plant is given as Figure 12.17.

The petrochemical refinery

The petrochemical refinery processes crude oil to produce feedstock for chemical plants. The two most important processes are:

- The production of aromatics
- The production of olefins

Both of these types of petrochemical refineries have been described briefly in Chapter 1 of this Handbook. It is not proposed to describe these type of refineries further here. This chapter will deal with a typical example of the process configuration for the production of aromatics only. This is probably the more common integration of the petrochemical refinery because a normal energy refinery is more easily adapted to aromatics with the minimum of additional processes.

The production of aromatics

The aromatics referred to here are:

- Benzene
- Toluene
- Ethyl benzene
- Para-xylene
- Meta-xylene
- Ortho-xylene

The configuration described here begins with a mixed aromatic stream which has been obtained by catalytic reforming of a high naphthene content naphtha. This naphtha would probably have been a product of a hydrocracker producing energy products from heavy waxy distillate. There are refineries that do hydrocrack heavy distillates to extinction to produce this kind of high naphthene naphtha only. The more common though is the energy hydrocracker producing a range of products of which the naphtha is just one of them. The reformate from this high naphthene feed is very rich in the aromatics listed above. To increase the aromatic content as feed to the aromatic complex the aromatics are separated from the remaining paraffin's by an extraction process.

The aromatics recovery complex which takes as feed the mixed aromatic stream is shown in Figure 12.18.

In this particular scheme the objective is to produce and maximize the benzene product and the ortho-xylene products only. Many aromatic complexes also produce para-xylene as product via a crystallization or adsorption step. A description of these units and the process flow of the complex follows.

Feed fractionation

The fresh mixed aromatics is delivered from off plot to enter a 35 tray splitter tower. Benzene and toluene are removed as overhead product while the mixed xylene streams leave as the bottom product.

Xylene splitter and isomerization process

The mixed xylene stream leaves the splitter and is routed to a xylene splitter. This is a super fractionating tower containing at least 135 fractionating trays. The fractionation split is between the meta- and ortho-xylene components. A recycle stream from an isomerization plant rich in ortho-xylene is also fed to this xylene splitter. The overhead rich in ethyl benzene and the para, meta-xylenes, is routed from the splitter to an isomerization plant. These C_8 aromatics are isomerized over a catalyst and in the presence of a rich hydrogen stream to a product rich in ortho-xylene but containing also benzene, toluene, and some ethyl benzene with some light hydrocarbons and hydrogen in equilibrium. This isomerate enters a fractionator in which the light hydrocarbons and some ethylbenzene are removed as overheads while the bottom product, containing mostly ortho-xylene, with the other C_8's in equilibrium is returned to the xylene splitter. The light isomerate overhead product from the fractionator is stabilized in a separate stabilizing column before being routed to the benzene recovery section. The bottom product from the xylene splitter enters an ortho-xylene rerun tower from which commercially pure ortho-xylene leaves as the overhead product. The

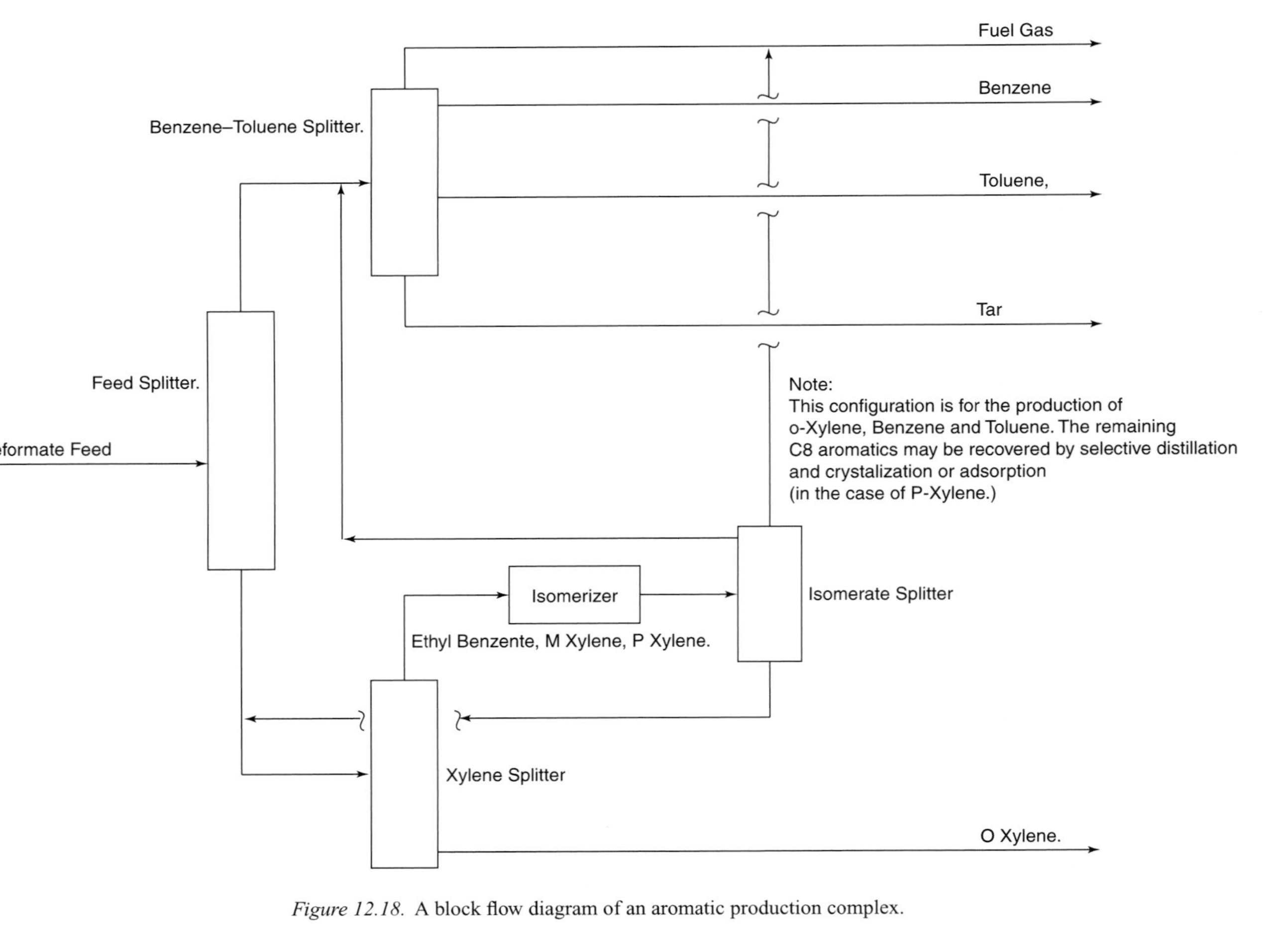

Figure 12.18. A block flow diagram of an aromatic production complex.

bottom product leaving this tower contains heavy aromatics (heavier than C_8's) and is routed to fuel, or, in some cases, to trans alleylation with toluene to yield more C_8s.

Dealkylation and benzene recovery

The overheads from the feed fractionator and the bottom product from the light isomerate stabilizer combine to form the fresh feed to a benzene fractionator. This feed stream will contain mostly benzene and toluene. A recycle stream from the toluene dealkylation plant, if present, containing a high proportion of benzene joins the fresh feed to enter the benzene fractionator. Benzene is removed from this fractionator as an overhead product, while toluene is removed as a re-boiled stripped side stream product. A small bottom make of polymer and tar is removed as a bottom product. The toluene stream may be routed to a dealkylation unit in which about 98 mol% of the toluene is converted to benzene. This dealkylated product forms the recycle stream to join the fresh feed to the benzene fractionator.

The data given in the block flow diagram. Figure 12.18 is based on processing the following mixed aromatic feed to the units:

Benzene	20,000 long tons/year
Toluene	50,000 long tons/year
Ethyl benzene	9,000 long tons/year
Para-xylene	11,400 long tons/year
Meta-xylene	27,000 long tons/year
Ortho-xylene	12,600 long tons/year

Make-up gas to the isomerization unit and the dealkylation unit has the following composition:

H_2	81.5 mol%
C_1	12.2 mol%
C_2	5.2 mol%
C_3+	1.1 mol%

The products required to be maximized are:
High purity benzene to an SG of 0.882 (Min) and 0.886 (Max) all at 60°F
Ortho-xylene will contain not less than 99.0 wt% of ortho-xylene

Process discussion

Feed fractionation and xylene splitter

Good separation of the light aromatics and the xylenes can be achieved in a 35 trayed column operating at about 35 psig in the reflux drum. At this pressure the tower overheads are at a temperature high enough to offer some preheating of the tower feed by the overhead condensers. Splitting the xylenes however requires a tower of about 136–140 trays. A reasonable economic overhead pressure (in the reflux drum) is about 98 psig. Again at this pressure, as in the case of the feed fractionator, the overhead temperatures offers some feed preheat from the overhead condensers and to reboil the ortho-xylene rerun tower.

The isomerization unit

This is a licensed process which converts the xylene streams to approach an equilibrium xylene composition. Some isomerization units convert ethylbenzene to xylenes while others essentially dealkylate the ethylbenzene. By removing one or more of the components (ortho-xylene in this case) the remaining xylenes in the feed are isomerized to recover the equilibrium distribution lost by the removal. The unselected isomers in the fresh feed are recycled to extinction by this method. The process shown here utilizes around 1.2 mmScf/D of 81.5 mol % hydrogen.

The dealkylation unit

This unit is also a licensed process. its purpose is to de-alkylate the toluene component in the feed to produce benzene.

This process utilizes a separate polymer reactor to increase benzene selectivity. It also uses a hydrogenation reactor to control impurities in the benzene such as unsaturates and sulfur products. The yield of benzene in this process is about 98.2 mole% on toluene in the feed. Except for very small units a cryogenic step is applied to remove the reaction products of methane and heavier. Where hydrogen from a catalytic reformer is used as make up the C_3's, and heavier paraffins and any H_2S contained in it must be removed. An absorber operating at about 250 psig is used to remove the hydrocarbons and a caustic wash used to remove the H_2S. The removal of the C_3's and heavier is advisable as these hydrocrack under the dealkylation reactor conditions causing additional reactor temperature rise and increase in hydrogen consumption.

Appendix 12.1: Sizing a bitumen oxidizer

Design specification

Product required 816 BPSD of 25 pen asphalt.
Crude source Pennington—West African offshore. (Assay data see Figures 12.19–21.)

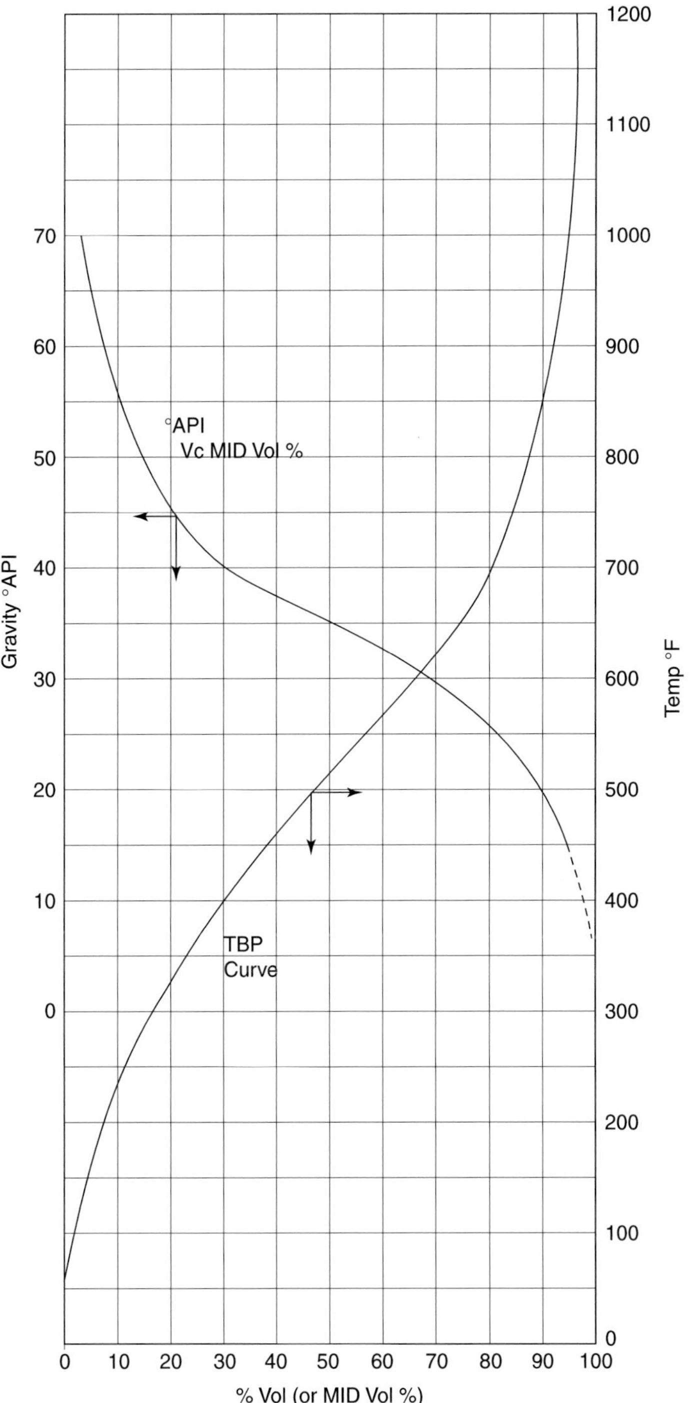

Figure 12.19. The TBP and gravity curves for Pennington crude.

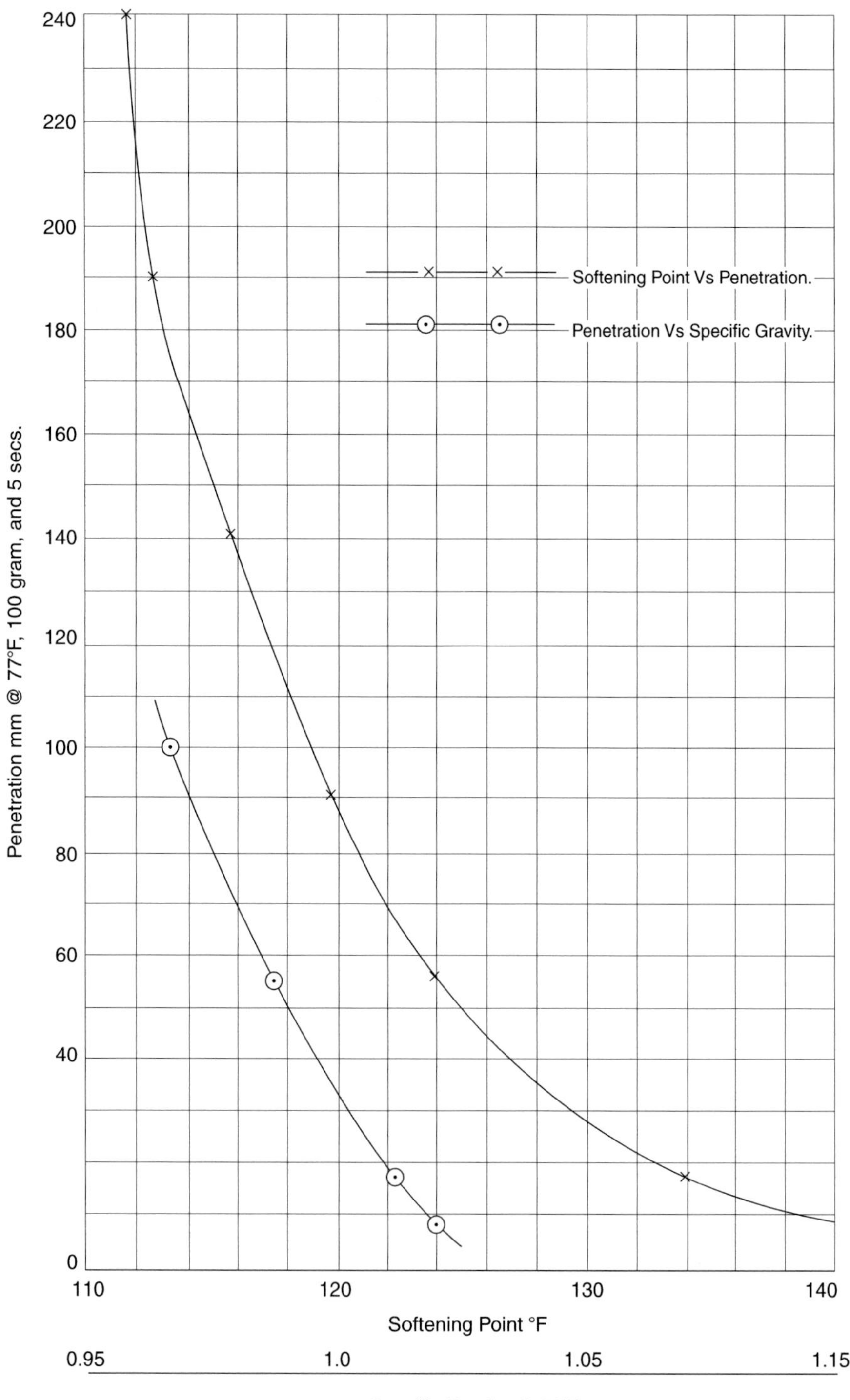

Figure 12.20. Plot of softening point versus penetration and penetration versus specific gravity for Pennington crude.

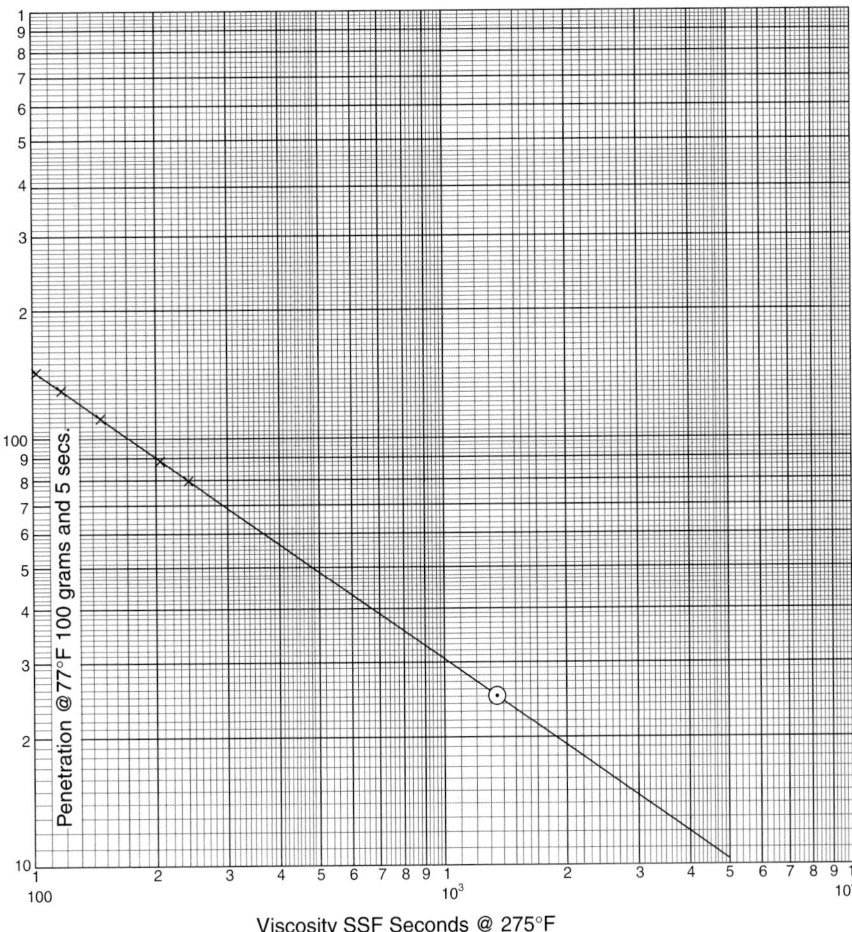

Figure 12.21. Plot of viscosity versus penetration for asphalt.

Vacuum unit feed = +650°F on crude = 75 vol%.
Required size of oxidizing reactor.

Step 1.0 Predict yield of asphalt
The Patel equation shall be used for this purpose.

Trial 1. Assume initial boiling point of asphalt is 1,015°F which equates to a cut 95 vol% on crude.

$$\text{vol\% cut point on crude} = 0.75 \times 0.95 = 71 \text{ vol\%}$$

using Figure 12.6 and data from the TBP curve (Figure 12.19):

$$\text{Slope A} = \frac{1{,}015 - 400}{95 - 31} = 9.6$$

$$\text{Slope B} = \frac{1{,}015 - 625}{95 - 71} = 16.25$$

$$\text{Slope C} = \frac{625 - 400}{71 - 31} = 5.6$$

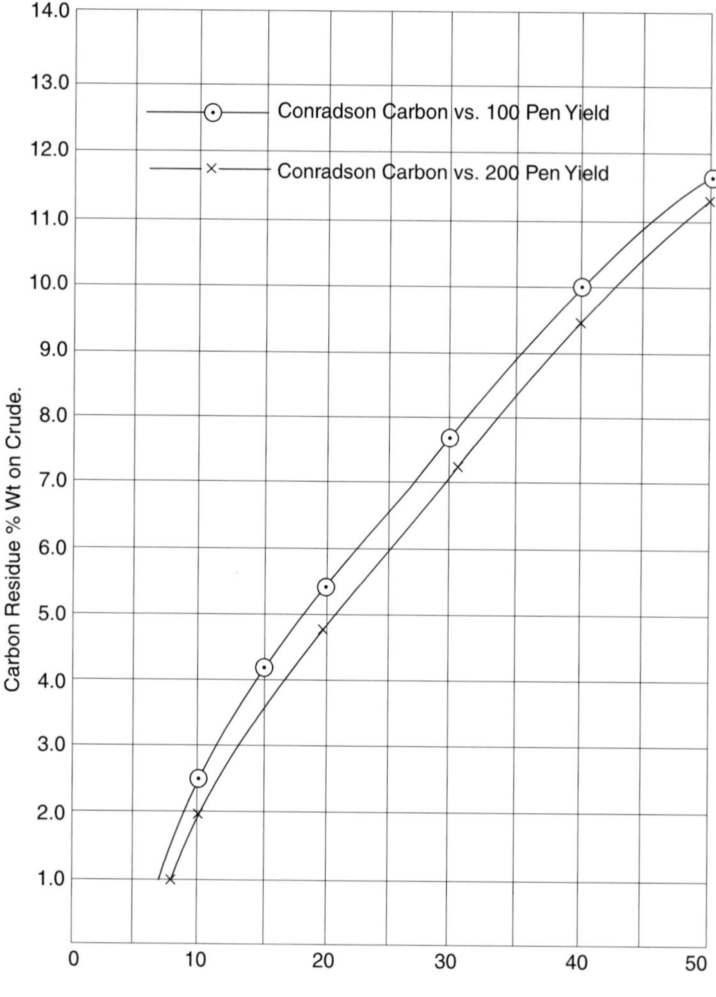

Figure 12.22. Asphalt yields versus carbon residue % weight on crude.

Conradson carbon content of +750°F residue = 4.6 wt% (see Figure 12.22)

$$K \text{ factor at } 750^\circ\text{F} = 11.2$$

$$\text{Exponent } `n' = 1.2$$

$$F = \frac{9.6 \times (4.6)^{1.2}}{(16.25/5.6) \times (11.2 - 10.4)} = 25.8$$

From Figure 12.7 percent asphalt yield = 4.3 (assumed was 5.0 vol%).

Trial 2. Assume boiling point of asphalt is 950°F which equates to a cut of 93.5 vol% on crude or a yield of 6.5 vol%.

Calculated '*F*' factor is 34.6 correlating to a yield of 7.6 vol%.

Trial 3. Assume boiling point of asphalt is 850°F which equates to a cut of 90 vol% on crude or a yield of 10 vol%.

Calculated '*F*' factor is 42.0 correlating to a yield of 10.0% which is that assumed, and is accepted as the yield of 100 pen asphalt from this crude.

Step 2.0 Calculate throughput of 100 pen and air to oxidizer
From Figure 12.23 the ratio of 100 pen asphalt to a 25 pen asphalt is read off at 0.89 for a characterization factor of 11.2. Then:

For 816 BPSD of 25 pen crude we require a throughput of 816/0.89 = 917 BPSD of 100 pen asphalt.

For a continuous process, the recommended air rate is 0.3 to 1.0 scfm per BPSD. A rate of 0.55 will be used here.

Then air rate will be $0.55 \times 917 = 504.4$ scfm.

Step 3.0 Calculate residence time and capacity of fresh feed.

Required final penetration = 25 mm @77°F, 100 g/sec.

Equivalent softening point = 131°F (Figure 12.20)

Initial pen = 100 and equivalent softening point is 119°F

Change in softening point = 12°F approximate residence time = $\dfrac{131 - 119}{12} = 1$ hr.

Capacity of fresh feed: Cuft of feed per hour = $\dfrac{917 \times 42}{24 \times 7.48} = 214.9$

Asphalt SG @ 60°F = 0.966

@ 500°F = 0.815

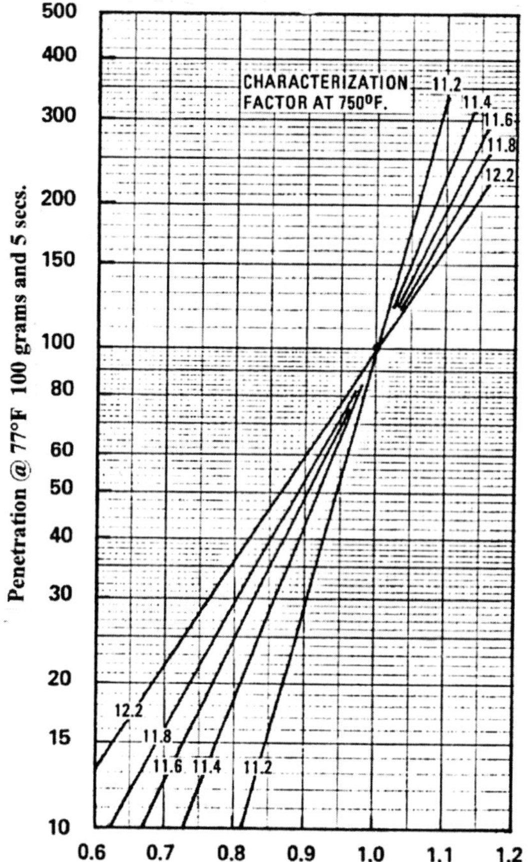

Figure 12.23. Fraction of yield of 100 PEN asphalts.

Volume of fresh feed at 500°F = 254.7 cuft/hr. Add 15% as contingency = 292.9 cuft/hr.

Step 4.0 Reactor material balance
Material balance over the reactor is as follows

$$\text{Feed of 100 pen in} = \frac{917\ \text{BPSD} \times 42}{24}$$
$$= 1{,}605\ \text{gals/hr}$$

$$\text{SG @ 60°F} = 0.966 = 8.044\ \text{lbs/gal}$$
$$\text{lbs/hr} = 12{,}908.$$

$$\text{Product 25 pen asphalt out} = \frac{816 \times 42}{24}$$
$$= 1{,}428 \text{ gals/hr}$$

$$\text{SG @ } 60^\circ\text{F} = 1.006 = 8.33 \text{ lbs/gal}$$
$$= 11{,}895 \text{ lbs/hr } (= 0.92 \text{ wt\% on feed})$$

$$\text{Air into the oxidizer} = 504.4 \text{ scft/min} = \frac{504.4 \times 60 \times 29}{378}$$
$$= 2{,}322 \text{ lbs/hr.}$$

approximately 90% of the oxygen in the air reacts with the asphalt.

$$\text{Lbs/hr of oxygen reacted} = 2{,}322 \times 0.232 \times 0.9$$
$$= 485 \text{lbs/hr.}$$

Approx 20% of oxygen reacted will be in the asphalt product = 97 lbs/hr.

$$\text{Unreacted oxygen leaves unit in the overhead vapor} = (2{,}322 \times 0.232) - 485$$
$$= 54 \text{ lbs/hr}$$

Nitrogen in o/head vapor = 2,322 − 539 = 1,783 lbs/hr

Hydrocarbon vapor in o/head vapor

= Feed + Air − (product + oxygen in product + nitrogen + unreacted Oxygen)

= 12,908 + 2,322 − (11,895 + 97 + 1,783 + 54)

= 1,401 lbs/hr

Total overhead vapor = 1,401 lbs/hr hydrocarbons + 1,783 lbs/hr N_2 + 54lbs/hr O_2

= 3,238 lbs/hr

Step 5.0 Heat balance over the reactor and amount of cooling stream
It is intended to maintain a reactor temperature of 500°F by recycling cold run down product.

Feed temperature to the oxidizer shall be 550°F. Heat of reaction in Btu/lb = 1.3 × diff in softening point.

The amount of recycle cooled product as reactor coolant is obtained from the following heat balance where x lbs/hr is the recycle.

Stream	V or L	° API	°F	lbs/hr	Btu/lb	mmBtu/hr
IN						
Feed	L	15	550	12,908	254	3.279
Air	V		86	2,322	21	0.049
Recycle	L	10	338	x	148	$238x$
Ht of reaction	–	–	–	–	–	0.201*
Total IN				$15{,}230 + x$		$3.529 + 148x$
OUT						
Product	L	10	500	11,895	238	2.831
Dis O_2	V	–		97	–	Neg
Recycle	L	10	500	x	238	$238x$
O/Head vap	V	–	300	3,238	124	0.402
Total OUT						$3.233 + 238x$

$$x = \frac{3{,}529{,}000 - 3{,}233{,}000}{238 - 148}$$

$= 3{,}289$ lbs/hr.

Volume of recycle @ 500°F (SG 0.872) = 7.26 lbs/gal = 453 gals/hr or 60 cuft /hr

*Heat of reaction as Btu/lb of asphalt feed is calculated as 1.3 × the difference in softening point. In this case it is 12,908 × (12 × 1.3) = 201,000 Btu/hr.

Step 6.0 Reactor sizing

The reactor size shall be based on hot volume of feed + contingency and the hot volume of recycle over the calculated residence time of 1 hr. The volume of feed and recycle shall be based on the reactor temperature (500°F).

Volume of fresh feed and contingency is 292.9 cuft. The recycle is 60 cuft and the total sizing volume is 352.9 cuft.

This amount shall occupy 65% of the oxidizer's total volume. This allows for the disengaging of vapor leaving the vessel.

Total volume of the oxidizer is $\frac{352.9}{0.65} = 543$ cuft.

The ratio length to diameter shall be 4.5.

$$\text{Thus} \frac{4.5\pi D_3}{4} = 543 \text{ cuft}$$

and

$$D = 5.36 \text{ ft say } 5.5 \text{ ft and } L = 4.5 \times 5.5 = 24.8 \text{ ft T-T.}$$

Check height of liquid to NLL $= \frac{352.9}{\text{Xsect area}} = 14.8$ ft $= 60\%$ of total vessel height which is acceptable.

Chapter 13

Support systems common to most refineries

D.S.J. Jones

This Chapter covers those systems which support the refining process, and is divided into the following parts:

- Control Systems
- Offsite Systems
- Utility Systems
- Safety Systems

These parts describe the systems and provides guidelines for the analysis and design of the more important sections of them. Finally the chapter will include description and discussion on the environmental aspect of these systems, with some reference to their environmental engineering.

13.1 Control systems

The proper operation and performance of any process depends as much on a properly designed control system as the correct design and specification of the equipment contained in the process. Indeed this statement can be extended to include the safe operation of the plant as being dependant to a large extent on the design of the control system.

Control systems in a process are aimed at maintaining the correct conditions of flow, temperature, pressure and levels in process equipment and piping. There are therefore four major types of controls which are:

- Flow control
- Temperature control
- Pressure control
- Level control

The principal objective of all these types of controls is to maintain a steady stable plant operation and to enable changes and emergencies to be handled safely. The system must also be designed to ensure that any process changes can be accommodated with minimum risk of damage to plant equipment.

Before looking at these in some detail it is necessary to define some of the more common terms found in control systems.

Definitions

Surge volumes. This is the volume of liquid between the normal liquid level (NLL) to the bottom (Tan Line) of a vessel.

Level control range. This is the distance between the high liquid level (HLL) and the low liquid level (LLL) in the vessel. When using a level controller the signal to the control valve at HLL will be to fully open the valve. At LLL the signal will be to fully close the valve.

Proportional band. This determines the response time of the controller. Normally a proportional band is adjustable between 5% and 150%. The wider the proportional band the less sensitive is the control. If a slower response time is required a wider proportional band is used.

Control valve response. The minimum time that should be allowed between HLL and LLL to permit the control valve to respond effectively to changes in level are:

CV size	Response, sec
1″	6
2″	15
3″	25
4″	35
6″	40
8″	45
10″	50

The above times allows for air signal lags and for operating at a proportional band of about 50%.

Surge volume. Surge volume is the volume retained in a vessel during operation at a set level. It is used for:

- Protecting equipment from damage caused by flow failures
- Protecting downstream processes from fluctuating flows which could cause poor process performance
- Protecting downstream processing from fluctuations in feed composition or temperature
- Protecting equipment from damage due to coolant failure

Types of surge volumes

There are basically two types of Surge Volumes. These are:

1. *Upstream protection surge.* This is a surge volume provided to protect the upstream equipment and its associated pump from feed failure.
2. *Downstream flow surge.* This is surge volume provided to protect downstream equipment from feed failures or fluctuations.

Examples of surge types

Process feed. Feed to process units is almost invariably on flow control. Many units also have a feed surge drum, particularly those units that are sensitive to flow fluctuations or where complete flow failure can cause equipment damage. This is an example of "downstream flow surge". The surge volume of the drum will depend on

- Source and reliability of source
- Type of control at source
- Variations and fluctuations in source rate

Column feed from an upstream column. This feed stream will usually be controlled by the level in the source column, hence it will be fluctuating. If surge volume is provided only in the source column it must be sufficient to cater for "upstream protection" and "downstream protection". The use of a surge vessel would be recommended for this case.

Feed to fired heaters or boilers. The failure of flow through the tubes of fired heaters or steam boilers can cause serious damage through overheating of the tubes. Consequently "downstream protection" is required in this case. Invariably flows to heaters are on flow control.

Reflux drums

1.0 When the drum only furnishes reflux or reflux and product to storage all that is needed in terms of surge volume is sufficient to provide "upstream protection". That is the surge volume required is only to protect the reflux pump from losing suction in the case of column feed failure. The pump will be required to circulate reflux and cool the column down during an orderly shut down period.

2.0 When the reflux drum furnishes reflux and the feed stream to another unit then the drum must furnish "upstream protection" surge and "downstream protection" surge.

3.0 If there is a vapor product from the drum additional volume must be provided in the drum to allow vapor/liquid disengaging. This will be such as to retain the same liquid surge capacity as described above.

4.0 Should the vapor phase from the reflux drum be routed to the suction of a compressor an even larger volume reflux drum will be required. This is to ensure complete disengaging of the vapor/liquid. Internal baffles or screens are also used in the drums vapor outlet section to ensure complete phase separation.

Quantity of surge volume

The amount of surge volume will vary with the various types given above and with the specific case in question. Sometimes this amount is set by company specifications or, in the case of engineering contractors, by the client. Generally however the process engineer will be responsible for setting a safe and economic surge volume. In doing this the engineer needs to analyze each case in terms of why the surge volume is being provided then deciding how much based on this answer. Figure 13.1 provides some guidelines to the amount of surge that should be applied.

Some useful equations used in setting and handling surge volumes.

For surge volume vol cuft = (GPM) (minutes)/7.48

For vessel size:

Diam, $D = \sqrt[3]{(\text{cuft}/2.35)}$ @ L/D = 3
Diam, $D = \sqrt[3]{(\text{cuft}/1.96)}$ @ L/D = 2.5
Diam, $D = \sqrt[3]{(\text{cuft}/1.57)}$ @ L/D = 2.0

For line size:

Diam, $D = \sqrt[3]{(\text{GPM}/25)}$ @ velocity = 10 ft/sec
Diam, $D = \sqrt[3]{(\text{GPM}/17)}$ @ velocity = 7 ft/sec.
Diam, $D = \sqrt[3]{(\text{GPM}/12)}$ @ velocity = 5 ft/sec.

For flow rate:

ft/sec = GPM/450

For approx control valve size:

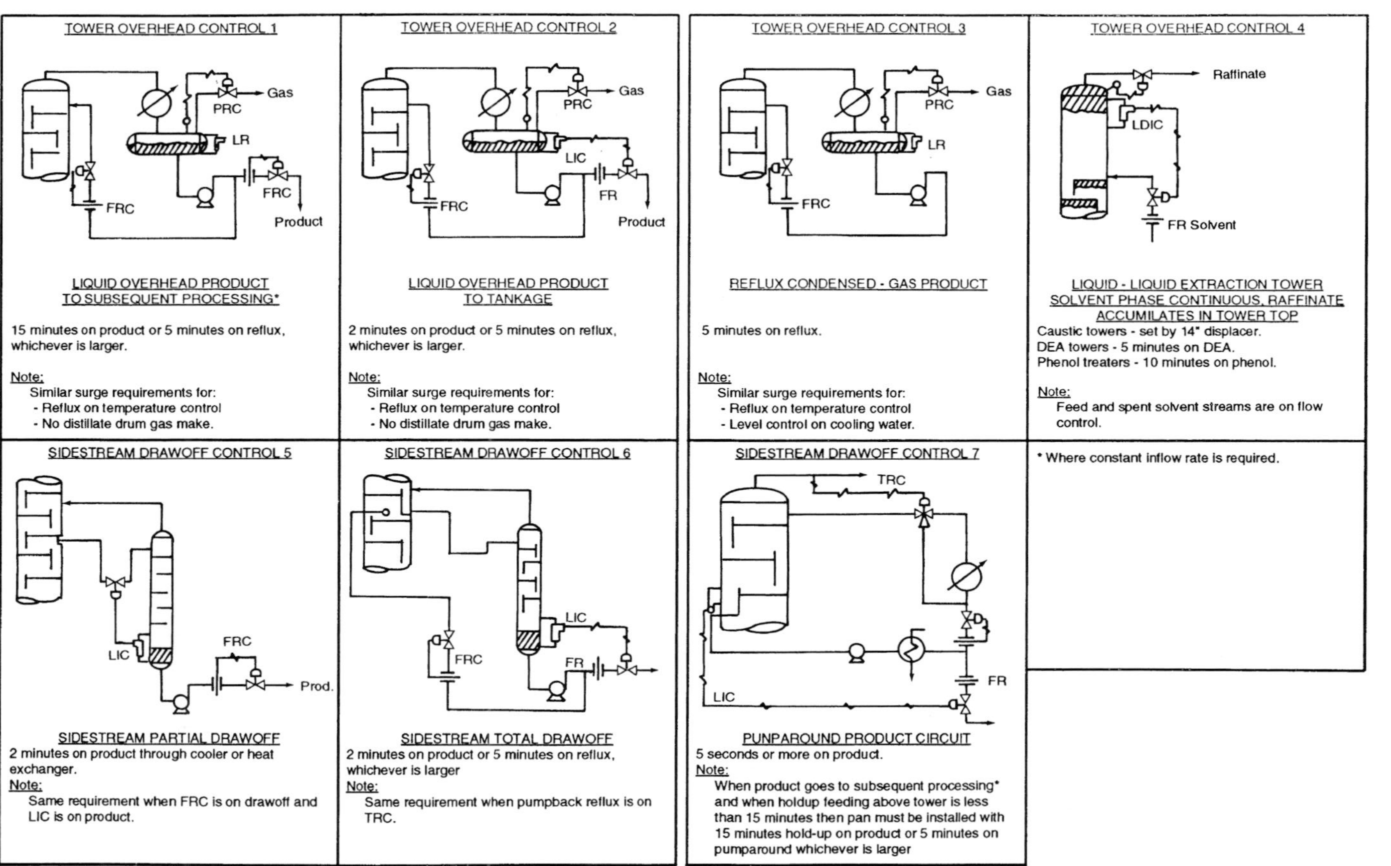

TOWER OVERHEAD CONTROL 1
Gas
PRC
LR
FRC
FRC
Product
LIQUID OVERHEAD PRODUCT TO SUBSEQUENT PROCESSING*
15 minutes on product or 5 minutes on reflux, whichever is larger.
Note:
Similar surge requirements for:
- Reflux on temperature control
- No distillate drum gas make.
TOWER OVERHEAD CONTROL 2
Gas
PRC
LIC
FRC
FR
Product
LIQUID OVERHEAD PRODUCT TO TANKAGE
2 minutes on product or 5 minutes on reflux, whichever is larger.
Note:
Similar surge requirements for:
- Reflux on temperature control
- No distillate drum gas make.
TOWER OVERHEAD CONTROL 3
Gas
PRC
LR
FRC
REFLUX CONDENSED - GAS PRODUCT
5 minutes on reflux.
Note:
Similar surge requirements for:
- Reflux on temperature control
- Level control on cooling water.
TOWER OVERHEAD CONTROL 4
Raffinate
LDIC
FR Solvent
LIQUID - LIQUID EXTRACTION TOWER SOLVENT PHASE CONTINUOUS, RAFFINATE ACCUMILATES IN TOWER TOP
Caustic towers - set by 14" displacer.
DEA towers - 5 minutes on DEA.
Phenol treaters - 10 minutes on phenol.
Note:
Feed and spent solvent streams are on flow control.
SIDESTREAM DRAWOFF CONTROL 5
FRC
LIC
Prod.
SIDESTREAM PARTIAL DRAWOFF
2 minutes on product through cooler or heat exchanger.
Note:
Same requirement when FRC is on drawoff and LIC is on product.
SIDESTREAM DRAWOFF CONTROL 6
LIC
FRC
FR
SIDESTREAM TOTAL DRAWOFF
2 minutes on product or 5 minutes on reflux, whichever is larger
Note:
Same requirement when pumpback reflux is on TRC.
SIDESTREAM DRAWOFF CONTROL 7
TRC
FR
LIC
PUNPAROUND PRODUCT CIRCUIT
5 seconds or more on product.
Note:
When product goes to subsequent processing* and when holdup feeding above tower is less than 15 minutes then pan must be installed with 15 minutes hold-up on product or 5 minutes on pumparound whichever is larger
* Where constant inflow rate is required.

Figure 13.1. Typical surge requirements.

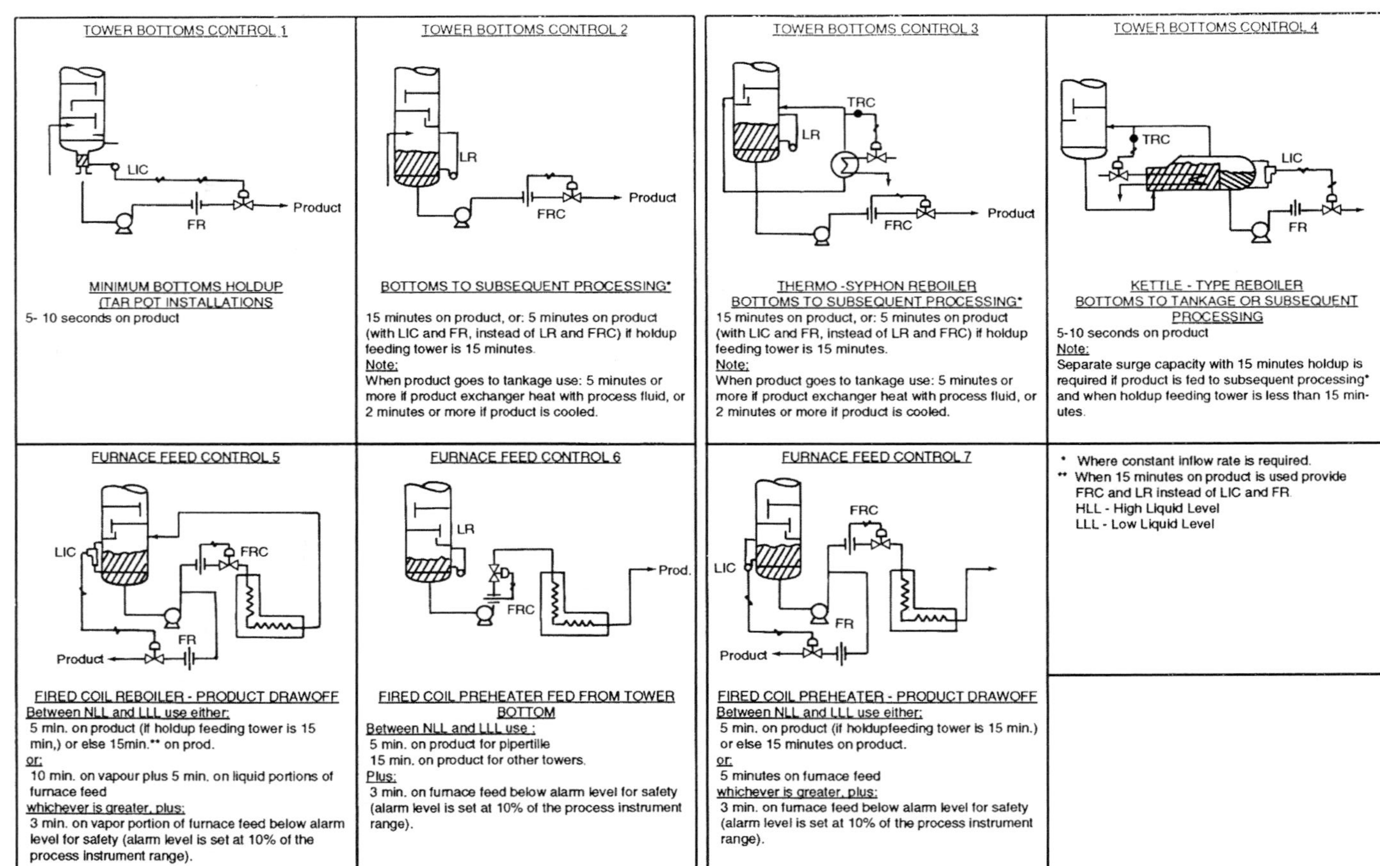

TOWER BOTTOMS CONTROL 1
LIC
FR
Product
MINIMUM BOTTOMS HOLDUP
(TAR POT INSTALLATIONS
5- 10 seconds on product
TOWER BOTTOMS CONTROL 2
LR
FRC
Product
BOTTOMS TO SUBSEQUENT PROCESSING*
15 minutes on product, or: 5 minutes on product (with LIC and FR, instead of LR and FRC) if holdup feeding tower is 15 minutes.
Note:
When product goes to tankage use: 5 minutes or more if product exchanger heat with process fluid, or 2 minutes or more if product is cooled.
TOWER BOTTOMS CONTROL 3
TRC
LR
FRC
Product
THERMO -SYPHON REBOILER
BOTTOMS TO SUBSEQUENT PROCESSING*
15 minutes on product, or: 5 minutes on product (with LIC and FR, instead of LR and FRC) if holdup feeding tower is 15 minutes.
Note:
When product goes to tankage use: 5 minutes or more if product exchanger heat with process fluid, or 2 minutes or more if product is cooled.
TOWER BOTTOMS CONTROL 4
TRC
LIC
FR
KETTLE - TYPE REBOILER
BOTTOMS TO TANKAGE OR SUBSEQUENT PROCESSING
5-10 seconds on product
Note:
Separate surge capacity with 15 minutes holdup is required if product is fed to subsequent processing* and when holdup feeding tower is less than 15 minutes.
FURNACE FEED CONTROL 5
LIC
FRC
FR
Product
FIRED COIL REBOILER - PRODUCT DRAWOFF
Between NLL and LLL use either:
5 min. on product (if holdup feeding tower is 15 min,) or else 15min.** on prod.
or:
10 min. on vapour plus 5 min. on liquid portions of furnace feed
whichever is greater, plus:
3 min. on vapor portion of furnace feed below alarm level for safety (alarm level is set at 10% of the process instrument range).
FURNACE FEED CONTROL 6
LR
FRC
Prod.
FIRED COIL PREHEATER FED FROM TOWER BOTTOM
Between NLL and LLL use :
5 min. on product for pipertille
15 min. on product for other towers.
Plus:
3 min. on furnace feed below alarm level for safety (alarm level is set at 10% of the process instrument range).
FURNACE FEED CONTROL 7
FRC
LIC
FR
Product
FIRED COIL PREHEATER - PRODUCT DRAWOFF
Between NLL and LLL use either:
5 min. on product (if holdupfeeding tower is 15 min.) or else 15 minutes on product.
or:
5 minutes on furnace feed
whichever is greater, plus:
3 min. on furnace feed below alarm level for safety (alarm level is set at 10% of the process instrument range).
* Where constant inflow rate is required.
** When 15 minutes on product is used provide FRC and LR instead of LIC and FR.
HLL - High Liquid Level
LLL - Low Liquid Level

Figure 13.1. (Cont.)

One size smaller than line size. Thus:

Line	CV
4″	3″
6″	4″
8″	6″
10″	8″

Level control

Surge volumes are maintained by controlling the amount of liquid entering or leaving the vessel in question. There are several means of accomplishing this and these are described and discussed in the following paragraphs:

(A) *Control the surge liquid outlet on level control.* This will give a close level control but a fluctuating outlet flow. The level control valve (LCV) will close completely at LLL. Thus flow through the outlet line will be completely shut off at LLL.

(B) *Control the surge liquid outlet on flow control and provide a low level alarm.* This will give a fluctuating level but will eliminate flow fluctuations. As there is no LCV to restrict flow at LLL then operators must physically reset the flow controller to maintain the surge volume. This has the disadvantage that the alarm condition could be missed or even ignored, resulting in possible damage to downstream equipment, and, in extreme cases, result in a fire or explosion hazard.

(C) *Control the surge liquid outlet on flow control reset by a level control.* This will give a wandering level but a smooth outlet flow. The LCV reset can still however cut off the flow completely on LLL.

(D) *Control the outlet flow by flow control and the system feed by surge volume level.* This will give close level control and also close outlet flow control. The outlet flow also will not be closed off by a LLL of the surge. In the case of the feed being that to a fractionating tower level control on the feed stream could cause tower upset conditions. This would be particularly undesirable on fractionators that operate close to critical conditions such as de ethanizer.

(E) *Control the surge liquid on level control to an intermediate surge vessel.* The liquid from the surge vessel may then be flow controlled. This is the ideal method for controlling feed to a fired heater. The only question in this case is one of economics.

Level control range

Should the decision now be to use an LC on the surge outlet (as in ‘A’ or ‘E’ above) the range of the instrument needs to be determined. The range is the vertical distance between the HLL and the LLL. Now if the liquid outlet is feeding another unit which requires a smooth flow it is possible to achieve this by using a wide proportional band and a large range. The larger the instrument range and the wider the proportional band the less sensitive is the level controller. Consequently the flow becomes smoother. However the larger the range, the more expensive is the controller and of course the larger is the tower or vessel in order to accommodate the greater distance between HLL and LLL.

The selection of level control system and the level control range depends therefore on:

1.0 How many outlet streams are there from the surge vessel.
2.0 Which streams cannot tolerate complete shut off before all the available surge is used.
3.0 The degree to which the outlet stream requires smooth flow.

Pressure control for gases

Pressure control for gases is similar to level control for liquids in that it is really a material balance control. If gas production rate varies, which is the usual case, a pressure controller will relieve the system of gas by holding a given pressure in a drum or tower. In this case the entire space above the liquid level actually constitutes a surge volume.

Summary

There are two very general rules to follow in selecting a control system. These are:

- If it is permissible for the product outflow rate to vary, use level control and a relatively small amount of surge capacity.
- If the product goes to a subsequent process where feed rate must be held constant, use flow control and considerably more surge.

The control valve

Control valves are used throughout the process industry to motivate the control of the operating parameters listed above.

Figure 13.2 shows a conventional control valve which in this case is taken as a double seated plug type valve. Like most control valves it is operated pneumatically by an air stream exerting a pressure on a diaphragm which in turn allows the movement of a spring loaded valve stem. One or two plugs (the diagram shows two) are attached to the bottom of the valve stem which, when closed, fit into valve seats thus providing

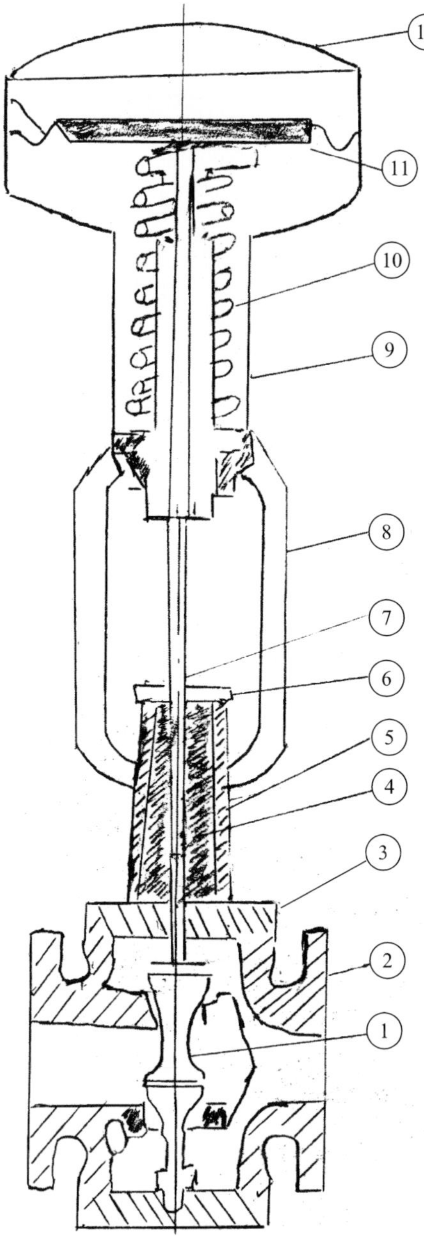

1 INNER VALVE

2 BODY

3 BONNET

4 PACKING

5 STUFFING BOX

6 GLAND

7 VALVE STEM

8 YOKE

9 SPRING BARREL

10 SPRINGS

11 DIAPHRAM AND PLATE

12 DIAPHRAM DOME

Figure 13.2. A conventional control valve.

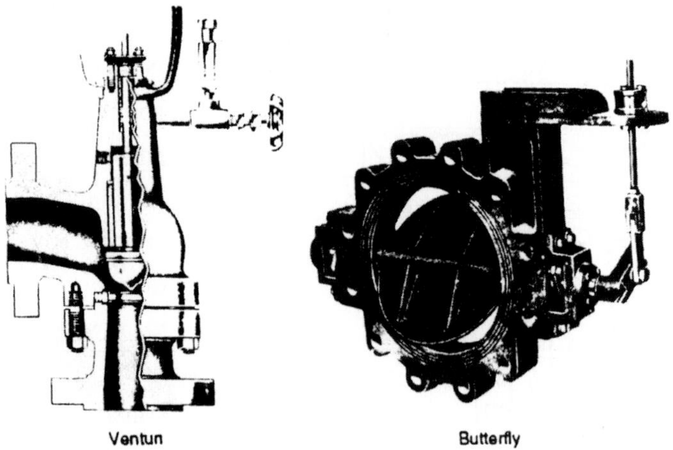

Figure 13.3. The venturi and butterfly control valves.

tight shut off. The progressive opening and closing of the plugs on the valve seats due to the movement of the stem, determines the amount of the controlled fluid flowing across the valve. The pressure of the air to the diaphragm controlling the stem movement is varied by a control parameter, such as a temperature measurement, or flow measurement, or the like. In many of the more modern refineries many control valves are operated electronically.

Figure 13.3 shows two other types of control valves in common service in the industry. These are the venturi type and the butterfly type. Both these types when pneumatically operated (which they usually are) work in the same way as described above for a plug type valve. The major difference in these two types are in the valve system itself. In the case of the venturi the fluid being controlled is subjected to a 90° angle change in direction within the valve body. The inlet and outlet dimensions are also different, with the inlet having the larger diameter. The valve itself is plug type but seats in the bend of the valve body. Venturi type or angle valves are used in cases where there exists a high-pressure differential between the fluid at the inlet side of the valve and that required at the outlet side.

Butterfly valves operate at very low-pressure drop across them. They can operate quite effectively at only inches of water gauge pressure drop, and where the operation of the conventional plug type valves would be unstable. The action of this valve is by means of a flap in the process line. The movement of this flap from open to shut is made by a valve stem outside the body itself. This stem movement, as in the case of the other pneumatically operated valves, is provided by air and spring loads onto the stem from a diaphragm chamber. The only major disadvantage in this type of valve is the fact that very tight shut off is difficult to obtain due to the flap type action of the valve.

The plug valve

There are two types of plug valves used for the conventional control valve function. These are:

- Single seated valves
- Double seated valves

Single seated valves are inherently unbalanced so that the pressure drop across the plug affects the force required to operate the valve. Double seated valves are inherently balanced valves and are the first choice in most services.

The conventional control valves predominantly in use can have either an equal percentage, or linear characteristics. The difference between these two is given in Figure 13.4.

With an equal percentage characteristic, equal incremental changes in valve stem lift result in equal percentage changes in the flow rate. For example if the lift were to

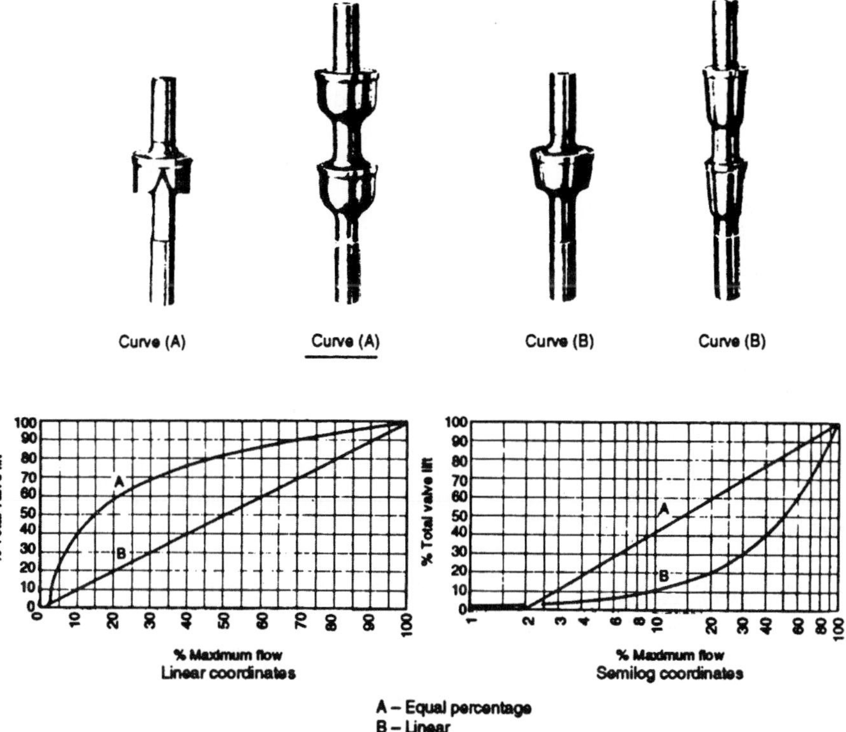

Figure 13.4. Control valve characteristics.

change from 20% to 30% of maximum lift the flow at 30% would be about 50% more than the flow at 20%. Likewise, if the lift increases from 40% to 50% the flow at 50% would be about 50% more than at 40%.

With a linear valve having a constant pressure drop across it, equal incremental changes in stem lift results in equal incremental changes in flow rate. For example, if the lift increases from 40% to 50% of maximum, the flow rate changes from 40% to 50% of maximum. Thus equal percentage is the more desirable characteristic for most applications and is the one most widely used.

Pressure drop across control valves

In sizing or specifying the duty of the control valve the pressure drop across the control valve must be determined for the design or maximum flow rate. In addition, if it is known that a valve must operate at a flow rate considerably lower than the maximum rate, the pressure drop at this lower flow rate must also be calculated. This will be required to establish the range ability of the valve.

As a general rule of thumb the sum of following pressure drops at maximum flow may be used for this purpose:

(a) 20% of the friction drop in the circuit[1] (excluding the valve).
(b) 10% of the static pressure of the vessel into which the circuit discharges up to pressures of 200 psig, 20 psig from 200 to 400 psig, and 5% above 400 psig.

The static pressure is included to allow for possible changes in the pressure level in the system (i.e., by changing the set point on the pressure controller on a vessel). The percentage included for static pressure can be omitted in circuits such as recycle and reflux circuits in which any change in pressure level in the receiver will be reflected through the entire circuit. In some circuits the control valve will have to take a much greater pressure drop than calculated from the percentages listed above. This occurs in circuits where the control valve serves to bleed down fluid from a high-pressure source to a low-pressure source. Examples are pressure control valves releasing gas from a tower or streams going out to tankage from vessels operating at high pressure. These are the circumstances where venturi or angle valves are used, as described earlier.

Valve action on air failure

In the analysis of the design and operation of any process or utility system the question always arises on the action of control valves in the system on instrument air failure.

[1] A circuit generally includes all equipment between the discharge of a pump, compressor or vessel and the next point downstream of which pressure is controlled. In most cases this latter point is a vessel.

Should the control valve fail open or closed is the judgment decision based principally on evaluating all aspects of safety and damage in each event. For example: control valves on fired heater tube inlets should always fail open to prevent damage to the tubes through low or no flow through them when they are hot. On the other hand control valves controlling fuel to the heaters should fail closed on air failure to avoid over heating of the heater during the air failure.

The failed action of the valve is established by introducing the motive air to either above the diaphragm for a failed open requirement or below the diaphragm for a failed shut situation. The air failure to the valve above the diaphragm allows the spring to pull up the plugs from the valve seats. Air failure to valves below the diaphragm forces the spring to seat the valves in the closed position.

Sizing a control valve

The sizing procedure for sizing control valves is given in Appendix 13.3 of this chapter.

13.2 Offsite systems

In most refineries the off sites facilities are often the major capital cost center second only to the process plants themselves. Indeed in some instances where the off sites include one or more complex jetties it can be the principal capital cost center. Among the major units found in most refineries are:

- Storage
- Product blending
- Road and rail loading
- Jetty facilities
- Waste disposal
- Effluent water treating

Many refineries consider that the flare is part of the safety system. To a large extent this is justifiable as most relief valves exhaust to the flare as does the process and other vent systems. For the purpose of this publication however the flare system is considered as part of the off sites and specifically comes under the section on waste disposal.

Storage facilities

The crude oil feed and the processed products are stored in storage tanks of various sizes and types. These tanks are usually collected and located together in the refinery area suitably defined as ‘The Tank Farm’. There will be many other tanks (usually

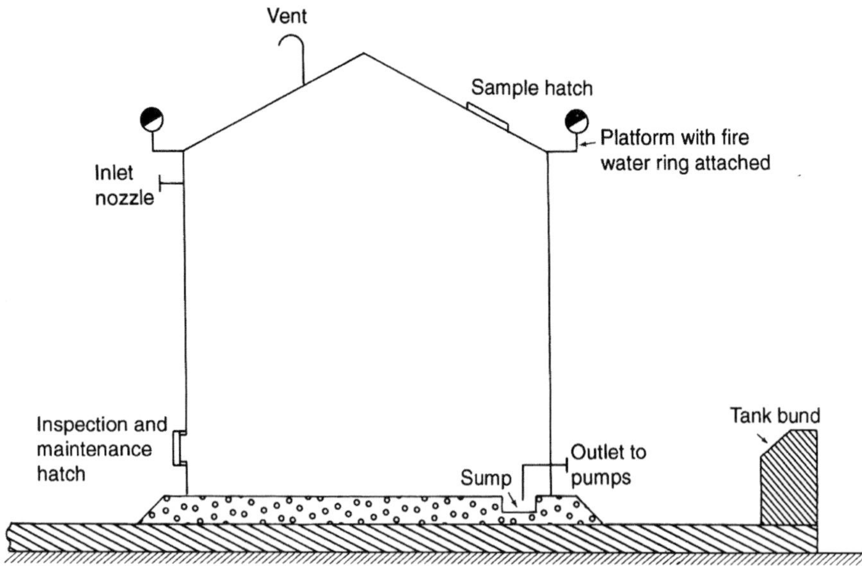

Figure 13.5. Cone roof tank.

much smaller than those in the tank farm) which will contain chemicals to be used in the processes and the slops or spent chemicals from the various processes and utility plants. Most refinery storage tanks fall into the following categories:

- Atmospheric storage
- Pressure storage
- Heated storage

Atmospheric storage

As the name implies, all atmospheric storage tanks are open to the atmosphere, or are maintained at atmospheric pressure by a controlled vapor blanket. These tanks fall into two categories:

- Cone roof tanks
- Floating roof tanks

Cone roofed tanks

Among the most common atmospheric storage tank is the cone roofed tank shown in Figure 13.5. This tank is used for the storage of non-toxic liquids with fairly low volatility. In its simplest form the roof of the tank will contain a vent, open to atmosphere, which allows the tank to "breathe" when emptying and filling. A hatch

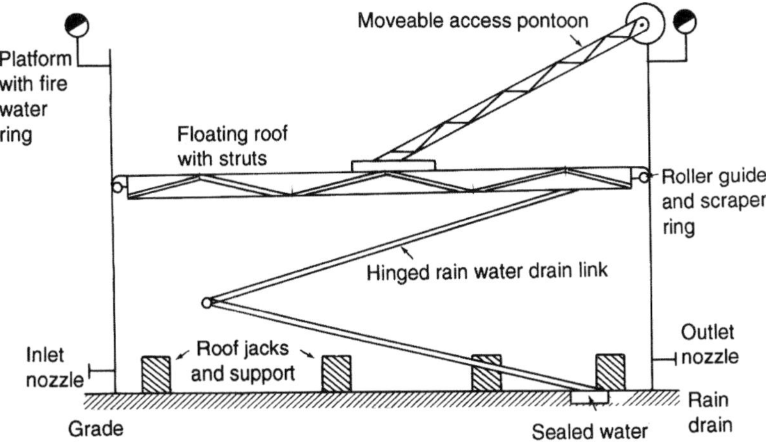

Figure 13.6. Floating roof tank.

in the roof also provides access for sampling the tank contents. In oil refining this type of tank is used for the storage of gas oils, diesel, light heating oil, and the very light lube oils (e.g., spindle oil).

In keeping with the company's fire protection policy tanks containing flammable material will be equipped with foam and fire water jets located around the base of the roof. All storage tanks containing flammable material and material that could cause environmental damage are contained within a dyked area or bund. The size of the bunded area is fixed by law and is usually such as to contain the total contents of one of the tanks included in the area. The number of tanks per bunded area is also fixed by legislation.

Floating roofed tanks

Light volatile liquids may also be stored at essentially atmospheric pressure by the use of "Floating Roof" tanks. A diagram of this type of storage tank is also given in Figure 13.6.

The roof of this tank literally floats on the surface of the liquid contents of the tank. In this way the air space above the liquid is reduced to almost zero, thereby minimizing the amount of liquid vaporization that can occur. The roof is specially designed for this service and contains a top skin and a bottom skin of steel plate, held together by steel struts. These struts also provides strength and rigidity to the roof structure. The roof moves up and down the inside of the tank wall as the liquid level rises when filling and falls when emptying. The roof movement is enhanced by guide rollers between the roof edge and the tank wall. A scraper ring around the edge of the roof top and pressed tightly against the tank wall ensures a seal between the contents and the atmosphere. It also provides additional guide to the roof movement and stability to the roof itself.

When the tank reaches the minimum practical level for the liquid contents the roof structure comes to rest on a group of pillars at the bottom of the tank. These provide the roof support when the tank is empty and a space between the roof and the tank bottom. This space is required to house the liquid inlet and outlet nozzles for filling and emptying the tank which, of course, must always be below the roof. The space is also adequate to enable periodic tank cleaning and maintenance.

A hinged drain line running inside the tank from the roof to a "below grade" sealed drain provides the facility for draining the roof of rain water. The hinges in the drain line allows the line to move up and down with the roof movement. A pontoon type access pier from the platform around the perimeter at the top of the tank provides access to the sample hatch located at the center of the roof. This "pontoon" also moves upward and downward with the roof movement. Automatic bleeder vents are provided on all floating roof tanks. They vent air from under a floating roof when the tank is being filled initially from empty. After the liquid rises high enough to float the roof off its supports, the vent automatically closes. Likewise when the tank is being emptied, the vent automatically opens just before the roof lands on its support, thereby preventing the development of a vacuum under the roof. Other accessories include rim vents, float gauges, anti rotation devices, and manholes.

Liquids stored in this type of tank have relatively high volatilities and vapor pressures such as gasoline, kerosene, jet fuel, and the like. In oil refining the break between the use of cone roof tank and floating roof is based on "flash point" of the material. Flash point is that temperature above which the material will ignite or "flash" in the presence of air. Normally this break point is 120°F.

Pressure storage

Pressure storage tanks are used to prevent or at least minimize the loss of the tank contents due to vaporization. These types of storage tanks can range in operating pressures from a few inches of water gauge to 250 psig. There are three major types of pressure storage. These are:

- *Low-pressure tanks*—These are dome roofed tanks and operate at a pressures of between 3 ins water gauge and 2.5 psig.
- *Medium pressure tanks*—These are hemispheroids which operate at pressures between 2.5 and 5.0 psig, and spheriodal tanks which operate at pressures up to 15 psig.
- *High-pressure tanks*—These are either horizontal "bullets" with elipsoidal or hemispherical heads or spherical tanks (spheres). The working pressures for these

types of tanks range from 30 to 250 psig. The maximum allowable is limited by tank size and code requirements. For a 1,000 bbl sphere, the maximum pressure is 215 psig, for a 30,000 bbl it is 50 psig. These pressure limits can be increased if the tank is stress relieved.

Although it is possible to store material in tanks with pressure in excess of 250 psig normally when such storage is required refrigerated storage is usually a better alternate.

Heated storage tanks

Heated storage tanks are more common in the petroleum industry than most others. They are used to store material whose flowing properties are such as to restrict flow at normal ambient temperatures. In the petroleum industry products heavier than diesel oil, such as heavy gas oils, lube oil, and fuel oil are stored in heated tanks.

Generally speaking tanks are heated by immersed heating coils or bayonet type immersed heaters. Steam is normally used as the heating medium because of its availability in petroleum complexes. Very often where immersed heating is used the tank is agitated usually by side located propeller agitators for large tanks. Where external circulating heating is used for tanks, the contents are mixed by means of jet mixing. Here the hot return stream is introduced into the tank via a specially designed jet nozzle as shown in Figure 13.7. External tank heating is used when there is a possibility of a hazardous situation occurring if an immersed heater leaks.

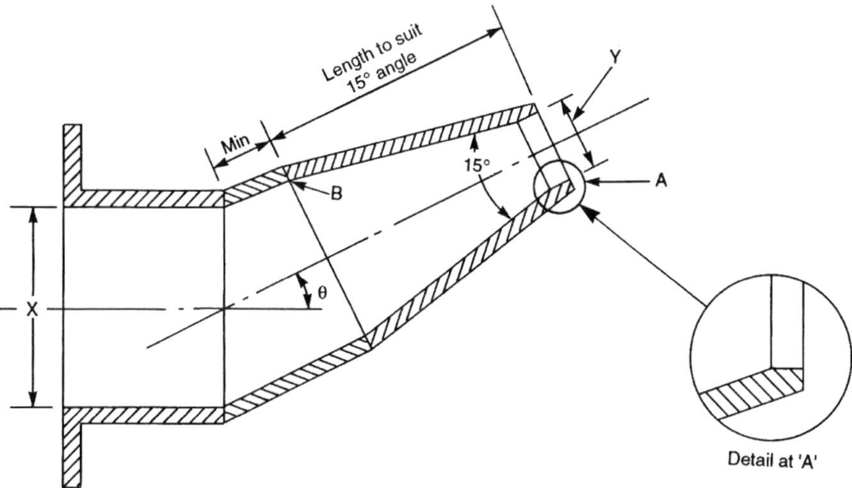

Figure 13.7. A jet mixing nozzle.

Calculating heat loss and heater size for a tank

Heat loss and the heater surface area to compensate for the heat loss may be calculated using the following procedure:

Step 1. Establish the bulk temperature for the tank contents. Fix the ambient air temperature and the wind velocity normal for the area in which the tank is to be sited.

Step 2. Calculate the inside film resistance to heat transfer between the tank contents and the tank wall. The following simplified equation may be used for this:

$$h_c = 8.5(\Delta t/\mu)^{0.25}$$

where

h_c = Inside film resistance to wall in Btu/hr sqft °F.
Δt = Temperature difference between the tank contents and the wall in °F.
μ = The viscosity of the tank contents at the bulk temperature in cps.

The heat loss calculation is iterative with assumed temperatures being made for the tank wall.

Step 3. Using the assumed wall temperature made in step 2 calculate the heat loss to atmosphere by radiation using Figure 13.8. Then calculate the heat loss from the tank wall to the atmosphere using Figure 13.9. Note the temperature difference in this case is that between the assumed wall temperature and the ambient air temperature. Correct these figures by multiplying the radiation loss by the emissivity factor given in Figure 13.8. Then correct the heat loss by convection figure by the factors as described in item 4 below.

Step 4. The value of h_{co} read from Figure 13.9 is corrected for wind velocity and for shape (vertical or horizontal) by multiplying by the following shape factors:

Vertical plates	1.3
Horizontal plates	2.0 (facing up)
	1.2 (facing down)

Correction for wind velocity use

$$F_w = F_1 + F_2$$

where

F_w = wind correction factor
F_1 = wind factor @ 200°F calculated from:
$F_1 = (\text{MPH}/1.47)^{0.61}$
F_2 = Read from Figure 13.10

Then the corrected h_{co} is:
$h_{co} \times$ shape correction $\times Fw$.

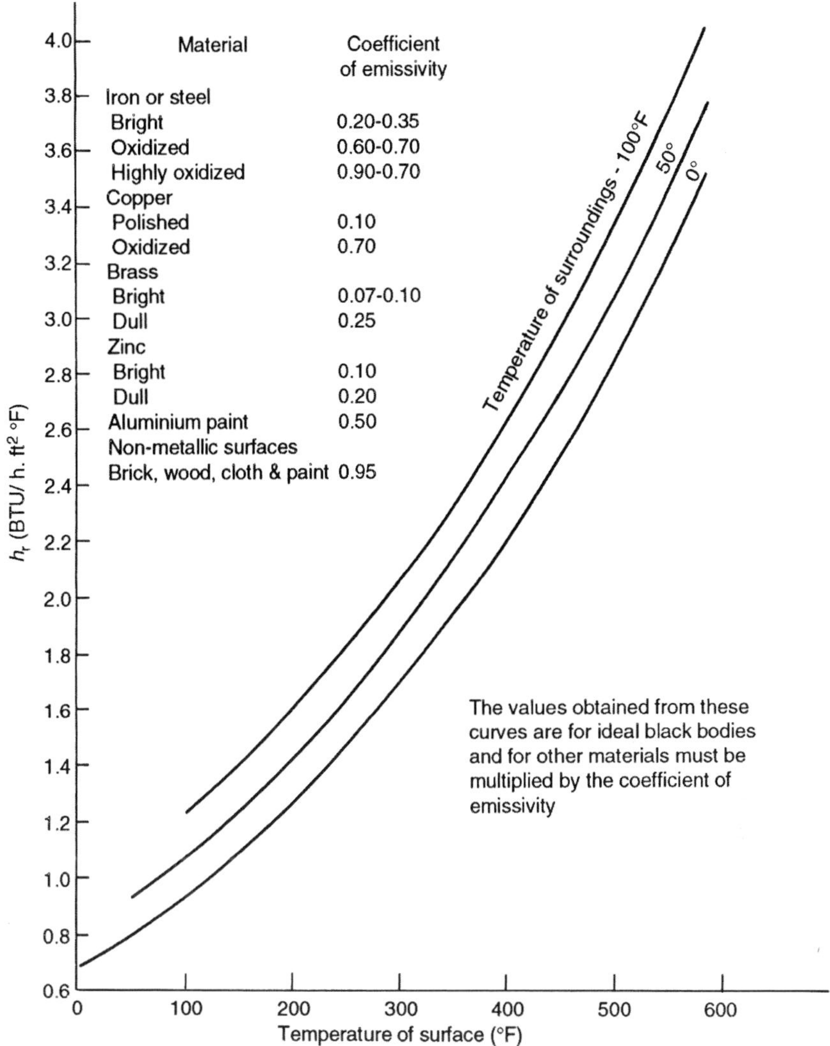

Figure 13.8. Heat loss by radiation.

Step 5. The resistance of heat transferred from the bulk of the contents to the wall must equal the heat transferred from the wall to the atmosphere. Thus:
Heat transferred from the bulk to the wall = '*a*'

$$= h_c \text{ from step 2} \times \Delta t \text{ in Btu/hr} \cdot \text{sqft.}$$

where Δt in this case is (bulk temp—assumed wall temp)

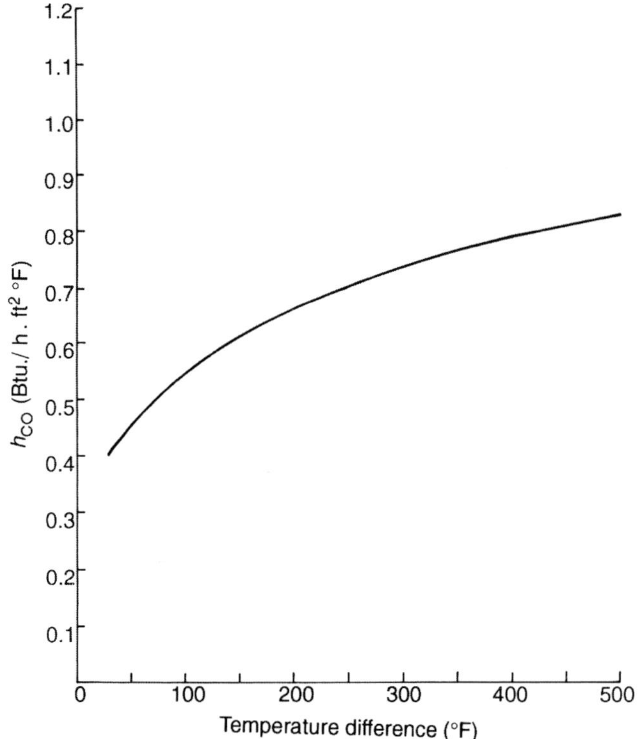

Figure 13.9. Heat loss to atmosphere by natural convection.

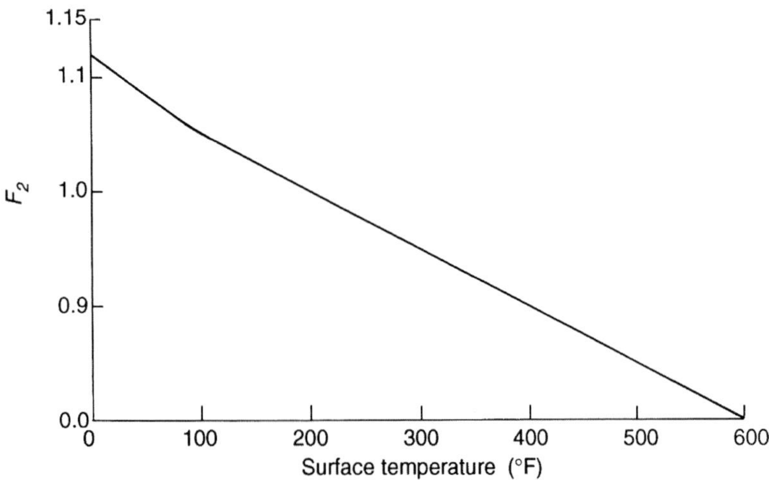

Figure 13.10. Plot of 'F_2' versus surface temperature.

Heat transferred from the wall to the atmosphere = 'b'

$$= (h_{co} + hr) \times \Delta t \text{ in Btu/hr sqft}$$

where Δt in this case is (assumed wall temp—air temp).

Step 6. Plot the difference between the two transfer rates against the assumed wall temperature. This difference ('a' − 'b') will be negative or positive but the wall temperature that is correct will be the one in which the difference plotted = 0. Make a last check calculation using this value for the wall temperature.

Step 7. The total heat loss from the wall of the tank is the value of 'a' or 'b' calculated in step 6 times the surface area of the tank wall. Thus:

$$Q_{wall} = h_c \times \Delta t \times (\pi D_{tank} \times \text{tank height}) \text{ in Btu/hr.}$$

Step 8. Calculate the heat loss from the roof in the same manner as that for the wall described in steps 2–7. Note the correction for shape factor in this case will be for horizontal plates facing upward, and the surface area will be that for the roof.

Step 9. Calculate the heat loss through the floor of the tank by assuming the ground temperature as 50°F and using;

$$h_f = 1.5 \text{ Btu/hr sqft°F}$$

Step 10. Total heat loss then is:

$$\text{Total heat loss from tank} = Q_{wall} + Q_{roof} + Q_{floor}.$$

Step 11. Establish the heating medium to be used. Usually this is medium pressure steam. Calculate the resistance to heat transfer of the heating medium to the outside of the heating coil or tubes. If steam is used then take the condensing steam value for h as 0.001 Btu/hr sqft °F. Take value of steam fouling as .0005 and tube metal resistance as 0.0005 also. The outside fouling factor is selected from the following:

Light hydrocarbon = 0.0013
Medium hydrocarbon = 0.002
Heavy Hc such as fuel oils = 0.005

The resistance of the steam to the tube outside $= \dfrac{1}{h + \mathrm{R}}$

where $R = r_{\text{steam fouling}} + r_{\text{tube metal}} + r_{\text{outside fouling}}$.

Step 12. Assume a coil outside temperature. Then using the same type of iterative calculation as for heat loss, calculate for 'a' as the heat from the steam to the coil outside surface in Btu/hr sqft. That is

$$\text{'}a\text{'} = h \times \Delta_{ti}$$

Calculate for 'b' as the heat from the coil outside surface to the bulk of the tank contents. Use Figure 13.11 to obtain ho and again 'b' is $h_o \times \Delta_{to}$ where the Δ_{to} is the temperature between the tube outside and that of the bulk tank contents. Make further assumptions for coil outside temperature until 'a' = 'b'.

Step 13. Use 'a' or 'b' from step 12 which is the rate of heat transferred from the heating medium in btu/hr sqft and divide this into the total heat loss calculated

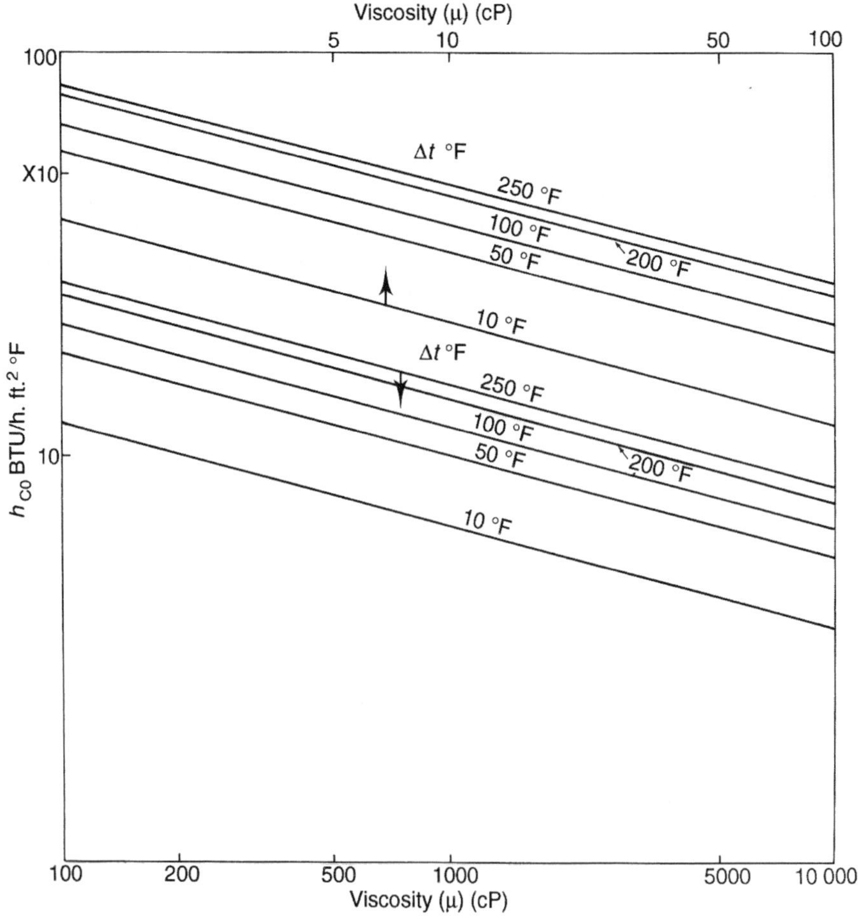

Figure 13.11. Convection heat transfer coefficient.

in step 10. The answer is the surface area of the immersed heater required for maintaining tank content's bulk temperature.

An example calculation using this technique is given as Appendix 13.1 at the end of this chapter.

Product blending facilities

Blending is the combining of two or more components to produce a desired end product. The term in refinery practice usually refers to process streams being combined to make a saleable product leaving the refinery. Generally these include gasolines, middle distillates such as: jet fuel, kerosene, diesel, and heating oil. Other blended finished products will include various grades of fuel oil and lube oil. The blending of the process streams is accomplished either by batch blending in blending tanks or in-line blending in the pipe line itself.

In-line versus batch blending

In batch blending the components are routed separately into a single receiver tank. They are mixed in this tank to meet the finished product specification. In the case on in-line blending the component streams are routed through automatically operated flow valves to a finished product tank. With modern computerized control technology in-line blending is becoming the more common form of blending process. In the case of gasolines and lube oils in particular in-line blending is extensively the accepted method. Middle distillates and residuum blending by batch has still some advantage because there are much fewer components to be handled, although the quantities involved are usually greater.

The in-line blender operation

An in-line blender is essentially a multiple stream controller with feed-back. The controller itself is a computer into which the recipe for the blend is keyed. Such blending recipes have been covered in some detail in Chapter 1 of this Handbook. The controller automatically starts the pumps for the blend components and motivates the flow control valves on the component lines to meet the required component quantities. In most cases the component lines join together to form the blended product which is then routed to the finished product tank. A series of on line analyzers located in the blend run down lines, monitor the finished product properties and in turn, reset the controller adjusting the component quantities to meet the end product specification.

The in-line blender design

Most refinery companies have their own proprietary component blending recipes for their finished products. These will be in computer program form and usually utilizing the linear programming techniques described in Chapter 8 of this Handbook. It is this program software that is installed into the blender controller to activate the respective component systems. The blender controller acts to start the selected pumps and the control valves. It also receives data from the on line analyzers located in the product run down line.

The design of the blending system as a whole is the combined effort of the process engineer, the instrument engineer, and the computer specialist. It will be the duty of the process engineer to develop the blending recipe in terms of component percentages and quantities to meet a particular product specification. The instrument engineer will ensure that the control valves, pump starting assembly and the on stream analyzers meet the process requirements. A typical list of the instrument engineer's responsibilities in this regard are as follows:

- The control panel, panel instruments, instrument power supply, annunciators pump start push buttons, indicating lights, graphic display and all panel wiring
- Turbine meters and pre-amplifiers
- Control valves with positioners
- Stream analyzers
- Field transmitters (flow, temperature, pressure)

Finally it will be the computer specialist's duty to translate these requirements into the software program for the controller computer.

Component tankage

Because the in line blender permits the rapid conversion of the components to finished products, the ratio of component to finished product tankage should be quite large. For good flexibility this ratio should be 4 or 5 to 1.

The most significant process requirement for successful blending is that the properties of the individual components do not change during the blending operation. Alkylates and catalytically cracked naphtha vary little unless feed or operation are changed. Reformates and straight run naphtha however have a greater variability due to changes in source crude feed quality. These are often stored in separate "running gage" tankage the quality of the contents of which are tested when full.

Lube oils are a particular problem in component storage, because of the tendency of the oils to 'stratify'. That is these oils tend to separate in storage with the heavier gravity and viscosity portion of the oil sinking to the tank bottom. In most cases the contents of these tanks are continually mixed using propeller type mixers. Many companies adopt this system to all heavy component tankage such as gas oils, and fuel oil product components.

Finished product tankage

Finished product tankage is needed even with the most efficient in line blender. This is because of required product disposal rates, and product certification. However in many design cases, in-line blending will still only require about one half the product storage required by batch blending.

Actual requirements of product tankage will usually be dictated by the manner in which the product is to be shipped from the refinery. The blend rate is usually sized to blend a day's production in a certain number of hours. The maximum rate of an economically sized blender will usually be too slow to blend into an ocean going tanker.

Thus, product tankage and loading pumps are needed to supplement the blender. Conversely, the minimum rate of a blender will usually be too fast to blend directly into a road tanker. In some cases however this is done by limited volume transfer pumps taking suction directly from component tankage.

Road and rail loading facilities

The extent of product shipping facilities required in a chemical or petroleum complex depends on the size of the complex, the local market, the number of different products to be shipped and the market to be supplied. Normally the shipping facilities installed in most plants is sufficient to cater for normal product handling and the flexibility required for seasonal demands. The capacity of these facilities will almost invariably exceed the plant's total production.

The most common method of shipping product is by road or rail in suitably designed tanker cars. In the case of large complexes located on coastal or river side sites shipping by barge or ships carry the bulk of the plant products. This section however will deal only with dispatch by road and rail.

Loading rates

Loading rates for road and rail tankers vary from as low as 150 GPM to as high as 1,000 GPM, but most terminals load at rates between 300 to 550 GPM. Road tankers have capacities from 1300 to 6500 gals and one tractor can haul two 6500 gals tanks. The number of loading arms required for each product to be loaded varies with:

- Truck size
- Number of loading hours per day
- Number of loading days per week
- Time for positioning, hook up, and depositioning of the truck.

Figure 13.12 gives the number of arms or spouts required for loading a 3,500 gal truck under various conditions.

The conditions shown in Figure 13.12 is for filling at a rate of 300 GPM (bottom curve) and for 500 GPM (top curve). The loading time is taken as the filling time per tank truck plus 10 minutes. Thus loading time is:

$$\frac{\text{Tank truck capacity gals}}{\text{GPM}} + 10 \text{ minutes}$$

Tank car capacity is taken as 3,500 Gallons. Thus for the lower curve loading time is 22 minutes per car, and for the upper curve 17 minutes per car. Assuming that a single product is loaded over 4 hr in an 8 hr day 5 days a week then number of trucks

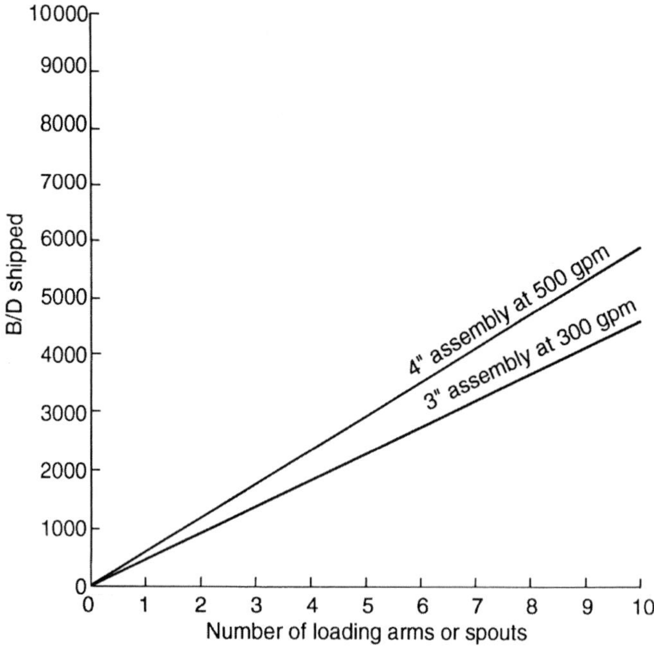

Figure 13.12. Number of loading arms for quantity shipment.

required per Barrel/Day is:

$$\frac{\text{B/D} \times 42 \times 5 \text{ days} \times 8 \text{ hr/day}}{3500 \text{ gals/truck} \times 20 \text{ hr loading per week}} = 0.024 \text{ trucks per Barrel/Day.}$$

Then for 1000 B/D number of trucks per working day = 0.024 × 1000

= 24 Trucks/day of 4 hr filling.

Time to load trucks at 300 GPM filling rate = 24 × 22 minutes = 528 minutes.

To complete loading in 4 hrs the number of arms required $= \frac{528}{240} = 2.2$

Continuing with this calculation for several more shipping capacities and for filling rates of 300 and 500 GPM produces the data given in Figure 13.12.

Loading equipment

Figure 13.13 is a schematic drawing for a typical road or rail loading facility.

The loading pumps which are located close to the product storage tanks take suction from these tanks. The loading pumps are high capacity, low head type with flat head/capacity characteristic. They operate at between 35 and 45 psi differential head.

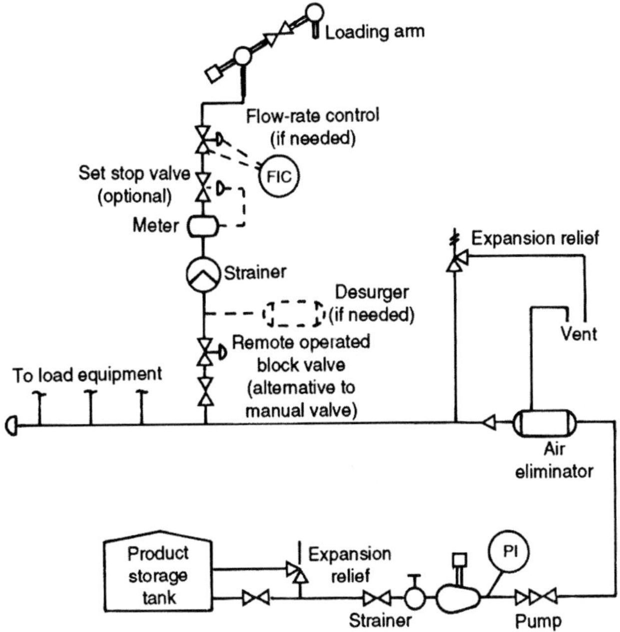

Figure 13.13. Schematic diagram of a loading facility.

These pumps discharge through an air eliminator drum into the loading header. Several loading arm assemblies are connected to the loading header. Each of these assemblies include a remote operated block valve followed by a desurger (optional), then a strainer located before the loading meter. The product flows from the meter into a swivel jointed loading arm and nozzle. Tank trucks and rail cars are loaded through their top hatches into which the nozzles of the loading arm fit.

Air eliminators are used to disengage air and other vapors which would interfere with the accuracy of the meters. Disengaging of the vapors is accomplished at about 3 psig. Should there not be sufficient static head at the disengaging vessel a back pressure valve must be provided to obtain this pressure. The meters are positive displacement type and desurgers are installed to decrease hydraulic shock resulting from quick shutoff.

Loading facilities arrangement

Figures 13.14 and 13.15 show the arrangement of loading facilities for truck and railcar, respectively.

The dimensions shown on the diagram are applicable to one world area and may not be applicable to other localities. The equipment and its arrangement shown in the diagrams however are standard for most of these facilities.

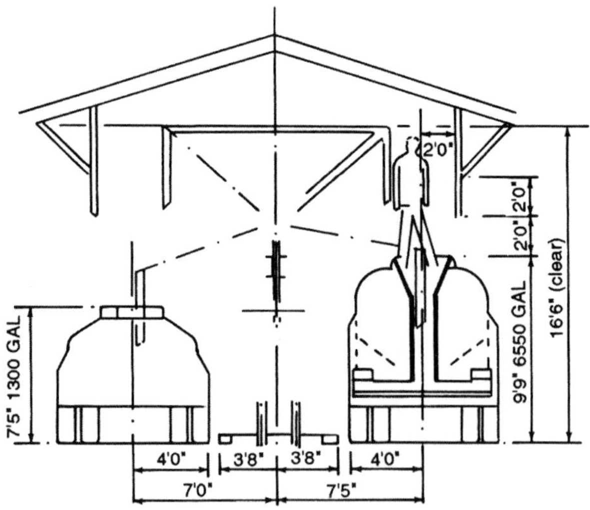

Figure 13.14. Tank truck loading facility.

In truck loading the meters and strainers are located at the loading station. The connection of the loading arm is made by an operator actually standing on the car itself. In the case of the truck loading facilities the loading arm is operated from an adjacent platform. As in the case of the truck loading the meter and strainer together with the on/off valve is located on the loading site.

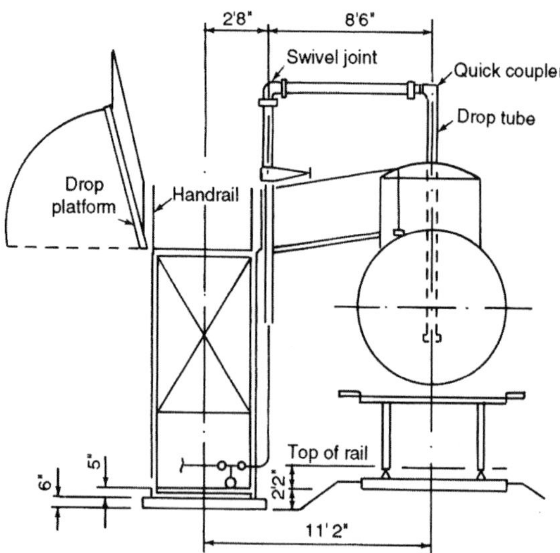

Figure 13.15. Rail car loading facility.

Jetty and dock facilities

Tankers and barges are loaded and unloaded at jetties or docks. In almost all circumstances these facilities for handling petroleum products are separate from those used for general cargo. Very often tankers, particularly the modern 'Super' tankers are loaded and unloaded by submarine pipelines at deep water anchorage. This section of the chapter however deals only with onshore docking facilities.

Jetty size, access, and location

Tanker sizes range from small coastal vessels of 10,000 bbl capacity to super tankers in excess of 250,000 bbl capacity. The more common tanker size is one of 140,000 bbl capacity and this size tanker is labeled a T2. This tanker is usually used for product carrying. It can carry as much as three different product parcels at the same time. The larger tankers are usually used for crude oil transportation.

Ideally the jetty size should be sufficient to cater for both these size tankers, and usually at least one of each size at the same time. In some refineries which have jetty facilities these usually include barge loading items also. The barge loading may however be located on remote docking facilities from the larger sea going tankers.

The location of the jetty itself must consider the following:

- There is sufficient deep water to cater for the larger crude carrying tanker
- It is located as close as possible to the refinery's tank farm
- It is in an area that has a good approach road and park
- There is sufficient room for a product/crude pipe-way
- There is sufficient waterway in which to maneuver and handle tanker docking

A fully loaded T2 tanker has a maximum draft of 30 ft. The larger crude carrying tanker would have a maximum draft around 45 ft. It is important to minimize rundown pipe lengths to and from the jetty loading area and the refinery tank farm. The first is the piping cost factor and the second is that often the pumping characteristics may have a negative static head during the pumping program.

A good onshore jetty approach road is mandatory for the operation of the jetty. This is required for safety reasons and the easy approach way for emergency vehicles (such as fire engines and ambulances). The approach road is also required for the transportation of the operating staff, ship's crew, and the ships chandler vehicles. Usually this approach road is dedicated for jetty use and will be quite independent of any adjacent refinery road.

There must be room on shore and on the jetty itself for the loading and unloading pipe way(s). This can be quite extensive depending on the refinery size and the number of products that are exported. There are several options for the location and size of this pipe way configuration. On the jetty itself it maybe carried on overhanging supports on both sides of the jetty pier or it can be supported by an independent pipe rack adjacent to the jetty pier (much more expensive however), and this pipe rack could be multitiered. On shore the pipe way can be located along the roadway at ground level or elevated with two or more tiers. It could also be elevated and run above the roadway. This does however restrict access by limiting vehicle height using the roadway.

Finally the location of the jetty must allow sufficient waterway room for tankers to be berthed properly. Tankers arriving from the open sea must have room so that tugs can handle and turn the ship around to face open sea before tying up at the jetty.

A layout plan of a typical tanker jetty is shown below as Figure 13.16.

Equipment

The equipment required for tanker loading include pumps, hose or flexible loading pipe, and handling cranes or structures. The loading pumps are located at the tank farm. These are centrifugal type with discharge pressures in excess of 100 psig. This is dependant on rundown pipe lengths, but pressure drop through the loading hoses or flexible loading pipes are within 10–25 psig. Also, and as mentioned earlier there is often a static head loss to the deck manifold of an empty tanker.

When a rubber hose is used, it is supported by a dockside derrick plus the tanker boom. Some installations employ a combination of hose and pipe or flexible assemblies of pipe and swivel joints supported by structures. Automatic adjustment for tide and tanker draft is incorporated in this equipment. Hose and various assemblies are available in sizes from 2 to 12 inches diameter with the 8 and 10 inch most frequently used for products.

Barges are usually loaded through hoses supported by dockside derricks or, in some cases by derricks on the barges. To conserve space barges are frequently moored two or three abreast with the loading hose being manhandled across the in board barges to the outboard barge. This hose is accordingly limited in size to a 6 inch diameter weighing about 8 lb/ft.

Loading rates

Tanker piping and pumping systems are designed for relatively high rates. The unloading pumps for the T2 tanker size can handle up to 8,000 bph. Super tankers unloading crude have pumps that can handle quantities of 22,000 bph or more. A

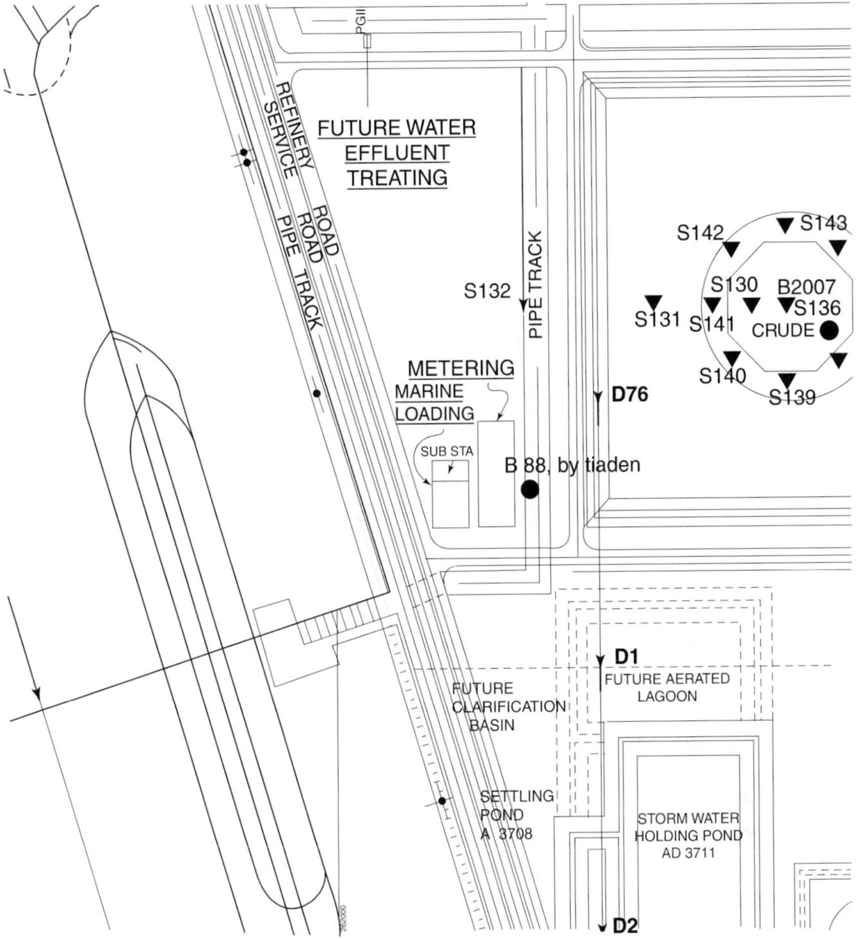

Figure 13.16. A typical tanker jetty layout.

desirable product loading rate for each product is 8,000–10,000 bph and it is general practice to load two products simultaneously. Loading rates for barges are usually limited to 2,000 bph. Barges have capacities ranging from 600 to 2,000 bbl. Barges are flat bottomed shallow draft vessels used to transport products short distances in canals, harbors, or inland waterways.

Quantities loaded aboard tankers are measured by metering storage tanks and tanker compartments. Products loaded into barges are measured by metering. Meters are located on the dock near the hose connections. These metering facilities include strainers, and flow controllers.

Other features

Some of the other features that are considered in the establishing of jetty facilities are:

- Ship ballast handling
- Tanker mooring facilities
- Slop and spill collection facilities
- Lighting and communication facilities

Ship ballast water is handled using specially allocated on shore tanks to collect the water which will of course be contaminated. This contamination will be the petroleum product residue remaining in the ships tanks after product unloading. Ships arriving at the refinery jetty under ballast are usually the smaller product tankers. These are the vessels that will load at the refinery with the finished products for shipment. The ballast water is pumped from the ships tanks to the onshore ballast water tankage by the ships pumps. From the ballast water tanks the content is drained off to the refinery's effluent treating facilities. The hydrocarbon contaminants are removed from the water in the treating plant and routed to the refineries slop system to finally enter the refinery processes. The treated effluent water is drained back to the sea.

The length of the jetty's loading/unloading wharf where the ships are moored are sized to accommodate two or more tankers of fixed length (say two T2's). The allocated space for these vessels must conform to standard conditions usually established by the particular Port Authority Regulations. One of these regulations which sets the length of the wharf is that the space between moored vessels should be such that the stern and aft mooring lines of adjacent vessels measured at an angle of 45° to the center line of the vessels do not overlap.

Slop and spill facilities around loading or unloading vessels at the wharf may be contained by a temporary boom installed around the vessel during these operations. Any spillage is contained by the boom and is subsequently disposed of by the same route as the ballast water.

Jetty lighting is based on the main refinery lighting code and practice. This means that all access ways and roads will have general street lighting. Areas where personnel are employed on a 24 hr basis will be flood lit between the hours of sun set and sun rise. This lighting will be supplemented by the ships lighting facilities as required for loading/unloading activities and for ship berthing and departure. Ship to onshore communication by means of telephone. radio, and company computer systems are activated as soon as the ship has been berthed.

Waste disposal facilities

All process plants including oil refineries produce large quantities of toxic and/or flammable material during periods of plant upset or emergencies. A properly designed flare and slop handling system is therefore essential to the plant operation. This section describes and discusses typical disposal systems currently in use in the oil refining industry where the hydrocarbon is immiscible with water. Where the chemical is miscible in water special separation systems must be used.

Figure 13.17 shows a completely integrated waste disposal system for the light end section of an oil refinery. The system shown here consists of three separate collection systems being integrated to a flare and a slops rerun system. A fourth system is for the disposal of the oily water drainage with a connection to the flare and a separate connection for any oil laden skimming. This later connection would be to route the skimmed oil to the refinery slop tanks. In the three integrated systems, the first collects all the vapor effluent streams from the relief headers. The contents of this stream will be material normally vapor at ambient conditions. It would be the collection of the vapors from the relief valve and the vapor venting on plant shut down or upset conditions. The second of the three systems is the liquid hydrocarbon drainage. The material in this system is liquid under normal ambient conditions and is collected from drain headers used to evacuate vessels during shutdown or upset conditions. Both the first and the second collection systems are routed to the flare knockout drum. The second (liquid system) may also be routed to the light ends slop storage drum. The liquid phase from the flare knockout drum is also routed to the slops storage drum. The third system is the light ends feed diversion. This allows the light ends unit to be bypassed temporarily by sending the feed to the slop drum for rerunning later.

Further description and discussion of these disposal systems is given in the following sections:

- Blow-down and slop disposal
- Flares

Blow-down and slop

This system generally consists of the following drums:

- Non-condensable blow-down drum
- Condensable blow-down drum
- Water disengaging drum

A typical non-condensable blow-down drum is shown in Figure 13.18. These types of drums are provided for handling material normally in the vapor state and high volatility liquids. These drums receive and disengage liquid from safety valve headers, and drain headers. These drums are often referred to as flare knockout drums as the disengaged vapor is routed directly to a flare. The drum is basically a surge drum and therefore should be sized as one using the following criteria:

1.0 Normal liquid surge is based on the daily liquid draw-off to drain per operating day of 24 hr. This includes spillage, sample point draining, etc.

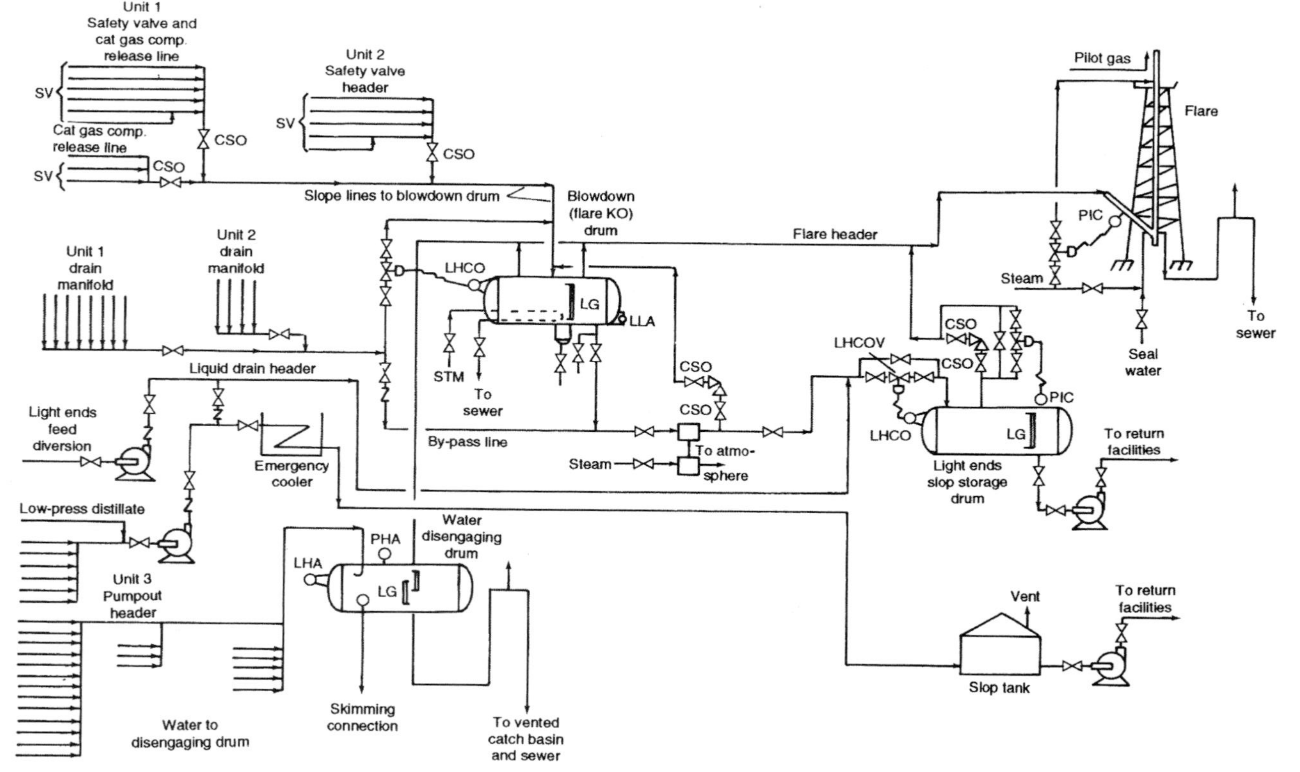

Figure 13.17. An integrated water disposal system.

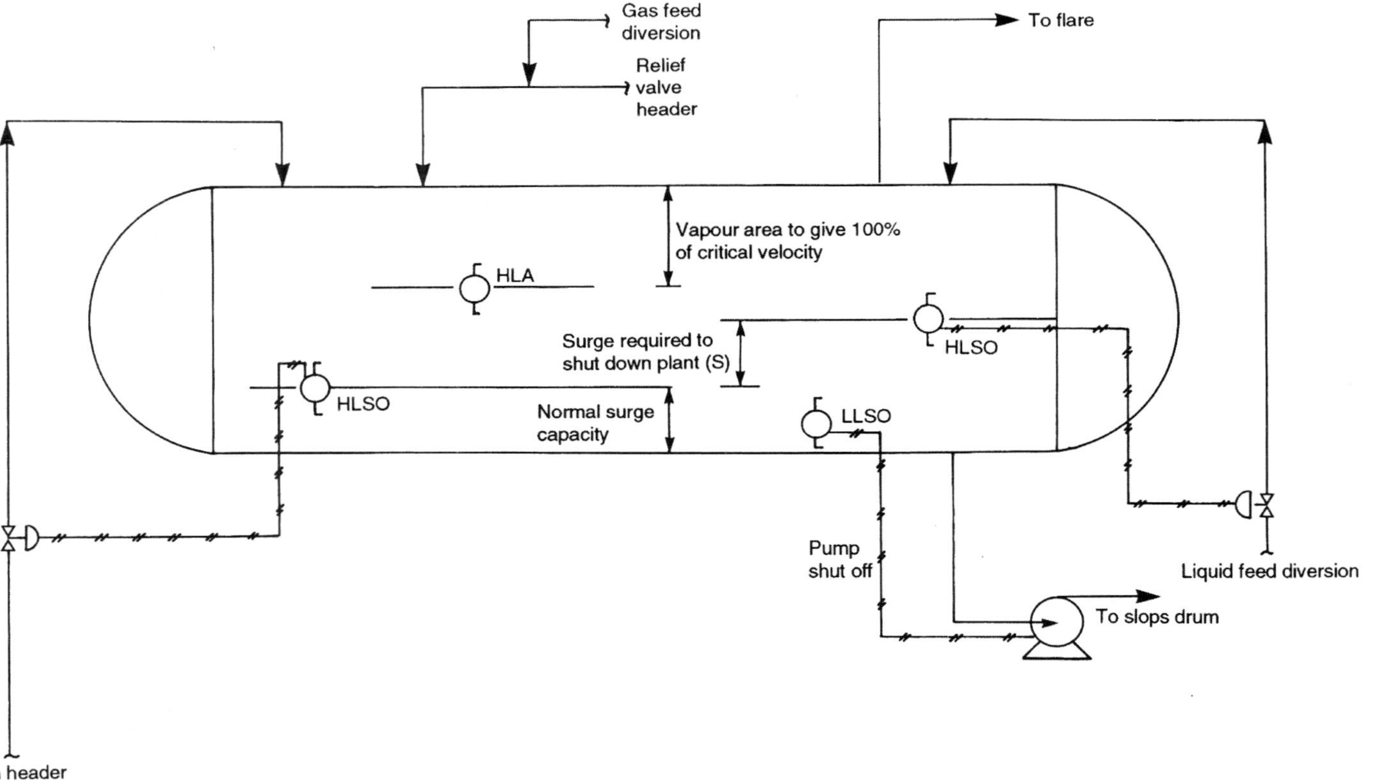

Figure 13.18. A non-condensable blow down drum.

2.0 The surge capacity between the HLSO (High Level Shut Off) for normal drainage and the HLSO for feed diversion should be such as to contain the total feed to a unit routed to this drum for a period sufficient to shutdown the unit producing the feed stream. Should there be more than one unit routed to the drum then this surge capacity should be for the largest of a single feed streams.
3.0 The capacity between the highest HLSO and the high level alarm (HLA) should be sized to handle the largest liquid volume that can be discharged in 30 minutes by the relief valves constituting any single risk.
4.0 The drum must be sized for a vapor velocity above the HLA at a maximum of 100% of the critical figure calculated by:

$$V = 0.157\sqrt{(\rho_l - \rho_v)/\rho_v}$$

where

V = critical velocity in ft/sec
ρ_l = liquid density in lbs/cuft
ρ_v = vapor density in lbs/cuft at drum conditions of pressure and temperature.

Drum pressure. The maximum operating pressure for this drum will be about 0.5 psig or that of the water disengaging drum tied to the same flare header.

Condensable blow-down drum and system is used for collection and containment of heavier hydrocarbons with low volatility. For example this would account for the middle and waxy distillates (kerosene, gas oils, and the like).

Figure 13.19 shows a typical blow-down drum and quench. The material entering this system is generally above ambient temperatures. Very often hot streams directly from operating units find their way into this system. To handle these materials the condensable blow-down drum is designed as a direct contact quench drum. The blow-down material leaves the unit in a drain collection system to enter the bottom section of the drum. Cooling water is introduced at the top of the drum and passes over a baffled tray section to contact the hot blow-down stream at the drum base. Any hot vapors rising from the blow-down stream are condensed in the baffle section of the drum and carried down to the bottom of the drum. Uncondensed material leaving the top of the vessel is routed to the flare. The aqueous mixture containing the condensed blow-down leaves the bottom of the drum through a seal system to enter the chemical or oily water sewer for separation and treatment.

The following criteria is used to size this vessel:

- The vapor load on the drum is based on the safety valve(s) constituting the largest single risk
- The maximum operating pressure for the drum is usually 1–2 psig
- The stack may vent to atmosphere rather than the flare if desired. However if vented to atmosphere the stack should vent at least 10 ft above the highest adjacent structure.

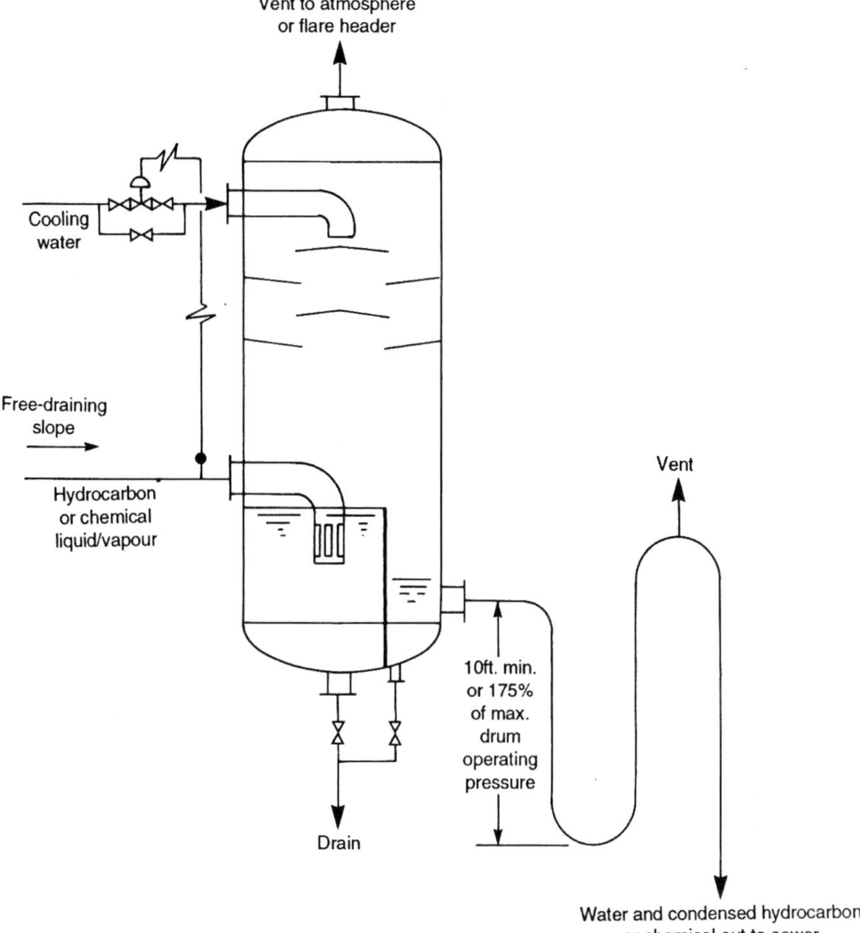

Figure 13.19. A typical blow down drum with quench.

In any case the vent should not release to atmosphere below 50 ft above grade. Snuffing steam should also be provided.

- The cooled effluent leaving the drum should be at 150°F or colder. The cold water supply should be controlled either by effluent temperature or inlet blow-down stream flow. There should however be a bypass flow of water entering the drum at all times.
- The drain system from the unit(s) to the drum should be free draining into the drum. The drum therefore should be located at a minimum height to grade. Where very waxy materials are likely to be handled, steam tracing of lines and a steam coil in the drum should be considered.

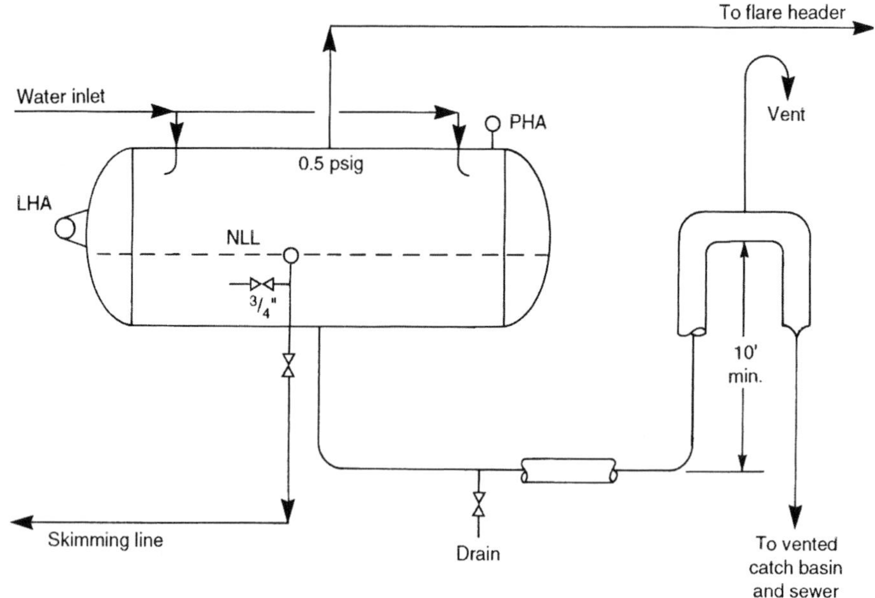

Figure 13.20. A water separation drum.

The water separation drum arrangement is shown in Figure 13.20. The purpose of this drum is to remove any volatile and combustible material from certain water effluent streams before they enter the sewer system.

Thus:

- All water from distillate drums which have been in direct contact with flammable material such as light hydrocarbons are sent to the disengaging drum before disposal to the sewer. The exception is where a sour water stripper is included in the plant; then these streams are sent to this stripper.
- Cooling water drainage from coolers and condensers which may have been contaminated with flammable high volatility material are sent to the disengaging drum. So too are steam condensate streams which fall in the same category.

The drum is located at a minimum height above grade. It operates at about 0.5 psig and vents into the flare system. The pressure and the liquid level in the drum are maintained by the free draining of the effluent through a suitable seal.

Design criteria used for the sizing of this drum are as follows:

- The vapor load on the drum will be the result of high volatile material flashing to equilibrium conditions at the drum pressure. This design load is based on the largest

amount of vapor arising from a single contingency. For exchangers this contingency will be due to a fractured tube. For liquid from a distillate drum this will be due to a failed open control valve on the water outlet.
- The liquid seal must be such as to eliminate air from the sewer system and to allow free drainage from the drum.
- An oil or chemical skimming valve is located at the water NLL. This allows for the draw off of the oil phase from time to time. A high interface level alarm is often included.

The flare

Vapors collected in a closed safety system are disposed of by burning at a safe location. The facilities used for this burning are called flares. The most common of these flares used in industry today are:

- The elevated flare
- The multi jet ground flare

The elevated flare is used where some degree of smoke abatement is required. The flare itself operates from the top of a stack usually in excess of 150 ft high. Steam is injected into the gas stream to be burnt to complete combustion and thereby reduce the smoke emission.

The multi-jet ground flare is selected where luminosity is a problem. For example at locations near housing sites. In this type of flare the vapors are burned within the flare stack thus considerably reducing the luminosity. Steam is again used in this type of flare to reduce the smoke emission.

Figure 13.21 shows a typical arrangement of an elevated flare and Figures 13.22 and 13.23 shows that for a multi jet ground flare.

The elevated flare

This type of flare is the normal choice in the larger process industries such as the petroleum refining industry. It consists of a flare stack over 150 feet in height that contains an ignitor system, a pilot flame, and the flare pipe itself. The flare header enters the stack through a water seal at the base of the stack immediately above an anchor of concrete plinth. The water seal maintains a back pressure of around 0.5 psig on the flare header. The waste gas to be flared moves up the stack to exit at the top. At the stack top there is an assembly of ignitor and pilot gas which ensures the safe burning of the waste flare gasses. This assembly is shown in Figure 13.24. It consists of three tubes all external to the stack itself and each supplied with the plant fuel gas. The first and largest of these tubes is the ignitor. Here the fuel gas supply

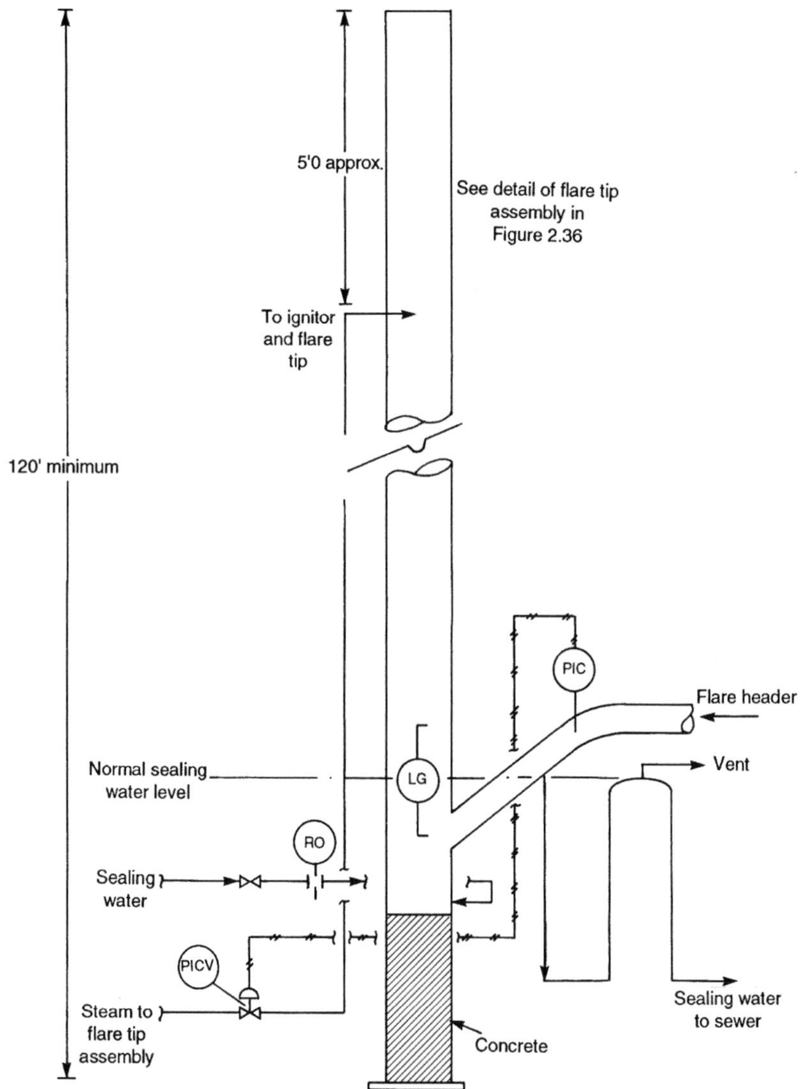

Figure 13.21. An elevated flare.

is mixed with air (plant or instrument air supply) before passing upwards through a venturi tube to an igniter chamber. A spark is induced in the ignitor chamber by an electric current of 15 Amp. The chamber and the venturi tube are located near grade and a sight glass on the ignitor chamber enables the operator to check on the ignitor's operation. The flame front from the ignitor travels up the ignitor tube to contact the waste gasses that are to be flared as they exit the top of the flare stack. The same flame

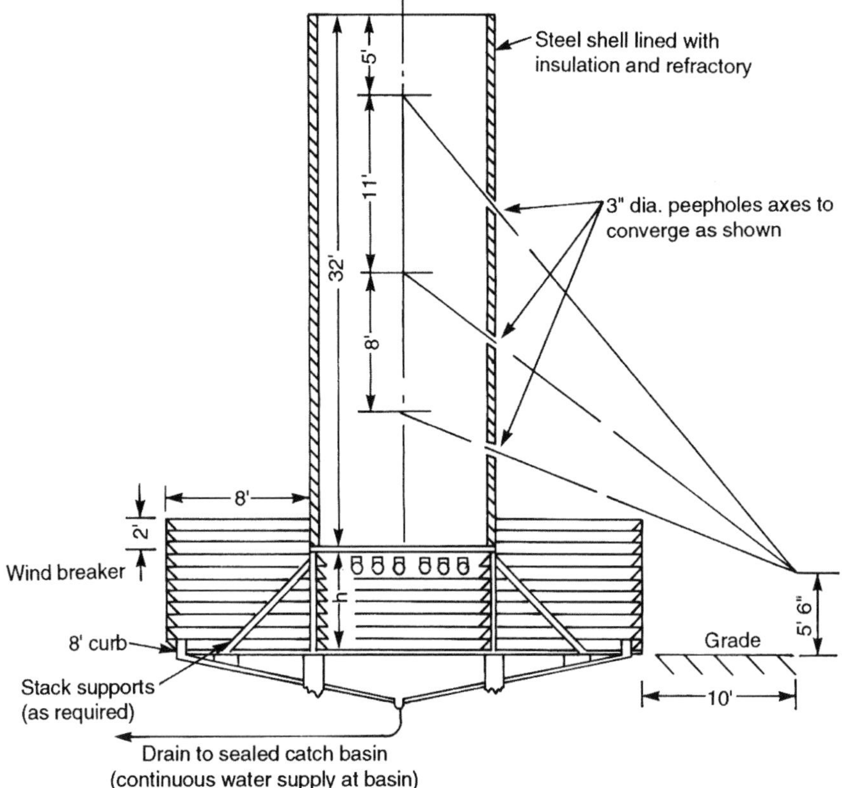

Figure 13.22. A multi jet ground flare and stack.

front ignites the "On and Off" pilot burner which is the center tube of the three and initially ignites the permanent pilot burner at the stack top. The outlets of these three tubes are located at the stack top such that the prevailing wind ensures that the flame from them is blown across the stack exit.

Steam is often injected into the stack at some point near the top to complete combustion and eliminate or at least reduce smoke emission. The amount of steam normally used for this purpose depends on the character or composition of the waste gasses. Aromatics and olefins when burnt produce a smokey flame: steam injection allows the free carbon which makes up the smoke to convert into CO and CO_2 which of course are invisible gasses. An estimate of the amount of steam required for smoke abatement is given in Figure 13.25.

A clear space around an elevated flare is required to allow for the effect of heat radiation from the flare to the ground. Flares which have a heat release of 300 million

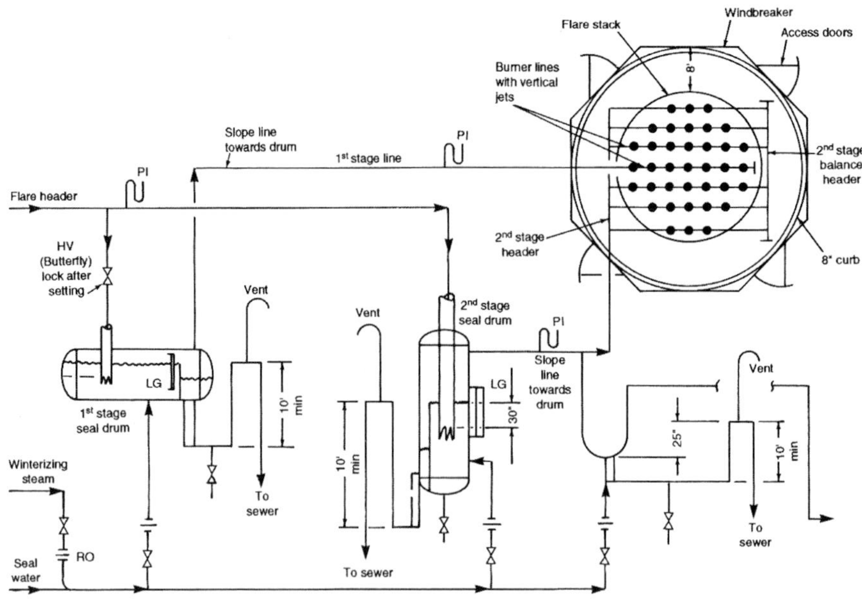

Figure 13.23. Multi jet flare plan with seal details.

to 1 billion btu/hr should be located at least 200 ft from the plant property line, or any pond, separator, tankage, or any equipment that could be ignited by a falling spark. The stack also must have a spacing of at least 500 ft from any structure or plant whose elevation is within 125 ft of the flare tip.

The elevated flare stack is designed to maintain a gas velocity of between 100 and 160 ft/sec during a major blow-down to flare. This rate is based on the maximum single emergency plus any steam added to improve the burning characteristics. Above a velocity of 160 ft/sec noise becomes a problem and the maintenance of ignition also is dubious unless multiple ignition tubes are used. Some proprietary flare tip designs however do claim ability to handle satisfactorily velocities up to 400 ft/sec.

The multi-jet ground flare

The multi-jet flare provides a completely noiseless, non-luminous flaring at a reasonable cost. At normal loads the flare is also essentially smokeless and is particularly useful where continuous flaring is required. Figures 13.22 and 13.23 show the elevation of the flare stack and the plan arrangement of a two stage multi-jet flare, respectively.

The two stage arrangement shown here shows the flare header being directed to one of two seal drums or to both. The fist stage seal drum operates at a back pressure of 20 ins of water at the first stage burners at its design capacity. The second stage burners

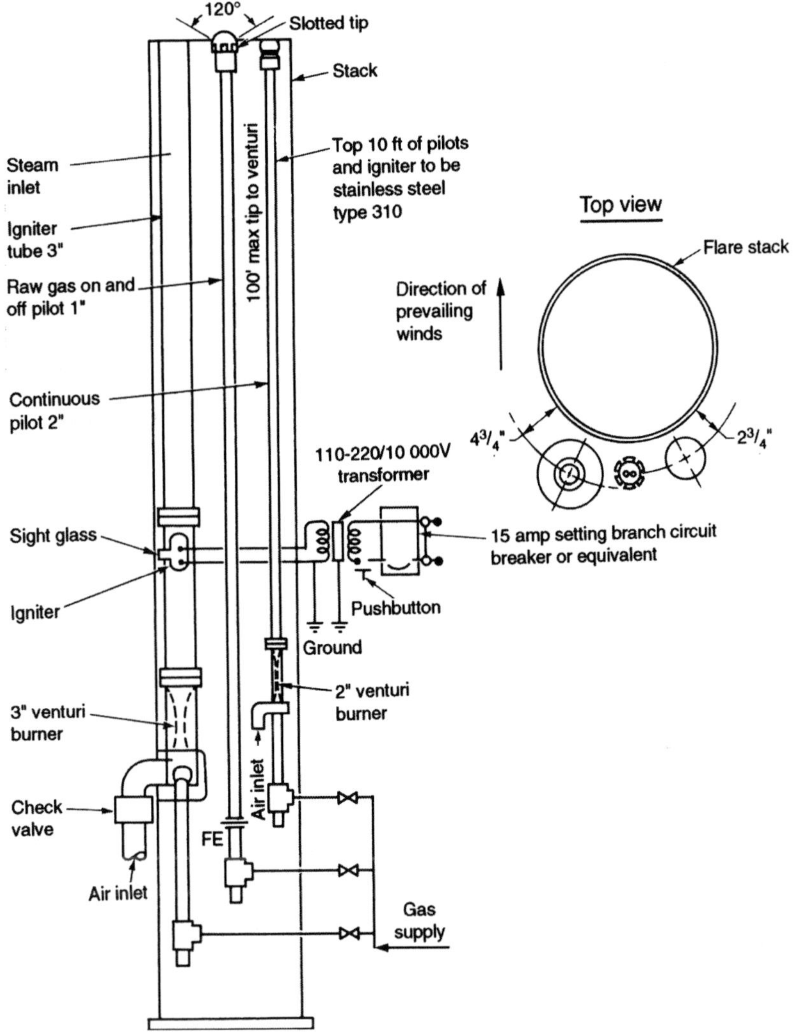

Figure 13.24. Flare top igniter assembly.

are activated when the pressure in the flare header reaches 30 ins of water gauge. Very often, particularly in large process complexes, the ground flare is designed to operate in conjunction with an elevated flare. The multi-jet flare takes a gas stream up to say 80% of its rated capacity additional flow is then diverted to an elevated flare system. Thus, if there is need for the continuous flaring of a reasonably small quantity, the ground flare caters for it. In an emergency or surge the elevated flare comes into operation automatically to take the additional load.

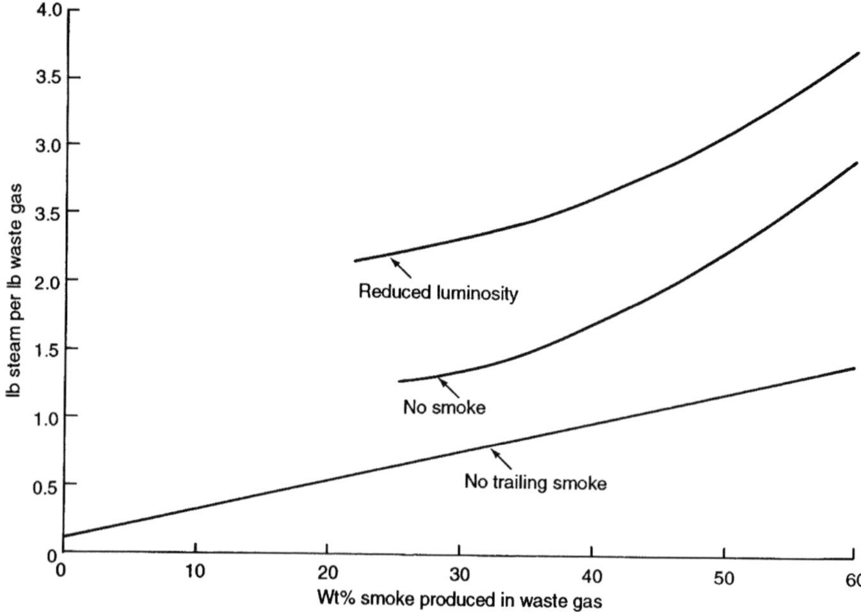

Figure 13.25. Approximate amount of steam for smoke abatement.

The burners of a multi-jet flare are jet nozzles approximately 15 ins in length of 1 ins diam stainless steel pipe. They discharge vertically from the horizontal burner lines which run across the bottom of the stack. The number of jets is based on gas velocity and is expressed by the equation:

$$N = 16.4\ \mathrm{V}$$

where

N = number of Jets
V = flare design capacity in mmScf/day.

The jets are placed on a square or rectangular pitch of 18 to 24 ins. A first estimate of the required pitch may be obtained from the expression:

$$P = \frac{100\ D^2}{NC}$$

where

P = pitch in ins
D = stack ID in feet
N = number of jets
C = distance between burner center lines in ins

No jet should be placed closer than 12 ins to the inside of the stack.

The inside diameter of the stack is based on the rate of heat release at design capacity. It is calculated using the following equation:

$$D = 0.826Q^5$$

where

D = stack inside diameter in feet
Q = heat release at max design in mm Btu/hr.

The stack height for diameters up to 25 ft is 32 ft and the steel shell of the stack is lined with 4″ of refractory material. A wind breaker completes the construction of the stack. This is necessary to prevent high wind gusts from extinguishing the flames.

Flame holders are installed above the burners to prevent the flames ‘riding’ up to the top of the stack. These are simply solid rods of 1″ refractory material supported horizontally above each burner line. The position of these flame holders relative to the bottom of the stack is critical to the proper operation of the burners. The stack itself is elevated to allow air for combustion to enter. The minimum space between grade and the bottom of the stack is set at 6 feet or 0.3D whichever is the larger.

As for any flare a continuous pilot burner is recommended. The proper operation of this pilot is important with respect to multi-jet type flares because of the danger of un-burnt flammable material escaping outside the flare at ground level. A gas pilot is provided at each end of the primary burner to minimize this risk.

Effluent water treating facilities

This section of the “Offsite Systems” deals with the treating of waste water accumulated in a chemical process complex before it leaves the complex. Over the years requirements for safeguarding the environment have demanded close control on the quality of effluents discharged from chemical and oil refining plants. This includes effluents that contain contaminants that can affect the quality of the atmosphere and those that can be injurious to plant and other life in river waters and the surrounding seas. Effluent management in the oil industry has therefore acquired a position of importance and responsibility to meet these environmental control demands.

Water effluents that are discharged from the process and other units are collected for treating and removal or conversion of the injurious contaminants. In most oil refineries imported water in the form of ship’s ballast water is also collected on shore for treatment before discharging back to the sea. Figure 13.26 is a schematic of the water effluent treating system for a major European oil refinery.

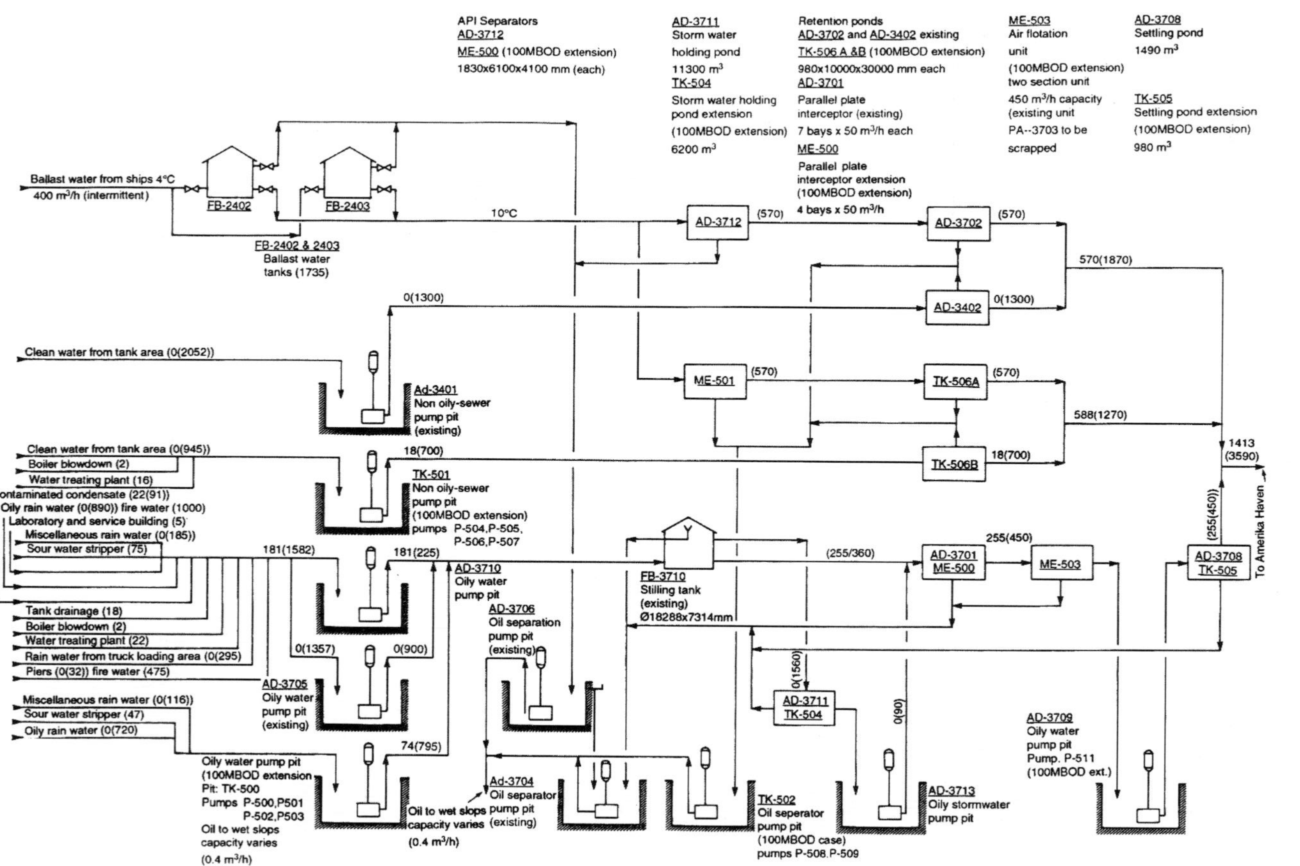

Figure 13.26. A schematic of a water effluent treating system.

Normally the water effluent treating facilities for a complex would be located at the lowest geographical point in the plant. In this way very little pumping is required to move the waste water to and from the treating plants. The schematic in Figure 13.26 is for a refinery that was sited below the sea level so more than the usual number of pumps are used.

The contaminant that is to be removed in the system shown in Figure 13.26 is, of course, oil. Five separate systems are used in this refinery's treatment plant. The first is that for handling ballast water from sea going tankers. The second is the handling of clean water. This is included because the system bypasses all the treating processes except the last "guard" process which, in this case, are the retention ponds. The third system is also for handling non-oily water but water that would be high in certain chemicals. This system also discharges into retention or storm water ponds. The water is held in these ponds to ensure that there is no contamination. Should there be then the water would be returned into one or the other treating processes for removal of the contaminants.

The last two systems shown are for the handling of contaminated water from the refineries paved area, various tank and process plant drainage and the like. These oily water systems and the ballast water stream are treated for oil removal. In the case of the ballast water the water drained from the bottom of the holding tanks is routed through an API separator. This is a specially designed pond that reduces the forward velocity of the water stream to allow the separation of oil from the water by settling or gravity.

The water/oil separation for the other refinery streams takes place in a series of settling pods. Final clean up in this case is accomplished by the use of parallel plate interceptors and an air flotation process. The principle of the parallel plate interceptor is to force the water stream to change direction several times in rapid sequence and thus "knocking out" any oil entrained in the stream. The air flotation unit causes the contaminated water stream to be agitated so as to force the lighter oil phase to the surface where it can be removed by skimming or by baffled overflow.

Other treating processes

Most chemical plants and indeed a few oil refining plants require more complex methods for clarifying their effluent water to meet environmental requirements for its disposal. The four more common methods are as follows:

- In-line clarification using coagulation, flocculation and filtration
- Plain filtration
- Sedimentation
- Chemically aided sedimentation using coagulation, flocculation, and settling

Clarification is a process that removes suspended (usually organic) matter that gives the stream color and turbidity. The removal of this matter especially in a colloidal form, requires the addition of chemicals to cause coagulation and flocculation to promote settling and separation of suspended solids. Coagulants and coagulant aids added to the influent stream chemically react with impurities to form precipitates. These, together with particles of enmeshed turbidity, are flocculated into larger masses that are then readily separated from the bulk liquid.

There are essentially three steps in the chemically-aided clarification process. These are:

- Mixing of the additives
- Flocculation
- Settling

Coagulation encompasses the process of mixing the first formation of agglomerates that form the floc. This is carried out in a series of separate compartments with the settling basin occupying the largest volume. Coagulation is the singular most important step in the clarifying process. Because it involves the build up of colloidal type particles, the chemicals and the process rate are specific to the material that is to be clarified. There are companies that specialize in the design construction and the operation of this type of effluent treating. These companies use their experience in handling the complex electrochemical kinetics associated with flocculation and coagulation principles.

Utility Systems

All oil refining processes require utilities in order to function. The more common utility systems are:

- A steam and condensate system
- A fuel system
- Water systems including cooling water, potable water, and boiler feed water (BFW)
- A compressed air systems
- A power supply system.

The engineering and design of the first four of these systems is usually the responsibility of a chemical engineer. On operating plants and processes the process engineer undertakes the responsibility for the correct and efficient operation of utility facilities. The duties associated with the power systems are usually left to the Electrical Engineer or department, although the process engineer does have an input in the sizing of the system be developing a list of power requirement for all electrically motivated equipment. This includes all motors and electrically operated equipment such as: the desalter, product dehydrators, and the like.

Brief descriptions of typical utility systems

The following paragraphs describe typical utility systems found in the oil refining industry. The details of these systems may vary from company to company but their format and general layout will be as described here. Only the first four systems are described and discussed here.

Steam and condensate systems

In most plants steam condensate accumulated in the various processes is collected into a single header and returned to the steam generating plant. It is stored separate to the treated raw water because condensates may contain some oil contamination. A stream of treated water and condensate are taken from the respective storage tanks and pumped to the deaerator drum. The pumps in this case are usually vertical pumps set in a pit near the storage tanks. The condensate stream passes through a simple filter on route to the deaerator to remove any oil contamination. The combined water and condensate streams enter the top of a packed section of the deaerator drum. Low-pressure steam is introduced immediately below the packing in the drum to flow upwards countercurrent to the liquid streams. Any air entrained in the water is removed by this counter current flow of steam to be vented to atmosphere.

The deaerated BFW is pumped by the BFW pumps into the steam drum of the steam generator. There will normally be three 60% pumps for this service. Two will be operational and one will be on standby. Those pumps normally operating are usually motor driven while the standby pump is very often driven by an automatically start up diesel engine. These pumps are quite large in capacity operating at high head and discharge pressure. The main steam lines in most plants are high-pressure (at least 700 psig at the generator coil outlet), so the pump discharge pressure will be much greater than the HP steam outlet. These pumps are the most important in any chemical plant. If they fail no steam can be generated and the whole complex is in danger of total shutdown or worse. Therefore three separate pumps are used to cater for the normal high head and high capacity, and a separate pump driver operating on a completely different power source than electrical power or steam is mandatory to minimize the danger of complete shut down.

The steam drum is located above the generator's firebox. The liquid in the drum flows through the generator's coils located in the firebox by gravity and thermo-syphon. A mixture of steam and water is generated in the coils and flows back to the steam drum. Here the steam and water are separated with the steam leaving the drum to enter the super-heater coil. The steam is heated to the plant's steam main temperature in this coil and enters the high-pressure steam header for distribution to the various users. The steam pressure is controlled be a pressure controller on the steam outlet to the header. Steam to the lower pressure headers is generated through turbines where possible. Where lower pressure steam is required and it is not possible to produce it through equipment then let down stations are located in suitable places in the system. When steam pressure is reduced to the lower pressure headers the associated increase of temperature above that specified for the lower pressure steam may need to be reduced. Desuperheaters are used for this purpose.

Desuperheaters consist of a chamber in the steam line into which cold condensate is injected. These items are purchased equipment with specially designed injection nozzles for the condensate. The amount of condensate delivered is controlled by the downstream temperature of the steam. Desuperheaters are located at critical locations of the plant where relatively large quantities of high-pressure steam are reduced to low pressures such as the discharge from turbines.

The condensate return header is usually operated at a positive pressure of between 5 and 10 psig. The condensate is stored at atmospheric pressure, and very often the small amount of steam flashed from the header pressure to the storage pressure is used in the deaerator instead of the low-pressure steam (the deaerator operates at or near atmospheric pressure).

Figure 13.27 is a schematic flow diagram of a typical steam generation unit.

Fuel systems

Most oil refineries have two separate fuel systems. They have:

- A fuel gas system
- And a fuel oil system

The user burners in these plants are of the dual purpose design. They fire either the fuel gas or the fuel oil stream and can be easily switched over from one to the other. The pilot burners however must be fuel gas, and the system is designed such that if the pilot burner is extinguished the whole burner system is shutdown. Generally the design of the burner system in most plants have many safety and shutdown features. After all in processes that handle flammable material the heater burners are the one feature in the plant design that are a major fire hazard source. Thus the design of the burner operation is such as to shutdown on:

- Low fuel pressure
- High process outlet temperature
- Pilot burner extinguished
- Atomizing steam failure (oil burners)
- Low process feed to heater

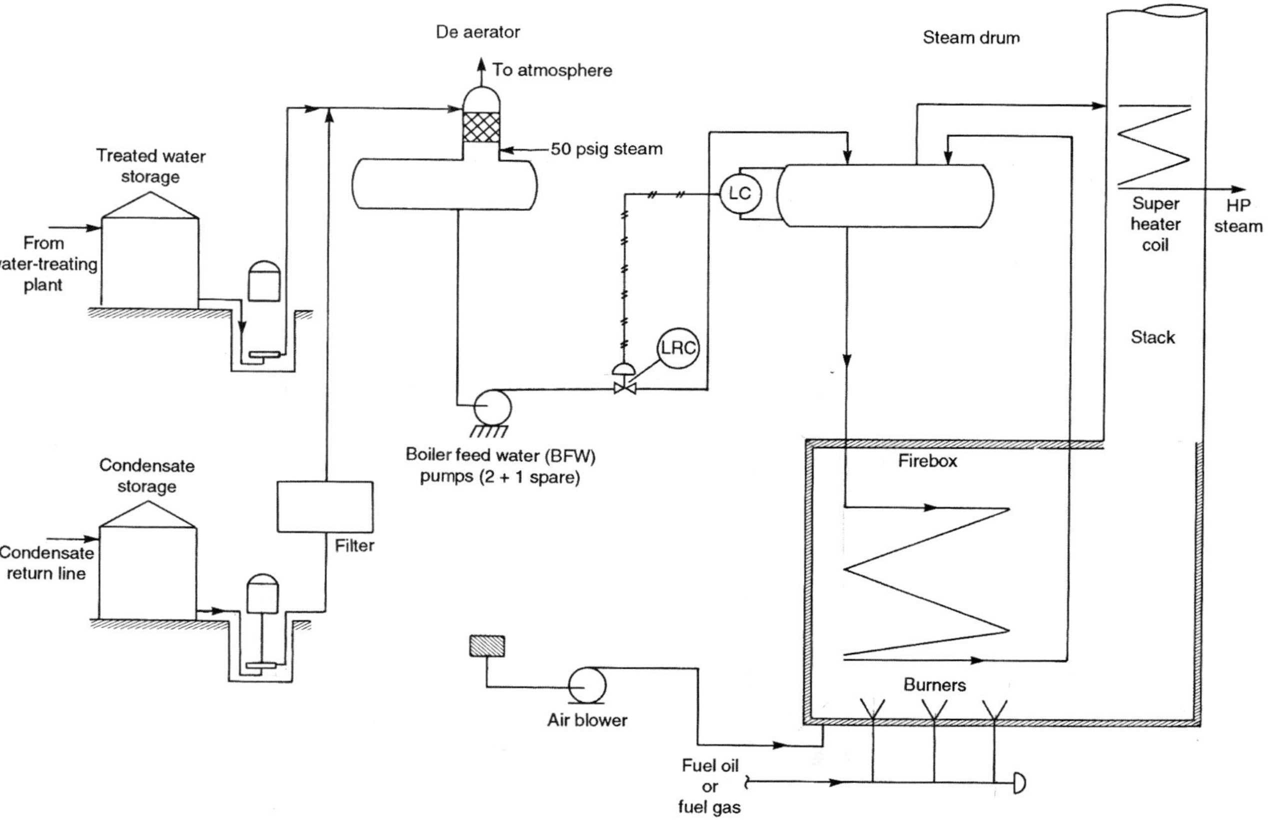

Figure 13.27. A typical steam generation unit.

Fuel gas system

This is the simplest of the two systems. Waste gas streams from the process plants are gathered and directed to the plant's fuel gas drum. This drum operates at 30 psig pressure and close to ambient temperature or 60°F. A small steam coil is usually installed in the drum to gasify any "below dew point" material that may have condensed out. The drum is held at the set pressure by pressure control valves which allows surplus gas to flow to flare and activates an emergency source of gas on low pressure. This emergency source may be in the form of imported liquefied petroleum gas (LPG) or some other plant gas stream that may be diverted to fuel. If LPG is used as a secondary fuel source it is routed to the fuel gas drum via a vaporizer. This item of equipment is a kettle type re-boiler heating and vaporizing the LPG at the drum pressure. Medium or low-pressure steam is used as the heating medium for the vaporizer.

The gas burner at the process fired heater operates at or close to atmospheric pressure. The burner draws fuel from a separate header than that used to supply gas to the pilot burner. Many heaters contain an automatic switch over from gas firing to fuel oil firing on low gas flows or when manually selected. The fuel gas system is "dead ended". That is there is no return system to the fuel gas drum, the gas header is pressured up and gas flows to the burners by means of this differential pressure and intermediate control valves. The fuel gas flow to the heater burners is controlled by a temperature control valve activated by a temperature controller on the heater coil process outlet line. The same controller also regulates the oil firing arrangement when the heater is operated on fuel oil.

Fuel oil system

Figure 13.28 is a schematic of a typical fuel oil system. Most plants use petroleum residues as fuel oil. These types of fuel are high in viscosity and very often have a high pour point. (Pour point temperature is the temperature that the oil ceases to flow.) For this reason the fuel oil is stored in insulated and heated cone roofed tanks. Heating may be accomplished by steam coils located in the base of the tank or by external steam heat exchanger through which the fuel oil is continually circulated.

Positive displacement pumps (usually rotary type) are used to deliver the fuel oil from the tank, through the distribution system to the heater burners. These pumps are always spared and the spare pump is driven by a steam turbine, while the operating pump is motor driven. The fuel oil passes through a duplex filter before entering the suction of the pumps. This filter is included to remove any solid contaminants that may be in the oil such as fine coke particles which would foul the fuel oil burner. The discharge pressure of the pumps is controlled by a slipstream routed back to the storage tanks through a pressure control valve. This valve is activated by a pressure control element on the pump discharge header.

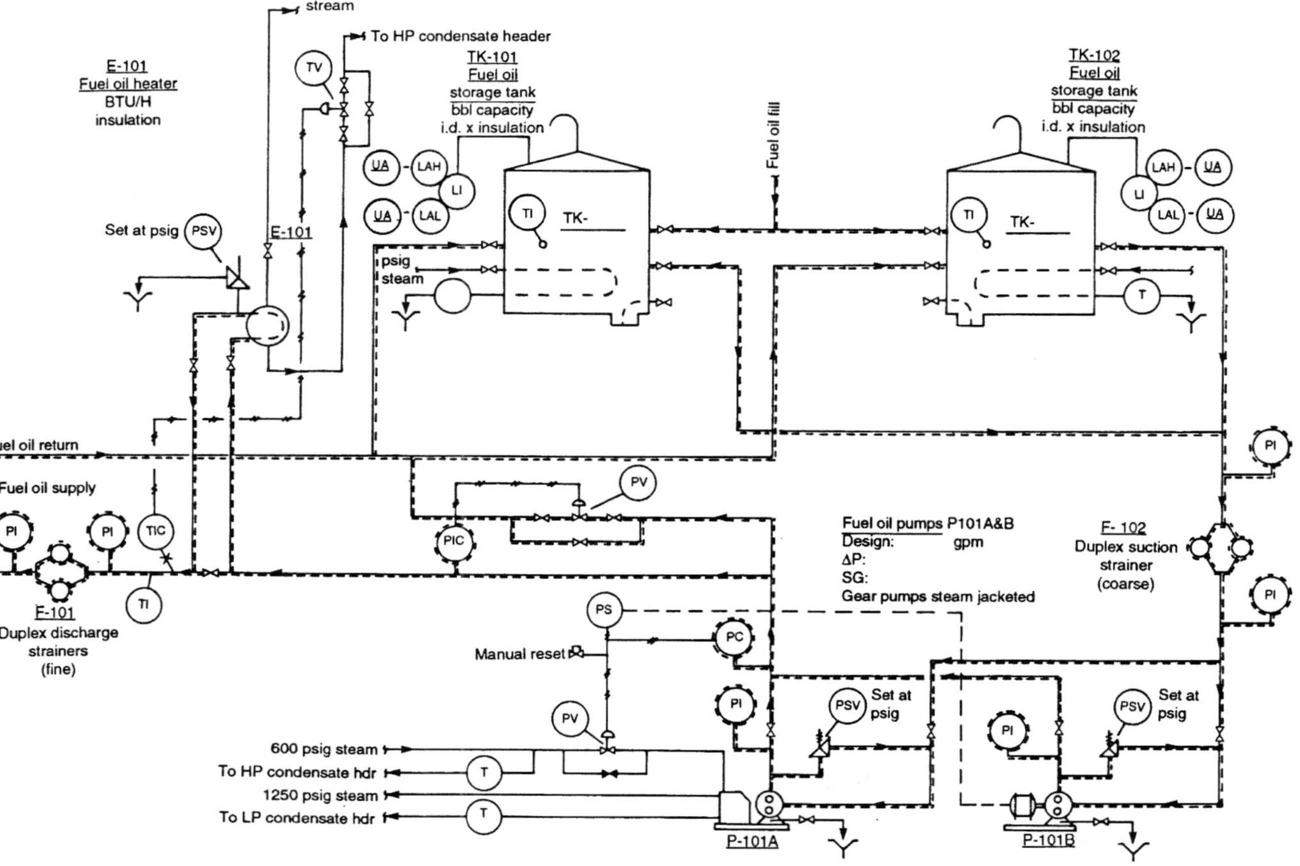

Figure 13.28. A typical fuel oil system.

The pumps discharge the fuel oil via the pressure controller to a pre-heater. This pre-heater may be a simple double pipe heat exchanger for relatively small units or regular shell and tube for the larger systems. Double pipe type exchangers are favored in this service when economical because they are easier to clean and maintain. The fuel oil leaves the pre-heater to enter the fuel oil distribution system at a temperature high enough to maintain a viscosity low enough for the oil to flow easily and to be easily atomized by steam at the fuel oil "gun" (or burner). All the piping associated with residual fuel systems are heavily insulated and steam traced. The distribution systems of residual fuel oils is usually the re-circulating type. That is the fuel leaves the pre-heater to circulate to all the user plants in a loop where the quantity used is taken off the stream and the remainder allowed to return to the system. The return header is routed back to the storage tanks. The circulation system handles between 1 to 3 times that amount of oil that is actually burnt.

Fuel oils are introduced into the fire box and ignited through a fuel oil burner sometimes called a fuel oil gun. In order to ensure combustion in a manner suitable for a fire box operation the fuel oil needs to be dispersed into small droplets or spray at the burner tip. In heavy residual oils this is almost always accomplished by steam. Compressed air is however sometimes used for this purpose. This atomizing stream is introduced into the gun chamber and comes into contact with the oil stream just before the burner tip. The kinetic energy in the atomizing medium forces the oil into suitable droplets as it leaves the burner. Steam is normally used as the atomizing material because it is usually cheaper, more readily available, and has a more reliable source than air from a compressor.

The steam pressure for atomizing should be 15–25 psig higher than the fuel oil pressure. The quantity of steam will range from 1.5 to 5 lbs per gallon of oil. Dry steam with a superheat of about 50°F is preferred for atomizing. In order to control the process heater operation oil burners require a turndown ability. That is they require to operate satisfactory over a prescribed range of flow. In keeping with an operating range for oil flow the atomizing medium must also have a similar operating range. Burner pressure is a critical requirement for turndown. The steam (or air) pressure should be 15 psi or 10% (whichever is the greater) higher than oil pressure. The fuel oil supply system should be 100 psi higher than the burner requirement.

The oil burner operation as with the fuel gas burner is controlled by the heater's process stream outlet temperature. The temperature control valve activated by the coil outlet temperature increases or decreases the oil flow from the circulating oil stream to the burner. A proportional control valve on the atomizing steam line regulates the flow of steam to the burner in keeping with the oil flow. Figure 13.29 shows a typical burner control system.

There are various methods for safety shutdown. The one shown here shuts down the oil flow on steam failure and on loss of the pilot burner. In some systems there is also an automatic change over to gas firing on low oil pressure.

Water systems

The major water systems generated in most chemical plants are:

- Cooling water
- Treated water for BFW

Potable water as raw water is usually drawn from municipal supply. Where water is required for cleaning or drinking this potable water is used without further processing.

Cooling water

Figure 13.30 is a mechanical flow sheet showing the arrangement around the base of a cooling tower for the collection and supply of cooling water.

The cooling water system is a circulating one. That is there is a cold supply line with an associated warmer return line from all its users. Figure 13.31 shows a section of this distribution system.

The water returned to the cooling tower by the return header enters the top of the tower and flows down across the tower internals counter current to an air flow either induced or forced by fans passing up through the tower. The water cooled by the air flow is collected in the cooling tower basin. Make up water (usually potable water) is added to the basin under level control. Vertical cooling water circulating water pumps take suction from the cooling tower basin sump to deliver the water into the plant's distribution header. These pumps are usually high capacity with a moderate differential head. Because of the critical nature of the water supply the pumps are each rated at around 60% of design capacity with two in operation and two on stand by. A mixture of motor and steam turbine drivers are quite common.

The supply header pressure is kept at around 30 psig and, very often in large plants covering long distances, booster stations are installed at predetermined locations to maintain the supply header pressure. These booster stations consist of pump pits with again high capacity vertical pumps rated smaller of course than the main supply pumps. The location of these booster stations is determined by a rigorous hydraulic analysis of the distribution system which also determines the header pipeline sizes. The return flow is collected from each user into the return header and flows back to the cooling tower under the users' outlet pressure.

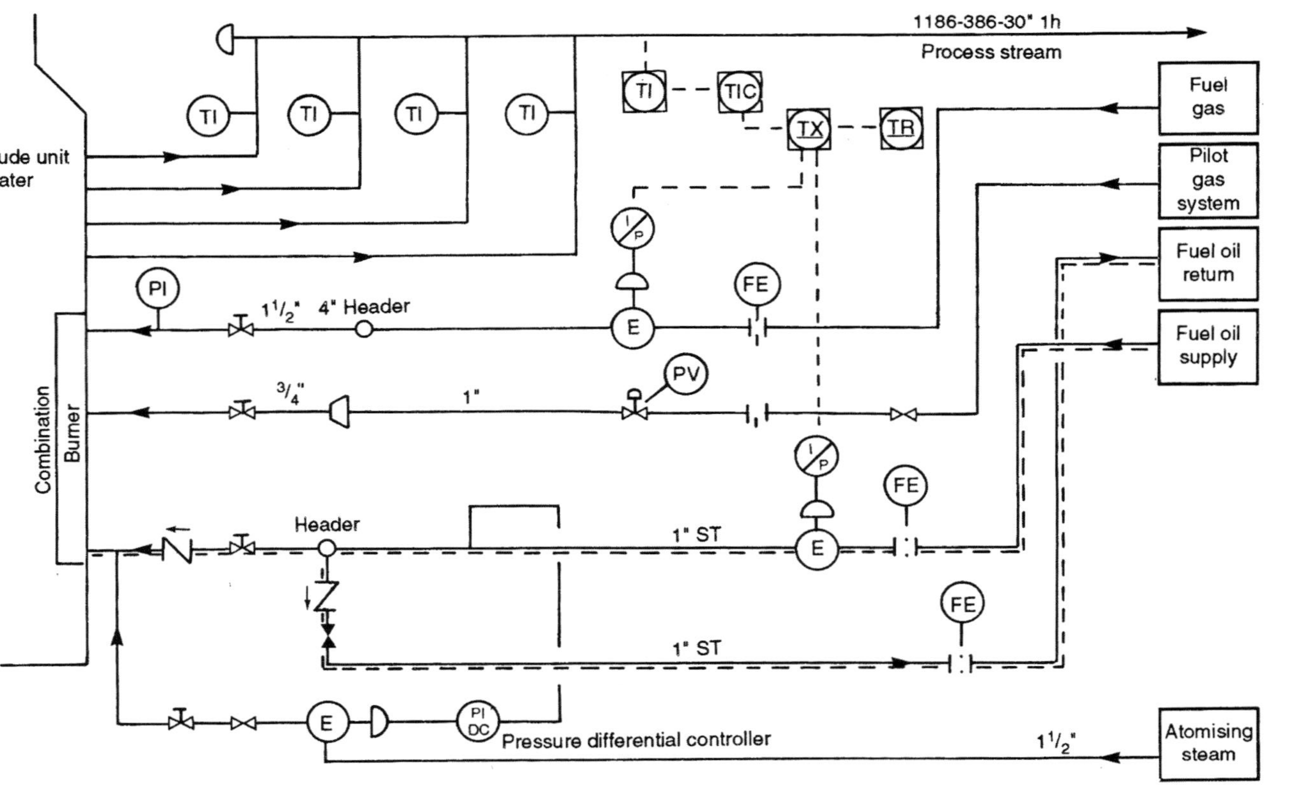

Figure 13.29. A burner control system.

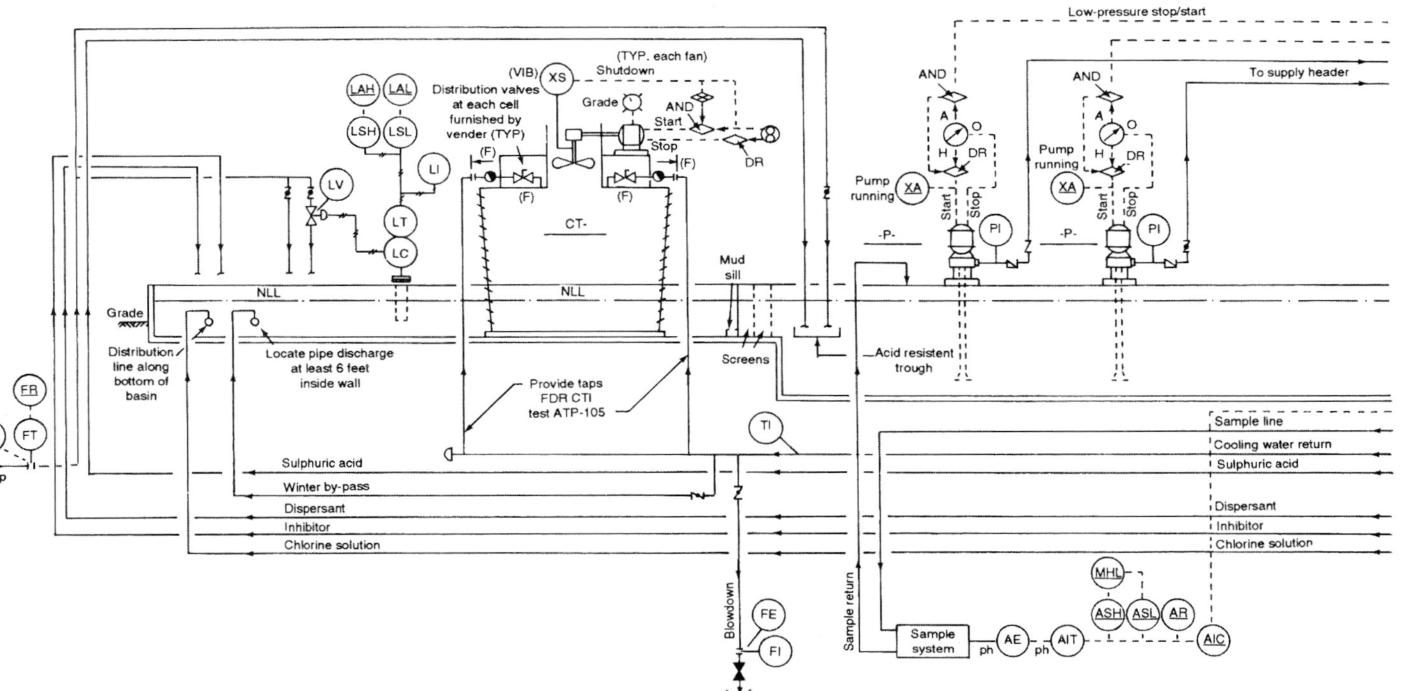

Figure 13.30. Mechanical flow diagram of a water cooling system.

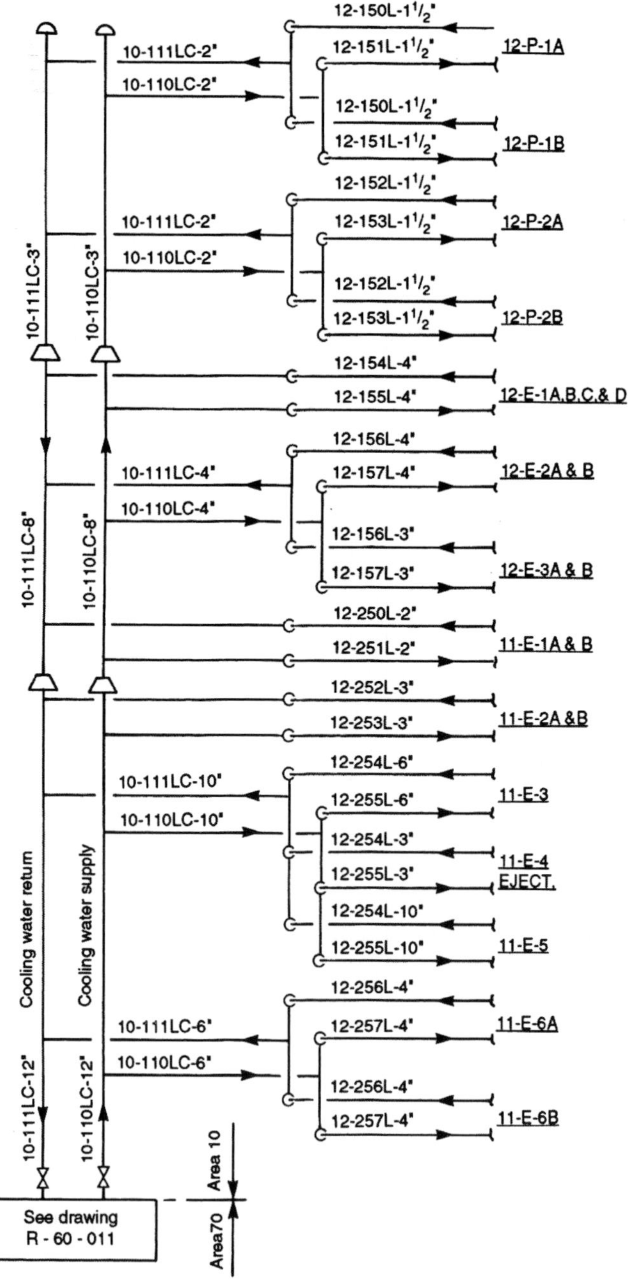

Figure 13.31. A diagram of a cooling water distribution system.

The water in the cooling tower basin and in the cooling tower itself requires some treatment to prevent the build up of algae and other undesirable contaminants. A separate small treatment plant is used for mixing the inhibiting chemicals and injecting them into the critical sections of the system.

Boiler feed water treating

All water contains impurities no matter from what source the water comes from. Appendix 13.2 gives a listing of the common impurities found in water for industrial use. Normally, in a chemical plant, water with most of these impurities can be used without treating of any kind. However when it comes to generating steam and particularly high-pressure steam these impurities become problematic. To operate steam generators effectively and to avoid serious damage to the unit, these impurities either have to be removed or be converted into compounds that can be tolerated in the system. Appendix 13.2 also provides a description of the effect of these impurities on steam generators and gives the normal means of treating.

In general there are three types of soluble impurities naturally present in water and which must be removed or converted in order to make the water suitable for boiler feed purpose. These are:

- *Scale forming impurities.* These are salts of calcium, magnesium, silica, manganese, and iron
- *Compounds that cause foaming.* These are usually soluble sodium salts
- *Dissolved gases.* These are usually oxygen and carbon dioxide. The soluble gases must be removed to prevent corrosion

Solid build up in the boiler itself is removed or kept at a low level by blow down. This is the mechanism of draining a prescribed amount of the boiling water from the boiler steam drum at regular intervals. This amount is calculated from the analysis of the solid content of the feed water and must equal the amount brought into the system by the feed water. Figure 13.32 gives an example of boiler blow down.

The American Boiler Manufacturers Association (ABMA) have developed limits for the control of various solids in BFW. These are given in Table 13.1.

Other considerations regarding the limits of solids in BFW are:

- *Sludge.* This is a direct measurement of feed water hardness (calcium and magnesium salts) since virtually all hardness comes out of solution in a boiler.
- *Total dissolved solids.* These consist of sodium salts, soluble silica, and any chemicals added. Total solids do not contribute to scale formation, but excessive amounts can cause foaming.

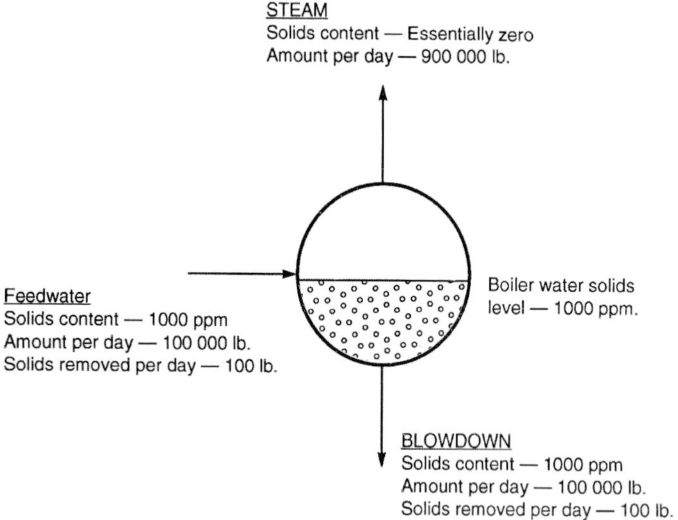

Figure 13.32. How blow down reduces the amount of solids build up.

- *Silica.* This may be the blow down controlling factor in pre-softened water containing high silica. At elevated pressures high silica content can cause foaming and carry over.
- *Iron.* High concentration of iron in BFW can cause serious deposit problems. Where concentration is particularly high blow down may be based on reducing this concentration.

There are two types of BFW treatment in use. There are the external type of treating and the internal processes. As the names suggest the external processes are those that treat the water before it enters the boiler. The internal treatments are those in the form

Table 13.1. ABMA limits of various solids in BFW

Boiler pressure (psig)	Total solids (ppm)	Alkalinity (ppm)	Suspended solids	Silica
0–300	3500	700	300	125
301–450	3000	600	250	90
451–600	2500	500	150	50
601–750	2000	400	100	35
751–900	1500	300	60	20
901–1000	1250	250	40	8
1001–1500	1000	200	20	2.5
1501–2000	750	150	10	1.0
Over 2000	500	100	5	0.5

of added chemicals that treat the water inside the boiler. Only the external processes are described here.

The "hot lime" process

This is a water softening process which uses a hot lime contact to induce a precipitate of the compounds contributing to the hardness. The sludge formed is allowed to settle out. Very often coagulation chemicals such as alum or iron salts are used to enhance the settling and the removal of the sludge formed. In most plants that use the "lime" process the reaction by the addition of lime and soda ash is carried out at elevated temperatures. However the reaction can be allowed to take place at ambient temperatures. The hardness of the water from the "cold" process will be about 17–35 ppm while that from the "hot" process will be 8–17 ppm. Clean up filters containing anthracite are often used to finish the treating process.

The ion exchange processes

As the name implies this process exchanges undesirable ions contained in the raw water with more desirable ones that produces acceptable BFW. For example, in the softening process, calcium and magnesium ions are exchanged for sodium ions. In dealkalization, the ions contributing to alkalinity (carbonates, bicarbonates, etc) are removed and replaced with chloride ions. Demineralization in this process replaces all cations with hydrogen ions (H^+), and all anions with hydroxyl ions (OH^-) making pure water ($H^+ + OH^-$).

The ion exchange material needs to be regenerated after a period of operation. The operating period will differ from process to process and will depend to some extent on the amount of impurities in the water and the required purity of the treated water. Regeneration is accomplished in three steps:

- Back washing
- Regenerating the resin bed with regenerating chemicals
- Rinsing

Figure 13.33 shows the internals of a typical ion exchange unit.

Under operating conditions the raw water is introduced through the top connection and distributor. The water flows through the resin bed where ion exchange takes place. The treated water is removed via the bottom connection. Under regeneration operation, raw water as backwash is introduced through the bottom connection and removed from the top connection. During its passage upward through the resin and support beds it "fluffs" the beds and removes any waste material that has adhered to them. The backwash water is then sent to the plants waste water disposal system.

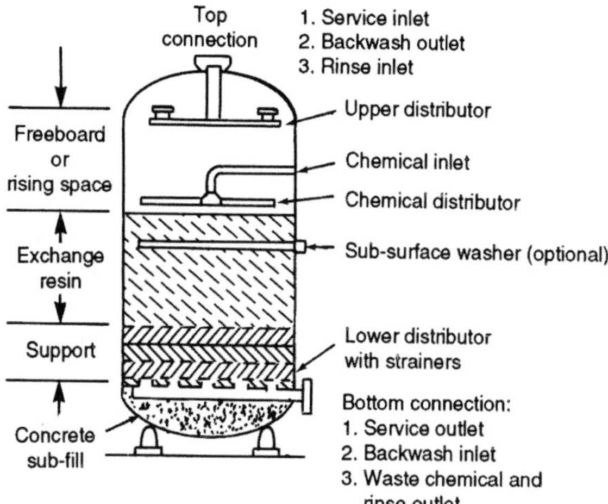

Figure 13.33. A typical ion exchange unit.

Regenerating exchange chemical is introduced directly above the resin bed through a chemical distributor and allowed to flow downward to be removed at the bottom water outlet. The regenerating cycle is completed with the rinsing of the bed to remove any surplus regenerating chemicals. This is done by introducing a stream of raw water at the top connection and removing it from the bottom connection. This water is also disposed to waste.

Normally ion exchange units are installed in pairs. When one is operating the other is being regenerated. An automatic switch over of electronically controlled valves takes the pair of units through the correct cycles at the prescribed time intervals, without disrupting the treating process. Figure 13.34 shows a typical "hook up" of an ion exchange unit.

Deaeration

The deaeration process is used in almost all BFW treatment to remove dissolved gases from the water. Normally treated water and returned condensate are routed to a deaerator immediately prior to entering the boiler steam drum. Figure 13.35 is a drawing of a typical deaerator drum layout.

The drum consists of a retention vessel surmounted by a degassing tower section. The degassing section contains a packed volume over which the treated water (and condensate) stream is passed. Low-pressure steam, usually let down saturated 50 psig steam or, if available 5 psig steam from condensate flash is introduced to the bottom of the degassing section. The steam flows upward through the packed section and

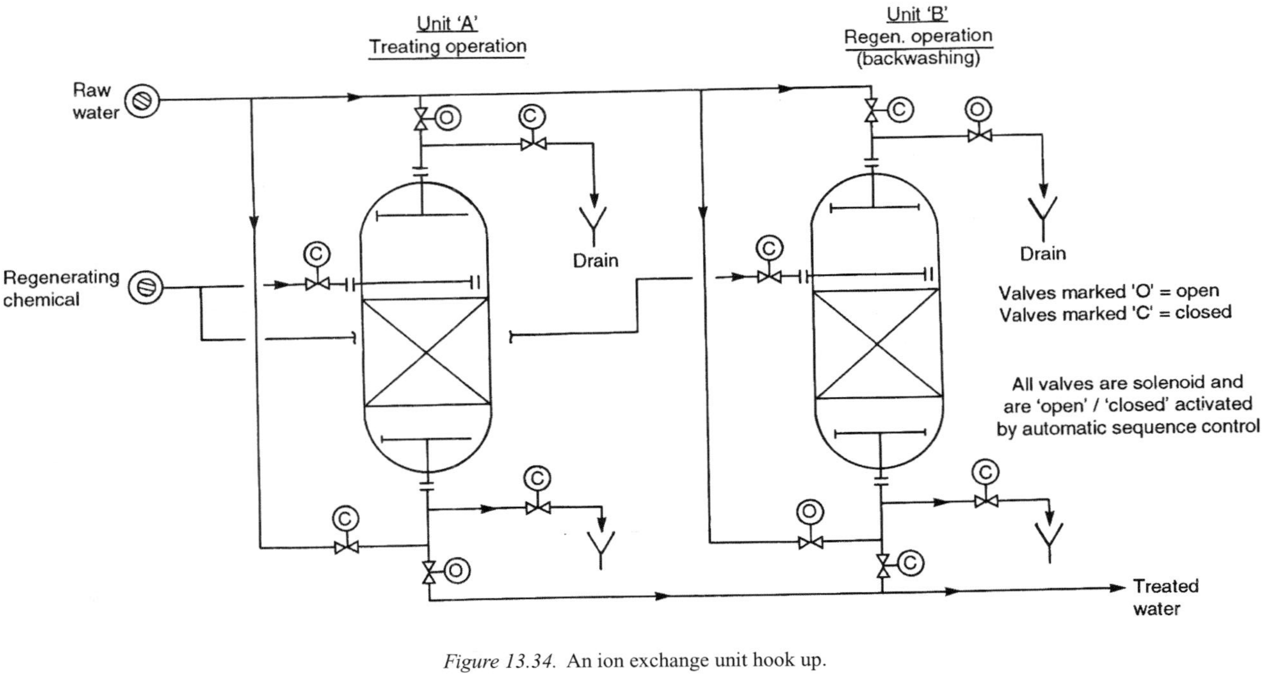

Figure 13.34. An ion exchange unit hook up.

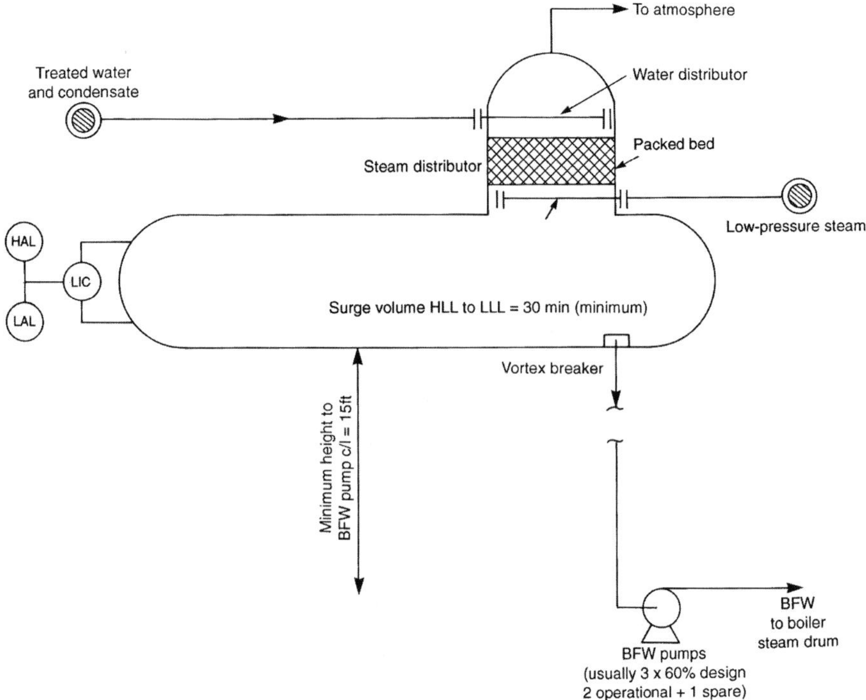

Figure 13.35. A deaerator drum.

counter current to the water. This action removes the dissolved gasses from the water. These gasses then leave with the steam from the top of the degassing section to be vented to atmosphere. The gas free water free flows into the main retaining section of the deaerator.

The treated water feed is introduced to the degassing section of the deaerator through atomizing spray/distributor. This reduces the water stream to fine droplets prior to entering the packed section. This enhances the removal of the gasses in the water. deaerators operate at about 2–3 psig and at this pressure all of the CO_2 contained in the water and most of the oxygen is removed.

The BFW pumps draw suction directly from the deaerator. To ensure that there is available sufficient NPSH for the pumps to operate properly deaerator drums are installed on a structure at least 15 ft above the center line of the pump suction. Most large pumps (as BFW pumps are) usually require a relatively high NPSH when handling hot water.

The retention section of the deaerator should have as a minimum 30 minutes of surge between HLL and LLL. The water feed to the deaerator is normally on retention section level control. The boiler feed from the BFW pumps will be on flow control with steam drum level reset. Very often boiler feed flow has low flow alarm and at very low flow has automatic boiler shut down device.

Compressed air system

All oil refining plants require a supply of compressed air to operate the plant and for plant maintenance. There are usually two separate systems and these are:

- Plant air system
- Instrument air system

Plant air is generally supplied by a simple compressor with an after cooler. Very often when plant air is required only for maintenance this is furnished by a mobile compressor connected to a distribution piping system. Air for catalyst regeneration and the like is normally supplied by the regular process gas compressor on the unit. Instrument air should always be a separate supply system. Compressed air for instrument operation must be free of oil and dry for the proper function of the instruments it supplies. This is a requirement which is not necessary for most plant air usage. A reliable source of clean dry instrument air is an essential requirement for plant operation. Failure of this system means a complete shut down of the plant.

Figure 13.36 shows a typical instrument air supply system.

Atmospheric air is introduced into the suction of one of two compressors via an air filter. The compressors are usually reciprocating or screw type non-lubricating. Centrifugal type compressors have been used for this service when the demand for instrument air is very high. The air compressors discharge the air at the required pressure (usually above 45 psig) into an air cooler before the air enters one of two dryers. One of the compressors is in operation while the other is on standby. The operating compressor is usually motor driven with a discharge pressure operated on/off start up switch. The standby compressor is turbine (or diesel engine) driven with an automatic start up on low discharge pressure switch.

The cooled compressed air leaves the cooler to enter the dryers. There are two dryer vessels each containing a bed of desiccant material. This material is either silica gel (the most common), alumina, or in special cases zeolite (molecular sieve). One of the two dryers is in operation with the compressed air flowing through it to be dried and to enter the instrument air receiver. The desiccant in the other dryer is being simultaneously regenerated. Regeneration of the desiccant bed is effected by passing through the bed a stream of heated air and venting the stream to atmosphere. This heated stream removes the water from the desiccant to restore its hygroscopic properties. At the end of this heating cycle cooled air is reintroduced to cool down the bed to its operating temperature. When cool the unit is ready to be

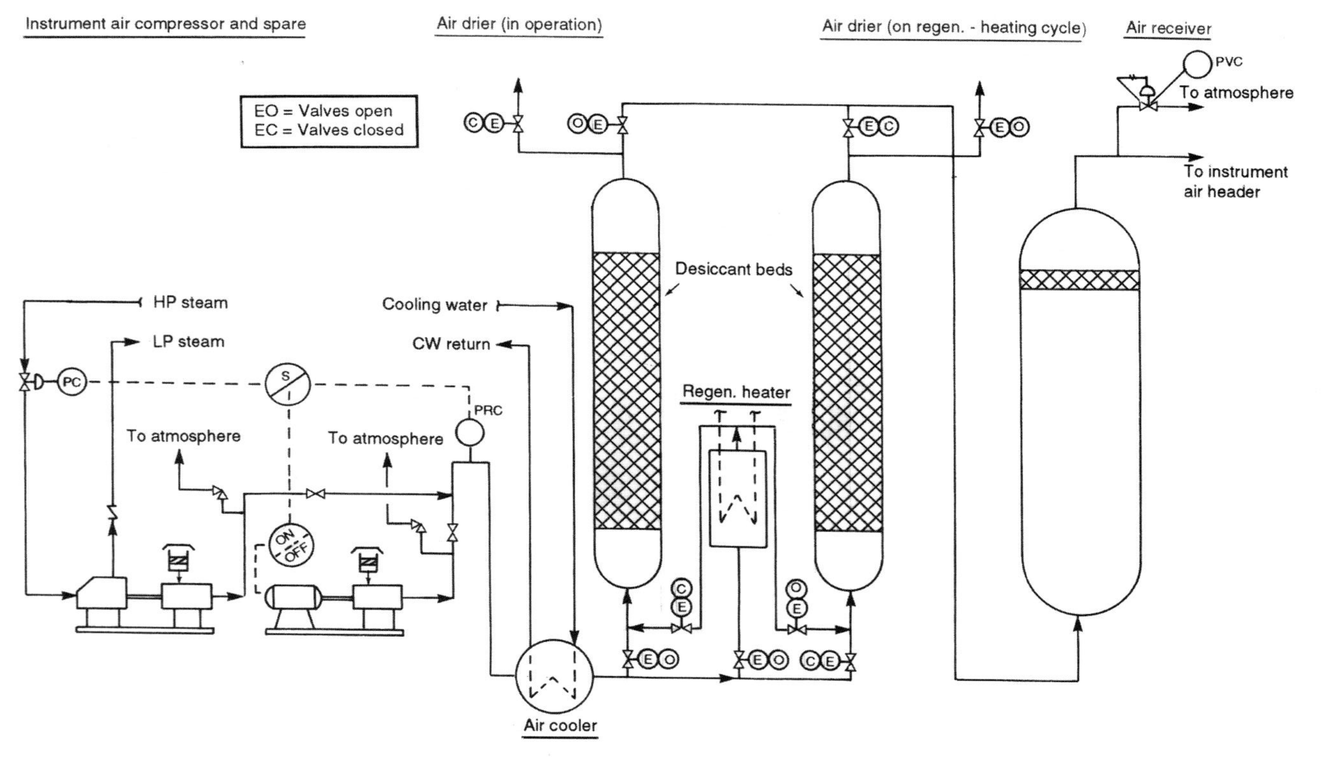

Figure 13.36. An instrument air supply system.

switched into operation for the first dryer to start its regeneration cycle. The various operating and regeneration phases are automatically obtained by a series of solenoid valves operated by a sequence timer switch control. These dryers (often including the compressor and receiver items) are packaged units supplied, skid mounted, and ready for operation.

The instrument air receiver vessel is a pressure vessel containing a crinkled wire mesh screen (CWMS) before the outlet nozzle. It is high-pressure protected by a pressure control valve venting to atmosphere, and of course is also protected by a pressure safety relief valve (not shown in diagram). The air leaves the top of this vessel to enter the instrument air distribution system servicing all the plants in the complex.

13.3 Safety systems

A major requirement in the design and engineering of a plant or system is to ensure its safe operation. Much of the effort in this respect is directed to determining the pressure limits of equipment and to protect that equipment from dangerous over pressuring. Pressure relief valves are normally used for this protection service, although under certain conditions bursting discs may be used. This section of Chapter 13 covers the various types of relief valves, and the procedure for calculating the correct orifice size of the valve and the valve selection.

Determination of risk

The cost of providing facilities to relieve all possible emergencies simultaneously is prohibitive. Every emergency arises from a specific cause, the simultaneous occurrence of two or more emergencies or contingencies is unlikely. Hence, an emergency which can arise from two or more contingencies (e.g., the simultaneous failure of a control valve and cooling water) is not considered when sizing safety equipment. Likewise, simultaneous but separate emergencies are not considered.

Every unit or piece of equipment must be studied individually and every contingency must be evaluated. The safety equipment for an individual unit is sized to handle the largest load resulting from any possible single contingency. If a certain emergency would involve more than one unit, then all must be considered as an entity. The equipment judged to be involved in any one emergency is termed "single risk". The single risk which results in the largest load on the safety facilities in any system is termed "largest single risk" and forms the design basis for the equipment.

Note: The emergency which results in the largest single risk on the overall basis may be different from the emergency(ies) which form the basis for individual pieces of equipment.

Contingencies generally fall into one of two categories—fire (external) or operating failure. Operating failure covers such contingencies as utility failure, mechanical failure, or mal-operation.

Definitions

The terms used and the descriptions given in this item are based on data given in two API publications. These are: API RP520 and 521. References are also made to: Part 1 ANSI Proposed Standard, "Terminology for Pressure Relief Devices" and to ASME PTC 25.2. These publications are the safety standards commonly in current use. The following definition of terms used in the design of safety systems helps to understand the design and criteria of safety systems.

Accumulation

Accumulation is the pressure increase over the maximum allowable working pressure of the vessel during discharge through the pressure relief valve expressed as a percent of that pressure, or in PSI.

Atmospheric discharge

Atmospheric discharge is the release of vapors and gases from pressure relief and de-pressurizing devices to the atmosphere.

Back pressure

Back pressure is the pressure existing at the outlet of the pressure relief device due to pressure in the discharge system.

Balanced safety relief valves

A balanced safety relief valve incorporates means for minimizing the effect of back pressure on the performance characteristics-opening pressure, closing pressure, lift, and relieving capacity.

Blow-down

Blow-down is the difference between the set pressure and the resealing pressure of a pressure relief valve, expressed as a percent of the set pressure, or in PSI.

Burst pressure

Burst pressure is the value of inlet static pressure at which a rupture disk device functions.

Conventional safety relief valve

A conventional safety relief valve is a closed bonnet pressure relief valve that has the bonnet vented to the discharge side of the valve. The performance characteristics-opening pressure, closing pressure, lift, and relieving capacity are directly affected by changes of the back pressure on the valve.

Design pressure

Design pressure is the pressure used in the design of a vessel to determine the minimum permissible thickness or physical characteristics of the different parts of a vessel.

Flare

A flare is a means for safe disposal of waste gases by combustion. With an *elevated flare* the combustion is carried out at the top of a pipe or stack where the burner and igniter are located. A *ground flare* is similarly equipped except that combustion is carried out at or near ground level. A *burn pit* differs from a flare in that it is normally designed to handle both liquids and vapors. Flare systems are described and discussed more fully in "Offsite Systems" of this chapter.

Lift

Lift is the actual travel of the disk away from closed position when the valve is relieving.

Overpressure

Overpressure is the pressure increase over the set pressure of the primary relieving device; it would be termed *accumulation* when the relieving device is set at the maximum allowable working pressure of the vessel.

Note: When the set pressure of the first (primary) pressure relief valve to open is less than the maximum allowable working pressure of the vessel, the overpressure may be greater than 10% of the set pressure of the valve.

Pilot-operated pressure relief valve

A pilot-operated pressure-relief valve is one that has the major flow device combined with and controlled by a self-actuated auxiliary pressure relief valve. This type valve does not utilize an external source of energy.

Pressure relief valve

Pressure relief valve is a generic term applied to relief valves, safety valves, or safety relief valves.

Relieving conditions

Relieving conditions pertain to pressure relief device inlet pressure and temperature at a specific overpressure. The relieving pressure is equal to the valve set pressure (or rupture disk burst pressure) plus the overpressure. The temperature of the flowing fluid at relieving conditions may be higher or lower than the operating temperature.

Set pressure

Set pressure in psig is the inlet pressure at which the pressure relief valve is adjusted to open under service conditions. In a safety or safety relief valve in gas, vapor, or steam service, the set pressure is the inlet pressure at which the valve pops under service conditions. In a relief or safety relief valve in liquid service, the set pressure is the inlet pressure at which the valve starts to discharge under service conditions.

Superimposed back pressure

Superimposed back pressure is static pressure existing at the outlet of a pressure relief device at the time the device is required to operate. It is the result of pressure in the discharge system from other sources. This type of back pressure may be constant or variable; it may govern whether a conventional or balanced-type pressure relief valve should be used in specific applications.

Vapor depressing system

A vapor depressing system is a protective arrangement of valves and piping intended to provide for rapid reduction of pressure in equipment by release of vapors. Actuation of the system may be automatic or manual.

Vent stack

A vent stack is the elevated vertical termination of a disposal system which discharges vapors into the atmosphere without combustion or conversion of the relieved fluid.

Types of pressure relief valves

The following is a list of those types of relief valves commonly used in industry. These have been approved according to ASME V111 "Boiler and Pressure Vessel" code.

Conventional safety relief valves

In a conventional safety relief valve the inlet pressure to the valve is directly opposed by a spring closing the valve, the back pressure on the outlet of the valve changes the inlet pressure at which the valve will open. A diagram of a conventional relief valve is shown below as Figure 13.37.

Balanced safety relief valves

Balanced safety valves are those in which the back pressure has very little or no influence on the set pressure. The most widely used means of balancing a safety relief

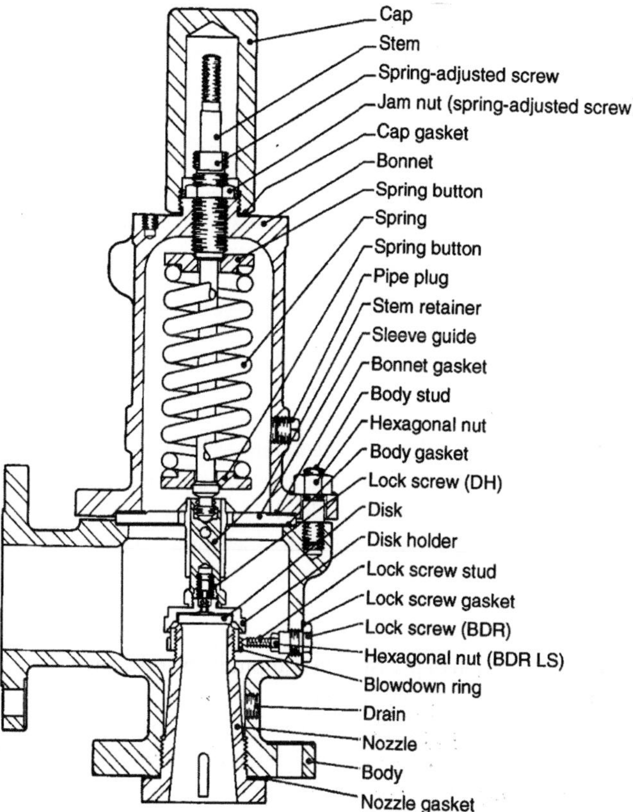

Figure 13.37. A diagram of a conventional safety relief valve.

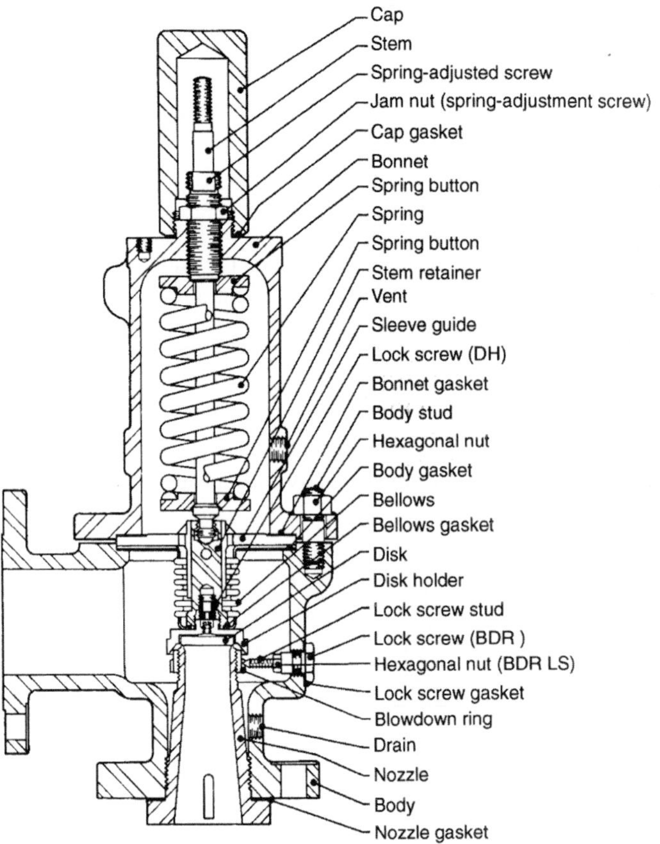

Figure 13.38. A diagram of a balanced safety relief valve.

valve is through the use of a bellows. In the balanced bellows valve, the effective area of the bellows is the same as the nozzle seat area and back pressure is prevented from acting on the top side of the disk. Thus the valve opens at the same inlet pressure even though the back pressure may vary. A diagram of a balanced safety relief valve is shown as Figure 13.38.

Pilot operated safety relief valves

A pilot-operated safety relief valve is a device consisting of two principal parts, a main valve and a pilot. Inlet pressure is directed to the top of the main valve piston, and with more area exposed to pressure on the top of the piston than on the bottom; pressure, not a spring, holds the main valve closed. At the set pressure the pilot opens reducing the pressure on top of the piston and the main valve goes fully open.

Resilient seated safety relief valves

When metal-to-metal seated conventional or bellows type safety relief valves are used where the operating pressure is close to the set pressure, some leakage can be expected through the seats of the valve (Refer to API Standard 527, "Commercial Seat Tightness of Safety Relief Valves with Metal-to-Metal Seats").

Resilient seated safety relief valves with either O-Ring seat seal or plastic seat such as Teflon provide seat tightness. Limitations of temperature and chemical compatibility of the resilient material must be considered when using these valves.

Rupture disk

A rupture disk consists of a thin metal diaphragm held between flanges. The disk is designed to rupture and relieve pressure within tolerances established by ASME Code.

Capacity

The maximum amount of material to be released during the largest single risk emergency determines the size of the safety relief valves in any given system. Any calculation to determine valve sizing must therefore be preceded by a calculation or some determination of the maximum amount. Among the most common sizing criteria is the event of fire and its effect on the contents of exposed vessels. There are also other criteria which can determine maximum release that are attributable to operational failure.

Capacity due to fire

The exact method of making this calculation must be established from the appropriate codes which apply, API RP-520, Part I, API Standard 2510, NFPA No. 58, or local codes which may apply. Each of the listed codes or standards approach the problem in a slightly different manner.

Liquid systems—A majority of the systems that are encountered will contain liquids or liquids in equilibrium with vapor. Fire relief capacity in this situation is calculated on the basis of heat energy from the fire translated in terms of vapor generated in the boiling liquid.

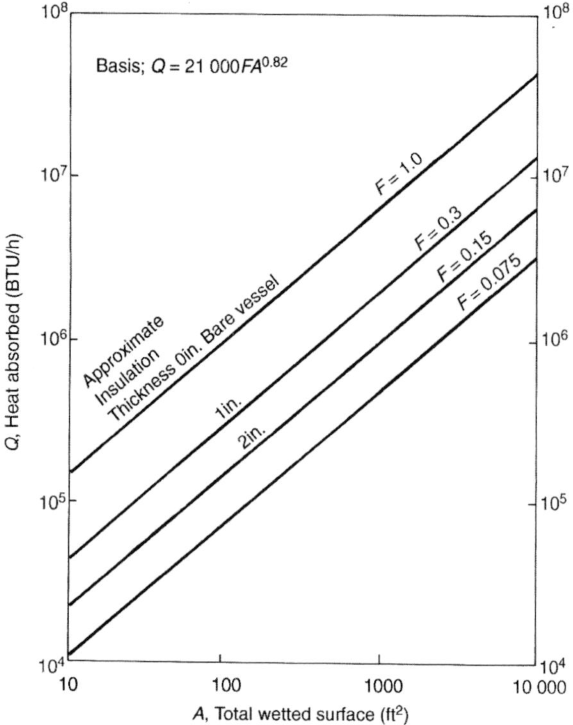

Figure 13.39. Insulation factors.

API RP-520, Part I, applies to refineries and process plants. It expresses requirements in terms of heat input.

$$Q = 21000FA^{0.82}$$

where

Q = Btu/hr heat input

A = area in sqft of wetted surface of the vessel up to 25 ft above grade. Wetted surface is calculated at the maximum fill level. Grade is the ground level under the vessel.

F = fireproof factor due to insulation becoming 1.0 for a bare vessel (see Figure 13.39).

The amount of vapor generated with this is then calculated from the latent heat of the material at the relieving pressure of the valve by the following equation. For fire relief only this may be calculated at 120% of maximum allowable working pressure. All other conditions must be calculated at 110% of maximum allowable working pressure thus:

$$Q/H_t = W$$

where

Q = Btu/hr heat input to the vessel
H_L = Btu/lb latent heat of the material being relieved
W = lb/hr of vapor to be relieved by the relief valve

The latent heat of pure and some mixed paraffin hydrocarbons materials may be estimated using the data given in API RP-520. A more accurate latent heat evaluation for mixed hydrocarbons will be found by utilizing vapor-liquid equilibrium K data and making a flash calculation. Mixed hydrocarbons will fractionate, beginning with the lowest temperature boiling mixture and progress to the highest temperature mixture; therefore, consideration must be given to the condition which will cause the largest vapor generation requirements from the heat input of a fire.

Latent heat will approach a minimum value near critical conditions; however, the effect does not go to zero. An arbitrary minimum value that may be used is 50 Btu/lb.

Sizing of required orifice areas

The safety relief valve manufacturers have standard orifice designation for area and the valve body sizes which contain these orifices (API Standard 526, "Flanged Steel Safety Relief Valves for Use in Petroleum Refineries"). The standard orifices available, by letter designation and area are:

D Orifice	0.110 inch^2
E Orifice	0.196 inch^2
F Orifice	0.307 inch^2
G Orifice	0.503 inch^2
H Orifice	0.785 inch^2
J Orifice	1.287 inch^2
K Orifice	1.838 inch^2
L Orifice	2.853 inch^2
M Orifice	3.60 inch^2
N Orifice	4.34 inch^2
P Orifice	6.38 inch^2
Q Orifice	11.05 inch^2
R Orifice	16.0 inch^2
T Orifice	26.0 inch^2

Note: however, many small safety relief valves are manufactured with orifice areas smaller than "D", and many pilot-operated types contain orifice areas larger than "T".

Sizing for gas or vapor relief, sonic or critical flow

Safety relief valves in gas or vapor service may be sized by use of one of these equations:

$$A = \frac{W\sqrt{T}\sqrt{Z}}{CKP_i K_b \sqrt{M}}$$

$$A = \frac{V\sqrt{T}\sqrt{M}\sqrt{Z}}{6.32\ CKP_i K_b}$$

$$A = \frac{V\sqrt{T}\sqrt{G}\sqrt{Z}}{1.175\ CKP_i K_b}$$

where

W = flow through valve, lb/hr
V = flow through valve, scfm
C = coefficient determined by the ratio of the specific heats of the gas or vapor at standard conditions
K = coefficient of discharge, obtainable from the valve manufacturer (usually 0.6–0.7).
A = effective discharge area of the valve, inch^2
P_i = upstream pressure, psia. This is set pressure plus overpressure plus the atmospheric pressure
K_b = capacity correction factor due to back pressure. This can be obtained from Figure 13.40 for conventional valves or pilot operated valves, and from Figure 13.41 for balanced bellows valves
M = molecular weight of gas or vapor
T = absolute temperature of the inlet vapor in °R (°F + 460)
Z = compressibility factor for the deviation of the actual gas from a perfect gas.
G = specific gravity of gas referred to air = 1.00 at 60° F and 14.7 psia

Sizing for liquid relief

Conventional and balanced bellows safety relief valves in liquid service may be sized by use of the following equation. Pilot-operated relief valves should be used in liquid service only after determining from the manufacturer that they are suitable for the service.

$$A = \frac{\text{gpm}\sqrt{G}}{38K(K_p K_w K_v)(1.25P - P_b)^{1/2}}$$

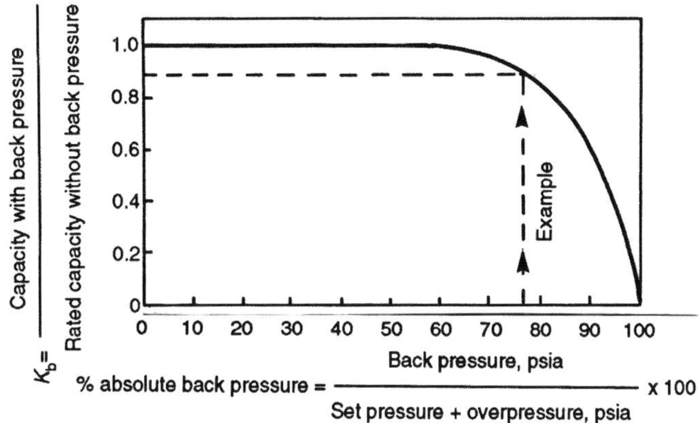

Figure 13.40. Back pressure sizing factor K_b for conventional valves.

Note: A coefficient of discharge of 0.62 is normally used for K.

where

gpm = flow rate at the selected percentage of overpressure, in US gals
A = effective discharge area, inch^2
K_p = capacity correction factor due to overpressure. For 25% overpressure $K_p = 1.00$. The factor for other percentages of overpressure can be obtained from Figure 13.42
K_w = capacity correction factor due to back pressure and is required only when balanced bellows valves are used. K_w can be obtained from Figure 13.43

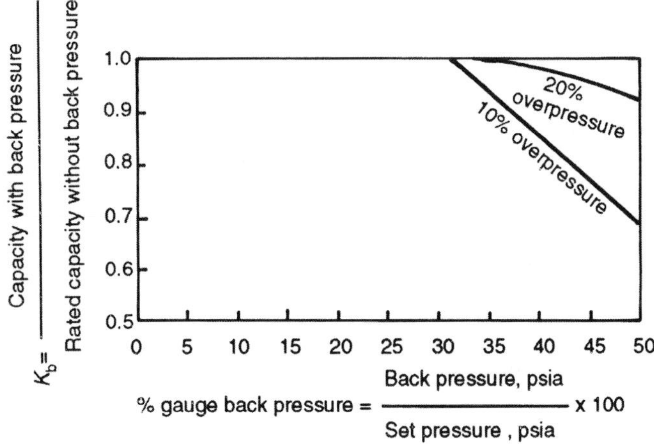

Figure 13.41. Back pressure sizing factor K_b for balanced valves.

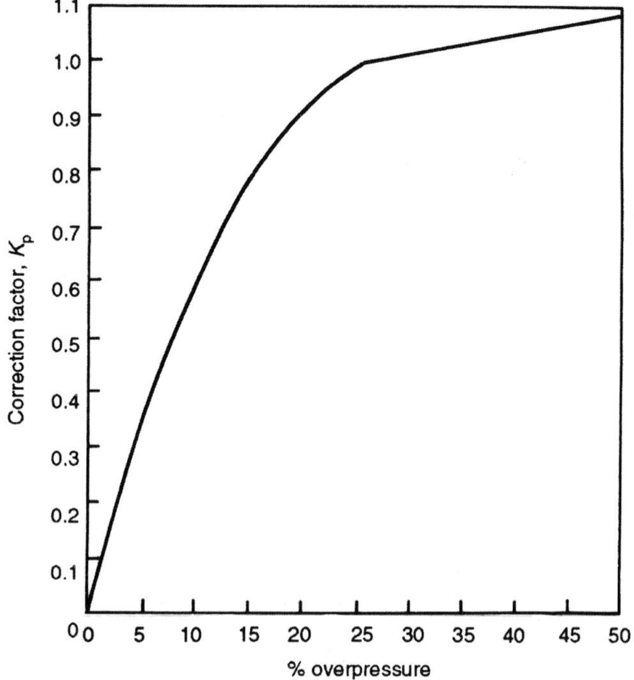

Figure 13.42. Capacity correction factor due to over pressure K_p.

K_v = capacity correction factor due to viscosity. For most applications, viscosity may not be significant, in which case $K_v = 1.00$. When viscous liquid is being relieved see method of determining K_v as described below.

P = set pressure at which relief valve is to begin opening, psig

P_b = back pressure, psig

G = specific gravity of the liquid at flowing temperature referred to water = 1.00 at 70°F.

When a relief valve is sized for viscous liquid service, it is suggested that it be sized first as for non-viscous-type application in order to obtain a preliminary required discharge area, A. From manufacturers' standard orifice sizes, the next larger orifice size should be used in determining the Reynold's number, R, from either of these relationships:

$$R = \frac{\text{gpm}\,(2800G)}{\mu\sqrt{A}}$$

$$R = \frac{12700\ \text{gpm}}{U\sqrt{A}}$$

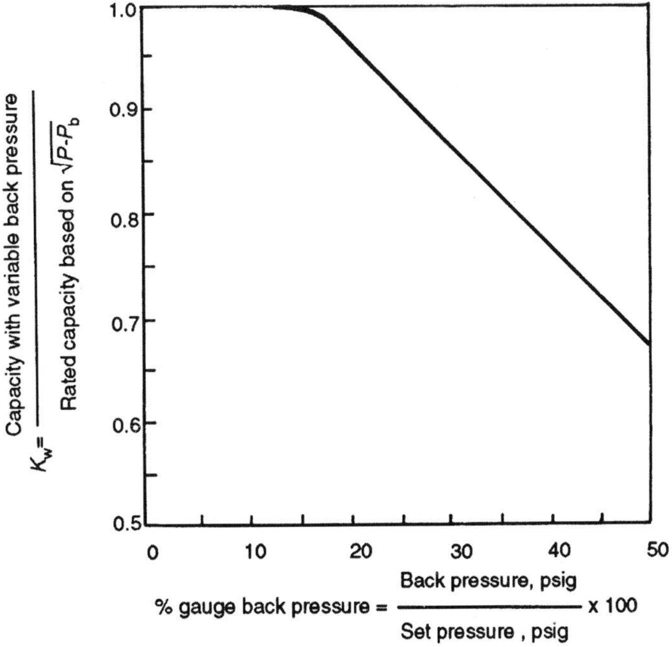

Figure 13.43. Capacity correction factor K_W.

where

gpm = flow rate at the flowing temperature, in US GPM
G = specific gravity of the liquid at the flowing temperature referred to water = 1.00 at 70°F
μ = absolute viscosity at the flowing temperature, in centipoise.
A = effective discharge area, in sqin (from manufacturers' standard orifice areas)
U = viscosity at the flowing temperature, in Saybolt Universal seconds

After the value of R is determined, the factor K_V is obtained from Figure 13.44. Factor K_V is applied to correct the "preliminary required discharge area." If the corrected area exceeds the "chosen standard orifice area," the above calculations should be repeated using the next larger standard orifice size.

Sizing for mixed-phase relief

When a safety relief valve must relieve both liquid and gas or vapor it may be sized by the following steps:

(a) Determine the volume of gas or vapor and the volume of liquid that must be relieved.

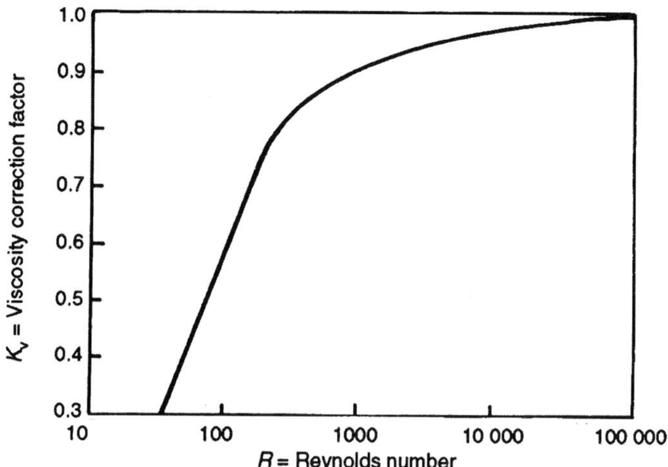

Figure 13.44. Capacity correction factor due to viscosity K_V.

(b) Calculate the orifice area required to relieve the gas or vapor as previously outlined.
(c) Calculate the orifice area required to relieve the liquid as previously outlined.
(d) The total required orifice area is the sum of the areas calculated for liquid and vapor.

Sizing for flashing liquids

The conventional method is to determine the percent flashing from a Mollier diagram or from the enthalpy values. Then consider the liquid portion and vapor portion separately as in mixed phase flow. A method to calculate the percent flashing is shown in the following equation:

$$\% \text{ Flash} = \frac{h_f(1) - h_f(2)}{h_{fg}(2)} \times 100$$

where

$h_f(1)$ = enthalpy in Btu/lb of saturated liquid at upstream temperature
$h_r(2)$ = enthalpy in Btu/lb of saturated liquid at downstream pressure
$h_{fg}(2)$ = enthalpy of evaporation in Btu/lb at downstream pressure

Sizing for gas or vapor on low-pressure sub-sonic flow

When the set pressure of a safety relief valve is very low, such as near atmospheric pressure, the K_b values obtainable from Figure 13.40 are not accurate. Safety relief

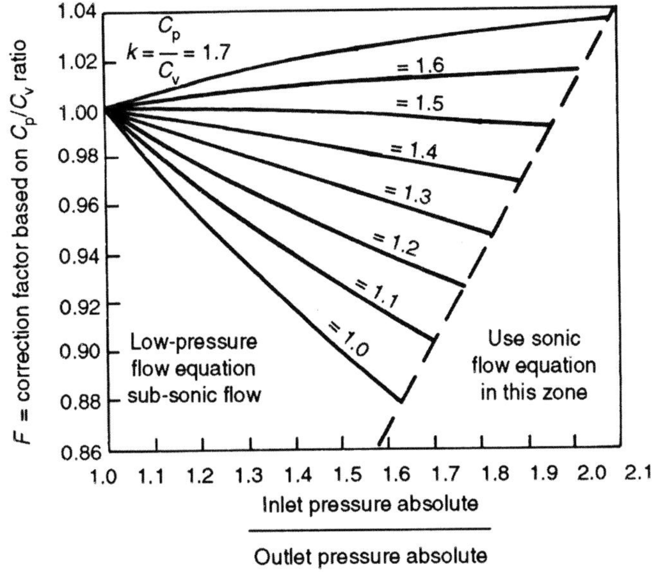

Figure 13.45. Flow correction factor '*F*' based on specific heats.

valve orifice areas for this low-pressure range may be calculated:

$$A = \frac{Q_V\sqrt{GT}}{863KF\sqrt{(P_1 - P_2)P_2}}$$

where

A = effective discharge area of the valve, inch^2
Q_V = flow through valve, scfm
G = specific gravity of gas referred to air = 1.00 at 60°F and 14.7 psia
T = absolute temperature of the inlet vapor, °R = ° F + 460
K = coefficient of discharge, is available from the valve manufacturer
F = correction faction based on the ratio of specific heats. This can be obtained from Figure 13.45.
P_1 = upstream pressure in psia = set pressure plus overpressure plus atmospheric pressure
P_2 = downstream pressure at the valve outlet in psia

An example calculation covering the sizing of a relief valve is given in Appendix 13.2 of this chapter.

Appendix 13.1: Example calculation for sizing a tank heater

Problem

It is required to calculate the surface area for a heating coil which will maintain the bulk temperature of fuel oil in a cone roofed tank at a temperature of 150°F. The ambient air temperature is an average 65°F and the wind velocity averaged over the year is 30 MPH. The fuel oil data is as follows:

$$\text{Viscosity}\,(\mu) = 36\,\text{cps} \,@\, 150°\text{F}$$

$$\text{Sg} \,@\, 150°\text{F} = 0.900$$

The tank is to be heated with 125 psig saturated steam. The tank dimensions are 60 ft diameter × 180 ft high it is not insulated but is painted with non metallic color paint.

Solution

1.0 Calculating the heat loss from the wall.

1st trial. Assume wall temperature is 120°F

$$h_i = 8.5(\Delta t_i/\mu)^{0.25}$$

where

$\Delta t_i = 150 - 120 = 30°\text{F}$
$h_i = 8.5 \times 0.955 = 8.12$ Btu/hr. sqft.°F
'a' $= 8.12 \times 30 = 243.6$ Btu/hr. sqft.

Δt_o is temperature difference between assumed wall temp and the ambient air = $120 - 65 = 55°\text{F}$.

$$h_{co} = 0.495 \times 1.3 \text{ (from Figure 13.9)}$$

Wind correction factor F_w is as follows:

$$\begin{aligned} F_1 &= (\text{MPH}/1.47)^{0.61} \\ &= 6.29 \\ F2 &= 1.04 \text{ (From Figure 13.10)} \\ F_w &= 6.29 + 1.04 \\ &= 7.33. \\ h_{co}\text{ (corrected)} &= 0.495 \times 1.3 \times 7.33 \\ &= 4.20 \text{ Btu/hr. sqft.°F} \end{aligned}$$

Heat loss from wall due to radiation h_{ro} is found from Figure 13.8 = 1.18 Btu/hr.sqft.°F.

Corrected for emissivity $h_{ro} = 1.18 \times .95 = 1.123$ Btu/hr.sqft.°F

$$\begin{aligned} \text{'}b\text{'} &= (h_{co} + h_{ro}) \times \Delta t_o \\ &= (4.21 + 1.12) \times 55 \\ &= 293 \text{ Btu/hr.sqft} \\ \text{'}a\text{'} - \text{'}b\text{'} &= 244 - 293 \\ &= -49 \text{ Btu/hr.sqft} \end{aligned}$$

2nd trial. Assume wall temperature is 110°F

Carrying out the same calculation procedure as for trial 1:

$$\text{'}a\text{'} - \text{'}b\text{'} = +172 \text{ Btu/hr.sqft}$$

3rd trial. Assume wall temperature is 115°F

Again carrying out the calculation procedure as for trial 1:

$$\text{'}a\text{'} - \text{'}b\text{'} = +35$$

The results of the above trials are plotted linearly below:

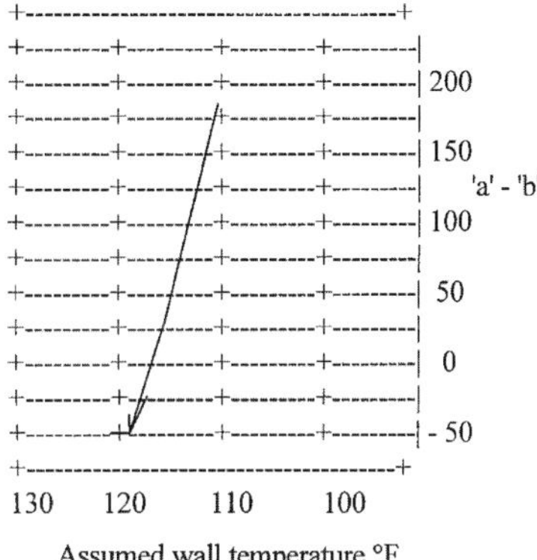

Final trial. At wall temperature of 117°F.

$$\begin{aligned} h_{io} &= 8.31 \text{ Btu/hr.sqrt.°F} \\ \text{'}a\text{'} &= 8.31 \times (150 - 117) = 274 \text{ Btu/hr.sqrt} \\ h_{co} \text{ (corrected)} &= 4.18 \text{ Btu/hr.sqft.°F} \\ h_{ro} \text{ (corrected)} &= 1.12 \text{ Btu/hr.sqft°F} \\ \text{'}b\text{'} &= (4.18 + 1.12) \times (117 - 65)\text{°F} \\ &= 275.7 \text{ Btu/hr.sqft} \end{aligned}$$

'*a*' and '*b*' are close enough call total heat loss 275.7 Btu/hr.sqft.

Surface area of wall = circumference × height.

$$= \pi D \times 180\,\text{ft}$$
$$= 33929\,\text{sqft}$$

Total heat loss through wall = 275.7 × 33929

$$= 9.35\,\text{mm Btu/hr}$$

2.0 Calculating heat loss through roof.

Trial 1. Assume roof temperature is 116°F.

$$h_i = 8.38\ \text{Btu/hr.sqft.°F}$$
$$\text{`}a\text{'} = 8.38 \times (150 - 116)\text{°F}$$
$$= 284.9\ \text{Btu/hr.sqft.}$$
$$h_{co}\ (\text{corrected}) = (.470 \times 2.0) \times 1.04 \times 6.29$$

(*Note*: the number read from Figure 13.9 is multiplied by 2.0 in this case as the roof is an upward facing plate.)

$$= 6.35\ \text{Btu/hr.sqft.°F}$$
$$h_{ro}\ (\text{corrected}) = 1.165 \times 0.95$$
$$= 1.11\ \text{Btu/hr.sqft.°F}$$
$$\text{`}b\text{'} = (6.35 + 1.11) \times (116 - 65)\text{°F}$$
$$= 380\ \text{Btu/hr.sqft.}$$
$$\text{`}a\text{'} - \text{`}b\text{'} = -95\ \text{Btu/hr.sqft}$$

Trial 2. Assume a wail temperature of 110°F

'*a*' – '*b*' in this case = +23 which is within acceptable limits.

The heat loss is taken as an average of '*a*' and '*b*'

$$= 338\,\text{Btu/hr.sqft.}$$

Total heat loss from the roof = area of roof × 338

$$= 2827\,\text{sqft} \times 338$$

$$= 0.956\,\text{mm Btu/hr}$$

3.0 Calculating the heat loss through the floor.

Assume the ground temperature is 50°F and the heat transfer coefficient is 1.5 Btu/hr.sqft.°F.

Then heat loss = 1.5 × (150 – 50) × 2827 sqft

$$= 0.424\ \text{mm Btu/hr}$$

4.0 Total heat loss from the tank.

Heat loss from the Walls = 9.350 mm Btu/hr
Heat loss from the Roof = 0.956 mm Btu/hr
Heat loss from the Floor = 0.424 mm Btu/hr
Total Heat Loss = 10.730 mm Btu/hr

5.0 Calculating the tank heater coil surface area required.

The heating medium is saturated 125 psig steam.

Temperature of the steam = 354°F

Steam side calculations.

Approx resistance of steam, $h_s = 0.001$ hr.sqft.°F/Btu
Fouling factor on steam side, $r_1 = 0.0005$ hr.sqft.°F/Btu
Tube metal resistance $r_2 = 0.0005$ hr.sqft.°F/Btu
Outside fouling factor $r_3 = 0.005$ hr.sqft.°F/Btu

$$\text{Heat transfer coefficient for the steam side} = \frac{1}{0.001 + 0.0005 + 0.0005 + 0.005}$$
$$= 143 \text{ Btu/hr.sqft.°F}$$

Oil side heat transfer coefficient is obtained from Figure 13.11.

1st trial. Assume a tube wall temperature of 310°F.

For steam side 'a' = 143 × (354 − 310)
= 6292 Btu/hr.sqft.
For oil side h_o = 31 Btu/hr.sqft.°F (Figure 13.11)
'b' = 31 × (310 − 150)
= 4960 Btu/hr.sqft
'a' − 'b' = +1332

2nd trial. Assume a tube wall temperature of 320°F.

'a' in this case calculated to be 4862

'b' was calculated to be 5355

'a' − 'b' = −493

Plotted on a linear curve the tube wall temperature to give 'a' = 'b' was 317°F.

Final trial. At a tube wall temperature of 317°F

Steam side = 143 × (354 − 317)
'a' = 5291 Btu/hr.sqft.
Oil side h_o = 31.2 (From Figure 13.11)
'b' = 31.2 × (317 − 150)
= 5210 Btu/hr.sqft
'a' − 'b' = +81 which is acceptable.

$$\text{Make rate of heat transfer} = \frac{5291 + 5210}{2}$$
$$= 5251 \text{ Btu/hr.sqft}$$

$$\text{Then surface area of coil required} = \frac{10.730 \text{ mm Btu/hr}}{5251 \text{ Btu/hr.sqft}}$$
$$= 2043 \text{ sq ft}$$

Appendix 13.2: Example calculation for sizing a relief value

A vessel containing naphtha C_5–C_8 range is uninsulated and is not fireproofed. The vessel is vertical and has a skirt 15′ in length. Dimensions of the vessel are I/D 6′0″ T-T 20′0 liquid height to HLL = 16′0. Calculate the valve size for fire condition relief. Set pressure is 120 psig.

Latent heat of naphtha at 200°F is 136 Btu/lb = H_L

$Q = 21000\ \text{FA}^{0.82}$
A = Wetted area and is calculated as follows:
Liquid height above grade = 15 + 16 ft
= 31 ft

Therefore wetted surface of vessel need only be taken to 25 ft above grade which is 25−15 = 10 ft of vessel height.

Wetted surface = $\pi D \times 10$ for walls
= 188.5 sq.ft
plus 28.3 sq.ft for bottom
= 216.8 sq.ft

$$Q = 21000 \times 1.0 \times (216.8)^{0.82}$$
$$= 1.729 \text{ Btu/hr} \times 10^6$$
$$Q/H_L = \frac{1.729 \times 10^6}{136} = 12713 \text{ lbs/hr} = W$$
$$A = \frac{W\sqrt{T}\sqrt{Z}}{CKP_1K_b\sqrt{M}}$$

where

A = effective discharge area in sq ins.
W = flow through valve in lbs/hr = 12713
T^* = absolute temp of inlet vapor = 460 + 200 = 660°R
Z = 0.98 (NC_5)
C = 356.06 (based on CA/CV = 1.4)
K = 0.65 (from Figure 13.40)
K_b = 0.9 M = 100 (use C_7)
P_1 = set pressure of valve = 134.7 psia.
*Bubble point of C_5 to C_8 and say 10 psig

$$A = \frac{12713 \times 25.7 \times 0.99}{356 \times 0.65 \times 134.7 \times 0.9 \times 1} = 1.153\,\text{inch}^2$$

nearest orifice size = 'J' at 1.287 inch^2

Appendix 13.3: Control valve sizing

Process flow coefficient (C_V) and valve sizing

Process flow coefficient C_V is defined as the water flow in GPM through a given restriction for 1 psi pressure drop. These C_Vs can be determined by the following equations:

$$C_v = Q_L\sqrt{\frac{G_l}{\Delta P}} \quad \text{for liquid}$$

$$C_v = \frac{Q_s}{82}\sqrt{\frac{T}{\Delta P - P_2}} \quad \text{for steam}$$

$$C_v = \frac{Q_G}{1360}\sqrt{\frac{\mu_2 ST}{\Delta P - P_2}} \quad \text{for gases}$$

where

C_v = Flow coefficient
Q_L = liquid flow in GPM at conditions
ΔP = pressure drop across valve, psi
G_L = specific gravity of liquid at conditions
Q_S = steam rate in lbs/hr
P_2 = pressure downstream of valve psia

Q_G = gas flow in SCFH (60°F, 14.7 psia).
T = temperature of gas in °R (°F + 460)
S = mol weight of gas divided by 29
μ_2 = compressibility factor at downstream conditions

The following are some special considerations that may have to be made in determining.

Process C_V values

Pressure drop
For compressible fluids the maximum usable pressure drop in equations (b) and (c) is the critical value. As a rule of thumb and for design purposes this value is 50% of the absolute upstream pressure. (The valve can take more than the critical pressure drop, but any pressure drop over the critical takes the form of exit losses).

Flashing liquids
In the absence of accurate information, it is recommended that for flashing service the valve body be specified as one nominal size larger than the valve port.

Two-phase flow
If two-phase flow exists upstream of the control valve experience has shown that for fluids below their critical point a sufficiently accurate process C_V value can be arrived at by adding the process C_V values for the gas and liquid portions of the stream. The calculation is based on the quantities of gas and liquid at upstream conditions. The valve body is specified to be one nominal size larger than the port to allow for expansion.

Valve rangeability
The rangeability of a control valve is the ratio of the flow coefficient at the maximum flow rate to the flow coefficient at the minimum flow rate. ($R = C_V$ Max/C_V Min). Valve rangeability is actually a criteria which is used to judge whether a given valve will be in a controlling position throughout its required range of operation (neither wide open nor fully closed). In practice the selection of the actual valve to be installed is the responsibility of the instrument engineer. As the process engineer is usually the person responsible for the correct operation of the process itself however he must be satisfied that the item selected meets the control criteria required. He must therefore satisfy himself that the valve will control over the range of the process flow.

Control valves are usually limited to a rangeability of 10 : 1. If R is greater than 10 : 1 then a dual valve installation should be considered in order to assure good control at the maximum and minimum flow conditions.

In some applications, particularly on compressor or blower suction, butterfly valves have been specified to be line size without considering that as a result the valve may operate almost closed for long periods of time. Under this condition there have been cases of erosion resulting from this. It is recommended therefore that butterfly valves be sized so that they will not operate below 10% open for any appreciable period of time and not arbitrarily be made line size.

Valve flow coefficient ($C_{V'}$)

In order to ensure that the valve is in a controlling position at the maximum flow rate, the valve $C_{V'}$ is the maximum process C_V value determined above, divided by 0.8. The reasons for using this factor are that:

(1) It is not desirable to have the valve fully open at maximum flow since it is not then in a controlling position.
(2) The valves supplied by a single manufacturer often vary as much as 10–20% in C_V.
(3) Allowance must be made for pressure drop, flow rate, etc., values which differ from design.

Control valve sizing

Control valve sizes are determined by the manufacturers from the process data submitted to them. However there are available some simple equations to give a good estimate of the required valve sizes to meet a process duty. Three of these are given below:

$$S\,(\text{inches}) = \left[\frac{\text{Valve } C_V}{9}\right]^{1/2}$$

Double seated control valves sizes may be estimated by:

$$S\,(\text{inches}) = \left[\frac{\text{Valve } C_V}{12}\right]^{1/2}$$

Butterfly valve sizes may be estimated by:

$$S\,(\text{inches}) = \left[\frac{\text{Valve } C_V}{20}\right]^{1/2}$$

The constants (9, 12 or 20) in these equations can vary as much as 25% depending on the valve manufacturer.

A control valve should be no larger than the line size. A control valve size that is calculated to be greater than line size should be carefully checked together with the calculation used for determining line size. Ideally a control valve size should be one size smaller than line size.

Once the valve size is estimated and the valve C_V known, then the percent opening of the valve at minimum flow and maximum flow can be obtained by dividing the respective process C_V values conditions by the valve C_V. This information is normally required to check the percent opening of a butterfly at minimum flow. It is not normally necessary to calculate it for any other type of valve.

Valve action on air failure

In the analysis of the design and operation of any process or utility system the question always arises on the action of control valves in the system on instrument air failure. Should the control valve fail open or closed is the judgement decision of the process engineer after evaluating all aspects of safety and damage in each event. For example, control valves on fired heater tube inlets should always fail open to prevent damage to the tubes through low or no flow through them when they are hot. On the other hand control valves controlling fuel to the heaters should fail closed on air failure to avoid over heating of the heater during the air failure.

The failed action of the valve is established by introducing the motive air to either above the diaphragm for a failed open requirement or below the diaphragm for a failed shut situation. The air failure to the valve above the diaphragm allows the spring to pull up the plugs from the valve seats. Air failure to valves below the diaphragm forces the spring to seat the valves in the closed position.

Chapter 14

Environmental control and engineering in petroleum refining

D.S.J. Jones

Introduction

Certainly during the last three decades of the 1900s control of contaminants from the refining of crude oil has become a significant consideration in the design and operation of the oil refinery. Chapter 2 of this book provides the changes that have taken place in the specification of motive fuels to meet legislative environmental control. These products have been singled out because they impact more prominently on the daily life style and health of the general public, particularly in the more advanced industrial countries. Similar changes and legislative control have been adopted on the levels of effluents leaving the refinery plants in the water and air. These changes have become more restrictive as greater knowledge of their effect on the environment emerges. This chapter then deals with some of the measures adopted in process design and operation of processes to meet these environmental protection measures. These are described and discussed in the following parts of the chapter:

Part 1. Aqueous wastes
Part 2. Emission to the atmosphere
Part 3. Noise abatement

The aqueous and atmospheric contaminants that most refineries emit are shown in Table 14.1. The table also shows the process plant source of these more common contaminants.

14.1 Aqueous wastes

This part is divided into a description of the pollutants in aqueous wastes, the processes that generate the wastes, and the treatment of the various aqueous waste streams.

Table 14.1. Sources of waste water and air contaminants in oil refining

Process	Waste water	Air
Atmospheric and vacuum distillation	Sour water (NH_3 and H_2S) Desalter water Spent caustic Process area waste water (pump glands, area drains, etc.)	Furnace Flue gases—SO_2
Thermal cracking delayed coking	Sour water Decking water (Oil) Process Area Waste Water	Furnace Flue gases
Fluid cat cracking unsat gas plant	Sour water (NH_3, H_2S, Phenols) Spent caustic Process area waste water	Furnace Flue gases SO_2, CO Particulates.
Hydrocracker hydrogen plant	Sour water (inc Phenols) Process area waste water	Furnace Flue Gases SO_2
Sat gas plant alkylation	Spent caustic Process area waste water	Nil
Naphtha hydrotreater cat reformer	Sour water Process area waste water	Furnace Flue Gases
Sulfur plant	Nil	Incinerator Flue Gas—SO_2 Hydrocarbons—Flare
Tankage area	Tank dike area drains Non contaminated rain runoff	Tank Vents Hydrocarbons

Pollutants in aqueous waste streams

The most undesirable pollutants in aqueous waste streams are:

- Those that deplete the dissolved oxygen content of the waterways into which they discharge
- Those contaminants that are toxic to all forms of life, such as arsenic, cyanide, mercury and the like
- Those contaminants that impart undesirable tastes and odors to streams and other waterways into which they discharge

A description and discussion on these contaminants are expressed in the following paragraphs.

The effect of contaminants on the oxygen balance of natural waterways

Natural waterways have a complex and delicate oxygen balance. The water contains an amount of dissolved oxygen which has an equilibrium of between 14 ppm in winter and 7 ppm in summer. Aquatic life (fish and tadpoles) continually consume oxygen from the water, but aquatic plants produce oxygen naturally maintaining the balance required to sustain fish life. Any oxidizable contaminant introduced into the natural waterways consumes the dissolved oxygen to be oxidized. The dissolved oxygen will be depleted below saturation point and will be replenished only by re-aeration. The relative rate of replenishment to its saturation level will depend on a time factor equivalent to the stream distance or the stream flow rate.

Oxygen depletion occurs by the introduction of one or more of the following contaminants entering the waterway:

- Natural pollution by surface run-off rainwater, or melting snow, in the form of soluble salts leached from the earth
- Natural pollution caused by decay of organic plants from swamps or other sources
- Human and animal life excretion
- Chemical pollution from reducing agents in industrial plant wastes. Such as sulfides, nitrites, ferrous salts, etc.
- Biochemical pollution from such industrial wastes as phenols, hydrocarbons, carbohydrates, and the like

The degree of oxygen depletion from the pollution sources described above may be catalogued by the following terms:

BOD—Biological Oxygen Demand.
COD—Chemical Oxygen Demand.
IOD—Immediate Oxygen Demand.

BOD. Since all natural waterways will contain bacteria and nutrient, almost any waste compound introduced into the waterway will initiate biochemical reactions. These reactions will consume some of the dissolved oxygen in the water. This is illustrated in Figure 14.1.

The depletion of oxygen due to biological pollution is not very rapid. It follows the laws of first order reaction. Because of this the effect of BOD is measured in the laboratory on a five day basis, and has been universally adopted as the measure of pollution effect. The "Ultimate" BOD is a measure of the total oxygen consumed when the biological reaction proceeds to almost completion. The "5 day" BOD is believed to be approximately the Ultimate. Figure 14.2 shows the "5 day" BOD versus the "Ultimate".

In summary BOD measures organic wastes that are biologically oxidizable.

COD. The COD is a measure of the oxygen depletion due to organic and inorganic wastes which are chemically oxidizable. There are several laboratory methods

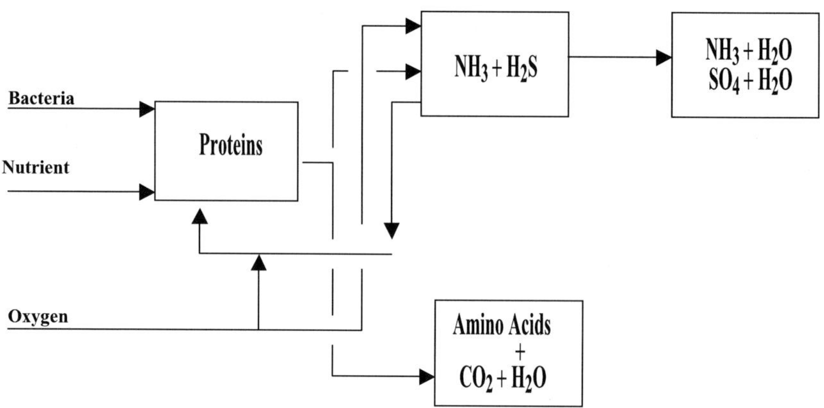

Protein Biochemical Oxidation.

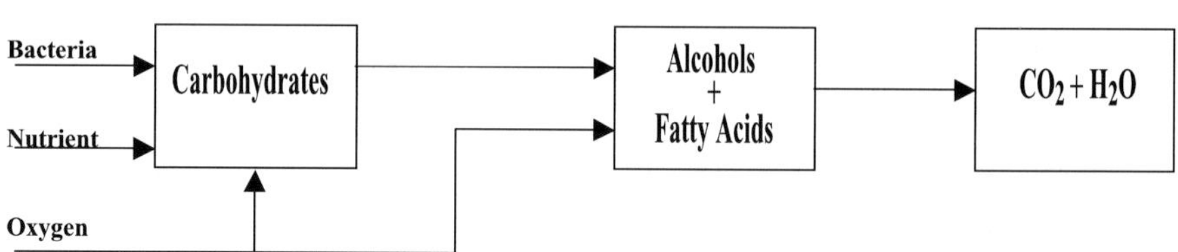

Carbohydrate Biochemical Oxidation.

Figure 14.1. Typical BOD chemistry.

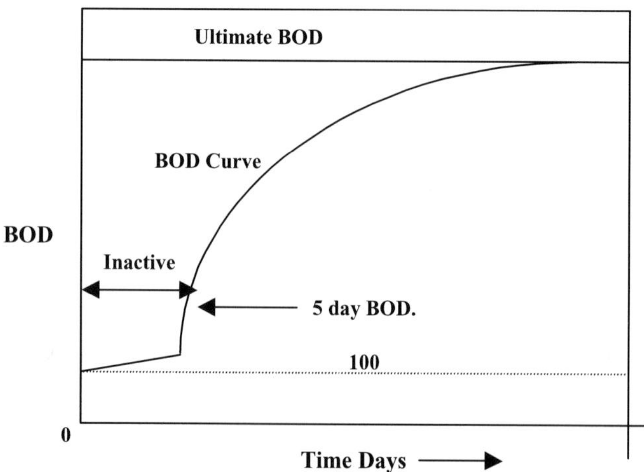

Figure 14.2. Lab determination of 5-day BOD versus ultimate BOD.

Table 14.2. Relationship between BOD and COD

Source	Pollutants	BOD "5 day"	COD "2 hr dichromate"
Brewery	Carbohydrates, Proteins	550	–
Coal gas	Phenols, Cyanides, Thiocyanates, Thio sulphates	6,500	16,400
Laundry	Carbohydrates, Soaps	1600	2700
Pulp mill	Carbohydrates, Lignins, Sulfates	25,000	76,000
Domestic sewage	Solids, Oil and Grease, Proteins, Carbohydrates	350	300
Petroleum refinery (Sour water)	Phenols, Hydrocarbons, Sulfides	850	1,500
Petroleum refinery	Phenols, Sulfides, Hydrocarbons, Mercaptans, Chlorides	125	2,600

accepted to measure the oxygen depletion effect of these pollution. The two most widely accepted are the "4 hour permanganate" method or the "2 hour dichromate" method. Although there is no generalized correlation between BOD and COD, usually the COD will be larger than the BOD. The following table illustrates how different wastes exhibit a different relationship between COD and BOD (Table 14.2).

IOD. Oxygen consumption by reducing chemicals such as sulfides and nitrates is typified as follows:

$$S^{2-} + 2O_2 \longrightarrow SO_4^{2-}$$

$$NO_2^- + {}^1\!/\!_2O_2 \longrightarrow NO_3^-$$

These types of inorganic oxidation reactions are very rapid and create what is measured in the laboratory as immediate oxygen demand. If waste contaminants contain these inorganic oxidizers the "5 day" BOD test will include the consumption of the oxygen due to IOD also. A separate test to determine IOD must be made and this result subtracted from the "5 day" BOD to arrive at the true BOD result.

Toxic pollutants common to oil refining

Toxic pollutants that are most commonly contained in untreated refinery aqueous wastes are:

Oil. Heavy oil and other hydrocarbons are the most problematic pollution to be found in refinery water effluent. All refineries exercise the most stringent methods to control and remove these undesirable pollutants. Indeed in many cases the treated effluent streams leaving the refinery may well be purer than incoming portable water used in the processes.

Phenols. These chemicals often formed in refinery processes such as catalytic and thermal crackers, are highly toxic to aquatic life in concentrations of 1 to 10 ppm. Apart from their toxicity phenols also unpleasant taste and odor to drinking water in the range of 50 to 100 ppb. In concentrations of 200 ppm and more, these chemicals can also deactivate water treatment plants such as trickle filters and activated sludge units.

Caustic soda and derivatives. Solutions containing sodium hydroxide are used in a number of refinery processes. Inevitably some of this chemical enters the refinery's waste water system. This contaminant is toxic to humans and marine life in even low concentrations. The spent caustic (compounds leaving the process) such as sodium sulfide is even more injurious.

Aqueous solutions of ammonium salts. The most common of these are ammonium sulfide NH_4SH, and ammonium chloride. Both these salts are present in effluent water from the crude distillation unit overhead accumulator, however the sulfide salt is present in all aqueous effluents from the cracking processes, and the hydro treaters. Other ammonium salts are also present in hydro cracking and deep oil hydro treating.

Acids in aqueous effluents. The most common of these are from the alkylation processes which use either hydrofluoric acid or sulfuric acid. In some isomerization processes hydrochloric acid is used to promote the aluminum chloride catalyst. In some older processes sulfur dioxide is used to remove aromatics. This effluent usually leaves these plants as dilute sulfuric acid.

Ketones, furfural, and urea. These compounds are used in the refining of lube oils. MEK and Urea are used in the dewaxing processes while furfural is used in the extraction processes for finished lube oil stock. All of these compounds are toxic.

Treating refinery aqueous wastes

The treatment of aqueous wastes from oil refineries fall into three categories. These are:

- In-plant treatment, these are onsite processes usually sour water strippers, spent caustic oxidizers, and spent caustic neutralizers

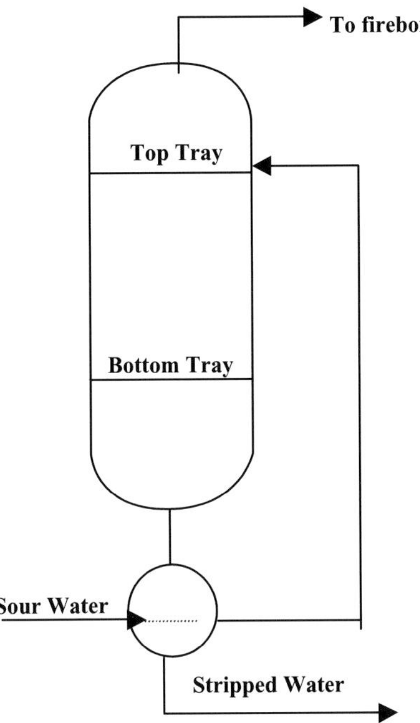

Figure 14.3. Sour water stripper with no reflux.

- The API separator, or similar oil/water separating device
- Secondary treatment, which includes chemical coagulation, activated sludge processes, trickle filters, air flotation and aerators

Most energy refineries contain only the first two of the above categories.

Sour water strippers. This is one of the most common treating process. Its purpose is to remove the pollutant gases included in process plant effluents. The more common pollutants are ammonia, ammonium salts, and hydrogen sulfide. The sour water stripper is almost always located in the process area of the refinery. The feed to the stripper is the effluent water from the crude unit overhead condenser, the water phase from the desalter, the condensed water from the vacuum unit's hot well, and all the water condensate from the hydro-treater product steam strippers. The sour water stripper is usually a single trayed tower with no reflux or a similar tower with an overhead reflux stream. Figure 14.3 shows a simple stripper tower with no reflux and Figure 14.4 shows a stripper with an overhead condenser and reflux.

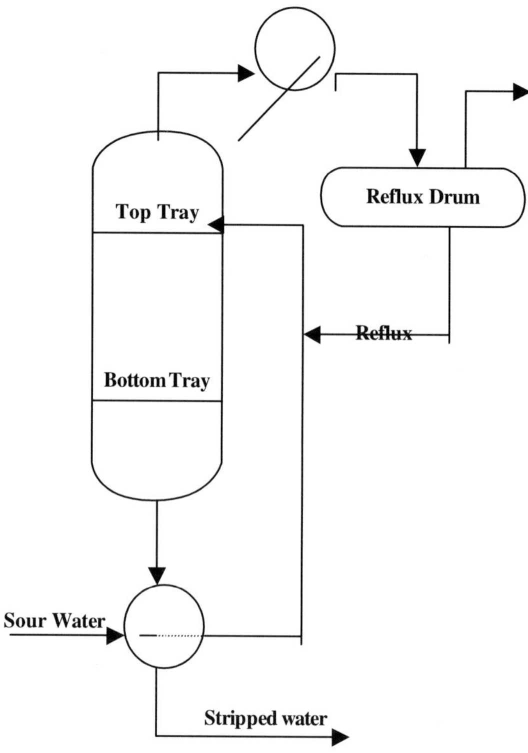

Figure 14.4. Sour water stripper with reflux.

Both these type of sour water stripping lend themselves to tray by tray mass and heat transfer. The sour water feed is introduced on the top tray of the tower while steam usually at a rates of 0.5–1.5 lbs/gallon of feed is introduced below the bottom tray. In the case of a tower with reflux the reflux enters the tower with the fresh feed. The design of both towers utilize the partial pressure relationship of NH_3 and H_2S in aqueous solution. A series of graphs giving these relationships are given in Appendix 14.1 of this chapter as Figures 14.A.1–14.A.10. An example of the design for a sour water stripper with no reflux is given in Appendix 14.2.

Spent caustic disposal. The other major effluent from oil refining is the spent caustic streams from hydrogen sulfide removal and also the removal of phenols. Refiners usually have the following options in the disposal of these streams. In order of preference these can be listed as follows:

Phenolic spent caustic.

- Disposal by sales
- Disposal by dumping at sea
- Neutralizing with acid
- Neutralizing with flue gas

Sulfidic spent caustic.

- Disposal by sales
- Oxidation with air and steam
- Neutralization with acid and stripping
- Neutralization with flue gas and stripping

Both the phenolic and sulfidic spent caustic have a commercial value in industrial areas and where transportation costs make sales an attractive economic option. In the case of the phenolic spent caustic, processors recover, separate, and purify various cresylic acid fractions for commercial use and sales. In the case of the sulfidic spent caustic this can be sold as sodium sulfide with some additional processing.

Neutralizing phenolic spent caustic. As listed above phenolic spent caustic can be neutralized using acid or flue gas. When neutralized the mixture separates into two liquid phases. The upper phase contains the acid oils while the lower phase is an aqueous solution of sodium sulfate or sodium carbonate. The neutralization using either acid or flue gas can be accomplished in a batch or continuous operation. Batch neutralization using acid is considered to be the preferable route in this case. Flue gas neutralization, although practical and commercially used in many installations is more complex in design than the acid application and does not lend itself readily to batch operation. It may require in some instances large pipe-work, switch valves, and a blower.

A diagram of a phenolic caustic batch neutralizer is shown in Figure 14.5.

The process is fairly simple involving the following steps:

1. Charging the spent caustic batch into the neutralizer vessel.
2. The addition and thorough mixing of the acid neutralizer into the batch. This has to be accomplished slowly and carefully to avoid lowering the pH of the batch too far. A pH of between 4 and 5 will be required to free the phenols completely.
3. A settling period is necessary to allow the sprung acids to separate from the sprung water.
4. This step is the pumping out and disposal of the sprung water.

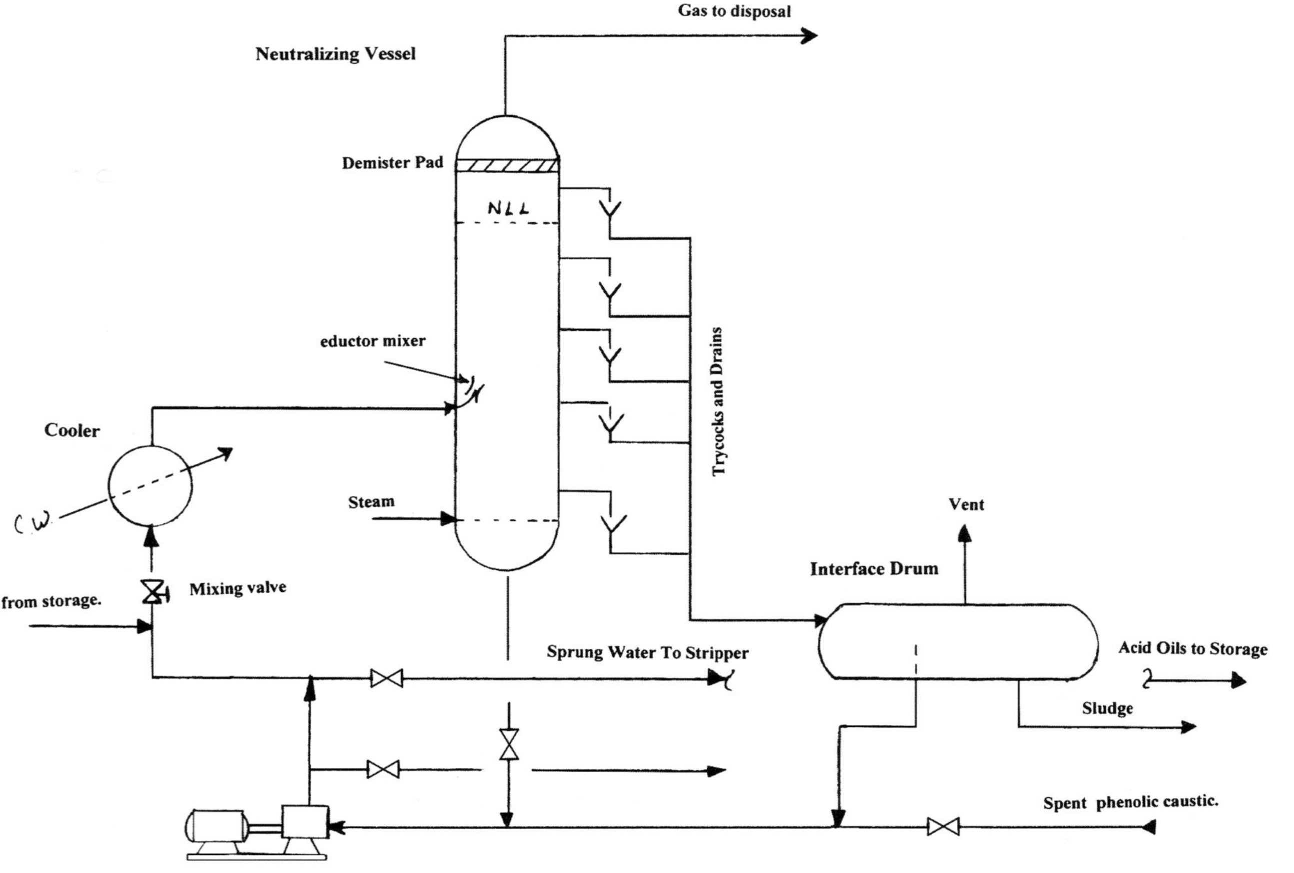

Figure 14.5. Batch neutralization of spent caustic with acid.

5. The sprung acid is steam stripped to remove any residual H_2S and light mercaptans.
6. Pump out the interface material which will be included in the next spent caustic batch.
7. Finally pumping out of the sprung acid.

The neutralization step is exothermic giving out around 125 Btu/lb of sprung acid. As the objective of the process is to produce a phenol-free sprung water for disposal the system temperature should be kept as low as possible until the sprung water is removed. For example the solubility of phenol in pure water at 120°F is about 11%wt while its solubility at 100°F is 8.5 wt%. For this reason therefore steam stripping should not begin until the sprung water has been removed. Routing the sprung water to the sour water stripper ensures the removal of any entrained H_2S in that stream.

Mixing of the acid/spent caustic is usually achieved using a mixing valve at the inlet of the neutralizer. An educator or a jet nozzle inlet to the vessel itself would ensure complete mixing if this is required.

Spent caustic oxidation

Spent caustic cannot be steam stripped to remove the sulfides contained in it due to the H_2S removal process in which the caustic was used. This is because sodium sulfide does not hydrolyze even when heated. Acids could be used, of course, to neutralize the spent caustic which would release gaseous H_2S. This would however be a costly procedure and causes a potential air pollution problem. The alkaline sulfide can be economically oxidized to form thiosulfates and sulfates. This is the process most commonly used in refineries where only sulfides are the pollutants in the spent caustic and the release of gaseous H_2S is a problem. The oxidation process also applies to the presence of ammonia sulfide in the stream. The process has not been applied to wastes containing a higher percentage of phenols, because the phenols interfere with the oxidation process.

Oxidation of sulfides to thiosulfates

The oxidation of Na_2S or $(NH_4)_2S$ to the corresponding thiosulfates may be expressed in ionic terms as follows:

$$
\begin{array}{lccccccccc}
 & 2S^{2-} & + & 2O_2 & + & H_2O & \rightarrow & S_2O_3^{2-} & + & 2OH^- \\
\text{lb} & 2(32) & & 2(32) & & 18 & & 112 & & 2(17) \\
H_f & 2(-10) & & 2(0) & & +68 & & +154 & & 2(+55)
\end{array}
$$

where H_f is the theoretical heat of formation in kcals/gm-mole in dilute solution. The plus sign denotes heat evolved while the minus sign denotes heat absorbed.

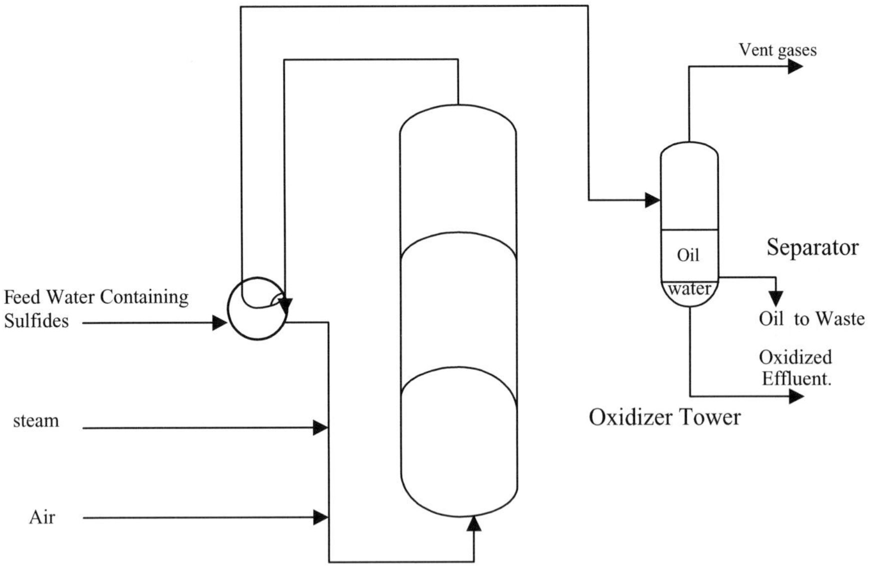

Figure 14.6. Schematic of an oxidizer plant.

Similarly the oxidation of NaSH or NH_4SH to the corresponding thiosulfate can be expressed as follows:

	$2SH^-$	+	$2O_2$	$\rightarrow$	$S_2O_3^{2-}$	+	H_2O
lb	2(33)		2(32)		112		18
H_f	2(+3)		2(0)		+154		+68

In both cases the theoretical oxygen required is 1 lb/lb of S, and the theoretical air required would be 4.33 lbs/lb of S. The theoretical heat of reaction in both cases is 216 kcals/gm-mole of S_2O_3 or 6,100 Btu evolved per lb of sulfur.

Figure 14.6 shows a typical oxidizing unit. Briefly the unit consists of an oxidizing tower, and an overhead separator. The sour water feed enters the oxidizing tower after preheating by exchange with the hot oxidizer overhead effluent. Heating steam and air is injected into the sour water feed stream before entering the oxidizer tower. The steam flow is controlled by the temperature of the inlet mixture while the air is flow controlled to meet the reactor condition premise. The oxidizing tower itself is divided into three sections with mixing nozzles connecting the inlet to each section. The reactants (air and the sour water feed) enter the base of the oxidizer also through a mixing nozzle and flow upwards through the sections of the tower. The oxidized effluent leaves the top of the tower to be cooled and partially condensed by exchange with the sour water feed. The cooled effluent enters a separator operating under pressure control. The vent

Table 14.3. Typical design criteria for sulfide oxidizers

Temperature at bottom, °F	165–225
Pressure at bottom, psia	75–100
Inlet air, (lbs/gal)/(1,000 ppm·S)	0.05–0.075
Inlet air, lb/lbs	1.4–2.1
Superficial air, ft/sec	0.08–0.1
Sulfide oxidation rate, (lbs/hr)/cuft	0.35

gases leave through this controller to flare or other suitable disposal equipment (such as a heaters firebox). The oil phase of the effluent is skimmed off as a side stream while the water phase after settling leaves the bottom of the separator as the oxidized effluent.

High temperature and low pressures tend to vaporize some of the water in the top section of the oxidizer. This results in more sulfides being stripped out into the vapor phase. Thus it is recommended that the oxidizer top pressure be at least 25–40 psi above the vapor pressure of the water at the oxidizer top temperature. Typical design criteria for these oxidizing units are given in Table 14.3.

Oxidation of mercaptans

If caustic soda has been used to remove H_2S there will almost certainly be mercaptans present in the sour stream. Caustic soda will also react with and remove these from the process stream. Certainly in treating these light streams from sour crudes the mercaptan content of the spent caustic stream could be as high as 3,000 ppm of sodium mercaptide sulfide. The greater part of the mercaptides will be oxidized to hydrocarbon disulfides according to the equation:

$$2RSNa + {}^{1}\!/_{2}O_2 + H_2O \longrightarrow RSSR + 2NaOH.$$

Almost all the mercaptides will be oxidized in a concurrent flow reactor. However some of the mercaptides will be partially hydrolyzed and it can be assumed that there will be a certain amount of free mercaptans present in accordance with the equilibrium:

$$RSNa + H_2O \longleftrightarrow RSH + NaOH.$$

The free mercaptan formed by this reaction will be stripped out of solution and leave the reactor in the overhead vapor phase. Now if the separator is operating at 62 psia and 125°F, the vapor pressure of water at 125°F is 2 psia, then the partial pressure acting to condense the overhead vapor is 60 psia. Methyl mercaptan may condense under these conditions but the heavier mercapans and the disulfide certainly will condense. The disulfides are immiscible in aqueous alkaline effluent and will form the upper oil layer to be skimmed off for disposal. The mercaptans that are condensed will form mercaptides with the NaOH present. Since this will really occur in the feed/effluent exchanger when the reformed mercaptides are still in contact with the reactor air supply and most if not all will be oxidized to disulfides again.

The theoretical oxygen requirement for oxidizing mercaptides to disulfides is 0.25 pounds per pound of mercaptide sulfur, and the theoretical air requirement is 1.09 pounds per pound of sulfur. For convenience, the following value of the theory in this case can be taken as:

$$1 \text{ Theory} = 0.0091 \text{ (lb air/gal)/(1,000 ppm RSH-S)}$$

Oxidation of sulfide to sulfate

The oxidation of sulfides to thiosulfides only reduces the ultimate oxygen demand of the sulfides from two pounds of oxygen per pound of sulfides to one pond per pound of sulfide. The oxidation to thiosulfate will remove the short term or immediate oxygen demand. Thiosulfate limits are not usually stipulated in water pollution regulations while sulfides are always stipulated. In those cases where thiosulfates are a regulated limit it will be necessary to convert the thiosulfate to the sulfate. The chemistry relating to this may be written as follows:

$$S^{2-} + 2O_2 \longrightarrow SO_4^{2-}$$

The above equation applies equally to aqueous ammonium sulfide and to sodium sulfides.

The theoretical oxygen requirement is 2 pounds per pound of S, and the theoretical air requirement is 8.66 ponds per pound of S. The theoretical heat of reaction is 12,700 Btu evolved/lb of S. That is one theory of air is:

$$0.072 \text{ (lb air/gal)/(1,000 ppm S)}$$

The following table compares those units oxidizing to thiosulfates to those oxidizing to sulfates (Table 14.4). Although many units processing the sulfides to sulfates do not use a catalyst but convert the sulfide to a mixture of thiosulfates and the sulfate. To convert the sulfide completely to the sulfate a copper chloride catalyst is used to accelerate the reaction rate. The reaction of this conversion is quite slow without the catalyst.

Oil–water separation

Most aqueous effluents from a refinery will contain oil. This oil content has to be reduced to at least 10 ppm before it can be deposited into a river, lake or ocean. The oil contamination sources are from process water run down, paved area drainage,

Table 14.4. Comparison of sulfide oxidation processes

	Oxidizing to thiosulfate only	Partial oxidizing to sulfate	Total oxidizing to sulfate
Anions produced			
SO_4 %	Nil	46	100
S_2O_3 %	100	54	Nil
% Sulfide oxidized to:			
SO_4	Nil	34	100
S_2O_3	100	66	Nil
Sulfide oxidation rate, (lbs/hr)/cuft.	0.33	0.033	0.035
Air rate in theories			
Oxidation to S_2O_3	2.05	1.55	6.53
Oxidation to actual anions	2.05	1.17	3.27
Air–water flow	Concurrent	Countercurrent	Countercurrent
Catalyst used	None	None	$CuCl_2$

storm catch-pots, tanker ballast pump out, and tank farm diked areas. All the water from these sources is treated in oil separation processes. The most important of these oil–water separation process is the API Separator.

The API oil–water separator

The design of an API separator is based on the difference in gravity of oil and water in accordance with the general laws of settling. These laws are given by the following equations:

$$V = 8.3 \times 10^5 \times \frac{d^2 \Delta S}{\eta} \quad \textit{(Stokes Law)}$$

$$V = 1.04 \times 10^4 \times \frac{d^{1.14} \times \Delta S^{0.71}}{S_c^{0.29} \times \mu^{0.43}} \quad \textit{(Intermediate Law)}$$

$$V = 2.05 \times 10^3 \left[\frac{d \times \Delta S}{S_c}\right]^{1/2} \quad \textit{(Newton's Law)}$$

where

V = settling rate in ins per min.
d = droplet diameter in ins
ΔS = specific gravity differential between the two phases.
η = viscosity of the continuous phase (water in this case) in centipoises
S_c = continuous phase specific gravity.

When the Re number is $<$ 2.0 use Stokes Law. When Re number is between 2 and 500 use the Intermediate Law. When the Re number is $>$500 use Newton's Law.

$$\text{The Re number in this case is given as} \frac{10.7 \times d\, V\, S_c}{\eta}$$

As a guide the minimum droplet size of the oil in water can be taken (as a minimum) 0.008 ins when the oil gravity is 0.850 or lighter and 0.005 ins when the gravity of oil is greater than 0.850.

The above laws of settling in oil refining processes usually apply to the hydrocarbon being the continuous phase with water (the heavier phase) being the one that settles out. In the case of the API separator the reverse is true. That is the continuous phase is water and the lighter oil phase is the one that is separated to be disposed as the product skimmed from the surface. In the design of this system *Stokes Law* is used to reflect the rate at which the oil rises to the surface through the water. This modified Stokes law may be written as follows:

$$V_r = 6.69 \times 10^4 \frac{d^2 \Delta S}{\eta}$$

where

V_r = rising rate of the oil phase in ft/min
d = droplet diameter in ins.
ΔS = difference in Specific gravity of the phases.
η = viscosity of the continuous phase (water) in centipoises.

An example of the application of the equation in the design of an API separator is given in Appendix 14.3.

The oil phase from the separator is removed using specially designed skim pipes and an oil sump. A simplified diagram of a typical API separator is given in Figure 14.7.

Ancillary equipment used in the design and operation of the API separator are described in the following items.

Oil skim pipes. As a good part of the oil will separate from the water on leaving the inlet flume and two oil retention baffles with skim pipes are provided. One is located ahead of the flow baffle and the other ahead of the overflow weir. The distance between these baffles will be the function 'L' as calculated in Appendix 14.3. The API manual recommends 10 ins diameter skim pipes where the horizontal run is less than 40 ft. Larger diameter skim pipes will be installed in greater horizontal runs.

The oil sump. The oil recovery sump is located at one edge mid way along the horizontal run. The size of the sump is based on the oil content of the inlet stream to

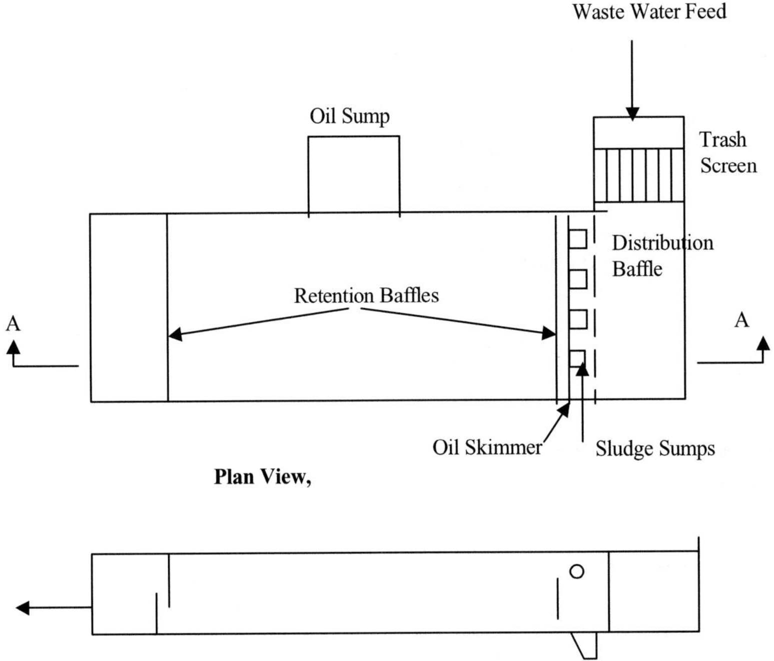

Figure 14.7. Schematic diagram of a typical api separator.

the separator plus four times that amount as water delivered by the skim pipes. For example if the inlet stream contains 200 ppm by volume of oil and the water outlet contains 50 ppm Then the oil removal from the separator will be 150 ppm or 0.015% vol which in the case of the specification given in Appendix 5.3 will equate to:

$$0.00015 \times 600 \times 1{,}440 = 130\,\text{gals/day}.$$

Use the API recommendation to add to this figure four times this figure for the amount of water contained with the oil delivered by the skim pipes. Then the total liquid to the sump will be 650 gals/day. For ease of operation a sump size is often based on a predicted weekly accumulation which amounts to 4,550 gals. Assume the sump cross sectional area is 144 sqft (12 ft $\times$ 12 ft) then the liquid depth to NLL should be $4{,}550/(7.48 \times 144) = 4.2$ ft and the HLL can be set at 5 ft. To ensure adequate gravity head for the oil removed by the skim pipes the bottom of the sump will be 8 ft below the NLL of the separator.

The sump pump. The sump pump will be a standard vertical pump with a capacity to pump out the entire sump in about 1 hr or less. Normally refineries pump the sump oil to the 'Bottom Settling Tanks' of the refinery. These are tanks used to store the bottoms of crude oil tanks which usually contain high water and sediment content not suitable for the normal process plants. The contents are allowed to sit for some time and then returned as feed to the normal refinery processes. The bottom settling from these tanks are routed to the API separator.

Oil retention baffles. API recommends that oil retention baffles have a depth below the level in the separator of 0.6–0.8 of the liquid level. Further the baffle should extend 1 ft above the normal liquid level of the separator.

Distribution baffles. An important feature in the design of the API separator is a means of distributing the influent evenly. API recommends either 'V' shaped baffles or pipes installed at the inlet of the separator to be used for this purpose. It is also recommended that the the open area between slots be 3–7% of the chamber cross sectional area.

Sludge sumps. Sludge sumps are located immediately downstream of the distribution baffles. API recommends that these sumps should be at least 30 ins deep. That is 30 ins below the floor of the separator. The sludge pump suction is to extend to 6 ins of the sump bottom and the suction line will be 3 ins diameter. Normally the sludge bottom will be 4 sqins in cross sectional area or 2 ins × 2 ins. API recommends that these sumps slope a minimum of 1.7 ins vertical to 1.0 ins horizontal. In the case of the example design in Appendix 5.3 an 18 ins × 18 ins sumps sloping to the bottom of 2 ins × 2 ins would give a slope of 1.9 which would meet the API recommendation. Such a layout could accommodate 6 sumps.

Trash rack. API recommends that a trash rack made up of series of 3/8 ins rods on 1.5 ins centers be installed as a trash trap.

Storm surge ponds

These ponds are installed to provide storage for maximum rainfall conditions. There are several forms of these surge ponds, some requiring pumps, some located upstream of the API separator, and some downstream of the separator. In most cases the storm drain system is directed to the storm surge pond. Thus in a storm the excess rainfall is held in this pond and fed to the API separator, over a period of time and at a rate that will not exceed the separator's capability to handle the water effectively. In this way the refinery ensures that any oily water will not by-pass the separator under the worst condition. The size of the surge pond must be able to handle the maximum rainfall and the flow from all catch basins and open culverts that form part of the refinery drainage system.

Surge ponds are constructed with a shallow depth over a large surface area. Its location is usually only slightly elevated above the API separator but near the low point of the refinery. Thus, even when the pond is full the water, it will not 'back-up' to the process area drains.

Other refinery water effluent treatment processes

Oxidation ponds

Oxidation ponds are usually used as a secondary effluent cleanup after the API separator. There are three types of these ponds which are:

Aerobic. Where the oxidation of the water utilizes oxygen from the atmosphere plus oxygen produced by photosynthesis.

Anaerobic. Where oxidation of the wastes does not utilize oxygen.

Aerated. Where oxidation of the wastes utilizes oxygen introduced from the atmosphere by mechanical aeration.

Aerobic oxidation. Consider a shallow pond containing bacteria and algae. The bacteria will utilize oxygen to oxidize bio-chemically the incoming water. In so doing it will produce H_2O, CO_2, and perhaps NH_3. The algae will utilize sunlight plus the H_2O and CO_2 to produce oxygen. Oxygen in turn produces additional bacteria and algea growth. This cycle is shown below:

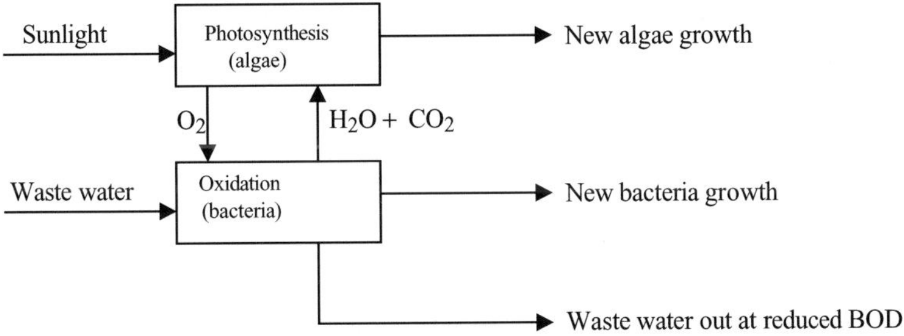

Air flotation

The purpose of the air flotation process is the clarification of waste water by the removal of suspended solids and oil. This is achieved by dissolving air in waste water under pressure, and then releasing it at atmospheric pressure. The released air forms bubbles which adhere to the solid matter and oil in the waste water. The bubbles cause the adhered matter to float in the froth on the surface of the water bulk. The dissolved air in the water also achieves a reduction in the BOD of the effluent stream. Figure 14.8 shows the principle of a typical air flotation process.

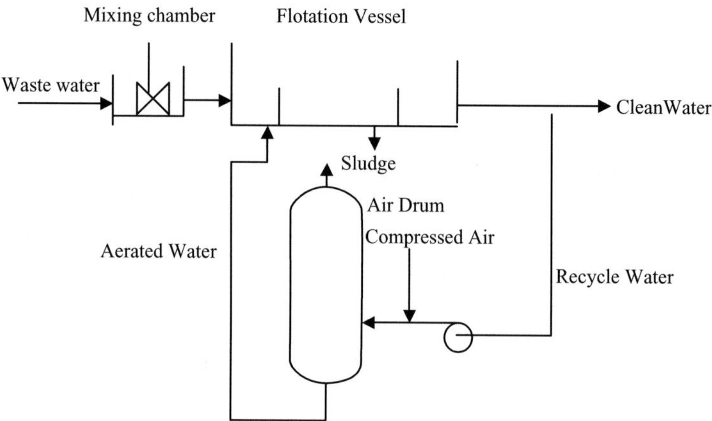

Figure 14.8. A typical air flotation process.

The process shown is the preferred recycle type. The process can also be designed for a once through operation. Typical design criteria for the recycle process is given in the Table 14.5.

Referring to Figure 14.8 the flocculating chemical and pH control chemical are mixed in the mixing tank with the raw water feed before the feed enters the flotation vessel.

The processes described above are those found in oil refining most often. Indeed of the processes described above most refineries only use the API separator and the surge ponds to meet the oil/water separation required.

Table 14.5. Design criteria for a typical recycle air flotation unit

Recycle rate	50% of raw feed rate
Air drum pressure	35–55 psig
Air drum retention time	2 min of recycle flow (from Liquid level to drum bottom)
Flotation tank retention	15–20 min of total flow (raw feed plus recycle)
Flotation tank rise rate	3.0 gal/min of total flow per sqft of liquid surface
pH	7.5–8.5
Flocculating chemical	25 ppm of alum in total flow.
Air rate	0.25 to 0.5 Scft/100gals of total flow
Flotation tank liquid depth	6–8 ft (to meet requirements above)

Reference

M.R. Beychok. *Aqueous Wastes from Petroleum and Petrochemical Plants*, John Wiley & Sons, New York, 1967.

14.2 Emission to the atmosphere

Oil refiners have two responsibilities with regard to atmospheric emissions. The first is to produce products that minimize the release of toxic or unacceptable emissions when used or stored. Much of this subject has been dealt with in Chapter 2 of this volume. The other responsibility, of course is to identify and control those unsavory emissions from the refining operation and processes. This is in keeping with all other industries that contain processes that have atmospheric emissions. It is this second responsibility that is addressed in this part of the chapter. It begins with a synopsis of the Clean Air Act. This Act is the basis of all standards and limits of control in the atmospheric emission from industrial and other sources. The sections of the Act given in the following sections refer to the USA although most other countries follow these items quite closely.

Features of the Clean Air Act

The Clean Air Act reflects a numbers of revisions of a law first passed in 1970. This summary covers some of the important provisions of the Clean Air Act.

This summary is only a brief introduction to the Clean Air Act. The document itself is much more extensive and provides details of all the most complicated aspects of air pollution. The reader should consult the latest versions and amendments to the Clean Air Act.

The role of the federal government and the role of the states

Although the Clean Air Act is a federal law covering the entire country, the states do much of the work to implement the Act. For example, a state air pollution agency holds a hearing on a permit application by a power or chemical plant or fines a company for violating air pollution limits.

Under this law, EPA sets limits on how much of a pollutant can be in the air anywhere in the United States. This ensures that all Americans have the same basic health and environmental protections. The law allows individual states to have stronger pollution controls, but states are not allowed to have weaker pollution controls than those set for the whole country.

The law recognizes that it makes sense for states to take the lead in carrying out the Clean Air Act, because pollution control problems often require special understanding of local industries, geography, housing patterns, etc.

States have to develop state implementation plans (SIPs) that explain how each state will do its job under the Clean Air Act. A SIP is a collection of the regulations a state will use to clean up polluted areas. The states must involve the public, through hearings and opportunities to comment, in the development of each SIP.

EPA must approve each SIP, and if a SIP isn't acceptable, EPA can take over enforcing the Clean Air Act in that state.

The United States government, through EPA, assists the states by providing scientific research, expert studies, engineering designs and money to support clean air programs.

Interstate air pollution

Air pollution often travels from its source in one state to another state. In many metropolitan areas, people live in one state and work or shop in an-other; air pollution from cars and trucks may spread throughout the interstate area. The Clean Air Act provides for interstate commissions on air pollution control, which are to develop regional strategies for cleaning up air pollution. The Clean Air Act includes other provisions to reduce interstate air pollution.

International air pollution

Air pollution moves across national borders. The law covers pollution that originates in Mexico and Canada and drifts into the United States and pollution from the United States that reaches Canada and Mexico.

Permits

One of the major breakthroughs in the Clean Air Act is a *permit* program for larger sources that release pollutants into the air [2].

[2] A source can be a power plant, factory or anything that releases pollutants into the air. Cars, trucks and other motor vehicles are sources, and consumer products and machines used in industry can be sources too. Sources that stay in one place are referred to as stationary sources; sources that move around, like cars or planes, are called mobile sources.

Requiring polluters to apply for a permit is not a new idea. Approximately 35 states have had state-wide permit programs for air pollution. The Clean Water Act requires permits to release pollutants into lakes, rivers or other waterways. Now air pollution is also going to be managed by a national permit system. Under the new program, permits are issued by states or, when a state fails to carry out the Clean Air Act satisfactorily, by EPA. The permit includes information on which pollutants are being released, how much may be released, and what kinds of steps the source's owner or operator is taking to reduce pollution, including plans to *monitor* (measure) the pollution. The permit system is especially useful for businesses covered by more than one part of the law, since information about all of a source's air pollution will now be in one place. The permit system simplifies and clarifies businesses' obligations for cleaning up air pollution and, over time, can reduce paperwork. For instance, an electric power plant may be covered by the acid rain, hazardous air pollutant and non-attainment (smog) parts of the Clean Air Act; the detailed information required by all these separate sections will be in one place-on the permit.

Permit applications and permits are available to the public; contact your state or regional air pollution control agency or EPA for information on access to these documents.

Businesses seeking permits have to pay *permit fees* much like car owners paying for car registrations. The money from the fees will help pay for state air pollution control activities.

Enforcement

The Clean Air Act gives important new *enforcement* powers to EPA. It used to be very difficult for EPA to penalize a company for violating the Clean Air Act. EPA has to go to court for even minor violations. The law enables EPA to fine violators, much like a police officer giving traffic tickets. Other parts of the law increase penalties for violating the Act and bring the Clean Air Act's enforcement powers in line with other environmental laws.

Public participation

Public participation is a very important part of the Clean Air Act. Throughout the Act, the public is given opportunities to take part in determining how the law will be carried out. For instance, you can take part in hearings on the state and local plans for cleaning up air pollution. You can sue the government or a source's owner or operator to get action when EPA or your state has not enforced the Act. You can request action by the state or EPA against violators.

The reports required by the Act are public documents. A great deal of information will be collected on just how much pollution is being released; these monitoring (measuring) data will be available to the public. The Clean Air Act ordered EPA to set up clearinghouses to collect and give out technical information. Typically, these clearinghouses will serve the public as well as state and other air pollution control agencies. See the list at the end of this summary for organizations to contact for additional information about air pollution and the Clean Air Act.

Market approaches for reducing air pollution: economic incentives

The Clean Air Act has many features designed to clean up air pollution as efficiently and inexpensively as possible, letting businesses make choices on the best way to reach pollution cleanup goals. These new flexible programs are called *market or market-based* approaches. For instance, the acid rain clean-up program offers businesses choices as to how they reach their pollution reduction goals and includes pollution allowances that can be traded, bought and sold.

The Clean Air Act provides economic incentives for cleaning up pollution. For instance, gasoline refiners can get credits if they produce cleaner gasoline than required, and they can use those credits when their gasoline doesn't quite meet clean-up requirements.

The major effects of air pollution and the most common pollutants

Air pollution can make people sick. It can cause burning eyes and nose and an itchy, irritated throat, as well as trouble in breathing. Some chemicals found in polluted air cause cancer, birth defects, brain and nerve damage and long-term injury to the lungs and breathing passages. Some air pollutants are so dangerous that accidental releases can cause serious injury or even death.

Air pollution can damage the environment. Trees, lakes and animals have been harmed by air pollution. Air pollutants have thinned the protective ozone layer above the Earth; this loss of ozone could cause changes in the environment as well as more skin cancer and cataracts (eye damage) in people. Air pollution can damage property. It can dirty buildings and other structures. Some common pollutants eat away stone, damaging buildings, monuments and statues. Air pollution can cause haze, reducing visibility in national parks and sometime interfering with aviation.

The Clean Air Act is aimed at improving air quality and therefore to protect the health of the general public. The Act changed the way people work or do business, and it could, in some ways, change the way people live.

Some of the major and more serious pollutants are described in the following sections (see also Chapter 2 on 'Motive Fuels').

Smog

Often, wind blows smog-forming pollutants away from their sources. The smog-forming reactions take place while the pollutants are being blown through the air by the wind. This explains why smog is often more severe miles away from the source of smog-forming pollutants than it is at the source.

The smog-forming pollutants literally cook in the sky, and if it's hot and sunny, smog forms more easily. Just as it takes time to bake a cake, it takes time to cook up smog-several hours from the time pollutants get into the air until the smog gets really bad. Weather and geography determine where smog goes and how bad it is. When temperature inversions occur (warm air stays near the ground instead of rising) and winds are calm, smog may stay in place for days at a time. As traffic and other sources add more pollutants to the air, the smog gets worse.

Since smog travels across county and state lines, the respective local governments and air pollution control agencies must cooperate to solve their problem.

The Clean Air Act is applied to reduce pollution from criteria air pollutants, including smog, in this way.

First, EPA and state governors cooperate to identify non-attainment areas for each criteria air pollutant. Then, EPA classifies the non-attainment areas according to how badly polluted the areas are. There are five classes of non-attainment areas for smog, ranging from *marginal* (relatively easy to clean up quickly) to *extreme* (will take a lot of work and a long time to clean up).

The Clean Air Act uses this new classification system to tailor clean-up requirements to the severity of the pollution and set realistic deadlines for reaching clean-up goals. If deadlines are missed, the law allows more time to clean-up, but usually a non-attainment area that has missed a clean-up deadline will have to meet the stricter clean-up requirements set for more polluted areas.

Not only must non-attainment areas meet deadlines, states with non-attainment areas must show EPA that they are moving on clean-up before the deadline-making reasonable further progress.

States will usually do most of the planning for cleaning up criteria air pollutants, using the permit system to make sure power plants, factories and other pollution sources meet their clean-up goals.

The comprehensive approach to reducing criteria air pollutants taken by the Act covers many different sources and a variety of clean-up methods. Many of the smog clean-up requirements involve motor vehicles (cars, trucks, buses). Also, as the pollution gets worse, pollution controls are required for smaller sources.

Acid rain

In addition to the commonly accepted term "acid rain", there are also acid snow, acid fog or mist, acid gas and acid dust. All of these "acids" are related air pollutants, that can harm health, cause hazy skies and damage the environment and property. The Clean Air Act includes an innovative program to reduce acid air pollutants (all referred to here as "acid rain").

The acid rain that has received the most attention is caused mainly by pollutants from big coal burning power plants. Coal from most sources contains sulfur. Sulfur in coal becomes sulfur dioxide (SO_2) when coal is burned. Big power plants burn large quantities of coal, so they release large amounts of sulfur dioxide, as well as NO_x (nitrogen oxides). These are acid chemicals, related to two strong acids: sulfuric acid and nitric acid. Refineries too which burn heavy hydrocarbons such as fuel oils are subject to the same scrutiny and restrictions that may be in force.

The sulfur dioxide and nitrogen oxides released from the burning of fossil fuels etc are routed high into the air by the heater stacks and are carried by winds to areas remote from the actual source. When winds blow the acid chemicals into areas where there is wet weather, the acids become part of the rain, snow or fog. In areas where the weather is dry, the acid chemicals may fall to Earth in gases or dusts.

Lakes and streams are normally slightly acid, but acid rain can make them very acid. Very acid conditions can damage plant and animal life. In the United States acid lakes and streams have been found all over the country. For instance, lakes in Acadia National Park on Maine's Mt. Desert Island have been very acidic, due to pollution from the Midwest and the East Coast industries. Streams in Maryland and West Virginia, lakes in the Upper Peninsula of Michigan, and lakes and streams in Florida have also been affected by acid rain. Heavy rainstorms and melting snow can cause temporary increases in acidity in lakes and streams in the eastern and western United States. These temporary increases may last for days or even weeks.

Acid rain does more than environmental damage; it can damage health and property as well. Acid air pollution has been linked to breathing and lung problems in children and in people who have asthma. Even healthy people can have their lungs damaged by acid air pollutants. Acid air pollution can eat away stone buildings and statues.

Health, environmental and property damage can also occur when sulfur dioxide pollutes areas close to its source. Sulfur dioxide pollution has been found in towns where paper and wood pulp are processed and in areas close to some power plants. The Clean Air Act's sulfur dioxide reduction program complements health-based sulfur dioxide pollution limits already in place to protect the public and the environment from both nearby and distant sources of sulfur dioxide.

The Act as originally enacted in 1990 took a new nationwide approach to the acid rain problem. The law set up a *market based system* designed to lower sulfur dioxide pollution levels. Beginning in the year 2000, annual releases of sulfur dioxide were to be about 40% lower than the 1980 levels. Reducing sulfur dioxide releases caused a major reduction in acid rain.

Phase I of the acid rain reduction program went into effect in 1995. Major users of high sulfur fuels had to reduce releases of sulfur dioxide. In 2000, *Phase II* of the acid rain program went into effect, further reducing the sulfur dioxide releases from the traditional sources, and covering other smaller polluters.

Reductions in sulfur dioxide releases are obtained through a program of *emission* (release) *allowances*. EPA issues allowances to those industries covered by the acid rain program; each allowance is worth one ton of sulfur dioxide released from the smokestack. To obtain reductions in sulfur dioxide pollution, allowances are set below the current level of sulfur dioxide releases. Plants may only release as much sulfur dioxide as they have allowances. If a plant expects to release more sulfur dioxide than it has allowances, it has to get more allowances, perhaps by buying them from another plant that has reduced its sulfur dioxide releases below its number of allowances and therefore has allowances to sell or trade. Allowances can also be bought and sold by "middlemen," such as brokers, or by anyone who wants to take part in the allowances market. Allowances can be traded and sold nationwide. There are stiff penalties for plants that release more pollutants than their allowances cover.

A summary of other major atmospheric pollutants

Ozone (ground-level ozone is the principal component of smog)

- *Source*—chemical reaction of pollutants; VOCs and NO_x
- *Health effects*—breathing problems, reduced lung function, asthma, irritates eyes, stuffy nose, reduced resistance to colds and other infections, may speed up aging of lung tissue
- *Environmental effects*—ozone can damage plants and trees; smog can cause reduced visibility
- *Property damage*—Damages rubber, fabrics, etc.

*VOCs** (volatile organic compounds); smog-formers

- *Source*—VOCs are released from burning fuel (gasoline, oil, wood coal, natural gas, etc.), solvents, paints glues and other products used at work or at home. Cars are

* All VOCs contain carbon (C), the basic chemical element found in living beings. Carbon-containing chemicals are called organic. Volatile chemicals escape into the air easily. Many VOCs, are also hazardous air pollutants, which can cause very serious illnesses. EPA does not list VOCs as criteria air pollutants, but are included because efforts to control smog target VOCs for reduction.

an important source of VOCs. VOCs include chemicals such as benzene, toluene, methylene chloride and methyl chloroform

- *Health effects*—In addition to ozone (smog) effects, many VOCs can cause serious health problems such as cancer and other effects
- *Environmental effects*—In addition to ozone (smog) effects, some VOCs such as formaldehyde and ethylene may harm plants.

Nitrogen Dioxide (One of the NO_x); smog-forming chemical

- *Source*—burning of gasoline, natural gas, coal, oil etc. Cars are an important source of NO_2
- *Health effects*—lung damage, illnesses of breathing passages and lungs (respiratory system)
- *Environmental effects*—nitrogen dioxide is an ingredient of acid rain (acid aerosols), which can damage trees and lakes. Acid aerosols can reduce visibility
- *Property damage*—acid aerosols can eat away stone used on buildings, statues, monuments, etc.

Carbon Monoxide (CO)

- *Source*—burning of gasoline, natural gas, coal, oil etc.
- Health effects—reduces ability of blood to bring oxygen to body cells and tissues; cells and tissues need oxygen to work. CO may be particularly hazardous to people who have heart or circulatory (blood vessel) problems and people who have damaged lungs or breathing passages.

Particulate Matter (PM-10); (dust, smoke, soot)

- *Source*—burning of wood, diesel and other fuels; industrial plants; agriculture (plowing, burning off fields); unpaved roads
- *Health effects*—nose and throat irritation, lung damage, bronchitis, early death
- *Environmental effects*—particulates are the main source of haze that reduces visibility
- *Property damage*—ashes, soots, smokes, and dusts can dirty and discolor structures and other property, including clothes and furniture.

Lead

- *Source*—leaded gasoline (being phased out), paint (houses, cars), smelters (metal refineries); manufacture of lead storage batteries
- *Health effects*—brain and other nervous system damage; children are at special risk. Some lead-containing chemicals cause cancer in animals. Lead causes digestive and other health problems
- *Environmental effects*—Lead can harm wildlife

Carbon Dioxide (CO_2)

- *Source*—Main combustion product

- Not monitored by EPA or by the Clean Air Act
- Blamed for climatic changes and subject of the Kyoto protocol

Mercury (Hg)

- *Source*—coal-fired power plants, mercury cell chlorine facilities
- subject to reductions under recent versions of the Clean Air Act
- *Health effects*—neurological damage

Monitoring atmospheric emission

Monitoring to develop an improved understanding of human activity on the local environment is the primary activity of most industrialized countries today. In a great many cases independent consultants are used for this purpose. Normally these consultants will set up a number of monitoring stations. For example in the highly industrial section of Eastern Canada (Ontario) and the neighboring cities in Eastern United States several of these stations are in operation. All instrumentation associated with these stations is frequently cross-calibrated by the Ontario Ministry of Environment and Energy, the local regulatory agency. The existence of this monitoring network provides a major cost benefit to members, in that the regulatory agency accepts the co-operative monitoring rather than requiring fenceline monitoring at individual facilities. Monitoring results are provided via telemetry to a central computer accessible through remote terminals by each member. In this way, all members have real-time access to current ambient conditions, and can respond quickly to changing ambient conditions.

The air monitoring stations are located within, as well as north and south of the industrialized area, including the United States. Local winds are predominantly north or south, generally aligned with the river. Not all sites are completely instrumented; however, most contaminants of concern are well represented among the stations, including:

- Sulfur dioxide, due to the many sources of heavy fuel combustion in the area;
- Ground level ozone, as it is a chronic problem in much of urbanized North America, resulting in high levels during the summer months;
- Nitrogen oxides, having been recognized as major contributors to ozone formation;
- Non-methane hydrocarbons, as they provide a simple measure of VOCs, important in a large petrochemical complex, and are also contributors to ground level ozone problems;
- Ethylene, which is monitored at most stations, is a major local commodity with high levels of local production and consumption;
- Total reduced sulfur, which provides a general measure of potential odourants, important in such a concentration of industries;

- Particulates, measured as both Total Suspended Particulates and as Inhalable Particulates on a six day cycle corresponding to the cycle used on a national basis in Canada.
- Wind speed and direction at 10 meters above ground level as this has proven to be valuable component in assessing the atmospheric emission data.

Trend data, such as average annual sulfur dioxide levels are available over many years. Data from the early 1960s showed sulfur dioxide routinely exceeding an annual average of 0.020 ppm. Data through the 1970s and 1980s demonstrated acceptable levels and more recent data showed continuing reductions.

Reducing and controlling the atmospheric pollution in refinery products

As stated earlier much of the atmospheric pollution from oil refining is centered in the pollution emitted by the products of that industry when they are utilized. Thus a major effort in refinery design and operation is geared to produce products that have a minimum effect on the environment and in particular atmospheric pollution. This results in the research into refining technology and product composition to change the character of many of the traditional refinery products. This is detailed in Chapter 2 of this volume. The following items included below are some additional requirements that are or have been adopted to minimize atmospheric pollution from motive fuels:

Each of today's cars produces 60–80% less pollution than cars in the 1960s. More people are using mass transit. Leaded gas was phased out, resulting in dramatic declines in air levels of lead, a very toxic chemical. Despite this progress, most types of air pollution from mobile sources have not improved significantly.

At present in the United States:

- Motor vehicles are responsible for up to half of the smog-forming VOCs and nitrogen oxides (NO_x)
- Motor vehicles release more than 50% of the hazardous air pollutants
- Motor vehicles release up to 90% of the CO found in urban air

What went wrong?

- More people are driving more cars more miles on more trips. In 1970, Americans traveled 1 trillion miles in motor vehicles; it is estimated that today they drive around 5 trillion miles.
- Many people live far from where they work; in many areas, buses, subways, and commuter trains are not available. Also, most people still drive to work alone, even when van pools, high-occupancy vehicle (HOV) lanes and other alternatives to one-person-per-car commuting are available.

- Buses and trucks, which produce a lot of pollution, haven't had to clean up their engines and exhaust systems as much as cars.

Auto fuel has become more polluting. As lead was being phased out, gasoline refiners changed gasoline formulas to make up for octane loss, and the changes made gasoline more likely to release smog-forming VOC vapors into the air.

Although cars have had pollution control devices since the 1970s, the devices only had to work for 50,000 miles, while a car in the United States is usually driven for 100,000 miles. The Clean Air Act took a *comprehensive* approach to reducing pollution from motor vehicles. The Act provided for cleaning up fuels, cars, trucks, buses and other motor vehicles. Auto inspection provisions were included in the law to make sure cars are well maintained. The law also includes transportation policy changes that can help reduce air pollution.

Cleaner fuels

It will be very difficult to obtain a significant reduction in pollution from motor vehicles unless fuels are cleaned up. The phase out of lead from gasoline was completed by January 1, 1996. Diesel fuel refining must be changed so that the fuels contain less sulfur, which contributes to acid rain and smog. Gasoline refiners will have to *reformulate* gasoline sold in the smoggiest areas; this gasoline will contain less VOCs such as benzene (which is also a hazardous air pollutant that causes cancer and aplastic anemia, a potentially fatal blood disease). Other polluted areas can ask EPA to include them in the reformulated gasoline marketing program. In some areas, wintertime CO pollution is caused by people starting their cars. In these areas, refiners have to sell *oxyfuel*, gasoline with oxygen added, usually in the form of 10% ethanol, to make the fuel burn more efficiently, thereby reducing CO release.

All gasolines have to contain *detergents*, which, by preventing build-up of engine deposits, keep engines working smoothly and burning fuel cleanly.

The Clean Air Act encouraged development and sale of *alternative fuels* such as alcohols, liquefied petroleum gas (LPG) and natural gas.

Gas stations in many areas have installed *vapor recovery nozzles* on gas pumps. These nozzles cut down on vapor release when you put gas in your car.

Cleaner cars

The Clean Air Act required cars to have under-the-hood systems and dashboard warning lights that check whether pollution control devices are working properly. Pollution control devices must work for 100,000 miles, rather than the older 50,000 miles.

Electric cars, which are low-pollution vehicles, are also available in California and, more recently, hybrid cars that improve mileage by using a combined gasoline/electric motor have been gaining in acceptance. Hydrogen-powered fuel cell cars are also being used, but mostly experimentally in fleet applications (e.g., some city buses).

Inspection and maintenance programs

Under the Clean Air Act, auto manufacturers are required to build cleaner cars, and cars to use cleaner fuels. However, to get air pollution down and keep it down, a third program was implemented, vehicle *inspection and maintenance* (I/M), which makes sure cars are being maintained adequately to keep pollution emissions (releases) low. The Clean Air Act includes very specific requirements for I/M programs.

Before the Clean Air Act went into effect, seventy United States cities and several states already had auto emission inspection programs. The law required that some areas that already had I/M programs enhance (improve) their emission inspection machines and procedures. Enhanced I/M machines and procedures give a better measurement of the pollution a car releases when it is actually being driven, rather than just sitting parked at the inspection station. Enhanced I/M programs resulted in changes in where and how cars are inspected in local areas. The added expense for the new machinery more than made up for by air pollution reductions: *emission I/M programs are expected to have a big payoff in reducing air pollution from cars.*

Cleaner trucks and buses

Starting with model year 1994, engines for new big diesel trucks had to be built to reduce *particulate* (dust, soot) releases by 90%. Buses had to reduce particulate releases even more than trucks. To reduce pollution, companies and governments that own buses or trucks had to buy new clean models. Introduction of ultra low-sulfur diesel (ULSD) is expected to greatly assist in these efforts. Small trucks will eventually be cleaned up by requirements similar to those for cars.

Non-road vehicles

Locomotives, construction equipment and even riding mowers may be regulated under the Clean Air Act. Air pollution from locomotives must be reduced. For the other non-road vehicles, EPA can issue regulations if a study shows that controls would help cut pollution.

Transportation policies

The smoggiest metropolitan areas had to change their transportation policies to discourage unnecessary auto use, and to encourage efficient commuting (van pools, HOV lanes, etc.).

Controlling emission pollution from the refining processes

Most oil refineries do not have the very hazardous emissions usually met in petrochemical or chemical plants. In most refineries the pollution that is met with is in the burning of fuels that may contain high levels of sulfur or nitrogen compounds. The emission of VOCs is usually restricted to storage tanks and in some cases relief valves open to atmosphere. The release of particulates from certain process plants such as the FCCU and the Coking units is an emission that refiners can handle quite well.

In all cases the pollution control in a refinery starts with the design of the processes and systems themselves, and extends to their proper operation. Some of the measures undertaken to meet the Clean Air Act in the refining industry are as follows:

Fired heaters

Almost all refinery heaters have double service burners. That is the heater burners can fire fuel gas, or fuel oil. The fuel gas stream will usually have been treated for the removal of sulfur, in the gas treating plant. So normally this fuel source is not a major pollutant problem. However all refineries have a 'waste liquid' pool, and this is used as the fuel oil stream to the heaters. Unless properly treated either in terms of the streams that are routed to this pool or indeed the fuel oil pool itself, it becomes the source for SO_2 pollution. The treatment must reduce the fuel oils sulfur content to acceptable levels that, when burnt in the fire box, the flue gases satisfy the SO_2 levels called for by the Act. Almost all refineries today hydrotreat most product streams for the removal of sulfur and nitrogen except perhaps the very heavy residue stream. Where these residue streams are used as part of the fuel oil pool they are usually blended with hydrotreated middle distillate streams (gasoil, or even kero) to reduce the total stream sulfur. The individual refinery planning schedules will be tailored to meet this refinery fuel criteria.

The emission of NO_x is quite another problem and this is usually controlled by the minimum amount of excess air that is necessary for proper heater operation. The solution to this problem often starts in the fired heater manufacturer's design shop where the best heater fire box geometry can assist in lowering or at least maintaining the required excess air criteria and in the use of low NO_x burners. Modern refinery design engineering takes note of the stack height requirements so that the emission fall out avoids populated areas as much as possible. It is now common refining engineering practice among major contractors and heater manufacturers to utilize developed computer programs which define the fallout contours at ground level distances from fired heater stacks. These contour map programs are based on fuel type and burning data, the proposed stack height, the prevailing wind direction, and speed. The maps are used in the refinery layout studies (for new refineries being engineered) in terms of locating processes and their fired heaters on the plant site. This assists in avoiding wherever possible particulate and other pollutants contaminating high populated and ecologically sensitive areas.

Particulate emission from the fluid catalytic cracking unit

The process most vulnerable to this type of pollutant emission is the Fluid Catalytic Unit (FCCU). In this case the emission of the particulates is either a problem in operation or less frequently a break down of protection equipment—in particular the cyclone separators located in the critical areas of the process (for example, in the reactor outlet and the regenerator flue gas outlet). The most common problem in fluidized bed processes are air surges that cause a disruption of the fluidized bed. In the case of the FCCU this fluidized bed is the catalyst bed being regenerated in the regenerator. This disruption results in loss of the catalyst to atmosphere from the regenerator exit stack. The incorporation of a CO boiler however does minimize this emission to some extent. Usually these air surges are minimized by good flow control systems with some anti surge system. The malfunction of the cyclones is a much more serious problem which results in a plant shut down for repair.

VOC emission from refinery facilities

The major sources of atmospheric pollution by VOC emission in refinery are as follows:

- Relief and vent valves open to atmosphere
- Leaks from poorly maintained control valves and flanged joints
- Storage tanks
- Rail and road tanker filling facilities
- Ship loading and unloading facilities (jetty area).

Relief and vent valves normally discharge to a closed relief header which is routed to flare system. However in some cases, because of the location of the relief and/or vent valve and the quantity of the discharge material the valve(s) are open to atmosphere. A typical example is the crude distillation unit relief valves. These are located at the top of the distillation towers at about 200 ft above grade. Steam injection facilities are installed at the valve discharge to facilitated atomization and dispersion. Nevertheless when these valves do open there will be some considerable VOC emission to atmosphere. Any unscheduled event that would cause these valves to discharge are extremely rare and do not warrant the extensive costs of increased flare header and flare design that would be necessary to cater for these valve discharges.

Leaks from control (and other valves), pump seals, packing, and flanged joints are related entirely to the refinery's maintenance policy and program.

Perhaps the biggest source of VOC from a refinery and its operation is from the storage tanks. This emission occurs during the tank filling, particularly in the case of fixed roof tanks and LPG bullets and spheres. The emissions from floating roof tanks are generally quite low and the function of a properly installed roof should suffice in minimizing the emission from this source. Many of today's refineries have installed a closed inert gas circulating system. This system circulates an inert gas (mostly nitrogen) throughout the refinery for many purposes which requires the absence of oxygen in the air. One of these is the use of inert blanketing of fixed roof tanks. Blanketing is often required in intermediate tanks whose contents are feed to processes which need to eliminate oxygen. Extension of this inert gas to all fixed roof tanks as a blanket prevents the emission of VOC during their filling. The storage of LPG is also a potential for emitting VOC to the atmosphere. One method used to combat this is to float the make/break valves of the spheres and bullets on the refinery fuel gas main.

Spillages from truck and railcar loading facilities have always been problematic. Most spillages occur when filling the vessels. Most loading facilities nowadays have automatic shut off on the loading arm nozzles, very similar to those commonly used in public filling stations. These operate using a level sensitive device which closes the filling valve at a prescribed filling level in the tank. The other source of VOC emission is a poorly designed slop system that allows the draining of bottoms from road tankers and railcars. Such a system should allow for steam out facilities on vessels carrying heavy petroleum cuts. Vents on this system should be routed to the flare header.

The jetty and ship loading and unloading is also a potential source of VOC emission. Again as in the case of the tank farm and product loading facilities the most vulnerable product in this respect is the handling of the light hydrocarbon streams. In the loading of LPG a flash recycle system has been found to be effective. In this system the flashed LPG which occurs as the LPG first enters the ships tank which when empty would be close to atmospheric pressure and temperature. This flashed vapor is then routed under pressure control to a compressor and cooler assembly where it is liquefied and returned to the LPG feed stream. The system continues to operate throughout the loading activity acting then as the loading relief system.

Other sources of emissions of atmospheric pollutants

The most notable of these is the refinery sulfur recovery plant, and in particular the sulfur storage pit. Most refineries have a sub-surface tiled pit for storing the liquid sulfur product from the sulfur plant. This pit is covered by concrete or other suitable slabs. These are sealed to prevent the emission of the SO_2 that is always present above the liquid sulfur. The SO_2 that does accumulate is normally vented off to a small bullet partially filled with water. The ensuing sulfuric acid formed is either used in the refinery for caustic neutralization or sold to other users.

In bitumen production the heavy vacuum residue, which is the basis for most bitumen grades, may be air blown. This becomes necessary to meet certain bitumen grades. The air used in this air blowing procedure must be perfectly dry to avoid an explosion in the reactor vessel. (Refer to Chapter 3.9 for details on the air blowing process.) Under normal operating conditions the vapors exiting the reactor vessel are routed to the flare header. Occasionally however an explosion does occur in the reactor. When this does happen the explosion doors (located at the top of the vessel) blow open. This emits the reactor vapors which are high in SO_2 and VOCs. The plant usually returns to normal pretty quickly, and the explosion doors close as the pressure inside the vessel returns to normal. There is no way to eliminate the effect of this occurrence except, perhaps, to ensure that the air used be dry.

14.3 Noise pollution

Noise problems and typical in-plant/community noise standards

Noise has been widely organized as a major industrial/environmental problem in most processing plants because of the risk of hearing loss involved when workers are exposed to high noise levels.

The high noise levels in process plants can be attributed to a great number of sources. Major noise sources are compressors, fans, pumps, motors, furnaces, control valves, steam and gas turbines, and piping systems. The noise generating mechanism for each piece of equipment is complex but, in most cases, the noise levels can be reduced to desired limits through the implementation of proper noise control measures.

With increasing awareness of the noise problem and its effect on the general public, regulations on noise standards have been adapted in many countries throughout the world. In the United States, the Occupational Safety and Health Act (OSHA) of 1970 (29 CFR 1910.95) and the Noise Control Act of 1972 and later amendments have served as basic guidelines for noise control requirements. OSHA contains maximum permissible sound pressure levels for each daily time of exposure. These guidelines are presented in Table 14.6 and serve as a basis for in-plant noise criteria for any process plant constructed in the United States.

Conversely, community noise criteria are more variable and depend on a number of factors including local ordinances, existing noise levels and the site of the plant with respect to the community. Some typical community noise limits are shown in Table 14.7.

The art of acoustics and noise control is beyond the scope of this paper. Only the most important concepts necessary for an analysis of process plant noise will be considered.

Table 14.6. OSHA noise exposure limits

Duration per day, hr	Sound level, DBA
8	90
6	92
4	95
3	97
2	100
1.5	102
1	105
0.5	110
0.5 or less	115

The classification of different areas of the community in terms of environmental noise zones shall be determined by the Noise Control Office, based upon assessment of community noise survey data.

The Noise Control Act of 1972, administered by the Environmental Protection Agency, was intended to establish federal noise emission standards. This act served a broader scope of coordinating all noise control efforts and places the primary responsibility for noise control on the states. Please refer to current and local regulations.

Fundamentals of acoustics and noise control

Several factors contribute to this problem.

Table 14.7. Typical community noise limits

		Noise levels, DBA		
		Noise Zone Classification		
Receiving land use category	Time period	Rural Suburban	Suburban	Urban
One and two family residential	10 pm–7 am	40	45	50
	7 am–10 pm	50	55	60
Multiple dwelling residential	10 pm–7 am	45	50	55
	7 am–10 pm	50	55	60
Limited commercial some dwellings	10 pm–7 am		55	
	7 am–10 pm		60	
Commercial	10 pm–7 am		60	
	7 am–10 pm		65	
Light industrial,	Any time		70	
heavy industrial			75	

(1) *Sound pressure level*
Sound is a fluctuation in the pressure of the atmosphere at a given point. Sound pressure level is expressed as a ratio of the particular sound pressure and a reference sound pressure:

$$\mathrm{SPL} = 10 \log \frac{p^2}{P_{\mathrm{ref}}^2}$$

where

SPL = sound pressure level in dB (decibel)
P_2 = mean square amplitude of the pressure variation
$P_{\mathrm{ref}} = 2 \times 10^{-5}\ \mathrm{N/m^2}$

(2) *Sound power level*
Sound power level is defined as the ratio of the particular sound power and the reference power:

$$\mathrm{PWL} = 10 \log \frac{W}{W_{\mathrm{ref}}}$$

where

W = sound power (rate of acoustic energy flow) in acoustic watt
$W_{\mathrm{ref}} = 10^{-12}$ Watt.

The relationship between SPL and PWL is:

$$\mathrm{SPL} = \mathrm{PWL} + K$$

where

K = a constant dependent upon geometry and other aspects of the situation.

(3) *Wavelength*
Consideration of the wavelength is important to noise control.
It is defined as:

$$\lambda = \frac{C}{f}$$

where

λ = wavelength in feet
C = speed of sound in feet per second
f = frequency, cycles per second (Hz)

(4) *Octave band*
An Octave refers to a doubling of frequency. Generally, the audible frequency range consists of ten preferred octave bands with following center frequencies: 31.5, 63, 125, 250, 500, 1,000, 2,000, 4,000, 8,000, and 16,000.

Table 14.8. A-weighting network band correction

Octave band (Hz)	Band correction (dB)
63	−25
125	−16
250	−9
500	−3
1,000	−0
2,000	+1
4,000	+1
8,000	−1

(5) *A-weighted sound pressure level (dBA)*
Most noise regulations set the maximum allowable noise limits based on the use of the "A" weighting network which provides a popular means of rating noise. This network is designed to account for the response of the human ear. The other "B" and "C" weighting networks are no longer in common use. Table 14.8 shows A-weighting network band correction for each octave band to convert the sound pressure level (dB) to an A-weighted sound pressure level (dBA).

(6) *Adding decibels*
The noise levels expressed in decibels cannot be added arithmetically but the addition should be performed on the basis of energy addition. Therefore, the combined dB level is determined by:

$$\mathrm{dB}_{\mathrm{total}} = 10 \log \sum_{i}^{N} 10\,\mathrm{dB_i}/10$$

where

$\mathrm{dB}_{\mathrm{total}}$ = the combined dB level
$\mathrm{dB_i}$ = the individual dB level
N = the total number of dB levels

(7) *Sound fields*
A sound field is a description of the relationship between the PWL of the source and the SPL at different points in the surrounding space.

(a) *Idealized sound field:*
The sound field for an idealized sound sources which can be considered as a very small, uniformly pulsating sphere is given by the following equation:

$$\mathrm{SPL} = \mathrm{PWL} - 20 \log r + K \text{ dB}$$

where

r = distance from source to measurement point
K = a constant

(b) *Non-idealized sound fields:*

(i) *Outdoor:*

The sound field of a directional source radiating over a plane (hemispherical radiation) is given by:

$$\mathrm{SPL} = \mathrm{PWL} + 10\log\frac{Q}{\mathrm{r}^2} + 2.5\ \mathrm{dB}$$

where

Q = directivity factor in the direction of interest
r = distance from source in feet.

(ii) *Enclosed space:*

If source is radiating in an enclosed space, the field equation becomes:

$$\mathrm{SPL} = \mathrm{PWL} + 10\log\left(\frac{Q}{4r^2} + \frac{4}{R}\right) + 10.5\ \mathrm{dB}$$

where

R = room constant in sqft.

(8) *Directivity*

The directivity factor may be defined as the ratio of the mean square sound pressure at a given distance in a particular direction to the value which would exist if the source were non-directional. The directivity is defined by:

$$DI = 10\log Q$$

where

Q = directivity factor in the direction of interest

Some typical directivity indices are shown in Table 14.9.

(9) *Sound propagation*

The propagation of sound waves can be affected by a number of factors. The factors important to noise control consists of sound absorption, transmission loss, barriers, atmospheric, and terrain effects.

Table 14.9. Some typical directivity indices

Location	Q	DI
Near a single plane surface	2	3 dB
Near the intersection of 2 plane surfaces	4	6 dB
Near a corner formed by 3 plane surfaces	8	9 dB

(a) *Sound absorption:*
Sound waves traveling in an enclosed space are affected by the absorptive quality of the incident surface. The amount of absorption is expressed as:

$$R = \frac{S \cdot ab}{1 - ab}$$

where

R = room constant in sqft.
S = total interior surface area in sqft.
ab = average absorption coefficient

(b) *Transmission loss:*
The sound isolating capability of a wall is defined as:

$$TL = 10 \log \frac{1}{\tau} \text{ dB}$$

where

τ = transmission coefficient (ratio of transmitted sound intensity to incident sound intensity)

(c) *Barriers:*
Appreciable sound attenuation can often be obtained by interposing a barrier or acoustical shield between the source and receiver. The sound attenuation of a barrier is given approximately by:

$$B = 10 \log \frac{20H^2}{\lambda r}$$

where

B = reduction of the sound pressure level at a given frequency
H = effective barrier height
r = distance from source to barrier
λ = wavelength of sound at the frequency being considered

(d) *Atmospheric and terrain effects:*
The propagation of sound outdoors at long distances may be significantly influenced by atmospheric and terrain effects. Sound propagated through the atmosphere is subject to small energy losses due to molecular effects. This loss is dependent upon air temperature and relative humidity. The molecular effects for 72°F and 50% RH> are shown in Table 14.10.
The other effects resulting in noise reduction include attenuation due to substantial vegetation, the effects of uneven terrain and tall buildings, and the effects of wind and temperature gradients. These effects are more complex and, in most cases, can be neglected in a process plant noise analysis.

Table 14.10. Molecular effects T = 72°F; RH = 50%

Frequency (Hz)	Attenuation (dB per 1,000 ft)
500	1
1,000	2
2,000	3
4,000	8
8,000	15

Coping with noise in the design phase

Because of local, state and/or federal noise regulations which establish the maximum noise reception limits, permission to build or expand any significant industrial facility may be dependent on predicting that the reception limits set by the controlling agency will not be exceeded.

Consequently, noise control engineering must be commenced early in the design stage. A typical noise control program adopted by some major engineering companies is shown as Figure 14.9:

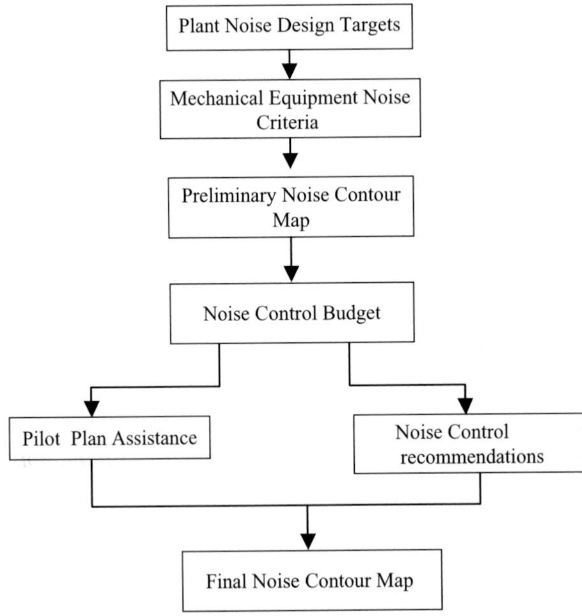

Figure 14.9. A typical noise control program flowchart.

(a) *Plant noise design targets:*
Design criteria should be developed early in the design stage with considerations given to federal, state and local laws, client standards, proximity and type of adjacent communities as well as anticipated community growth patterns.
(b) *Development of mechanical equipment and control valve noise criteria:*
Criteria for individual items of equipment will be developed to meet plant noise design objectives and will be made part of inquiry specifications for all noise-generating equipment.
(c) *Preparation of preliminary noise contour map:*
Noise reception levels will be predicted using the engineering company's in house estimate data bank and noise prediction computer programs.
(d) *Preparation of noise control budget:*
A budget is prepared which identifies funds necessary to implement noise control measures.
(e) *Plot plan assistance:*
Equipment location can be optimized based on projected noise levels which will minimize the need of attenuation treatment.
(f) *Noise control recommendations:*
Recommendations for noise attenuation are prepared based on the contribution of all major equipment to the composite noise level, of a given plant area.
(g) *Preparation of final noise contour map:*
After final noise level data have been obtained, the plot plan has been finalized and any noise control measures have been implemented, a noise contour map can be prepared. The noise contour map identifies that the plant noise design objectives have been met. This contour map is normally prepared using the company's own modeling program. Alternatively a proprietary program can be leased from an appropriate soft ware company.

An effective noise control program requires an analysis in the early stage of plant design when no equipment has been purchased. The most efficient and economical approach to nose control is to include noise control features as an integral part of equipment design through equipment specifications. An optimal plot plan arrangement can also minimize the need for attenuation treatment by strategically locating noisy equipment or positioning process areas or known noise sources at maximum distances from sensitive areas. The task of predicting noise levels to be used in the design phase can be overwhelming without the aid of the computer. A typical project may involve hundreds of thousands of noise sources, and attempting to do the noise level predictions by hand computer program to predict community/in-plant noise levels. This program is discussed below in "*a typical community/in-plant noise program*".

A typical community/in-plant noise program

The noise pollution cycle is one of emission, propagation and reception. A computer program is usually developed to simulate the noise propagation from several types of noise sources with different configurations. This section will describe the capabilities of such a program and its application to plant design.

(1) *Capabilities*
The computer program calculates sound pressure levels generated by single or multiple noise sources at specified grid points or special receptors. This program should utilize a simple algorithm to simulate the propagation of four different source models. These models include point, line, discrete points on a line, and plane sources.

(2) *The Mathematical model*
The basic equation used in the program is:

$$\text{SPL} = \text{PWL} + 10\log[F(R)] + DI + K - AE$$

where

SPL = sound pressure at any receptor in dB
PWL = source sound power level in dB
$10\log[F(R)]$ = distance attenuation factor for various types of source which will be defined in the following paragraphs in dB
DI = directivity index in dB
K = Characteristic resistance of air in dB
AE = total excess attenuation factor (molecular absorption, Ground absorption, screening effect, barrier effect, etc.) in dB.

- *Distance attenuation* $F(R)$

(A) *For a point source:*

$$F(R) = \frac{1}{R^2}$$

where

R = distance between source and receptor

(B) *For a continuous line source:*

$$F(R) = \frac{\alpha_2 - \alpha_1}{R_{\text{Od}}}$$

All terms are defined in Figure 14.10.

(C) *For discrete sources on a line:*

$$F(R) = \sum_{r=1}^{N} \frac{1}{R_n^2}$$

where

N = the number of sources on the line
R = the distance between receptor and each source

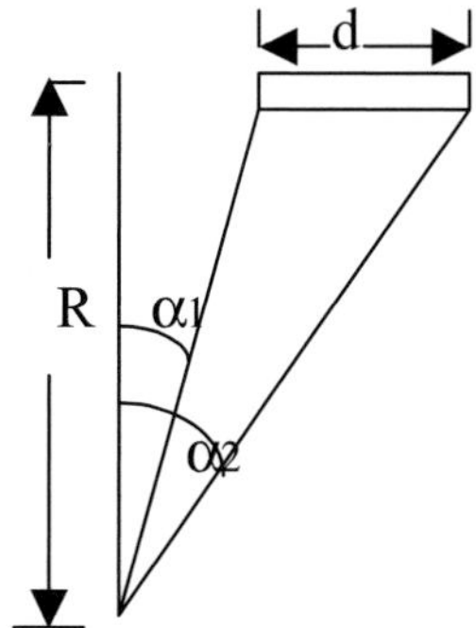

Figure 14.10. Continuous line source.

(D) *For a plane source:*

$$F(R) = 1/A \int_{-L/2}^{L/2} \int_{-L/2}^{L/2} R\, dx dy$$

where

R = distance from the receptor to a differential area *dxdy* on the plane.

All terms are defined on Figure 14.11.

- *Excess attenuation A*
  Excess attenuation due to ground and molecular absorption can be entered as input data. When these data are not entered, the default values shown in Table 14.11 should be used. In-plant shielding corrections may be included in the ground absorption correction. Corrections due to the effect of wind, temperature gradients, rain, sleet and barriers will be added at a future date.

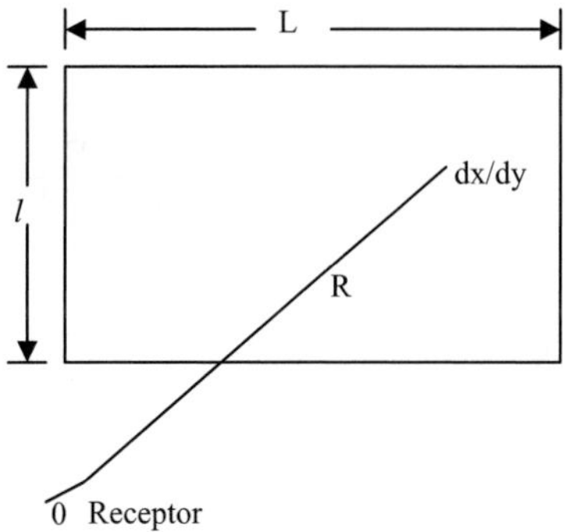

Figure 14.11. Plane source.

Table 14.11. Default values for excess attenuation

Frequency, Hz	Ground absorption, dB	Molecular absorption, dB/1,000 ft
63	↑	0
125		0
250		0.3
500	0	1.0
1,000		2.0
2,000		3.0
4,000		6.0
8,000	↓	11.0

(i) *Input data requirements*

The requirements for each noise source are basically the source sound power spectrum, the source location and the desired noise level prediction model. Additionally, excess attenuation factors including molecular and ground absorption data can be input if available.

(ii) *Output*

In the program described here, all the input data will be summarized and the calculation results will be printed in the computer print out. The calculated octave band sound pressure level, overall sound pressure levels, and A-weighted sound pressure levels will be tabulated for all the special receptors specified. The calculated A-weighted sound pressure levels for all grid points will also be listed in the computer printout.

Special routines are available to draw configurations of equipment normally present in a process plant, such as round-headed vessels, cooling towers, air coolers, and fired heaters.

(iii) *Applications*

The community/in-plant noise program described here can be used to perform the following applications:

- Estimating the net noise impact due to a prospective industrial activity
- Checking plant design for compliance with noise regulations
- Establishing noise emission levels and plant design necessary to comply with the applicable environmental regulations

As with other computer programs, the accuracy of computer-calculated noise levels depends on the accuracy of the input data as well as the application of the model selected for a particular noise source.

Appendix 14.1: Partial pressures of H_2S and NH_3 over aqueous solutions of H_2S and NH_3

Refer to Appendix 14.2 for the interpretation and use of these charts

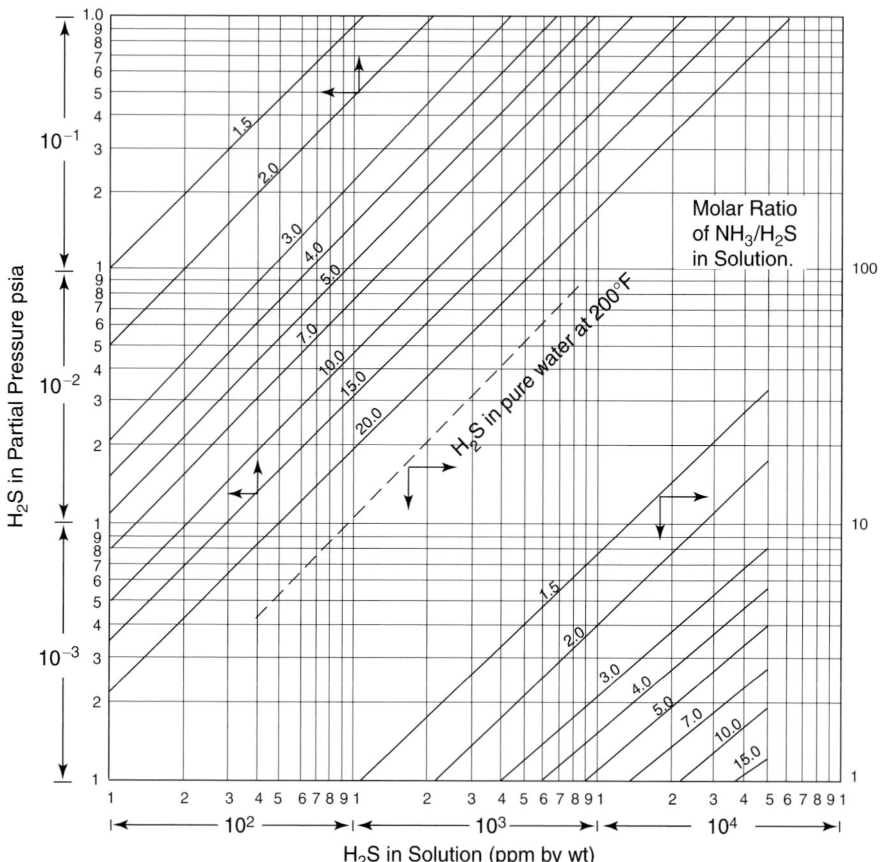

Figure 14.A.1. Partial pressure of H_2S over aqueous solutions of H_2S and NH_3 at 200°F.

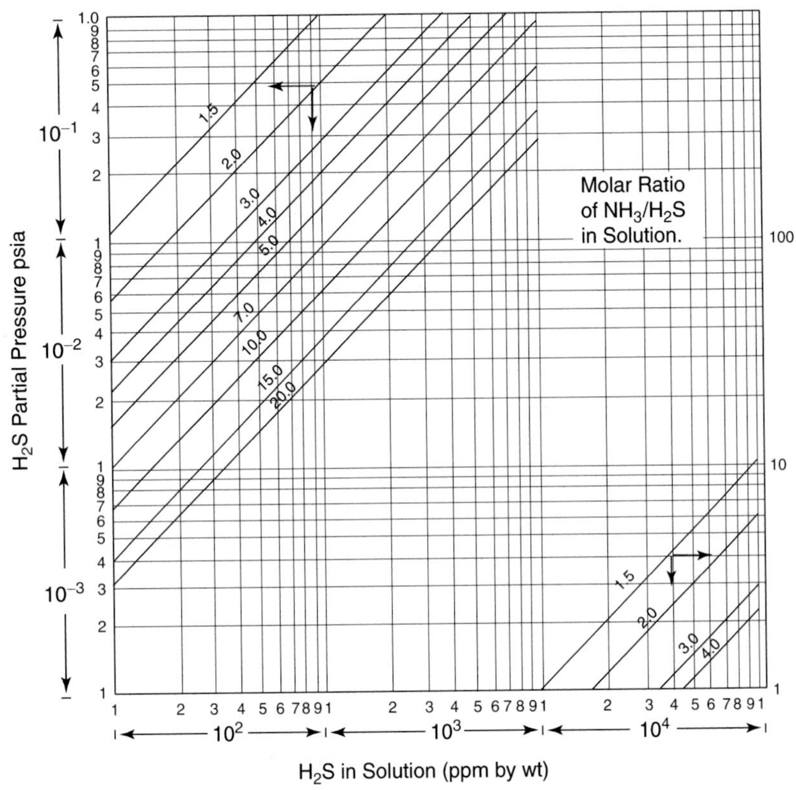

Figure 14.A.2. Partial pressure of H_2S over aqueous solutions of H_2S and NH_3 at 210°F.

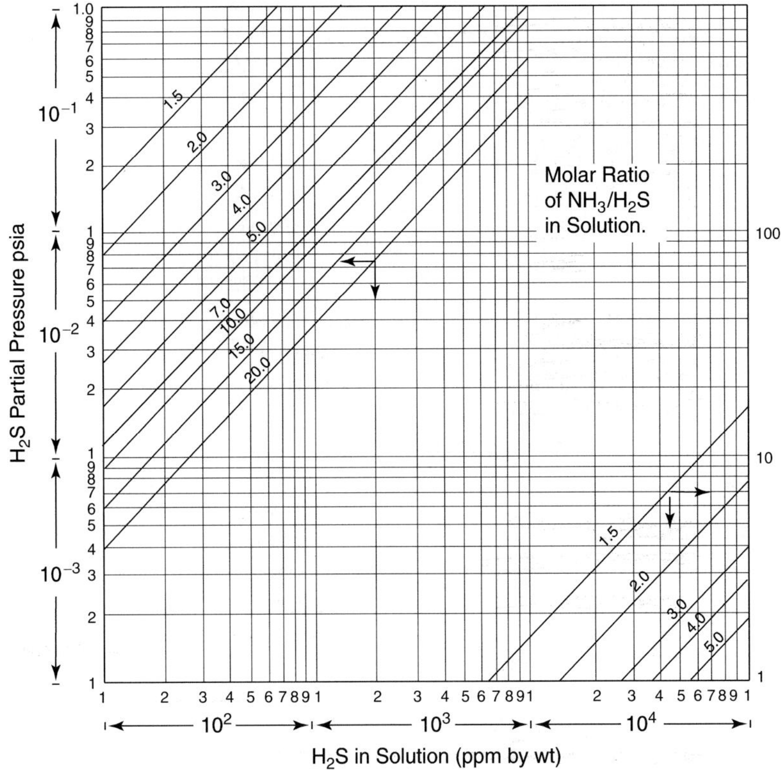

Figure 14.A.3. Partial pressure of H_2S over aqueous solutions of H_2S and NH_3 at 220°F.

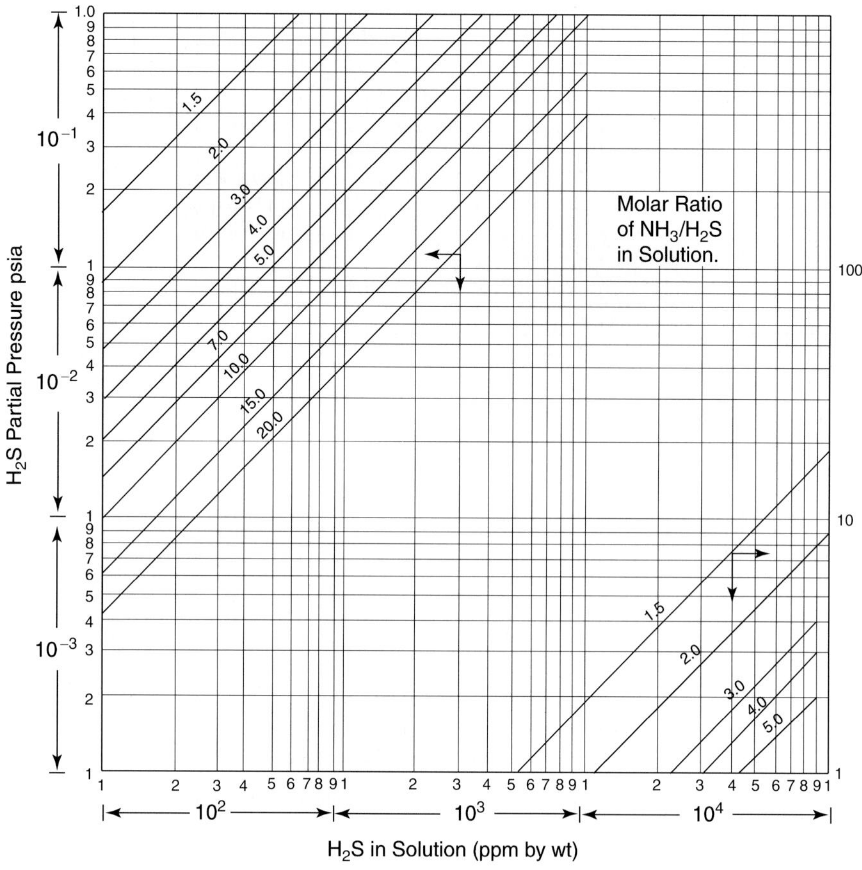

Figure 14.A.4. Partial pressure of H_2S over aqueous solutions of H_2S and NH_3 at 225°F.

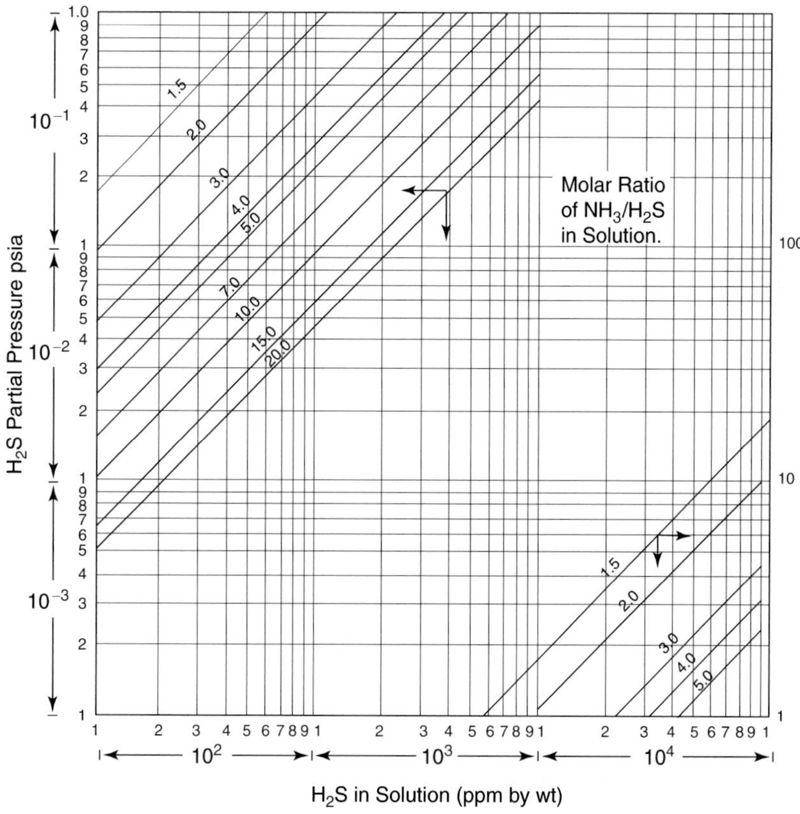

Figure 14.A.5. Partial pressure of H_2S over aqueous solutions of H_2S and NH_3 at 230°F.

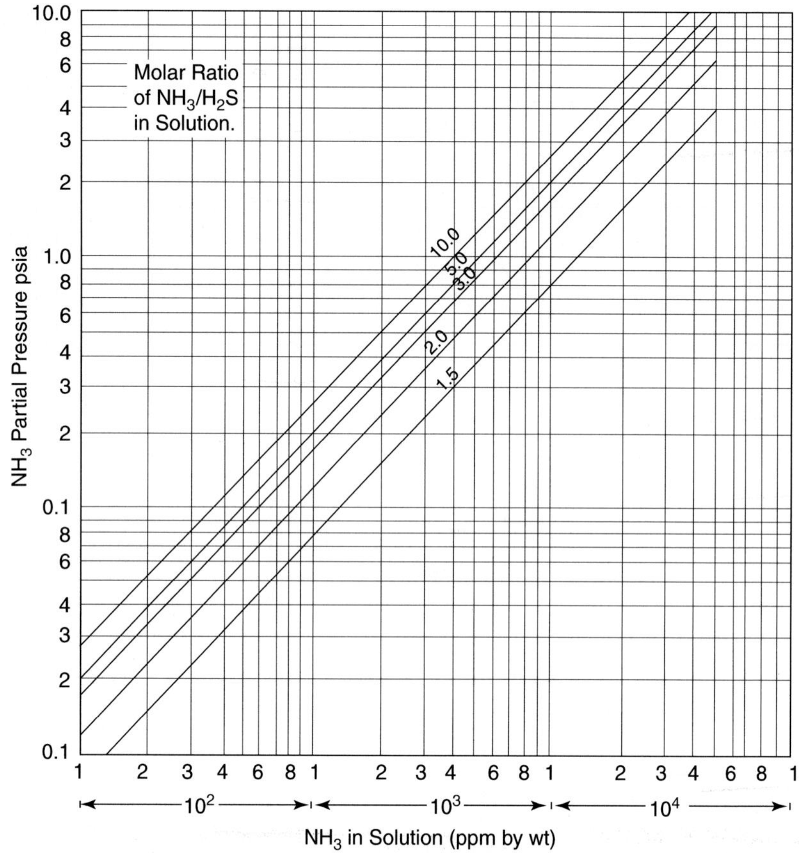

Figure 14.A.6. Partial pressure of NH_3 over aqueous solutions of H_2S and NH_3 at 200°F.

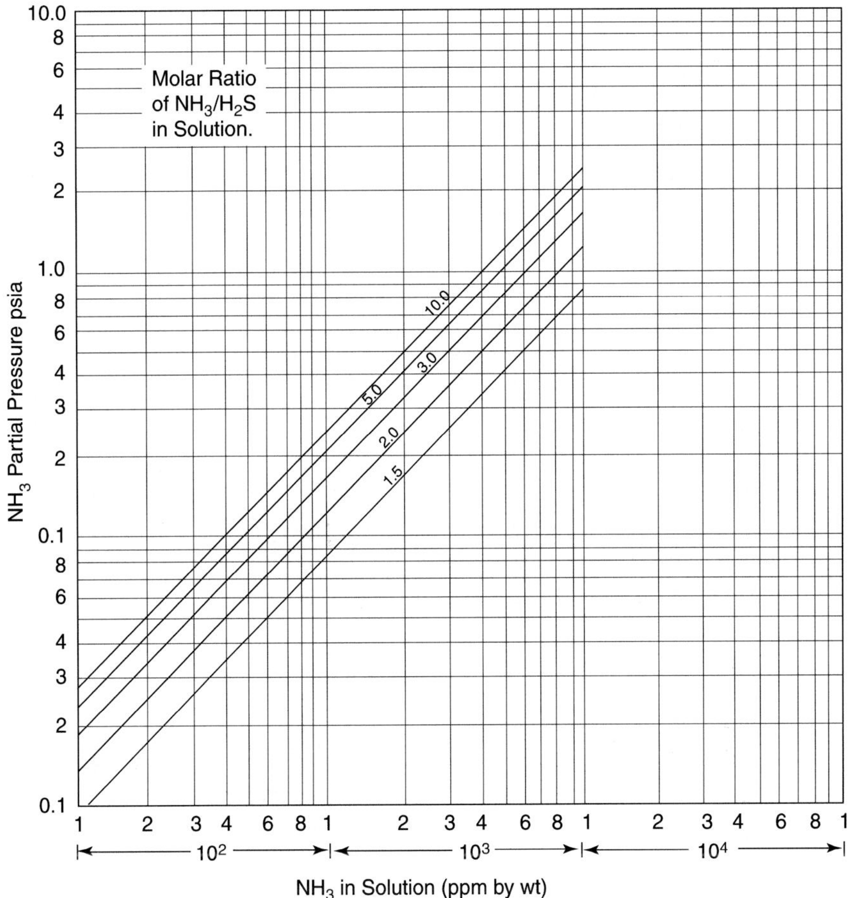

Figure 14.A.7. Partial pressure of NH_3 over aqueous solutions of H_2S and NH_3 at 210°F.

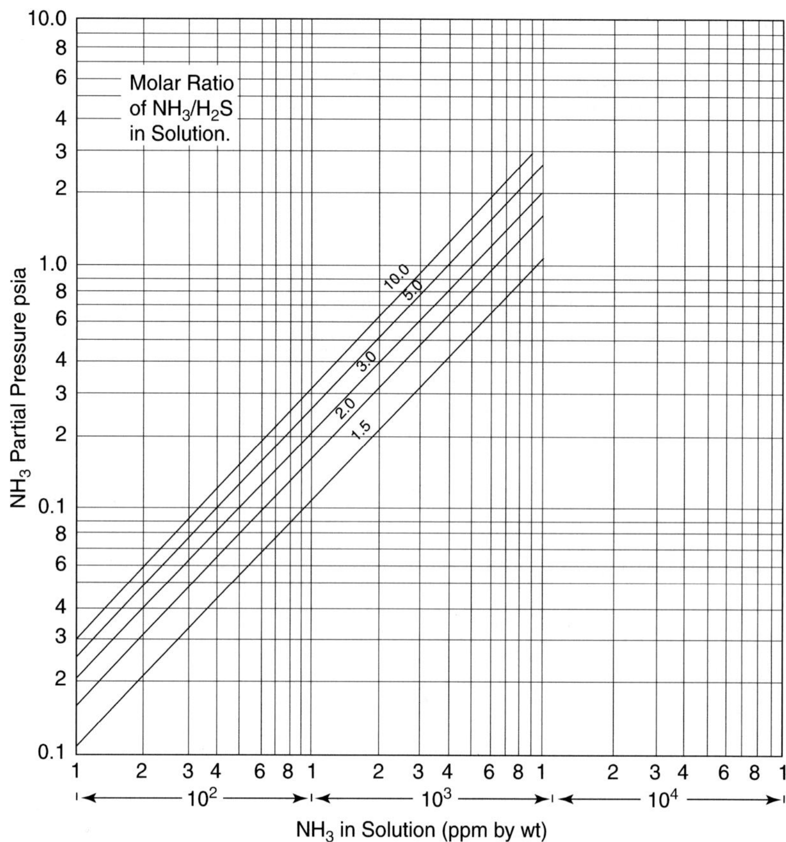

Figure 14.A.8. Partial pressure of NH_3 over aqueous solutions of H_2S and NH_3 at 220°F.

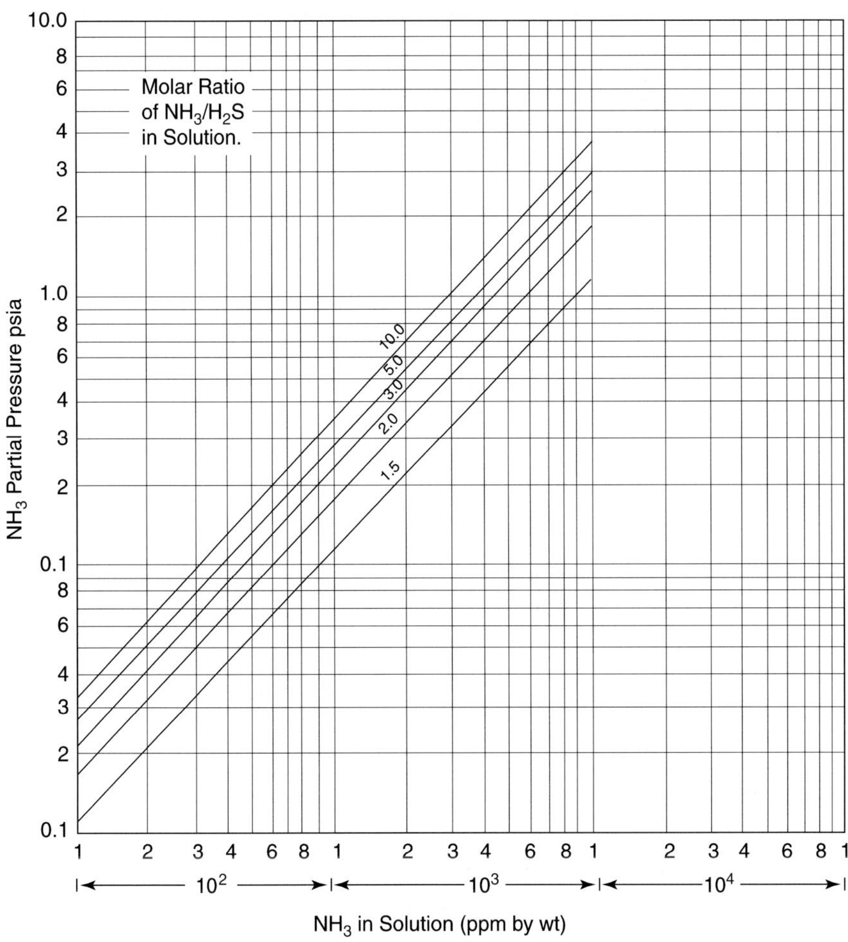

Figure 14.A.9. Partial pressure of NH_3 over aqueous solutions of H_2S and NH_3 at 225°F.

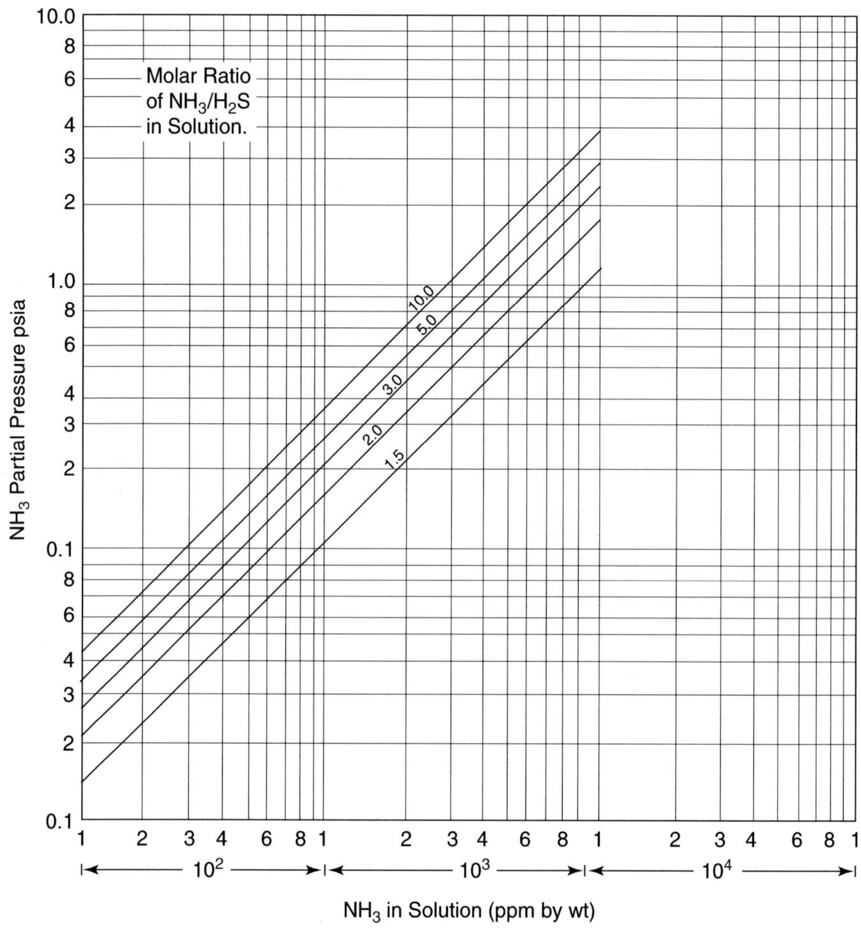

Figure 14.A.10. Partial pressure of NH_3 over aqueous solutions of H_2S and NH_3 at 230°F.

Appendix 14.2: Example of the design of a sour water stripper with no reflux

Specification

Feed: The tower is to be designed to handle 200 gpm (at 100°F) of sour water containing 10,000 ppm of H_2S and 7,500 ppm of NH_3 (by weight).

Unit: The unit shall be a trayed column using sieve trays (efficiency of 0.5) with no reflux.

Steam rate: Refinery 50 psig saturated steam shall be used at a rate of 1.3 lbs/gpm of feed.

Tower pressure and temperature: The vapors leaving the tower top shall have sufficient pressure to enter a 'Rat Tail' burner in a nearby heater. The tower top pressure shall be 20 psia. The feed entering the top tray shall be preheated to a temperature of 200°F. Total tower pressure drop shall be 2 psi.

Stripped water specification: The tower shall be designed and operated to remove 99.0% of the H_2S in the feed and 95% of the NH_3.

The design

Assume the stripping will be accomplished using four theoretical trays. Then at an efficiency of 0.5 the number of actual trays will be 8. The pressure drop per tray will be 0.25 (that is 2 psi/8 trays).

Calculate feed mass per hour:

Water @ 100°F has a specific volume of 0.1207 gals/lb.
Then 200 gpm of water $= \frac{200 \times 60}{0.1207} = 99{,}420$ lbs/hr.

Feed will be:

	lbs/hr	moles/hr	
Water	99,420	5,523.3	
NH_3	745.65	43.86	(7,500 ppm by wt)
H_2S	994.2	29.24	(10,000 ppm by wt)

Calculate stripping steam:

Tower bottom pressure will be 20 psia $+ (8 \times 0.25) = 22$ psia.
From steam tables tower bottom temperature will be water at 22 psia $= 233$°F

Feed temperature = 200°F
Then steam used for heating = 99,420(33/924) = 3,550 lbs/hr.
And steam used for stripping = 15,550 − 3,550 = 12,000 lbs/hr.

Calculate stripped water quantity and composition:

Stripped water shall contain feed water plus condensate. 99,420 + 3,550 = 102,970 lbs/hr
NH_3 in stripped water shall be 5% of total = 37.28 lbs/hr = 2.19 moles/hr = 361 ppm by wt
H_2S in stripped water shall be 1% of total = 9.94 lbs/hr = 0.29 moles/hr = 100 ppm by wt

Calculate the overhead vapor partial pressures:

Overhead vapor leaving the tower V_o will be as follows:

	moles/hr	lbs/hr	PP psia
NH_3	41.67	708	1.13
H_2S	28.95	984	0.78
Steam	667	12,000	18.09
Total	737.62	13,692	20.00

Temperature of top theoretical tray is steam at a saturation pressure of 18.09 psia = 223°F.

Top tray calculation:

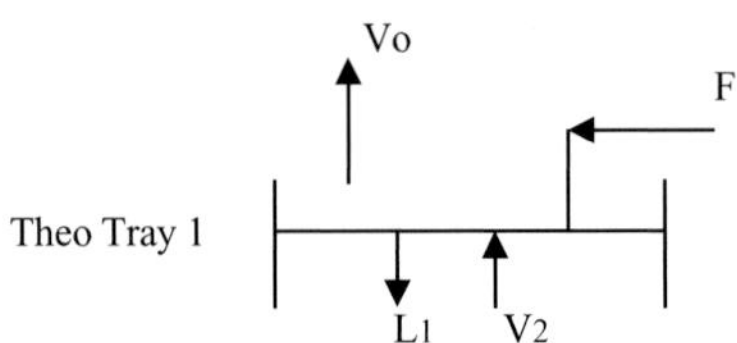

Assume a ratio of NH_3/H_2S as 3.8 moles that is (17/34) × 3.8 = 1.9 by wt.
From Figure A5.9 NH_3 at a partial pressure of 1.13 will be 4,600 ppm by wt H_2S ppm by wt will be 4,600/1.9 = 2,421 ppm
From Figure A5.4 partial pressure of H_2S = 0.75 which is acceptably close to 0.78, which was established.
Note: Should the Partial Pressure of H_2S be substantially different to that for V_o then a different ratio of the two components would have to be chosen and the calculation repeated.

Calculate liquid from top tray L_1:

	ppm	lbs/hr	moles/hr
NH_3	4,600	457	26.9
H_2S	2,421	241	7.08

Calculate vapor from Theo tray 2 V_2

Moles vapor for NH_3 and H_2S will be those moles in $V_o + L_1 - F$
Pressure on tray Theo tray 2 will be $(1/0.5) = 2$ actual trays @ 0.25 psi pressure drop = 20 psia + 0.5 = 20.5 psia.

	moles/hr	PP psia	
NH_3	24.71	0.73	
H_2S	6.79	0.2	
Steam	667	19.57	
Total	698.5	20.5	Tray temp (from Steam tables = 226°F)

Calculate balance over theoretical tray 2:

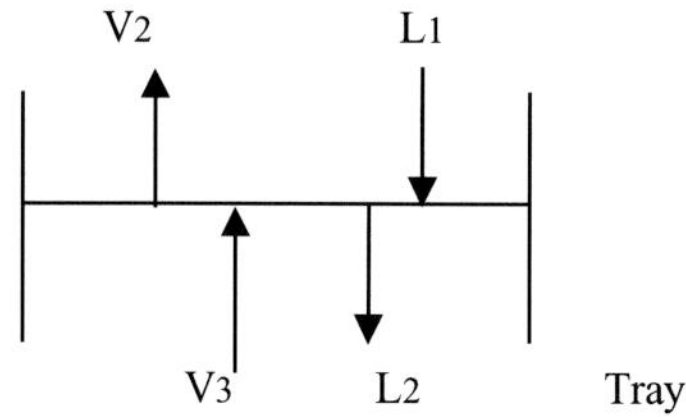

Assume NH_3/H_2S ratio is 5.0 molar and 2.5 ppm by weight.

NH_3 @ a PP of 0.73 and 5.0 molar ratio = 2,400 ppm by wt (from Figure 14.A.10).
H_2S ppm is 2,400/2.5 = 960.
From Figure A5.5 H_2S ppm = 0.19, which is a satisfactory match.

Liquid from tray 2. L_2:

	ppm	lbs/hr	moles/hr
NH_3	2,400	278.4	16.37
H_2S	960	95.5	2.81

Vapor from tray 3 V_3:

Total tray Pressure = 21 psia and temperature is 231°F
Vapor from tray 3 = $V_2 + L_2 - L_1$

	moles/hr	PP psia
NH_3	14.18	0.44
H_2S	2.52	0.08
Steam	667	20.52
Total	683.7	21.00

Calculate balance over theoretical tray 3:

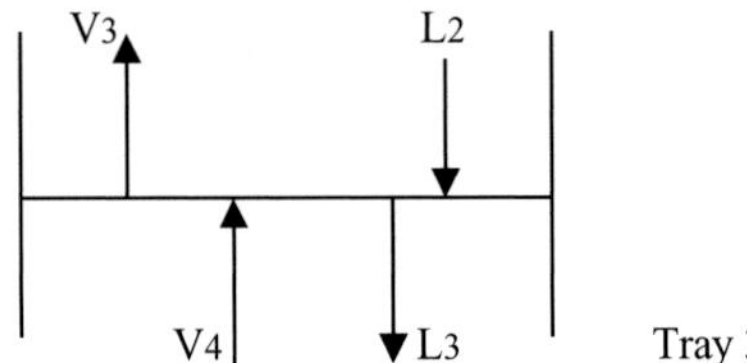

Assume NH_3/H_2S ratio is 6.5 Molar and 3.25 by weight.

NH_3 ppm from Figure 14.A.10 = 1,600 ppm by weight.
H_2S is 493 ppm from Figure 14.A.6 PP of H_2S is 0.085 which is a satisfactory match.

Liquid from theoretical tray 3, L_3:

	ppm	lbs/hr	moles/hr
NH_3	1,600	159.1	9.35
H_2S	493	49	1.3

Vapor from theoretical tray 4, V_4:

Tray pressure 21.5 psia Temperature 232°F

$$V_4 = V_3 + L_3 - L_2$$

	Moles/hr	PP psia
NH_3	7.16	0.23
H_2S	1.01	0.032
Steam	667	21.24
Total	675.17	21.5

Calculate balance over theoretical tray 4:

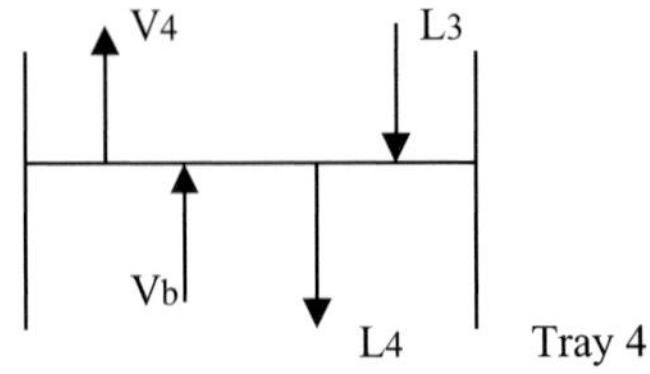

Assume NH_3/H_2S ratio is 6.6 molar and 3.3 by weight

NH_3 ppm from Figure 14.A.10 is 610. H_2S ppm is 185 PP psia of H_2S from Figure 14.A.6 is 0.032 which is a satisfactory match.

Liquid from theoretical tray 4, L_4

	ppm	lbs/hr	moles/hr
NH_3	610	60.6	3.57
H_2S	184.8	18.4	0.54

Vapor from tower bottom V_b

Pressure 22 psia Temperature 233°F

$$V_b = V_4 + L_4 - L_3$$

	moles/hr	PP psia
NH_3	1.38	0.045
H_2S	0.25	0.008
Steam	667	21.947
Total	668.83	22.0

Calculate the bottom of the tower:

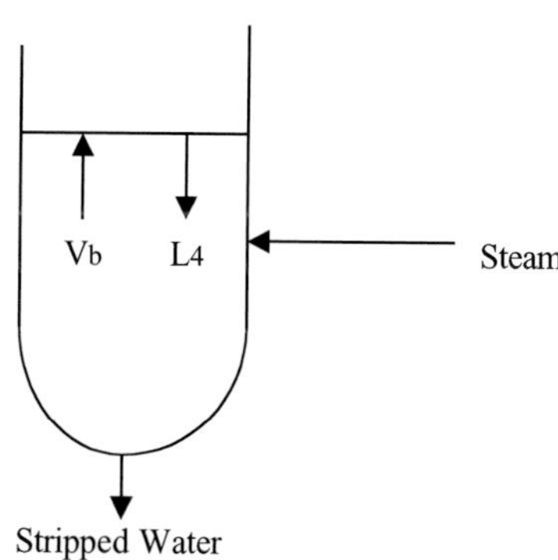

Assume NH_3/H_2S ratio is 7.0 molar and 3.5 by weight

NH_3 ppm from fig A5.10 is 180
H_2S ppm is 51.4 PP psia of H_2S from Figure A5.6 is 0.008 (extrapolated) which is a satisfactory match.

Contaminants in Stripped Liquid Product

	ppm	lbs/hr
NH_3	180	17.9
H_2S	51	5.1

Conclusion

The tower will handle 200 gpm of sour water and remove 97.6 % weight of NH_3 and 99.49% weight of H_2S using 8 actual trays and 1.3 lbs/hr per gpm of feed. Feed will be preheated to 200°F by heat exchange with tower bottoms before entering on the top tray of the tower.

Appendix 14.3: Example design of an API separator

Specification

It is required to design an oil/water separator to handle the normal quoted rainfall and process waste from a 4,500 BPSD hydro-skimming refinery. The quantity of in flow to the separator is estimated to be 600 gpm. The normal rundown temperature for this stream is taken as 100°F. The design of the separator shall be in accordance with the appropriate section of the *API Manual*—sixth edition. The following data shall be used in the design:

Specific gravity of water	0.995
Specific gravity of the oil	0.890
Viscosity of water	0.7 Centipoise.
Diameter of the oil globules	0.006 ins

The oil shall be removed from the separator by means of oil skimming pipes and an oil sump designed to meet the oil influent content of 400 ppm by volume.

The design

The rising rate of the oil is calculated from the equation:

$$V_r = 6.69 \times 10^4 \times \frac{d^2 \times \Delta S}{\mu}$$

where

V_r = rising rate of oil in ft/min.
d = diameter of oil globule in ins = 0.006
ΔS = difference in the SGs of oil and water phases = 0.105
μ = Viscosity of the water phase in centipoise = 0.7

Then

$$V_r = \frac{6.69 \times 0.36 \times 0.105}{0.7}$$
$$= 0.361 \text{ ft/min.}$$

V_h = the horizontal velocity of the effluent = $15 \times V_r$ but not to exceed 3ft/min.
= 3 ft/min.

$$\text{Volumetric rate of flow} = Q = \frac{600}{7.48} = 80 \text{ cuft/min}$$

$$\text{Minimum cross sectional area of flow} = \frac{80}{3} = 26.7 \text{ sqft say 27 sqft.}$$

API recommend the following limits:

Depth 3 to 6 ft.
Width 6 to 20 ft
Ratio of depth to width 0.3 to 0.5

Width of section $= 27/3 = 9$ ft which meets the API recommended depth to width ratio.

Calculating the effective separation length of unit:

The API manual gives the following factors to include in the determination of the unit's length.

Turbulence factor $F_t = 1.22$
Short circuit factor $F_s = 1.2$

The length is then calculated by the expression :

$$L = (F_t + F_s)(V_H/V_r) \times \text{depth.}$$
$$= 2.42 \times 8.31 \times 3$$
$$= 60 \text{ ft.}$$

This 60 ft is the length between the oil retention baffles (see Figure 14.7).

Recommendations regarding the types and sizes of the internal equipment is found in the appropriate section of the API manual. A brief description of these are given in the text of this chapter.

Chapter 15

Refinery safety measures and handling of hazardous materials

D.S.J. Jones

Introduction

From the very nature of crude oil, its refining and the processes relating to its operation provide an extremely hazardous situation. Above all, of course, is the inherent danger of fire. Considerable steps are taken therefore to prevent this, and if a fire does occur, to combat and restrain it in the most effective manner.

Although the fire hazard is always the primary concern in the refining of petroleum, there are other hazards that are present and always need to be addressed. Among these are the handling of toxic and dangerous chemicals that are used in the refining processes. There is also always present the danger to life of toxic products that are produced in some of the refining processes. Perhaps the most notable is hydrogen sulfide, which is common to all modern refineries.

This chapter deals first of all with the nature and handling of these common hazardous materials (Part 1). The chapter then continues with the description and discussion of those basic fire protection methods used in refinery design and operation (Part 2).

15.1 Handling of hazardous materials

Anhydrous hydrofluoric acid

Because of anhydrous hydrofluoric acid's (AHFs) highly toxic and corrosive nature, this item is included to highlight its characteristics and the safe handling of the acid.

Anhydrous hydrofluoric acid is a colorless, mobile liquid which boils at 67°F at atmospheric pressure, and therefore requires pressure containers. The acid is also hygroscopic therefore its vapor combines with the moisture of air to form "fumes". This tendency to fume provides users with a built-in detector of leaks in AHF storage

Table 15.1. The physical properties of AHF

Boiling point at 1 atm, °F	66.9
Freezing point, °F	–117.4
Specific gravity at 32°F	1.00
Weight per gallon at 32°F, lb	8.35
Viscosity at 32°F, cp	0.31

and transfer equipment. On the other hand, care is needed to avoid accidental spillage of water into tanks containing AHF. Dilution is accompanied by a high release of heat. The physical properties of AHF are given in Table 15.1.

AHF vapor, even at very low concentrations in air, has a sharp penetrating odor that is an effective deterrent to willful overexposure by operating personnel. Both the vapor and liquid forms of AHF cause severe and painful burns on contact with the skin, eyes, or mucous membranes.

Hydrofluoric acid is very corrosive. It attacks glass, concrete, and some metals—especially cast iron and alloys which contain silica (e.g., Bessemer steels). The acid also attacks such organic materials as leather, natural rubber, and wood, but does not promote their combustion.

Although AHF is non-flammable, its corrosive action on metals, particularly in the presence of moisture, can result in hydrogen forming in containers and piping to create a fire and explosion hazard. Potential sources of ignition (sparks and flames) should be excluded from areas around equipment containing hydrofluoric acid.

Despite its corrosive nature, AHF can be handled with relative safety if the hazards are recognized and the necessary precautions taken. This item describes certain procedures for the safe handling of large bulk quantities of AHF.

Safe handling

The safe handling of AHF requires that well-designed equipment be properly operated and maintained by well-trained, adequately protected, responsible personnel.

Tanks and other containers of AHF should be protected from heat and the direct rays of the sun. Storage-area temperatures should preferably remain below 100°F. If they reach or exceed 125°F, means for cooling the containers must be applied.

Acid-transfer lines between the unloading station and the storage tank should tilt toward the latter to insure free drainage. Relief valves should be installed in those sections of acid-transfer lines where acid may be entrapped between two closed valves in the line, because expansion of the liquid might create excessive pressure and rupture the line.

No open fires, open lights, or matches should be allowed in or around acid containers or lines. The possibility of acid acting on metal to produce hydrogen gas is ever present. Only non-sparking tools and spark-proof electrical equipment should be used in the AHF storage and handling areas.

Safety showers should be readily accessible at the unloading station, in the storage area, and at other locations where acid is handled. The showers should be capable of supplying volume flows of 30 GPM through quick-opening valves in 2-inch water lines. Handles at hip level should actuate the valves which, with a 0.25-inch weep hole directly above the valve, should be positioned below the frost line and surrounded by crushed rock or gravel to provide drainage.

A water hydrant and hose should also be available in the unloading area to flush away spilled acid. Good drainage should be provided, and also a supply of dry soda ash, ground limestone, or hydrated builders lime. Accidental spills of acid on walkways or equipment should be washed off immediately with large volumes of water and, if necessary, neutralized with one of the agents mentioned.

Personal protective equipment

Personal protective equipment is not a substitute for good, safe working conditions. Its purpose is to protect the wearer in the event of an accident—major or minor. The extent of protection needed depends upon the degree of exposure attending the particular job at hand. Protective equipment should not be worn or carried beyond the operating area. It should be thoroughly washed with sodium bicarbonate solution immediately after each use.

The minimum protection required for operating and maintenance personnel includes the following items:

- Coveralls with sleeves to the wrists
- Face shield or chemical safety goggles
- Hard hat
- Poly(vinyl chloride)—or neoprene-dipped gauntlets
- Poly(vinyl chloride)—or neoprene-soled rubber shoes

When taking acid samples, opening equipment which may contain hydrofluoric acid, or performing similar hazardous duties, operators should wear the following:

- Poly(vinyl chloride) or neoprene overalls
- Poly(vinyl chloride) or neoprene boots
- Lightweight poly(vinyl chloride) or neoprene gloves under poly(vinyl chloride)—or neoprene-dipped gauntlets

- Poly(vinyl chloride) or neoprene jumper
- Airline hood

Air should be applied to the hood until the absence of fumes in the work area has been fully established.

Unloading and transfer of AHF

AHF is shipped in rail tank cars having capacities ranging from approximately 5,400 gallons to 25,000 gallons and in road tank trucks of approximately 5,250 gallons AHF capacity. Compressed dry gas (air, hydrocarbon, or nitrogen) is the preferred means for transferring bulk quantities of AHF, but a centrifugal, rotary or positive-pressure pump can be used if necessary.

The unloading of AHF tank cars or tank trucks, with transfer of the acid to plant storage, consists of five steps:

(1) Spotting the tank car or tank truck at the unloading station.
(2) Connecting the plant compressed-gas (or vapor) and AHF-unloading lines to the proper valves on the carrier tank.
(3) Transferring the AHF from the carrier tank to the storage tank.
(4) Disconnecting the plant compressed-gas (or vapor) and AHF-unloading lines from the carrier tank valves.
(5) Releasing the tank car or tank truck for return to the shipper.

Equipment
Mild steel is satisfactory for storing and handling AHF at temperatures up to 150°F maximum. Type 300 stainless steels are useful up to 200°F. "Monel" nickel-copper alloy and "Hastelloy" C nickel steel are suitable for higher temperatures. TEFLON TFE fluorocarbon resin is completely resistant to all concentrations of hydrofluoric acid at temperatures up to 500°F.

Steel should not be used for movable parts because the corrosion-product film will cause movable parts to "freeze". Cast iron, type 400 stainless steel, and hardened steels are unsatisfactory for AHF handling. Copper is velocity-sensitive. Stressed Monel may stress crack if exposed to moist vapors or aerated acid containing water. Welds in Monel corrode rapidly.

The selection of construction materials used for AHF equipment depends very much on such corrosion-affecting variables as moisture, temperature, aeration, fluid velocity, and impurities. Each storage and handling situation requires separate study to evaluate these factors before selecting materials which must meet the requirements of the installation.

Storage tank. The capacity of the storage system should be approximately $1^1/_2$ times the maximum quantity normally ordered to insure against running out of acid between receipt of shipments. As a rule, too large a storage system is preferable to too small a system. The additional investment required for the larger installation is not great. The larger installation permits further expansion, less precise scheduling of shipments, and larger inventories when desired.

The horizontal cylindrical storage tank should be manufactured according to the current ASME Code for Unfired Pressure Vessels or other equivalent codes which meet State or local mandatory requirements. It is further recommended that the wall thickness of the tank be at least 1/8 inch in excess of the ASME Code requirements. The tank should be double-welded, butt-joint construction, the welds to be slag-free (conforming to ASME Code, Section 8) and ground smooth inside to facilitate inspection. X-ray inspection of welds is recommended.

The storage tank should be suitably supported above ground level. Structural steel supports or concrete saddles (protected with an acid-resistant paint) are satisfactory.

Safety devices for relieving abnormal internal tank pressures should be obtained from qualified manufacturers who are familiar with AHF. The maximum working pressure of the storage system should not exceed 2/3 the rated relief or bursting pressure of the safety devices. The dual relief system is recommended which has a 2-way valve and rupture discs ahead of the relief valves and also a separate rupture-disc line in case of relief valve failure.

Piping. All pipe lines should be installed so that they drain toward the storage tank, or toward the point of consumption. This will prevent the accumulation of acid in low points, thereby eliminating possible safety hazards when repairs are necessary. Relief valves should be installed in the various sections of the lines in case acid becomes confined between two closed valves in the line. All flanges in the lines should preferably be coated with an acid-indicating paint such as Mobil #220-Y-7 hydrofluoric acid-detecting paint, which changes in color from orange to yellow in the presence of AHF liquid or vapor.

The line from the unloading station to the storage tank should be equipped with a gate valve so acid flow can be stopped at any time. The line should also be securely anchored to the storage tank as considerable vibration may occur, especially when unloading by means of compressed gas.

Extra heavy (Schedule 80) or, better, triple extra-heavy black seamless or welded steel pipe which is free from non-metallic inclusions is satisfactory.

Fittings. Larger lines (2 ins and over) should preferably be welded to conform to ASME Code, Section 8. Alternatively, properly gasketed forged steel flanges can be used.

On smaller lines, extra-heavy forged steel, screw-type unions with steel-to-steel seats can be used for pipe joints. Graphite-and-oil is satisfactory as lubricant.

Gaskets. Gaskets made of TEFLON TFE fluorocarbon resin are recommended.

Valves. Jamesbury "Double-Seal" ball valves have given excellent service to AHF manufacturing operations. The valve seats are preferably of TEFLON TFE; the balls and bodies of 316 stainless steel, "Durimet" 20 austenitic stainless steel alloy, or equivalent. Gate valves should be of the O, S, and Y flange type, with a ring of TEFLON TFE or "Kel-F" fluorocarbon thermoplastic material on the plug seat and packing of either TEFLON TFE or Kel-F.

Globe valves can be of Monel nickel-copper alloy or have a forged steel body and trim of Monel.

Good service has been reported for Hills–McCanna diaphragm valves with body of Durimet 20 or equivalent, or Monel alloy; diaphragm of polyethylene, Kel-F, or a laminate of neoprene and TEFLON TFE; and a wheel closure.

Plug valves of Monel with a sleeve of TEFLON TFE have been found satisfactory.

Check valves should be of the forged ball and body type—made of Monel metal.

Pressure gauges. Pressure gauges should be constructed of 316 stainless steel or Monel metal Bourdon tubes. The bottom connection of the gauge should be ½-inch. The case should have a "blow-out" back.

Pumps. Centrifugal, rotary or positive-pressure types of pumps are satisfactory. The 300 series stainless steels; Durimet 20 or equivalent, Hastelloy C, Monel alloys; nickel, bronze, and acid bronze have been recommended as construction materials.

Level gauge. ANF Manufacturers suggest the use of a magnetic-type level gauge, such as a Fischer & Porter Model 13 C 2265W Liquid "Levelrator" with donut-type float.

An alternative means for monitoring storage-tank content is to set the tank on load cells or strain gauges.

Filters. Where the critical nature of the process has warranted, cartridge-type filters in the storage tank-to-process line have been recommended. Two such filters are normally mounted in parallel to permit replacing the cartridge in one line while diverting the acid flow through the other. Construction materials used in fabricating the filters are the same or similar to those described above for other auxiliary equipment.

Poly-propylene can be used as filter material for AHF alone if the liquid temperature remains below 200°F.

When a fouled filter is removed for replacement, it should be promptly flushed with water, neutralized with a solution of soda ash, and rinsed before discard.

Personal safety

Liquid AHF causes immediate and serious burns to any part of the body on contact.

Dilute solutions of hydrofluoric acid often do not cause an immediate burning sensation where they came in contact with skin. Several hours may pass before the solution penetrates the skin sufficiently to cause redness or a burning sensation.

Wearing clothing which may have absorbed small amounts of hydrofluoric acid (such as leather shoes or gloves) can result in painful delayed effects similar to those caused by dilute acid solutions.

Hydrofluoric acid vapor causes skin irritation and inflammation of the mucous membranes; the burns become apparent a few hours after exposure. Inhaling the vapor in high concentrations may cause lung damage (pulmonary edema).

The American Conference of Governmental Industrial Hygienists recommends a Threshold Limit Value of 3 parts (by volume) AHF vapor (hydrogen fluoride) per million parts air. This value refers to a time weighted concentration for a seven or 8 hr workday and 40-hr work week.

The 3 ppm figure is based on both experimental and occupational evidence; however, nosebleeds and sinus troubles have reportedly occurred among metal workers exposed to even lower concentrations of a fluoride or fluorine in air. Therefore, for protection against acute irritation, 3 ppm should be considered a ceiling limit.

Anyone who knows or even suspects he has come in contact with hydrofluoric acid should immediately seek first aid.

In the event of an accident, the plant nurse or physician should be called as soon as possible; however, all plant supervisors should be aware of first aid procedures for HF burns. All affected persons should be referred to a physician even when the injury seems slight.

The amines used in gas treating

Amine solvents are used in petroleum and natural gas refining to remove hydrogen sulfide from the various streams. In petroleum refining, the monoethanolamine

Table 15.2. Properties of amines used in petroleum refining

Type	MEA	DEA	DGA	ADIP	SULFINOL
Mole weight	61.1	105.1	105.14	133.19	120.17
Boiling Pt °F	338.5	515.1	405.5	479.7	545
Boiling range 5–95% °F	336.7–341.06	232–336.7	205–230	–	–
Freezing Pt °F	50.5	77.2	9.5	107.6	81.7
S.G. @ 77°F	1.0113	1.0881	1.0572	–	1.256
Visc @ 77°F	18.95 cps	351.9 cps	40 cps	870 cps	12.1 cps
140°F	5.03 cps	–	6.8 cps	86 cps	4.9 cps
Flash Pt °F	200	295	260	255	350
Fire Pt °F	205	330	285	275	380

(MEA) compound of the homologue is the more common in the treating processes. Diethanolamine (DEA) however is a close second with proprietary compounds of amine such as Sulfinol and ADIP following closely in usage. All of these amines however are similar in respect to their hazard in the refining operation, their handling and their effect on people's health. For the purpose of this work, only MEA will be considered here. However as a point of reference Table 15.2 gives the physical properties of those amines used in petroleum refining (see also Chapter 10). MEA is corrosive and a combustible liquid and requires special handling and personnel protection considerations.

Personal safety

All Amines are injurious to personnel. The effects of exposure to MEA are as follows:

Target organs: Kidneys, central nervous system, liver.

Potential health effects

The eyes: MEA causes severe eye irritation and burning.

The skin: May be absorbed through the skin in harmful amounts. Causes moderate skin irritation.

Ingestion: Causes gastrointestinal tract burns.

Inhalation: Inhalation of high concentrations may cause central nervous system effects. This is characterized by headaches, dizziness, unconsciousness, and coma. Also causes respiratory tract irritation.

Chronic: May cause liver and kidney damage.

Safe handling

MEA is transported by road or rail tanker in its concentrated form. It is transferred in the normal way to an onsite storage bullet, which is blanketed by an inert gas. MEA is degraded on exposure to air. The use of this, and other enthanolamines, in the refinery processes is in a dilute form. This dilution and its onsite storage is very often in a suitably constructed pit usually in the proximity of the user plant. In some cases, a cone roof tank may be used for onsite storage. In all cases, though, the product must be kept free from exposure to air by inert gas blanketing. The dilution of MEA for use in the refinery process is between 15 and 20 wt%. The water for this dilution is usually treated boiler feed water, which is essentially free of impurities.

Vessels and piping in the process in which the amine is used should be of a suitable grade of carbon steel. The temperatures at which the amine is exposed to is below 300°F throughout the process. Personnel likely to be exposed to the amine must wear protective clothing, including eye pieces. As in the case of handling AHF, the minimum protection for operating and maintenance personnel should be:

- Coveralls, with sleeves to the wrist
- Chemical safety goggles
- Hard hat
- Gauntlets (polyvinyl chloride)
- Standard safety footwear

In addition to the standard, protective clothing listed certain operating and maintenance work requires the use of a respirator. Such an instance would be in the changing of the amine filter (see Chapter 10) cartridge. Although the filter will have been steam cleaned prior to opening the filter vessel respiratory protection is essential until the presence of amine and H_2S is certified to be absent. The certification is established by gas testing and the special processes to determine the absence of the sulfide (Lead Acetate test).

Equipment and piping

As is the case with all alkaline substances Amines cause stress corrosion. Consequently, all vessels and piping (welds) are stress relieved. Valves and piping are in carbon steel as are pumps and heat exchanger tubes.

Caustic soda

Caustic soda solution is used in oil refining mostly for the absorption of hydrogen sulfide or light mercaptans from light petroleum products from LPG thro the Kerosene cut. Very often the compound is delivered to the refinery in a strong aqueous solution to be further diluted on site to the strength required by a specific process. Sodium hydroxide in a solution is a white, odorless, non-volatile solution. It will not burn but it is highly reactive. It can react violently with water and numerous commonly encountered materials, generating enough heat to ignite nearby combustible materials. Contact with many organic and inorganic chemicals may cause fire or explosion. Reaction with metals releases flammable hydrogen gas.

Sodium hydroxide is produced mainly in three forms: 50% and 73% aqueous solutions, and anhydrous sodium hydroxide in the form of solid cakes, flakes or beads. The major impurities include sodium chloride, sodium carbonate, sodium sulfate, sodium chlorate, potassium, and heavy metals such as iron and nickel. This record reviews the information relevant to solutions. Chemical data are as follows:

Molecular weight:	40.00
Melting point:	12°C (53.6°F) (50% solution; freezing point); 62°C (143.6°F) (70–73% solution)
Boiling point:	140°C (284°F) (50% solution)
Specific gravity:	1.53 (50% solution); 2.0 at 15.5°C (70–73% solution)
Solubility in water:	Soluble in all proportions.
Solubility in other liquids:	Soluble in all proportions in ethanol, methanol, and glycerol
pH value:	12 (0.05% solution); 13 (0.5% solution); 14 (5% solution).

Hazards associated with handling sodium hydroxide

Skin. Sodium hydroxide is extremely corrosive and is capable of causing severe burns with deep ulceration and permanent scarring. It can penetrate to deeper layers of skin and corrosion will continue until removed. The severity of injury depends on the concentration of the solution and the duration of exposure. Burns may not be immediately painful; onset of pain may be delayed minutes to hours. Several human studies and case reports describe the corrosive effects of sodium hydroxide. A 4% solution of sodium hydroxide, applied to a volunteer's arm for 15–180 min, caused damage which progressed from destruction of cells of the hard outer layer of the skin within 15 min to total destruction of all layers of the skin in 60 min. Solutions as weak as 0.12% have damaged healthy skin within 1 hr. Sodium hydroxide dissolved the hair and caused reversible baldness and scalp burns when a concentrated solution (pH 13.5) dripped onto a worker's head and treatment was delayed for several hours.

Eyes. Sodium hydroxide is extremely corrosive. The severity of injury increases with the concentration of the solution, the duration of exposure, and the speed of penetration into the eye. Damage can range from severe irritation and mild scarring to blistering, disintegration, ulceration, severe scarring, and clouding. Conditions which affect vision such as glaucoma and cataracts are possible late developments. In severe cases, there is progressive ulceration and clouding of eye tissue which may lead to permanent blindness.

Inhalation. A worker, exposed for 2 hr daily over 20 years to mists from boiling a solution of sodium hydroxide in two large containers in a small room with inadequate ventilation, developed severe obstructive airway disease. It was concluded that the massive and prolonged exposure induced irritation and burns to the respiratory system eventually leading to the disease. It was noted that chronic exposure had not previously been reported, probably since the strong and immediate irritation would normally deter workers from further exposure. Actual exposures to sodium hydroxide aerosols were not measured and it could not definitely exclude late onset asthma as a cause of the man's condition.

A report of workers exposed to sodium hydroxide aerosol for at least 16 months, was confounded by the presence of high concentrations of Stoddard solvent and other solvent vapors, as well as other chemicals.

There was no trend of increased mortality in relation to duration (up to 30 years) or intensity of exposure (0.5–1.5 mg/m^3) among 291 workers exposed to sodium hydroxide dust during the production of flakes or beads of concentrated sodium hydroxide from chlorine cell effluent. This study is limited by the small population size.

Ingestion. There are no reported cases of industrial workers ingesting sodium hydroxide solutions. Non-occupational ingestion has produced severe corrosive burns to the esophageal tissue, which has in some cases progressed to stricture formation. Should ingestion occur, severe pain; burning of the mouth, throat, and esophagus; vomiting; diarrhea; collapse, and possible death may result.

Long term effects. Sodium hydroxide has been implicated as a cause of cancer of the esophagus in individuals who have ingested it. The cancer may develop 12–42 years after the ingestion incident. Similar cancers have been observed at the sites of severe thermal burns. These cancers may be due to tissue destruction and scar formation rather than the sodium hydroxide itself.

A case-control study reported an association between renal cancer and history of employment in the cell maintenance area of chlorine production. The major exposures in this work were presumed to be to asbestos and sodium hydroxide. An association was made between renal cancer and sodium hydroxide exposure. This study is limited by factors such as small numbers of exposed workers, multiple exposures, reliance on work histories and is not considered sufficiently reliable.

Skin. Owing to its corrosive nature, repeated or prolonged skin contact would be expected to cause drying, cracking, and inflammation of the skin (dermatitis).

First aid and personal safety

Contact with the skin. Avoid direct contact with this chemical. Wear chemical resistant protective clothing, if necessary. As quickly as possible, remove contaminated clothing, shoes, and leather goods (e.g., watchbands, belts). Flush contaminated area with lukewarm, gently flowing water for at least 60 min, by the clock. DO NOT INTERRUPT FLUSHING. If necessary, keep emergency vehicle waiting. Transport victim to an emergency care facility immediately. Discard contaminated clothing, shoes, and leather goods.

Contact with the eyes. Avoid direct contact. Wear chemical resistant gloves, if necessary. Quickly and gently blot or brush away excess chemical. Immediately flush the contaminated eye(s) with lukewarm, gently flowing water for at least 60 min, by the clock, while holding the eyelid(s) open. Neutral saline solution may be used as soon as it is available. DO NOT INTERRUPT FLUSHING. If necessary, keep emergency vehicle waiting. Take care not to rinse contaminated water into the unaffected eye or onto the face. Quickly transport victim to an emergency care facility.

Suffering from inhalation. Avoid direct contact. Wear chemical resistant gloves, if necessary. Quickly and gently blot or brush away excess chemical. Immediately flush the contaminated eye(s) with lukewarm, gently flowing water for at least 60 min, by the clock, while holding the eyelid(s) open. Neutral saline solution may be used as soon as it is available. DO NOT INTERRUPT FLUSHING. If necessary, keep emergency vehicle waiting. Take care not to rinse contaminated water into the unaffected eye or onto the face. Quickly transport victim to an emergency care facility.

Ingestion. DO NOT INDUCE VOMITING! Give large quantities of water or milk if available. Never give anything by mouth to an unconscious person. Get medical attention immediately.

Protective clothing

Personal respirators (NIOSH Approved). If the exposure limit is exceeded and engineering controls are not feasible, a half-face piece particulate respirator (NIOSH type N95 or better filters) may be worn for up to 10 times the exposure limit or the maximum use concentration specified by the appropriate regulatory agency or respirator supplier, whichever is lowest. A full-face piece particulate respirator (NIOSH type N100 filters) may be worn up to 50 times the exposure limit, or the maximum use concentration specified by the appropriate regulatory agency, or respirator supplier, whichever is lowest. If oil particles (e.g., lubricants, cutting fluids, glycerin, etc.) are present, use a NIOSH type R or P filter. For emergencies or instances where the exposure levels are not known, use a full-face piece positive-pressure, air-supplied respirator. WARNING: Air-purifying respirators do not protect workers in oxygen-deficient atmospheres.

Skin Protection. Wear impervious protective clothing, including boots, gloves, lab coat, apron or coveralls, as appropriate, to prevent skin contact.

Eye protection. Use chemical safety goggles and/or a full face shield where splashing is possible. Maintain eye wash fountain and quick-drench facilities in work area.

Materials of construction

Carbon steel can be used throughout at moderate temperatures. At temperatures in excess of 350°F nickel or nickel alloy is recommended. All carbon steel piping, flanges, welds, and vessel must be stress relieved. Caustic soda solution also attacks glass and dissolves it to some extent. As caustic soda is used in cleaning process plants during commissioning site, glasses and level gauges are removed and replaced with silica ones for this cleaning phase of the commissioning.

Furfural

Furfural is used in petroleum refining for the production of lube oils. It is a solvent in the extraction of undesirable compounds such as naphthenes and aromatics from lube oil stocks to improve the color of the lube oil product (see Chapter 12). Furfural or furfuraldehyde, C_4H_3OCHO, is a viscous, colorless liquid that has a pleasant aromatic odor; which upon exposure to air it turns dark brown or black. It boils at about 160°C. It is commonly used as a solvent; it is soluble in ethanol and ether and somewhat soluble in water. Furfural is the aldehyde of pyromucic acid; it has properties similar to those of benzaldehyde. A derivative of furan, it is prepared commercially by dehydration of pentose sugars obtained from cornstalks and corncobs, husks of oat and peanut, and other waste products. It is used in the manufacture of pesticides, phenolfurfural resins, and tetrahydrofuran. Tetrahydrofuran is used as a commercial solvent and is converted in starting materials for the preparation of nylon. Its chemical properties are as follows:

Appearance:	Colorless to yellowish liquid.
Odor:	Almond odor.
Solubility:	8 gm/100 gm water @ 70°F.
Specific gravity:	1.16 @ 77°F
% Volatiles by volume @ 70°F:	100
Boiling point:	324°F
Melting point:	–38°F
Vapor density (Air = 1):	3.3
Vapor pressure (mm Hg):	1 @ 64°F

Hazards associated with handling furfural

Inhalation. Causes irritation to the mucous membranes and upper respiratory tract. Symptoms may include sore throat, labored breathing, and headache. Higher concentrations act on the central nervous system and may cause lung congestion. Inhalation may be fatal.

Ingestion. Highly toxic. May cause gastrointestinal disorders. Can cause nerve depression and severe headache. May be fatal. Other effects are not well known.

Skin contact. Irritant to skin. May cause dermatitis and possibly eczema, allergic sensitization, and photosensitization. May be absorbed through the skin with possible systemic effects.

Eye contact. Vapors irritate the eyes, causing tearing, itching, and redness. Splashes may cause severe irritation or eye damage.

Chronic exposure. Can cause numbness of the tongue, loss of sense of taste, headache. Other effects are not well-known.

Aggravation of pre-existing conditions. Persons with pre-existing skin disorders or eye problems, or impaired liver, kidney or respiratory function may be more susceptible to the effects of the substance.

First aid and personal safety

Inhalation. Remove to fresh air. If not breathing, give artificial respiration. If breathing is difficult, give oxygen. Get medical attention immediately.

Ingestion. If swallowed, give large quantities of water to drink and get medical attention immediately. Never give anything by mouth to an unconscious person.

Skin contact. Immediately flush skin with plenty of soap and water for at least 15 min while removing contaminated clothing and shoes. Get medical attention, immediately. Wash clothing before reuse. Thoroughly clean shoes before reuse.

Eye contact. Immediately flush eyes with plenty of water for at least 15 min, lifting lower and upper eyelids occasionally. Get medical attention immediately.

Protective clothing and equipment

Airborne exposure limits. Between 2 and 5 ppm.

Ventilation system. A system of local and/or general exhaust is recommended to keep employee exposures below the Airborne Exposure Limits. Local exhaust ventilation is generally preferred because it can control the emissions of the contaminant at its source, preventing dispersion of it into the general work area. Please refer to the ACGIH document, *Industrial Ventilation, A Manual of Recommended Practices*, most recent edition, for details.

Personal respirators (NIOSH approved). If the exposure limit is exceeded, a full-face piece respirator with organic vapor cartridge may be worn up to 50 times the exposure limit or the maximum use concentration specified by the appropriate regulatory agency or respirator supplier, whichever is lowest. For emergencies or instances where the exposure levels are not known, use a full-face piece positive-pressure, air-supplied respirator. WARNING: Air purifying respirators do not protect workers in oxygen-deficient atmospheres.

Skin protection. Wear impervious protective clothing, including boots, gloves, lab coat, apron or coveralls, as appropriate, to prevent skin contact.

Eye protection. Use chemical safety goggles and/or a full face shield where splashing is possible. Maintain eye wash fountain and quick-drench facilities in work area.

Fire prevention and fighting

Fire

Flash point: 140F Pensky Marten.
Autoignition temperature: 601°F
Flammable limits in air % by volume:
lel: 2.1; uel: 19.3
Flammable Liquid and Vapor!

Explosion. Above flash point, vapor-air mixtures are explosive within flammable limits noted above. Reacts violently with oxidants. Reacts violently with strong acids and bases causing fire and explosion hazards. Sealed containers may rupture when heated. Sensitive to static discharge.

Fire extinguishing media. Water spray, dry chemical, alcohol foam, or carbon dioxide. Water spray may be used to keep fire exposed containers cool. Water may be used to flush spills away from exposures and to dilute spills to non-flammable mixtures.

Special information. In the event of a fire, wear full protective clothing and NIOSH-approved self-contained breathing apparatus with full-face piece operated in the pressure demand or other positive pressure mode.

Materials of construction and storage

There are no specific materials of construction, a suitable grade of carbon steel is adequate. Storage and handling of furfural must exclude air. Furfural polymerizes readily on exposure to air. In refinery practices, start up of furfural extraction plants usually requires that the equipment which handles furfural be first filled with either the lube oil feed or a suitable middle distillate to eliminate air.

Hydrogen sulfide H_2S

Hydrogen sulfide in a refinery is usually formed during the desulfurizing processes used to sweeten distillate product streams. Hydrogen sulfide (H_2S) is a colorless, extremely poisonous gas that has a very disagreeable odor, much like that of rotten eggs. It is slightly soluble in water and is soluble in carbon disulfide. Dissolved in water, it forms a very weak dibasic acid. Hydrogen sulfide is flammable, and in excess air it burns to form sulfur dioxide and water, where not enough oxygen is present, it forms elemental sulfur and water. It may be made by reacting hydrogen gas with molten sulfur or with sulfur vapors, or by treating a metal sulfide (e.g., ferrous sulfide, FeS) with an acid. Hydrogen sulfide reacts with most metal ions to form sulfides; the sulfides of some metals are insoluble in water and have characteristic colors that help to identify the metal during chemical analysis. Hydrogen sulfide also reacts directly with silver metal, forming a dull, gray-black tarnish of silver sulfide (Ag_2S). One method of detecting small concentrations of hydrogen sulfide is to expose it to a filter paper impregnated with lead acetate. The paper turns black (due to the precipitation of lead sulfide. The degree of H_2S concentration is measured by the shade of 'Blackness' of the lead acetate paper compared with standard colors. The chemical data on hydrogen sulfide is given in Table 15.3.

Table 15.3. Properties of hydrogen sulfide

Chemical symbol	H_2S
Relative density	1.189 (air = 1.0)
Autoignition temperature	260°C
Flammability	Very flammable
Lower explosive limit	4.3% in air by volume
Upper explosive limit	46% in air by volume
Color	Colorless, invisible
Odor	Strong rotten egg
Vapor pressure	17.7 atm at 20°C
Boiling point	−60°C
Melting point	−83°C
Reactivity	Dangerous with acids and oxidizers
Solubility	Yes in water, hydrocarbons, alcohol

Quoted from National Safety Council Data Sheet 1-284-67.

The hazards and toxicity of hydrogen sulfide

Table 15.4 indicates the toxicity of hydrogen sulfide.

At 1 ppm, most people can smell the gas. A strong smell does not necessarily mean a high concentration and a slight smell does not mean a low concentration. A person could work in a 10 ppm concentration of H_2S for 8 hr. If the concentration exceeds 10 ppm for a short period of time, then the time must be reduced.

A concentration of 15 ppm can be tolerated for a period of time not exceeding 15 min. There can be no more than 4 exposures of 15 ppm in an 8 hr shift with 1 hr

Table 15.4. The toxicity of hydrogen sulfide

PPM	Percent	Comment
1	0.0001	Most people can smell the gas
10	0.001	Occupational exposure limit. Maximum continuous exposure for 8 hr
15	0.0015	Occupational exposure for 15 min
20	0.002	Ceiling occupational exposure limit. This level of exposure cannot be exceeded at any time without respiratory protection
100	0.01	Dulls sense of smell. Causes burning sensation in the eyes and throat
500	0.05	Attacks the respiratory center of brain; causes loss of reasoning and balance
700	0.07	Victim quickly loses consciousness; breathing will stop, and death will result if not rescued promptly
1,000	0.1	Unconscious immediately; permanent brain damage or death occurs if victim is not rescued and resuscitated immediately

between exposures. If the concentration of H_2S exceeds 20 ppm, a worker must wear approved breathing apparatus. If the concentration is not known, a worker must wear breathing apparatus until the concentration is determined.

If exposed to a concentration of 100 ppm (1/100 of 1%), the sense of smell will be lost or become ineffective within 2–15 min. The H_2S might cause a burning sensation to the eyes, throat, and lungs, and could cause headache or nausea.

A 200 ppm concentration will cause immediate loss of smell and a burning sensation in the eyes, throat, nose, and lungs. (The hydrogen sulfide combines with alkali in body fluids to form caustic sodium sulfide.)

At a concentration of 500 ppm, the victim will appear to be intoxicated, and will lose his sense of balance and reasoning. In this state, the victim may attempt to continue with the job he was doing when he encountered the gas. For this reason, a person *must* know the people he works with, and be able to detect any unusual behavior of a coworker. Obviously, persons under the influence of alcohol, or any other mind-altering drugs, should never be allowed in an area which may contain sour gas. A victim must be watched very closely and may require resuscitation. A victim should be taken for medical attention and not allowed to return to work for at least 8 hr.

At 700 ppm, the victim will be rendered unconscious very quickly, and may develop seizures similar to those caused by epilepsy. Loss of bladder and bowel control can be expected. Breathing will stop, and death will result, if not rescued and resuscitated promptly. At a concentration of 1,000 ppm (1/10 of 1%) the victim will be rendered unconscious immediately. THE VICTIM WILL NOT BEGIN BREATHING VOLUNTARILY IF BROUGHT TO FRESH AIR. ARTIFICIAL RESUSCITATION MUST BE COMMENCED WITHIN THREE MINUTES OF EXPOSURE TO HYDROGEN SULPHIDE!

Protective clothing and personal safety

Much of the personal protection regarding hydrogen sulfide has been discussed in item 1 above. Essentially when working in an area which is or may be exposed to hydrogen sulfide an appropriate respirator must be at hand if not actually worn. In confined areas such as enclosed compressor or pump houses which handle sour gas or liquids constant monitoring for H_2S concentration in the atmosphere must be made. Preferably this will be accomplished by an automatic air analyzer with an alarm attachment. Failing this a routine analysis using a lead acetate paper should be made. In addition all such buildings should always be properly vented using an exhauster fan system.

Materials of construction in H_2S service

Atmospheres containing hydrogen sulfide and completely or almost free of oxygen give rise to rapid corrosion of unalloyed steel by forming a sulfide film on its surface. The corrosion rate in hydrogen/hydrogen sulfide increases as the content of hydrogen sulfide increases up to about 5 vol%, while increases beyond that point generally only have a slight effect on the corrosion rate. Steels alloyed with chromium and aluminum have improved resistance to hydrogen sulfide, while nickel has no deleterious effect.

Moist and aqueous solution of hydrogen sulfide cause some minor pitting in unalloyed steel and there is a risk of stress corrosion. This pitting corrosion is about 1 mm/year. This rate can be considerably reduced to about 0.1%/year using an alloyed steel of 18% chrome and 9% nickel. Vessels and piping should all be stress relieved.

Methyl ethyl ketone

Methyl ethyl ketone (MEK) is used in oil refining for the removal of wax from lube oil stock (see Chapter 12). Methyl ethyl ketone is a colorless liquid with a sweet/sharp, fragrant, acetone-like odor. It is extremely flammable in both the liquid and vapor phase. The vapor is heavier than air and may spread long distances and distant ignition and flashback are possible. MEK is highly volatile. Its chemical data is given in Table 15.5. MEK has a boiling point of 176°F and a vapor pressure of 3 ins Hg at 68°F.

Hazards associated with MEK

Inhalation. Causes irritation to the nose and throat. Concentrations above 200 ppm may cause headache, dizziness, nausea, shortness of breath, and vomiting. Higher concentrations may cause central nervous system depression and unconsciousness.

Ingestion. May produce abdominal pain, nausea. Aspiration into lungs can produce severe lung damage and is a medical emergency. Other symptoms expected to parallel inhalation.

Table 15.5. Chemical data for methyl ethyl ketone

Appearance	Clear, colorless, stable liquid
Purity, % minimum	99.5
Water content, % maximum	0.30
Acidity, % maximum (as acetic acid)	0.003
Color, Pt–Co maximum	10
Specific gravity, 20/20°C	0.805–0.807
Non-volatile matter (g/100 ml), maximum	0.002

Skin contact. Causes irritation to skin. Symptoms include redness, itching, and pain. May be absorbed through the skin with possible systemic effects.

Eye contact. Vapors are irritating to the eyes. Splashes can produce painful irritation and eye damage.

Chronic exposure. Prolonged skin contact may defat the skin and produce dermatitis. Chronic exposure may cause central nervous system effects.

Aggravation of pre-existing conditions. Persons with pre-existing skin disorders or eye problems or impaired respiratory function may be more susceptible to the effects of the substance.

First aid and personal protection

Inhalation. Remove to fresh air. If not breathing, give artificial respiration. If breathing is difficult, give oxygen. Get medical attention.

Ingestion. Aspiration hazard. If swallowed, vomiting may occur spontaneously, but DO NOT INDUCE. If vomiting occurs, keep head below hips to prevent aspiration into lungs. Never give anything by mouth to an unconscious person. Call a physician immediately.

Skin contact. Immediately flush skin with plenty of soap and water for at least 15 min while removing contaminated clothing and shoes. Get medical attention. Wash clothing before reuse. Thoroughly clean shoes before reuse.

Eye contact. Immediately flush eyes with plenty of water for at least 15 min, lifting upper and lower eyelids occasionally. Get medical attention.

Clothing and protective equipment

Airborne exposure limits.

Permissible exposure limit (PEL): 200 ppm (TWA)
Threshold limit value (TLV): 200 ppm (TWA), 300 ppm (STEL)

Ventilation system. A system of local and/or general exhaust is recommended to keep employee exposures below the Airborne Exposure Limits. Local exhaust ventilation is generally preferred because it can control the emissions of the contaminant at its source, preventing dispersion of it into the general work area.

Personal respirators. If the exposure limit is exceeded and engineering controls are not feasible, a full-face piece respirator with organic vapor cartridge may be worn up to 50 times the exposure limit or the maximum use concentration specified by the appropriate regulatory agency or respirator supplier, whichever is lowest. For emergencies or instances where the exposure levels are not known, use a full-face piece positive-pressure, air-supplied respirator. WARNING: Air purifying respirators do not protect workers in oxygen-deficient atmospheres.

Skin protection. Wear impervious protective clothing, including boots, gloves, lab coat, apron or coveralls, as appropriate, to prevent skin contact. Butyl rubber is a suitable material for personal protective equipment.

Eye protection. Use chemical safety goggles and/or a full face shield where splashing is possible. Maintain eye wash fountain and quick-drench facilities in work area.

Fire prevention and fighting

Fire

Flash point: −9°C (16°F) CC
Auto-ignition temperature: 404°C (759°F)
Flammable limits in air % by volume:
  lel: 1.4; uel: 11.4
Extremely Flammable.

Explosion. Above flash point, vapor-air mixtures are explosive within flammable limits noted above. Vapors can flow along surfaces to distant ignition source and flash back. Contact with strong oxidizers may cause fire. Sealed containers may rupture when heated. Sensitive to static discharge.

Fire extinguishing media. Dry chemical, foam or carbon dioxide. Water spray may be used to keep fire exposed containers cool, dilute spills to non flammable mixtures, protect personnel attempting to stop leak and disperse vapors.

Special information. In the event of a fire, wear full protective clothing and NIOSH-approved self-contained breathing apparatus with full-face piece operated in the pressure demand or other positive pressure mode. This highly flammable liquid must be kept from sparks, open flame, hot surfaces, and all sources of heat and ignition.

Accidental release measures. Ventilate area of leak or spill. Remove all sources of ignition. Wear appropriate personal protective equipment as specified in Section 8. Isolate hazard area. Keep unnecessary and unprotected personnel from entering. Contain and recover liquid when possible. Use non-sparking tools and equipment. Collect liquid in an appropriate container or absorb with an inert material (e.g., vermiculite, dry sand, earth), and place in a chemical waste container. Do not use combustible materials, such as saw dust. Do not flush to sewer! If a leak or spill has not ignited, use water spray to disperse the vapors, to protect personnel attempting to stop leak, and to flush spills away from exposures.

Storage and handling

MEK is usually delivered to a refinery by road or rail truck. It maybe stored in small bullets or a cone roof tank under an inert gas. The materials of construction is an appropriate grade of carbon steel.

15.2 Fire prevention and fire fighting

Fire prevention and fire protection is a paramount requisite in the operation of any hydrocarbon facility. It is more important perhaps in oil refining than any other related facility because of the relative size of most refineries compared with petrochemical or chemical facilities. Refinery prevention and protection begins at the early stages of the refinery design, and engineering. This section of the chapter begins with the refinery company's development of the design and engineering specification, which is the document that instructs the Engineering and Construction company in details of the refinery standards that are to be implemented in the building of the refinery. This section begins with a list of the items usually contained in the design specification. It continues with more details of those items that pertain to fire prevention and fire fighting.

The design specification

Client companies each have their specific design specification in terms of format and indeed details of the various specifications. However most design specifications will contain the following subject as a minimum.

The duty specification. This is of interest to the design process and mechanical engineer. It states exactly what facilities are to be built and the duty of each item in terms of throughput and in some cases composition of the various streams. It will also give specific detail of the local meteorological data and the parameters to be considered in economic decisions. The duty specification will include the products required and the composition of the products. It will also give full details of utilities required or available for the process. In cases of 'grass roots' facilities the duty specification will cover off sites such as tankage, blending, loading, and unloading facilities, etc.

Mechanical specification. This deals with the requirement and standards (including the codes to be used) the client requires for the equipment that will be installed. The narrative specifications that the mechanical engineer will produce based on this, will form part of the package to be used for procuring these items of equipment. These equipment items will be at least the following:

- Pumps
- Compressors and Turbines
- Heat Exchangers (including Air coolers, double pipe, etc.)
- Fired Heaters
- Other miscellaneous equipment such as ejectors, blenders, and the like

The electrical and instrument specification. This part will set the standards to be used for all electrical equipment, and the bulk materials associated with the equipment. It will deal with the 'Area Classification Code' which sets the parameters for equipment in terms of fire proofing (i.e., whether the item is to be spark proof, etc.) that will be located in the various areas of the refinery. For instrumentation it specifies the type of instruments to be used and the basic control and measuring systems to be used.

Piping and layout specification. This is usually the largest section of a design specification. It will detail the piping codes to be used and the material break points. It will proceed to establish the criteria for equipment and tankage layout with respect to:

- Maintenance accessibility
- Fire prevention (e.g., distance of fired heaters from other equipment)
- Tank area layout and size of tank bunds
- Degree of fire protection piping for plant units and tanks
- Under ground piping corrosion protection, etc.

Other sections of the design specification. These include detail requirements for vessels, civil, and structural equipment and associated materials.

Fire prevention with respect to equipment design and operation

Fired heaters. Design of fired heaters must incorporate snuffing steam facilities for the fire box. With respect to the maintenance and operation of the heater the following points should be included into operating procedures:

1. Implement a formal, regularly scheduled preventive maintenance program for burners and the clean-up of any refractory debris to prevent flame impingement.
2. Introduce a mechanism to monitor and record, more accurately and thoroughly, tube skin temperature to prevent hot spots.
3. Implement a thorough and complete procedure to deal with hot spots once detected and reported. This includes the recording of descriptions and locations of any suspected hot spot detected in a visual inspection.

4. Introduce an accurate method to determine the impact on tube life once a hot spot is detected.
5. The refinery personnel should be trained to recognize and respond to the changes in metallurgical properties and characteristics demonstrated by different material in the furnaces that are subjected to high temperature and pressure. Documentation to be updated to reflect this information.

Pressure vessels. These include towers, horizontal and vertical process vessels all of which must be protected properly with pressure relieving devices (see Chapter 13 for pressure relief facilities). In the design of vessels the correct wall thickness is based on the API and ASME codes covering the design temperatures and pressures. Fireproofing of vertical vessel skirts and horizontal vessel supports must be specified. In the operation and maintenance of vessels the following items must be included as standard procedures:

- Vessel alarms must be recognized and acted upon.
- Action must be taken immediately when excessively high temperatures and high or low liquid levels are observed.
- In the case of a fire on the unit or adjacent unit the towers must be shut down according to the plant's emergency shut down procedures.
- In the event of an emergency shut down the vessel must be purged free of hydrocarbons either by steam or inert gas.
- No entry or work must be done on the vessel before it is certified 'Gas Free'.
- All vessels must be inspected before re commissioning after a shut down. The inspection must ensure that all flanges are secure and the correct gaskets have been properly installed.
- All instruments and relief systems on the vessels must be checked and re-calibrated on every scheduled or unscheduled shut down of the plant.
- All vessels must be pressure tested before commissioning and after any work has been completed on them.
- All drains and vents on the vessel must be checked during scheduled plant 'turnaround' or un scheduled shut down before re commissioning the vessel.

Heat exchangers and coolers. Shell and tube heat exchangers are designed and fabricated to ASME and TEMA codes. Their shell design follows closely to that of vessels with respect to design temperatures and pressures. The thickness of tubes and tube sheets are also calculated using the ASME code. In the case of air coolers the tube sheets and tubes are also calculated to ASME codes as are the inlet and outlet manifolds. Air coolers tubes are usually finned to enhance the heat exchanger properties of the units. The operation and maintenance of shell and tube exchangers follow many of those conditions stated for vessels with respect to fire (or explosion) prevention. In the case of air coolers however some different rules and procedures apply. For example a number of air coolers in a refinery are installed above an elevated pipe rack. The pipe rack often runs through the center of a process plant, that is there will be equipment on both sides of the rack. It is usual for the pipe rack structure to be encased in a fire proof concrete as a fire prevention measure. This fireproofing should extend upwards to cover the air cooler structures as well. Fire water and foam sprays should be installed above the tube nests such as to snuff out any outbreak of tube leaks and subsequent fires.

Rotary equipment. This group includes all pumps, and compressors. Next to fired heaters the compressor units, which may include gas driven turbine or turbo expanders, are the items most vulnerable as fire hazards. The reason in this case is the high pressures of the gas they handle, and this is more so in those units handling hydrogen. All reciprocal compressors must have pressure relief facilities on the compressor discharge. Most also have an emergency shut down system in the case of high discharge pressure. Centrifugal compressors and their drivers require a more sophisticated emergency shut down procedures however. Here critical condition can arise by poor suction conditions as well as the high-pressure discharge. All centrifugal compressors therefore require protection against 'Surge' (or 'Pumping') condition. For example this condition can occur when the compressor suction pressure is so low that the unit cannot pull any of the feed gas into its impellers. Severe vibration of the unit follows which, in extreme cases, will cause considerable structural damage to the unit and even an explosion and fire. Most modern day units do have anti surge devices, which if properly maintained will close the unit down long before any severe damage.

Pumps, even those handling high-pressure liquefied petroleum gases are less hazardous. Overheating due to defective seals or packing is the more common hazard that can cause a fire. Pump cavitating due to NPSH problems being a much lesser second hazard.

In most present day refineries rotary equipment is installed in the open or in an area which has only a roof as a cover. This helps a great deal to fight any rotary equipment fire. Usually the area containing rotary equipment will have both fire water and foam facilities installed. Close monitoring of the rotary equipment during operation and regular maintenance of the items are essential for the prevention of fires from these sources.

Tanks and the tank farm. Tank fires are probably the worse type of fire in a refinery. This is because tanks hold a high inventory of hydrocarbons. The one thing that lowers the degree of hazard in the tank farm is that there is usually no direct source of ignition. There are no fired heaters or large compressor units present in tank farms as a rule. Fires can occur however from accidents remote from the area such as a fire on a jetty or a ship nearby. There was a tank farm fire in New York during February 2003, that was caused by an explosion aboard a barge unloading gasoline at a nearby jetty. Large fuel oil tanks on shore were set ablaze by hot debris from the explosion. Another source of fire is an explosion in a product loading station while loading a light product such as gasoline. These explosions are usually caused by static electricity. Most petroleum companies do have strict procedure in place to prevent such occurrences, but accidents do happen from time to time.

Middle distillates and fuel oil tanks present the most difficult fires to fight and extinguish. The large inventory of these tanks and their low volatility but high heating value can negate for quite a time even using the most up to date foam extinguishing techniques once the fire has really established itself. The major effort then is to keep adjacent tanks cool by extensive water spray. Fires on tanks storing lighter liquid products are a little easier to combat. Foam is used to smother the oil inventory. As these light products are stored in floating roof tanks, it provides the means to build a depth of foam on the roof itself directly on the surface of the liquid if the roof has been damaged by an earlier explosion. Fire water sprays are installed on LPG spheres and bullets. In almost every case the initial fire on LPG tankage causes an explosion. The material remaining can be snuffed out by foam and again the main purpose of the fire fighting is to activate the fire water sprays on adjacent tanks to keep them cool. Incidentally all refineries these days have in place fire water and foam spray rings on all storage tanks (see Chapter 13). Water on its own should only be used to keep the tank cool from adjacent fire sources. The danger of using water alone to extinguish a fire is that it may cause the fire to spread by the hydrocarbon continuing to burn on the surface of the water stream flowing from the tank area.

The design of storage tanks and their installation shall comply with the local Fire Regulations. In general crude oil and petroleum products with a flash point of $<130^\circ$F shall be stored in floating roof tanks, or in the case of LPG in bullets or spheres. Middle distillates and heavier shall be stored in cone roof tanks. Storage tanks, both floating roof or cone roof, shall be installed on concrete pads which will be piled if required. The space between tanks shall not be less than 10 ft (or according to local regulations). The tank area shall be dyked to hold at least 110% of a single tank inventory. This dyked or bunded area shall contain adequate drainage leading to the API separator or holding pond. LPG storage generally does not require a dyked area but both bullets and spheres shall be installed with proper fireproofed support structures.

Jetty and on shore loading stations. One of the biggest fire hazards in areas of loading rail/road cars and ship loading is fire by spark caused by static electricity. The proper earthing of the items being loaded is essential in preventing this fire hazard. The same applies to the unloading of hydrocarbons such as crude oil feed or other product streams transferred from other refineries or depots. In small refinery installations the jetty facilities and the on shore loading islands are protected by an extension of the off sites fire main and foam systems. In larger refineries the jetty may have its own fire main using sea water as the main water medium.

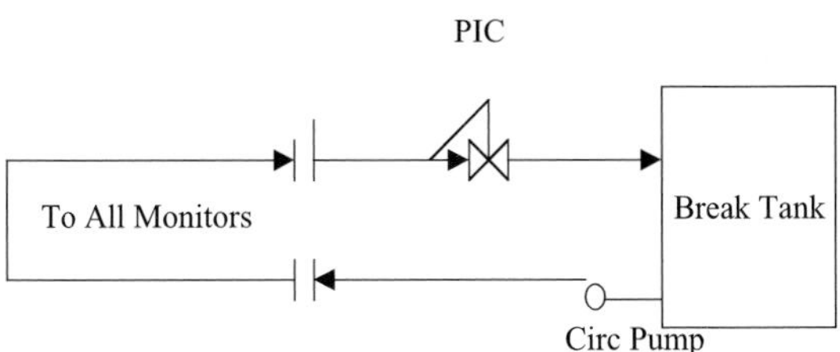

Figure 15.1. Schematic fire main system.

The fire main

Fire mains are usually designed as a pressurized circulating water system with an atmospheric break tank. A schematic sketch of such a system is given as Figure 15.1. The fire main is maintained at a pressure of around 100 psig by the circulating pump. This pump is automatically switched on or off to maintain the line pressure when monitors are opened to deliver the water to fire sources or foam system or for any other application for which they are opened to atmosphere. The pump is usually electric motor driven with a spare which is usually driven by a diesel motor or steam turbine. The spare pump drive is automatically activated on electrical failure and the inability of the normal fire circulating pump to maintain line pressure. Make up water to the break tank is maintained under break tank level control from the refinery's fresh (or salt water) main.

On smaller refinery facilities and smaller systems, such as a jetty fire main, the system may be 'dead ended'. That is the fire water pump would be used only to maintain the fire main pressure. The water would not be circulated to a break tank. A water source tank or pit would be used as the fire water source reservoir. In the case of the jetty fire main usually sea water is used and this would be delivered from a suitable seabed pit by a vertical pump. Again an electric motor driver would be used for the normal pump with usually a steam turbine driver for the spare pump.

Fire foam and foam systems

Fire fighting foam is simply a stable mass of small, air-filled bubbles with a lower density than oil, gasoline, or water. The foam is made up of three ingredients:

- Water;
- A foam concentrate; and
- Air

The water is mixed with the concentrate (proportioned) to form a foam solution. This solution is then mixed with air (aspirated) to produce the foam which is very fluid and flows readily over liquid surfaces.

Balanced pressure proportioning is the most common method used for foam system applications. The foam concentrate pressure is balanced with the water pressure at the proportioner inlets allowing the proper amount of foam concentrate to be metered into the water stream. With an aspirating discharge device, foam solution passes through an orifice, past air inlets, and into an expansion area to produce an expanded foam. With non-aspirating devices, foam solution passes through the orifice and discharge outlet where it mixes with air en route to the fire.

Fire fighting foam is used in a variety of applications to extinguish flammable and combustible liquid fires, to control the release of flammable vapors and to cool fuels and sources of ignition. Typical foam applications include:

- Loading racks
- Refineries
- Pumping stations
- Power plants
- Airports
- Heliports
- Marine applications
- Manufacturing plants
- Storage tanks
- Chemical plants
- Flammable liquid storage
- Offshore platforms
- Aircraft hangars
- Crash rescue vehicles
- Mining facilities
- Warehouses
- Hazardous material spill control.

Fire fighting foam agents suppress fire by separating the fuel from the air (oxygen). Depending upon the type of foam agent, this is done in several ways:

- Foam blankets the fuel surface, smothering the fire and separating the flames from the fuel surface
- The fuel is cooled by the water content of the foam
- The foam blanket suppresses the release of flammable vapors that can mix with air

A film forming foam (AFFF) agent forms an aqueous film on the surface of a hydrocarbon fuel. An alcohol-resistant concentrate (ARC) will form a polymeric membrane on a polar solvent fuel. Essentially there are six general types of foam agents:

Class B fire foams

Film-forming foams

Film-forming foams (AFFF) are based on combinations of fluorochemical surfactants, hydrocarbon surfactants, and solvents. These agents require a very low energy input to produce a high-quality foam. Consequently, they can be applied through a wide variety of foam delivery systems. This versatility makes AFFF an obvious choice for airports, refineries, manufacturing plants, municipal fire departments, and any other operation involving the transportation, processing, or handling of flammable liquids. These Foam Forming agents are available as 1% and 3% freeze-protected concentrates.

Alcohol-resistant concentrates

Alcohol-resistant concentrates are based on AFFF chemistry to which a polymer has been added. ARCs are the most versatile of the foam agents in that they are effective on fires involving polar solvents like methanol as well as hydrocarbon fuels like gasoline. When used on a polar solvent type fuel, the ARC concentrate forms a polymeric membrane which prevents destruction of the foam blanket. When used on hydrocarbon fuels, the ARC produces the same rugged aqueous film as a standard AFFF agent. ARCs provide fast flame knockdown and good burn back resistance when used on both types of fuels.

Protein Foam Concentrates

Protein foam concentrates are recommended for the extinguishment of fires involving hydrocarbons. They are based on hydrolyzed protein, stabilizers, and preservatives. Protein foams produce a stable mechanical foam with good expansion properties and excellent burn back resistance characteristics. Protein foam concentrates are available in 3% and 6% concentrations.

Fluoroprotein foam concentrates

Fluoroprotein foam concentrates are based on hydrolyzed protein, stabilizers, preservatives, and synthetic fluorocarbon surfactants. When compared to protein foams, fluoroproteins provide better control and extinguishment, greater fluidity, and superior resistance to fuel contamination. Fluoroprotein foams are useful for hydrocarbon vapor suppression and have been recognized as very effective fire suppressing agents for sub-surface injection into hydrocarbon fuel storage tanks. Fluoroprotein foam is available in a 3% concentrate.

High expansion foam concentrates

High expansion foam concentrates are based on combination of hydrocarbon surfactants and solvents. They are used with foam generators for applying foam to large areas in total flooding and three-dimensional applications such as warehouses, ship cargo holds, and mine shafts. They are especially useful on fuels such as liquefied natural gas (cryogenic fuels) for vapor dispersion and control. In certain concentrations, high expansion foams are effective on hydrocarbon spill fires of most types and in confined areas.

Class A fire foams

Class A foams are typically formulated from a combination of specialty hydrocarbon surfactants, stabilizers, inhibitors, and solvents. They reduce the surface tension of water for improved wetting and penetrating characteristics and create a clinging foam blanket that suppresses combustible vapors while cooling the fuel. Class A foams can be applied using a variety of proportioning/discharge devices and have proven to be highly effective for use in structural and forest firefighting, coal mines, tire, and rubber manufacturing, lumber mills, coal bunkers, paper warehouses, and other hazards involving ordinary combustible materials. Please note that Class B foams are acceptable for use on Class A fires, however they are not designed for use on Class A fires as such. The foam concentrate percentage refers to the amount of concentrate that is proportioned or pre-mixed with water to give the resulting foam solutions.

Environmental impact and toxicity. Most foam concentrates are formulated to maximize performance and minimize environmental impact and human exposure hazards. All concentrates are readily biodegradable—both in the natural environment and in sewage treatment facilities. However, all foam agents should be metered into the facility to prevent overloading the plant due to foam formation. They are not considered skin irritants; however, prolonged contact may cause some dryness of the skin. For this reason, we recommend that areas of the skin which have come in contact with the foam concentrate be flushed with fresh water.

Shelf life. Shelf life is the length of time over which foam concentrates remain stable without significant changes in performance characteristics. Many AFFF, high-expansion and Class A foam concentrates—if stored in accordance with recommended guidelines—have a normal shelf life of 20–25 years. Other foam agents—those which are not totally synthetic—have a normal shelf life of 7–10 years.

Chapter 16

Quality control of products in petroleum refining

D.S.J. Jones

Introduction

This chapter is concerned mostly with the laboratory testing for the control of petroleum products. It will be concerned with those tests that establish the quality of refinery streams and in some finished and saleable products. What is not covered are those specialized tests such as Mass Spectrometry, motor road tests and the like. Where possible sketches of test apparatus are included. Most of these tests are performed in the refinery laboratory, and generally follow the methods provided by the respective ASTM numbered tests.

The most common tests to check the quality of finished products cover the following:

- Specific gravity
- ASTM distillations
- Pensky Marten closed cup flash point
- Kinematic viscosities
- Octane numbers
- Sulfur content tests
- LPG weathering tests
- Reid vapor pressure
- Bromine number

The chapter begins with some details of typical product specifications. More discussion and description of modern day trends in product quality (such as the effect of Reformulated Gasoline) are given in Chapter 2 of this book.

16.1 Specifications for some common finished products

This chapter is concerned only with those energy products that are common to everyday living. They include, the LPGs, gasolines, burning oil (including kerosene), diesel fuel, jet fuels, and the black oils including fuel oil and marine diesel.

The LPG products

This category includes the normal specification for propane and butane LPG. They are given in Table 16.1.

The gasolines

Two grades of gasoline are given here. They represent present day and near future quality requirements meeting the environmental parameters already discussed in Chapter 2 of this volume. The two grades are:

Table 16.1. Specifications covering saleable LPG

Propane LPG (Note 1.0)		Method of test
Vapor pressure at 113°F, psig	255 max	IP 161
C_1 Hydrocarbons, mol%	0.1 max	ASTM 2163
C_2 Hydrocarbons, mol%	5.0 max	ASTM 2163
C_3 Hydrocarbons, mol%	95.0 min	ASTM 2163
C_4 and heavier hydrocarbons, mol%	4.0 max	ASTM 2163
Total unsaturated hydrocarbons, mol%	1.0 max	ASTM 2163
Total sulfur content, wt%	0.1 max	ASTM D1266
Mercaptan sulfur grains per 100 cuft @ STP	3.0 max	IP 104
H_2S content	Absent	IP 103
Butane LPG (Note 1.0)		
Vapor pressure at 113°F, psig	70–85	IP 161
C_1 Hydrocarbons, mol%	Nil	ASTM 2163
C_2 Hydrocarbons, mol%	0.5 max	ASTM 2163
C_5 Hydrocarbons (as nC_5), mol%	2.0 max	ASTM 2163
Total sulfur content, wt%	0.01 max	ASTM D1266
Mercaptan sulfur grains per 100 cuft @ STP	2.0 max	IP 104
H_2S content	Absent	IP 103

Note 1.0. The above products shall not contain harmful quantities of toxic or nauseating substances and shall be free from entrained water. The odor of the gases shall be distinctive, unpleasant and non persistent, and shall indicate the presence of the gas at concentrations in air down to 1/5th of the lower limit of inflammability.

- Regular grade
- Premium grade (100 octane)

The specifications given in Table 16.2 cover present day quality requirements. Note only the major quality items and tests are listed here.

Table 16.2. Gasoline quality requirements

	Regular gasoline	Premium gasoline	
Effective date: September 2002			
Research octane number (RON)	91.0 minimum	100 minimum	ASTM D2699
Motor octane number (MON)	82.0 minimum	91.0 minimum	ASTM D2700
Color	Off white	Orange	Visual
Distillation:			
vol% distilled @ 70°C			
Minimum	20	22	ASTM D86
vol% distilled @ 100°C		Not applicable	ASTM D86
Maximum	50	50	
Minimum	60	60	
vol% distilled @ 180°C			
Minimum	90	90	
End point °C	220°C maximum	210 maximum	ASTM D86
Residue vol%	2 maximum	2 maximum	ASTM D86
Vapor pressures			
Psig @ 100°F	◄——Summer 10 max 7.0 min——►		ASTM D323
	◄——Winter 13.5 max 8.5 min——►		
Copper strip corrosion (3 hr at 50°C)	◄——Class 1 maximum——►		ASTM D130
Sulfur wt%	◄——0.035 max——►		IP 336 or ASTM D5453
Existent gum (solvent washed) (mg/100 ml)	◄——4 maximum——►		ASTM D381
Oxidation stability induction period (minutes)	◄——360 minimum——►		ASTM D525
Other specifications for regular and premium grade gasoline are:			
Lead, mg ms/litre		5 maximum	IP 224
Benzene, vol%		4 maximum	ASTM D5580
Total aromatics		42 maximum (regular) 48 maximum (premium)	ASTM D 5580
Oxygenates, vol%		1 maximum (except ethanol)	ASTM D 4815
Ethanol, vol%		10% maximum	
Olefins, vol%		18 maximum	ASTM D 1319

Table 16.3. Specifications for kerosene products

	Odorless kerosene	Regular grade	Premium grade	Test methods
Specific gravity @ 15	0.820 max	0.820 max	0.820 max	ASTM D 4052
Appearance @ 15°C	◀———	Clear & Bright	———▶	Visual
Color saybolt	◀———	+ 20	———▶	ASTM D156
Odor	Merchantable	N/A	N/A	
Distillation IBP °C	◀———	175	———▶	ASTM D86
50%	240 max	235 max	N/A	
90%	210 min	N/A	N/A	
FBP	280 max	N/A	280 max	
Residue vol%	N/A	2.0 max	N/A	
Flash point °C	75 min (PMCC)	20 min (Abel)	43 min (Abel)	
Smoke point mm	N/A	20 min	35 min	IP 57
Sulfur content Wt%	0.01 max	0.2 max	0.04 max	IP 336
Copper 3 hrs 100°C	1 max	1 max	1 max	ASTM D 130
Aromatics vol%	1 max	N/A	N/A	
Doctor test	◀———	Negative	———▶	IP 30
Burning test mg/kg	N/A	N/A	10 max	

The kerosenes

There are three grades of kerosenes which are common in most refineries. These are:

- Odorless kerosene
- Regular grade kerosene
- Premium kerosene

Specifications for each of these are given in Table 16.3.

Aviation turbine gasoline (ATG) and jet fuels

This product is among the most important item produced in a refinery. Its quality must conform with stringent requirements particularly for use in commercial aircraft. In this chapter the description and discussion on ATG will be confined mainly to that for civil purposes, although reference will be made to the military grades of ATG. The specification that sets the standard for commercial jet fuel quality is the ASTM D 1655. A summary of the specification is given in Table 16.4.

IATA (International Air Transport Association) publishes a document for the guidance of international users of ATG. This document is titled *Guidance Material for Aviation Turbine Fuels Specifications.* This guidance contains specifications for four aviation turbine fuel types. Three of these are Kerosene type fuels (Jet A, Jet A1, and TS1) and one wide cut fuel Jet B. Jet A meets the ASTM requirements, Jet A1 meets the Joint Checklist requirements, TS1 meets the Russian GOST requirements, and Jet B meets the Canadian CGSB requirements. These specifications are summarized in Table 16.5. Jet A1 is listed as the Defense Standard 91–91 which is close to the UK JetA1 standard.

Table 16.4. Commercial jet fuel specification

Composition			
Acidity, total, mg KOH/g, max	0.10	–	D 3242
Aromatics, vol%, max	25	25	D 1319
Sulfur, mercaptan, % mass, max	0.003	0.003	D 3227
Sulfur, total, % mass, max	0.30	0.3	D 1266, D 1552, D 2622 D 4294, or D 5453
Volatility			
Distillation, °C			D 86
Volume percent recovered			
10, max	205	–	
20, max	–	145	
50, max	Report	190	
90, max	Report	245	
Final boiling point, max	300	–	
Distillation yields, vol%:			
Residue, max	1.5	1.5	
Loss, max	1.5	1.5	
Flash point, °C, min	38	–	D 56 or D 3828
Density, 15°C, kg/m^3	775 to 840	751 to 802	D 1298 or D 4052
Vapor pressure at 38°C, kPa, max	–	21	D 323 or D 5191
Fluidity			
Freezing point, °C, max	−40 (Jet A) −47 (Jet A-1)	−50	D 2386, D 4305, D 5501, or D 5972
Viscosity at −20°C, mm^2/sec max	8.0	–	D 445
Combustion			
Net heat of combustion, MJ/kg, min	42.8	42.8	D 4529, D 3338, or D 4809
One of the following requirements:			
1. Luminometer number, min	45	45	D 1740
2. Smoke point, mm, min	25	25	D 1322
3. Smoke point, mm, min and	18	18	D 1322
Naphthalenes, vol%, max	3.0	3.0	D 1840
Corrosion			
Copper strip, 2 hr. at 100°C, max	No. 1	No. 1	D 130
Stability			
Thermal stability, 2.5 hr. at 260°C:			
Filter pressure drop, mm Hg, max	25	25	D 3241
Tube deposit, less than Contaminants	Code 3	Code 3	
Existent gum, mg/100 ml, max	7	7	D 381
Water reaction, interface rating, max	1b	1b	D 1094
	Jet A or A1	Jet B	

Table 16.5. Summary of specifications for civil jet fuels

Specification	ASTM D 1655	DEF STAN 91–91	GOST 10227*	CGSB-3.22
Aromatics, vol%, max	25	25.0	22 (% mass)	25.0
Distillation, °C:				
Initial boiling point	–	Report	150	Report
10% recovered, max	205	205	165	Report
50% recovered, max	Report	Report	195	Min 110; max 190
90% recovered, max	Report	Report	230	245
End point	300	300	250	Report
Vapor pressure, kPa, max	–	–	–	21
Flash point, °C, min	38	38.0	28	–
Density, 15°C, kg/m^3	775–840	775–840	–	750–801
Density, 20°C, kg/m^3, min	–	–	775	–
Freezing point, °C, max	−40	−47.0	–	−51
Chilling point, °C, max	–	–	−50	–
Viscosity, −20°C, mm^2/sec, max	8	8.0	–	–
Viscosity, 20°C, mm^2/sec, min	–	–	1.25	–
Viscosity, −40°C, mm^2/sec, max	–	–	8.0	–

*Some of the GOST test methods are significantly different from those used in other specifications, so values for these properties may not agree.

Military jet fuel. The governments of the United States and many other countries maintain separate specifications for jet fuel for military use. The reasons for separate specifications include the operational and logistical differences between the military and civilian systems and the additional demands high performance jet fighter engines place on the fuel.

There are currently two fuels in widespread use by the US military. They are JP-5 used by the Navy, and JP-8 used by the Air Force. Both are kerosene type fuels and the major difference between them is their flash point temperature. The minimum flash point for JP-8 is 38°C while that for JP-5 is 60°C. The higher flash point for JP-5 is for safer handling of the fuel on aircraft carriers.

The gas oils

Three grades of gas oils are considered here. They are:

- Heating oil (regular grade gas oil)
- Automotive grade (diesel fuel)
- Marine diesel

Table 16.6. General specification for regular grade gas oil

Properties		Method of test
Specific gravity @ 60°F	0.820–0.845	ASTM D 1298
Color NPA	$1^1/_2$	ASTM D 155
Pour point		ASTM D 97
Summer, °F	14	
Winter, °F	5	
Cloud point		ASTM D 97
Summer, °F	25	
Winter, °F	20	
Sulfur, wt%	0.5	ASTM D 129
Diesel index	57 min	IP 21
Cetane number	50 min	ASTM D 613
Distillation:		ASTM D 158
Recovered at 230°C, vol%	10 min	
Recovered at 240°C, vol%	50 max	
Recovered at 300°C, vol%	70 min	
Recovered at 357°C, vol%	90 max	
FBP, °C	385 max	
Flash point (PM), °F	150 min	ASTM D 93
Copper strip (3 hr @ 100°C)	1 A	ASTM D 130
Viscosity @ 100°F SUS	37.5 max	ASTM D 88
Carbon residue (Condradson) on 10% Btms, wt%	0.1 max	ASTM D 189
Ash, wt%	0.01 max	ASTM D 482
Water by distillation, wt%	0.05 max	ASTM D 95
Sediment by extraction, wt%	0.01 max	ASTM D 473
Calorific value (Gross), Btu/lb	19,300 min	ASTM D 240

Heating oil is used predominately as domestic heating oil in many countries. Its most important properties are the pour point and sulfur content, A full specification of the gas oil is given as Table 16.6.

The specification for automotive grade gas oil (Diesel fuel) is given as Table 16.7. Certain properties of this product are more stringent than those for the heating oil to meet the present environmental requirements and the present design features of the diesel engine. The one item that stands out as being very stringent is the sulfur content of the diesel. This specification became effective in 2002. All indications are that it will be even more stringent in 2006 (see Chapter 2 for more details).

The third gas oil product is really a blend of the gas oil distillate and atmospheric residuum. It's color is black and has the following, much smaller specification. This is given as Table 16.8.

Table 16.7. The specification for diesel fuel

Properties		Method of test
Specific gravity @ 60°F	0.820–0.860	ASTM D 1298
Color NPA	3 max	ASTM D 155
Pour point		ASTM D 97
Summer, °F	15	
Winter, °F	7	
Cloud point		ASTM D 97
Summer, °F	N/A	
Winter, °F	N/A	
Sulfur, wt%	0.1*	ASTM D 129
Diesel index	54 min	IP 21
Cetane number	47 min	ASTM D 613
Distillation:		
Recovered at 230°C, vol%	10 min	ASTM D 158
Recovered at 240°C, vol%	50 max	
Recovered at 347°C, vol%	50 min	
Recovered at 370°C, vol%	95 max	
FBP, °C	385 max	
Flash point (PM), °F	14 0 min	ASTM D 93
Copper strip (3 hr @ 100°C)	1 max	ASTM D 130
Viscosity @ 100°F, SUS	35.9 max	ASTM D 88
Carbon residue (Condradson) on 10% Btms, wt%	0.25 max	ASTM D 189
Ash, wt%	0.1 max	ASTM D 482
Water by distillation, wt%	0.02 max	ASTM D 95
Sediment by extraction, wt%	0.005 max	ASTM D 473
Calorific value (Gross), Btu/lb	Not specified	ASTM D 240

(*) Down to 50 wt-ppm (0.005 wt%) for ultra.low.sulfur diesel

The fuel oil products

The single most common fuel oil marketed today is the No 6 grade. Of the lighter grades No 1, 2, 3 were distillate fuels (mostly regular grade kerosene and gas oil

Table 16.8. The specification for marine diesel fuel

Properties		Method of test
Specific gravity @ 60°F	0.840 min	ASTM D 1298
Pour point—Winter, °F	15 max	ASTM D 97
Sulfur, wt%	N/A	ASTM D 129
Diesel index	50 min	IP 21
Flash point (PM), °F	150 min	ASTM D 93
Viscosity @ 100°F SUS	32.0–43.9	ASTM D 88
Carbon residue (Condradson) on 10% Btms, wt%	0.2 max	ASTM D 189

Table 16.9. Specification for No 6 fuel oil product

Properties		Method of test
Specific gravity @ 60°F	0.990 max	ASTM D 1298
Pour point, °F	65 max	ASTM D 97
Sulfur, wt%	2.0 max*	ASTM D 129
Viscosity kinematic @ 122°F, Cs	370 max	ASTM D 445
Flash point (PM), °F	160 min	ASTM D 93
Ash, wt%	0.1 max	ASTM D 482
Water (by distillation), vol%	1.0 max	ASTM D 95
Sediment (by extraction), wt%	0.1 max	ASTM D 473

*Sulfur content is probably closer to 1.5% wt for most areas now.

types). These pretty much disappeared from the markets during the 1980s. No 5 fuel was a light residual fuel used mostly in the steel industry. It is no longer widely used. No 6 fuel is a heavy residual fuel normally a mixture of atmospheric and vacuum distillation residues cut back to adjust for viscosity with kerosene or gas oil cutter stock. This fuel oil product is almost entirely used as ship bunker oil, with some market outlet to stationary power plants in producing steam for their turbines. For the last 20 years or so these markets have been in decline. Power plants are converting to the use of coal or natural gas, and ships are leaning more to the use of marine diesel or just diesel fuel. Environmental restrictions are calling for a continual decrease in the sulfur content of this particular product. As a result no 6 fuel is the least valued refinery product and is worth less than the crude feed itself. Refiners are now installing processes (such as 'deep oil cracking', and residue hydrocracking) to upgrade this product to the more valued distillates such as gasoline and the middle distillates. The specification for a No 6 fuel oil is given as Table 16.9.

The lube oils

Most marketable lube oils produced in a refinery are graded by their viscosity. Proprietary brands of these lube oil grades will contain additives as provided by their particular marketing outlets. These basic grades and their specification are given in Table 16.10. Other specifications apply for multi-grade lube oils

These grades and their processing are described in some detail in Chapter 12.

The asphalts

The production of the asphalt grades and their specification has been described and discussed in Chapter 12 of this book.

Table 16.10. Specification for basic (single grade) lube oil grades

Grade	Max viscosity @ 0°F SUS	Max viscosity @ 210°F SUS	Min viscosity @ 210°F SUS
5 W	6,000	–	–
10 W	12,000	–	–
20 W	48,000	–	–
SAE 20	–	58	45
SAE 30	–	70	58
SAE 40	–	86	70
SAE 50	–	110	85

Petroleum coke

Petroleum coke is not found in crude oil, but is the carbon compound formed from the thermal conversion of petroleum containing resins and asphaltenes. Coke is formed in the processes to convert the residuum fuels to the more desirable distillate products of naphtha and lighter through to the middle distillates. There are two routes by which this coking process proceeds. The first and the most common is the delayed coking route. The second is the fluid coking method, and this has been made more attractive to many refiners with the development of Exxon Mobil's proprietary process of *Flexicoking*. This proprietary process eliminates the coke completely by converting it to low Btu fuel gas. By far the largest production of coke is the sponge coke from the delayed coking process. Uncalcined sponge coke has a heating value of about 14,000 Btu/lb and is used primarily as a fuel. High sulfur sponge coke however is popular for use in cement plants since the sulfur reacts to form sulfates. Sponge coke is calcined to produce a coke grade suitable for anodes in the aluminum industry. Details of the specifications for green (uncalcined) coke and calcined coke are given as Table 16.11.

Table 16.11. Specification for sponge coke

Parameter	Green coke	Calcined coke
Fixed carbon %	86–92	99.5
Moisture %	6–14	0.1
Volatile matter %	8–14	0.5
Sulfur %	<2.5	<2.5
Ash %	0.25	0.4
Silicon %	0.02	0.02
Nickel %	0.02	0.03
Vanadium %	0.02	0.03
Iron %	0.01	0.02

Table 16.12. A typical specification for sulfur

Purity	99.8 wt% sulfur, on a dry basis.
Ash	500 ppm by weight maximum
Carbon	1,000 ppm by weight maximum
Color	Bright Yellow
Hydrogen sulfide	10 ppm by weight maximum. This is particularly important for international transport and sales
State	Shipped either liquid or solid

Sulfur

Sulfur is a byproduct of modern day refineries. By the processes that make up the modern hydrogen skimming refinery a significant portion of the sulfur contained in the crude is removed as elemental sulfur for marketing as a product. It is true to say that as the environmental requirements for reduced sulfur levels in products and emissions from the refinery decreases the sulfur produced as a product increases. In most countries today over half of the required sulfur is produced from the petroleum and gas industries. Sulfur is stored and transported from the refinery as a molten product or as solid sulfur. Almost half the world's sulfur production is used in making sulfuric acid and phosphate fertilizers using the sulfuric acid produced from the sulfur. A typical specification for sulfur is given as Table 16.12.

16.2 The description of some of the more common tests

The tests described here are usually carried out in the refinery laboratory. The results these tests are used in plant control and for the quality control of finished products. There are many more laboratory and other tests that are carried out in the refinery company's research and development establishment. These later tests are carried out to improve product quality parameters or to establish design data for processes and development work. These later tests would include mass spectrometry for crude oil data and assay development, pilot plant tests to establish optimum operating data, and items such as motor vehicle road tests. Those tests described briefly in the following sections are those listed earlier.

Specific gravity (D1298)

Density ASTM D 1298—Density, relative density (specific gravity), or API gravity of crude petroleum and liquid petroleum products by hydrometer method. Fuel is transferred to a cylindrical container and a hydrometer is carefully lowered into the cylinder and allowed to settle. After the temperature of the sample has equilibrated, the value on the hydrometer scale positioned at the surface of the sample and the

Table 16.13. Relationship of specific gravity @ 60°F, °API, and lbs/US gal

°API	10	20	30	40	50	60
Spec grav	1.000	0.934	0.878	0.825	0.780	0.739
Lbs/Gall	8.328	7.778	7.296	6.870	6.490	6.151

sample temperature are recorded. The hydrometer value is converted to density at 15.6°C or API gravity at 60°F using standard tables. The API gravity which is always quoted in Degrees API can be calculated from the hydrometer test at 60°F using the equation as follows:

$$\text{Specific Gravity} = \frac{141.5}{131.5 + {}^{\circ}\text{API}}$$

The calculation of the weight per unit volume from the specific gravity is based on the US measure of volume (i.e. US Gallons). A summary table of the relationship between specific gravity at 60°F, °API, and lbs per gallon is shown as in Table 16.13.

ASTM distillations

There are two types of ASTM distillations that are used in the refinery for plant control and finished product quality. These are the ASTM D86 for naphtha and equivalent and for the kerosenes. D156 is used for the ASTM distillation of atmospheric gasoils (heating oil and diesel). The major difference between the two tests are the volume of sample used. In the case of the D86 the sample will be 100 ml while that for D156 the sample will be 200 ml. There are other differences and these will be noted in their descriptions which follow:

ASTM D86

The diagram in Figure 16.1 is the apparatus used for both ASTM distillations. Only the apparatus item sizes will change and the temperature levels in the condenser bath. The measured sample is introduced into an Engler glass flask (A) at 100 ml for the D86 test. The liquid fills about two thirds of the flask leaving the space above the liquid to the cork in the vessel neck as vapor space. About half way along the neck of the flask there is a vapor offtake tube. The open end of this tube is connected to the condenser tube which is routed through the condenser bath (B). This condenser tube emerges from the 'Bath' and the open end is directed into a measuring cylinder. For the lighter boiling range samples (i.e. naphthas) this cylinder is placed in a cold water (slightly below room temperature) bath. The Engler flask rests on an asbestos or similar plate (F) which has a hole 1" in diameter exposing the bottom of the flask

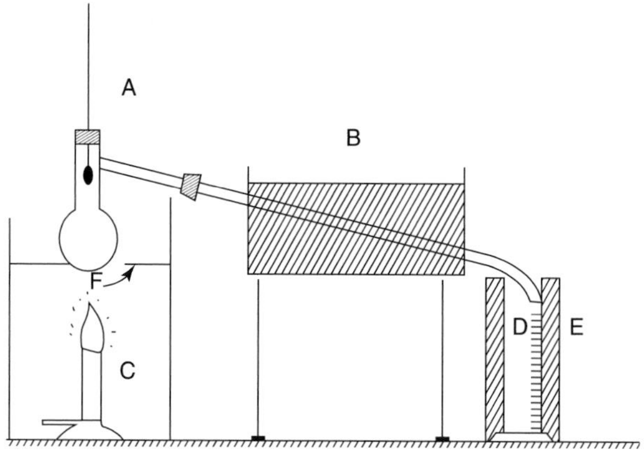

Figure 16.1. Diagram of a typical ASTM distillation apparatus (included in text).

to the heat source. In this case the heat source is shown as a Bunsen burner (C). A thermometer is introduced into the top of the flask and is positioned so that the bulb is directly in line with the vapor offtake. The condenser bath is filled with water and ice and allowed to reach 32°F before heat is applied to the sample flask and the test begun. The temperature of the sample is increased slowly until the liquid begins to boil. The initial boiling point is read as the temperature measured by the thermometer located in the flask when the first condensate drop enters the receiving cylinder (D) at the end of the condensate bath (B). For light boiling point samples (i.e. naphthas) this receiver is cooled in a water bath (E). For kerosenes the water bath (E) is not required. The test is allowed to proceed at a constant rate and temperature readings are taken at predetermined recovery levels of condensate. (usually these temperatures will be at 10 vol% recovered, 30, 50, and 90 vol% recovered). When the flask (A) has been boiled apparently dry, the temperature shown by the thermometer will rise sharply and then begin to fall. The highest temperature observed in this rise and fall is the final boiling point of the sample.

ASTM D156

The same equipment arrangement is used for this test as for the D86. In this case however 200 ml of gas oil or diesel sample is used and the flask will be an Engler 200 ml standard. The other differences are that the hole in the screen (F) has a diameter of 2″. The condenser bath contains water at room temperature and the Receiver (D) is a 200 ml measuring vessel, an no water bath is required.

The ASTM distillation curves from all of the methods described above and those in the determination of vacuum ASTM distillation (not described here as these are not normal marketing or process operation measures) can be converted to TBP and EFV curves. The methods for such conversions are given in Chapters 1 and 3 of this book. Such conversions are used primarily for process design and planning.

Flash point test method (D93)

There are two methods used for determining the flash point of an intermediate and finished petroleum products. These are the ASTM D56 The Tag Closed Cup method (commonly known as the ABEL flash point) and the ASTM D93 The Pensky Marten Closed Cup method. The D56 method is used for material which has a flash of between 68°F and 148°F while the D93 method is used for all other distillates and residuum products with flash points above 148°F. Based on this premise the D56 is used almost exclusively for the kerosene cut range materials. Only the Pensky Marten D93 will be described here.

ASTM D93

A brass test cup is filled to an inside mark with the test specimen. A cover is fitted of specified dimensions, see Figure 16.2. The specimen is heated and stirred at specified rates. An ignition source in the form of a small flame is directed into the cup at regular intervals. When the specimen is seen to flash, the temperature of the specimen is noted as the flash point of the sample.

Significance and use

The flash point is a measure of the tendency of the, material to form a flammable mixture with air under controlled laboratory conditions. It is however only one of several properties that must be considered in assessing the overall flammable hazard of the material. The flash point is used to establish the flammable criteria in transporting the material. Generally shipping and safety regulations will be based on the flash point criteria. The flash point should NOT however be used to describe or appraise the fire hazard or risk under actual fire conditions. This test method (D93) provides the only closed cup flash point test procedures for temperatures up to 698°F.

Pour point and cloud point (D97)

Pour points are determined initially by heat treating the petroleum specimen above it's expected pour point and then to cool the specimen in controlled stages until the pour point is observed. The pour point is the temperature of the material that it ceases to flow.

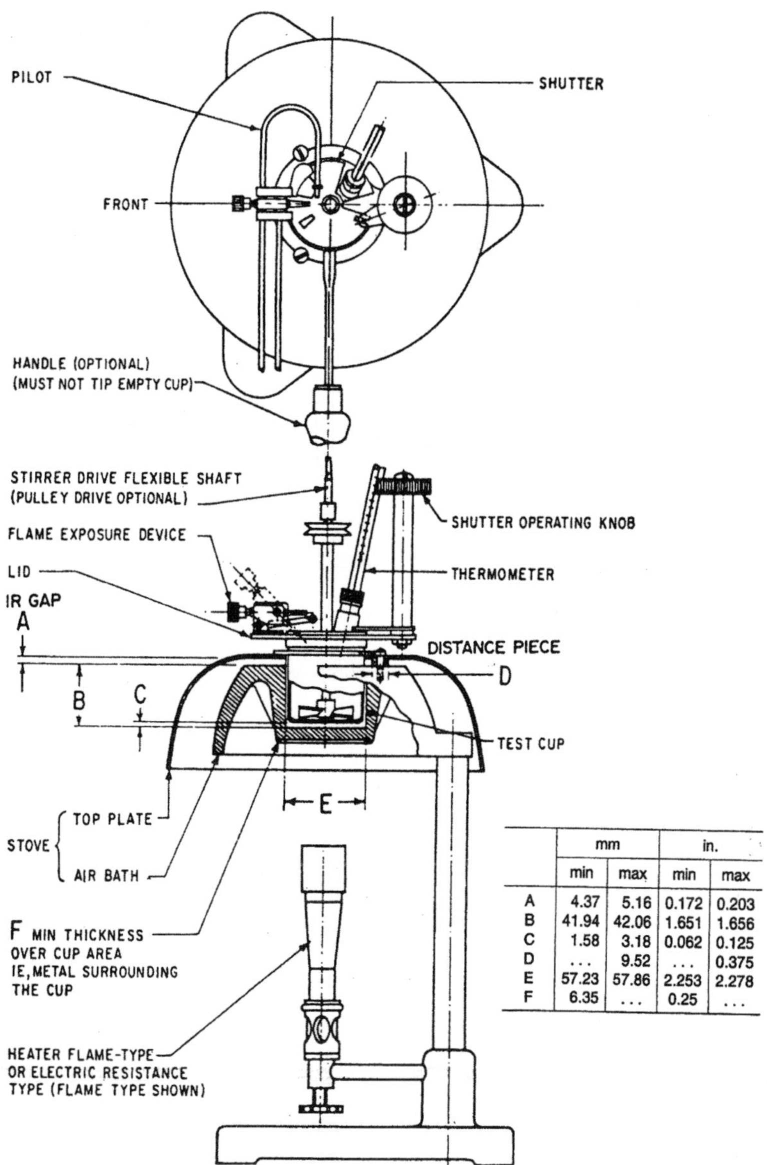

	mm		in.	
	min	max	min	max
A	4.37	5.16	0.172	0.203
B	41.94	42.06	1.651	1.656
C	1.58	3.18	0.062	0.125
D	...	9.52	...	0.375
E	57.23	57.86	2.253	2.278
F	6.35	...	0.25	...

Figure 16.2. The Pensky Marten closed cup flash point apparatus.

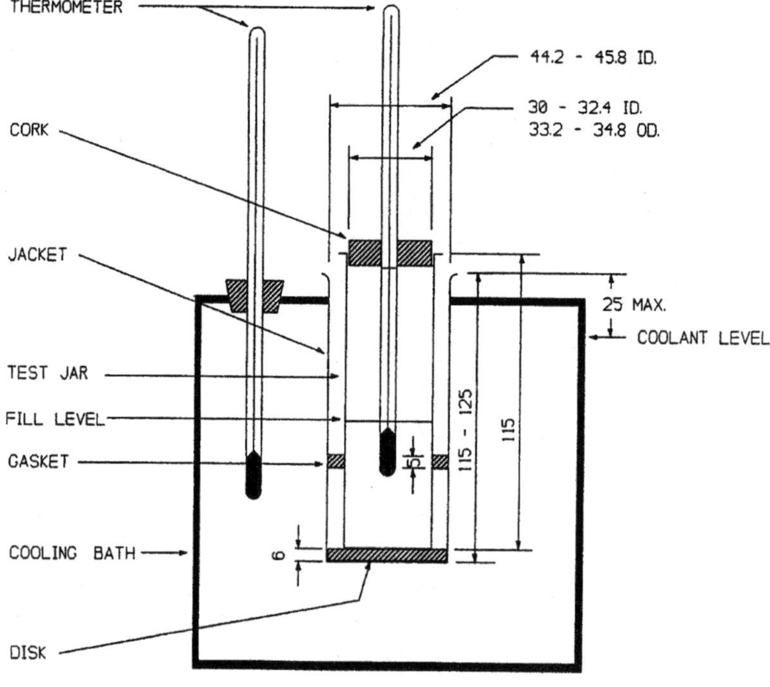

Note 1—Dimensions are in millimetres (not to scale).

Figure 16.3. Apparatus for pour point tests.

ASTM D97

The apparatus used to determine pour point (and cloud point) consists of a cylindrical glass jar with a flat bottom a coke stopper at it's top. A thermometer is inserted through the coke stopper so that the bulb is immersed up to 3 mm of the capillary in the specimen. The sample is inserted into the jar up to a prescribed level. The apparatus is shown in the following Figure 16.3 and is self-explanatory.

The sample is first heated as follows:

Material with an estimated pour point above 33°C—heat to a temperature of expected pour point plus 9°C but at least 45°C in a bath controlled to at least 48°C. Material with an expected pour point of below 33°C—heat to at least 45°C and cool to 15°C in a bath controlled at 6°C.

Commence the test then by sequential cooling and observing the flow of the specimen using cooling baths as follows:

	For temperatures down to
Bath 1—Ice and water	9°C
Bath 2—Crushed ice and sodium chloride	−12°C
Bath 3—Crushed ice and calcium chloride	−27°C
Bath 4—Acetone and solid carbon dioxide	−57°C.

The specimen is checked at regular intervals of cooling (every 3–5°C) for flow. This checking must be done with great care by just slowly tilting the jar to the horizontal position for not more than 5 seconds, and observing if there is flow. The moving of the sample from bath to bath should be at the following schedule:

Specimen at 27°C move to bath 1
Specimen at 9°C move to bath 2
Specimen at −6°C move to bath 3
Specimen at −24°C and below move to bath 4.

The reporting of the pour point is the temperature where no flow is observed plus 3°C.

Kinematic viscosity (D446)

The kinematic viscosity of an oil is obtained by measuring the time required for a sample of the oil to flow, under gravity through a capillary. The capillary is a part of the calibrated viscometer and the flow through it of the oil is at a known temperature. These viscometers are in several sizes and differing design. The configuration and size of the capillary are calibrated and tested to provide a constant value for that particular viscometer.

ASTM D446

The viscometer shown in Figure 16.4 is one of many types. It is used for obtaining the viscosity of transparent liquids. The viscometer shown in Figure 16.4 is designated a SIL. Its important features and its calibration constant are given in Table 16.14.

The viscometer(s) shown may be one of many. These are suspended using a specially designed holder in a bath containing water with some glycol added to reduce vaporization or prevent boiling. The bath is maintained at a constant temperature at which

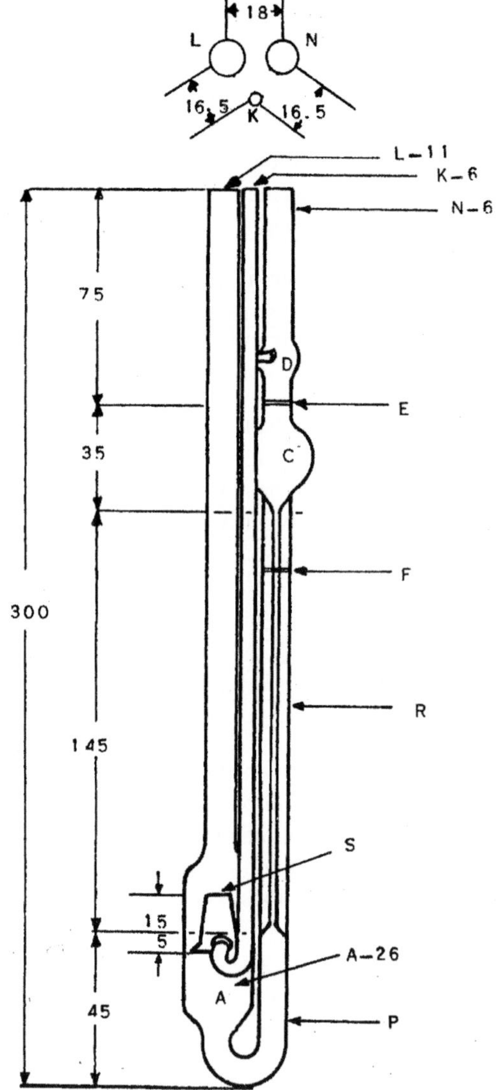

NOTE 1—All dimensions are in millimetres.

Figure 16.4. Viscometer for transparent liquids.

Table 16.14. Viscometer SIL details

Size number	Approximate constant*	Kinematic viscosity range, cst	Inside diam of tube R, mm	Inside diam of tubes E & P, mm	Volume of bulb C, ml
0C	0.003	0.6–3	0.41	4.5–5.5	3.0
1	0.01	2.0–10	0.61	4.5–5.5	4.0
1C	0.03	6–30	0.73	4.5–5.5	4.0
2	0.1	20–100	1.14	4.5–5.5	5.0
2C	0.3	60–300	1.5	4.5–5.5	5.0
3	1.0	200–1,000	2.03	4.5–5.5	5.0
3C	3.0	600–3,000	2.68	4.5–5.5	5.0
4	10.0	2,000–10,000	3.61	4.5–5.5	5.0

*Constant is in $(\text{mm}^2/S)/S$.

the viscosity is to be reported. The test temperature may be as high as 100°C thus the use of glycol in the bath water to prevent actual boiling.

The oil sample is introduced into the viscometer through tube L by tilting the viscometer to about 30° from the vertical with bulb A below capillary R. After introducing the sample attach the viscometer into the holder and insert the viscometer into the bath so that it is vertical. Allow the sample to reach the bath temperature usually about 30 min before starting the test. Using suction (all laboratories should have a vacuum system), draw up the sample through bulb C to about 5 mm above the upper timing mark E. Release the vacuum and allow the sample to flow by gravity. Measure the time for the sample to flow from the timing mark E to the lower timing mark F. This is the time t in the equation:

$$V = C \times t$$

where

V = the kinematic viscosity in centistokes
C = the approximate constant in $(\text{mm}^2/t)/t$ (from Table 16.14)
T = the flow time from the test.

Reid vapor pressure (D323)

This test is the standard test for low boiling point distillates. It is used for naphthas, gasolines, light cracked distillates and aviation gasolines. For the heavier distillates with vapor pressures expected to be below 26 psig at 100°F the apparatus and procedures will be different. Only the Reid vapor pressure for those distillates with vapor pressures above 26 psig at 100°F are described here. The apparatus used for this test is given as Figure 16.5.

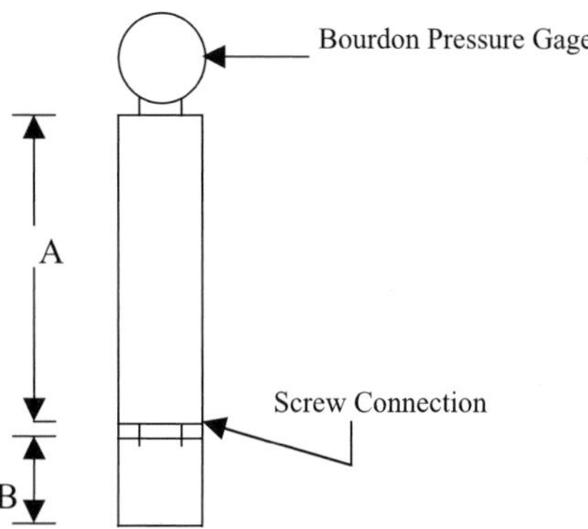

Figure 16.5. A Reid vapor pressure apparatus (included in text).

ASTM D323

In the above diagram A represents the vapor chamber which has a length of 254 mm and a diameter of 51 mm. The ratio of the volume of the vapor chamber A to the liquid chamber B shall be between 3.8 and 4.2. The diameters of both the vapor chamber and the liquid chamber shall be the same. The liquid test sample is placed in the liquid chamber to fill the vessel. The liquid chamber is then connected to the vapor chamber and the pressure gage inserted as shown in the diagram. The assembled apparatus is then immersed vertically in a water bath up to a level of 190 mm of the vapor chamber. The water bath is kept at a temperature of 100°F for 30 min and then removed tilted once or twice and replaced for a further 15 min. At the end of this time the vapor pressure of the test sample is read from the pressure gage to the nearest 0.2 psi.

Weathering test for the volatility of LPG (D1837)

This test is a measure of the relative purity of liquefied petroleum gases (LPG). The test results give the indication of the pentane content of butane LPG or the amount of butane in a sample of propane LPG. Volatility expressed in terms of 95% evaporated temperature of the test sample indicates the measure of least volatile fuel components present. The apparatus used for this test is shown as Figure 16.5.

The weathering tube is a centrifuge cone-shaped tube conforming to the dimensions given in Figure 16.6. The shape of the lower tip is especially important—the

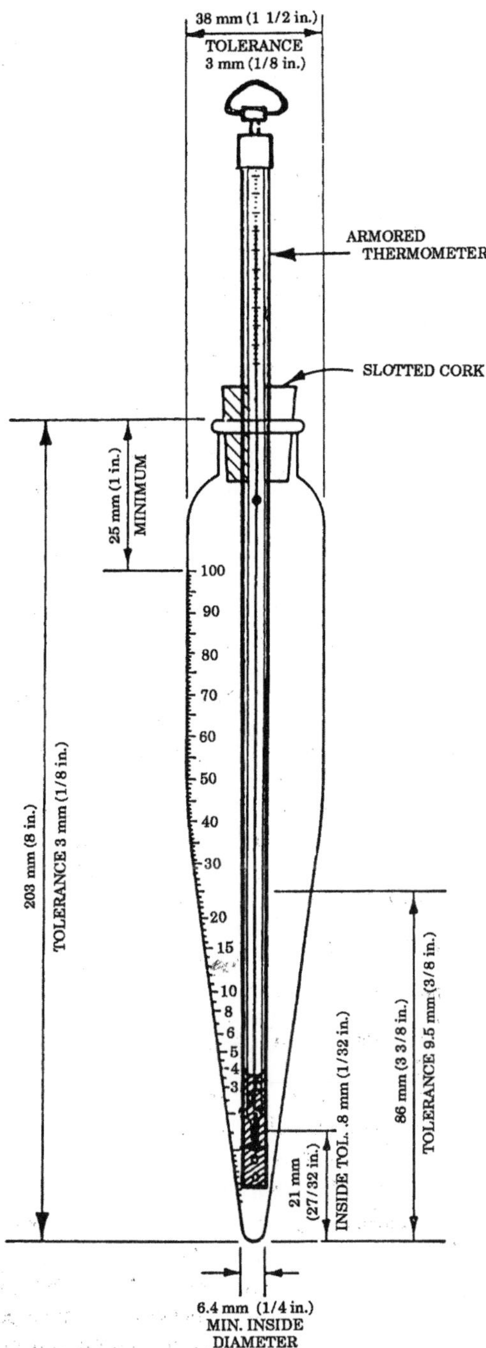

Figure 16.6. Weathering tube.

Table 16.15. Weathering tube graduation tolerances

Range, ml	Scale division, in ml	Limit of error, ml
0.0–0.1	0.05	0.02
0.1–0.3	0.05	0.03
0.3–0.5	0.05	0.05
0.5–1.0	0.1	0.05
1.0–3.0	0.1	0.1
3.0–5.0	0.5	0.2
5.0–25.0	1.0	0.5
25.0–100.0	1.0	1.0

taper shall be uniform and the bottom rounded as shown in the Figure 16.6. The tube is calibrated as shown and the calibration tolerances are given in the following Table 16.15.

ASTM D1837

Testing a butane or a butane—propane mixture

If the sample is at a temperature of 10°F or below allow it to weather in the atmosphere to 10°F then introduce the thermometer and assemble the apparatus as shown in Figure 16.6. Place the weathering tube vertically in a water bath (temperature between 60°F and 70°F) submerging it to the 1.5 ml mark on the tube. Record the temperature when 95% of the sample has evaporated.

Weathering propane type LPG

Allow the test sample to weather in the atmosphere. Clean off the frost that will accumulate on the vessel outer surface with a swap containing acetone or alcohol. Read off the temperature at 95% as in the case for butane.

Smoke point of kerosenes and aviation turbine fuels (D1322)

The sample is burned in an enclosed wick fed lamp that is calibrated daily against pure aromatic blends of known smoke point. The smoke point of the test sample is quoted as the maximum height of flame that can be achieved without smoking. The apparatus that is used for this test is shown as Figure 16.7.

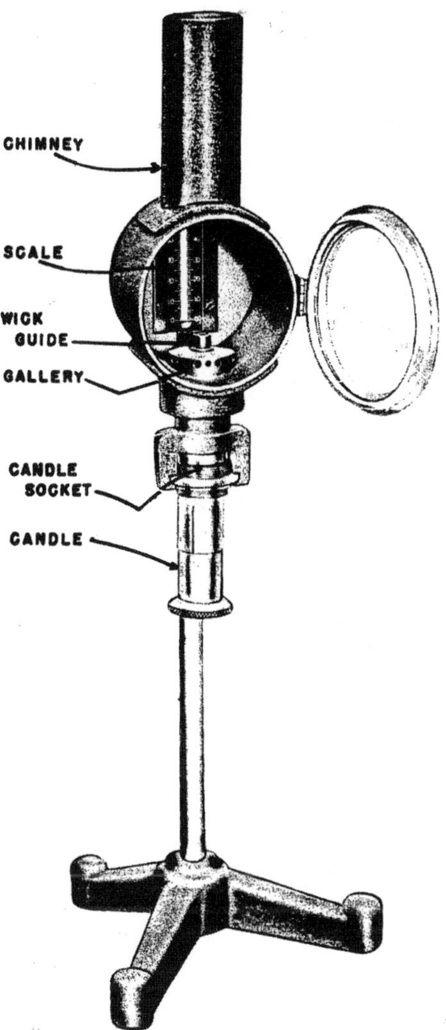

Figure 16.7. A smoke point lamp.

ASTM—D1322

Calibrating the apparatus

The apparatus is calibrated daily against known aromatic compound smoke points to establish a correction factor. The two reference fuel blends are: Toluene and 2,2,4-methyl pentane and details of these fuels are given in Table 16.16.

Table 16.16. Reference fuel blends

Standard smoke point at 101.3 kPa mm	Toluene % (v/v)	2,2,4-Methyl pentane % (v/v)
14.7	40	60
20.2	25	75
22.7	20	80
25.8	15	85
30.2	10	90
35.4	5	95
42.8	0	100

The smoke point of the blends are determined in the same way as the test procedure given below. From the calibrating readings the correction factor f is determined from the equation:

$$f = \frac{(A_s/A_d) + (B_s/B_d)}{2}$$

where

A_s = the standard smoke point of the first reference fuel blend
A_d = the actual smoke point determined for the first fuel blend
B_s = the standard smoke point of the second reference fuel blend
B_d = the actual smoke point of determined for the second fuel blend

The test procedure is as follows:

Figures 16.8 and 16.9 are shown below for detailed reference:

1.0 Soak a piece of extracted and dried wick about 125 mm in length in the test sample, and place it in the wick tube of the candle. Carefully ease out any twists arising from this operation.
2.0 Introduce about 20 ml of the test sample into the clean dry candle place the wick tube into the candle and screw and secure. Then insert the candle into the lamp.
3.0 Light the candle and adjust the wick so that the flame is approximately 10 mm high. Allow the lamp to burn for 5 min. Raise the candle until a smoky tail appears, then lower the candle slowly until the following stages occur:
   - First stage: An elongated pointed as sown as Flame A in Figure 16.10
   - Second stage: The pointed tip just disappears, leaving a slightly blunted flame shown as Flame B in Figure 16.10
   - Third stage: A well rounded flame tip as shown in Flame C in Figure 16.10

   The correct flame profile is Flame B. Determine the height of this flame and record to the nearest 5 mm.
4.0 Make three separate observations using the same Flame sequence as given in item 3.0 above. If these values vary by more than 1.0 mm repeat the test with a fresh sample.

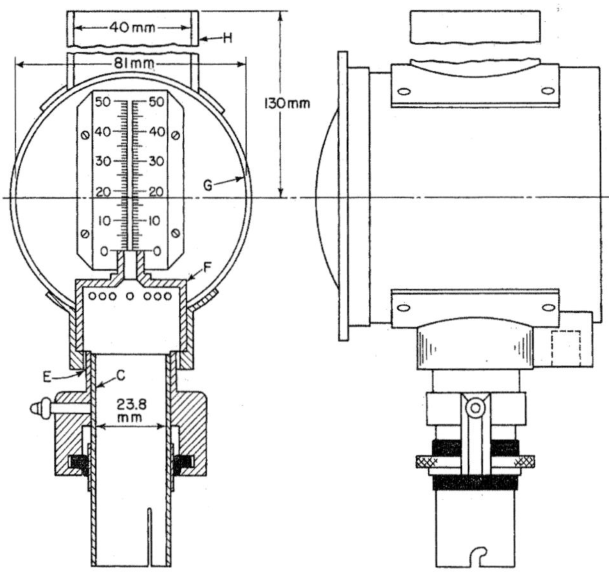

Critical Dimensions of the Lamp Body.

	Dimensions in mm
Candle Socket C internal diameter	23.8
Wick Guide C internal diameter	6.0
Air inlets 20 in number E diameter	2.9
Galary F internal diameter	35.0
Air inlets 20 in number	3.5
Lamp Body G internal diameter	81.0
Internal depth	81.0
Chimney H internal diameter	40
Top of chimney to center of lamp body	130

Figure 16.8. The lamp body.

5.0 The smoke point is calculated to the nearest 1.0 mm using the following equation:

$$\text{Smoke point} = L \times f$$

where

L = the average value of Flame B over the 3 individual tests readings (item 4).
f = the correction factor rounded to the nearest 0.01.

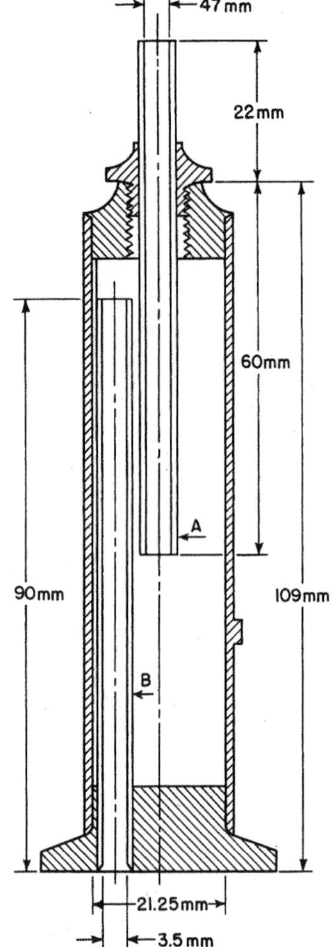

Critical Dimensions of the Candle.

		Dimensions in mm
Candle Body:	**Internal diameter**	**21.25**
	External diameter	**Sliding fit in holder**
	Length without cap	**109**
	Thread on cap	**9.5 mm dia 1.0 mm screw pitch**
Wick Tube A :	**internal diameter**	**4.7**
	External diameter	**Close fit in flame guide**
	Length	**82.0**
Air Vent B ;	**Internal diameter**	**3.5**
	Length	**90.0**

Figure 16.9. The candle.

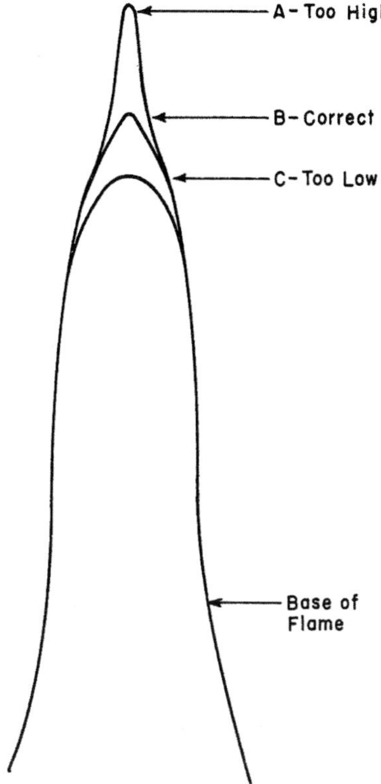

Figure 16.10. Typical flame appearance.

Conradson carbon residue of petroleum products (D189)

This test is to determine the amount of carbon residue left after complete evaporation and pyrolysis of an oil. The degree of this carbon content is indicative of the coke forming properties of the oil. The value of the Conradson carbon content is important in most heavy oil hydrotreating, and most cracking processes. A number of modern catalysts used in heavy oil cracking have been developed to withstand the adverse effect of carbon lay-down. Still excessive coke formation remains a problem in most cracking processes. The apparatus used for this test is shown as Figure 16.11.

ASTM—D189

The apparatus consists of a porcelain crucible containing a weighed sample of the oil. This crucible is placed in the center of an iron crucible which is itself placed in a larger iron crucible. Both iron crucibles are fitted with covers and are vented through

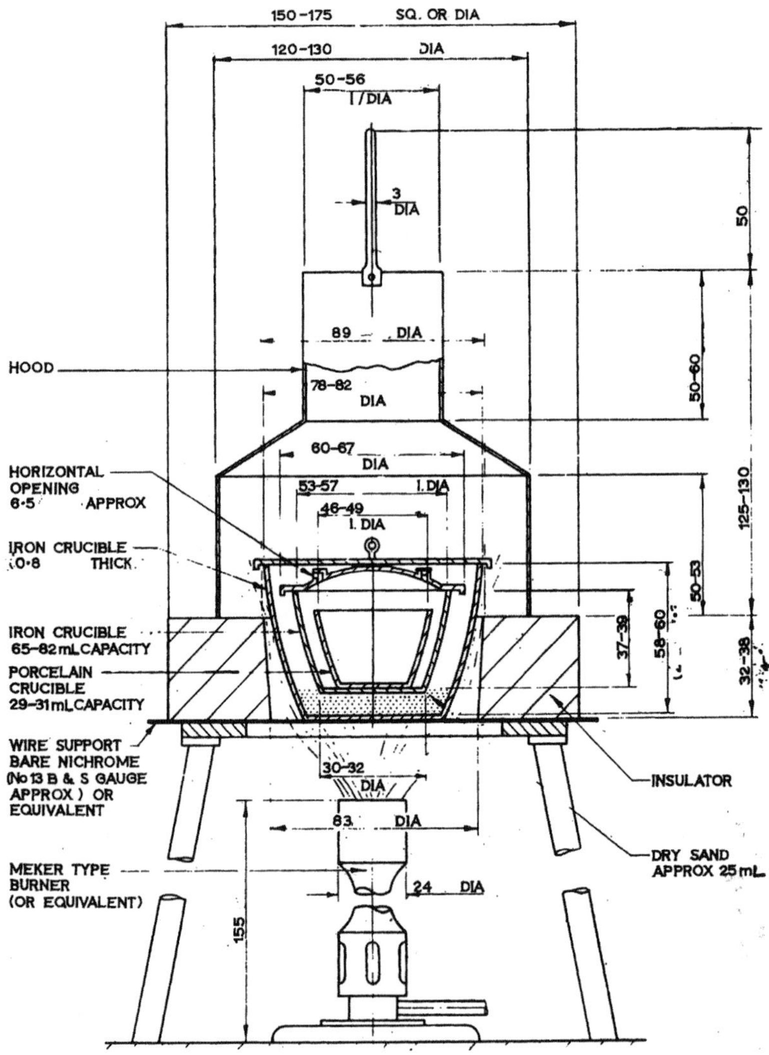

Figure 16.11. The apparatus for determining Conradson carbon residue.

a metallic hood. This assembly is placed on a stand with a supporting wire mesh. The crucibles are partially surrounded by insulating brick work on which the metallic hood rests. Heat is supplied to the crucibles by a burner located below the wire mesh support. The sample is heated for a specified period of time until all the volatile content of the oil has evaporated leaving the carbon as residue. The carbon content of the oil is calculated as the weight of the carbon left as a percentage of the weight of the original sample.

Table 16.17. Weight of test specimen vs. predicted bromine number

Predicted br number	Specimen weight grams
0–10	20–16
Over 10–20	10–8
Over 20–50	5–4
Over 50–100	2–1.5
Over 100–150	1–0.8
Over 150–200	0.8–0.6

Bromine number of petroleum distillates (D1159)

The bromine number is a measure of olefins in a petroleum distillate boiling up to 600°F. The number is supporting evidence of the purity of the product with respect to its olefin content.

D1159

The apparatus used in this test is any apparatus that can perform titration to a pre-set end point. Such an apparatus may include an electrical meter with two polarized electrodes of sufficient voltage to indicate the end point. The titration vessel is a jacketed glass vessel about 120 mm high and 45 mm internal diameter. The sample contained in the vessel will be kept at 32–45°F. The burette will be a normal titration burette capable of measuring the titration reagent in graduations of 0.05 ml or smaller. Place 10 ml of 1,1,1-trichloromethane or dichloromethane in the titration vessel and introduce the test specimen as indicated in Table 16.17.

Fill the titration vessel to the mark with the selected solvent. Mix the contents well. Titrate the sample with the standard bromine–bromate solution—stir the specimen well during the titration. The end point is noted by a sudden change in potential on an electrometric end point indicator due to the presence of bromine. The bromine number is calculated by the following equation:

$$\text{Bromine Number} = \frac{(A - B)(M_1)(15.98)}{W}$$

where

A = milliliters of bromide–bromate solution required for titration of the test sample.
B = milliliters of bromide–bromate solution required for titration of the blank.
M_1 = molarity of the bromide–bromate solution, as Br_2
W = grams of the test specimen.

Table 16.18. Boiling range correction for olefins

Boiling range correction f	Boiling range IBP to FBP °C
1.0	0
0.975	7
0.950	14
0.925	21
0.900	28
0.875	38
0.850	43
0.825	53
0.800	62
0.775	72
0.750	95
0.725	99
0.700	125 or greater

The olefin content of the test sample can be calculated from the bromine number by the equation

$$\text{Olefin mass\%} = f(BM/160)$$

where

f = Boiling range correction from Table 16.18.
B = Bromine number expressed as grams of bromine per 100 grams of sample.
M = Molecular weight (relative molecular mass of olefin as given in Table 16.19.

Sulfur content by lamp method (D1266)

This test method determines the total sulfur in liquid petroleum products in concentrations of 0.01–0.4 wt%. This method is applicable to all petroleum liquids that can be burned in a wick lamp. It therefore covers gasoline, kerosene, and naphtha. A blending procedure can be applied to determine the sulfur content in gas oils and heavier

Table 16.19. Relation of average molecular weight to 50% boiling point

50% boiling point °C	Average molecular weight
30	72
66	83
93	96
121	110
149	127
177	145
204	164
232	186

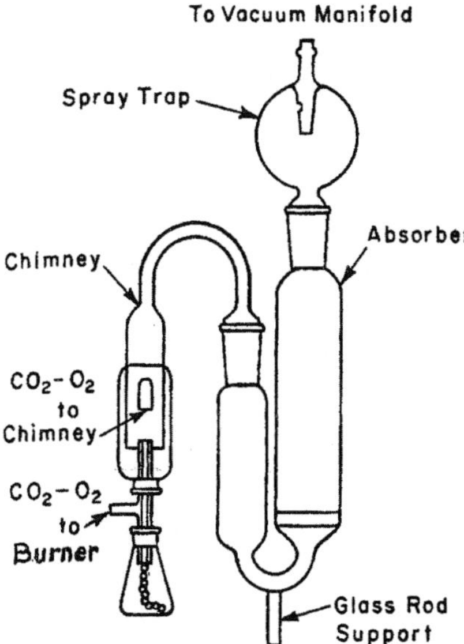

Figure 16.12. Details of assembled lamp unit.

which can not be burned directly in the wick burner. Figure 16.12 shows the details of the assembled lamp unit. Figure 16.13 is a schematic diagram of the CO_2–O_2 supply manifold and lamp system.

D1266

The sample is burned in a closed system using a suitable lamp (Figure 16.12) and an artificial atmosphere of 70% carbon dioxide and 30% oxygen to prevent the formation of nitrogen oxides (Figure 16.13). The oxides of sulfur are absorbed and oxidized to sulfuric acid by means of a hydrogen peroxide solution. This solution is flushed with air to remove carbon dioxide. The sulfur as sulfate in the absorbent is determined by titration of the sulfuric acid with standard sodium hydroxide. The sample size for direct combustion is 10–15 grams if the predicted sulfur content is below 0.05% for predicted sulfur content 0.05–0.4% the sample size should be 5–10 g.

The calculation of the sulfur content from the titration is given by the following equation.

$$\text{Sulfur Content wt\%} = 16.03\ M \times (A/10W)$$

where

A = milliliters of NaOH titrated
M = molarity of the NaOH solution (Note 1 below)
W = grams of sample burned.

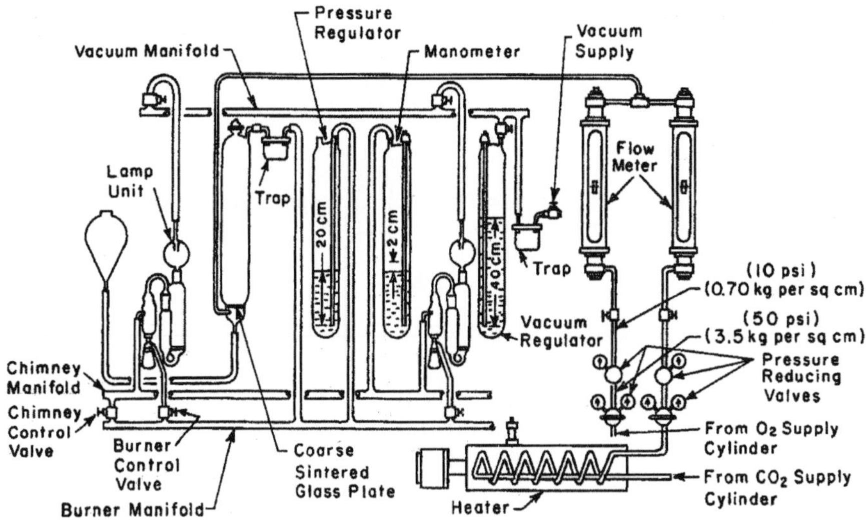

Figure 16.13. Schematic diagram of the CO_2–O_2 supply manifold.

Note 1: The calculation can be simplified by adjusting the molarity of the NaOH solution to 0.624. Then 1 ml of the solution will be equivalent to 0.001 g of sulfur. In this case the factor 16.03M in the calculation becomes 1.0.

Octane number research and motor *(details of this test are given in ASTM standards Part 7)*

By definition, an octane number is that percentage of isooctane in a blend of isooctane and normal heptane that exactly matches the knock behavior of the gasoline. Thus, a 90 octane gasoline matches the knock characteristic of a blend containing 90% isooctane and 10% *n*-heptane.

The knock characteristics are determined in the laboratory using a standard single cylinder test engine equipped with a super sensitive knock meter. The reference fuel (isooctane and C_7 blend) is run and compared with a second run using the gasoline sample. Two octane numbers are usually given: the first is the research octane number (RON) and the second is the motor octane number (MON). The same basic equipment is used for both octane numbers, but the engine speed is increased for the motor octane number. The actual octane number obtained in a commercial engine would be somewhere between the two numbers. The significance of the two octane numbers is to evaluate the sensitivity of the gasoline to the severity of the operating conditions in the engine. Invariably the research octane number is higher than the motor octane number, the difference between them is quoted as the 'sensitivity' of the gasoline.

Conclusion

The specifications and the selected tests given here represent the control measures used to establish product quality control. Much of the testing described here are used for plant control while all the tests are necessary to ensure that product quality of finished products is met. In modern refinery practice much the plant control is achieved using 'in line' analyzers tied to control system computers. These computers automatically adjust plant conditions to meet set product quality criteria. A similar 'in line' analysis is used for blending rundown streams for finished products. Invariably, though, laboratory testing is still used to 'spot check' plant intermediate product quality, and laboratory testing is always used as final check on finished product quality.

Chapter 17

Economics

17.1 Refinery planning, economics and handling new projects

D.S.J. Jones

This chapter is divided into three parts. The first deals with the planning of a refinery's operation, which includes its optimized crude runs, product slate, and any process expansion or debottlenecking that may be required to meet this optimized operation. The second part deals with the process economic evaluation of a proposed new refinery or new facilities within an existing refinery. The third part is a brief description and discussion of a refining company's participation in the execution of an approved engineering and construction project.

This second part of the chapter is closely related to the first part in so far that process studies and the ultimate selection of any process or processing route is based on the approved marketing strategy for the refining company. This in turn is based on the conclusions following the work undertaken in Part 1. The final phase in this refinery operation is described in Part 3. This outlines the preparation of contracts to third parties for the installation of new or expanded facilities. It includes the choice of the type of contract to be solicited and follows with the selection of a suitable engineering company to execute the work. Finally, it discusses the company's activities associated with the execution of the project by contractors and the subsequent commissioning and acceptance of the completed facilities.

17.1.1 Refinery operation planning

The basic organization of an oil company consists essentially of three main departments:

1.0 The Supply Department
2.0 The Refining Department
3.0 The Marketing Department

The function of these three main departments are co-ordinated by a supply department. This department undertakes this role in accordance with the following sequence:

1.0 Marketing department informs the supply department with the quantities of products they can sell.
2.0 Supply department, after making adjustments for stock levels, advise the refinery of the quantity of each product required.
3.0 The refining department advise the supply department of the actual quantities they can produce and how much crude they will require to do so.
4.0 The supply department arranges with the marine department for the necessary shipping to:
    - Provide the crude
    - Lift the products
    - Import the quantities that cannot be produced

In the above sequence of events Item 3 is carried out by the planning section of the refinery department. Their objective is to translate the production requirements given by the supply department into a workable system, which specifies the actual operation of each plant to ensure that the final result is the most economical one practicable. This is termed the 'Running Plan' for the operation.

Running plans

Running plans are developed and issued on a monthly basis. They usually cover periods varying from 11 to 18 months ahead. Long term periods are broken down into quarters, short term (up to 6 months ahead) are broken down into months. The main uses of the running plan are to provide data for:

- Keeping the supply department and the company executives informed
- Provide information for long term chartering of ships
- Arrangement of import programs
- Pin point future product quality and equipment difficulties

The running plan is used in a refinery as the basis for an Operating Program. Such a program is developed in the refinery by converting the plan into a day to day operation program for each refinery plant. At the end of every month the plant operation for that month is summarized and compiled into an Operations Summary which compares the actual plant operation with the program. Usually the refinery organization will include a planning department and the production of the both the Operating Program and the subsequent summary is their responsibility.

The rules of planning

Before looking at the development of running plans in some detail it is necessary to understand some fundamental rules. These are listed below and are the "Planning Data":

- Equipment Availability
- Plant Performance
- Service Factors
- Product Quality
- Planning Properties
- Blending Properties
- Crude Availability
- Crude Quality
- Tankage and Lines Availability
- Economic Rules

Most of these rules are self explanatory, but notes on a few of those items, which are not so apparent, are as follows:

Plant performance

This is an important requirement in planning the refinery daily and long term operation. The exact operating capability of the various refinery plants must be known. Such data that is essential are:

- Yields under any operating condition and any feed quality
- Product quality under any operating condition and feed quality
- Maximum and minimum plant throughput
- Fuel consumption

Blending properties

Most finished products leaving the refinery are blends of one or more basic process streams. For example all gasoline products will be blends of straight run naphtha's and some high octane streams from a cracking unit and/or streams from a catalytic reformer. Full description and discussion on blending and blended products are given in Chapter 1 of this Handbook. In planning the refinery's day to day operation these blending techniques and, when applicable, blending indices are used.

Crude availability

With most major oil refining companies the company's central supply department issues the availability of crude slates (quantities and crude source) when they issue the requirement data. In most cases the respective refining company has contract for the purchase of a few specific crudes and their quantities. Of these basic crude slates one or two of the crude oils may be increased or decreased if necessary. These are called marginal crude oils. Most often there are two types of crude oils in the slate which are marginal. One of these marginal crude oils would be used if more crude than is shown on the slate is required. The other would be backed out of the slate if the entire projected slate was not required. Thus marginal crude oils play an important part in the economic running plans for a refinery operation.

Economic rules

Using the planning rules, it is possible to devise an infinite number of schemes that would meet the majority of the market requirements. Only one such scheme however would be the most profitable. It is the most profitable plan or scheme that has to be determined and implemented, and to do this it is necessary to follow some basic economic rules. These rules may be developed with answers to the following typical questions:

1.0 If all requirements cannot be met, in which priority order should the various products be placed?
2.0 Is it profitable to run the refinery to meet just a fuel oil product requirement. That is not to produce waxy distillates at all?
3.0 What is the optimum octane number severity operation for the reforming and/or cracking units to meet gasoline requirements?
4.0 Which is the cheapest middle distillates to import if this becomes necessary to meet a product slate?
5.0 Is it economical to utilize a surplus unit capacity to produce feed to another? For example to use surplus deasphalting capacity to produce feed for the catalytic cracker.
6.0 Does it pay to divert straight run naphtha to refinery fuel in a situation where naphtha outlet is limiting?

The following table is a simplified example of the use of these economic rules to determine the changing of fuel oil ‘length’ in a middle distillate outlet limiting situation. The only units considered here are the crude distillation unit (atmospheric and vacuum), cat reformer unit, and the cat cracker (FCCU). Being a middle distillate outlet limiting there will be spare capacity in the units considered, and all requirements are met except for fuel oil. The economic rule in this case is that the most profitable plan will be when the crude and fuel oil are at a minimum while still meeting the requirements of the other products in the slate.

Developing the running plan

In most refining companies with more than one refinery installation running plans are developed in the company's corporate offices. In this way access to the required data (such as crude availability, product slate etc.) are readily available to the planner. Normally running plans are published on a monthly basis. Again normally these plans cover a period of 18 months. The early part of the period, say the first six months, will be studied on a monthly basis, and the remainder on a quarterly basis.

The use and purpose of the refinery running plans

The running plans are designed to provide corporate management with data with respect to the following:

1. The anticipated pattern of production i.e. the surpluses and the deficits that are likely to occur.
2. The likely operations of the refineries in the company. This covers crude runs and refinery throughputs.
3. Any inadequacies in refining capacities to meet projected profit margins. This may form the basis for refinery expansion program. (This is dealt with in a separate item of this section of the chapter).
4. Indication as to where and when extra business can be accommodated in the company's operating plan.
5. The basis for financial forecasting and estimates of profit and loss.

The planning schemes

The content of the running plan will conform to the rules of planning listed and discussed earlier. The development of all running plans begin with schematic diagrams giving as many options as possible of the routes by which a required product slate can be achieved from a particular crude availability slate. The mechanism to arrive at this is not unlike the process configuration study described in Chapter 2 of this Encyclopedia. In the case of the plan schemes however the objective is to fit the process streams to a fixed refinery configuration in order to meet a product slate. In these schemes excess products are corrected by either reducing crude throughput or using the stream as blend stock or feed to other unit(s). Deficits in the product slate may be solved by importing the necessary product components to complete a scheme. An example of a typical planning scheme to meet middle distillates (Jet Fuel in this case) and specialties (White Spirit in this case) is given as Figure 17.1.1.

Several of such schemes may be developed to meet this particular product slate criteria. Each of these will be subject to an economic analysis, and this may extend to one or more of the company's several refineries. These economic analyses are made using

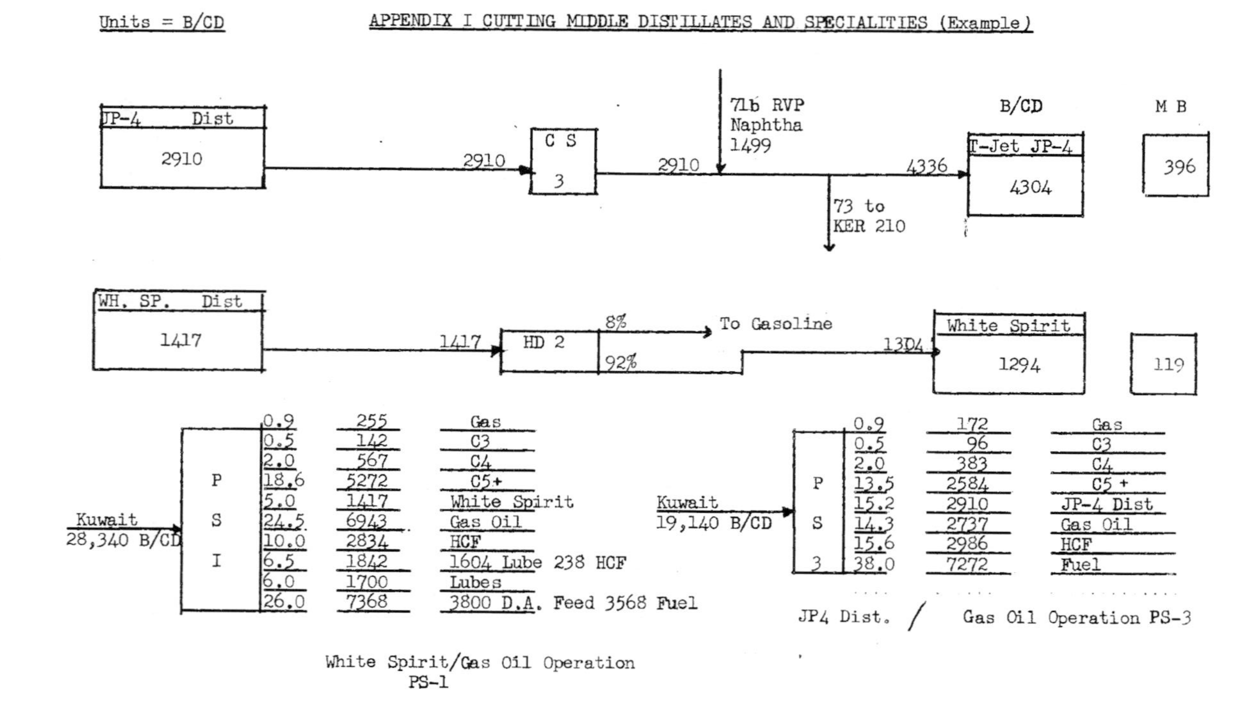

Figure 17.1.1. An example of a planning scheme diagram.

Table 17.1.1. Changes to running plan to maximize profit

Change Base case	Crude MB/CD 200	Fuel MB/CD 83	Mid dist MB/CD 12	Naphtha MB/CD 40	Cum profit over base case $/D
Reduce viscosity of cat feed to fuel from 46.9 RI to 40.2 RI@ 100°F	198.267	81.370	11.65	40.667	595
Reduce residue length from 24 to 22%	189.798	73.406	9.685	43.886	3,558
Decrease octane from 50 to 45 res.	189.702	73.276	9.179	42.870	4,000
Lower cat plant conversion by 3%	187.129	71.074	8.667	46.186	5,024

the same criteria of cumulative product profits (including plant operating costs) as described above in *economic rules* and the example provided by Table 17.1.1. Computer technology is used extensively in planning these days. Indeed most refinery companies possess their proprietary programs similar to the linear program described in Part 2 of this chapter to evaluate the economics of the schemes and to determine the optimum one. However, the development of the scheme itself still remains with the planning engineer or development engineer for most companies.

The application of running plans in identifying refining inadequacies

This item illustrates a function of the running plan in identifying lack of refining capacity under certain conditions. An example of a work sheet for the report is given in Appendix 17.1 of this chapter. The scenario surrounding this exercise is as follows:

Background

A large international oil company Concord Petroleum Company (Fictional) has requested two of its European refining affiliates to submit plans for low cost debottle necking of distillation capacity which can be 'on stream' by next winter. The two de-bottle neck cases that are to be considered are the company's Kent refinery and the company's South Wales refinery. The expansion for the Kent refinery crude unit will be 34,000 BPSD and that for the South Wales Refinery crude unit will be 40,000 BPSD. The study will determine whether either or both of these debottlenecking routes represents an economic investment route for Concord Petroleum Company to meet the long range expansion program.

Table 17.1.2. The base running plan

MB/CD	←——— Kent	Summer S. Wales	———→ Total	←——— Kent	Winter S. Wales	———→ Total
Input (Crude)						
Aramco	66.0	–	66.0	66.0	–	66.0
Brega	113.5	46.5	160.0	130.0	30.0	160.0
Safaniyah	20.0	–	20.0	–	20.0	20.0
Bachequero	18.5	41.5	60.0	22.0	–	22.0
Iraq	–	–	–	–	28.0	28.0
Kuwait	–	37.0	37.0	–	47.0	47.0
Tia Juana	12.0	–	12.0	12.0	–	12.0
Total Crude	230.0	125.0	355.0	230.0	125.0	355.0
T. J. Dist	1.1	–	1.1	–	–	–
Output	Prod + / ( )					
Gas	3.8 –	–	3.8 –	4.4 –	–	4.4 –
LPG	9.9 (2.3)	4.9 (2.3)	14.8 –	11.0 (6.6)	4.9 (1.5)	15.9 (5.1)
Chemicals	9.7 –	–	9.7 –	9.7 –	–	9.7 –
Naphtha Lt	5.9 (2.9)	3.0 (6.7)	8.9 (9.6)	– (15.2)	5.3 (11.6)	5.3 (26.8)
Hy Naphtha	53.3 (9.3)	24.7 (3.1)	78.0 (12.4)	54.8 7.6	22.8 (10.2)	77.6 (2.6)
Light Dist	26.4 2.7	4.0 –	30.4 2.7	27.6 (2.7)	5.2 –	32.8 (2.7)
Gas oils	36.2 –	17.7 –	53.9 –	50.1 1.9	22.0 (1.9)	72.1 –
Fuel Blend stk	5.7 –	3.7 –	9.4 –	7.7 –	4.9 –	12.6 –
Heavy Fuel Oil	– (41.3)	43.8 (19.8)	43.8 (21.5)	– (43.8)	– (31.4)	– (75.2)
Special FO	54.4 (7.6)	21.6 (7.6)	76.0 –	43.7 (49.9)	53.1 (31.9)	96.8 (18)

Basis for assessing requirements

The following table is the Base Running Plan. It highlights the two UK refinery's crude availability and their respective requirements with surplus and deficit highlights:

Table 17.1.2 represents the short range status. Other running plans for other years, are summarized below. This Table 17.1.3 gives the product deficiencies/(surplus) versus crude requirement to satisfy product deficiencies over those years.

See the following notes for Table 17.1.3:

1.0 *Swing capacity* is defined as the sum of international bunkers plus volume of seasonal storage. It is assumed that crude is run for international bunkers in the summer if necessary, and an equal volume of fuel oil is imported in the winter against inland fuel oil requirements. Thus, *swing capacity* represents the volume of crude which can be run in the summer against winter requirements.

2.0 Net crude is the volume of crude running capacity required in summer and winter when maximum use is made of *swing capacity*.

Table 17.1.3. Crude requirements to satisfy deficiencies

	year 1			year 2			year 3			year 4		
Year	Min	Max	Ave	Min	Max	Ave	Min	Max	Ave	Min	Max	Ave
Deficiencies					MB/CD							
Naphtha	–	8.8	4.4	—	17.0	8.5	9.8	13.4	11.7	9.6	26.8	18.2
Distillates	–	–	–	–	–	–	–	–	–	(2.7)	2.7	–
Fuel Oil	(15.9)	32.5	8.2	3.5	48.2	25.9	8.5	70.0	39.2	21.5	93.2	57.4
Total	(15.9)	41.3	12.6	3.5	65.2	34.4	18.3	83.4	50.9	35.8	115.3	75.6
Crude equiv	(16.3)	42,4	13,0	3,6	67.0	35.2	10.0	85.7	52.2	36.8	118.3	77.6
Swing capacity	40.8	(40.8)	—	39.4	(39.4)	--	37.0	(37.0)	—	35.0	(35.0)	—
Net Cap Req	24.5	1.6	13.0	43.0	27.6	35.2	55.8	48.7	52.2	71.8	83.3	77.6
Mon Cap		13.0			35.2			52.2			77.6	

3.0 Minimum crude capacity is the volume of distillation capacity which must be available just to meet deficiencies and making maximum practical use of the *swing capacity*.

Eight basic schemes were studied to meet the minimum capacities for satisfying the deficiencies. These were as follow:

Case 1. Expand Kent distillation by 82 mbsd—Debottleneck
Case 2. Expand Kent distillation by 34 mbsd on CDU 101 and 50 mbsd on CDU 102
Case 3. Expand Kent distillation by 50 mbsd on CDU 101 and 34 mbsd on CDU 102
Case 4. Expand S. Wales distillation by 100 mbsd—Grass Roots
Case 5. Expand Kent distillation by 100 mbpsd—Grass Roots
Case 6. Expand Kent distillation by 34 mbsd and S. Wales by 40 mbpsd—De-bottle neck
Case 7. Expand Kent distillation by 40 mbsd and S. Wales by 34 mbpsd – De-bottle neck
Case 8. Expand S. Wales distillation by 75 mbsd—Grass Roots.

There were in addition one sub case for each of cases 4, 6, 7, and 8. These sub cases took into account special depreciation allowances for the S. Wales refinery.

The results

The discounted cash flows at 10% for the 8 cases and the four sub cases were calculated for the entire period as shown in Table 17.1.4 (This took into account the company's long range plan and forecast). The preferred case was case 6 with the cognizance of the S. Wales refinery depreciation allowance. This cash flow for this item is as follows:

Table 17.1.4. Study cash flow and result

M$ out/ (in)											
Year	base	1	2	3	4	5	6	7	8	Total	Preference
Case 6A	–	5,991	1,054	2,953	380	(1,252)	(923)	(859)	(58)	7,286	1
Cash flow of some other cases are as follows:											
Case 5	6,973	7,100	1,395	1,871	779	(1,760)	(1,300)	(1,196)	(81)	13,781	12
Case 7A	–	1,352	6,706	804	(401)	(1,070)	(886)	(409)	(27)	8,069	2
Case 6	–	5,991	1,054	2,953	1,841	(1,352)	(1,090)	(948)	(64)	8,385	4
Case 4	5,212	8,191	1,467	1,040	534	(1,614)	(1,233)	(1,138)	(77)	12,382	11

Author's Note: The names and location of the company and its refineries are, of course, fictitious. The study though is not. The author actually participated in the study itself, and was later the lead engineer in the process design of one of the selected refinery (Case 6A) debottlenecking expansion.

A copy of one worksheet used in the study is given in the appendix to this section. Most calculations at that time were done by hand, although the discounted cash flow calculations were computerized.

The refinery operating program

One of the major uses of the running plan is to enable the refineries of the company to develop their own operating program. The running plan sets the objective for the refinery production over a short term and provides a long term forecast. The operating program translates these requirements into a day by day program of work. Operating programs are concerned about the short term plans for the operation, and are prepared for a monthly period. Thus while running plans are calculated in terms of Barrels per Calendar Day, operating programs are always in terms of Barrels per Stream Day.

The content of the operating program

The actual format of operating programs will vary from company to company, however, they will follow a similar pattern. They will consist of the following items:

- A crude oil program
- An intermediate stock report
- A table of programmed production
- A blending schedule
- An operating schedule
- Operating notes

A crude oil program is a table of the refinery's current crude oil inventory with projected imports, feed outlets, and the storage tank reference. An example of this table is given as Appendix 17.2 of this Chapter.

An intermediate stock report is similar to the crude oil program but it refers to the refineries intermediate stocks, which will usually be blended to finished products. An example of this table is also given as Appendix 17.3.

The table of programmed production is simply the scheduled finished product make from the refinery. The blending schedule is self-explanatory and is usually in a form of simple block flow diagrams. A typical middle distillate schedule is given in Figure 17.1.2.

Operating notes include any critical requirements or operational changes. Note operating programs can be updated on a weekly or even daily basis if the Company's Supply Department initiates a change in running plans, that affect a particular refinery's operating program. An example of an operating note is as follows:

Current program

Unit F101	Light gas oil	Marine diesel	White spirit	Turbo Jet JP4
Supply dept requirement	750 m bpsd	235 m bpsd	40 m bpsd	181 m bpsd
Refinery make	668	202	40	181

Supply must have absolute minimum of 725 mbpsd of light gas oil i.e. +57 MB's

Can stand only 152 MB's of Marine diesel i.e. −50 MB's

Action: Change kero/gas oil operation to kero/diesel operation.

Preparing an operating program

In present day refining industry both the running plan and the subsequent operating program are computerized. However, their preparation requires the same input as the old manual system. In the case of the operating program these would follow the sequence listed below:

1.0 In order to begin assembling the operating program the planner will need to know the following details:

- The running plan for the refinery for the month in question
- The latest crude slate data
- What plants are scheduled for shut down during the life of the program
- Estimate of the intermediate product stocks (Appendix 17.3)

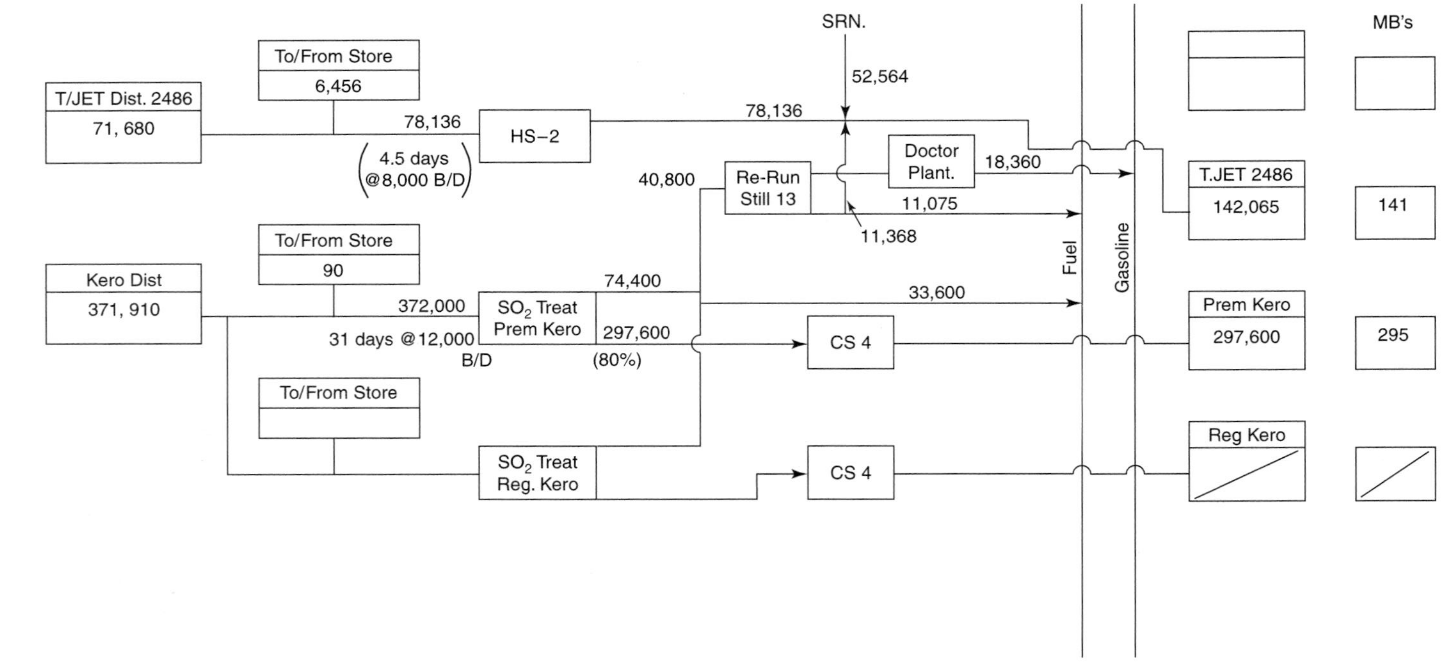

Figure 17.1.2. A typical middle distillate schedule.

2.0 Prepare an outline operating schedule to meet intermediate product requirements and timed to maintain adequate intermediate products in tankage (Appendix 17.4).
3.0 Cut the crude(s) TBP into the required product cuts, for the crude unit operation.
4.0 Cut the product from the other units (such as FCCU, and cat reformer) to meet intermediate products.
5.0 Balance the crude run against production. From the cutting scheme and schedule outline develop the detailed and balanced refinery flow plan.
6.0 Develop the detailed operating program (Appendix 17.5).

Monitoring the operating program

The planning section in a refinery organization is usually delegated the duty of monitoring the Program. The following activities are associated with this monitoring function:

1.0 A check is maintained on the actual product produced, by control room charts, tank dips, and the like. This is done daily and deviation from program noted with explanation of cause for the deviation.
2.0 Weekly operating schedules are issued to allow for any deviation to program. These would include such causes as:
   - Delay in tanker arrivals
   - Changes in crude slate
   - Changes in product requirements
   - Changes in crude fractionator yields
   - Plant operational problems

   Example is provided in Appendix 17.6.
3.0 Daily meeting with the department responsible for oil movements and shipping in the refinery. These meetings are to review shipping and oil movement status to and from tankage.
4.0 Frequent meetings with the operating and the refinery process department to review plant operations.
5.0 Contact the company's supply department with regard to:
   - Crude shipment and changes
   - Lifting products out of refinery schedule (see note below)
   - Changes in requirements

Note: At times of high inventories in the refinery prompt lifting is essential to keeping the refinery operating as programmed.

Operating summary

An operating summary is a document giving details of the refinery's performance against plan. This is issued at the end of each month. A summary of throughputs and yields are usually supplied by the accounts department of the refinery. Comments on production versus program are made by the refinery's planning department.

17.1.2 Process evaluation and economic analysis

Among the principal functions of a refinery's development engineer is to generate ideas for implementing the supply department's planned requirements (both short term and long term) with the most economic process configuration. In carrying out this function the engineer is required to study various processing routes or changes to existing process routes and configurations. When satisfied that the technical aspect of any proposed changes or additions are feasible and sound he or she must now satisfy the company management that it is economically attractive. The proposal must also be shown to be the best of any possible alternatives that may be available to achieve the same objective.

This chapter sets out to describe some basic techniques used in carrying out this function.

Study approach

Process studies are usually initiated after a very precise definition by the company's management of the company's immediate and long term production objectives. Apart from this definition being a premise of the company's annual operating budget it also aims to fulfill its marketing strategy. Often such a definition results in changes being required to be made to the process facilities either in the short term or long term to meet these objectives. The company will look to the Development Engineers to provide definition of these changes and to support the definition(s) with whatever technical and economic data necessary for management to make their "GO/NO GO" decisions. Figure 17.1.3 shows the steps in a typical study.

This particular study route is to define an expansion of an existing process configuration by adding a new group of plants. The steps shown in Figure 17.1.3 are described as follows:

Step 1. Receive and review the supply departments requirements.
This step is extremely important. The development engineer or engineers working on the study must understand completely what is required and what the end product of the study must fulfill. There is usually also a time element for the completion of the work. This too must be understood and planned for. Normally the study premise is presented in a written document with copies to all study participants. The contents of the document must be studied by all concerned. A study 'Kick Off' meeting is always beneficial where participants and management review the study premises together before commencing work on the study.

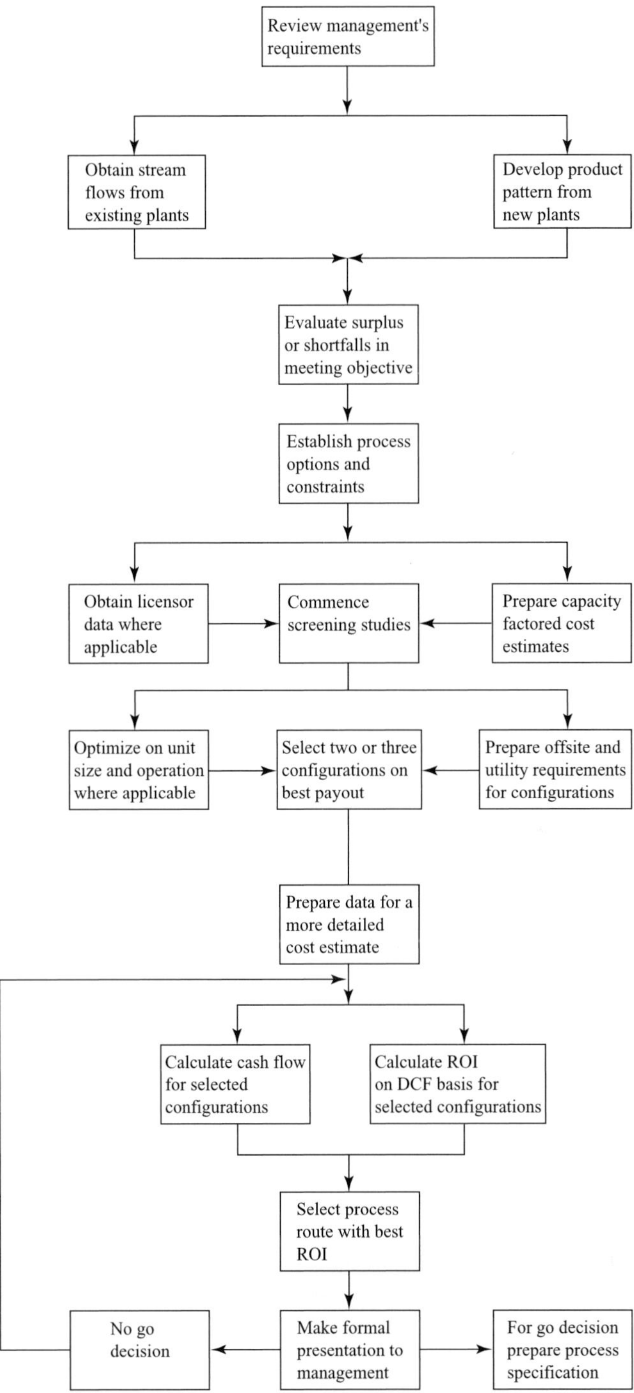

Figure 17.1.3. Steps in a typical process study.

Step 2. Obtain stream flows from existing plants.

In cases where existing plants are to form part of the study premise a detail analysis of the performance of the plants must be made. This almost certainly will involve performance test runs under pre-calculated operating conditions. The test run results are evaluated and included in the analysis of products to meet the study's product slate.

Step 3. Develop product pattern from plant data for new processes.

This will be executed in parallel with step 2. Here new processes that may be included in the study to meet the product requirements are examined and their product pattern developed. Process licensors may be involved at this stage to provide data on their proprietary processes. As may the refining company's own research department.

Step 4. Evaluate surplus and shortfalls in meeting the objective.

Steps 2 and 3 are now used to determine the possible product slates achievable using plant yields and blending recipes. Any surplus or shortfalls in the product slate that may now exist must be eliminated by further work in step 2 or 3 or both.

Step 5. Establish process options and constraints.

With a firm product slate work may now commence to map processing routes that will achieve the product objective. This will include very rough plant configurations irrespective of cost or other constraints that will fulfill a processing route. This step should end with the elimination of the more obvious undesirable routes. For example those routes which contain very small process units or those that contain unit processes that have undesirable by products. This step should provide 5 or 6 reasonable process options for further study.

Step 6. Commence screening studies.

The purpose of this item is to reduce the number of possible options to two or at the most three process configurations that appear most economically attractive. To arrive at the selected options those alternates arrived at in step 5 are allocated a capital and operating costs. The capital costs in this case are very rough estimates arrived at by applying factors to the installed cost for similar processes. Details of this technique in cost estimating are given later in this chapter in item 4 "Capital Cost Estimates". Licensors are usually requested to participate in the development of these cost items in the case of their proprietary processes. The operating costs for nonproprietary processes are also factored from similar processes. With these cost data in place an incremental pay-out time is calculated for each option. The option with the lowest capital cost is used as a base case and the others compared against it. Details of this calculation are given later in this chapter.

The two (or three) options with the lowest positive incremental pay-out time against the base case are selected for further study.

Step 7. Prepare data for a more accurate cost estimate.

The credibility of any economic study depends to a large extent on the accuracy of the data used in predicting the various cost items. The principle cost item will be the capital cost of the processes to be installed and/or revamped. The next step in the study therefore is for the development engineer to provide basic engineering for a "Semi Definitive" capital cost estimate to be prepared. (Very often the refining company may solicit third party Engineering Contractors to undertake this step.) Should the project proceed from study stage to installation, this estimate may be used as the first project budget estimate for project control. Details of this type of estimate is given later in this chapter. Briefly for this estimate development engineers (and/or third party engineering contractors) will:

- Prepare process flow sheets (complete with mass balances).
- Prepare preliminary mechanical flow sheets (complete with line and instrument Sizes and specifications).
- Initiate all equipment data sheets sufficient for obtaining manufacturers firm prices.
- Review and select manufacturers offers for equipment.
- Using selected manufacturer's equipment data, prepare a preliminary utility balance and size by factoring new or expanded utilities facilities.
- Factor additional offsite facilities as required.

Step 8. Calculate cash flow for selected configurations.

The selected manufacturers equipment offer will also include a delivery time for the equipment item. Using these and other experience factors prepare a preliminary schedule for the installation of the facilities. This is executed by the experienced schedule engineers of the Company or by accepted consultants.

Using this schedule the development engineer participates with others (Usually members of the company's project management department or finance departments) to split the budget down to annual payments over the construction period as scheduled. The exercise continues to allocate pay-out of the capital costs on an annual basis over a prescribed period after plant start-up. This is also discussed later in this chapter.

Step 9. Calculate the return on investment based on a discounted cash flow basis.

The development engineer proceeds to develop the return on investment using the cash flow developed in step 8 for the selected configurations. A technique using the discounted cash flow principle is used for this purpose. This provides a more accurate and more credible result than a ROI arrived at by the simple method where no account is taken of the construction time and cash flow. Details of ROI calculations are given later in this chapter.

Step 10. Prepare the study report and presentation to management.

Much of the report should be written as the study project proceeds. In this way facts which are still fresh in ones mind are recorded accurately. The final report must be concise but complete with all salient points. Techniques in report writing is not included as a subject here. Sufficient to say that communicating the study and its results effectively is as important as any technique and good engineering "know how" used in the work. The same applies to the final oral presentation to management. The study, the methodology, and finally, the results must be presented in a manner that maintains interest and stimulates the decision making mechanism.

This completes a typical study approach. It needs to be pointed out that should the screening study reveal an option which far surpasses any of the others studied in terms of pay-out this could be the only option that then needs the further study as described in steps 7–10. Such a situation saves a considerable time and effort.

Building process configurations and the screening study

(See also Chapter 2 of this Handbook).

From item 1 above it can be seen that the first major event, that involves in depth examination in a process study, is the screening of options. Here a very simple and preliminary economic comparison of the options is carried out. This comparison is viable because the basis for the economic criteria is the same for each of the options. It must be borne in mind that a prerequisite for carrying out any of these process studies is a good working knowledge of the processes involved. This example has been chosen because the ultimate product slate is achieved through the addition of new processes, changing operating conditions of existing units, and of course blending base stocks.

The example then is a requirement by a petroleum refinery to upgrade its fuel oil product to more financially lucrative lighter products. This is a fairly common program in this particular industry because of the fluctuating needs of the fuel market. The example calculation given here to illustrate the study is the base configuration published in an earlier book "*Elements of Petroleum Processing" by the author*. This may be found at the end of the first chapter of that book, and the block flow sheet is given here as Figure 17.1.4. The objective of this refinery configuration is to maximize gasoline and middle distillate but still retaining sufficient of the atmospheric residue from the crude distillation as fuel oil.

In the example given here it is intended to reduce the amount of fuel oil and to produce as much gasoline and diesel as possible to give the best return on investment. For this purpose six well proven processing routes have been selected for preliminary study. These are as follows:

- Thermally crack all of the atmospheric residue. This will be a single stage unit.
- Vacuum distill all of the atmospheric residue. The waxy distillate will be cracked in a fluid catalytic cracker producing gas, gasoline, naphtha, and light cycle oil as a diesel blend component. The vacuum residue from the vacuum unit will be thermally cracked to give additional gas, LPG, naphtha (for gasoline production), and gas oil for diesel blending.
- The same as option 2. except there will be no thermal cracker for the vacuum residue.
- Vacuum distill all the atmospheric residue and hydrocrack the waxy distillate.

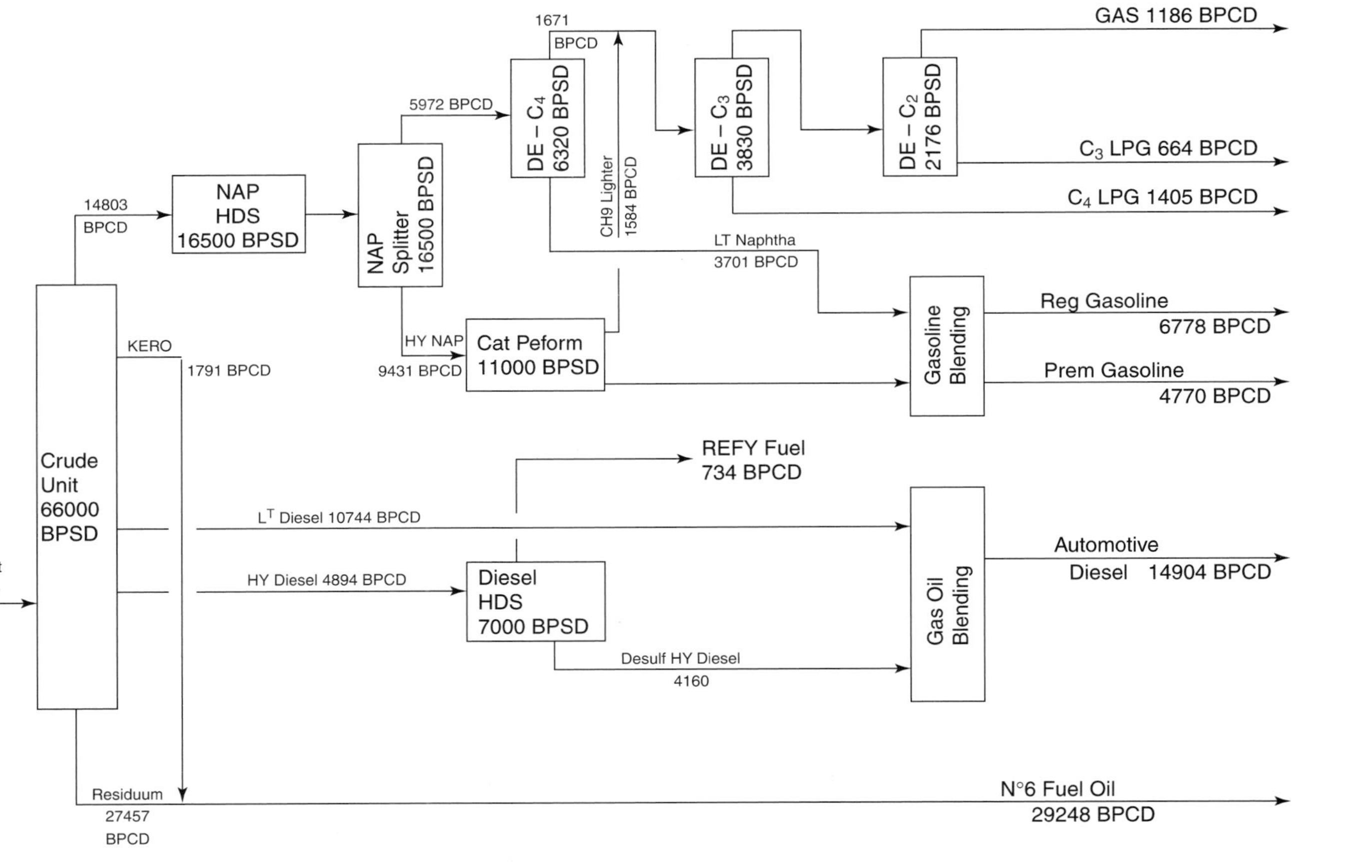

Figure 17.1.4. The existing refinery.

- Process the atmospheric residue through a delayed coking unit.
- Finally, use a proprietary fluid coking unit for the atmospheric residue that eliminates the coke by converting it into low Btu gas.

The last two options are eliminated after a preliminary study because the is no economically attractive outlet for the coke and no effective use for the low Btu gas in this refinery.

The example calculation that follows provides the detail calculation for Scheme 1 and provides the results of the same calculations for Schemes 2, 3, and 4. The configuration for Scheme 1 is given in the block flow sheet Figure 17.1.5. Similar block flow sheets would be prepared for the other study schemes but these are not shown here but a flow description is given for these three schemes.

Example calculation

It is required to develop a processing scheme in which the gasoline and diesel products are increased at the expense of the fuel oil product. The capacity of the present units, the flow configuration and flow rates are given in Figure 17.1.4. The refinery fence pricing of the products are historical and do not reflect recent escalation

	\$/bbl
LPG (C3 & C4)	20.2
Premium gasoline	35.2
Regular gasoline	27.6
Automotive diesel	28.6
No 6 Fuel oil	17.2

The cost of Kuwait Crude Oil is posted at 17.0 \$/bbl delivered to refinery tankage.

Cost of utilities and labor is as follows:

Power (purchased)	\$0.042/Kwh.
Water	\$ 0.5/1,000 US gals
Steam	\$ 0.83/1,000 lbs
Fuel	\$ 0.01/lb.
Labor	\$15/hr salary + 40 burdens. Three 8 hr shift plus one shift off. There will be 8 staff on each Shift. Total man hours per month is 185.

The calculation

Scheme 1. Thermal cracking

The first step is to determine the yields and the quality of the products that can be obtained by the thermal cracking of the atmospheric residuum. These are obtained

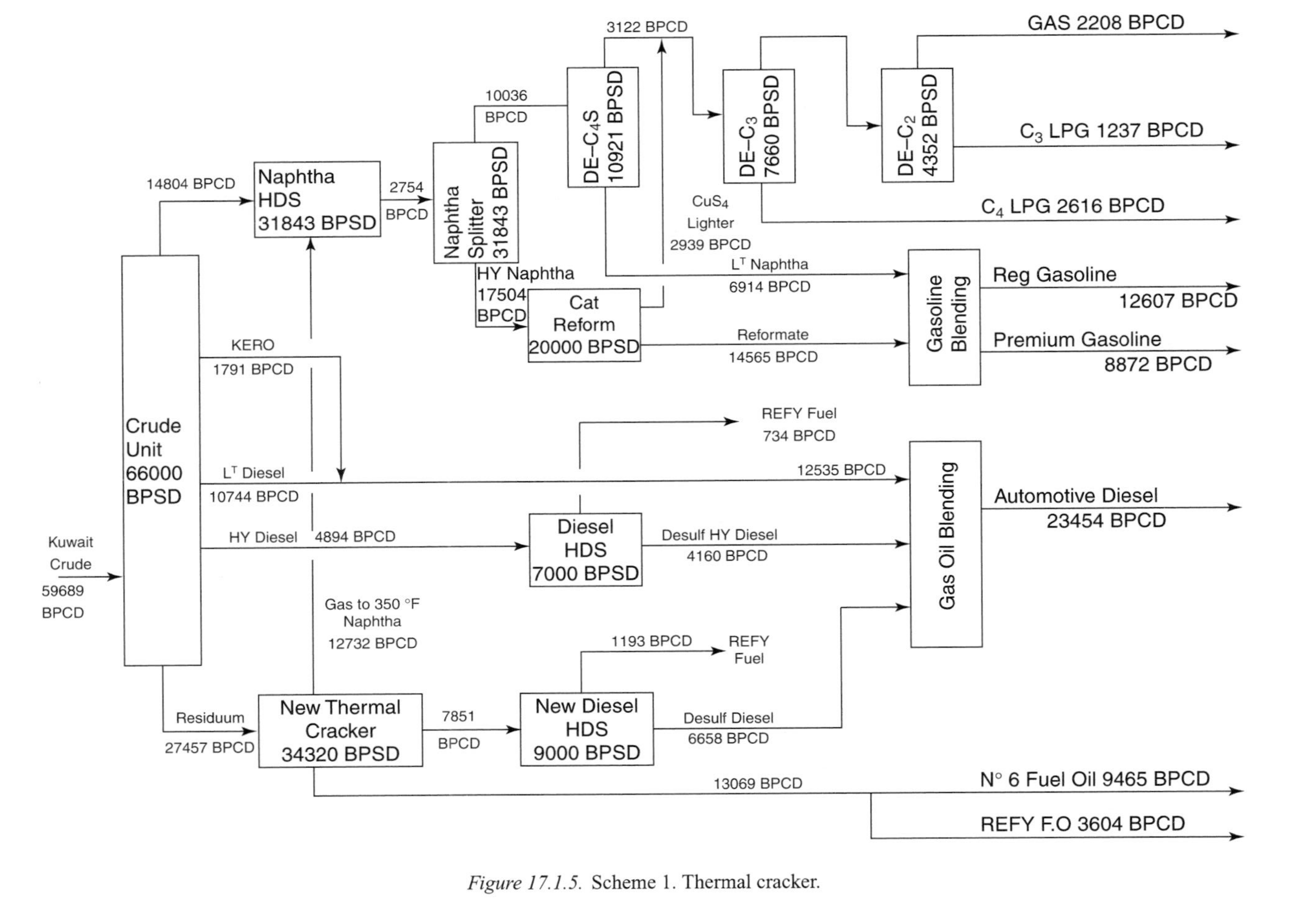

Figure 17.1.5. Scheme 1. Thermal cracker.

by methods given in the authors publication "*Elements of Petroleum Processing*" Chapter 17 and also in Chapter 1 of this Handbook. These are:

Stream	wt%	lbs/hr	lbs/gal	Gph	BPCD
Gas to C5	9.02	35,079	4.67	7,506	4,289
Naphtha to 390°F	20.29	78,899	5.34	14,775	8,443
Gas oil 390°F to 622°F	24.66	95,878	6.89	13,916	7,952
Residue +622°F	46.03	179,013	7.73	22,870	13,069
Total	100.00	388,869	6.58	59,067	33,753

Note: There is a volumetric increase in the total products over the feed. This is due to the changes in gravity of the products of cracking. The total weight and moles will still be the same between feed and products.

Handling the products

Figure 17.1.5 shows the routing of the product streams from unit to unit. The gas from the thermal cracker (Gas to C5) and the naphtha are routed to the existing naphtha hydrotreater which will be expanded in size to accommodate this. The hydrotreated material will subsequently go to the splitter and debutanizer, respectively. The existing reformer and light ends unit will also be expanded to cater for the thermal cracker light products. The gas oil from the thermal cracker will be hydrotreated in a new or expanded hydrotreater to meet diesel pool requirements. Thus, new unit sizes will be:

Naphtha hydrotreater: Total new flow = 14,809 BPCD + 12,732 BPCD
= 27,541 BPCD

$$\text{New size @ 86\% service factor} = \frac{27{,}541}{0.86} = 32{,}024 \text{ BPSD}$$

which is an increase of 72% above the existing unit. This is too large to consider a revamp therefore build a new unit 13,500 BPSD in parallel. The same applies to the splitter.

Catalytic reformer

New feed to the reformer is 16,191 BPCD. At a 86% service factor the new size will be:

$$\frac{17{,}504}{0.86} = 20{,}353 \text{ BPSD.}$$

Revamp the catalytic reformer by 9,000 BPSD.

Light ends

New total feed to the debutanizer is 9,283 BPCD. With a service factor of 90% this gives a new capacity of 10,000 BPSD, an increase of 58%. This unit will be revamped at a cost of 10% of a new unit.

The depropanizer and deethanizer will be expanded with parallel units the same size as the existing ones.

The diesel hydrotreater

A new hydrotreater of 9,000 BPSD will be designed to cater for the additional gas oil from the thermal cracker. The new total amount of gas oil to be treated is 4,894 BPCD + 7,851 BPCD = 12,745 BPCD. There is 706 BPCD spare capacity in the existing hydrotreater through improved catalyst and service factor. Taking this into consideration with a service factor of 80% then the additional capacity required will be 9,000 BPSD.

Operating costs

The following tables (Tables 17.1.5 and 17.1.6) give approximate factors for utilities and labor for some refinery plants. Note the use of these is acceptable for the screening analysis given here. For detailed economic analysis of the selected configuration a more accurate operating costs using certified vendor and licensor data must be used.

Looking at the configuration under study here, the operating cost for the extension of the refinery is as follows:

Total cost of utilities per stream year using 0.86 ave service factor
$= 28{,}227$ \$/day $\times .86 \times 365 = 8.840$ mm \$/year

Labor salaries at 15 \$/hr $= 15 \times 8 \times 4 \times 185$ hr/month $\times 1.4$.

Table 17.1.5. Utilities requirement per bbl of feed.

	Steam l	Water	Power	Fuel	Labor
	lbs	gals	Kw hr	mmBtu	Pers/Shift
Light end dist	25	12	0.4	–	1
Splitter	0	9	0.4	0.1	"
Cat reformer	50	15	12	0.35	3
Nap HDS	–	4	6	0.12	1
Diesel HDS	30	8	1.5	0.12	1
Thermal cracker	70	20	3	0.5	2
FCCU	20	20	7	–	4
Wash plants	1	2.5	0.4	–	1
VDU	15	20	0.2	0.07	3

Table 17.1.6. $/Calendar day for utilities (Exc Labor)

Unit	Steam	Water	Power	Fuel	Labor Pers/shift
Nap HDS (13,500 BPSD)	–	27	3,400	900	1
Nap splitter (1,350 BPSD)	–	60	227	750	1
Lt Ends (3,000 BPSD)	60	18	50	–	
Ther cracker (34,320 BPSD)	1,922	343	4,324	9,500	2
Diesel HDS (9,000 BPSD)	216	36	567	600	1
Cat reformer (9,000 BPSD)	360	68	4,536	263	3
Total	2,558	552	13,104	12,013	8

Notes on labor:

1. The salaries are increased by 40% to take up government benefits, paid days off, vacation etc.
2. There are four shifts three on duty and one off every 24 hr.
3. Hours worked per person per month is 185 hr.

$$\begin{aligned}\text{Then total labor cost per year} &= 124{,}320\ \$/\text{month} \times 12\\ &= 1.491\ \text{mm}\ \$/\text{year.}\\ \text{Total operating cost per year} &= 10.331\ \text{mm}\ \$.\end{aligned}$$

Investment costs for the new facilities

There are two costs which make up the overall investment outlay. These are:

- Capital cost of the plant
- Associated costs

The capital cost

For the screening study a capital cost of plant estimated on experience and factors are used. This and other types of estimates are described later in this chapter. A set of capacity cost data for plants of this size are listed below:- Note that these numbers are pure fiction and bear no relationship to any company's cost data.

Type of unit	*Capital cost factor*
Light end units	412 $/BPD
Catalytic reformer	3,120 $/BPD
Hydrotreater (naphtha)	780 $/BPD
Hydrotreater (Diesel)	1,200 $/BPD
Wash plants	170 $/BPD
Naphtha splitters	360 $/BPD

Thermal crackers	2,000 \$/BPD
Fluid catalytic crackers	3,960 \$/BPD
Vacuum distillation units	700 \$/BPD

Associated costs
To complete the full investment picture the following costs must be included as part of the investment. These are:

- Cost of first catalyst and chemical inventory.
- Cost of licensing fees for proprietary processes.
- Cost of additional utility and offsite facilities. This cost is included because additional processing plants will either require additions to utilities & offsite facilities or take up available spare capacity. This will result at some future date an expansion to these facilities which would not be necessary if this process expansion had not occurred.

For the purpose of a screening study a figure of 15% of the capital cost is used for these associated costs. In more definitive studies this item would be developed with more detail.

The capital cost of the process expansion considered here is as follows:

New naphtha hydrotreater
Capacity 13,500 BPSD @ 780 \$/bpsd = 10,530,000 \$

New naphtha splitter

Capacity 13,500 BPSD @ 360 \$/bpsd = 4,860,000 \$

Revamp debut and new light ends

$$\begin{aligned}\text{Debutanizer revamp} &= 10\%\text{ of new plant} = \frac{2{,}963 \times 412}{10}\\ &= 122{,}000\,\$\\ \text{New light ends unit} &= (3{,}830\,\text{bpsd} \times 412) + (2{,}176 \times 412)\\ &= 2{,}474{,}472\,\$\\ \text{Total} &= 2{,}596{,}500\,\$\end{aligned}$$

New thermal cracker
At a service factor of 80% the capacity of the unit is

$$\begin{aligned}\frac{27{,}457\ \text{BPCD}}{0.8} &= 34{,}320\,\text{bpsd}\\ \text{Capital cost} &= 34{,}320 \times 2{,}000\,\$\\ &= 68{,}600{,}000\,\$.\end{aligned}$$

New diesel hydrotreater

$$9{,}000\ \text{bpsd} \times 1{,}200\,\$ = 10{,}800{,}000\,\$.$$

Revamped catalytic reformer

$$9{,}000\ \text{bpsd} \times 3{,}120\,\$ = 28{,}100{,}000\,\$.$$
$$\text{Total capital cost} = 125.490\,\text{mm}\,\$.$$

Associated costs

$$\text{This is taken as } 15\% \text{ of capital cost} = 18.823\,\text{mm}\,\$.$$
$$\text{Total cost} = 144.313\,\text{mm}\,\$.$$

The product slate

The new refinery product slate is summarized below (all in BPCD):

Product	Existing	Expanded	Difference
C3 LPG	664	1,237	+ 573
C4 LPG	1,405	2,616	+ 1,211
Reg gasoline	6,778	12,607	+ 5,829
Prem gasoline	4,770	8,872	+ 4,102
Auto diesel	15,638	23,454	+ 7,816
Fuel oil	26,323	9,465	– 16,858
Refy fuel gas	1,186	4,135	+ 2,949
Refy fuel oil	2,925	3,604	+ 679
	*	*	
TOTAL	59,689	65,990	+ 6,301

*Note difference in BPCD between input and output is due to changes in SG's due to cracking. Mass (lbs/hr) In still equals mass Out.

Gross income from increase in product slate

Product	Increase BPCD	$/bbl	$/day
LPGs	1,784	20.2	36,037
Reg gasoline	5,829	27.6	160,880
Prem gasoline	4,102	35.2	144,390
Auto diesel	7,816	28.6	223,538
Fuel oil	−16,858	17.2	−289,906
Refy fuel	−3,628	17.1	−62,039
Gross income			212,900 $/CD

Table 17.1.7. Scheme 1. Addition of thermal cracker simple economic analysis

Item	$ mm
Capital cost of plant	125.490
Associated costs	18.823
Total investment	144.313 mm $/year
Gross income	77.709
Less	
Operating cost	10.331
Depreciation	8.030
Net income before Tax	59.348
Tax @ 42.5 %	25.223
Net income after Tax	34.125
Pay-out time	3.4 years
Return on investment	23.7%

Depreciation

In the economic analysis of the various processes the depreciation of the plant value must be taken into account. Details of depreciation will be described and discussed later in this chapter. For the purpose of the screening study the following depreciation factors may be used:

Crude distillation units	6.6% of capital cost per year
Cracking units	6.6% of capital cost per year
Light end units & gas plants	5.0% of capital cost per year
Hydrotreaters	6.0% of capital cost per year

The screening analysis based on a pay-out time and a Return on Investment (ROI) is given in the following table (Table 17.1.7). Note the ROI is given by

$$\frac{\text{Net Income After Taxes}}{\text{Net investment cost}} \times 100 = \%\,\text{ROI}$$

and Pay-out time is given by

$$\frac{\text{Net investment cost}}{\text{Net income after taxes} + \text{depreciation}} = \text{Years.}$$

Details of the other schemes
Scheme 2. This calls for the vacuum distillation of the atmospheric residue to produce a waxy distillate and a heavy vacuum residue. The waxy distillate is cracked in a FCCU (Fluid Catalytic Cracker) to give gas, LPG, naphtha for gasoline, gas oil for

diesel. The heavy vacuum residue is also cracked in a thermal cracker to give gas, LPG, naphtha feed to the cat reformer, gas oil for diesel.

The total capital cost for these additional facilities = 218.580 mm $.
Total additional product income = 108.202 mm $/year and
Total operating cost is an additional 13.856 mm $/year.

Scheme 3. This is the same as scheme 2 except the vacuum residue is not cracked. Details of this scheme are:

Total capital costs for additional facilities = 194.780 mm $.
Total additional product income = 96.509 mm $/year.
Total operating cost is an additional 8.122 mm $/year.

Scheme 4. This also includes the vacuum distillation of the atmospheric residue as in schemes 2 and 3.The wax distillate in this case however is routed to a hydrocracker where LPG, naphtha for reforming to gasoline, and diesel is produced. An additional hydrogen plant is required for this scheme. Details of the scheme are summarized below:

Total capital cost for additional facilities = 306.012 mm $
Total additional income 119.502 mm $/year
Total operating cost is an additional 12.6 mm $/year.

Table 17.1.8. Comparison of the economic analysis of the schemes studied

Scheme	1	2	3	4
	mm $			
Capital cost	125.490	218.580	194.780	306.012
Associated costs	18.823	32.828	29.217	46.080
Total investment	144.313	251.408	223.997	352.092
	mm $/year			
Gross income	77.709	108.202	96.509	119.502
Less				
Operating cost	10.331	13.856	8.121	12.600
Depreciation	8.030	15.007	13.237	18.361
Net income before tax	59.348	79.339	75.150	88.541
Tax @ 42.5%	25.223	33.719	31.939	37.630
Net income after tax	34.125	45.620	43.211	50.911
Payout time years	3.4	4.15	3.97	5.1
Return on investment%	23.7	18.1	19.2	14.4

The obvious selection in this case is scheme 1 with the lowest pay-out time and the highest ROI.

Preparing more accurate cost data

Once a selected configuration or study case or cases have been selected from the screening study it is now necessary to firm up as much as possible the cost data used in those studies. Three items of data fall into this category. These are:

- The capital cost estimate
- The associated costs
- The operating cost

Development engineers supply the basic data from which an update can be arrived at. This level of data is arrived at from a process design of the facilities making up the case studies. The process design will include a material balance and a firm process flow sheet. Sufficient data will be generated for all equipment to allow manufacturers to give a good and realistic equipment budget cost. Where applicable too the manufacturers will be requested to provide equipment efficiencies from which utility usage for the various items can be calculated. Where options exist within the process design itself these will be optimized for equipment sizing and layout. An example of such an optimization is the case of the process heat exchange system. Full and complete equipment data sheets need not necessarily be developed for this part of the study. Usually manufacturers are able to provide good budget costs from equipment summary sheets similar to those given in the Example Calculation below.

Where licensed processes are included in the case study the Development Engineer will need to develop a specification to a suitable licensor soliciting:

- A budget estimate for the plant (installed)
- The utility requirements and the operating cost
  (*Note:* In the case of a catalytic process part of the operating cost will be the catalyst usage or loss in operation.)
- First inventory of catalyst or chemicals where applicable
- The licensing fee

The specification must include the capacity of the plant and the required product quality. It must also of course give the licensor(s) details of the feedstock, utilities conditions, and in most cases local climatic conditions. The additional information concerning feed and product pricing together with utilities costs will enable the licensor to respond with a firmer and optimized cost data.

H301 - Thermal cracker heater
Heating duty 121.789 mm BTU/h
Heat of reaction 249.106 mm BTU/h
Total duty 370.895 mm BTU/h

T301 - Syn tower
6'0" i/d and 13'5" i/d 108' T-T

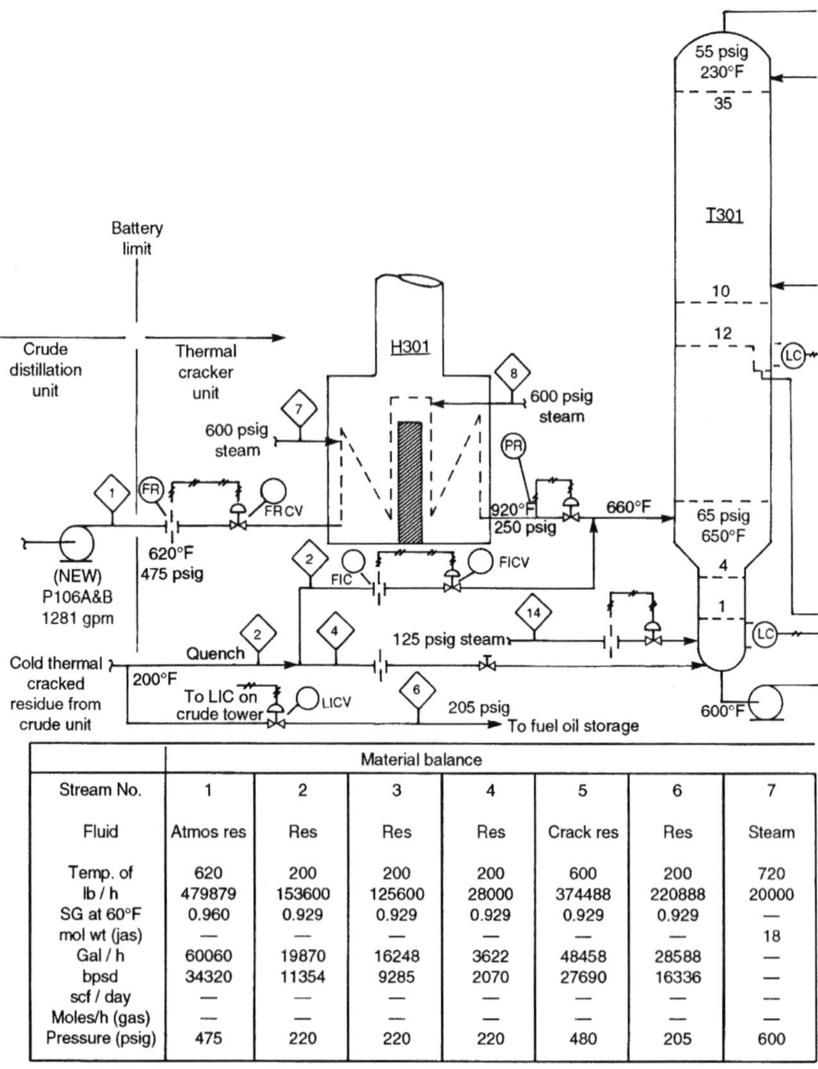

	Material balance						
Stream No.	1	2	3	4	5	6	7
Fluid	Atmos res	Res	Res	Res	Crack res	Res	Steam
Temp. of	620	200	200	200	600	200	720
lb / h	479879	153600	125600	28000	374488	220888	20000
SG at 60°F	0.960	0.929	0.929	0.929	0.929	0.929	—
mol wt (jas)	—	—	—	—	—	—	18
Gal / h	60060	19870	16248	3622	48458	28588	—
bpsd	34320	11354	9285	2070	27690	16336	—
scf / day	—	—	—	—	—	—	—
Moles/h (gas)	—	—	—	—	—	—	—
Pressure (psig)	475	220	220	220	480	205	600

Figure 17.1.6. The process flow sheet for a thermal cracker.

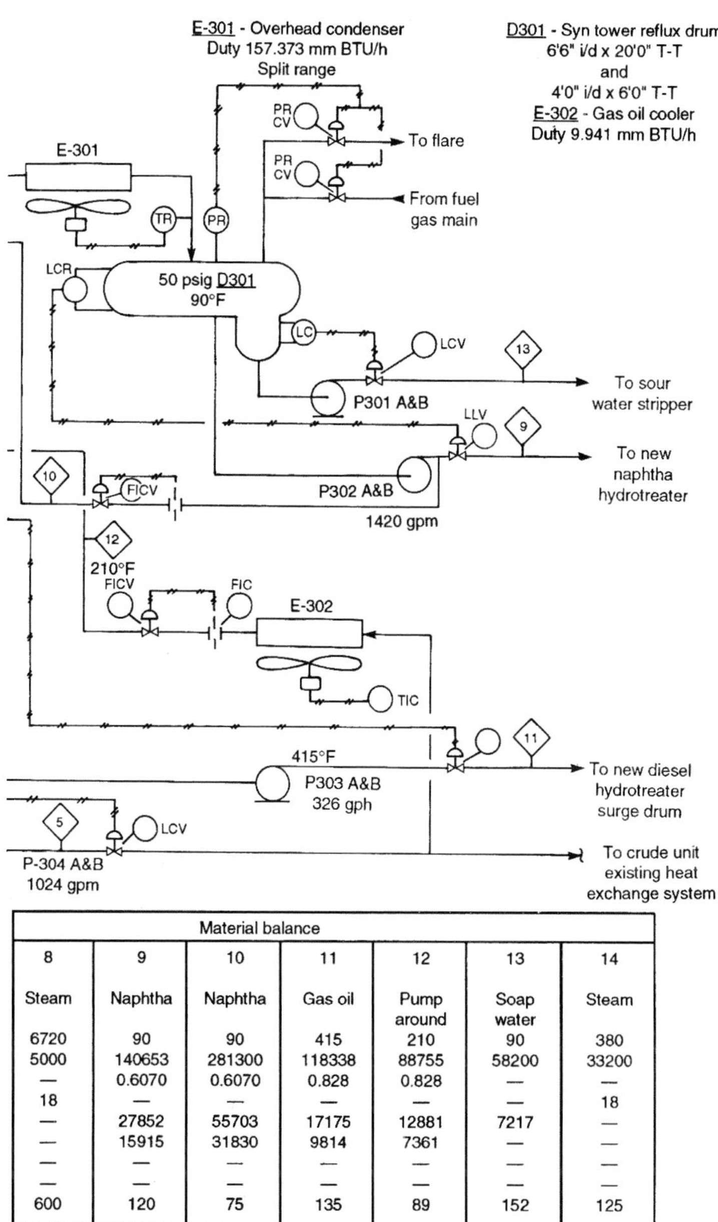

Material balance						
8	9	10	11	12	13	14
Steam	Naphtha	Naphtha	Gas oil	Pump around	Soap water	Steam
6720	90	90	415	210	90	380
5000	140653	281300	118338	88755	58200	33200
—	0.6070	0.6070	0.828	0.828	—	—
18	—	—	—	—	—	18
—	27852	55703	17175	12881	7217	—
—	15915	31830	9814	7361	—	—
—	—	—	—	—	—	—
—	—	—	—	—	—	—
600	120	75	135	89	152	125

Figure 17.1.6. (Cont.)

The following example follows the study screening given in Item 2.0 above.

Example calculation

From the screening study the new plants included in the selected case (which was case 1) were as follows:

- Naphtha hydrotreater (13,500 BPSD) licensed
- Naphtha splitter (13,500 BPSD)
- Catalytic reformer (9,000 BPSD) licensed
- De-propaniser (3,830 BPSD) and de-ethanizer (2,176 BPSD)
- Diesel hydrotreater (9,000 BPSD) licensed
- Thermal cracker (34,320 BPSD)

The existing debutanizer is to be revamped and a new sized equipment list will be prepared for this unit.

The units of normally licensed processes are indicated as such in the above list. The thermal cracker may be licensed or can be designed by some major engineering companies as a nonproprietary process. It is a major item of cost and for this example the degree of process engineering required at this stage of the study is demonstrated for this unit.

The process design of the thermal cracker

The process flow sheet for this unit is given in Figure 17.1.6. This process flow sheet contains the overall material balance for the unit together with the calculated reflux and quench flows. Major process stream routes are shown with stream temperatures where appropriate. The material balance table also gives the flow conditions (temperature and pressure) for those major streams. The pump capacities are shown on the flow sheet next to the respective pump. These quantities are flows in gallon per minute at the pump temperature.

The major item in a thermal cracker is the furnace or heater. This has two duties, the first being to heat the feed to its reaction temperature and the second to maintain the feed at its reaction temperature for the prescribed period of time. Note therefore that the duty of this heater is shown on the flow sheet as the duty to heating the oil (which also includes the steam in this case) and the duty to supply the heat of reaction.

Data for the remainder of the equipment are given in Tables 17.1.9, 17.1.10, and 17.1.11. Sufficient data for obtaining a cost and efficiency for the heater is provided in a "Process Specification" shown in Table 17.1.12. A special specification sheet

Summary data sheets

Table 17.1.9. Centrifugal pumps

Item No	P106 A&B	P301 A&B	P302 A&B	P303 A&B	P304 A&B
Service	Feed	S water	REF +PR	PA/PROD	Bot Prod
No of units	1 + 1 spare	1 + 1 spare	1 + 1 Spare	1 + 1 spare	1
Fluid	Resid	Water	Naphtha	Gas oil	Cr res
lbs/hr	479,879	58,200	421,953	207,093	374,488
Pump temp °F	620	90	90	415	600
Sg @ PT	0.758	0.995	0.592	0.692	0.732
Visc @ PT Cs	1.8	0.76	0.33	0.25	1.2
Rate (norm) @ PT GPM	1,281	117	1,420	599	1,024
Rate (Max) @ PT GPM	1,470	130	1,700	690	1,178
Norm suct press PSIG	39	63	63	75	72
Max suct press PSIG	89	132	132	135	122
Discharge press PSIG	475	152	85	135	480
Diff press PSI	436	89	22	60	408
Diff head Ft	1,332	207	86	200	1,290
NPSH avail Ft	>15Ft	>15Ft	>15Ft	>15Ft	>15Ft
Hyd horse power	323	6.1	18.3	20.9	244
Corr/erros	Sulph	H2S	H2S	Sulph	Sulph
Driver R (Note 2) S	Motor	Motor	Motor	Motor	Turb
	Turb	Motor	Turb	Motor	Note 1

Notes: 1.0. May be common spared with P106 A&B. 2.0. Driver R = Normally running, S = Normally spare.

Table 17.1.10. Vessels

Item No	T 301	D301		
Service	Syn Tower	Ref drum		
Type	Vert	Horiz		
Dimensions:	*Feet*	*Feet*	*Feet*	*Feet*
Top Sect I/D				
Top Sect T-T	13.5	6.5		
Bot Sect I/D	74.0	20.0		
Bot Sect T-	7.5	4.0		
Overall T-T	31.0	6.0		
Skirt	108.0(1)	20.0		
	15.0	In Struc		
Internals:				
Trays	Top Sect			
	30 sieve	None		
	Bot Sect			
	4Disc Don			
Operating conditions:				
Press PSIG				
Temp °F	55/67	50		
	230/650	90		

(Cont.)

Table 17.1.10. (Cont.)

Design conditions:		
Press PSIG		
Temp °F	125	125
Vacuum PSIA	920	810
Min Temp °F	7.0	7.0
Max Liquid Level Ft	90	90
	8.0	N/A
Materials:		
Shell	C.S (2)	C.S
(Corr allow)	0.125″	0.125″
Internals	11/13 Cr	
(Corr Allow)	0.125″	
Trays	Monel/Cr	None
(Corr Allow)	0.125″	
Packing.	N/A	
Insulation:		
Yes/No	YES	No
Stress relieved:		
Yes/No	No	Yes

Notes: 1. Swage Sect 3.0 ft 2. Clad with 11/13 Cr.

Table 17.1.11. Air condensers/coolers

Item No	E 301		E302	
Service	O/Head cond		Gas oil Cooler	
Type	Forced		Forced	
Duty mmBtu/hr	157.373		9.941	
Fluid	Nap + H_2O		Gas oil	
Temp °F	In	Out	In	out
	230	90	415	210
Liquid				
M lbs/hr HC		422	88.8	88.8
H_2O		58.2	nil	nil
Sg @ 60°F HC		.607		.828
Visc @ 60 Cs HC		.12		.52
Therm cond HC		.07		.06
Specific Heat Btu/lb °F HC		.73		.80
Vapour				
M lbs/hr HC	422			
H_2O	58.2			
Mol wt HC	96			
Therm Cond	.009			
Specific Heat Btu/lb °F	.41			
Fouling HC factor	—			
Ambient Air Temp °F	60			
Allowable Δpsi	5.0		10.0	

Table 17.1.12. Process specification for a thermal cracking heater/reactor

Item number: H 301 Thermal cracker heater
Overall heat duty: 370.895 mmBtu/hr.
Feedrate lbs/hr: 479,879 lbs/hr.
Temperature of feed in: 620°F temperature of effluent: 920°F.
Specific gravity of feed @ 60°F: 0.960 (+650°F TBP Kuwait Residue).
Effluent TBP and EFV curves are attached Figure 1.
Effluent Sg curves versus mid Bpt curve is attached Figure 2.
Feed conversion to gasoline of 390°F cut point = 30 vol%.
Coil temperature profile (oil + steam) versus coil volume given as Figure 3 attached.
Steam (600 psig) injected at coil volume of 50 cuft and 150 cuft, respectively. Total steam to be injected is 74,000 lbs/hr.
Outlet pressure of effluent to be 250 psig. This pressure to be maintained by downstream pressure control valve (supplied by others).
Coil pressure drop to be 225 psi max.

is required for this item because besides being an oil heater it is also a cracking reactor. In this function the design of the unit must be such as to allow the oil to be retained for a specific period of time at the reaction temperature.

Coil tubes should be standard 4″ 11/13 Cr with wide radius bends. The firebox should be constructed to heat the incoming feed to a temperature of about 870°F in the first 150 cuft of coil. This should constitute the HEATER SECTION. The SOAKER SECTION and crossover should be designed to heat the oil to 920°F and retain it at this temperature. The sections should be divided by a firebrick wall so that temperatures on both sides of the wall can be controlled independently by burner adjustments.

Table 17.1.13. Summary of accepted prices

			Price $/unit		
Pumps	HHP	BHP	Pump	Motor	Turbine
P 106 A	323	430	606,826	489,082	–
106 B			606,826	–	503,903
P 301 A	6.1	10.1	14,253	19,563	–
301 B			14,253	19,563	–
P 302 A	18.3	28.1	39,655	39,126	–
302 B			39,655	–	40,312
P 303 A	20.9	32.2	45,441	48,908	–
303 B			45,441	48,908	–
P 304 A	244	325	458,648	–	453,513
Total			1,870,998	665,150	997,728

The above are accepted vendor prices for equipment delivered to site, subject to updates
Total pumps and drivers = 3,533,884 $.

Vendor may recommend that the convection side of the heater be used for additional heater capacity or for steam generation or to remain unused if better temperature control may result.

Vessels

• T 301 syn tower

Shell and heads	1,766,810 $
Trays	1,590,129 $
Internals	176,681 $
Total	3,533,620 $

• D 301 Reflux drum

Shell and heads	561,222 $
Lining	37,415 $
Internals	24,943 $
Total	623,580 $

Total vessel account = 4,157,200 $

Coolers and condensers

• E 301 Overhead condenser

Unit including fans & structure = 1,627,540 $.

• E 302 Gas oil cooler

Unit including fans = 243,196 $ (unit located in pipeway, no stand alone structure).

Total cooler & condenser account = 1,870,740 $

Heater/reactor

• H 301 Thermal cracker heater

This unit will be field erected. Included in this account is the cost of fabricated material delivered to site.

Unit material cost = 8,314,400 \$.

Total equipment material account summary

	\$
Pumps and drivers	3,533,884
Vessels	4,157,200
Cond & coolers	1,870,740
Heater/reactor	8,314,400
Total	17,876,224

Capital cost estimates

Capital cost estimates are usually prepared by specialist engineers who are fully conversant with their company's estimating procedures and their company's material and labor cost records. Refinery development engineers are involved in cost estimating only so far as they develop and provide the basic technical data that is used to prepare the capital cost of a plant or process. To execute this part of their duty properly development engineers should know a little about cost estimates and the significance of the various items included in them. This section therefore sets out to define in broad terms the various levels of estimating, their degree of accuracy, and the scope of the engineering development that is required for each of them.

The accuracy of cost estimates will vary between that based on comparing similar plants or processes on a capacity basis to the actual cost of a plant when all the bills have been paid and the plant is operational. Obviously the accuracy based on similarity of the process or plant is very low while that based on project end cost must be 100%. That is it is no longer an estimate but is now a fact. In between these two levels however estimates may be prepared giving increasing degree of accuracy as engineering and construction of the process advances. The accuracy of estimates are enhanced as more and more money is committed to the project. The progress of a project uses milestone cost estimates coupled with completion schedules for its measurement and control. In the life of a construction project therefore about four "control estimates" may be developed. For want of better terms these may be referred to as:

- The capacity factored estimate
- The equipment factored estimate
- The semi definitive estimate
- The definitive estimate

Companies have their own terms for these estimates but the important point is to develop these estimates at a time in the project when they can be most useful and when sufficient information is available for their best accuracy. Figure 17.1.7 shows typical phases of a project when these estimates may be best developed.

The following paragraphs now describe these estimates in more detail and emphasize the process engineer's input.

The capacity factored estimate

In most process studies this is the first estimate to be prepared. It is the one that requires the minimum amount of engineering but is the least accurate. This is because the plant that is used to factor from will not exactly match the plant in question. It is good enough however when comparing different processes where the estimates are on the same basis.

Past installed costs of similar plants (definitive estimate) are used coupled with some experienced factors to arrive at this type of estimate. The cost of a higher or lower capacity plant is obtained from the equation:

$$C = K(A/B)^b$$

where

C = cost of the plant in question.
K = known cost of a similar plant of size B.
A = is the capacity of the plant in question (usually in vol/unit time).
B = capacity of known plant.
b = is an exponential factor ranging between 0.5 and 0.8.

The estimator's experience and records of the type of plant determines the exponent to be used. Usually this is 0.6; a listing of these factors are given in Appendix 17.7. Table 17.1.14 shows an example of a capacity factored estimate for the thermal cracking unit used in Item 3 above.

In this example the direct field cost is used as the cost figure to be capacity factored. This allows the cost estimator to adjust between the direct field cost and the total cost using current experience for field indirects, office costs, escalation and contingencies. In this case a multiplier of 1.5 times the direct field cost is used to collect these items.

The equipment factored estimate

This estimate requires a substantial amount of process engineering to define the specific plant or process that is to be estimated. Briefly the following process activity is required:

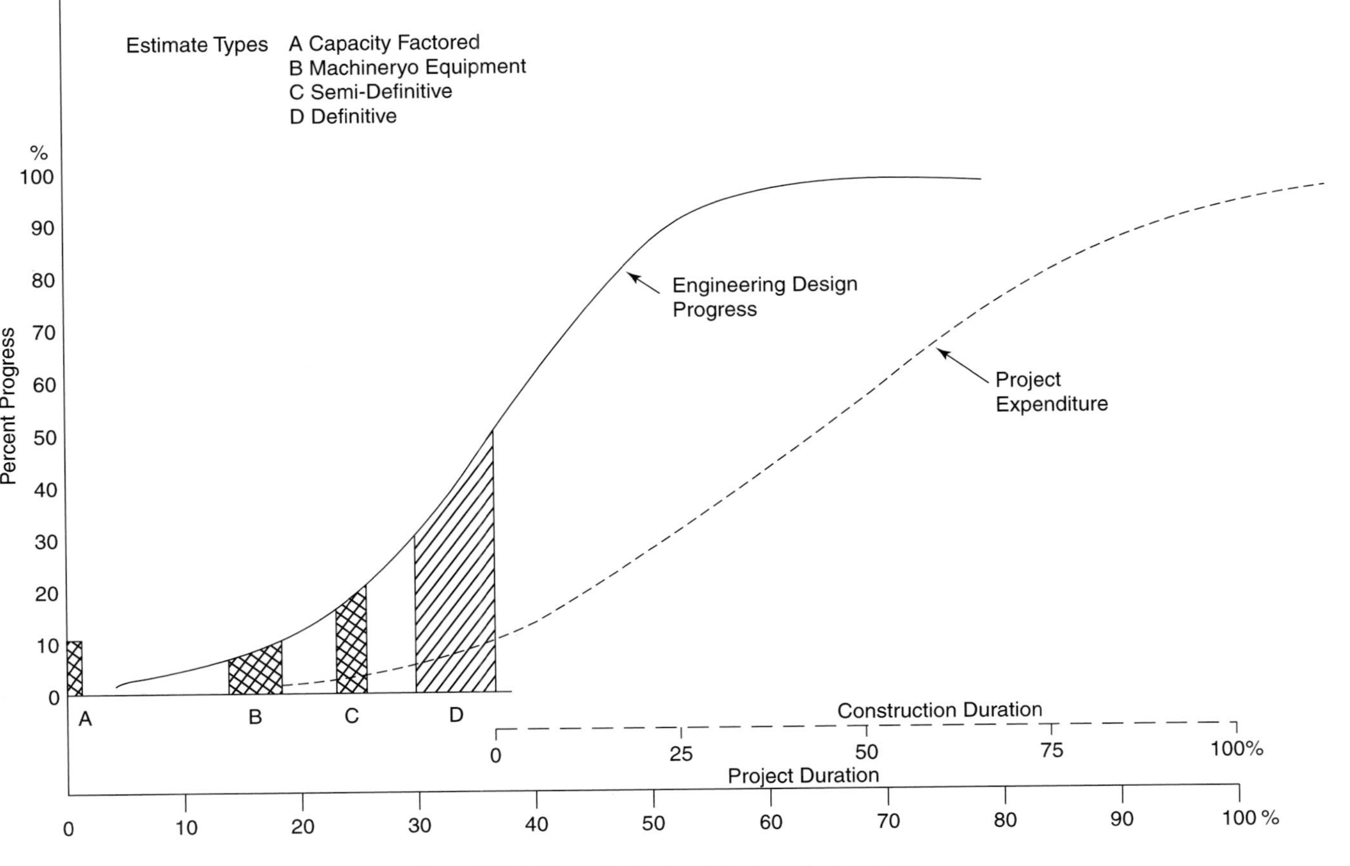

Figure 17.1.7. Type of estimate relative to project progress.

Table 17.1.14. The capacity factored estimate

Plant:		Description:		Date:		
A/C			Estimated cost, mm$			
No	Item	M/hr	Labor	Subcont	Material	Total
	Excavation		0.6			
	Concrete	$C =$	$K\,(A/B)$			
	Structural steel	Where	$K = 28.93$ m	m $		
	Buildings		$A = 34{,}320$	BPSD		
	Machinery/equip		$B = 16{,}000$	BPSD		
	Piping					
	Electrical	C	$= 1.57\,\times$	28.93 mm$	= 45.733	
	Instruments					
	Painting/scaffold					
	Insulation					
	Direct field cost					45.733
	Temp const facil					
	Constr expense					
	Craft benefits and burdens					
	Equip rental					
	Small tools					
	Field O/heads					
	Indirect field costs.					
	Total field costs					
	Engineering Salaries					× 1.5
	Purchasing Sal					
	Gen Office Sal					
	Office Expense					
	Payroll Burdens					
	Overhead Costs					
	Total office costs					
	Total field & office costs					
	Other Costs					
	Total cost					68.59

- Development of a firm heat and material balance
- An acceptable process flow sheet to be developed
- An approved equipment list
- Equipment summary process data sheet for all major equipment
- Process specification for specialty items
- A detailed narrative giving a process description and discussion. (This will be required for the management review and approval of the Estimate.)

This estimate is normally developed by the process engineer and the cost estimator only. There may be some participation from equipment vendors and licensors in the case of licensed processes. There is seldom any further input required from other engineering disciplines for this level of estimating.

The equipment summary process data sheets and any process specifications that may be required are used to solicit equipment prices from pre selected vendors. In addition to the pricing data that will be used for the estimate development vendors should also be asked for delivery time required, and sufficient details to enable the operating cost of the item to be calculated. The delivery time required is important to management to establish a project schedule and life span for the engineering and construction of the facilities. Later process engineers will also require this schedule to develop cash flow curves and data.

On receipt of the vendor's pricing data the cost estimator will apply factors to each of the equipment account data to arrive at a Direct Field Cost for each group of equipment. This factored cost will be the multiplier of the equipment price which will include the materials and labor to actually install those items on site. In Table 17.1.16 this factor is shown as a composite for the *whole* equipment prices to direct field cost. In this case the figure developed was a multiplier of 2.5. These factors are proprietary to respective companies but most major companies do have the statistics to develop their own factors.

The total cost is again obtained by factoring the DFC (direct field cost) as in the previous 'Capacity Factored' Estimate. Note in the case of this example of an equipment factored estimate a factor of 1.53 has been used slightly higher than that in Table 17.1.14. This is to account for the field erection of the heater, which was not known previously. This would impact on future home office engineering and field indirects to some extent (Table 17.1.15).

The semi-definitive and the definitive cost estimates

It is not proposed to describe these two estimates in detail here. Sufficient to say that these estimates will include input from all engineering disciplines that are involved in the engineering and construction of the plant or process. Indeed after the equipment factored estimate the input from process engineers is merely updates of the process design as it develops. However, Tables 17.1.16 and 17.1.17 shows the increasing amount of actual data as opposed to factored data that is included in these two estimates.

The numbers for these estimates are not included in these tables but the crosshatched items show where normally firm cost data would be included. The factored data are also indicated for both tables.

Table 17.1.15. The equipment factored estimate

Plant:		Description:	Date:			
A/C			Estimated cost, mm$			
No	Item	M/hr	Labor	Subcont	Material	Total
	Excavation					
	Concrete					
	Structural steel					
	Buildings					
	Machinery/Equip				17.876	
	Piping					
	Electrical					
	Instruments					× 2.5
	Painting/scaffold					
	Insulation					
	Direct field cost					44.690
	Temp Const Facil					
	Constr Expense					
	Craft Benefits and burdens					
	Equip Rental					
	Small Tools					
	Field O/heads					
	Indirect field costs					
	Total field costs					
	Engineering salaries					
	Purchasing sal					× 1.53
	Gen office sal					
	Office expense					
	Payroll burdens					
	Overhead costs					
	Total office costs					
	Total field & office costs					
	Fee (contractor)					
	Escal/contingency					
	Total costs					68.377

Although major manufacturing companies such as large oil companies and chemical companies have sufficient statistical data to develop semi definitive and definitive estimates these are normally left to the engineering and construction companies. These companies develop the estimates and use them during the course of the installation project as cost control tools for project management.

Accuracies of estimates and contingencies

For an estimate to be a meaningful basis of a process study its accuracy needs to be established and sufficient contingency be added to make the final cost figure as realistic

Table 17.1.16. Semi-definitive cost estimate summary

Plant: Description: Date:

A/C			Estimated cost, mm$			
No	Item	M/hr	Labor	Subcont	Material	Total
	Excavation				xxxxxxxxx	
	Concrete				xxxxxxxxx	
	Structural steel				xxxxxxxxxx	
	Buildings				xxxxxxxxxx	
	Machinery/equip				17.876	
	Piping				xxxxxxxxx	
	Electrical				xxxxxxxxx	
	Instruments				xxxxxxxxx	
	Painting/scaffold					
	Insulation					
	Direct field cost	Xxxxxxxx	xxxxxxxx		xxxxxxxxx	Xxxxxxxx
	Temp const facil					
	Constr expense					
	Craft benefits & burdens					
	Equip rental			FACTOR		
	Small tools					
	Field O/heads					
	Indirect field costs					Xxxxxxx
	Total field costs					Xxxxxxxx
	Engineering salaries					
	Purchasing sal					
	Gen office sal					
	Office expense					FACTOR
	Payroll burdens					
	Overhead costs					
	Total office costs					Xxxxxxxx
	Total field and office costs					Xxxxxxxx
	Fee (contractor)					Xxxxxxx
	Escal/contingency					Xxxxxxx
	Total costs					Xxxxxxxx

as possible. Unfortunately this is easier said than accomplished. Most companies have however developed analysis programs to meet this requirement and many of these are quite sophisticated. Engineering and Construction companies whose daily work depend on good cost estimates for project control are among the front runners in this exercise. Most of these programs revolve around principles of Risk Analysis. Figure 17.1.8 has been developed from a collection of statistical values over a long period of time and does provide some quick guidance to judging accuracy of the estimates and then setting a contingency value. Detailed software packages are available that can provide reasonable estimates, subject to the quality of the inputs.

Table 17.1.17. The definitive estimate cost estimate summary

Plant:		Description:		Date:		
A/C				Estimated cost, mm$		
No	Item	M/hr	Labor	Subcont	Material	Total
	Excavation		xxxxxxxx	xxxxxxx	xxxxxxxxx	Xxxxxxx
	Concrete		xxxxxxxx	xxxxxxx	xxxxxxxxx	Xxxxxxx
	Structural steel		xxxxxxxx	xxxxxxx	xxxxxxxxx	Xxxxxxx
	Buildings		xxxxxxxx	xxxxxxx	xxxxxxxxx	Xxxxxxx
	Machinery/equip		xxxxxxxx	xxxxxxx	17.876	Xxxxxxx
	Piping		xxxxxxxx	xxxxxxx	xxxxxxxxx	Xxxxxxx
	Electrical		xxxxxxxx	xxxxxxx	xxxxxxxxx	Xxxxxxx
	Instruments		xxxxxxxx	xxxxxxx	xxxxxxxxx	Xxxxxxx
	Painting/scaffold			Factored		
	Insulation			Factored		
	Direct field cost	xxxxxx	xxxxxxxx	xxxxxxxx	xxxxxxxxx	Xxxxxxxx
	Temp const facil		xxxxxxxx		xxxxxxxx	Xxxxxxxx
	Constr expense					Xxxxxxx
	Craft benefits and burdens					Xxxxxxx
	Equip rental				xxxxxxxx	Xxxxxxx
	Small tools				xxxxxxxx	Xxxxxxx
	Field O/heads					Xxxxxxx
	Indirect field costs		xxxxxxxx		xxxxxxxx	Xxxxxxx
	Total field costs					Xxxxxxx
	Engineering salaries	xxxxxxx	xxxxxxx			
	Purchasing sal	xxxxxxx	xxxxxxx			
	Gen Office sal	xxxxxxx	xxxxxxx			
	Office expense				xxxxxxxx	
	Payroll burdens				xxxxxxxx	
	Overhead costs				xxxxxxxx	
	Total office costs	xxxxxxxx	xxxxxxxxx		xxxxxxxx	Xxxxxxxx
	Total field and office costs					Xxxxxxxx
	Fee (contractor)					Xxxxxxxx
	Escal/contingency					Xxxxxxxx
	Total costs					Xxxxxxxx

Figure 17.1.8 shows two sets of curves. The first is a positive to negative range of accuracy against the type of estimate. This type of estimate scale is based on Figure 17.1.7 as the progress of the project. Thus, the widest accuracy range is for a capacity factored estimate at the beginning of a project. The narrowest range therefore is at definitive estimate stage around 30–40% of project duration. The second set of curves gives a range of contingency to be applied to the various estimate types. This is a plot of percent of total field and office cost against the estimate type used for the accuracy curves.

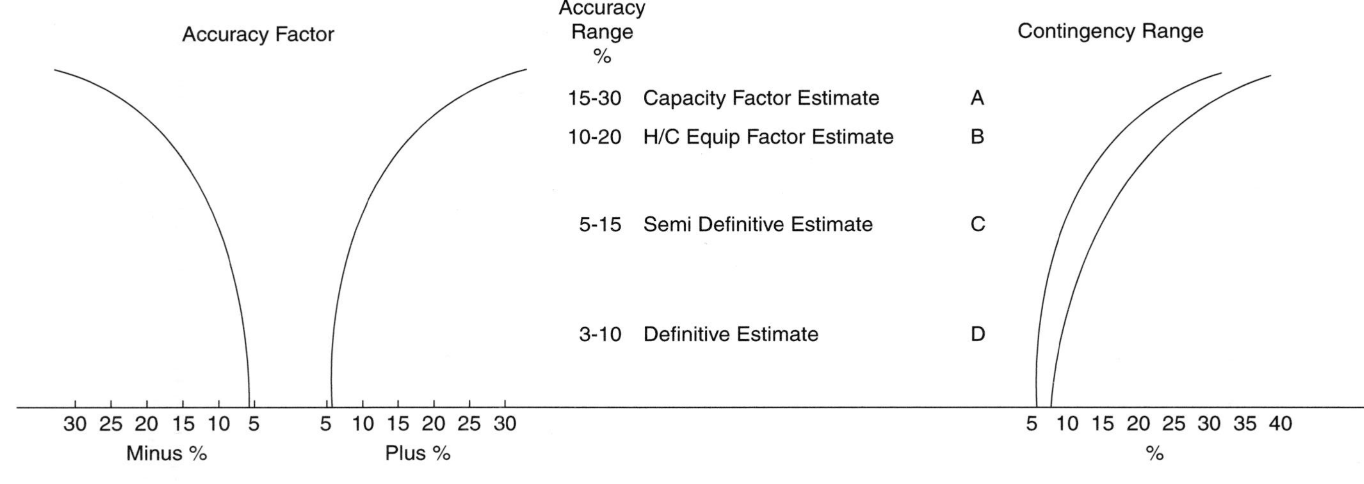

Figure 17.1.8. Accuracy of an estimate.

Discounted cash flow and economic analysis

Throughout this chapter a major role of a process engineer in conducting economic studies has been described and discussed. The purpose of these studies is to provide the company's management with sound technical data to enable them to improve the company's profitability at minimum financial risk to the company. Up to now the course of these studies have covered

- Identifying the viable options that will meet the study objectives
- Short listing and screening the options using simple return on investment techniques
- Providing process input into the preparation of a more detailed capital cost estimate of those options selected for further detailed economic evaluation

The cost estimates based on firm equipment costs are the most accurate possible without committing to more detail design and capital expenditure. Using these estimates the detailed and more reliable prediction of the profitability of the selected options can now proceed. This prediction is based on a Return on Investment calculated from a projected cash flow of the process over a prescribed economic life for the facility.

There are several methods of assessing profitability based on Discounted Cash Flow (DCF), but one such method is a return on investment method using the present worth (or net present value) concept. This concept equates the present value of a future cash flow as a product of the present interest value factor and the future cash flow. *Based on this concept, the Return on Investment is that Interest value or Discount Factor which forces the Cumulative Present Worth value to Zero over the economic life of the project.* Other methods are described and discussed in Chapter 17.2 which follows.

Whereas the development of capital cost of plants is usually a combined effort between the Development engineer and cost estimator so is the development of a DCF return on investment a combined effort between the Development engineer and the company's Finance specialist. The development engineer provides the technical input to the work such as operating costs, type of plants, construction schedules and cost, yield and refinery fence product prices, and the like. While the financial specialist provides the financial data based on statutory and company policies, such as the form of depreciation, tax exemptions, tax credits (if any), items forming part of the company's financial strategy, etc. The calculation itself is in two parts, which are:

- Calculation of cash flow
- Present worth calculation

These are described in the following paragraphs.

Calculating cash flow

Figure 17.1.9 gives a graph of the cumulative cash flow of a typical project in relation to its project life.

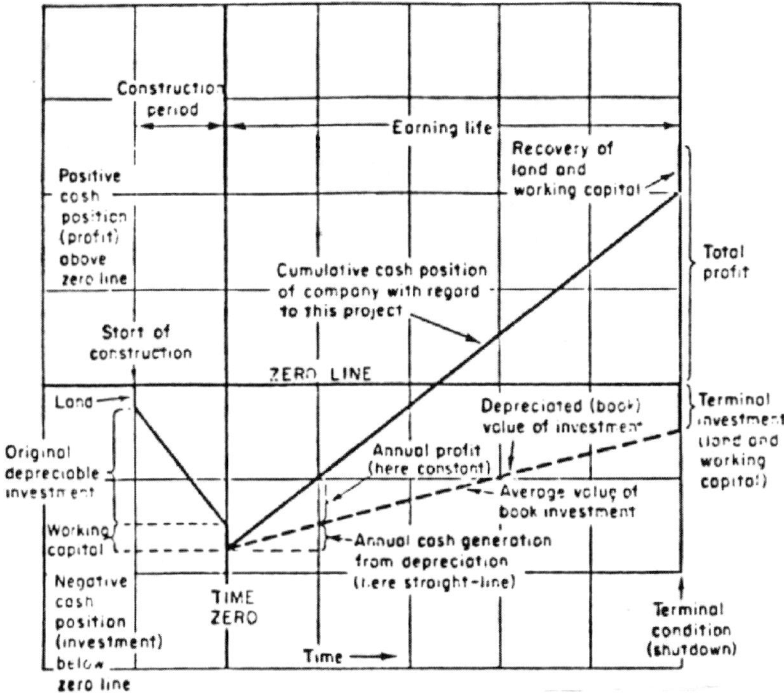

Figure 17.1.9. A diagram of cumulative cash flow.

Initially there is a financial loss when the company has to purchase land, equipment, pay contractors to erect equipment, and the like. To do this not unlike most private individuals, companies borrow money, on which of coarse they have to pay interest. In addition to the cost of construction the company must keep in hand some capital by which it can buy feedstock, chemicals, and pay the salary of its staff during the commissioning of the plant. This working capital must also be considered as debt at the project initiation.

At the end of construction and commissioning the cash flow into the system moves upward toward a positive value and after a prescribed number of years moves into a positive value. During this period of positive cash flow the money recovered from the operation must be sufficient to:

- Repay the loan and its interest
- Pay all associated taxes
- Pay all the operating costs of the project
- Make an acceptable profit

When these conditions have been met over the prescribed period of time called the "Earning Life" or "Economic Life", there still remains the plant hardware, the land, and the working capital. These are considered as the projects terminal investment and are added to the final project's net cost recovery.

The cash flow calculation recognizes the financial milestones shown in Figure 17.1.9 and the contents of a typical cash flow calculation are discussed below:

Economic life of the project
This is the number of years over which the project is expected to yield the projected profit and pay for its installation. These are the number of years starting at year 0 which indicates the end of construction and the commissioning of the facilities. The last year (usually year 10) is the year in which all loans and other project costs are repaid and the "Terminal Investment" released.

Construction period
This is the period before year 0 during which the plant is constructed and commissioned. Assume this period is three years then this is designated as end of year −2,−1, 0. During this period the construction company will receive incremental payments of the total capital cost of the plant with final payment at the end of year 0. The construction cost may be paid from the company's equity alone or from equity and an agreed loan or entirely from a loan. In the case of a loan to satisfy this debt the payment of the loan interest commences in this period. The interest payment over this period however is usually capitalized and paid over the economic life of the project.

Net investment
The net investment for the project includes the capital cost of the plant, which is subject to depreciation, and the associated costs, which are not subject to depreciation. The capital cost of the plant is the contractor's selling price for the engineering, equipment, materials and the construction of the facilities. In a process study using a DCF return on investment calculation the capital cost should be an estimate with an accuracy at least that based on an equipment factored type.

The associated costs include the following elements:

- Any licensor's paid up royalties
- Cost of land
- First inventory of chemicals and catalyst
- Cost of any additional utilities or offsite facilities incurred by the project
- Change in feed and product inventory
- Working capital
- Capitalized construction period loan interest

Revenue

This is the single source of income into the project. In most cases it is the income received for the sale of the product(s). This is calculated from projected process yields of products multiplied by the marketed price of the products. A market survey should already have been completed to ensure that the additional products generated by the project are in demand and the price is in an acceptable range. Later a sensitivity analysis of the DCF return on investment may be conducted changing the revenue recovered by price escalation or other means.

Expenses

This is the major cost of carrying out the project. It consists of the following items:

- *Plant operating cost.* This includes the cost of utilities used in the process, such as power, steam, and fuel. It also includes the cost of plant person in salaries, burdens, and indirects and the cost of chemicals and catalyst used.
- *Maintenance.* There are two kinds of maintenance cost included in this item. These are the preventive maintenance carried out on a routine basis, and those costs associated with incidental breakdowns and repair.
- *Loan repayment.* The loan principle is paid back in equal increments over the economic life of the plant. This item includes the payback increments and the associated interest on a declining basis.

These are incurred running costs and as such are considered tax free.

Depreciation

The second cost to the project which is considered as a deductible from the gross profit for tax purposes is the depreciation of the plant value. This is calculated over the PLANT LIFE as the plant capital cost divided by the plant life. The term Plant Life is the predicted life of the facility before it has to be dismantled and sold for scrap. Usually this is set at 20 years and indeed all specifications relating to Engineering and Design of the facilities will carry this requirement. So that all material and design criteria such as corrosion allowances associated with the plant will meet this plant life parameter.

Ad valorem tax

This is the fixed tax levied in most countries payable to local municipal authorities, provincial, or state authorities to cover property tax, municipal service costs etc.

Taxable income

Taxable income is revenue less operating cost, depreciation, and Ad Valorem Tax. This of course is simply put as in most countries, States, or Provinces there will probably be certain local tax relief principles and tax credits that will affect the final taxable income figure. The company's financial specialist will be in the position to apply these where necessary.

Tax
This is quite simply the tax rate applied to the "Taxable Income" figure. This will vary from location to location but will be taken as one rate over the economic life of the project for the purpose of a process study, unless there are legislative changes already in place.

Profit after tax
This is the gross profit less the tax calculated in the previous item.

Net cash flow
This item is calculated for each year of the project's economic life. It commences with year 0 with the net investment shown as a negative net cash flow item. Then for each successive year until the end of the last year of the economic life, the net cash flow is calculated as the sum of profit after tax PLUS the depreciation.

The depreciation is added here because it is not really a cost to the project. It is a "Book" cost only and is used specifically for tax calculation. The cash flow item for the last year of the economic life must now include the "Terminal Investment Item". This item is the sum of the salvage or net scrap value of the plant (scrap value less cost of dismantling), the estimated value of the land and the Working Capital initially used as part of the "Associated Costs".

Thus the final cash flow item will be the sum of profit after tax, plus depreciation, plus terminal investment.

With the net cash flow in place the second part of the calculation which is the determination of the return on investment based on the present worth concept for the project can be carried out.

Calculating the cumulative present worth

This is an iterative calculation using three or more values as the discounting percentage. The result of the cumulative values of the present worth at the end of the economic life from these calculations are plotted against the discount percentages used for each case. The discount value when the cumulative present worth is zero is the return on investment for the project.

For each value of the discount percentages selected first calculate the discount factor for each year of the economic life starting at year 0 where the discount factor is always 1.0. Then divide the discount factor for year 0 which is 1.0 by $(1.0 + d)$ where d = discount percent divided by 100. Do the same for each successive year. Thus, if the discount percent selected is 10% then the discount factor for year 0 = 1.0, for year 1 = 1/1.1 = 0.909, and for year 2 = 0.909/1.1 = 0.826, and for year 3 = 0.826/1.1 = 0.751, and so on until the last year.

The present worth value for each year is then calculated as the net cash flow value for that year multiplied by the discount factor for that year. Remember the cash flow value for year 0 is always negative. Next determine the cumulative present worth value by adding the value for each successive year to the value of the year before. Starting with that value for year 0 which is negative add that value for year 1 which is positive to give a positive or negative cumulative value for year 1. Continue by adding the present worth value for year 2 to the cumulative value for year 1 to give the cumulative value for year 2, and so on through the last year. The cumulative value of the present worth for the last year in each of the discount percentages selected is plotted linearly against the discount percentage. This cumulative value can either be positive or negative. Indeed to be meaningful the discount percentages selected must be such that the calculated present worth values for the last economic years be a mixture of negative and positive values. In this way the resulting curve plotted must pass through zero.

The following example calculation is based on Scheme 1 in Item 8.2. Similar calculations may also be carried out on one or more of the other schemes that were screened to confirm their comparative profitability. In this case, it is important to remember that the economic parameters used for each case are identical to enable a proper comparison and analysis to be made.

Example calculation

This scheme includes the addition of the following units into an existing oil refinery.

- New naphtha hydrotreater (13,500 BPSD)
- New naphtha splitter (1,350 BPSD)
- Revamped debutanizer and new light end units (3,830 BPSD)
- New thermal cracker (34,320 BPSD)
- New diesel hydrotreater (9,000 BPSD)
- New catalytic reformer (9,000 BPSD)

The capacity factored capital cost estimate used in the screening studies for this scheme was 125.490 mm \$. A subsequent estimate based on a more definitive design and equipment definition (equipment factored estimate) gave a capital cost figure of 119.216 mm \$ for this configuration. This latter figure will be used as the capital cost in this example calculation.

Net investment cost

This will consist of:

- The capital cost
- Associated costs
- Capitalized construction loan interest

Associated costs

1.0 Paid up royalties. This is a once off licensing fee paid to the licensors of the hydrotreater processes, and the catalytic reformer process. There may also be a running licensing fee for these processes, but this will be included into the operating cost.
Paid up royalties = 2.026 mm \$.
2.0 First catalyst inventory = 4.052 mm \$.
3.0 Cost of land = 1.0 mm \$.
4.0 Cost of incremental utility / offsite facilities = 2.501 mm \$.
5.0 Cost of increased product / feed inventory (only product inventory is considered here as there is no change in crude oil throughput). Statutory requirements for this refinery location is a mandatory inventory of 14 days feed and product.
Additional inventory cost = $14 \times ((564{,}845 + 162{,}798) - (452{,}756))$
= 3.848 mm \$. (see this chapter Item 2)
6.0 Working capital. This is taken as 5.0% of the capital cost = 5.835 mm \$.

Total associated costs = 19.262 mm \$.

Construction cost and payment

End of year	−2	−1	0
Construction schedule of payments mm \$.	11.922	47.686	59.608

The construction costs will be paid out of equity up to the limit of the equity. The remainder will be paid by a loan at 8.0% interest. Initial equity is 60% of Capital + associated costs = 0.6 (119.216 + 19.262) = 83.087 mm \$.

The loan to complete the construction schedule of payments is raised during year 0 and interest on this is 2.89 mm \$. This is capitalized but will be paid out of an increased equity.

Net Investment:

	mm \$.
Capital cost of plant	119.216
Associated costs	19.262
Construction loan interest	2.89
Net investment	141.368

Table 17.1.18. Schedule of operating and maintenance costs

End of year	Operating costs mm $	Maintenance mm $
1	11.01	4.76
2	11.34	4.81
3	11.681	4.86
4	12.031	4.904
5	12.392	4.953
6	12.764	5.003
7	13.147	5.053
8	13.541	5.103
9	13.947	5.154
10	14.366	5.206

Operating and maintenance costs
Operating cost for year 1 is made up as follows:

	mm $
Operating labor	= 1.49
Utilities	= 8.32
Chemicals, catalyst, running royalties	= 1.20
Total	= 11.01

Operating costs escalate at a rate of 3.0% per year. The yearly operating cost schedule is given in Table 17.1.18.

The maintenance cost for year 1 is taken as 4% of capital cost which is $0.04 \times 119.216 = 4.76$ mm \$. Maintenance costs are escalated at a rate of 1.0% per year. The annual schedule for this item is also given in Table 17.1.18.

Loan repayments and interest
The total loan for the project is 55.390 mm $ and is repaid over 10 years at an interest rate of 8.0% per annum discounted annually.

The schedule of repayments and interest is given in Table 17.1.19 assuming uniform returns in principal.

Revenue
There is a single source of revenue which is from the sale of all products at the refinery price given in Item 2 of this section of the chapter. For the base case given in this example there is no escalation of this figure which remains at 77.7 mm $ per year.

Sensitivity analysis performed later gives the change in ROI for escalated product pricing of 3% and 4% per year, respectively.

Table 17.1.19. Schedule of debt repayments and interest

End of year	Principal, mm $	Principal repayments, mm $	Interest, mm $	Total, repayments Total, repayments mm $
0	55.390			
1	49.851	5.539	4.431	9.970
2	44.312	5.539	3.988	9.527
3	38.773	5.539	3.545	9.084
4	33.234	5.539	3.102	8.641
5	27.695	5.539	2.659	8.198
6	22.156	5.539	2.216	7.755
7	16.617	5.539	1.772	7.311
8	11.078	5.539	1.329	6.868
9	5.539	5.539	0.886	6.425
10	0	5.539	0.443	5.982

Schedules of increased pricing are given in Table 17.1.20.

Depreciation

Depreciation is normally the capital cost of the plant divided by the plant life. The plant life in this example is 20 years but the capital cost in this case has been taken as the original capacity factored estimate of 125.490 mm $. This allows for such items in the associated costs such as the precious metal content of catalysts which are subject to depreciation. For this example therefore depreciation is taken as 7.7 mm $ per year throughout.

Ad Valorem Tax

This item includes plant insurance and is set at 2.0% of capital cost per year.

Tax

This is 40% of taxable income. For the purpose of this study it is assumed there are no tax credits.

Table 17.1.20. Schedule of escalated revenue (for sensitivity analysis)

End of year	Escalated @ 3.0%, mm $/year	Escalated @ 4.0%, mm $/year
1	77.709	77.709
2	80.040	80.817
3	82.441	84.050
4	84.915	87.412
5	87.462	90.901
6	90.086	94.545
7	92.789	98.327
8	95.572	102.260
9	298.439	106.351
10	101.392	110.601

Table 17.1.21. Consolidation of net cash flow

	mm $										
End of year	0	1	2	3	4	5	6	7	8	9	10
Investment											
Cap cost	*119.2*										
Assoc cost	*22.2*										
Net investment	*141.4* (1)										18.7 (2)
Revenue @ 0% Esc		77.7	77.7	77.7	77.7	77.7	77.7	77.7	77.7	77.7	77.7
Expenses (1)											
Operating		*11.0*	*11.3*	*11.7*	*12.0*	*12.4*	*12.7*	*13.1*	*13.5*	*13.9*	*14.4*
Maintenance		*4.8*	*4.8*	*4.8*	*4.9*	*4.9*	*5.0*	*5.1*	*5.1*	*5.2*	*5.2*
Loan repayment (Table 8.17)		*10.0*	*9.5*	*9.1*	*8.6*	*8.2*	*7.8*	*7.3*	*6.9*	*6.4*	*6.0*
Total expense (1)		*25.8*	*25.6*	*25.6*	*25.5*	*25.5*	*25.5*	*25.5*	*25.5*	*25.5*	*25.6*
Depreciation (1)		*7.5*	*7.5*	*7.5*	*7.5*	*7.5*	*7.5*	*7.5*	*7.5*	*7.5*	*7.5*
Ad Valorem Tax (1)		*2.4*	*2.4*	*2.4*	*2.4*	*2.4*	*2.4*	*2.4*	*2.4*	*2.4*	*2.4*
Taxable income		42.0	42.2	42.2	42.3	42.3	42.3	42.3	42.3	42.3	42.2
Tax @ 40%		*16.8*	*16.9*	*16.9*	*16.9*	*16.9*	*16.9*	*16.9*	*16.9*	*16.9*	*16.9*
Profit after Tax		25.2	25.3	25.3	25.4	25.4	25.4	25.4	25.4	25.4	25.3
Net cash flow	141.4 (1)	32.7	32.8	32.8	32.9	32.9	32.9	32.9	32.9	32.9	51.5

Note: (1) These are costs and therefore negative values in cash flow. (Shown here in Italics.)
(2) These are cash recoveries and therefore have positive values in cash flows.

Results

Consolidation of the net cash flow is given in Table 17.1.21.

Figure 17.1.10 gives a plot of the cumulative net present worth versus percent discount from the results of the calculations given in Table 17.1.22.

From Figure 17.1.9 the DCF return on investment for this scheme is 20.5% per year. Based on the plant location and the current investment environment Return on Investment above 18% DCF makes the venture economically attractive. Similar calculations for scheme 3 will also be conducted to verify its ROI on a DCF basis for comparison with scheme 1.

The sensitivity of scheme 1 to escalation of refinery fence product costs are as follows:

- At an escalation rate of 3.0% per year the ROI becomes 23.5%
- At an escalation rate of 4.0% per year the ROI becomes 24.8%

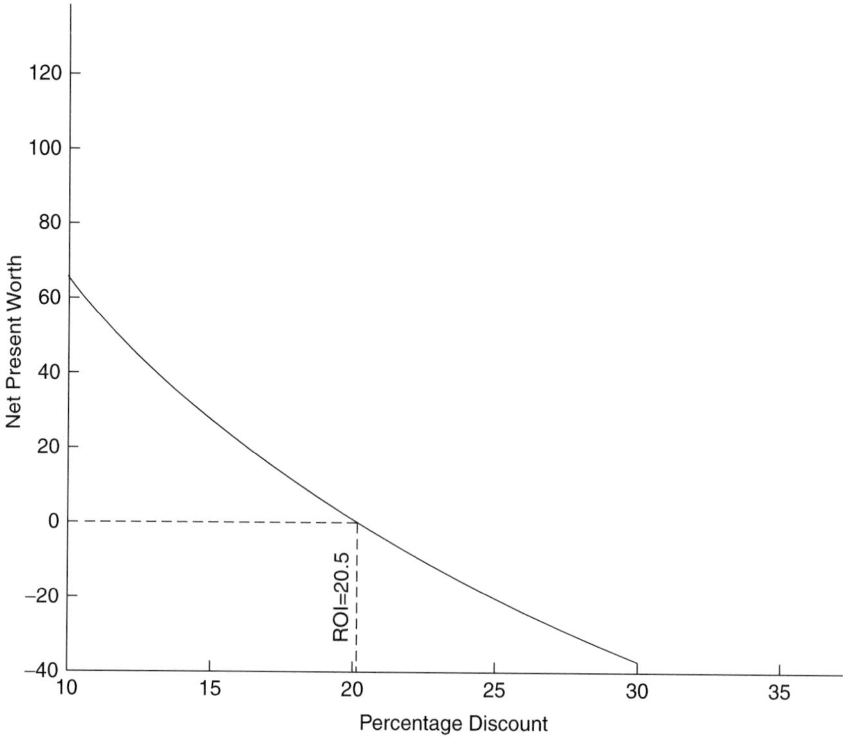

Figure 17.1.10. A plot of net present worth versus percent discount.

Using linear programs to optimize process configurations

Linear programming is a technique to solve complex problems having multi variable conditions by the use of linear equations. The equations are developed to define the inter relationship of the variables. Computers are used to solve these equations and to select from a matrix of these equations the solution or solutions to the problem.

This is not a new technique. It has been used for many years in industry and particularly the oil industry to plan and optimize its operation. Indeed in oil refining today there is considerable development in process control using linear programming "on line" to optimize the units operation. In its use as a management tool it can refine the calculations described in item 8.2 to a very fine degree. It is possible also by using this technique to examine many more options than could be examined in a manual operation.

Table 17.1.22. Present worth calculation

Net cash flow	*141.4*	32.7	32.8	32.8	32.9	32.9	32.9	32.9	32.9	32.9	51.5
discounted @ 10%											
Discount factor	1.0	0.909	0.826	0.751	0.683	0.621	0.564	0.513	0.467	0.424	.385
Present worth	*141.4*	29.7	27.1	24.6	22.5	20.4	18.6	16.9	15.4	13.9	19.8
Cumulative PW	*141.4*	*111.7*	*84.6*	*60.0*	*37.5*	*17.1*	1.5	18.4	33.8	47.7	67.5
Discounted @ 15%											
Discount factor	1.0	0.870	0.756	0.658	0.572	0.497	0.432	0.376	0.327	0.284	.247
Present worth	*141.4*	28.4	24.8	21.6	18.8	16.4	14.2	12.4	10.8	9.3	12.7
Cumulative PW	*141.4*	*113.0*	*88.2*	*66.6*	*47.8*	*31.4*	*17.2*	4.8	6.0	15.3	28.0
Discounted @ 25%											
Discount factor	1.0	0.800	0.640	0.512	0.410	0.328	0.262	0.210	0.168	0.134	.107
Present worth	*141.4*	26.2	21.0	16.8	13.5	10.8	8.6	6.9	5.5	4.4	5.5
Cumulative PW	*141.4*	*115.2*	*94.2*	*77.4*	*63.9*	*53.1*	*44.5*	37.6	32.1	27.7	22.2
Discounted @ 30%F											
Discount factor	1.0	0.769	0.592	0.455	0.350	0.269	0.207	0.159	0.123	0.094	.073
Present worth	*141.4*	25.1	19.4	14.9	11.5	8.9	6.8	5.2	4.0	3.1	3.8
Cumulative PW	*141.4*	*116.3*	*96.9*	*82.0*	*70.5*	*61.6*	*54.8*	*49.6*	*45.6*	*42.5*	*38.7*

The first objective using this technique is to build a mathematical model that fully defines the problem. The main model constitutes the "objective function." The purpose of linear programming is to either maximize or minimize the objective function (e.g., maximize overall profit or minimize overall costs) subject to a number of constraints. The model itself will consist of many submodels which will be inter-related by linear equations. In the case of an oil refinery configuration study the data for the model development will include:

- All the processing plants to be considered
- The yield from each plant
- The product and feed stream properties
- All the possible routing of the streams
- All the possible final blending recipes
- Utility and operating requirements for each process on a unit throughput basis
- Investment cost for each process

These data are coded and the coded items used in equations that represent all the relationships to one another. The coded data items are shown graphically in Figure 17.1.11. The relationship equations representing physical and financial data are input to the computer program. Proprietary sub routines included in the program resolve nonlinear relationships to linear. Such routines as the DCF calculation are

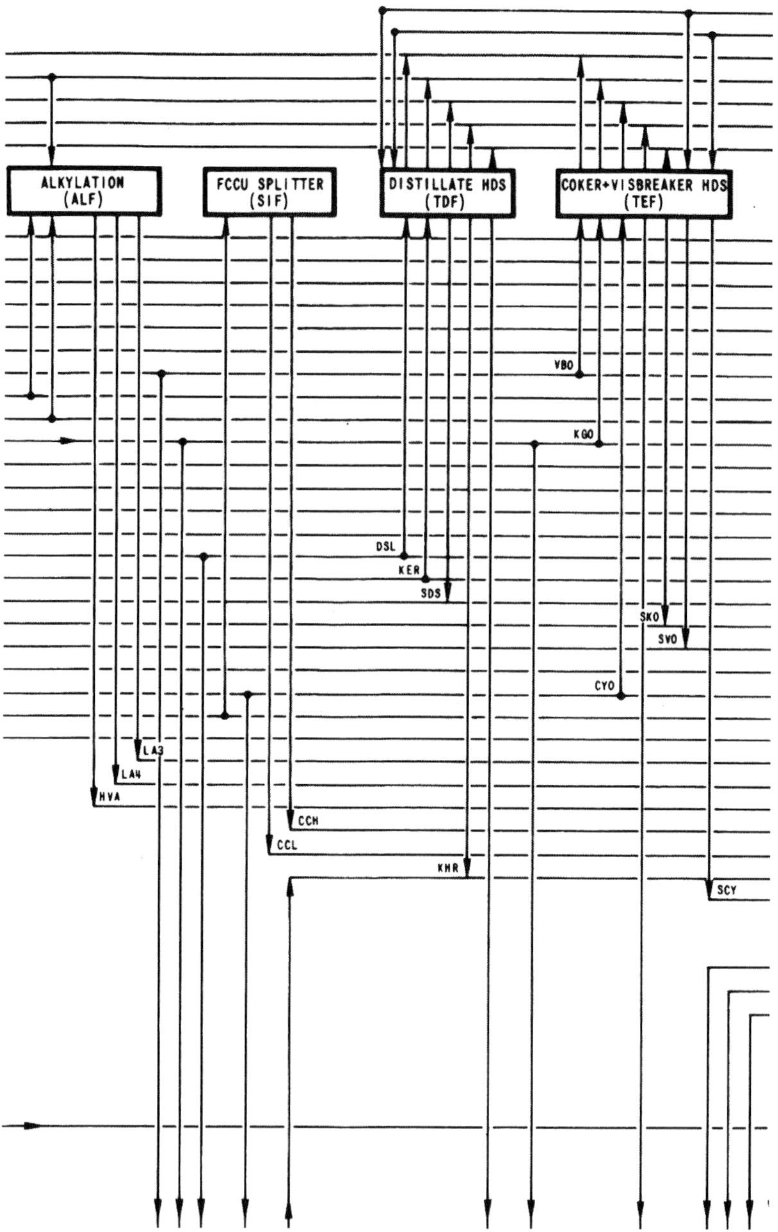

Figure 17.1.11. An example of a mathematical model used in linear program.

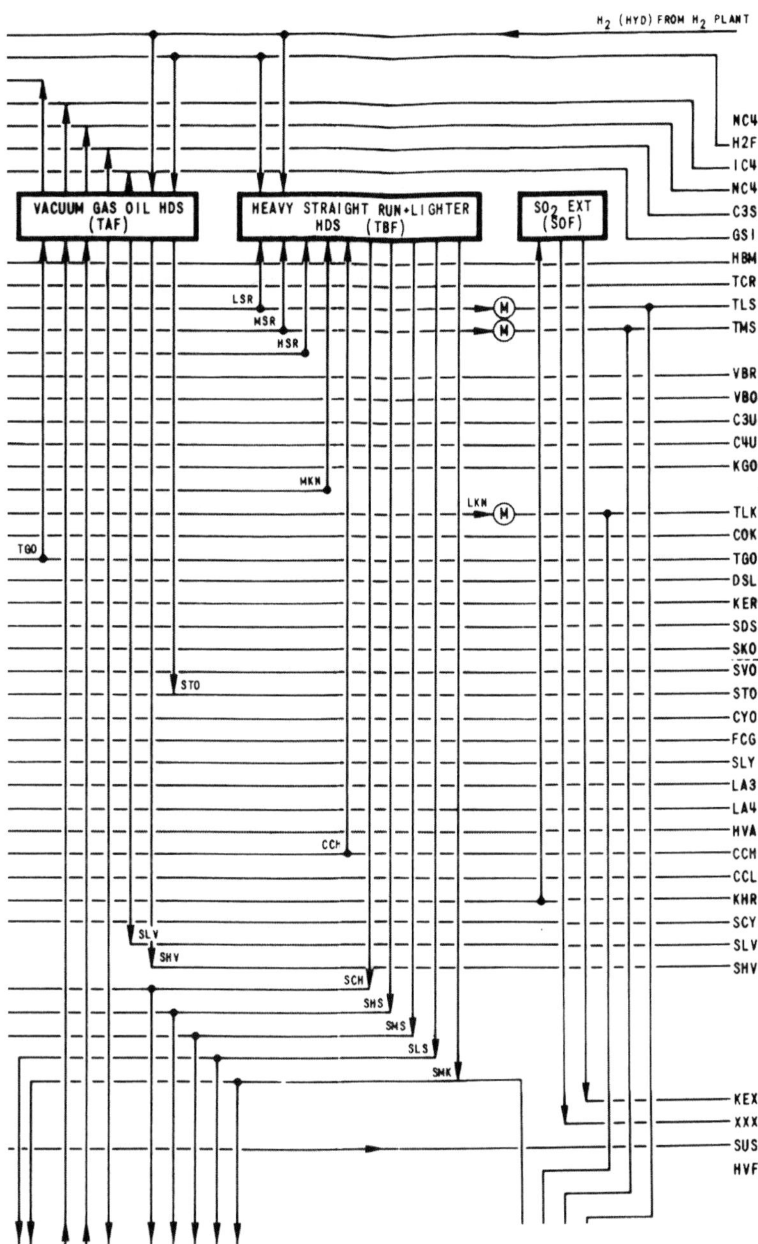

Figure 17.1.11. (Cont.)

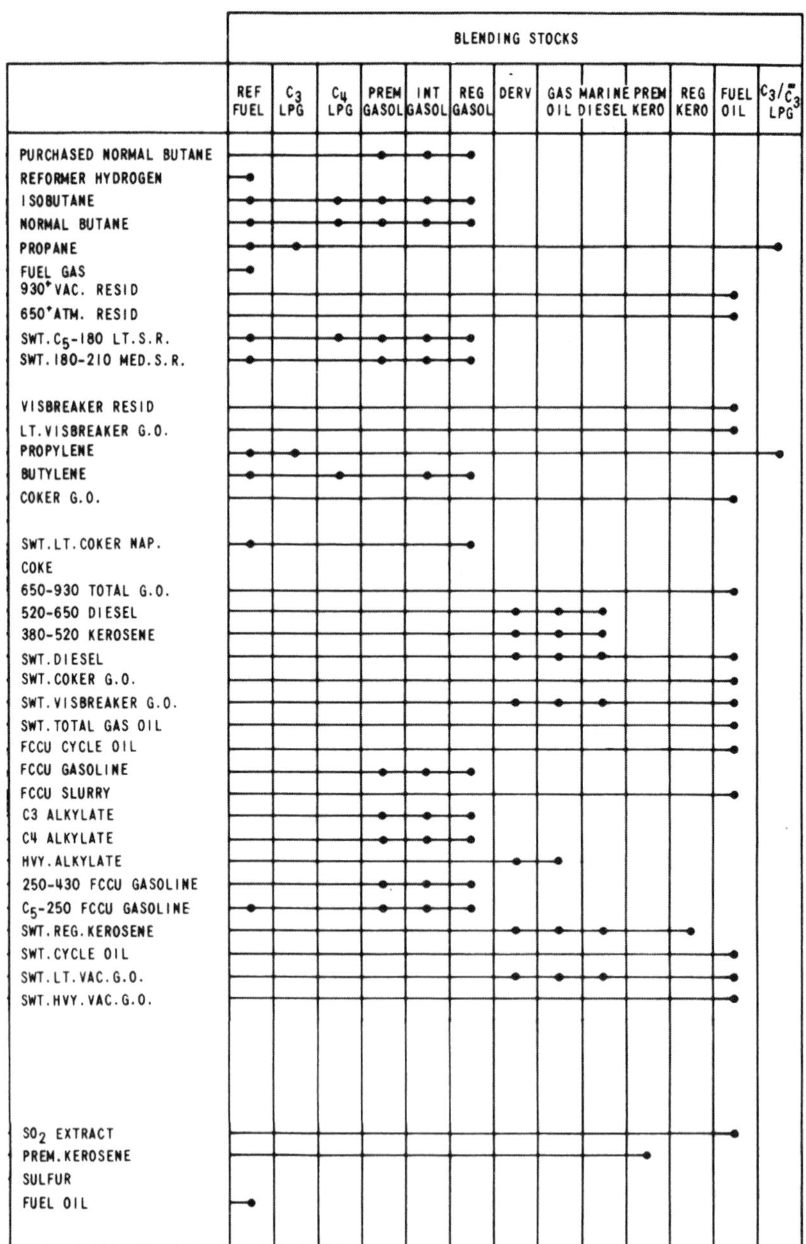

	BLENDING STOCKS												
	REF FUEL	C_3 LPG	C_4 LPG	PREM GASOL	INT GASOL	REG GASOL	DERV	GAS OIL	MARINE DIESEL	PREM KERO	REG KERO	FUEL OIL	$C_3/\bar{C}_3$ LPG
PURCHASED NORMAL BUTANE				•	•	•							
REFORMER HYDROGEN	•												
ISOBUTANE	•		•	•	•	•							
NORMAL BUTANE	•		•	•	•	•							
PROPANE	•	•											•
FUEL GAS	•												
930⁺ VAC. RESID												•	
650⁺ ATM. RESID												•	
SWT. C_5-180 LT. S. R.	•		•	•	•	•							
SWT. 180-210 MED. S. R.	•			•	•	•							
VISBREAKER RESID												•	
LT. VISBREAKER G.O.												•	
PROPYLENE	•	•											•
BUTYLENE	•		•		•	•							
COKER G.O.												•	
SWT. LT. COKER NAP.	•					•							
COKE													
650-930 TOTAL G.O.												•	
520-650 DIESEL							•	•	•				
380-520 KEROSENE							•	•	•				
SWT. DIESEL							•	•	•			•	
SWT. COKER G.O.												•	
SWT. VISBREAKER G.O.							•	•	•			•	
SWT. TOTAL GAS OIL												•	
FCCU CYCLE OIL												•	
FCCU GASOLINE				•	•	•							
FCCU SLURRY												•	
C3 ALKYLATE				•	•	•							
C4 ALKYLATE				•	•	•							
HVY. ALKYLATE							•	•					
250-430 FCCU GASOLINE				•	•	•							
C_5-250 FCCU GASOLINE	•			•	•	•							
SWT. REG. KEROSENE							•	•	•		•		
SWT. CYCLE OIL												•	
SWT. LT. VAC. G.O.							•	•	•			•	
SWT. HVY. VAC. G.O.												•	
SO_2 EXTRACT												•	
PREM. KEROSENE										•			
SULFUR													
FUEL OIL	•												

Figure 17.1.11. (Cont.)

also added into the main program. The computer solves the many hundreds of equations meeting the problem premises and any other applied constraints. The selected configurations are optimized within the program and the final printout shows product quantities, product quality, stream flows, unit capacities, investment costs, operating cost, and the ROI. The graphical representation of the model, part of which is shown here as Figure 17.1.11, contains 511 variables and over 400 equations. By present linear program standards this is a family small model.

17.1.3 Executing an approved project

This part of the chapter deals with the activities that are carried out by the refinery in installing a new or revamped facility which has been approved by the mechanism defined in Part 2 of this Chapter. Much of the initial work is carried out by the company's development engineers. Later, during the early part of the work being executed in a selected engineering office, the company will set up a project team headed by a project manager to monitor and oversee that the company's interests and requirements are met.

Developing the duty specification

Among the first activities to initiate a refinery project is the development of the "Project Duty Specification". The process major input to this most important document is supplied by the company's process development engineers.

As soon as a project and the funds have been approved for implementation work must begin to appoint an engineering and construction contractor to do the work. Before this can be accomplished however it is necessary to complete the following:

1.0 Prepare the duty specification for the plant.
2.0 Develop the contractor's scope of work.
3.0 Develop the contractual terms and the contract itself.
4.0 Assemble the document inviting contractor bids.
5.0 Decide on the selection procedure for the contractor appointment.

Normally the operating company will assemble a team of people who will be dedicated to carry out these activities. This team will usually be headed by a member of the company's middle management who will be responsible for these precontract activities. The team itself will consist of senior personnel from its engineering disciplines, and other parties such as the company's purchasing and legal departments.

The development engineer will almost invariably participate in some or all of the activities listed above. His/her main participation however will be in the development of the duty specification. The importance of this document is that it forms the basis of the selected contractor's scope of work. It also establishes the standard of quality required by the operating company as the client. A well drafted project duty specification will contain at least the following sections and information:

- The process specification
- General design criteria
- Any preliminary flow sheets (duly labeled "Preliminary")
- Utilities specification
- Basis for economic evaluations
- Materials of construction
- Equipment standards required
- Instrument standards required

These are now discussed in the following paragraphs:

The process specification

This document is developed entirely by the senior development engineer assigned to the project team. It gives in precise terms, the plant required, the number of units, its throughput, the product yields and quality, the required test standards, and any salient process requirements. An example of a typical process specification is given in Appendix 17.8.

General design criteria

This section of the project duty specification is usually compiled by the refinery's project engineers assigned to the team. Some input is required from the development engineer to ensure that technical documents and data developed by others will conform with the company's usual format. This section, as the name implies, supplies the general data associated with any work done in the company's plant site. The following topics make up this section:

Scope
This is a brief statement covering the objective of the project. This is followed by a list of the company's standards that are to be used in the implementation of the work.

Climatic data
A list of the following data is given here. These are the data that will be used in the various specifications and calculations developed during the course of the project:

- Dry and wet bulb temperatures
- Winter design dry bulb temperature
- Temperature extremes
- Barometric pressure

General design considerations
This section lists *all* of the legislative criteria associated with the building of an industrial plant in the area. It will include the environmental requirements, and safety and quality standards to be adhered to by contractors and licensor.

Units of measurement
The units of measurement that will be used in all calculations, flow sheets, and specifications are given in this item. It begins by stating the units in terms of:

- English units
- or SI units
- or any others

The next section then defines these units in more detail such as for example:

Linear—Millimeters (mm)
Mass—Kilograms (kg)
Flow gas—Normal cubic meters per hour (Nm^3/h).
Flow liquid—(large) cubic meters per hour (m^3/hr).
—(small) liters per minute (l/min).

There follows then a list of conversion units that should be used on the project.

Engineering flow diagrams
Many companies include those flow diagrams they have developed during the appropriation stage of the project. These are usually the process flow diagram and the mechanical flow diagram. The inclusion of these adds to the description of the work scope by supporting the process specification. In most contracts however the client company will expect the contractor and/or the licensor to take responsibility for the process performance and to guarantee it. Under this circumstance these engineering diagrams are released into the project duty specification as "Preliminary" issues. The contractor is expected to check and, if necessary, revise them. Thereafter, the contractor must accept responsibility for the technical content of the diagrams as a basis for future normal design development.

Utilities specification
Full details of the utility streams available in the clients plant is given in this item. This item is generally prepared by the process engineers and must include data concerning all steam and condensate systems, water systems, air, and fuel. Such data will include at least the temperature and pressure of these streams. In the case of circulating systems such as fuel oil and cooling water both supply and return conditions will be required. Raw water available should also have a complete analysis of its impurities. The analysis of the normal fuel oil and fuel gas should be included. These would contain at least the following:

Fuel oil: Supply temperature.

Sg @ supply temp
Visc @ 120°F cs
Visc @ 210°F cs
Supply and return pressure.

Fuel gas: Supply temperature and pressure.

Molecular weight.
Approximate molar analysis.

The dew point of instrument air required at the air supply header must be stated. If the client has any preference as to the type of desiccant to be used in the drying of the air this too must be stated.

All other preferences or standards that the client company requires to be utilized in the utilities design must be given in this section. For example most companies have a standard burner control system for their plants. This should be fully described here together with some appropriate sketches.

The electrical engineers usually add details of the plant power supply and distribution in this section. If some existing switch boards and substations are to be utilized this should be noted together with a list of drawings that should be referenced.

Other systems, although not strictly utilities, may be included here. Among these would be:

- The fire main
- The flare(s)
- Water effluent treating and disposal
- Other environmental protection systems
- Boiler blow-down systems

Basis for economic evaluations

The process engineer completes this section with the criteria used in the evaluation studies and the appropriation design. The section should begin with a statement that contractors are encouraged to review all flow and equipment arrangements so that all possible alternatives are considered. The incremental cost of any alternative arrangements must however yield a minimum of the company's stipulated return on investment. This return on investment is stated here.

This section continues with the detail costs of labor, utilities, feedstock, and products to be used for any such economic analysis.

Materials of construction
This section is usually compiled by the team's project engineers with some input by the process engineers. It should begin with a paragraph on the references to be used in selecting the materials used in handling the corrosive streams. For example:

- Selection of steels exposed to hydrogen service shall be based on "API-941."
- Corrosion allowances for steels subject to hydrogen sulfide environment shall be in accordance with "Computer correlations to estimate corrosion of steels" by Couper and Gorman, NACE paper No 67.

The next paragraph or sub section should detail the client's requirement for corrosion allowances in terms of equipment life. For example:

Equipment	Life in years
Columns, drums, and reactors	20
Heat exchanger-shell	20
Heat exchanger-tubes	5
Pumps	10
Compressors	10
Heater tubes	100,000 hours.
Atmospheric tankage	20
Piping material	10

These data may also be supplemented by a table giving the company's accepted minimum corrosion allowance in mm or ins (mils) for the various equipment items.

Equipment standards to be used for sizing and design
There follows in this section data to be used for the sizing and the specifying of the major equipment items, which are:

- Vessels and columns
- Heat exchangers
- Pumps
- Compressors
- Heaters

Examples of such data included in this section are as follows:

Vessels and columns
Columns to be sized on a specified type of tray (e.g. valve trays).
Design for a percentage of flood (usually 80%).
Design pressure and temperature criteria for all vessels.
Minimum diameter for trayed columns.

Heat exchangers

Design standards to be used (e.g. TEMA, ASME, etc.).
Rating procedures.

General design criteria (e.g. use of fixed and floating head tube bundles, expansion bellows, tube side flows where special considerations are required, use of kettle or thermo-syphon reboilers etc.).

Tube sizes and pitch relative to fouling factors.
Table of fouling factors to be used.
Allowable velocities and pressure drops.
Approach temperatures to be used.
Design temperature and pressures.

Pumps
The design standard to which the pumps are to be designed. For example API 610 "Centrifugal Pumps for General Refinery Services."

Rated capacity of the pump over normal. For example:

Service	*Rated Capacity over Normal %.*	
	Flow control	Level or temp control
Feed	5	10
Product	10	15
o/hd reflux.	15	20
Inter reflux.	20	20
Reboiler feed	15	20

1. Sparing requirement.
2. NPSH calculation criteria—Source pressure
   - Rate for suction line losses
   - Level for static head. (e.g. bottom tangent line for columns)
   - Pump centre line etc.
3. Pump selection preferences
4. Minimum flow criteria
5. Casing design conditions
6. Type of drivers
7. Piping hook-ups (company standard)

Compressors
Standard to be used for the design of compressors.
(e.g. centrifugal compressors designed to API-617)

Compressor type selection preferences
Rating of compressors. Includes % over normal capacity
Sparing requirements
Mode of control
Type of drivers
Type of cooling
Piping hook-ups

Fired heaters
Types and selection preferences
Any company standards
Heater design criteria, such as:

- Burners & flues to be designed to 125% of normal heat release
- Environmental constraints
- Acceptable average flux, coil pressure drops, and mass velocities
- Surge volumes and acceptable % vaporization for fired reboilers
- The rated duty of the heater as a % of normal duty

Allowable maximum coil film temperature (where coking is a problem).

Coil design temperature and pressure criteria
Fuel system details (company standard)
Burner types preferred and burner control (company standards)

Instrument standards required
This section of the specification is usually compiled by the instrument and the process engineers. Its objective is to convey to the contractors the instrument, safety, and control philosophy required for the company's plants. It may begin with the sizing criteria for control and relief valves that the company wishes to use. For example:

- Criteria to set the control valve pressure drop as a % of line losses, etc.
- Design capacities for control valves
- Design standards required for relief valve sizing (e.g. API-520 and 521)

This item should then continue to define the company's preferences in piping design for control valve hook-up, relief valve venting, and relief valve locations. Some basic criteria concerning the relief header and its condition should be described. Finally, the section should give a list of instrument drawing symbols that the company prefers or, in the case of a plant being build in an existing complex, the symbols already in use.

This completes a review of a typical project duty specification. This document which usually forms part of a request for contractor's quotes may be smaller and less detailed than described here. Sometimes the client company may wish to depend solely on the contractor's standards and criteria. However to ensure good competitive quotations a document such as described here or even more detailed is desirable.

The project team

Immediately after the appropriation approval for a new project the company usually appoints a Project Manager to look after the execution of the project as a whole. The Project Manager is chosen from senior engineers in the company. He will be a person with wide experience in the management of engineering and installation of petroleum refining facilities and ideally very familiar with his company's codes and standards.

One of the first activities of the Project Manager will be to form a team of experienced people who will assist him in directing the effort of his company and third parties who will be engaged in executing the project. In the early stages of the project the team will be relatively small, consisting of the senior development engineer, a scheduler, and a cost estimator. Later a contracts specialist and legal advisor will join the team on a part time basis. As the project continues with the selection of the engineering and construction company the team will include additional development/process engineers, procurement or purchasing specialists, and one or more assistant project engineers. This team will probably move into the contractor's offices, so that a day by day interface with the contractor is established.

At some time soon after the contractor's field 'move in' the company's project team will move into the contractor's field offices. At this stage of the project the team structure will be changed. In all probability the senior Development Engineer and the assisting process and project engineers will return to their home office and will participate only in addressing any process problems that may arise. Their place on the team will be taken up by one or more experienced construction supervisors who will assist in monitoring progress both with respect to time and to budget during the construction period.

Most oil refining companies carry their own commissioning and start-up specialists. The duty of these specialists commence when the facilities or part of the facilities have been declared by the contractor to be complete. Their first activity is to check the plant out for completeness and that it conforms to the company's standards and requirements. In doing this they develop a 'Punch List' of those items that do not conform or require more work. After the contractor has worked through the 'Punch List' and the work has been accepted the plant is handed over to the 'care, custody, and control' of the company. At this point commissioning of the facilities can commence with the 'lead' specialist taking charge. Part of the duties of the commissioning specialist is to orientate and/or train plant operating personnel to run the facilities. The senior commissioning specialist usually remains on site to monitor the guarantee test run and interpret its results.

Primary activities of the project team

The function of the project team has been touched on in the last section. The activities that satisfy these functions are listed as follows:

Pre contract award:

1.0 Preparation of the project specification
2.0 Determining the type of contract
3.0 Preparing the enquiry document
4.0 Issuing the enquiry document
5.0 Tender evaluation
6.0 Recommendation of the award contract

The first of these activities 'The preparation of the project specification' has already been discussed at the beginning of the third part of this chapter.

Determining the type of contract
Types of contracts fall into three main categories. These are:

"Lump Sum". With this type of contract the contractor undertakes to provide the complete plant ready for operation for a fixed lump sum of money.

"Fixed Fee". Here the contractor undertakes to execute some of the services associated with the plant installation for a fixed sum of money. All the other costs would be reimbursable at net cost to the contractor. For example: The contract may require the contractor to fix his 'Home Office' engineering cost plus his profit fee with all other costs being reimbursable (this is called an Omnibus Fee contract). Other forms of this type of contract may require the contractor to fix just his profit fee, or perhaps part of the office engineering cost.

"Cost Plus". In this type of contract the contractor is reimbursed for all costs applicable to the work plus a percentage of these cots as the contractor's profit fee and overheads.

There are several factors that determine the choice of these types of contracts. Among the more important of these are the following considerations:

- Is the project subject to process design changes?
- Can the scope of work be accurately defined?
- Is time an essential factor?
- The current and anticipated trend in material and labor prices
- The amount of competition among the contracting organizations

A "Lump Sum" contract would be the preferable type of contract if the scope of work can be accurately described in the project specification and it was certain there would not be any major changes to the process. However, contractors would require a reasonable time to respond to the request for a "lump sum" bid. After all they would require the time to properly analyze the project specification, obtain equipment quotes, make preliminary layouts, design, and material take off for cost estimating. The standard of the cost estimate in this case would need to be at least a semi definitive type (see second part of this chapter).

The advantage of "Lump Sum" contract is that the owner knows from day one the cost of the project. However, the owner is also faced with the need to closely monitor the project to ensure the contractor does not effect savings at the expense of quality to increase his, the contractor's, profitability.

Inquiry documents for "Fixed Fee" contracts can be prepared in a much shorter time and with less definition than for "Lump Sum" contracts. If time is an all important element and the project itself is so complicated that even a fairly accurate scope of work cannot be made, then a "Cost Plus" type contract is the more preferable one. In both these types of contract the owner has complete control over material and labor costs. This is a big advantage when material prices and labor costs are falling, but it imposes a significant onus on the owner.

The types of contracts described here are the "basic" types. There are many variations or small changes to these that are often used. For example a contract may call for a "Cost Plus" up to, say, the stages where a semi definitive estimate can be developed. It may then require the contract to be converted to "Lump Sum" for the remainder of the project.

Preparing the enquiry document

"The Project Specification" is the instruction to the contractor as to what facility is required to be built and to what standards and codes. The enquiry document must now include the exact scope of the work that has to be undertaken by the contractor and what work has to be provided by the owner. It is this scope of work that will become the basis for the contractor's program and ultimately his master control plan for executing the work. It will also be the basis for negotiations with respect to changes and the cost/schedule implication of such changes.

The next important item in the enquiry document must define in clear terms the type of contract. This is most important in the case of "Fixed Fee" type contract. Contractors' bidding the work must understand fully what is included as a fixed portion and what is to be reimbursed. As mentioned earlier the contract type may be a variation of the three major categories discussed earlier. Such terms must be clearly defined and if changes are to be made during the project such as conversion to a different type or an addition of target pricing this too has to be clearly defined. If there is a time element for completion a Bonus/Penalty clause is often included.

The enquiry document must include information regarding the procedures to be adhered to during the life of the project and toward the latter part when the plant is considered ready for operation. Such instructions, which may affect the contractor's bid, are as follows:

- List of approvals required by the owner
- Attendance by the owner for critical reviews and conferences
- Correspondence and communication to/from the owner
- Preferred vendor's list (particularly in the case of "Lump Sum" contracts)
- Guarantees-process and mechanical
- Environmental considerations
- Site data (such as: elevation, location, ambient conditions etc.)
- Obtaining licenses to build
- Hand over procedures

Finally, the enquiry document gives instructions as to how the contractor's bid is to be submitted. This includes the number of copies required, when and in what form the bids are presented.

Issuing the enquiry document

Prior to issuing the enquiry document and in the time span in which the document is compiled, the owner's company should review the contractor market. This review should include a limited number of suitable engineering companies. Preparing bids with the subsequent follow-up work does cost contracting companies a significant amount of time and money; therefore the selection of contractors to be invited to bid should be carefully made.

In this task the first question to be resolved is: Which contractors are qualified to undertake this work? For example if the project requires engineering, design, and installation of a hydrocracker then ideally the qualifying contractor should have installed a hydrocracker or a similarly complex unit before. It would not be appropriate for example to solicit a bid from a civil contractor for this kind of project or indeed a contractor who had only installed smaller petroleum plants.

The next task is to meet with potential contractors at their respective establishments. The purpose here is to determine the extent of their home office facilities for engineering and to establish the method they use for executing a project. The other factor during this meeting would be to determine the spare capacity they had available for undertaking the work and their program for growth or completing their existing 'in house' work.

Finally and during this meeting at the potential contractors' offices the owner should meet the key people they would assign to the project. Certainly meeting the proposed project manager is a must. In the case of a major refinery project meeting the lead process engineer is equally as important. Present day contractor evaluation must include the extent of this company's computer applications. This will be a significant factor in selection.

Normally three to four bidders should be selected from this pre enquiry review to receive the invitation to bid. These selected bidders should be informed by phone or fax and confirmed by letter. In this way the bidders are requested if they are interested to receive the bid documents and subsequently submit a bid. The bid documents when complete should be delivered by hand if possible or by courier at a predetermined date.

Contractors' bid evaluation, and recommendation for award

The first activity after receipt of the bids is to carefully review their contents to ensure that they comply with the enquiry documents. In the case of Lump Sum bids careful review of the equipment offered is made to ensure they meet the owner's vendor list and approval. The bids are reviewed regarding the commercial offer. The selection is never made on price alone but on the complete offer. While the price at first glance may seem very attractive, in some cases an evaluation of the other contractual offer may prove more costly in the end.

In most companies approval of major contracts is made by an executive group or person. It is usually the duty of the Project Team to compile the bidders' offers into a succinct form for review by the approval board. The Project Manager of the team should include a recommendation for approval in the report to the company's approval executive.

A suggested content of a "Contractors' Bid Summary" for a Lump Sum Contract is as follows:

Contractor's name:

Total price mm $
Engineering $
Design $
Procurement $
HO overheads $

Materials
Equipment $
Piping $
Instruments $
Electrical $
Civil Structural $
Others $

Construction
Direct field costs $
Indirect field costs $
Total construction $

Deviation from the duty specification
Compliance with contractual terms.
Completion time.

There may be further items depending on the enquiry content.

Project manager's recommendation
Project manager's notes and justification.

The bid summary is given top priority and early completion is usually of major importance. Most bids do carry a limit on the time that the quotes are valid. It is in the owner's interest to make the award within those valid dates if possible. Many owner companies prefer to short list the bids to two or three and to interview the companies so listed to discuss their bids before making the final selection. This is very usual in the case of Lump Sum or Fixed Fee type contracts. Such reviews certainly help to minimize "Change Order" negotiations further down the line.

Monitoring the execution of new projects

Project initiation
As soon as possible after the award of the project the owner's project team moves into the successful contractor's offices. The content and size of the team will vary with the type of contract and the status of the work in the contractor's home office. In general though the Project Team has more responsibility when administrating a Cost Plus contract than either of the other two types. The major activities of the team in this

section of the work refers to those required for the Cost Plus contract. Initially the team should consist of the Project Manager, the Senior Process or Development Engineer, one or two Project Engineers, possibly a procurement (purchasing) specialist, and a Cost/Scheduling engineer. The activities of this team during the initial stages of the work will be:

- Conduct the client's "kick off" meeting
- Review and approve preliminary schedules and budgets for the project
- Review and comment on the contractors project procedure manual

These are described in some detail in the following paragraphs:

The kick off meetings

One of the most important conferences held during the course of the project life is 'The kick off' meeting. This takes place as soon as possible after the award of the project by the client to the contractor. As the name implies this meeting formally releases the contractor to begin work on the client's plant and describes again in detail the client's requirements. Its purpose is to communicate to the contractor's project team, the scope of work, the time span required, the budget (approximate or firm), and details of quality requirements and specifications as outlined in the client's 'project specification'. Discussion should then follow on the project management aspect of the work. The completion date required for the project is tabled together with major milestones to be met with during the project life. These milestones usually indicate when cost estimates (and schedules) are to be updated or when critical overall project decisions are to be made.

This first or formal 'kick off meeting' conducted usually by the client's project manager is invariably followed by a second, less formal, meeting of the contractor's project team. The client's team may or may not be invited to attend this meeting. However at the end of both these kick off meetings each key member of the client's and the contractor's project teams should be absolutely sure of what the project requirements are and what role they are to play in achieving them.

Preliminary schedule and budget

Soon after the "kick off" meetings a preliminary schedule for the engineering work and a budget will be developed by the contractor. The contractor will be able to produce a fairly accurate schedule for the front end work based on their experience in similar projects. The remainder of the schedule to 'job end' will be far less accurate. Much of this forward scheduling will depend on equipment vendor quotes and if applicable licensor data, and later sub contractor quotations. The same applies to the accuracy of this first budget. However, both this schedule and budget will form the first control parameter for the project execution plan. On "Cost Plus" projects the clients approval for these items is usually mandatory.

Review and comment on the contractor's project procedure manual
Most contractor project managers develop a procedure manual on major projects that they work on. This project manual contains details of the particular lists, directory of the project team, and specific procedures that will be adopted on the project. This eliminates the need for a great number of correspondence defining instructions during the course of the project. Although this is contractors internal document the client's project team is always invited to review and often participate in it's development.

A typical project procedure manual is divided into several sections. An example of these sections are as follows:

- Introduction
- Project organization and directory
- The master schedule
- Man hour budgets (restricted issue)
- The cost code of accounts
- Project control procedures
- Correspondence and communication
- Conferences and meetings
- Filing index
- Individual engineering, procurement and design interface procedures
- Drawing index
- Field organization and directory
- Hand over procedures, and close out reports

The items given above may be increased or decreased depending on the wishes of the contractor's project manager. After all this is the project manager's document although it is compiled by the members of the contractor's project team. A brief description of the contents of some of these sections now follows:

Introduction
This section begins with the general information as to who the client is, the official address of the client and the phone, fax, or telex numbers, e-mail address etc. It continues with a short history of the project up to the award, and concludes with a synopsis of the process(es) involved. This section should not be more than three or four pages in length.

Project organization and directory
The project organization chart for the project, is included in this section. Following the chart is a table of the key positions (and those of the client's project team) with the individual's name and office phone number or extension, and e-mail addresses.

The master schedule
This is the master schedule in bar chart form which shows the scheduled progress by discipline and activity. Initially and of necessity this will be preliminary but will indicate critical milestone dated clearly. It may be followed by a list of these milestone dates with explanatory notes. This will be subject to updates as the project progresses.

Manhour budgets
This includes the allocated manhour budgets by discipline for the project. This item is restricted to the copies of the manual issued to key personnel who are responsible for their discipline's budget control. As in the case of the master schedule, initially these manhour estimates will be preliminary, and will be subject to updates as the project proceeds.

The cost code of accounts
This is a summary of the cost coding that will be used on the project to identify all the various cost centers. This coding will also be used to identify and code such items as purchase orders, client billing, change orders and the like. More often than not the contractor's normal cost codes are used for this purpose.

Project control procedures
This section outlines the reporting procedures the project manager will adopt to control the project in terms of progress and budget. It will detail the data required from the key members of the project and the timing of the data for each reporting period.

Correspondence and communication
All correspondence leaving the project does so under the project manager's signature. Similarly all correspondence originating outside the project is addressed to the project manager. All correspondence are given a coded reference number and this coded reference number also identifies the origin of the correspondence. This reference code is given in this section, together with instructions on the routing of the correspondence. Detail procedures as to the communication by other means, such as telephone, e-mail, and fax is also given here. A list giving the required distribution of all correspondence to the project team personnel and possibly others outside the project is given in this section.

Conferences and meetings
Routine meetings and conferences are scheduled in this section. This will also include a list of permanent attendees to these meetings. As time taken up in meetings is a large utilizer of manhours, the project manager may elect to outline certain recommended procedures for unscheduled meetings in this section. All meetings on the project are minuted, and the minutes distributed within the team and other interested parties according to the project's correspondence distribution list.

The filing index
The project secretary organizes the project filing system. The files generated on the project often become legal documents which may be used for such purposes as 'job close out' negotiations, any possible litigation, or inquiries at later dates into industrial accidents and the like. Maintaining the filing therefore is an important function on the project, and to accomplish this effectively a filing index is initiated and developed. This is given in the manual to enable members of the project team to utilize the files effectively when required.

Individual engineering, procurement, and design interface procedures
During the life of a project there will a considerable amount of data and information generated and distributed to the various disciplines in the project team. To ensure the correct movement of these data etc each discipline develops it's interface procedures. For example in the case of a process engineer who has now completed the data sheet for a pump he needs to ensure that this document is sent to the correct discipline for further work and ultimately purchasing. In this particular project the mechanical engineer is designated to be the interface between process and the purchasing department. Then this section of the procedure manual will detail such interface with instructions of how the transmittal of the data sheet is to be done.

Drawing index
This is initiated and maintained by the design coordinator. In developing the index as a list of all the drawings he also demonstrates what the items in the drawing reference numbers signify. This is updated as the project proceeds.

The remaining items of the manual are self explanatory. These are usually added only in later editions of the manual. For example just before or immediately after field 'move in' etc.

Monitoring the engineering, procurement, and design phase

With the project now underway the work continues in the home office of the contractor to detail the engineering of the plants, to purchase the equipment and materials to be installed, and to arrange the 'lay out' the facility on the site provided by the client. Each of the various disciplines proceed with their function to meet the progress milestones defined by the 'master schedule'. Some of these milestones are finalized by conferences of all the contractor's participants and the client's project team. The first of these, and perhaps the most important is the mechanical flow sheet conference.

The mechanical flow sheet conference and approval
One of the major objectives of the process engineering's early activities is to provide sufficient data to enable other disciplines to commence effective work as early as possible. A major milestone in achieving this objective is to conference the MFD

(mechanical flow diagram) and have it approved by the client. This approval allows a considerable amount of design work (which is the major man hour user on most projects) to commence. The development of the equipment data sheets has already allowed the mechanical engineering and the procurement functions to commence work but it is the release of the approved mechanical flow diagram that really puts the activities of the project into top gear.

Participation of the client company in the flow sheet conference depends on the type of contract and the relative location of the client's offices to that of the contractor. In the case of a cost plus contract however participation by the client is mandatory. Whether there is client participation or not the flow sheet conference follows some basic guidelines, and these are described in the following paragraphs:

- *Preparing for the conference.* The contractor's lead process engineer is responsible for organizing and conducting the flow sheet conference. The flow sheet that will be used in the conference will be 'clean' and free from any pencil or other marks on the diagram itself. It will be given the revision number '0', and clearly marked. 'FLOW SHEET CONFERENCE MASTER'.

The diagram will be complete with all lines sized and specified. Instrumentation will be correctly shown with control valves sized, relief valves will also be shown but not necessarily sized at this time. All equipment will be properly titled and labeled in accordance with the agreed format for flow sheet production.

Accompanying the 'FLOW SHEET CONFERENCE MASTER' will be the current revision of the process flow diagram. This will also be 'clean' with no pencil marks other than the label "Flow sheet Conference Copy". The lead process engineer will also ensure that copies of all the equipment data sheets pertaining to the equipment shown on the flow sheet are available at the conference. The current equipment list will also be made available at the conference. Finally, the lead process engineer or the process engineer he has assigned to the particular process that is to be conferenced should have available a copy of the process calculations developed. This often helps to resolve minor issues that may arise.

- *Attendees.* After notice of the conference has been issued by the lead engineer, the project manager or his assignee will designate the location where the conference is to be held. He will also issue directives to attendees giving the date, time, and location of the conference. In cases of major conferences where client representatives are present professional secretaries may be employed to take notes at the meeting. Otherwise project engineers will be assigned this task. The notes taken at this meeting are most important to the final approval mechanism. The attendees are usually:
  - ➢ Client's representatives (if applicable)
  - ➢ Project manager (usually part time)
  - ➢ Lead process engineer (convener)

- ➢ Responsible process engineer(s)
- ➢ Engineering manager
- ➢ Area project engineer(s)
- ➢ Lead discipline engineers (except civil & structural—called for as required)
- ➢ Design coordinator
- ➢ Design section heads (piping, instruments, electrical)
- ➢ Others as required, such as environmental engineer, cost & schedule control engineers etc.

- *Conducting the conference.* The process engineer conducting the conference commences with a brief outline of the process and its objective. He proceeds to describe the process using the process flow sheet to illustrate the process flow and the equipment. He then commences with the description of the mechanical flow diagram which is attached to a wall or suitable frame from which it can be easily read by the attendees. Now most companies have a color code depicting 'additions to', 'deletions from', etc. Colored pencils are used therefore to cover all such points made during the conference by marking up the flow sheet as the conference proceeds. Assume the color coding in this case is as follows:
  - ➢ Yellow—Shows all items that have been conferenced
  - ➢ Red—All changes and additions
  - ➢ Blue—All deletions
  - ➢ Green—Numbers in green refer to note items taken during the conference

The process engineer conducting the conference starts at a suitable point on the left of the flow sheet. He selects a major line, checks and 'yellows out' the source label, and traces the line using the yellow pencil to the first equipment item. In tracing the line he highlights the line number, size, and specification shown on the line. The design coordinator or piping designer checks this against his line list. It must match exactly or it is marked as requiring further review. All valves and instrumentation along the line are similarly checked against piping lists or the instrument register.

The equipment item into which the line is connected is checked for correct labeling and data. These shown on the flow sheet are compared with equipment list, data sheets, and the process flow diagram for an exact match. The process engineer continues with a second line in a similar manner. He may elect to go back to the left side of the flow sheet to select this line or take a line leaving the equipment item. However, the same exercise is carried out for all lines and equipment until the flow diagram is completely yellowed out with the discussed notes marked, and with all deletion and changes shown.

This flow sheet conference is a necessary activity but is nevertheless a high consumer of man hours. The process engineer should therefore use his experience and knowledge to minimize the time taken in this activity. For example: where there is some considerable contention or where there is an area of concern to the client highlighted in the conference, the process engineer should attempt to defer a solution until after the conference. Appropriate note should be made however of the problem and the item clearly marked with a "HOLD" sign. Clients will normally approve flow sheets with some "HOLDS" providing there are not too many. In further development of the flow diagram, priority is given to resolving the issue(s) and removal of the "HOLD".

Immediately after the conference the lead process engineer and the project manager formally solicits the client's approval and the release of the project to detail design work. The approved flow sheet is marked as Rev 1. "Approved for Construction".

Approval of equipment selection

On a cost plus contract the client's project manager must approve the selection of all pieces of equipment that will be purchased and installed. Next to the approval of the mechanical flow sheet the commitment to purchase equipment is certainly the next most important. Certified equipment vendor data are essential for updating the budget and schedule. It is also important in obtaining correct outline drawings for certain items to initiate piping lay out and also plot plan development. This certified and guaranteed data will only be made available to the project after the issue of the purchase order.

In addition to releasing the project to essential design work in piping, civil (foundations), and electrical engineering, it also enables the contractor's process engineers to build an accurate utility balance. This in turn initiates the engineering of the utility facilities if this is part of the project.

The contractor's process engineers initiate this function by developing data sheets that define each equipment item. The data in these sheets define the equipment size and performance required to meet the facilities heat and material balances. On completion of theses data sheets the contractor's procurement department complete the commercial conditions associated with the purchase, and send these documents to a small number of vendors for each type of equipment soliciting their quotes for supplying the items.

The receipt of the quotations initiates the selection process. Usually the contractor's procurement department develops a bid summary list of the quotes from each vendor for a specific equipment item. The enter all the associated price and delivery, and together with the actual quotations, this bid summary is then sent to the process and other engineering disciplines who analyze the technical content of the quotes giving their comments on the bid summary. The summary is then reviewed by the cost/scheduling engineers who compare the offers against the control budget and the master schedule. They add their comments and route the summary to the contractor's project manager for recommended selection. Finally, the client's project manger receives the package and the recommended selection. He usually solicits input from his project team before approving the selection.

If the approval mechanism meets with some problems the client's project team through the project manager may request a conference of members of the contractor's team to resolve the problem. This is sometimes requested anyway. After all the equipment cost is a significant portion of the total budget, and more than that an incorrect or poor selection could well adversely affect the detail design work depending on it, and even, much later, the construction effort (late deliveries to the field etc.).

Purchase orders

With the selection of the equipment vendors, the contractor's procurement develops the purchase orders for the client's approval (for cost plus or omnibus fee contracts). Prior to this activity usually the successful vendor is invited to the contractor's offices for a pre-award meeting to discuss his offer in detail. The contractor's purchasing department prepares the purchase order for the approval signature of the contractor's project manager, and, when appropriate the client's project manager.

Plot plan development and approval

Shortly after the commitment of the major equipment purchase orders, certified vendor data in the form of outline drawings, completed data sheets, and the like will begin to be received by the contractor. These data are received by the purchasing department whose duty is then to route the data to all the engineering and design departments for checking and data acquisition. The receipt of equipment dimensioned outline drawings enable the design department to initiate the plot layout, showing the exact location of the items relative to one another. This is a combined operation involving most of the engineering and design disciplines. The key members of this group will be the process engineer and the lead piping designer. Most contractors will conduct this activity using small scale wooden models of the items on a scaled board representing the total site. Computerised piping isometrics and 3D visualizations can be of great help at this stage.

When satisfied with a reasonable layout, the contractor's project manager will arrange a conference with all interested parties including the client's project team, to view the layout. Comments made at this review will be incorporated and the layout will be the basis the plot plan drawing. This plot plan will then become the basic tool for routing pipe ways, electrical and underground piping, and the civil foundation and structural design.

On cost plus and fixed fee contracts the plot plan requires client approval before use is made of it for further design work.

Routine progress metings and conferences

Although there is a continual interface between the two project managers during this phase, there need to be formal meetings between the two teams. This is usually on a set weekly basis where key members of both teams sit down together and solve the problems that have arisen or are anticipated to occur in the near future. These routine meetings are minuted and become part of the project record, although they may be conducted in an informal manner.

A more formal meeting takes place once per month. Again the participants are the two project managers and their key staff members. Very often senior executives of both companies may attend, particularly if there is a question affecting either company policy. This monthly meeting is convened primarily to discuss and review progress and expenditures. Obviously these topics more often than not involve some technical interface. Minutes of this meeting form the basis of the contractor's project manager's monthly written progress report. The distribution of this report includes the client's project manager, and members of his home office senior management, and the contractor's home office senior management.

The master construction plan
At some point toward the end of the engineering phase and when all equipment is defined, and the bulk of material quantities are known, the master plan for the construction is developed. As in the case of the project master plan this item is detailed both in terms of time and in terms of cash flow. In all contracts where the field cost is reimbursable, the client's project team, including a representative of their field specialist, must be involved. In all projects field cost (and, of necessity, schedule) are significant items in the overall project cost. Indeed in many parts of the world where climatic conditions or accessibility to site is a problem field cost are the major item in the overall project cost.

Among the first items in the plan that has to be decided is: What method of construction is to be adopted? Some of these options are:

- Subcontracts
- Direct hire
- Field erected
- Remote modularization

There can be a mixture of these options in the overall plan for the construction, but the decision has to be made in plenty of time for a smooth field move in.

Monitoring the construction phase

With the development of an approved construction plan the client's project team in the contractor's home office can be substantially reduced. Most of the engineering work should be close to completion at this stage. There will still be some design work, such as detail pipe spooling, electrical, and instrument hook-up details, to be completed. These do not require the monitoring and input from the client's team however. The only two major engineering items at this stage that will require some client input and approval will be:

- The operating instruction manual
- The mechanical catalogue compilation
- Plant guarantees

These items continue to be developed through most of the construction phase by a small contractor's team consisting of one or two process engineers, a mechanical engineer, and/or a project engineer. These will be discussed later.

The field move in

The extent of the field 'move in' will depend on the type of construction decided on, the location of the site, and the proximity of the labor source to site. The most extensive effort in this respect is the situation where it is decided to utilize direct labor and this labor source is a considerable distance from site. In this case the contractor must set up a camp for the labor and field staff complete with canteen, social facilities and the like. Developing labor contracts also becomes a major activity for both the contractor and the client in approving these contracts. In this situation there would be a definite advantage in totally subcontracting the work. This field aspect of the 'move in' would become part of the subcontract.

Monitoring the field activities

The activities of the client's project team during the construction phase is reduced to monitoring the cost and schedule with respect to the master construction plan. The team itself will probably be reduced to the project manager and possibly a project engineer with the process engineer on a part time basis. It is also likely that this smaller project team will remain in the contractor's office rather than move permanently to site. Usually monthly site meetings are conducted and the project manager among others would attend these meetings. During such meetings physical review of progress in the field would accompany an in depth written and verbal report on the schedule and the cost aspect.

Although the project manager would not be present full time on site, there would be a client presence there continually. Heading this team would be the client's construction specialist, with some project engineering assistance. As the construction work progresses this team would expand to include specialists in all disciplines who will follow the installation and testing of equipment relating to these various disciplines. These would include mechanical engineers, electrical, and instrument engineers. The follow-up for the piping installation would usually be undertaken by piping specialists (to check such details as piping layout conforming to plan, pipe stress analysis and the like). Most clients also include the proposed permanent operators and maintenance staff of the facility in this team.

Final activities of the project team-mechanical completion

The erection of the facilities is considered complete when all the equipment, piping, cables, buildings (if any) have been installed, and such testing that is necessary (such as line and vessel pressure testing) has been done. The contractor is also responsible for completing all insulation, such refractory lining as may be required, and the disposal of debris from the site. Most vessels may be left open for final inspection by the client.

When satisfied that all activities required to be done by him according to contract have been completed, the contractor will issue a formal notice to the client that the facilities are 'Mechanically Complete'. With the receipt of such a notice the client usually has a fixed period according to contract to check the plant out and to inform the contractor of any defects or omissions to the plant. The contractor will remedy these items notified by the client and then issue a formal notice handing over the plant to the client in terms of 'Transfer of Custody and Control'. On acceptance of this notice the client takes over the facilities and prepares to commission them.

Developing the operating manual and plant commissioning

The development of the operating manual occurs toward the end of the engineering and design phase in the contractors home office. Although normally compiled by the contractor's process engineers the client's operating staff invariably participate.

Very often too the process engineers who have been most closely associated with particular process plant(s) on the project may be required to assist in the commissioning of those plants. This section of Chapter 17 describes the activities associated with both these functions.

Developing the operating manual

The operating manual is a compilation of instructions and data reflecting recommended procedures for:

- Prestart-up conditioning of the plant
- Plant start-up
- Normal operation, and trouble shooting
- Emergency action and shut down
- Normal shut down
- Catalyst regeneration and de-coking (where applicable)

These procedures include equipment manufacturers' recommended handling, conditioning, and operation of their items of equipment and the experience of the process engineer who designed the plant. To the process engineer the writing of the operating manual provides him with the final 'In Depth' review of the mechanical flow diagram of the plant. During the course of writing the manual he will use the MFD continually to make reference to the operating procedures he proposes. Anomalies or missing valves and piping become obvious as the process engineer develops the procedure logic.

A typical operating manual has the following table of contents:

Introduction and process description

This will include a brief statement as to the contents of the manual and its purpose. This will be followed by a process description which will include the "Approved for Construction" revision of the process flow diagram as reference.

Prestart-up conditioning

This item describes the cleaning out of the plant after construction and its reassembly following the clean out. For example this item will designate those lines and equipment that will be flushed out by water and those that will be blown out by air. It will also describe the equipment and materials that will be used for this clean out such as piping spool pieces to replace control valves and nozzles and lines that will be blanked off during this activity. Finally, this item will draw attention to the manufacturer's procedures for the various equipment clean out and conditioning. Copies of the actual manufacturer's documents relating to this activity should be included in the appendices of the manual. It is important in order to maintain the warranties that accompany most equipment that the manufacturer's instructions be carefully adhered to.

Plant start-up

This section of the manual presupposes that the plant has been properly cleaned, all utilities to the plant have been commissioned, all equipment is ready for operation, all drains are free, and fire fighting equipment and procedures are in place. The process engineer then begins his detailed description of the activities to be carried out in their proper sequence to bring the plant on line and producing the products intended. To accomplish this he refers to the MFD. Using this document he describes each action in the start-up sequence by referring to line numbers, control valve numbers, equipment item numbers and titles as they appear on the diagram. Normally all these actions start with introducing the cold feed. For example in the case of the Thermal Cracker project, cold feed from storage would be used for start-up. Sufficient lines and valves would be provided for this purpose and the manual text would be somewhat as follows:

"Open block valves at tank TK 101 and at the plant battery limits on line 1234 CS 12″ ST. Open the suction valves on pump P 106 A and start the pump (Note: Only the term "Start the pump" need be made as instructions on how to start the pump are included in the manufacturer's documents found in the appendices.). Set the flow control valves 106 FRCV 1 A, B, and C to full open. Set back pressure controller 300 PRCV 1 at the outlet of heater H 301 to 250 psig. Commission the level control indicator 301 LCI 1 and when the desired level in the bottom of the fractionator is reached (NLL) open the suction valves to bottoms pump P304 A on line 1235 10″ Cr ST. Control discharge flow from pump P304 by activating control valve 301 LCIV 1 on line 1236 8″ Cr ST, etc".

Normal Operation
This section carries on from the end of 'start up' with the unit lined out and in stable operation. It then describes a series of procedures during normal operation to fine tune certain parameters, and to maintain the plant on set conditions when minor changes in feed composition, temperature changes, etc. occur. It must be noted here that in modern control systems associated with plant operation these type of adjustments are made automatically with very little need for operator intervention.

The other adjustments that may still require operator action are those associated with changing the product grade or specification. This section therefore still includes a selected list of operating changes that may be required to be made under normal operating conditions and the procedures used to make these changes.

Normal shutdown procedures
Normal shutdown procedures are usually divided into two forms, which are:

- Short duration shutdown
- Shutdown for an extended period

This first type of shutdown is that associated with a minor mechanical problem, or temporary loss of feed, or a minor instrumentation problem and the like. In these cases the feed to the plant is diverted but some or all of the product streams are rerouted back to take the place of the normal feed. The feed heater, if it is a fired heater, may continue to be fired on a lower level of operation but sufficient to enable a quick resumption to normal operation when the fault causing the shutdown has been rectified. The unit under these conditions is said to be "Boxed In".

The second type of normal shut is the more common and occurs when a major fault has occurred requiring equipment to be taken out of service for repair. This type of shutdown is also placed on a predefined schedule for routine maintenance. Such a shutdown requires the unit feed to be withdrawn and the unit itself allowed to cool down to ambient temperature, de-pressured, cleaned out usually by steam, and drained free of any hazardous material.

This section of the manual describes both these procedures in the same detail as described earlier for 'Normal Start up'. Again the MFD is used as a reference in these detail procedures.

Emergency procedures
This section is the most important in the manual, and requires careful development and presentation. Its importance is reflected in the fact that very often the preparation of this section becomes a combined effort of both the client process engineers and the contractor's process engineer. Its obvious importance to the safety of all personnel

working on the plant is often coupled with the requirements of insurance underwriters and, in the case of a major mishap, its legal significance.

The section should contain at least the following sub sections:

- Emergency shutdown
- Emergency action by personnel
- Plan showing location of fire fighting equipment
- Emergency telephone numbers (by client)
- Location and setting of pressure relief valves

The contractor's process engineer is primarily concerned with the writing of the emergency shutdown procedures. In most process unit this follows the same basic principles, which are:

1.0 If there is a fired heater either as a feed pre-heater or reboiler, shut down the burners immediately and introduce steam into the firebox.
2.0 Again, in the case of fired heaters, take out the fresh feed and recycle products through the coil until the coil is cool enough not to be damaged. Use vendor data to fix this temperature.
3.0 Cool down fractionating towers by increasing reflux flow and wherever possible reducing return temperatures.
4.0 In the case of reactors, remove the feed stream immediately and purge with inert gas such as nitrogen. In the case of reactors in exothermal service and where there is a coolant, increase the coolant duty.
5.0 Divert all product streams to slop, but maintain any cooling cycle until safe conditions are reached.
6.0 De-pressure all pressure vessels to flare or to atmosphere at a safe location. Note: The process engineer must make sure that there exists a de-pressuring system on all these vessels and that they are clearly shown on the MFD.
7.0 As soon as conditions allow, introduce purge steam into all vessels, towers, and, where applicable, reactors. In some cases steam would be injurious to catalysts in reactors, under these circumstances the reactors should be purged with inert gas.
8.0 If the emergency is the result of a fire in a heater firebox due to a fractured coil, shutdown the burners and take out the feed immediately. Introduce snuffing steam into the firebox, and do not recycle product or introduce steam into the fractured coil.

As in the case of start-up and normal shutdown procedures the same kind of detail covering the basic actions given above would be written for the emergency shutdown. Again the MFD would be used as the reference document.

Appendices

The appendices to the manual should contain as much data as possible to assist the operator of the plant in the safe operation of the plant. As a minimum it should contain:

- The "As Built Revision" of the MFD
- An up to date equipment list
- Equipment manufacturers operating instructions and data
- A list of hazardous materials used in the process, with a summary of recommended handling procedures

Plant commissioning

Process engineers from the contractor's organization are often called upon to assist in supervising plant commissioning. Process engineers from the client's organization are invariably called to supervise the commissioning. This task begins at the point where the contractor's field organization has completed all their installation activities and have handed the units over to the client's "care, custody and control". At this point all contractor's debris has been removed, pressure testing has been completed, and the unit has been reassembled after the initial flushing out.

The commissioning activities fall into the following sequence of events:

- Pre energizing activities
- Energizing the plant
- Conditioning equipment, calibrating instruments, and setting relief valves
- Final check out, and closing up all vessels
- Preparation for "start up"
- Start-up
- Lining out
- Performance test runs and guarantee test run

Much of this work is carried out by the future operators of the plant to enhance their familiarity with the process, and as mentioned earlier, the team is supervised and the activities planned by the responsible process engineer(s). Further description of the major commissioning activities now follow:

Pre-energizing activities
When the plant is handed over by the contractor to the client no hazardous material such as fuel oil, fuel gas or permanent electrical power has been introduced into the plant area. Obviously before much of the work to check out and condition equipment utility services must be established, and this will be the objective of this phase of the work. Normally while the plant is in this "safe" condition many client companies take advantage of this to carry out their final physical check of the plant on a line by line, item by item basis.

Although the contractor has flushed the plant out before hand over a further and more thorough flush out is advisable before start-up. This is done at this "safe plant" stage when utility lines and underground lines can also be flushed out. If the plant is a unit in an existing complex oily or chemical drains would at this point be blocked off from the complex's main drain systems. Immediately prior to the plant being energized these drain systems must be unplugged and checked that they are free of any obstruction.

Energizing the plant
As soon as the responsible process engineer is satisfied that all pre energizing activities are complete he will instruct that the utility systems be commissioned. Normally this starts with the introduction of permanent power and the checking out of the circuits by the electrical technicians. When this has been completed to the responsible process engineer's satisfaction, instructions for commissioning the steam, condensate, and fuel systems will follow. Note: the water system is usually commissioned before the final flush out and the system used for this flushing activity.

The commissioning of the fuel systems indicates that the plant is now a 'Hazardous Area' and all regulations pertaining to this type of area comes into effect.

Conditioning equipment
Certain new process equipment will require conditioning before being used in the process. Some of these are:

Reactors: In some processes these will have refractory lining which will need to be 'Cured'. Curing is the subjection of the refractory to a controlled increasing temperature environment until the curing temperature is reached. The refractory is held at this temperature for a prescribed period of time, before cooling back to ambient temperature. The cooling is also undertaken in a controlled fashion. Many reactors and reactor systems also require drying out. This too is accomplished during this period using heated air or inert gas and the recycle gas compressor if there is one in the circuit.

Heaters: All heaters will have refractory lining and these will need to be 'Cured' as described above. If the heater contains coils that are to be used in steam generation service, these coils need to be treated with hot caustic soda to remove all traces of grease or other undesirable contaminants.

Fractionating columns and vessels generally: These usually need to be dried. This is usually done at the same time that the associated heaters are being conditioned. The hot air or inert gas stream passing through the heater coils to protect them during the refractory curing is routed to the vessels and used for the drying activity.

Piping in sour service: Piping and pots etc in high H_2S and hydrogen concentration service often need to be acid treated to protect them against local corrosion or embattlement under these conditions. This is called 'Pickling' and is usually accomplished by setting up a temporary system. This contains a reservoir for the fresh acid, one or two small skid mounted reciprocating pumps and a receptacle for the used acid. The pumps deliver the acid through the piping and equipment for a prescribed period of time.

All of these conditioning procedures are provided by the respective equipment manufacturers and would be included as part of the operating manual.

The opportunity is normally taken at this time to calibrate as much instrumentation as possible. Flow meter orifice plates are installed and, where possible, flow meters and control valves used in these procedures. Relief valves are set, and this is always done by the client's organization, and the settings certified.

Final check out, and closing up of all vessels
This will be the last opportunity to check such items as the internals of towers, fractionation trays, condition of refractory and other linings, hold up grids, distributors, and the bottom of the tower baffling system (to and from the re-boiler). A final check-out of the piping layout also needs to be carried out at this point. When satisfied that all is satisfactory the process engineer will authorize the following final prestart-up activities to be completed:

- Catalyst loaded into the respective reactors. This is often supervised by the licensor's representative. The reactor closed up by installing the correct operating gaskets. This later point is made because during the prestart-up activities and conditioning temporary gaskets are used on manways, nozzle connections and the like. These are replaced by the correctly specified gaskets for the operating conditions. The temporary gaskets are never used again and are thrown away.
- All towers containing random packing are loaded with the packing and closed up. Permanent specified gaskets are used to replace temporary gaskets as described above.
- All other vessels and towers are closed up using the permanent specified gaskets.
- In cases where equipment has been subject to a caustic wash the temporary silica level gauges used during the wash are replaced by the specified operational ones.

With the completion of these final checks and vessel close up, the plant is now ready for start-up.

Start-up and lining out
The activities and their sequence for starting up the plant are carried out as described in the operating manual. In the case of an oil refining plant for example the first activity is to eliminate air from the plant systems. This is done by using water or steam or inert gas or a combination of all three. The use of steam or water is prohibited for most reactor system where the catalyst would be irreparably damaged by such contact. Thus inert gas is circulated using the recycle compressor for this purging. Thus the recycle compressor is commissioned and will then continue to function until plant shut down. In the case of the fractionating units of the crude distillation and thermal

cracking systems, water and steam are used for purging. The water is used to purge the heaters and tower bottom systems and steam is used for purging the upper sections of the fractionating towers.

After the purging comes the introduction of the cold feeds. In the case of units that contain reactors and use hydrogen under pressure a leak testing program is required. First of all the plant is subject to leak testing at operating pressure using the inert gas. Then as the inert gas is replaced by hydrogen further leak tests are required. All leaks must be repaired before start-up of these units can begin.

Where water has been used for purging, the water is replaced by the oil feed. This is termed the ‘oil squeeze’. The steam is not replaced by anything. It continues to flow through the tower until after start-up where it may then be replaced by the vapor phase of fractionation.

Start-up may be defined as beginning when the purge program shows conditions to be safe to apply heat into the plant. In the case of catalytic plants using a hydrogen stream, the hydrogen stream is circulated and heated up to its operating temperature first. When the plant operating conditions are on gas circulation the oil feed is introduced and heated.

In the case of the noncatalytic units where only fractionation is being carried out heat may by applied to the oil feed directly. This is done when the oil squeeze is complete and the water content of the oil flowing out of the plant is at an acceptable level. The plant conditions are obtained in accordance with the procedures described in the operating manual by adjusting heat input, reflux rates, pressures and respective circulating rates in the case of the catalytic plants. Up to the point when the operating conditions have been reached the total products have been routed to slop. They may be returned later as feed or used in the refinery fuel system. As soon as the plant has reached its operating conditions product streams are required to be diverted to storage. This is done however only after the product quality has been confirmed by laboratory tests.

Performance and guarantee test runs

Usually as soon as possible after the plant has been lined out and is in production a performance test run may be carried out. The purpose of this test run would be to establish as far as possible the limits of the various systems included in the process. Normal feedstock would be used and the performance of the plants under various operating conditions examined. The other objective of this exercise is to familiarize the operators with the plant control responses to the various changes.

The guarantee test run is a more formal requirement. At the end of the test run carried out over a fixed period and under specific conditions, the client will accept the plant completely or require remedial work and further test runs to meet the guarantees. More details on this test run is given in the next section of this chapter.

An example of a commissioning plan used in the commissioning of a 'grass roots' refinery on sites facilities is given in Figure 17.1.12. This plan reflects the program for the onsite units and tank farm only. All the utility plants had already been commissioned and started up before this plan begins. The program was successful and apart from some minor problems the refinery was brought on stream in the time span predicted.

Process guarantees and the guarantee test run

Among the last activities that a contractor's process engineer performs on a project is the preparation of the process guarantees that are usually required by the client, and the procedure for testing the plant to meet the guarantees. The process guarantees may begin to be developed as soon as a firm process has been established and manufacturer's guarantees obtained for the performance of the various manufactured items of equipment. The process engineer may be required to guarantee the performance of any plant that he has calculated and specified equipment, piping, instruments etc for. He will not be expected to guarantee the performance of any item of plant or piece of equipment he has not specifically designed. He will also not be expected to guarantee any criteria that cannot be measured during a test run. Thus if a process engineer calculates a fractionating unit to meet a specified separation, he will be expected to guarantee that the design will make the separation at the design feed rate and composition. He will not however be expected to guarantee the performance of individual items of equipment contained in the unit, and which have been properly specified by him. This guarantee is carried by the respective manufacturers who have designed and fabricated the items.

The process performance is tied also to a guarantee of its efficiency. This will be in terms of a guarantee of the utility consumption in the plant whilst operating on the design throughput and conditions. Of course the guarantees as written will differ from process to process but will usually follow a similar pattern or format. This is as follows:

- Description of the feed in terms of throughput, composition, or source (in the case of crude oil for example).
- Design operating conditions and the guarantee of the product specification at these conditions. Alternatively a guarantee of the fractionation or separation performance of the process in terms of key component separation.
- A guarantee of the hydraulic capacity of the process system. This would only be in the terms that the process system will handle a design quantity per unit time. Note, this hydraulic guarantee would not normally be combined with the stream composition specification guarantee. This is discussed later.

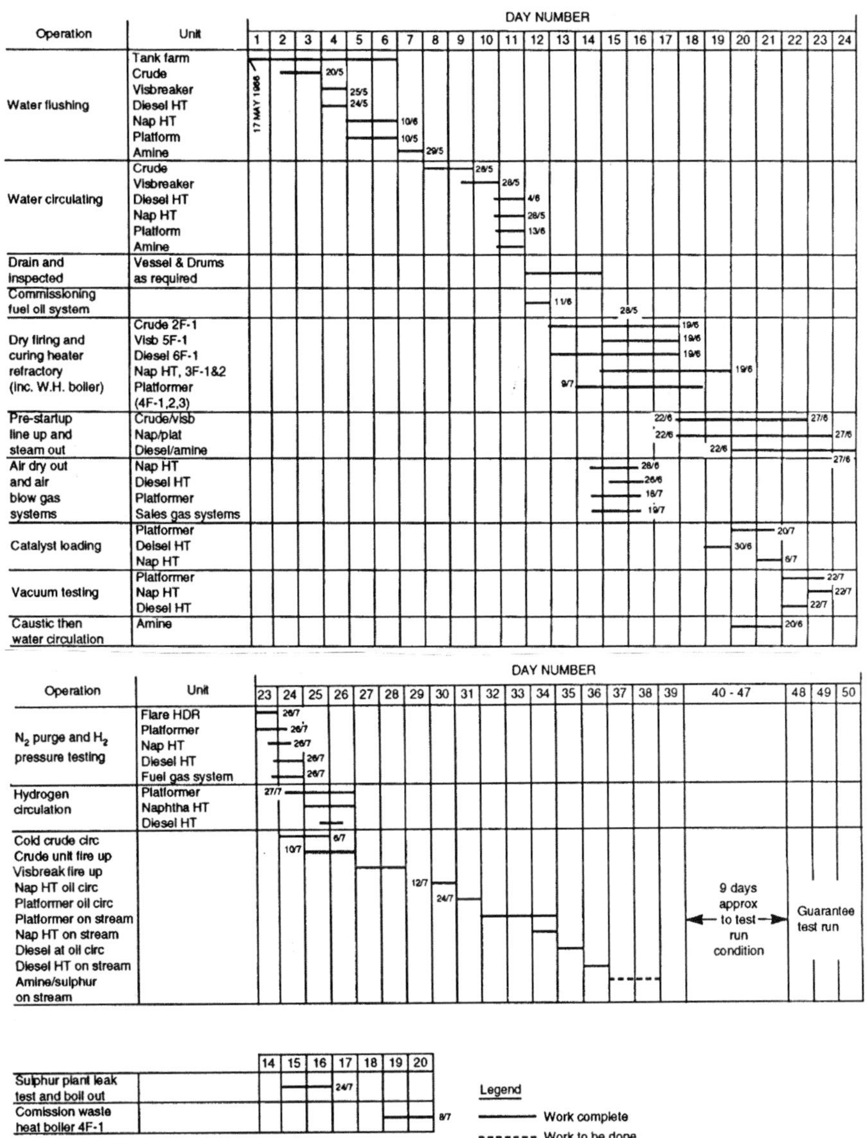

Figure 17.1.12. A typical commissioning plan.

- The utility consumption guarantees. These are usually taken for the total plants in the new or revamped complex. Rarely are utility guarantees written for individual units making up the new or revamped complex.
- A list of the accepted test procedures that will be used.
- The guarantee test run procedures that are written in some detail. Provision is normally made in them for the contractor's process engineer to instruct operating changes to be made as necessary before or during the test run.
- Description of the notices to be given in event that:
  1. The test run was successful and met all the requirements of the guarantee.
     or
  2. The test run was not successful and requires to be run again.
     or
  3. Some or all of the guarantee requirements cannot be met and a limited liability clause is evoked.

The following paragraphs describe and discusses the contents of a performance guarantee in more detail than given in the list above.

Description of the design feedstock

Although the plant may have been designed on the basis of handling more than one feedstock of slightly differing composition, the test run is conducted using one of the design feedstocks as a parameter. This practice is more common in the oil refining industry than in the production of chemicals or even petrochemicals. This is so because most refineries handle crude oil from various sources around the world, and these can vary considerably in composition. The intermediate products from the various crude oils therefore can also vary in composition.

The choice and proper definition of the test run feedstock is absolutely basic in developing the guarantees and assessing the result of the guarantee test run(s). This begins by providing a statement of the design feed throughput. This should be as simple as possible and preferably in the units of measurement used in the client's original "Duty Specification". If the measurement units used in the design or the units of measurement used in the operation of the process differs from that in the "Duty Specification", then the conversion factors used to convert from one unit of measurement to the others must be noted here. There must be no ambiguity concerning this quantity.

The second item of importance in this section is the composition of the design feedstock. Where possible this should be a breakdown in the quantity of each component in either weight or molar terms. Occasionally volume composition may be used, however, the composition and the units it is quoted *must* be able to be measured using accepted normal test equipment and/or reagents.

When it is not possible to give the feed composition as described above other acceptable criteria for establishing the feed quality may be used. For example in defining the quality or composition of a crude oil or the products of crude oil distillation boiling points and boiling curves are used (See *"Elements of Petroleum Processing"* also by the author[1]). Again however these characteristics of the feedstock must be measurable by accepted means. The test methods to be used in evaluating the guarantee test run results are listed in this document. This item is discussed later.

Test run conditions and the guarantee of product quality

Following the description covering the design feedstock comes the heart of the performance guarantee—the guarantee of the end product. This guarantee however is always tied to the feedstock and to the conditions that the plant is operated on to meet the product specification. Where there is a catalytic or any other type of reactor involved in the process this guarantee may extend to cover the yield of product promised by the contractor or the licensor.

The pertinent process conditions to obtain the guarantee with respect to the end product is defined here. Obviously these conditions will reflect the basic parameters used in the design of the facility. The process engineer developing the guarantee document however must be sure that the conditions described here *are pertinent* to meeting the end product specifications and yield. Over or incomplete definition of the process conditions may lead to ambiguity and conflict in operating the plant under test run conditions.

The product specification, and its yield under certain circumstances should reflect exactly those given in the client's "Duty Specification". Yields and or product quality which have been left open or to be confirmed by a licensor's pilot plant tests, during the compiling of the "Duty Specification" must be included here. These should be qualified however with reference to the client's acceptance of them and his approval for their use in the plant design. They become part of the client's licensor agreement. The contractor may then develop the remaining process guarantees on the design carried out by him based on this approved licensor data.

Performance of the plant may also be judged by the performance of some engineering principle used in the design. For example in oil refining the performance of some of the complex fractionators used (e.g., the atmospheric crude distillation unit) is most often defined in terms of fractionating efficiency. This is measured by the difference in the ASTM distillation temperature reading between a 95 vol% recovery of a light product and the 5 vol% reading of the adjacent heavier product. The test used to determine these readings is the accepted ASTM distillation test carried out routinely

[1] Published by John Wiley & Sons Ltd.

in all refineries. The measure of these test temperatures indicates how much of the lighter product is contained in the heavier one and vice versa. This of course is also a measure of the respective products composition and the contaminants they contain.

The hydraulic guarantee

Although this is tied to the plant design and its ability to handle the feed and product streams as stated earlier it is guaranteed separately to the performance guarantee described above. This is so because very often it is not possible for a contractor to guarantee the quantity of product that the plant will make. An example of this is again seen in the oil industry. When a contractor's process engineer designs the unlicensed units of an oil refinery he does so using the design crude's assay data. All the product yields and composition therefore are based on the assay data which the contractor could not have developed—he is therefore not expected to guarantee it. From his design calculation however he has sized the various systems in terms of piping instruments and of course specified the various pumps. He is expected to guarantee these based on the design flows he has used. Thus, the hydraulic guarantee should start with the following statement:

"When the unit is operated on the design feedstock and at the design feed rate the following systems shall be capable of handling:

- Stream One 450 gpm measured at 60°F
- Stream Two 600 gpm measured at 60°F

Utility consumption guarantees

Normally contractors offer utility guarantees as a summation of the cost of all utilities consumed by the process. These are also determined as a summary of the utilities consumed by all the plants considered in the test run. The utility guarantee figure is developed from the utility balances compiled during the process design. Much of these figures are based on manufacturers guaranteed equipment efficiencies for electric motors and turbines. Others are based on the process heat and material balances The calculated utility consumption for each utility is multiplied by the respective utility unit cost as provided by the client in his "Duty Specification". The utility guarantee figure is then quoted as the sum of all these cost items as a single cost figure.

In preparing the utility guarantee the following should appear in the statement of the guarantee:

- Unit cost of each utility
- Expected normal consumption of each utility
- Either the *guaranteed* total daily cost of all the utilities as one figure (This would include a contingency factor agreed to by the client.)

Or

- A guaranteed daily cost of all the utilities as a percentage of the expected normal consumption

The description of the remaining items that make up the guarantee document are self explanatory and are illustrated in the example of a typical process guarantee given in Appendix 17.1.9. This example is based on the study case for the thermal cracker illustrated in the various sections of this Chapter.

In this example it is assumed that the contract has been awarded on a lump sum basis for engineering, all other home office functions and the contractor's profit fee. The other costs such as equipment, materials and the installation of the plant is on a cost payable directly by the client basis.

APPENDICES

CASE 6

40 MB/SD (39 MB/CD) ON S. WALES ; 34 MB/SD (33 MB/CD) on KENT.

NORMAL DEPRECIATION — Debits/(Credits)

	1967	1969	1969	1970	1971	1972	1973	1974
Crude Run MB/CD	-	35	39	72	-	-	-	-
Imports MB/CD	-	-	13	6	-	-	-	-
Investments $M								
Onsite	3,456	-	2,943	-	-	-	-	-
Offsite & Utilities	1,285	-	-	-	-	-	-	-
Expense	62	-	221	-	-	-	-	-
Inventory	907	-	188	-	-	-	-	-
Tankage	-	-	332	-	-	-	-	-
Jetty Debit	874	-	-	395	-	-	-	-
Utilities	-	-	-	128	-	-	-	-
Services, etc.	-	-	-	594	-	-	-	-
Total Net Costs (Cumulative)	6,584	6,584	10,268	11,385	-	-	-	-
Operating Cost								
Utilities Costs	-	654	729	1,504	-	-	-	-
Ship Waiting Time	-	(266)	(106)	(394)	-	-	-	-
Milford Marine Credit	-	(132)	(52)	(291)	-	-	-	-
Maintenance, etc.	-	361	361	958	-	-	-	-
Sub-Total	-	617	932	1,777	-	-	-	-
Net Feedstock Costs	-	906	767	1,162	-	-	-	-
Gross Annual Costs	-	1,523	1,414	2,939	-	-	-	-
Tax Relief	-	(253)	(1,161)	(1,349)	(2,180)	(1,946)	(1,858)	(136)
Net Cash Outgoings	6,584	1,270	3,937	2,707	(2,180)	(1,946)	(1,858)	(136)
10% Discounted Cash Outgoings	5,991	1,054	2,953	1,841	(1,352)	(1,090)	(948)	(64)

Total Outgoings Discounted at 10% = $M8.385

CASE 6A

SPECIAL DEPRECIATION RATES AT S. WALES.

	1967	1968	1969	1970	1971	1972	1973	1974
Net Cash Outgoings	6,584	1,270	3,937	559	(2,019)	(1,648)	(1,684)	(124)
10% Discounted Cash Outgoings	5,991	1,054	2,953	380	(1,252)	(923)	(859)	(58)

Total Outgoings Discounted at 10% = $M7.286

Appendix 17.1.1. Refinery plant inadequacies report—example work sheet.

SFR-22

CRUDE OIL PROGRAME

TOTAL CRUDE STOCK M.B.	DATE	KUWAIT				IRAN IRAQ ARAMCO, BASRAN				TIA JUANA.			
		VESSEL	TOTAL STOCK BARRELS	IMPORTS	RUN TO STILLS	VESSEL	TOTAL STOCK BARRELS	IMPORTS	RUN TO STILLS	VESSEL	TOTAL STOCK BARRELS	IMPORTS	RUN TO STILLS
1782	1st		1341		117		441		69		276		-
1596	2nd		1224		186		372		-		276		-
1410	3rd		1038		186		372		-		276		-
1224	4th	OXFORD E. MUNCHEN	852	185 186	186		372		-		276		-
1408	5th		1036		117		372		69		276		-
1222	6th		919	204	119		303		69		276		-
1244	7th	JOHN P.G	1010	215	119		234		69		276		-
1271	8th	CHARIOT	1106	240	119	LARGO (ARAMCO)	165	270 210	69		276		-
1803	9th		1227		119	FERN ROSE (IRAN) CANTERBURY (IRAN)	576	185	69		276		-
1800	10th	BRITISH ENGINEER	1108		119		692		69		276		-
1612	11th		989		119	WORLD GRANDEUR GUILD FORD (IRAN)	623	270 270	69		276		-
1964	12th		870	131	119		1094		69		276		-
1906	13th	SAN. NICOLA	881		92		1025		96		276		-
1718	14th		789		64		929		122		276		-
1532	15th		725		64		807	270	122		276		-
1616	16th		661	189	64	CRYSANTHY ' L' (ARAMCO)	955	115	122		276		10
1730	17th	EXETER.	782		64	STEENSMOUNTAIN (IRAN)	948	115	122		266		10
1659	18th		718	185	64	SAGUARO (BASRAN)	946		122		256	130	10
1658	19th	OLYMPIC WIND	839	150	64	N.COMMANDER (IRAQ)	819	270	122		246	TIA JUAN	10
1892	20th	NEDERLAND RAGNA GORTHEN	925	135	117		967		69		306		10
1841	21st		943		117		898		69	WORLD INSPIRATION	296		10
1655	22nd	HAV KONG WESTMINSTER.	826	105 185	186		829	210	-		286		10
1979	23rd		940		133	MASTER PETER (IRAN)	1039		53		276		10
1793	24th	CHEMAWA	807	115	133		986		53		266		10
1722	25th		789		64		933		122		256		10
1536	26th		725		64		811		122		246		10
1350	27th		661	185 130	64	AARHUS (IRAN)	689	175	122		236		10
1524	28th	TENACIA HOEGH SHIELD	782		64	SALISBORY (IRAQ)	742	270	122		226		10
1608	29th	HOEGH GRACE	718	220	64		890		122		216		10
1642	30th		874		64		768		122		206		10
1448	31st	REIN.	810	240	64	(TIA JUAN) EORLD GRANDEUR (IRAQ)	646	270	122		196		10

FIG. 1.

Appendix 17.1.2. Example of a crude oil inventory schedule.

INTERMEDIATE STOCK SHEET

______ Refinery   Distribution ______

Month of ______

DAY OF MONTH	TANKS 335/336 WHITE SPIRIT PRODN.	RUN	DIFF.	MAX. 26.5 MR'S MIN. 1.5 " EST. STOCKS	ACT. STOCKS	TANKS 450/457 KEROSINE PRODN.	RUN	DIFF	MAX. 82 MR'S MIN. 4 " EST. STOCKS	ACT. STOCKS	TANK 353 T/JET 2486 PRODN.	RUN	DIFF	MAX. 58 MR'S MIN. 5 " EST. STOCK	ACT. STOCK	TANK 332 T/JET 2482 PRODN.	RUN	DIFF.	MAX. 16 MR'S MIN. 1 " EST STOCK
1	2.9	2.0	+0.9	9.2		15.1	12	+3.1	46.0										
2	2.9	2.0	+0.9	10.1		15.1	12	+3.1	49.1										
3	2.9	2.0	+0.9	11.0		15.1	12	+3.1	52.2										
4	2.9	2.0	+0.9	11.9		15.1	12	+3.1	55.3										
5	2.9	2.0	+0.9	12.8		15.1	13	+2.1	57.4										
6	2.9	2.0	+0.9	13.7		15.1	13	+2.1	59.5							6.3	–	+6.3	6.9
7	2.9	2.0	+0.9	14.6		7.5	13	+2.1	61.6							6.3	–	+6.3	13.2
8	2.9	2.0	+0.9	15.5		–	13	−6.5	55.1							6.3	–	+6.3	19.5
9	2.9	2.0	+0.9	16.4		–	13	−13.0	42.1							6.3	8	−1.7	17.8
10	2.9	2.0	+0.9	17.3		–	13	−13.0	19.1							6.3	8	−1.7	16.1
11	2.9	2.0	+0.9	18.2		15.1	13	−13.0	6.1							6.3	8	−1.7	14.4
12	2.9	2.0	+0.9	19.1		15.1	13	+2.1	8.2							6.3	8	−1.7	12.7
13	–	2.0	+0.9	20.0		15.1	13	+2.1	10.3							4.0	8	−4	8.7
14	–	2.0	+2.0	18.0		15.3	13	+2.1	12.4								7	−7	1.7
15	–	2.0	−2.0	16.0		15.5	12	+2.3	14.7										
16	–	2.0	−2.0	14.0		15.5	12	+3.5	18.2										
17	–	2.0	−2.0	12.0		15.5	12	+3.5	21.7					31.0					
18	–	2.0	−2.0	10.0		15.1	12	+3.5	25.2		9.6	8	−8	23.0					
19						15.1	12	+3.1	28.5		9.6	8	+1.6	24.0					
20						15.1	13	+3.1	31.4		9.6	8	+1.6	26.2					
21						–	13	+3.1	34.5		9.6	8	+1.6	27.8					
22						–	13	−13.0	21.5		9.6	8	+1.6	28.4					
23						15.1	13	−13.0	8.5		9.6	8	+1.6	30.0					
24						15.1	12	+2.1	10.6		9.6	8	+1.6	31.6					
25						15.1	12	+3.1	13.7		9.6	8	+1.6	33.2					
26						15.3	12	+3.1	16.8		–	8	+1.6	34.8					
27						15.5	12	+3.3	20.1		–	8	−8	26.8					
28						15.5	12	+3.5	23.5										
29						15.5	12	+3.5	27.0										
30						15.1	12	+3.5	30.5										
31						15.1	12	+3.1	33.6										
TOTALS																			

Appendix 17.1.3. Example of a product inventory and schedule.

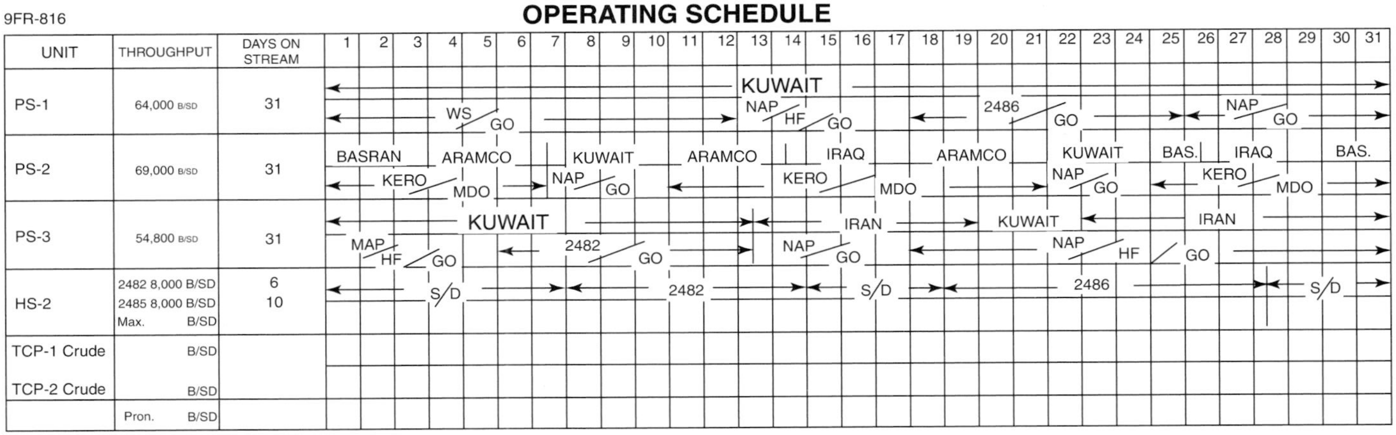

9FR-816

OPERATING SCHEDULE

UNIT	THROUGHPUT	DAYS ON STREAM	1	2	3	4	5	6	7	8	9	10	11	12	13	14	15	16	17	18	19	20	21	22	23	24	25	26	27	28	29	30	31
PS-1	64,000 B/SD	31														KUWAIT																	
							WS/GO									NAP/HF/GO							2486/GO							NAP/GO			
PS-2	69,000 B/SD	31	BASRAN				ARAMCO			KUWAIT				ARAMCO				IRAQ			ARAMCO				KUWAIT		BAS.		IRAQ				BAS.
					KERO/MDO					NAP/GO							KERO/MDO							NAP/GO				KERO/MDO			
PS-3	54,800 B/SD	31						KUWAIT										IRAN					KUWAIT					IRAN					
				MAP/HF/GO							2482/GO					NAP/GO							NAP/HF/GO								
HS-2	2482 8,000 B/SD 2485 8,000 B/SD Max. B/SD	6 10				S/D							2482					S/D							2486						S/D		
TCP-1 Crude	B/SD																																
TCP-2 Crude	B/SD																																
	Pron. B/SD																																

Appendix 17.1.4. An outline operating schedule.

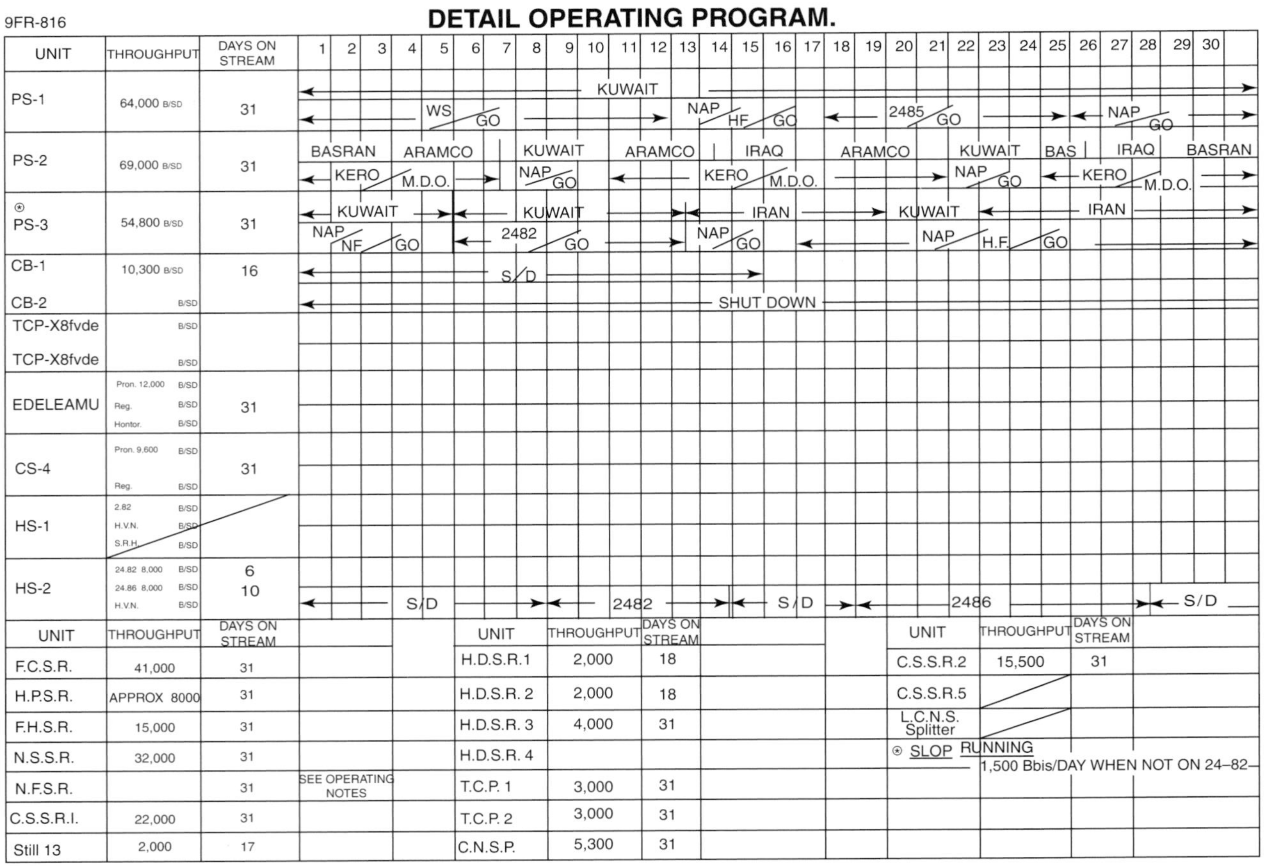

9FR-816

DETAIL OPERATING PROGRAM.

UNIT	THROUGHPUT	DAYS ON STREAM	Days 1–30
PS-1	64,000 B/SD	31	KUWAIT (1–30) / WS/GO (1–12); NAP/HF/GO (13–17); 2485/GO (18–24); NAP/GO (25–30)
PS-2	69,000 B/SD	31	BASRAN, ARAMCO, KUWAIT, ARAMCO, IRAQ, ARAMCO, KUWAIT, BAS, IRAQ, BASRAN / KERO/M.D.O. (1–7); NAP/GO (8–9); KERO/M.D.O. (10–21); NAP/GO (22–24); KERO/M.D.O. (25–30)
⊛ PS-3	54,800 B/SD	31	KUWAIT (1–5); KUWAIT (5–13); IRAN (13–19); KUWAIT (20–23); IRAN (23–30) / NAP/NF/GO (1–5); 2482/GO (5–13); NAP/GO (13–15); NAP/H.F./GO (16–30)
CB-1	10,300 B/SD	16	S/D (1–15)
CB-2	B/SD		SHUT DOWN (1–30)
TCP-X8fvde	B/SD		
TCP-X8fvde	B/SD		
EDELEAMU	Pron. 12,000 B/SD Reg. B/SD Hontor. B/SD	31	
CS-4	Pron. 9,600 B/SD Reg. B/SD	31	
HS-1	2.82 B/SD H.V.N. B/SD S.R.H. B/SD		
HS-2	24.82 8,000 B/SD 24.86 8,000 B/SD H.V.N. B/SD	6 10	S/D (1–7); 2482 (8–14); S/D (14–18); 2486 (19–27); S/D (28–30)

UNIT	THROUGHPUT	DAYS ON STREAM	
F.C.S.R.	41,000	31	
H.P.S.R.	APPROX 8000	31	
F.H.S.R.	15,000	31	
N.S.S.R.	32,000	31	
N.F.S.R.		31	SEE OPERATING NOTES
C.S.S.R.I.	22,000	31	
Still 13	2,000	17	

UNIT	THROUGHPUT	DAYS ON STREAM
H.D.S.R.1	2,000	18
H.D.S.R. 2	2,000	18
H.D.S.R. 3	4,000	31
H.D.S.R. 4		
T.C.P. 1	3,000	31
T.C.P. 2	3,000	31
C.N.S.P.	5,300	31

UNIT	THROUGHPUT	DAYS ON STREAM
C.S.S.R.2	15,500	31
C.S.S.R.5		
L.C.N.S. Splitter		

⊛ SLOP RUNNING 1,500 Bbis/DAY WHEN NOT ON 24–82

Appendix 17.1.5. A typical detailed operating program and schedule.

9 FR-1086 ISSUE No. 2

WEEKLY OPERATING PROGRAM.

UNIT	THROUGHPUT	DAYS ON STREAM	WED 11	THUR 12	FRI 13	SAT 14	SUN 15	MON 16	TUES 17	WED 18
P. S.-1	64,000 B/SD	8	←			KUWAIT				→
			307	307	310	311	311		306	307
			2482 GO	←	WS	G.O.	→		2486 GO	2486 GO
P. S.-2	67,000 B/SD	8	ARAMCO	ARAMCO	←	IRAQ	→		ARAMCO	ARAMCO
			303	301		304		309	301	301
			KERO	KERO MDO		KERO	KERO GO		KERO	KERO MDO
P. S.-3 (1)	55,000 B/SD	8	←		IRAN	→		KUW	KUW	IRAN
				313		314		312		315
			←		NAP	GO		→		
HS-2	8,000 B/SD	4	2482		SHUT DOWN					2486
E.P.S.R. (2)	MAX	8	←			PREMIUM				→

SHIPPING

DEC 11th	"BR. ENGINEER"	(KUW)	306, 312
" 11th	"GUILDFORD"	(IRAQ)	309, 304, 303
" 11th	"WORLD GRANDEUR"	(ARAM.)	301, 313, 303
" 13th	" SAN NICOLA"	(KUW)	307, 312
" 16th	"EXETER"	(KUW)	310, 311
" 16th	"CRYSANTNY 'L'"	(ARAMCO)	303, 304, 314
" 17th	"STEENSMOUNTAIN"	(IRAN)	312
" 18th	"SAGUARO"	(BAS)	309

NOTES.

(1) Stop to be run on Nap/Go operations.

(2) Awaiting decision on Prem. Kerosene specification. If Edeleanu plant runs at only 10,000 BU1/D the Kero run on Ps-2 willl be curtailed.

Rev. 1

Appendix 17.1.6. A typical weekly program.

Appendix 17.1.7. Typical factors used in capacity factored estimates

Type of plant	Factor (*b*)
Atmospheric and vacuum distillation	0.6
Catalytic reforming	0.6
Fluid cat cracking	0.7
Naphtha splitter	0.7
Thermal cracker	0.7
Delayed coking	0.6
Fluid coking	0.65
Gas compression—Recip	0.9
Gas compression–Centrif	0.75
Hydrogen plant	0.55
Sulfur production	0.55
Steam generation	0.55
Utilities (general)	0.55

All other type of plants including tankage and off sites generally may be taken as 0.6.

Appendix 17.1.8. An example of a process specification

XYZ Refinery Project

Process specification for a thermal cracker

Number of units. One

Capacity. The unit shall have an input capacity of 34,500 barrels per stream day of a residue from the atmospheric distillation of Kuwait crude oil.

Charge. The normal feedstock will be an atmospheric residuum boiling above 700°F TBP cut point on Kuwait crude.

Duty. The duty required from this unit will be to thermally crack the feedstock to produce gas, naphtha distillate, gas oil, and fuel oil.

Yields. The unit shall be designed to make the required products in the following relative proportions from the feedstock specified above. The process licensor shall confirm these proportions by pilot plant tests on samples of the feedstock provided by the owner.

	wt% on feed
Conversion to 340°F TBP cut point	25.0
Products:	
Gas to C_5	9.0
Naphtha distillate to 390°F TBP cut point	20.3
Gas oil 390 to 622°F cut points	24.7
Fuel oil + 622°F cut point	46.0

Products

1.0 The gas shall include C_3 an C_4's and shall be routed to the crude unit overhead distillate drum.
2.0 The overhead naphtha distillate product shall also be routed to the crude unit overhead naphtha drum. The naphtha product shall have an ASTM distillation end point of not more than 387°F.
3.0 The gas oil side stream product shall have an ASTM 90 vol% distilled at a temperature no higher than 645°F.
4.0 The fuel oil as residue from the cracker primary tower shall have a minimum Pensky Martin flash point of 200°F and shall be thermally stable.

Process conditions

1.0. Thermal cracker furnace

It is required that the furnace and transfer line be capable of effecting a crack of 25 wt% on feed of gas and naphtha to a TBP cut point of 340°F. The transfer line outlet temperature from the furnace shall not be higher than 920°F at a pressure of 250 psig. The injection of HP steam into the furnace coils may be considered to increase turbulence and minimize the lay-down of coke. Should steam be used it's volume under the furnace conditions must be accounted for in the design of the furnace coil(s).The temperature of the feed to the inlet of the furnace soaking section must be controllable.

2.0. The main fractionator

The feed to the fractionator from the furnace must be quenched to a suitable temperature to meet the residue cut point requirement. This temperature must also be sufficient to produce enough over flash for proper fractionation between the distillate streams.

The column will be operated at a pressure in the overhead distillate drum of 50 psig. Naphtha distillate stream shall be maximized and steam stripping of the gas oil side stream should be considered in this respect. Fractionation criteria in terms of gaps (and overlaps) between the distillate streams shall be:

ASTM Dist 95% temp of naphtha and 5% temp of gas oil to be not less than +15°F.

The residue leaving the bottom of the tower shall be steam stripped for flash point control. It shall also be quenched in the well of the tower to prevent further cracking. A cold residue stream is recommended for this quench.

Appendix 17.1.9. An example of a process guarantee

XYZ Refinery Project Contract 1234

The process performance guarantees are handled on an individual process unit basis as set forth in subsequent sections of this exhibit. The utility guarantee covers all of the process units associated with this contract and is lumped into an aggregate utility cost as set forth in Section 4.0 of this exhibit.

Where licensed or proprietary processes not owned by contractor are involved or proprietary catalysts are supplied by others, such as in the case of the thermal cracker, naphtha hydrotreater, diesel hydro-treater, and catalytic reformer, contractor offers only the guarantees set forth herein and identified as the responsibility of the contractor. Any other guarantees shall be negotiable between Client and Licensors and are not part of this exhibit.

3.0. Thermal cracker

3.1. Feed

The thermal cracking unit shall be designed to process 34,320 BPSD of Kuwait reduced crude to produce gas to C_5, gasoline, gas oil, and fuel oil as residue. The feedstock shall be material boiling above 700°F TBP cut point from an atmospheric crude distillation unit. The feedstock shall be substantially in accordance with the Kuwait assay titled "ASSAY of KUWAIT 31.2°API CRUDE Dated xxxxx" and shall have the following properties:

Gravity °API 15.8
ASTM D—1,160

	°F
IBP	650
5%	750
10%	775
20%	800
30%	850

Pour point (Max) °F	85
Visc . CS @ 122°F	55
Flash point (Pensky Martin)	>250°F

3.2. Performance guarantees

3.2.1. When owner operates the thermal cracker at conditions defined by the contractor with the feedstock and rates given in 3.1 above, the estimated (but not guaranteed) yield structure will be as set forth in Table 17.9.1 attached.

3.2.2. The thermal cracker will be designed so that it will have the hydraulic capacity to process 34,320 BPSD of the feedstock defined in 3.1 above, with the design yield structure defined in 3.2.1 above.

3.2.3. Based on the yield structure defined in 3.2.1 above, the contractor guarantees that the fractionation section of the unit shall be designed so that the numerical difference between the 5% point of the thermal cracker gas oil and the 95% point of the thermal cracked naphtha, expressed as ASTM gap, shall be no less than 10°F, as measured by ASTM D-86.

3.2.4. If the feedstock does not meet the assay defined in Section 3.1 or if the thermal cracker is not operated in accordance with contractor's instructions, then the performance guarantees set forth herein will be modified in accordance with sound engineering practice, as appropriate, based on actual conditions during the test run and using the same data sources and calculation methods as used in the design so as to obtain a true measure of the unit's performance.

3.2.5. The test methods that shall be used in evaluating all aspects of this guarantee shall be as listed below:

Feed flash point—ASTM D-93
Feed distillation—ASTM D-1,160
Product distillation—ASTM D-86.
Feed viscosity—ASTM D-445
Feed pour point
Gravity—ASTM D-287

Process performance and utility guarantees

4.0. Utilities

4.1. The process units will be designed so that the daily aggregate cost of utilities consumed for all facilities covered in Section 4.2 herein shall not exceed $xxxxxxxxx

Table 17.9.1. Design yield structure

	wt% on Feed
Conversion to 340°F TBP in furnace & transfer line	25.0
Products:	
Gas to C5	9.0
Naphtha to 390°F cut point	20.3
Gas oil 390°F to 622°F cut point	24.7
Fuel oil +622°F cut point	46.0

when the process units are operated at the rates and conditions summarized in Section 4.2 herein and with the individual unit utility costs as summarized in Section 4.3 herein.

4.2. The utilities guarantee is based on operating the process units at the rate and conditions summarized below and defined in more detail in Sections 2.0–5.0 of this exhibit:

(Sec 2.0) New naphtha hydrotreater	13,500 BPSD fresh feed.
(Sec 3.0) New naphtha splitter	13,500 BPSD fresh feed.
(Sec 4.0) New thermal cracker	34,320 BPSD fresh feed.
(Sec 5.0) New light end unit	3,830 BPSD C_4's, C_3, C_2 mixed feed
(Sec 6.0) New diesel hydrotreater	9,000 BPSD fresh feed.
(Sec 7.0) New catalytic reformer	9,000 BPSD fresh feed.

This utility guarantee covers only those units listed above and does not include the utilities of any revamped units (such as the existing debutanizer). All conditions regarding feedstock composition and operating conditions defined in Section 2.0 to 7.0 of this exhibit shall extend to the utilities guarantee.

4.3. The following unit costs for the individual utilities shall be used in computing the daily aggregate cost of utilities:

Fuel gas	$0.47 per mmBtu, LHV.
Fuel oil	$0.23 per BPSD
Power	$0.042 per KWH.
Water	$0.5/1,000 US Gals.
Steam	$0.83 / 1,000 Lbs

4.4. In the event that the daily aggregate cost quoted in Section 8.1 herein is exceeded, contractor shall have the right to make any alterations it deems necessary in order that the utilities guarantee can be met in a subsequent test run. Contractor shall make alterations or pay a penalty as defined in Section 8.5 herein at his sole option.

4.5. The penalty that the contractor shall pay in the event that the utilities guarantee is not met ,and the contractor elects not to modify the plant, shall be the difference between the average daily aggregate cost of utilities for the best test run made under Section 10.0, and the guaranteed daily cost of utilities in Section 4.1 herein multiplied by 700.

5.0. *Qualifications for guarantees*

5.1. Notwithstanding any other sections or statements in this exhibit, the guarantees in Sections 2.0 to 4.0 inclusive, above are subject to change if Licensor information, data, or designs for the licensed units should be revised at any time so as to differ from the Licensor's information, data, or designs in the Contractor's possession on –Date–, or if licensed units require more utilities than specified by the aforesaid Licensor information, data or designs.

5.2. In no event shall the contractor be liable for contingent or consequential damages, including damages for loss of products or profit or for plant downtime.

6.0. Performance test procedures.

To determine whether the guarantees defined in Sections 2.0 to 4.0 are met, the following test run procedure will be used:

6.1. The contractor will notify the owner in writing when the plants or any portion thereof is ready for initial operation. Within 30 days thereafter, the owner will perform a series of test runs unless delays are caused by deficiencies that are the contractor's responsibility. If the performance test run for the plants or any portion thereof is not conducted by the owner within 30 days after notification by the contractor to owner of availability for test run, it shall be conclusively presumed that the performance guarantees have been met and that contractor's obligations covered herein have been satisfied and owner agrees to pay any sums due to the contractor as if the test runs had been successfully met. The time of these test runs may be initiated at any hour of day or night. Test periods shall be of 72 hr duration or less as mutually agreed upon by owner and contractor and may be interrupted as follows:

6.1.1. For minor alterations, repairs, failure of feedstock, utility supply, or other condition beyond the contractor's reasonable control, each of which do not exceed 24 hr. the test run shall proceed promptly after the interruption and as soon as the contractor deems the plant operation has leveled out. The sum of normal operating periods before, between, and after such interruption shall be considered the required test period when it totals 72 hr.

6.2. Owner shall be responsible for supplying all the necessary operating labor, feed stocks, utilities, catalysts, chemicals, sampling laboratory analyses, and other supplies for operating and testing the process units. Owner shall maintain the process units in accordance with good practice, and all catalysts shall be in essentially new condition to the end that the process units will be in proper condition for the performance tests provided for herein.

6.3. The contractor shall furnish observers and test engineers, excluding licensor personnel, to technically advise owner during the performance tests. Contractor's observers shall have the right to issue instructions regarding the manner in which the plant is to be operated during the test run. Owner shall comply with these instructions unless the instructions are contrary to generally accepted safety practices or expose equipment conditions of temperature, pressure, or stress greater than their maximum allowable operating conditions.

6.4. Analyses of the streams and products for these tests shall be determined by methods mutually agreed to by owner and contractor. Analysis will be conducted by owner and may be observed or witnessed by contractor. In case of disagreement, a referee laboratory may be selected by approval of both parties and paid for by the owner.

6.5. Samples shall be spot and/or composite and sampling procedure shall be by methods as mutually agreed upon by owner and contractor. Samples shall be taken at uniform intervals. Elapsed time between samples or sample increments shall not exceed 4 hr, and composite samples increments shall be of equal volume when flow rate is essentially uniform. All flow rates, product rates, and analysis shall be averaged over the test period. Where possible, tanks shall be gauged at frequent intervals to substantiate meter readings.

6.6. The performance guarantees shall be considered satisfactorily met when the average of the performance results during the period meets or exceeds the performance specified in Sections 2.0 to 4.0 of this exhibit. Owner shall be responsible for the security of the unit operating log sheets, charts, laboratory test results, gauging records, and other pertinent information for the test period. Within 10 days after the completion of each performance test, owner shall submit a written statement for each unit indicating whether the guarantee has been met. If the test is not acceptable to the owner, then the owner shall specify in writing to the contractor in what respect the performance has not been met .On request the owner shall submit to contractor all records and calculations for review.

6.7. A performance test run shall be stopped when in the judgement of the contractor alterations, adjustments, repairs, and/or replacements which cannot be made safely with the equipment in operation must be made to enable the plant to meet and fulfill the performance guarantees, or the data obtained during the test will not be sufficient to establish the actual performance of the unit within desired limits of error, or it becomes obvious that such performance test cannot be satisfactorily concluded in the current attempt.

6.8. If a performance test is stopped as provided above, or if the unit or units do not meet their guaranteed performance during the test run, owner, when requested, shall make the unit or units available to the contractor as soon as possible to make such alterations or additions thereto as in contractor's judgment are required to enable the plant to fulfill the aforesaid guarantees. If the owner does not make the unit or units available within three months, the performance guarantee will be considered to have been met in it's entirety. Contractor will make such alterations as it shall deem necessary to make the unit or units perform as guaranteed and a further performance test shall be conducted in accordance with the above procedure.

6.9. Process performance tests for the units described in Sections 2.0 to 3.0 along with the appropriate pro rata of the utilities guarantee as set forth in Section 4.0, all of this exhibit, may be carried out collectively and/or on an individual basis as may be mutually agreed to by owner and contractor. If a test run involving more than one process unit or section of a process unit indicates that any unit or section meets it's guarantees as set forth in this exhibit, then that unit or section shall be accepted.

6.10. Notwithstanding any other provision hereof, contractor's liability for making the changes units to make them perform as guaranteed in Sections 2.0 to 3.0 of this exhibit shall be limited to $xxxxxx. In determining the total cost expended by contractor there shall be included the engineering costs to the contractor of making the necessary design changes, and the labor and material costs of implementing these changes.

Chapter 17

17.2 Economic analysis

P. R. Pujadó

Introduction

In this introductory economic analysis we cover two main subjects:

Economic analysis at one point in time

This "instantaneous" analysis is the most common since it entails the least difficulty and requires the least amount of information. This is the type of analysis seen in Chem Systems' PERP reports and in SRI's PEP reports. If properly done and properly used, the results derived from this analysis are excellent. In this write-up we devote most of the attention to this analysis; in particular, we try to analyze each of the terms in detail. Although this analysis requires the least information, in order to do a proper job the amount of information required is usually still beyond what is commonly available in the initial screening stages of a project. However, the insight provided by examining the various entries is of great value as it helps to identify missing data. Also, even with incomplete information, this type of analysis is especially useful to determine the sensitivity of a project relative to a number of parameters.

Economic analysis over an extended period of time

This analysis, usually known as "cash flow analysis," is performed by analyzing the (expected) performance of a project over an extended period of time, usually a number of years at least equal to the capital depreciation or amortization period. The value of this type of analysis is that it recognizes the "time value of money."

A cash flow analysis does not require more information than an instantaneous analysis, but it does introduce some additional concepts in the calculation of the discounted cash flow rate or return or internal rate of return (IRR) and other discounting parameters.

Calculation of the IRR used to be difficult when slide rules prevailed, since it entails a trial-and-error calculation; it is quite trivial with just about any computer or even a small programmable calculator. A cash flow analysis is an absolute necessity to account for the effects of taxation and inflation since these are usually spread out over a number of years and the effects are carried over from year-to-year, especially when deferred taxes accrue out of the different depreciation and/or amortization methods used for tax accounting as opposed to financial accounting. However, in order to keep it simple, this Introductory Economic Analysis will not cover multiple depreciation schedules, tax accounting, and the like.

Indirectly, the cash flow analysis can also be used to decide what the prospective sale price of a product should be. The price can be calculated so as to yield a desired discounted cash flow rate of return.

Accounting for inflation

The effects of inflation, though important in real life, tend to be irrelevant and misleading for economic evaluations and can best be dismissed by always performing a constant-dollar analysis.

If inflation needs to be taken into account, it is easy to incorporate inflationary factors that adjust the numerical values on an annual basis. The formulas provided in the appendices in general do not apply when variable annual cash flows have to be accounted for. The calculations are then performed by creating a series of estimated annual cash flows and conducting the calculations by trial-and-error. Some of the trial-and-error calculations are already programmed in the functions usually available in most PC software packages. Other calculations may require some simple program coding; half interval splitting (dichotomous search) is usually suitable for convergence.

The interpretation of the results when inflation is accounted for tends to be muddied by the additive effect of inflation. The gross returns will increase if inflation leads to increasing cash flows but may decrease if inflationary pressures raise the costs faster than the revenues.

Analysis at one point in time

This section examines what can be regarded as an analysis at one given point in time. This "instant" analysis is normally confined to a typical period in the life of the project, usually one year. This is the usual type of economic analysis and is also the type found in Chem Systems' reports, SRI's reports, etc.

An analysis of this type normally entails three parts:

- Capital summary
- Production cost summary
- Reporting parameters

In the first part, capital summary, we tabulate those items that may be candidates for depreciation, amortization, or interest charges for both long-term and short-term capital. Typical items are:

- Erected plant cost
- Erected offsites cost
- Interest during construction
- Catalyst and metals inventory
- Working capital

If the economic analysis is intended to account for tax effects, a far more detailed breakdown is required for capital items since, for tax purposes, the depreciation schedules may vary considerably from item to item, principally as concerns the treatment of machinery, equipment, and offsites.

The second part, production cost summary, provides a condensed tabulation of the various items that contribute toward the final cost of the manufactured product. There are two basic ways of conducting an "instantaneous" economic analysis as described herein: (a) to arrive at a cost of production, or (b) to arrive at a given return-on-capital (ROI). Both formats are similar in concept and execution, the only difference being that the latter, **b**, requires the assignation of cash values to the main product or products.

The following items are normally included:

Variable credits (often shown as credits against variable costs)
Materials

Product revenue (only shown if required to set margins)
By-product credits

Utilities

Utility credits

Variable costs
Materials

Raw materials costs
Catalysts and chemicals

Utilities

Utility requirements

Operating costs

Labor
Maintenance

Operating expenses

Plant insurance
Property taxes
Direct overhead
General plant overhead
Sales and administration (often accounted for as overhead)

Capital items

Depreciation
Amortization
Interest on capital
Interest on working capital

Taxes

(only if the analysis is to consider tax effects)

Following is an itemized summary of each section.

The final section of an economic analysis, instantaneous, or over time, may be referred to as "reporting parameters." In this section the project performance is measured by using some commonly acceptable yardsticks: return-on-investment (ROI), pay-back or pay-out time, internal rate of return (IRR), discounted cash flow rate of return (DCFRR), etc. They will be discussed in more detail later.

Capital items

The first part of an economic analysis requires an exposition of the capital items. This includes those items that can be either depreciated, amortized, or that enter indirectly in the cost of production through either a long-term or a short-term interest charge. Typical entries are:

- Erected plant cost
- Erected offsites cost

- Interest during construction
- Royalties
- Catalyst and metals inventory
- Working capital

Erected plant cost

By erected plant cost we mean the estimated cost of construction inclusive of materials, equipment and labor, plus design, engineering, and contractor's fees within battery limits as usually provided by cost estimating [Estimated Erected Cost or EEC]. It is usually assumed that all the elements included within battery limits qualify for the same depreciation rate; this need not be the case in real life.

Erected offsites cost

An estimated erected cost of offsites usually is not available at the early stages of a project. Estimating the cost of offsites is difficult, but not impossible. Two approaches are available:

- A gross estimation approach
- An itemized estimation approach

The first approach provides only a very rough estimate and makes use of some simple but often unreliable guidelines. Essentially, these guidelines dictate that, if a plant is a grass-roots facility, the cost of the offsites is about the same as the cost of the plant within battery limits; if the plant is built on a developed site, the cost of the offsites is about 30–50% of the cost of the plant within battery limits. These rules are good only to show an allowance for offsites, but are inadequate for quantification purposes. The scope of offsites to be built for a grass-roots facility can vary enormously from case to case: tank age, flare, feed water treatment, steam plant, power distribution center, cooling water facilities, effluent treatment facilities, road terminals, rail terminals, marine terminals (jetties, etc.), canteen, shops, administration buildings, etc. just to give a few examples; units in remote locations may include "offsites" such as workers' housing, sports, religious and recreational facilities, etc.

The second approach entails more work, but yields results that, at least, are more justifiable. In this approach the various offsites and their capacities are itemized and cost estimating rules are applied. An example can be found in the guidelines provided by Chem Systems.

Interest during construction

This concept is often ignored or overlooked. Unfortunately, it is a significant amount that greatly contributes to the final erected cost of a unit or complex.

A unit is not built overnight. Expenditures normally accrue over the procurement, fabrication, and construction period that, for a plant of average complexity, can usually take anywhere from 2 to 4 years. Not knowing further details, 2 years may be assumed for relatively simple units and 3 years or more for more complex ones; very simple units may be assumed built in one to one-and-a-half years.

It is common practice to assume that the cost of the capital expenditures that accrue over the construction period is covered by a short-term loan on a 100% debt basis. This assumption is not unreasonable: even if the funds are provided on an equity basis, the expenditures represent a short-term opportunity cost that can be accounted for through the short-term loan concept. It is true that, often, funds are provided through a long-term loan, but this is of relatively little consequence over the construction period; assumption of a short-term loan at worst provides for a more conservative estimate.

How are the capital expenses allocated over the construction period? Unless actually known, there is no common school of thought as to how construction costs ought to be spread. Equal shares (i.e., one-third each over a three-year period) can be used for simplicity; others might prefer to use a 30%–50%–20% distribution over a 3-year period to account for the fact that expenses tend to occur up-front.

Regardless of the way used to allocate capital over the construction period, the interest charges accrued over this period are accumulated and capitalized (usually on a year-end basis) as part of the erected cost of the unit. If properly accounted for, separate allocations should be made for items within battery limits and for offsite items. In fact, because different individual items such as columns, heat exchangers, etc. may have to be retired and replaced at different times over the life of the unit, a running count of interest expenses should be maintained for each individual piece of equipment. To provide an example in a similar, but unrelated, situation in a plane the jet engines have a different depreciation schedule and tax treatment than the main body within the fuselage. It is therefore always advisable to account for the capitalization of as many individual items as possible; it is not an impossible task with a relatively small computer to keep track of events and charges.

Royalties

Royalties are normally added to and capitalized with the plant investment even though a different depreciation schedule may be used for royalties.

Adsorbent, catalyst, and metals inventories

Catalyst base and metal make-up, additions and reprocessing costs should be handled as an operating expense, along with chemicals and other consumables, in the production cost summary.

A question arises as to how to handle a catalyst base inventory or an adsorbent inventory, principally when the life of the catalyst or adsorbent is expected to span several years.

Essentially there are two ways:

- We can treat the catalyst as a consumable with the annual consumption calculated by prorating the inventory over the expected life span of the catalyst. The catalyst inventory itself can still be treated as a capital item for the purposes of calculating interest on capital, if the investment is leveraged.
- We can treat the catalyst inventory purely as a capital item. The catalyst inventory can then appear in the production cost summary as (a) an amortization term *and* (b) interest on capital.

Tax law considerations may favor or dictate which approach should be taken. In general, however, the latter approach is preferred; namely, the catalyst or adsorbent inventory is capitalized and amortized over its expected useful life. While, in principle, both approaches would seem to be numerically equivalent, the amortization approach has a more favorable impact on the cash flow since amortization is an add-back item. Whenever allowable, this approach also makes it possible to use accelerated amortization schedules.

Catalysts should *never* be amortized and treated as a consumable at the same time!

Those metals that are not recovered and are disposed of with the catalyst base are treated as an integral part of the catalyst and accounted for as the catalyst base.

Metals—principally noble metals—that are recovered at the end of the useful life of each catalyst batch are carried only as working capital and, therefore, appear in the production cost summary only as an interest item.

Small metal losses incurred when the spent catalyst is reprocessed and the cost of reprocessing itself can also be amortized over the useful life of the catalyst, but are most often handled as operating expenses.

Catalyst stockpiled as inventory for future use is part of the working capital.

Working capital

Working capital items are those items that:

- require an additional front-end investment even though they are not part of the physical plant
- accrue additional interest charges, usually in the form of short-term loans
- can be recovered at the end of the useful life of the project

For timing purposes, working capital is usually assumed to be available upon completion of construction. The following items comprise the working capital in most economic studies:

Additive items

Feedstock inventory
Supply of raw materials valued at delivered prices. Excluded are any materials that are not normally stored (gases, for example). Unless specified otherwise, a 1/2 month supply or equivalent liquid and solid feedstock storage is assumed.

Finished product inventory
Supply of principal products and by-products, if any, valued on a gross cost of production basis. Again, a 1/2 month storage is assumed unless differently specified. Excluded too are items that are not normally stored (e.g., hydrogen or light ends).

Accounts receivable
Typically it is assumed to consist of one-month's worth of production. Production includes main products, by-products, and any other items that might be sold on a regular basis.

Cash on hand
Typically estimated as the equivalent of one week's cash flow.

Adsorbent, catalyst, and metals
Normally, only those items that are fully recoverable, and thus not amortizable, are part of the working capital. Typical is the inventory of noble metals in the plant.

Warehouse and spare parts inventory
Any spare items kept in storage on a regular basis are included as part of the working capital. Excluded are items ordered and kept in storage on a short-term temporary basis until installed. Typical working capital items would be spare loads of adsorbents or catalysts and spare parts such as compressor rotors, rotary valve gaskets, replacement pumps, etc. Since a detailed inventory of spare parts is not usually available for a preliminary economic analysis, it is common practice to assume that the total amounts to, say, 2 or 3% of the erected cost within battery limits.

Subtractive items

Accounts payable
Typically assumed as one month's worth of raw materials and other consumables at delivered prices.

Cost of production

The concepts usually included as part of the Cost of Production summary were itemized in the opening section. Essentially, they were grouped under five major sections:

- Variable costs
  - Materials
  - Utilities
- Operating costs
- Operating expenses
- Capital items
- Taxes

Variable costs

Variable costs are those that can be directly associated with the production process as concerns the consumption of materials and services or utilities. A characteristic of variable costs is that they are roughly proportional to the amounts produced. Thus, if the total annual production were to increase by, say, 20%, the annual variable costs would be expected to also increase by roughly 20% (any differences would be due to changes in production efficiencies only) so as to maintain a uniform variable cost of production per unit item. Typical variable costs of production are:

- Raw materials costs
- Catalysts and chemicals
- Utility requirements

and, in general, any other consumables expended in the production process.

Although not part of the costs, it is customary practice to subtract here any variable credits such as:

- By-product credits
- Utility credits

More appropriately, these concepts should be entered under a separate heading of variable credits.

If the establishment of a cost of production is the main objective of the analysis, all the various costs are added up to the final grand total. If, on the other hand, it is desired to calculate an income or cash flow result, the various costs will be added up and the result subtracted from a base term that will comprise the revenue generated by the production of all major products or co-products assumed sold at prevailing market prices estimated at an equivalent production level from other manufacturers.

Operating costs and operating expenses

These are typically *fixed* operating charges unless there is a major revision of the plant's operating conditions. Typical items are:

- Labor
- Maintenance
- Plant insurance
- Property taxes
- Direct overhead
- General plant overhead
- Sales and administration expenses

and, possibly, many others depending on an individual plant's characteristics and accounting basis. In general, all these expenses are *fixed* in that they remain about the same on an annual basis irrespective of the plant's actual running capacity. Naturally, if business is poor and production rates are low on a consistent or extended basis, we can expect that measures will be taken to lower these charges to the extent that it may be feasible (e.g., reduction in labor through lay-offs or reduction in overheads through some other belt-tightening measures) but, in general, changes in these concepts will always be limited, short of shutting down the plant.

Labor

Labor includes the personnel required at various skill levels to operate the unit on a 24-hr basis. Since shifts usually run for 8 hr, a minimum of three operators are required to staff each position. Actually, because of holidays, vacation, leave time, etc. the staffing requirements are considerably higher.

An average year of 365.25 days consists of 8766 hr. The basic work load per individual at 8 hr/day and 5 days/week is about 2090 hr/year, corresponding to an average of 4.2 operators/shift. This value is often seen in economic estimates, but is not quite correct. If we instead estimate the time off as follows:

	days/year
Weekends	104
Vacation (3 week average)	15
Holidays	10
Sick leave (estimate)	10
	139

The corresponding annual workload is 1,810 hr. Therefore, the number of individuals required to fill each shift position is 4.84 (say 4.8) rather than 4.2.

Labor rates are those prevailing within a particular geographical area. Base labor rates are typically used. Fringe benefits, often estimated at 35–45% of base labor rates, are more commonly accounted for as direct overhead.

Supervision, including overall supervision and shift foremen, is frequently estimated as 25% of labor. Otherwise, one foreman per shift may be assumed together with one single overall supervisor on a day shift basis only.

Maintenance

Maintenance labor and maintenance materials are always accounted for as a percentage of the investment cost within battery limits (ISBL). This percentage may be as high as 4 or 7%, but it should be estimated based on past experience. Often an allowance of 3% and in many cases as low as 2% is often sufficient for most of refinery or petrochemical units built according to heavy duty specifications (e.g., API). The periodic replacement of parts due to corrosion, wear, etc. should be accounted for as part of the overall annual maintenance allowance.

Property taxes and plant insurance

It is customary to estimate the total of these charges at about 2% of the plant erected cost within battery limits (EEC). (Others may use, say, 1.5% of fixed investment, including battery limits, and offsites.)

Direct overhead

Direct overhead in its usual interpretation consists of fringe benefits and other labor related concepts (personnel department, etc.). It is often allowed for at 35–45% of the total charges for labor and supervision. Some estimates include direct overhead as part of labor expenses.

General plant overhead

This allowance is best estimated based on past experience; it may vary considerably depending on each company's practice. A common practice is to assign to it, say, 65% of the total of labor and maintenance. Others may use, say, 80% of labor.

The general plant overhead concept is meant to include the costs of operating shops, labs, etc. as they are allocated to the operation of the plant. Research and analytical expenses are often allocated also as part of plant overhead unless accounted for separately.

Sales and administration

In order to avoid further complexity it is best to assume that these expenses are included as part of the general plant overhead. At times, however, this is not practical, principally if the plant is part of a more extensive organization. A corporate overhead entry may be included in these cases. If not part of the general plant overhead, sales, and administration can then be assumed to be part of corporate overhead. At other times a cost of sales, say 5%, is allocated as a percentage of product revenue.

The actual accounting for this item is largely immaterial except for those few cases when one wishes to distinguish a cost of production from a cost of manufacturing, thus making a distinction between plant-related costs and other extraneous costs.

Capital items

Capital items are those directly related to the plant investment and include:

- Depreciation
- Amortization
- Interest on capital
- Interest on working capital

Depreciation and amortization expenses are significant in that they are substantial, but are not real. Thus, while both depreciation and amortization charges are subtracted in order to calculate a net (pre-tax or post-tax) income, they in no way affect the cash flow out of the unit and, therefore, are added back for the purpose of determining the cash flow.

Depreciation

Depreciation represents a capital recovery. As such, it is a certain amount of money related to the capital investment spent up-front that is set aside on a regular basis, without interest, over a certain period of time to represent the recovery of the original capital.

Depreciation is accounted for separately, apart from any return-on-capital (ROI). The reasoning behind this practice is that, as the plant ages, provision should be made to accumulate capital for its eventual replacement. Obviously, in actual practice, financing of an actual replacement, if any, will rarely depend on capital accumulated in such a manner. If viewed as a capital build-up for plant replacement, depreciation would be equivalent to a zero-interest sinking fund.

Apart from philosophical interpretations, the real value of depreciation is that it is almost universally tax deductible. On the basis that depreciation represents a gradual recovery of the original capital, it is regarded as not being part of the profits generated by the plant and, therefore, not subject to taxation.

An interesting—and legal—dichotomy arises at this point. From an actual operating viewpoint one wishes to show as much profit as possible and, therefore, as little depreciation as possible. However, from a tax viewpoint, one wishes to maximize depreciation so as to minimize the immediate tax liability. This is the only place where two perfectly legal and different accounting methods can be applied that lead to different results.

For financial reporting purposes, the Financial Accounting Standards Board (FASB) in the United States dictates what minimum depreciation levels should be used. If not subject to any particular requirements, it is customary to use 10-year straight-line depreciation for plant equipment and 20-year straight-line depreciation for (most) offsites. (Sometimes, however, in order to bring depreciation schedules closer to those used for tax accounting 5- and 10-year straight-line depreciations are used for plant and offsites, respectively.)

For tax accounting purposes it is advisable to use whichever fastest rate of depreciation is allowable under the local current tax laws. (Whenever allowed, it may be advisable to start with the fastest rate of accelerated depreciation allowable and later, when the amount to be depreciated is relatively small, switch to, say, straight-line so as to prolong the depreciation benefits.) The golden rule in tax accounting is

> pay the least and the latest . . . within the law!

Since tax effects are ignored in most of our representations, this discussion is largely irrelevant. However, when tax effects are included it becomes apparent that the tax liability incurred under financial accounting principles is much higher than the tax actually paid. This gives rise to a deferred tax liability. The reasoning is that eventually this deferred tax will have to be paid; in actual practice, if sufficient capital items are added or replaced over the life of the plant, depreciation can be rolled over so as to largely neutralize the effects of deferred tax liabilities.

Amortization

Amortization is akin to depreciation. Amortization is the general term for the process of allocating the acquisition cost of assets to the periods of benefits as expenses. It is called depreciation for plant assets, depletion for wasting assets or natural resources, and amortization for intangibles.

We use the general term amortization to allocate expenses for the initial catalyst inventory costs. (In the sense that a catalyst life is relatively brief we could think of this amortization as a depletion allowance.)

It is customary to amortize the catalyst inventory over its estimated useful life, irrespective of what the actual life might be.

Interest on capital

Though very important in real life, interest on capital should not be used in a preliminary economic analysis.

Most commercial plants are heavily leveraged; namely, most plant owners choose to borrow as much capital as possible instead of investing their own money. Usually, the final investment will consist of a certain amount of debt (amount borrowed as long-term loans) and the rest equity (investors' capital). If the return-on-capital is calculated relative to equity, its value is distorted by the relative percentage of debt. For example, if the plant is financed through 100% debt, a net return of $1.00 will represent an infinite rate of return for the investors who have exposed none of their own capital.

Also, the interest paid on capital ultimately is part of the plant return-on-investment and should be added back to the usual return-on-investment calculation in a traditional analysis.

Finally, if interest on capital must be used, it should be used on an average interest basis as explained in Appendix 17.2.4. The base interest rate, typically the prime rate applicable to a large long-term loan, will be averaged over the pre-established life of the loan or calculated on a year-to-year basis if a cash flow analysis is performed over time.

Interest on working capital

Working capital is normally assumed carried as 100% debt. Because of its nature, working capital revolves somewhat like a line of credit and, therefore, is more representative of a short-term loan for which a higher rate of interest applies than for a long-term loan. Also, there is no interest averaging for working capital loans as they are assumed renewed on a year-to-year basis.

Reporting parameters

Several terms are often used for reporting purposes:

Cost basis

The summation of all costs without capital items leads to what is called the cash cost of production. The cash cost of production represents the total of raw materials costs, chemicals and catalysts costs, utility costs, operating costs, and overheads.

The net cost of production is obtained by taking the cash cost of production and adding capital items: depreciation and amortization. As indicated above, interest on capital should not be used; interest on working capital may or may not be included at this point depending on individual practice. It is customary to include it, but others prefer to calculate cash flows on a total equity basis, including working capital.

Return basis

The simplest parameter is the return-on-investment (ROI). This simply is the difference between product revenue (product value) and net production cost divided by the total investment. For this purpose, total investment is normally viewed as total fixed investment or total plant investment. Total fixed investment is the plant investment (EEC) within battery limits (ISBL) plus the plant investment outside battery limits (OSBL). EEC may be used instead of total fixed investment if clearly understood and agreed upon that this is the case. ROI's are normally calculated on a pre-tax basis; they can also be estimated on a post-tax basis, if desired. Royalties and interest during construction should be included as part of the total plant investment.

A second simple return parameter often used is pay-out time (or pay-back time). This represents the ratio of total fixed or plant investment to operating cash flow. (This type of pay-out time will not agree with the pay-out times calculated on a discounted cash flow (DCF) basis.)

Typical rules of thumb are that a new smallish project is attractive if a, say, 25% pre-tax ROI can be obtained. If this is coupled with overall depreciation estimated at 10% (10-year straight-line), the corresponding simple pay-out time is about 3 years. Very large, expensive projects may be justified with longer pay-out times of about 5 or even 7 or more years.

Higher pre-tax ROIs and shorter pay-out times are usually required for revamps or retrofits. Typical 1–2 years maximum pay-out times are often required for revamp projects, again depending on their relative magnitude.

Other parameters

Other common reporting parameters are:

- discounted cash flow (DCF) rate of return (internal rate of return)
- required sales price for a given DCF rate of return
- cash flow pay-out time

These parameters are associated with the use of discounting over a time series of cash flows.

Cash flow analysis

The previous section examined the economic summary prepared at one particular point in time, usually a "typical" production year. In this section we examine the economic analysis extended over an extended period of operation, usually 10 or 20 years.

The basis for the analysis is identical to what has been said before, so there is no need to repeat it here. All that needs to be discussed is the discounting effect over time.

A *cash flow analysis* is a *net present value analysis*. If we assume an initial outflow or investment at time zero, the total of the net present values of all future cash flows must exceed the initial investment for the project to be attractive.

Obviously, the result will be highly dependent on the rate of discounting used to bring future cash flows to time zero. What rate of discounting should be used? We should use a rate of discounting equivalent to the money market rate; namely, the rate of return that we would obtain if we were to grant a loan to others. On this basis, then, the appropriate rate of discounting is equivalent to the long-term interest rate that we would be charged for any debt incurred in connection with the construction of the plant. The rate of discounting should be increased above this value if we feel that the project involves risks such as market potential, future obsolescence of the project, etc.

If, when we use this rate of discounting, the net present value of all future cash flows exceeds the net present value of all investments (initial and future, if any), the project is regarded as profitable relative to a passive interest generating activity. If the net present value of all future cash flows is less than the net present value of the investments, the project is relatively less attractive than a passive interest generating activity.

One may ask, if a project is attractive at a certain rate of discounting, what would be the rate of discounting that renders the project unattractive? The break-even rate of return at which the summation of the net present values of all future cash flows (inflows) just equals the net present value of the outflows (investments, etc.) is the so-called internal rate of return (IRR). If, therefore, the IRR exceeds the discounting rate, the project may be regarded as profitable and economically attractive. If it does not, the project is not attractive or less attractive. The IRR has no real meaning of its own; it must always be viewed relative to the appropriate discount rate.

The IRR is also called the discounted cash flow rate of return or DCF rate of return or DCFRR. Everything else being the same, one can raise or lower the IRR by adjusting the sale price of the product and, therefore, the projected revenue. If we then fix the desired IRR, we can calculate the necessary sale price. The desired IRR should eventually be higher than the discounting rate as explained above.

There is a further complexity. Sometimes, one wishes to examine the overall return over an extended period—say, 10 or 20 years. Whether we consider an active activity (i.e., a plant investment) or a passive activity (i.e., a loan to others) the capital is always returned in discrete parcels over this period. What becomes of this capital once it is returned? Logically, one can expect that it will be reinvested but, at what rate? Treatments found in the literature often assume that the capital that is returned can be reinvested at the same rate of return as the original project. This tends to be unrealistic, principally for highly profitable projects; a more conservative approach is to assume that the capital that is returned can be reinvested at a then prevailing rate (constant or varied, depending on the desired complexity) so as to generate a fund at the end of the economic analysis timeframe. This fund can then be discounted to time zero to add to the overall cash flow. Depending on the choice of rates and discount rates, this adjustment can have a positive or negative effect on the overall analysis.

It is often customary to assume a gradual sales build-up. For example, 60% of capacity the first year, 80% the second year, and 100% the third year and thereafter. A simple computer program can easily keep track of this and any other assumptions. For the sake of simplicity, however, we normally assume a uniform production capacity over the entire period of 10 or 20 years.

Tax accounting, whenever desired, should be done on an extended discounted cash flow basis. In this format it is easier to keep track of such terms as financial depreciation, tax depreciation, tax liability, taxes paid, carryover losses, investment tax credits, etc. so as to be able to determine the actual cash flow every year. Even if done on a constant-dollar basis, the annual cash flows can be expected to vary from year-to-year. A discounted cash flow analysis can easily keep track of such variations. Again, for simplicity, the concept can be explained equally well if we assume a uniform stream of cash flows.

It is helpful to visualize expenses and cash flows as a series of arrows: positive arrows for positive cash flows (net inflows) and negative arrows for outflows such as capital investment and negative cash flows. We start with a fully capitalized investment figure at time zero, the start of production.

Let n be the total number of periods. For simplicity, let us assume that n are years and that we consider either 10 or 20 years.

Let I represent the initial investment at time zero and CF the annual cash flow that is assumed uniform on a constant-dollar basis (i.e., no inflation).

The internal rate of return (IRR) is that at which the cash flows over n years just pay back the invested capital, or, equivalently, when the capital equals the stream of net present values:

$$I = \frac{\mathrm{CF}}{1+r} + \frac{\mathrm{CF}}{(1+r)^2} + \cdots + \frac{\mathrm{CF}}{(1+r)^n} \tag{a}$$

Then, by using the summation formula (Appendix 17.2.2), we can write:

$$I = \mathrm{CF}\frac{(1+r)^n - 1}{r(1+r)^n} \quad \text{or} \quad \mathrm{CF} = I\frac{r(1+r)^n}{(1+r)^n - 1} \tag{b}$$

Internal rate of return

In the formulas above, given I, n, and CF, find r. The calculated rate of return, r, is the "internal rate of return" of the project over a period of n years. Note that it varies with the number of years and it increases with the number of years. Thus, the IRR calculated for 20 years will be slightly higher than the IRR calculated for 10 years.

Computers use an iterative built-in function to calculate the IRR, often in the form:

$$@\mathrm{irr}(0.15, -I, \mathrm{CF}_1, \mathrm{CF}_2, \mathrm{CF}_3, \ldots, \mathrm{CF}_n)$$

Where "0.15" is an initial guess that should be modified if the function does not converge in some 20 iterations. –I is the initial investment expressed as a negative number and CFs are the values of the annual cash flows.

Some software packages require that the cash flow numbers be entered in a separate array of cells; others allow for direct input.

The summation formula (b) derived above only applies if the annual cash flows are uniform. The net present value summation (a) applies however even if the CFs vary from year-to-year. The @irr function also accepts variable annual CFs.

Discounted pay-back time

Given I, r, and CF, find n. This calculated n is a discounted pay-back time. The simple pay-back time (I/CF) is equivalent to a discounted pay-back time when $r = 0$ (i.e., with no discounting). Likewise, if we calculate IRR for, say, $n = 20$, and then we set $r = \mathrm{IRR}$, we will just back-calculate $n = 20$. Therefore, for $0 < r < \mathrm{IRR}$, the calculated pay back time will be $(I/\mathrm{CF}) < n < 20$. Likewise, if $r > \mathrm{IRR}$, then $n > 20$.

If the cash flows, CF, are uniform, we can solve for n as follows:

$$n = \frac{\ln\left[\dfrac{\mathrm{CF}}{\mathrm{CF} - Ir}\right]}{\ln(1+r)}.$$

Sales price required to obtain a given discounted cash flow rate or return

Given I, n, and a desired rate of return, r, we can calculate the required cash flow, CF, according to formula (b). This cash flow can be used in the economic analysis to calculate the product sales price that will be necessary to generate this cash flow. If this cash flow is obtained, then we have the desired discounted cash flow rate of return, r.

This calculated sales price is a function of both r and n.

For the calculation of the cash flow, CF, the appropriate computer function, if available, is:

@pmt(I, r, n)

where r must be used in the decimal form for the corresponding period.

APPENDIX 17.2.1

Background for economic calculations

The following summary provides the background necessary for most economic calculations and should be used for reference as needed.

First of all, let us consider the following terms:

P present value of a sum of money or a series of sums
F future value of a sum of money or a series of sums
C capital (investment or loan) at one given time
A sum of money every compounding period
n number of compounding periods
i annual rate of interest
r rate of interest per compounding period
L salvage value

The rate of interest *per compounding period* (r) should be used in all cases.

The interest rate stated legally is the nominal annual rate (i). If the compounding period is less than one year, the rate of interest per compounding period is obtained by dividing the nominal annual rate, i, by the number of compounding periods in a year. It is done this way by law.

A commercial year is understood to consist of 360 days. Likewise, a generic commercial month consists of 30 days. Thus, if the compounding is done daily, the annual rate of interest, i, will have to be divided by 360; if monthly, the annual rate will be divided by 12, irrespective of the month. Note, however, that the amount compounded over a period of time will be calculated on the basis of the actual days. Thus, 28 days for a normal February, 31 days for May, 365 days for a normal year or 366 days for a leap year.

Future value

$$F_n = P(1+r)^n$$

F_n, or simply F, represents the value *at the end* of the nth period of a sum of money, P, at time zero (e.g., present time).

The factor, $(1+r)^n$, is sometimes called the (F/P) factor or the "compound amount factor for a single sum," (CAFS). Thus,

$$F = P(F/P)_n^r$$

or

$$F = P(\mathrm{CAFS})_n^r$$

Present value

The present value, P, of a future sum of money, F_n, represents the reciprocal of the future value calculation.

$$P = F_n \frac{1}{(1+r)^n}$$

The reciprocal factor is also called the (P/F) factor or the "present worth factor for a single sum," (PWFS). Thus,

$$P = F(P/F)_n^r$$

or

$$P = F(\mathrm{PWFS})_n^r$$

In all these formulas the compounding becomes effective at *the end of the period.* This is true in all the expressions used in economic calculations; modified formulas are needed if the compounding becomes effective at the beginning of the period, but this is seldom, if ever, the case. The formula modification, if needed, involves a $(1+r)$ factor.

Future value of a uniform series of payments (made at the end of each period)

See Appendices B and C for justification.

If the amount paid at the end of each period is A, the future value, F_n, of the sum is given by:

$$F_n = A\frac{(1+r)^n - 1}{r}$$

The multiplier is also known as the (F/A) factor or the "compound amount factor for a uniform series," (CAFUS).

$$F = A(F/A)_n^r$$

or

$$F = A(\text{CAFUS})_n^r$$

Sinking funds
A sinking fund consists of a series of payments made for the purpose of accumulating a given amount of money, F, at some future time. As usual, we assume a uniform series of payments, A being the amount paid at the end of each period. The calculation assumes that the rate of interest does not change over time; trial-and-error would be required otherwise.

In order to accumulate an amount F_n at the end of n periods, the amount A to be deposited at the end of each period is:

$$A = F_n\frac{r}{(1+r)^n - 1}$$

This is the reciprocal of the future value of a series of uniform payments.

The multiplier is at times called the (A/F) factor or the "sinking fund factor" (SFF). Thus,

$$A = F(A/F)_n^r$$

or

$$A = F(\text{SFF})_n^r$$

As used in economic estimates, depreciation and amortization are essentially sinking funds at zero rate of interest. Again, to repeat, no item should ever be treated both as a sinking fund amortization and as an expense at the same time.

Loan repayments
The loan repayment calculation, or "mortgage formula," is one of the most useful expressions in economic calculations. It is a combination of the previous cases. It represents the series of uniform payments, A, to be made *at the end of each period* to repay a loan, P, taken out at time zero. In this situation, the present value, P, often is just denoted by "capital" C.

The derivation of the formula is given in Appendices B and C.

$$A = P\frac{r(1+r)^n}{(1+r)^n - 1}$$

The multiplier is also variously known as the (A/P) factor or the "capital recovery factor" (CRF).

$$A = P(A/P)_n^r$$

or

$$A = P(\text{CRF})_n^r$$

It ought to be very clear what these formulas mean. A, the uniform series of payments, represents a constant amount paid to the lender at the end of each compound period. This quantity, A, is made up of two variable parts: a capital return part and an interest payment part. The capital return part represents a very small fraction at the beginning of the loan repayment period, but it increases slowly (and nonlinearly) until it becomes the dominant term toward the end. Conversely, the interest payment term is proportionately very large at the beginning of the loan repayment period and decreases progressively until it becomes small at the end.

This formula is predicated on the idea of making uniform payments, as is often done with mortgages; hence the name. Naturally, there are many other ways of repaying loans and expressions could be derived for other situations too, if desired.

An offshoot of the mortgage formula is the concept of an "average rate of interest," sometimes used in economic calculations to reflect the average cost of long-term debt. See Appendix 17.2.4 for this concept.

Present value of a uniform series
Logically, this represents the reciprocal of the loan repayment schedule.

$$P = A\frac{(1+r)^n - 1}{r(1+r)^n}.$$

The multiplier is also called the (P/A) factor or the "present worth factor for a uniform series," (PWFUS).

$$P = A(P/A)_n^r$$

or

$$P = A(\text{PWFUS})_n^r$$

The above expressions represent most of the situations encountered in economic evaluations. There are other expressions available for other situations (e.g., when the series of payments increases or decreases progressively at a uniform rate—the "gradient" method), but these are far rarer and need not be discussed here.

APPENDIX 17.2.2

Progressions

Though trivial, the concept of arithmetic and geometric progressions is useful in economic calculations. These concepts are outlined in as a refresher.

Arithmetic progressions

An arithmetic progression (or an arithmetic series) is a series of terms such that each is made up of the previous term plus an additive (positive or negative) constant.

The following series is an arithmetic progression:

Term:	1	2	3	4	...	n
Value:	a	$a+q$	$a+2q$	$a+3q$	...	$a+(n-1)q$

where a is the first term and q is the constant addendum.

As seen, the nth term of an arithmetic progression is given by:

$$b = a + (n-1)q$$

The sum of the first n terms of an arithmetic progression is:

$$S = \frac{(a+b)n}{2}$$

There is a spurious anecdote attributed to child Gauss who, having been punished by his teacher along with his school mates to add the first 100 numbers, he immediately came up with the answer, 5,050. Supposedly, he had just observed that $1 + 100 = 101$, $2 + 99 = 101$, and so on to $50 + 51 = 101$. Since there are 50 such pairs, the answer is immediate.

Geometric progressions

Geometric progressions (or geometric series) are of more interest for economic calculations.

A geometric progression is a series of terms such that each term is the previous one multiplied by a constant factor.

Term:	1	2	3	4	...	n
Value	a	aq	aq^2	aq^3	...	$aq^{(n-1)}$

Therefore, the nth term of a geometric progression is given by:

$$b = aq^{(n-1)}$$

More importantly, the sum of the first n terms is:

$$S = \frac{bq - a}{q - 1}$$

or

$$S = \frac{aq^n - a}{q - 1} = a\frac{q^n - 1}{q - 1}$$

Often $q < 1$. In this case, both the numerator and the denominator will be negative and, for comfort, we can write:

$$S = a\frac{1 - q^n}{1 - q}$$

These expressions are the basis for the mortgage formula and many other economic calculations.

APPENDIX 17.2.3

Loan repayments (mortgage formula)

Let us assume that we have borrowed a certain amount of money, C, that we plan to return over a total of n periods such that at the end of each period we repay the same amount, A. We further assume that the rate of interest remains constant and is r for each period.

Time zero

Take the loan out. We owe C.

End of the first period
We owe $C(1+r)$. We repay an amount A. Therefore, we now owe $C(1+r) - A$.

End of the second period
We owe $[C(1+r) - A](1+r) = C(1+r)^2 - A(1+r)$. Again we repay an amount A. Therefore, we now owe $C(1+r)^2 - A(1+r) - A$.

End of the third period
After repaying the amount A, we owe $[C(1+r)^2 - A(1+r) - A](1+r) - A$, or $C(1+r)^3 - A[(1+r)^2 + (1+r) + 1]$.

Keep repeating the reasoning for all other intermediate periods.

End of the nth period
After making the last payment, A, we owe $C(1+r)^n - A[(1+r)^{(n-1)} + (1+r)^{(n-2)} + \cdots + (1+r) + 1]$. But since we are repaying the loan in n periods, at this time the balance must be zero!

By applying the summation formulas in Appendix 17.2.2, we can rewrite this as:

$$C(1+r)^n - A\frac{(1+r)^n - 1}{(1+r) - 1} = 0$$

or

$$C(1+r)^n = A\frac{(1+r)^n - 1}{r}$$

or, solving for A:

$$A = C\frac{r(1+r)^n}{(1+r)^n - 1}$$

which is the "mortgage" formula as seen in Appendix 17.2.1. All other formulas in Appendix A can be derived in a like manner.

APPENDIX 17.2.4

Average rate of interest

The "average rate of interest" is a concept that arises when a loan is repaid over a number of time periods, but we wish to conduct a preliminary economic analysis over one of the time periods only. We saw in Appendix 17.2.1, Item 5, that, if a uniform series of payments is used, the actual amounts paid as interest vary from a high at the beginning to relatively small amounts toward the end. The question then is: what is the "average" amount of interest paid over a typical period? This value is normally calculated by adding up all the interest payments and simply prorating them by dividing the sum over the total number of periods. Calculation of this value has interest in cases when the capital may be strongly leveraged (i.e., a high percentage of debt) since, obviously, the *average* interest paid will be less than the interest paid over the first period on the basis of the total amount of debt outstanding.

How much interest is paid over the life of the loan? Since we have borrowed a capital C and we have made a total of n payments of A dollars each, the difference between the amount paid and the capital borrowed must be the amount paid as interest.

$$\text{Total interest paid} = nA - C$$

Therefore, the average amount paid out as interest every compounding period is:

$$\text{Average interest payment} = A - \frac{C}{n}$$

or, on a percentage basis relative to C, we obtain:

$$\text{Average rate of interest} = \frac{A}{C} - \frac{1}{n}$$

This rate is per compounding period; it should be annualized by multiplying it by the number of compounding periods in one year.

What is the effect of using interest averaging? Just to give an example, a 10 year loan at 10% compounded annually ($A = 0.1627454$, if $C = 1$) represents an average annual interest of only 6.27%.

Chapter 18

Process equipment in petroleum refining

D.S.J. Jones

Introduction

This chapter deals with the items of equipment normally met with in the petroleum refining industry. Indeed, many of the items that will be described and discussed here are also common to many other process industries. Knowledge of these equipment items are essential for good refinery design, operation, and troubleshooting when necessary. The equipment described here falls into the following categories, and will be presented in the following parts:

Part 1 Vessels
Part 2 Pumps
Part 3 Compressors
Part 4 Heat Exchangers
Part 5 Fired Heaters

These sections will include a description of the various types, an in depth discussion, and design features. They will also provide an example of the data sheet usually forwarded to manufacturers for the items required. Invariably in refinery technical libraries these data sheets are included as part of the 'Mechanical Catalogues' and supported by narrative specifications which give details of metallurgy and fabrication codes etc. These catalogues are provided by the equipment supplier and are part of all the information dossier on each item. Included also are such items as installation details, start-up procedures, routine maintenance procedures and the like. In most refineries today the catalogues are kept on computer discs or microfilm.

18.1 Vessels

This section address the pressure vessels that are common to most refineries. These include:

- Columns and Towers
- Knock out Drums and Separators
- Accumulators and Surge vessels

Storage tanks have been dealt with in Chapter 13 of this Handbook.

Fractionators, trays, and packings

Trayed towers

Columns normally constitute the major cost in any chemical process configuration. Consequently it is required to exercise utmost care in handling this item of equipment. This extends to the actual design of the vessel or evaluating a design offered by others. Normally columns are used in a process for fractionation, extraction or absorption as unit operations. Columns contain internals which may be trays, or packing. Both types of columns will also contain suitable inlet dispersion nozzles, outlet nozzles, instrument nozzles, and access facilities (such as manholes or handholes). This item deals with the trayed towers.

Tray types

There are three types of trays in common use today. These are:

- Bubble cap
- Sieve
- Valve

Bubble cap trays
This type of tray was in wide use up until the mid to late 1950s. Their predominance was displaced by the cheaper sieve and valve trays. The bubble cap tray consists of a series of risers on the tray which are capped by a serrated metal dome. Figure 18.1 shows two types of caps. One is used in normal fractionation service while the other is designed for vacuum distillation service. Vapor rises up through the risers into the bubble cap. It is then forced down through the serrated edge or, in some cases, slots at the bottom of the cap. A liquid level is maintained on the tray to be above the slots or serrations of the cap. The vapor therefore is forced out in fine bubbles into this liquid phase thereby mixing with the liquid. Mass and heat transfer between vapor and liquid is enhanced by this mixing action to effect the fractionation mechanism.

Capacity. Moderately high with high efficiency.

Efficiency. Very efficient over a wide capacity range.

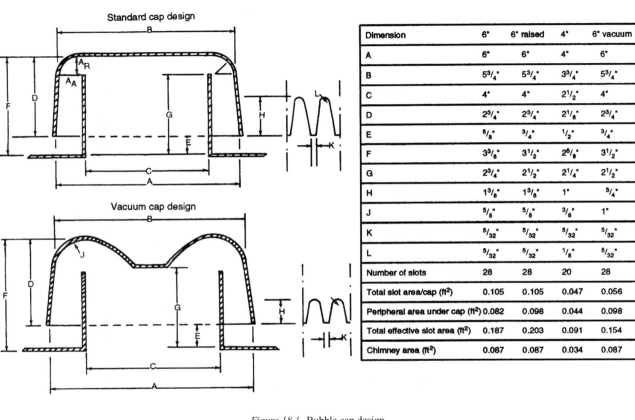

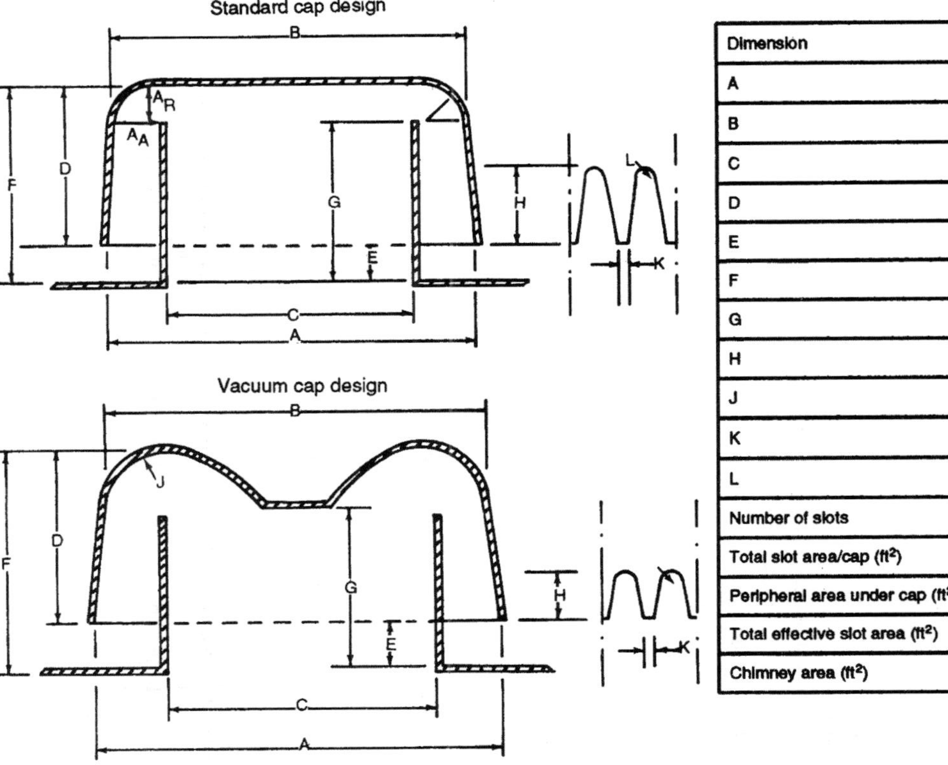

Dimension	6"	6" raised	4"	6" vacuum
A	6"	6"	4"	6"
B	$5^3/_4$"	$5^3/_4$"	$3^3/_4$"	$5^3/_4$"
C	4"	4"	$2^1/_2$"	4"
D	$2^3/_4$"	$2^3/_4$"	$2^1/_8$"	$2^3/_4$"
E	$^5/_8$"	$^3/_4$"	$^1/_2$"	$^3/_4$"
F	$3^3/_8$"	$3^1/_2$"	$2^5/_8$"	$3^1/_2$"
G	$2^3/_4$"	$2^1/_2$"	$2^1/_4$"	$2^1/_2$"
H	$1^3/_8$"	$1^3/_8$"	1"	$^3/_4$"
J	$^5/_8$"	$^5/_8$"	$^3/_8$"	1"
K	$^5/_{32}$"	$^5/_{32}$"	$^5/_{32}$"	$^5/_{32}$"
L	$^5/_{32}$"	$^5/_{32}$"	$^1/_8$"	$^5/_{32}$"
Number of slots	28	28	20	28
Total slot area/cap (ft^2)	0.105	0.105	0.047	0.056
Peripheral area under cap (ft^2)	0.082	0.098	0.044	0.098
Total effective slot area (ft^2)	0.187	0.203	0.091	0.154
Chimney area (ft^2)	0.087	0.087	0.034	0.087

Figure 18.1. Bubble cap design.

Entrainment. Much higher than perforated type trays due to the "Jet" action that accompanies the bubbling.

Flexibility. Has the highest flexibility both for vapor and liquid rates. Liquid heads are maintained by weirs.

Application. May be used for all services except for those conditions where coking or polymer formation occur. In this case Baffle or Disc & Donut trays should be used.

Note: Because of the relatively high liquid level required by this type of tray it incurs a higher pressure drop than most other types of trays. This is a critical factor in tray selection for vacuum units.

Tray Spacing. Usually 18″ to 24″. For Vacuum service this should be about 30″ to 36″.

Sieve trays

This is the simplest of the various types of trays. It consists of holes suitably arranged and punched out of a metal plate. The vapor from the tray below rises through the holes to mix with the liquid flowing across the tray. Fairly uniform mixing of the liquid and vapor occurs and allows for the heat/mass transfer of the fractionation mechanism to occur. The liquid flows across a weir at one end of the tray through a downcomer to the tray below.

Capacity. As high as or higher than bubble cap trays at design vapor/liquid rates. Performance drops off rapidly at rates below 60% of design.

Efficiency. High efficiency at design rates to about 120% of design. The efficiency falls off rapidly to around 50–60% of design rate. This is due to "weeping" which is the liquid leaking from the tray through the sieve holes.

Entrainment. Only about one-third of that for bubble cap trays.

Flexibility. Not suitable for trays operating at variable loads.

Application. In most mass transfer operations where high capacities in vapor and liquid rates are required. Handles suspended solids and other fouling media well.

Tray spacing. Requires less tray spacing than bubble cap. Usually spacing is rarely less than 15″ although some services can operate at 10″ and 12″. In vacuum service a spacing of 20″ to 30″ is acceptable.

Valve trays

These trays have downcomers to handle the liquid traffic and holes with floating caps that handle the vapor traffic. The holes may be round or rectangular and the caps over the holes are moveable within the limits of the length of the "legs" which fit into the

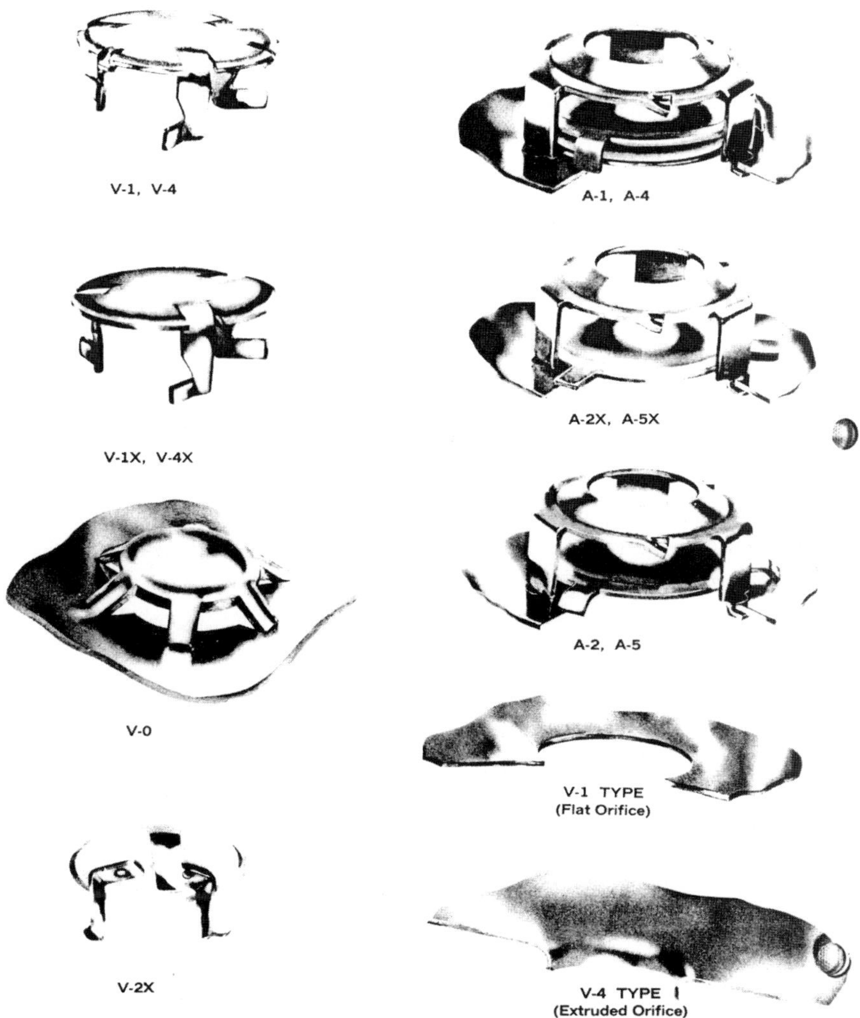

Figure 18.2. Valve unit types—Glitsch Ballast.

holes. Figures 18.2 and 18.3 show the type of valves and valve trays offered by Glitsch as their "Ballast" trays.

Valve trays are by far the most common type of tray used in the chemical industry today. The tray has good efficiency and a much better flexibility in terms of turn down than the sieve or bubble cap trays. Its only disadvantage over the sieve tray is that it is slightly more expensive and cannot handle excessive fouling as well as the sieve tray. The remainder of this item will now be dedicated to the sizing and analysis of the valve tray tower.

Description of Ballast® Units

The various types of Ballast units are shown on the page 881. A description of each unit follows:

V-0 A non-moving unit similar in appearance to the V-1 in a fully open position. It is used in services where only moderate flexibility is required and minimum cost is desired.

V-1 A general purpose standard size unit, used in all services. The legs are formed integrally with the valve for deck thicknesses up to ⅜″.

V-2 The V-2 unit is similar to the V-1 unit except the legs are welded-on in order to create a more leak-resistant unit. The welded legs permit fabrication of Ballast units for any deck thickness or size. Large size units are frequently used for replacement of bubble caps.

V-3 A general purpose unit similar to the V-2 unit except the leg is radial from the cap center.

V-4 This signifies a venturi-shaped orifice opening in the tray floor which is designed to reduce substantially the parasitic pressure drop at the entry and reversal areas. A standard Ballast unit is used in this opening normally, although a V-2 or V-3 unit can be used for special services. The maximum deck thickness permissible with this opening is 10 gage.

V-5 A combination of V-0 and V-1 units. It normally is used where moderate flexibility is required and a low cost is essential.

A-1 The original Ballast tray with a lightweight orifice cover which can close completely. It has a separate Ballast plate to give the two-stage effect and a cage or travel stop to hold the Ballast plate and orifice cover in proper relationship.

A-2 The same as A-1, except the orifice cover is omitted.

A-4 An A-1 unit combined with a venturi-shaped orifice opening in order to reduce the pressure drop.

A-5 An A-2 unit combined with a venturi-shaped opening.

The diameter of the standard size of the V-series of Ballast units is 1⅞″. The V-2 and V-3 units are available in sizes up to 6″.

Photographs of several Ballast trays are shown on page 8 and 9.

Figure 18.2. (Cont.)

Trayed tower sizing

The height of a trayed tower is determined by the number of trays it contains, the liquid surge level at the bottom of the tower, and the tray spacing. The number of trays is a function of the thermodynamic mechanism for the fractionation or absorption duty required to be performed. This is described in Chapters 3 and 4 of this Handbook.

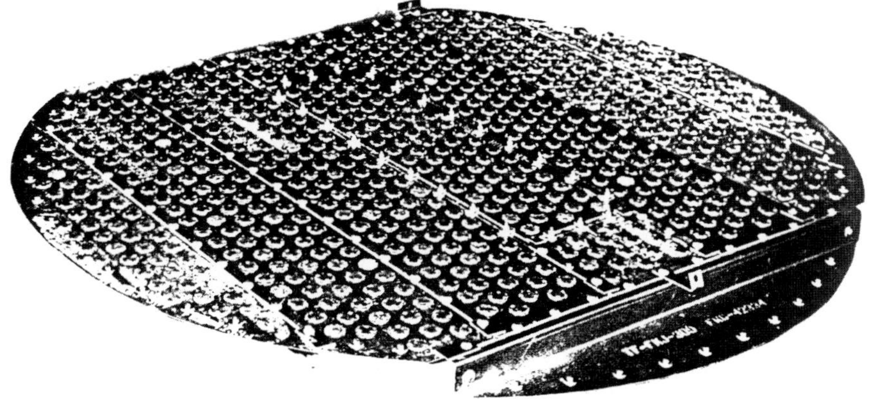

V-1 BALLAST TRAY
(with Recessed Inlet Sump)
10′-0″ DIA.

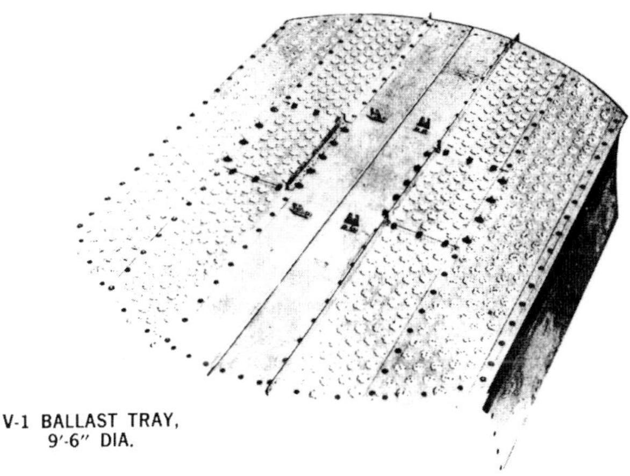

Figure 18.3. Valve trays—Glitsch Ballast.

The diameter of the tower is based on allowable vapor and liquid flow in the tower and the type of tray. This section now deals with determining the tower diameter using valve trays.

The "quickie" method

This method is good enough for a reasonable estimate of a tower diameter which can be used for a budget type cost estimate or initial plant layout studies. The steps used for this calculation are as follows:

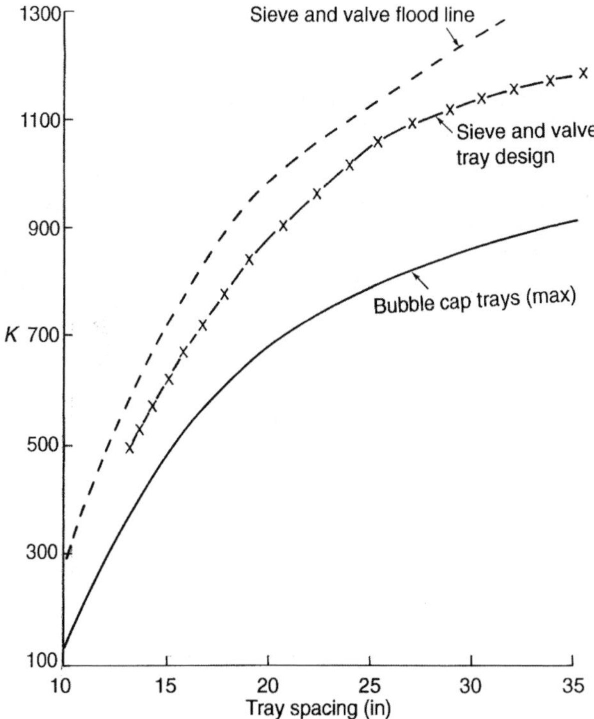

Figure 18.4. Tray spacing versus 'K' factor.

Step 1. Establish the liquid and vapor flows for the critical trays in the section of the tower that will give the maximum values. These are obtained by heat balances (see Chapters 3 and 4 of this Handbook). The critical trays are:

The top tray.
Any side stream draw off tray.
Any intermediate reflux draw off tray.
The bottom tray.

Step 2. Calculate the actual cuft/hr at tray conditions of the vapor. Then using the total mass per hour of the vapor calculate its vapor density in lbs/cuft.

Step 3. From the heat balance determine the density of the liquid on the tray at tray temperature in lbs/cuft.

Step 4. Select a tray spacing. Start with a 24″ space. Read from Figure 18.4 a value for 'K' on the flood line. Using the equation:

$$G_f = K\sqrt{((\rho_v \times (\rho_l - \rho_v))}$$

where

G_f = mass vapor velocity in lbs/hr sqft at flood,
K = the constant read from the flood curve in Figure 18.4,
ρ_v = density of vapor at tray conditions in lbs/cuft,
ρ_L = density of liquid at tray conditions in lbs/cuft.

Step 5. Multiply G_f by 0.82 to give mass velocity at 82% of flood which is the normal recommended design figure. Divide the actual vapor rate in lbs/hr by the vapor mass velocity to give the area of the tray. Calculate tray diameter from this area.

An example calculation now follows.

Example calculation
Calculate the diameter of the tower to handle the liquid and vapor loads as follows:

Vapor to tray	Liquid from tray
lbs/hr = 47,700	GPH @ 60 = 119.7
moles/hr = 929.7	Hot GPH = 153.0
ACFS = 7.83	Hot CFS = 0.339
lbs/cuft ρ_V = 1.69	lbs/hr = 33,273
Temp °F = 167	lbs/cuft ρ_L = 27.3
Pressure PSIA = 220	Temp °F = 162

Tray spacing is set at 24″ and the trays are valve type.

From Figure 18.4 'K' = 1,110 at flood.

$$\rho_v = 1.69 \text{ lbs/cuft at tray conditions.}$$

To calculate ρ of vapor at tray conditions use:

$$\rho = \frac{\text{wt/hr} \times 520 \times \text{pressure (psia)}}{378 \times 14.7 \times \text{moles/hr} \times \text{Temp °R}}$$

where

Press is tray pressure.
Temp °R is tray temperature in °F + 460.

$$\rho_L = 27.3 \text{ lbs/cuft.}$$

$$\text{Then } G_f = 1{,}110\sqrt{1.69 \times (27.3 - 1.69)}$$
$$= 7{,}302 \text{ lbs/hr.sqft}$$

$$\text{Area of tray @ 82\% of flood} = \frac{47{,}700}{0.82 \times 7{,}302} \text{ sqft}$$
$$= 7.97 \text{ sqft}$$
$$= 3.18 \text{ ft. say } 39''$$

The tray dimensions and configuration for design purposes are subject to a much more rigorous examination. This is normally undertaken by the tray manufacturer from data supplied by the purchaser's process engineer. However this process engineer needs to be able to check the manufacturer's offer before committing to purchase. The following calculation procedure offers a rigorous method for this purpose which establishes tray size and geometry. This calculation is based on a method developed by Glitsch Inc., a major manufacturer of valve and other types of trays and packing.

The rigorous method

A rigorous method used in the design of valve trays is described by the following calculation steps:

Step 1. Establish the liquid and vapor flows as described earlier for the "quickie" method.

Step 2. Calculate the down comer Design Velocity V_{dc} using the following equations (or by Figure 18.5).

(a) $V_{dc} = 250 \times$ system factor
(b) $V_{dc} = 41 \times \sqrt{(\rho_L - \rho_v)} \times$ system factor.
(c) $V_{dc} = 7.5 \times \sqrt{TS} \times \sqrt{(\rho_L - \rho_v)} \times$ system factor

where TS = Tray spacing, in inches
Use the lowest value for the design velocity in gpm/sqft.
Down comer System Factors are given in Table 18.1.

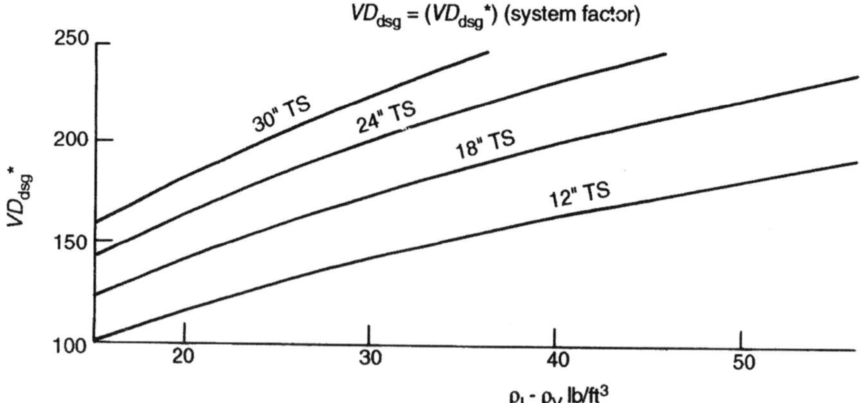

Figure 18.5. Down-comer design velocity.

Table 18.1. Down comer system factors

Service	System factor
Non foaming, regular system	1.0
Fluorine systems	0.9
Moderate foaming (amine units)	0.85
Heavy foaming (glycol, Amine)	0.73
Severe foaming (MEK units)	0.6
Foam stable systems (caustic regen)	0.3

Step 3. Calculate the Vapor Capacity Factor CAF using Figure 18.6.

$$\text{CAF} = \text{CAFo} \times \text{system factor.}$$

System factors used for this equation are given in Table 18.2.
Step 4. Calculate the vapor load using the equation:

$$V_1 = \text{CFS}\sqrt{\rho_v/(\rho_L - \rho_v)}$$

where CFS = actual vapor flow in cuft/sec.
Step 5. Establish tower diameter using Figure 18.7. Tray spacing is usually 18″, 24″, or 30″ for normal towers operating at above atmospheric pressures. Large vacuum towers may have tray spacing 30 to 36″. Note this diameter may be increased if other criteria of tray design are not met.
Step 6. Calculate the approximate Flow Path Length (FPL) based on tower diameter from Step 5 using the equation:

$$\text{FPL} = 9 \times \text{DT/NP}$$

where

FPL = Flow Path Length in ins.
DT = Tower Diameter from step 5 in ft
NP = Number of passes. For small towers with moderate liquid flows this will be 1. For larger towers this will depend on liquid velocities in the down comer. The highest number of passes is usually 4.
Step 7. Calculate the *minimum* active area (AA_m) using the expression:

$$AA_m = \frac{V_1 + (L \times \text{FPL}/13{,}000)}{\text{CAF} \times \text{FF}}$$

where

AA_m = Minimum active area in sqft.
V_1 = Vapor load in CFS.
L = Liquid flow in actual gpm.

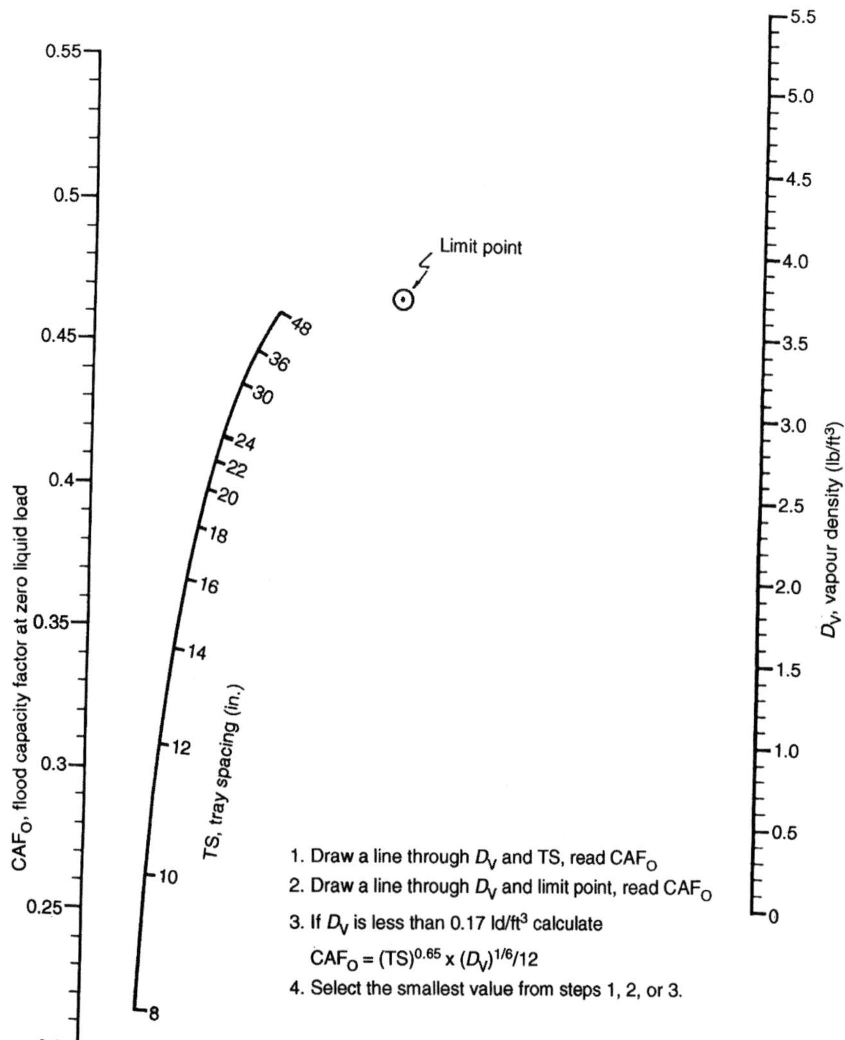

Figure 18.6. Flood capacities of valve trays.

FPL = Flow path length in ins.
CAF = Capacity Factor from Step 3.
FF = Flood Factor usually 80–82%.

Step 8. Calculate *minimum* down comer area (AD_m) using the equation

$$AD_m = \frac{L}{V_{dc} \times 0.8}$$

Table 18.2. Vapor system factors

Service	System factor
Non-foaming, regular	1.0
Fluorine systems	0.9
Moderate foaming	0.85
Heavy foaming	0.73
Severe foaming	0.6
Foam stable system	0.3–0.6

where

AD_m = minimum down comer area in sqft
L = Actual liquid flow in gpm
V_{dc} = Design down comer velocity from step 2.

Note: The down comer liquid velocity using the calculated minimum down comer area should be around 0.3 to 0.4 ft/sec.

Step 9. Calculate the *minimum* tower cross-sectional area using the following equations:

$$AT_m = AA_m + 2AD_m$$

or

$$AT_m = \frac{V_l}{0.78 \times CAF \times 0.8}$$

where

AT_m = minimum tower cross-sectional area in sqft.

Step 10. Calculate actual down comer area using the following equation:

$$AD_c = \frac{AT \times AD_m}{AT_m}$$

where

AD_c = Actual down comer area in sqft.
AT = Tower area in sqft from the diameter calculated in step 5.

Step 11. Determine down comer width (H_1) (from Table 18.A.1 in the Appendix of this Chapter) for side down comers. For multipass trays use the following equation with

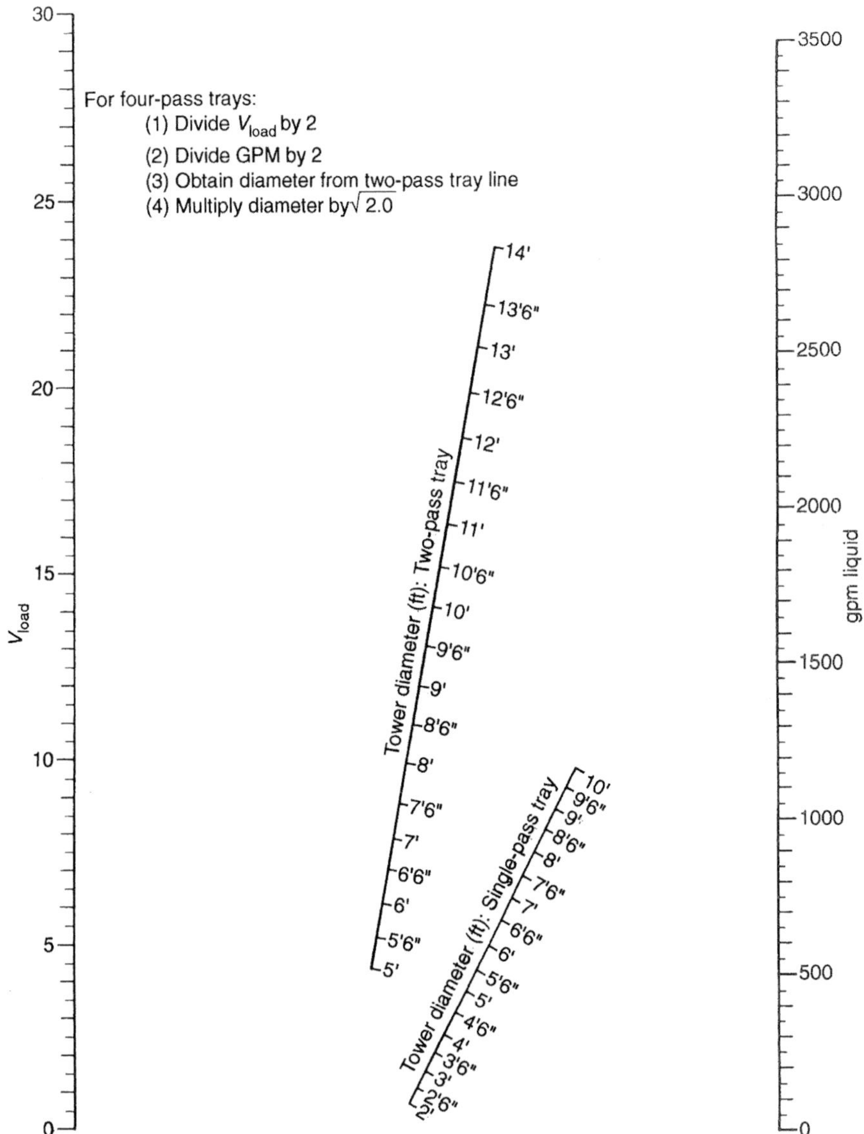

Figure 18.7. Tray diameter versus vapor loads.

Table 18.3. Allocation of down comer area and width factors

		← Fraction of Total Down ComerArea →		
No of passes	AD1	AD2	AD3	AD4
1	0.5% each	1.0	–	–
2	0.34% each	–	0.66	–
3	0.25% each	0.5	0.5 each	–
4	0.2	–	0.4	0.4
		Width Factors (WF)		
Passes	H4	H5	H7	
1	12.0	–	–	
2	–	8.63	–	
3	6.0	6.78 each pass	–	
4	–	5.66	5.5	

the width factors given in Table 18.3:

$$H_i = WF \times \frac{AD}{DT}$$

where

H_i = width of individual down comers in ins
WF = width factor from Table 18.3
AD = total down comer area in sqft
DT = actual tower diameter in ft

See Figure 18.8 for allocation of down comers in multi pass trays.

Step 12. Calculate the actual FPL from the equation:

$$FPL = \frac{12 \times DT - (2H_1 + H_3 + 2H_5 + 2H_7)}{NP}$$

where

H_{1-7} are individual down comer widths in ins (see Figure 18.8)
NP = number of passes.

Step 13. Calculate actual active area (AA) using values for H calculated in step 12 and Table 18.A.1 in the Appendix to establish inlet areas of multi pass tray down comers.

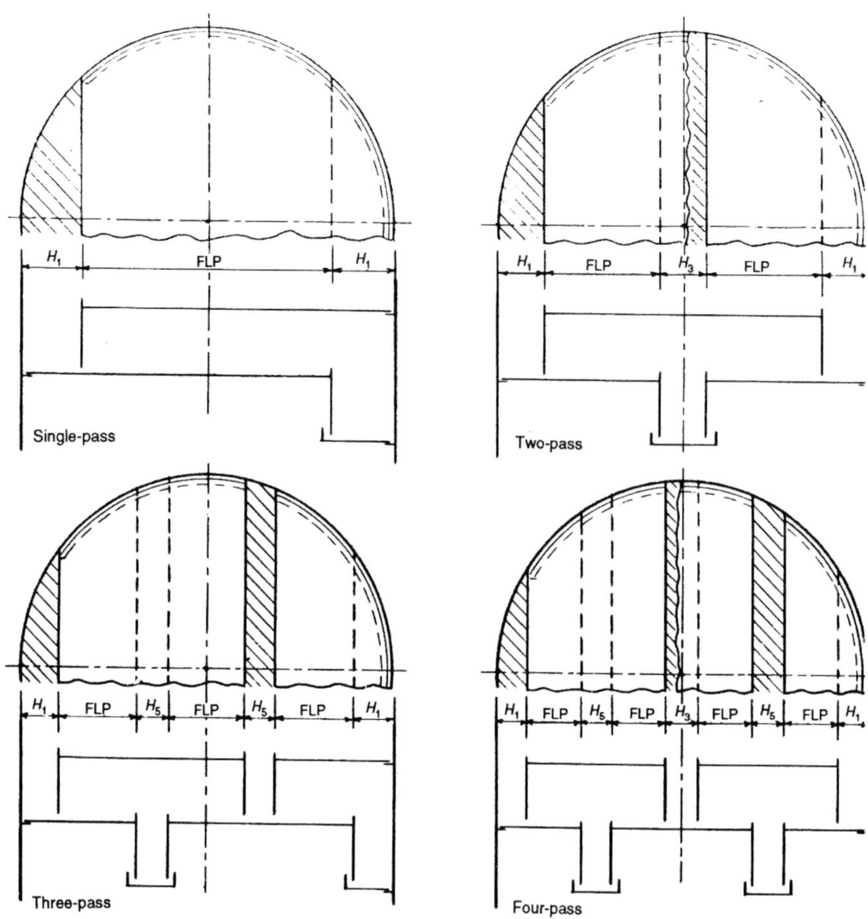

Figure 18.8. Types of tray.

$$AA = AT - (2AD_1 + AD_3 + 2AD_5 + 2AD_7)$$

where

AA = Actual active area in sqft.
AT = actual tower area in sqft.
AD_{1-7} are individual down comer areas in sqft for multi pass trays corresponding to H_{1-7}.

Step 14. From the data now developed calculate the actual percent of flood or flood factor (FF).

The following expression is used for this:

$$\%\,\text{Flood} = \frac{V_l + (L \times FPL/13{,}000) \times 100}{AA \times CAF}$$

Step 15. Calculate vapor hole velocity V_h assume 12–14 units (holes) per sqft of AA. Then

$$V_h = \frac{CFS \times 78.5}{NU}$$

where

V_h = Hole velocity in ft/sec.
CFS = Actual cuft/sec of the vapor.
NU = Total number of units.

Step 16. Calculate dry tray pressure drops from:

$$\text{Valves partly open:}\quad \Delta P_D = 1.35\, t_m \rho_m/\rho_l + K_1(V_h)(\rho_v/\rho_l)$$

where

ΔP_D = dry tray valve pressure drop in ins liquid
t_m = valve thickness in ins (see Table 18.4)
ρ_m = valve metal density in lbs/cuft (see Table 18.4)
K_1 = pressure drop coefficient (see Table 18.4)

Valves fully open: $\Delta P_D = K_2(V_h)^2 \rho_v/\rho_l$

where K_2 = pressure drop coefficient (see Table 18.4).

Table 18.4. Pressure drop coefficients (Glitch Ballast Type Trays)

		← K_2 for deck thickness (ins) →			
Type of Unit	K1	0.074	0/104	0.134	0.25
V1	0.2	1.05	0.92	0.82	0.58
V4	0.1	0.68	0.68	0.68	—

← Thickness →		← Densities of valve material →	
Gauge	tm (ins)	Metal	Density (lbs/cuft)
20	0.37	CS	400
18	0.5	SS	500
16	0.6	Ni	553
14	0.074	Monel	550
12	0.104	Titanium	283
10	0.134	Hasteloy	560
		Aluminum	168
		Copper	560
		Lead	708

Step 17. Calculate total tray pressure drop.

$$\Delta P = \Delta P_D + 0.4(L/L_{wi})^{0.87} + 0.4H_w$$

where

ΔP = total tray pressure drop in ins of liquid.
L_{wi} = weir length in ins (from Chapter 3 Appendix Figure A4.0).
H_w = weir height in ins (usually 1–2).

Step 18. Calculate height of liquid in down comer.
First calculate the head loss under the down comer H_{UD}, where $H_{UD} = 0.65(V_{UD})^2$. V_{UD} is calculated from the liquid velocity in CFS or gpm/450 divided by the area under the down comer. Use weir length times weir height for the area. This velocity should be around 0.3–0.6 ft/sec for most normal towers.

Then:

$$H_{dc} = H_w + 0.4(L/L_{wi})^{.67} + (\Delta P + H_{UD})(\rho_l/(\rho_l - \rho_v)).$$

where

H_{dc} = height of liquid in down comer in ins
For normal design this should not exceed 50% of tray spacing.

An example calculation now follows:

Example calculation
In this example the same Liquid and Vapor flows and data will be used as used in the 'Quickie Calculation'. The objective of this calculation will be to determine the tower diameter, tray pressure drop and configuration and the percent flood for the design flow rates given.

Calculating the down comer design velocity V_{dc}
System factor in this case is 1.

$$\begin{aligned}
& V_{dc} = 250 \times 1.0 = 250 \\
\text{or}\quad & V_{dc} = 41 \times (\rho_l - \rho_v) \times 1.0 \\
& \quad = 41 \times 5.06 \times 1.0 = 207 \\
\text{or}\quad & V_{dc} = 7.5 \times \text{TS} \times (\rho_l - \rho_v) \times 1.0 \\
& \quad = 7.5 \times 4.9 \times 5.06 = 186\ (\text{TS} = \text{tray spacing} = 24'') \\
\text{or}\quad & V_{dc} = 188 \text{ from Figure 18.5, use } V_{dc} = 186 \text{ gpm/sqft.}
\end{aligned}$$

Vapor capacity factor CAF.

System factor in this case is 1.
From Figure 18.6 CAF = 0.43.

Actual vapor load V_l.

$$
\begin{aligned}
V_l &= \mathrm{CFS}\sqrt{\rho_v/(\rho_l - \rho_v)} \\
&= 7.83\sqrt{1.69/(27.3 - 1.69)} \\
&= 2.01.
\end{aligned}
$$

Approximate tower diameter AT.

$$
\begin{aligned}
\text{Using Figure 9.7 } V_l &= 2.01 \\
\mathrm{TS} &= 24'' \\
L &= 153\ \text{gpm} \\
\text{Tower id} &= 3.25\ \text{ft} = 39'' \\
\text{Area} &= 8.30\ \text{sqft.}
\end{aligned}
$$

Calculate approx flow path length FPL.

$$
\mathrm{FPL} = \frac{9 \times 39}{\text{No of passes}} \times 12 = 29.25\ \text{ins.}
$$

Calculate minimum active area AA_m.

$$
\begin{aligned}
\mathrm{AA_m} &= \frac{V_l + (L \times \mathrm{FPL}/13{,}000)}{\mathrm{CAF} \times \mathrm{FF}} \\
&= \frac{2.01 + (153 \times 29.25/13{,}000)}{0.43 \times 0.8} \qquad \text{(using 80\% flood)} \\
&= 3.01\ \text{sqft.}
\end{aligned}
$$

Calculating minimum down comer area AD_m.

$$
\begin{aligned}
\mathrm{AD_m} &= \frac{\mathrm{L}}{V_{dc} \times 0.8} \\
&= \frac{153}{186 \times 0.8} = 1.028\ \text{sqft.}
\end{aligned}
$$

Calculating the minimum tower cross-sectional area.

$$
\begin{aligned}
\text{Either } \mathrm{AA_m} + 2\mathrm{AD_m} &= 3.01 + 2.056 \\
&= 5.066\ \text{sqft.}
\end{aligned}
$$

$$
\text{or} \qquad \frac{V_l}{0.78 \times \mathrm{CAF} \times 0.8} = 7.49\ \text{sqft}
$$

Use the larger which is 7.49 sqft.

Min diameter therefore is 3.09 ft say 37″.

Calculating actual downcomer area AD_c.

$$\begin{aligned} AD_c &= \frac{AT \times AD_m}{AT_m} = \frac{8.3 \times 1.028}{7.49} \\ &= 1.14\,\text{sqft.} \end{aligned}$$

Down comer width $H = AD/AT = 1.14/8.30 = 0.137$.

From Chapter 3 Appendix Figure A4.0 H/D = 0.197 then

$$H = 0.197 \times 3.25 = 0.633\,\text{ft} = 76''.$$

Recalculating flow path length FPL.

$$\begin{aligned} FPL &= 12 \times D_t - (2H) \\ &= 12 \times 3.25 - (2 \times 7.6) \\ &= 23.8\,\text{ins.} \end{aligned}$$

Recalculating active area based on actual down comer area.

$$\begin{aligned} AA &= AT - (2AD) \\ &= 8.3 - 2.28 \\ &= 6.02\,\text{sqft. Which is greater than min allowed.} \end{aligned}$$

Checking percent of flood.

$$\begin{aligned} \%\,\text{flood} &= \frac{V_l + (L \times FPL/13{,}000)}{AA \times CAF} \times 100 \\ &= \frac{2.01 + (153 \times 23.8/13{,}000)}{6.02 \times 0.43} \times 100 \\ &= 88.0\% \quad \text{Which is a little high for design but is acceptable.} \end{aligned}$$

Check down comer velocity.

CFS of liquid = 0.34 cfs
Area of down comer = 1.14 sqft

$$\begin{aligned} \text{Velocity of liquid in down comer} &= \frac{0.34}{1.14} \\ &= 0.3\,\text{ft/sec.} \end{aligned}$$

Calculating pressure drops and down comer liquid height.

- *Dry tray pressure drop.*
  Partially open valves.

$$\Delta P_D = 1.35\, t_m \rho_m/\rho_l + K_1 (V_h)^2 (\rho_v/\rho_l)$$

$$V_h = \frac{7.83 \times 78.5}{72} \qquad \text{(assumes 12 units per sqft of AA).}$$

$$= 8.5 \text{ ft/sec}$$

$$\Delta P_D = 1.35 \times 0.74'' \times (490/27.3) + (0.2 \times 72.25 \times 0.062)$$

$$= 2.69'' \text{ liquid.}$$

Fully Open valves.

$$\Delta P_D = K_2\,(V_h)^2 \cdot (\rho_v/\rho_l)$$

$$= 0.92 \times 72.25 \times 0.062$$

$$= 4.12'' \text{ liquid. This will be used.}$$

- *Total tray pressure drop.*

$\Delta P = \Delta P_D + 0.4\,(L/L_{wi})^{0.67} + (0.4 \times H_w)$
H_w (weir height) is fixed at 2″.
L_{wi} (down comer length) is calculated from Appendix A, Chapter 3 as 30.9″.
$\Delta P = 4.12 + 0.4\,(153/30.9)^{0.67} + 0.8$
$= 6.09''$ of liquid.

Height of liquid in down comer.

$$H_{dc} = H_w + 0.4\,(L/L_{wi})^{0.67} + (\Delta P + H_{UD})(\rho_l/\rho_l - \rho_v)$$

$$H_{dc} = 2 + 0.4 \times 1.16 + (6.09 + 0.405)(27.3/25.61)$$

$$= 10.08'' \text{ liquid. This is 42.0\% of tray spacing which is acceptable.}$$

(H_{UD} was calculated using a down comer outlet area of $L_{wi} \times 2''$ giving a velocity of 0.339 CFS divided by 0.429 sqft which is 0.79 ft/sec. H_{UD} is then $0.65\,(0.79)^2 = 0.405$).

Calculating the actual number of valves for tray layout. With truss lines parallel to liquid flow.

$$\text{Rows} = \left[\frac{\text{FPL} - 8.5}{0.5 \times \text{Base}} + 1\right] \text{NP}$$

where

Base = spacing of units usually 3.0″, 3.5″, 4.0″, 4.5″, or 6.0″.

$$\text{Units/row} = \frac{\text{WFP}}{5.75 \times \text{NP}} - (0.8 \times \text{number of Beams}) + 1$$

With truss lines perpendicular to liquid flow.

$$\text{Rows} = \left[\frac{\text{FPL} - (1.75 \times \text{N}_\text{o}\ \text{trusses} - 6.0)}{2.5}\right]\text{NP}$$

$$\text{Units/row} = \frac{\text{WFP}}{\text{Base} \times \text{NP}} - (2 \times \text{N}_\text{o}\ \text{Major Beams}) + 1$$

where

WFP = Width of flow path in ins.
= AA × 144/FPL.

Using a base pitch of 3.5″ the number of rows on the trays with trusses parallel to flow were calculated to be 9.7. Units per row and were then calculated to be 8.73. This gives total number of valves over the active area as 84.7. Thus number of valves per sqft of AA is 14. The assumption of 12 in the calculation (item 14) gives a more stringent design therefore the assumption is acceptable.

Calculation Summary

Tower diameter = 3.25 ft or 39″
Down comer Area (ea) = 1.14 sqft (single pass)
Active area = 6.02 sqft.
Percent of flood = 88
Tray spacing = 24″
Down comer backup = 42.0% of tray spacing.
Number of valves = 85
Number of rows = 10
Valve pitch = 3.5″

Packed towers and packed tower sizing

Although trayed towers are generally the first choice for fractionation and absorption applications, there a number of instances where packed towers are preferable. For example on small diameter towers (below 3ft diameter) packed towers are generally cheaper and more practical for maintenance, fabrication, and installation. At the other end of the spectrum packing in the form of grids and large stacked packed beds have superceded trays in vacuum distillation towers whose diameter range up to 30 ft in some cases. This is because packing offers a much lower pressure drop than trays.

The packing in the tower itself may be stacked in beds on a random basis or in a defined structured basis. For towers up to 10–15 ft the packing is usually dumped or random packed. Above this tower size and depending on its application the packing may be installed on a defined stacked or structured manner. For practical reasons and to avoid crushing the packing at the bottom of the bed the packing is installed in beds. As a rule of thumb packed beds should be around 15 ft in height. About 20 ft should be a maximum for most packed sections.

Properties of good packing are as follows:

- Should have high surface area per unit volume
- The shape of the packing should be such as to give a high percentage of area in active contact with the liquid and the gas or in the two liquid phases in the case of extractors
- The packing should have favorable liquid distribution qualities
- Should have low weight but high unit strength
- Should have low pressure drop, but high coefficients of mass transfer

Some data on the various common packing available commercially are given in Tables 18.5–18.7. Figure 18.9 shows a sectional layout of a typical packed tower. Note this tower has bed supports designed for gas distribution and includes intermediate weir liquid distributors between some of the beds.

Other salient points concerning packed towers are as follows:

1.0 Reflux ratios, flow quantities, and number of theoretical trays or transfer units are calculated in the same manner as for trayed columns.
2.0 Internal liquid distributors are required in packed towers to ensure good distribution of the liquid over the beds throughout the tower.
3.0 The packed beds are supported by grids. These are specially designed to ensure good flows of the liquid and the gas phases.
4.0 Every care must be taken in the design of the packed tower that the packing is always properly "wetted" by the liquid phase. Packing manufacturers usually quote a minimum wetting rate for their packing. This is usually around 2.0–2.5 gpm of liquid per sqft of tower cross section. Most companies prefer this minimum to be around 3.0–3.5 gpm/sqft (Tables 18.5 and 18.6).

Sizing a packed tower

The height of the tower is determined by the methods used to calculate the number of theoretical trays required to perform a specific separation. These have been discussed earlier in Chapter 1. A figure equivalent to the height of a theoretical tray is then calculated to determine the height of packing required. This is used as the basis to determine the overall height of the tower by adding in the space required for distributors, support trays and the like.

Table 18.5. Physical properties of some common packing

Packing type	Size (in.)	Wall thickness (in.)	OD and length	Approx. no./per ft^3	Approx. wt/per ft^3	Approx. surface area (ft^2 ft^3)	% void volume
Raschig rings (ceramic)	1/4	1/32	1/4	88 000	46	240	73
	5/16	1/16	5/16	40 000	56	145	64
	3/8	1/16	3/8	24 000	52	155	68
	1/2	1/32	1/3	10 600	54	111	63
	1/2	1/16	1/2	10 600	48	114	74
	5/8	3/32	3/8	5 600	48	100	68
	3/4	3/32	3/4	3140	44	80	73
	1	1/3	1	1350	40	58	73
	1 1/4	3/16	1 1/2	680	43	45	74
	1 1/2	1/4	1 2/3	375	46	35	68
	1 1/2	3/16	1 1/2	385	42	38	71
	2	1/4	2	162	38	28	74
	2	3/16	2	164	35	29	78
	3	3/8	3	48	40	19	74
Raschig rings (metal) (1)	1/4	1/32	1/4	88 000	150	236	69
	5/16	1/32	5/6	45 000	120	190	75
	5/16	1/16	5/16	43 000	198	176	60
	1/2	1/32	4/2	11 800	77	128	84
	1/2	1/16	1/2	11 000	132	118	73
	19/32	1/32	19/32	7 300	66	112	86
	19/32	1/32	19/32	7 000	120	106	75
	3/4	1/32	3/4	3400	55	84	88
	3/4	1/16	3/4	3190	100	72	78
	1	1/32	1	1440	40	63	92
	1	1/16	1	1345	73	57	85
	1 1/4	1/16	1 1/2	725	62	49	87
	1 1/2	1/16	1 1/2	420	50	41	90
	2	1/16	2	180	38	31	92
	3	1/16	3	53	25	20	95
Raschig rings (carbon)	1/4	1/16	1/4	85 000	46	212	55
	1/2	1/16	1/2	10 600	27	114	74
	3/4	1/8	3/4	3140	34	75	67
	1	1/3	1	1325	27	57	74
	1 1/4	1/16	1 1/4	678	31	45	69
	1 1/2	1/2	1 1/2	392	34	37	67
	2	1/5	2	166	27	28	74
	3	5/16	3	49	33	19	78
Berl saddles (ceramic)	1/4	–	–	113 000	56	274	60
	1/2	–	–	16 200	54	142	63
	3/4	–	–	5 000	48	82	66
	1	–	–	2200	45	76	69
	1 1/2	–	–	580	38	44	75
	2	–	–	250	40	32	72
Intalox saddles (ceramic)	1/4	–	–	117 500	54	300	75
	1/2	–	–	20 700	47	190	78
	3/4	–	–	6500	44	102	77
	1	–	–	2385	42	78	77
	1 1/2	–	–	709	37	60	81
	2	–	–	265	38	36	79

Table 18.6. Coefficients for use in the HETP equation

Packing type	Packing size (ins)	K_1	K_2	K_3
Raschig rings	0.375	2.10	– 0.37	1.24
	0.500	0.853	– 0.24	1.24
	1.000	0.57	– 0.10	1.24
	2.000	0.52	0.00	1.24
Saddles	0.500	5.62	– 0.45	1.11
	1.000	0.76	– 0.14	1.11
	2.000	0.56	– 0.02	1.11

The diameter of the tower is calculated using a method which allows for good mass and heat transfer while minimizing entrainment. The same principle of tower flooding is applicable to packed towers as for trayed towers. A calculation procedure for determining a packed tower diameter and the height of packed beds now follows:

Step 1. From examination of the flows of vapor and liquid in the tower determine the critical section of the tower where the loads are greatest. Usually this is at the bottom of an absorption unit and either the top or bottom of a fractionator.

Step 2. Determine the conditions of temperature and pressure at the critical tower section. This is usually accomplished by bubble and dew point calculations as described in Chapter 1. That is bubble point of the bottoms liquid (either in a fractionator or an absorber) determines the bottom of the tower conditions and dew point calculation of the overhead vapor determines the tower top conditions.

Step 3. Establish the liquid and vapor stream compositions at the critical tray conditions. See Chapter 3 for determining vapor/liquid streams in absorption and fractionation towers. Calculate the properties of these streams such as densities, mass/unit time, moles/unit time, viscosity, etc at the conditions of the critical tower section. Next select a packing type and size. Use Table 18.7 for this.

Table 18.7. Recommended packing sizes

Packing type	← Tower diameter ft →				
	1.0	2.0	3.0	4.0	+ 5.0
Raschig rings	0.5	0.75	1.00	1.5	3.00
Berl saddles	0.75	1.50	2.00	2.00	3.00
Interlox saddles	0.75	1.50	2.00	2.00	3.00
Pall rings	1.00	1.50	2.00	2.00	3.00

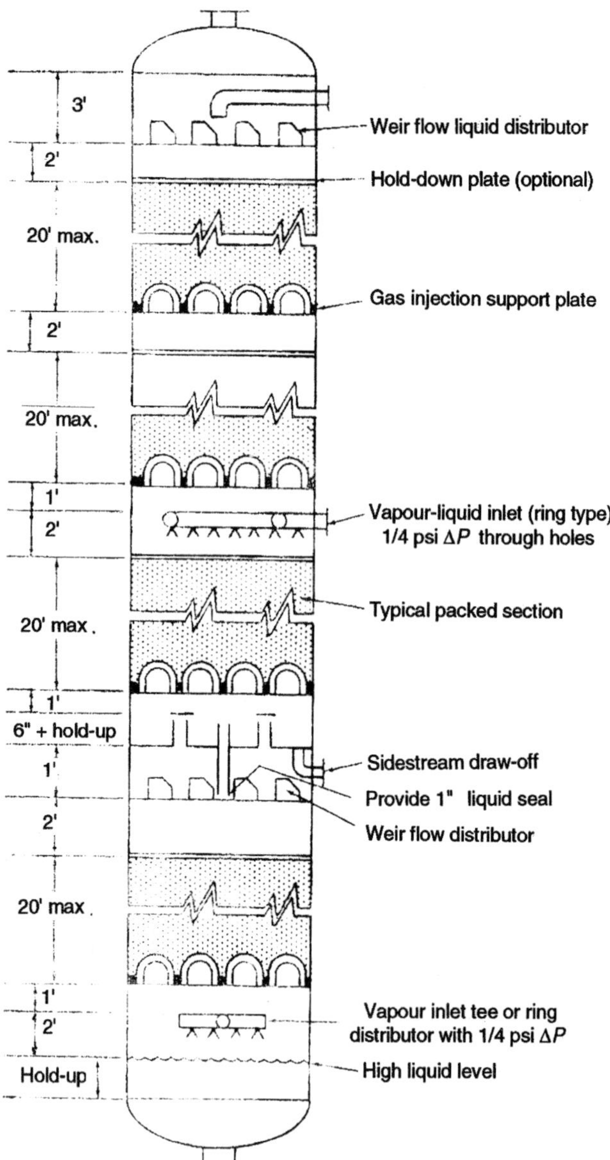
3'
2'
20' max.
2'
20' max.
1'
2'
20' max.
1'
6" + hold-up
1'
2'
20' max.
1'
2'
Hold-up
Weir flow liquid distributor
Hold-down plate (optional)
Gas injection support plate
Vapour-liquid inlet (ring type)
1/4 psi ΔP through holes
Typical packed section
Sidestream draw-off
Provide 1" liquid seal
Weir flow distributor
Vapour inlet tee or ring
distributor with 1/4 psi ΔP
High liquid level

Figure 18.9. A typical packed tower.

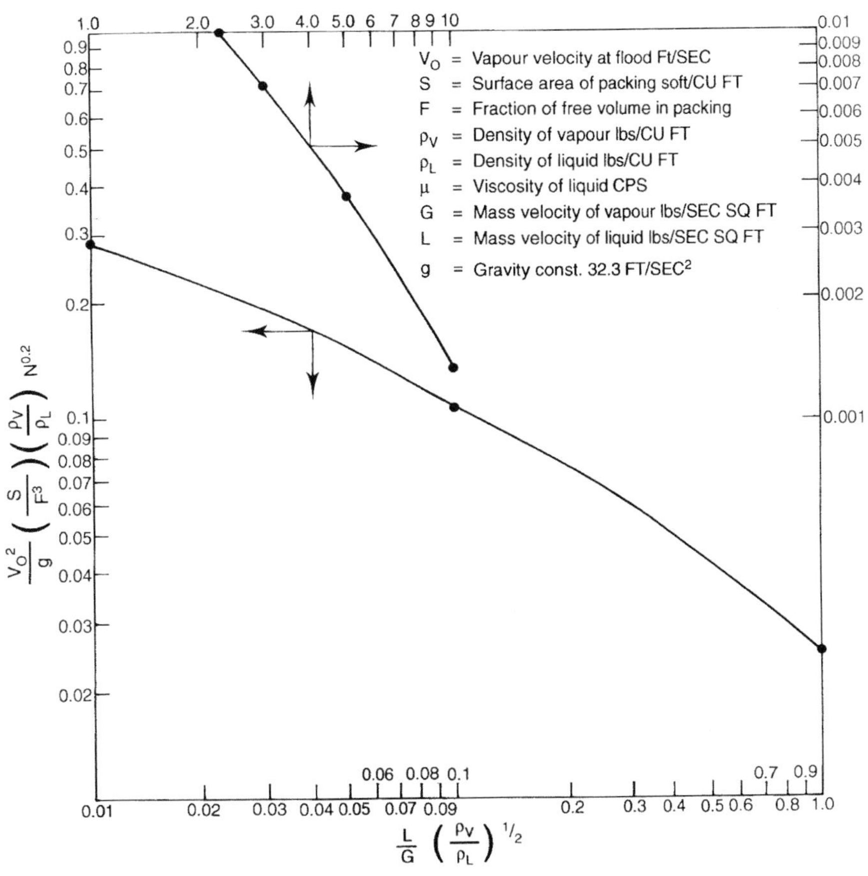

Figure 18.10. Packed tower flooding criteria.

Step 4. Commence the tower sizing by calculating the diameter. First calculate a value for

$$(L/G)\sqrt{(\rho_v/\rho_1)}$$

where

L = mass liquid load in lbs/sec sqft
G = mass vapor load in lbs/sec sqft
ρ_v = density of vapor in lbs/cuft @ tower conditions
ρ_1 = density of liquid in lbs/cuft @ tower conditions

Then using Figure 18.10 read off a value for the equation:

$$\left[\frac{L\rho_v}{G\rho_l}\right]^{1/2} = \frac{V^2 \cdot S \cdot \rho_v \cdot \mu^{0.2}}{g\,(F)^3 \rho_1}$$

where

V = the vapor velocity at flood in ft/sec
g = 32.2 ft/sec/sec
S = Surface area of packing in sqft/cuft of packing. (see Table 18.5)
F = Fraction of void (see Table 18.5)
μ = Viscosity of liquid in cps.

Step 5. Solve the equation from step 4 to give a value for V. This is the superficial velocity of the vapor at flood. Designing for 80% of flood multiply V by 0.8.

Step 6. Divide the total cuft/sec of the vapor flowing in the tower by 0.8V to give the tower cross-sectional area in sqft. Calculate the tower diameter from this area.

Step 7. The next part of the calculation is to determine the height of the tower. The number of theoretical trays has been determined by either the fractionation or absorption calculation described in Chapter 3. It is now required to establish either the actual number of trays for a trayed tower or the height of packing in the case of a packed tower. This calculation deals with the second of these.

The next step sets out to establish the HETP which is the height equivalent to a theoretical tray.

Step 8. The HETP is calculated from the following equation:

$$\text{HETP} = K_1 \cdot G_{\text{h}}^{K_2} \cdot D^{K_3} \cdot \frac{62.4 \times \alpha \times \mu^2}{\rho_1}$$

where

$K_{1,2,3}$ = factors from Table 18.6
D = tower diameter in ins
α = relative volatility of the more volatile component in the liquid phase (see step 9 below)
μ = viscosity in cps
G_{h} = mass velocity of the vapor in lbs/hr sqft

Step 9. To determine the relative volatility α, select a key light component that is in the lean liquid and the wet gas. The relative volatility is the equilibrium constant of the lightest significant component in the rich liquid divided by the equilibrium constant of the light key component. Solve for a value of HETP and multiply this by the number of theoretical stages to give the total packed height.

Step 10. Determine the number of beds to accommodate the packed height. Allow space between the beds for vapor/liquid redistribution and holdup plates. Use Figure 18.9 as a guide for this. The tower height will be the sum of beds, internal distributors packing support trays, liquid hold up and vapor disengaging space.

An example calculation now follows.

Example calculation

In this example the number of theoretical trays for an absorption unit has been fixed as 4. The compositions of the "wet" gas and the lean liquid have been given and used to determine the composition and quantities of the rich liquid and the lean gas. The quantities to be used in the following calculation are as follows:

Rich liquid leaving the bottom of the absorber = 452.66 moles/hr
Wet gas entering the bottom of the tower = 1,018.35 moles/hr

Their respective composition and conditions are as follows:

Wet gas

	Mole Frac	Mole wt	Weight
H_2	0.467	2.0	0.93
C_1	0.190	16.0	3.40
C_2	0.059	30.0	1.77
H_2S	0.242	34.0	8.24
C_3	0.604	44.0	1.32
iC_4	0.006	58.0	0.35
nC_4	0.006	58.0	0.35
Total	1.000	16.0	16.00

$$\text{Temperature} = 95^\circ\text{F} \quad \text{Pressure} = 175\,\text{psia}$$

$$\text{Cuft} = \frac{378 \times 14.7 \times 555}{175 \times 520} = 33.89$$

$$\rho_v = 16/33.89 = 0.473\,\text{lbs/cuft.}$$

Rich liquid

	Mole Frac	Mole wt	wt Fact	lbs/Gal	Vol Fact
H_2	0.004	2.0	0.002	–	–
C_1	0.013	16.0	0.208	2.5	0.083
C_2	0.016	30.0	0.480	2.97	0.162
H_2S	0.092	34.0	3.128	6.56	0.477
C_3	0.028	44.0	1.232	4.23	0.291
iC_4	0.011	58.0	0.638	4.68	0.136
nC_4	0.013	58.0	0.754	4.86	0.155
C_9	0.823	128.0	105.344	6.02	17.499
Total	1.000		111.786	5.95	18.803

$$\text{Temperature} = 95\ ^\circ\text{F}$$

$$\begin{aligned} \text{Sg}\,@\,60 &= 0.715 \\ \text{Sg}\,@\,95 &= 0.696 \\ \rho_1\,@\,95^\circ\text{F} &= 43.3\ \text{lbs/cuft.} \end{aligned}$$

$$\frac{L}{G}\left[\frac{\rho_v}{\rho_1}\right]^{1/2} = \frac{V^2 \cdot S \cdot \rho_v \cdot \mu^{0.3}}{g\,F^3\,\rho_1}$$

where

$$\begin{aligned} V &= \text{the vapor velocity at flood in ft/sec.} \\ g &= 32.2\ \text{ft/sec.} \\ S &= \text{Surface area of packing in sqft/cuft of packing. (see Table 18.5)} \\ &= 36\ \text{sqft/cuft} \\ F &= \text{Fraction of void (see Table 18.5)} = 0.79 \\ \mu &= \text{Viscosity of liquid in cps.} = 0.56\ \text{cps} \end{aligned}$$

$$\frac{L}{G}\frac{[\rho_v]^{1/2}}{[\rho_1]} = \frac{50{,}670}{16{,}325} \times \left[\frac{0.473}{43.3}\right]^{1/2} = 0.324$$

From Figure 18.10 = 0.055.

Then:

$$\frac{V^2 \times S \times \rho_v \times \mu}{g \times F^3 \times \rho_1} = 0.055$$

$$\begin{aligned} V^2 &= \frac{0.055}{0.022} = 2.5 \\ V &= 1.58\ \text{ft/sec.} \\ @\,80\%\ \text{of flood}\ V &= 1.58 \times 0.8 \\ &= 1.26\ \text{ft/sec.} \\ \text{Total vapor flow} &= 16{,}325\ \text{lbs/hr} \\ &= \frac{16{,}325}{0.473} = 34{,}514\ \text{cuft/hr.} \\ &= 9.59\ \text{cuft/sec} \\ \text{cross-sectional area} &= \frac{9.59}{1.26} = 7.6\ \text{sqft.} \\ \text{Tower diameter} &= 3.1\ \text{ft say } 3.25\ \text{ft or } 39''. \end{aligned}$$

To calculate HETP. Use 2″ Berl Saddles.

$$\text{HETP} = K_1 \cdot G_H^{K_2} \cdot D^{K_3} \cdot \frac{62.4 \times \alpha \times \mu_1}{\rho_1}$$

where

$K_1 = 0.56$
$K_2 = -0.2$
$K_3 = 1.11$
α = relative volatility (neglect H_2 and C_1 in liquid composition as non-condensable). The Key component is C_3 then is KC_2/KC_3, which is 4.6/1.0 = 4.6.
$G_H = 16{,}325/7.6 = 2{,}148$ lbs/hr·sqft
D = tower diam = 39″

$$\text{HETP} = \frac{0.56 \times 39 \times 62.4 \times 4.6 \times 0.56}{(2{,}148)^{0.02} \times 43.3}$$
$$= 25.83 \text{ ft per theoretical tray}$$

The number of theoretical trays was fixed at 4 for this separation. Then using 4 theoretical trays the total height of packing = 4 × 25.83 ft = 103 ft call it 100 ft. Five packed beds each 20 ft would satisfy the required duty. Using Figure 18.9 as a guide the tower height is developed as follows:

Bottom tan to HLL (holdup). Liquid is feed to a stripping column, therefore let the holdup time be 3 mins to NLL. Then NLL = 6.9 ft say 7.0 ft and HLL = 10 ft.

HLL to vapor inlet distributor. This will be set at 2.0 ft.

Distributor to bottom bed packing support. This will be set at 1.0 ft.

Bottom packed bed support to top of packed bed. Packed height which is 20 ft.

Top of bottom bed to bottom of next bed. Set this at 3.0 ft to allow for a liquid weir type distributor.

Height to top of top bed.

Packed height which is 4 × 20 ft = 80
4 distributors 12 ft
total = 92 ft

Top of top bed to top tan (tangent). Make this 5 ft to allow for liquid distribution tray and liquid inlet pipe.

Total height tan to tan = 130 ft.

Drums and drum design

Drums may be horizontal vessels or vertical. Generally drums do not contain complex internals such as fractionating trays or packing as in the case of towers. They are used however for removing material from a bulk material stream and often use simple baffle plates or wire mesh to maximize efficiency in achieving this. Drums are used in a process principally for:

- Removing liquid droplets from a gas stream (knockout pot) or separating vapor and liquid streams
- Separating a light from a heavy liquid stream (separators)
- Surge drums to provide suitable liquid hold up time within a process
- To reduce pulsation in the case of reciprocating compressors

Drums are also used as small intermediate storage vessels in a process.

Vapor disengaging drums

One of the most common examples of the use of a drum for the disengaging of vapor from a liquid stream is the steam drum of a boiler or a waste heat steam generator. Here the water is circulated through a heater where it is risen to its boiling point temperature and then routed to a disengaging drum. Steam is flashed off in this drum to be separated from the liquid by its superficial velocity across the area above the water level in the drum. The steam is then routed to a super-heater and thus to the steam main. The performance of the steam super-heater depends on receiving fairly "dry" saturated steam. That is steam containing little or no water droplets. The separation mechanism of the steam drum is therefore critical. The design of a vapor disengaging drum depends on the velocity of the vapor and the area of disengagement. This is expressed by the equation:

$$V_c = 0.157\sqrt{\frac{\rho_l - \rho_v}{\rho_v}}$$

where

V_c = critical velocity of vapor in ft/sec
ρ_l = density of liquid phase in lbs/cuft
ρ_v = density of vapor phase in lbs/cuft.

The area used for calculating the linear velocity of the vapor is:

- The vertical cross-sectional area above the high liquid level in a horizontal drum
- The horizontal area of the drum in the case of vertical drums

The allowable vapor velocity may exceed the critical, and normally design velocities will vary between 80% and 170% of critical. Severe entrainment occurs however above 250% of critical. Table 18.8 gives the recommended design velocities for the various services. The minimum vapor space above the liquid level in a horizontal drum should not be less than 20% of drum diameter or 12″, whichever is greater.

Table 18.8. Some typical drum applications

Service	Liquid surge and distillate	Settling drums	Compressor suction	Fuel gas KO drums	Steam drums	Water disengaging drums
			Cent Recip			
Allowable vapor velocity without CWMS % V_c	170		80 80	170	–	170
Allowable vap velocity with 1CWMS % V_c			150 120		100	
Allowable vap velocity with 2CWMS % V_c			– –		150	
Liquid hold up set by	Water settling	Settling Requirements	10 min liquid spill	Should be at least volume of a 20 ft slug of condensate.	1/3 the heater and steam piping volume	50 ins per minimum settling rate for Hydrocarbon vapors from water.
	Minimum instrument	Minimum instrument	When taking suction from absorbers	Following an absorber—5 mins on total lean oil circulation		Minimum height to low level 1.5 ft
	Controlling Process	Controlling process	For refrigerators—5 mins based on largest cooling unit			
	Inventory requirement					
Normal drum position	Horizontal	Horizontal	Vertical	Vertical	Vertical	Horizontal
Type of nozzle inlet	90° bend	90° bend	Tee Dist	Flush	Tee Dist	90° bend
Outlet vapor	Flush	–	Flush	Flush	Flush	Flush
Outlet liquid	Flush	Flush	Flush	Flush	Flush	Flush

Crinkled wire mesh screens (CWMS) screens are effective entrainment separators and are often used in separator drums for that purpose. When installed they improve the separation efficiency so vapor velocities much above critical can be tolerated. They are also a safeguard in processes where even moderate liquid entrainment cannot be tolerated.

CWMS are now readily available as packages that include support plates and installation fixtures. Normally for drums larger than 3 ft in diameter 6″ thick open mesh type screen is normally used.

Liquid separation drums
The design of a drum to perform this duty is based on one of the following laws of settling:

Stokes law

$$V = 8.3 \times 10^5 \times \frac{d^2 \Delta S}{\mu}$$

When the Re number is < 2.0.

Intermediate law

$$V = 1.04 \times 10^4 \times \frac{d^{1.14} \Delta S^{0.71}}{S_c^{0.29} \times \mu^{0.43}}$$

When the Re number is 2–500.

Newtons law

$$V = 2.05 \times 10^3 \times \left[\frac{d \Delta S}{S_c}\right]^{1/2}$$

When the Re number is > 500.

where

$$\text{Re number} = \frac{10.7 \times d \cdot V \cdot S_c}{\mu}$$

V = settling rate in ins per minute
d = droplet diameter in ins
S = droplet specific gravity
S_c = continuous phase specific gravity
ΔS = specific gravity differential between the two phases
μ = viscosity of the continuous phase in cps

The following may be used as a guide to estimating droplet size:

Lighter phase	Heavy phase	Minimum droplet size
0.850 SG and lighter	Water	0.008 ins.
Heavier than 0.850	Water	0.005 ins.

The holdup time required for settling is the vertical distance in the drum allocated to settling divided by the settling rate. Some typical applications of drums for this service are given in Table 18.8.

Settling baffles, are often used to reduce the holdup time and the height of the liquid level.

Surge drums
This type of drum, the calculation of holdup time and surge control has been described fully in Chapter 4.0 under "Control Systems".

Pulsation drums or pots
This type of drum will be described in some detail in Part 3 of this chapter in the section on reciprocating compressors.

An example calculation on drum sizing now follows.

Example calculation
It is required to provide the dimensions and process data for the design of a reflux drum receiving the hydrocarbon distillate, water, and uncondensed hydrocarbon vapor from a distillation column. Details of flow and drum conditions are as follows:

Vapor: 12,000 lbs/hr, 40 mole wt, 300 moles/hr.
Distillate product: 76,650 lbs/hr, Sg @ 100°F 0.682.
Reflux liquid: 61,318 lbs/hr, Sg @ 100°F, 0.682.
Water: 17,381.
Temperature of drum: 100°F
Pressure of drum: 30 psia.

The drum is to be a horizontal vessel located on a structure 45 ft above grade. The liquid product is to feed another fractionating unit and therefore requires a holdup time of 15 min between LLL and HLL. The vapor leaving the drum is to be routed to fuel gas via a compressor, therefore complete disengaging of liquid droplets is required. Complete separation of water from the oil is required. However as the water is routed to a de-salter separator from the drum separation of oil from the water is not critical.

In all probability the surge volume required by the product will be the determining feature of this design. Setting the liquid levels in the drum will depend on the settling out of the water from the hydrocarbon phase. The design will be checked for satisfactory vapor disengaging.

The design

1.0 Calculating the surge volume for the distillate product.

Holdup time = 15 min.

$$\text{Product rate} = \frac{76,650\ \text{lbs/hr}}{0.682 \times 62.2} = 1,807\ \text{cuft/hr}$$

$$\text{Holdup volume} = \frac{1,807 \times 15}{60} = 452\ \text{cuft.}$$

Then volume of liquid between HLL and LLL is 452 cuft. Let this be 60% of total drum volume. Then drum volume 452/.60 = 753 cuft.

Using a length to diameter ratio (L/D) of 3, diameter and length are calculated as follows:

$$753\ \text{cuft} = \frac{\pi \cdot D^2}{4} \times 3D$$

$$D = \sqrt[3]{\frac{753 \times 4}{3\eta}}$$

$$= 6.8\ \text{ft make it } 7.0\ \text{ft}$$

$$L = 3 \times 7.0\ \text{ft} = 21.0\ \text{ft.}$$

2.0 Calculating water settling rate.

Using "intermediate law" then:

$$V = 1.04 \times 10^4 \times \frac{d^{1.14} \Delta S^{0.71}}{Sc^{0.29} \times \mu^{0.43}}$$

V = settling rate in ins/min.
d = droplet size in ins = 0.008″
Sc = Sg of continuous phase = 0.682
Sw = Sg of water = 0.993
$\Delta S = 0.311$

$$V = 1.04 \times 10^4 \times \frac{0.004 \times 0.44}{0.895 \times 0.78}$$

$$= 29.2\ \text{ins/min}$$

Check Re number:

$$\text{Re} = \frac{10.7 \times 0.008 \times 29.2 \times 0.682}{0.56}$$

= 3.0 so use of "intermediate law" is correct.

3.0 Setting the distance between bottom tan and LLL

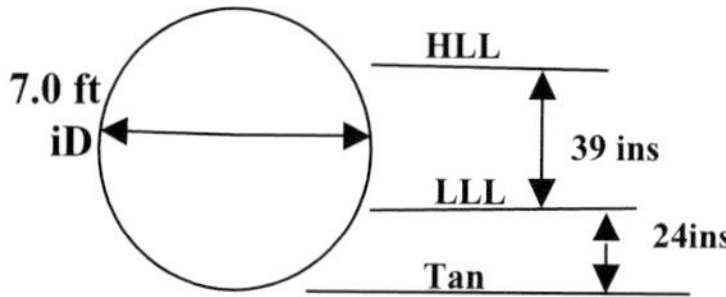

Sufficient distance or surge should be allowed below LLL to provide a LLL alarm at a point about 10% below LLL and bottom tangent. The remaining surge should be sufficient to provide the operator with some time to take emergency action (such as shutting down pumps).

Let LLL be 2 ft above bottom tan. Then the surge volume in this section is as follows:

R = 2/7 = 0.286. From Chapter 3 Appendix Figure A4.0 = 0.237 area of section = 0.237 × 38.48 = 9.1 sqft and volume = 21 × 9.1 = 191 cuft.

$$\text{Total flow rate} = \text{Product} + \text{Reflux} + \text{Water.}$$
$$= 3{,}531.4 \text{ cuft/hr} = 58.86 \text{ cuft/min.}$$

Minutes of hold up below LLL = 3.25 mins

By the same calculation holdup after alarm = 2.9 mins, which is satisfactory.

4.0 Checking settling time for the water

At the LLL a distance of 2 ft from Tan
Residence time for liquid below LLL = 3.25 min.

Minimum settling time required:

$$\frac{\text{Vert distance to bottom of drum}}{\text{Settling rate}}$$
$$= \frac{24''}{29.2 \text{ inch/min}}$$
$$= 0.82 \text{ min,}$$

which is adequate.

5.0 Calculating height of HLL above LLL

$$\text{Total volume to HLL} = 191 + 452 = 643 \text{ cuft}$$
$$\text{Area above HLL} = \frac{808 \text{ cuft} - 643 \text{ cuft}}{21 \text{ ft}}$$
$$= 7.58 \text{ sqft.}$$

Using table in Appendix of Chapter 3

$$\frac{A_D}{A_s} = \frac{7.58}{38.48} = 0.197 \quad R = 0.251$$

$$r = 0.251 \times 7.0 = 1.76 \text{ ft.}$$

$$\text{Height of HLL above LLL} = 7 - (1.76 + 2.0)$$

$$= 3.25 \text{ ft. (39 ins)}$$

6.0 Checking the vapor disengaging space.

$$V_c = 0.157\sqrt{\frac{\rho_l - \rho_v}{\rho_v}}$$

where

V_c = critical velocity of vapor in ft/sec
ρ_l = density of liquid phase in lbs/cuft = 42.42
ρ_v = density of vapor phase in lbs/cuft = 0.216

$$V_c = 2.28 \text{ ft/sec.}$$

Actual velocity of vapor is as follows:

Cross-sectional area of vapor space above HLL = 7.58 sqft

$$\text{Vapor linear velocity} = \frac{59{,}840 \text{ cuft/hr}}{7.58 \times 3{,}600} = 2.19 \text{ ft/sec}$$

which is 96% of critical.

The drum design meets all necessary criteria and will be used.

Specifying pressure vessels

Process engineer's responsibility extends to defining the basic design requirements for all vessels. These data include:

- The overall vessel dimensions
- The type of material to be used in its fabrication
- The design and operating conditions of temperature and pressure
- The need for insulation for process reasons
- Corrosion allowance and the need for stress relieving to meet process conditions
- Process data for internals such as trays, packing, etc.
- Skirt height above grade
- Nozzle sizes, ratings and location (not orientation)

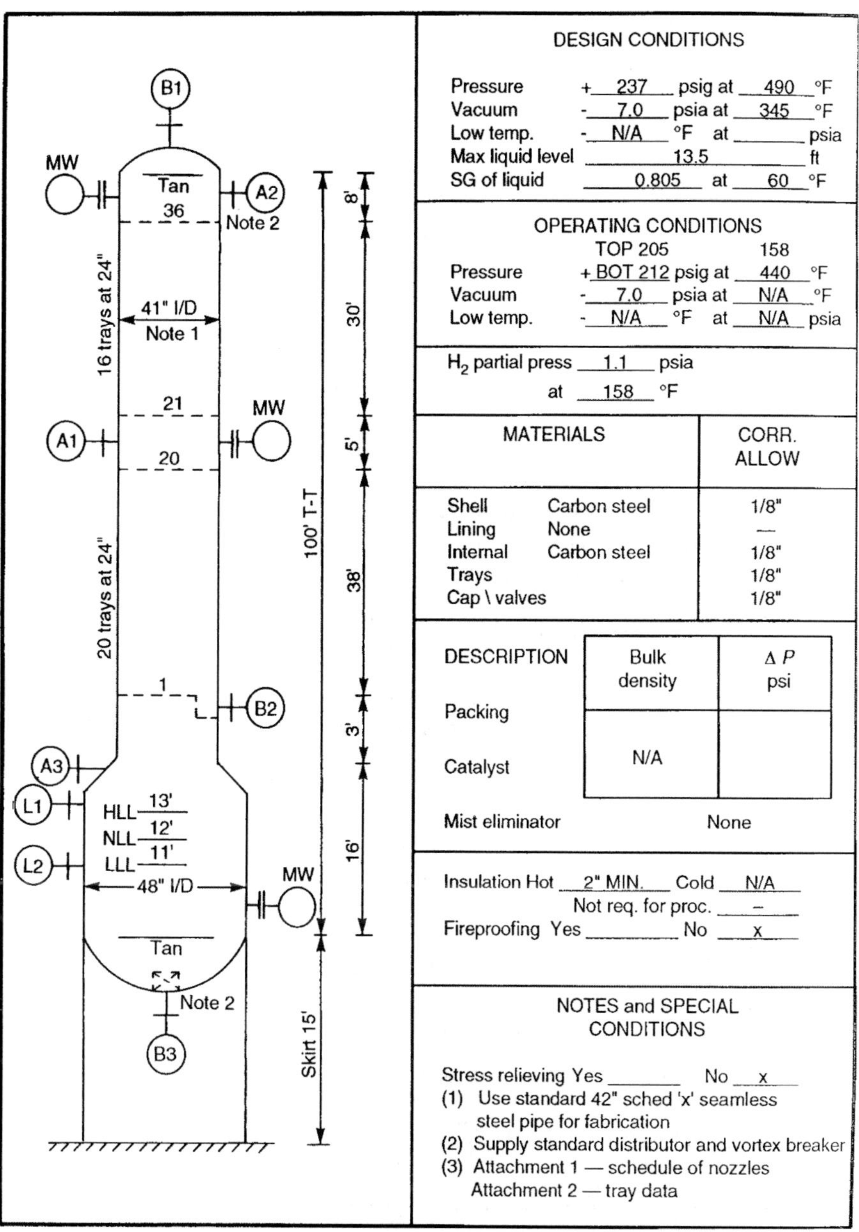

Figure 18.11. A typical process data sheet for columns.

Typical process data sheets used for specifying columns and horizontal drums are given in Figures 18.11 and 18.12 (with their attachments) respectively. These data sheets have been completed to reflect the examples calculated in this chapter. The following paragraphs describe and discusses the contents of these data sheets.

The attachments to Figure 18.11 are as follows:

Figure 18.11. Attachment 1. Nozzle Schedule

Ref	Description	Size, in inch	RTG
A 1	Feed inlet nozzle	6	150 RF
A 2	Reflux inlet nozzle	4	150 RF
A 3	Inlet from reboiler	6	300 RF
B 1	O/Head vapor outlet	8	150 RF
B 2	Outlet to reboiler	4	300 RF
B 3	Bottom product outlet	3	300 RF
L1 L2	Instrument nozzles	3/4	300 RF
MW	Manways	24	150 RF

Figure 18.11. Attachment 2. Tray Data Sheet

Vessel no	C401	
Vessel name	Reformate stabilizer	
Description of material	Un-stabilized light hydrocarbons	
Section	**Top trays 21–36**	**Bottom trays 1–20**
Total trays in section	16	20
Max ΔP per tray, psi	0.25	0.25
Conditions on tray	Top Tray No 30	Bottom Tray No 1
Vapor		
Temp, °F	167	440
Pressure, psig	205	212
Density, lbs/cuft	1.69	2.0
Rate, lbs/hr	47,700	71,021
ACFS	7.83	9.81
Liquid		
Temp, °F	162	430
Viscosity, Cps	0.3	0.85
Density, lbs/cuft	27.3	38.2
Mole weight	57	100
Rate, lbs/hr	33,273	104,950
Rate, cuft/min	20.34	45.79
Tower diameter, ft	3′ 5″	
Number of tray passes,	One	
Type of tray	Valve	
Tray spacing, ins	24	

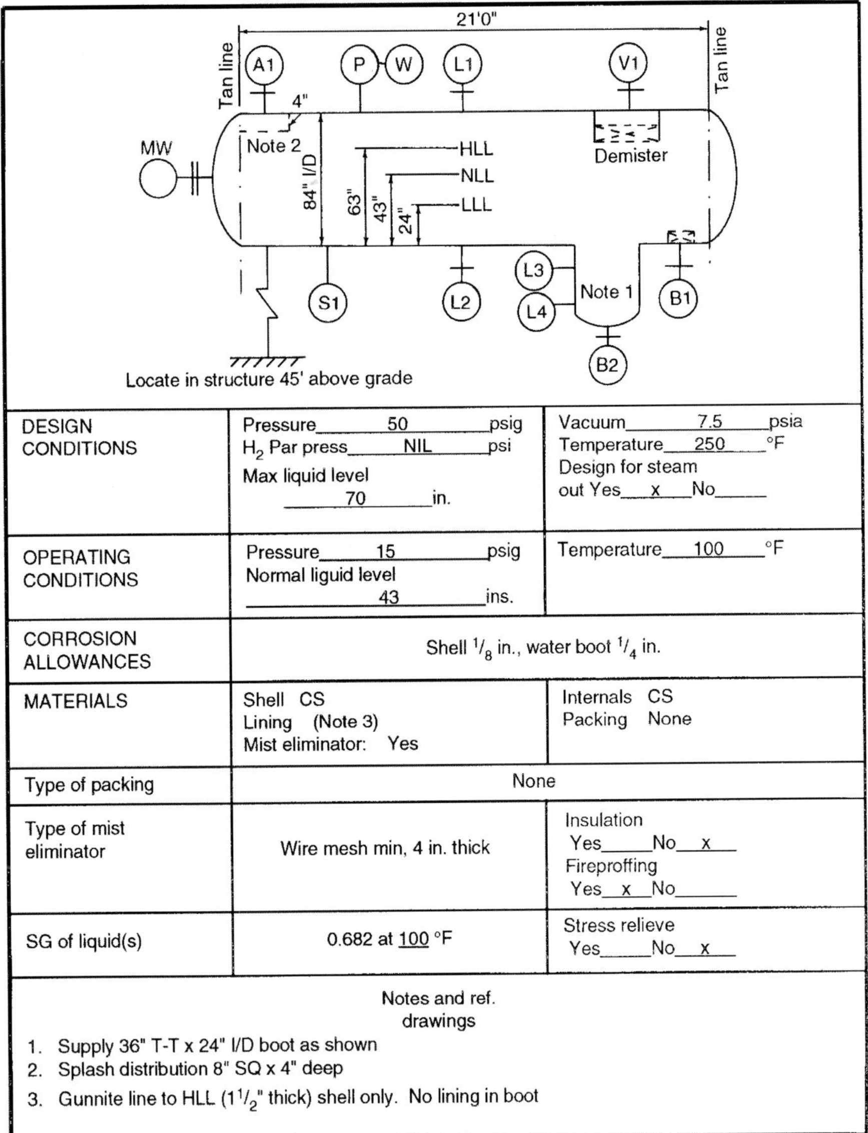

DESIGN CONDITIONS	Pressure 50 psig H_2 Par press NIL psi Max liquid level 70 in.	Vacuum 7.5 psia Temperature 250 °F Design for steam out Yes x No
OPERATING CONDITIONS	Pressure 15 psig Normal liguid level 43 ins.	Temperature 100 °F
CORROSION ALLOWANCES	Shell $^1/_8$ in., water boot $^1/_4$ in.	
MATERIALS	Shell CS Lining (Note 3) Mist eliminator: Yes	Internals CS Packing None
Type of packing	None	
Type of mist eliminator	Wire mesh min, 4 in. thick	Insulation Yes No x Fireproffing Yes x No
SG of liquid(s)	0.682 at 100 °F	Stress relieve Yes No x

Notes and ref. drawings

1. Supply 36" T-T x 24" I/D boot as shown
2. Splash distribution 8" SQ x 4" deep
3. Gunnite line to HLL ($1^1/_2$" thick) shell only. No lining in boot

Figure 18.12. A typical process data sheet for horizontal vessels (drums).

The vessel sketch

This particular vessel is a light ends fractionator and has a single tray diameter (i.e., it is not swaged). The tower contains 36 valve trays on a 24″ tray spacing and has a calculated diameter of 39″ for the trayed section. This diameter will be specified as 41″ i/d however. This can be met by a standard 42″ schedule "X" pipe and this will reduce the cost of the vessel. The overall dimension for the tower is completed by setting the height of the tower from Tan to Tan. In the example here this has been done as follows:

$$\text{Height of trayed section} = (\text{No of trays} - 1.0) \times \text{tray spacing.}$$

The fractionation calculation (see Chapter 3.2 for this type of calculation) has determined 36 actual trays for this tower. Thus the trayed height is $(36 - 1) \times 24'' = 70$ ft.

Add another 3 ft to accommodate the feed inlet distributor on tray 20. Then total trayed height is 73 ft.

Bottom of the tower must accommodate the liquid surge requirement. As the tower diameter is relatively small a swaged section of 4 ft diameter will be considered below the bottom tray for surge. The liquid product goes to storage, therefore the surge requirement need not be more than 3 min on product.

From the unit's material balance the bottom product is as follows:

$$\begin{aligned} \text{Weight per hour} &= 117{,}513 \text{ lbs} \\ \text{Temperature} &= 440^\circ\text{F} \\ \text{Density @ } 440^\circ\text{F} &= 40 \text{ lbs/cuft} \end{aligned}$$

Then hot cuft/min of product = 48.96.

The product goes to storage, therefore only 2–3 min surge is required. This will be set at 3 min surge to NLL.

$$\begin{aligned} \text{Total surge to NLL} = 48.96 \times 3 &= 146.9 \text{ cuft.} \\ \text{Cross-sectional area of surge section} &= \eta/4 \times 4^2 \\ &= 12.6 \text{ sqft.} \\ \text{Height of NLL above Tan} &= \frac{146.9}{12.6} \\ &= 11.7 \text{ ft make it 12 ft.} \end{aligned}$$

level range will be 24″ then HLL will be 12 +1 = 13 ft above NLL.

Let the reboiler inlet nozzle be 3 ft above HLL = 16 ft above Tan.

Allow another 3 ft between reboiler inlet nozzle to bottom tray. This provides adequate space for vapor separation from the high liquid level.

At the top of the column space must be provided between the top tray and the top vapor outlet nozzle to accommodate the reflux return distributor and vapor disengaging from the top tray. Let this be 8 ft from top tray to the tower top Tan line. Then

Total tower height:

From bottom Tan to Bottom tray = 19.0 ft
The trayed section = 73.0 ft
From top tray to top Tan = 8.0 ft
Total = 100.0 ft Tan to Tan

These overall dimensions are now inserted on the vessel diagram as shown in Figure 18.11. An attached sheet will be included to give the nozzle description, size and flange rating referring to those shown in the small circles on the sketch. A schedule of flange ratings for carbon steel is given in Table 18.9.

The only other dimension that will be shown on the sketch is that of the Skirt. Now the vessel is installed onsite supported by a metal skirt fixed to the concrete foundation of the vessel. The height of this skirt is fixed by the following criteria. If the product

Table 18.9. Schedule of flange ratings for carbon steel

Flange ratings psig Serv temp, °F	150	300 Max	400 operating	600 pressures	900 Psig
100	275	720	960	1,440	2,160
150	255	710	945	1,420	2,130
200	240	700	930	1,400	2,100
250	225	690	920	1,380	2,070
300	210	680	910	1,365	2,050
350	195	675	900	1,350	2,025
400	180	665	890	1,330	2,000
450	165	650	870	1,305	1,955
500	150	625	835	1,250	1,875
550	140	590	790	1,180	1,775
600	130	555	740	1,110	1,660
650	120	515	690	1,030	1,550
700	100	470	635	940	1,410
750	100	425	575	850	1,275
800	92	365	490	730	1,100
850	82	300	400	600	900
900	70	225	295	445	670
950	55	155	205	310	465
1,000	40	85	115	170	255

from the bottom of the tower is to be pumped (as is usual), the skirt height must be such as to accommodate the suction conditions for the pump. The most important of these conditions is the head required to meet the pump's net positive suction head (NPSH). See Part 2 of this chapter "PUMPS" for details. Usually a skirt height of 15 ft meets most NPSH requirements. The second consideration dictating skirt height is the head required by a thermosyphon reboiler. If the vessel is new and being designed the skirt height of 15 ft remains adequate with properly designed piping to and from the reboiler.

Design conditions

This particular tower operates at 212 psig and 440°F in the bottom and 205 psig & 158°F at the top. It is therefore classed as a pressure vessel and will be fabricated to meet a pressure vessel code. The most common of these codes is the ASME code either Section 1 or Section 8. Most vessels in the chemical industry are fabricated to ASME Section 8.When fabricated, inspected, and approved it will be stamped to certify that its construction conforms to this pressure vessel code.

Among the data required by the code and for complete vessel engineering and fabrication are the design conditions of temperature and pressure for the vessel.

Design pressure

The design pressure is based on the maximum operating pressure at which the relief valve will open plus a suitable safety increment. The following table provides a guide to this increment:

Maximum: Operating pressure, psig	Design pressure, psig
Full or partial vacuum	50
0–5	50
6–35	50
36–100	Operating + 15
101–250	Operating + 25
251–500	Operating + 10%
501–1,000	Operating + 50
over 1,000	Operating + 5%

In cases where vessels relieve to a flare header it may be necessary to add a little more to the differential between operating and design pressures to accommodate for the flare back pressure.

Design temperatures
The following table may be used as a guide to the max and min design temperatures. *Note:* Very often companies will have their own standards for these design criteria. The table given here may be used if there are no company standards.

Maximum: Operating temperature, °F	Design temperature, °F
Ambient –200	250
201–450	Operating + 50
Over 450	Divide into zones add 50 to each operating zone.
Vessels	
up to 225	250
226–600	Operating + 25
Over 600	Operating + 50
Minimum: Operating temperatures, °F	**Minimum: Design temperatures, °F**
15 to Ambient	Operating − 25
14 to 10	− 20
−10 to 80	Operating − 10
Below −80	Operating

A vacuum condition can exist in a tower during normal steam out if the tower is accidentally shut in and the steam valve closed. Normally a design vacuum pressure of 7 psia is specified at the steam saturated temperature to cover this contingency.

Low temperature
This applies to towers in cryogenic services (such as de methanizers, and LNG plants). There may be a situation in a non cryogenic service where rapid de-pressurizing causes sub zero temperatures to exist. If this is a situation that can exist for several hours and occurs frequently this condition should be entered. Otherwise make an appropriate remark in “Notes and Special Conditions”.

Max liquid level
This is the liquid level under operating conditions that will:

Either—Activate the high liquid level alarm.
Or—Shut down the feed pump.

whichever system is applicable to protect the plant operation. Usually this is quoted as the HLL and 1–2% of surge.

SG of liquid
Quote this as the SG of the liquid on which the surge volume was based. This SG is usually quoted at 60°F.

Operating conditions

In most fractionation towers there will be two distinctly different conditions of temperature and pressure—those for the tower top and those for the bottom of the tower. Both these conditions must be quoted in this case. The same situation may not necessarily arise in an absorption column.

Operating temperatures and pressures

Quote the calculated data as they will appear also on the process flow diagram. Show the tower top pressure and temperature first, followed by the bottom set of conditions. If the tower has been sized on data for more than one design case, show the highest numbers calculated for top and bottom. Also make a note in the "notes and Special Conditions" section of the cases the data was based on.

Other operating data

Vacuum conditions in this case only apply if the tower operates normally at sub atmospheric pressure. In this case quote the lowest pressure the tower will be operated on together with the normal operating temperature(s). Note in many vacuum fractionators there will be a spectrum of these conditions along the tower, these should be quoted for critical locations in the tower. Such locations would be feed inlet (flash zone), side stream and pump-around, draw off, tower top.

Low temperatures and the associated pressure apply only to cryogenic plants in this case.

Hydrogen partial pressure

This item is important to the metallurgist who will select the grade of metal to be used in the fabrication of the vessel. Generally the hydrogen partial pressure that will be quoted will be the one that exists at the tower top under normal operating conditions. For example the dew point calculation used in the sizing of the tower given in Figure 18.11 was based on the following tower top vapor composition.

	Mole fraction
H_2	0.005
C_1	0.021
C_2	0.117
C_3	0.378
iC_4	0.207
nC_4	0.268
iC_5	0.004
Total	1.000

The tower top pressure is 220 psia and the temperature is 158°F.

$$\text{Hydrogen partial pressure} = \frac{\text{Moles } H_2}{\text{Total Moles Gas}} \times \text{System Pressure}$$

$$= \frac{0.005}{1.0} \times 220 = 1.1 \text{ psia}$$

Materials and corrosion allowance

The process engineer will state the type of materials required to meet the process condition. For example where carbon steel only is to be used the process engineer indicates "CS". He is not normally required to state the grade of steel to be used, this is the responsibility of the vessel specialist or the metallurgist. However if the process engineer has a special knowledge of the material to be used and its specifics he should note it on the data sheet.

The same applies to the corrosion allowance. Normally 1/8″ is used for this allowance, however there maybe some mild corrosive condition existing which may justify using a higher number.

Description of internals

This is self explanatory when it refers to packed columns. In the case of trayed towers a separate data sheet giving sufficient data for tray rating and sizing is attached to the process data sheet front page. This is shown in the attachments to Figure 18.11.

Other common internals, such as distributors, vortex breakers, and the like are not normally shown on this data sheet. These are normally standard to a particular design office and will be added to the engineering drawings developed from this process data sheet later.

Insulation and fireproofing

The insulation requirement for heat conservation is specified by the process engineer as required. An approx thickness is shown. This will be checked later by the vessel specialist. In the case of fireproofing the process engineer indicates whether or not it is needed. The process engineer's relief valve sizing based on a fire condition takes into consideration the inclusion or not of fireproofing.

Notes and special conditions

This item is a "catch all" and is used to make note of whatever other information the process engineer may wish to add to the data sheet to ensure the equipment item will meet the process requirements. The question of stress relieving of the vessel is an item which is most important to the proper fabrication of the vessel and to its cost. The process engineer usually has knowledge whether this is needed or not to handle the process material at the conditions specified. He must therefore indicate this in this section of the data sheet. Other entries in this item should be a list of the attachments to the data sheet.

Most of the process data used to define the requirements for a horizontal drum are the same as those applied to a column, and these have already been discussed for Figure 18.11. The data included in the example given in Figure 18.12 have been calculated earlier in this chapter. In the data sheet however a "boot" measuring 2 ft i/d × 3 ft high has been added to the outlet end of the vessel to accumulate the water phase for better control of its level and to allow the disengaging of the hydrocarbon from the water.

18.2 Pumps

Pumps in the petroleum and other process industries are divided into two general classifications which are

- Variable head-capacity
- Positive displacement

The variable head-capacity types include centrifugal and turbine pumps whilst the positive displacement types cover reciprocating and rotary pumps.

The centrifugal pump

Centrifugal pumps comprise a very wide class of pumps in which pumping of liquids or generation of pressure is effected by a rotary motion of one or several impellers. The impeller or impellers force the liquid into a rotary motion by impelling action, and the pump casing directs the liquid to the impeller at low pressure and leads it away under a higher pressure. There are no valves in centrifugal type pumps (except, of course, isolation valves for maintenance, etc.), flow is uniform and devoid of pulsation. Since this type of pump operates by converting velocity head to static head, a pump impeller operating at a fixed speed will develop the same theoretical head in feet of fluid flowing regardless of the density of the fluid. A wide range of heads can be handled. The maximum head (in ft of fluid) that a centrifugal pump can develop is determined primarily by the pump speed (rpm), impeller diameter and number of impellers in series. Refinements in impeller design and the impeller blade angle primarily effect the slope and shape of the head-capacity curve and have a minor effect on the developed head. Multistage pumps are available which will develop very high heads; up to 5,000 ft. and up to 120 gpm. This versatility in handling high pressure head makes the centrifugal pump the most commonly used type in the process industry.

The turbine pump

Turbine pumps are a type of centrifugal pumps designed to recover power in systems of high flow and high differential pressure. These pumps transmit some of the kinetic energy in the fluid into brake horsepower. The actual energy recovery is about 50% of the hydraulic horsepower available. This type of pump is expensive and is therefore not as widely used as the centrifugal pump.

The rotary pumps

Rotary pumps are positive displacement pumps. Unlike the centrifugal type pump these types do not throw the pumping fluid against the casing but push the fluid forward in a positive manner similar to the action of a piston. These pumps however do produce a fairly smooth discharge flow unlike that associated with a reciprocating pump. The types of rotary pumps commonly used in a process plant are:

The gear pump. This pump consists of two or more gears enclosed in a closely fitted casing. The arrangement is such that when the gear teeth are rotated they are unmeshed on one side of the casing. This allows the fluid to enter the void between gear and casing. The fluid is then carried around to the discharge side by the gear teeth, which then push the fluid into the discharge outlet as the teeth again mesh.

Screw pumps. These have from one to three suitably threaded screwed rotors of various designs in a fixed casing. As the rotors turn, liquid fills the space between the screw threads and is displaced axially as the threads mesh.

Lobular pumps. The Lobular pump consists of two or more rotors cut with two, three, or more lobes on each rotor. The rotors are synchronized for positive rotation by external gears. The action of these pumps is similar to that of gear pumps, but the flow is usually more pulsating than that from the gear pumps.

Vane pumps. There are two types of vane pumps: those that have swinging vanes and those that have sliding vanes. The swinging vane type consists of a series of hinged vanes which swing out as the rotor turns. This action traps the pumped fluid and forces it into the pump discharge. The sliding vane pump employs vanes that are held against the casing by the centrifugal force of the pumped fluid as the rotor turns. Liquid trapped between two vanes is carried around the casing from the inlet and forced out of the discharge.

Reciprocating pumps

These are positive displacement pumps and use a piston within a fixed cylinder to pump a constant volume of fluid for each stroke of the piston. The discharge from reciprocating pumps is pulsating. Reciprocating pumps fall into two general categories. These are the simplex type and the duplex type. In the case of the simplex pump there is only one cylinder which draws in the fluid to be pumped on the back stroke and discharges it on the forward stroke. External valves open and close to enable the pumping action to proceed in the manner described. The duplex pump has a similar pumping action to the simplex pump. In this case however there are two parallel cylinders which operate on alternate stroke to one another. That is, when the first cylinder is on the suction stroke the second is on the discharge stroke.

Reciprocating pumps may have direct acting drives or may be driven through a crankcase and gear box. In the case of the direct acting drive the pump piston is connected to a steam drive piston by a common piston rod. The pump piston therefore is actuated by the steam piston directly. Reciprocating pumps driven by electric motors, turbines, etc are connected to the prime mover through a gearbox and crankcase.

Other positive displacement pumps

There are other positive displacement pumps commonly used in the process industry for special services. Some of these are:

Metering or proportioning pumps. These are small reciprocating plunger type pumps with an adjustable stroke. These are used to inject fixed amount of fluids into a larger stream or vessel.

Diaphragm pumps. These pumps are used for handling thick pulps, sludge, acid or alkaline solutions, and fluids containing gritty solid suspensions. They are particularly suited to these kind of service because the working parts are associated with moving the diaphragm back and forth to cause the pumping action. The working parts therefore do not come into contact with this type of fluid which would be harmful to them.

Characteristic curves

Pump action and the performance of a pump are defined in terms of their *Characteristic Curves*. These curves correlate the capacity of the pump in unit volume per unit time versus discharge or differential pressures. Typical curves are shown in Figures 18.13–15. Figure 18.13 is a characteristic curve for a reciprocating simplex pump which is direct driven. Included also is this reciprocating pump on a power drive.

Figure 18.14 gives typical curves for a rotary pump. Here the capacity of the pump is plotted against discharge pressure for two levels of pump speed. The curves also show the plot of brake horsepower versus discharge pressure for the two pump speed levels.

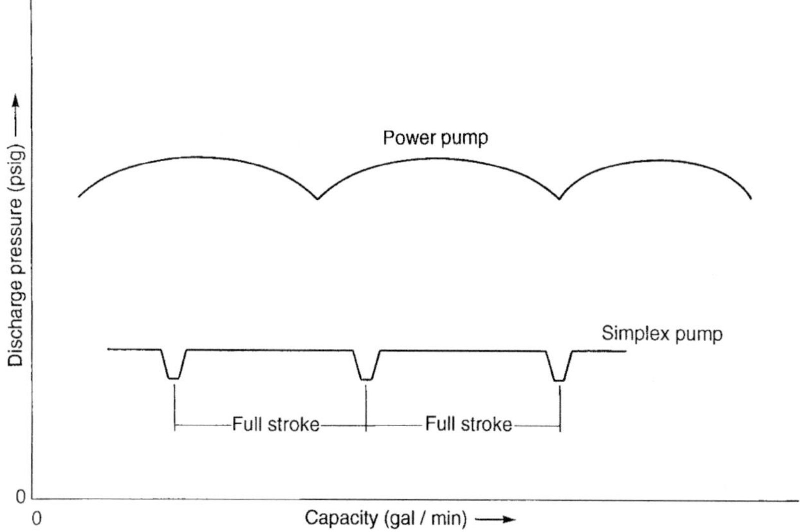

Figure 18.13. Characteristic curves for a reciprocating pumps.

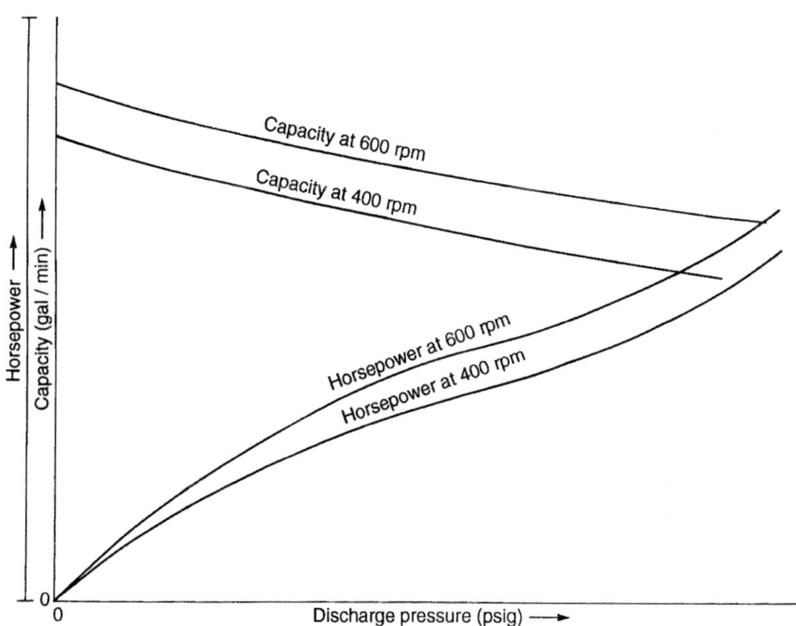

Figure 18.14. Characteristic curves for a rotary pump.

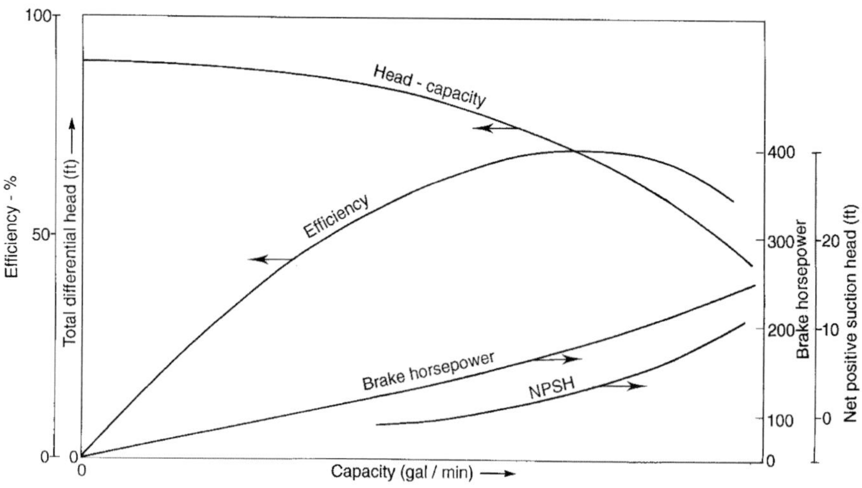

Figure 18.15. Characteristic curves for a centrifugal type pump.

Figure 18.15 is a typical characteristic curve for a centrifugal pump. This curve usually shows four pump relationships in four plots. These are:

- A plot of capacity versus differential head. The differential head is the difference in pressure between the suction and discharge
- The pump efficiency as a percentage versus capacity
- The brake horsepower of the pump versus capacity
- The net positive head (NPSH) required by the pump versus Capacity. The required NPSH for the pump is a characteristic determined by the manufacturer

Pump selection

Most industrial pumping applications favor the use of centrifugal pumps. The prominence of this type of pump stems from its ability to handle a very wide spectrum of fluids at a large range of pumping conditions. It is fitting that in considering pump selection the first choice has to be the centrifugal pump and all others become a selection by exception. Centrifugal pumps are generally the simplest in construction, lowest initial cost and simplest to operate and to maintain. This item therefore begins with the selection characteristics of the centrifugal type pump.

The centrifugal pump

Before looking at the selection of the centrifugal pump it is necessary to define the following terminology associated with pumps in general. These are:

- Capacity
- Differential Head
- Available NPSH
- Required NPSH

Capacity. This can be defined as the amount of fluid the pump can handle per unit time and at a differential pressure or head. This is usually expressed as gallons per minute at a differential head of so many pounds per square inch or so many feet.

Differential head. This is the difference in pressure between the suction of the pump and the discharge. It is usually expressed as PSI and FEET in specifying a pump. The following formula is the conversion from PSI to Feet:

$$\Delta H = \frac{2.31 \times \Delta P}{\text{SG}}$$

where

H = the differential head in feet of fluid being pumped.
P = the differential pressure of the fluid across the pump measured in pounds per square inch.
SG = the specific gravity of the fluid at the pumping temperature.

Available NPSH. The available NPSH is the static head available (in feet or meters) above the vapor pressure of the fluid at the pumping temperature. This is a feature of the design of the system which includes the pump.

Required NPSH. Is the static head above the vapor pressure of the fluid required by the pump design to function properly. The required NPSH must always be less than the available NPSH.

Selection characteristics

Selection of any pump must depend on its ability to handle a particular fluid effectively, and the efficiency of the pump under normal operating conditions. The second of these primary requirements can be determined by the pump's characteristic curves. These have already been described earlier in this part of the chapter, and a further discussion on these now follows:

Capacity range

Normal
Figures 18.16 and 18.17 show the normal capacity range for various types of centrifugal pumps in two different speed ranges, 3,550 rpm, 2,950 rpm.

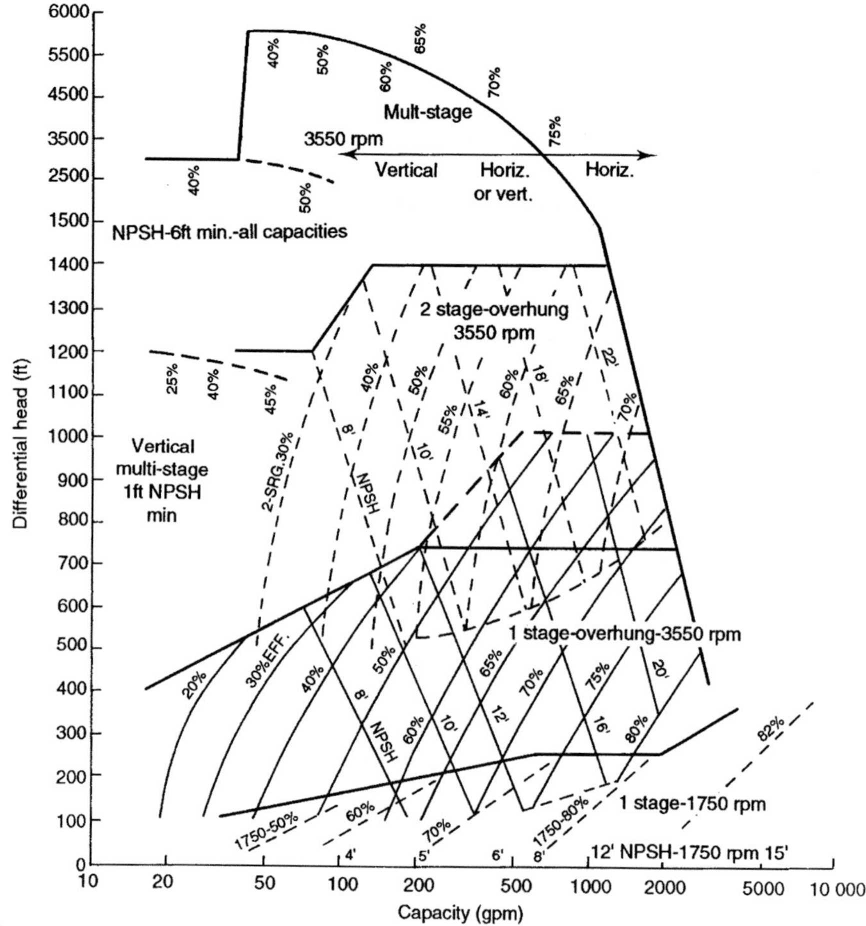

Figure 18.16. Centrifugal pumps at 3,550 rpm.

These values correspond to motor full load speeds available with current at 60 and 50 cycles, respectively. Most process applications call for these speed ranges. Lower speeds are for low or medium head and high capacity requirements, and for special abrasive slurries or corrosive liquids. Low capacity centrifugal pump applications may require special recirculation provisions in the process design to maintain a minimum flow through the pump. Because of practical consideration in impeller construction, the smallest available process type centrifugal pumps are rated at about 50 gpm.

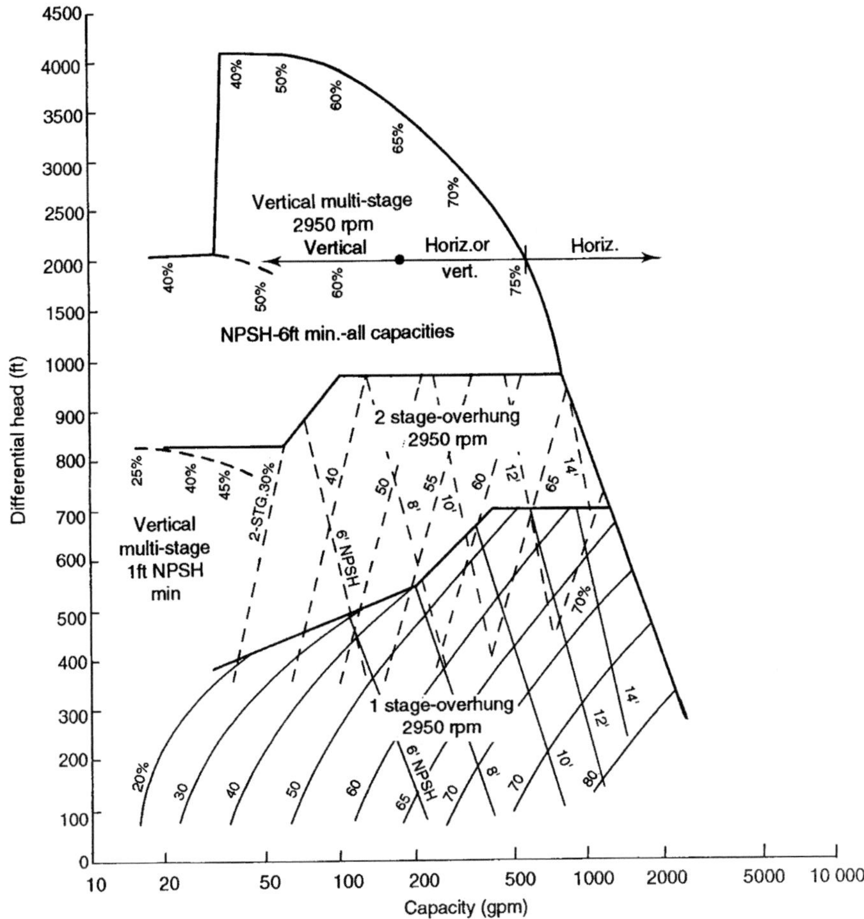

Figure 18.17. Centrifugal pumps at 2,950 rpm.

High and low capacity ranges

Pumps above the limits shown in Figure 18.16 and 18.17 will normally require large horsepower drivers. Special investigation of efficiency, speed, NPSH requirements, etc., will normally be justified. As an example, heads at or above the limits shown for multistage pumps at standard motor speed may be obtained by speed increasing gears (motor drive) or turbines to give pump operating speeds above maximum motor speeds, (NPSH requirements increase with speed).

In general, centrifugal pumps should not be operated continuously at flows less than approximately 20% of the normal rating of the pump. The normal rating for the pump is the capacity corresponding to the maximum efficiency point. The following table lists minimum desirable flow rates which should be maintained by continuous recirculation, if the required process flow conditions are of lower magnitude:

Head range feet	Pump type	Minimum continuous capacity rating, gpm	Normal rating of pump, gpm
60 Cycle speed (3,550 rpm)			
To 100	1 stg	10	60
100–350	1 stg	15	75–100
350–650	2 stg	30	150
650–1,100	2 stg	40	160
400–1,200	Multistg	15	50
1,200–5,500	Multistg	40	100–120
50 Cycle Speed (2,950 rpm)			
To 75	1 stg	10	50
75–250	1 stg	15	60–80
250–450	2 stg	25	120
450–775	2 stg	30	130
250–850	Multistg	10	40
850–3,800	Multistg	30	80–100

Care must be exercised in the design of any recirculation system to insure that the recirculated flow does not increase the temperature of pump suction and cause increased vapor pressure and reduction of available NPSH.

For low head pumps that can operate at 1,750 or 1,450 rpm, the above normal and minimum continuous capacities are reduced by 50%.

Effect of liquid viscosity

When suitably designed, centrifugal pumps can satisfactorily handle liquids containing solids, dirt, grit and corrosive compounds. Though fluids with viscosities up to 20,000 SSU (440 centistokes) can be handled, 3,000 SSU (650 centistokes) is usually the practical limitation from an economical operating standpoint.

Effect of suction head

An important requirement is that there be sufficient NPSH at the eye of the first stage impeller. This is static pressure above the vapor pressure of the fluid handled to prevent vaporization at the impeller eye. Flashing of the fluid produces a shock or cavitation effect at the impeller which results in metal loss, noise, lowered capacity and discharge pressure, and rapid damage to the pump. NPSH requirements for various centrifugal pumps will normally vary from 6 to over 20 (in feet of fluid) depending on type, size and speed. Vertical pumps can be built for practically no NPSH at all at the nozzle. These will have extended barrels in order to provide the required NPSH at the eye of the first stage impeller.

Efficiency

The efficiency of centrifugal pumps varies from about 20% for low capacity (< 20 gpm) pumps to a range of 70–80% for high capacity > 500 gpm pumps. Extremely large capacity pumps (several thousand gpm) may have efficiencies up to nearly 90%.

The rotary pumps

Rotary pumps deliver constant capacity against variable discharge pressure. This is a feature for all positive displacement pumps. Rotary pumps are available for process application over a range of 1 to 5,000 gpm capacity and a differential pressure up to about 700 psi. Displacement of the pump varies directly as the speed except in the case where the capacity may be influenced by a the viscosity of the fluid. In this case thick viscous liquids may limit capacity because they cannot flow into the cylinder fast enough to keep it completely filled. Rotary pumps are used mostly in low capacity service where the efficiency of a centrifugal pump would be very low.

Reciprocating pumps

The liquid discharge from reciprocating pumps is pulsating. The degree of pulsation is higher for SIMPLEX pumps than for the DUPLEX type. The pulsation is also higher for direct driven pumps than for those driven by motor or turbine through a gear box.

Pulsation is generally not a problem in the case of small low speed pumps of this type. It affects only the associated instrumentation which can be compensated for by local dampeners. However as the pump speed is increased the pulsation effect becomes more serious affecting the piping design of the system. Under these conditions of high pulsation instantaneous piping pressures may often exceed the design pressure of the piping. The piping must then be designed to meet this higher pressure requirement which invariably results in a higher cost. As an alternative, discharge dampeners are often considered but these too add to the cost of the pump installation.

Reciprocating pumps are used mostly in situations where their low piston speeds will withstand corrosive and abrasive conditions. They are ideal also for pumping at low capacity against high differential head and where it is necessary to maintain a constant flow rate against a gradually increasing discharge pressure.

Evaluating pump performance

Many process engineers are involved with the day to day operation of process plants. Much of their duties in this respect is concerned with maintaining plant efficiency, locating trouble areas and solving operational problems. This Item is directed to these engineers and presents the calculation methods to check pump performance in terms of the pump horsepower and the available NPSH for a pump.

Brake horsepower efficiency

Actual running efficiency can be calculated from plant data. This may be compared with a typical expected efficiency to evaluate the pump performance. The steps below are followed to arrive at this efficiency figure.

Step 1. Obtain flow rate from plant readings. Also read discharge pressure and if available suction pressure. If this later reading is not available calculate it from source pressure, height of liquid above pump suction and frictional loss.

Step 2. Read stream temperature and obtain Sg of stream from lab data. Calculate Sg at flow conditions.

Step 3. Calculate differential head which is discharge pressure PSIA—suction pressure PSIA. Convert to differential head in feet by

$$\frac{\Delta P \times 144}{62.2 \times \text{Sg (at flow cond)}}$$

Step 4. Convert feed rate to pounds per hour. Then calculate hydraulic horsepower from the expression

$$\frac{\text{Head (ft)} \times \text{rate (lbs/hr)}}{60\ \text{(mins)} \times 33{,}000\ \text{(ft-lbs/min)}}$$

Step 5. From motor data sheet obtain pump motor running efficiency from plant data read power usage in kW.

Step 6. Convert pump power to HP by dividing kW by 0.746. Multiply this by pump efficiency expressed as a fraction. This is the brake horsepower.

Step 7. Divide hydraulic horsepower by brake horsepower and multiply by 100 to give efficiency as a percentage.

Step 8. Check against Figure 18.16 or 18.17 to evaluate pump performance. That is, the calculated efficiency within a reasonable agreement with the expected efficiency given by Figure 18.16 or 18.17. Should there be a large discrepancy then the appropriate mechanical or maintenance engineers should be informed.

Checking available NPSH

This needs to be done if the pump is showing signs of vibration and losing suction under normal operations.

Step 1. Obtain details of the fluid being pumped (temperature, Sg, flow rate, source pressure, and vapor pressure of fluid).
Step 2. Calculate the frictional pressure drop in the suction line.
Step 3. Calculate the suction pressure of the pump by taking source pressure and adding in the static head. For this calculation take static head as being from the bottom of the vessel (not from the liquid level).
Step 4. From the suction pressure calculated in Step 3 take out loss through friction.
Step 5. Calculate NPSH available as being net suction pressure less vapor pressure. This is usually quoted in feet (or meters) so convert using:

$$\frac{\text{PSI} \times 144}{62.2 \times \text{SG}}$$

Step 6. Check against manufacturer's data sheet for required NPSH. If the available NPSH is less than required the pump will continue to cavitate. Fill tank to a level that the vibration stops and maintain it at that level.

If the problem is really troublesome and maintaining a liquid level as suggested in step 6 above is not practical contact the pump manufacturer. Very often he is able to make some minor changes to the pump design that will solve the problem.

Example calculation
Example calculation No. 1—Pump brake HP efficiency.

Flow capacity of pump at flow conditions = 200 gpm
(from plant data) = 82,632 lbs/hr

$$\begin{aligned}
\text{Suction pressure (plant data)} &= 120 \text{ psig}\\
\text{Discharge pressure} &= 450 \text{ psig}\\
\Delta\text{Pressure} &= 330 \text{ psi}\\
\text{Sg at flow conditions} &= 0.827\\
\text{Differential head ft} &= \frac{330 \times 144}{62.2 \times 0.827} = 924 \text{ ft}\\
\text{Hydraulic horsepower} &= \frac{82,632 \times 924}{60 \times 33{,}000}\\
&= 38.56
\end{aligned}$$

From motor rating, motor efficiency is 92%
Plant readings show motor power usage = 48.1 kW

$$\text{Then motor HP input} = \frac{48.1}{0.746} = 64.5 \text{ HP}$$

$$\begin{aligned} \text{Motor output} &= 64.5 \times 0.92 \\ &= 59.3 \end{aligned}$$

This is brake HP

$$\text{Then pump efficiency} = \frac{38.56 \times 100}{59.3} = 65\%$$

Motor is 50 cycle speed. From Figure 4.5 for multistage 2,950 rpm. This efficiency figure is about right.

Example calculation No. 2—Checking available NPSH
Fluid is gas oil from surge drum at 250°F and 15 psig drum is 12 ft above grade (bottom of drum which is horizontal).

Boiling point at 34 psia is 488°F.

Vapor pressure of gas oil at 250°F = 0.8 psia (VP curves)

$$\begin{aligned} \text{Source pressure} &= 29.7 \text{ psia} \\ \text{Sg @ conditions} &= 0.815 \\ \text{Head above grade} &= 12 \text{ ft (to bottom of drum)} \\ \text{Pump center line} &= 2 \text{ ft above grade} \\ \text{Liquid head to pump} &= 10 \text{ ft} = 3.5 \text{ psi} \\ \text{Friction pressure drop} &= 0.4 \text{ psi (calculated from 0.2 psi/100)} \\ \text{Less friction} &= 0.4 \\ \text{Less vapor press} &= 0.8 \end{aligned}$$

$$\begin{aligned} \text{Available NPSH} = 33.2 - (0.4 + 0.8) &= 32 \text{ psia} \\ &= 91 \text{ ft} \end{aligned}$$

Most pumps only require 10 ft or less.

Specifying a centrifugal pump

In all engineering companies and in most petroleum refineries two disciplines are responsible for correctly specifying a pump. These are the Mechanical Engineer and the Process Engineer. Most pumps are designed and built in accordance with set and accepted industrial codes, such as the API codes. The mechanical engineer ensures that the mechanical data supplied to the manufacturer for a particular pump meets the requirements of the code and standards to which the pump is to be built. The process engineer develops and specifies precisely the performance required of the pump in meeting the process criteria of the plant. To accomplish this the mechanical engineer develops a mechanical specification, and the process engineer initiates the pump specification sheet.

The mechanical specification

This specification is in a narrative form and will contain at least the following topics:

- Scope—Introductory paragraph which gives the code the pump manufacturer is to conform to (such as API 610).
  A list of other standards the pump shall conform to, if these are required.
- Main Body of the Specification—This covers all additions and any exceptions to the selected code. It provides for the type of drive shaft acceptable if different to code. Items such as impeller size as a percent of maximum allowable by code is given. The need for special bearing arrangements in the case of multi stage pumps are detailed in this document.
- Ancillary equipment and piping arrangements—The specification describes in detail the type of cooling medium that shall be used. It provides a guide also to the piping requirements that is required to satisfy the cooling system(s).
- Seal or Packing requirements—The mechanical specification details the type of seal or packing that will be installed. It also provides details of the seal arrangement required if this is different to the standard code.
- Pump Mounting—Some installation guide is provided by this specification. The method by which the pump is mounted on the base plate is detailed. It also details under what conditions the manufacturer is to provide pedestal cooling facilities.
- Metallurgy—Although the process engineer will specify the general material of construction for the pump(such as carbon steel or cast iron etc.) it is the mechanical engineer who details this. This detail includes the specific grade of the material and in many cases its pre-operational treatment.
- Inspection—The mechanical specification will provide details of the inspection that the company will carry out during the manufacture of the pump and before its delivery. This will include dimensional checking during manufacture and some checks on the metallurgy. Prior to shipping the purchaser may require a running test of the pump and will witness this test. For this purpose the pump is run in the workshop under specified process conditions.

The mechanical specifications may continue to detail other requirements that the purchaser may wish. Its objective is to ensure that the pump when delivered is mechanically robust, is safe and easily maintainable. The mechanical specification must also be aware of the cost implication of the requirements on the pump, and to keep them as low as possible.

The process specification

The data provided by the process engineer must be sufficient to ensure that the pump delivered for the process purpose will meet the duty required of it. These data are furnished to the pump manufacturer in the form of a data sheet similar to the one shown here as Figure 18.18. The data sheet collects the essential input from the process engineer, the mechanical engineer and, later, by the manufacturer to describe fully what is required of the pump and what the manufacturer has supplied. All the data given here will be unique to this pump.

The process input to the pump specification shown on Figure 13.6 are those items marked with the 'P'. Input by other disciplines and the manufacturer are not indicated on the form. The process engineer compiles much of these data from an 'Hydraulic Analysis' of the piping system. A calculation sheet given as Figure 18.19 shows the development of this and is described as follows:

Compiling the pump calculation sheet

The pump number, title, and service

This first section of the calculation sheet is important because it identifies the pump and what it is intended to do. The item number and service description will be unique to this item and will remain as its identification throughout its life. All the data below this section will refer only to this pump and to no other. The item number may contain the suffix 'A', 'B', 'C', etc. This indicates identical pumps in parallel service or as spare or both. This section also shows how many of these pumps are motor driven and how many are turbine (steam) driven. Usually spare pumps in critical service will be turbine driven. The remark column in this section should give any information that will be of benefit to the pump manufacturer or future operators of the pump. For example, if the spare pump is turbine driven, the process engineer may require an automatic start up of the turbine on a 'low flow' of the pumped stream. This should be noted here.

Operating conditions for each pump

The details of the fluid to be pumped and a summary of the calculations given below are entered here. Starting on the left of this section:

Liquid: This is a simple definition of the pumped material. In the example given here this will simply be 'vacuum gas oil'.

Pumping temperature (PT): This is the temperature of the gas oil at the pump. There are two temperatures called for 'normal' and 'max'. The normal temperature is that shown

SHEET NO. P REV.
JOB NO. P DATE P
BY P CHK'D.
P.O. NO.

NOTE: ○ INDICATES INFORMATION TO BE COMPLETED BY PURCHASER; □ BY MANUFACTURER

FOR P SITE P
UNIT P SERVICE P
NO. PUMPS REQ'D P NO. MOTORS REQ'D P ITEM NO. P PROVIDED BY MTD BY
NO. TURBINES REQ'D P ITEM NO. P PROVIDED BY MTD BY
PUMP MFR SIZE AND TYPE (P-Type only) SERIAL NO.

OPERATING CONDITIONS, EACH PUMP

LIQUID P m^3/h at PT, NOR. P RATED P
DISCH. PRESS., kg/cm^2g P
PT, °C NOR. P MAX. P SUCT. PRESS., kg/cm^2g MAX. P RATED P
SP.GR. at PT P DIFF. PRESS., kg/cm^2 P
VAP. PRESS. at PT, kg/cm^2a P DIFF. HEAD, m P
VIS. at PT, Ssu P CP P NPSHA, m P
CORR/EROS. CAUSED BY P HYD. HP(metric) P

PERFORMANCE

PROPOSAL CURVE NO.
RPM NPSHR (WATER) m
EFF. metric BHP RATED
MAX. metric BHP RATED IMP
MAX. HEAD RATED IMP m
MIN. CONTINUOUS m^3/h
ROTATION (VIEWED FROM CPLG END)

CONSTRUCTION

NOZZLES	SIZE	RATING	FACING	LOCATION
SUCTION				
DISCHARGE				

CASE-MOUNT: □ CENTERLINE □ FOOT □ BRACKET □ VERT. (TYPE)
-SPLIT: □ AXIAL □ RAD; TYPE VOLUTE □ SGL □ DBL □ DIFFUSER
-PRESS: □ MAX. ALLOW. kg/cm^2g °C, □ HYDRO TEST kg/cm^2g
-CONNECT: □ VENT □ DRAIN □ GAGE
IMPELLER DIA.: □ RATED □ MAX. □ TYPE:
MOUNT: □ BETWEEN BRGS □ OVERHUNG
BEARINGS-TYPE: □ RADIAL □ THRUST
LUBE: □ RING OIL □ FLOOD □ OIL MIST □ FLINGER □ PRESSURE
COUPLING: □ MFR □ MODEL
DRIVER HALF MTD BY: ○ PUMP MFR ○ DRIVER MFR ○ PURCHASER
PACKING: □ MFR & TYPE □ SIZE/NO. OF RINGS
MECH. SEAL: □ MFR & MODEL API CLASS. CODE
□ MFR CODE

SHOP TESTS

○ NON-WIT. PERF. ○ WIT. PERF.
○ NON-WIT. HYDRO ○ WIT. HYDRO
○ NPSH REQ'D. ○ WIT. NPSH
○ SHOP INSPECTION
○ DISMANT. & INSP. AFTER TEST
○ OTHER

MATERIALS

PUMP: CASE/TRIM CLASS ○
BASEPLATE: □

AUXILIARY PIPING

○ C.W. PIPE PLAN ○ CU; ○ S.S.; ○ TUBING; ○ PIPE
□ TOTAL COOLING WATER REQ'D, m^3/h ○ SIGHT F.I. REQ'D
○ PACKING COOLING INJECTION REQ'D: □ TOTAL m^3/h □ kg/cm^2g
○ SEAL FLUSH PIPE PLAN ○ C.S. ○ S.S. ○ TUBING ○ PIPE
○ EXTERNAL SEAL FLUSH FLUID □ m^3/h □ kg/cm^2g
○ AUXILIARY SEAL PLAN ○ C.S. ○ S.S. ○ TUBING ○ PIPE
○ AUX. SEAL QUENCH FLUID

VERTICAL PUMPS

PIT OR SUMP DEPTH ○
MIN. SUBMERGENCE REQ'D. □
COLUMN PIPE: □ FLANGED □ THREADED
LINE SHAFT: □ OPEN □ ENCLOSED
BRGS: □ BOWL □ LINE SHAFT
BRG. LUBE □ WATER □ OIL □ GREASE
FLOAT & ROD ○ C.S. ○ S.S. ○ BRZ ○ NONE
FLOAT SWITCH □
PUMP THRUST, kg □ UP □ DOWN

MOTOR DRIVER

HP(metric) RPM P FRAME VOLTS/PHASE/CYCLES P
MFR BEARINGS LUBE
TYPE INSUL FULL LOAD AMPS
ENC TEMP RISE, C LOCKED ROTOR AMPS
○ VHS ○ VSS VERT. THRUST CAP., kg MTR. ITEM NO.

APPROX. WT. PUMP & BASE
MOTOR TURBINE

API STANDARD 610 GOVERNS UNLESS OTHERWISE NOTED. APPLICABLE TO: PROPOSALS ○ PURCHASE ○ AS BUILT ○

P = Specified by Process

Figure 18.18. A centrifugal pump specification sheet.

PUMP CALCULATION.

Item No P 103- A- Unit Crude Vacume Unit Sheet No 1 Rev 0
Service HGO Product and BPA Motor Drive 1
Turbine Drive 1 Remarks Turbine to have Auto Start By PSTT App. J.S

OPERATING	CONDITIONS (Each Pump)	TURBINE CONDITIONS
Liquid VAL GAS OIL	US GPN &Pt.Min __ MOR 13Q Rated 1585	Inlet Stean psig 600
PTP NOR 545 MAX 740	Dish press Psig 85.5	Temp P 670
SP GR & PT 0.755	Suct Press Psig Max 50 Rated −0?	Exhaust Psig 50
Vap Press & Pt.Psig 0.29	Diff Press Psi 86.2	PUMP MATERIALS
Vis & PT Cp 0.906	Diff Bead FT 264.3	Casing C.S.
Corrosion\Erosion None	NPSH Available, P 740 Ayd HP 79.7 (1)	Internal Parts C.S.

ALTERNATES		B.L		
DESTINATION :				
Destination Press	psig	50		
Static Head	psi	5.4		
Line Loss	psi	10.7		
Meter Loss	psi	0.2		
11-E-9/10Δ Ht Exchangers	psi	14.0		
Δ Control Valves	psi	5.20		
TOTAL DISCHARGE PRESS	psig	85.50		
SUCTION.				
Surce Press	psig	−14.4		
↓ Static Head	psi	14.7		
- Ststem Losses	psi	1.0		
TOTAL SUCTION PRESS	psig	−0.7		
NPSH AVAILABLE.		0.29		
Source Press	psia	−1.29		
-(Vap Press ↓ suct losses) sub total Psia\Pt	psia	−1.00		
Elev of liquid - pump CL	Pt	45.00		
NPSH Available	Pt	40.40		

SKETCH

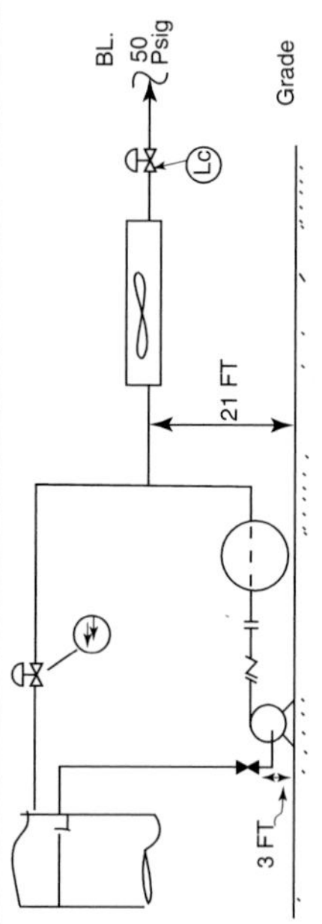

DATED 24.3.92 REV 0 DATED ____ REV ____.
NOTES (1) Based on Rated Flow.

Figure 18.19. A centrifugal pump calculation sheet.

on the process flow diagram, while the max temperature is that used for the pump design conditions. It should be the same as the design temperature of the vessel the fluid is pumped from.

Specific gravity @ PT: This is self explanatory. Note the item also calls for the SG @ 60°F.

Vapor pressure @ PT: This is read from the vapor pressure curves given in the appendix in Chapter 3 Appendix Figure 1. First locate the vapor pressure of the stream at atmospheric pressure. (This is the material's normal boiling point). Follow the temperature line down or up to the PT and read off the pressure at that point.

Viscosity @ PT: This too is self explanatory. Note this calculation sheet requires the viscosity to be in *Centipoise*. This is Centistokes × SG.

US gpm @ PT: This is the pump capacity and three rates are asked for. These are:

- *Minimum rate:* The anticipated lowest rate the pump will operate at for any continuous basis. This rate sets the control valve range.
- *Normal rate:* This is the rate given in the material balance and the basis for the hydraulic analysis.
- *Maximum rate:* This is normally set based on the type of service that the pump will undertake. For example: Pumps used only as rundown to storage will have a max rate about 10% above normal. Those used for reflux to towers will have between 15% and 20% above normal.

Discharge pressure, psig: This figure is calculated in the column below. It will also have been determined by the hydraulic analysis of the system.

Suction pressure, psig: Two pressures are asked for in this item. Rated pressure is that calculated in the column below and in the hydraulic analysis given in item 1. It is based on the 'Norm' Rate. The 'Max' suction pressure is based on a source pressure at the *design* pressure rating of the vessel the pump is taking suction from.

Differential pressure, psi: This is the discharge pressure minus the rated suction pressure.

Differential head feet: The head is determined from the differential pressure by the equation:

$$\frac{\text{Diff Press (psi)} \times 144}{62.2 \times \text{SG @ PT}}.$$

NPSH feet: Is calculated in the column below. This is the suction head available greater than the fluid vapor pressure (at the PT) at the pump impeller inlet.

Hydraulic horsepower: This is calculated from the weight per unit time (usually minutes or seconds) of fluid being pumped times the differential head in feet divided by 550 ft-lbs/sec or 33,000 ft-lbs/min. The differential head is always based on the rated suction pressure and the weight on the rated capacity (gpm) for this calculation.

Corrosion/erosion: The process engineer notes any significant characteristic of the fluid regarding its corrosiveness or abrasiveness here.

Turbine conditions

Although this item is not strictly part of a pump definition it should be included for completeness in the case of turbine drives. The data required to complete this item is self explanatory.

Pump material

The process engineer indicates here the acceptable material for the pump in handling the fluid. For example: carbon steel, or cast iron, etc; it is not necessary to specify grade of steel, etc.

The calculation columns

The objective of this section of the calculation sheet is to itemize all the data that are used to provide the figures given in the operating conditions described in Section "Operating conditions for each pump" above. The first column lists those items while the other three columns are available for entering the corresponding numbers. These three columns are provided to cater for alternate conditions that may need to be studied. A space is left on the right of the form to sketch the pumping system (it is very advisable to do a sketch).

The first column starts with the destination pressure, and continues down with the list of the pressure drops in the system to the pump discharge. This section of the column ends with the sum of the pressure drops giving the pump discharge pressure. The items that make up the pump suction pressure are listed next. This starts with he source pressure (usually a vessel) and its static head above the pump. All the pressure drops in the suction side are listed and deducted from the sum of the source pressure and static head to give the pump suction pressure.

The last section in the column itemizes the data that gives the *available* NPSH for the pump. The development of the NPSH is self explanatory.

Centrifugal pump seals

A pump seal is any device around the pump shaft designed to prevent the leakage of liquid out of or air into a pump casing. All industrial pumps have shafts protruding through the casings which require sealing devices. Pump sealing devices are usually either a "packed box" with or without a lantern ring, or a mechanical seal. Controlled leakage is a system sometimes used.

A flushing stream must be introduced into the pump seals for one or more of the following reasons:

- To effect a complete seal
- To provide cooling, washing or lubrication to the seal
- To keep grit from the seal
- To prevent corrosive liquid from reaching the seal

The facilities for accomplishing this are called "flushing system", and there are two types of these in general use.

- A dead-end system
- A through system

In a "dead-end" system the flushing liquid enters the casing through the stuffing box and combines with the pumped fluid (see Figure 18.20a). A "through" system is one in which the flushing liquid is re circulated between a double seal arrangement and does not enter the pump (see Figure 18.20b). The liquid source may be external to the pump or as on most mechanical seals is a self-flushing system in which the pumped liquid is used as the flushing fluid.

A description of each of the types of sealing devices is presented below and illustrated in Figure 18.21.

Packed boxes (without lantern ring) Figure 18.21a.

This is the simplest type of pump seal. Its principal components are a stuffing box, rings of packing, a throat bushing, and a packing gland. A slight leakage through the packing is required at all times to lubricate the packing. A water quench is used at the packing gland if the packing "leakage" is considered flammable or toxic.

Packed box (with lantern ring) Figure 18.21b

When a packed box pump seal is used in conjunction with a flushing oil system, a lantern ring is usually provided. This metallic ring provides a flow path for the flushing oil to reach the pump shaft. For very erosive or corrosive services, the lantern ring is

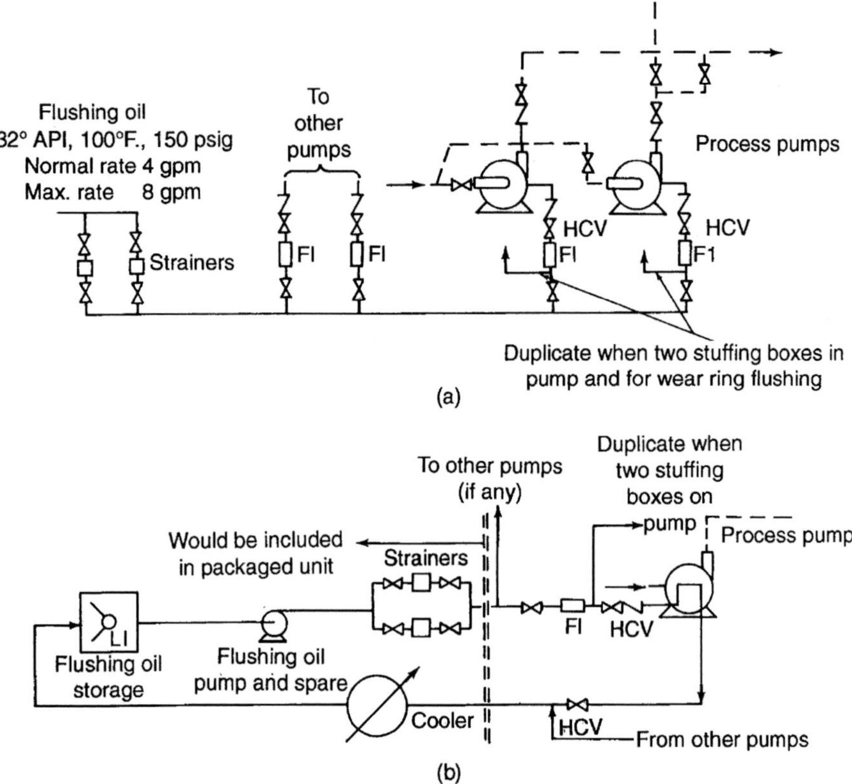

Figure 18.20. Typical flushing systems (a) a dead end system, (b) a through—re-circulating system.

often located next to the throat bushing and a liquid is injected into the throat bushing to prevent the pumped fluid from reaching the packing area. For a pump operation with vacuum suction conditions, the lantern ring is installed at the middle of the box and liquid is injected to prevent air entering the system. This type also operates with positive leakage with the same comments as the packed box without lantern ring.

Mechanical seals Figure 18.21c and d.

Typical basic elements of a single seal are shown in Figure 18.21c. Sealing is affected between the precision-lapped faces of the rotating seal ring and stationary seal ring. The stationary seal ring is usually carbon, and is mounted in the seal plate by an "O" ring. The two "O" ring packing serves the dual purpose of sealing off any liquid tending to leak behind the seal rings, and also to provide flexibility in allowing the seal faces to align themselves exactly so as to compensate for any slight "wobble" of the rotating seal face caused by shaft whip.

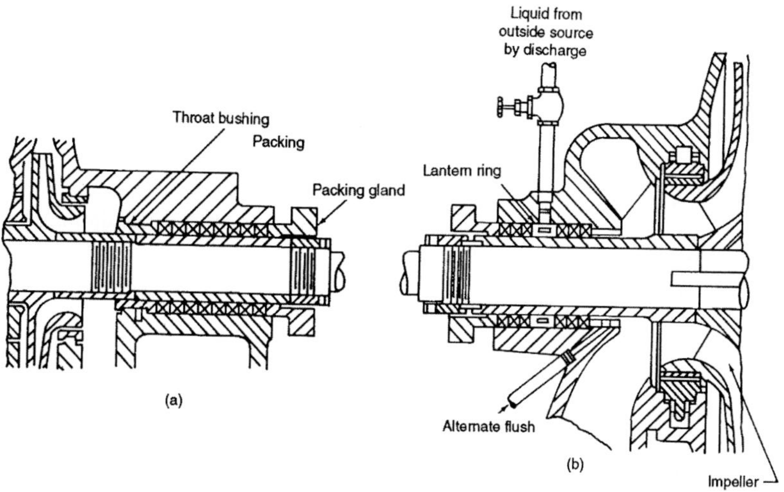

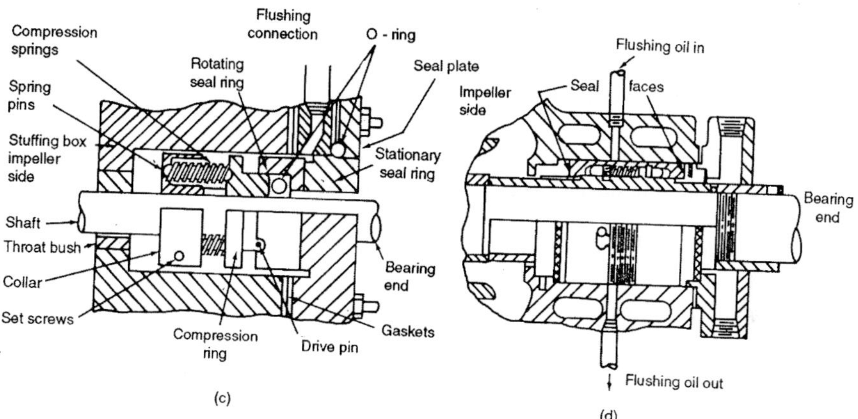

Figure 18.21. Pump shaft packing and seals. (a) Stuffing box completely filled with packing no lantern ring. (b) Externally sealed stuffing box. (c) Single Mechanical Seal. (d) Double Mechanical Seal.

The rotating seal ring is usually stainless steel with a Stellite face. The springs furnish the necessary force to set the "O" ring and hold the seal faces closed under low stuffing box pressures. Any pressure in the box exerts additional force on the rotating sealing ring. The seal is frequently "balanced" so that the face pressure is in correct ratio to the liquid pressure to ensure adequate sealing without excessive loading of the faces. Flushing oil enters the stuffing box through a connection in the seal plate.

A double seal consists of two single seals back to back. See Figure 18.21d. As a double seal is more expensive and requires a complicated seal-oil system, it is used only where single seals are not practical.

Table 18.20 summarizes the application of the various sealing systems in the refinery operation.

Pump drivers and utilities

Most pumps in the process industry are driven either by electric motors, or by steam, usually in the form of steam turbines. This item deals with the calculation of the driver requirements and its specification.

Electric motor drivers

Electric motors are by far the most common pump driver in industry. They are more versatile and are cheaper than a comparable size of steam turbine. The electric motors used for pump drivers are the induction type motor. They range in size from fractional horsepower to 500 and higher horsepower. Sizing the required motor for a pump driver takes into consideration the pump brake horsepower, the energy losses occurring in the coupling device between the pump and the motor, and a contingency factor of about 10%. These are expressed by the equation:

$$\text{Minimum driver BHP} = \frac{\text{Maximum pump BHP} \times 1.1}{\text{Mechanical efficiency of coupling}}$$

If the pump is driven through a direct coupling the efficiency will be 100%. With gears or fluid coupling the efficiency will be between 94% and 97%.

Specifying motor driver requirements

Process engineers are called upon very often to specify pump driver requirements or to check the already existing. In doing this two items of data need to be obtained or calculated. These are:

- The actual required horsepower of the pump motor to drive the pump at its specified duty
- What is actually installed in terms of horsepower

These data are tabulated in terms of power load as follows:

Operating load, kW—This is power input to the motor a normal operating horsepower. Connected load, kW—This is power input to the motor at motor rated horsepower.

If the pump is spared by another motor driven pump then the connected load will be the sum of *both motors*.

$$\textit{Operating load} = \frac{\textit{Minimum required driver HP} \times .746}{\textit{Efficiency of the motor @ its operating HP}}$$

$$\textit{Connected load} = \frac{\textit{Rated motor HP} \times .746 \times \textit{number of motors}}{\textit{Efficiency of the motors @ 100\% Full load}}.$$

Table 18.20. Application of pump sealing systems

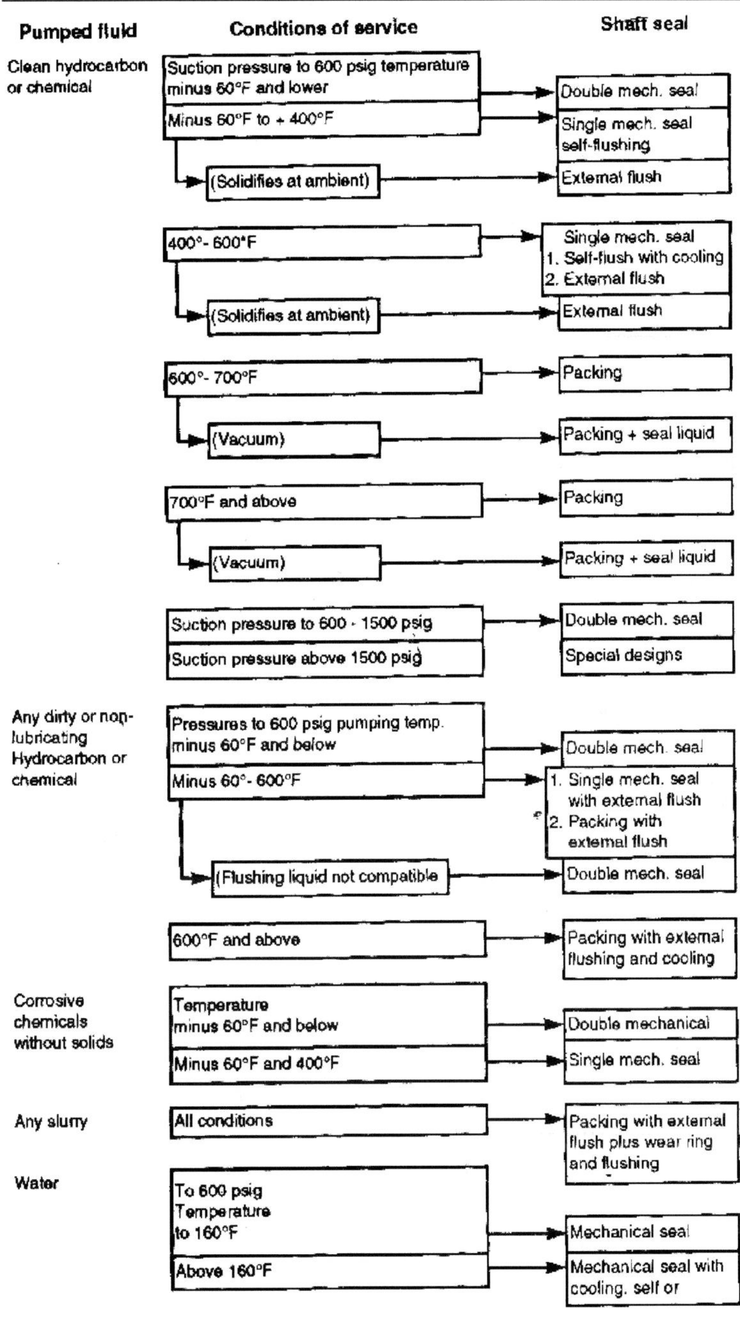

Pumped fluid	Conditions of service	Shaft seal
Clean hydrocarbon or chemical	Suction pressure to 600 psig temperature minus 60°F and lower	Double mech. seal
	Minus 60°F to + 400°F	Single mech. seal self-flushing
	(Solidifies at ambient)	External flush
	400°- 600°F	Single mech. seal 1. Self-flush with cooling 2. External flush
	(Solidifies at ambient)	External flush
	600°- 700°F	Packing
	(Vacuum)	Packing + seal liquid
	700°F and above	Packing
	(Vacuum)	Packing + seal liquid
	Suction pressure to 600 - 1500 psig	Double mech. seal
	Suction pressure above 1500 psig	Special designs
Any dirty or non-lubricating Hydrocarbon or chemical	Pressures to 600 psig pumping temp. minus 60°F and below	Double mech. seal
	Minus 60°- 600°F	1. Single mech. seal with external flush 2. Packing with external flush
	(Flushing liquid not compatible	Double mech. seal
	600°F and above	Packing with external flushing and cooling
Corrosive chemicals without solids	Temperature minus 60°F and below	Double mechanical
	Minus 60°F and 400°F	Single mech. seal
Any slurry	All conditions	Packing with external flush plus wear ring and flushing
Water	To 600 psig Temperature to 160°F	Mechanical seal
	Above 160°F	Mechanical seal with cooling. self or

Table 18.21. Electric motor size and efficiency

Motor rating BHP	Motor connected load, kW	Motor efficiency at a % of full load 50	75	100
1	0.98	68	74	76
1.5	1.42	72	76.5	79
2	1.86	73	78	80
3	2.76	77.5	81.5	82
5	4.39	83	82	85
7.5	6.65	81	83.5	84
10	8.78	84	85	85
15	13.0	85	86	86
20	17.05	86.5	87.5	87.5
25	21.0	87.5	88.5	88.5
30	25.1	88	89	89
40	33.5	88	89	89
50	41.7	88	89.5	89.5
75	62.1	89	90	90
100	82.0	84	89	91
125	102.0	85	89.5	91.5
150	123.0	86	89	91
200	161.0	88	91	92.5
250	201.0	90.5	92.5	92.5
300	241.0	90.5	92.5	92.8
350	281.0	90.9	92.6	92.9
400	320.0	91.1	92.8	93.1
450	360.0	91.2	93	93.2

Table 18.21 gives the motor sizes, and standard efficiencies at % of full load. (Higher-efficiency and variable speed motors may also be available.)

Example calculation
Calculate the operating and connected loads for pump 11-P-3 A&B as specified in Figure 18.19 of this chapter.

From the pump calculation sheet the hydraulic horsepower is calculated:

$$\text{HHP} = \frac{\text{lbs/min} \times \text{diff head in feet}}{33{,}000}$$

$$= \frac{8{,}665 \times 264.3}{33{,}000}$$

$$= 69.4\ \text{HP}$$

From Figure 18.19 and assuming 60 cycle pump speed pump efficiency is 79%

$$\text{Then brake horsepower} = \frac{\text{Hydraulic Horsepower}}{.79}$$

$$= 87.8\ \text{HP}$$

This will be a direct driven pump thus coupling efficiency is 100%

$$\text{Minimum motor size} = \text{BHP} \times 1.1 = 96.6 \text{ HP}$$

The closest motor size to this requirement is 100 HP (Table 18.21). This is a little too close so a motor size of 125 HP will be selected.

$$\text{Operating Load} = \frac{\text{Rated HP} \times 0.746}{\text{Efficiency @ \% of full load.}}$$

$$\text{\% of Full load} = \frac{87.8}{125} = 70.2\%$$

$$\text{Efficiency} = 89\% \text{ (from Table 18.21)}$$

$$\text{Operating load} = \frac{87.8 \times .746}{0.89}$$
$$= 73.6 \text{ kW}$$

$$\text{Connected load} = \frac{125 \times .746}{0.915}$$
$$= 101.9 \text{ kW say } 102 \text{ kW}$$

Note if both regular and spare pumps were motor driven then the connected load would be $2 \times 102 = 204$ kW.

Reacceleration requirement

To complete the specification for the motor requirement a degree of process importance of the pump must be established and noted. Voltage drops that can occur in any system may be sufficient to stop the pump. The process engineer must determine how important it is to the process and the safety of the process to be able to restart and reaccelerate the particular pump quickly. The following code of importance has been adopted:

- Re acceleration absolutely necessary.—A
- Re acceleration desirable.—B
- Re acceleration unnecessary.—C

The 'A' category involves any pump critical to keeping the process on stream safely and with no possibility of equipment damage. The 'B' category applies to those pumps that in operation with the 'A' category will maintain the unit 'on spec'. The 'C' category refers to those pumps that can be started manually without any problems.

In the case of the example pump 11-P-3 A&B given here, the service required of the pump is so critical to the operation and orderly shutdown of the process in the case of power failure that the spare pump is Turbine driven. Thus the motor driven regular pump need only be coded 'C' for reacceleration.

Steam turbine drivers

Steam turbines are the second most common pump drivers in modern day process industry. Although more expensive than the electric motor they offer an excellent stand-by to retain the maximum process 'on stream' time. The one big disadvantage with power driven pumps is the reliability of power availability. Steam turbines therefore offer a good alternative in cases of power failure. Another alternative means of pump drivers is the diesel engine or gas engine, but these require their own fuel storage etc. and are certainly not as reliable as the steam turbine.

Most process plants therefore spare the critical pumps in the process with a turbine driven unit which may be started automatically on low process flow.

The principle of the turbine driver

Turbines are the most flexible of prime movers in today's industry. Their horsepower output can be varied by the number and size of the steam nozzles used, speeds can be changed readily, and high speeds without gearing are possible. They have a very wide range of horsepower applications. The operation of the steam turbine is analogous to that of a water wheel where buckets are attached to the wheel which collect the water. The wheel is moved downwards by the weight of the water collected and thus cause the rotation of the wheel. In steam turbines the buckets are replaced by vanes which are impinged by the motive steam to cause the rotating motion. Turbines may consist of one set of vanes keyed to the shaft in the case of a single stage machine or several sets of vanes in the case of multi stage machines. These sets of vanes are called simply 'wheels' and the number of stages are referred to as the number of wheels.

In the case of multistage turbines, the steam leaving the first wheel is directed towards a set of stationary vanes attached to the casing. These stationary vanes reverse the steam flow and serve as nozzles directing the steam toward the second wheel attached to the same shaft.

Most turbines used on a regular basis in a process plant are single stage. Multi stage machines are more efficient but are also much more expensive. Their use therefore is for drivers requiring horsepower in excess of 300. The power industry is a good example for the use of large multi stage turbine drivers. Single or multi stage turbines may be operated either condensing or non condensing. However pump drivers should not be made *condensing* without a rigorous review to see if other types of drives can be used. The complexity of condensing is hardly worth the small savings in utilities that are made.

The performance of the steam turbine

The salient factors in the performance of the steam turbine are:

- Horsepower output
- Speed
- Steam inlet and outlet conditions
- Its mechanical construction (e.g., number of wheels, size of the wheel, etc.)

These factors are interrelated and their effect on the performance of the turbine is reflected by a change in over-all efficiency. The *overall efficiency may be defined as the ratio of the energy output to the energy of the steam theoretically available at constant entropy as obtained from a Mollier diagram.* This over-all efficiency is the product of mechanical and thermal efficiencies. The losses in turbines are partly due to friction losses of the rotating shaft and partly to thermodynamic losses and turbulence. Figure 18.22 gives the overall efficiencies plotted against delivered horsepower.

The steam required by a turbine for a given horsepower application is called its "Water Rate". The actual water rate for a turbine is supplied by the manufacturer from test runs carried out on the actual machine in the workshop. Plant operators and other engineers very often need to be able to estimate these water rates for their work. A typical situation arises when determining the best steam balance for a plant. Such estimates may be obtained from Figures 18.23a and b. This and the accompanying notes are self explanatory.

Cooling water requirements

Many pumps in process service require water cooling to various parts of the pump. This cooling water is applied to bearings, stuffing boxes, glands, and pedestals. The application of the cooling water is determined by the manufacturer in accordance with his standard for the service and conditions that the pump must satisfy. Most of the cooling water may be recovered in a closed cooling water system. However gland cooling water is never recovered but is routed to the waste water drain. The following lists the approximate cooling water requirements for pumps and steam turbines:

Pumps	*To 1,000 gpm*	*Above 1,000 gpm*
Up to 350 F	0 gpm	0 gpm
350 F–500 F	2 gpm	4 gpm
Above 500 F	3 gpm	6 gpm

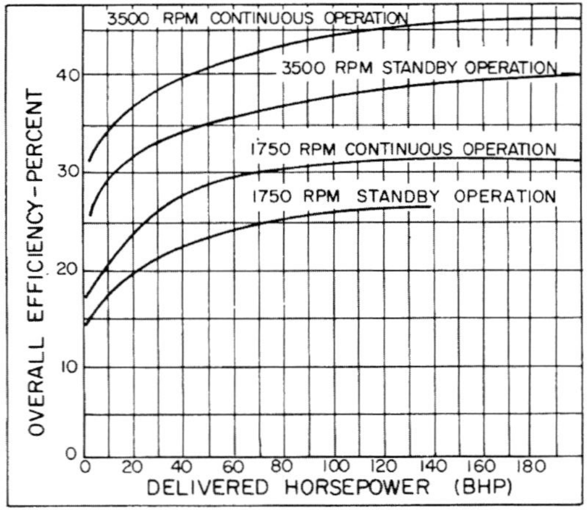

3500 RPM CONTINUOUS OPERATION
3500 RPM STANDBY OPERATION
1750 RPM CONTINUOUS OPERATION
1750 RPM STANDBY OPERATION
OVERALL EFFICIENCY-PERCENT
DELIVERED HORSEPOWER (BHP)

GRAPH 1
EFFICIENCY OF TURBINES AT 110 PSIG SAT. STEAM INLET WITH 20 PSIG STEAM EXHAUST

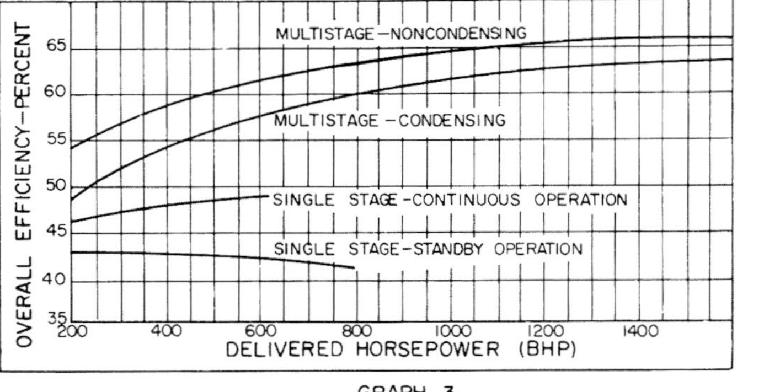

GRAPH 3
EFFICIENCY OF HIGH HORSEPOWER TURBINES

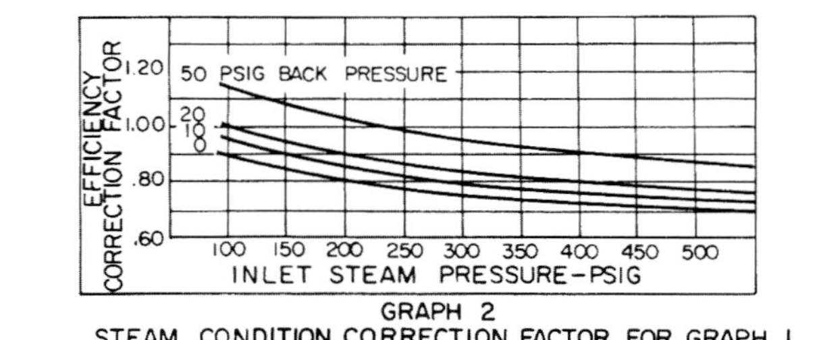

GRAPH 2
STEAM CONDITION CORRECTION FACTOR FOR GRAPH 1

Figure 18.22. Steam turbine efficiencies.

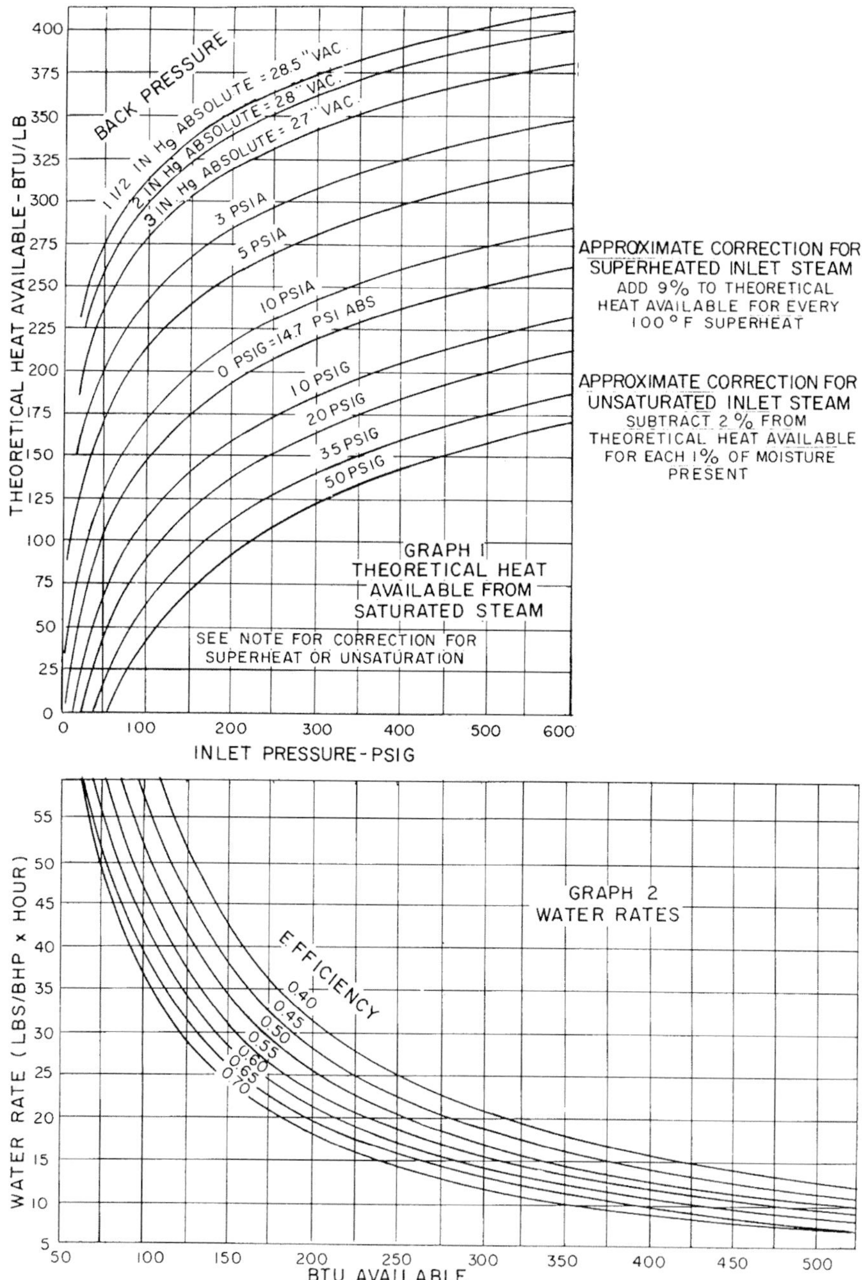

Figure 18.23. Water rates for condensing and non-condensing turbines.

Steam Turbines.

450 F	0 gpm	0 gpm
Above 450 F	3 gpm	3 gpm

18.3 Compressors

Types of compressors and selection

Compressors are divided into four general types, these are:

- Centrifugal
- Axial
- Reciprocating
- Rotary

The name given to each type is descriptive of the means used to compress the gas and comparison of the different types of compressors and typical applications is shown in Table 18.22. A brief description of each of the types now follows:

Centrifugal

This type of compressor consists of an impeller or impellers rotating at high speed within a casing. Flow is continuous and inlet and discharge valves are not required as part of the compression machinery. Block valves are required for isolation during maintenance.

Centrifugal compressors are widely used in the petroleum, gas, and the chemical industries primarily due to the large volumes of gas that frequently have to be handled. Long continuous operating periods without an overhaul make centrifugal compressors desirable for use for petroleum refining and natural gas applications. Normally they are considered for all services where the gas rates are continuous and above 400 ACFM (actual cubic feet per minute) for a clean gas, and 500 ACFM for a dirty gas. These rates are measured at the discharge conditions of the compressor. Dirty gases are considered to be gases similar to those from a catalytic cracker, which may contain some fine particles of solid or liquid material.

The slowly rising head-capacity performance curves make centrifugal compressors easy to control by either suction throttling or variable speed operation.

The main disadvantage of this type compressor is that it is very sensitive to gas density, molecular weight and polytropic compression exponent. A decrease in density or molecular weight results in an increase in the polytropic head requirement of the compressor to develop the required compression ratio.

Axial flow

These compressors consist of bladed wheels that rotate between bladed stators. Gas flow is parallel to the axis of rotation through the compressor. Axial flow compressors become economically more attractive than centrifugal compressors in applications where the gas rates are above 70 000 ACFM at *suction* conditions. The compressors are extremely small relative to capacity and have a slightly higher efficiency than the centrifugal. Axial flow compressors are widely used as air compressors for jet engines and gas turbines.

Reciprocating

Reciprocating compressors are widely used in the petroleum and chemical industries. They consist of pistons moving in cylinders with inlet and exhaust valves. They are cheaper and more efficient than any other type in the fields in which they are used. Their main advantages are that they are insensitive to gas characteristics and they can handle intermittent loads efficiently. They are made in small capacities and are used in applications where the rates are too small for a centrifugal. Reciprocating compressors are used almost exclusively in services where the discharge pressures are above 5,000 psig.

When compared with centrifugal compressors, the reciprocating compressors require frequent shutdowns for maintenance of valves and other wearing parts. For critical services this requires either a spare compressor or a multiple compressor installation to maintain plant throughput. In addition, they are large and heavy relative to their capacity.

Rotary

Recent developments in the rotary compressor field have opened up areas of application in the process industry with the use of the following types of rotary compressors:

(1) *High pressure screw*
These compressors have been developed into heavy duty type machines. They consist of two rotating helices that rotate in a casing without actual contact. Rotary compressors are lower in cost and have a higher efficiency than centrifugal compressors. They are not sensitive to gas characteristics since they are positive displacement machines. Parts are standardized production items so that a spare rotor is not generally required to be stocked for emergency replacement.

This compressor is noisy and sensitive to temperature rise along the screws due to the close clearances involved. They are good for fouling services where the fouling material forms a soft deposit. This decreases the clearances and leakage along the screws and casing. They are not recommended for use in fouling services in which the deposits are hard.

Variation in speed or a discharge bypass to suction are the only types of control that can be used.

(2) *Low pressure screw, lobe and sliding vane*

These compressors should be used only for low pressure, light duty, non-critical applications. They operate on the same principle as the high pressure screw type but have different mechanical designs. The same advantages and disadvantages apply as those for rotary high pressure screw compressors. They are even lower in cost than the high pressure screw compressors but contain parts having limited life, thus requiring more maintenance.

Only centrifugal and reciprocating compressors will be discussed further in this book.

Calculating horsepower of centrifugal compressors

Centrifugal compressors are used in process service where high capacity flows are required. A typical example is the recycle compressor for handling a hydrogen rich stream in some oil refining and petrochemical processes. The following table gives some idea of the centrifugal compressors capacity range.

Centrifugal compressor flow range

Nominal flow range (inlet acfm)	Average polytropic efficiency	Average adiabatic efficiency	Speed to develop 10,000 ft head/wheel
500–7,500	0.74	0.70	10,500
7,500–20,000	0.77	0.73	8,200
20,000–33,000	0.77	0.73	6,500
33,000–55,000	0.77	0.73	4,900
55,000–80,000	0.77	0.73	4,300
80,000–115,000	0.77	0.73	3,600
115,000–145,000	0.77	0.73	2,800
145,000–200,000	0.77	0.73	2,500

In general the head or differential pressure levels served by centrifugal compressor is considerably lower than that for reciprocal. The following diagram illustrates this feature.

Table 18.22. Comparison of compressors and typical applications

Type and control lable range	Percent availability	Operating speed volumetric capacity compression ratio per stage (6)	Compression eff.	Advantages	Disadvantages	Usual drivers	Common applications
Centrifugal 70–100%	99.5 to 100% (1)	3,000–15,000 RPM (2) 400–500 ACFM minimum @ discharge 150,000 ACFM max. Suction volume (5) 80–100,000 ft. polytrophic head/casing	70–78%	1. Long continuous operating periods 2. Low maintenance costs 3. Small size relative to capacity 4. Ease of capacity control	1. Pressure ratio is sensitive to gas density and molecular weight 2. Spare rotor required	Steam turbine Gas turbine Electric motor Waste gas Expander	Large refrigeration system Cat cracker air Large catalytic reformer recycle gas
Axial flow 80–100%	99.5 to 100%	4,000–12,000 RPM (2) 70,000 min. ACFM 2–4 compression ratio per casing	75–82%	1. Very high throughputs possible 2. Extremely small size relative to capacity 3. Higher efficiency than centrifugals 4. Good for parallel operation with other axials or centrifugals	1. Capacity flexibility limited by steep head-capacity curve and short stable operating range except when variable pitch stators are used 2. Performance and efficiency are sensitive to fouling 3. Spare rotor and spare stator blading are required	Steam turbine Gas turbine Electric motor Waste gas Expander	Cat cracker air (large)
Reciprocating See Sect. D for Range	98% clean gas (3) 95% dirty gas (3) 95% clean gas (4) 93% dirty gas (4)	300–1,000 RPM 5 max. compression ratio or 330–380°F max. discharge temp.	75–85%	1. Handles intermittent loads efficiently 2. Lower cost for small capacities 3. Used for very high discharge pressures (up to 50,000 psig) 4. Higher efficiency than centrifugal in lower capacity ranges 5. Insensitive to gas characteristics	1. Short continuous operating periods require spare or multiple machine installations if service is critical 2. Higher maintenance costs than centrifugal 3. Pulsation and vibration require engineered piping arrangement 4. Availability decreases when non-lubricated machines required to avoid lubricating oil in gas discharge	Synchronous motor Coupled or integral electric motor Coupled or integral engine	Instrument air Refinery air Fuel gas Synthesis gas Crude gas Small catalytic reformer recycle gas. Small refrigeration system Low Mole. Wt. gas

Rotary High Pressure Screw 55–100%	99–99.5	2,500–10,000 RPM 1,000–20,000 ACFM @ suction 4 to 7 max. compression ratio per casing but not exceeding 100 psi ΔP	75–80%	1. Lower cost than centrifugals 2. Higher efficiency than centrifugals 3. Not sensitive to gas characteristics 4. Parts are standardised production items and no spare rotor required	1. Noisy, require inlet and discharge silencers 2. Sensitive to temperature rise due to close clearances 3. Not recommended for use where fouling produces hard deposits 4. Speed or bypass control are only type applicable	Electric motor Steam turbine Gas turbine Waste gas expander	Refinery air Fuel gas Cat. Cracker Air (small)
Low Pressure Screw, Lobe and Vane type Rotaries Fixed Capacity	Not recom-mended for continuous service	1,500–3,600 RPM 100–12,000 ACFM suction 2 compression ratio per stage 50 psi max. discharge pressure	75–80%	1. Low first cost 2. Low maintenance cost 3. Parts are standardised production items and no spare rotor is required	1. Limited life 2. Speed or bypass control are only type applicable 3. Very noisy	Electric motor Steam turbine	Low pressure, light duty, non-critical services

Note: Clean service machines have the highest availability.
Large machines run at lower speeds.
Between turnarounds of 3 days every 8–12,000 hr with electric drive. 95–98% includes 8 hr shutdowns every few months for valve maintenance.
Between turnarounds of 2 weeks every 8–12,000 hr with engine drive. 93–95% includes 8 hr shutdowns every month for maintenance checks on compressor valves and engine driver.
Axial flow compressors should be considered at gas rate above 70,000 ACFM @ suction conditions.
Stages can be compounded in series for higher rates.

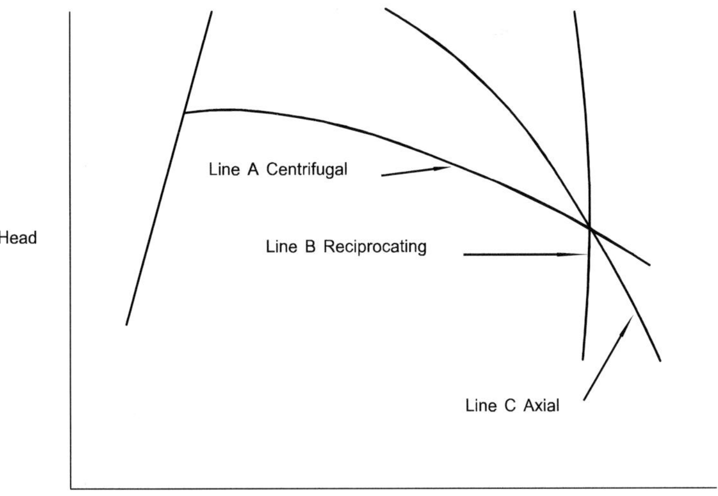

Process Engineers are often required to establish the capability of a centrifugal compressor in a particular service or to assess the machine's capability to handle a different service. In conducting these studies it is necessary to determine the machine's horsepower under the study conditions. This item provides a procedure where the gas horsepower (and thereafter the brake horsepower) of a compressor can be calculated.

This procedure is as follows:

Step 1. Establish the duty required from the compressor in terms of:

- Capacity in CF/min at inlet conditions
- Design inlet temp
- Design inlet pressure
- The mole wt of the gas to be handled
- Compression ratio (P2/P1) P2 being the discharge pressure and P1 the inlet pressure

Step 2. Establish the '*K*' value for the gas. If this is a pure gas (such as Oxygen) the '*K*' value can be read from data books. Otherwise the '*K*' value is the ratio CP/CV. See Figure 18.A.3 in the Appendix. (*Note*: Do not confuse this '*K*' factor with Equilibrium Constants.)

Step 3. Calculate volume of the gas in SCF/min. This is the inlet CFM times inlet pressure times 520 divided by 14.7 psia times inlet temperature in °R, thus

$$\text{SCFM} = \frac{1\ \text{CFM} \times \text{inlet press} \times 520}{14.7 \times \text{inlet temp}\ ^\circ\text{R}}$$

Step 4. Calculate number of moles gas/min by dividing SCF/m by 378. Multiply number of moles by mole wt for lbs/min of gas.

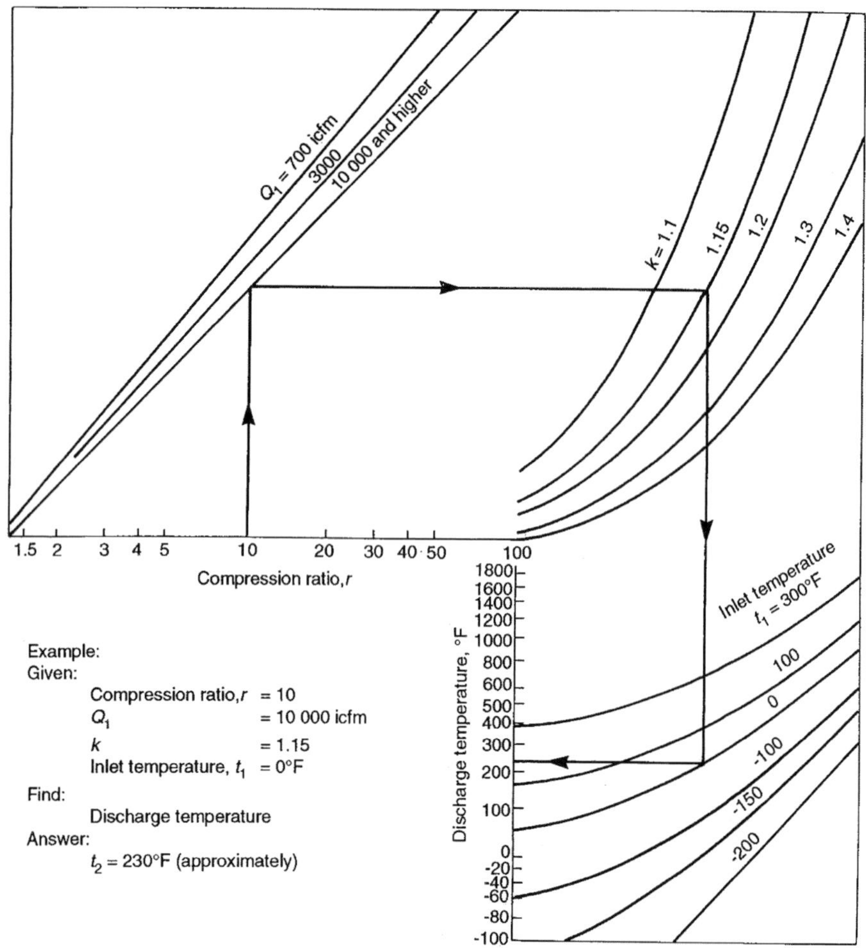

Figure 18.24. Estimated discharge temperatures for centrifugal compressors.

Step 5. Read off the estimated discharge temperature from Figure 18.24 using this and the discharge pressure calculate the volume in CF/min at discharge.

Step 6. Calculate density of gas at suction and discharge using the weight calculated in Step 4 and the CFM for suction and the CF/min calculated in Step 5 for discharge. This density will be in lbs/cuft.

Step 7. The average value for Z is taken as Z at suction + Z at discharge divided by 2. Z (compressibility factor) is calculated by the expression.

where

$$Z = \frac{\text{MP}}{T\rho_{\text{v}} \times 10.73}$$

M = mole weight
P = pressure @ psia
T = °R (°F + 460)
ρ_V = density in lbs/cuft

Step 8. Calculate the adiabatic head in ft lbs/lb using the expression

$$H_{ad} = \frac{Z_{ave} \times R \times T_i}{(K-1)/K} \left[\left(\frac{P_2}{P_1} \right)^{(K-1)/K} - 1 \right]$$

where

H_{ad} = the adiabatic head in ft lbs/lb
Z_{ave} = average compressibility factor
R = gas constant = 1,545/mole wt
K = adiabatic exponent CP/CV
P_2 = discharge pressure psia
P_1 = suction pressure psia
T = inlet temperature °R

Step 9. The gas HP is obtained using the expression

$$\text{HP} = \frac{W \times H_{ad}}{\eta_{ad} \times 33{,}000}$$

where

W = weight in lbs/min of gas
H_{ad} = adiabatic head in ft lbs/lb
η = adiabatic efficiency (0.7–0.75)

Step 10. Check GHP using Figure 18.25.

A example calculation now follows:

Example calculation
To determine the Gas HP of a centrifugal compressor assuming isentropic compression:

Compression ratio = 10.0
Capacity (actual inlet CF/min) = 10,000
K_{ave} = 1.15
T_1 °F = 100
P_1 psia = 100
Mole wt = 30
lbs/min = 5,013

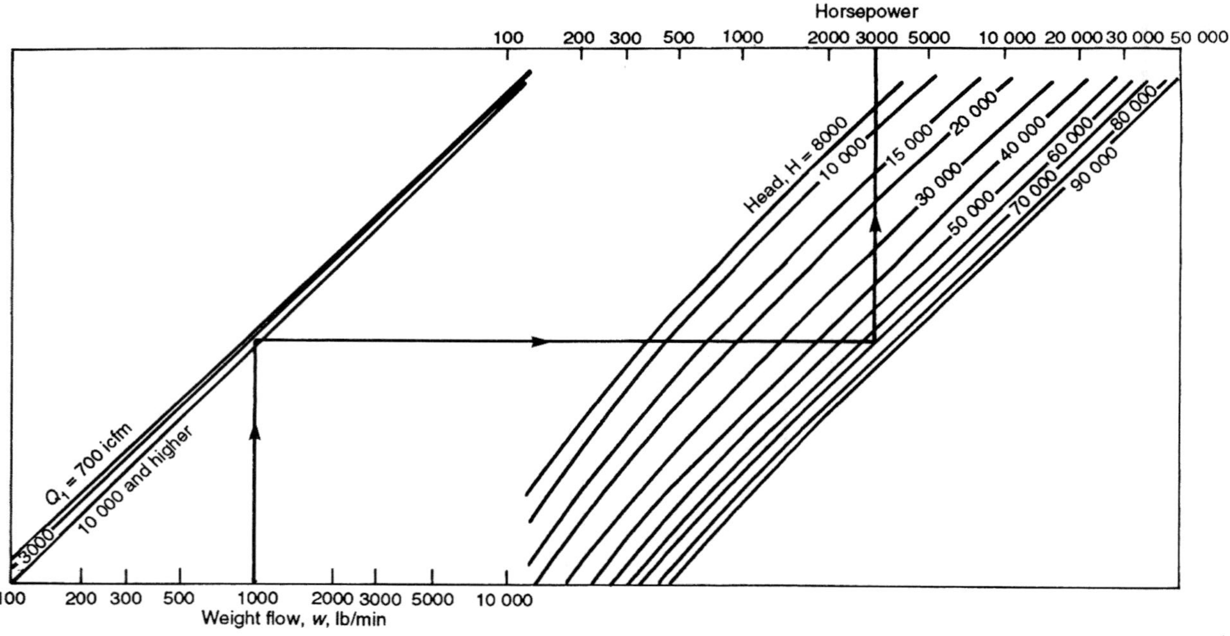

Figure 18.25. Determination of centrifugal compressor horsepower.

For isentropic compression

$$H_{ad} = \frac{Z_{ave} \times R \times T_i}{(K-1)/K}\left[\left(\frac{P_2}{P_1}\right)^{(K-1)/K} - 1\right]$$

where

H_{ad} = adiabatic head in ft lbs/lb
Z = compressibility factor (ave)
R = gas constant = 1,545/MW
K = adiabatic exponent $CP/CV = 1.15$
T = temperature in °R = °F + 460°F
P_2 = discharge pressure psia
P_1 = suction pressure psia

Z at inlet conditions

$$\rho_V = 0.5 \text{ lbs/cuft}$$

$$Z = \frac{MP}{T\rho_v} \times 10.73 = \frac{30 \times 100}{560 \times 0.5} \times 10.73 = 0.998$$

Estimated discharge temp (Figure 18.24) = 400°F

$$Z_{dis} = \frac{30 \times 1{,}000}{860 \times 3.26} = 0.997$$

Use 0.998

$$H_{ad} = \frac{0.998 \times 51.5 \times 560}{\frac{0.15}{1.15}}\left[\left(\frac{1{,}000}{100}\right)^{0.13} - 1\right]$$

$$= 77{,}004 \text{ ft lbs/lb}$$

$$= 220{,}664\,(1.349 - 1)$$

$$\text{Gas HP} = \frac{W \times H_{ad}}{\eta_{ad} \times 33{,}000}$$

Let η_{ad} be 0.75

$$\frac{5{,}013 \times 77{,}004}{0.75 \times 33{,}000} = 15{,}598 \text{ ghp}$$

This compares well with the estimate based on Figure 18.25.

Centrifugal compressor surge control, performance curves and seals

Centrifugal compressors can be counted on for uninterrupted run lengths of between 18 and 36 months after the initial shakedown run. The 18 month run corresponds to a compressor handling dirty gas, such as furnace gas, and the 36 month run corresponds to a clean gas service, such as refrigerant.

Spare compressors are not usually provided. A spare rotor, however, is required to be stocked as insurance against an extended downtime. Since this rotor is part of the capital cost of the equipment, it is not accounted for as spare parts. Only reliable drivers such as an electric motor, steam or gas turbine can be used where long continuous run lengths are required. In the case of steam and gas turbines, the drivers will probably dictate the maximum possible run length. The high operating speed of a centrifugal compressor also favors the selection of these type of high-speed drivers. The speed of these drivers can be specified to be the same as those of the compressor. For electric motor drives, a speed increasing gear is normally required. Centrifugal compressors can be broadly classified with regards to head and capacity as follows:

	Speed, RPM	Suction, ACFM	Polytropic head ft #/#
Small standard multistage	3,000–3,600	100–1,000	to 8,500
Standard single stage	3,000–3,600	700–60,000	1,000–6,700
Special single stage	3,000–15,000	1,000–60,000	6,700–11,500
Special multistage casing, uncooled	3,000–15,000	1,000–140,000	6,700–100,000
Special multistage, multi-casing, inter-cooled	3,000–15,000	2,500–140,000	37,000 up

As a guide, the maximum head per impeller is about 10,000 ft. Normally, about 8 impellers can be used in a casing.

The minimum allowable volume of gas at the compressor discharge is about 400 ACFM for a clean gas and 500 ACFM for a dirty gas. Dirty gases are considered to be similar to the gas from a steam or catalytic cracking unit.

The discharge temperature is limited to about 250°F for gases that may polymerize and 400°F for other gases. Normally inter-coolers will be used to keep the discharge temperature within these limits. These temperature limitations do not apply to special centrifugal flue gas re-circulator which can be obtained to operate at over 800°F. There is also a temperature rise limitation of 350°F per casing. This is the maximum temperature rise that can be tolerated due to thermal expansion considerations.

Use of cast iron as a casing material is limited to 450°F maximum. Temperatures of −150°F to −175°F can be tolerated in conventional designs. Lower temperatures are not common and will require consulting on individual design features.

Surge

A characteristic peculiar to centrifugal and axial compressors is a minimum capacity at which the compressor operation is stable. This minimum capacity is referred to as the surge or pumping point. At surge, the compressor does not meet the pressure of the system into which it is discharging. This causes a cycle of flow reversal as the compressor alternately delivers gas and the system returns it.

The surge point of a compressor is nearly independent of its speed. It depends largely on the number of wheels or impellers in series in each stage of compression. Reasonable reductions in capacity to specify for a compressor are shown below.

Wheels/compression stage	% of normal capacity at surge—maximum
1	55
2	65
3 or greater	70

An automatic re-circulation bypass is required on most compressors to maintain the minimum flow rates shown. These are required during start-up or when the normal load falls below the surge point. Cooling is required in the recycle circuit if the discharge gas is returned to the compressor suction.

Performance curves

The rise of performance curves should be specified for a compressor. This is normally done by specifying the pressure ratio rise to surge required in each stage of compression. A continuously rising curve from normal flow rate to surge flow is required for stable control.

The pressure ratio rise to surge is largely a function of the number of impellers per compression stage. Reasonable pressure ratio rises to specify are shown below:

Wheels/compression stage	Minimum % of rise in pressure ratio from normal to surge flow
1	$3\frac{1}{2}$
2	6–7
2 or greater	$7\frac{1}{2}$

Frequently, the performance curves for a compressor have to be plotted to determine if all anticipated process operations will fit the compressor and its specific speed control. Three points on the head-capacity curve are always known. These are the normal, surge and maximum capacity points. The normal capacity is always considered to be on the 100% speed curve of the compressor. The surge point and the compression ratio rise to surge have been specified. From this the head produced by the compressor at the surge point can be back-calculated using the head-pressure ratio relationship. The maximum capacity point is specified to be at least 115% capacity at 85% of normal head.

The head-capacity curve retains its characteristic shape with changes in speed. Curves at other speeds can be obtained from the three known points on the 100% speed curve by using the following relationships:

1. The polytropic head varies directly as the speed squared.
2. The capacity varies directly as speed.
3. The efficiency remains constant.

Figure 18.26 shows a typical centrifugal compressor performance curve.

Control

Speed
Speed control is the most efficient type of control from an energy consideration. It requires, however, that a variable speed driver such as a steam turbine or gas turbine, or a variable-speed electric motor be used. The compressor is controlled by shifting its performance curve to match the systems requirement.

Suction throttling

- *Adjustable inlet guide vanes.* Adjustable inlet guide vanes are the most efficient method of adjusting the capacity of a constant speed compressor to match the system characteristics. They consist of a venetian blind device that is positioned by a rack and pinion linkage. While the guide vanes do some throttling, their main effect is to change the velocity of the gas to that of the impeller vane by changing the direction of flow. This changes the head produced and in effect changes the characteristic of the machine.
- *Suction throttle.* This control consists of a control valve located in the compressor suction which regulates the suction pressure to the compressor. The control valve results in a greater power loss compared to adjustable inlet guide vane control since it is a pure throttling effect. Suction throttle valves are lower in cost than adjustable inlet guide vanes.
- *Discharge throttling.* This control consists of a control valve located in the compressor discharge. Discharge throttle valves are seldom used since they offer relatively little power reduction at reduced capacity. The effect is simply to “push” the compressor back on its curve.

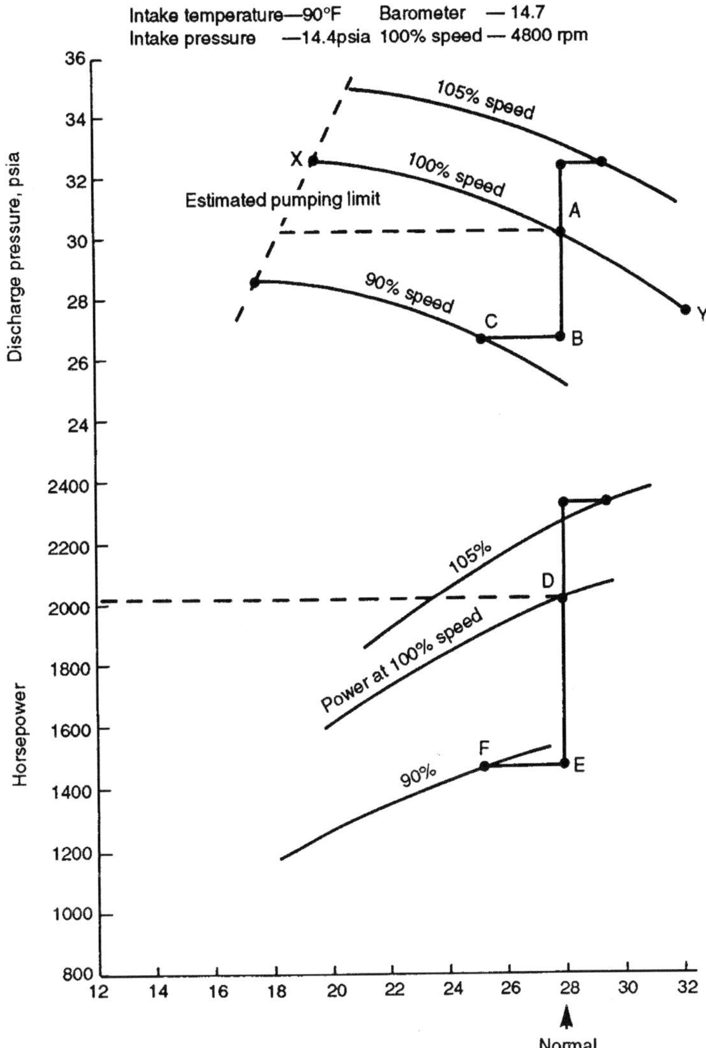

Figure 18.26. An Example of a centrifugal compressor performance curve.

Seals

Table 18.23 shows the types of seals that are commonly used in centrifugal compressors. The start-up as well as the operating conditions of the compressor should be considered in selecting a seal. Often the system is evacuated when hydrocarbons are handled prior to its start-up. This requires that the seal be good for vacuum conditions.

Table 18.23. Centrifugal compressor seals.

No	Application	Gas being handled	Inlet pressure, psia	Seal arrangement
1	Air compressor	Atmospheric air	Any	Labyrinth
2	Gas compressor	Non-corrosive Non-hazardous Non-fouling inexpensive	Any	Labyrinth
3	Gas compressor (note 1)	Non-corrosive or Corrosive Non-hazardous or Hazardous Non-fouling or Fouling	10–25	Labyrinth with injection or ejection of fluid being handled.
4	Gas compressor	Non-corrosive Non-hazardous Non-fouling	All pressures	Oil seal combined with Lube oil system.
5	Gas compressor	Corrosive Non-hazardous or Hazardous	All pressures	Oil seal with seal oil Separate from lube oil System.

Note 1: Where some gas loss or air induction is tolerable.

Specifying a centrifugal compressor

The process specification must give all the information concerning the gas that is to be handled, its inlet and outlet conditions, the utilities that are available and the service that is required of the compressor. The process specification sheet given here shows the minimum that a process engineer should provide in approaching manufacturers. An explanation of this specification now follows covering each of the items in the specification.

Title block

This requires the item to be identified by item number and its title. The number of units that the specification refers to is also given here. For a centrifugal compressor this will normally be just one as very seldom is a spare machine required.

Normal and rated columns

More often than not the conditions and quantities required to be handled will vary during the operation of the machine. The two columns therefore will be completed showing the average normal data in the first column and the most severe conditions and duty required by the compressor in the second. The severe conditions in column two are for a continuous length of operation not instantaneous peaks (or troughs) that may be encountered.

Gas

The composition and gas stream identification must be included as part of the process specification. Usually the composition of the gas is listed on a separate sheet as shown in the example. Note in many catalytic processes that utilize a recycle gas the composition of the gas will change as the catalyst in the process ages. Thus it will be necessary to list the gas composition at the start of the run (SOR) and at the end of the run (EOR).

The compressor may also be required to handle an entirely different gas stream at some time or other. This too must be noted. For example in many petroleum refining processes a recycle compressor normally handling a light predominately hydrogen gas is also used for handling air or nitrogen during catalyst regeneration, purging, and start-up.

Volume flow

This is the quantity of gas to be handled stated at 14.7 psia and 60 F.

Weight flow

This is the weight of gas to be handled in either lbs/min or lbs/hr.

Inlet conditions

Pressure: This is the pressure of the gas at the inlet flange of the compressor in psia.
Temperature: This is the temperature of the gas at the inlet flange of the compressor.
Mole weight: The mole weight of the gas is calculated from the gas composition given as part of the specification.
C_p/C_v: This is the ratio of specific heats of the gas again obtained from the gas mole wt and Table 18.A.1 in the Appendix.
Compressibility factor (z): use the value at inlet conditions calculated as shown in step 7 of the chapter 'Calculating the Horsepower of Centrifugal Compressors'.
Inlet volume: This is the actual volume of gas at the conditions of temperature and pressure existing at the compressor inlet. Thus:

$$\text{ACFM} = \frac{\text{SCFM} \times 14.7 \times (\text{inlet temp F} + 460)}{(60\text{ F} + 460) \times \text{inlet press psia}}$$

Discharge conditions

Pressure: This is the pressure at the compressor outlet flange and is quoted in either psia or psig.
Temperature: This is estimated using Figure 18.24.

C_p/C_v: This will be the same as inlet.
Adiabatic efficiency: Will be as given in Figure 18.27.

Approximate driver horsepower

This item will include the adiabatic (or gas) horsepower as calculated in the section on calculating the horsepower of centrifugal compressors in this chapter plus the following losses:

Leakage loss—1% of Adiabatic HP
Seal losses—35 HP for all HP ranges.
Bearing loss—5 HP for all HP ranges.

The reminder of the spec sheet contains all the essential data and requirements that may affect the duty and performance of the compressor. Much of this is self explanatory however there are some items that require comment. These are:

1. Most compressor installations today are under an open sided shelter with a small overhead gantry crane assembly for maintenance.
2. Usually the lube and seal oil assemblies have their own pump and control systems. Consequently even if the compressor itself is to be steam driven there may still be need to give details of utilities for the ancillary equipment.
3. Details of the gas composition is essential for any development of the compressor. This is listed on the last page of the specification together with any notes of importance concerning the machine and its operation.

An example calculation for a specification sheet follows:

Example calculation
Prepare a process specification sheet for a compressor to handle the hydrogen recycle stream in an o-xylene isomerization plant. Details are as follows:

Fresh feed rate : 5,000 bpsd of m-xylene.
Recycle gas rate : 7,000 Scf of hydrogen per Bbl of Fresh feed.
Gas composition mole % :

	Start of Run (SOR)	End of Run (EOR)
H_2	85.00	68.78
C_1	4.4	9.17
C_2	4.2	8.86
C_3	3.6	7.36
iC_4	0.78	1.62
nC_4	0.99	2.05
C_5s	1.03	2.16

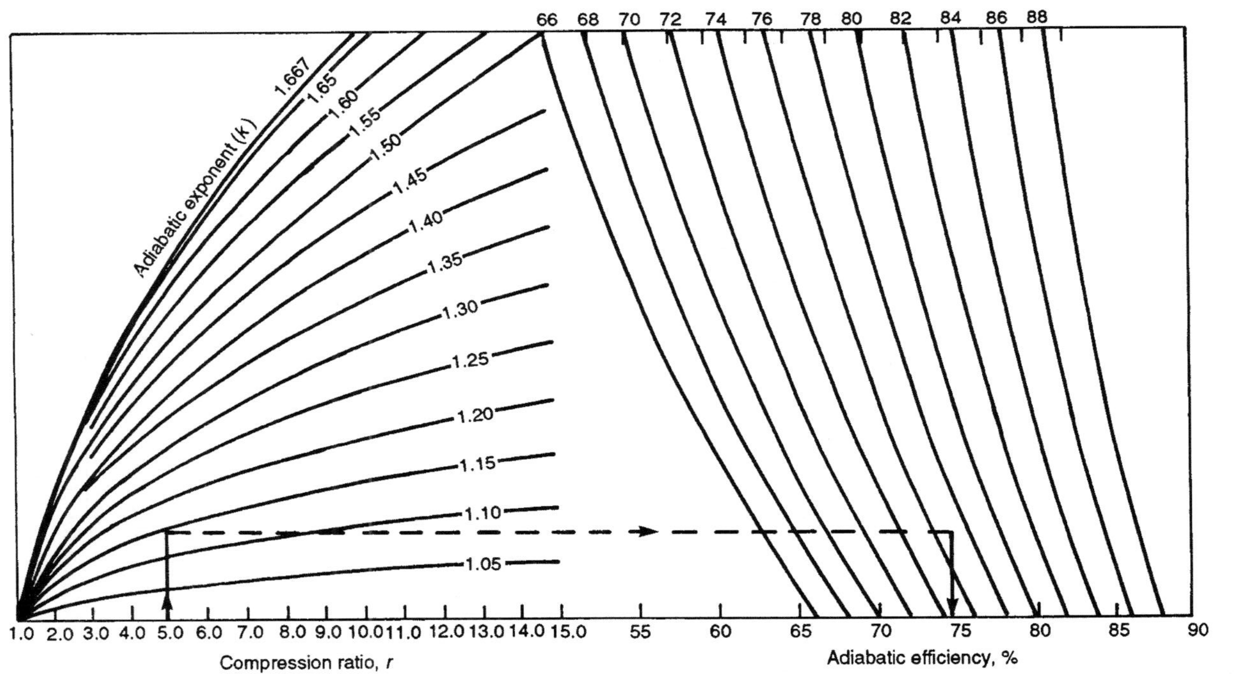

Figure 18.27. Adiabatic efficiencies for centrifugal compressors.

Suspection pressure : 150 psig
Reactor pressure : 500 psig.
Suction temperature : 100 F

Step 1. Calculate the mole weight of the gas.

	SOR			EOR	
	mole%	MW	wt Factor	mole%	wt Factor
H_2	85.0	2	170	68.78	138
C_1	4.4	16	70	9.17	147
C_2	4.2	30	126	8.86	266
C_3	3.6	44	158	7.36	324
iC_4	0.78	58	45	1.62	94
nC_4	0.99	58	57	2.05	119
C_5s	1.03	72	74	2.16	156
Total	100.00	700	100.00	1,244	
MW	7.0			12.44	

Step 2. Calculate volume flow of gas in SCF/min.

$$\begin{aligned}\text{Total volume of } H_2 \text{ required} &= 5{,}000\,\text{BPSD} \times 7{,}000\,\text{SCF} \\ &= 35.00\,\text{mmScf/day} \\ &= 24{,}306\,\text{Scf/min.}\end{aligned}$$

$$\begin{aligned}\text{For SOR volume gas Flow} &= \frac{24{,}306}{0.85} \\ &= 28,595\,\text{Scf/min.}\end{aligned}$$

$$\begin{aligned}\text{For EOR volume gas flow} &= \frac{24{,}306}{0.6878} \\ &= 35,339\,\text{Scf/min.}\end{aligned}$$

Step 3. Calculate weight flow in lbs/min.

$$\text{moles/min of gas} = \frac{\text{Scf/min}}{378}$$

$$\begin{aligned}\text{For SOR moles/min} &= 75.6 \\ \text{For EOR moles/min} &= 93.5 \\ \text{lbs/min for SOR} &= 75.6 \times 7.0 = 529\,\text{lbs/min} \\ \text{lbs/min for EOR} &= 93.5 \times 12.44 = 1{,}163\,\text{lbs/min}\end{aligned}$$

Step 4. Calculate ACFM at inlet conditions.

$$\begin{aligned}\text{Compressor inlet pressure} &= 165\,\text{psia} \\ \text{'' \quad '' Temp} &= 100\,\text{F}\end{aligned}$$

$$\text{For SOR ACFM} = \frac{28{,}597 \times 14.7 \times 560}{520 \times 165}$$
$$= 2{,}744 \text{ cft/min}$$

$$\text{For EOR ACFM} = \frac{35{,}339 \times 14.7 \times 560}{520 \times 165}$$
$$= 3{,}391 \text{ cft/min}$$

Step 5. Estimate the C_p/C_v ratio.

The molal proportions will be used for this purpose. The ratio for each component will be taken from Table 18.A.1 in the Appendix.

		SOR		EOR	
	C_p/C_v	mole%	C_p/C_v fact	mole%	C_p/C_v fact
H_2	1.40	85.0	119	68.78	96.29
C_1	1.30	4.4	5.7	9.17	11.92
C_2	1.22	4.2	5.12	8.86	10.81
C_3	1.14	3.6	4.10	7.36	8.39
iC_4	1.11	0.78	0.87	1.62	1.80
nC_4	1.11	0.99	1.10	2.05	2.28
C_5S	1.09	1.03	1.12	2.16	2.35
Total		100.00	131.9	100.00	133.84

Then C_p/C_v for the gas is:

SOR = 1.319
EOR = 1.338

Step 6. Calculate compressibility factors.

$$Z = \frac{MW \times P_1}{T \times \rho_v \times 10.73}$$

$$\text{For SOR flows} \quad Z = \frac{7.0 \times 165}{560 \times 0.193 \times 10.73}$$
$$= 0.996$$
$$\rho_v = \frac{\text{wt lbs/min}}{\text{ACFM}}$$
$$\text{For EOR flows} \quad Z = \frac{12.46 \times 165}{560 \times 0.343 \times 10.73}$$
$$= 0.998$$

Step 7. Calculate outlet temperature.

Approx discharge temperature is read from Figure 18.24 in this chapter using the following:

$$\begin{aligned}
\text{ACFM for SOR} &= 2{,}744 \\
\text{ACFM for EOR} &= 3{,}391 \\
\text{Compression ratio} &= \frac{515}{165} \\
&= 3.12 \\
\text{Inlet temp F} &= 100
\end{aligned}$$

Then

$$\begin{aligned}
\text{Discharge temp for SOR} &= 370\ \text{F} \\
\text{'' '' '' EOR} &= 340\ \text{F}
\end{aligned}$$

Step 8. Calculate approx driver HP.

$$H_{ad} = \frac{Z_{ave} \times R \times T_i}{(K-1)/K}\left[\left(\frac{P_2}{P_1}\right)^{(K-1)/K} - 1\right]$$

where

$$\begin{aligned}
H_{ad} &= \text{adiabatic head in ft lbs/lb} \\
Z &= \text{compressibility factor (ave)} \\
R &= \text{gas constant} = 1{,}545/\text{MW} \\
K &= \text{adiabatic exponent } C_P/C_V = 1.15 \\
T &= \text{temperature in } {}^\circ\text{R} = {}^\circ\text{F} + 460{}^\circ\text{F} \\
P_2 &= \text{discharge pressure psia} \\
P_1 &= \text{suction pressure psi}
\end{aligned}$$

Then for SOR:

$$\begin{aligned}
\text{Had for SOR conditions} &= 161{,}063 \\
\text{Had for EOR conditions} &= 91{,}546
\end{aligned}$$

Step 9. The gas HP is obtained using the expression

$$\text{Gas HP} = \frac{W \times H_{ad}}{\eta_{ad} \times 33{,}000}$$

where

$$\begin{aligned}
W &= \text{weight in lbs/min of gas} \\
H_{ad} &= \text{adiabatic head in ft lbs/lb} \\
\eta &= \text{adiabatic efficiency (0.7–0.75)}
\end{aligned}$$

Let η_{ad} be 0.73

$$\begin{aligned}
\text{For SOR Gas HP} &= 3{,}536 \\
\text{For EOR Gas HP} &= 4{,}420
\end{aligned}$$

Step 10. The driver HP is as follows:

	SOR	EOR
Gas HP	3,536	4,420
Leakage losses	35	44
Bearing losses	35	35
Seal losses	35	35
DRIVER HP	3,641	4,534

Calculating reciprocating compressor horsepower

Reciprocating compressors are used extensively in the process industry. They vary in size from small units used for gas recovery (such as those on a crude distillation overhead system) to fairly large complex machines used for recycle gas streams and for transporting natural gas. Engineers are frequently required therefore to assess the horsepower of these machines and their capability to handle various streams. This item describes a method used to determine horsepower and proceeds with the following steps:

Step 1. Obtain the capacity and the properties of the gas to be handled. Fix the ultimate (discharge) pressure level.
Step 2. From the machine data sheet ascertain the number of stages.
Step 3. Estimate the brake horsepower from the expression:

$$\text{BHP} = 22 \times (\text{compression ratio/stage}) \times \text{number of stages} \times \text{capacity (in cuft/day} \times 10^6) \times \text{F}$$

where

F = 1.0 for 1 stage
1.08 for 2 stages
1.10 for 3 stages

Ratio/Stage = $\sqrt{\ }$ ratio for two stages and $\sqrt[3]{\ }$ ratio for three stages.
Step 4. Check the estimate with Figure 18.28. The use of these graphs is self explanatory.
Step 5. Confirm actual suction conditions and compression ratio required (discharge pressure).
Step 6. Calculate compression ratio/stage.
Step 7. Calculate 1st stage discharge pressure. This will be suction pressure times compression ratio per stage from Step 6.
Step 8. Allow about 3% for inter-stage pressure drop then calculate second stage discharge pressure. Check that overall compression ratio/stage is close to that calculated for Step 6.

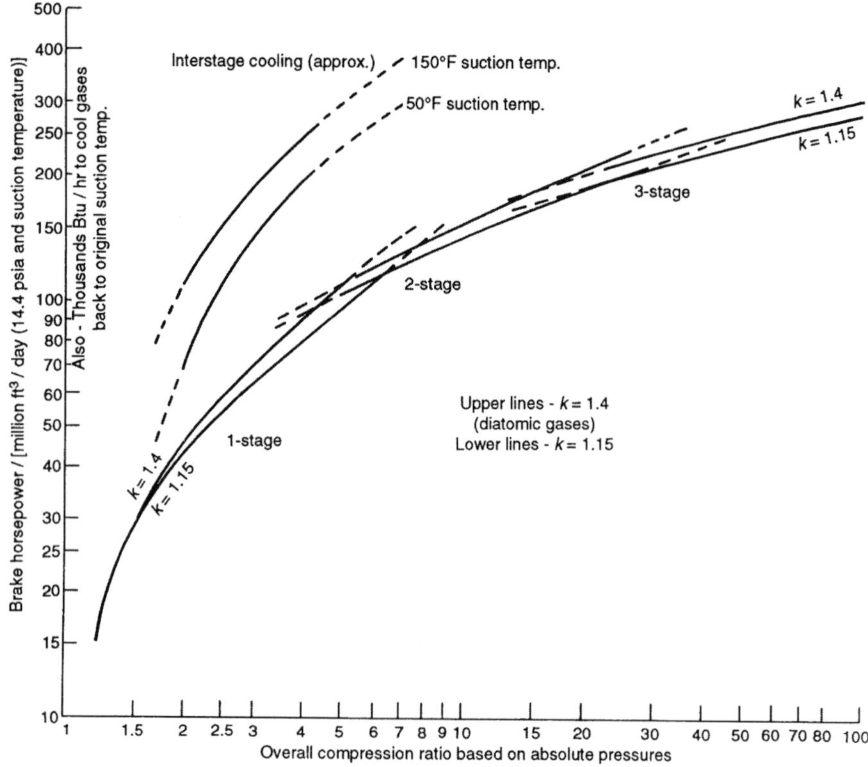

Figure 18.28. An estimate of brake horsepower per mm cfd for reciprocating compressors.

Step 9. Calculate the 'K' value of the gas. 'K' value is C_p/C_v of the gas. If the gas is a mixture of components, 'K' value may be calculated as the sum of each component multiplied be each of their 'K' values given in Table 18.A.1 in the Appendix. Alternatively for a good approximation data in Figure 18.29 may be used.

Step 10. Calculate discharge temperature from 1st stage using Figure 18.30. Assume some inter-cooling (or calculate inter-cooling from plant data) and fix 2nd stage discharge temperature using also using Figure 18.30.

Step 11. Calculate the compressibility factor Z at suction and discharge from the expression

$$\rho_v = \frac{MP}{T \times Z \times 10.73}$$

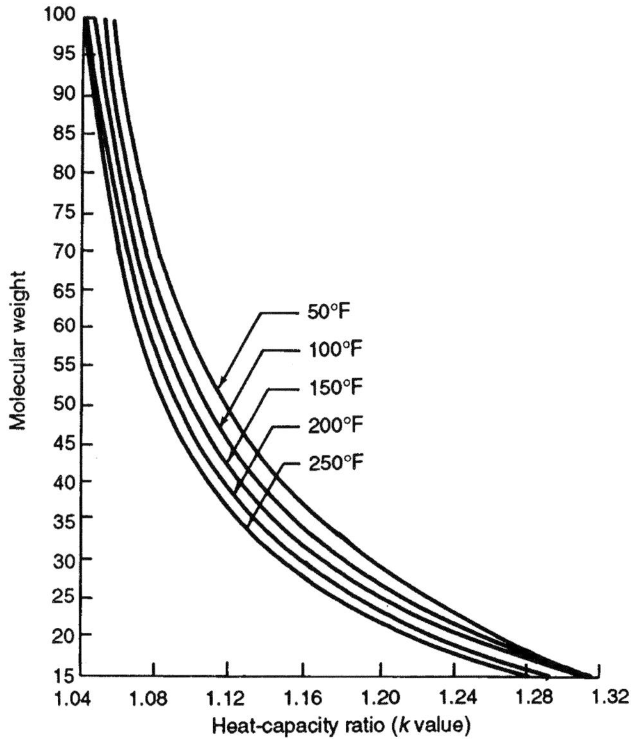

Figure 18.29. Approximation of 'K' from mole weights.

where

ρ_v = gas density in lbs/cuft at condition
T = °Rankine (°F + 460°F)
Z = compressibility factor
M = mole weight
P = pressure in psia

Use average value at suction and discharge for each stage.

Step 12. Read off BHP/mm cfd at the compression ratio/stage (from Step 7) and '*K*' from Step 9 for each stage.

Step 13. Calculate BHP per stage from the expression:

$$\text{Bhp} = (\text{bhp/mmcfd}) \times \frac{\text{P}_\text{L}}{14.4} \times \frac{T_\text{S}}{T_\text{L}} \times Z_{\text{ave}} \times \text{mmScf/D}$$

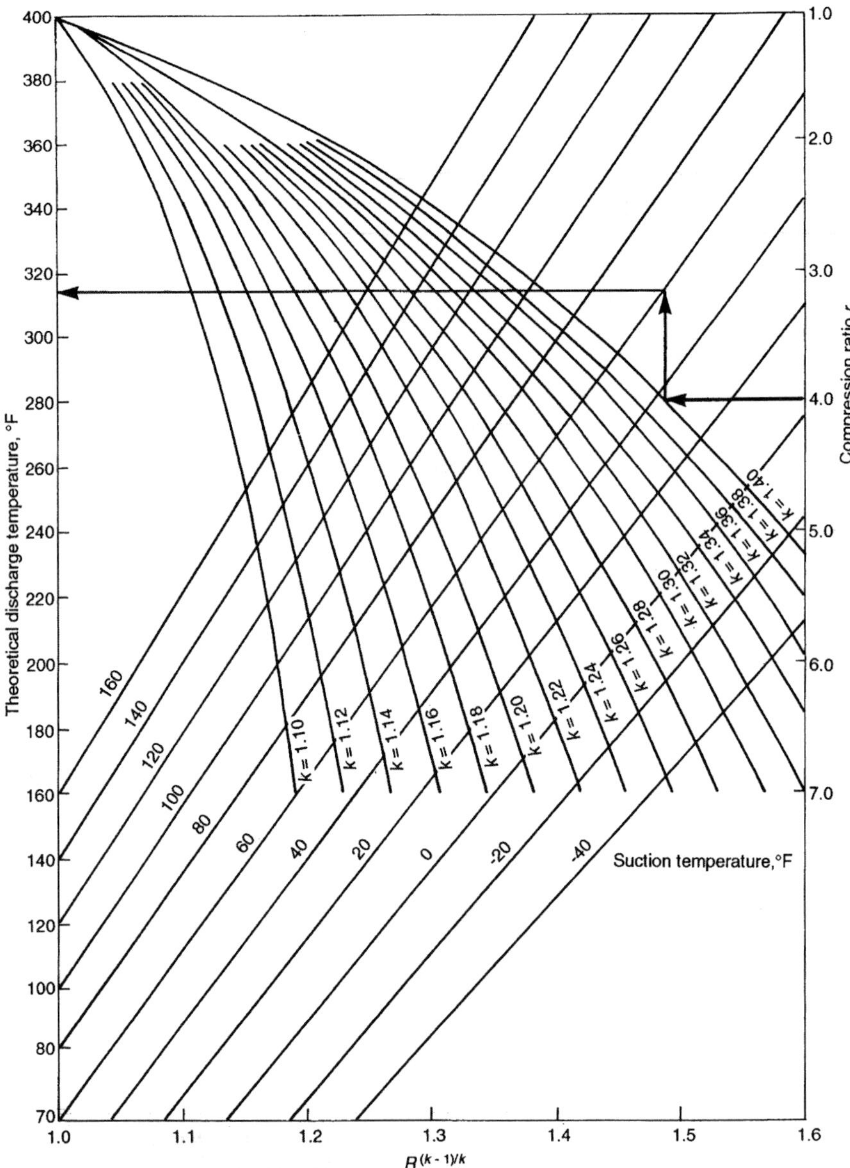

Figure 18.30. Determination of discharge temperature for reciprocating compressors.

where

Bhp/mmcfd = From Figure 18.29
P_L = pressure base used in contract psia
T_S = intake temperature °R
T_L = temperature base used in contract °R (usually 520°R)

Step 14. Brake horsepower for the machine is the sum of the BHP calculated for each stage in Step 13 above.

Reciprocating compressor controls and inter-cooling

A reciprocating compressor is a constant displacement type compressor. It compresses the same volume of gas to the same pressure level without regard to whether the gas is hydrogen or butane. This characteristic makes them desirable for use in services where the gas will have a widely varying composition. In some cases when an extremely low density gas will be compressed, a reciprocating compressor may be more economical than a centrifugal compressor, even though the flow rate may be very high, due to the large number of stages required for the centrifugal.

Reciprocating compressors are widely used in process services where the flow rates are too small for centrifugal compressors. These units can be obtained with integral or coupled electric motor in sizes from a few HP to 12,000 HP and separate or integral gas drivers varying in sizes from 100 HP to 5,500 HP.

A range of air-cooled light duty compressors is available for intermittent service. They range in size from 1/4 to about 100 HP at pressures up to 300 psig and are usually single acting. A primary process use of such equipment is for starting air compressors on gas engine driven machines. Reciprocating compressors can be designed to handle intermittent loads efficiently. This is done by using cylinder unloaders such as clearance pockets or suction valve lifters. Power losses are low at part load operation with these devices.

Reciprocating parts and pulsating flow present several engineering problems. The foundation and piping system must be constructed to withstand the vibrations produced by the compressor. The pulsating flow produced by the compressor must be dampened by the use of properly engineered suction and discharge bottles. These problems do not arise with the use of other types of compressors.

Reciprocating compressor control

Control of the compressor to prevent driver overload can be accomplished with clearance pockets, suction valve lifters, a throttling valves in the suction line, or a control valve in a bypass around the compressor. A hand operated bypass without cooler is usually furnished inside the block valves for start-up purposes.

1. Clearance pockets

Of the above types of regulation, control by opening fixed clearance pockets gives the smoothest and most efficient control within its range of application. It has the following advantages:

- Minimizes the intake pulsation as the gas flow is not reversed in the intake lines to the cylinder.
- Results in lower bearing loads as all inertia loads are cushioned.
- Results in very efficient part load operation. When the gas compressed into the pockets is expanded, it follows the adiabatic line of compression and results in little power loss.

Clearance control has the following disadvantages which sometimes completely eliminates it from consideration:

- When low ratios of compression are combined with high suction pressure, clearance pockets of sufficient size to unload the compressor cannot physically be installed in the machine.
- Clearance control is designed for one set of pressure conditions and any variation in either suction or discharge pressure affects the amount of unloading accomplished by a given pocket.
- Condensable corrosive gases sometimes cause corrosion and liquid slugging problems.

2. Suction valve lifters

Suction valve lifters are the other type of internal unloading devices for compressor cylinders and have characteristics that make them applicable when clearance control is not. Suction valve lifters completely unload their end of the cylinder whenever they are opened, regardless of the pressure. They do result in increased bearing loads due to unbalanced inertia forces. Suction line pulsation may increase because the single acting cylinder may excite a different frequency in the gas.

3. Suction throttle valve

A throttle valve in the suction line should be considered only for small reciprocating compressors. For large size machines, the suction valve cannot give tight enough shut off to permit unloading the compressor for starting.

4. Bypass control

External bypass control around the compressor is applicable to all sizes of compressors. It results in a loss of power since the full compressor capacity must be compressed to and delivered at the full discharge pressure before being bled back to suction pressure. Care must be taken with this type of control to ensure that the bypassed gas is cooled sufficiently to prevent increasing the discharge temperature. This type of control is preferred for installations up to several hundred horsepower because of its smoothness and lack of complexity. Individual machines can be shutdown for large process variations.

5. Variable speed reciprocating compressors
With a variable speed driver, cylinder control can usually be eliminated and speed control used to obtain desired process conditions. However, start-up unloading must be furnished, usually consisting of a hand operated bypass within the machine, or cylinder block valves. On turbine driven reciprocating compressors economics usually dictate that the compressor be run at constant speed and that cylinder controls or system bypasses be used to obtain the required control.

Reciprocating compressor inter-cooling
Inter-cooling for multistage compressors is advisable whenever there is a large adiabatic temperature rise within the cylinder and the cylinder discharge temperature would exceed 350°F. When inter-cooling is employed, the inlet temperature to the higher stage should be as close to the cooling water temperatures as practical. On standard commercial air inter-coolers, approach temperatures of 15–20°F are commonly used. Cooling to first stage inlet temperature is usually economical on process gas compressors.

Inter-cooling is employed for two basic reasons:

1. For mechanical reasons whereby discharge temperature must be limited to 350°F for lubrication purposes.
2. An economic reason as inter-cooling will save from 3% to 5% of the required BHP.

In general, on process compressors handling low "*n*" value gases inter-cooling is not employed unless the temperature limitation is exceeded. On high "*n*" value di-atomic gas mixtures, such as air, inter-cooling is the rule above about a 4 compression ratio and ambient temperature at suction.

In general, cooling water for electric driven compressors can be any water available, including salt water. (If the compressor is tied into a plant having gas engine driven compressors, the electric driven machine should be tied into the closed system). The cooling water should be available at a minimum pressure of 25 psig.

For estimating purposes, cooling water temperature rise across the cylinders can be taken as 15°F and that across the inter and after coolers can be taken as 15°F. For estimating purposes, cooling water requirements are as follows:

Jacket water cooling	500 Btu/BHP/hr
Inter cooler	1,000 Btu/BHP/hr
After cooler	1,000 Btu/BHP/hr

Use motor rating for number of horsepower required.

If the discharge temperature of the gas does not exceed 180°F, it is common practice to eliminate cooling water on the cylinder and operate with cooling passages which are filled with oil. Any jacketed cylinder must be filled with some fluid to ensure even temperature distribution.

Specifying a reciprocating compressor

As in the case of the centrifugal compressor all data necessary to give a precise requirement for the duty and performance required of a reciprocating compressor must be given in the specification sheet. Much of this data is the same as that given in a specification for a centrifugal compressor (discussed earlier in this chapter). For completeness however all the items in a reciprocating compressor are included below:

Title block

This requires the item to be identified by item number and its title. The number of units that the specification refers to is also given here. For a centrifugal compressor this will normally be just one as very seldom is a spare machine required. This may not be so in the case of a reciprocating compressor.

Normal and rated columns

More often than not the conditions and quantities required to be handled will vary during the operation of the machine. The two columns therefore will be completed showing the average normal data in the first column and the most severe conditions and duty required by the compressor in the second. The severe conditions in column two are for a continuous length of operation not instantaneous peaks (or troughs) that may be encountered.

Gas

The composition and gas stream identification must be included as part of the process specification. Usually the composition of the gas is listed on a separate sheet as shown in the example. Note in many catalytic processes that utilize a recycle gas the composition of the gas will change as the catalyst in the process ages. Thus it will be necessary to list the gas composition at the SOR and at the EOR.

The compressor may also be required to handle an entirely different gas stream at some time or other. This too must be noted. For example in many petroleum refining processes a recycle or make up compressor normally handling a light predominantly hydrogen gas is also used for handling air or nitrogen during catalyst regeneration, purging, and start-up.

Volume flow

This is the quantity of gas to be handled stated at 14.7 psia and 60 F.

Weight flow

This is the weight of gas to be handled in either lbs/min or lbs/hr.

Inlet conditions

In the case of multistage compressors the conditions for each stage must be shown. Where inter stage cooling is used the effect must be reflected in the conditions specified.

Pressure: This is the pressure of the gas at the inlet of the compressor stage in psia. Note if inter cooling is used this pressure must include the inter cooler pressure drop.

Temperature: This is the temperature of the gas at the inlet of the compressor stage—after the inter cooler if applicable.

Mole weight: The mole weight of the gas is calculated from the gas composition given as part of the specification.

C_p/C_v: This is the ratio of specific heats of the gas again obtained from the gas mole wt and Table 18.A.1 in the Appendix.

Compressibility factor (z): use the value at inlet conditions calculated as shown in step 7 of item on horsepower calculation for reciprocating compressors.

Inlet volume: This is the actual volume of gas at the conditions of temperature and pressure existing at the compressor stage inlet. Thus:

$$\text{ACFM} = \frac{\text{SCFM} \times 14.7 \times (\text{inlet temp F} + 460)}{(60\text{ F} + 460) \times \text{inlet press psia}}$$

Discharge conditions

Pressure: This is the pressure at each stage outlet and is quoted in either psia or psig.
Temperature: This is estimated for each stage using Figure 18.30.
C_p/C_v: This will be the same as in the inlet.

Approximate driver horsepower

The brake horsepower for the reciprocating compressor is calculated using the method described earlier. This is *Brake Horsepower* and includes an allowance for mechanical inefficiencies. The approximate minimum driver horsepower is 1.1 × Brake

Horsepower, but the approximate driver HP will be calculated using the inefficiencies for leakage, seals, etc as for centrifugal compressors.

The reminder of the spec sheet contains all the essential data and requirements that may affect the duty and performance of the compressor. Much of this is self explanatory; however there are some items that require comment. These are:

1. Most compressor installations today are under an open sided shelter with a small overhead gantry crane assembly for maintenance.
2. Usually the lube and seal oil assemblies have their own pump and control systems. Consequently even if the compressor itself is to be steam driven there may still be need to give details of utilities for the ancillary equipment.
3. Details of the gas composition is essential for any development of the compressor. This is listed on the last page of the specification together with any notes of importance concerning the machine and its operation.

An example calculation for a specification sheet follows:

Example calculation

A hydrotreater make up compressor is required to handle a gas stream such as to provide the unit with 260 SCF per barrel of feed of pure hydrogen. The composition of the gas varies as follows:

Mole %	Start of Run	End of Run
H_2	74.9	65.80
C_1	14.17	19.31
C_2	5.85	7.97
C_3	2.43	3.31
iC_4	1.13	1.54
nC_4	1.00	1.36
C_5S	0.52	0.71
Total	100.00	100.00

The fresh feed throughput is fixed at 30,000 BPSD (barrels per stream day). It is proposed to use 3 × 60% machines of which one will be standby and turbine driven. The inlet pressure of the gas is 50 psig at a temperature of 80 F. The gas is to be delivered at a pressure of 600 psig and 100 F. Prepare a process specification for reciprocating compressors to meet these requirements.

1.0 Calculating volume flows.

$$\text{SOR conditions Total flow required} = \frac{260}{0.749}$$

$$= 347 \text{ Scf of GAS per Bbl of feed.}$$

$$= \frac{347 \times 30{,}000}{24 \times 60} = 7{,}229 \text{ Scf/Min}$$

$$\text{EOR conditions Total flow required} = \frac{260}{0.658}$$
$$= 395 \text{ Scf/Bbl}$$
$$= \frac{395 \times 30{,}000}{24 \times 60} = 8{,}229 \text{ Scf/Min}$$

Volume flow per machine:

$$\text{SOR} = 7{,}229 \times 0.6 = 4{,}337$$
$$\text{EOR} = 8{,}229 \times 0.6 = 4{,}937$$

2.0 Calculate mole wt of gas.

	MW	SOR		EOR	
		mole%	wt Factor	mole%	wt Factor
H_2	2	74.9	149.8	65.8	131.6
C_1	16	14.17	226.7	19.31	309.0
C_2	30	5.85	175.5	7.97	239.1
C_3	44	2.43	106.9	3.31	145.6
iC_4	58	1.13	65.5	1.54	89.3
nC_4	58	1.00	58.0	1.36	78.9
C_5s	72	0.52	37.4	0.71	51.1
Total		100.0	644.3	100.00	1,044.6

SOR gas mole wt = 6.44 EOR gas mole wt = 10.44

3.0 Weight of gas lbs/min per machine

One mole of any gas occupies 378 cf at 60 F and 14.7 psia. Then

$$\text{For SOR Conditions moles/min of gas per machine} = \frac{4{,}337}{378} = 11.5$$

and

$$\text{lbs/min} = 11.5 \times 6.44$$
$$= 74.06 \text{ lbs/min}$$

$$\text{For EOR Conditions moles/min of gas per machine} = \frac{4{,}937}{378} = 13.06$$

and

$$\text{lbs/min} = 13.06 \times 10.44$$
$$= 136.30 \text{ lbs/min}$$

4.0 Inlet conditions

Inlet pressure = 50 psig = 65 psia.

Required outlet pressure = 600 psig = 615 psia.

$$\text{Overall compression ratio} = \frac{615}{65} = 9.46$$

This will be a two stage compressor. *Note:* At this level of compression in reciprocating compressors the compression ratio should not exceed 4:1 for any stage.

$$\text{Compression ratio per stage} = \sqrt{9.46} = 3.07.$$
$$\text{Discharge pressure stage 1} = 65 \times 3.07 = 199.6 \text{ psia}$$

Allowing 2 psi for the pressure drop across the inter cooler the suction pressure of stage 2 is 197.6 call it 198 psia.

Check the compression ratio of stage 2:

$$\text{Required discharge pressure} = 615 \text{ psia}$$
$$\text{compression ratio} = \frac{615}{198} = 3.1$$

which is close to the originally predicted of 3.07.

5.0 Calculate ratio C_p/C_v.

	C_p/C_v	SOR		EOR	
		mole%	Factor	mole%	Factor
H_2	1.4	74.9	1.049	65.8	0.921
C_1	1.3	14.17	0.184	19.31	0.251
C_2	1.22	5.85	0.071	7.97	0.097
C_3	1.14	2.43	0.028	3.31	0.038
iC_4	1.11	1.13	0.013	1.54	0.017
nC_4	1.11	1.00	0.011	1.36	0.015
C_5's	1.09	0.52	0.006	0.71	0.008
Total		100.0	1.362	100.00	1.347

C_p/C_v SOR gas = 1.362
C_p/C_v EOR gas = 1.347

6.0 Calculate inlet ACFM per stage

SOR.

Inlet volume for 1st stage:

$$\text{ACFM} = \frac{\text{Scf/min} \times 14.7 \times \text{ inlet temp R}}{(60 + 460) \times \text{ Inlet press psia}}$$
$$= \frac{4{,}337 \times 14.7 \times 540}{520 \times 65}$$
$$= 1{,}019 \text{ cf/min}$$

Inlet volume for 2nd stage: (Inter cooled to 100 F)

$$\text{ACFM} = \frac{4{,}337 \times 14.7 \times 560}{520 \times 198}$$
$$= 347 \text{ cf/min}$$

EOR.

Inlet volume for 1st stage:

$$\text{ACFM} = \frac{4{,}937 \times 14.7 \times 540}{520 \times 65}$$
$$= 1{,}159 \text{ cf/min}$$

Inlet volume for 2nd stage:

$$\text{ACFM} = \frac{4{,}937 \times 14.7 \times 560}{520 \times 198}$$
$$= 488.5 \text{ cf/min}$$

7.0 Calculate inlet compressibility factor (Z).

$$Z = \frac{\text{MW} \times P_i}{T_i \times \rho \times 10.73}$$

where

$$\rho = \frac{\text{wt/min}}{\text{ACFM}}$$

SOR conditions

$$\text{1st stage } Z = \frac{6.44 \times 65}{540 \times 0.0727 \times 10.73}$$
$$= 0.994$$

$$\text{2nd stage } Z = \frac{6.44 \times 198}{560 \times 0.213 \times 10.73}$$
$$= 0.991$$

EOR conditions

$$\text{1st stage } Z = \frac{10.44 \times 65}{540 \times 0.1176 \times 10.73}$$
$$= 0.996$$

$$\text{2nd stage } Z = \frac{10.44 \times 198}{560 \times 0.345 \times 10.73}$$
$$= 0.997$$

8.0 Determine discharge temperature

From Figure 18.30.

For SOR Conditions: 1st stage Comp Ratio = 3.07
$C_p/C_v = 1.362$
Suction Temp = 80 F
Discharge Temp read as 270 F
2nd stage Comp Ratio = 3.1
$C_p/C_v = 1.362$
Suction Temp = 100 F
Discharge Temp read as 299 F

For EOR Conditions: 1st stage Comp Ratio = 3.07
$C_p/C_v = 1.347$
Suction Temp = 80 F
Discharge Temp read as 262 F
2nd stage Comp Ratio = 3.1
$C_p/C_v = 1.347$
Suction Temp = 100 F
Discharge Temp read as 292 F

9.0 Compressibility factors for discharge conditions

SOR Conditions:

1st stage ACFM on discharge (before inter cooler).

$$= \frac{4{,}337 \times 14.7 \times 730}{520 \times 200\,(\text{neglect IC pressure drop})}$$

$$= 447.5 \text{ cf/min}$$

$$\rho = \frac{74.1}{447.5} = 0.166$$

$$Z = \frac{6.44 \times 200}{730 \times 0.166 \times 10.73}$$

$$= 0.991$$

2nd stage ACFM on discharge.

$$= \frac{4{,}337 \times 14.7 \times 759}{520 \times 615}$$

$$= 151.3 \text{ cf/min}$$

$$\rho = \frac{74.1}{151.3} = 0.490$$

$$Z = \frac{6.44 \times 615}{759 \times 0.49 \times 10.73}$$
$$= 0.992$$

EOR conditions.

These are calculated in the same way as those above and give the following results:

1st stage $Z = 0.995$
2nd stage $Z = 0.995$

10.0 Approximate driver horsepower.

Use the expression:

BHP = 22 × (Comp Ratio/stage) × No of stages × Capacity × Factor F
Comp ratio/stage = 3.07
No of stages = 2
Capacity per machine in mm cf/day at suction temperature
Factor for 2 stage machine = 1.08

For SOR conditions:

$$\text{BHP} = 22 \times 3.07 \times 2 \times (0.6 \times 10.81) \times 1.08$$
$$= 946$$

For EOR conditions:

$$\text{BHP} = 22 \times 3.07 \times 2 \times 7.38 \times 1.08$$
$$= 1{,}077$$

Use the efficiency factors as given in Figure 18.27 for centrifugal compressors. In this case there will be a gear assembly between compressor and driver. Use the efficiency of this as 97%. Thus:

	SOR	EOR
BHP	946	1,077
Gear losses (3% of BHP)	29	33
leakage	10	10
seal	35	35
Bearings	35	35
Driver HP	1,055	1,190

ITEM No. C101 A&B TITLE Hydrotreater Recycle Gas Compressor

Number of units required 3 (2 + 1 spare)

	Normal				Rated			
GAS (see attached composition)	75% H_2				65.8% H_2			
VOL. flow scf/min	4337				4937			
WEIGHT lb/min	74.1				136.3			
INLET CONDITIONS (each stage)	stage 1	stage 2	stage 3	stage 4	stage 1	stage 2	stage 3	stage 4
Pressure psia	65	198*			65	198*		
Temperature °F	80	100*			80	100*		
Mol weight	6.44				10.44			
C_p/C_v	1.362				1.347			
Compressibility factor	0.994	0.996			0.996	0.997		
Inlet vol acf/min	1019	347			1159	394		
DISCHARGE CONDITIONS (each stage)								
Pressure psia	198	615			198	615		
Temperature °F	270	299			262	292		
C_p/C_v	1.362				1.347			
Compressibility factor	0.991	0.992			0.995	0.995		
APPROX. DRVIER HORSEPOWER	1055				1190			

*After intercooling

COMPRESSOR SERVICE REQUIRED:

Length of uninterrupted service: 8000 hours

Type of compressor: Lubricated Yes

Non Lubricated

Discharge RV setting (each stage) 250 and 750 psig

Figure 18.31. An example of a process specification for a reciprocating compressor.

Compressor drivers, utilities, and ancillary equipment

This item covers details on various compressor drivers, the utilities associated with operating the compressors and their ancillary equipment.

ITEM No. C 101 A&B TITLE Hydrotreater Recycle Gas Compressor

General Data and Requirements

Enclosure:

Open to weather ______

Under shelter Yes

In building ______

Corrosiveness and remarks concerning gas ______

None

Type of driver: Motor: 2 normal operating

Steam turbine:

Condensing ______

Non-condensing Spare machine

Utilities:

Power:

Voltage ______

Cycle ______

Phase ______

Steam:

Inlet:	Pressure (psig)	600
	Temperature (°F)	710
Condensing exhaust		
	Pressure (psia)	N/A
	Temperature (°F)	N/A
Non-condensing exhaust		
	Pressure (psig)	50
	Temperature (°F)	(by vendor)

Cooling water: Pressure 60 psig: Temperature 40 °F

Allowable temperature rise 30 °F

Materials of construction:

Cylinder	CS	Type	By vendor
Piston	Ni.Cr	Type	By vendor

Type of shaft seal: Labyrinth

Figure 18.31. (Cont.)

Compressor drivers

Table 18.24 gives a listing of the more common types of compressor drivers. It provides some of the data that would influence the choice of the driver. The most common drivers by far in a process plant are the electric motor and the steam turbine.

ITEM No. C 101 A&B TITLE Hydrotreater Recycle Gas Compressor

General Data and Requirements (cont.)

Gas Composition

Mol %	Start of run	End of run
H_2	74.9	65.80
C_1	14.17	19.31
C_2	5.85	7.97
C_3	2.43	3.31
iC_4	1.13	1.54
nC_4	1.00	1.36
c_5s	0.52	0.71
Total	100.00	100.00

Remarks and notes:

1. Vendor to provide:

   Intercoolers: Yes

   Aftercooler: Yes

   Dampeners: Yes

   Flushing and sealing oil systems: Yes

2. If motor driven Re acceleration required class A

3. Suction line: Size 6" RTG 300# RF

   Discharge line: Size 4" RTG 600# RF

Figure 18.31. (Cont.)

For very large machines as encountered in handling natural gas the gas turbine or gas engine become the more prominent prime mover.

Sizing drivers

As a basic rule drivers are sized for the most severe duty required of the compressor plus a factor as an operating contingency. In general the most severe duty is that design case which has the highest suction temperature, the maximum ratio of specific heats, the lowest suction pressure, the highest required discharge pressure, and the gas molecular weight which gives the highest HP. The driver rated horsepower shall

Table 18.24. Comparison of compressor drivers

Driver	HP range	Available speed, RPM	Efficiency %	Common applications
Synchronous motor	100–20,000	3,600	90–97	Reciprocating compressors
Induction motor	1–15,000	3,600	86–94	All types of compressors
Wound rotor induction motor	–	–	–	Normally not used
Steam engine	10–4,000	400–140	60–80	All types of rotary equip.
Steam turbine	10 to 2,000	2,000 to 15,000	50 to 76	Centrif, axial and recip.
Combustion gas turbine	3,000 to 35,000	10,000 to 3,600	19 to 24 see note 1 below	All types of compressors (except recip)
Gas and oil engines	100 to 5,000	1,000 to 300	35 to 45	Reciprocating compressors

Note 1: The efficiency given here does not include for waste heat recovery. With WHR the efficiency can be increased to between 28 and 35%.

therefore be greater than:

$$\text{Driver Brake HP} = \frac{\text{Max compressor BHP @ the most severe duty}}{\text{Mechanical efficiency of the power transmission}}.$$

The mechanical efficiency in this case includes for energy losses for bearings, seals, lube oil etc, in the case of centrifugal compressors and gears in the case of reciprocating compressors.

Electric motor drivers

Squirrel cage motors are preferred for this type of duty. These may be drip proof open type where the location is not a fire or explosion hazard. Where it is required that the units must be explosion or fire proof these motors must be totally enclosed type. In sizing the motor, efficiencies for squirrel cage motors up to 450 HP given in item 4.6 for pumps may be used. The following Table 18.25 is used for motors above 500 HP.

The driver rated brake horsepower is the compressor horsepower times a load factor divided by a service factor. Normally the load factor is 10% and a service factor for an enclosed squirrel cage motor is 1.0 and 1.15 for an open type.

Example calculation

Calculate the Operating Load and the Connected Load for the driver of a 4,000 HP centrifugal compressor (includes Leakage, Seal, and Bearing losses). A gear is used and this has a 97% efficiency. The load factor is 10% and the motor is open type squirrel cage with a service factor of 1.15. There will be a normal operating unit and a spare, both motor driven.

Table 18.25. Motor efficiencies

Motor Rated HP	Motor efficiencies full load @ percent of		
	50	75	100
500	91.4	93.1	93.4
1,000	92.1	93.8	94.1
	92.4	94.1	94.4
2,000	92.7	94.4	94.7
2,500	92.9	94.6	94.9
3,000	93.0	94.7	95.0
3,500	93.0	94.7	95.0
4,000	93.1	94.8	95.1
4,500	93.1	94.8	95.1
5,000	93.2	94.9	95.2

$$\text{Minimum required driver HP} = \frac{4{,}000 \times 1.1}{0.97}$$
$$= 4{,}536$$
$$\text{Driver nameplate rating} = \frac{4{,}536}{1.15} = 3,944$$

call it 4,000 HP.

Connected load for the motor is:

$$\begin{aligned}\text{Motor nameplate rating} \times 1.15 &= 4{,}000 \times 1.15\\ &= 4{,}600 \text{ HP (rated HP)}\\ &= \frac{4{,}600 \times .746}{0.951 \text{ (@ 100\% load)}}\\ &= 3{,}608 \text{ kW}\end{aligned}$$

$$\begin{aligned}\text{There are two units then Total connected load} &= 3{,}608 \times 2\\ &= 7{,}216 \text{ kW}\end{aligned}$$

Operating load for the motor is:

$$\frac{4{,}000}{0.97} = 4{,}124 \text{ HP}$$
$$\% \text{ load} = \frac{4{,}124}{4{,}600} = 90\% \text{ (use 75\% eff)}$$
$$\text{Operating load} = \frac{4{,}124 \times 0.746}{0.948}$$
$$= 3{,}245 \text{ kW.}$$

Table 18.26. Steam Turbine Efficiencies

	Adiabatic efficiencies %	
	Inlet pressures (psig.)	
Driver BHP	900	100
500	48	59
800	53	64
1,000	56	67
1,200	58	68
1,500	60	71
2,000	63	73
2,500	65	74
3,000 and up	67	76

Steam turbine drivers. Next to the motor drivers steam turbines are the most common form of drivers for rotary equipment in general and compressors in particular. The two most common types of theses are turbines that exhaust to a lower pressure but the exhaust steam is not condensed and those in which the exhaust steam is condensed. Normally the latter is only used in the case of large driver horsepower 5,000 and above. It is far more expensive than the non condensing type as the exhaust is normally sub atmospheric in pressure and of course the cost of the condenser must be included.

Steam turbine approximate efficiencies are listed in Table 18.26.

These efficiencies are based on the exhaust pressure of the steam being 50 psig for non condensing type and 2″ Hg Abs for the condensing type.

Determining the rated horsepower of the steam turbine driver follows closely to the method for motor horsepower. First determine the minimum horsepower required of the turbine. Thus:

Step 1. Determine the *minimum* driver horsepower by multiplying the compressor BHP by 1.1.

Step 2. Now the turbine will deliver the normal HP at the normal speed. A contingency in the form of additional speed is added to the driver capability. This will be controlled in practice by a steam governor. This contingency is usually 5% above normal speed.

Step 3. Horsepower capability varies as the cube of the speed. Thus the Rated horsepower of the turbine will be:

$$\text{Rated HP} = \text{Minimum HP} \times (1.05)^3$$

Step 4. The amount of steam that will be used calculated by the change in enthalpy of the inlet steam to the outlet steam at constant entropy. The change in enthalpy for the two conditions is read from the steam Mollier diagram.

Step 5. The theoretical steam rate is:

$$\frac{2{,}544}{\text{Inlet enthalpy} - \text{Outlet enthalpy (in Btu/lb)}}$$

This figure divided by the turbine efficiency gives the steam rate in lbs/BHP/hr.

Example calculation

Calculate the turbine horsepower requirements and the theoretical steam rates to drive a 4,000 BHP centrifugal compressor. No gears are included in this case. Steam is available at 650 psig and 760 F. The steam will exhaust into the plant's 125 psig header.

$$\begin{aligned}\text{Minimum driver horsepower} &= 4{,}000 \times 1.1\\ &= 4{,}400 \text{ BHP}\\ \text{Rated turbine HP @ 105\% speed} &= 4{,}400 \times (1.05)^3\\ &= 5{,}094 \text{ HP.}\end{aligned}$$

$$\begin{aligned}\text{Enthalpy of steam @ 650 psig and 760 F} &= 1{,}390 \text{ (entropy 1.62)}\\ \text{Enthalpy of steam @ 125 psig} &= 1{,}225 \text{ (entropy 1.62)}\end{aligned}$$

Difference in Enthalpy = 165

Efficiency of turbine (from Table 18.26) = 67%.

$$\begin{aligned}\text{Theoretical Steam Rate} &= \frac{2{,}544}{165 \times 0.67}\\ &= 23 \text{ lbs/BHP/hr.}\end{aligned}$$

Gas turbine drivers. These items of equipment are the most expensive and because they require a high capital investment their use can only be justified as compressor drivers where the continual load on the compressor is also very high. These drivers therefore are met mostly in the natural gas industry. They are used extensively in recompressing natural gas after treating for dew point control or desulfurizing.

The thermal efficiencies of gas turbines are low (about 16–20%) but it is common practice to use the exhaust gases which are usually at a temperature of above 800 F in waste heat recovery. This involves exchanging the waste heat of the exhaust gases

Table 18.27. Gas turbine sizes and data

HP rating @ 80, and 1,000 ft	Fuel consumption LHV lbs/Hp-hr	Exhaust Flow #/sec	Exhaust Temp F	RPM
430	1.25	10.3	950	19,250
1,000	0.66	11.1	960	19,500
1,080	0.63	13.7	860	22,300
1,615	0.84	23.6	1,000	13,000
2,500	0.76	43.0	795	9,000
3,800	0.75	53.0	900	8,500
5,500	0.76	77.3	945	5,800
7,000	0.8	101.0	938	5,500
8,000	0.65	102.0	935	5,800
9,000	0.70	130.0	850	5,000
10,000	0.59	123.5	805	6,000
12,000	0.65	160.0	720	4,750
13,500	0.62	187.0	800	4,860
15,000	0.61	188.0	835	4,860
24,000	0.61	258.0	850	3,600

with boiler feed water to generate steam or to preheat a process stream, for example distillation. Table 18.27 gives some turbine sizes and data. It should be noted that considerable development work is continuing in the field of gas turbines and consequently the data given here may be subject to revision or updating.

To obtain the gas turbine rated horsepower for a specific compressor duty follows closely the same calculation route as the steam turbine. Thus:

Step 1. Obtain the *minimum driver horsepower* by multiplying the compressor Brake horsepower by 1.05 (BHP includes Seals, Leakage, etc).

Step 2. The rated turbine horsepower is the minimum driver HP divided by the gear efficiency.

Step 3. The horsepower of the turbine selected must equal or slightly exceed the horsepower calculated in Step 2. This HP must be corrected for site conditions as shown in Step 4.

Step 4. The HP's given in Table 18.27 are at an ambient temperature of 80 F and at an elevation of 1,000 ft. correction for any specific site is given by the following expression:

$$\text{SITE HP} = \text{Quoted HP}\,(1.00 + \text{A} \times 10)^{-2}(1.00 - \text{B} \times 10)^{-2}$$

$$\times\, (1.00 - \text{C} \times 10)^{-2} \times \frac{\text{Site Atmos Press}}{14.7}$$

where

A = Temp adjustment of % per F
B = Inlet press loss % per ins water gauge.
C = Discharge press loss % per ins water gauge.

'A' will be plus for ambient temperatures above 80°F and minus for ambient temperatures below 80°F.

Ancillary equipment

Reciprocating compressor dampening facilities. Dampening facilities are used in conjunction with reciprocating compressors to smooth out the pulsation effect of the compressor action. These facilities are simply in line bottles sized larger than the gas line which cushion the gas motion. These are essential to minimize expensive piping designs that would be necessary without them. Calculating the size of these bottles is important in the design of the compressor facilities.

The following calculation technique is used to determine the size of new dampers or to evaluate the adequacy of an existing facility. This calculation is described by the following steps:

Step 1. From compressor data sheet obtain cylinder diameter and stroke dimensions.
Step 2. Calculate the swept volume per cylinder using the expression:

$$\frac{\pi D^2}{4} \times S$$

where

D = cylinder diameter
S = stroke

Step 3. Knowing the suction and discharge pressures the pulsation bottle capacity (both suction and discharge) is obtained from Figure 18.32 in terms of a multiple of swept volume.

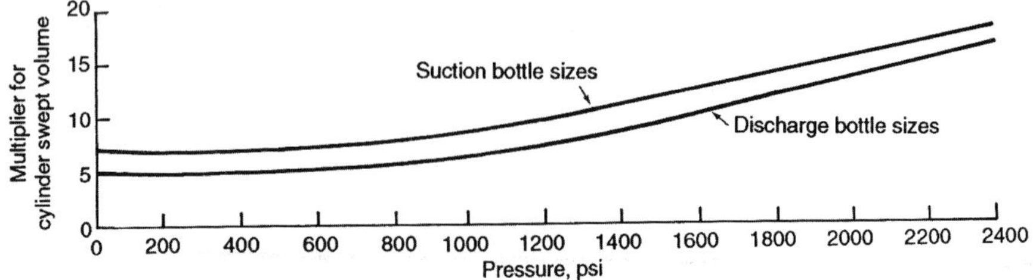

Figure 18.32. Dampener bottle sizing.

Step 4. Use the rule of thumb that pulsation bottle diameter equals 1½ times the compressor cylinder diameter. Calculate the suction and discharge bottle length.

Example calculation

To determine the dimensions of the compressor pulsation bottle of a reciprocating compressor having a 6″ diameter cylinder and a stroke of 15″. The compressor delivers 3.0 mm Scf/D gas at a suction pressure of 100 psia and 100°F, and a discharge pressure of 1,200 psia.

The cylinder diameter is 6″ and stroke is 15″

$$
\begin{aligned}
\text{then swept volume} &= \pi/4 \times 6^2 \times 15 \\
&= 424 \text{ cu.ins} \\
\text{capacity of machine} &= 3 \text{ mmScf/D} \\
\text{in a mm CF/D} &= \frac{3 \times 14.7 \times 560}{520 \times 100} \\
&= 0.475 \text{ mm ACFD} \\
&= 330 \text{ ACF/min} \\
&= 570240 \text{ AC ins/min} \\
\text{machine speed} &= \frac{570240}{424} = 1{,}345 \text{ RPM}
\end{aligned}
$$

From Figure 18.32:

Suction bottle size should be 7 × swept volume (at 100 psia) discharge bottle size should be 7 × swept volume (at 1,200 psia) = 2,968 cu ins or 1.718 cu.ft

As a rule of thumb diameter of bottle should be 1½ × cylinder diameter = 0.75 ft (9″) length = 2,968/63.6 = 47 ins or 4 ft.

18.4 Heat exchangers

Type and selection of heat exchangers

Heat exchange is the science that deals with the rate of heat transfer between hot and cold bodies. There are three methods of heat transfer, they are:

- Conduction
- Convection
- Radiation

In a heat exchanger heat is transferred by conduction and convection with conduction usually being the limiting factor. The equipment used in heat exchanger service is designed specifically for the duty required of it. That is, heat exchange equipment cannot be purchased as a stock item for a service but has to be designed for that service.

The types of heat exchange equipment used in the process industry and their selection for use are as follows:

The shell and tube exchanger

This is the type of exchanger most commonly used in a process plant. It consists of a bundle of tubes encased in a shell. It is inexpensive and is easy to clean and maintain. There are several types of shell and tube exchangers and some of these have removable bundles for easier cleaning. The shell and tube exchanger has a wide variety of services that it is normally used for. These include vapor condensation (condensers), process liquid cooling (coolers), exchange of heat between two process streams (heat exchangers), and reboilers (boiling in fractionator service). Most of this chapter will be dedicated to the uses and design specification of the shell and tube exchanger.

The double pipe exchanger

A double pipe exchanger consists of a pipe within a pipe. One of the fluid streams flows through the inner pipe while the other flows through the annular space between the pipes. The exchanger can be dismantled very easily and therefore be easily cleaned. The double pipe exchanger is used for very small process units or where the fluids are extremely fouling. Either true con-current or counter current flows can be obtained but because the cost per square foot is relatively high it can only be justified for special applications. The following table gives the heat transfer area for various pipe lengths and diameters:

No of tubes	Shell size, ins	Tube size, ins	Surface area, sqft		
			10 ft	20 ft	30 ft
1	2	1	5.8	11.0	16.3
1	3	1.5	10.9	20.9	30.9
1	4	2	13.7	26.1	38.5

Extended surface or fin tubes

This type of exchanger is similar to the double pipe but the inner pipe is grooved or has longitudinal fins on its outside surface. Its most common use is in the service where one of the fluids has a high resistance to heat transfer and the other fluid has a low resistance to heat transfer. It can rarely be justified if the equivalent surface area of a shell and tube exchanger is greater than 200–300 sqft.

Finned air coolers

These are the more common type of air coolers used in the process industry. Air cooling for process streams gained prominence during the early 1950s. In a great many applications and geographic areas they had considerable economic advantage over the conventional water cooling. Indeed today it is uncommon to see process plants of any reasonable size without air coolers.

Air coolers consist of a fan and one or more heat transfer sections mounted on a frame. In most cases these sections consist of finned tubes through which the hot fluid passes. The fan located either above or below the tube section induces or forces air around the tubes of the section.

The selection of air coolers over shell and tube is one of cost. Usually air coolers find favor in condensing fractionator overheads to temperatures of about 90–100°F and process liquid product streams to storage temperatures. Air coolers are widely used in most areas of the world where ambient air temperatures are most times below 90°F. At atmospheric temperatures above 100°F humidifiers are incorporated into the cooler design and operation. The cost under these circumstances is greatly increased and their use is often not justified.

In very cold climates the air temperature around the tubes is controlled to avoid the skin temperature of the fluid being cooled falling below a freezing criteria or in the case of petroleum products its pour point. This control is achieved by louvers installed to recirculate the air flow or by varying the quantity of air flow by changing the fan pitch.

Box coolers

These are the simplest form of heat exchange. However they are generally less efficient, more costly and require a large area of the plant plot. They consist of a single coil or "worm" submerged in a bath of cold water. The fluid flows through the coil to be cooled by the water surrounding it. The box cooler found use in the older petroleum refineries for cooling heavy residuum to storage temperatures. Modern day practice is to use a tempered water system where the heavy oil is cooled on the shell side of a shell & tube exchanger against water at a controlled temperature flowing in the tube side. The water is recycled through an air cooler to control its temperature to a level which will not cause the skin temperature of the oil in the shell & tube exchanger to fall below its pour point.

Direct contact condensers

In this exchanger the process vapor to be condensed comes into direct contact with the cooling medium (usually water). This contact is made in a packed section of a small tower. The most common use for this type of condenser is in vacuum producing equipment. Here the vapor and motive steam for each ejector stage is condensed in a packed direct contact condenser. This type has a low pressure drop which is essential for the vacuum producing process.

General design considerations

Basic heat transfer equations

The following equations define the basic heat transfer relationships.

These equations are used to determine the overall surface area required for the transfer of heat from a hot source to a cold source.

The overall heat transfer equation
The usual heat transfer mechanism are conduction, natural convection, forced convection, condensation, and vaporization. When heat is transferred by these means the overall equation is as follows:

$$Q = UA(\Delta t_{\mathrm{m}})$$

where

Q = Heat transferred in Btu/hr
U = Overall heat transfer coefficient, Btu/hr/sqft/°F
A = Heat transfer surface area, sqft.
Δt_{m} = Corrected log mean temperature difference, °F

The overall heat transfer coefficient U
This coefficient is the summation of all the resistances to the flow of heat in the transfer mechanism. These resistances are the resistance to heat transfer contained in the fluids, the resistance caused by fouling, and the resistance to heat transfer of the tube wall. The resistance to the flow of heat from the liquid outside the tube wall is measured by the film coefficient of that fluid. The resistance of the flow of heat from the fluid inside the tube is similarly the film coefficient of the inside fluid. These film coefficients are products of dimensionless numbers which include:

- The Reynolds Number
- The Graetz Number
- The Grashof Number

- The Nusselt Number
- The Peclet Number
- The Prandtl Number
- The Stanton Number

The format of these numbers and their use are found in all standard text books on heat transfer. For example: Kern's *Process Heat Transfer*, and McAdams *Heat Transmission*.

These resistances are defined therefore by the following expression:

$$\frac{1}{U_o} = \frac{1}{h_o} + \frac{1}{h_i} \times \frac{A_o}{A_i} + \frac{1}{h_w} + (rf)_o + (rf)_i \times \frac{A_o}{A_i}$$

where

U_o = overall heat transfer coefficient based on outside tube surface, in Btu/hr/sqft/°F.
h = The film coefficient in Btu/hr/sqft/°F.
rf = fouling factors in $\frac{1}{\text{Btu/hr/sqft/°F}}$
h_w = Heat transfer rate through tube wall in Btu/hr/sqft/°F.
A = Surface area in sqft

subscripts "o" and "i" refer to outside surface and inside surface, respectively.

Flow arrangements

The two more common flow paths are Con-current and Counter-current. In Con-current flow both the hot fluid and the cold fluid flow in the same direction. This is the least desirable of the flow arrangement and is only used in those chemical processes where there is a danger of the cooling fluid congealing, subliming, or crystallizing at near ambient temperatures.

Counter-current flow is the most desirable arrangement. Here the hot fluid enters at one end of the exchanger and the cold fluid enters at the opposite end. The streams flow in opposite directions to one another. This arrangement allows the two streams exit temperatures to approach one another.

Logarithmic mean temperature difference Δt_m

In either counter-current or con-current flow arrangement the log mean temperature difference used in the overall heat transfer equation is determined by the following expression:

$$\Delta t_{\mathrm{m}} = \frac{\Delta t_1 - \Delta t_2}{\ln \dfrac{\Delta t_1}{\Delta t_2}}$$

The Δt's are the temperature differences at each end of the exchanger and Δt_1 is the larger of the two. In true counter-current flow the Δt_{m} calculated can be used directly in the overall heat transfer equation. However such a situation is not common and true counter-current flow rarely exists. Therefore a correction factor needs to be applied to arrive at the correct Δt_{m}. These are given in Figure A9.1 in the Appendix.

Fluid velocities and pressure drops

Film coefficients are a function of fluid velocity, density (vapor), and viscosity (liquids). Within limits increasing the velocity of a fluid reduces its resistance to heat transfer (i.e., it increases its heat transfer coefficient). Increasing the fluid velocity however increases its pressure drop. An economic balance needs to be sought therefore between the cost of heat transfer surface and pumping cost. This exercise should be undertaken to find a pay-out balance of 2–4 years. This exercise has been done many times and the following data is considered a reasonable balance between velocity and pressure drop for some common cases:

	Tube side		Shell side	
	Velocity, ft/sec	Press drop, psi	Velocity, ft/sec	Press drop, psi
Non-viscous liquids	6–8	10	1.5–2.5	10
Viscous liquids	6	20	3.0 max	15–20
Clean cooling water	6–8	10–15	–	–
Dirty cooling water	3 min	10+	–	–
Suspended solids in	2–3 min	10	1.5 min	15 liquids. (Note 1)
Gases and Vapors	$\frac{100}{\sqrt{\text{Gas density}}}$ max	3–5	–	3
Condensing vapors	–	–	–	3–5

Note 1: Normally erosion by suspended solids in liquids occurs at velocities of above 6 ft/sec.

For condensing steam pressure drop is usually not critical but a minimum steam pressure drop is desirable. Allowable steam velocities in tubes are as follows:

Pressure	Velocity ft/sec
Below atmospheric	225
Atmos to 100 psig	175
Above 100 psig	150

Choice of tube side versus shell side

There are no hard and fast rules governing which fluid flows on which side in a heat exchanger. Much is left to the discretion of the individual engineer and his or her experience. There are some guidelines and these are as follows:

(i) *Tube side flow*

Fouling liquids.
Tube cleaning is much easier than cleaning the outside of the tubes. Also fouling can be reduced by higher tube side velocities.

Corrosive fluids. It is cheaper to replace tubes than shells and shell baffles so as a general rule corrosive fluids are put on the tube side. There are exceptions and a major one are those corrosive fluids that become more corrosive at high velocities. An example of this are naphthenic acids, which are present in some crude oils and their products.

High pressure. Fluids at high pressure are usually put on the tube side as only the tubes, tube sheet, and channel need to be rated for high pressure in the unit design. This cheapens the overall cost of the exchanger.

Suspended solids. Fluids containing suspended solids should whenever possible be put to flow tube side. Shell side flows invariably have "dead spaces" where solids come out of suspension and build up to cause fouling.

Cold boxes. These are exchangers used in cryogenic processes where condensing of a vapor on one side of the exchanger is accompanied by boiling of a liquid on the other side. The condensing fluid is preferred on the tube side. Better control of the refrigerant flow is accomplished by the level control across the shell side.

(ii) *Shell side flow*

Available pressure drop. Shell side flows generally require lower pressure drop than tube side. Therefore if a system is pressure drop limiting it should be routed to the shell side.

Condensers. Condensing vapors should flow on the shell side wherever possible. The larger free area provided by the shell side space permits minimum pressure drop and higher condensate loading through better film heat transfer coefficients.

Large flow rates. In cases where both streams are of a similar nature with similar properties the stream with the largest flow rate should be sent to the shell side where the difference in flow rates are significant. The shell side provides more flexibility in design by baffle arrangements to give the best heat transfer design criteria.

Boiling service. The boiling liquid as in the case of reboilers, waste heat recovery units and the like should be on the shell side of the exchanger. This allows space for the proper disengaging of the vapor phase and provides a means of controlling the system by level control of the liquid phase.

Types of shell and tube exchangers

Figure 18.33 gives some of the more common arrangements in shell and tube exchanger design. The arrangements shown here are all one shell pass and one or two tube passes. Equipment with more than two tube passes (up to five) is also fairly common particularly in petroleum refining. Shell arrangements are however left at one if at all possible. Where multi-pass shell side is required companies prefer to use complete exchangers in series or in parallel or both rather than making two or more shell passes using horizontal baffling in one exchanger.

Estimating shell and tube surface area and pressure drop

There are many excellent computer programs available that calculate exchanger surface area and pressure drops from simple input. The actual calculation when done manually is tedious and long. However to understand a little of the importance of the input required by these computer programs it does well to at least view a typical manual calculation. The one given here is for a shell and tube cooler with no change of phase for either tube side or shell side fluids.

The calculation follows these steps:

Step 1. Establish the following data by heat balances or from observed plant readings:

- The inlet and outlet temperatures on the shell side and on the tube side.
- The flow of tube side fluid and that for the shell side. It may be necessary to calculate one or the other from a heat balance over the exchanger.
- Calculate the duty of the exchanger in heat units per unit time (usually hours).
- Establish the stream properties for tube side and shell side fluids. The properties required are: SG, Viscosity, Specific heats, Thermal conductivity.

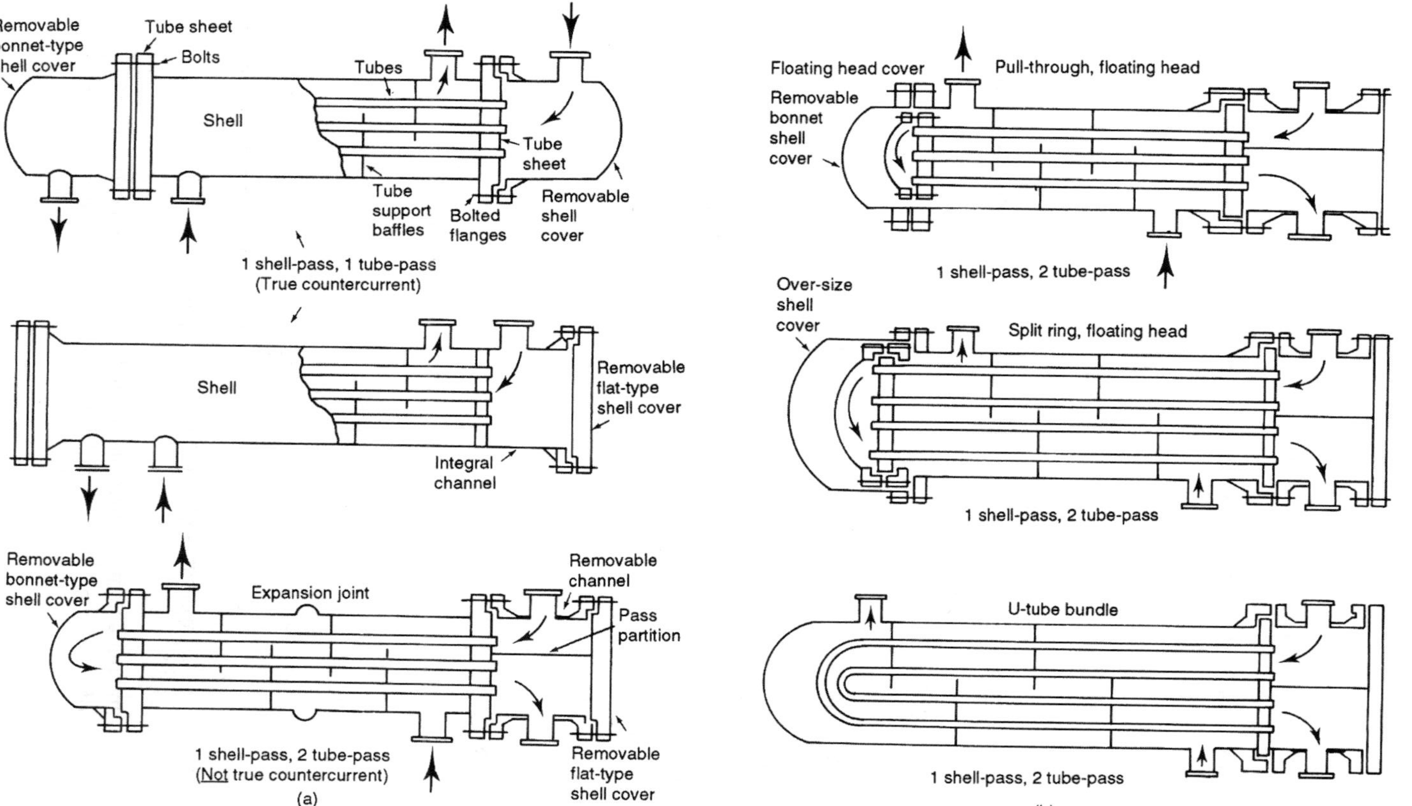

Figure 18.33. Some Common types of shell and tube exchangers. (a) Fixed tube sheet. (b) Removable bundle.

Step 2. Calculate the log mean temperature difference (Δt_m).

Assume a flow pattern (i.e. either co current or counter-current). Most flows will be a form of counter-current. Then show the temperature flow as follows:

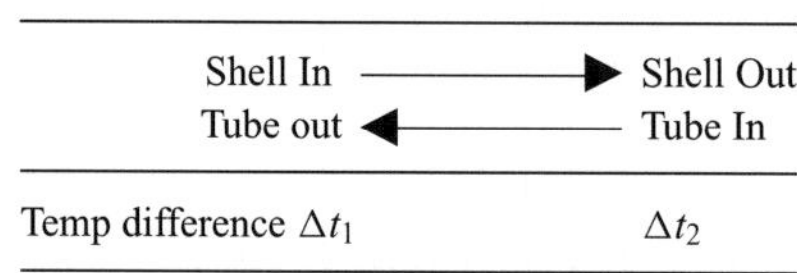

The log mean temperature difference is then calculated using the expression:

$$\Delta t_m = \frac{\Delta t_1 - \Delta t_2}{\log_e \dfrac{\Delta t_1}{\Delta t_2}}$$

This temperature needs to be corrected for the flow pattern, and this is done using the correction factors given in Figure 18.A.1. The use of these are self explanatory and are given in the figure.

Step 3. Calculate the approximate surface area.

From Table 18.A.2 in the Appendix select a suitable overall heat transfer coefficient U in Btu's/hr. sqft. F. Use the expression to calculate for 'A':

$$Q = UA\Delta t_m$$

where

Q = Heat transferred in Btu/hr. (the exchanger duty)
U = Overall heat transfer coefficient.
A = Exchanger surface area in sqft.
Δt_m = Log mean Temperature difference (corrected for flow pattern) in °F.

From the surface area calculated select the tube size and pitch. Usually ¾ ins on a triangular pitch for clean service and 1″ on a square pitch for dirty or fouling service. A single standard shell will hold about 4,100 sqft of surface per pass. Now most companies do not use multi-pass shells and prefer sets of shells in series if this becomes necessary. The 'norm' therefore are single pass shells each containing up to 4,100 sqft of surface.

Step 4. Calculate the tube side flow and the number of passes.

If it cannot be read from plant data calculate tube side flow in cuft/hr by heat balance. Select the tube gauge and length. The tube data are given in Table 18.A.3 in the Appendix and standard lengths of tubes are 16 and 20 ft. Calculate the number of selected tubes per pass from the expression:

$$Np = \frac{Ft \times 144}{3{,}600 \times At \times Vt}$$

where

Np = number of tubes per pass
Ft = tube side flow in cuft/hr
At = cross-sectional area of 1 tube
Vt = linear velocity in tube in ft/sec

See earlier item on 'Fluid Velocities and Pressure Drop' for recommended fluid velocities. The number of tube passes is arrived at by dividing the total surface area required by the total (external) surface area of the number of tubes per pass calculated above.

Step 5. Calculate tube side film coefficient corrected to outside diameter (h_{io}).
The tube side film coefficient may be calculated for water by the expression:

$$h_{io} = \frac{300 \times (V_t \times \text{tube i/d ins})^{0.8}}{\text{tube o/d ins}}$$

where

h_{io} = Inside film coefficient based on outside tube diameter in Btu/hr sqft °F.
V_t = linear velocity of water tube side in ft/sec.

For fluids other than water flowing tube side use the expression:

$$h_{io} = \frac{K}{D_o}(C\mu/K)^{1/6}(\mu/\mu_w) \cdot \phi(DG^{.14}/\mu)$$

where

h_{io} = inside film coefficient based on outside diameter in Btu/hr sqft °F.
K = thermal conductivity of the fluid in Btu/hr sqft (°F per ft). See Maxwell *Data Book on Hydrocarbons* or Perry *Chemical Engineers Handbook*
D = Inside tube diameter in ins.
D_o = Outside tube diameter in ins.
C = Specific heat in Btu/lb/°F.
G = mass velocity in lbs/sec sqft.
μ = Absolute viscosity Cps at average fluid temp.
μ_w = Absolute viscosity Cps at average tube wall temp.
$\phi(DG/\mu)$ = from Figure 18.34.

Step 6. Calculate shell side dimensions.
First determine the shell side average film temperature as follows:

$$\text{Inlet ave} = \frac{T_1 + T_2}{2} \qquad \text{Outlet ave} = \frac{T_3 + T_4}{2}$$

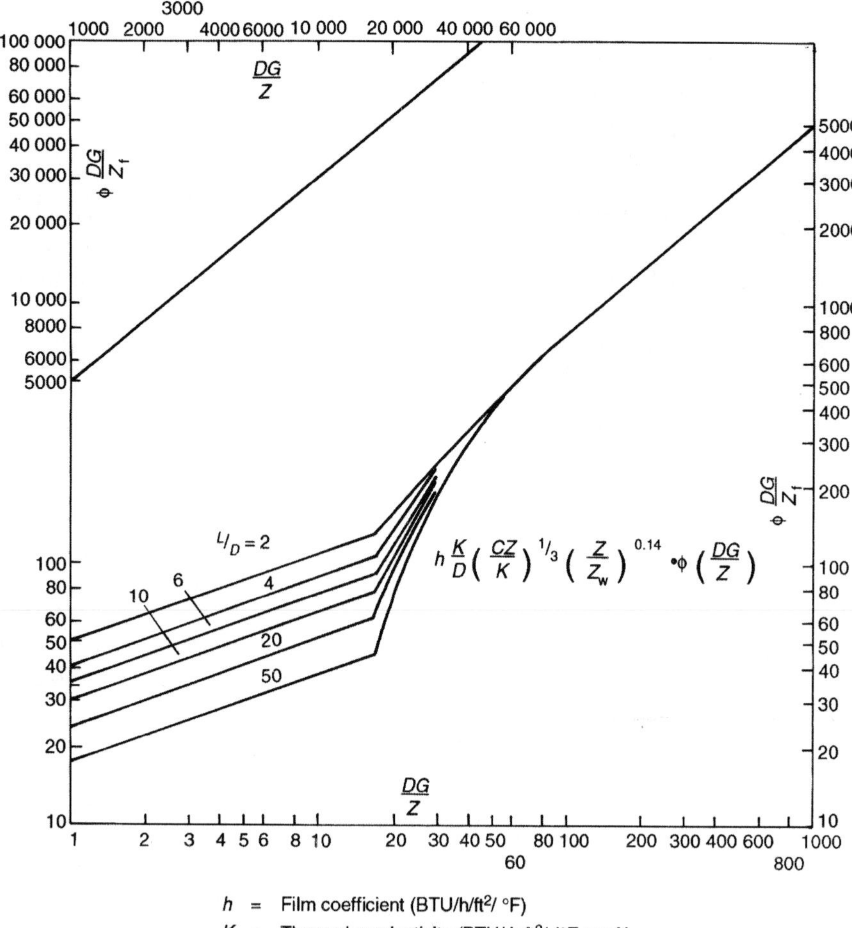

h = Film coefficient (BTU/h/ft^2/ °F)
K = Thermal conductivity (BTU/ h ft^2/ (°F per ft)
L = Heated tube length (ft)
D = Inside tube diameter (in)
G = Mass velocity (lb/ s / ft^2)
C = Specific heat of fluid at av. fluid temp. (BTU/ lb/ °F)
Z = Absolute viscosity (cp) at av. temp. of fluid
Z_W = Absolute viscosity (cp) at av. tube wall temp.

Figure 18.34. Heat transfer inside tubes.

where

T_1 = Shell fluid inlet temperature.
T_2 = Tube outlet temperature.
T_3 = Shell outlet temperature.
T_4 = Tube inlet temperature.

Average shell side film temperature:

$$\frac{\text{Inlet ave} + \text{outlet ave}}{2}$$

Use this temperature to determine density and viscosity used in the shell side film coefficient calculations.

The shell diameter. Next calculate the diameter of the tube bundle and the shell diameter. For this use one of the following equations to calculate the number of tubes across the center line of the bundle:

1. For square pitch tube arrangement:
   $T_{cl} = 1.19\ (\text{number of tubes})^{0.5}$
2. For triangular pitch tube arrangement:
   $T_{cl} = 1.10\ (\text{number of tubes})^{0.5}$

Note these are *total* number of tubes; namely, those calculated in Step 4 times number of tube passes.

Set number of baffles and their pitch. The type of baffles usually used are shown in Figure 18.35. Disc and donut type baffles are only used where pressure drop available is very small and there is a pressure drop problem. Baffles on the bias are used in sq pitch tube arrangement and baffles perpendicular to the tubes are usual for triangular tube arrangements.

The minimum baffle pitch should not be less than 16% of the shell diameter. Pitch in this case is the space between two adjacent baffles. Normally 20% of shell i/d is used for the baffle pitch. The number of baffles is calculated from the expression:

$$N_B = \frac{10 \times \text{tube length}}{\text{baffle pitch}\,\% \times \text{diameter of shell.}}$$

Free area of flow between baffles. The space available for flow on the shell side is calculated as:

$$W = D_i - (d_o \times T_{cl})$$

where

W = space available for flow in sq ins.
D_i = shell inside diameter in ins.
d_o = tube outside diameter in ins.
T_{cl} = number of tubes across centerline.

The free area of flow between baffles is now calculated as follows:

$$A_f = W \times (B_p - 0.187)$$

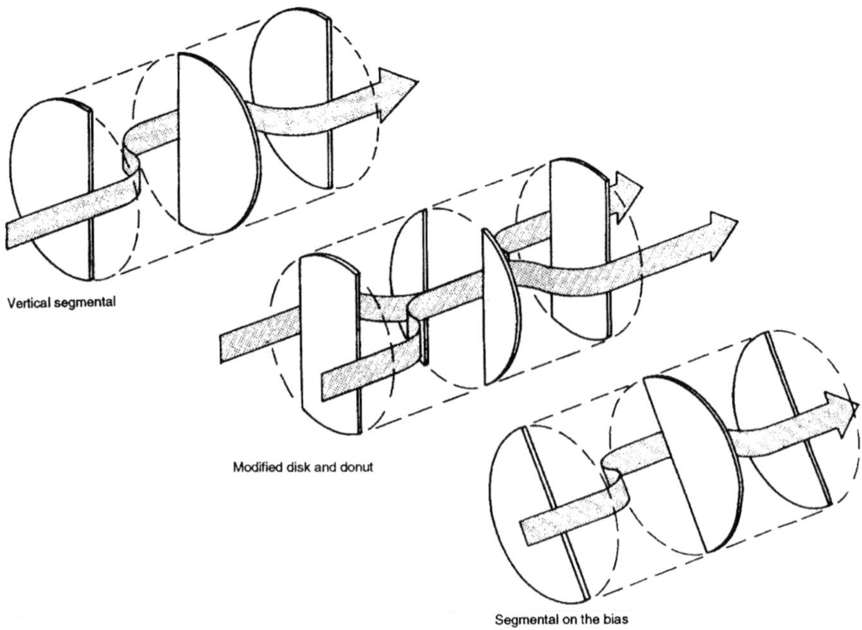

Figure 18.35. Types of baffles.

where

A_f = free flow area between baffles in sq ins.
B_p = baffle pitch in ins.

Step 7. Calculate the shell side film coefficient h_o.

The following expression is used to determine the outside film coefficient:

$$h_o = \frac{K}{d_o}(C\mu/K)^{1/3} \cdot \phi(d_o G_m/\mu_f) \cdot \frac{4P_b}{D}$$

where

h_o = outside film coefficient in Btu/hr sqft °F
G_m = maximum mass velocity in lbs/sec. sqft
d_o = outside tube diameter in ins
K = thermal conductivity.
C = specific heat of fluid in Btu/lb/F
μ_f = viscosity at mean film temperature in Cps
P_b = baffle pitch in ins
D = shell internal diameter in ins

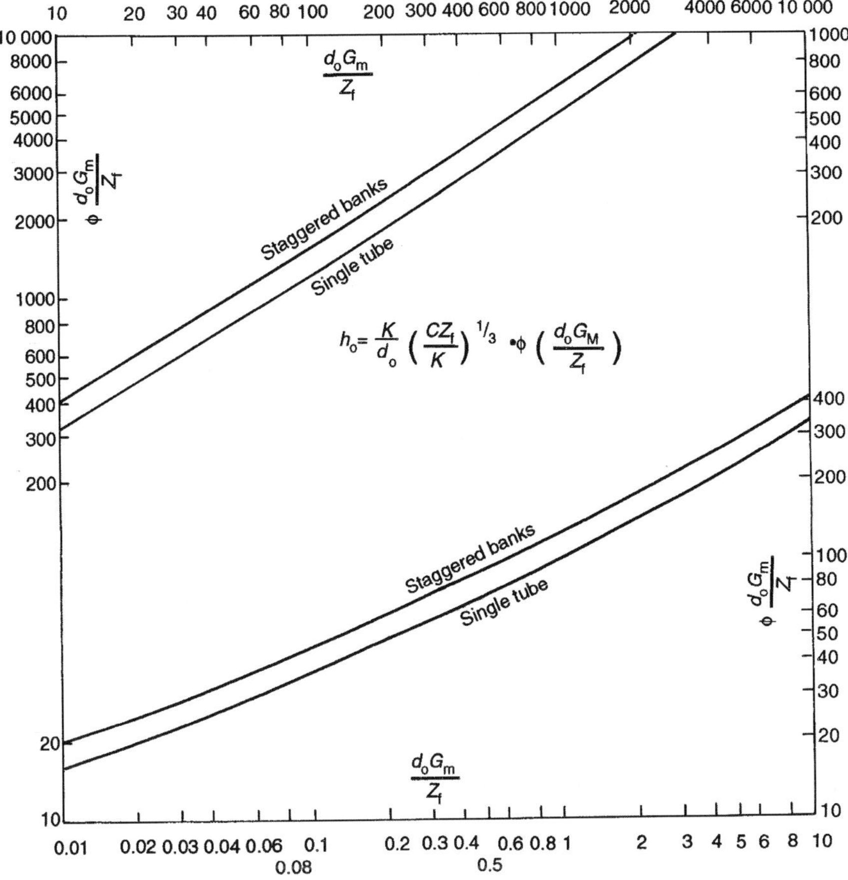

Figure 18.36. Heat transfer to fluids outside tubes.

$\phi(d_o G_m/\mu_f)$ is a function of the Reynolds number read from Figure 18.36. The Reynolds Number is:

$$\mathrm{Re} = \frac{d_o G_m}{\mu_f}$$

where G_m = lbs/sec sqft.

This film coefficient is corrected for the type of baffle and tube arrangement by multiplying it by one of the following factors:

For square pitch vertical to tube rows	0.50
square pitch on the bias	0.55
triangular tube pitch	0.70

Table 18.28. Thermal conductivity of tube metals

	K, Btu/hr.sqft. F/ft
Admiralty brass	64
Aluminum brass	58
Aluminum	117
Brass	57
Carbon steel	26
Copper	223
Cupronickel	41
Lead	20
Monel	15
Nickel	36
Red Brass	92
Type 316 alloy steel	9
Type 304 alloy steel	9
Zinc	65

Step 8. Calculate the *overall heat transfer coefficient* U_o.

The film coefficients calculated in steps 5 and 7 are now used in the expression:

$$\frac{1}{U_{\text{o}}} = \frac{1}{h_{\text{io}}} + r_{\text{io}} + \frac{1}{h_{\text{o}}} + r_{\text{o}} + r_{\text{w}}$$

where

U_{o} = overall heat transfer coefficient in Btu/hr.sqft.°F.

r_{io} and r_{o} = tube side and shell side fouling factors respectively in hr. sqft.°F/Btu. For clean tubes this is 0.001 as a sum of both factors.

r_{w} = Tube wall resistance to heat transfer in hr. sqft. °F/Btu, which is expressed as:

$$r_{\text{w}} = \frac{t_{\text{w}} \cdot d_{\text{o}}}{12 \times K \times (d_{\text{o}} - 2t_{\text{w}})}$$

where

t_{w} = Tube wall thickness ins.
d_{o} = Outside tube diameter ins.
K = Thermal conductivity Btu's/hr sqft °F/ft. See Table 18.28.

Compare the calculated value of the overall heat transfer coefficient with the assumed one in step 3.

If there is agreement within ±10% then the calculated one will be used for revising the calculation for surface area and the other dimensions. If there is no agreement repeat the calculation using a new value for the assumed U.

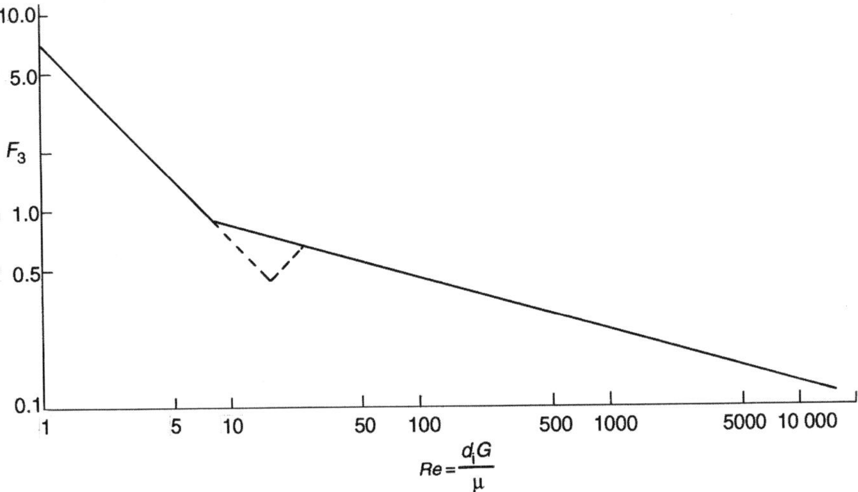

Figure 18.37. Pressure drop factor F_3 for flows inside tubes.

Step 9. Calculate tube side pressure drop.

Using the adjusted dimensional values from the calculated U_o, calculate the tube side pressure drop using one of the following equations:

$$\Delta P_t = .02 F_t \times N_p \times (V^2 + (0.158 L V^{1.73}/d_i)^{1.27})$$

For water only.

For fluids other than water use:

$$\Delta P_t = F_t \times N_p \times (\Delta P_{tf} + \Delta P_{tr})$$

where

$\Delta P_{tf} = F_3 \cdot \frac{L}{d_i} \cdot (\rho_m \times V^2/9,270) \cdot (\mu_w)^{.14}$

$\Delta P_{tr} = 3 \times (\rho_m \times V^2/9,270)$

F_3 = factor based on Reynolds number see Figure 18.37

ρ_m = density in lbs/cuft at mean fluid temperature.

μ_w = viscosity of fluid at tube wall temperature in Cps (use mean film temperature)

V = linear velocity in ft/sec.

F_t = pressure drop fouling factor as follows (dimensionless)

Tube OD	Tube metal	F_t
0.75	Steel	1.50
1.00	Steel	1.40
1.50	Steel	1.20
0.75	Ad Brass	1.20
1.00	Copper	1.15

The pressure drop figure calculated by these equations are for one unit. Where there are more than one shell in series multiply the figures by the number of shells.

Step 10. Calculate the shell side pressure drop.

Using the revised dimensions calculated in Step 8 the total shell side pressure drop is calculated using the following equation:

$$\Delta P_s = F_s(\Delta P_{sr} + \Delta P_{sf})$$

where

$\Delta P_{sf} = B_2 F_{sp} N_{tc} N_b (m \times V^2/9{,}270)$

$\Delta P_{sr} =$ pressure drop due to turns given by:

$$(N_b + 1) \cdot (3.5 - 2P_b/D) \cdot \frac{(m \times V^2)}{9{,}270}$$

$B_2 =$ Factor as follows:

Baffle position	*Tube layout*	B_2
Vertical	square	0.30
Bias @ 45°	square	0.40
Vertical	triangular	0.50

$F_{sp} =$ Factor based on Reynolds number. See Figure 18.38.
$N_{tc} =$ Number of tubes on center line.
$N_b =$ Number of shell baffles.
$P_b =$ Space between baffles ins.
$D =$ Shell i/d in ins.

The pressure drop calculated here is for one shell. If there are more than one shell in series then multiply these pressure drops by the number of shells.

Air coolers and condensers

Air cooling of process streams or condensing of process vapors is more widely used in the process industry than cooling or condensing by exchange with cooling water. The use of individual air coolers for process streams using modern design techniques has economized in plant area required. It has also made obsolete those large cooling towers and ponds associated with product cooling. This item of the chapter describes air coolers in general and outlines a method to estimate surface area, motor horsepower and plant area required by the unit.

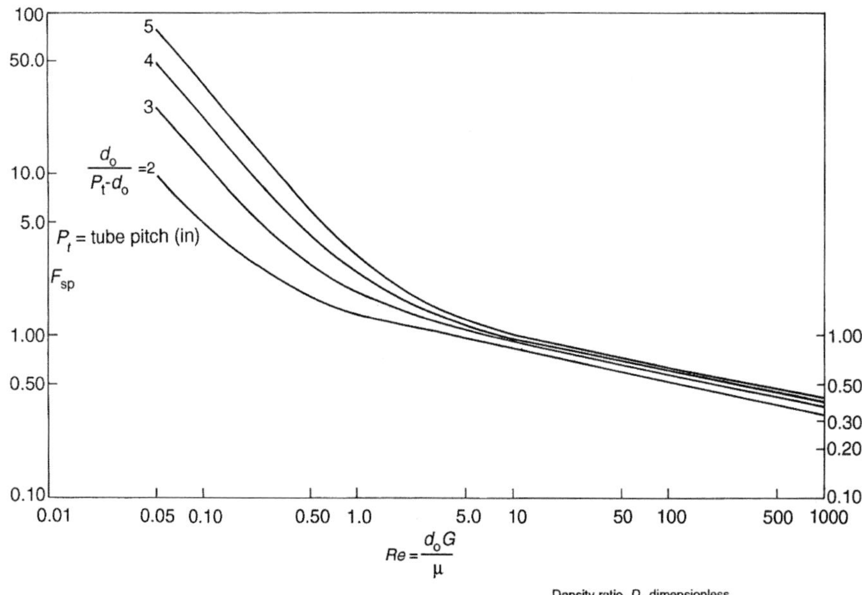

Figure 18.38. Pressure drop factor F_{SP} for flows across banks of tubes.

As in the case for shell and tube exchangers there are many excellent computer programs that can be used for the design of air coolers. The method given here for such calculation may be used in the absence of a computer program or for a good estimate of a unit. The method also emphasizes the importance of the data supplied to manufacturers for the correct specification of the units.

General description of air coolers/condensers

Figures 18.39 show the two types of air coolers used in the process industry. Both units consist of a bank of tubes through which the fluid to be cooled or condensed flows. Air is passed around the tubes either by a fan located below the tubes forcing air through the tube bank or a fan located above the tube bank drawing air through the tube bank. The first arrangement is called 'forced draft' and the second 'induced draft'.

Air in both cases is motivated by a fan or fans driven by an electric motor or a steam turbine or in some cases a gas turbine. The fan and prime driver are normally connected by a 'V' belt or by a shaft and gear box. Electric motor drives are by far the most common prime drivers for air coolers.

The units may be installed on a structure at grade or as is often the case on a structure above an elevated pipe rack. Most air coolers in condensing service are elevated above pipe racks to allow free flow of condensate into a receiving drum.

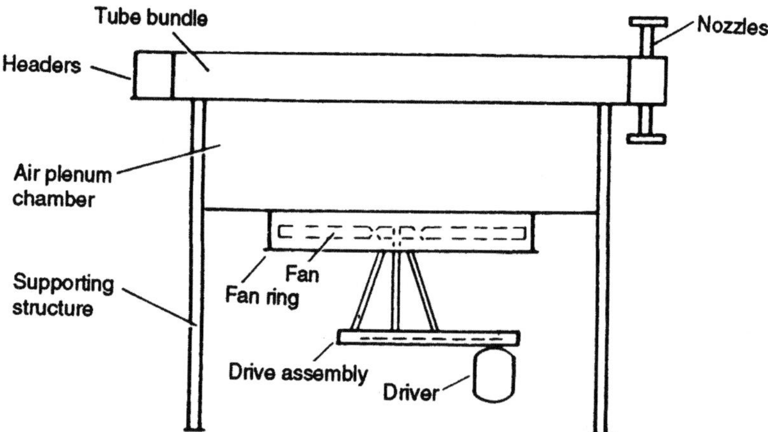

Fig A A Forced Air Flow Arrangement

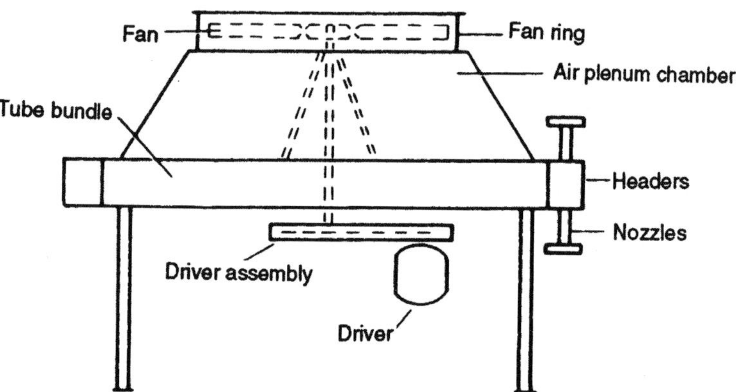

Fig B An Induced Air Flow Arrangement

Figure 18.39. Air coolers—(a) Forced draught. (b) Induced draught.

Thermal rating

Thermal rating of an air cooler is similar in some respects to that for a shell and tube described in the previous item. The basic energy equation

$$Q = U \Delta T A$$

is used to determine the surface area required. The calculation for U is different in that it requires the calculation for the air side film coefficient. This film coefficient is usually based on an extended surface area which is formed by adding fins to the bare surface of the tubes. Thermal rating, surface area, fan dimensions and horsepower are calculated by the following steps:

Step 1. Calculate the heat duty and the tube side material characteristics.
Step 2. Calculate the log mean temperature for the exchanger.
Using the following equation determine the temperature rise for the air flowing over the tubes:

$$\Delta t_m = ((U_e + 1)/10)) \cdot ((\Delta t_m/2) - t_1))$$

where

Δt_m = air temperature rise °F
U_e = overall heat transfer coefficient assumed. (from Table 18.29).
Δt_m = mean tube side temperature °F
t_1 = inlet air temperature °F.

Calculate the log mean temperature difference (LMTD) as in Step 2 of previous item on shell and tubes.

Step 3. Determine an approximate surface area using the expression:

$$A_E = \frac{Q}{U_E \cdot \Delta t_m}$$

where

A_E = extended surface area in sqft.
Q = exchanger duty in Btu/hr.
U_E = overall heat transfer coefficient based on extended surface from Table 18.29.
Δt_m = log mean temperature difference corrected for number of passes in °F.

Step 4. Calculate the number of tubes from the expression:

$$N_t = \frac{A_E}{A_f \times L}$$

where

N_t = total number of tubes.
A_E = extended surface area in sqft.
A_f = extended area per ft of fin tube read from Table 18.30.
L = length of tube (30 ft is standard).

Step 5. Fix the number of passes (usually 3 or 4) and calculate the mass flow of tube side fluid using the expression:

$$G = \frac{\text{lbs/hr of tube side fluid} \times N_p \times 144}{N_t \times A_t \times 3{,}600}$$

Table 18.29. Some common overall transfer coefficients for air cooling

	1/2″ by 9 Fin ht by Fin/ins		5/8′ by 10 Fin ht by Fin/ins	
Service	U_e	U_o	U_e	U_o
Process water	95	6.5	110	5.2
Hydrocarbon Liquids				
Visc @ ave temp cps				
0.2	85	5.9	100	4.7
1.0	65	4.5	75	3.5
2.5	45	3.1	55	2.6
6.0	20	1.4	25	1.2
10.0	10	0.7	13	0.6
Hydrocarbon gasses				
@ Pressures psig				
50	30	2.1	35	1.6
100	35	2.4	40	1.9
300	45	3.1	55	2.6
500	55	3.8	65	3.0
1,000	75	5.2	90	4.2
Hydrocarbon condensers				
Cooling range 0°F	85	5.9	100	4.7
10°F	80	5.5	95	4.4
60°F	65	4.5	75	3.5
100+°F	60	4.1	70	4.2
Refrigerants				
Ammonia	110	7.6	130	6.1
Freon	65	4.5	75	3.5

U_e is transfer coefficient for finned surface.
U_o is transfer coefficient for bare tubes.

Table 18.30. Fin tube to bare tube relationships based on 1″ O/D tubes

Fin Ht by Fins/ins	1/2″ by 9		5/8″ by 10	
Area/ft Fin tube	3.8		5.58	
Ratio of Areas Fin/Bare Tube	14.5		21.4	
Tube Pitch ins	2Δ	2¼Δ	2¼Δ	2½Δ
Bundle Area sqft/ft (Note 1)				
3 Rows	68.4	60.6	89.1	80.4
4 Rows	91.2	80.8	118.8	107.2
5 Rows	114.0	101.0	148.5	134.0
6 Rows	136.8	121.2	178.2	160.8

Note 1: Bundle area is the external area of the bundle face area in sqft/ft.

where

G = Mass velocity in lbs/sec sqft.
N_p = Number of tube passes.
A_t = inside cross-sectional area of tube in sq ins

Step 6. Calculate the Reynolds Number for tube side using the expression:

$$\text{Re} = \frac{d_i \cdot G}{\mu}$$

where

R_e = Reynolds Number (dimensionless)
d_i = Tube i/d in ins.
μ = Tube side fluid viscosity at average temperature in Cps.

Step 7. Calculate the inside film coefficient from the expression:

$$h_{io} = \frac{K}{D}(C\mu/K)^{1/3} \cdot (\mu/\mu_w)^{.14} \cdot \phi(DG/Z)$$

where

h_i = inside film coefficient in Btu/hr · sqft °F.
K = thermal conductivity of the fluid in Btu/hr · sqft (°F per ft). See Maxwell *Data Book on Hydrocarbons*.
D = inside tube diameter in ins.
C = specific heat in Btu/lb/°F.
G = mass velocity in lbs/sec sqft.
μ = absolute viscosity Cps at average fluid temp.
μ_w = absolute viscosity Cps at average tube wall temp.
$\phi(DG/\mu)$ = from Figure 18.34

Step 8. Calculate the mass velocity of air and the film coefficient on the air side thus:

$$\text{Weight of air} = \frac{Q}{C_{Air} \times \Delta t_{Air}}$$

where

Q = exchanger duty in Btu/hr
C_{Air} = specific heat of air (use 0.24)
Δt_{Air} = temperature rise of the air °F.

Face area of tubes A_f is calculated as follows:
Set the O/D of the tubes (usually 1″), length, fin size (usually 5/8″ @ 10 to the ins or 1/2″ @ 9 to the ins), Pitch (see Table 18.30), and number of tube rows (start with 3 or 4). Then face area is:

$$A_f = \frac{\text{Total extended surface area } A_E}{\text{External area per ft of bundle (from Table 18.30)}}$$

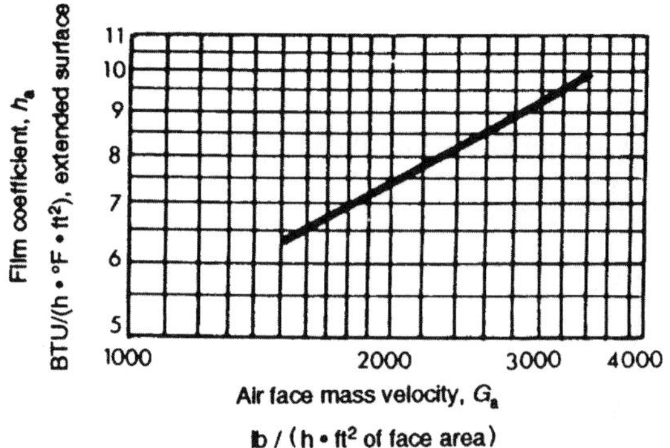

Figure 18.40. Air film coefficients.

Mass velocity of air is calculated from the expression:

$$G_a = \frac{\text{lbs per hour of air flow}}{\text{face area } A_f}$$

The film coefficient for the air side is read from Figure 18.40.

Step 9. Calculate the overall heat transfer coefficient as follows:

Area ratio of bare tube outside to finned outside is read from Table 18.30. Then factor to convert all heat flow resistance to outside tube diameter basis is:

$$F_t = \frac{A_r \times \text{tube o/d}}{A_t}$$

where

A_r = Area ratio
A_t = inside tube cross-sectional area sq ins.

Then:

$$\frac{1}{U_o} = \frac{1 + (r_t \times F_t) + r_w + 1}{h_i h_o}$$

where

r_t = inside fouling factor.
r_w = tube metal resistance (normally ignored.)

If the calculated U is within 10% of the assumed there will be no need to recalculate with a new assumed value for the U. The dimensions and data are adjusted however using the calculated value for U.

Step 10. Calculate the required fan area and the fan diameter as follows:

$$\text{Fan area} = \frac{0.4 \times \text{Face Area } A_f}{\text{Assumed Number of Fans}}$$

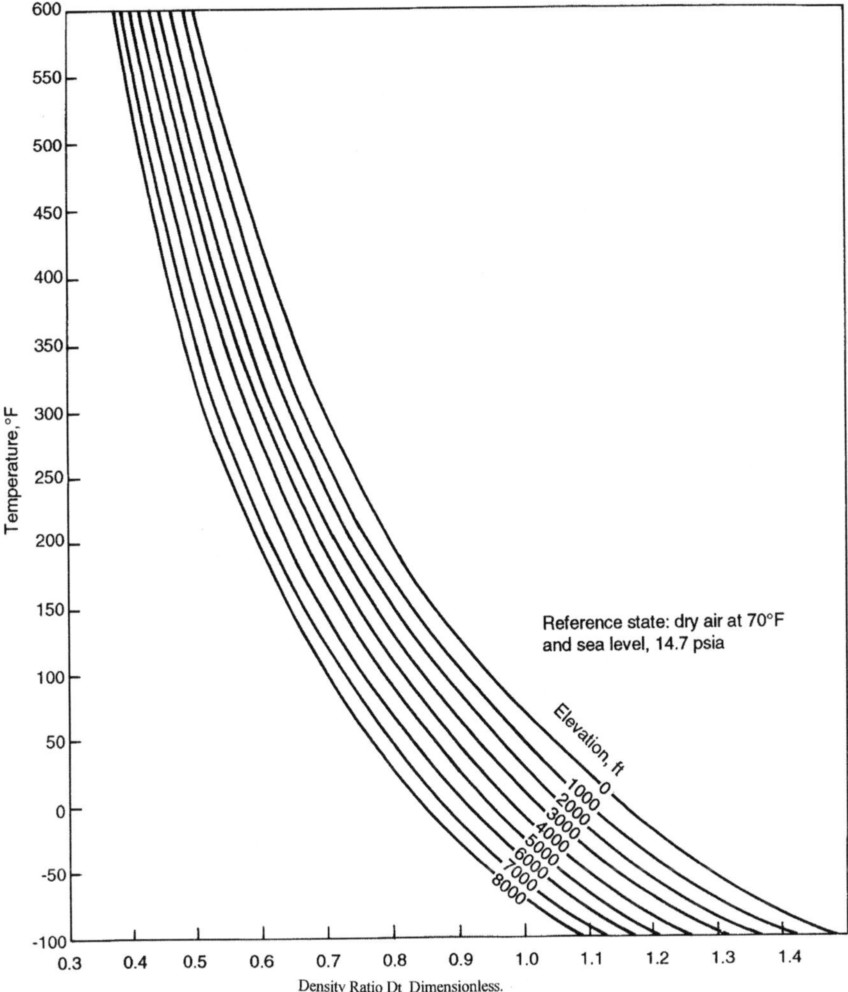

Figure 18.41. Relative density of air.

Begin by assuming 2 fans and continue with multiples of 2 until a reasonable fan diameter (about 10 to 12 ft) is obtained. On very large units fans can be maximized at 16 ft.

$$\text{Fan diameter} = \sqrt{(\text{Fan area} \times 4/\pi)}$$

Step 11. Calculate air side pressure drop and actual air flow in cuft/min.

$$\text{Average air temperature} = \frac{t_1 + t_2}{2}$$

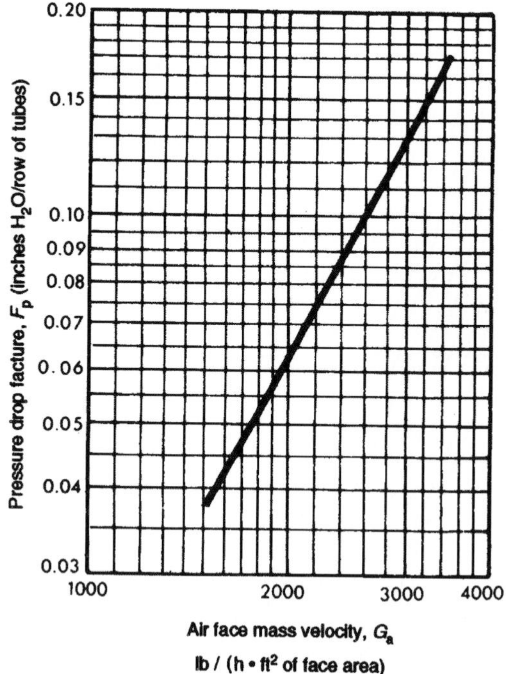

Figure 18.42. Pressure drop air side in inches of water.

From Figure 18.41

$$D_r = \text{Relative density factor for air at elevations of site.}$$

From Figure 18.43, ΔP_a = Pressure drop of air in ins of H_2O

$$\Delta P_a \text{ corrected} = \frac{\Delta P_a \times \text{No of rows}}{D_r}$$

Density of air at corrected ΔP_a

$$\frac{29}{(378 \times 14.7 \times T_2)/(T_1 \times (\text{Corr } \Delta P_a + 14.7))}$$

where $T_{1\&2}$ are absolute temperatures.
ACFM of air therefore is:

$$\frac{\text{lbs/hr of air}}{\text{Density} \times 60}$$

ΔP of air at the fan is obtained by the expression:

$$\Delta P_m = \left[\frac{\text{ACFM}}{(4{,}000(\pi d)/4)}\right]^2$$

in ins of water gauge.

Step 11. Calculate the fan horsepower as follows:

$$\text{Hydraulic HP} = \frac{\text{ACFM} \times \text{Density of air} \times \text{Diff head in Ft}}{33{,}000}$$

$$\text{Differential head} = \frac{\text{Total } \Delta P @ \text{ fan in ins } H_2O \times 5.193}{\text{Density}}$$

$$\text{Bhp} = \frac{\text{Hydraulic HP}}{\eta_r}$$

where η_r is the fan efficiency (usually 70%).

Condensers

In petroleum refining and most other chemical process plants vapors are condensed either on the shell side of a shell and tube exchanger, the tube side of an air cooler, or by direct contact with the coolant in a packed tower. By far the most common of these operations are the first two listed. In the case of the shell and tube condenser the condensation may be produced by cooling the vapor by heat exchange with a cold process stream or by water. Air cooling has overtaken the shell and tube condenser in the case of water as coolant in popularity as described in the previous item.

In the design or performance analysis of condensers the procedure for determining thermal rating and surface area is more complex than that for a single phase cooling and heating. In condensers there are three mechanisms to be considered for the rating procedure. These are:

- The resistance to heat transfer of the condensing film
- The resistance to heat transfer of the vapor cooling
- The resistance to heat transfer of the condensate film cooling

Each of these mechanisms is treated separately and along pre-selected sections of the exchanger. The procedure for determining the last two of the mechanisms follows that described earlier for single phase heat transfer. The following expression is used to calculate the film coefficient for the condensing vapor:

$$h_c = \frac{8.33 \times 10^3}{(M_c/L_c \cdot N_s)^{.33}} \times k_f \times \left[\frac{Sg_c^2}{\mu_f}\right]^{0.33}$$

where

h_c = condensing film coefficient.
M_c = mass condensed in lbs/hr
L_c = tube length for condensation.
$= \frac{A_{zone}}{A} \times (L - 0.5)$

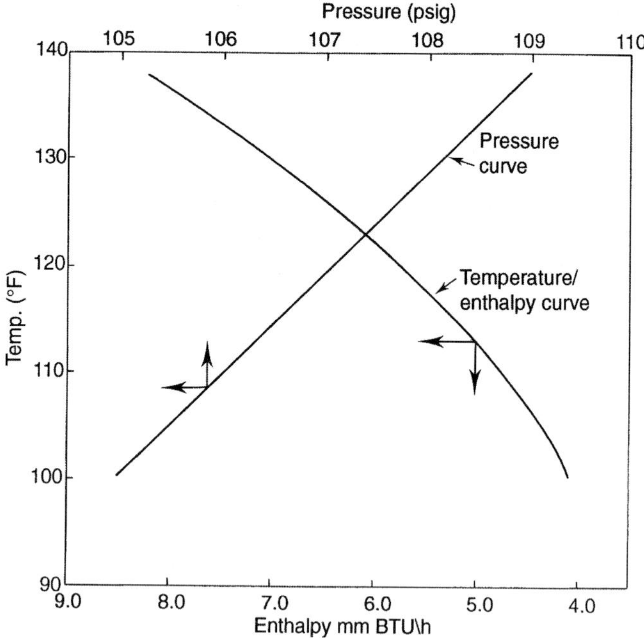

Figure 18.43. Enthalpy curve for a de-butanizer overhead condenser.

$N_s = 2.08\ N_t^{0.495}$ for triangular pitch.
k_f = thermal conductivity of condensate at film temperature.
Sg = specific gravity of condensate.
μ_f = viscosity of condensate at film temperature in Cps

Again there are many excellent computer programs that calculate condenser thermal ratings, and these of course save the tedium of the manual calculation. However no matter which method of calculation is selected there is required one major additional piece of data over that required for single phase heat exchange. That item is the enthalpy curve for the vapor.

Enthalpy curves are given as the heat content per lb or per hour contained in the mixed phase condensing fluid plotted against temperature. An example of such a curve is given in Figure 18.43. These Enthalpy curves are developed from the vapor/liquid or flash calculations described in Chapter 3 of this volume.

Briefly the calculation for the curve commences with determining the dew point of the vapor and the bubble point of the condensate. Three or more temperatures are selected between the dew and bubble points and the V/L calculation of the fluid at these temperatures carried out. Enthalpy for the vapor phase and the liquid phase are added for each composition of the phases at the selected temperatures. These together with the enthalpy at dew point and bubble point are then plotted.

As in the case of the shell and tube exchanger and the air cooler a manual calculation for condensers is described here. Again this is done to provide some understanding of the data required to size such a unit and its significance in the calculation procedure. Computer aided designs should however be used for these calculations whenever possible.

The following calculation steps describe a method for calculating the film coefficient of a vapor condensing on the shell side of a S&T exchanger. The complete rating calculation will not be given here as much of the remaining calculation is simply repetitive.

Step 1. Calculate the dew point of the vapor stream at its source pressure. Estimate the pressure drop across the system. Usually 3–5 psi will account for piping and the exchanger pressure drop. Calculate the bubble point of the condensate at the terminal pressure. Select three or more temperatures between dew point and bubble point and calculate the vapor/liquid quantities at these conditions of temperature and pressure.

Step 2. Calculate the enthalpy of the vapor and liquid at these temperatures. Plot the total enthalpies against temperature to construct the enthalpy curve. Establish the properties of the vapor phase and liquid phase for each temperature interval. The properties mostly required are Sg, viscosity, Mole wt, thermal conductivity, and specific heats.

Step 3. In the case of a water cooler calculate the duty of the exchanger and the quantity of water in lbs/hr. Commence the heat transfer calculation by assuming an overall heat transfer coefficient (use the data given in Table A9.1). Calculating the corrected LMTD, and the surface area.

Step 4. Using the surface area calculated in step 3 define the exchanger geometry in terms of number of tube passes, number of tubes on the center line, shell diameter, baffle arrangement and the shell free flow area. Calculate also the water flow in feet per sec.

Step 5. Divide the exchanger into 3 or 4 zones by selecting the zone temperatures on the enthalpy curve. Calculate the average weight of vapor and the average weight of condensate in each zone. Using these averages calculate the average heat transferred for:

Cooling of the vapor Q_v

Cooling of the condensate Q_L

Condensing of the vapor which will be:

Total heat in the zone (from the enthalpy curve) less the sum of Q_v and Q_L.

Step 6. Calculate the film coefficient for the tube side fluid. See previous item 'Estimating Shell and Tube Surface Area and Pressure Drop'.

Step 7. Starting with zone 1 and knowing the outlet temperature of the coolant fluid, the total heat duty of the zone, and the shell side temperatures calculate the coolant inlet temperature. Using this calculate the LMTD for the zone and, assuming a zone overall heat transfer coefficient U, calculate a surface area for the zone. Using this and the total exchanger area estimated in step 4 establish L_c in feet.

Step 8. Calculate the condensing film coefficient from the equation given earlier. This will be an uncorrected value for h_c. This will be corrected to account for turbulence by the expression:

$$h_{c\,(corr)} = h_c \times (G_v/5)$$

where

$$G_v = \text{average vapor mass velocity in lbs/hr} \cdot \text{sqft}$$

Step 9. Calculate the value of G_v using the free flow area allocated to the vapor γ_v. The following expressions are used for this:

$$\gamma_v = 1 - \gamma_L$$

$$\frac{1}{\gamma_L} = 1 + \frac{\text{Ave mass vapor}}{\text{Ave mass liquid}} \times (\mu_v/\mu_L)^{0.111} \times (\rho_L/\rho_v)^{0.555}$$

$$G_v = \frac{\text{Ave mass Vapor}}{25 \times \text{Free flow area} \times \gamma_v}$$

Step 10. Calculate the film coefficient hv for the vapor cooling mechanism. This will be the procedure used for a single phase cooling given in a previous item. This is corrected to account for resistance of the condensate film by the expression:

$$\frac{1}{h_{v\,corr}} = -\frac{1}{h_c} + \frac{1}{1.25 h_v}$$

Step 11. Calculate the film coefficient for the condensate cooling mechanism. Again this is the procedure described in the Item for single phase cooling on the shell side. This is corrected for drip cooling that occurs over a tube bank.

$$\text{Drip cooling } h_{dc} = 1.5 \times h_c$$

$$\text{and } h_L \text{ corrected} = \frac{2 \times h_{dc} \times h_L}{h_{dc} + h_L}$$

Step 12. Calculate the total zone film coefficient ho using the following expression:

$$h_o = \frac{Q_{zone}}{\dfrac{Q_c}{h_c} + \dfrac{Q_v}{h_v} + \dfrac{Q_L}{h_L}}$$

where

Q_c, Q_v, Q_L are the enthalpies for condensing, vapor cooling, and condensate cooling, respectively.

Step 13. Calculate the overall heat transfer coefficient neglecting the shell side coefficient from Step 12. Thus:

$$\frac{1}{U_x} = r_o + r_w + r_{io} + R_{io}$$

where

r_o, r_w, r_{io} are fouling factors for shell fluid, wall, and tube side fluid respectively. R_{io} is the tube side film coefficient calculated in Step 6.

Step 14. Calculate the overall heat transfer coefficient U_{zone} for the zone using the expression:

$$U_{zone} = \frac{h_o \times U_x}{h_o + U_x}$$

Check the calculated U against that assumed for the zone. Repeat the calculation if necessary to make a match.

Step 15. Calculate the zone area using the acceptable calculated U. Repeat steps 7 through 14 for the other zones. The total surface area is the sum of those for each zone.

Reboilers

Reboilers are used in fractionation to provide a heat source to the system, and to generate a stripping vapor stream to the tower. Reboilers are operated either by the natural circulation of a fluid or by forced circulation of the fluid to be reboiled. This chapter deals only with natural circulation reboilers.

There are three common types of reboilers and these are:

- The kettle type reboiler
- The once through thermosyphon reboiler
- The re-circulating thermosyphon reboiler

The kettle reboiler

This type of reboiler (Figure 18.44) is extremely versatile. It can handle a very wide range of vaporization loads. (e.g., when used as LPG vaporizer for fuel gas purposes it vaporizes 100% of the feed). The equipment consists of a large shell into which is fitted a tube bundle through which the heating medium flows. The liquid to be reboiled enters the bottom of the shell at the end adjacent to the tube inlet/outlet chamber. The liquid is boiled and partially vaporized by flowing across the tube bundle. The diameter of the shell is sized such that there is sufficient space above the tube bundle and the top of the shell to allow some disengaging of the liquid and vapor. A baffle weir is installed at the end of the tube bundle furthest from the inlet. This baffle weir establishes a liquid level over the tube bundle in the shell. The boiling liquid flows over this weir to the shell outlet nozzle, while the vapor generated is allowed to exit from the top of the shell through one or two nozzles.

The space downstream of the weir is sized for liquid holdup to satisfy the surge requirements for the product. Thus it is not necessary to provide space in the bottom of the tower for product surge. If the heating medium is non fouling it is permissible to use U tubes for the tube bundle. Otherwise the tube bundle must be of the floating head type. The kettle type reboiler should always be the first type to be considered if there are no elevation constraints to pumping the bottoms product away.

Once through thermosyphon reboiler

This type of reboiler and its location relative to the tower is given in the sketch below:

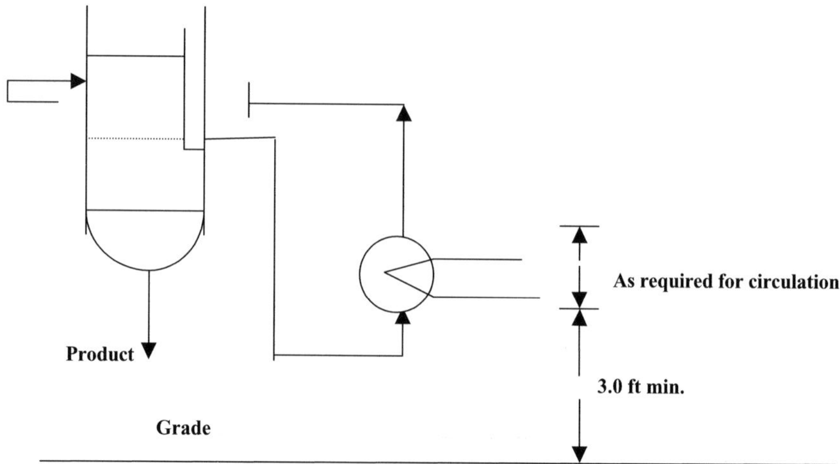

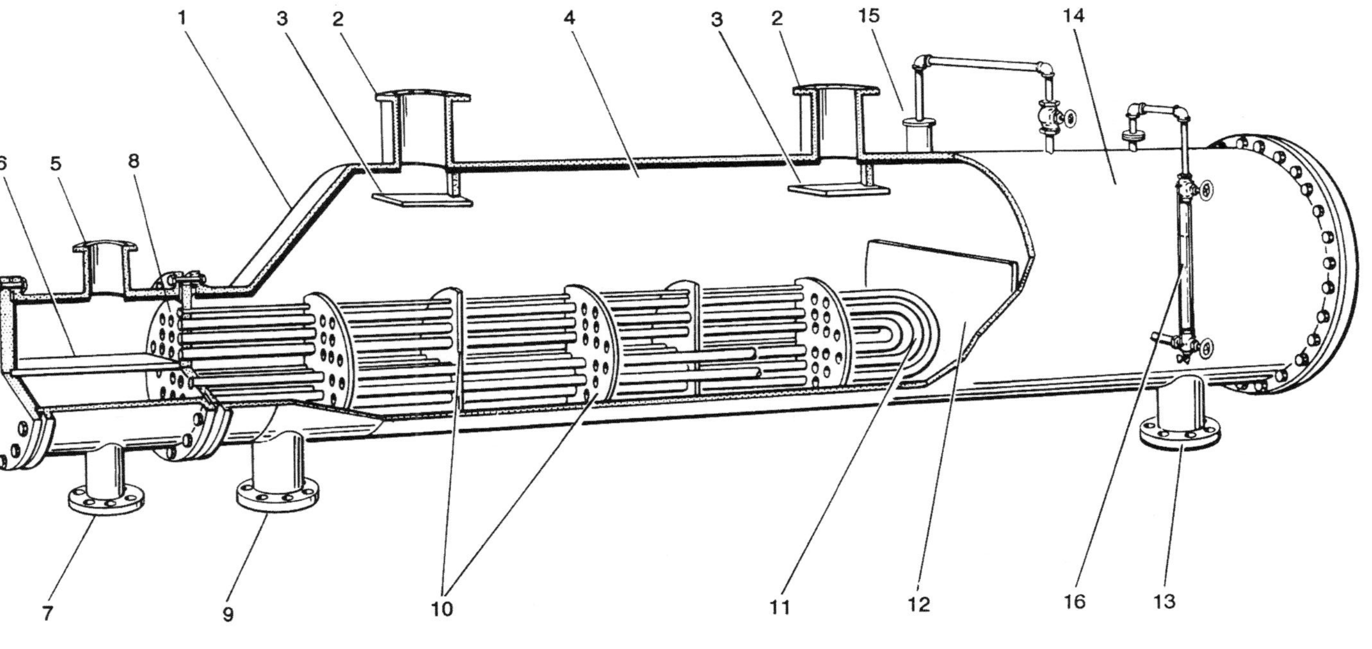

1 shell ; 2 Shell outlet nozzles Vapor ; 3. Entrainment Baffles ; 4 Vapor Disengaging Space. 5. Channel Inlet Nozzle; 6. Channel Partition. ; 7.Channel outlet nozzle. ; 8. Tube Sheet; 9. Shell inlet nozzle.; 10. Tube support sheets.; 11 U Tube returns.; 12 Weir. ; 13 Shell outlet nozzle (liquid),; 14. Liquid hold up (Surge) section.; 15Top of level – instrument housing (external displacer).; 16. Liquid level gauge.

Figure 18.44. The components of a kettle reboiler.

This type of reboiler should be considered when a relatively high amount of surge is required for the bottom product and when it is necessary to provide head for the product pump (NPSH requirement).

This type of reboiler takes the liquid from the bottom tray of the fractionator as feed. This stream enters the shell side of a vertical single tube pass shell and tube exchanger by gravity head to the bottom of the shell. The heating medium flows tube side to partially vaporize the liquid feed. A siphoning effect is caused by the difference in density between the reboiler feed and the vapor/liquid effluent. This allows the reboiler effluent to exit from the top of the shell side and reenter the tower where the vapor disengages from the liquid phase. The liquid is the bottom product of the fractionator and is discharged from the bottom of the tower.

Both the kettle type and the once through thermo-siphon type constitute a theoretical tray as regards fractionation. Unlike the kettle reboiler the once through thermo-siphon is limited to a vaporization of not more than 60% of the feed. The low holdup of the feed from a tray results in severe surging through the reboiler at high vaporization rates.

The re-circulating thermo-siphon reboiler

When vaporization rates higher than 60% of reboiler feed is required and a kettle re-boiler is unsuitable a recirculating Thermo-siphon type reboiler should be considered. A sketch showing this type of reboiler is given below:

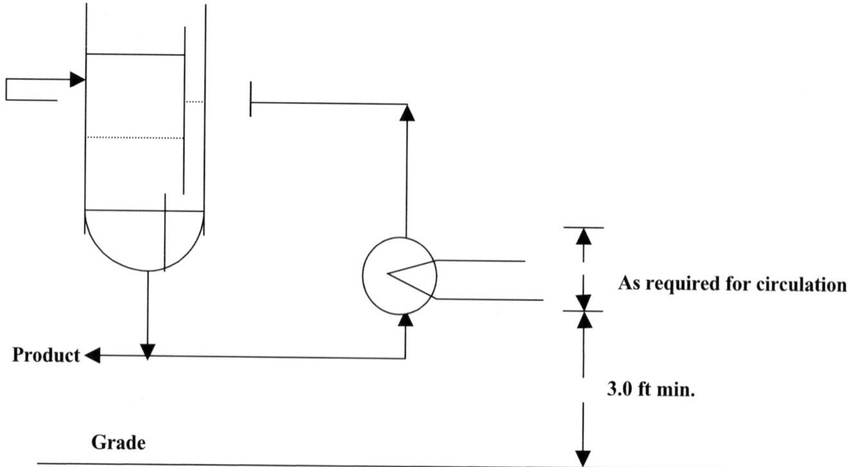

This reboiler is similar to the once through thermo-siphon in as far as it operates by flowing a liquid feed through the shell side of the vertical reboiler by the siphon mechanism. In the case of the recirculating reboiler however the feed to the reboiler is a stream of the bottom product from the fractionator. This is vaporized as described earlier and the liquid/vapor effluent returned to the tower. The vaporization by this reboiler can exceed 60% without danger of surging. However vaporization in this type of reboiler should not exceed 80%. Its action is directed solely to imputing heat to the tower and, because it recycles the same composition stream to the tower bottom, it cannot be considered as a theoretical fractionating tray. (Although some amount of fractionation does occur in this system).

Note in the description of both the thermo-siphon type reboilers the heating fluid is shown as flowing tube side. There may be cases where this stream will be routed shell side and the reboil fluid directed tube side. Some guidance to this selection is provided by the following preference for tube side fluid:

1. Corrosive or fouling fluids.
2. The less viscous of the two fluids.
3. The fluid under the higher pressure.
4. Condensing steam.

Reboiler sizing

As in the case of most heat exchangers the sizing calculation is quite rigorous and complex. Normally process engineers rarely need to compute this in detail. There will be need however to estimate the size of these items for cost purposes or for plot layout studies. This sizing is greatly simplified by applying heat flux quantities to the predetermined reboiler duty. Heat flux is the value of heat transferred per unit time per sqft of surface. The following list gives a range of heat fluxes that have been used in design and observed in operating units.

	Design	Observed
	(Btu/hr · sqft)	
Kettle type	12,000	15,000–20,000
Once through	15,000	17,500+
Recirculating	15,000	up to 20,000
Forced circulation	20,000	–

The duty of the reboiler is obtained by the overall heat balance over the tower. This is accomplished by equating the total heat out of a fractionating tower to the heat supplied, making the reboiler duty the unknown in the heat supplied statement. Now the heat out of the fractionator is the total heat in the products leaving plus the condenser duty. The heat supplied to the tower is the heat brought in with the feed, and the heat supplied by the reboiler.

Example calculation
The feed to a fractionator is 87,960 lbs/hr of mixed hydrocarbons. It enters the tower as a vapor and liquid stream and has a total enthalpy of 15.134 mm Btu/hr. The overhead products are a distillate and a vapor stream at 95°F. The vapor is 1,590 lb/hr with an enthalpy of 320 Btu/lb. The distillate is 8,028 lbs/hr with an enthalpy of 170 Btu/hr. The bottom product from the tower is 78,342 lbs/hr and leaves as a liquid at its boiling point at 440°F. Its enthalpy is 370 Btu/lb. The overhead condenser duty is 4.278 mm Btu/lb. Calculate the reboiler duty.

Calculation
Calculate the reboiler duty from the overall tower heat balance as in the following:

	V/L	°API	°F	lbs/hr	Enthalpy, Btu/lb	mm Btu/hr
In						
Feed	VL	–	300	87,960	–	15.134
Reboiler duty						x
Total In				87,960		$15.134 + x$
Out						
Bottom Prod	L	–	440	78,342	370	28.986
O/head Dist	L	–	95	8,028	170	1.364
O/head vapor	V	–	95	1,590	320	0.508
Condenser duty						4.278
Total out				87,960		35.136

$$\text{Heat In} = \text{Heat Out}$$

Then

$$15.134 + x = 35.136$$
$$\text{Reboiler duty } x = 20.002 \text{ mm Btu/hr.}$$

Using a heat flux of 15,000 Btu/hr sqft the surface area for the reboiler becomes

$$\frac{20002000}{15{,}000} = 1{,}333.5 \text{ sqft.}$$

Estimating the liquid and vapor flow from the reboiler
It is necessary to know the vapor and liquid flow leaving the reboiler and entering the tower for the following reasons:

- To establish that there is sufficient vapor rising in the tower to strip the bottom product effectively
- To establish the vapor loading to the bottom tray for calculating the tray loading

- To be able to calculate the driving force for flow through the exchanger in the case of thermo-siphon reboilers

The calculation of this flow is again based on a heat balance. In this case it is the heat balance across the reboiler itself. With the duty of the reboiler now established by the overall tower heat balance, as described above, the balance over the reboiler can proceed as follows:

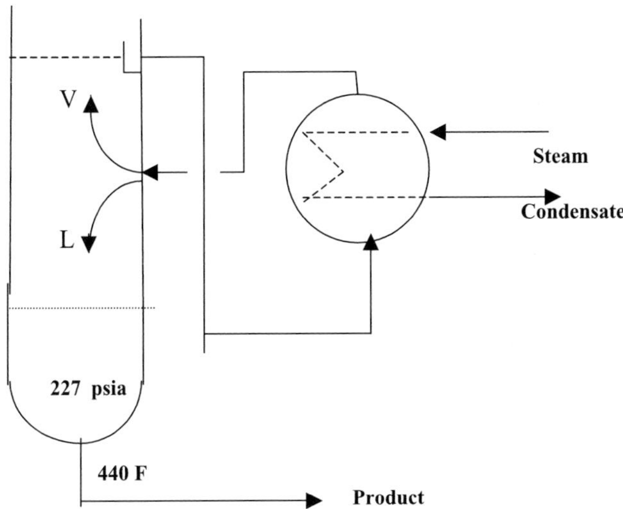

	V/L	°API	°F	lbs/hr	Enthalpy, Btu/lb mm Btu/hr
In					
Liquid from tray 1	L	–	430	78,342 + V	369 28.908 + 369 V
Reboiler duty					20.002
Total In				7,834 + V	48.910 + 369 V
Out					
Bottom prod	L	–	440	78,342	370 + 28.986
Vapor to tray 1	V	–	440	V	458 + 458 V
Total Out				78,342 + V	28.986 + 458 V

The temperature of the bottom tray (430°F) is estimated from a straight line temperature profile of the tower. As a rule of thumb—for a 30 to 40 tray tower the bottom tray will be about 10°F lower than the bottom temperature.

$$\text{Again Heat In} = \text{Heat Out}$$

Then
$$48.910 + 369\text{ V} = 28.986 + 458\text{ V}$$
$$\text{V} = 223{,}865\text{ lbs/hr}$$

Now the mole weight of the vapor is determined from the bubble point calculation of the bottom product used to determine the tower bottom temperature (see Chapter 1). In the case of the calculation example given above the bubble point calculation for the bottom product was as follows:

$$\text{Pressure at bottom of tower} = 220 + (30 \times 0.25)$$
$$= 227\,\text{psia}$$

		1st Trial = 400°F		2nd Trial = 435°F					
	X_W	K	$Y = XK$	K	$Y = XK$	MW	Weight factor	lbs/gal	Vol factor
nC_4	0.017	2.7	0.046	3.1	0.053	58	3.1	4.86	0.64
iC_5	0.047	1.9	0.089	2.2	0.103	72	7.4	5.20	1.42
									(°API 53.1)
nC_5	0.055	1.7	0.094	1.9	0.105	72	7.6	5.25	1.45
C_6	0.345	0.96	0.331	1.3	0.449	81	36.4	6.83	5.33
C_7	0.322	0.44	0.142	0.59	0.190	102	19.4	6.84	2.84
C_8	0.214	0.17	0.036	0.25	0.054	128	6.9	6.94	0.99
Total	1.000		0.738		0.954	84.7	80.8	6.38	12.67

Actual temp = 440°F

The mole weight of the vapor is that calculated for the "y" column in the above table which is 84.7.

Calculating the pressure head driving force through a thermo-siphon reboiler

The big advantage of thermo-siphon reboilers is that there are no working parts, such as pumps, that can go wrong and cause failure. However a major cause of failure in a thermo-siphon reboiler is loss of driving force to move the fluid over or through the tube bundle. This problem mostly occurs during commissioning when the reboiler has been incorrectly positioned relative to the tower nozzles, or during start up where debris left after maintenance blocks one or other of the nozzles. In both these cases the problem is really the loss of pressure head that drives the fluid to be reboiled through the exchanger. The calculation to determine the theoretical driving force is based on the density of the incoming liquid, the head of that liquid to the inlet nozzle, the density of the out flowing liquid/vapor fluid, and its head. An example of a pressure driving force calculation based on a once through thermo-siphon (as shown in the diagram below) is as follows. The flow data is based on the heat balance given earlier in this item.

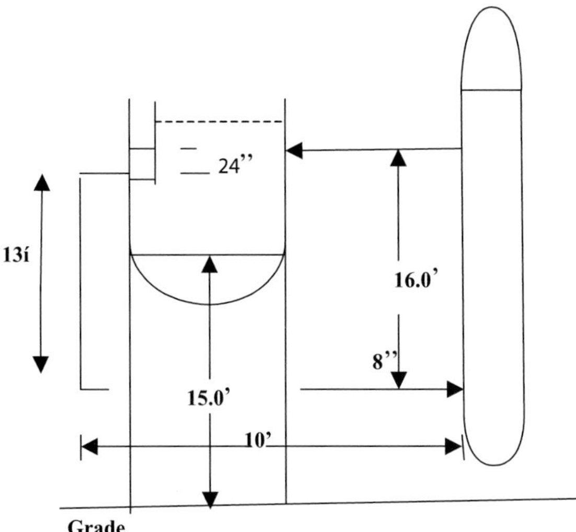

The density of the liquid to the reboiler is 38.2 lbs/cuft at 430°F and the total flow is 302,207 lbs/hr.

$$\text{Hot cuft/hr} = 7,911$$
$$\text{Hot gpm} = 966$$

The transfer line from the tower to the bottom nozzle of the reboiler is a 8″ schedule 40 seamless steel pipe. The head between the bottom of the tower draw off pot and the reboiler nozzle is 13 ft. The equivalent horizontal line length including fittings is 15 ft.

From the friction loss tables in the Appendix, head loss due to friction = 66.4 ft/100 ft of line (viscosity is taken as 1.1 cs).

$$\begin{aligned}
\text{Total line length to the reboiler is } 13 + 15\text{ ft} &= 28\text{ ft.}\\
\text{Head loss due to friction} = \frac{28 \times 66.4}{1,000} &= 1.85\text{ ft.}\\
\text{Head of liquid in draw off pot} &= 24''\\
\text{Head of liquid to the reboiler inlet nozzle} = 13 + 2\text{ ft.} &= 15\text{ ft.}\\
\text{Pressure head at the reboiler inlet} = 15 - 1.85\text{ ft} &= 13.15\text{ ft}\\
&= \frac{13.15 \times 38.2}{144}\\
&= 3.49\text{ psi}
\end{aligned}$$

The density of the vapor/liquid stream leaving the reboiler is calculated as follows:

$$\frac{\text{Total mass of fluid}}{\text{cuft liquid} + \text{cuft vapor}}.$$

lbs/cuft of liquid (this is bottom product) = 39.4 @ 440°F
mole wt of vapor (see bubble point calculation above) = 84.74

$$\text{cuft/hr of liquid} = \frac{78{,}342 \text{ lbs/hr}}{39.9} = 1{,}963.5$$

$$\text{cuft/hr of vapor} = \frac{223{,}865}{84.5} \text{lbs/hr} = 2{,}643 \text{ moles/hr}$$

$$= \frac{2{,}643 \times 378 \times 14.7 \times (460 + 440°\text{F})}{227 \times 520}$$

$$= 111{,}974.6 \text{ cuft/hr}$$

$$\text{density of fluid from reboiler} = \frac{302{,}207}{1{,}963.5 + 111{,}974.6}$$

$$= 2.65 \text{ lbs/cuft.}$$

In this case the fluid to be reboiled flows on the shell side of the exchanger. The manufacturer's certified shell side pressure drop based on all vapor flow is 1.5 psi. The mixed phase pressure drop is calculated using Figure 18.46 thus.

$$m_v = \text{average mass of vapor} = \frac{223{,}865}{2} = 111{,}933 \text{ lbs/hr}$$

$$m_l = \text{average mass of liquid} = \frac{302{,}207 + 78{,}342}{2} = 190{,}275 \text{ lbs/hr}$$

$$\rho_v = \text{average density of vapor} = \frac{2.0}{2} = 1.0 \text{ lbs/cuft.}$$

$$\rho_l = \text{average density of liquid} = \frac{38.8 + 39.9}{2} = 39.35 \text{ lbs/cuft}$$

Referring to Figure 18.45

$$R_m = \frac{1}{\frac{m_v}{m_l} + \frac{\rho_v}{\rho_l}}$$

$$= \frac{1}{\frac{111{,}933}{190{,}275} + \frac{1.0}{39.32}}$$

$$= 1.63$$

From Figure 18.45

$$\alpha = 0.42$$

$$\begin{aligned}\Delta P_{\text{mixed phase}} &= \alpha \times \Delta P_{\text{gas}} \\ &= 0.42 \times 1.5 \text{ psi} \\ &= 0.63 \text{ psi.}\end{aligned}$$

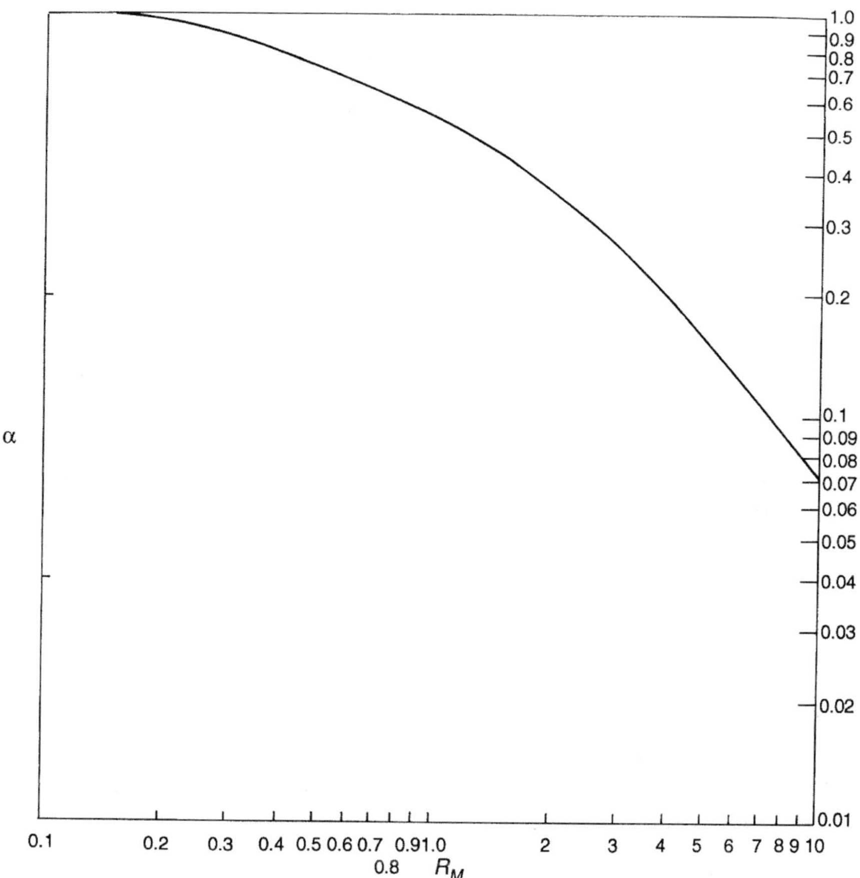

Figure 18.45. Two phase pressure drop factor for flow across staggered tubes.

To calculate the total head of liquid from exchanger inlet to outlet nozzle.

Tube length is standard 16 ft.

Assume bottom 20% of length is all liquid phase at a density of 38.8 lbs/cuft.

Then head is $\dfrac{16 \times 0.2 \times 38.8}{144} = 0.862$ psi

Remaining head is mixed phase at a density of 2.65 lbs/cuft.

Then head is $\dfrac{16 \times 0.8 \times 2.65}{144} = 0.24$ psi

Neglecting the small pressure drop due to friction in the 4 ft long return line to tower,

$$\begin{aligned} \text{Friction loss through exchanger} &= 0.630 \text{ psi} \\ \text{Lower section head} &= 0.862 \text{ psi} \\ \text{Upper section head} &= 0.240 \text{ psi} \\ \text{TOTAL} &= 1.732 \text{ psi} \end{aligned}$$

$$\begin{aligned} \text{Then driving force} &= \text{Pressure head available} - \text{pressure head required.} \\ &= 3.49\,\text{psi} - 1.732\text{psi} \\ &= 1.758\,\text{psi which is satisfactory for good flow.} \end{aligned}$$

Note the height of the tower above grade is usually fixed by pump suction requirements in the first place. It may be adjusted upwards if necessary to accommodate a head to a reboiler. However this necessity is quite rare. The transfer line to the reboiler should have its horizontal section at least 3.0 ft above grade to allow for maintenance, etc.

18.5 Fired Heaters

Types of fired heaters

This chapter provides some features and detail of fired heaters.

Most chemical plants and all petroleum refineries contain fired heaters as a means of providing heat energy into a system. Because the equipment utilizes an outside source of fuel it is usually supported and enhanced by a heat exchange system to minimize the quantity of fuel required.

Generally fired heaters fall into two major categories:

- Horizontal type
- Vertical type

The horizontal type heater usually means a box type heater with the tubes running horizontally along the walls. Vertical type is normally a cylindrical heater containing vertical tubes. Figures 18.46 and 18.47 show examples of these two types of heaters.

These figures also give some nomenclature used in describing these items of equipment. Other terms used in connection with fired heaters are as follows:

Headers & Return Bends: are the fittings used to connect individual tubes.

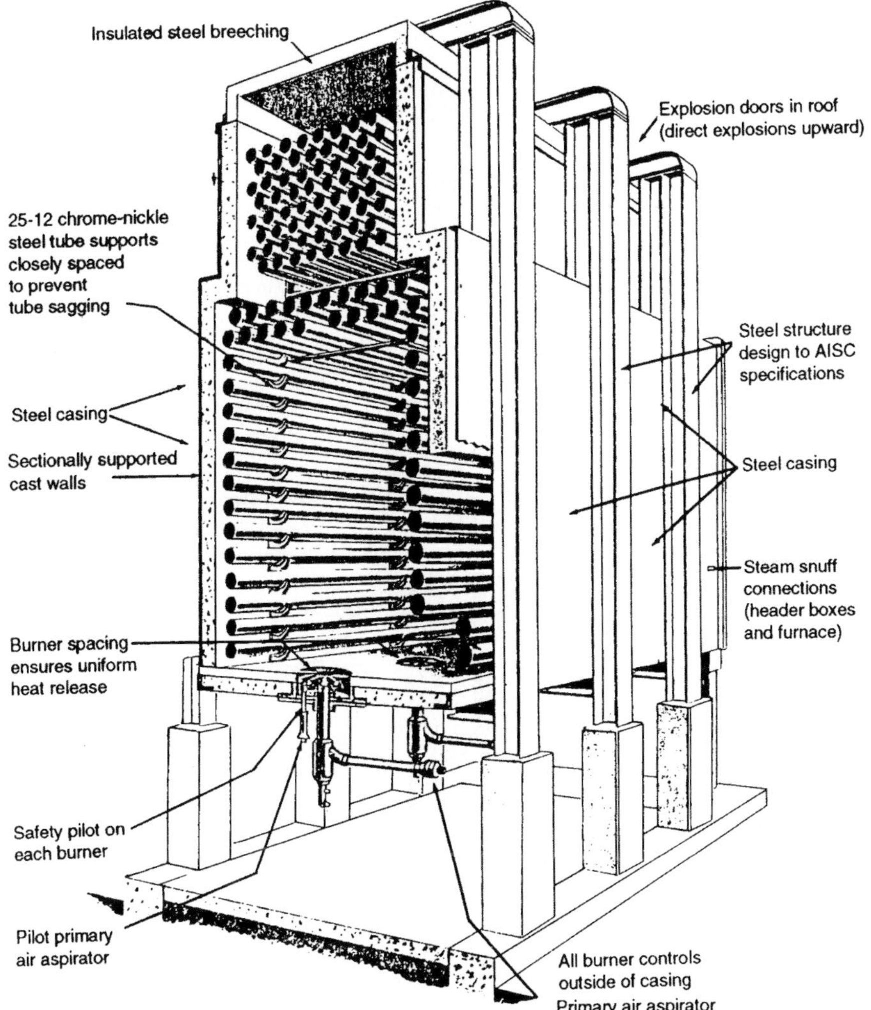

Figure 18.46. Horizontal type heater.

Terminals: are the inlet and outlet connections.

Crossovers: are the piping used to connect the radiant with the convection section; usually external to the heater.

Manifold: is the external piping used to connect the heater passes to the process piping; may be furnished with the heater.

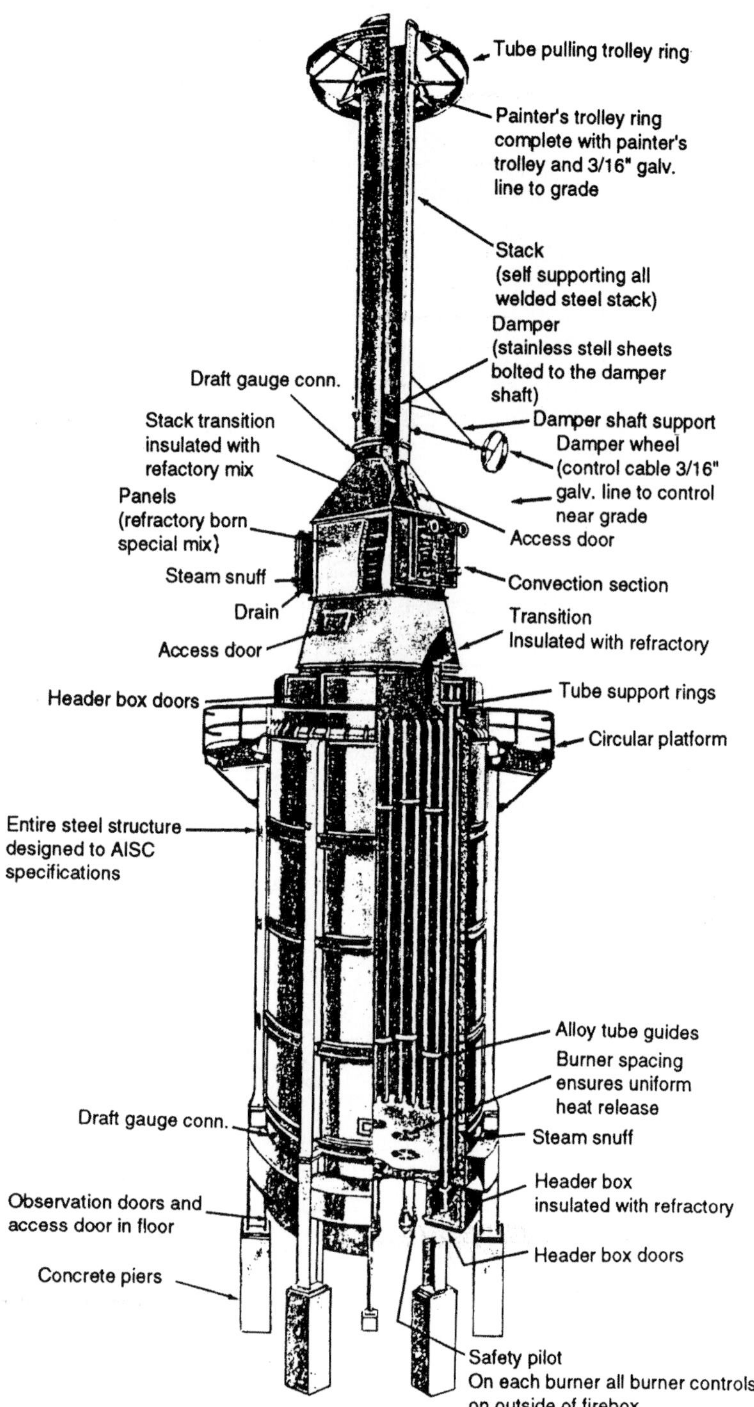

Tube pulling trolley ring
Painter's trolley ring complete with painter's trolley and 3/16" galv. line to grade
Stack (self supporting all welded steel stack)
Damper (stainless stell sheets bolted to the damper shaft)
Draft gauge conn.
Damper shaft support
Stack transition insulated with refactory mix
Damper wheel (control cable 3/16" galv. line to control near grade
Panels (refractory born special mix)
Access door
Steam snuff
Convection section
Drain
Transition Insulated with refractory
Access door
Header box doors
Tube support rings
Circular platform
Entire steel structure designed to AISC specifications
Alloy tube guides
Burner spacing ensures uniform heat release
Draft gauge conn.
Steam snuff
Header box insulated with refractory
Observation doors and access door in floor
Header box doors
Concrete piers
Safety pilot On each burner all burner controls on outside of firebox

Figure 18.47. Vertical type heater.

Setting: any and all parts that form:

(a) coil supports
(b) enclosure (housing)

Casing: Is the steel shell which encloses the heater.

Bridge wall or partition wall: are the refractory walls inside the heater that divide the radiant section into separately fired zones.

Shield tubes or shock tubes: are the first 2 or 3 rows of tubes in the convection section. They protect or shield the convection tubes from direct radiant heat and must have the same metallurgy as the radiant tubes and have no fins.

Air plenum: is the chamber enclosing burners under the heater and having louvers to control the air flow.

Cylindrical heaters require less plot space and are usually less expensive. They also have better radiant symmetry than the horizontal type.

Horizontal box types are preferred for crude oil heaters, although vertical cylindrical heaters have been used in this service. Vacuum unit heaters should have horizontal tubes to eliminate the static head pressure at the bottom of vertical tubes and to reduce the possibility of two-phase slugging in the large exit tubes.

Occasionally, several different services ("coils") may be placed in a single heater with a cost saving. This is possible if the services are closely tied to each other in the process. Catalytic reforming pre-heater and re-heaters in one casing is an example. Reactor heater and stripper reboiler in one casing is another example. This arrangement is made possible by using a refractory partition wall to separate the radiant coils. The separate radiant coils may be controlled separately over a wide range of conditions by means of their own controls and burners. If a convection section is used, it is usually common to the several services. If maintenance on one coil is required, the entire heater must be shut down. Also, the range of controllability is less than with separate heaters.

Each of these types may be shop fabricated if size permits. Shop fabrication reduces costs. However, shop fabrication should not be forced to the extent of getting an improperly proportioned heater.

Codes and standards

Fired heaters have a "live" source of energy. That is, they use a flammable material in order to impart heat energy to a process stream. Because of this the design, construction and operation of process fired heaters and boilers are strictly controlled by legislative and other codes and standards. This item outlines some of the more important of these codes and standards which need to be recognized by engineers dealing with fired heaters in any way.

Codes and standards directly applicable to fired heaters are listed below. In addition, there are many codes and standards covering such factors as materials, welding, refractories, structural steel, etc., which apply to fired heaters.

API RP-530

Calculation of heater tube thickness' in refineries

This recommended practice sets forth procedures for calculating the wall thickness of heater tubes for service at elevated temperatures in petroleum refineries.

API Standard 630

Tube and heater dimensions for fired heaters for refinery service

This standard establishes certain standard dimensions for heater tubes and for cast and wrought headers.

Tube sizes and header centre-to-centre dimensions covered by the standard are as follows:

Tube OD, inches		Header C-TO-C, inches	
Primary	Secondary	Group A	Group B
2.375	–	4.00	4.75
2.875	–	5.00	5.25
3.50	–	6.00	–
4.00	–	7.00	6.50
4.50	–	8.00	7.25
–	5.00	9.00	7.75
5.563	–	10.00	8.50
–	6.00	11.00	9.00
6.625	–	12.00	10.00
–	7.625	14.00	12.00
8.625	–	16.00	14.00

Groove dimensions and tolerances for rolled headers are also given. Much of this standard is also used in chemical and petrochemical plants.

API RP-2002 Fire protection in natural gasoline plants

This practice contains a brief statement about the use of snuffing steam. This system provides the piping of a steam source to the heater fire box which in an emergency can introduce steam into the box to quench any uncontrolled fire. This system may be automatically controlled or activated manually.

API Guide for Inspection of Refinery Equipment, Chapter IX, Fired Heaters and Stacks.

This reference gives a general description of fired heaters and describes how to inspect them, what damage to look for, and how to report the results of the inspection.

ASME boiler code and boiler codes of the USA

These are applicable in process plants if steam is generated, superheated or boiler feed water preheated in the convection section. Special materials are required according to ASME Section I. The external piping and pressure relieving devices must also be in accordance with ASME Section I.

Contractors' standards

These will be covered in the "Narrative Specification" for the particular job and/or heater. The narrative specification is written by the heat transfer engineer specialists in the Contractors' Mechanical Equipment Group. These specifications detail all of the pertinent aspects required in the manufacture of the equipment. It will encompass all of the requirements of the applicable.

Thermal rating

Refinery process engineers are seldom if ever required to thermal rate a fired heater or indeed check the thermal rating. This is a procedure that falls in the realm of specialist mechanical engineers with extensive experience in heater design and fabrication. Process engineers are however required to specify the equipment so that it can be designed and installed to meet the requirements of the process heat balance. To do this effectively it is desirable to know something about the mechanism of heater thermal rating.

A fired heater is essentially a heat exchanger in which most of the heat is transferred by radiation instead of by convection and conduction. Rating involves a heat balance between the heat releasing and heat absorbing streams, and a rate relationship.

Fuel is burned in a combustion chamber to produce a “flame burst”. The theoretical flame burst temperature may vary from 4,000°F when burning refinery gases with 20% excess air preheated to 460°F, down to 2,300°F when burning residual fuel oils with 100% excess air at 60°F. Heat is transferred from the flame burst to the gases in the firebox by radiation and mixing of the products of combustion. Heat is then transferred from the firebox gases to the tubes mainly by radiation.

The common practice is to assume a single temperature for the firebox gases for the purpose of radiation calculations. This temperature may be the same as the exit gas temperature from the firebox to the convection section (bridge wall temperature), or it may be different due to the shape of the heater and to the effect of convection heat transfer in the radiant section. Experience with the particular type of heater is required in order to select the effective firebox temperature accurately.

Whilst this chapter does not detail the rating procedure or give an example calculation the following steps summarize the rating procedure:

1.0 Calculate net heat release and fuel quantity burned from the specified heat absorption duty and an assumed or specified efficiency.
2.0 Select excess air percentage and determine flue gas rates.
3.0 Calculate duty in the radiant section by assuming 70% of total duty is radiant. This is a typical figure and will be checked later in the calculations. For very high process temperatures, such as in steam-methane reforming heaters, the radiant duty may be as low as 45% of the total.
4.0 Calculate the average process fluid temperature in the radiant section and add 100°F to get the tube wall temperature. The figure of 100°F is usually a good first guess and can be checked later by using the calculated inside film coefficient and metal resistance.
5.0 Calculate the radiant surface area using the average allowable flux. Convection surface is usually about equal to the radiant surface.
6.0 Select a tube size and pass arrangement that will give the required total surface and meet specified pressure drop limitations.
7.0 Select a center-to-center spacing for the tubes from the API 630 Standard or from dimensions of standard fittings, and calculate firebox dimensions. Long furnaces minimize the number of return bends and thus reduce cost. Shorter and wider fireboxes usually give more uniform heat distribution and lessen the probability of flame impingement on the tubes. For vertical cylindrical heaters, the ratio of radiant tube length to tube circle diameter should not exceed 2.7.
8.0 The remainder of the calculation involves determining the firebox exit temperature from assumption (3) above, applying an experience factor for the type of furnace to obtain the average firebox temperature, and then checking if this temperature will transfer the required radiant heat.

9.0 The average heat flux (proportional to radiant surface) and the percent of total duty of the radiant section (which affects average tube wall temperature) are varied until a balance is obtained. The convection section surface and arrangement can now be calculated.

10.0 The heater is normally designed to allow adequate draft at the burners with at least 125% of design heat release and an additional 10% excess air at the maximum and minimum ambient air temperatures.

Heat flux

Although a process engineer is not normally required to thermally rate a process heater he is often required to estimate the heater size, for preliminary cost estimates or plot layout and the like. This can be accomplished quite simply by the use of "Heat Flux". Whilst this figure is quoted as Btu/hr sqft, it is not however an overall heat transfer coefficient, it lacks the driving force ΔT for this. Heat flux is the rate of heat transmission through the tubes into the process fluid.

The maximum film temperature and tube metal temperature are a function of heat flux and the inside film heat transfer coefficient.

The heat flux varies around the circumference of the tube, being a maximum on the side facing the firebox. The value depends upon the sum of the heat received directly from the firebox radiation and the heat re-radiated from the refractory.

Single fired process heaters are usually specified for a maximum average heat flux of 10,000–12,000 Btu/hr sqft. The maximum point heat flux is about 1.8 times greater.

Double fired heaters are usually specified for about 13,500 × 18,000 Btu/hr sqft *average* heat flux with the maximum point flux being about 1.2 times greater.

The following are typical flux values for heaters in hydrocarbon service:

Horizontal, fired on one side	8,000–12,000
Vertical, fired on one side from bottom	9,000–12,000
Vertical, single row, fired on both sides	13,000–18,000

Heater efficiency

The efficiency of a fired heater is the ratio of the heat absorbed by the process fluid to the heat released by combustion of the fuel expressed as a percentage.

Heat release may be based on the lower heating value (LHV) of the fuel or higher heating value (HHV). Process heaters are usually based on LHV and boilers on HHV. The HHV efficiency is lower than the LHV efficiency by the ratio of the two heating values.

Heat is wasted from a fired heater in two ways:

- with the hot stack gas,
- by radiation and convection from the setting.

The major loss is by the heat contained in the stack gas. The temperature of the stack gas is determined by the temperature of the incoming process fluid unless an air pre-heater is used. The closest economical approach to process fluid is about 100°F. If the major process stream is very hot at the inlet, it may be possible to find a colder process stream to pass through the convection section to improve efficiency, provided plant control and flexibility are adequately provided for. A more common method of improving efficiency is to generate and/or superheat steam and preheat boiler feed water.

The lowest stack temperature that can be used is determined by the dew point of the stack gases. See the section on stack emissions.

Figures 18.48 and 18.49 may be used to estimate flue gas heat loss.

The loss to flue gas is expressed as a percentage of the total heat of combustion available from the fuel. These figures also show the effect of excess air on efficiency. Typically excess air for efficiency guarantees is 20% when firing fuel gas and 30% when firing oil.

Heat loss from the setting, called radiation loss, is about 1½ to 2% of the heat release.

The range of efficiencies is approximately as follows:

Very high	—	90%+. Large boilers and process heaters with air pre-heaters.
High	—	85%. Large heaters with low process inlet temperatures and/or air pre-heaters.
Usual	—	70–80%.
Low	—	60% and less. All radiant.

Engineers are often required to check the efficiencies of the process heaters on operating units assigned to them. This can be done using the heats of combustion given in Figures 18.A.2 and 18.A.3 in the Appendix to this Chapter. The steps used to carry out these calculations are as follows:

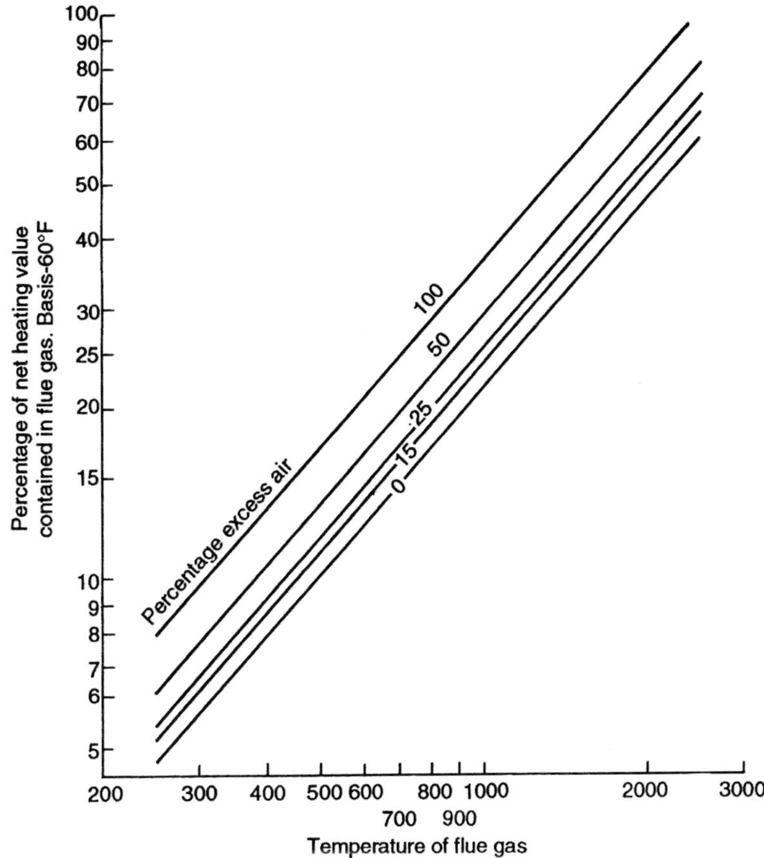

Figure 18.48. Percentage of net heating values contained in flue gas when firing fuel gas.

Step 1. Obtain details of the heater from the manufacturer's data sheet or drawings. The data required are:

- Tube area
- Layout (is it vertical and how are the burners located?)

Step 2. From plant data obtain coil inlet and outlet temperature pressures and flow. Calculate the outlet flash (i.e., vapor or liquid or a mixture of vapor and liquid) condition, then calculate its enthalpy. Do the same for the inlet flow. Usually this will be a single phase either liquid or vapor.

Step 3. Again from plant data obtain the quantity of fuel fired and its properties (API Gravity in particular).

Step 4. The difference between the enthalpies calculated in Step 2 is the enthalpy absorbed by the feed in Btu per hour.

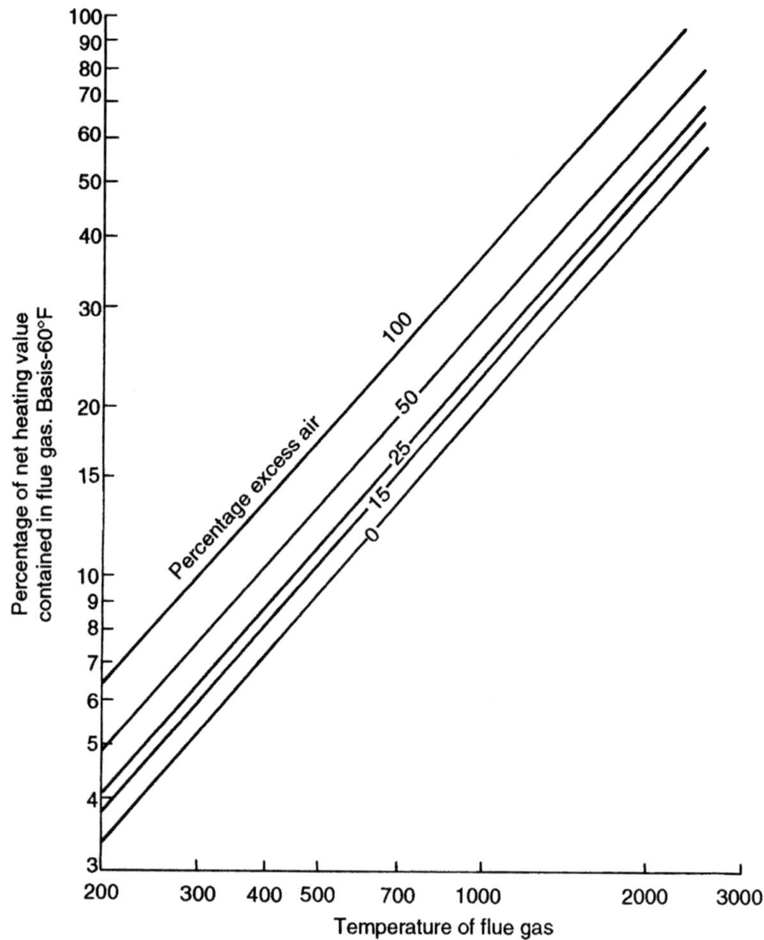

Figure 18.49. Percentage of net heating values contained in flue gas when firing fuel oil.

Step 5. Divide this absorbed enthalpy by the tube area to give the heat flux in Btu/hr/ft^2. Heat flux's are generally as follows:

Horizontal, fired one side	8,000 to 12,000 Btu/hr/ft^2
Vertical, fired from bottom on one side	9,000 to 12,000 Btu/hr/ft^2
Vertical, single row, fired on both sides	13,000 to 18,000 Btu/hr/ft^2

If the heat flux falls outside this range given above, there could be excessive fouling. Check the pressure drop—if this is far above manufacturer's calculated value then fouling is certainly present.

Step 6. Check the thermal efficiency of the heater by giving the fuel fired a heating value. This is provided by figures in the appendix. Use the LHV (lower heating valve) in Btu/lb and multiply it by the lbs/hr of the fuel.

Step 7. Divide the heat absorbed by the heat released calculated in Step 7 to give the thermal efficiency. For most heaters this should be between 70% and 80%. If it falls below this range note should be taken of burner operation and the amount of excess air being used.

Burners

The purpose of a burner is to mix fuel and air to ensure complete combustion. There are about 12 basic burner designs. These are:

Direction	—	vertical up fired
		vertical down fired
		horizontally fired
Capacity	—	high
		low
Fuel type	—	gas
		oil
		combination
Flame shape	—	normal
		slant
		thin, fan-shaped
		flat
		adaptable pattern
Hydrogen content	—	high
Excess air	—	normal
		low
Atomization	—	steam
		mechanical
		air assisted mechanical
Boiler types		
Low NO_x		
High intensity		

Various combinations of the above types are available.

Gas burners

The two most common types of gas burners are the “pre-mix” and the “raw gas” burners.

Pre-mix burners are preferred because they have better “linearity”, i.e., excess air remains more nearly constant at turndown. With this type, most of the air is drawn in through an adjustable “air register” and mixes with the fuel in the furnace firebox. This is called secondary air. A small part of the air is drawn in through the “primary air register” and mixed with the fuel in a tube before it flows into the furnace firebox. A turndown of 10:1 can be achieved with 25 psig hydrocarbon fuels. A more normal turndown is 3 : 1.

Oil burners

An oil burner “gun” consists of an inner tube through which the oil flows and an outer tube for the atomizing agent, usually steam. The oil sprays through an orifice into a mixing chamber. Steam also flows through orifices into the mixing chamber. An oil-steam emulsion is formed in the mixing chamber and then flows through orifices in the burner tip and then out into the furnace firebox. The tip, mixing chamber and inner and outer tubes can be disassembled for cleaning.

Oil pressure is normally about 140–150 psig at the burner, but can be lower or higher. Lower pressure requires larger burner tips, the pressure of the available atomizing steam may determine the oil pressure.

Atomizing steam should be at least 100 psig at the burner valve and at least 20–30 psi above the oil pressure. Atomizing steam consumption will be about 0.15–0.25 lbs steam/lb oil, but the steam lines should be sized for 0.5.

Combination burners

This type of burner will burn either gas or oil. It is better if they are not operated to burn both fuels at the same time because the chemistry of gas combustion is different from that of oil combustion. Gases burn by progressive oxidation and oils by cracking. If gas and oil are burned simultaneously in the same burner, the flame volume will be twice that of either fuel alone.

Pilots

Pilots are usually required on oil fired heaters. Pilots are fired with fuel gas.

Pilots are not required when heaters are gas fired only, but minimum flow bypasses around the fuel gas control valves are used to prevent the automatic controls from extinguishing burner flames.

Excess air and burner operation

The excess air normally used in process fired heaters is about 15–25% for gas burners and about 30% for oil burners. These excess air rates permit a wide variation in heater firing rates which can be effectively controlled by automatic controls without fear of 'Starving' the heater of combustion air. There has been considerable work lately to reduce this excess air considerably mostly to minimize air pollution. This practice has not been used in process heaters to date. It has however been adopted in the operation of large power station type heaters with some success.

Normally companies specify that burners be sized to permit operation at up to 125% of design heat release with a turndown ratio of 3 : 1. This gives a minimum controllable rate of 40% of design without having to shut down burners.

Burner control

Burner controls become very important from safety and operation considerations. Most systems include an instrumentation system with interlocks that prohibit:

- Continuing firing when the process flow in the heater coil fails
- The flow of fuel into the firebox on flame failure

Under normal operating conditions the amount of fuel that is burnt is controlled by flow controllers operated on the coil outlet temperature. With combination burners the failure of one type of fuel automatically introduces the second type. Such a switch over can also be effected manually. This aspect is usually activated on pressure control of the respective fuel system. That is, on low pressure being sensed on the fuel being fired automatically switches to the second fuel.

Most companies operate their own specific controls for the heater firing system.

Heater noise

All heaters are noisy and this noise is the result of several mechanisms. Among these are the operation of the burners. Gas burners at critical flow of fuel emit a noise. This can be minimized by designing for low pressure drop in the system. Intake of primary and secondary air is another source of noise. Forced draft burners are generally quieter than natural draft if the air ducting is properly sized and insulated. The design of the fan can also reduce noise in this mechanism. Low tip speed fan favors low noise levels.

Refractories, stacks, and stack emissions

Refractories are used on the inside walls of the heater fire box, floor and through the convection side of the heater. The purpose of this refractory lining is to conserve the heat by limiting its loss to atmosphere by convection. It is also necessary for personal safety of those working on or about the heater who may accidentally touch the heater walls.

Good insulation has the following qualities:

- It has good high temperature strength
- It is resistant to abrasion, spalling, chemical reaction, and slagging
- It has good insulating properties

Among the most common refractories that meet some if not all of the above criteria are silica refractories, high alumina and fire clay brick. These have high resistance to spalling and to thermal shock. Their insulating qualities are also good.

Silica refractories tend to form slag with metal oxide dust and ashes. Compounds of sodium and potassium attack most refractories while refractories containing magnesium react with acids and acid gasses. Carbon monoxide and other reducing chemicals that may be present in the firebox reduce the life of refractories, particularly fire clay brick and silica.

Dense refractories with low porosity are the strongest but have the poorest insulation qualities. Castable refractories containing a mixture of cement and refractory aggregate are the cheapest and the easiest to install. They are not very rugged however. Normally in process heaters conditions are such that the use of a light insulating refractory will satisfy all that is required from a refractory lining.

The ASTM standard part 13 gives more detail on refractories. This standard provides the classification of refractories and describes their characteristics and composition. It also offers a procedure for calculating the heat loss through the insulation and thus its thickness.

Preparing refractories for operation

All new refractories need to be 'cured' after installation and before use. Refractories contain moisture, some due to the installation procedure and some in the form of water of crystallization. Curing the refractory means removing this moisture by applying a slow heating mechanism. New heater manufacturers will usually provide details of the curing procedure they recommend. The following procedure may however be used as a guide to refractory curing when manufacturers' procedures are not available:

1.0 Rise temperature at 50°F per hour to a temperature of 400°F and hold for 8 hr at that temperature.
2.0 Then rise the temperature again at 50°F per hour from 400°F to 1,000°F and hold for another 8 hr.
3.0 If necessary continue heating at 100°F per hr to the operating temperature if higher than 1,000°F. Hold at the operating temperature for a further 8 hrs.
4.0 Cool at 100°F per hr to about 500°F and hold ready for operation.

5.0 On start up the heater can be heated up to its operating temperature at a rate of 100°F per hr.

6.0 During the curing of the refractory it will be necessary to pass some fluid through the heater coils to protect them from overheating. Steam or air may be circulated through the coils for this purpose. In certain cases such as in the catalytic reforming of petroleum stock it may be necessary to circulate the nitrogen or the hydrogen that was used to purge and pressure test the unit. Air and steam in this process would not be desirable.

Stacks

Stacks are used to create an updraft of air from the firebox of a heater. The purpose of this is to cause a small negative pressure in the firebox and thus enable the introduction of air from the atmosphere. This negative pressure also allows for the removal of the products of combustion from the firebox. The stack therefore must have sufficient height to achieve these objectives and overcome the frictional pressure drop in the firebox and the stack itself.

The height required for a stack to achieve good draft can be estimated from the following equation:

$$D = 0.187H(\rho_{\mathrm{a}} - \rho_{\mathrm{g}})$$

where

D = draft in ins of water.
H = Stack height in feet.
ρ_{a} = Density of atmospheric air in lbs/cuft.
ρ_{g} = Density of stack gasses in lbs/cuft at stack conditions.

For stack gas temperature use 100°F lower than gasses leaving the convection section. Stack gasses have a molecular weight close to that of nitrogen. For this calculation use 28 as the mole weight. Burner draft requirements range from 0.2 to 0.5 ins of water. Use 0.3 ins of water as a good design value.

Stacks must also be designed to handle and disperse stack emissions. This usually results in having to build stack heights greater than that required for obtaining draft. There are available specific computer programs relating plume height of the stack gasses above the stack outlet to the probable ground level fallout of the impurities in the gas. These programs also produce a map of the relative concentrations of these impurities at ground level. Such data is usually available from government authority offices in most countries. The predicted ground concentration and map calculated are checked against local legal requirements. The results usually form part of the government approval to build and/or operate a facility.

The stack diameter is based on an acceptable velocity of stack gasses in the stack. This is generally taken as 30 ft/sec. Some allowance must be made for frictional losses these are:

1.5 velocity heads for inlet and outlet losses.
1.5 velocity head for damper.
1.0 velocity head for each 50 ft of stack height.

Stack emissions

The obnoxious compounds in stack gas emissions arise from:

- Impurities in the fuel
- Chemical reactions resulting from the fuel combustion with air

The three major impurities in oil or gas fuels which produce undesirable emissions are:

- Sulfur
- Metals
- Nitrogen

Sulfur

All gas and oil fuels contain sulfur at some level of concentration. When these fuels are burned the sulfur reacts with air to form SO_2 and SO_3. These compounds are objectionable because they cause:

- Air pollution in the form of smog
- They contribute to the corrosion of heater tubes and stack
- SO_3 lowers the dew point of the flue gasses resulting in an objectionable visible plume at the stack exit

Figure 18.50 shows the effect of sulfur in the feed on the flue gas dew point. This dew point also of course varies with the amount of excess air and the relative partial pressure of the combustion gasses.

There is no precise way of determining the amount of SO_3 that is formed in the flue gas. Nor can it be determined with any degree of accuracy where the SO_3 is formed in the system. The total amount of the sulfur oxides is of course determined simply from the sulfur content of the feed. Because of this uncertainty it is important to restrict the minimum allowable convection side metal temperatures to 350°F when firing fuels containing sulfur. Also the minimum temperatures to the stack should be 320°F when firing fuel gas and 400°F when firing fuel oil. In the case of metal stacks the use of a non-corrosive lining must be used in the colder section of the stack if the flue gas temperature falls below those stated above.

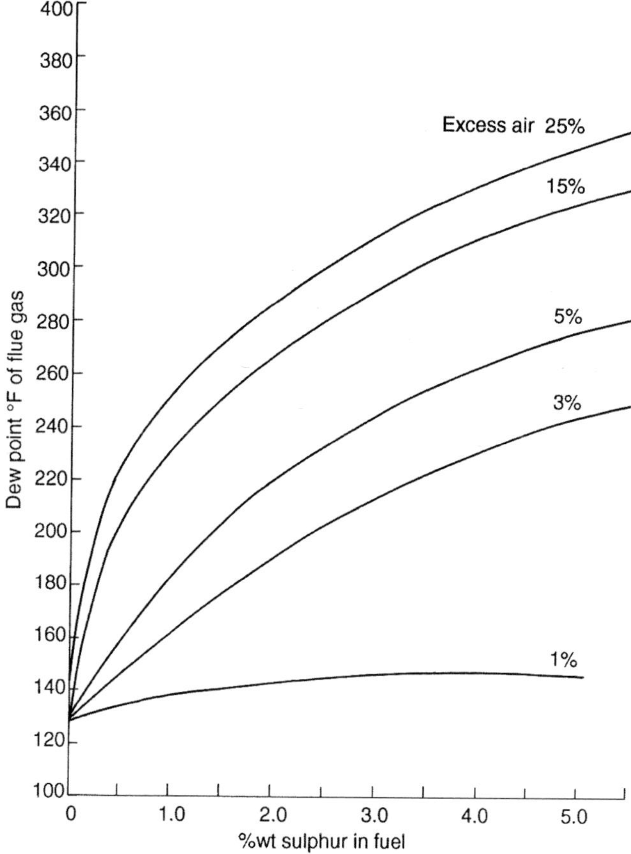

Figure 18.50. Dew point of flue gases versus sulfur content.

Metals

The most objectionable metal impurities in fuel oil are sodium and vanadium. These can cause severe corrosion of tubes and refractory lining. By high vanadium content is meant concentrations between 200 ppm and 400 ppm, above 400 ppm is considered very high and should not be used as a fuel. Vanadium in the presence of sodium and oxygen readily forms a corrosive compound:

$$Na_2O \cdot 6V_2O_5$$

This compound attacks iron or steel to form ferric oxides according to the equation:

$$Na_2O \cdot 6V_2O_5 + Fe \longrightarrow Na_2O\ V_2O_4 \cdot 5V_2O_5 + FeO$$

This vanadium product is oxidized back to the original corrosive compound according to the reaction:

$$Na_2OV_2O_4 \cdot 5V_2O_5 + O_x \longrightarrow Na_2O \cdot 6V_2O_5$$

These reactions occur at temperatures between 1,070°F and 1,220°F with the vanadium compound being continually regenerated to its corrosive state.

Sodium and sulfur also combine to form undesirable corrosive compounds without vanadium being present. This reaction forms sodium iron tri-sulfate $Na_3Fe(SO_4)_3$ at temperatures of around 1,160°F. The critical temperature for both vanadium and sodium corrosion is around 1,100°F. Vanadium is not a major problem at this temperature but becomes so at temperatures above this level.

Nitrogen

Some nitrogen oxides will be formed in combustion gasses even if there are no nitrogen compounds in the fuel. The presence of the nitrogen compounds in the fuel significantly increases the nitrogen oxide content of the flue gas. There are six oxides of nitrogen present in flue gasses but only two are present in any appreciable amount. These are:

NO Nitrogen Oxide.
NO_2 Nitrogen Dioxide.

Nitrogen oxide is a poisonous gas which readily oxidizes to nitrogen dioxide on entering the atmosphere. Nitrogen dioxide is a yellowish gas which readily combines with the moisture in the air to form nitric acid. In the presence of sunlight and oxygen nitrogen dioxide also contributes to the formation of other air pollutants collectively labeled NO_x. The production of NO_x can be reduced by limiting the amount of excess air, by washing the flue gasses with aqueous ammonia, or by catalytically reducing NO_x in the presence of ammonia.

Specifying a fired heater

Some basic data concerning a fired heater must be made known before the equipment can be designed for fabrication or even costed. These data are provided by a specification sheet or sheets. In the case of a fired heater as in the case for compressors this specification can run into several sheets or forms. Such sheets will define the unit in terms of:

- Process Requirement
- Mechanical Detail
- Civil engineering Requirement
- Operational Requirement
- Environmental Requirement

FIRED HEATER.

Design Duty 40886000 Btu\hr Service Hydrocarbons

No Heaters 1 Unit Vac Unit Preheater

Item No H 201 Type Horizontal

DESIGN DATA :-	Radiant sec	Conv sec
Service	Red Crude	Steam s.heat
Heat Absorption mm Btu\hr	38.501	2.385
Fluid	Hydrocarbon	Sat Steam
Flow Rate lbs / hr	197085	30040
Allowable Pressure drop PSI	300	30
Allowable average Flux Btu / hr. sqft	15000	___
Maximum Inside Film Temp °F	800	___
Fouling Factor °F. sqft. hr / Btu	0.004	0.001
Residence Time sec	N/A	N/A
Inlet Conditions		
Temperature °F	554	368
Pressure PSIG	270	155
Liquid Flow lbs / hr	197085	nil
Vapor Flow lbs/ hr	nil	30040
Liquid Density lbs / cuft	48.4	___
Vapour Density lbs\cuft	___	Steam
Visc Liq\Vap Cps	2.31 / ___	__ / __
Specific Heats liq\vap Btu\lb	0.65 / ___	/
Thermal Cond liq\vap Btu\hr.sqft.°F\Ft	0.0671 / ___	/

Figure 18.51. Specification sheet for fired heaters.

This item will deal only with the "Process Requirement" and the duty of the heater. A typical process originated specification sheet is shown here as Figure 18.51. The data provided on this sheet are the minimum that will be necessary for a heater manufacturer or a heater specialist to begin to size the heater or price out the item. Usually this data would also be supported with a vaporization curve of the feed if there is a change of phase taking place in the heater coils. The process engineer developing this specification would also provide details of services required in addition to the main duty of the heater. For example if it is intended that the convection side of

FIRED HEATER. (Cont)

DESIGN DATA (Cont)	Rad Sec	Conv Sec
Outlet Conditions.		
Temperature °F	750	500
Pressure (Psia) PSIG	(0.68)	125
Liquid Flow lbs\hr	69641	nil
Vapour Flow lbs\hr	127444	30040
Liquid Density lbs\cuft	47.3	____
Vapour Density lbs\cuft	0.023	0.246
Visc of Liq\Vap Cps	1.4 /0.002	__ /__
Specific Heat Liq\Vap Btu\lb	0.64 /0.69	__ /__
Thermal Cond Liq\Vap Btu\hr.sqft.°F\ Ft	0.062 /0.023	/
FUEL DATA		
Type (Gas or Oil)	OIL	GAS
LHV Gas Btu\cuft Oil Btu\lb	17560	2320
HHV Gas Btu\cuft Oil Btu\lb	18580	2520
Pressure at Burner PSIG	82	25
Temp at Burner °F	176	68
Mol Wt of Gas		44
Visc of Oil @ Burner Cps	23.3	
Atomising Steam Temp°F Press PSIG	500 125	
Composition of Gas Mol %		
H2		0.2
C1		12.0
C2		28.2
C3		31.3

Figure 18.51. (Cont.)

the heater is to be used for steam generation, preheating, or steam superheating, the system envisaged must be properly described by additional diagrams and data.

The example specification sheet Figure 18.51 is completed for a crude oil vacuum distillation unit heater with a steam super heater coil located in the convection side of the heater. The following paragraphs describe the content on a line by line basis:

FIRED HEATER (Cont).

FUEL DATA (Cont)		
Composition of Gas (Cont)		
C4's		
C5's		
C6+		
Properties of Fuel oil		
° API	15.2	
Visc @ 100°F cs	175	
Visc @ 210°F cs	12.5	
Flash Pt °F	200	
Vanadium ppm	12	
Sodium ppm	32	
Sulphur %wt	2.4	
Ash %wt	< 1.0	

REMARKS

1.0 Flash curve of the reduced crude feed versus % volume distilled is attached for atmos pressure and at 35mm Hg Abs.

2.0 Gravity and mole wt curves for the reduced crude versus mid volume % are attached.

3.0 Soot blowers are to be considered in this package. Steam is available at 600 psig and 750°F.

4.0 Studded tubes to be considered for the convection side.

Figure 18.51. (Cont.)

Line 1. *Design duty* refers to the total duty required of the unit. In the case of the example given here it includes the duty required to heat and partially vaporize the oil in the radiant section and the duty required to superheat the steam. The data sheet will be split into two sections to reflect each of these duties.

Service describes the main purpose for which the heater will be used. In this case it will be used to heat and vaporize hydrocarbons.

Line 2. *No of heaters.* This is self explanatory. In this case there is only one heater required. Should there have been more than one identical unit this would be reflected here.

Unit. This is the title given to this unit of equipment as it appears on an equipment list. It will correspond to the item number also given in the equipment list. In the case of the example it is "Vac unit pre-heater."

Line 3. *Item No.* This is the reference number given to the item in the equipment list. This reference number and unit title identifies the equipment on all drawings where it appears, and all documents used in its purchase, costing, maintenance, etc.

Type. The type of heater (if decided on) is given here. In the case of the example heater a cabin (Horizontal Tubes) type has been selected for vacuum services considerations.

Line 4. This is the first line of the specification sheet proper. It commences with the service of the heater or section of the heater. In the case of this example only two columns have been provided. These are designated for the 'Radiant' section and 'Convection' section. On most preprinted specification forms there would be at least 4 columns.

Line 5. *Heat absorption.* This line divides the duty of the heater into that required from the radiant coils and that required from the convection side. In this example the oil is routed through the radiant section only while saturated steam from a waste heat boiler is superheated in the convection coils. Both are measured in million Btu/hr.

Partial vaporization of the oil occurs in the radiant coil. Therefore a flash curve or a phase diagram of the oil must accompany this specification sheet. In the example the duty to the oil is calculated from data developed in the material balance and heat balance of the process. Thus:

Temperature of the feed into the heater is 554°F and that of the coil outlet is 750°F. From the flash curve at the outlet pressure of 35 mm Hg abs (0.68 psia) and the material balance the weight per hr of vapor is calculated to be 127,444 lbs/hr and the liquid portion is 69,641 lbs/hr. The heat absorbed in the radiant section is:

$$\begin{aligned}
\text{Heat in with feed} &= 197{,}085 \text{ lbs/hr} \times 268 \text{ Btu/lb (all liquid)} \\
&= 52.819 \text{ mm Btu/hr.} \\
\text{Heat out in feed:} & \\
\text{Liquid portion} &= 69{,}641 \text{ lbs/hr} \times 378 \text{ Btu/lb} \\
&= 26.324 \text{ mm Btu/hr.} \\
\text{Vapor portion} &= 127{,}444 \text{ lbs/hr} \times 510 \text{ Btu/llb} \\
&= 64.996 \text{ mm Btu/hr.} \\
\text{Total heat out} &= 91.320 \text{ mm Btu/hr.} \\
\text{Duty of radiant sect} &= 91.320 - 26.324 \\
&= 38.501 \text{ mm Btu/hr.}
\end{aligned}$$

For the convection section the duty is calculated as follows:

Weight of saturated steam at 155 psig is 30,040 lbs/hr
Temperature of 155 psig steam = 368°F
From steam tables steam enthalpy = 1,195.9 Btu/lb
Temperature of steam out = 500°F
Pressure of steam out = 125 psig.
From steam tables steam enthalpy = 1,275.3 Btu/lb.
Duty of convection side = 30,040 (1,275.3 − 1,195.9)
= 2.385 mm Btu/hr.

Line 6. *Fluid.* This refers to the material flowing in the coil. In the case of this example it will be hydrocarbons in the radiant coil and steam in the convection coil.

Line 7. *Flow rate.* This is the total flow rate in lbs per hour entering the respective section of the heater. Thus for the radiant side the figure will be 197,085 lbs/hr, and for the convection side it will be 30,040 lbs/hr.

Line 8. *Allowable pressure drop.* The process engineer enters the required pressure drop calculated from the hydraulic analysis of the system. This pressure drop is measured from the heater side of the inlet manifold down stream of the balancing control valves and the coil outlet down stream of the outlet manifold.

Line 9. *Allowable average flux.* This is usually a standard set by the company for its various heaters. In the example here this value would be between 13,500 and 18,000 Btu/hr sqft. (a horizontal heater fired on both sides). It is specified as 15,000 Btu/hr sqft for this example and refers only to the radiant section.

Line 10. *Maximum inside film temperature.* It is important to notify the heater manufacturer of any temperature constraint that is required by the process. In the case of this example temperatures of the oil above 800°F may lead to the oil cracking. Such a situation could adversely affect the performance of the downstream fractionation equipment and therefore high temperatures in excess of 800°F must be avoided. There is no constraint on the convection coil.

Line 11. *Fouling factor.* The fouling factors used in heat exchanger rating can be used here also. In this example therefore the radiant side would have a fouling factor of about 0.004°F sqft hr/Btu for the oil and 0.001 for the steam.

Line 12. *Residence time.* This becomes important when a chemical reaction of any kind takes place in the heater tubes. In the case of this example this item does not apply. If the example were a thermal cracking heater or a visbreaker the appropriate kinetic equations and calculations would be attached to the specification sheet to support this item.

Line 13–30. These are self explanatory. The only comment here is that the data are quoted at the inlet or outlet conditions of temperature and pressure.

Line 31. *LHV.* The section of the specification sheet that follows deals with the characteristics of the fuel that will be used in the heater. This section is divided into oil and gas which are the usual fuels used in modern day processes. This item requires the "Lower heating value" of the fuel. This can be read off charts such as Figures A9.2 and A9.3.

Line 32. *HHV.* This is the other heating value data required by the heater designer. This "higher heating value" data can also be read off charts such as Figures A9.2 and A9.3.

Line 33. *Pressure at burner.* This normally refers to the oil fuel and is measured at the heater fuel oil manifold.

Line 34. *Temperature at the burner.* This item too is self explanatory and these measurements are also taken at the respective manifolds.

Line 35. *Mol wt of gas.* This refers to the gas stream normally expected to be used. Obviously in practice this will vary with the process operation from day to day.

Line 36. *Viscosity of the oil at the burner.* This viscosity is quoted at the burner temperature and may be arrived at from the two viscosity figures given later in the specification sheet. This is important for the best design or selection of the burner itself.

Line 37. *Atomizing steam.* This item calls for the temperature and pressure of the steam that will be used for atomizing the oil fuel. This is also required for the best design or selection of the oil burner.

Line 38. *Composition of gas.* This section requires the composition of the gas fuel in terms of mol percent. This is the normal expected fuel gas that will be used in the heater. If there is likely to be a wide variation in the quality of the fuel gas that will be used this should be noted here as a range of two or even three compositions. This situation is particularly common in petroleum refining.

Line 46. *Properties of fuel oil.* This final section of the specification sheet requires details of the fuel oil that will be used. These details are:

Gravity of the oil @ 60°F (in °API).
Viscosity of the oil @ 100°F and at 210°F
Flash point in °F.

From the two viscosities the 'Refutus' Graph can be used to determine the viscosity at any other temperature. This graph can be found in most data books that carry viscosity data. Note the viscosity requested in this section is in centistokes (kinematic viscosity). This is because most suppliers quote in centistokes. To convert to centipoise multiply by the specific gravity (grams/cm^3).

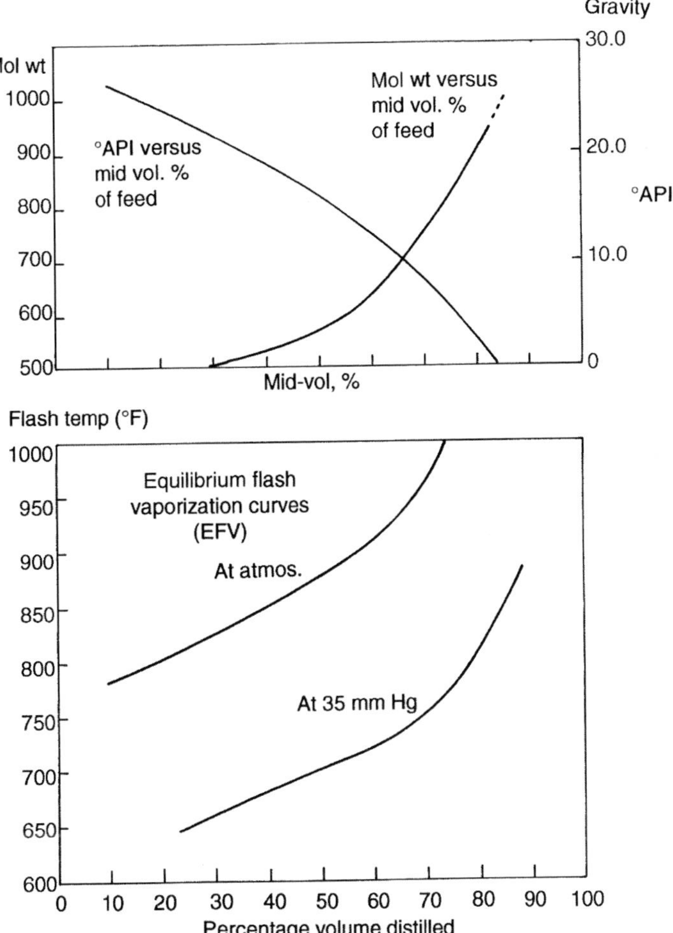

Figure 18.52. Specification sheet attachment.

The flash point required is that determined by the Pensky Marten method and is a measure of the oil's flammability.

This completes the explanation of the main body of the specification sheet. It represents the minimum data required to commence the sizing of the item. The last part of the specification sheet is given to "REMARKS". In this section the engineer should provide all the other data that may influence the design of the heater, e.g. Figure 18.52.

APPENDICES

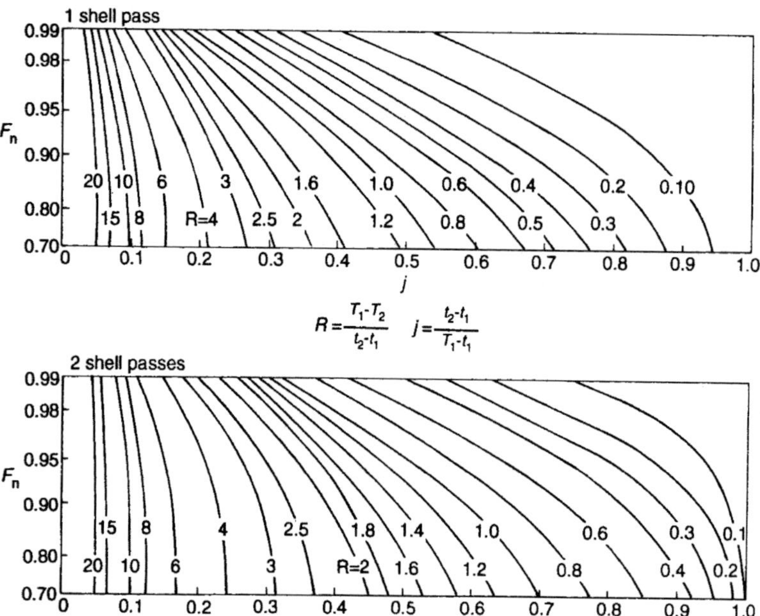

Figure 18.A.1. LMTD correction factors.

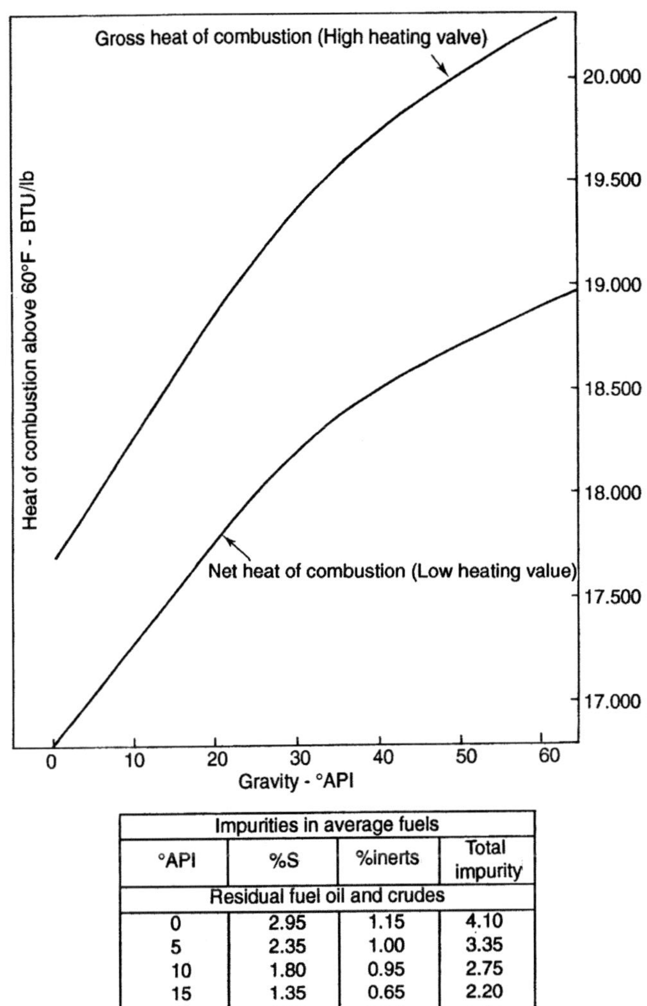

Impurities in average fuels			
°API	%S	%inerts	Total impurity
Residual fuel oil and crudes			
0	2.95	1.15	4.10
5	2.35	1.00	3.35
10	1.80	0.95	2.75
15	1.35	0.65	2.20
20	1.00	0.75	1.75
Crude oils			
25	0.70	0.70	1.40
30	0.40	0.65	1.10
35	0.30	0.60	0.90

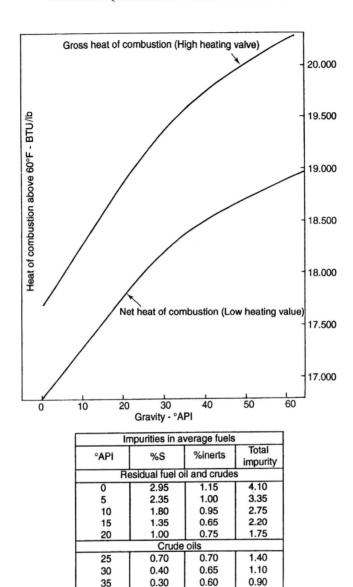

Figure 18.A.2. Heat of combustion of fuel oils.

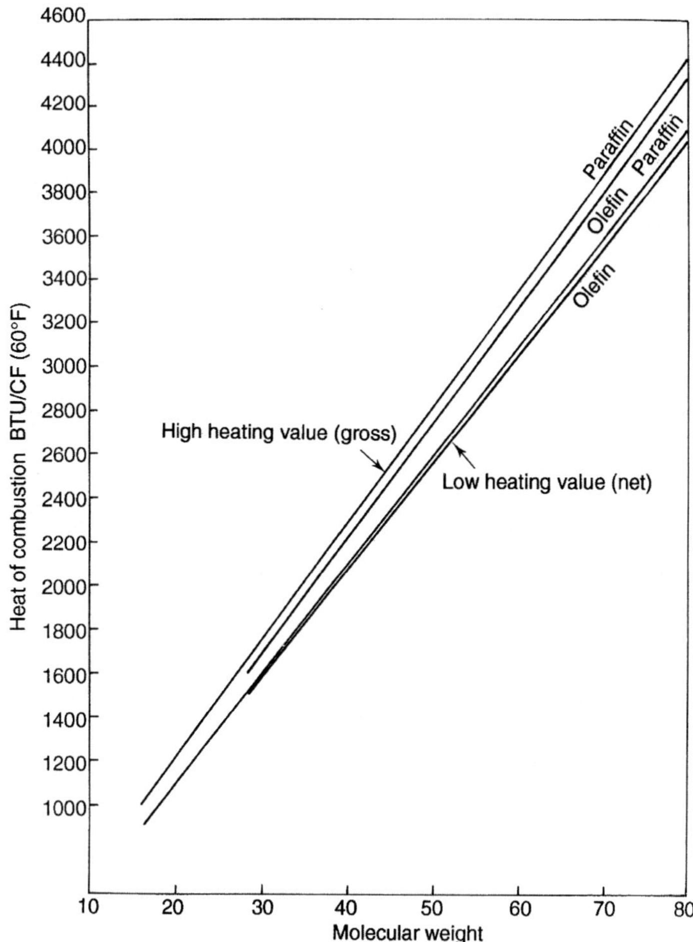

Figure 18.A.3. Heat of combustion of fuel gasses.

Table 18.A.1. Values for coefficient C.

Values of coefficient $C = 520\sqrt{k\left(\frac{2}{k+1}\right)^{\frac{k+1}{k-1}}}$; $k = \frac{C_p}{C_v}$

k	C	k	C	k	C	k	C	k	C	k	C
0.41	219.28	0.71	276.09	1.01	316.56*	1.31	347.91	1.61	373.32	1.91	394.56
0.42	221.59	0.72	277.64	1.02	317.74	1.32	348.84	1.62	374.09	1.92	395.21
0.43	223.86	0.73	279.18	1.03	318.90	1.33	349.77	1.63	374.85	1.93	395.86
0.44	226.10	0.74	280.70	1.04	320.05	1.34	350.68	1.64	375.61	1.94	396.50
0.45	228.30	0.75	282.20	1.05	321.19	1.35	351.60	1.65	376.37	1.95	397.14
0.46	230.47	0.76	283.69	1.06	322.32	1.36	352.50	1.66	377.12	1.96	397.78
0.47	232.61	0.77	285.16	1.07	323.44	1.37	353.40	1.67	377.86	1.97	398.41
0.48	234.71	0.78	286.62	1.08	324.55	1.38	354.29	1.68	378.61	1.98	399.05
0.49	236.78	0.79	288.07	1.09	325.65	1.39	355.18	1.69	379.34	1.99	399.67
0.50	238.83	0.80	289.49	1.10	326.75	1.40	356.06	1.70	380.08	2.00	400.30
0.51	240.84	0.81	290.91	1.11	327.83	1.41	356.94	1.71	380.80	2.01	400.92
0.52	242.82	0.82	292.31	1.12	328.91	1.42	357.81	1.72	381.53	2.02	401.53
0.53	244.78	0.83	293.70	1.13	329.98	1.43	358.67	1.73	382.25	2.03	402.15
0.54	246.72	0.84	295.07	1.14	331.04	1.44	359.53	1.74	382.97	2.04	402.76
0.55	248.62	0.85	296.43	1.15	332.09	1.45	360.38	1.75	383.68	2.05	403.37
0.56	250.50	0.86	297.78	1.16	333.14	1.46	361.23	1.76	384.39	2.06	403.97
0.57	252.36	0.87	299.11	1.17	334.17	1.47	362.07	1.77	385.09	2.07	404.58
0.58	254.19	0.88	300.43	1.18	335.20	1.48	362.91	1.78	385.79	2.08	405.18
0.59	256.00	0.89	301.74	1.19	336.22	1.49	363.74	1.79	386.49	2.09	405.77
0.60	257.79	0.90	303.04	1.20	337.24	1.50	364.56	1.80	387.18	2.10	406.37
0.61	259.55	0.91	304.33	1.21	338.24	1.51	365.39	1.81	387.87	2.11	406.96
0.62	261.29	0.92	305.60	1.22	339.24	1.52	366.20	1.82	388.56	2.12	407.55
0.63	263.01	0.93	306.86	1.23	340.23	1.53	367.01	1.83	389.24	2.13	408.13
0.64	264.72	0.94	308.11	1.24	341.22	1.54	367.82	1.84	389.92	2.14	408.71
0.65	266.40	0.95	309.35	1.25	342.19	1.55	368.62	1.85	390.59	2.15	409.29
0.66	268.06	0.96	310.58	1.26	343.16	1.56	369.41	1.86	391.26	2.16	409.87
0.67	269.70	0.97	311.80	1.27	344.13	1.57	370.21	1.87	391.93	2.17	410.44
0.68	271.33	0.98	313.01	1.28	345.08	1.58	370.99	1.88	392.59	2.18	411.01
0.69	272.93	0.99	314.19*	1.29	346.03	1.59	371.77	1.89	393.25	2.19	411.58
0.70	274.52	1.00	315.38*	1.30	346.98	1.60	372.55	1.90	393.91	2.20	412.15

Values of C for gases

	Mol wt	$k = C_p/C_v$	C	$C/356$		Mol wt	$k = C_p/C_v$	C	$C/356$
Acetylene	26	1.28	345	0.969	Hydrochloric acid	36.5	1.40	356	1.000
Air	29	1.40	356	1.000	Hydrogen	2	1.40	356	1.000
Ammonia	17	1.33	351	0.986	Hydrogen sulphide	34	1.32	348	0.978
Argon	40	1.66	377	1.059	Iso-butane	58	1.11	328	0.921
Benzene	78	1.10	327	0.919	Methane	16	1.30	346	0.972
Carbon disulphide	76	1.21	338	0.949	Methyl alcohol	32	1.20	337	0.947
Carbon dioxide	44	1.28	345	0.969	Methyl chloride	50.5	1.20	337	0.947
Carbon monoxide	28	1.40	356	1.000	N-butane	58	1.11	328	0.921
Chlorine	71	1.36	352	0.989	Natural gas	19	1.27	345	0.969
Cyclohexane	84	1.08	324	0.910	Nitrogen	28	1.40	356	1.000
Ethane	30	1.22	339	0.952	Oxygen	32	1.40	356	1.000
Ethylene	28	1.20	337	0.947	Pentane	72	1.09	325	0.913
Helium	4	1.66	377	1.059	Propane	44	1.14	331	0.930
Hexane	86	1.08	324	0.910	Sulphur dioxide	64	1.26	342	0.961

* Interpolated values, since C becomes indeterminate as k approaches 1.00. (Reproduced by permission of Gas Processors Suppliers Association)

Table 18.A.2. Some common heat transfer coefficients Uo.

Fluid being cooled	Fluid being heated	$\frac{U_0}{(\text{BTU/h.f}^2.^\circ\text{F})}$
Exchangers		
C_4s and lighter	Water	75–110
C_4s and lighter	LPG	75
Naphtha	Naphtha	75
Naphtha	Water	80–100
BPT 450 (kero)	Hy oil (crude)	70–75
Gas oils	Crude	40–50
Gas oils	Water	40–70
Light fuel oil	Crude	20–30
Waxy distillates	Hy oils	30–40
Slurries	Waxy distillate	40
MEA or DEA	Water	140
MEA or DEA	MEA or DEA	120–130
Water	Water	180–200
Air	Water	20–30
Lt HC vapour	H_2-rich stream	35–40
Lt HC vapour	Naphtha	38
Condensers		
Full-range naphtha	Water	70–80
Amine stripper O/heads	Water	100
C_4s and lighter	Water	90
Reformer effluent (Lt HC)	Water	65

Table 18.A.3. Standard exchanger tube sheet data.

d_0 = o.d. of tubing (in.)	BWG gauge	l = thickness (ft)	d_i =i.d. of tubing (in.)	Internal area (in.^{2})	External surface per foot length (ft^2)
$\frac{3}{4}$	10	0.0112	0.482	0.1822	0.1963
$\frac{3}{4}$	12	0.00908	0.532	0.223	0.1963
$\frac{3}{4}$	14	0.00691	0.584	0.268	0.1963
$\frac{3}{4}$	16	0.00542	0.620	0.302	0.1963
$\frac{3}{4}$	18	0.00408	0.652	0.334	0.1963
1	8	0.0137	0.670	0.355	0.2618
1	10	0.0112	0.732	0.421	0.2618
1	12	0.00908	0.782	0.479	0.2618
1	14	0.00691	0.834	0.546	0.2618
1	16	0.00542	0.870	0.594	0.2618
1	18	0.00408	0.902	0.639	0.2618
$1\frac{1}{2}$	10	0.0112	1.232	1.192	0.3927
$1\frac{1}{2}$	12	0.00908	1.282	1.291	0.3927
$1\frac{1}{2}$	14	0.00691	1.334	1.397	0.3927
$1\frac{1}{2}$	16	0.00542	1.37	1.474	0.3927

Chapter 19

A dictionary of terms and expressions

D.S.J. Jones and P.R. Pujadó

A

Abel flash points

This is a test procedure for determining the flash point of light distillate such as kerosene. The flash points determined by this method will be in the range of 85–120°C. The apparatus consists of a water bath with a heating source into which is suspended a cup containing the material to be tested. The water bath is heated and retained at a fixed temperature. The temperature of the sample being tested in the cup is measured by a thermometer. The lid of the cup contains a shutter and a small gas burner from which a flame of determined length exits. The rate of temperature rise of the test material is noted and at predetermined temperatures the shutter is opened and the burner flame exposed to the space above the test liquid in the cup. The temperature at which a flame is observed crossing the surface of the oil sample when the burner is dipped into the cup is the flash point. The ASTM name and number for this test is D56-01 Standard Test Method for Flash Point by Tag Closed Tester.

Absorption units

Absorption units usually consist of a trayed or packed tower in which a gas stream is contacted with a lean solvent in a counter current flow. Usually the gas stream enters below the bottom tray or packed bed and rises up the tower meeting the solvent liquid stream which enters the tower above the top tray or packed bed flowing down the tower. Undesirable material in the gas stream is selectively absorbed into the liquid stream. This liquid stream then enters a stripping column (usually a Steam Stripper), where the absorbed material is stripped off and leaves as an overhead product. The stripped solvent stream is then rerouted to the absorber tower to complete the cycle. These processes are used in petroleum refining to remove heavy hydrocarbons from a light

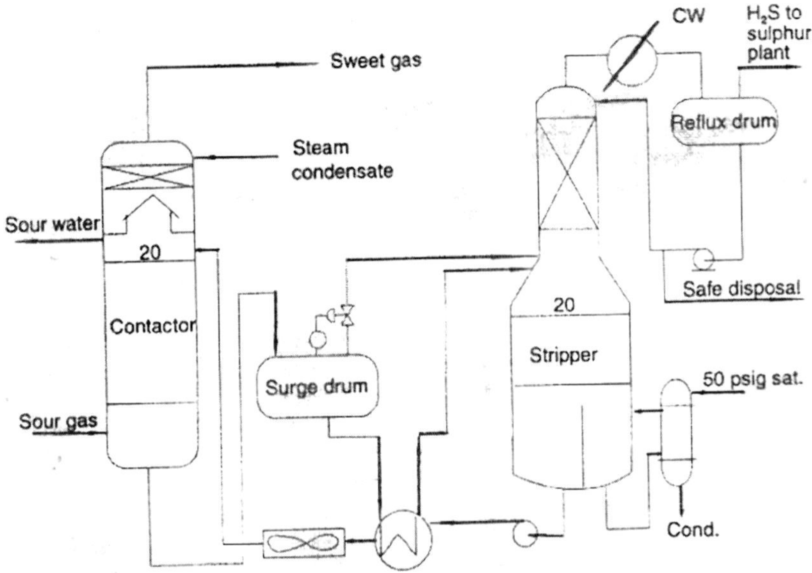

Figure 19.A.1. A schematic of a typical absorption unit.

gas stream. More commonly this type of unit is used in gas treating for the removal of H_2S from a gas stream (see Chapter 10 of this book). A schematic flowsheet of an absorption (gas treating unit) is shown as Figure 19.A.1.

Anhydrous hydro fluoric acid

Anhydrous hydrofluoric acid (AHF) is a colorless, mobile liquid that boils at 67°F at atmospheric pressure, and therefore requires pressure containers. The acid is also hygroscopic therefore its vapor combines with the moisture of air to form "fumes". This tendency to fume provides users with a built-in detector of leaks in AHF storage and transfer equipment. On the other hand, care is needed to avoid accidental spillage of water into tanks containing AHF. Dilution is accompanied by a high release of heat. Hydrofluoric acid is very corrosive. It attacks glass, concrete, and some metals—especially cast iron and alloys that contain silica (e.g., Bessemer steels). The acid also attacks such organic materials such as leather, natural rubber, and wood, but does not promote their combustion (see Chapter 15 for details).

Air condensers and coolers

Air cooling of process streams or condensing of process vapors is more widely used in the process industry than cooling or condensing by exchange with cooling water. The

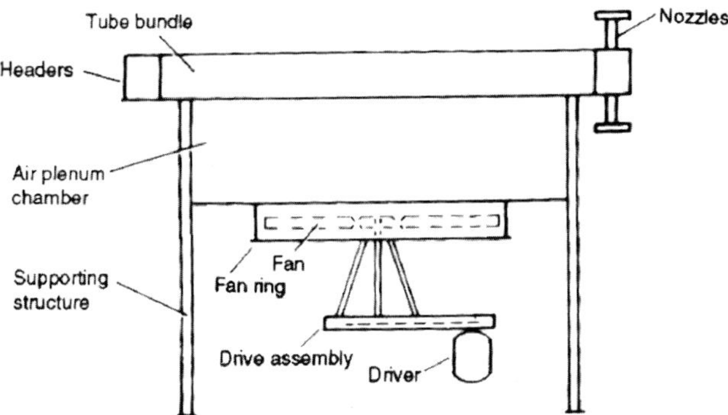

Figure 19.A.2. A forced air flow arrangement.

use of individual air coolers for process streams using modern design techniques has economized in plant area required. It has also made obsolete those large cooling towers and ponds associated with product cooling. This item in Chapter 18 describes air coolers in general and outlines a method to estimate surface area, motor horsepower, and plant area required by the unit.

As in the case for shell and tube exchangers there are many excellent computer programs that can be used for the design of air coolers. The method given in Chapter 18 for such calculation may be used in the absence of a computer program or for a good estimate of a unit. The method also emphasizes the importance of the data supplied to manufacturers for the correct specification of the units.

Figures 19.A.2 and 19.A.3 show the two types of air coolers used in the process industry. Both units consist of a bank of tubes through which the fluid to be cooled or condensed flows. Air is passed around the tubes either by a fan located below the tubes

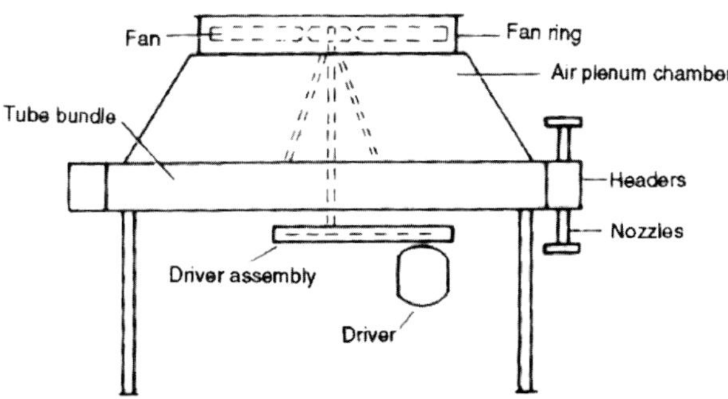

Figure 19.A.3. An induced air flow arrangement.

(Figure 19.A.2) forcing air through the tube bank or a fan located above the tube bank drawing air through the tube bank (Figure 19.A.3). The first arrangement is called 'Forced draft' and the second 'Induced draft'.

Air in both cases is motivated by a fan or fans driven by an electric motor or a steam turbine or in some cases a gas turbine. The fan and prime driver are normally connected by a 'V' belt or by a shaft and gear box. Electric motor drives are by far the most common prime drivers for air coolers.

The units may installed on a structure at grade or as is often the case on a structure above an elevated pipe rack. Most air coolers in condensing service are elevate above pipe racks to allow free flow of condensate into a receiving drum.

Air systems

All chemical and petroleum plants require a supply of compressed air to operate the plant and for plant maintenance. There are usually two separate systems and these are:

- Plant air system
- Instrument air system

Plant air is generally supplied by a simple compressor with an after cooler. Very often when plant air is required only for maintenance this is furnished by a mobile compressor connected to a distribution piping system. Air for catalyst regeneration and the like is normally supplied by the regular gas compressor on the unit. Instrument air should always be a separate supply system. Compressed air for instrument operation must be free of oil and dry for the proper function of the instruments it supplies. This is a requirement which is not necessary for most plant air usage. A reliable source of clean dry instrument air is an essential requirement for plant operation. Failure of this system means a complete shut down of the plant.

Figure 13.36 in Chapter 13 of this Handbook shows a typical instrument air supply system. Atmospheric air is introduced into the suction of one of two compressors via an air filter. The compressors are usually reciprocating or screw type non-lubricating. Centrifugal type compressors have been used for this service when the demand for instrument air is very high. The air compressors discharge the air at the required pressure (usually above 45 psig) into an air cooler before the air enters one of two dryers. One of the compressors is in operation while the other is on standby. The operating compressor is usually motor driven with a discharge pressure operated on/off start up switch. The standby compressor is turbine (or diesel engine) driven with an automatic start up on low/low discharge pressure switch.

The cooled compressed air leaves the cooler to enter the dryers. There are two dryer vessels each containing a bed of desiccant material. This material is either silica gel (the most common), alumina, or in special cases zeolite (molecular sieve). One of the two dryers is in operation with the compressed air flowing through it to be dried and to enter the instrument air receiver. The desiccant in the other dryer is being simultaneously regenerated. Regeneration of the desiccant bed is effected by passing through the bed a stream of heated air and venting the stream to atmosphere. This heated stream removes the water from the desiccant to restore its hygroscopic properties. At the end of this heating cycle cooled air is reintroduced to cool down the bed to its operating temperature. When cool, the unit is then shut in ready to be switched into operation for the first dryer to start its regeneration cycle. The various operating and regeneration phases are automatically obtained by a series of solenoid valves operated by a sequence timer switch control. These dryers (often including the compressor and receiver items) are packaged units supplied, skid mounted, and ready for operation.

The instrument air receiver vessel is a pressure vessel containing a crinkled wire mesh screen (CWMS) before the outlet nozzle. It is high-pressure protected by a pressure control valve venting to atmosphere, and of course is also protected by a pressure safety valve. The air leaves the top of this vessel to enter the instrument air distribution system servicing all the plants in the complex.

Alkylation

Alkylation as utilized in the petroleum industry was developed independently by UOP, Shell, the Anglo Iranian Company, and Texaco in 1932–1936. Motor fuel alkylation in the petroleum refining industry refers to the acid catalyzed conversion of C_3–C_5 olefins with isobutane into highly branched C_5–C_{12} isoparaffins collectively called alkylate, a valuable gasoline blending component. A major constituent of alkylate is 2,2,4-trimethyl pentane which is defined as 100 on the octane scale.

Alkylation reactions are catalyzed by liquid and solid acids, including H_2SO_4, $AlCl_3$–HCl, HF, HF–BF_3, H_2SO_4–HSO_3F (fluorosulfuric acid), trifluoromethane sulfonic acid chlorided Pt alumina, BF_3 on alumina, zeolites, and ion exchange resins. However, the catalysts and associated processes commercialized during World War II for aviation gasoline, were HF alkylation and sulfuric acid alkylation. These are the process met with in most of today's refineries (see Chapter 9 for a fuller historical description).

The chemistry of the alkylation process is quite complex and is given in some detail in Chapter 9 of this Handbook. Briefly it can be summarized by the following reaction stages:

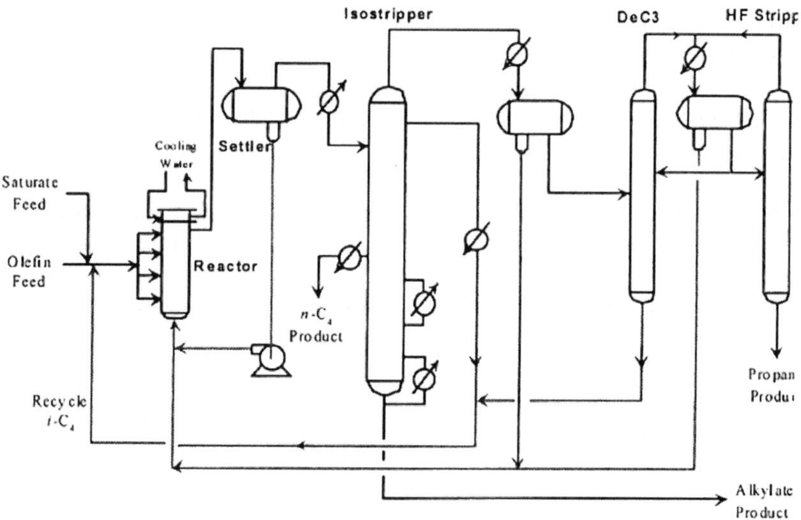

Figure 19.A.4. AHF alkylation process.

- Initiation $C_4 = + HF \text{ (or } H_2SO_4) \rightarrow C_4^+ + F^-$
- Alkylation $C_4^+ + C_4 \rightarrow C_8^+$
- Saturation and Continuation $C_8^+ + iC_4 \rightarrow C_8 + iC_4^+$
- Polymerization $C_8^+ + C_4 = \rightarrow C_{12}^+$
- Cracking $C_{12}^+ \rightarrow iC_5^+ + C_7$

The last two reactions explain the variety of side products occurring typically at either end of the boiling range of primary product.

Alkylation unit HF

Figure 19.A.4 is a simplified process flow diagram for the UOP propylene–butene HF alkylation process.

After pretreatment to remove H_2S and mercaptans the olefin feeds are combined with a large excess of recycle isobutane. This mixture provides an 6–14 isobutane/olefin molar ratio which is fed with a circulating HF acid catalyst to the shell side of a water-cooled reactor. The alkylation reaction is very fast with 100% olefin conversion. The excess isobutane, alkylate product, non-reactive hydrocarbons (propane, n-butane) in the feeds and the acid catalyst pass on to the settler vessel. The dense acid phase separates from the hydrocarbons rapidly by gravity and is then pumped back to the reactor. The hydrocarbons containing dissolved HF flow off the top of the settler to the isostripper. This is a large tower with two sidedraws. Its primary function is to produce the recycling isobutane stream to maintain the high isobutane/olefin molar ratio of the reactor feed. The tower typically has two reboilers. Alkylate is drawn off the bottom of the tower, cooled in exchangers, and sent to product storage. The next product draw up the tower is the n-butane sidedraw and above that is the isobutane recycle draw. The isostripper overhead vapor is a propane-enriched isobutane stream and HF which is condensed and separated in a settling drum. The HF phase is pumped back to the reactor section. The HF saturated hydrocarbon phase is charged to the depropanizer.

This tower and its associated HF stripper remove propane from the isobutane recycle. The depropanizer bottoms is returned to the reactors as part of the recycle isobutane. The depropanizer overhead containing the propane product and HF are condensed and separated in the overhead receiver. The acid phase is returned to the reactor section and the acid-saturated propane is stripped free of acid in the HF Stripper column. The HF stripper bottoms is an acid-free propane product which is treated with hot alumina to remove organic fluorides, cooled and treated with KOH pellets to remove traces of HF and water.

The UOP HF Alkylation process contains an acid regenerator. This unit takes a small sidestream of the recycle HF and strips out the acid leaving the hydrocarbon to join the isobutane recycle. (A full process description and discussion is given in Chapter 9.)

Alkylation unit H_2SO_4

Figure 19.A.5 is a simplified process drawing of a H_2SO_4 alkylation unit. Today there are two processes for H_2SO_4 alkylation, the cascade process licensed by Exxon Mobil and MWKellogg and the Stratco effluent refrigerated process. The one shown as Figure 19.A.5 is the cascade process. Full details of this process and the Stratco process are given in Chapter 14.

Amine solvents

Amines are used as solvents in the removal of hydrogen sulfide from refinery gas streams. There are many amine compounds used for this purpose. The more common of these are listed as follows

- Monoethanol amine (MEA)
- Diethanol amine (DEA)
- Diglycol amine (DGA)

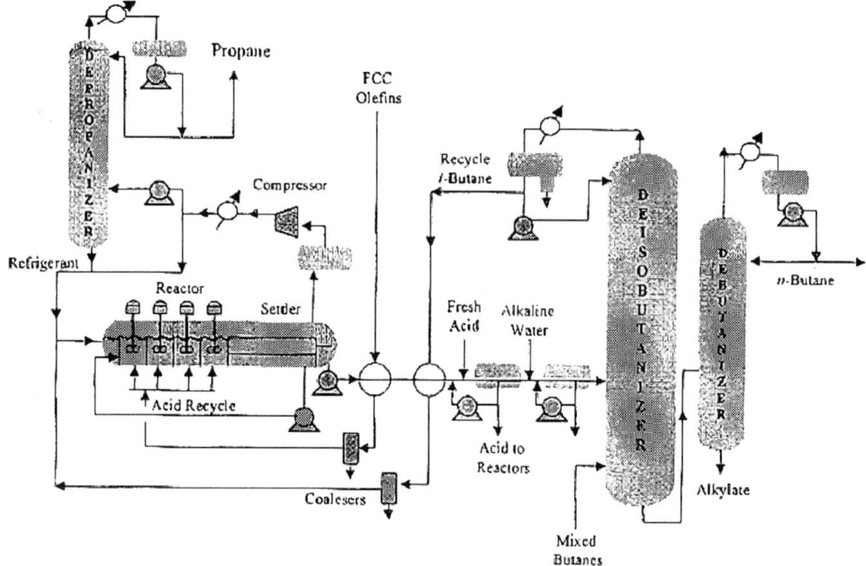

Figure 19.A.5. H_2SO_4 alkylation process.

There are also two proprietary processes licensed by Shell Petroleum which are used extensively world wide. These are the ADIP process and the Sulfinol process.

Monoethanol amine

MEA is the most basic (and thus reactive) of the ethanol amines. MEA will completely sweeten natural gases removing nearly all acid gases if desired. The process is well proven in refinery operations.

Like all of the amine solvents used for acid gas removal, MEA depends upon its amino nitrogen group to react with the acidic CO_2 and H_2S in performing its absorption. MEA is considered a chemically stable compound. If there are no other chemicals present it will not suffer degradation or decomposition at temperatures up to its normal boiling point.

The process reactions are given below.

$$HOCH_2CH_2NH_2 = RNH_2$$

$$2RNH_2 + H_2S \rightarrow (RNH_3)_2S$$

$$(RNH_3)_2S + H_2S \rightarrow 2RNH_3HS$$

Some of the degradation products formed in these systems are highly corrosive. They are usually removed by filtration or reclaimer operations. Filtration will remove corrosive byproducts such as iron sulfide. Reclaiming is designed to remove heat stable salts formed by the irreversible reaction of MEA with COS, CS_2 (Carbonyl Sulfide and Carbon Disulphide). The reclaimer operates on a sidestream of 1–3% of the total MEA circulation. It is operated as a stream stripping kettle to boil water and MEA overhead while retaining the higher boiling point heat stable salts. When the kettle liquids become saturated at a constant boiling point with the degradation products it is shut in and dumped to the drain.

Diethanol amine

DEA does not degrade when contacted with CS_2, COS, and mercaptans as MEA does. Because of this, DEA has been developed as a preferred solvent when these chemicals are present in the stream to be treated.

DEA is a weaker base (less reactive) than MEA. This has allowed DEA to be circulated at about twice the solution strength of MEA without corrosion problems. DEA systems are commonly operated at strengths up to 30 wt% in water and it is not unusual to see them as high as 35 wt%. This results in the DEA solution circulation rate usually being a little less than MEA for the same system design parameters.

The process reactions are shown below.

$$HOCH_2CH_2NHCH_2CH_2OH = R_2NH = DEA$$

$$2R_2NH + H_2S \rightarrow (R_2NH_2)_2S$$

$$(R_2NH_2)_2S + H_2S \rightarrow 2R_2NH_2HS$$

Because the system has much fewer corrosion problems and removes acid gases to nearly pipeline specifications it has been installed as the predominant system in recent years.

Diglycol amine

This process uses 2-(2-amono ethoxy) ethanol at a recommended solution strength of 60 wt% in water. DGA has almost the same molecular weight as DEA and reacts mole for mole with acid gases. DGA seems to tie up acid gases more effectively so that the higher concentration of acid gas per gallon of solution does not cause corrosion problems as experienced with the usual amine processes.

Table 19.A.1. A summary of the common amines

Amine	MEA	DEA	DGA
Molecular weight	61.1	105.1	105.14
Boiling point °F	339	514	405
Freezing point °F	51	77	−9.5
Sg @ 77°F (25°C)	1.0113	1.0881	1.0572
Visc @ 77°F Cp	18.95	352	40
Visc @ 140°F Cp	5.03	53.9	6.8
Flash point °F	200	295	260

The system reactions are given below.

$$HOCH_2CH_2OCH_2CH_2NH_2 = RNH_2 = DGA$$

$$2RNH_2 + H_2S \rightarrow (RNH_3)_2S$$

$$(RNH_3)_2S + H_2S \rightarrow 2RNH_3HS$$

DGA does react with COS and mercaptans similarly to MEA but forms bis (hydroxy, ethoxy ethyl) urea, BHEEU. BHEEU can only be detected using an infra-red test rather than chromatography. Normal operating levels of 2–4% BHEEU are carried in the DGA without corrosion problems. BHEEU is removed by the use of a reclaimer identical to that for an MEA system but operating at 385°F (196°C). Materials of construction are the same as those for MEA systems.

DGA allows H_2S removal to less than 1/4 grain per 100 Scf (about 0.006 kg per 1,000 cubic meters) and removes CO_2 to levels of about 200 ppm using normal absorber design parameters.

A summary of these amines is given in Table 19.A.1.

Amine units

The use of chemically "basic" liquids to react with the acidic gases was developed in 1930. The chemical used initially was tri-ethanolamine (TEA). However as mono-ethanolamine (MEA) became commercially more available it became the preferred liquid reactant due to its high acid gas absorptivity on a unit basis.

Since 1955, numerous alternative processes to MEA have been developed. These have fewer corrosion problems and are to a large extent more energy efficient. Inhibitor systems have however been developed which have eliminated much of the MEA corrosion problems. Some of these newer processes also are designed to selectively remove the H_2S, leaving the CO_2 to remain in the gas stream.

A process diagram of a typical amine gas treating unit is given in Chapter 10 in of this Handbook and is shown as Figure 19.A.1. A brief description of such a unit is as follows:

Referring to the above flow sheet, sour gas (rich in H_2S) enters the bottom of the trayed absorber (or contactor). Lean amine is introduced at the top tray of the absorber section to move down the column. Contact between the gas and amine liquid on the trays results in the H_2S in the gas being absorbed into the amine. The sweet gas is water washed to remove any entrained amine before leaving the top of the contactor.

Rich amine leaves the bottom of the contactor to enter a surge drum. If the contactor pressure is high enough, a flash stream of H_2S can be routed from the drum to a trayed stripper. The liquid from the drum is preheated before entering a 20 tray stripping column on the top tray. This stripper is re-boiled with 50 psig saturated steam. High temperatures cause amines to break down. The H_2S is stripped off and leaves the reflux drum usually to a sulfur production plant. Sulfur is produced in this plant by burning H_2S with a controlled air stream, and then reacting H_2S with SO_2 over a catalyst.

The lean amine leaves the stripper and is cooled. The cooled stream is routed to the contactor.

Amine absorber

Amine absorbers do not have a high-tray efficiency. Generally, the efficiency of a contactor will range between 10% and 20%. This can be determined on an operating plant using plant data to determine the number of theoretical trays required to achieve the plant's operating performance. There are several accepted methods to calculate theoretical trays in this service. Among these are the MCabe Thiele graphical Method and the calculation method described by the following equation. This later calculation method is considered by many to be the sounder and more accurate of the methods available. The equation used for determining the theoretical trays for the absorption process is as follows:

$$N = \left[\frac{\text{Log } 1/q\,(A - 1)}{\text{Log } A}\right] - 1$$

where

N = Number of theoretical trays.

$q = \dfrac{\text{Moles } H_2S \text{ in Lean gas}}{\text{Moles } H_2S \text{ in Feed gas}}$

A = Absorption Factor LK/V

The absorption factor is obtained from the equation:

$$A = \frac{a(1 + R - r)(1 - q)}{pp/P}$$

where

A = The absorption factor.
a = Mole fraction of H_2S in gas feed.
R = Moles MEA/moles H_2S absorbed.
r = Residual H_2S in lean MEA solution.
pp = Partial pressure of H_2S in rich amine solution.
P = Tower pressure psia.

Note: The above equations are suitable for all absorbents.

The Tower: The conventional Amine Contactor is divided into two parts:

The Absorption Section
The Water wash section

This is shown in Figure 19.A.6.

The lean amine enters the tower above the top absorber tray, through a distributor. It flows down the absorber section trays counter current to the rich gas moving up the tower. The rich gas enters the tower via a distributor under the bottom absorber tray. The H_2S contained in the gas is absorbed by the amine solvent by mass transfer on the trays. The rich amine leaves the bottom of the tower to the amine stripper column. The lean gas leaves the top of the of the absorber section to enter the water wash section of the tower. This wash section contains about 4–5 trays or a packed bed. Water is introduced above the top wash tray and flows down counter current to the lean gas to remove any entrained amine in the gas. The water is collected in a chimney tray and pumped to the oily water sewer. The washed gas leaves the top of the tower into the gas system.

Anhydrous ammonia

Ammonia is used in the refinery to neutralize vapor containing HCL at the overhead section of the atmospheric crude distillation tower. The ammonia is injected into the vapor spaces above the top four (or five) trays of the tower and into the overhead vapor line. The ammonia may be in the anhydrous form and introduced directly from cylinders or may be in the form of an aqueous solution. The aqueous form is injected from a storage bullet by means of metering pumps.

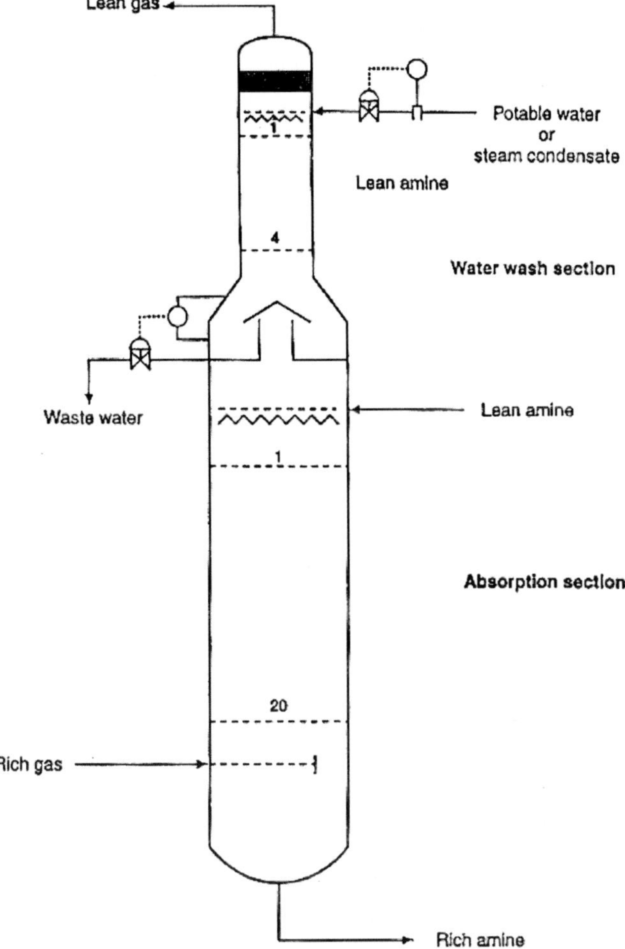

Figure 19.A.6. A typical amine absorber for MEA or DEA unit.

API codes

API stands for the American Petroleum Institute. The purpose of this body is to provide the industry with a set of standards which defines the design and measurement parameters that will be used in the petroleum industry. These codes cover such items as: vessel design, oily water separators, boiler design, safety items etc, and a number of laboratory test procedures for feed and petroleum products.

API gravity

This item is used in the compilation of most crude assays (see Chapter 1 of this Handbook). Although not a laboratory test as such it is derived from the standard test to determine the specific gravity of a liquid. The correlation between specific gravity and degrees API is as follows:

$$\text{Sp Gr} = \frac{141.5}{131.5 + {}^{\circ}\text{API}}$$

The specific gravity and the API are at 60°F. Note API is always quoted in degrees.

Aromatics

Aromatics are present throughout the entire boiling range of crude oil above the boiling point of benzene, the compound with the lowest boiling point in the homologue. These compounds consist of one or more closed conjugated rings with one or more alkyl groups attached. The lighter aromatics such as benzene, toluene, and the xylenes are removed as products in the petroleum chemical plants (see Chapter 12 of this Handbook). In the energy petroleum refinery these lighter aromatics are included in the finished gasoline products to enhance the octane rating of the products. Indeed the refinery process of catalytic reforming is aimed at converting the lower octane compounds (predominately naphthenes) into the high octane light aromatics. The heavier aromatic compounds however are often undesirable compounds in many products, such as kerosene, jet fuel, and many lube oils. In these cases the aromatic compounds are either converted (de-aromatizing hydrotreater for kerosenes) or removed by solvent extraction as in the case of lube oils.

The production of benzene, toluene, ethyl benzene, and the xylenes in the petrochemical refinery commences with the catalytic reforming of the naphtha product produced in the normal energy refinery. This reformate is treated to remove the residual aliphatic compounds of the naphtha by solvent extraction. The rich aromatic stream is then subjected to a series of distillation processes and selective con version to maximize the BTX products required. A typical configuration for an aromatic complex is shown in Figure 19.A.7.

This is one of many configurations for aromatic production. In Figure 19.A.7, the production maximizes benzene and ortho xylene at the expense of some toluene and all of ethyl benzene. This is accomplished by a cryogenic de-alkylation unit to produce more benzene, and a catalytic isomerization unit to convert ethyl benzene to ortho xylene.

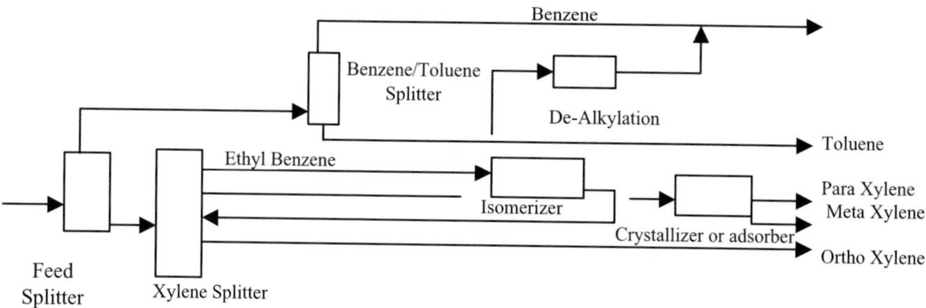

Figure 19.A.7. A typical aromatic plant configuration.

Ash content (Petroleum)

Petroleum ash content is the non-combustible residue of a lubricating or fuel oil determined in accordance with ASTM D582—also D874 (sulfated ash).

ASME (American Society of Mechanical Engineers)

The American Society of Mechanical Engineers has been a world leader in codes, standards, accreditation, and certification for over a century. These programs have now been extended to include the registration (certification) of quality systems in conformance with the standards set by the International Organization of Standardization (ISO). In the petroleum industry, this organization sets the quality requirements for vessel fabrication, piping and in particular the boiler code among many other standard definitions.

Asphalt

Asphalt is a group of products produced from the vacuum distillation residue of crude oil. The products that make up this group have distinct properties that must be met by the treating of the vacuum residue.

There are two major categories of asphalt products. These are:

- Paving and Liquid Asphalt
- Roofing Asphalt

The paving grades will have a penetration specification of 300 (30.0 mm) or less @ 77°F and 100 g weight, while the softer liquid grades will have a penetration of 300 and higher.

The liquid asphalt grades are typified as, RC (rapid curing), MC (medium curing), and SC (slow curing) each of these grades will also have 4 viscosity grades as given below:

70 grade—RC/MC/SC	70–140 Cs @ 140°F.
250 grade—RC/MC/SC	250–500 Cs
800 grade—RC/MC/SC	800–1,600 Cs
3,000 grade—RC/MC/SC	3,000–6,000 Cs.

Rapid curing cutbacks are penetration asphalt and naphtha blends having a viscosity range from 75 SSF @ 77°F to 600 SSF @ 180°F. The naphtha content may be as high as 75 vol%. Cutbacks are hot sprayed onto existing roads as a binding medium for new wearing surfaces.

Medium curing cutbacks are penetration grade asphalt and kerosene blends with four grades having the same viscosities as the RC cutbacks. These are used in road building in the same way as the RC cutbacks.

Slow curing cutbacks are penetration asphalt and gas oil blends normally produced directly from the crude oil atmospheric or vacuum distillation. The volume produced is small and they are used mainly as a gravel dust layer or mixed with aggregates for cold patching of asphalt surfaces.

Roofing asphalt

The second largest asphalt use is that for roofing. Most of these asphalts are produced by air blowing. There are three major roofing grades and may be classified by penetration and softening points. These are:

	Pen @ 77°F	Softening point °F
Type 1	25–50	140–150
Type 2	20–30	166–175
Type 3	15–25	190–205

Type 1 grade is used on 'Dead Flat' roof while the other two are used on intermediate and steep slope roofs, respectively.

Asphalt blowing process

Blowing air into a vessel containing asphalt from vacuum distillation or the de asphalting process will change its penetration and softening properties. Among the variables that affect the manufacture of air blown asphalt are:

- The rate of the air injected
- The temperature of the asphalt
- Retention time of the asphalt
- The system pressure

Description and details of this process are given in Chapter 12 in of this Handbook.

Asphaltenes

Asphalt is basically a colloidal dispersion of asphaltenes in oil with resins as the stabilizing agent. The quantities of these can vary widely with the type of crude.

Asphaltenes can be hydrogenated to resins, resins to oils. Resins can be oxidized to asphaltenes, oil to resins. Thus:

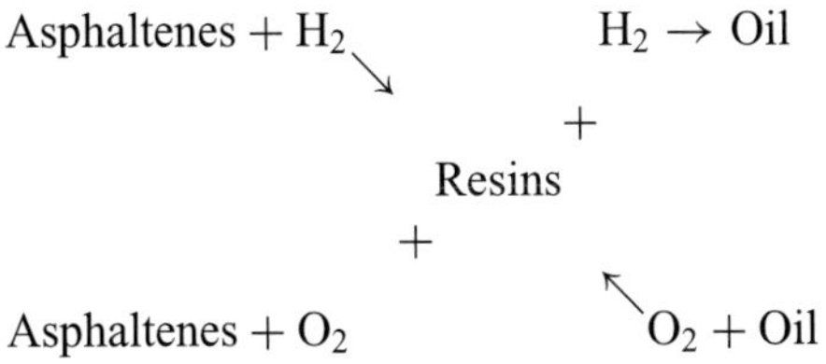

Asphaltenes + H_2 ↘ Resins + H_2 → Oil

Asphaltenes + O_2 ← Resins + ↖ O_2 + Oil

Oxidation is really a misnomer, as air blown asphalt has essentially the same oxygen content as the charge. Air blowing asphalt increases the asphaltene content, hardens it, decreases penetration, increases softening, and reduces ductility. Basically air blowing is a polymerization process following the route below:

1. Addition of O_2 to form unstable compounds from which H_2O is eliminated leaving unsaturated compounds which polymerize.
2. Addition of O_2 to form carboxyl derivatives from which CO_2 is eliminated followed again by polymerization.
3. Formation and elimination of volatile oxidation products other than H_2O and CO_2 followed again by polymerization.

It is worth noting, that the best crudes for asphalt air blowing are those with high percentage of asphaltene fraction and low in paraffin hydrocarbons, and waxes. Resins can be oxidized to asphaltenes relatively easily, so crudes rich in resin are good raw materials for asphalt manufacture. Oils can also be oxidized to asphaltenes, but they must be oxidized to resins first, which requires a more severe operation. Cracked residua simply do not make good asphalt.

Assay: The crude

The crude oil assay is a compilation of laboratory and pilot plant data that define the properties of the specific crude oil. At a minimum, the assay should contain a distillation curve for the crude and a specific gravity curve. Most assays however contain data on pour point (flowing criteria), sulfur content, viscosity, and many other properties. The assay is usually prepared by the company selling the crude oil, it is used extensively by refiners in their plant operation, development of product schedules, and examination of future processing ventures. Engineering companies use the assay data in preparing the process design of petroleum plants they are bidding on, or having been awarded the project, they are now building.

The data normally contained in a well produced assay will be:

- A TBP curve
- A specific gravity curve
- A sulfur content curve
- A pour point and cloud point curve
- Product tables of some of the lighter products. These should contain at least the following:
  - ➢ The product boiling range
  - ➢ The product yield on crude
  - ➢ The cuts gravity (usually in °API)
  - ➢ A PONA (paraffin, olefin, naphthene, and aromatic content)
  - ➢ In the case of the naphthas the cuts octane values
  - ➢ In the case of kerosene cuts, their sulfur and smoke point data
  - ➢ In the case of diesel and gas oil cuts their diesel index or cetane value

Full details of a typical Assay and a description of the tests and data are given in Chapter 1 of this Handbook.

ASTM (American Standard Testing Methods)

The tests and test methods provided by this body define and establish the quality of petroleum products and provide data on petroleum intermediate streams. This later provision is used as a basis for refinery planning, operation, and engineering work associated with the refinery. Among the more important tests are those given below:

- D56—Tag Closed Flash Point (The Abel Flash)
- D86—Standard Test For Distillation of Petroleum Products
- D93—The Pensky Marten Flash Point Closed Cup Test
- D97—Cloud & Pour Points
- D129—Sulfur Content (Bomb Method)
- D189—Condradson Carbon Content
- D323—Reid Vapor Pressure
- D445—Kinematic Viscosity
- D613—Cetane Number
- D908—Octane Number Research
- D1160—Gas oil Distillation (Sub Atmospheric)

- D1298—Specific Gravity (by hydrometer)
- D1837—Weathering Test for LPGs
- D2163—Component Analysis of LPGs (by gas chromatography)

These are the most common tests to define marketable quality and in plat operation control. Further details are given in Chapter 7 of this Handbook.

Atmospheric crude distillation unit

History of the process

The distilling of petroleum products from crude oil to some extent or other has long been practiced. Certainly, the ancient Egyptians, Greeks, and Romans had some form of extracting a flammable oil from, probably, weathered crude oil seepage. It wasn't though until the turn of the nineteenth and twentieth century that crude oil well drilling was first discovered and commercialized. Originally the crude oil was refined to produce essentially Kerosene (Lamp Oil), and a form of gasoline known then as benzene (as opposed to benzene already being produced from coal) and the residue used as pitch for calking and sealing. The lamp oil or kerosene was produced to provide a means of illumination, later a lighter cut known as Naphtha was produced for the same purpose but used in special pressurized lamps.

The production of these early distillates was by made by cascading the crude oil through successive stills each operating at successively higher temperatures. This is shown in diagram Figure 19.A.8. This is described more fully in Chapter 3 of this Handbook.

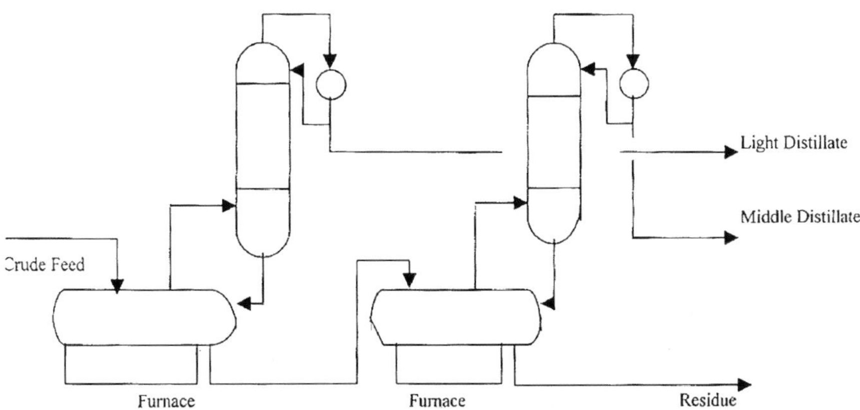

Figure 19.A.8. Schematic of a continuous atmospheric crude still.

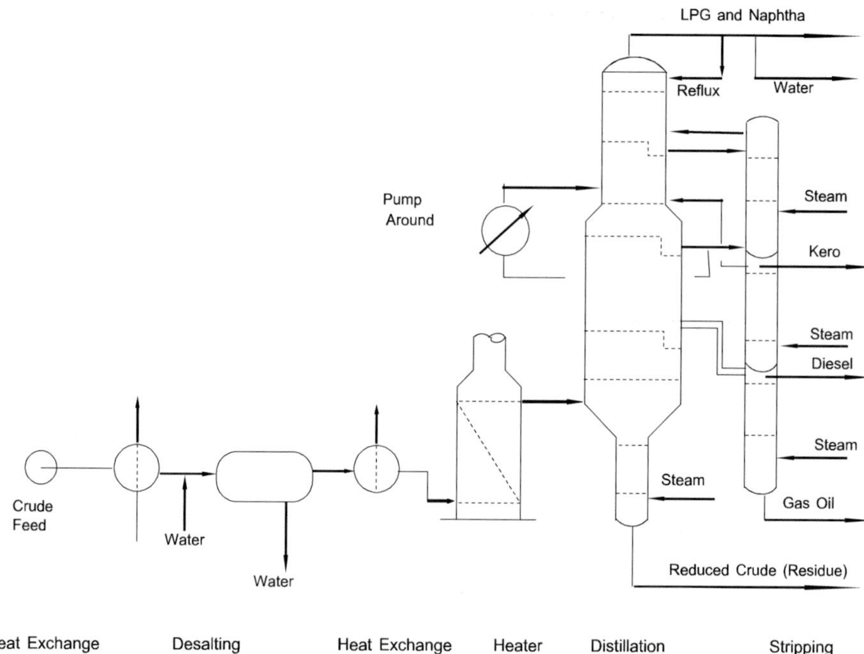

Figure 19.A.9. A Typical atmospheric crude distillation unit.

The major breakthrough in this process was the development of the single fractionating column. This came in the late 1920s and 1930s. In many refineries however, even in some modern ones built after say 1960, the atmospheric crude unit contains a primary fractionator which removes the light ends and some naphtha, before the main tower. These cases are rare and are only used when the crude feed contains a great deal of light ends (spiled crude) or a primary tower is added to increase throughput.

The process

Crude oil is pumped from storage to be heated by exchanges against hot overhead and product streams in the crude unit. At a temperature of about 200–250°F, the crude enters the desalter. Free salt water contained in the crude is washed out and separated by means of an electrostatic precipitator contained in the desalter drum. The water phase from the drum is sent to a sour water stripper to be cleaned before disposal to the oily water sewer.

The oil phase is further heated by exchange with hot sidestream and bottom products before entering a fired heater. The crude oil is heated in the heater to a temperature that will vaporize the distillate products in the crude tower. Some additional heat is added to the crude to vaporize about 5% more than required for the distillate streams. This is called overflash and is used to ensure good reflux streams in the tower.

The un-vaporized portion of the crude leaves the bottom of the tower via a steam stripper section as the bottom product. The distillate vapors move up the tower counter current to the cooler liquid reflux streams coming down. Heat and mass transfer take place on the trays contained in the tower. Distillate products are removed from the various sections of the tower, they are stream stripped and sent to storage. The full naphtha vapor is allowed to leave the top of the tower to be condensed and collected in the overhead drum. A portion of this stream is returned as reflux while the remainder is delivered to the light end processes for stabilizing and further distillation.

The side-stream distillates shown in Figure 19.A.9 above are:

- Heavy Gas Oil (has the highest Boiling Point)
- Light Gas Oil (will become Diesel)
- Kerosene (will become Jet Fuel)

A "Pump around" section is included at the light gas oil draw-off. This is simply an internal condenser which takes heat out of that section of the tower. This in turn ensures a continued reflux stream flow below that section.

B

Barrels

In the petroleum industry, the barrel is a standard form of measuring liquid volume. A barrel of oil is defined as 42 U.S. gallons (one U.S. gallon equals 231 cubic inches). It is still used extensively in most countries, but is being replaced particularly in European countries by the metric measures of cubic meters or liters, or by weight measures like kg or tonnes (1000 kg).

BPCD and BPSD

BPCD is the measure of throughput or stream flow based on an operation over one year of 365 days. BPSD is the rated throughput of a plant or the rate of a stream over the total operating days in the year. BPCD is barrels per calendar day, and BPSD is barrels per stream day. BPSD is defined as BPCD divided by the service factor as a fraction. The service factor is the percentage time over a calendar year that the unit is operating. Each type of unit is allocated a service factor depending on the amount of scheduled shut down time the particular unit requires over the year for maintenance. The following is an example of some service factors:

Unit	Service Factor Percent
Crude distillation unit	95–98
Light ends distillation	98
Vacuum crude distillation	95–98
Visbreaker and thermal cracker	90
Cat reformer	90–92
Naphtha hydrotreaters	90–92
Gas oil hydrotreaters	90
Fluid catalytic crackers	85–90
Hydrocrackers (distillate feed)	90
Hydrocrackers (residue feed)	80–82 (includes residue hydrotreating)
Cokers	85

These service factors are based on modern day plant operating controls and state of the art catalyst quality and regenerating frequency.

Base lube oils

This refers to a lube product that meets all lube specifications and is suitable for blending to meet performance specifications.

Bitumen

Bitumen is the term often given to untreated asphalt from the vacuum distillation of crude and the extract from the de-asphalting unit. This is before the stream has been treated, with cut backs or by air blowing to make the various asphalt product grades.

Boiling points and boiling range

It is not possible to separate the components of crude oil into individual chemical compounds. However groups of these component mixtures are grouped together and identified by the boiling point at atmospheric pressure of the lightest component in the group and the boiling point of the heaviest component in the group. The group itself is called a *Cut* and the range of temperatures that identify it is called the boiling range or cut range. These can be related to the crude TBP curve to determine its yield on that particular crude (see Chapter 1 of this Handbook).

Bright stocks (lube oils)

These are processed from the raffinate of the vacuum residue de-asphalting unit. (See Chapter 12 of this Handbook)

Bromine number

The Bromine number of petroleum distillates is determined by Electrometric Titration in accordance with ASTM laboratory test D 1159. It is measure of olefins in the sample according to the equation:

$$\%\,\text{Olefins} = \frac{\text{Bromine number} \times \text{Mole weight of olefins}}{160}$$

Bubble points

Bubble point—is the temperature and pressure at which a hydrocarbon begins to boil.

The overhead from a fractionator has the following composition:

Mole fraction composition:

C_2	0.008
C_3	0.054
iC_4	0.021
nC_4	0.084
C_5s	0.143
C_6	0.155
C_7	0.175
Comp 1	0.124
Comp 2	0.124
Comp 3	0.075
Comp 4	0.037
Total	1.000

The components 1–4 are pseudo components that have properties similar to real components with the same boiling point. The mid boiling points of these components are:

	°F
Comp 1	260
Comp 2	300
Comp 3	340
Comp 4	382

The reflux drum temperature will be fixed at 100°F. The pressure will be calculated at *the bubble point* of this material at 100°F.

Bubble point is defined as the sum of all y's = sum of all Kx's.

x is the mole fraction of a component in the liquid phase and y is the mole fraction of the component in the vapor phase at equilibrium. At bubble point $\Sigma y = \Sigma Kx$.

K is the equilibrium constant and can be read from the curves found in text books such as Maxwell's *Data book on Hydrocarbons* or can be considered (rough and not be used for definitive design) as:

$$\frac{\text{Vapor Pressure}}{\text{Total Pressure}} = K$$

This relationship will be used for this calculation.

The calculation is iterative (trial and error) as follows:

AT 100°F

		1st Trial @ 5 psig		2nd Trial @ 10 psig	
	Mol fract. X	K	$Y = KX$	K	$Y = KX$
C_2	0.008	40.6	0.325	32.4	0.259
C_3	0.054	9.3	0.502	7.42	0.401
iC_4	0.021	3.55	0.075	0.38	0.05
nC_4	0.084	2.54	0.213	0.03	0.171
C_5	0.143	0.89	0.127	0.71	0.102
C_6	0.155	0.254	0.039	0.20	0.031
C_7	0.175	0.084	0.015	0.067	0.011
Comp 1	0.124	0.023	0.003	0.020	0.002
Comp 2	0.124	NEG	NEG		
Comp 3	0.075	NEG	NEG		
Comp 4	0.037	NEG	NEG		
	1.000		1.299		1.027

For 2nd trial (Estimate)

Take the K value of the highest fraction of y (in this case C_3) where $K = 9.3$.
Take this $K = 7.16$ (new K)
1.299

Make the 2nd trial with KC_3 at 7.16 which gives a systems pressure P as follows:

VP C3 = 7.1 where VP C3 @ 100°F is 190 psia

$$\text{Then } P = \frac{190}{7.1} = 26.5 \text{ psia.}$$

2nd trial pressure = 26.5 psia = 11.8 psig
Lets set it at 12 psig.

The second trial gives sum of Y's = 1.027 and this is considered close enough to 1.000. Then the drum will be operated at 100°F and at 12 psig.

Burners

The purpose of a burner is to mix fuel and air to ensure complete combustion. There are about 12 basic burner designs. These are:

Direction	—	vertical up fired vertical down fired horizontally fired
Capacity	—	high low
Fuel type	—	gas oil combination
Flame shape	—	normal slant thin, fan-shaped flat adaptable pattern
Hydrogen content	—	high
Excess air	—	normal
low		
Atomization	—	steam mechanical air assisted mechanical

Boiler types
Low NO_x
High intensity

Various combinations of the above types are available.

Gas burners

The two most common types of gas burners are the "pre-mix" and the "raw gas" burners.

Pre-mix burners are preferred because they have better "linearity", i.e., excess air remains more nearly constant at turndown. With this type, most of the air is drawn in through an adjustable "air register" and mixes with the fuel in the furnace firebox. This is called secondary air. A small part of the air is drawn in through the "primary air register" and mixed with the fuel in a tube before it flows into the furnace firebox. A turndown of 10 : 1 can be achieved with 25 psig hydrocarbon fuels. A more normal turndown is 3 : 1.

Oil burners

An oil burner "gun" consists of an inner tube through which the oil flows and an outer tube for the atomizing agent, usually steam. The oil sprays through an orifice into a mixing chamber. Steam also flows through orifices into the mixing chamber. An oil-steam emulsion is formed in the mixing chamber and then flows through orifices in the burner tip and then out into the furnace firebox. The tip, mixing chamber and inner and outer tubes can be disassembled for cleaning.

Oil pressure is normally about 140–150 psig at the burner, but can be lower or higher. Lower pressure requires larger burner tips, the pressure of the available atomizing steam may determine the oil pressure.

Atomizing steam should be at least 100 psig at the burner valve and at least 20–30 psi above the oil pressure. Atomizing steam consumption will be about 0.15–0.25 lbs steam/lb oil, but the steam lines should be sized for 0.5.

Combination burners

This type of burner will burn either gas or oil. It is better if they are not operated to burn both fuels at the same time because the chemistry of gas combustion is different from that of oil combustion. Gases burn by progressive oxidation and oils by cracking. If gas and oil are burned simultaneously in the same burner, the flame volume will be twice that of either fuel alone.

Pilots

Pilots are usually required on oil fired heaters. Pilots are fired with fuel gas.

Pilots are not required when heaters are gas fired only, but minimum flow bypasses around the fuel gas control valves are used to prevent the automatic controls from extinguishing burner flames.

Excess air and burner operation

The excess air normally used in process fired heaters are about 15–25% for gas burners and about 30% for oil burners. These excess air rates permit a wide variation in heater firing rates which can be effectively controlled by automatic controls without fear of 'Starving' the heater of combustion air. There has been considerable work lately to reduce this excess air considerably mostly to minimize air pollution. This practice has not been used in process heaters to date. It has however been adopted in the operation of large power station type heaters with some success.

Normally, companies specify that burners be sized to permit operation at up to 125% of design heat release with a turndown ratio of 3 : 1. This gives a minimum controllable rate of 40% of design without having to shut down burners.

Burner control

Burner controls become very important from safety and operation considerations. Most systems include an instrumentation system with interlocks that prohibit:

- Continuing firing when the process flow in the heater coil fails
- The flow of fuel into the firebox on flame failure

Under normal operating conditions the amount of fuel that is burnt is controlled by flow controllers operated on the coil outlet temperature. With combination burners the failure of one type of fuel automatically introduces the second type. Such a switch over can also be affected manually. This aspect is usually activated on pressure control of the respective the fuel system. That is on low pressure being sensed on the fuel being fired automatically switches to the second fuel.

(See Chapter 18 for more details on burners and heaters.)

Heater noise

All heaters are noisy and this noise is the result of several mechanisms. Among these are the operation of the burners. Gas burners at critical flow of fuel emit a noise. This can be minimized by designing for low pressure drop in the system. Intake of primary and secondary air is another source of noise. Forced draft burners are generally quieter than natural draft if the air ducting is properly sized and insulated. The design of the fan can also reduce noise in this mechanism. Low tip speed fan favors low noise levels.

C

Catalytic reforming unit

History. Without doubt, the development of catalytic reforming was a major evolution of the petroleum refining industry. It, in fact signified the birth of the 'Hydroskimming

Refinery'. The process produces hydrogen as a by-product in the upgrading of the naphtha octane value, and this hydrogen provided the means to upgrade a number of products by hydro treating.

Catalysts had been used during World War II for the upgrading of aviation fuels in terms of their octane number. Catalytic reforming of a kind had also been developed from a coal usage plant to process the aromatics used for war time explosives, etc. It was in the late 1950s and early 1960s that the fixed bed of bi-metallic catalyst was first developed. This process accomplished what the modern day catalytic reformers do in converting the cycloparaffins to their corresponding aromatics with the release of hydrogen molecules.

The process. Catalytic reforming is a process for improving the octane quality of straight-run naphthas and of mixed naphthas containing cracked naphtha. The principal reaction is dehydrogenation of naphthenes to aromatics. Contributing to the high octane of the product are side reactions such as hydrocracking of high-boiling hydrocarbons to low-molecular weight paraffins, isomerization of paraffins to branched-chain structures, and dehydrocyclization of paraffins and olefins to aromatics. Hydrogen recycle reduces the formation of carbon. Figure 19.C.1 is a process sketch of a typical catalytic reformer.

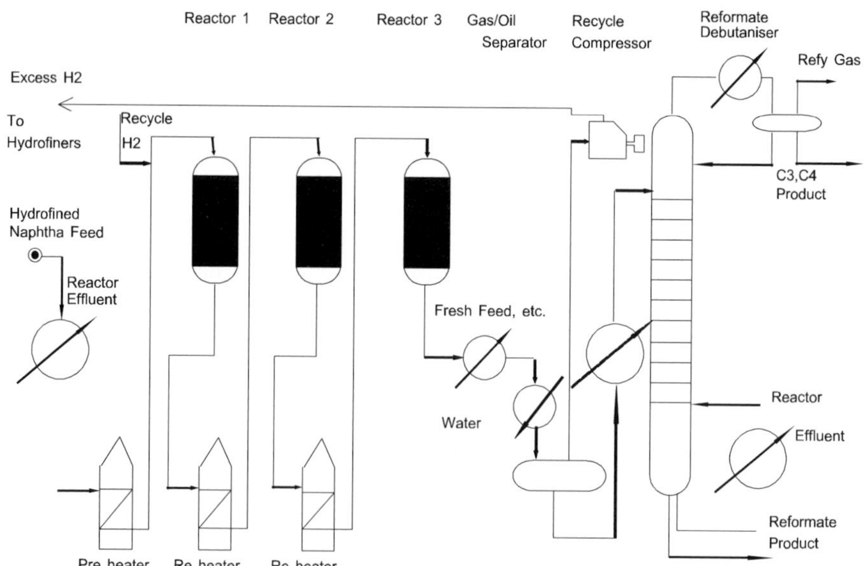

Figure 19.C.1. A typical catalytic reforming unit.

Feedstocks preferably are within the 200–430°F (TBP) boiling range. Feeds with lower initial boiling-point contain C_5 naphthenes which do not aromatize well. Feeds with higher final boiling point:

(a) Cause excessive lay-down of carbon on catalyst.
(b) Necessitate product re-running because reforming may increase the final boiling-point beyond gasoline specification.

The water content of the feedstock must be under 10 ppm to prevent rapid stripping of chloride from the platinum catalyst. Poisoning of the platinum catalyst is prevented by hydrofining the feed to a sulfur content below 10 ppm. Recycle hydrogen is added to the reformer feed and the mixture is preheated and charged to the first of a series of reactors. As the reactions are highly endothermic, each reactor effluent is reheated before being charged to the following reactor.

The effluent from the final reactor is separated into hydrogen rich gas and reformate. With heavy feeds, the reformate may contain 430°F + polymer.

The catalyst, 0.6 wt% platinum and 0.6 wt% chloride on alumina base, is regenerable. The type of regeneration system and the cycle length between regeneration periods depends on feed quality and reforming severity. Good feedstock quality and reforming severity favor the use of the semi-regeneration system, in which the plant is shut down after three months of operation to regenerate all reactors simultaneously. When reforming poorer feed and/or going into higher octane, the cyclic regenerative system is used in which there is a spare or "swing" reactor. A deactivated reactor can be regenerated in one day without process interruption. Many units now operate at high severity with continuous catalyst regeneration (CCR). Further details at this process are given in Chapter 5.

Centrifugal compressors

Compressors are divided into four general types, these are:

- Centrifugal
- Axial
- Reciprocating
- Rotary

This item refers only to the centrifugal compressor a description of which now follows.

This type of compressor consists of an impeller or impellers rotating at high speed within a casing. Flow is continuous and inlet and discharge valves are not required as part of the compression machinery. Block valves are required for isolation during maintenance.

Centrifugal compressors are widely used in the petroleum, gas, and chemical industries primarily due to the large volumes of gas that frequently have to be handled. Long continuous operating periods without an overhaul make centrifugal compressors desirable for use in petroleum refining and natural gas applications. Normally they are considered for all services where the gas rates are continuous and above 400 ACFM (actual cubic feet per minute) for a clean gas, and 500 ACFM for a dirty gas. These rates are measured at the discharge conditions of the compressor. Dirty gases are considered to be gases similar to those from a catalytic cracker, which may contain some fine particles of solid or liquid material.

The slowly rising head capacity performance curves make centrifugal compressors easy to control by either suction throttling or variable speed operation.

The main disadvantage of this type of compressor is that it is very sensitive to gas density, molecular weight, and polytropic compression exponent. A decrease in density or molecular weight results in an increase in the polytropic head requirement of the compressor to develop the required compression ratio.

The flow range of centrifugal compressors is as shown in Table 19.C.1.

In general, the head or differential pressure levels served by centrifugal compressors is considerably lower than that for reciprocal. The following diagram illustrates this feature (Figure 19.C.2).

Surge. A characteristic peculiar to centrifugal and axial compressors is a minimum capacity at which the compressor operation is stable. This minimum capacity is referred to as the surge or pumping point. At surge, the compressor does not meet the pressure of the system into which it is discharging. This causes a cycle of flow reversal as the compressor alternately delivers gas and the system returns it.

Table 19.C.1. Centrifugal compressor flow range

Nominal flow range (inlet acfm)	Average polytropic efficiency	Average adiabatic efficiency	Speed to develop 10,000 ft head/wheel
500–7,500	0.74	0.70	10,500
7,500–20,000	0.77	0.73	8,200
20,000–33,000	0.77	0.73	6,500
33,000–55,000	0.77	0.73	4,900
55,000–80,000	0.77	0.73	4,300
80,000–115,000	0.77	0.73	3,600
115,000–145,000	0.77	0.73	2,800
145,000–200,000	0.77	0.73	2,500

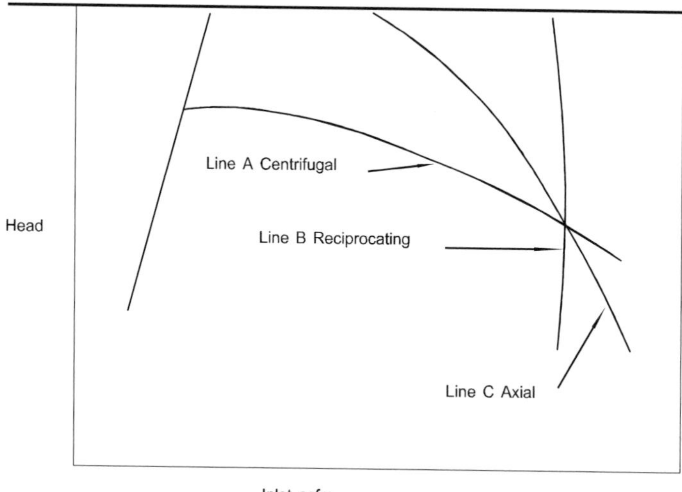

Figure 19.C.2. Comparison of differential compressor heads.

The surge point of a compressor is nearly independent of its speed. It depends largely on the number of wheels or impellers in series in each stage of compression. Reasonable reductions in capacity to specify for a compressor are shown in Table 19.C.2.

An automatic re-circulation bypass is required on most compressors to maintain the minimum flow rates shown. These are required during start-up or when the normal load falls below the surge point. Cooling is required in the recycle circuit if the discharge gas is returned to the compressor suction.

Performance curves for centrifugal compressors

The rise of performance curves should be specified for a compressor. This is normally done by specifying the pressure ratio rise to surge required in each stage of compression. A continuously rising curve from normal flow rate to surge flow is required for stable control.

Table 19.C.2. Effect of capacity reduction

Wheels/compression stage	% of Normal capacity at surge-maximum
1	55
2	65
3 or greater	70

The pressure ratio rise to surge is largely a function of the number of impellers per compression stage. Reasonable pressure ratio rises to specify are shown below:

Wheels/compression stage	Minimum % of rise in pressure ratio from normal to surge flow
1	$3^1/_2$
2	6–7
2 or greater	$7^1/_2$

Frequently, the performance curves for a compressor have to be plotted to determine if all anticipated process operations will fit the compressor and its specific speed control. Three points on the head capacity curve are always known. These are the normal, surge, and maximum capacity points. The normal capacity is always considered to be on the 100% speed curve of the compressor. The surge point and the compression ratio rise to surge have been specified. From this the head produced by the compressor at the surge point can be back-calculated using the head-pressure ratio relationship. The maximum capacity point is specified to be at least 115% of capacity at 85% of normal head.

The head capacity curve retains its characteristic shape with changes in speed. Curves at other speeds can be obtained from the three known points on the 100% speed curve by using the following relationships:

1. The polytropic head varies directly as the speed squared.
2. The capacity varies directly as speed.
3. The efficiency remains constant.

Figure 19.C.3 below shows a typical centrifugal compressor performance curve.

Cloud points

The cloud point of a transparent or semi transparent oil sample is the temperature at which a cloud or mist forms in the sample. The method is ASTM D 2500, and is carried out in the laboratory where a sample of the oil is reduced in temperature by submerging the sample in its container first in ice then in iced salt and finally in a solid CO_2 bath. The container is a cylindrical glass vessel about 19 mm diameter by 100 mm deep. The vessel is filled to a mark about 80 mm deep and a low range thermometer inserted so the bulb is about 10–12 mm below the sample surface. Temperature readings are taken at every stage. Readings are more frequently taken in the last cooling stage and that temperature at which the sample becomes misty is taken as its cloud point.

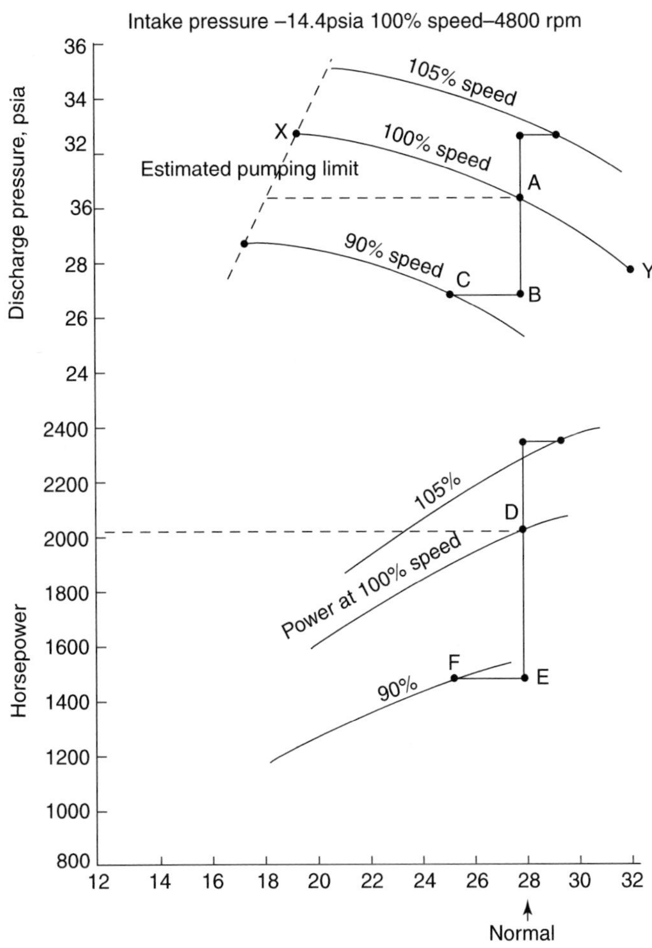

Figure 19.C.3. Typical performance curves for a centrifugal compressor.

Coke

Coke is formed in the processes to convert the residuum fuels to the more desirable distillate products of naphtha and lighter through to the middle distillates. By far the largest production of coke is the sponge coke from the delayed coking process. Uncalcined sponge coke has a heating value of about 14,000 Btu/lb and is used primarily as a fuel. High sulfur sponge coke however is popular for use in cement plants since the sulfur reacts to form sulfates. Sponge coke is calcined to produce a coke grade suitable for anodes in the aluminum industry.

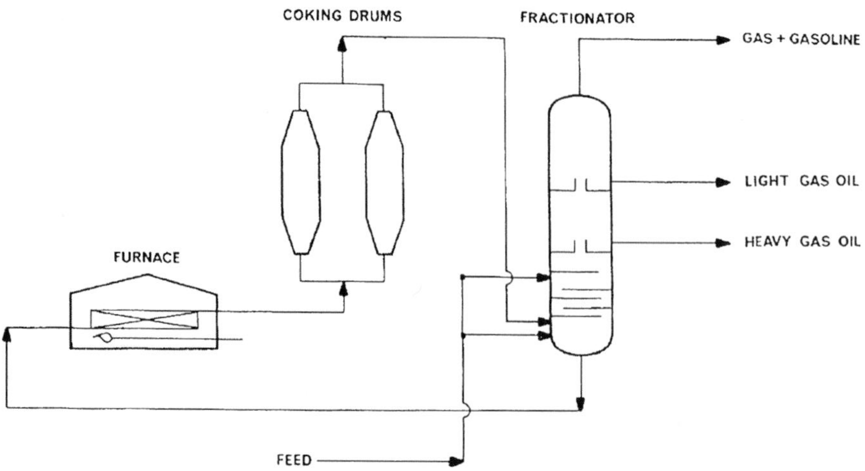

Figure 19.C.4. A delayed coker.

Coking processes

Coke is formed in the processes to convert the residuum fuels to the more desirable distillate products of naphtha and lighter through to the middle distillates. There are two routes by which this coking process proceeds. The first and the most common is the delayed coking route. The second is the fluid coking method, and this has been made more attractive to many refiners with the development of ExxonMobil's proprietary process of *Flexicoking*. This proprietary process eliminates the coke completely by converting it to low Btu fuel gas (see Chapter 11).

The delayed coking process

The delayed coking process is illustrated in Figure 19.C.4.

The fuel oil or heavy oil feed is routed to a cracking furnace similar to other thermal cracking processes. The effluent from the furnace is sent to one of a set of several coking drums without quenching. The effluent is normally at a coil outlet temperature of around 920°F and at a pressure of 30–50 psig and the coking drum is filled in about 24 hr. The vapors leaving the top of the drum whilst filling are routed to the fractionator, where they are fractionated to the distillate products. The very long residence period of the effluent in the coking drums results in the complete destruction of the heavy fractions, and a solid residue rich in carbon is left in the drum as coke. When the drum is full of coke it is cooled and the top and bottom of the drum are opened. Special high-pressure water jets are used to cut and remove the coke through the drum's bottom opening.

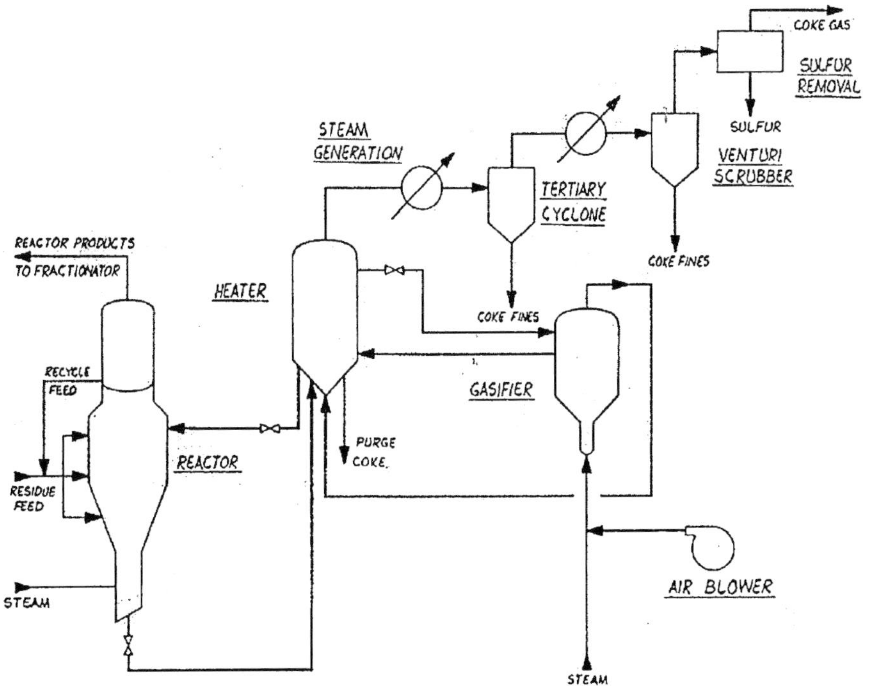

Figure 19.C.5. A flexi (fluid) coking process.

Flexi coking (Fluid coker)

Flexi-coking is a thermal conversion process licensed by Exxon Research and Engineering Company. The process itself is an extension of the traditional fluid coking process. The extension allows for the gasification of the major portion of the coke make to produce a low-Btu gas. The gasification step follows closely to the concept used in the coal gasification processes.

Figure 19.C.5 is a simplified flow diagram of the Flexi-coking process.

Heavy residuum feed is introduced into the reactor vessel where it is thermally cracked. The heat for cracking is supplied by a fluidized bed of hot coke transferred to the reactor from the heater vessel.

The vapor products of the reaction leave the reactor zone to enter the scrubber section. Fine coke and some of the heavy oil particles are removed from the cracked products in the scrubber zone and returned to mix with the fresh feed entering the reactor. The reactor products subsequently leave the scrubber and are routed to a conventional fractionating facility. Steam is introduced to the bottom of the reactor to maintain a fluid bed of coke and to strip the excess coke leaving the reactor free from entrained oil. The coke leaving the reactor enters the heater vessel, where sufficient coke is converted into CO/CO_2 in the presence of air. This conversion of the coke provides the heat for cracking which is subsequently transmitted to the reactor by a hot coke stream. The net coke make leaves the heater and enters the gasifier vessel. Air and steam are introduced into the gasifier to react with the coke producing a low-Btu gas consisting predominately of hydrogen, CO, CO_2, and nitrogen. This gas together with some excess air is transferred to the heater, and leaves this vessel to be suitably cleaned and cooled.

Flexi-coking is an extinctive process. By continuous recycle of heavy oil stream all the feed is converted into distillate fractions, refinery gas, and low Btu gas. There is a very small coke purge stream which amounts to about 0.4–0.8 wt% of fresh feed. When suitably hydrotreated, the fractionated streams from the Flexi-coker provide good quality products. Hydrotreated coker naphtha provides an excellent high naphthene feed to the catalytic reformer.

Cold flash separator

In many high temperature and high pressure hydrocracker or hydrotreater units the reactor effluent is reduced in pressure and temperature in several stages. The last of these stages is the cold flash separation stage. Figure 19.C.6 shows a typical residuum hydrocracker with a cold flash separator.

The reactor effluent leaves the reactor to enter a hot flash drum which is near to the reactor pressure and temperature conditions. Here the heavy bituminous portion of the effluent leaves from the bottom of the drum while the lighter oil and gas phase leaves as a vapor from the top of the drum. This vapor is subsequently cooled by heat exchange with the feed and further cooled and partially condensed by an air cooler. This cooled stream then enters a cold separator operating at a pressure only slightly lower than that of the reactor. A rich hydrogen gas stream is removed from this drum to be amine treated and returned as recycle gas to the process. The distillate liquid leaves from the bottom of the separator to join a vapor stream from the hot flash surge drum (in this case a thermal cracker feed surge drum). Both these streams enter the *cold flash drum* which operates at a much lower pressure than the upstream equipment. A gas stream is removed from the drum to be routed to the absorber in a light ends unit. The liquid distillate from the drum is routed to the debutanizer in the light ends unit.

Component balances

Component balances are derived from the TBP of the material requiring the balance. These component balances are derived by splitting the TBP into mid-boiling point

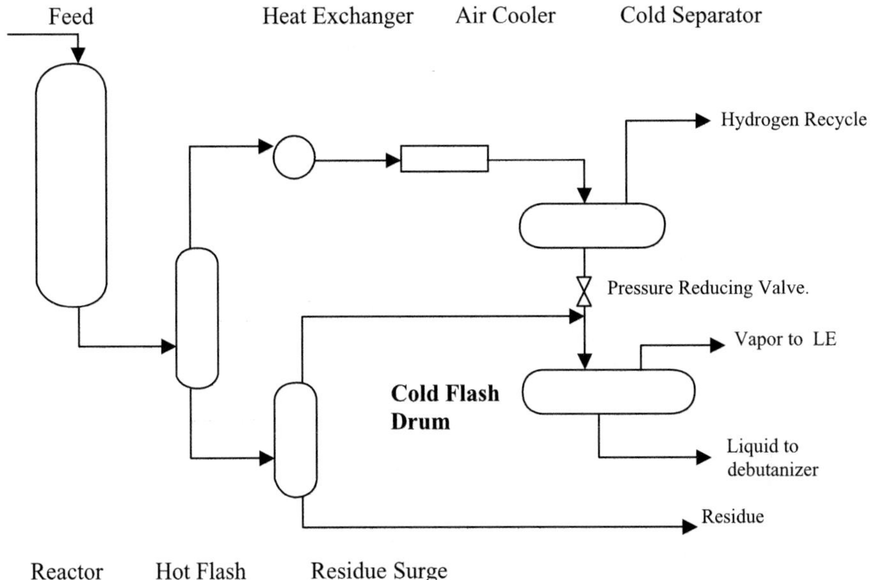

Figure 19.C.6. Typical residue hydrocracker with cold flash drum.

pseudo components (see Chapter 1 for definition of pseudo components and mid-boiling points). The purpose of component balances is to calculate more accurately fractions of the feed material properties in terms of specific gravity, sulfur content, mole weight, cloud and pour points and the like. Figure 19.C.7 shows a hypothetical TBP curve of a middle distillate fraction. This fraction has been broken up into 6 pseudo components, with mid boiling points of 410°–591°F. By referencing the crude feed assay from which this fraction originates, the SG of each component can be read off as °API. If it is a well-produced assay the component sulfur, cloud and pour points can also be read off for each pseudo component. The volume of each pseudo component forming the fraction is shown as the 'x' of the TBP curve.

Some of the components shown in the diagram are tabled in order to determine the fraction's properties. Table 19.C.3 is an example of the purpose of the component breakdown In this case only the specific gravity is determined, but by applying the respective assay data for other properties a reasonable definition of the fraction can be obtained (see Chapter 1).

Condensers (shell and tube)

In the chemical process plants vapors are condensed either on the shell side of a shell and tube exchanger, the tube side of an air cooler, or by direct contact with the coolant

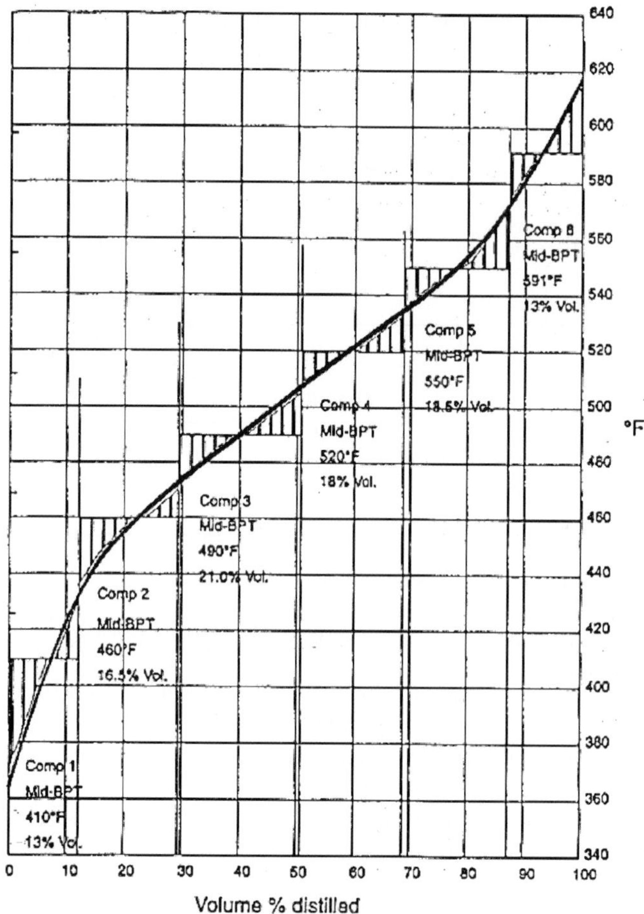

Figure 19.C.7. A Typical component breakdown.

in a packed tower. By far the most common of these operations are the first two listed. In the case of the shell and tube condenser, the condensation may be produced by cooling the vapor by heat exchange with a cold process stream or by water. Air cooling has overtaken the shell and tube condenser in the case of water as coolant in popularity as described in item 6.4.

In the design or performance analysis of condensers the procedure for determining thermal rating and surface area is more complex than that for a single phase cooling and heating. In condensers there are three mechanisms to be considered for the rating procedure. These are:

Table 19.C.3. Determination of SG using components

Component	Volume %	Mid BPt °F	SG @ 60°F (From assay)	Weight factor
1	13.0	410	0.793	10.3
2	16.5	460	0.801	13.2
3	21.0	490	0.836	17.6
4	18.0	520	0.844	15.2
5	18.5	550	0.846	15.7
6	13.0	591	0.850	11.1
Totals	100.0			83.1

$$\text{SG @ }60°\text{F} = \frac{83.1}{100} = 0.831 \text{ or } 38.8°\text{API}.$$

- The resistance to heat transfer of the condensing film
- The resistance to heat transfer of the vapor cooling
- The resistance to heat transfer of the condensate film cooling

Each of these mechanisms is treated separately and along pre-selected sections of the exchanger. The procedure for determining the last two of the mechanisms follows that for single-phase heat transfer. The following expression is used to calculate the film coefficient for the condensing vapor:

$$h_c = 8.33 \times 10^3 \times k_f \times \left[\frac{\text{Sg}_c}{\mu_f}\right]^{2.33} \times (M_c/L_c \cdot N_s)^{33}$$

where

h_c = Condensing film coefficient.
M_c = Mass condensed in lbs/hr
L_c = Tube length for condensation.
$= \frac{A_{zone}}{A} \times (L - 0.5)$
$N_s = 2.08\ \text{N}_t^{0.495}$ for triangular pitch.
k_f = Thermal conductivity of condensate at film temperature.
Sg = Specific gravity of condensate.
μ_f = Viscosity of condensate at film temperature (cP).

Again there are many excellent computer programs that calculate condenser thermal ratings, and these of course save the tedium of the manual calculation. However, no matter which method of calculation is selected there is required one major additional piece of data over that required for single-phase heat exchange. That item is the Enthalpy Curve for the vapor.

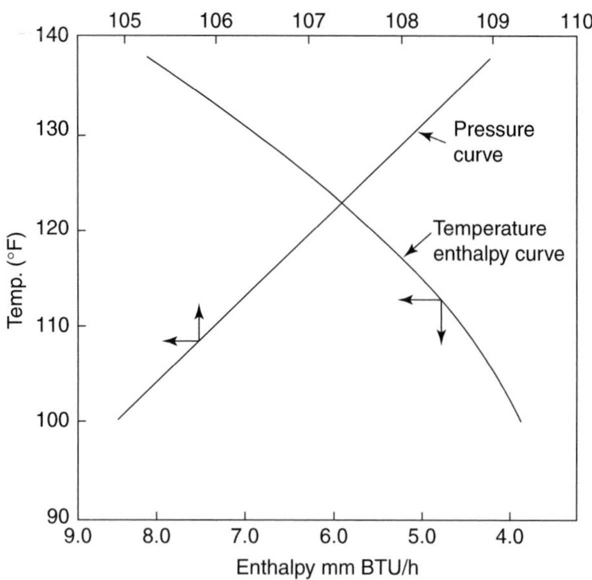

Figure 19.C.8. Typical enthalpy curves.

Enthalpy curves are given as the heat content per lb or per hour contained in the mixed phase condensing fluid plotted against temperature. An example of such a curve is given in Figure 19.C.8. These Enthalpy curves are developed from the vapor/liquid or flash calculations described in Chapter 1 and 3.

Briefly, the calculation for the curve commences with determining the dew point of the vapor and the bubble point of the condensate. Three or more temperatures are selected between the dew and bubble points and the V/L calculation of the fluid at these temperatures carried out. Enthalpy for the vapor phase and the liquid phase are added for each composition of the phases at the selected temperatures. These together with the enthalpy at dew point and bubble point are then plotted.

A manual calculation for condensers is described here. Again this is done to provide some understanding of the data required to size such a unit and its significance in the calculation procedure. Computer aided designs should however be used for these calculations whenever possible.

The following calculation steps describe a method for calculating the film coefficient of a vapor condensing on the shell side of a S&T exchanger. The complete rating calculation will not be described here as much of the remaining calculation is simply repetitive.

Step 1. Calculate the dew point of the vapor stream at its source pressure. Estimate the pressure drop across the system. Usually 3–5 psi will account for piping and the exchanger pressure drop. Calculate the bubble point of the condensate at the terminal pressure. Select three or more temperatures between dew point and bubble point and calculate the vapor/liquid quantities at these conditions of temperature and pressure.

Step 2. Calculate the enthalpy of the vapor and liquid at these temperatures. Plot the total enthalpies against temperature to construct the enthalpy curve. Establish the properties of the vapor phase and liquid phase for each temperature interval. The properties mostly required are specific gravity, viscosity, Mole wt, thermal conductivity and specific heats.

Step 3. In the case of a water cooler calculate the duty of the exchanger and the quantity of water in lbs/hr. Commence the heat transfer calculation by assuming an overall heat transfer coefficient, calculating the corrected LMTD, and the surface area.

Step 4. Using the surface area calculated in Step 3 define the exchanger geometry in terms of number of tube passes, number of tubes on the center line, shell diameter, baffle arrangement and the shell free flow area. Calculate also the water flow in feet per sec.

Step 5. Divide the exchanger into 3 or 4 zones by selecting the zone temperatures on the enthalpy curve. Calculate the average weight of vapor and the average weight of condensate in each zone. Using these averages calculate the average heat transferred for:

Cooling of the vapor Q_v
Cooling of the condensate Q_L

Condensing of the vapor which will be:
Total heat in the zone (from the enthalpy curve) plus the difference of Q_v and Q_L.

Step 6. Calculate the film coefficient for the tube side fluid.

Step 7. Starting with zone 1 and knowing the outlet temperature of the coolant fluid, the total heat duty of the zone, and the shell side temperatures calculate the coolant inlet temperature. Using this calculate the LMTD for the zone and, assuming a zone overall heat transfer coefficient U, calculate a surface area for the zone. Using this and the total exchanger area estimated in Step 4 establish L_c in feet.

Step 8. Calculate the condensing film coefficient from the equation given earlier. This will be an uncorrected value for h_c. This will be corrected to account for turbulence by the expression.

$$h_{c\,(corr)} = h_c \times (G_v/5)^{0.7}$$

where

G_v = average vapor mass velocity in lbs/hr sqft

Step 9. Calculate the value of G_v using the free flow area allocated to the vapor γ_v. The following expressions are used for this:

$$\gamma_v = 1 - \gamma_L$$

$$\frac{1}{\gamma_L} = 1 + \frac{\text{Ave mass vapor}}{\text{Ave mass liquid}} \times (\mu_v/\mu_L)^{.111} \times (\rho_L/\rho_v)^{.555}$$

$$G_v = \frac{\text{Ave mass Vapor}}{25 \times \text{Free flow area} \times \gamma_v}$$

Step 10. Calculate the film coefficient h_v for the vapor cooling mechanism. This will be the procedure used for a single phase cooling. This is corrected to account for resistance of the condensate film by the expression:

$$\frac{1}{h_{v\,corr}} = \frac{1}{h_c} + \frac{1}{1.25\,h_v}$$

Step 11. Calculate the film coefficient for the condensate cooling mechanism for single phase cooling on the shell side. This is corrected for drip cooling that occurs over a tube bank.

$$\text{Drip cooling } h_{dc} = 1.5 \times h_c$$

and

$$h_{L\,corrected} = \frac{2 \times h_{dc} \times h_L}{h_{dc} + h_L}$$

where h_t = The inside film coefficient based on outside diameter as follows

$$h_t = \frac{300 \times (V_t \times \text{tube i/d in ins})^{0.8}}{\text{tube o/d in ins}}$$

and V_t = Linear velocity of tube side flow in ft/sec.

Step 12. Calculate the total zone film coefficient h_o using the following expression:

$$h_o = \frac{Q_{zone}}{\frac{Q_c}{h_c} + \frac{Q_v}{h_v} + \frac{Q_L}{h_L}}$$

where Q_c, Q_v, Q_L, are enthalpies for condensing, vapor cooling, and condensate cooling, respectively.

Step 13. Calculate the overall heat transfer coefficient neglecting the shell side coefficient from step 12. Thus:

$$\frac{1}{U_x} = r_o + r_w + r_{io} + R_{io}$$

where

r_o, r_w, r_{io} are fouling factors for shell fluid, wall, and tube side fluid, respectively. R_{io} is the tubeside film coefficient calculated in Step 6.

Step 14. Calculate the overall heat transfer coefficient U_{zone} for the zone using the expression:

$$U_{zone} = \frac{h_o \times U_x}{h_o + U_x}$$

Check the calculated U against the assumed for the zone. Repeat the calculation if necessary to make a match.

Step 15. Calculate the zone area using the acceptable calculated U. Repeat Steps 7–14 for the other zones. The total surface area is the sum of that for each zone.

Conradson carbon (ASTM D-189)

See Chapter 16.

Control valves

Control valves are used throughout the process and oil refining industries to control operating parameters. These parameters are:

- Flow
- Pressure
- Temperature
- Level

Figure 19.C.9 shows the conventional control valve which in this case is taken as a double seated plug type valve. Like most control valves it is operated pneumatically by an air stream exerting a pressure on a diaphragm which in turn allows the movement of a spring loaded valve stem. One or two plugs (the diagram shows two) are attached to the bottom of the valve stem which, when closed, fit into valve seats thus providing tight shut off. The progressive opening and closing of the plugs on the valve seats due to the movement of the stem, determines the amount of the controlled fluid flowing across the valve. The pressure of the air to the diaphragm controlling the stem movement is varied by a control parameter, such as a temperature measurement, or flow measurement, or the like.

There are two types of plug valves used for the conventional control valve function. These are:

- Single seated valves
- Double seated valves

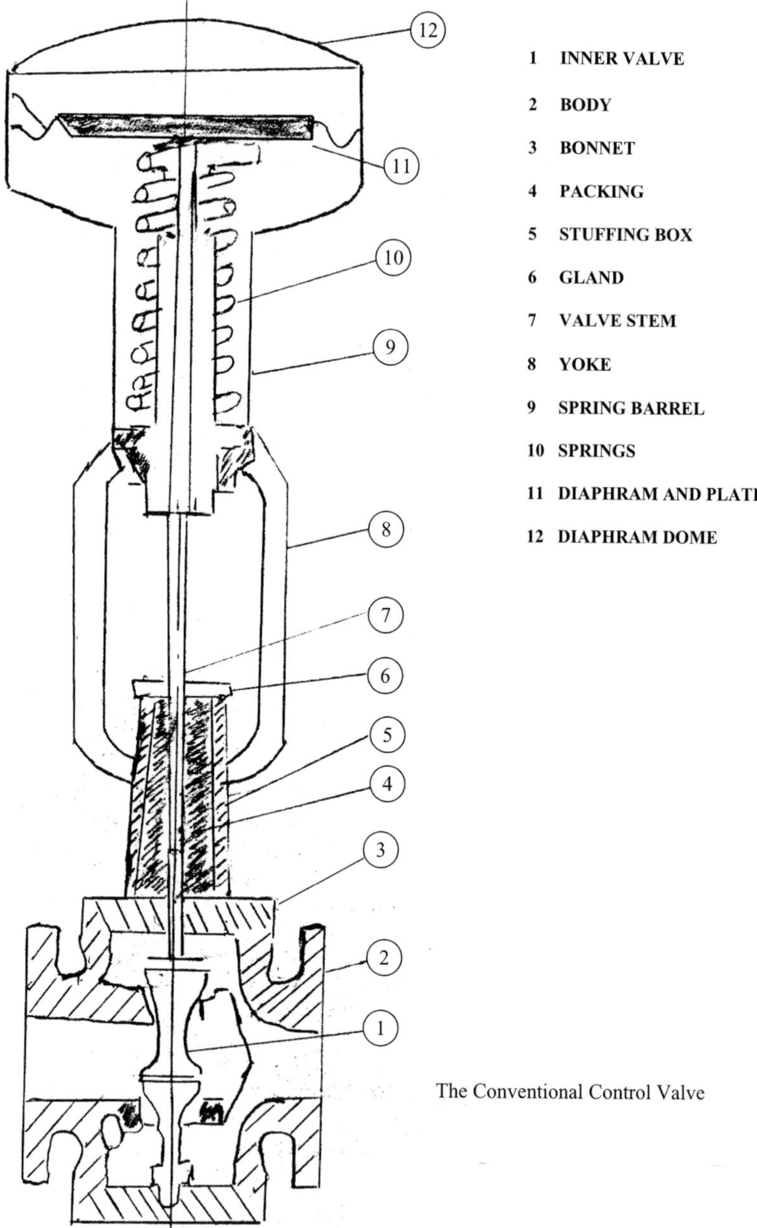

Figure 19.C.9. A doubled seated control valve.

Single seated valves are inherently unbalanced so that the pressure drop across the plug affects the force required to operate the valve. Double seated valves are inherently balanced valves and are the first choice in most services.

There are two other types of control valves in common service in the industry. These are the Venturi Type and the Butterfly type. Both these types when pneumatically operated (which they usually are) work in the same way as described above for a plug type valve. The major difference in these two types are in the valve system itself. In the case of the venturi the fluid being controlled is subjected to a 90° angle change in direction within the valve body. The inlet and outlet dimensions are also different, with the inlet having the larger diameter. The valve itself is plug type but seats in the bend of the valve body.

Venturi type or angle valves are used in cases where there exists a high-pressure differential between the fluid at the inlet side of the valve and that required at the outlet side. Details of these valves may be found in Chapter 13 of this book.

Butterfly valves operate at very low-pressure drop across them. They can operate quite effectively at only ins of water gauge pressure drop, and where the operation of the conventional plug type valves would be unstable. The action of this valve is by means of a Flap in the process line. The movement of this flap from open to shut is made by a valve stem outside the body itself. This stem movement, as in the case of the other pneumatically operated valves, is provided by air and spring loads onto the stem from a diaphragm chamber. The only major disadvantage in this type of valve is the fact that very tight shut off is difficult to obtain due to the flap type action of the valve. Details of valve characteristics and valve sizing are given in Chapter 13 of this book.

Cost estimating

Details of cost estimating as it applies to process plants including oil refineries are given in Chapter 17 of this book. The following is a synopsis of the cost estimate sections of Chapter 17.

The progress of any construction or development project in a refinery uses milestone cost estimates coupled with completion schedules for its measurement and control. In the life of a construction project therefore about four "control estimates" may be developed. For want of better terms these may be referred to as:

- The Capacity Factored Estimate
- The Equipment Factored Estimate
- The Semi Definitive Estimate
- The Definitive Estimate

These are described briefly as follows:

The capacity factored estimate

In most process studies this is the first estimate to be prepared. It is the one that requires the minimum amount of engineering but is the least accurate. This is because the plant that is used to factor from will not exactly match the plant in question. It is good enough however when comparing different processes where the estimates are on the same basis. Past installed costs of similar plants (Definitive Estimate) are used coupled with some experienced factors to arrive at this type of estimate. The cost of a higher or lower capacity plant is obtained from the equation:

$$C = K(A/B)^b$$

where

C = Cost of the plant in question.
K = Known cost of a similar plant of size B.
A = Is the capacity of the plant in question. (usually in vol/unit time).
B = Capacity of known plant.
b = Is an exponential factor ranging between 0.5 and 0.8.

The equipment factored estimate

This estimate requires a substantial amount of process engineering to define the specific plant or process that is to be estimated. Details of the degree of engineering that is to be performed is given in item 8.4. Briefly the following process activity is required:

- Development of a firm Heat and Material Balance
- An acceptable Process flowsheet to be developed
- An Approved Equipment list
- Equipment summary Process Data sheet for all major equipment
- Process Specification for Specialty Items

A detailed narrative giving a process description and discussion (This will be required for the management review and approval of the estimate).

The semi definitive and the definitive cost estimates

These estimates will include input from all engineering disciplines that are involved in the engineering and construction of the plant or process. Although major manufacturing companies such as large oil companies and chemical companies have sufficient statistical data to develop semi definitive and definitive estimates these are normally left to the engineering and construction companies.

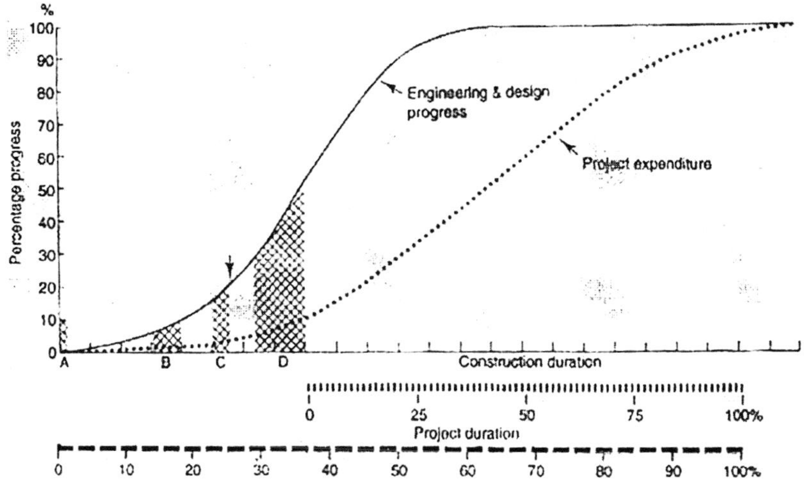

Figure 19.C.10. Time phasing of estimates.

These companies develop the estimates and use them during the course of the installation project as cost control tools for project management.

Phasing of the estimates

The phasing of the estimates in the project schedule is reflected in Figure 19.C.10. The estimates are as follows:

A) Capacity Factored
B) Machinery and Equipment
C) Semi Definitive
D) Definitive

Crude oil

Crude oil is a mixture of hydrocarbon compounds. These compounds range in boiling points and molecular weights from methane as the lightest compound to those whose molecular weight will be in excess of 500. Chapter 1 of this book describes these hydrocarbon families in some detail. Briefly the major groups are:

Paraffins
Olefins
Naphthenes
Aromatics

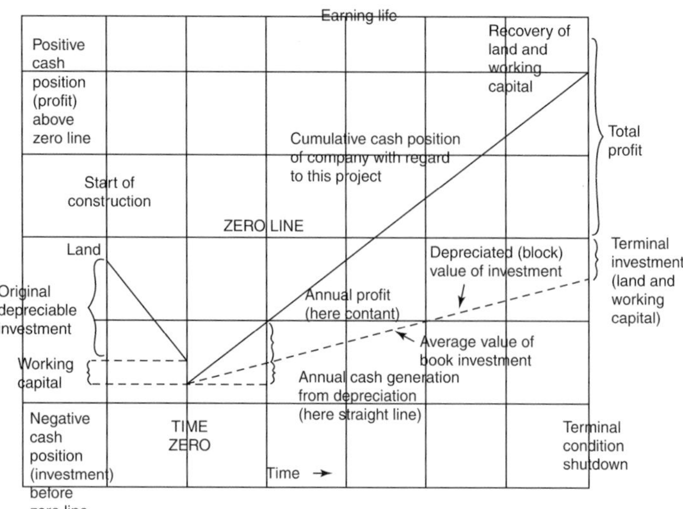

Figure 19.C.11. Graph of cumulative cash flow.

In addition to these basic compounds there will be impurities. The major non-hydrocarbon impurity are the sulfur compounds. In the lighter boiling range of the crude oil sulfur appears mostly in the form of mercaptans, while in the higher boiling point section of the crude the sulfur is in the form of complex thiophenes. In addition to sulfur, nitrogen, and heavy metal are usually present in small quantities, but are sufficient to cause problems to some processes. Olefins are usually present in small quantities in the crude oil, but are produced in the processes of refining the crude. Most notably are the thermal and fluid catalytic process that produce olefins.

The nature and the properties of crude oil vary considerably. This variation is usually connected to the area from which the crude oil is produced. The properties of crude oils from respective sources are recorded in the Crude Assay which is produced by the vendors of the individual crude oils.

Summary assays of the more common crude oils are produced as Appendix 3 to this part of the book.

Cumulative cash flow and present worth

There are several methods of assessing profitability based on discounted cash flow (DCF), but the most reliable yard stick is a return on investment method using the Present Worth (or Net Present Value) concept. This concept equates the present value of a future cash flow as a product of the present interest value factor and the future cash flow. *Based on this concept, the Return on Investment is that Interest value or*

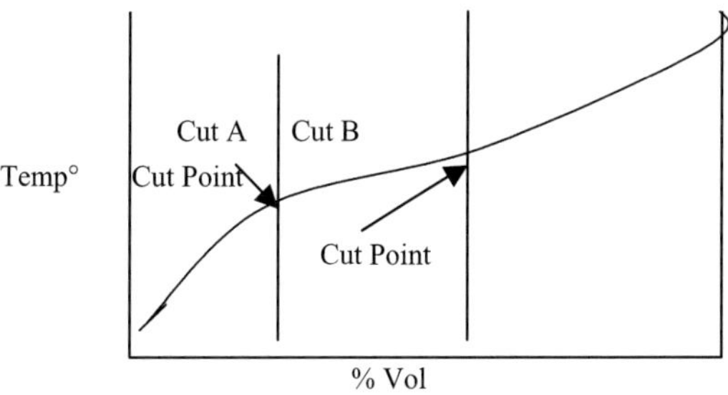

Figure 19.C.12. Typical cut points on a TBP curve.

Discount Factor which forces the Cumulative Present Worth value to Zero over the economic life of the project.

The calculation itself is in two parts, which are:

- Calculation of Cash Flow
- Present Worth Calculation

Both these calculations are described in detail in Chapter 17 of this book. The phases and summary of Cumulated Cash Flow is provided by Figure 19.C.11.

The calculation for present worth is iterative, and is fully explained in Chapter 17 with an example calculation.

Cut points

A cut point is defined as that temperature on the whole crude TBP curve that represents the limits (upper and lower) of a fraction to be produced. Consider the curve shown in Figure 19.C.12. Below of a typical Crude oil TBP curve.

A fraction with an upper cut point of 100°F produces a yield of 20 vol% of the whole crude as that fraction. The next adjacent fraction has a lower cut point of 100°F and an upper one of 200°F this represents a yield of 50–20% = 30 vol% on crude.

D

De-aromatization process

The de-aromatization process is a hydrotreating process that converts aromatic compounds to naphthenes and paraffins. Its most common use is to lower the smoke

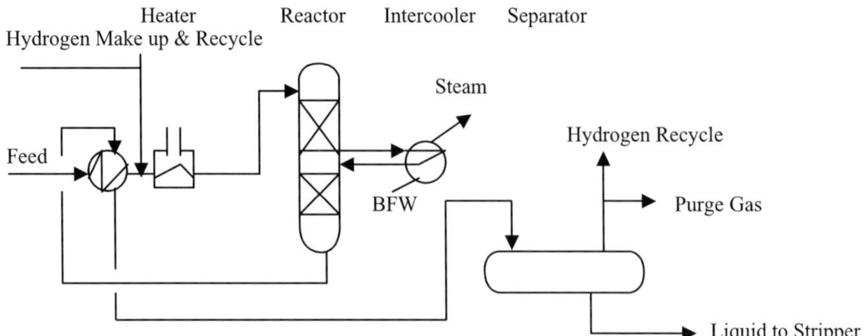

Figure 19.D.1. Typical hydro de-aromatizer process.

point of kerosene, particularly if the product is to be blended to make aviation turbine gasoline. Initially the reduction of kerosene smoke point was effected by the removal of aromatic compounds by extraction using liquid SO_2. A typical kerosene de-aromatizing unit is shown as Figure 19.D.1.

De-sulfurized kerosene feed is introduced and preheated with hot reactor effluent. Hydrogen make-up and the hydrogen recycle stream are mixed with the preheated feed to enter the reactor heater which rises the feed temperature to reaction temperature. Leaving the heater the feed enters the top catalyst bed of the reactor. It flows down through the bed consisting of a nickel type catalyst. The reaction is exothermal and temperature is controlled by an inter-cooler between the top bed and a second catalyst bed. Steam is generated in the inter-cooler as the cooling medium. The reactor effluent leaves the second catalyst bed to be condensed by heat exchange with incoming feed, and, if necessary, by a trim air cooler. This condensate enters a separator drum the pressure of which is set to remove a light hydrocarbon and hydrogen stream at a temperature of 100°F. The pressure in this drum also determines the reactor pressure and, depending on the catalyst used, will be between 350 and 450 psig. The liquid from the separator will be stripped free of light ends in a conventional trayed and reboiled stripper column. The purity of the hydrogen recycle stream is maintained by removing a purge stream that reduces the total hydrocarbons in the recycle. The recovery of kerosene product by this process is between 85% and 90% volume.

De-asphalting process

De-asphalting the heavy end of crude, that is the vacuum residue, is not entirely reserved for the production of lube oils. In energy refineries it is used to remove the asphaltene portion of the residue to prepare a suitable feedstock for catalytic conversion units. In most of these conversion units the performance of the catalyst is greatly impaired by the presence of heavy metals and the high Conradson carbon content of the residue feed. Most of the metals are entrained in the asphaltene compounds which make up most of the asphalt portion of the residue. These asphaltenes are also high in Conradson carbon content, so the removal of these serves to eliminate both the heavy metal content and the high Conradson Carbon. In the case of lube oil production, the light liquid phase resulting from the extraction of the asphalt makes excellent lube base oil. This de-asphalted oil is termed 'Bright Stock' and can now be further refined in the same way as neutral base stock, which are vacuum distillates, to meet the specifications for blend stocks. Details of the process are given in Chapter 12 of this book.

De-butanizer

De-butanizers are used in refineries to remove butanes and lighter compounds from product streams. The most notable of the feed streams is the overhead distillate from the atmospheric crude distillation tower. Other uses of this process are in the removal of these light ends from hydrotreaters (kero & heavier products), and the reactor effluent liquids from catalytic reforming and hydrocracking. The overhead from the fluid catalytic cracker main tower is also debutanized, but the butane compounds from this process will also contain the olefinic compounds from the cracking process. A typical process configuration for a debutanizer is shown below as Figure 19.D.2.

The debutanizer operates at an overhead accumulator pressure of about 125 psia at 100°F. Usually the whole column overhead is condensed under these conditions

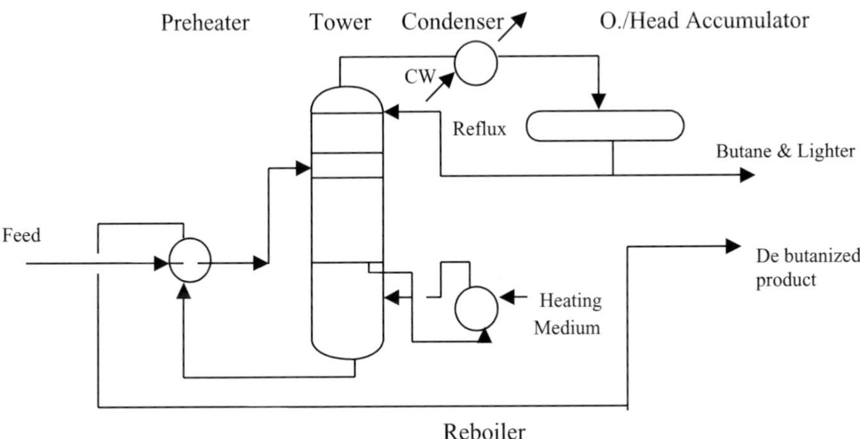

Figure 19.D.2. A typical de-butanizer configuration.

unless there is hydrogen present. Thus, in the case of hydotreaters, reformers, and hydrocrackers there will be partial condensation and there will be a vapor phase from the drum as well as the liquid distillate. The tower will consist of between 30 and 35 distillation trays and the external reflux will be about 2 : 1. More details and design procedures are given in Chapter 4 of this Handbook.

De-ethanizer

The purpose of the de-ethanizer is to remove ethane from the product stream of LPG. Normally it will be the last tower in a light end distillation configuration (see the item 'Light End Distillation' under 'Distillation' in this Part 2 of the book. The de-ethanizer tower operates at an overhead accumulator of below 450 psia at 100°F. This ensures that the operating pressure at the bottom of the tower, which is propane LPG, will be below its critical pressure. In the design example given in Chapter 4 the pressure in the accumulator was calculated to be 350 psia at 100°F, and this was the dew point of the overhead vapor product. De-ethanizers operating without overhead refrigeration facilities have partial condensers and in this case only sufficient overheads from the tower are condensed to meet the reflux required in the tower. Thus the accumulator becomes a theoretical tray itself and there will be no liquid distillate product as such.

De-propanizer

In the refinery configuration of light ends distillation this unit is usually located between the de-butanizer and the de-ethanizer. The process flow is similar to the de-butanizer, with total overhead product and reflux being condensed. The tower operates at a reflux drum pressure of between 200 and 250 psia at 100°F. The tower contains 35–40 actual trays, and the bottom product will be butane LPG. The overhead distillate will be feed to the de-ethanizer whose bottom product will be propane LPG.

Dew points

Is the temperature and pressure condition at which a hydrocarbon vapor begins to condense. That is in a calculation the sum of the mole fraction composition of the vapor divided by the equilibrium constant of each compound must equal the sum of the sum of the mole fraction of the liquid phase at the dew point condition of temperature and pressure. The following example illustrates this concept. (This example is based on a tower top condition for an atmospheric crude distillation unit.)

The following dew point calculation will be carried out at 8.3 psia which is the partial pressure of the hydrocarbons in the overhead vapor.

COMP	Mole Frac Y	1st Trial @ 220 K	$X = Y/K$	2nd Trial @ 225°F K	$X = Y/K$	MOL wt	Weight factor	SG	Vol factor	Liquid prop
C_2	0.008	–	NEG		NEG					
C_3	0.054	84.3	0.001	93.9	0.001	44	0.044	0.508	0.009	
iC_4	0.021	38.6	0.001	39.8	0.001	58	0.058	0.563	0.010	Mol wt = 130.7
nC_4	0.084	29.52	0.003	30.1	0.003	58	0.174	0.584	0.030	
C_5	0.143	12.53	0.011	12.65	0.011	72	0.792	0.629	0.126	SG = 0.766
C_6	0.155	4.70	0.033	4.94	0.031	85	2.635	0.675	0.390	
C_7	0.175	2.17	0.081	2.19	0.080	100	8.00	0.721	1.110	°API = 53
Mid-BP 260	0.124	1.00	0.124	1.16	0.107	114	12.198	0.743	1.642	
Mid-BP 300	0.124	0.506	0.245	0.518	0.239	126	30.114	0.765	3.936	K = 12
Mid-BP 340	0.075	0.229	0.328	0.253	0.293	136	39.848	0.776	5.135	
Mid-BP 382	0.037	0.108	0.343	0.126	0.294	152	44.688	0.788	5.671	
Totals	1.000		1.170		1.06	130.7	138.551	0.767	18.059	

$$K = \frac{\text{Vapor press @ selected temperature}}{\text{Total systems pressure}}$$

2nd trial = 0.108 × 1.170 ('K' for mid-BP 362 component)

New K = 0.126 then VP = 8.3 psia × 0.126 = 1.05 psia ≡ 225°F

2nd trial is close enough to $\sum x_i = 1.00$

Notes:

a. In estimating for the 2nd trial and final temperature the 'K' of the highest X component is multiplied by the total value of X function. Then vapor pressure curves are used to give the component temperature corresponding to this new vapor pressure.

b. The molar composition of the final 'X' is the composition of the liquid in equilibrium with the product vapor.
The tower top conditions are 229°F @ 15 psig. (This is total pressure and includes the steam effect)

Diesel oil

Diesel oil (sometimes called automotive gas oil) is used as fuel for heavy internal combustion engines such as heavy lorries and rail locomotives. Its main component is the light gas oil cut from the atmospheric crude distillation unit. This product has

Table 19.D.1. A shortened diesel fuel specification

				Method of Test
Specific gravity @ 60°F			0.820–0.860	ASTM D 1298
Color NPA			3 Max	ASTM D 155
Pour point	Summer	°F	15	ASTM D 97
	Winter	°F	7	
Cloud point	Summer	°F	N/A	ASTM D 97
	Winter	°F	N/A	
Sulfur % wt			0.1	ASTM D 129
Diesel index			54 Min	IP 21
Cetane number			47 Min	ASTM D 613
Distillation:				
Recovered at 230°C		% Vol	10 Min	ASTM D 158
" 240°C		% Vol	50 Max	
" 347°C		% Vol	50 Min	
" 370°C		% Vol	95 Max	
FBP		°C	385 Max	
Flash point (PM)		°F	14 0 Min	ASTM D 93

a boiling range of around 480°–610°F. This cut is hydrotreated to remove sulfur (and the process does reduce pour and cloud points to some extent), and blended with kerosene and some heavier middle distillate stock to meet the diesel oil specification. A summarized specification is given below as Table 19.D.1.

Because diesel oil is a motive fuel and therefore emissions are in direct contact with the general public, extensive emission controls are in place. These include the composition of the fuel itself and the engine design in which it is used. Full details of these controls are described and discussed in Chapters 2 and 16 of this book.

Distillate hydro cracking

Hydrocracking is a versatile catalytic refining process that upgrades petroleum feedstocks by adding hydrogen, removing impurities, and cracking to a desired boiling range. Hydrocracking requires the conversion of a variety of types of molecules and is characterized by the fact that the products are of significantly lower molecular weight than the feed. Hydrocracking feeds can range from heavy vacuum gas oils and coker gas oils to atmospheric gas oils. Products usually range from heavy diesel to light naphtha.

While the first commercial installation of a unit employing the type of technology in use today was started up in Chevron's Richmond, California, refinery in 1960, hydrocracking is one of the oldest hydrocarbon conversion processes. Hydrocracking technology for coal conversion was developed in Germany as early as 1915 designed to secure a supply of liquid fuels derived from domestic deposits of coal. The first plant for hydrogenation of brown coal was put on stream in Leuna, Germany, in 1927, applying what may be considered the first commercial hydrocracking process.

In the mid-1950s the automobile industry started the manufacture of high-performance cars with high-compression ratio engines that required high-octane gasoline. Thus catalytic cracking expanded rapidly and generated, in addition to gasoline, large quantities of refractory cycle stock that was a material that was difficult to convert to gasoline and lighter products. This need to convert refractory stock to quality gasoline was filled by hydrocracking. Furthermore, the switch of railroads from steam to diesel engines after World War II and the introduction of commercial jet aircraft in the late 1950s increased the demand for diesel fuel and jet fuel. The flexibility of the newly developed hydrocracking processes made possible the production of such fuels from heavier feedstocks. The early hydrocrackers used amorphous silica alumina catalysts. The rapid growth of hydrocracking in the 1960s was accompanied by the development of new, zeolite based hydrocracking catalysts. They showed a significant improvement in certain performance characteristics as compared with amorphous catalysts: higher activity, better ammonia tolerance, and higher gasoline selectivity.

There are several process flow schemes for the Hydrocracker process. These are discussed in Chapter 7. The following recycle flow scheme is one of the more common flow configuration (Figure 19.D.3).

Hydrocracking reactions proceed through a bifunctional mechanism. A bifunctional mechanism is one that requires two distinct types of catalytic sites to catalyze separate steps in the reaction sequence. These two functions are the acid function, which

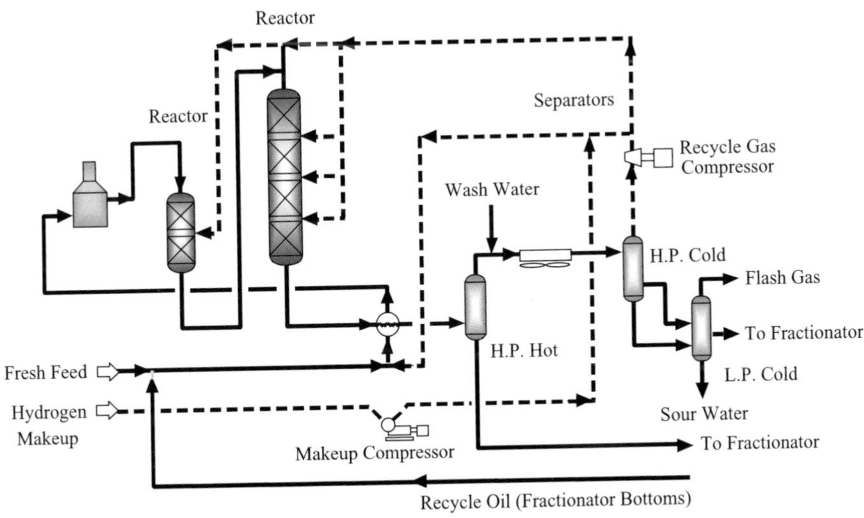

Figure 19.D.3. A distillate hydrocracker reactor section.

provide for the cracking and isomerization and the metal function, which provide for the olefin formation and hydrogenation. The cracking reaction requires heat while the hydrogenation reaction generates heat. Overall, there is heat release in hydrocracking, and just like in treating, it is a function of the hydrogen consumption (the higher the consumption, the more important the exotherm). Generally, the hydrogen consumption in hydrocracking (including the pre-treating section) is 1,200–2,400 SCFB/wt% change (200–420 Nm^3/m^3/wt% change) resulting in a typical heat release of 50–100 Btu/Scf H_2 (2.1–4.2 kcal/m^3 H_2) which translates into a temperature increase of about 0.065°F/Scf H_2 consumed (0.006°C/Nm^3/m^3 H_2). This includes the heat release generated in the treating section. In general, the hydrocracking reaction starts with the generation of an olefin or cycleolefin on a metal site on the catalyst. Next, an acid site adds a proton to the olefin or cyclo-olefin to produce a carbenium ion. The carbenium ion cracks to a smaller carbenium ion and a smaller olefin. These products are the primary hydrocracking products. These primary products can react further to produce still smaller secondary hydrocracking products. Some of these reactions are given in Figure 19.D.4.

Full details and discussion on Flow schemes and hydrocracking reactions are given in Chapter 7.

(A) Formation of olefins

$$R\text{-}CH_2\text{-}CH_2\text{-}\underset{\displaystyle CH_3}{\underset{|}{CH}}\text{-}CH_3 \xrightarrow{\text{Metal}} R\text{-}CH=CH\text{-}\overset{\displaystyle CH_3}{\overset{|}{CH}}\text{-}CH_3$$

(B) Formation of tertiary carbenium ions

$$R\text{-}CH=CH\text{-}\underset{\displaystyle CH_3}{\underset{|}{CH}}\text{-}CH_3 \xrightarrow{\text{Acid}} R\text{-}CH_2\text{-}CH_2\text{-}\overset{\displaystyle CH_3}{\overset{|}{\underset{\oplus}{C}}}\text{-}CH_3$$

(C) Isomerization and cracking

$$R\text{-}CH_2\text{-}CH_2\text{-}\underset{\displaystyle CH_3}{\underset{|}{\overset{\oplus}{C}}}\text{-}CH_3 \xrightarrow{\text{Acid}} R\text{-}\overset{\oplus}{CH_2} + CH_2=\underset{\displaystyle CH_3}{\underset{|}{C}}\text{-}CH_3$$

(D) Olefin hydrogenation

$$CH_2=\overset{\displaystyle CH_3}{\overset{|}{C}}\text{-}CH_3 \xrightarrow[H_2]{\text{Metal}} CH_3\text{-}\overset{\displaystyle CH_3}{\overset{|}{CH}}\text{-}CH_3$$

Figure 19.D.4. Hydrocracking reactions of paraffins.

Distillation

Part 1: Laboratory distillation tests

- *The ASTM distillation curve.* While the TBP curve is not produced on a routine basis the ASTM distillation curves are. Rarely however is an ASTM curve conducted on the whole crude. This type of distillation curve is used however on a routine basis for plant and product quality control. This test is carried out on crude oil fractions using a simple apparatus designed to boil the test liquid and to condense the vapors as they are produced. Vapor temperatures are noted as the distillation proceeds and are plotted against the distillate recovered. Because only one equilibrium stage is used and no reflux is returned, the separation of components is poor and mixtures are distilled. Thus the initial boiling point for ASTM is higher than the corresponding TBP point and the final boiling point of the ASTM is lower than that for the TBP curve. There is a correlation between the ASTM and the TBP curve, and this is dealt with later in Chapter 1 of this Handbook.
- *The true boiling point curve (TBP).* This is a plot of the boiling points of almost pure components, contained in the crude oil or fractions of the crude oil. In earlier times this curve was produced in the laboratory using complex batch distillation apparatus of a hundred or more equilibrium stages and a very high reflux ratio. Nowadays this curve is produced by mass spectrometry techniques much quicker and more accurate than by batch distillation.
- *The equilibrium vaporization curve (EFV).* The EFV curve of an oil is determined in a laboratory using an apparatus that confines liquid and vapor together until the required degree of vaporization is achieved. The percentage vaporized is plotted against temperature for several runs to produce the EFV curve. Separation is poorer for this type of distillation than for an ASTM or TBP. Therefore the initial boiling point will be higher for the EFV than for the ASTM. The final boiling point of the EFV will be lower than that for the ASTM. This test is rarely done but the EFV curve is calculated either from a TBP curve or an ASTM curve. These methods are given in Chapters 1 and 3.

Part 2: The distillation processes

Next to heat exchange the unit operation of distillation is the most utilized operation in oil refining. The separation of products by distillation fall into three major categories. These are:

- Total vapor condensation such as the atmospheric crude distillation
- Vacuum distillation processes. Such as the vacuum distillation of atmospheric residue
- Light ends distillation

- *Total vapor condensation processes.* The best example of this type of distillation units in modern refining is the atmospheric crude distillation unit. In these type of units the feed is heated and vaporized to a temperature above the total product cut point. This mixed vapor/liquid is produced by an external heater (and heat exchanger system) and the mixed stream flashed in the lower section of a tower. The vapor rises in the tower and condensed by cooled reflux stream at various stages up the tower according to the various distillate boiling points. Full details of this type of distillation is given in Chapter 3 of this book. Similar distillation systems are also used in the primary separation of Fluid Catalytic Cracking Unit effluent, and the reactor effluent from hydrocrackers.
- *Vacuum distillation processes.* These processes are designed to operate as total vapor condensation similar to the atmospheric units. However the feed to these units is usually heavy residual oils, which if heated to vaporize the product distillate required at atmospheric pressure (or near atmospheric pressure) would cause the feed to crack and coke. The system is therefore set at low vacuum pressure so that the vaporizing temperature is well below the feed's cracking properties. The distillates are produced in the same way as that for the atmospheric units. That is by selective cooled reflux streams. Full details of a crude oil vacuum process are also given in Chapter 3 of this book. Vacuum distillation is also used for the distillation of light residues from hydrocrackers to produce light vacuum distillates and heavy residuum for further thermal cracking.
- *Light ends distillation.* The most common of this type of distillation is the crude unit light ends process. The feed to this unit is the overhead distillate from the atmospheric crude unit. This feed may also include overhead distillate from other processes (such as the catalytic reformer). The product from the light end s process is the production of light and heavy naphtha, butane LPG, propane LPG, and refinery gas. These processes are discussed fully in Chapter 4 of this book.

Drums

Drums may be horizontal or vertical vessels. Generally drums do not contain complex internals such as fractionating trays or packing as in the case of towers. They are used however for removing material from a bulk material stream and often use simple baffle plates or wire mesh to maximize efficiency in achieving this. Drums are used in a process principally for:

- Removing liquid droplets from a gas stream (knockout pot), or separating vapor and liquid streams
- Separating a light from a heavy liquid stream (separators)
- Surge drums to provide suitable liquid hold up time within a process
- To reduce pulsation in the case of reciprocating compressors

Drums are also used as small intermediate storage vessels in a process.

Vapor disengaging drums

One of the most common example of the use of a drum for the disengaging of vapor from a liquid stream is the steam drum of a boiler or a waste heat steam generator. Here the water is circulated through a heater where it is risen to its boiling point temperature and then routed to a disengaging drum. Steam is flashed off in this drum to be separated from the liquid by its superficial velocity across the area above the water level in the drum. The steam is then routed to a super-heater and thus to the steam main. The performance of the steam super-heater depends on receiving fairly "dry" saturated steam. That is steam containing little or no water droplets. The separation mechanism of the steam drum is therefore critical.

The design of a vapor disengaging drum depends on the velocity of the vapor and the area of disengagement. This is expressed by the equation:

$$V_e = 0.157\sqrt{\frac{\rho_l - \rho_v}{\rho_v}}$$

where

V_c = Critical velocity of vapor in ft/sec.
ρ_l = Density of liquid phase in lbs/cuft.
ρ_v = Density of vapor phase in lbs/cuft.

The area used for calculating the linear velocity of the vapor is:

- The vertical cross sectional area above the high liquid level in a horizontal drum
- The horizontal area of the drum in the case of vertical drums

The allowable vapor velocity may exceed the critical, and normally design velocities will vary between 80% and 170% of critical. The use of crinkled wire mesh screens (CWMS) screens are an effective entrainment separators and are often used in separator drums for that purpose.

Liquid separation drums

The design of a drum to perform this duty is based on one of the following laws of settling:

Stoke's law

$$V = 8.3 \times 10^5 \times \frac{d^2\,\Delta S}{\mu}$$

When the Re number is < 2.0.

Intermediate law

$$V = 1.04 \times 10^4 \times \frac{d^{1.14} \Delta S^{0.71}}{S_c^{0.29} \times \mu}$$

When the Re number is 2–500.

Newton's law

$$V = 2.05 \times 10^3 \times \left[\frac{d \cdot \Delta S}{S_c}\right]^{1/2}$$

When the Re number is > 500,

where

$$\text{Re number} = \frac{10.7 \times d_v S_c}{\mu}$$

V = Settling rate in ins per minute.
d = Droplet diameter in ins.
S = Droplet specific gravity.
S_c = Continuous phase specific gravity.
ΔS = Specific gravity differential between the two phases.
μ = Viscosity of the continuous phase in cps.

The following may be used as a guide to estimating droplet size:

Lighter Phase.	*Heavy Phase.*	*Minimum Droplet Size.*
0.850 SG and lighter.	water	0.008 ins.
Heavier than 0.850.	water	0.005 ins.

The holdup time required for settling is the vertical distance in the drum allocated to settling divided by the settling rate.

Settling baffles, are often used to reduce the holdup time and the height of the liquid level.

Surge drums

This type of drum, the calculation of holdup time and surge control is used in the control the movement of liquids from one process to another. It's purpose is to smooth out the flow to meet the process criteria. For example, when a liquid feed leaves the first process item (such as an overhead reflux drum) under level control a surge drum may be used to collect the feed and deliver it under flow control to the next process. This is particularly desirable if the feed enters a fired heater as the first item of the second process.

Pulsation drums or pots

This type of drum is used almost entirely in the operation of a reciprocating compressors. It is used to smooth out the compressed gas flow leaving the compressor cylinders.

E

Economic evaluations

Economic evaluation is used for most aspects of the refinery planning, and its operation. The methods used in this evaluation may differ from company to company, but the end product must reflect the profitability of the present and often the future profitability of a proposed venture or operation. This measure is reflected in terms of the Return on Investment of the item. This is described and discussed in Chapter 8.

The following definitions and items are used in the economic evaluation exercises:

Ad volorum tax. This is the fixed tax levied in most countries payable to local municipal authorities, provincial, or state authorities to cover property tax, municipal service costs etc.

Calculation of cash flow method. There are several methods of assessing profitability based on discounted cash flow (DCF), but the most reliable yardstick is a return on investment method using the Present Worth (or Net Present Value) concept. This concept equates the present value of a future cash flow as a product of the present interest value factor and the future cash flow. *Based on this concept, the Return on Investment is that Interest value or Discount Factor which forces the Cumulative Present Worth value to Zero over the economic life of the project.*

Net investment costs. The Net Investment for the project includes the capital cost of the plant, which is subject to depreciation, and the Associated Costs, which may not be subject to depreciation.

The capital cost of the plant is the contractor's selling price for the engineering, equipment, materials, and the construction of the facilities. In a process study using a DCF return on investment calculation, the capital cost should be an estimate with an accuracy at least that based on an equipment factored type.

The associated costs include the following elements:

- Any licensor's paid up royalties
- Cost of land
- First inventory of chemicals and catalyst
- Cost of any additional utilities or offsite facilities incurred by the project
- Change in feed and product inventory
- Working capital

Capitalized construction period loan interest

Construction period. This is the period before year 0 during which the plant is constructed and commissioned. Assume this period is 3 years, this is designated as end of year –2, –1, 0. During this period, the construction company will receive incremental payments of the total capital cost of the plant with final payment at the end of year 0. The construction cost may be paid from the company's equity alone or from equity and an agreed loan or entirely from a loan. In the case of a loan to satisfy this debt, the payment of the loan interest commences in this period. The interest payment over this period, however, is usually capitalized and paid over the economic life of the project.

Depreciation. Part of the cost to a project or venture which is considered as a deductible from the gross profit for tax purposes is the depreciation of the plant value. This is calculated over the PLANT LIFE as the plant capital cost divided by the plant life. The term Plant Life is the predicted life of the facility before it has to be dismantled and sold for scrap. Usually this is set at 20 years and indeed all specifications relating to engineering and design of the facilities will carry this requirement. So all material and design criteria, such as corrosion allowances, associated with the plant will meet this plant life parameter.

Discounted cash flow, definition of. The development of a DCF return on investment is a combined effort between the Technical Disciplines and the company's Finance Specialist. The engineers provide the technical input to the work such as operating costs, type of plants, construction schedules, and cost, yield and refinery fence product prices, and the like. While the financial specialist provides the financial data based on statutory and company policies, such as the form of depreciation, tax exemptions, tax credits (if any), items forming part of the company's financial strategy, etc. The calculation itself is in two parts, which are:

- Calculation of Cash Flow
- Present Worth Calculation. (or any other method for calculating the Return on Investment the Company may use)

Economic life. This is the number of years over which the project is expected to yield the projected profit and pay for its installation. These are the number of years starting at year 0 which indicates the end of construction and the commissioning of the facilities. The last year (usually year 10) is the year in which all loans and other project costs are repaid and the "Terminal Investment" released.

Plant Operating Cost. This includes the cost of utilities used in the process, such as power, steam, and fuel. It also includes the cost of plant personnel in salaries, burdens, and indirects and the cost of chemicals and catalyst used.

- *Maintenance.* There are two kinds of maintenance costs included in this item These are the preventive maintenance carried out on a routine basis, and those costs associated with incidental breakdowns and repair.
- *Loan repayment.* The loan principal is paid back in equal increments over the economic life of the plant. This item includes the payback increments and the associated interest on a declining basis.

Net cash flow. This item is calculated for each year of the project's economic life. It commences with year 0 with the net investment shown as a negative net cash flow item. Then for each successive year until the end of the last year of the economic life, the net cash flow is calculated as the sum of Profit After Tax *PLUS* Depreciation.

The depreciation is added here because it is not really a cost to the project. It is a "Book" cost only and is used specifically for tax calculation.

The cash flow item for the last year of the economic life must now include the "Terminal Investment Item". This item is the sum of the net scrap value of the plant (scrap value less cost of dismantling), the estimated value of the land and the Working Capital initially used as part of the "Associated Costs".

Thus, the final cash flow item will be the sum of Profit After Tax, plus Depreciation, plus Terminal Investment.

With the Net Cash Flow in place, the second part of the calculation which is the determination of the Return on Investment for the project can be carried out.

Pay out time. The pay out time for a project or any venture is given by the expression:

$$\frac{\text{Net Investment Cost}}{\text{Net Income after taxes} + \text{depreciation}} = \text{Years}$$

Simple return on investment. This calculation is used for screening purposes of several options that are apparent in any venture. It does not consider construction time or the discounting of the cash flow over the economic life of a plant. Its definition is given by the equation:

$$\frac{\text{Net Income after Taxes}}{\text{Net Investment Cost}} \times 100 = \%\ \text{ROI}$$

Where

Net Investment is Capital Cost plus Associated Cost.
Net Income is Gross income less operating cost, depreciation, and tax.

Revenue. This is income received for the sale of the product(s). This is calculated from projected process yields of products multiplied by the market price of the products. A market survey should already have been completed to ensure that the additional products generated by the project are in demand and the price is in an acceptable range. Later, a sensitivity analysis of the DCF return on investment may be conducted changing the revenue recovered by price escalation or other means.

Taxable income. Taxable income is revenue less operating cost, depreciation, and Ad Valorem Tax. This of course is simply put as in most countries, states, or provinces there will probably be certain local tax relief principals and tax credits that will affect the final taxable income figure. The company's financial specialist will be in the position to apply these where necessary.

Tax. This is quite simply the tax rate applied to the "Taxable Income" figure. This will vary from location to location but will be taken as one rate over the economic life of the project for the purpose of a process study, unless there are legislative changes already in place.

Working capital. In all capital projects there is initially a financial loss when the company has to purchase land, equipment, pay contractors to erect equipment, and the like. To do this not unlike most private individuals, companies may need to borrow money, on which of coarse they have to pay interest. In addition to the cost of construction, the company must keep in hand some capital by which it can buy feedstock, chemicals, and pay the salary of its staff during the commissioning and the initial operation of the plant. During this initial operation, the working capital must also be considered as debt. At the end of this period of the project initiation, the operation must recover the working capital and make its prescribed profit.

Edmister correlations

A series of correlations which relates an ASTM distillation test to the TBP of a petroleum cut is presented in W. Edmister's publication titled "*Applied*

Thermodynamics". The correlation relating to the ASTM and TBP curves from this book is used in this Handbook and is given in Chapter 1. Similar correlations have also been prepared by Edmister for the relation between TBP and the equilibrium flash vapor curve (EFV).

Effluent water

This item is described and discussed in Chapter 14. Pollutants in waterways from refineries are listed in the following table.

Process	Waste water	Air
Atmospheric and vacuum distillation	Sour water (NH_3 & H_2S) Desalter water Spent caustic Process area waste water (pump glands, area drains etc)	Furnace Flue gases—SO_2
Thermal cracking	Sour water	Furnace
Delayed coking	Decking water (Oil) Process area waste water	Flue gases
Fluid cat cracking	Sour water (NH_3, H_2S, Phenols)	Furnace
Unsat gas plant	Spent caustic Process area waste water.	Flue gases SO_2, CO particulates.
Hydrocracker hydrogen plant	Sour water (inc Phenols) Process area waste water	Furnace Flue gases SO_2
Sat gas plant alkylation	Spent caustic Process area waste water	Nil
Naphtha hydrotreater Cat reformer	Sour water Process area waste water	Furnace Flue gases
Sulfur plant	Nil	Incinerator Flue gas—SO_2 Hydrocarbons—Flare
Tankage area	Tank dike area drains Non contaminated rain runoff	Tank vents Hydrocarbons.

The most undesirable pollutants in aqueous waste streams are:

- Those that deplete the dissolved oxygen content of the waterways into which they discharge.
- Those contaminants that are toxic to all forms of life, such as arsenic, cyanide, mercury, and the like.
- Those contaminants that impart undesirable tastes and odors to streams and other waterways into which they discharge.

Oxygen depletion occurs by the introduction of one or more of the following oxidizable contaminants entering the waterway:

- Natural pollution by surface run-off rainwater, or melting snow, in the form of soluble salts leached from the earth.
- Natural pollution caused by decay of organic plants from swamps or other sources.
- Human and animal life excretion.
- Chemical pollution from reducing agents in industrial plant wastes. Such as sulfides, nitrites, ferrous salts, etc.
- Biochemical pollution from such industrial wastes as phenols, hydrocarbons, carbohydrates, and the like.

The degree of oxygen depletion from the pollution sources described above may be catalogued by the following terms:

BOD—Biological Oxygen Demand
COD—Chemical Oxygen Demand
IOD—Immediate Oxygen Demand

BOD. Since all natural waterways will contain bacteria and nutrients; almost any waste compound introduced into the waterways will initiate biochemical reactions. These reactions will consume some of the dissolved oxygen in the water.

The depletion of oxygen due to biological pollution is not very rapid. It follows the laws of first order reaction. Because of this, the effect of BOD is measured in the laboratory on a five day basis, and has been universally adopted as the measure of pollution effect. The "Ultimate" BOD is a measure of the total oxygen consumed when the biological reaction proceeds to almost completion. The "5 day" BOD is believed to be approximately the Ultimate.

In summary, BOD measures organic wastes which are biologically oxidizable.

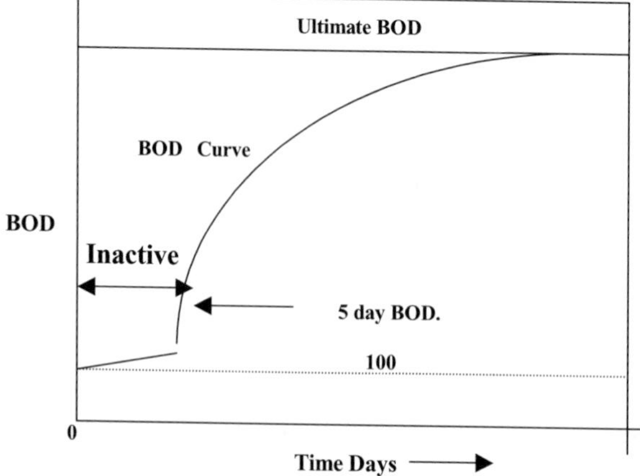

COD. The COD is a measure of the oxygen depletion due to organic and inorganic wastes which are chemically oxidizable. There are several laboratory methods accepted to measure the oxygen depletion effect of these pollution. The two most widely accepted are the "4 hour Permanganate" method or the "2 hour Dichromate" method. Although there is no generalized correlation between BOD and COD, usually the COD will be larger than the BOD. The following table illustrates how different wastes exhibit a different relationship between COD and BOD.

Source	Pollutants	BOD "5 day"	COD "2 hr dichromate"
Brewery	Carbohydrates Proteins	550	–
Coal gas	Phenols, Cyanides Thiocyanates Thio sulphates	6,500	16,400
Laundry	Carbohydrates Soaps	1,600	2,700
Pulp mill	Carbohydrates Lignins, Sulfates	25,000	76,000
Domestic sewage	Solids, Oil, and Grease, Proteins Carbohydrates	350	300
Petroleum refinery (Sour water)	Phenols Hydrocarbons Sulfides	850	1,500
Petroleum refinery	Phenols, Sulfides Hydrocarbons, Mercaptans, Chlorides	125	2,600

IOD. Oxygen consumption by reducing chemicals such as sulfides and nitrates is typified as follows:

$$S^{--} + 2O_2 \rightarrow SO_4^{--}$$

$$NO_2^- + \tfrac{1}{2}O_2 \rightarrow NO_3^-$$

These types of inorganic oxidation reactions are very rapid and create what is measured in the laboratory as immediate oxygen demand. If waste contaminants contain these inorganic oxidizers, the "5 day" BOD test will include the consumption of the oxygen due to IOD also. A separate test to determine IOD must be made and this result subtracted from the "5 day" BOD to arrive at the true BOD result.

Toxic pollutants common to oil refining. Toxic pollutants that are most commonly contained in untreated refinery aqueous wastes are:

Oil. Heavy oil and other hydrocarbons are the most problematic pollution to be found in refinery water effluent. All refineries exercise the most stringent methods to control and remove these undesirable pollutants. Indeed in many cases the treated effluent streams leaving the refinery may well be purer than incoming portable water used in the processes.

Phenols. This chemical often formed in refinery processes such as catalytic and thermal crackers, are highly toxic to aquatic life in concentrations of 1–10 ppm. Apart from its toxicity phenols also unpleasant taste and odor to drinking water in the range of 50–100 ppb. In concentrations of 200 ppm and more, these chemicals can also deactivate water treatment plants such as trickle filters and activated sludge units.

Caustic soda and derivatives. Solutions containing sodium hydroxide are used in a number of refinery processes. Inevitably some of this chemical enters the refinery's waste water system. This contaminant is toxic to humans and marine life in even low concentrations. The spent caustic (compounds leaving the process) such as sodium sulfide is even more injurious.

Aqueous solutions of ammonium salts. The most common of these are ammonium sulfide NH_4SH, and ammonium chloride. Both these salts are present in effluent water from the crude distillation unit overhead accumulator, however the sulfide salt is present in all aqueous effluents from the cracking processes, and the hydro treaters. Other ammonium salts are also present in hydrocracking and deep oil hydrotreating.

Acids in aqueous effluents. The most common of these are from the alkylation processes which use either hydrofluoric acid or sulfuric acid. In some isomerization processes, hydrochloric acid is used to promote the aluminum chloride catalyst. In some older processes sulfur dioxide is used to remove aromatics. This effluent usually leaves these plants as dilute sulfuric acid.

Ketones, furfural, and urea. These compounds are used in the refining of lube oils. MEK and Urea are used in the dewaxing processes while furfural is used in the extraction processes for finished lube oil stock. All of these compounds are toxic.

Treating refinery aqueous wastes

The treatment of aqueous wastes from oil refineries fall into three categories. These are:

- In-plant treatment. These are onsite processes usually sour water strippers, spent caustic oxidizers, and spent caustic neutralizers.
- The API separator, or similar oil/water separating device.
- Secondary treatment, which includes chemical coagulation, activated sludge processes, trickle filters, air flotation, and aerators.

Most energy refineries contain only the first two of the above categories. These processes are described in detail in this Handbook.

Effluent water treating

Sour water is usually treated in petroleum refineries by one or more of the following processes:

- Steam Stripping
- Caustic/Acid Neutralization
- Caustic Oxidization
- Oil Removal by Settling

All three of these processes with brief description of the more uncommon processes such as coagulation are also described fully in Chapter 14 of this Handbook. A brief summary of these are as follows:

Sour water strippers. This is one of the most common treating processes. Its purpose is to remove the pollutant gases included in process plant effluents. The more common pollutants in this case are ammonia, ammonium salts, and hydrogen sulfide. There are two types of strippers in this service; they are shown as Stripper A and B, respectively, below:

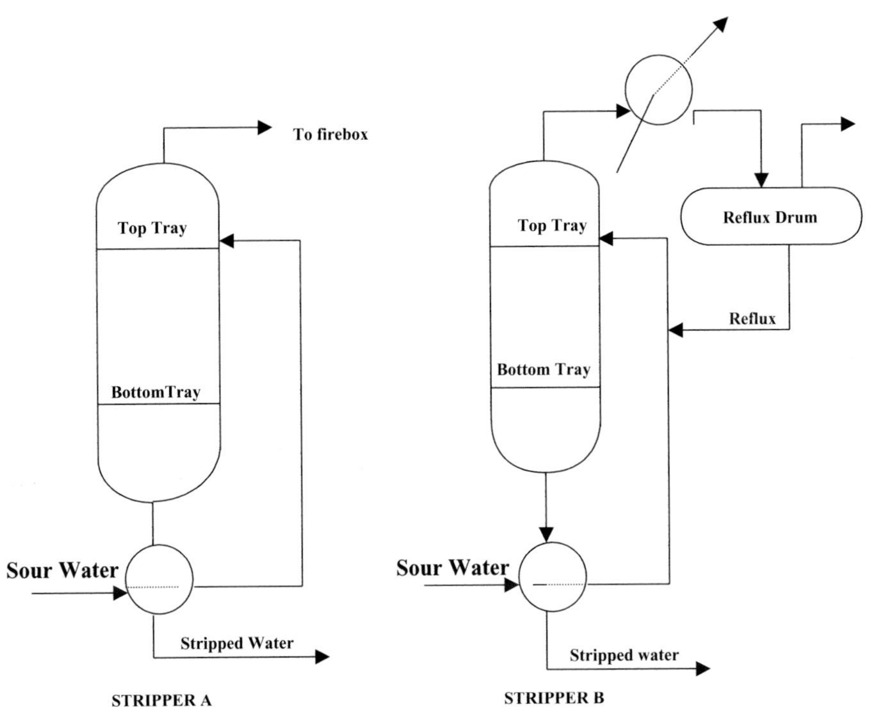

Both these type of sour water stripping lend themselves to tray by tray mass and heat transfer. The sour water feed is introduced on the top tray of the tower while steam usually at a rates of 0.5–1.5 lbs/gallon of feed is introduced below the bottom tray. In the case of a tower with reflux the reflux enters the tower with the fresh feed. The design of both towers utilizes the partial pressure relationship of NH_3 and H_2S in aqueous solution. These relationships are given by a series of graphs to be found in Appendix 1 of Chapter 14.

The sour water stripper is almost always located in the process area of the refinery. The feed to the stripper is the effluent water from the crude unit overhead condenser, the water phase from the desalter, the condensed water from the vacuum unit's hot well, and all the water condensate from the hydro-treater product steam strippers. The sour water stripper may be a single tower with no reflux as in Type A or a single trayed tower with an overhead reflux stream (Type B). The amount of steam used in both cases will be between 0.5 lbs steam per lb of feed to 1.5 lbs of steam.

Spent caustic disposal. The other major effluent from oil refining is the spent caustic streams from hydrogen sulfide removal and also the removal of phenols. Refiners usually have the following options in the disposal of these streams. In order of preference these are listed as follows:

Phenolic spent caustic

- Disposal by sales
- Disposal by dumping at sea
- Neutralizing with acid
- Neutralizing with flue gas

Sulfidic spent caustic

- Disposal by sales
- Oxidation with air and steam
- Neutralization with acid and stripping
- Neutralization with flue gas and stripping

These processes are also described and discussed in full in Chapter 14 of this Handbook. A brief summary is as follows:

Neutralizing phenolic spent caustic. As listed above phenolic spent caustic can be neutralized using acid or flue gas. When neutralized the mixture separates into two liquid phases. The upper phase contains the acid oils while the lower phase is an aqueous solution of sodium sulfate or sodium carbonate. The neutralization using either acid or flue gas can be accomplished in a batch or continuous operation. The neutralization step is exothermic giving out around 125 Btu/lb of sprung acid. As the objective of the process is to produce a phenol free sprung water for disposal the system temperature should be kept as low as possible until the sprung water is removed. Routing the sprung water to the sour water stripper ensures the removal of any entrained H_2S in that stream.

Spent caustic oxidation. Spent caustic cannot be steam stripped to remove the sulfides contained in it due to the H_2S removal process in which the caustic was used. This is because sodium sulfide does not hydrolyze even when heated. Acids could be used, of course, to neutralize the spent caustic which would release gaseous H_2S. This would however be a costly procedure and causes a potential air pollution problem. The alkaline sulfide can be economically oxidized to form thiosulfates and sulfates. This is the process most commonly used in refineries where only sulfides are the pollutants in the spent caustic and the release of gaseous H_2S is a problem. Details of this process together with a process schematic drawing are given in Chapter 14.

Oil–water separation. Most aqueous effluent from a refinery will contain oil. This oil content has to be reduced to at least 10 ppm before it can be deposited into a river, lake, or ocean. The oil contamination sources are from process water run down, paved area drainage, storm catch-pots, tanker ballast pump out, and tank farm diked areas. All the water from these sources are treated in oil separation processes. The most important of these oil–water separation process is the API Separator.

The API oil–water separator. The design of an API separator is based on the difference in gravity of oil and water in accordance with the general laws of settling. In the design of an API separator a modified version of Stoke's Law is used. In this law the rate of settling is given by the equation:

$$Vr = 6.69 \times 10^4 \frac{d^2 \Delta S}{\eta}$$

where

Vr = Rising rate of the oil phase in ft/min
d = Droplet diameter in ins.
ΔS = Difference in specific gravity of the phases.
η = Viscosity of the continuous phase (water) in centipoises.

The above modified Stoke's Law for the API separator recognizes that the continuous phase is water and the lighter oil phase is the one that is separated to be disposed as the product skimmed from the surface.

An example of the application of the equation in the design of an API separator is given in Chapter 14 Appendix 14.3 of this Handbook.

The oil phase from the separator is removed using specially designed skim pipes and an oil sump. A simplified diagram of a typical API Separator is given below.

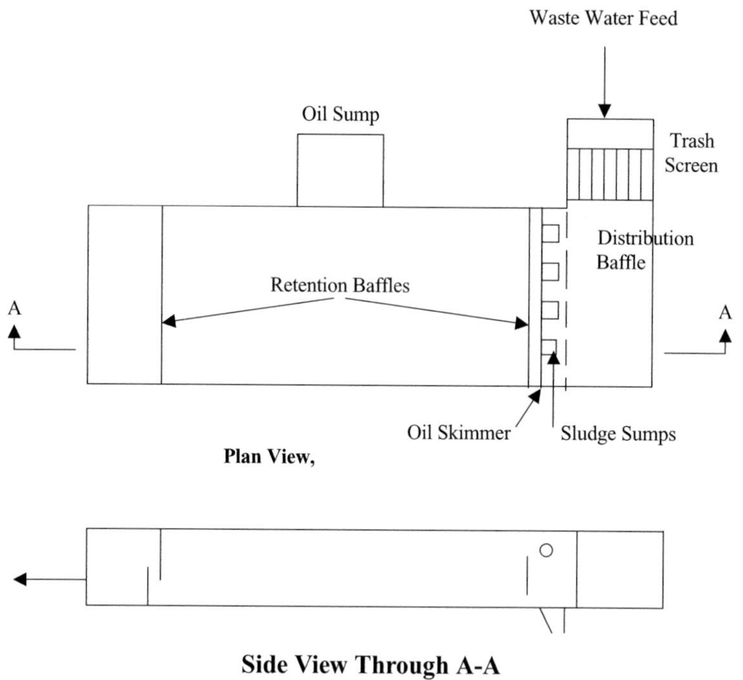

Side View Through A-A

A process description of the separator together with its ancillary equipment is given in Chapter 14.

Storm surge ponds. These ponds are installed to provide storage for maximum rainfall conditions. There are several forms of these surge ponds, some requiring pumps, some located upstream of the API separator, and some downstream of the separator. In most cases, the storm drain system is directed to the storm surge pond. Thus in a storm, the excess rainfall is held in this pond and fed to the API separator, over a period of time and at a rate that will not exceed the separator's capability to handle the water effectively. In this way, the refinery ensures that any oily water will not by-pass the separator under the worst condition. The size of the surge pond must be able to handle the maximum rainfall and the flow from all catch basins and open culverts that form part of the refinery drainage system.

Other refinery water effluent treatment processes. These are summarized below, but are described more fully in Chapter 14.

Oxidation ponds. Oxidation ponds are usually used as a secondary effluent clean up after the API separator. There are three types of these ponds which are:

- Aerobic—Where the oxidation of the water utilizes oxygen from the atmosphere plus oxygen produced by photosynthesis.
- Anaerobic—Where oxidation of the wastes does not utilize oxygen.
- Aerated—Where oxidation of the wastes utilizes oxygen introduced from the atmosphere by mechanical aeration.

The processes described above are those met with in oil refining most often. Indeed of the processes described above most refineries only use the API separator and the surge ponds to meet the oil/water separation required.

Air flotation. The purpose of the air flotation process is the clarification of waste water by the removal of suspended solids and oil. This is achieved by dissolving air in waste water under pressure, and then releasing it at atmospheric pressure. The released air forms bubbles which adhere to the solid matter and oil in the waste water. The bubbles cause the adhered matter to float in the froth on the surface of the water bulk. The dissolved air in the water also achieves a reduction in the BOD of the effluent stream. The figure below shows the principal of a typical air flotation process.

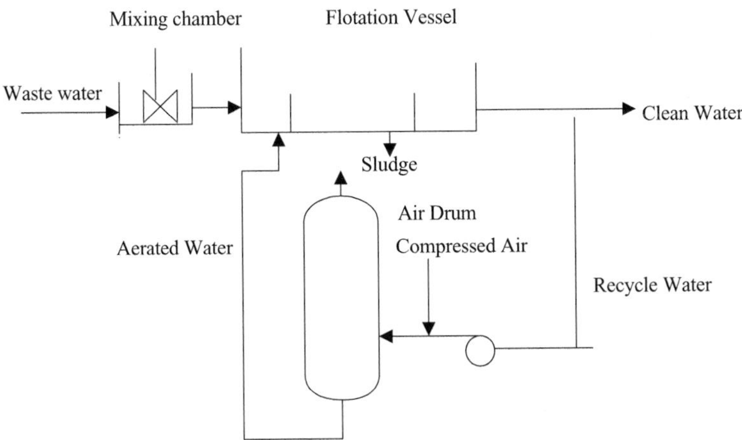

A typical Air Flotation Process

These processes are also described and discussed more fully in Chapter 14.

End points

Whereas a cut point is an ideal temperature on a TBP curve to define the yield of a fraction, the end points define the shape of the fraction when produced commercially. In actual process, the initial boiling point of a fraction will be much lower than its front end cut point. The final fraction boiling point will be higher than the corresponding cut point. This is demonstrated by the figure below:

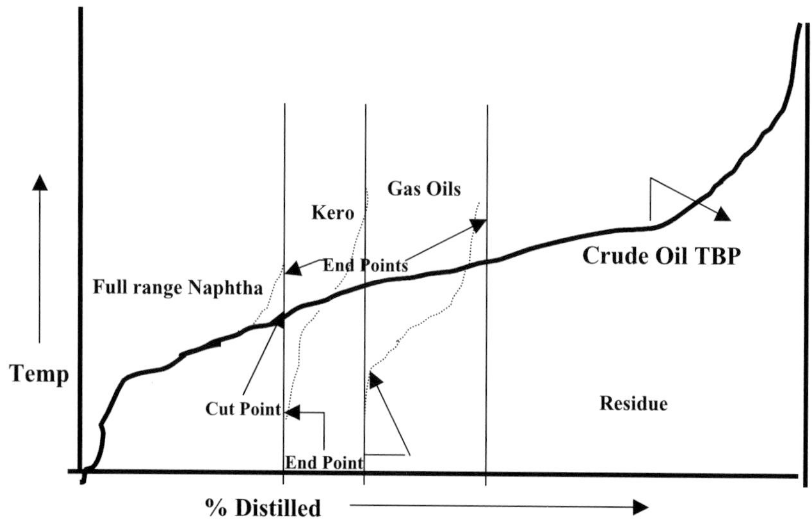

There is a correlation between the TBP cut point and the ASTM end point. This is described in Chapter 1 of this Handbook. A relationship also exists for the 90% ASTM point and the TBP 90% cut point. With these two ASTM data and using an ASTM probability graph (see Chapter 1) a full ASTM curve with its end points can be drawn. This converted to a TBP curve is used to define the cut's properties.

Engineering flow diagrams

Diagrams are used extensively by all disciplines of engineers to convey ideas and data.

Process engineers use and in some cases develop three types of flow diagrams to project their work and responsibilities.

These are:

- The Process Flow Diagram
- The Mechanical Flow Diagram (sometimes called the P&I Diagram)
- The Utilities Diagram

The process flow diagram is always originated by the process engineer and he retains sole responsibility for its future development and update. The Mechanical flow diagram may also be initiated by the process engineer or it may be developed by others from the process flow diagram. In many companies, however, the process engineer remains responsible for its technical content, development, and its completeness. The utilities diagram shows the routing, sizing, and specification of all the utility flow lines between units and within units of a process. This diagram is usually superimposed on the plot layout diagram. The piping engineers or those engineers who are responsible for initiating the plot plan drawing usually initiate the utilities drawing and administer it. The process engineer in this case is responsible for sizing the flow lines, establishing flow conditions of temperature and pressure, and for ensuring that all lines have been included.

The process flow diagram

This diagram is usually the first drawing that will be produced for an engineering or development project. In some cases, it may be preceded by a process block flow diagram but it is the process flow diagram that is the basis for:

- A process definition
- A budget cost estimate
- An equipment list
- A mechanical flow diagram
- Process equipment data sheets

The flow diagram supports the material and energy balances for the process and establishes the sequence and direction of the process flow. The diagram also shows the control philosophy that will be adopted for the process and the salient temperature/pressure conditions within the process. As a minimum therefore process flow diagrams should contain the following:

Vessels. The outlines of all major vessels, such as towers, drums, tanks are shown. Their equipment item number and their overall dimensions are indicated on the diagram. Where vessels contain special internals, such as trays, demisters, packing, etc these too should be simply indicated on the vessel drawing. For example, the number of trays in a tower may be indicted by showing the top, feed, and bottom trays only but including their respective tray numbers. The main temperature and pressure conditions are also shown on the vessel drawing. For example on fractionation towers the tower top and bottom operating temperatures will be shown but only the top pressure is normally shown.

Heat exchangers. All heat exchangers are shown as single shells on the process flow diagram. That is, no attempt is made to show the number of passes or the type (i.e., shell & tube, or double pipe, etc.) on this diagram. The process flows to and from these items are shown as flowing through the shell side or the tube side. The exchanger item number only is indicated on the flow diagram adjacent to the equipment drawing. Its equipment name is normally not shown. The heat duty of the exchanger is also shown on the flow diagram again adjacent to the equipment drawing and below the equipment number. The temperature conditions for the exchanger are shown on the process lines in and out of the equipment. No pressures are normally shown for this equipment.

Air coolers. Air coolers are shown simply as a narrow rectangular block in a process line with a fan symbol shown inside the rectangle as a dotted outline. The item number appears directly below the rectangle and below that is given the operating duty of the item in mm Btu/hr or kcal/hr, etc.

Heaters. Fired heaters are shown as box type with a stack outlet. The specific type of heater or the number of tube passes are not shown on the process flow diagram. Again, only the equipment item number and the heater duty is given in the flow diagram adjacent to the item drawing. The heating medium is shown as a single line entering the bottom of the equipment and marked only as 'fuel'. The line would contain a control valve with an instrument control line to the process coil outlet showing the firing control philosophy. The temperatures connected with the heater are shown on the process line in and out of the equipment. Normally no pressure levels are shown for this item.

Pumps. Pumps are shown in as simple a manner as possible. Most companies carry their own symbols for equipment. Many however show a centrifugal pump simply as a circle with the suction line proceeding to the center of the circle and the discharge line leaving tangentially from the top. Other symbols are used for positive displacement pump types. Only these two types are differentiated on the process flow diagram. The various types within these categories are not indicated. The pump item number and an indication as to whether the pump is spared is shown adjacent to the pump drawing on the flow diagram. If the pump is spared then the item number will be followed by "A + B". If it is not spared it will be followed by the letter "A" only. The capacity of the pump as gallons per minute (GPM) or cubic meters per hour will be shown under the item number. This is the normal or operating capacity.

Compressors. These equipment are shown as either centrifugal or reciprocating machines. The centrifugal type are shown similar to the centrifugal pump or as a trapezoid and the reciprocating type is shown as two small square boxes connected by a double line. The item number appears below the item and in some cases indicates the number of stages. The capacity in standard or normal units is given below the item number while the temperature/pressure conditions to and from the item are shown on the process lines to and from.

Process lines. All major process lines interconnecting equipment, recycle, or bypasses are shown on the process flow diagram. These lines will show direction of flow as black arrows on the lines. Control valves and major block valves are also shown on these lines. The temperature and pressure conditions of the flowing material are given on the lines at appropriate positions in the drawing.

Instruments. Only major control instrumentation are shown and the instrument symbols are kept as simple as possible. Details of instrument 'hook ups' are confined to showing the control valve being activated from either a level, flow, pressure, or temperature elements by a dotted line to the valve. The measurement instruments that affect control are shown as circles with the type of instrument (e.g., FC—for flow controller, TC—for temperature controller, etc.) printed in the circle.

The material balance. The material balance for the process represented by the process flow diagram is either shown in table form on the bottom of the flow sheet or on an attached but separate table. Preferably it should appear on the flow sheet itself. The table should contain at least the following:

- The stream number—This is the number given to the process stream and referenced on the respective process line in the diagram. This initiates the columns that will make-up the table.
- The stream identification—This next line should identify the stream, such as "Debutanizer bottoms" for each of the columns.
- The items following down the title column consists of, the flow rate for each stream as wt per unit time, the stream temperature, the specific gravity at a standard temperature for liquids, the mole wt for gas streams, volume flow at standard temperature (and pressure for gasses), stream pressure for gasses.

The mechanical flow diagram

The mechanical flow diagram (MFD) is developed from the process flow diagram. The detail provided by the MFD is sufficient for other engineering disciplines to:

- Initiate a plot layout
- Prepare a line list. (Piping design)
- Initiate piping arrangement drawings (Piping design)
- Prepare a preliminary piping material take off
- Prepare an instrument register (Instrument engineering)
- Initiate instrument "hook up drawings" (Instrument design)
- Prepare electrical "one line drawings" (Electrical engineering)
- Prepare preliminary Instrument and electrical material take off
- Initiate civil and structural design (Civil engineering)
- Develop the project execution plan (Planning engineers)
- Prepare a semi definitive cost estimate (Cost estimators)

To meet these objectives, the Mechanical Flow Diagram will contain more detail of the process and the equipment included in it than the Process Flow Diagram. This is described briefly in the following paragraphs:

Vessels. Vessels are shown in approximate relative size to one another whereever possible. Again, only the top and bottom trays need be shown, unless other trays are required to indicate the location of side draws, instruments, sample points or changes in type of tray layout. Trays are numbered sequentially either from top to bottom or bottom to top. Catalyst beds, packing, demisters, and the type of tray (such as single pass or double pass, etc.) are shown. The height of packing etc is shown adjacent to the vessel and the height of all vessels above grade is also shown on the vessel. The following detail is usually shown on the top of the flow sheet directly above the vessel:

- Vessel Item Number (this should also appear in or near the vessel drawing).
  - ➢ Title
  - ➢ Size (Inside diameter, and length tan to tan)
  - ➢ Design temperature and pressure
  - ➢ Insulation (i.e., Indication is the vessel insulated or not)
  - ➢ Trim Number (Line specification for miscellaneous vessel connections)

Heat exchangers. The actual arrangement and type of heat exchangers are shown on the MFD. Shell and tube exchangers are still shown as circles with tube side flows in dotted lines as in the process flow diagram. Here, however the actual number of shell passes is shown. Double pipe and reboilers are shown in their specific format, and again the actual number of shell passes. The following data for each exchanger is shown at the top of the flow diagram:

- The heat exchanger item number. (This is also shown adjacent to the exchanger drawing)
- The title
- The duty of the exchanger (In Btu/hr, or Kcal/hr, etc.)
- Insulation

Air coolers. As in the case of the heat exchangers, air coolers are shown in more detail on the MFD than on the process flow diagram. Usually on the MFD the air cooler is shown as a narrow rectangle. The fan in this case is drawn below the rectangle for a forced draft cooler or above the rectangle in the case of an induced air cooler. Only one process line is shown to and from the cooler but the number of passes is shown on these lines. Any temperature control by louvres or variable pitch fans is shown on the diagram together with the appropriate instrumentation. The following data is given at the top of the MFD above the cooler:

- The cooler item number (this is also shown adjacent to the drawing)
- The title
- Duty of the cooler

Fired heaters. The outline of the furnace is shown as being a cylindrical or a cabin type heater. Most companies carry their own symbols to portray this feature. The actual number of passes is shown on an inlet and outlet header together with the control system for the process flow. Only one pass is shown as entering and leaving the heater however and flowing through the item. The internal flow is shown as a dotted line in the fire box. All instrumentation for the heater is shown in detail—such as temperature points on the heater coil and in the fire box, oxygen analyzers in the chimney and the chimney damper control. Snuffing steam and de-coking manifolds are usually shown as separate details. The firing control system is shown in detail. Usually this follows a standard adopted by the company for all the heaters. All the instrumentation interlocks and safety features are detailed in the MFD for all heaters. The following data is provided for this item at the top of the flow diagram and above each heater:

- The item number
- Title
- Duty of the heater

Pumps. Pumps are shown in much greater detail on the MFD than on the process flow diagram. The pump drawing itself shows the type of pump and the type of driver. For example, a centrifugal pump is shown as a narrow vertical parallel lines curved at both ends and resting on a base plate. A motor driver is shown as a small horizontal rectangle curved at both ends also resting on the base plate and connected to the pump by a short line. Process suction lines enter the center of the pump case and the discharge line leaves from the top of the casing. All pumps are shown in the same detail, that is both normal operating pump and its spare are shown. Process (and utility) lines connecting the pumps in a set are shown together with the valve configuration. Isolating block valves, non-return valves, etc on the pumps are shown together with cooling systems where required. Instrumentation for automatic pump start and stop is also detailed together with electrical switches for manual start/stop facilities. Pumps and other rotary equipment are normally drawn on the bottom of the flow diagram. The data supplied on the MFD for pumps are given below the item on the bottom of the diagram and are as follows:

- Pump item number (followed by “A” for normally operating and “B” for spare pump).
  - ➢ Title
  - ➢ Flow rate (gpm or m^3/hr, etc)
  - ➢ Differential pressure (psi, kPa, etc)

- ➢ SG of pumped fluid at pumping temperature
- ➢ Miscellaneous auxiliary piping (cooling water, flushing oil, seal oil, etc.)

Piping. The MFD is among the most important documents that are developed in the course of a fully engineered project that can proceed to construction and finally to operation. Its importance is probably highest in the case of the piping engineering and design function of a project. For this discipline, the MFD is the basis for all their work. Any piping detail that is required to define the work must appear on this flow diagram if it is to be included at all in the constructed facility. The diagram then becomes a major communication tool for multi discipline interfaces.

In the layout of a MFD, process feed lines should originate at the left hand end of the drawing and the process product lines terminate at the right hand of the drawing. Where this is impractical, origin and terminus of the lines are located for clarity and convenience. The origin and terminus of each process line in any case are identified by a box which shows the descriptive title of the line, the drawing number, and section number of any reference drawing. Where possible process lines between equipment drawings are located either above the line of equipment or below the line of equipment. Every effort in the layout should be made to avoid breaking lines around equipment. Piping high point vents or low point drains are not shown on the MFD unless they have some significant process requirement. Any pertinent requirement for process reasons on any line must be noted. This includes such requirements as no pocketing, sloping, etc. These type of notes are clearly marked on the respective lines.

Utility lines originate and terminate adjacent to the equipment involved. Only the length of line necessary for valves, instrumentation, and line numbering is shown. The utility line origin and terminus are identified by a descriptive title only. (e.g., "LP Steam" and "LP Condensate"). Main utility headers are not shown—these will appear on the "Utility Flow Diagram".

All line sizes are shown on the lines they refer to. Where there is a change in a line size this is also indicated by a "swage" up or down. All valve sizes are indicated on the MFD even if they are line size. Flow direction on all lines (whether process or utility) are clearly shown by directional arrows on the lines. Steam or electrical heat tracing are also indicated on the lines that require it.

Instrumentation. All instrumentation is shown on the MFD in detail. This detail includes:

- Size of control valves
- Instrument hookup method
- Vessel surge levels and level range
- Type of instrument activation (i.e., pneumatic or electronic)

- Computer interface if it applies
- Instrument identification number
- Position of the control valve on air failure (i.e., failed open or closed)

Instrument symbols are usually to ISA standard with some minor modifications to meet the respective company's needs.

The utility flow diagram

Process input to the utility flow diagram development is minor compared with that for the PFD and MFD. It involves the sizing of the utility lines and valves only. The process engineer is responsible to ensure that all the necessary utility lines to satisfy the process have been included. The UFD itself is based on the approved plot layout of the plant and/or plants and is usually prepared by others. It shows the direction of flow and sequence of equipment geographically just as they appear on the plot plan. Although process input to the UFD as such is minor, there will be considerable detail necessary to complete the entire utility picture. For example, the UFD will show an instrument air header serving all the units in the process. This header will originate at the instrument air compressor set and dryers. This origin will only be indicated on the UFD with reference to a detail drawing for the instrument air compressor set. This detail drawing will be a MFD and the process engineer will have the same input and responsibility for this diagram as for any other MFD.

Drawings of sections of all these engineering drawings are given in Appendix A of this volume to illustrate this topic.

Equilibrium flash vaporization (EFV)

When a mixture of compounds vaporizes or condenses, there is an unique relationship between the composition of the mixture in the liquid phase and that in the corresponding phase at any condition of temperature and pressure. This relationship is termed the equilibrium flash vaporization for the mixture. It can be calculated using the composition of the feed mixture and the equilibrium constant of the components in the mixture. This is expressed by the equation:

$$L = \frac{x_\mathrm{f}}{1 + (V/L)K}$$

where

L = Total moles/hr of a component in the liquid phase.
x_f = Moles/hr of the component in the feed.

V/L = The ratio of total moles vapor to total moles liquid.
K = The equilibrium constant for each component at the flash condition of temperature and pressure.

There are several publications giving values for K. Among these are the charts in Maxwell's "Data Book on Hydrocarbons" which are based on fugacities. Others may be found in engineering data books such as "Gas Processors Suppliers Association" which are based on convergence pressures. A rough and ready substitute for K factors is to use the vapor pressure of the component divided by the system pressure. This, however, should not be used for any definitive design work nor in systems which have azeotropes or are near their critical conditions. A method for calculating equilibrium flash vaporization is given by the following steps.

Step 1. Prepare a table with the first column giving the components making up the feed. The second column will be the composition of the feed in mol/hr. The third column is a listing of the equilibrium constant K for each component at the temperature and pressure of the flash condition. Allow up to three columns following for assumptions of V/L. Each of these three columns should be subdivided into two. The first giving the product of $(V/L)K$ and the second for listing the "L" for each component. Other columns may be added to calculate mole wt of vapor and SG of the liquid phase.

Step 2. Assume a value for V/L. This is a judged assumption but start with 1.0 or 0.1 whichever seems to be the more realistic. Calculate $(V/L)K$ for each component.

Step 3. Calculate "L" for each component from the equation:

$$L = \frac{X_{\mathrm{f}}}{1 + (V/L)K}$$

Step 4. The calculated V/L is now obtained by adding the "L" column and subtracting this value from the total moles of feed in column 1. This subtraction is the vapor moles as calculated. Then the calculated V/L is arrived at by dividing the total V by the total of the "L" column.

Step 5. The calculation is complete when V/L calculated is equal to V/L assumed. An answer within 5% is usually acceptable. If the calculated V's assumed is not within this limit make another assumption for V/L and repeat steps 2, 3, 4. For this second assumption try 5, or 0.5, or 0.05, whichever is more appropriate.

Step 6. If there is still no agreement between assumed and calculated V/L plot the two trial points (assumed Vs calculated) on log graph paper. Draw a straight line through these two points and note where on this line assumed V/L = calculated V/L. This value is the next assumed V/L. Repeat the calculation Steps 2–4 using this value; this usually completes the calculation. If it does not then check that the conditions for the flash are within the boiling point and condensing point for the feed.

In this example, it is required to determine the amount of vapor and liquid and their composition in a feed to a fractionator at 112 psig and 300°F.

			1st Trial $V/L = 0.5$		2nd Trial V/L = 0.2		3rd Trial V/L = 0.1					
	Moles/ hr °F	127 psia K 300°F	V/LK	$L = 1 + \frac{F}{V/LK}$	V/LK	$L = 1 + \frac{F}{V/LK}$	V/LK	$L = 1 + \frac{F}{V/LK}$	MW	Liquid lbs/hr	lbs/gal	GPH
C_2	6.4	9.1	4.55	1.15	1.82	2.27	0.9	3.37	30	101	2.97	34
C_3	43.5	5.0	2.50	12.43	1.00	21.75	0.5	29.0	44	1,276	4.69	301.7
iC_4	16.9	3.3	1.65	44.79	0.66	10.18	0.33	12.71	58	737	4.69	157.1
nC_4	67.6	2.9	1.45	30.04	0.58	42.78	0.29	52.40	58	3,039	4.87	624.0
iC_5	80.5	1.8	0.90	42.37	0.26	59.19	0.18	68.22	72	4,912	5.21	942.8
nC_5	34.6	1.6	0.80	19.22	0.32	26.21	0.16	29.83	72	2,148	5.26	408.4
C_6	124.9	0.85	0.425	87.65	0.17	106.75	0.085	115.12	86	9,900	5.54	1,787.0
C_7	140.9	0.48	0.240	113.63	0.096	128.56	0.048	134.45	100	13,445	5.74	2,342.3
NP260	99.8	0.212	0.106	90.24	0.042	95.74	0.021	97.75	114	11,144	6.18	1,803.2
NP300	99.8	0.116	0.058	94.33	0.023	97.54	0.012	98.62	126	12,426	6.37	1,950.7
NP340	60.4	0.063	0.032	58.53	0.126	59.55	0.006	60.04	136	8,165	6.46	1,263.9
NP382	29.8	0.035	0.0175	29.29	0.007	29.60	0.004	29.80	152	4,530	6.56	690.5
Total	805.1			623.67		680.07		731.31	98.2	71,823	5.4	12,305.6
Calculated V/L				0.29		0.184		0.1				°API = 70

lbs/hr liquid = 71,823 lbs/gal = 5.4
lbs/hr vapor = 4,998 mol wt = 67.7
lbs/hr feed = 76821

Note the components NP260 to NP382 are psuedo components having mid boiling points of 260, 300, 340, and 382°F, respectively. K for these components are based on their vapor pressure and system pressure relationship.

Predicting the EFV curve from TBP data. For crude oils and complex mixtures such as the heavy products the equilibrium vaporization curve can be calculated from the TBP curve using empirical methods given by Edmister or Maxwell. In this work the EFV is based on the method by Maxwell in his book "*Hydrocarbon Data*". This method has been described in Chapter 1 of this Handbook.

F

FCCU (The fluid catalytic cracking unit)

Introduction and a brief history of the process. The Fluid Catalyst Cracking Unit (FCCU) is a process for converting middle and heavy distillates to high-octane gasoline and olefin-rich light gases. It was developed during World War II to provide the high-octane fuel required by the war effort. The first unit of its kind was brought on stream in the USA in 1942. These first models of the process were a side-by-side reactor/regenerator configuration. The transfer line for the catalyst between the two items was relatively short and the design as such was aimed at concentrating the reaction in the reactor's fluid bed. To achieve effective transfer of the catalyst between the two items the reactor was operated at a slightly higher pressure than the regenerator and the regenerator elevated to provide the hydraulic head. A series of slide valves provided the control of catalyst flow between the two units.

The catalysts used in these early units were simply finely ground silica alumina. These had poor fluidization characteristics and were fragile producing a high quantity of fines. The conversion rate with this type of catalyst was also low at about + or −50LV%. After the war and during the 1950's the development of catalytic reforming of straight run naphtha with its additional bonus of cheap hydrogen reduced the popularity of the FCCU considerably. Its revival as the accepted 'work horse' of refining came with the development of riser cracking and the use of zeolite catalysts during the early 1960s. These two developments projected the process back to its prominence principally because it could upgrade low quality products to high quality gasoline and LPG cheaply and at high conversion. Today it still remains a principal process in heavy oil processing. Further advances in catalyst management and in the process itself provides a unit which can convert certain atmospheric and vacuum residues directly to more valuable products. A complete description of the process with details of its chemistry and reactor mechanism is given in Chapter 6 of this Handbook. The following is a summary of the process reactor side mechanism.

Process description

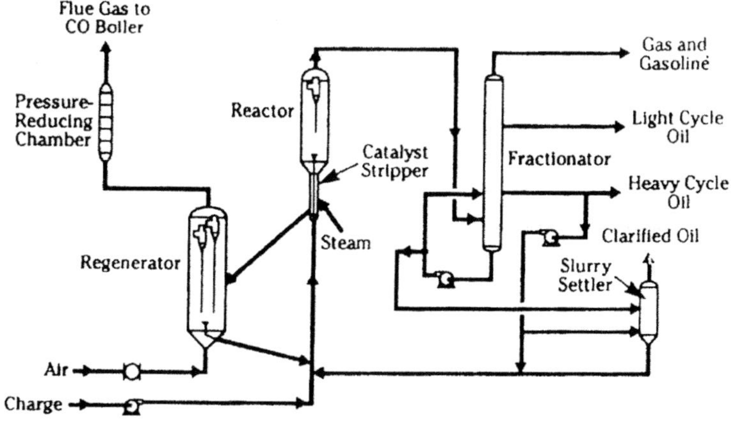

"A typical 'side by side' fluid catalytic cracking unit".

The "heart" of the process consists of reactor vessel and a regenerator vessel interconnected to allow the transfer of spent catalyst from the reactor to the regenerator and of regenerated catalysts back to the reactor. The heat for the oil cracking is supplied by exothermic heat of the catalyst regeneration. This heat is transferred by the regenerated fluid catalyst stream itself. The oil streams (feed and recycle) are introduced into this hot catalyst stream on route to the reactor. Much of the cracking occurs in the dispersed catalysed phase along this transfer line or riser.

The final contact with catalyst bed in the reactor completes the cracking mechanism. The vaporized cracked oil from the reactor is suitably separated from entrained catalyst particles (by cyclone) and routed to the recovery section of the unit. Here it is fractionated by conventional means to meet the product stream requirements. The spent catalyst is routed from the reactor to the regenerator after separation from entrained oil. Air is introduced into the regenerator and the fluid bed of the catalyst. The air reacts with the carbon coating on the catalyst to form CO/CO_2. The hot and essentially carbon free catalyst completes the cycle by its return to the reactor. The flue gas leaving the regenerator is rich in CO. This stream is often routed to a specially designed steam generator where the CO is converted to CO_2 and the exothermic heat of reaction used for generating steam (the CO boiler). Alternatively, CO combustion promoters may be used within the regenerator.

Feed stocks to the FCCU are primarily in the heavy vacuum gas oil range. Typical boiling ranges are 640°F (10%) to 980°F (90%). This gas oil is limited in end point by maximum tolerable metals, although the new zeolite catalysts have demonstrated higher metals tolerance than the older silica–alumina catalysts. The process has considerable flexibility. Apart from processing the more conventional waxy distillates to produce gasoline and other fuel components, feed stocks ranging from naphtha to suitably pretreated residuum are successfully processed to meet specific product requirement.

A summary of the mechanism of fluid catalytic cracking. The feed stock to FCCUs are usually the higher distillates of the crude barrel. The feed stock in this range of material therefore contains compounds of complex structure, some of which are contaminated by inorganic molecules such as sulfur, metals (vanadium, sodium, nickel, and the like). The amounts of these contaminants will vary for different crudes and the boiling range of the feed. The feed stock will also demonstrate a differing ease to its ability to crack under the conditions of the FCCU.

The mechanism of the cracking itself is extremely complex, and many theories have been offered to account for this. Certainly under the high temperature conditions of the FCC reactor, one can expect thermal cracking to occur, and to some extent this happens. Thermal cracking however results in the random fracture of the hydrocarbon compounds and there is very little selectivity in the resulting product content and yields. This is not the case in Fluid Catalytic cracking, indeed one of the process's main attraction is its high selectivity in its product make. The most generally accepted theory for this is the carbonium ion mechanism. It is believed that the carbonium ion in this cracking mechanism is initiated from an olefin's early contact with the catalyst at the high riser temperature condition. Once formed the ion reacts to:

- Crack larger molecules to smaller ones
- React with other molecules
- Isomerize
- React with the catalyst to terminate a chain

The cracking reaction then tends to follow the following route where the C—C bond in the β position relative to the carbonium ion is severed to form the olefin and the reactive positively charged ion. This is illustrated as follows:

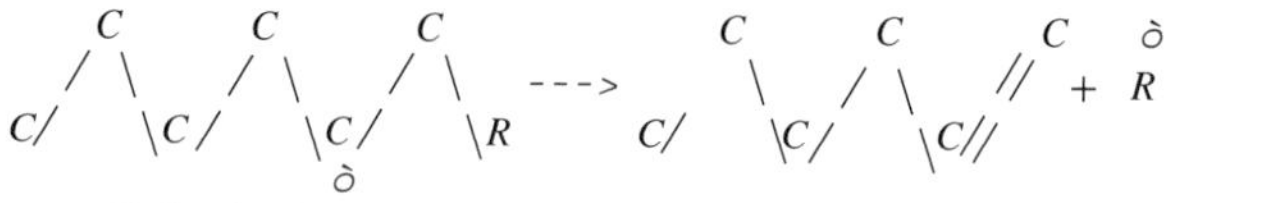

```
     C       C        C                   C        C         C   ò
    / \     / \      / \     --->          \      / \       //  + R
  C/   \C /    \C /    \R          C/       \C/     \C//
                ò
     Carbonium ion                               Olefin     Carbonium ion
```

The olefin may form new carbonium ions while the R ion reacts to form secondary and tertiary carbonium ions. The tertiary ion is the more stable, and is probably the more active in this cracking mechanism. A more detailed account of this mechanism is provided by Venuto P.B. and Habib E.T. in their paper "Catalyst–feed stock—engineering interactions in Fluid Catalytic Cracking".

Formation of coke on the catalyst. The formation of coke on the catalyst is an unavoidable occurrence of the catalytic cracking process. The mechanism of the coke formation is also very complex as was the case in the cracking mechanism itself. It is likely that coke is formed by the dehydrogenation and condensation of poly aromatics and olefins. Its deposition on the catalyst however tends to block the active pores which will reduce the effectiveness of the catalyst. The only recourse to this is to burn off the coke with air to form CO and CO_2. The heat of reaction for this coke burn off becomes important to the process as it provides the heat of reaction on its transfer from the regenerator to the reactor riser.

Typical yields and the effect of feedstock types. This item gives typical yield structures for various FCCU feed stocks and severities of operation. These were based on 100% zeolite catalyst processes and are included here as a guide to conventional FCCU process capability.

Note while these data may be used in study work and comparative process analysis, they should not be used for any definitive design work. For this definitive work specific licensor data from pilot plant test runs on the design feed stock must be used.

Hydrotreated Gas Oil (520–630°F)

Conversion vol%	50	60	70
Yields vol%:			
Gas	3.8	4.0	4.2
Propylene	4.3	6.0	8.0
Butylene	6.9	10.0	17.5
Propane	2.5	2.7	3.0
i-Butane	3.6	5.0	7.5
n-Butane	1.5	1.6	1.8
C_5-Naphtha	28.9	31.4	34.6
Light cycle oil	44.5	35.9	26.0
Heavy cycle oil	2.2	2.2	2.2
H_2S wt%	1.2	1.2	1.2

Straight Run Diesel (520–650)

Conversion vol%	30	40	50
Yields vol%:			
Gas	3.0	3.3	3.9
Propylene	3.0	4.2	6.0
Butylene	5.5	6.1	8.4
Propane	2.0	2.4	2.8
i-Butane	2.4	3.8	5.2
n-Butane	1.3	1.4	1.6
C_5-Naphtha	17.5	20.0	23.0
Light cycle oil	68.5	54.5	49.0
Heavy cycle oil	2.8	2.8	2.8
H_2S wt%	1.4	1.4	1.4

Hydrocracker Gas Oil (650–750°)

Conversion vol%	70	80	90
Yields vol% :			
Gas	4.2	4.8	5.0
Propylene	8.0	10.0	13.0
Butylene	15.0	24.0	41.0
Propane	3.1	3.5	3.9
i-Butane	7.0	9.0	12.0
n-Butane	1.5	1.8	2.3
C_5-Naphtha	40.0	49.5	52.0
Light cycle oil	23.0	16.0	9.0
Heavy cycle oil	3.0	3.0	3.0
H_2S wt%	1.1	1.1	1.1

Hydrotreated Visbreaker Gas Oil (380–650°F)

Conversion vol%	60	70	80
Yields vol% :			
Gas	3.0	3.6	4.0
Propylene	3.8	5.0	6.8
Butylene	5.0	6.8	8.0
Propane	2.0	2.3	3.0
i-Butane	2.6	4.0	6.0
n-Butane	1.6	1.8	1.9
C_5-Naphtha	50.0	53.0	60.0
Light cycle oil	37.0	28.5	18.5
Heavy cycle oil	2.0	2.0	2.0
H_2S wt%	←	Neg	→

Fired heaters

Generally fired heaters fall into two major categories:

- Horizontal type
- Vertical type

The horizontal type heater usually means a box type heater with the tubes running horizontally along the walls. Vertical type is normally a cylindrical heater containing vertical tubes. Full details of both these heater types are given in Chapter 18.

Horizontal box types are preferred for crude oil heaters, although vertical cylindrical have been used in this service. Vacuum unit heaters should have horizontal tubes to eliminate the static head pressure at the bottom of vertical tubes and to reduce the possibility of two-phase slugging in the large exit tubes. Occasionally, several different services ("coils") may be placed in a single heater with a cost saving. This is possible if the services are closely tied to each other in the process. Catalytic reforming pre-heater and re-heaters in one casing is an example. Reactor heater and stripper reboiler in one casing is another example. This arrangement is made possible by using a refractory partition wall to separate the radiant coils. The separate radiant coils may be controlled separately over a wide range of conditions by means of their own controls and burners. If a convection section is used, it is usually common to the several services. If maintenance on one coil is required, the entire heater must be shut down. Also, the range of controllability is less than with separate heaters.

Each of these types may be shop fabricated if size permits. Shop fabrication reduces costs. However, shop fabrication should not be forced to the extent of getting an improperly proportioned heater.

As fired heaters have a live source of energy they are designed and manufactured to strict codes. Details of these are also given in Chapter 18. These codes set the parameters for heater tube thickness, tube, and heater dimension, Inspection requirements, and fire protection.

Thermal rating. Thermal rating of fired heaters is a complicated and specialized procedure. While this procedure is described in Chapter 18 it is not detailed. For most basic design and evaluation studies the following rules of thumb may be used:

	Btu/hr sqft
Horizontal, fired on one side	8,000–12,000
Vertical, fired on one side from bottom	9,000–12,000
Vertical, single row, fired on both sides	13,000–18,000

Heater efficiency. The efficiency of a fired heater is the ratio of the heat absorbed by the process fluid to the heat released by combustion of the fuel expressed as a percentage.

Heat release may be based on the LHV (Lower Heating Value) of the fuel or HHV (Higher Heating Value). Process heaters are usually based on LHV and boilers on HHV. The HHV efficiency is lower than the LHV efficiency by the ratio of the two heating values.

Heat is wasted from a fired heater in two ways:

- with the hot stack gas
- by radiation and convection from the setting

The major loss is by the heat contained in the stack gas. The temperature of the stack gas is determined by the temperature of the incoming process fluid unless an air preheater is used. The closest economical approach to process fluid is about 100°F. If the major process stream is very hot at the inlet, it may be possible to find a colder process stream to pass through the convection section to improve efficiency, provided plant control, and flexibility are adequately provided for. A more common method of improving efficiency is to generate and/or superheat steam and preheat boiler feed water. The lowest stack temperature that can be used is determined by the dew point of the stack gases. The figures in Chapter 18 may be used to estimate flue gas heat loss. Heat loss from the setting, called radiation loss, is about 1½ to 2% of the heat release.

The range of efficiencies is approximately as follows:

Very high –90%+. Large boilers and process heaters with air pre-heaters.
High –85%. Large heaters with low process inlet temperatures and/or air pre-heaters.

Usual—70–80%.
Low—60%.

Burners. The purpose of a burner is to mix fuel and air to ensure complete combustion. There are about 12 basic burner designs. These are:

Direction — vertical up fired
vertical down fired
horizontally fired
Capacity — high
low
Fuel type — gas
oil
combination
Flame shape — normal
slant
thin, fan-shaped
flat
adaptable pattern
Hydrogen content — high
Excess air — normal
low
Atomization — steam
mechanical
air assisted mechanical
Boiler types
Low NO_x
High intensity

Various combinations of the above types are available.

Details of gas, oil, and combination burners are given in Chapter 18, together with burner controls and heater noise.

Chapter 18 under the section on fired heaters continues with a description and discussion on refractories, stacks, and stack emission. Finally the chapter describes the specification for a typical fired heater.

Flash points

The flash point is a measure of the tendency of the, material to form a flammable mixture with air under controlled laboratory conditions. It is however only one of several properties that must be considered in assessing the overall flammable hazard of the material. The flash point is used to establish the flammable criteria in transporting the material. Generally shipping and safety regulations will be based on the flash point criteria. The flash point should not however be used to describe or appraise the fire hazard or risk under actual fire conditions. Test method (D93 provides the closed cup flash point test procedures for temperatures up to 698°F). Details of the apparatus used and the laboratory method for determining a flash point is given in Chapter 16 of this Handbook. The flash point of two or more blended components is determined by the volume composition of the components in the blend and the use of the flash point index curve. This method and the index curve is given in Chapter 1 of this Handbook. A simple estimate of a material's flash point can be calculated from its ASTM distillation by the equation:

$$\text{Flash Point}\,^{\circ}\text{F} = 0.77(\text{ASTM } 5\% - 150^{\circ}\text{F}).$$

Flash zone

A flash zone is associated with the distillation of crude oil, both atmospheric and vacuum, the main fractionating towers of the fluid catalytic cracking unit, visbreaker or thermal cracking units. The flash zone is the area in these distillation towers where the distillate vapors are allowed to separate from the un-vaporized liquid. The transfer line from the heater enters the flash zone. The vapors rise up through the tower to be condensed by cold reflux streams coming down. Steam enters the flash zone from the bottom product stripper section located below the flash zone.

Flash zone conditions, particularly flash zone temperature is difficult to measure accurately. This is due to the profiles set up in the flash zone by turbulence of the feed entering and vapor disengaging. Knowledge of flash zone conditions is, however, essential when designing these units, evaluating their performance or trouble shooting them. The item in Chapter 3 of this Handbook deals with a calculation procedure to establish these flash zone conditions for atmospheric and vacuum distillation crude distillation towers.

Fractionation

This is a unit operation in chemical engineering which separates components from mixtures in which they are contained. In petroleum refining this separation process is a major means of separating precise groups of petroleum components from the crude oil feed and other intermediate refining processes. Separation by fractionation is accomplished by heat and mass transfer on successive stages represented by carefully designed trays. These trays are designed to enhance the heat and mass transfer by good mixing of hot vapors rising through the trays with colder liquid entering the tray. The mixing tends to achieve a phase equilibrium between the liquid and vapor traffic.

Fractionation as a separating tool is used extensively throughout the process industry. It can be a simple process separating just two components (such as water and alcohol) of very complex separation of multi components. It is the later that is required in the petroleum processing industry. The calculation of the number of stages required for the simple binary fractionation can be accomplished graphically using the McCabe Thiele diagram, the number of stages required for the multi-component separation is achieved by using a number of complex equations, and the individual equilibrium constants for the various components. The rigorous method for calculating the number of fractionation stages in multi component separation is by stage to stage calculation with 'trial and error' determination of the temperature and pressure conditions on each stage. The calculation is complete when convergence of phase equilibria is met on the feed input stage. This is the method used in developing the computer simulation program for multi-component fractionation.

The method is simplified for manual computation by a correlation developed by Gilliland (reference Maxwell's Data Book on Hydrocarbons). This is a correlation which relates minimum stages at total reflux to minimum reflux at infinite number of stages. Figure 4.2, Chapter 4, is reproduced below to show this correlation.

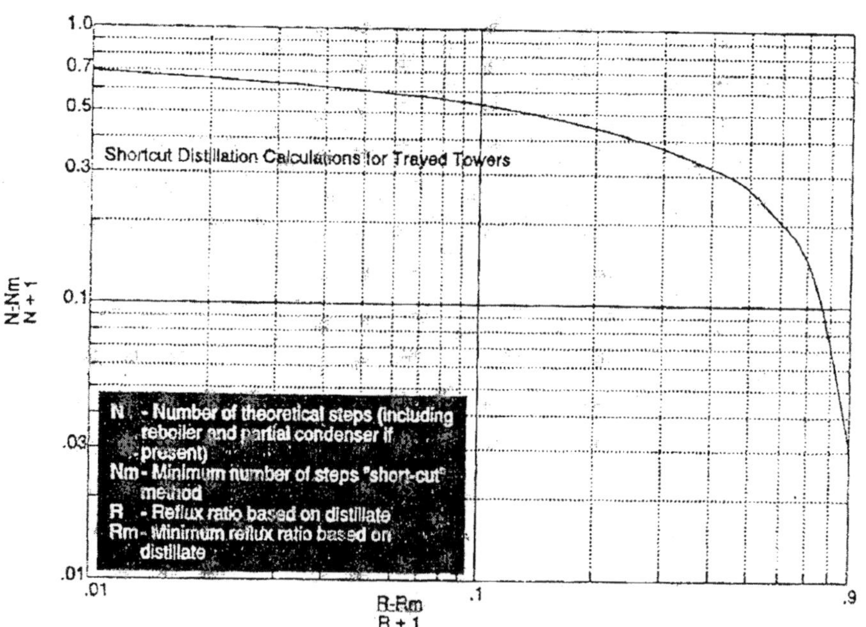

The Gilliland correlation for calculating theoretical trays.

Where:

N = actual number of trays.
N_m = minimum number of stages at total reflux.
R = actual reflux
R_m = minimum reflux at infinite number of stages.

There are two equations that define the items R_m and N_m these are:

- The *Fenske equation* which is used to calculate the minimum trays at total reflux.

$$N_{m+1} = \text{Log}\left[\left(\frac{\text{Lt Key}}{\text{Hy Key}}\right)_D \cdot \left(\frac{\text{Hy Key}}{\text{Lt Key}}\right)_W\right] \div \text{Log}\,\phi$$

Where:

N_M = minimum number of theoretical strays at total reflux. The + 1 is the reboiler which is counted as a theoretical tray
LT Key = is the mole fraction of the selected light key
HY Key = is the mole fraction of the selected heavy key
D = fractions in the distillate product
W = fractions in the bottom product
ϕ = KLt Key/Khy key
KLt key = the equilibrium constant of the light key at mean system condition of temperature and pressure
KHy key = the equilibrium constant of the heavy key again at mean system conditions

- *The Underwood equation* which establishes the minimum reflux at infinite stages. The Underwood equation is more complex than the Fenske equation and requires trial and error method for its solution. The equation itself is in two parts. The first looks at the vapor volatility (ratio of K's) of each component in the feed to one of the selected keys. Then by trial and error arriving at an expression for a factor 'B' that forces the equation to zero.
  This first equation is written as follows:

$$\frac{\Sigma(\phi \text{i})(X_i F)}{X_i F - B} = 0$$

ϕ_i = is the relative volatility of component i
$X_i F$ = is the mole fraction of component i in the feed
B = is the factor that forces the sum of the expression for each component to zero

The second equation uses the Factor B calculated from the first equation to determine the minimum reflux. This equation is

$$R_{\mathrm{m}+1} = \frac{\Sigma(\phi_i)(X_i D)}{X_i D - B}$$

where

ϕi = is the relative volatility of component i
$X_i D$ = is the mole fraction of component i in the distillate
B = is the factor obtained from first equation
R_{m} = minimum reflux at infinite trays (R_{m+1} include the reboiler)

R is taken as $1.5 \times R_{\mathrm{m}}$ while N is taken as $1.5 \times N_{\mathrm{m}}$.

Fractionation in the atmospheric and vacuum crude distillation towers

The method of Gilliland is impractical for use in the complex fractionation that is experienced in the main refinery towers handling the crude feed and other similar streams in the refining process. Packie J.W. developed an empirical method based on ASTM distillation gaps to define the fractionation between adjacent cuts. This degree of fractionation is measured by the temperature difference between the ASTM 95% of the lighter cut and the 5% temperature of the adjacent heavier cut. These differences can be positive or negative. That is the 5% temperature of the heavier cut can be higher than the 95% temperature of the lighter cut. This difference is called an *ASTM Gap.* On the other hand some adjacent cuts may demonstrate the reverse that is the 5% temperature of the heavier cut may be lower than the 95% temperature of the lighter cut. This difference is called an *ASTM Overlap.* Thus in a typical crude unit specification the fractionation may be quoted as follows:

The fractionation in the crude atmospheric distillation tower shall be:

95% naphtha ASTM temperature and the 5% temperature of the kero shall have a gap of 25°F.
95% kero ASTM temperature and the light gas oil 5% temperature shall have a gap of 0°F.
95% LGO ASTM temperature and the heavy gas oil 5%temperature shall have an overlap of no more than 15°F (a –15°F gap).

The number of the fractionation trays required to meet the respective fractionation gaps (overlaps) between cut draw-off in these main columns is based on a correlation. This correlation is a series of curves based on the 50% distillation temperature difference between the total vapor entering the draw off tray and the liquid leaving the tray. There are two sets of these curves, the first with steam present and the second based on no

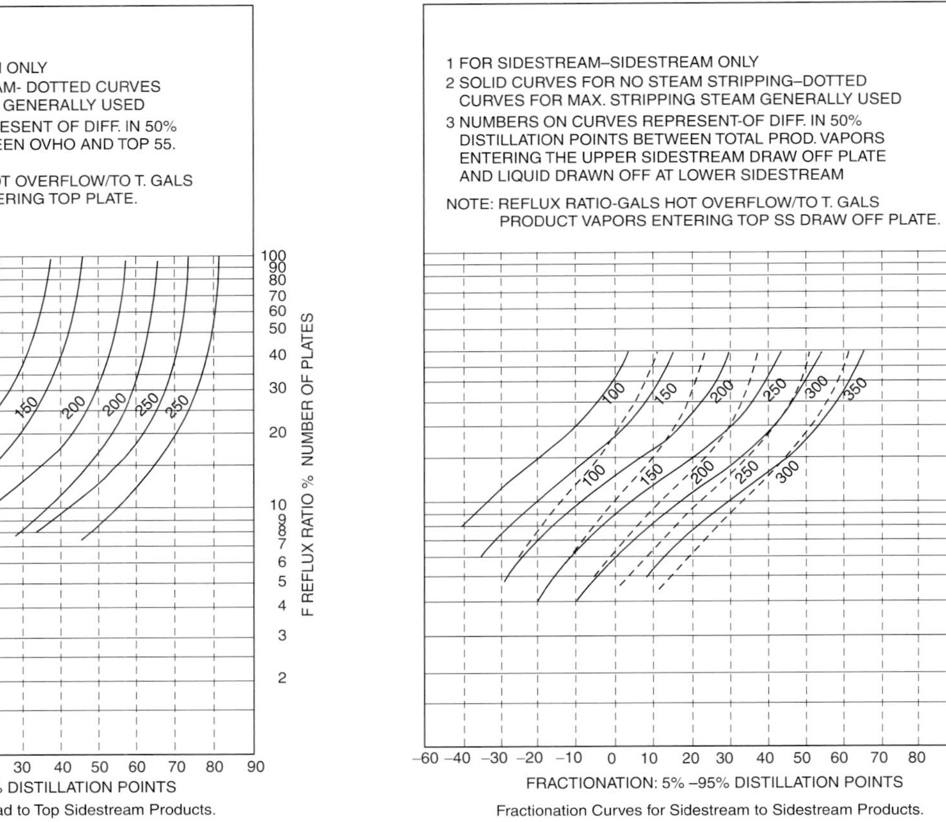

Fractionation Curves for Overhead to Top Sidestream Products.

Fractionation Curves for Sidestream to Sidestream Products.

ASTM gaps and overlaps.

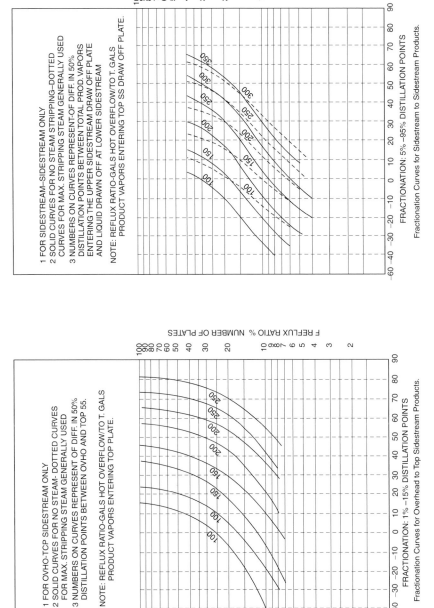

steam. On the basis of these curves the reflux rate times the number of stages are read off for the required ASTM gaps.

A copy of this correlation is given in Chapter 3 Appendix 1. For convenience it is also printed on page 1169.

G

Gaps and overlaps

Gaps and overlaps refer to the difference between the temperatures of the 95 vol% recovered in a lighter cut and the 5 vol% recovered of the adjacent heavier cut based on their ASTM distillation. The following diagrams illustrate this concept:

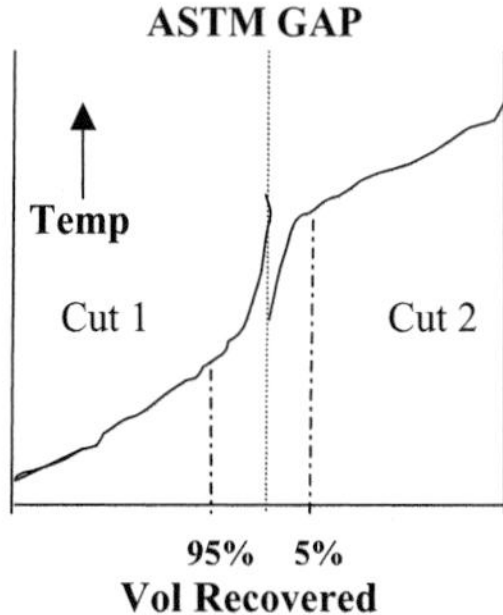

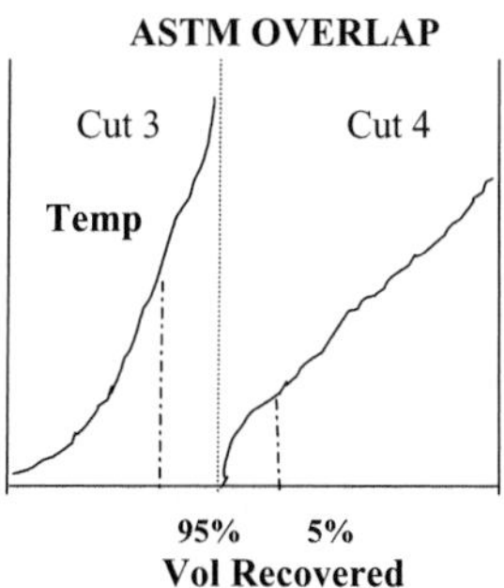

The ASTM Gap shown above would be typical in the separation between naphtha and kero in the atmospheric distillation unit. The 5% temperature of the kero (cut 2) is higher than the naphtha (cut 1) 95% temperature. This is a good separation because there are few kero components in the heavy end of the naphtha. The overlap illustration is typical of the separation between the heavier products of the atmospheric distillation of crude oil such as between light gas oil cut and the heavy gas oil cut. A GAP then is when the numeric difference between the 5% of the heavier cut and the 95% of the lighter cut is positive. An OVERLAP is when this difference is negative.

The ASTM gap and overlaps are used as a measure of fractionation. Packie (*Am Inst Chem Eng Transactions Vol 317*) *1941* has related a series of Gaps and overlaps to the 50% temperature difference between the vapors rising through a distillation tray and the liquid leaving. These series of curves are related to gaps and overlaps to give a product of reflux times number of trays. The series of curves have also been constructed for a 'No steam system' and 'Maximum steam system'. Two of these curves are shown under section "F" preceding.

Gas-refinery fuel

The gas fractions in petroleum refining may be taken as that fraction on crude or produced in a process as boiling below propane. This vapor fraction is usually used as a fuel to the refinery's fired heaters. The heaters themselves have burners on dual heating source, either fuel oil or fuel gas. The pilot flames on all the heater burner assemblies are on fuel gas.

The fuel gas is the collection of the vapors from the appropriate overhead distillation condensers, purge gas streams (from the hydro-treating processes), and any other continuous hydrocarbon gas stream meeting the fuel gas criteria. Emergency venting and start up purge streams are routed to the refinery flare system. The fuel gas streams are stored in bullets with heating coils to vaporize the heavier components in the mixture (i.e., butanes and occasionally pentanes). With the introduction of the 'clean air act' the fuel gas stream must be treated for the removal of sulfur. This is accomplished by absorption of the sulfur components (hydrogen sulfide, and possibly mercaptans) into amine solutions or other absorbents. See the item on gas treating in Chapters 10 and 15.

Gas oil

Usually there will be two gas oil side streams, a light gas oil side stream and, below this take off, a heavy gas oil side stream. Both these side streams are steam stripped to meet their respective flash point specification (usually 150°F minimum). The lighter side stream cut of about 480–610°F on crude) is the principal precursor for the automotive diesel grade finished product, this side stream is desulfurized to meet the diesel sulfur specification in a hydrotreater (see Chapter 8). The lower gas oil stream is really a guard stream to correct the diesel distillation end point. This heavy gas oil may also be hydrodesulfurized and routed to either the fuel oil pool (as a precursor for marine diesel for example) or to a finished heating oil product from the gas oil pool. See also Chapter 2.

Gasolines

These are probably among the most important and controversial refinery products. This is due to the fact that they are readily and widely used by the general public and are growing more and more in demand as the automobile industry expands. With this expansion in demand and their use as fuels, augmented are those problems associated with their effect on the environment and the health hazards due to emission from vehicles. These have become major concerns in most highly developed countries of the world.

Two major gasoline products are produced in the petroleum refinery. Their specification and standard quality are fully described and discussed in Chapters 2, 9, and 15. The two grades shipped from refineries are usually a regular grade with an octane number of 87 and a premium grade with an octane number of 93. These octane levels may differ slightly from country to country, but these are the key quality for North America, with octane numbers defined as (RON + MON)/2.

Gasolines are processed from the catalytic reforming of a heavy straight run naphtha 190–360°F cut. This cut is the bottom product of a naphtha splitter which takes as fee the de-butanized overhead distillate from the crude unit. The top distillate product from this splitter will be a light straight run naphtha. This will be blended with other components (such as the reformate, catalytic cracked naphtha, and other octane) enhancement cuts to make the specified two gasoline refinery products.

The catalytic reformer is run to make a 91–100 octane number research reformate after the removal of butanes and lighter. Prior to the clean air of the 1960s, tetra ethyl lead was used extensively as an octane enhancer, and the reformer operating severity in terms of octane number was much reduced. The clean air act prohibits the use of the lead compound and now only lead free gasoline is used in all vehicles. More stringent controls are becoming to the fore in environmental controls in most of the developed countries. Among these are the further reduction of sulfur compounds in gasoline and perhaps even more important the reduction of aromatic compounds in the products. The development of higher blending stock such as the oxygenated gasoline and the increased production and use of paraffin isomers make this a very possible achievement. These are detailed in Chapter 9 and Chapter 15.

Gas treating processes

Refinery gas treating usually refers to the process used to remove the so called 'Acid Gasses' which are hydrogen sulfide and carbon dioxide from the refinery gas streams. These acid gas removal processes used in the refinery are required either to purify a gas stream for further use in a process or for environmental reasons associated with the use of the gas for fuel. Clean air legislation now being practiced through most industrial countries requires the removal of these acid gases to very low concentrations in all gaseous effluent to the atmosphere. Hydrogen sulfide combines with the atmosphere to form very dilute sulfuric acid and carbon dioxide forms carbonic acid both of which are considered injurious to personal health. These compounds also cause excessive corrosion to metals and metallic objects and may contribute to "global warming."

The use of chemically "basic" liquids to react with the acidic gases was developed in 1930. The chemical used initially was tri-ethanolamine (TEA). However in more recent times as mono-ethanolamine (MEA) has become commercially the more available and preferred liquid reactant due to its high acid gas absorbency on a unit basis. Numerous alternative processes to MEA have been developed. These have fewer corrosion problems and are to a large extent more energy efficient. Inhibitor systems have however been developed which have eliminated much of the MEA corrosion problems. Some of these newer processes also are designed to remove the H_2S leaving the CO_2 to remain in the gas stream.

In an amine treating unit sour gas (rich in H_2S) enters the bottom of the trayed absorber (or contactor). Lean amine is introduced at the top tray of the absorber section to move down the column. Contact between the gas and amine liquid on the trays results in the H_2S in the gas being absorbed into the amine. The sweet gas is water washed to remove any entrained amine before leaving the top of the contactor.

Rich amine leaves the bottom of the contactor to enter a surge drum. If the contactor pressure is high enough a flash stream of H_2S can be routed from the drum to a trayed stripper. The liquid from the drum is preheated before entering a stripping column on the top stripping tray. This stripper is reboiled with 50 psig saturated steam. Saturated 50 psig steam is used because higher temperatures cause amines to break down. The H_2S is stripped off and leaves the reflux drum usually to a sulfur production plant. Sulfur is produced in this plant by burning H_2S with a controlled air stream. The lean amine leaves the stripper bottom and is cooled. The cooled stream is routed to the contactor.

There are several liquid solvents in commercial use for the removal of H_2S and CO_2 from refinery gases. Among the more common are the amines. These include:

MEA — Mono-ethanolamine
DEA — Di-ethanolamine
DGA — Di-glycolamine

In addition to these amine base solvents there are also the hot potassium carbonate process (Benfield), Sulfinol and ADIP. These latter two processes are marketed by the Shell company and are quite common in world wide usage. Full details of the more common of these compounds together with the reaction mechanism are given in Chapter 10. A comparison of these compounds and their properties are given in Table 19.G.1.

Grids

Grids are used as low pressure drop packing in certain fractionation towers. They came into prominence with the development of the crude oil 'dry vacuum' units. See Chapter 3 and Chapter 18 for details of this type of packing and its use.

Table 19.G.1. A comparison of gas treating absorbents

	*MEA	*DEA	*DGA	*DIPA	Sulfinol	*Sulfolane
Molecular wt	61.1	105.1	105.14	133.19		120.17
Boiling point, °F	338.5	515.1	405.5	479.7		545
Boiling range, 5–95%,	336.7–341.06	232–336.7	205–230	–		–
Freezing point, °F	50.5	77.2	9.5	107.6		81.7
Sp. gr., 77°F	1.0113	1.0881	1.0572	–		1.256
140°F	0.9844	(86°F) 1.0693	1.022	0.981 (129°F)		(86°F) 1.235
Pounds per gallon, 77°F	8.45	9.09 (86°F)	8.82	8.3 (86°F)		10.46 (86°F)
Abs. visc., cps., 77°F	18.95	351.9 (86°F)	40	870 (86°F)		12.1 (86°F)
140°F	5.03	53.85	6.8	86 (129°F)		4.9
Flash point, °F	200	295	260	255		350
Fire point, °F	205	330	285	275		380
Sp. ht. Btu/lb, °F	0.663	0.605	0.571	0.815		0.35
Critical–temp., °F	646.3	827.8	765.6	–		982.4
Critical–press., atm.	44.1	32.3	37.22	–		52.2
Ht. of vaporize., Btu/lb	357.94	267.00	219.14	202.72		225.7
Ht. of reaction– CO_2 Btu/lb (Approx)	825	620	850		580	
Ht. of reaction– H_2S, Btu/lb (Approx)	650	550	674		500	

*MEA = $HOC_2H_4NH_2$
*DEA = $(HOC_2H_4)_2NH$
*DGA = $HOCH_2OCH_2C_2H_4NH_2$
*DIPA = $(HOC_3H_6)_2NH$
*SULFOLANE = $(CH_2)_4SO_2$

H

Heaters

Heaters are used extensively in petroleum refining to provide heat energy to the process plants utilizing an independent energy source namely fuel oil or fuel gas. Generally fired heaters fall into two major categories:

- Horizontal type
- Vertical type

The horizontal type heater usually means a box type heater with the tubes running horizontally along the walls. Vertical type is normally a cylindrical heater containing vertical tubes. Figures 18.47 and 18.48 show examples of these two types of heaters.

Cylindrical heaters require less plot space and are usually less expensive. They also have better radiant symmetry than the horizontal type.

Horizontal box types are preferred for crude oil heaters, although vertical cylindrical have been used in this service. Vacuum unit heaters should have horizontal tubes to eliminate the static head pressure at the bottom of vertical tubes and to reduce the possibility of two-phase slugging in the large exit tubes.

Occasionally, several different services ("coils") may be placed in a single heater with a cost saving. This is possible if the services are closely tied to each other in the process. Catalytic reforming preheater and reheaters in one casing is an example. Reactor heater and stripper reboiler in one casing is another example. This arrangement is made possible by using a refractory partition wall to separate the radiant coils. The separate radiant coils may be controlled separately over a wide range of conditions by means of their own controls and burners. If a convection section is used, it is usually common to the several services. If maintenance on one coil is required, the entire heater must be shut down. Also, the range of controllability is less than with separate heaters.

Full details of the two types of heaters are given in Chapter 18 of this book. This includes the mandatory codes that apply to all fired heaters for their fabrication and operation.

Heater burners

Gas burners. The two most common types of gas burners are the "pre-mix" and the "raw gas" burners. Premix burners are preferred because they have better "linearity", i.e., excess air remains almost constant at turndown. With this type, most of the air is drawn in through an adjustable "air register" and mixes with the fuel in the furnace firebox. This is called secondary air. A small part of the air is drawn in through the "primary air register" and mixed with the fuel in a tube before it flows into the furnace firebox. A turndown of 10:1 can be achieved with 25 psig hydrocarbon fuels. A more normal turndown is 3 : 1.

Oil burners. An oil burner "gun" consists of an inner tube through which the oil flows and an outer tube for the atomizing agent, usually steam. The oil sprays through an orifice into a mixing chamber. Steam also flows through orifices into the mixing chamber. An oil-steam emulsion is formed in the mixing chamber and then flows through orifices in the burner tip and then out into the furnace firebox. The tip, mixing chamber, and inner and outer tubes can be disassembled for cleaning.

Oil pressure is normally about 140–150 psig at the burner, but can be lower or higher. Lower pressure requires larger burner tips, the pressure of the available atomizing steam may determine the oil pressure.

Atomizing steam should be at least 100 psig at the burner valve and at least 20–30 psi above the oil pressure. Atomizing steam consumption will be about 0.15–0.25 lbs steam/lb oil, but the steam lines should be sized for 0.5.

Combination burners. This type of burner will burn either gas or oil. It is better if they are not operated to burn both fuels at the same time because the chemistry of gas combustion is different from that of oil combustion. Gases burn by progressive oxidation and oils by cracking. If gas and oil are burned simultaneously in the same burner, the flame volume will be twice that of either fuel alone.

Pilots. Pilots are usually required on oil fired heaters. Pilots are fired with fuel gas and are not required when heaters are gas fired only, but minimum flow bypasses around the fuel gas control valves are used to prevent the automatic controls from extinguishing burner flames.

More details on fired heater burners are given in Chapter 18 of this Handbook.

Heater efficiencies

The efficiency of a fired heater is the ratio of the heat absorbed by the process fluid to the heat released by combustion of the fuel expressed as a percentage. Heat release may be based on the LHV (Lower Heating Value) of the fuel or HHV (Higher Heating Value). Process heaters are usually based on LHV and boilers on HHV. The HHV efficiency is lower than the LHV efficiency by the ratio of the two heating values.

Heat is wasted from a fired heater in two ways:

- with the hot stack gas
- by radiation and convection from the setting

The major loss is by the heat contained in the stack gas. The temperature of the stack gas is determined by the temperature of the incoming process fluid unless an air preheater is used. The closest economical approach to process fluid is about 100°F. If the major process stream is very hot at the inlet, it may be possible to find a colder process stream to pass through the convection section to improve efficiency, provided plant control, and flexibility are adequately provided for. A more common method of improving efficiency is to generate and/or superheat steam and preheat boiler feed water.

The lowest stack temperature that can be used is determined by the dew point of the stack gases. Figures 18.49 and 18.50 may be used to estimate flue gas heat loss.

The loss to flue gas is expressed as a percentage of the total heat of combustion available from the fuel. These figures also show the effect of excess air on efficiency. Typically excess air for efficiency guarantees is 20% when firing fuel gas and 30% when firing oil.

Heat loss from the setting, called radiation loss, is about 1½–2% of the heat release.

The range of efficiencies is approximately as follows:

Very high — 90%+. Large boilers and process heaters with air preheaters.
High — 85%. Large heaters with low process inlet temperatures and/or air preheaters.
Usual — 70–80%.
Low — 60% and less. All radiant.

More detailed discussion on this subject is given in Chapter 18 of this Handbook.

Heat exchangers

Heat exchange is the science that deals with the rate of heat transfer between hot and cold bodies. There are three methods of heat transfer, they are:

- Conduction
- Convection
- Radiation

In a heat exchanger heat is transferred by conduction and convection with conduction usually being the limiting factor. The equipment used in heat exchanger service is designed specifically for the duty required of it. That is, heat exchange equipment cannot be purchased as a stock item for a service but has to be designed for that service.

The types of heat exchange equipment used in the process industry and their selection for use are as follows:

The shell and tube exchanger. This is the type of exchanger most commonly used in a process plant. It consists of a bundle of tubes encased in a shell. It is inexpensive and is easy to clean and maintain. There are several types of shell and tube exchangers and some of these have removable bundles for easier cleaning. The shell and tube exchanger has a wide variety of service that it is normally used for. These include vapor condensation (condensers), process liquid cooling (coolers), exchange of heat between two process streams (heat exchangers), and reboilers (boiling in fractionator service).

The double pipe exchanger. A double pipe exchanger consists of a pipe within a pipe. One of the fluid streams flows through the inner pipe while the other flows through the annular space between the pipes. The exchanger can be dismantled very easily and therefore be easily cleaned. The double pipe exchanger is used for very small process units or where the fluids are extremely fouling. Either true concurrent or countercurrent flows can be obtained but because the cost per square foot is relatively high it can only be justified for special applications.

Extended surface or fin tubes. This type of exchanger is similar to the double pipe but the inner pipe is grooved or has longitudinal fins on its outside surface. Its most common use is in the service where one of the fluids has a high resistance to heat transfer and the other fluid has a low resistance to heat transfer. It can rarely be justified if the equivalent surface area of a shell and tube exchanger is greater than 200–300 sqft.

Finned air coolers. These are the more common type of air coolers used in the process industry. In a great many applications and geographic areas they have considerable economic advantage over the conventional water cooling. Indeed today it is uncommon to see process plants of any reasonable size without air coolers.

Air coolers consist of a fan and one or more heat transfer sections mounted on a frame. In most cases these sections consist of finned tubes through which the hot fluid passes. The fan located either above or below the tube section induces or forces air around the tubes of the section.

The selection of air coolers over shell and tube is one of cost. Usually air coolers find favor in condensing fractionator overheads to temperatures of about 90–100°F and process liquid product streams to storage temperatures. Air coolers are widely used in most areas of the world where ambient air temperatures are mostly below 90°F. At atmospheric temperatures above 100°F humidifiers are incorporated into the cooler design and operation. The cost under these circumstances is greatly increased and their use is often not justified.

In very cold climates the air temperature around the tubes is controlled to avoid the skin temperature of the fluid being cooled falling below a freezing criteria or in the case of petroleum products its pour point. This control is achieved by louvers installed to recirculate the air flow or by varying the quantity of air flow by changing the fan pitch.

Box coolers. These are the simplest form of heat exchange. However, they are generally less efficient, more costly and require a large area of the plant plot. They consist of a single coil or "worm" submerged in a bath of cold water. The fluid flows through the coil to be cooled by the water surrounding it. The box cooler found use in the older petroleum refineries for cooling heavy residuum to storage temperatures. Modern day practice is to use a tempered water system where the heavy oil is cooled on the shell side of a shell and tube exchanger against water at a controlled temperature flowing in the tube side. The water is recycled through an air cooler to control its temperature to a level which will not cause the skin temperature of the oil in the shell and tube exchanger to fall below its pour point.

Direct contact condensers. In this exchanger the process vapor to be condensed comes into direct contact with the cooling medium (usually water). This contact is made in a packed section of a small tower. The most common use for this type of condenser is in vacuum producing equipment. Here the vapor and motive steam for each ejector stage is condensed in a packed direct contact condenser. This type has a low pressure drop which is essential for the vacuum producing process.

Details of these heat exchange equipment are given in Chapter 18 of this Handbook.

Basic heat transfer equations. The following equations define the basic heat transfer relationships.

These equations are used to determine the overall surface area required for the transfer of heat from a hot source to a cold source.

The overall heat transfer equation;
The principal equation for heat transfer is given as:

$$Q = UA\,(\Delta t_m)$$

where

Q = Heat transferred in Btu/hr
U = Overall heat transfer coefficient, Btu/hr/sqft/°F
A = Heat transfer surface area sqft.
Δt_m = Corrected log mean temperature difference °F

The overall heat transfer coefficient U is defined by the expression:

$$\frac{1}{U_o} = \frac{1}{h_o} + \frac{1}{h_i} \times \frac{A_o}{A_i} + \frac{1}{h_w} + (rf)_o + (rf)_i \times \frac{A_o}{A_i}$$

where

U_o = overall heat transfer coefficient based on outside tube surface, in Btu/hr/sqft/°F.
h = The film coefficient in Btu/hr/sqft/°F.
r_f = fouling factors in $\frac{1}{\text{Btu/hr/sqft/°F}}$
h_w = Heat transfer rate through tube wall in Btu/hr/sqft/°F.
A = Surface area in sqft

subscripts "o" and "i" refer to outside surface and inside surface, respectively.

Flow arrangements. The two more common flow paths are concurrent and counter-current. In concurrent flow both the hot fluid and the cold fluid flow in the same direction. This is the least desirable of the flow arrangement and is only used in those chemical processes where there is a danger of the cooling fluid congealing, subliming, or crystallizing at near ambient temperatures.

Countercurrent flow is the most desirable arrangement. Here the hot fluid enters at one end of the exchanger and the cold fluid enters at the opposite end. The streams flow in opposite directions to one another. This arrangement allows the two streams exit temperatures to approach one another.

Logarithmic mean temperature difference Δt_m. In either countercurrent or concurrent flow arrangement the log mean temperature difference used in the overall heat transfer equation is determined by the following expression:

$$\Delta t_m = \frac{\Delta t_1 - \Delta t_2}{\log_e\left[\frac{\Delta t_1}{\Delta t_2}\right]}$$

The Δt's are the temperature differences at each end of the exchanger and Δt_1 is the larger of the two. In true countercurrent flow the Δt_m calculated can be used directly in the overall heat transfer equation. However such a situation is not common and true countercurrent flow rarely exists. Therefore a correction factor needs to be applied to arrive at the correct Δt_m. These are given in Figure 18.A.1 in the appendix to Chapter 9. Details of the shell and tube exchangers including the following:

- Choice of tube side or shell side fluid
- Calculation methods
- Shell and tube baffling
- Types of shell and tube exchangers (fixed sheet, floating head etc.)

are also given in Chapter 18.

Air coolers and condensers. Air cooling of process streams or condensing of process vapors is more widely used in the process industry than cooling or condensing by exchange with cooling water. The use of individual air coolers for process streams using modern design techniques has economized in plant area required. It has also made obsolete those large cooling towers and ponds associated with product cooling.

As in the case for shell and tube exchangers there are many excellent computer programs that can be used for the design of air coolers. A method given in Chapter 9 for such calculation may be used in the absence of a computer program or for a good estimate of a unit.

A general description and diagrams of the two types of air coolers which are forced air flow and induced air flow has already been given under item “Air Coolers” earlier in this part of the Handbook.

Thermal rating. Thermal rating of an air cooler is similar in some respects to that for a shell and tube described in the previous item. The basic energy equation

$$Q = U \, \Delta T \, A$$

is used to determine the surface area required. The calculation for U is different in that it requires the calculation for the air side film coefficient. This film coefficient is usually based on an extended surface area which is formed by adding fins to the bare surface of the tubes. Thermal rating, surface area, fan dimensions, and horsepower are calculation are given in full in Chapter 18 of this Handbook.

Condensers. In petroleum refining and most other chemical process plants vapors are condensed either in the shell side of a shell and tube exchanger, the tube side of an air cooler, or by direct contact with the coolant in a packed tower. By far the most common of these operations are the first two listed. In the case of the shell and tube condenser the condensation may be produced by cooling the vapor by heat exchange with a cold process stream or by water. Air cooling has overtaken the shell and tube condenser in the case of water as coolant in popularity as described in the previous item.

In the design or performance analysis of condensers the procedure for determining thermal rating and surface area is more complex than that for a single phase cooling and heating. In condensers there are three mechanisms to be considered for the rating procedure. These are:

- The resistance to heat transfer of the condensing film
- The resistance to heat transfer of the vapor cooling
- The resistance to heat transfer of the condensate film cooling

Each of these mechanisms is treated separately and along preselected sections of the exchanger. The procedure for determining the last two of the mechanisms follows that described earlier for single phase heat transfer. The following expression is used to calculate the film coefficient for the condensing vapor:

$$h_c^{.33} = \left[\frac{8.33^3 \times 10^2}{\mu_f}\right]^{0.33} \times k_f \times \frac{Sg_c}{(M_c/L_c \cdot N_s)}$$

where

h_c = Condensing film coefficient.
M_c = Mass condensed in lbs/hr
L_c = Tube length for condensation.
$= \frac{A_{zone} \times (L - 0.5)}{A}$
$N_s = 2.08\ N_t^{0.495}$ for triangular pitch.
k_f = Thermal conductivity of condensate at film temperature.
S_g = Specific gravity of condensate.
μ_f = Viscosity of condensate at film temperature in Cps

Again there are many excellent computer programs that calculate condenser thermal ratings, and these of course save the tedium of the manual calculation.

As in the case of the shell and tube exchanger and the air cooler a manual calculation for condensers is described in Chapter 18 of this Handbook. Again this is done to provide some understanding of the data required to size such a unit and its significance in the calculation procedure. Computer aided designs should however be used for these calculations whenever possible.

Heavy oil cracking

Up to the late 1980s feedstock to FCCU were limited by characteristics such as high Conradson carbon and metals. This excluded the processing of the "bottom of the barrel" residues. Indeed, even the processing of vacuum gas oil feeds was limited to

Conradson carbon < 10 wt%
hydrogen content < 11.2 wt%
metals N1 + V < 50 ppm

During the late 1980s significant research and development breakthroughs have produced a catalytic process that can handle residuum feed.

Feed stocks heavier than vacuum gas oil in conventional FCCU tend to increase the production of coke and this in turn deactivates the catalyst. This is mainly the result of:

- A high portion of the feed that does not vaporize. The unvaporized portion quickly cokes on the catalyst choking its active area.

- The presence of high concentrations of polar molecules such as poly-cyclic aromatics and nitrogen compounds. These are absorbed into the catalyst's active area causing instant (but temporary) deactivation.
- Heavy metals contamination that poison the catalyst and affect the selectivity of the cracking process.
- High concentration of poly-naphthenes that dealkylate slowly.

In FCCU that process conventional feedstocks cracking temperature is controlled by the circulation of hot regenerated catalyst. With the heavier feedstocks an increase in Conradson carbon will lead to a larger coke formation. This in turn produces a high regenerated catalyst temperature and heat load. To maintain heat balance therefore catalyst circulation is reduced leading to poor or unsatisfactory performance. Catalyst cooling or feed cooling is used to overcome this high catalyst heat load and to maintain proper circulation.

The extended boiling range of the feed as in the case of residues tends to cause an uneven cracking severity. The lighter molecules in the feed are instantly vaporized on contact with the hot catalyst, and cracking occurs. In the case of the heavier molecules vaporization is not achieved so easily. This contributes to a higher coke deposition with a higher rate of catalyst deactivation. Ideally the whole feed should be instantly vaporized so that a uniform cracking mechanism can commence. The *mix temperature* (which is defined as the theoretical equilibrium temperature between the uncracked vaporized feed and the regenerated catalyst) should be close to the feed dew point temperature. In conventional units this is about 20–30°C above the riser outlet temperature. This can be approximated by the expression:

$$T_M = T_R + 0.1 \text{ Delta } H_C$$

where

T_M = the mix temperature
T_R = riser outlet temperature °C
Delta H_C = heat of cracking in kJ/kg

This mix temperature is also slightly dependent on the catalyst temperature.

Cracking severity is affected by poly-cyclic aromatics and nitrogen. This is so because these compounds tend to be absorbed into the catalyst. Rising the mix temperature by increasing the riser temperature reverses the absorption process. Unfortunately, a higher riser temperature leads to undesirable thermal cracking and production of dry gas.

The processing of resid feedstocks therefore requires special techniques to overcome:

- Feed vaporization
- High concentration of polar molecules
- Presence of metals

Some of the techniques developed to meet heavy oil cracking processing are as follows:

- Two stage regeneration
- Riser mixer design and mix temperature control (for rapid vaporization)
- New riser lift technology minimizing the use of steam
- Regen catalyst temperature control (catalyst cooling)
- Catalyst selection for:
  - Good conversion and yield pattern
  - Metal resistance
  - Thermal and hydrothermal resistance
  - High gasoline RON

Conventional fluid catalytic crackers can be revamped to incorporate the features necessary for heavy oil (residual) cracking.

An important issue in the case of deep oil (residue) cracking is the handling of the high coke lay down and the protection of the catalyst. One technique that limits the severe conditions in regeneration of the spent catalyst is the two stage regeneration. Figure 11.11 in Chapter 11 shows this configuration.

The spent catalyst from the reactor is delivered to the first regenerator. Here the catalyst undergoes a mild oxidation with a limited amount of air. Temperatures in this regeneration remain fairly low around 700–750°C range. From this first regeneration the catalyst is pneumatically conveyed to a second regenerator. Here excess air is used to complete the carbon burn off and temperatures up to 900°C are experienced. The regenerated catalyst leaves this second regeneration to return to the reactor via the riser.

The technology that applies to the two stage regeneration process is innovative in that it achieves burning off the high coke without impairing the catalyst activity. In the first stage the conditions encourage the combustion of most of the hydrogen associated with the coke. A significant amount of the carbon is also burned off under mild condition. These conditions inhibit catalyst deactivation.

All the residual coke is burned off in the second stage with excess air and in a dry atmosphere. All the steam associated with hydrogen combustion and carry over from the reactor has been dispensed within the first stage. The second regenerator is refractory lined and there is no temperature constraint. The catalyst is allowed to come to equilibrium. Even at high regen temperatures under these conditions lower catalyst deactivation is experienced. The two stage regeneration technique leads to a better catalyst regeneration as well as a lower catalyst consumption. Typically the clean catalyst contains less than 0.05 wt% of carbon. This is achieved with an overall lower heat of combustion. Full details of this concept are given in Chapter 11. Since the unit remains in heat balance coke production stays essentially the same. The circulation rate of catalyst adjusts itself to any changes in coke deposition on the catalyst according to the expressions:

$$\text{coke make} = \text{delta coke} + c/o$$

and

$$\text{regenerator temperature} = \text{riser temperature} + C \times \text{delta coke}$$

where

delta coke = difference between the weight fraction of coke on the catalyst before and after regeneration.
c = unit constant (typically 180–230)
c/o = catalyst to oil ratio

In this regard a small circulation of extremely hot catalyst may not be effective as a large circulation of cooler catalyst. It has been found that there is a specific catalyst temperature range that is desirable for a given feed and catalyst system. A unique dense phase catalyst cooling system provides a technique through which the best temperature and heat balance relationship can be maintained.

Consider the enthalpy requirements for a FCC reactor given in the following table:

	Per pound of feed	
	Btu	%
Feed heating/vaporizing	530.0	69.00
Stripping steam enthalpy	5.0	0.65
Feed steam for dispersion	12.7	1.65
Feed water for heat balance	18.4	2.40
Heat of reaction	200.0	26.04
Heat loss	2.0	0.26
Total	768.1	100.00

It can be seen from this table that 69% of the enthalpy contained in the heat input to the reactor is required just to heat and vaporize the feed. The remainder is essentially available for conversion. To improve operation it would be desirable to allow more of the heat available to be used for conversion. The only variable that can be changed to achieve this requirement is the feed inlet enthalpy. That is through preheating the feed. Doing this, however, immediately reduces the catalyst circulation rate to maintain heat balance. This of course has an adverse effect on conversion. The preheating of the feed can, however, be compensated for by cooling the catalyst. Thus the catalyst circulation rate can be retained and in many cases can be increased. Indeed, by careful manipulation of the heat balance the net increase in catalyst circulation rate can be as high as 1 unit cat/oil ratio. The higher equilibrium catalyst activity possible at the lower regeneration temperature also improves the unit yield pattern. This is demonstrated in the following table:

Feedstock

°API	24.5
Conradson carbon	1.6

Yields

	Without catalyst cooling	With catalyst cooling
H_2S wt%	0.1	0.19
C_2- wt%	3.4	2.00
C_3 LV%	9.9	10.34
C_4 LV%	13.9	14.51
C_5+ (430:EP) LV%	58.2	60.87
LCO (650:EP) LV%	17.1	15.54
CLO LV%	8.6	8.10
Coke wt%	5.9	6.07
Conversion LV%	74.3	76.36

In summary, catalyst cooling will:

- Slightly increase unit coke
- Give a higher plant catalyst activity
- Be able to handle more contaminated feeds
- Improve conversion and unit yield
- Provide better operating flexibility

In residue cracking commercial experience indicates that operations at regenerated catalyst temperatures above 1,350°F result in poor yields with high gas production. Where certain operations require high regenerator temperatures the installation of a catalyst cooler will have a substantial economic incentive. This will be due to improved yields and catalyst consumption.

There are two types of catalyst coolers available. These are shown as Figure 11.14 in Chapter 11. They are:

- The back mix type
- Flow through type

Both coolers are installed into the dense phase section of the regenerator.

The back mix cooler. Boiler feed water flows tube side in both cooler types. The catalyst in the back mix cooler circulates around the tube bundle on the shell side.

The heat transfer takes place in a dense low velocity region so erosion is minimized. The back mix cooler can remove approximately 50 million Btu's/hr.

The flow through cooler. As the name suggests the catalyst flows once through on the shell side of this cooler. Again erosion is minimized by low velocity operation in the dense phase. This type of cooler is more efficient than the back mix. This unit can achieve heat removal as high as 100 million Btu's per hour.

Mix temperature control and lift gas technology. The equilibrium temperature between the oil feed and the regenerated catalyst must be reached in the shortest possible time. This is required in order to ensure the rapid and homogeneous vaporization of the feed. To ensure this it is necessary to design and install a proper feed injection system. This system should ensure that any catalyst back mixing is eliminated. It should also ensure that all the vaporized feed components are subject to the same cracking severity.

Efficient mixing of the feed finely atomized in small droplets is achieved by contact with a preaccelerated dilute suspension of the regenerated catalyst. Under these conditions feed vaporization takes place almost instantaneously. This configuration is shown in the diagram below:

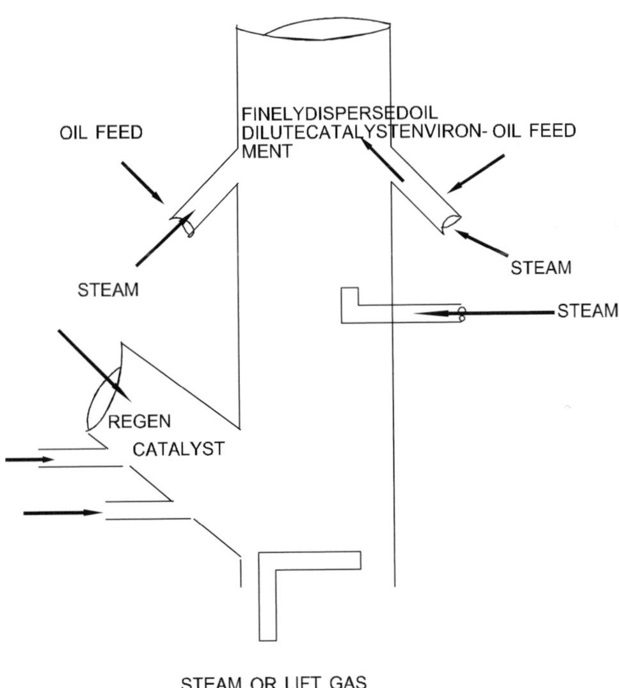

The regen catalyst stream from the regenerator is accelerated by steam or lift gas injection to move up the riser. The oil feed is introduced atomized by steam into the catalyst environment. The main motive steam into the riser is introduced below the feed inlet point. Good mixing occurs in this section with maximum contact between oil, catalyst, and the steam.

In residue cracking the proper selection of catalyst enables even the most bulky molecules to reach the active catalyst zone. Such zeolite catalyst have a high silica to alumna ratio which cracks the heavy molecules into sizes that can enter the active zone.

Efficient mixing of the catalyst and feed together with the catalyst selection ensures:

- Rapid vaporization of the oil
- Uniform cracking severity of the oil

Another problem that is met within residue cracking is the possibility of the heavier portion of the oil being below its dew point. The presence of poly-cyclic aromatics also affects cracking severity. Increasing the mix temperature to the riser temperature reverses the effect of poly-cyclic aromatics. In so doing, however, thermal cracking occurs which is undesirable. To solve this problem it is necessary to be able to control riser temperature independently of mix temperature. Mix Temperature Control (MTC) is achieved by injecting a suitable heavy cycle oil stream into the riser above the oil feed injection point. This essentially separates the riser into two reaction zones. The first is between the feed injection and the cycle oil inlet. This zone is characterized by a high mix temperature, a high catalyst to oil ratio and a very short contact time.

The second zone above the cycle oil inlet operates under more conventional catalytic cracking conditions. The riser temperature is maintained independently by the introduction of the regenerated catalyst. Thus an increase in cycle oil leads to a decrease in riser temperature, which introduces more catalyst, increases the mix temperature, and the catalyst to oil ratio, and decreases the regenerator temperature.

The lift gas technology. As described earlier it is highly desirable to achieve good catalyst / oil mixing as early and as quickly as possible. The method described to achieve this requires the preacceleration and dilution of the catalyst stream. Traditionally steam was the medium used to maintain catalyst bed fluidity and movement in the riser. Steam, however, has deleterious effects on the very hot catalyst that is used in residue cracking processes. Steam under these conditions causes hydrothermal deactivation of the catalyst.

Much work has been done in reducing the use of steam in contact with the hot catalyst. Some of the results of the work showed that if the partial pressure of steam is kept low, the hydrothermal effects are greatly reduced, in the case of relatively metal free catalyst. A more important result of the work showed that light hydrocarbons imparted favorable conditioning effects to the freshly regenerated catalyst. This was even more pronounced in catalysts that were heavily contaminated with metals.

Light hydrocarbon gases have been introduced in several heavy oil crackers since 1985. They have operated either with lift gas alone or mixed with steam. The limitations to the use of lift gas rests on the ability of downstream units to handle the additional gas. The following table compares the effect of lift gas in resid operation with the use of steam.

Feed: Atmospheric residue 4.3 wt% con carb.

Product distribution	Lift gas	Steam
C_2- wt%	3.2	4.0
C_3/C_4 LV%	11.4/15.1	11.6/15.4
C_5-Gasoline LV%	56, 9	55, 0
LCO + Slurry LV%	23, 9	24, 4
Total C_3 + LV%	107, 3	106, 4
Coke wt%	8, 6	8, 5
H_2 SCFB	70	89
H_2/C_1 Mol	0, 74	0, 85
Catalyst		
Material	←——same——→	
*SA, M_2/G	91	90
Ni + V WT ppm	7, 100	7, 300

*SA—surface area of equilibrium FCC catalyst, M_2/Gram.

As can be seen the use of lift gas as an alternative to steam gives:

- Lower hydrogen production
- Lower hydrogen/methane ratio
- Increase in liquid yield

Hot and cold flash separators

Hot and cold flash separation are used in high pressure, high temperature hydro cracking processes. The principal purpose is to recover high purity hydrogen for recycle in as economically a manner as possible. Secondly, as the high temperature and pressure cracker effluent is to be further processed, it is desirable to route the flashed liquid at as high a temperature as possible to the subsequent process unit.

Figure 11.15 of Chapter 11 of this Handbook shows the general configuration of a residue hydrocracker and visbreaking combination unit.

Bitumen feed from a crude vacuum distillation unit enters the reaction section of the hydro-cracker to be preheated by hot flash vapors in a shell and tube exchanger. A recycle and make up hydrogen stream is similarly heated by exchange with hot flash vapors. The hydrogen stream is mixed with the hot bitumen stream before entering the hydro-cracker heater. The feed streams are risen to the reactor temperature in the heater and leave to enter the top of the reactor vessel. The feed streams flow downwards through the catalyst beds contained in the reactor. Additional cold hydrogen is injected at various sections of the reactor to provide temperature control as the hydro-cracking process is exothermal.

The reactor effluent leaves the reactor to enter a hot flash drum. Here the heavy bituminous portion of the effluent leaves from the bottom of the drum while the lighter oil and gas phase leaves as a vapor from the top of the drum. This vapor is subsequently cooled by heat exchange with the feed and further cooled and partially condensed by an air cooler. This cooled stream then enters a cold separator operating at a pressure only slightly lower than that of the reactor. A rich hydrogen gas stream is removed from this drum to be amine treated and returned as recycle gas to the process. The distillate liquid leaves from the bottom of the separator to join a vapor stream from the hot flash surge drum (visbreaker feed surge drum). Both these streams enter the cold flash drum which operates at a much lower pressure than the upstream equipment. A gas stream is removed from the drum to be routed to the absorber in a distillate hydro-cracking unit. The liquid distillate from the drum is routed to the de-butanizer in the distillate hydro-cracking unit.

The visbreaker section of the unit takes as feed the heavy bituminous liquid from the hot flash drum. This enters the visbreaker furnace via a surge drum. The visbreaker heater has two parallel coils. The oil feed enters these coils to be thermally cracked to form some lighter products. The stream leaving the heater is quenched before entering a flash chamber. This vessel contains some baffled trays and a light gas and oil vapor stream leaves overhead. This stream is subsequently cooled and the distillate formed routed to the cold flash drum. The bottoms from the flash chamber may be fed to a visbreaker vacuum distillation unit where vacuum gas oil can be removed as feed to a fluid catalytic cracker unit. Alternatively this bottom product may be simply routed to an asphalt plant for suitable feed to an asphalt air blower or simple blended with cut-back material for marketable asphalt. Full details of this subject including the development of flashed stream compositions are given in Chapter 11.

Hydraulic analysis of process systems

This item deals with calculating a pressure profile of a process system. Such a calculation is used to size pipelines and to determine the pumping requirements for the system. The following calculation is an example of a typical system process engineers encounter in a design of a plant or in checking out an operating plant's process flow. The attached diagram H5 is used as the basis for this example.

Example calculation

$$\begin{aligned}
\text{Total flow to P-103 A\&B} &= 519{,}904 \text{ lbs/hr}\\
\text{API gravity} = 20.7 &= \text{Sg of } 0.930\ @\ 60^\circ\text{F}\\
\text{Stream temperature} &= 545^\circ\text{F}\\
\text{Sg } @\ 545^\circ\text{F} = 0.755 &= 6.287 \text{ lbs/gallon.}\\
\text{Gallons/min } @\text{ stream temp} &= \frac{519{,}904}{6.287 \times 60}\\
&= 1{,}378 \text{ gpm.}
\end{aligned}$$

viscosity of the oil @ 545°F = 1.2 cs
The suction line to P-103 A&B is 10″ sched 40.
From Table 19.A.1 in the Appendix Part 2.
Friction loss in feet /1,000 ft of pipe is 9.3 ft (equivalent to .30 psi/100 ft).

Suction line equivalent length

The following information would normally be obtained from a piping general arrangement drawing (a piping GA). For this calculation this information is fictional:

$$\begin{aligned}
\text{Number of standard elbows in the line} &= 8.\\
\text{Number of gate valves (all open) in the line} &= 3\\
\text{Total straight length of 10'' line} &= 85 \text{ ft.}
\end{aligned}$$

From Figure 19.B.2 in the appendix:

$$\begin{aligned}
\text{The equivalent length for 10'' elbows is 22 ft per elbow.} &= 22 \times 8 = 176 \text{ ft.}\\
\text{The equivalent length for 10'' gate valves is 5 ft per valve.} &= 5 \times 3 = 15 \text{ ft.}\\
\text{Total equivalent line length: } 85 + 176 + 15 &= 276 \text{ ft.}
\end{aligned}$$

Head loss to pump suction due to friction

$$= \frac{9.30 \times 276}{1{,}000} = 2.57 \text{ ft.}$$

In terms of pounds per square inch this is

$$= \frac{2.57 \times 62.2 \times .755}{144} = 0.83 \text{ psi.}$$

Pump suction pressure

$$\text{Source pressure at vacuum tower draw off} = 15\ \text{mmHg} \\ = 0.29\ \text{psia}$$

$$\text{Static head} = 45\ \text{ft.} = \frac{45 \times 62.2 \times .755}{144}$$
$$= 14.7\ \text{psia.}$$
$$\text{line loss} = 0.83\ \text{psi}$$

$$\text{Total pressure at the pump suction flange} = 0.29 + 14.7 - 0.83$$
$$= 14.21\ \text{psia. (which is} - 0.49\ \text{psig).}$$

Calculating the pressures at the pump discharge

Destination pressure at battery limits = 50 psig.
Temperature of the oil at battery limits = 140°F
Viscosity of the oil @ 140°F = 20 cs
S G of the oil @ 140°F = 0.900; lbs/ gal = 7.495.

From a material balance or from plant data: Flow of oil = 191,121 lbs/hr

$$\text{Flow rate} = \frac{191{,}121}{7.495 \times 60}$$
$$= 425\ \text{gpm.}$$

1.0 Line pressure drop from E-109 to battery limits
Equivalent length of line:

Straight line = 126 ft.
Number of elbows = 18 equiv length = 18 × 16 = 288 ft.
Number of gate valves = 6 equiv length = 6 × 3.5 = 21 ft.
Number of TEE's = 1 equiv length = 1 × 30 = 30 ft.
Total equivalent length = 465 ft.

Line to battery limits (BL) is a 6″ schedule 40. Then from Table 19.A.2 loss due to friction is 21.4 ft/ 1,000 ft which is equivalent to 0.83 psi/100 ft.

Then line friction loss is

$$\frac{21.4 \times 465}{1{,}000} = 9.95\ \text{ft.}$$

or

$$\frac{9.95 \times 62.2 \times 0.900}{144} = 3.86\,\text{psi.}$$

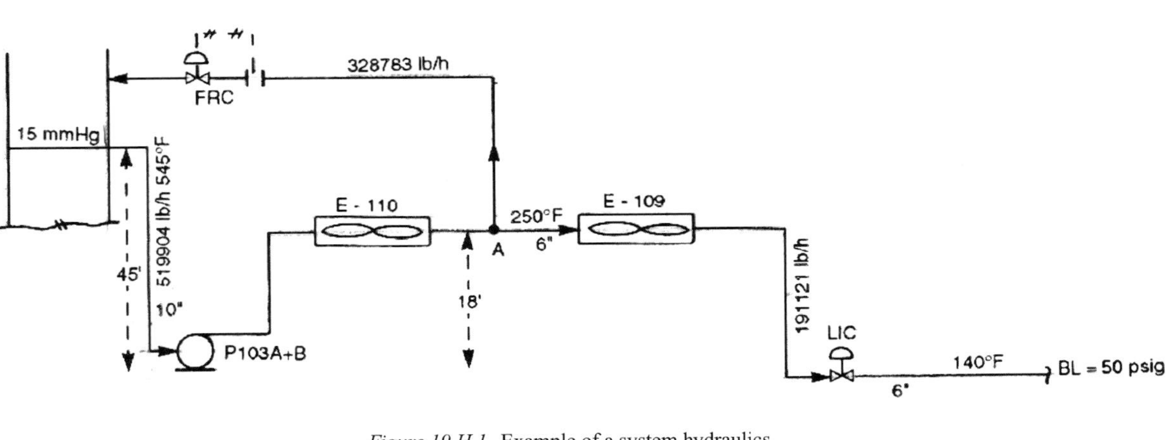

Figure 19.H.1. Example of a system hydraulics.

2.0 Control valve pressure drop

There is a battery limit Level Control Valve (LICV) between E-109 and the BL It is required to calculate the pressure drop for this valve at design flow. The rule of thumb given in Chapter 1.9 will be used to determine this. Thus, the pressure drop will be estimated as 20% of the circuit frictional pressure drop plus 10% of the static head of the receiving vessel.

The oil discharges into a surge drum of another downstream unit. This drum is pressurized by a blanket of inert gas. The net static head to this drum is 15 ft above valve outlet flange. This is equivalent to 6 psi.

The total line pressure drop for the whole circuit is estimated at 3 times that calculated above. (This will be checked and may be revised when the analysis of the whole circuit is complete). This pressure drop therefore is $3 \times 3.86 = 11.58$ psi for line losses. In addition to this line loss there are also two air coolers which have pressure drops as follows:

E-109 = 6 psi (from data sheets).
E-110 = 8 psi

Then total system pressure drop is estimated as:

$$11.58 + 14 = 25.58 \text{ psi.}$$

then control valve pressure drop is:

$$(0.1 \times 6.0) + (0.2 \times 25.58) = 5.2 \text{ psi}$$

3.0 Calculate pressure at point 'A' which is the reflux stream take off

Legnth of line between point 'A' and inlet to E-109 is as follows:

$$\begin{aligned} \text{Straight line} &= 121 \text{ ft} \\ \text{4 Elbows} = 4 \times 16 &= 64 \text{ ft} \\ \text{2 Valves} = 2 \times 3.5 &= 7 \text{ ft} \\ \text{Total 6'' sched 40 line equiv} &= 192 \text{ ft} \end{aligned}$$

Temperature of stream at this point is 250°F.

$$\text{Viscosity @ } 250^{\circ}\text{F} = 3.5 \text{ Cs}$$

SG @ 250°F is 0.860 which is 7.16 lbs/gal.

$$\begin{aligned} \text{Rate of flow into E-109} &= \frac{191{,}121}{7.16 \times 60} \\ &= 445 \text{ gpm.} \end{aligned}$$

Friction loss in 6″ pipe = 16.1 ft/1,000 ft

Total line loss:

$$\frac{192 \times 16.1}{1{,}000} = 3.09 \text{ ft or } 1.15 \text{ psi.}$$

Pressure at point 'A' therefore is the sum of:

Destination pressure at BL = 50 psig
Pressure drop for E-109 = 6 psi
Lossin line from E-109 to battery limits. = 3.86 psi

Loss in line from point 'A' to E-109 = 1.15 psi
Control valve pressure drop = 5.2 psi
Flow meter (not shown) say = 0.2 psi
Total pressure at point 'A' = 66.41 psig

4.0 Calculating the pressure at the pump discharge flange

Total flow from the pump is 519,904 lbs/hr

Oil temperature at outlet of E-110 is 250°F.

gpm of flow from E-110 is

$$\frac{519{,}904}{7.16 \times 60} = 1{,}210 \text{ gpm}$$

Line size at this point is 8″ sched 40.
Head loss in this line is 31.1 ft/1,000 ft.

Equivalent length of line:

Straight line = 82 ft
1 Tee = 30 ft
Total equiv legnth = 112 ft.

$$\text{Total head loss in line} = \frac{112 \times 31.1}{1{,}000}$$
$$= 3.5 \text{ ft or } 1.3 \text{ psi.}$$

Pressure at E-110 inlet will be:

Pressure at point 'A' = 66.41 psig.
Head loss in line = 1.3 psi
Pressure drop across 11-E-10 = 8 psi
= 75.7 psig

This pressure is 18 ft above grade as these air coolers are located above the pipe-rack. Allowing 1.5 ft from grade to pump center line, the static head at pump discharge flange is 16.5 ft.

Equivalent length of line from pump to E-110 is:

$$\begin{aligned}\text{Straight length} &= 155 \text{ ft}\\ \text{12 elbows} &= 12 \times 20.5 = 246 \text{ ft.}\\ \text{3 gate valves} &= 3 \times 4.8 = 14.4 \text{ ft}\\ \text{1 non return valve} &= 1 \times 51 = 51 \text{ ft}\\ \text{Total equivalent length} &= 466.4 \text{ ft}\end{aligned}$$

Head loss in 8″ sched 40 pipe at a flow rate of 1,378 gpm.
Pump temperature is 545°F
Viscosity at pump temperature is 1.2 Cs
SG at pump temperature is 0.755
Head loss = 28.8 ft/1,000 ft.

$$\begin{aligned}\text{Total line head loss} &= \frac{28.8 \times 466.4}{1{,}000}\\ &= 13.4 \text{ ft or } 4.38 \text{ psi}\end{aligned}$$

Then the pump discharge pressure is the sum of:

$$\begin{aligned}\text{Pressure at E-110 inlet} &= 75.7 \text{ psig}\\ \text{Line pressure drop} &= 4.38 \text{ psi}\\ \text{Static head (16.5 ft)} &= 5.38 \text{ psi}\\ \text{Total discharge pressure} &= 85.46 \text{ psig.}\end{aligned}$$

5.0 Calculating the pressures in the reflux line from point ‘A’

At point ‘A’ the pressure has been calculated as 66.41 psig.
Flow of the reflux stream (from the material balance) is:

$$519{,}904 - 191{,}121 = 328{,}783 \text{ lbs/hr.}$$

Temperature of the stream is 250°F.

$$\begin{aligned}\text{Viscosity @ 250°F} &= 3.5 \text{ Cs.}\\ \text{SG @ 250°F} &= 0.860 \text{ and lbs/gal is } 7.16\\ \text{Rate of flow} &= \frac{328{,}783}{7.16 \times 60}\\ &= 765 \text{ gpm}\end{aligned}$$

Line to tower from point ‘A’ is a 6″ sched 40 and head loss in this line is found to be 40.2 ft/1,000 ft.

Equivalent line lengths:
To the flow controller inlet flange:

$$\begin{aligned}\text{Straight line} &= 16 \text{ ft}\\ \text{2 elbows} &= 2 \times 16 = 32 \text{ ft}\\ \text{1 valve} &= 1 \times 3.5 = 7 \text{ ft}\\ \text{Total equivalent length} &= 81.5 \text{ ft}\end{aligned}$$

From the flow controller to tower:

$$\begin{aligned} \text{Straight line} &= 71 \text{ ft} \\ 6 \text{ elbows} &= 6 \times 16 = 96 \text{ ft} \\ 1 \text{ valve} &= 1 \times 3.5 = 3.5 \text{ ft} \\ \text{Total equivalent length} &= 171 \text{ ft} \end{aligned}$$

Total line length from point 'A' to tower:

$$81.5 + 171 = 252.5 \text{ ft}$$

Total head loss in line due to friction:

$$\frac{252.5 \times 40.2}{1{,}000} = 10.15 \text{ ft or } 3.77 \text{ psi}$$

The pressure required to deliver 765 gpm of reflux to the tower excluding the pressure drop across the control valve at this rate is the sum of the following:

$$\begin{aligned} \text{Destination pressure} &= 0.29 \text{ psia} \\ \text{Static head} &= 14.11 \text{ psi} \\ \text{Distributor (tower internals)} &= 2.0 \text{ psi (from data sheet)} \\ \text{Flow meter pressure} &= 0.5 \text{ psi (from data sheet)} \\ \text{Head loss in line} &= 3.77 \text{ psi} \\ \text{Total required} &= 20.67 \text{ psia} \\ &\quad \text{or } 5.97 \text{ psig} \end{aligned}$$

Then valve pressure drop at design flow of 765 gpm is:

Pressure at point 'A'—required pressure

$$\begin{aligned} &= 66.41\text{–}5.97 \text{ psig} \\ &= 60.44 \text{ psi} \end{aligned}$$

Note: Line pressure drops are given in Part 2 Appendix B.1.

Hydro-cracking (distillate)

Details of this process are given in Chapter 7. The following section is a description of a typical distillate hydro-cracker having a crude vacuum distillate as feed.

The diagram (Figure 19.H.2) shows a typical hydro-cracking unit in which the fresh feed is pumped through a series of feed/effluent exchangers to the inlet of the first reactor. The preheated feed is mixed with hot recycle gas at the inlet to the first reactor. In this case only the recycle gas passes through the fired heater although in some processes the combined feed and recycle gas are heated. The first reactor is

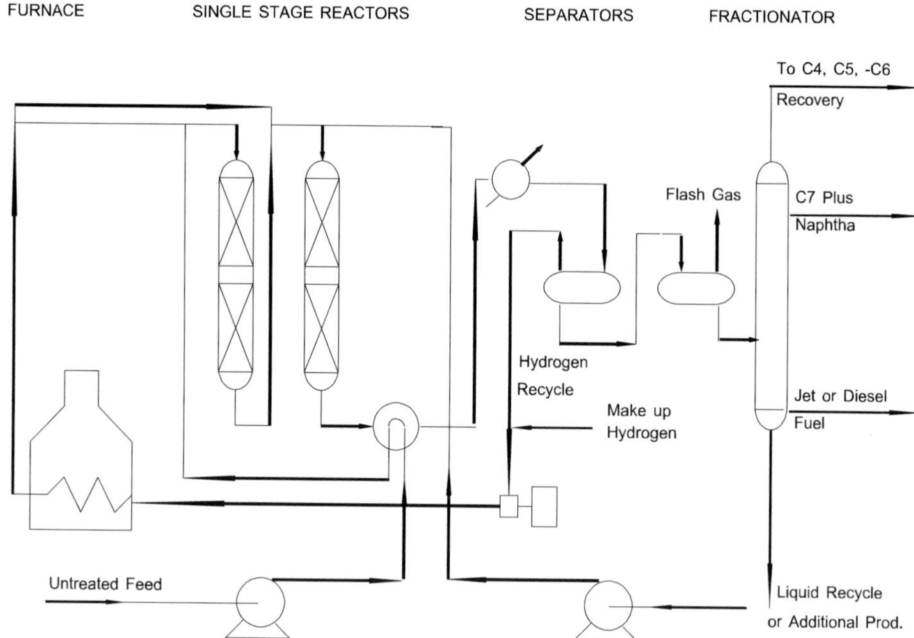

Figure 19.H.2. A distillate hydro cracker.

usually filled with a hydro-treating catalyst for the partial de-sulfurization and de-nitrogenation of the fresh feed. The catalyst employed for the hydro-treating is an alumna based type containing cobalt and molybdenum. To protect the catalyst from fouling by iron compounds and any salt in the feedstock, the first few feet of the reactor is usually used as a guard bed—some processes utilize an external guard reactor.

The hydrogenated feedstock is then mixed with additional hydrogen and passes to the hydro-cracking reactor, where cracking to the desired products takes place. The number of hydro-cracking reactor stages will either be one or two depending upon the products required.

The hydro-cracker reactor effluent is cooled by exchange with fresh feed and recycle hydrogen and is then flashed in the high pressure separator. The liquid from the high pressure separator passes to a further low pressure separator whilst the gas stream, which is rich in hydrogen, is recycled to the reactors. Make up hydrogen is added as required.

The liquid from the low pressure separator is then pumped to the fractionation train where products are separated. The unconverted portion of the fresh feed may then either be recycled for further cracking or used as product. Flash gas from the low pressure separator is usually treated to remove the acidic components before being sent to the refinery fuel gas system.

Typical feeds include atmospheric and vacuum virgin gas oils, catalytic cycle oils, deasphalted oil, coker gas oil, thermal gas oil, and paraffin raffinates. Heavy feed stocks such as vacuum gas oils are usually limited to an end point of about 1,050°F.

Products range from LPG, naphtha, gasoline through to lube oils; the more usual being naphtha, gasoline, and diesel fuel. It is possible to vary the products from a hydro-cracking unit by merely changing the reactor operating conditions and the fractionation cut point. This flexibility of operation is one of the major factors in favor of hydro-cracking.

The overall hydro-cracking reaction is exothermal. Temperature control of the reactor is accomplished by the introduction of cold hydrogen quench streams at the appropriate locations in the reactor.

The hydro-cracker reactor side

This side of the hydro-cracker is licensed to oil refinery clients. The data requirements therefore are proprietary to the licensor and are provided to clients under the licensing agreement. Use and disclosure of these data are limited to the various clients and/or client's engineering contractor personnel who require such data to design or operate the process.

Operating variables are available to client personnel in the form of graphs or curves. For example and referring to the process diagram H6 some of these would be:

R1 reactor temperature response

- Effect of feed end point and density on required temperature
- Effect of feed nitrogen content on required temperature
- Effect of temperature on R1 product nitrogen content
- Effect of feed rate on required temperature
- Effect of recycle hydrogen purity on required temperature

R2 reactor temperature response

- Effect of feed end point and density on required temperature
- Effect of feed nitrogen content on required temperature
- Effect of R1 product nitrogen on required temperature
- Effect of feed rate on required temperature
- Effect of temperature on conversion to diesel and lighter
- Effect of temperature on diesel/gas oil cut point

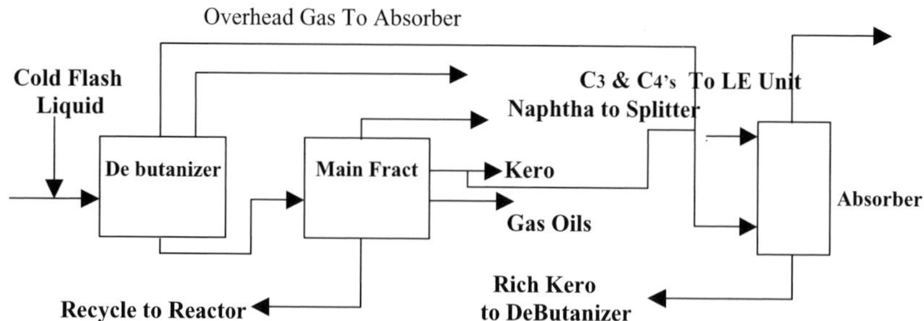

Figure 19.H.3. Block flow diagram of a typical hydro cracker recovery side.

- Effect of recycle hydrogen purity on required temperature
- Effect of gas to oil ratio on required temperature

These curves are made available from the section supervisors as required and under strict security. Discussion on the effect of certain process variables and the mechanism of hydro-cracking are provided in Chapter 7.

Hydro-cracker recovery side

A block flow diagram of a typical hydro-cracker product recovery is given in Figure 19.H.3.

The feed to the main fractionator contains the de-butanized effluent from the hydro-cracker reactor side and the cold flash liquid. The absorber rich liquid is also fed to the de-butanizer. This absorber rich liquid is kerosene saturated with some light gas and most of the propanes and butanes in the de-butanizer overhead gas stream.

The de-butanizer bottom stream is preheated in a fired heater before entering the flash zone of the main fractionator. In this fractionator a full range naphtha stream is taken off overhead. Kero and gas oils are taken off as the two side streams. Kero is stripped by reboiling while the gas oil is steam stripped. A small stream of unconverted oil leaves as the tower bottom product. This is returned to the reactors as recycle.

The ASTM distillation lab data for all streams, together with plant data giving their respective flows are available and can be used to develop a TBP curve for the main fractionator feed. The procedure is given by the following steps:

Step 1. Calculate the TBP curve for light naphtha, heavy naphtha, kero, diesel, and UCO from their respective ATSM distillation data.
Step 2. Plot each of the TBP curves as shown in Figure 19.H.4.
Step 3. Starting with the naphtha divide the TBP curve into about six boiling point fractions.

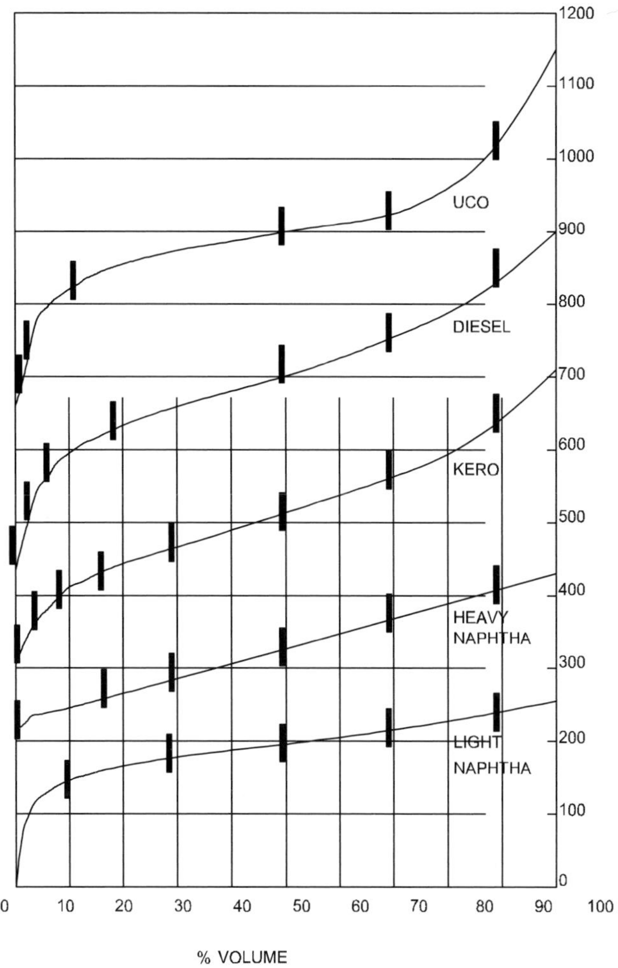

Figure 19.H.4. The TBP of the cuts.

Step 4. Divide the heavy naphtha curve in a smaller manner. In this case, however, the cut points of the fractions in the light naphtha that are also in the heavy naphtha must be identified. These must also be a fraction in the heavy naphtha. For example if the penultimate fraction of the light naphtha has a cut range 220–240°F, and the heavy naphtha starts with a TBP IBP of 220°F, then the first fraction of the heavy naphtha must have an IBP of 240°F.

Step 5. Divide the kero, diesel, and UCO TBP curves in the same manner as the heavy naphtha described in Step 4. Make sure that all common fractions are identified and measured in each product.

Step 6. From the plant flow data calculate the yield of each stream as a volume percent on total feed.

Step 7. List the cut points for each fraction developed in Steps 3 through 5. Against each cut point calculate its percentage on total feed that will be in all 5 product streams. That is, if cut points 180–190°F in light naphtha is 20 vol% on its TBP and light naphtha is 5.06 vol% on total feed then this component with cut range to 180–190°F will be 0.2 g× 5.06 vol% on total feed.

Step 8. Step 8 List the percent of each fraction on feed for all 5 products. Then by adding them horizontally the total of each cut point fraction on total feed is established.

Step 9. Plot the cumulative volume percent from Step 8 against cut point. This is the fractionator feed TBP curve given as Figure 19.H.5.

An example calculation now follow:

Example calculation

The following are lab and plant data from a test run on the recovery of the DHC.

Lab results

	ASTM distillation °F			All converted to D86 °F		
	Light nap	Hy nap	Kero	Diesel	UCO	Total
IBP	27.9	131.5	193.3	261	365.5	
10	79.7	139.2	231.3	324	433	
30	88.8	152.8	247.0	346	447.5	
50	94.1	164.9	264.0	361	455.1	
70	100.7	176.9	283.3	381	463.0	
90	109.2	194.8	317.7	424	510.0	
FBP	114.8	210.3	354.8	455	N/A	
°API	72.6	60.7	46.3	28.3	16.5	
Plant data flows m3/hr	3.38	13.06	24.74	22.78	2.90	66.86

Calculate liquid TBP curve

LV% Yield

LY naphtha	5.06
HY naphtha	19.53
Kero	37.00
Diesel	34.07
UCO	4.34
Total	100.00

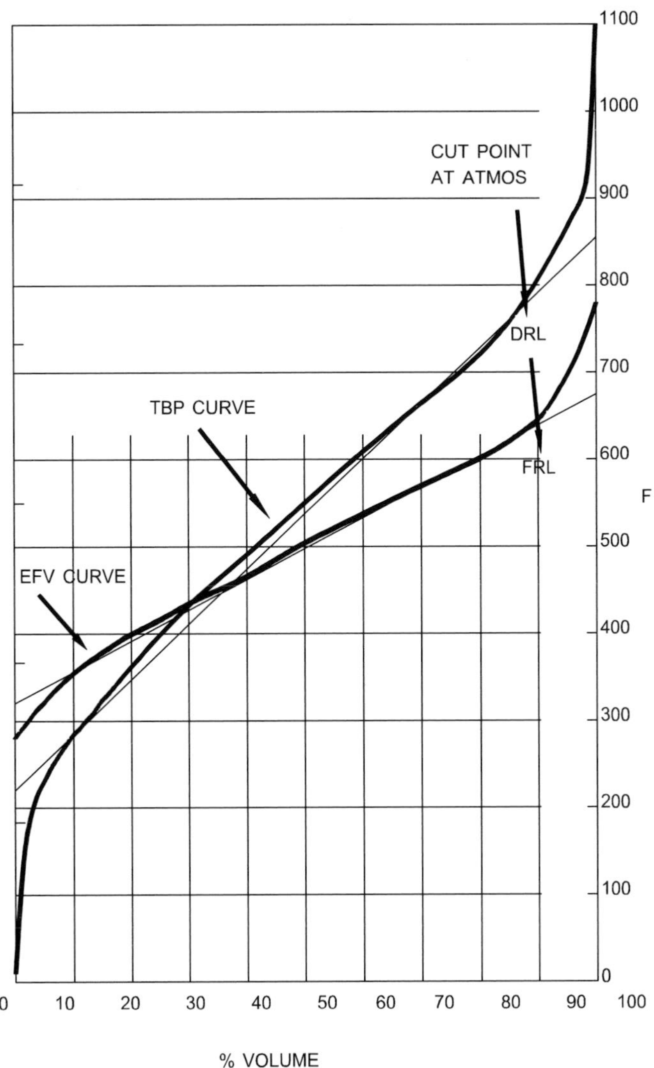

Figure 19.H.5. TBP and EFV curves for the main fractionator feed.

Light naphtha

	ASTM		TBP	
	°F	ΔT	ΔT	°F
IBP	82			12
10	176	94	134	146
30	192	16	32	178
50	201	9	17	195
70	214	13	22	217
90	228	14	20	237
FBP	239	11	14	251

Heavy naphtha

	ASTM		TBP	
	°F	ΔT	ΔT	°F
IBP	270			216
10	282	12	27	243
30	307	25	46	289
50	329	22	37	326
70	351	22	34	360
90	383	32	42	402
FBP	410	27	31	433

Kero

	ASTM		TBP	
	°F	ΔT	ΔT	°F
IBP	379			310
10	448	69	104	414
30	477	29	51	465
50	507	30	47	512
70	541	34	48	560
90	604	63	73	633
FBP	671	67	74	707

Diesel

	ASTM		TBP	
	°F	ΔT	ΔT	°F
IBP	502			439
10	615	113	154	593
30	655	40	65	658
50	682	27	43	701
70	718	36	51	752
90	795	77	85	837
FBP	851	56	61	898

UCO

	ASTM		TBP	
	°F	ΔT	ΔT	°F
IBP	690			662
10	811	121	162	824
30	838	27	49	873
50	851	13	24	897
70	865	14	22	919
90	950	85	93	1012
FBP	00			00

Composite curve for the main fractionator feed

Cut points	% Volume on feed						Cumulative
°F	LT nap	HY nap	Kero	Diesel	UCO	Total	total
12 to 149	0.51					0.51	0.51
to 180	1.01					1.01	1.52
to 190	1.01					1.01	2.53
to 220	1.01					1.01	3.54
to 240	1.01	0.59				1.60	5.14
to 255	0.51	2.72				3.23	8.37
to 285		2.54				2.54	10.91
to 320		3.91	0.37			4.28	15.19
to 365		3.91	1.11			5.02	20.21
to 405		3.91	1.85			5.76	25.97
to 432		1.95	2.59			4.54	30.51
to 465			5.18	0.34		5.52	36.03
to 510			7.40	0.34		7.74	43.77
to 560			7.40	1.36		8.76	52.53
to 635			7.40	4.43		11.83	64.36
to 710			3.70	11.25	0.04	14.99	79.35
to 755				6.13	0.09	6.22	85.57
to 835				6.81	0.39	7.20	92.77
to 900				3.41	1.65	5.06	97.83
to 920					0.87	0.87	98.70
to 1,010					0.87	0.87	99.57
to 1,160					0.43	0.43	100.00
Totals	5.06	19.53	37.00	34.07	4.34	100.00	

Hydrogen

The development of the catalytic reformer and the significant amount of hydrogen provided by the process gave rise to the hydro skimming refinery that is most common in present day petroleum refining. In this type of refinery the produced reformer hydrogen stream is used as the major component for the following treating processes:

- Hydro-desulfurizing
- Kerosene de-aromatization
- Olefin saturation
- Lube oil de colorization

Details of catalytic reforming are given in Chapter 5 while the major hydro-treating processes are discussed in Chapter 8. The availability of hydrogen from the catalytic reformer continued to support the hydro-skimming refinery comfortably until the development of the hydro-cracker process. The availability of hydrogen from the traditional source of catalytic refining fell far short both in quantity and in sustained purity to satisfy this new process. To overcome this shortfall in refineries whose process configuration included hydro-cracking, Hydrocarbon conversion to hydrogen and CO were installed. The most common of these is the deep conversion of naphtha or light hydrocarbon gases. These processes contained a fired reactor whose tubes contain a catalyst. The feed with a quantity of steam are introduced in this reactor. The hydrocarbon is reduced to its basic components of carbon as CO and hydrogen. A series of shift reactors and a methanation reactor produces the rich hydrogen stream required by the hydro-cracker process. As some sulfur removal is required, this is usually accomplished by traditional means (i.e., amine absorption, or hot potassium carbonate, or in some cases molecular sieves).

Hydro-treating

Naphtha hydro-desulfurization. This uses cat reformer hydrogen or similar on a once through basis. Heavy naphtha feed to the cat reformer is fed to the naphtha hydro-desulfurizer from storage. The feed stream and the hydrogen gas stream are preheated by exchange with the hot reactor effluent stream. The feed then enters the fired heater which brings it up to the reactor temperatures (about 450°F) and leaves the heater to enter the reactor which operates at about 400–450 psig. Sulfur is removed from the hydrocarbon as hydrogen sulfide in this reactor and the reactor effluent is cooled to about 100°F by heat exchange with the feed. The cooled effluent is collected in a flash drum where the light hydrogen rich gas is flashed off. This gas enters the suction side of the booster compressor which delivers it to other hydro-treaters. The liquid phase from the drum is pumped to a reboiled stabilizer. The overhead vapor stream from the stabilizer is routed to fuel while the bottom product, cat reformer feed, is pumped to the cat reformer.

Gas oil hydro-desulfurizer. This process uses a recycled hydrogen stream to de-sulfurize a gas oil feed. The flow sheet, Figure 19.H.6, shows the gas oil feed entering

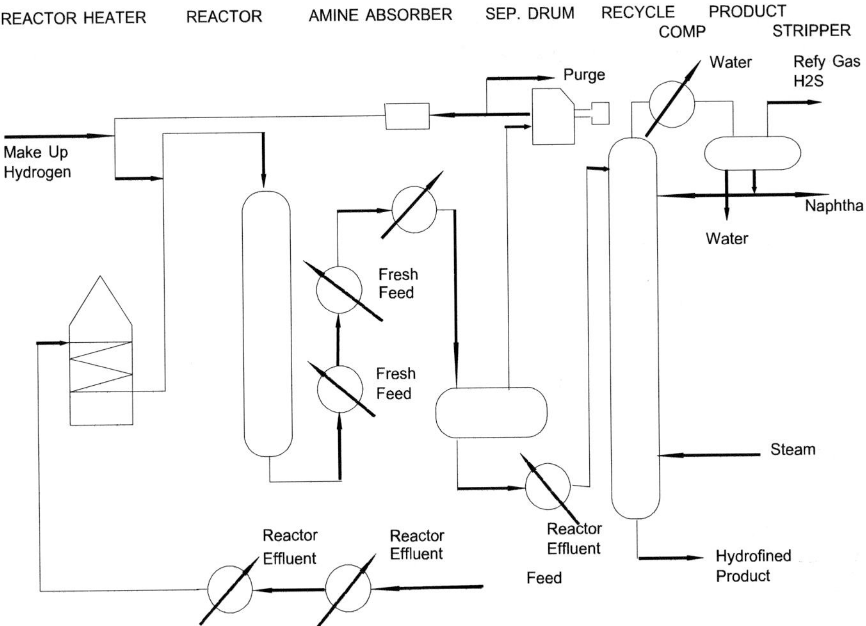

Figure 19.H.6. A typical gas oil hydro de-sulfurizer.

the unit to be preheated with hot effluent stream before entering a fired heater. Its temperature is increased to the reactor temperature of about 750°F in this heater. A hydrogen rich stream is introduced at the coil outlet prior to the mixed streams entering the reactor. The reactor contains a bed of cobalt molybdenum on alumna catalyst and de-sulfurization takes place over the catalyst with 70–75% of the total sulfur in the oil being converted to H_2S.

The reactor effluent is cooled by the cold feed stream, water, or air. This cooled effluent enters a flash drum where the gas phase and liquid phase are separated. The gas phase rich in H_2S and hydrogen enters the recycle compressor. The gas stream then enters an amine contactor where the H_2S is absorbed into the amine and removed from the system. Although the diagram shows a purge stream before the amine absorber in most cases the purge is down stream after the amine cleanup. The purged gas is replaced by fresh hydrogen-rich make up stream thus maintaining the purity of the recycle gas.

The liquid phase leaving the flash drum is preheated before entering a stream stripping column where the light ends created in the process are removed as overhead products. The bottom product leaves the tower to be cooled and stored.

Reactor conditions at start of run. Reactor conditions required for the proper operation of catalytic units are dependent on the type of catalyst used. In hydro-treating therefore reactor side conditions and operation are proprietary to the licensor of the process. However, the following data is an average and does not reflect any particular licensors' process.

The variables for hydro-treater operation are:

- Reactor inlet temperature
- Reactor pressure
- Recycle rate required
- Purity of the inlet hydrogen stream
- Space velocity

These change with the degree of de-sulfurization required, feed cut and quality, and the catalyst age.

There are typical conditions for the reactor side which are in the public domain:

Straight run naphtha de-sulfurization

Reactor inlet temperature — 650°F.
Reactor pressure — 500–600 psig
Space velocity v/v/hr — 4–6
Hydrogen purity at inlet — 75% mole
Recycle rate scf/bbl — 500–550

De-sulfurization to $<$ 5 ppm total sulfur.

Diesel de-sulfurization

Reactor temperature — 650°F
Reactor pressure — 750–1,000 psig
Space velocity v/v/hr — 1.5–2
Recycle rate scf/bbl — 600–700
De-sulfurization: — remove 85% sulfur in feed.

Hydrogen consumption in these units can be calculated and this depends to a large extent only on feed quality. That is the amount of sulfur (and nitrogen in some cases) and the presence of olefinic material.

Predicting hydrogen consumption in naphtha hydro de-sulfurization

This item describes a method of predicting the hydrogen consumed in hydro-treating straight run naphtha. Emphasis is based on SR naphtha. Treating cracked naphtha and gas oils need approximately three to five times the amount of hydrogen than straight run.

The method for predicting hydrogen consumption now follows with the steps described below:

Step 1. From the naphtha TBP and the assay estimate the weight percent sulfur in the naphtha feed (see Item 1.4).
Step 2. Establish the throughput of the naphtha. Assume the feed will be completely desulfurized in the process.
Step 3. Calculate the moles of sulfur removed by the chemical equation

$$H_2 + S = H_2S$$

Fix the moles of hydrogen required to satisfy the reaction given above. That is moles hydrogen equals moles sulfur removed.
Step 4. Estimate the amount of hydrogen required to saturate the hydrocarbon chain or ring *after* the sulfur molecule is removed. In naphtha fractions the sulfur compounds are relatively simple in structure. A figure of 2 or 3 times the hydrogen used for sulfur removal is reasonable for resaturation. Remember this only applies to straight run feeds.
Step 5. The remaining consumption of hydrogen is to replace the hydrogen lost from the system in solution with the liquid product leaving. This can be a very significant quantity. Calculate using Steps 6 through 10.
Step 6. Establish the component analysis of the make up gas. This is usually catalytic reformer off gas. Calculate the amount in moles/hr of each component that satisfies the chemical reaction quantity calculated in Steps 3 and 4.
Step 7. Let x moles/hr be the hydrogen that leaves in solution with the product. Calculate in terms of x the proportion of the other components related to the hydrogen that also leave in solution with the product. This is each component's moles divided by the hydrogen component and multiplied by x.
Step 8. Add the C_1 through C_5's portion of the make up gas to the x component's calculated in Step 7.
Step 9. The quantity calculated in Step 8 plus the moles of naphtha product is the liquid phase that leaves the separator (i.e., the unstabilized product). By definition this is in equilibrium with the gas phase that leaves the separator drum. Calculate, using the liquid phase composition, its bubble point at the separator drum conditions.

Step 10. Again by definition the y factor calculated in Step 9 (i.e., $y = Kx$) is the composition of the vapor phase leaving the separator drum. As $\Sigma y = \Sigma Kx$ then equate and solve for x as moles hydrogen leaving in solution with the hydro-treated product.

Step 11. The gas phase composition is calculated by substituting the calculate value for x in Step 10. If the unit has a recycle gas stream, this is the composition of the recycle gas.

Step 12. The total hydrogen and hydrogen gas stream make up to the unit can now be completed with the addition of the moles in solution calculated in Step 10.

Predicting the hydrogen consumption in gas oil hydro-treating—with gas purge

Diesel hydro-treating is in many ways similar to naphtha hydro-treating. However, because of a more complex molecular structure in the diesel fraction and more complex sulfur compounds some additional consideration must be made in this case. There will be a higher quantity of sulfur and these will contain disulphides and thiophenes which are complex ring compounds of sulfur. More light ends are made in the process when these compounds are broken to release the sulfur.

Because of the quantity of sulfur released as H_2S there is every probability that the recycle gas will require amine treating to remove this H_2S and thus retain its purity. There is also a need to purge off some of the recycle gas and replace it with fresh catalytic reformer (or high hydrogen content gas). This will be so if the light end is high and the subsequent purity of the recycle gas diminished.

In this item consideration has been given to amine treating and purging the recycle. The purging is an added item to be satisfied by the make up gas consumption. The method of predicting hydrogen consumption and recycle gas purity is given by the following steps:

Step 1. From the feed TBP and assay calculate its sulfur content.

Step 2. Set the degree of de-sulfurization required. This will depend on catalyst and reactor conditions. Most well designed modern diesel hydro-treaters can de-sulfurize heavy gas oil to remove at least 85 wt% of its sulfur content. In the case of automotive diesel, current requirements for ultra-low sulfur diesel (ULSD) are expected to bring sulfur specifications down to <50 ppm. This, however, varies from country to country, with many still in the 300–500 ppm range or even higher.

Step 3. Establish the feed throughput. This will depend on the space velocity required for Step 2.

Step 4. Estimate the light ends produced in the process. This can be done by referring to plant records. As a rule of thumb this can be in the form of C_5+ naphtha at about

6–10% by vol of feed. Some C_5's and lighter are also formed but these are usually in small quantities.

Step 5. As in the method used for naphtha hydro-treating, calculate the hydrogen required to remove the sulfur molecules. Again add 2 times this quantity to saturate the compounds that contained the sulfur.

Step 6. The light ends formed through the minor cracking to release thiophenes and disulfides will need to be saturated. This consumes hydrogen. Approximately 2 moles of hydrogen will be required for this purpose per mole of light ends formed. Using past lab tests on the hydro-treater naphtha develop the TBP and split to pseudo components. Allocate mole weights and gravities to these components to arrive at a number of moles/hr for the light ends. Calculate the hydrogen consumption.

Step 7. The remaining hydrogen requirements will be to replace hydrogen lost in solution with the liquid product and of course that lost in the purge stream. Commence with the calculation to determine solution loss as follows.

Step 8. Establish the component analysis of the make up gas. This is usually catalytic reformer off gas or if the naphtha hydro-treater is operating on a once through gas basis it will be the off gas from that unit.

Step 9. Calculate the amount in moles/hr of each component associated in the make up gas with the hydrogen required for the chemical reactions calculated in Steps 5 and 6.

Step 10. Let x moles/hr be the hydrogen that leaves in solution with the liquid product from the separator and the purge gas. Calculate in terms of x the proportion of the other components in the make up gas associated with the hydrogen.

Step 11. Add the C_1–C_5 portion of the make up gas to the x components calculated in Step 10. Add also the C_5+ naphtha components which were made in the process and of course a guess at the number of moles of H_2S that will be in the liquid phase of the separation drum. To do this look at the "K" (equilibrium constant) for H_2S at drum conditions. Use this to estimate its proportion in liquid. For example if $K = 1$ then the split will be close to 50% in liquid and 50% in vapor.

Step 12. Set the amount of purge in terms of its proportion to the liquid product (that is set the V/L for flash vaporization). This will be such as to provide a recycle gas hydrogen content of above 63% mole after H_2S removal. This figure is trial and error. Start with $V/L = 0.1$.

Step 13. Carry out a flash calculation in terms of x and using V/L set in Step 12. Solve for x. The vapor steam from this calculation is the purge gas in terms of moles/hr and composition. It also is the composition of the recycle gas.

Step 14. Complete the calculation for hydrogen consumption and make up gas using the value for x above.

An example calculation now follows.

Example calculation

Feed to diesel hydro-treater in this case will be heavy gas oil.

Gas oil cut = 610°F → 690°F Kuwait crude
Mid boiling point 650°F.
Unit throughput = 5,500 BPSD (blocked operation)
From assay sulfur content = 2.1 wt%

De-sulfurization shall be 85%.

lbs/hr gas oil = 70,078
Sulfur in feed = 1,472 lbs/hr
Sulfur removed = 1,251 lbs/hr

$$\text{Hydrogen for sulfur removal} = \frac{1{,}251}{32} = 39 \text{ moles/hr}$$

7 vol% of C_5+ naphtha is produced in the reaction.

Say this has the following composition:

	Vol%	BPSD	GPH	lbs/hr	MW	Moles/hr
C_6	18.8	72.4	127	702	86	8.2
C_7	23.6	90.9	159	911	100	9.1
C_8	27.1	104.3	183	1077	114	9.4
C_9	13.2	50.8	89	535	128	4.2
C_{10}+	17.3	66.6	117	715	142	5.0
	100.0	385	675	3940		35.9

Approximately 2 moles of H_2 per mole will be required to saturate light ends as they are produced.

Then this hydrogen = 35.9 × 2 = 71.8
Then total hydrogen make up will be
Sulfur removal = 39 moles/hr
Saturating after de-sulfurizing = 78 moles/hr
Light ends = 71.8
Total = 188.8 moles/hr

In addition there will be losses out of the system by H_2 in liquid solution and a purge stream. See following diagram

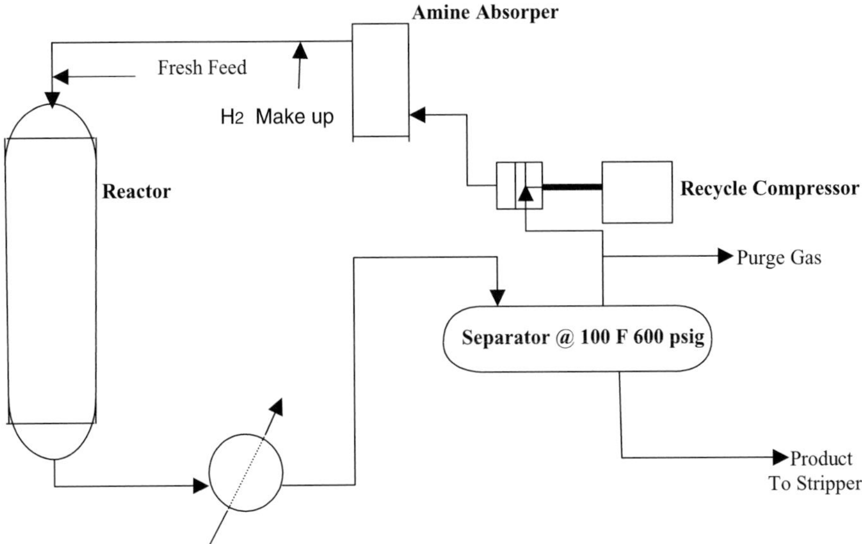

Total make up gas for chemical reaction:

	Mole fract*	Moles/hr
H_2	0.697	188.8
C_1	0.093	25.2
C_2	0.089	24.1
C_3	0.074	20.0
iC_4	0.016	4.3
nC_4	0.020	5.4
iC_5+	0.011	3.0
	1.000	270.8

* From a catalytic reformer.

Let x moles/hr be H_2 in solution lost from the system in liquid solution and purge. Total make up gas then is:

	Consumed in reaction	Loss in solution/purge	Total make up		
H_2	188.8	X	188.8	+	X
C_1	25.2	0.134 X	25.2	+	0.133 X
C_2	24.1	0.128 X	24.1	+	0.128 X
C_3	20.0	0.106 X	20.0	+	0.106 X
iC_4	4.3	0.023 X	4.3	+	0.023 X
nC_4	5.4	0.029 X	5.4	+	0.028 X
C_5+	3.0	0.016 X	3.0	+	0.016 X
	270.8	1.436 X	270.8	+	1.434 X

Set the purge to be 15% mole of liquid to stripper. Thus V/L = 0.15 (1st trial).

Calculate flash for effluent in terms of x at V/L = 0.15. Solve for x. Thus

		615 PSIK	V/L = 0.15				
	x_F effluent	K 100°F	V/L K	$L = \frac{X_F}{1+VLK}$	Liquid to stripper L moles/hr	Purge gas V moles/hr	Recycle gas composition
H_2	x	32	4.8	0.172 x	7.15	34.42	62.85
C_1	25.2 + 0.133 x	3.8	0.51	16.05 + 0.085 x	19.58	11.15	20.36
C_2	24.1 + 0.128 x	1.2	0.18	20.42 + 0.108 x	24.91	4.51	8.23
H_2S	19.0	1.0	0.15	16.52	16.52	2.48	4.53
C_3	20.0 + 0.106 x	0.5	0.075	18.60 + 0.099 x	22.72	1.69	3.09
iC_4	4.3 + 0.023 x	0.25	0.0375	4.14 + 0.022 x	5.05	0.21	0.38
nC_4	5.4 + 0.028 x	0.19	0.0285	5.25 + 0.027 x	6.37	0.19	0.35
C_5	3.0 + 0.016 x	0.085	0.0128	2.96 + 0.016 x	3.63	0.04	0.07
C_6	8.2	0.044	0.0066	8.15	8.15	0.05	0.09
C_7	9.1	0.020	0.0030	9.07	9.07	0.03	0.05
Oil	242	–		242	242	Nil	–
Total	360.3 + 1.434 x			343.16 + 0.529 x	365.15	54.77	100.00

$$\frac{(360.3 + 1.434\,x) - (343.16 + 0.529\,x)}{(343.16 + 0.529\,x)} = 0.15$$

$$17.14 + 0.905\,x = 51.474 + 0.0794\,x$$

$$x = \frac{34.334}{0.826} = 41.57 \text{ moles/hr.}$$

Substituting in the table above total effluent (less recycle) = 419.92 moles/hr.

The recycle gas H_2 content will be 62.9 mole% before H_2S removal.

If there is an amine contactor in the system H_2 purity of the gas becomes 65.8 mole%, which is quite good. This gave a purge steam of 36.59 moles/hr.

And the recycle gas purity was 61.2 mole% with no H_2S removal. With H_2S removal this became 64.2 mole%, which is borderline.

Total hydrogen to the plant at a purge ratio of 0.15 will be:

Sulfur removal	= 39 moles/hr
Saturating	= 78 moles/hr
Light ends	= 71.8 moles/hr
Hydrogen in liquid	= 7.2 moles/hr
Hydrogen in purge	= 34.4 moles/hr
	= 230.4 moles/hr

$$\text{Make up gas stream from catalytic reformer} = \frac{230.4}{0.697} = 330.5 \text{ moles/hr}$$

or 545 SCF/136L of fresh feed

Estimating the hydrogen consumption for olefin separation

This item gives a simple method for estimating the hydrogen consumption when treating cracked stocks. This method uses the correlation

$$\text{BR number} = \frac{\text{Percent olefins} \times 160}{\text{MW olefins}}$$

Where BR number is the bromine number of the stock. This can be obtained by lab analysis. The mole weight of the olefins is estimated as 1.3 times the mole weight of the paraffin with the same mid boiling point of the cut.

The hydrogen consumption for saturation is taken at 6.5–8.0 SCF hydrogen per barrel for every unit of bromine number reduction that occurs in the process. Bromine reduction in hydro-treating is estimated as follows

LT cracked naphtha	— 90–100% reduction
FCCU cycle oil	— 12 unit reduction
Coker gas oil	— 40 unit reduction
Visbreaker gas oil	— 40–45 unit reduction

Use 6.5 SCF/Bbl per bromine unit reduction for naphtha and light oil and 8.0 for the heavier feeds.

A sample calculation now follows:

Example calculation

Consider a cracked naphtha 180–380°F cut

API = 51.0
% S = 0.89

Pona analysis given 30% olefins by vol.

Calculate bromine number from the following.

$$\text{BR no.} = \frac{\%\,\text{olefins} \times 160}{\text{MW olefins}}$$

Mole wt of Olefins 1.3 times mole weight paraffin with same boiling point as the Cuts mid BPT.

In this case mid BPT is 200°F = Heptane C_7 Mole wt 100

$$\text{Olefin mole wt} = 130$$

$$\text{Then bromine no.} = \frac{30 \times 160}{130} = 37$$

Olefin saturation is 6.5 SCF hydrogen/Bbl per unit of bromine number reduction. With naphtha this is usually about 95%.

$$\text{Then hydrogen required} = 6.5 \times 37 \times 0.95 \text{ SCF/Bbl}$$
$$= 228.5 \text{ SCF/Bbl of Feed}$$

Further details including reaction mechanism of hydro-treating are discussed and described in Chapter 7 of this Handbook.

I

i-component

The “i” as a prefix to a chemical component indicates that the compound is an isomer. There are several isomers in petroleum refining and the most common relate to the light components of the structure. Notably these are:

iso-Butane—iC_4
iso-Pentane—iC_5
iso-Hexane—iC_6

and so on through the homologue. When isomers exist and are quoted together with the normal compound this normal compound will be identified by a prefix n.

Impeller speeds (pumps)

Two types of centrifugal pumps are used in petroleum refining. One at an impeller speed of 3,550 rpm and the other type operating at an impeller speed of 2,950 rpm. Some difference in their operating characterization are given in Table 19.I.1.

In general, centrifugal pumps should not be operated continuously at flows less than approximately 20% of the normal rating of the pump. The normal rating for the pump is the capacity corresponding to the maximum efficiency point. The table lists the minimum desirable flow rates which should be maintained by continuous re-circulation, if the required process flow conditions are of lower magnitude: Care must be exercised in the design of any re-circulation system to insure that the re-circulated flow does not increase the temperature of pump suction and cause increased vapor pressure and reduction of available NPSH.

For low head pumps that can operate at 1,750 or 1,450 rpm, the above normal and minimum continuous capacities are reduced by 50%. Pump details, description and discussion are given in Chapter 18.

Table 19.I.1. Impeller speed characteristics

		60 Cycle Speed (3,550 rpm)	
Head range feet	Pump type	Minimum continuous capacity rating GPM	Normal rating of pump GPM
To 100	1 stg	10	60
100–350	1 stg	15	75–100
350–650	2 stg	30	150
650–1,100	2 stg	40	160
400–1,200	Multistg	15	50
1,200–5,500	Multistg	40	100–120
50 Cycle speed (2,950 rpm)			
To 75	1 stg	10	50
75–250	1 stg	15	60–80
250–450	2 stg	25	120
450–775	2 stg	30	130
250–850	Multistg	10	40
850–3,800	Multistg	30	80–100

Table 19.I.2. Example of viscosity blending

Component	Vol%	Mid BPt °F	Viscosity Cs 100°F	Blending index	Viscosity factor
	(A)			(B)	(A × B)
1	13.0	410	1.49	63.5	825.5
2	16.5	460	2.0	58.0	957
3	21.0	489	2.4	55.0	1,155
4	18.0	520	2.9	52.5	945
5	18.5	550	3.7	49.0	906.5
6	13.0	592	4.8	46.0	598
Total	100.0				5,387.0

Overall viscosity index $= \dfrac{5{,}387}{100} = 53.87$

From Figure 1.8 an index of 53.87 = 2.65 Cs
(Actual plant test data was 2.7 Cs)

Indices

Indices are used extensively in petroleum refining technology to correlate one set of data with another. Chapters 1 and 3 provide indices that relate the properties of various components to temperature, viscosities, flash point blending and the like. A number of these indices are used in the blending of petroleum fraction to give the properties of the blended product. For example, the components listed in Table 19.I.2 are to be blended in the proportions given and the viscosity of the blended product determined.

Initial boiling points

Initial boiling points or IBP's refer to the temperature at which a petroleum cut begins to boil. Usually this temperature is taken as that at atmospheric pressure. These are determined in the refinery's laboratory from the ASTM distillation carried out as a routine test. Details of these tests are given in Chapter 7. Briefly the initial boiling point is the temperature of the boiling liquid vapor whose first condensate drop enters the measuring cylinder at the condenser outlet. The IBP of the TBP curve is usually calculated from the ASTM test results in the refinery. TBP curves are not usually part of the refinery test schedules but belong to the company's research and development center. A similar comment applies also to the EFV curve. See Chapter 3 for the calculation of TBP and EFV curves. The final boiling point of the ASTM distillation test is the maximum temperature noted after the boiling flask has boiled dry. Sometimes the final boiling point (FBP) is called the ASTM end point.

Instrumentation

The proper operation and performance of any process depends as much on a properly designed control system and its supporting instruments as the correct design and specification of the equipment contained in the process.

Control systems in a process are aimed at maintaining the correct conditions of Flow, Temperature, Pressure, and Levels in process equipment and piping. These are described and discussed in detail in Chapter 13. The system covers four major types of controls which are:

- Flow control
- Temperature control
- Pressure control
- Level control

The principal objective of all these types of controls is to maintain a steady stable plant operation and to enable changes and emergencies to be handled safely. The system must also be designed to ensure that any process changes can be accommodated with minimum risk of damage to plant equipment. The instruments that support the plant control system are gauges and control valves. Not all gauges have a control function; many are for plant performance record or for indication only, for example a pressure gauge on a pump discharge is there to indicate that the pump is operating correctly. A flow gauge in the line as orifice plate assembly indicates the flow quantity (usually on a control room chart) and sets an associated control valve action. In the petroleum refinery the major instruments are:

Flow	— Orifice assembly
Temperature	— Thermocouples
Pressure	— Bourdon tube gauge, differential pressure
Level	— Float type, displacer, or differential pressure type

The addition of 'on stream' analyzers constitutes a major instrument system, particularly if the refinery operates on automated control. This is standard practice in all major refineries.

Internals (Vessels)

The purpose of vessel internals may be characterized as follows:

- To enhance heat and mass transfer (e.g., fractionating towers)
- To maintain proper residence time for settling (e.g., condensate drums)
- To promote good distribution of fluids (vessel inlet distributors)
- To prevent vortexing of fluids leaving vessels to pump suction

A brief description of each of the vessel internals are as follows:

Heat and mass transfer is required in the operation of distillation fractionators. These include steam (or inert gas) strippers. The method used to achieve this is to fit a tower with fractionating trays. With the use of a reboiler supplying heat to the tower botton and cold reflux stream introduced at the tower top heat and mass transfer occurs successively on these trays. Full details, discussion and examples are given in Chapters 3–12. For the same purpose of heat and mass transfer the trays may be replaced with a packing, again this is described in Chapter 3 and Chapter 18. These internals are also used where mass transfer only is required (e.g., some absorber units, and liquid/liquid extraction units).

A number of vessels in petroleum refining are used to collect and separate immiscible fluids. The most common are fractionator overhead distillate drums. In many cases the fluid entering the drum is a mixture of hydrocarbon and water. In these cases it is necessary to retain the mixed fluid in the drum until the phases separate out. A baffle arrangement is used for this purpose slowing the fluid flow from the drum inlet to outlet meeting the required residence time.

In many processes, particularly those that utilize trays or packing it is important that the fluid entering is evenly distributed over the tray or the packing surface. This is achieved by a carefully designed distributor pipe with slots or holes. In large packed towers there may be several rigs of sprays covering the packing surface. In the case of drums the fluid entering is impinged on a horizontal plate to effect a curtain of the fluid entering the body of the drum. This enhances the separation of any vapor that the fluid contains and separates the two phases. Most fractionator overhead condenser drums are fitted with these 'Splash Plates'.

All vessels whose outlet nozzles feed directly to pump suction are fitted with a vortex breaker. This is simply a plate placed directly above the outlet nozzle in the vessel. The plate is supported by legs keeping it above the nozzle itself. This breaks down any vortex that may tend to form owing to the velocity of the fluid being drawn into the pump suction.

Investment analysis

(See earlier item on economic analysis) Refer also to Chapter 17.

Isomerization

The primary commercial use of the branched isomers of C_4, C_5, and C_6 paraffins is in the production of clean-burning, high-performance transportation fuels. The elimination of tetraethyl lead over the last 30 years as a means of improving the antiknock properties of gasoline and more-recent regulations restricting motor fuel composition have led refiners to select alternative means of producing high-quality gasoline, (see also Chapter 2). As a result of benzene concentration restrictions, end-point and olefin content limitations, and potential limitations on total aromatics concentration, the choices of high-quality gasoline blending components available in the typical refinery are limited. Isomerate, the gasoline blending component from light paraffin isomerization, is an ideal choice. Another equally valuable blending components is alkylate resulting primarily from the acid-catalyzed reaction of iso-butene with an aliphatic olefin. Both isomerization and alkylation yield highly branched, high-octane paraffinic blending components that by themselves can satisfy the strictest environmental requirements. Often, n-butane isomerization is one of the sources for the iso-butane requirements in alkylation.

The process flow for a typical isomerization plant is give below as Figure 19.I.1.

The naphtha feed is pre-heated by the reactor effluent before entering the isomerization reactor. The feed mixed with the hydrogen stream flows down the reactor through the catalyst bed. The isomerization reaction takes place in the catalyst bed and the reactor effluent leaves the bottom of the reactor to be cooled first by heat exchange with the incoming feed and then by water or air cooling. The cooled effluent enters the middle of the separator vessel and is flashed with a hydrogen rich gas stream leaving the top and a liquid phase leaving the bottom of the separator. This liquid stream is routed to a reboiled stripper column where a debutanized isomerate liquid leaves as the bottom product. The butanes and lighter components that leave the top of the tower are partially condensed to provide reflux to the tower and a liquid product rich in butanes and propane. The uncondensed overhead leaves the stripper condenser drum to be routed to fuel gas or other processes. Details of the isomerization process

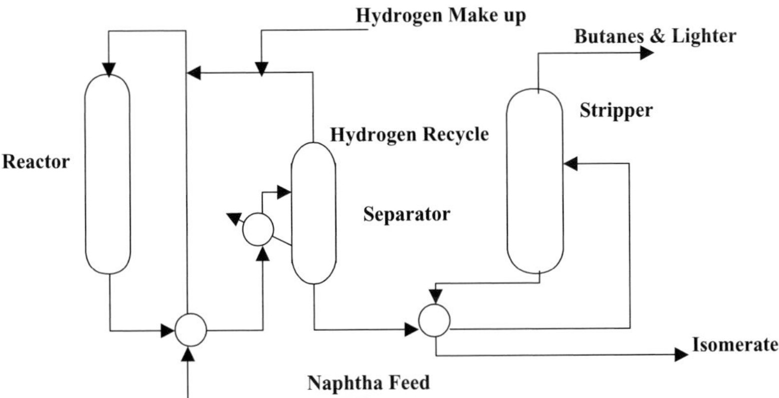

Figure 19.I.1. A process diagram of a typical isomerization unit.

together with a description of the catalyst and chemistry are given in Chapter 9 of this Handbook.

J

Jet fuels

Product specifications are a mechanism by which producers and users of a product identify and control the properties necessary for satisfactory and reliable performance.

Civilian jet fuel

Two organizations have taken the lead role in setting and maintaining specifications for civilian aviation turbine fuel (*jet fuel*): the American Society for Testing and Materials (ASTM) and the United Kingdom Ministry of Defence (MOD). The specifications issued by these two organizations are very similar but not identical. Many other countries issue their own national specifications for jet fuel; these are very nearly or completely identical to either the ASTM or MOD specifications. In the Commonwealth of Independent States (CIS) and parts of Eastern Europe, jet fuel is covered by GOST specifications.

ASTM D 1655 The *Standard Specification for Aviation Turbine Fuels* includes specifications for three commercial jet fuels: two kerosene-type fuels (Jet A and Jet A-1) and a wide-cut fuel (Jet B). Jet A is used for almost all domestic commercial aviation flights in the United States.

Defense Standard 91-91 The United Kingdom Ministry of Defense maintains this specification (formerly titled DERD 2494) for Jet A-1, which is used for most civil aviation fuels outside the United States and the CIS. There are minor differences between the DEF STAN 91-91 Jet A-1 and ASTM D 1655 Jet A-1 specifications.

CGSB-3.22 This Canadian General Standards Board specification covers wide-cut fuel (Jet B) used in parts of Canada and Alaska.

GOST 10227 This Russian specification covers the light kerosene-type fuel, TS-1, used in the CIS and parts of Eastern Europe, along with T-1, T-2, and RT grades of fuel.

Joint Checklist A group of oil companies that operate aviation fuel systems internationally have combined the most restrictive requirements from ASTM D 1655 and DEF STAN 91-91 into a single document: *Aviation Fuel Quality Requirements*

Table 19.J.1. Comparison of civil jet fuels

Specification	ASTM D 1655	Def Std 91–91
Aromatics Vol% Max	25	25
Distillation °C		
10% recovered Max	205	205
50% recovered Max	report	Report
90% recovered Max	report	Report
End point	300	300
Flash point °C Min	38	38
Density @ 15°C kg /M3	775–840	775–840
Freezing point °C Max	−40	−47

for Jointly Operated Systems. This publication is sometimes referred to as the *Joint Checklist*.

International Air Transport Association IATA publishes a document entitled *Guidance Material for Aviation Turbine Fuels Specifications*. The guidance material contains specifications for four aviation turbine fuels: three kerosene-type fuels (Jet A, Jet A-1, and TS-1) and one wide-cut fuel (Jet B). Jet A meets the ASTM requirements, Jet A-1 meets the Joint Checklist requirements, TS-1 meets the Russian GOST requirements, and Jet B meets the Canadian CGSB requirements.

Military jet fuel

The governments of the United States and many other countries maintain separate specifications for jet fuel for military use. The reasons for separate specifications include the operational and logistical differences between the military and civilian systems and the additional demands that high-performance jet fighter engines place on the fuel. Presently two fuels are in widespread use by the U.S. military: JP-5 by the Navy, and JP-8 by the Air Force. Both are kerosene-type fuels. The major difference between them is the flash point temperature, although there are also minor differences in other fuel properties. The minimum flash point temperature for JP-8 is 38°C, and for JP-5 is 60°C. The higher flash point for JP-5 affords an additional degree of safety in handling fuel on aircraft carriers.

The major difference between U.S. military fuels and commercial fuels is in the use of *additives*. Otherwise, JP-8 and Jet A-1 are very similar fuels. Table 19.J.1 provides a comparison between the civilian and military jet fuel specification.

Jetties

Tankers and barges are loaded and unloaded at jetties or docks. In almost all circumstances these facilities for handling petroleum products are separate from those used for general cargo. Very often tankers, particularly modern 'Super' tankers are loaded and unloaded by submarine pipelines at deep water anchorage. Only the 'Onshore' jetty facility is described here. Tanker sizes range from small coastal vessels of 10,000 bbl capacity to super tankers in excess of 250,000 bbl capacity. The more common tanker size is one of 140,000 bbl capacity and this size tanker is labeled a T2. This tanker is usually used for product carrying. It can carry as much as three different product parcels at the same time. The larger tankers are usually used for crude oil transportation.

Ideally the jetty size should be sufficient to cater for both these size tankers, and usually at least one of each size at the same time. In some refineries which have jetty facilities these usually include barge loading items also. The barge loading may however be located on remote docking facilities from the larger sea going tankers. A good onshore jetty approach road is mandatory for the operation of the jetty. This is required for safety reasons and the easy approach way for emergency vehicles (such as fire engines and ambulances). The approach road is also required foe the transportation of the operating staff, ship's crew and the ships chandler vehicles. Usually this approach road is dedicated for jetty use and will be quite independent of any adjacent refinery road.

Finally the location of the jetty must allow sufficient waterway room for tankers to be berthed properly. Tankers arriving from the open sea must have room so that tugs can handle and turn the ship around to face open sea before tying up at the jetty.

A layout plan of a typical tanker jetty is shown in Chapter 13 Figure 13.16. This chapter also includes a description of ideal jetty configuration, and location relative to the refinery site. The item also covers the environmental issue when loading and unloading hydrocarbons.

K

Kerosene

Most straight run kerosenes are desulfurized and routed directly to the finished product blending pools. The exception is in the case of the aviation turbine gasoline (ATG) which will require the kerosene precursor to be treated for aromatics reduction or removal to meet the strict smoke point specifications associated with this finished product. In present day refineries this is accomplished by hydrotreating the kerosene using a nickel catalyst. In older refineries the aromatics contained in the cut were removed as an extract in a process using SO_2 as a solvent. Some details

Table 19.K.1. Specification of some kerosene finished products

Parameters	Reg Kero	ATG	TVO	
Flash point °F	100	<66	100	D-56
Aromatics vol%	–	20	–	D-1319
Temperature @ 20% Max		293°	–	D-86
Temperature @ 50% Max		374°	–	D-86
Temperature @ 90% Max		473°	540	D-86
Final boiling point	572°F	572°	–	D-86
Sulfur Max wt%	0.04	0.04	0.3	D-1266
Smoke point Min	–	25 mm	25 mm	D-1322
Freeze point °C	–	−47	–	D-2386

of the de-aromatization hydro-treating process are given in Chapter 8 of this book. Table 19.K.1 gives the general spec of the kerosene products.

Kinematic viscosity

The Darcey–Weisbach and Colebrook relationships are based on using a Reynolds number which varies inversely with the *kinematic* viscosity. This kinematic viscosity is defined as the Dynamic (or Absolute) viscosity divided by its density, where dynamic or absolute viscosity is force × (time/length squared) and the unit for this is the Poise. The unit most frequently used for the kinematic viscosity is the metric unit—the Stoke. Both viscosities are however usually quoted in the hundredth unit. Thus, absolute viscosity would be the centipoise while the kinematic viscosity would be centistokes.

L

Leaded gasolines

Until the restrictions on lead compounds imposed by the 'Clean Air' acts of the 1960s, tetra ethyl lead was used extensively as a gasoline additive to improve Octane Number.

Tetra ethyl lead is a liquid with a gravity of 1.66 and a formula $Pb(C_2H_5)_4$. It is extremely toxic. The restriction of 'No Lead' in gasolines promoted further development of the catalytic reformer process to obtain higher conversion. It also influenced the use of the alkylation process and the development of isomerization and the oxygenated compounds as octane enhancers in gasoline. Coupled with this, motor manufacturers improved their respective auto engine design to operate efficiently on lower octane number fuel. Tetra Ethyl Lead has all but disappeared from the petroleum industry. Details of modern gasoline development are given in Chapter 2 and details of the Octane enhancer processes are given in Chapter 9.

Light end units

The light end units in the refinery produces the light and heavy naphtha cuts, the butane LPG, and the propane LPG products respectively. The straight run light end units take as feed the atmospheric overhead distillate, the overhead distillate from the catalytic reformer stabilizer, and the overhead distillate from a thermal cracker fractionator (if there is one in the configuration). A typical process flow schematic showing the sequence of the light ends unit is given as Figure 19.L.1.

The long range naphtha as the distillate from the atmospheric crude distillation overhead condensate drum is preheated, usually by heat exchange with the debutanizer bottom product before entering the debutanizer tower. Here the butanes and lighter

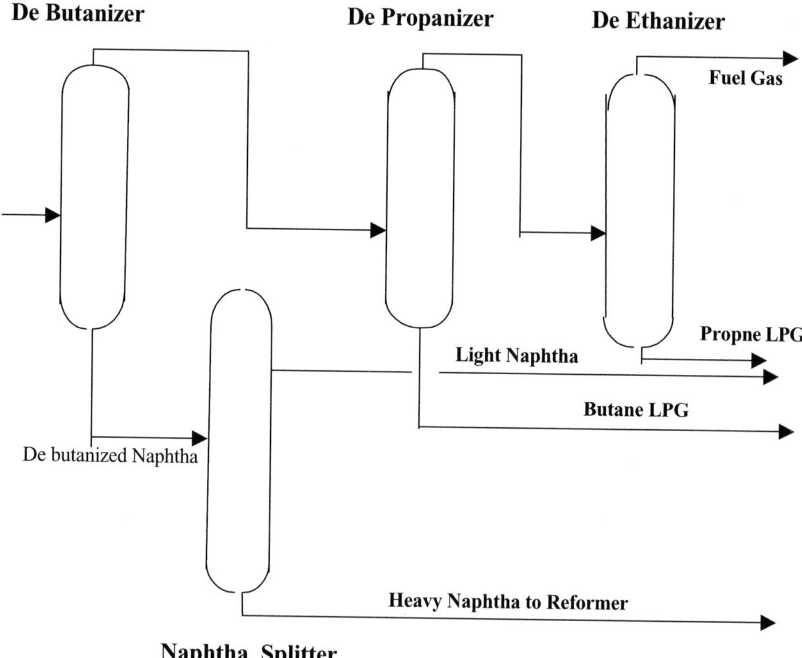

Figure 19.L.1. Process flow schematic for light ends plant.

fractions are removed as the overhead distillate. The bottom product from this tower is the long-range debutanized naphtha. This cut enters the naphtha splitter tower where it is fractionated to meet the specification of light naphtha as the overhead distillate product and heavy naphtha as the bottom product. The light naphtha fraction is routed to the gasoline pool for blending, while the bottom product is the feed to a catalytic reformer.

The overhead distillate from the debutanizer is preheated and enters the de-propanizer tower. Butane LPG is the bottom product from this tower and after treating to remove any sulfur products is routed to the LPG pool. The overhead distillate from the de-propanizer contains propane and lighter components. This is preheated and enters the de-ethanizer column. There is no overhead distillate product from this tower. The overhead product is simply the components lighter than propane and leaves as a gas to the refinery's fuel gas system. Some of the overhead vapor leaving the tower top is partialy condensed to produce the tower reflux. The bottom product from this tower is propane LPG and is routed to the LPG pool.

All three towers are reboiled and are operated with overhead reflux. Full details of this process is given in Chapter 4. A similar process is utilized for the light ends from the FCCU. This is usually kept separate from the straight run process described above. Wild distillates (i.e., containing the light ends) from the catalytic reformer stabilizer and the thermal cracker are routed to the straight run light end units.

Linear programming

This is a computerized technique that came into prominence during the late 1960s and early 1970s. It is used extensively now by most refiners to:

- Optimize new process configurations
- Plan the refinery operation
- Select crude oil feed slate and product slate

The technique uses equations (linear) that represent the properties of the crude feed and the resulting products. These equations also describe the blending characteristics of the components making up the finished product slate. Included also are the cost parameters such as the price of crude feed, the refinery fence selling price of products, operating cost, and any other relevant cost centers (such as licensing fees, interest on loans, etc.). These equations form a mathematical model and a suitable programmed computer is used to solve these equations to meet the objective function subject to the constraints of the analysis. Some further details describing linear programs is given in Appendix E.

LPG—liquefied petroleum gas

Refineries produce two types of LPG (liquefied petroleum gas) products, butane LPG and propane LPG. These two products are recovered in the straight run light ends units as described above. They are routed to finished product storage after treating for the removal of sulfur in the form of mercaptans. Their specifications are as follows:

Propane LPG

Vapor pressure psig	255 max
C_1 hydrocarbons mole%	0.1 max
C_2 hydrocarbons mole%	5.0 max
C_3 hydrocarbons mole%	95 min
C_4 hydrocarbons & heavier mole%	4.0 max (expressed as C_5)
Total unsaturated hydrocarbons mole%	1.0 max
Total sulfur content wt%	0.01 max
Mercaptans grains/100 cuft at STP	3.0 max
H_2S content	Nil.

The product shall not contain harmful quantities of toxic or nauseating substances and shall be free of entrained water

Butane LPG

Vapor pressure psig	70–85
C_1 hydrocarbons mole%	Nil
C_2 hydrocarbons mole%	0.5 max
C_5 hydrocarbons mole%	2.0 max
Total sulfur Wt%	0.01 max
Mercaptan grains/100 cuft at STP	2.0 max
H_2S content	Nil

The LPG's are utilized widely in industry and domestically as portable fuel source. The fact that they are liquids under pressure and are stored in moveable cylinder containers gives them prominence in the work place, at home, and in outside recreation areas. Propane LPG is used domestically for outdoor barbecues, in recreational vehicles, for cooking and heating, and on private boats, etc. Butane LPG is used mostly in industry as a portable heat source. LPGs are also used as automotive fuels. In the U.S. propane is often used in agricultural tractors. Butane LPG is often used in Europe for taxi fleets, etc.

Lube oils

Some, not all, refineries produce lubricating oils. These are considered non-energy products but are essential to modern living just as much as the energy products of

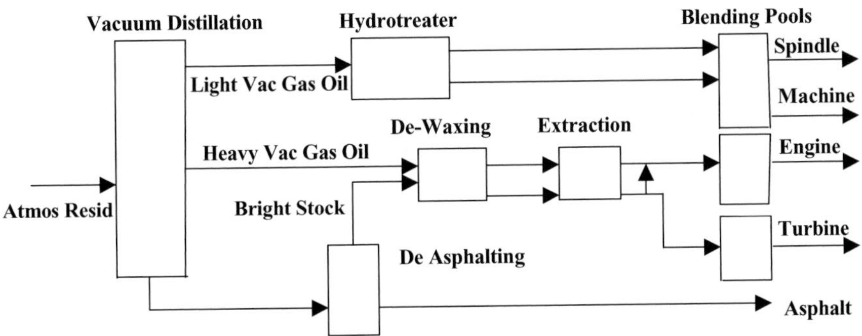

Figure 19.L.2. A schematic of a lube oil processing configuration.

petroleum. About 3–10 grades of basic lube oil components are produced and can be blended to meet the many grades of light lube oils, engine lubes and heavy turbine lube oils. The production of these basic lube products commences with the vacuum distillation of atmospheric residue. The processes to arrive at the blendable basic stocks follow the block flow diagram shown as Figure 19.L.2.

The atmospheric residue feed is vacuum distilled to provide two distillate side streams and a heavy bituminous residue. The light distillate side stream is hydro-treated to saturate any olefins present and, thus, improve the product color. It is then split in a vacuum distillation tower to produce the very light spindle oil lube and the slightly heavier machine oil lube. The bottom side stream distillate from the vacuum unit is routed to a de-waxing plant and then to a solvent extraction plant for color improvement. Both these units may be operating in a blocked operation with a raffinate from propane de-asphalting unit. This unit processes the vacuum residue to remove the asphalt leaving the de-asphalted oil, called 'bright stock' to be dewaxed and olefin extracted in the blocked operation through these two plants. The treated 'bright stock' is the base stock for the heavy turbine lube oils. Details of the lube oil production processes are given in Chapter 12.

M

Mass spectrometry

Mass spectrometry is concerned with the separation of matter according to atomic and molecular mass. It is most often used in the analysis of organic compounds of molecular mass up to as high as 200,000 Daltons, and until recent years was largely restricted to relatively volatile compounds. Continuous development and improvement of instrumentation and techniques have made mass spectrometry the most versatile, sensitive and widely used analytical method available today. One of its major uses in the petroleum refining industry is for the production of distillation curves such as TBP and EFV. This technique does away with the cumbersome distillation apparatus previously used for this purpose. It is also by far the more accurate method.

Material balances

Material balances form the basis for plant design, and are essential in refinery operation to account for gains and/or losses in the refinery's daily production. They are also essential in process plant audits and trouble shooting. The primary material balance in the design of a refinery is that for the atmospheric and vacuum crude distillation unit. For this balance the TBP distillation curve (usually from the crude assay) is split into real components up to C_5's or C_6's and then into mid boiling point or mid volume point pseudo components. The splitting of the crude in this way provides the component breakdown for the product cuts and the volume, mass, and the number of moles, together for the same information for the whole crude feed. Details on the development of the material balances are given in Chapter 3–12. The material balance in all cases is complete and correct when the quantities into the process equals the total quantities out when expressed in mass (weight) units. Volumes and mols in and out may differ because of chemical reactions and/or thermal and pressure effects.

Mechanical flow diagrams

This type of flow diagram is sometimes referred to the P&ID or very often the Engineering Drawing. Details of this type of drawing is given under the item flow sheets.

Metals in crude oil

Metallic organic compounds have a deleterious effect on some products, and are also usually poisonous to catalysts in some processes. The most common metals met with are:

- Sodium
- Nickel
- Vanadium

and sometimes mercury, etc. in the associated gas streams.

Most of these metallic compounds are found in the asphalt portion of the crude oil and are usually deeply imbedded in the asphaltene molecules. In the production of fuel oil the metal content of the fuel makes the product problematic to the steel production companies who use fuel oil in their processes. Secondly, in the upgrading of the 'bottom of the barrel' using a catalytic process (such as hydrocracking or fluid catalytic cracking) these metallic compounds deteriorate the catalyst life and performance of the processes. Much has been done in the last decade to improve the catalysts to withstand metals. However, in many cases the true solution to the presence of metals in the crude is to de-asphalt the heavy vacuum residue and route the de-asphalted oil as the cracker feed. Not all crude oils contain metal impurities in quantities that cause process problems. The deasphalting process (as propane deasphalting unit) is described in Chapter 11.

Mid boiling point and mid volume point components

These have been defined and described earlier in this Part 2, 'Component balances', and also in Chapter 1.

N

Naphtha

There are usually two straight run naphtha cuts produced from crude. These are:

- Light naphtha (sometimes called light gasoline)
- Heavy naphtha

Both these naphtha cuts form part of the atmospheric crude distillation overhead distillate product. This product also contains all of the light ends of the crude oil feed. The stream is therefore debutanized with the components butane and lighter being routed to the light ends unit for further fractionation.

The naphtha stream that leaves the bottom of the debutanizer is fed to a naphtha splitter where it is fractionated to produce the two naphtha cuts. The light naphtha as the splitter's overhead distillate product contains most of the crude's C_5's and much of the paraffin portion of the crude's C_6's. Heavy naphtha is recovered as the splitter's bottom product. This product will contain the heavier naphthenes and will be a suitable feed for a catalytic reformer. The light naphtha has a TBP distillation range of C_5 to around 190°F and is often used as feed to steams crackers for the production of light olefins. The heavy naphtha as the feed to the catalytic reformer consists of a cut of about 190°–360°F. This end point of 360°F can vary depending on the severity operation of the catalytic reformer, the volatility specification of the finished gasoline, and the refinery's production requirements.

The non energy product refineries

There are two major types of non-energy refineries. They are:

- The lube oil refinery
- The petrochemical refinery

These two production units are usually part of a conventional energy refinery, which makes it convenient for obtaining their respective feedstocks and for using common facilities such as loading stations and jetty. The lube oil refinery has been described in an earlier item. More information on this subject is found in Chapter 12. This chapter also details the production and properties of the asphalt product of this type of refining.

The petrochemical refinery

This refinery is concerned with processing products generated by the energy portion of the refining of crude oil into suitable feed streams to the:

- Production of aromatics
- Production of olefinic products

Both of these types of petrochemical refineries have been described briefly in Chapter 1. Only the production of aromatics is dealt with in detail.

The aromatics referred to here will be the more common with respect to crude oil refining. These are:

- Benzene
- Toluene
- Ethyl benzene
- Para-xylene
- Meta-xylene
- Ortho-xylene

The configuration described here begins with a mixed aromatic stream which has been obtained by catalytic reforming of a high naphthene content naphtha (usually from a hydro-cracking process). The reformate from this high naphthene feed is very rich in the aromatics listed above. To increase the aromatic content as feed to the aromatic complex the aromatics are separated from the remaining paraffins by an extraction process. The rich aromatic feed is separated into the required products by a series of distillations, a crystallization or adsorption process, an aromatic isomerization process, and α dealkylation process. The block flow diagram is shown as Figure 12.18 of Chapter 12.

In this particular configuration the purpose is to maximize benzene, toluene, and ortho xylene at the expense of ethyl benzene, meta and para-xylenes. To recover para-xylene all that would be required is an adsorption or crystallization unit before the isomerization unit.

O

Octane number

Octane numbers are a measure of a gasoline's resistance to knock or detonation in a cylinder of a gasoline engine. The higher this resistance is the higher will be the efficiency of the fuel to produce work. A relationship exists between the antiknock characteristic of the gasoline (Octane Number) and the compression ratio of the engine in which it is to be used. The higher the octane rating of the fuel then the higher the compression ratio of engine in which it can be used. By definition, an octane number is that percentage of iso-octane in a blend of iso-octane and normal heptane that exactly matches the knock behavior of the gasoline. Thus, a 90 octane gasoline matches the knock characteristic of a blend containing 90% iso-octane and 10% n-heptane.

The knock characteristics are determined in the laboratory using a standard single cylinder test engine equipped with a super sensitive knock meter. The reference fuel (iso octane and C7 blend) is run and compared with a second run using the gasoline sample. Two octane numbers are used: the first is the research octane number (RON) and the second is the motor octane number (MON). The same basic equipment is used for both octane numbers, but the engine speed is increased for the motor octane number. The actual octane number obtained in a commercial engine would be somewhere between the two numbers. The significance of the two octane numbers is to evaluate the sensitivity of the gasoline to the severity of the operating conditions in the engine. Invariably the research octane number is higher than the motor octane number, the difference between them is quoted as the 'sensitivity' of the gasoline.

The pump octane posted in the U.S. is (RON + MON)/2. In Europe often only RON is given.

Offsite systems

Among the major units found in most refinery offsite systems are:

- Storage
- Product blending
- Road and rail loading
- Jetty facilities
- Waste disposal
- Effluent water treating

Storage facilities

Crude oil feed and processed products are stored in storage tanks of various sizes and types. These tanks are usually located together in the refinery area suitably defined as 'the Tank Farm'. Most refinery storage tanks fall into the following categories:

- Atmospheric storage
- Pressure storage
- Heated storage

Details of all these type of tankage are given in Chapter 13.

Atmospheric storage
As the name implies, all atmospheric storage tanks are open to the atmosphere, or are maintained at atmospheric pressure by a controlled vapor blanket. These tanks fall into two categories:

- Cone roof tanks
- Floating roof tanks

Cone roofed tanks. Used for the storage of non-toxic liquids with fairly low volatility. In its simplest form the roof of the tank will contain a vent, open to atmosphere, which allows the tank to "breathe" when emptying and filling. A hatch in the roof also provides access for sampling the tank contents. In oil refining this type of tank is used for the storage of gas oils, diesel, light heating oil, and the very light lube oils (e.g., spindle oil).

Floating roofed tanks. Light volatile liquids may also be stored at essentially atmospheric pressure by the use of "Floating Roof" tanks. The roof of this type of storage tank literally floats on the surface of the liquid contents of the tank. In this way the air space above the liquid is reduced to almost zero, thereby minimizing the amount of liquid vaporization that can occur. The roof is specially designed for this service and contains a top skin and a bottom skin of steel plate, held together by steel struts. These struts also provide strength and rigidity to the roof structure. The roof moves up and down the inside of the tank wall as the liquid level rises when filling and falls when emptying. Liquids stored in this type of tank have relatively high volatilities and vapor pressures such as gasoline, kerosene, jet fuel, and the like. In oil refining the break between the use of cone roof tank and floating roof is based on the "flash point" of the material. Normally this break point is for materials with a flash point of 120°F or below. Floating roof tanks (very large in capacity) are used for storing the crude oil feed. Diagrams of a cone roof tank and a floating roof tank are shown (Figure 19.O.1) as Figures 13.5 and 13.6 in Chapter 13.

Pressure storage

Pressure storage tanks are used to prevent or at least minimize the loss of the tank contents due to vaporization. These types of storage tanks can range in operating pressures from a few inches of water gauge to 250 psig. There are three major types of pressure storage. These are:

- Low-pressure tanks—These are dome roofed tanks and operate at a pressures of between 3 ins water gauge and 2.5 psig.
- Medium-pressure tanks—These are hemispheroids that operate at pressures between 2.5 psig and 5.0 psig, and spheriodal tanks that operate at pressures up to 15 psig.
- High-pressure tanks—These are either horizontal "bullets" with ellipsoidal or hemispherical heads or spherical tanks (spheres). The working pressures for these types of tanks range from 30 to 250 psig. The maximum allowable is limited by tank size and code requirements. For a 1,000 bbl sphere, the maximum pressure is 215 psig, for a 30,000 bbl it is 50 psig. These pressure limits can be increased if the tank is stress relieved.

Although it is possible to store material in tanks with pressures in excess of 250 psig, normally when such storage is required refrigerated storage is usually a better alternate.

Heated storage tanks

Heated storage tanks are more common in the petroleum industry than most others. They are used to store material whose flowing properties are such as to restrict flow at normal ambient temperatures. In the petroleum industry products heavier than diesel oil, such as heavy gas oils, lube oil, and fuel oil are stored in heated tanks. Most often tanks are heated by immersed heating coils or bayonet type immersed heaters. Steam is normally used as the heating medium. Where immersed heating is used the tank is agitated usually by side located propeller agitators. External circulating heating is used for tanks if the contents are mixed by means of jet mixing. External tank heating is used when there is a possibility of a hazardous situation occurring if an immersed

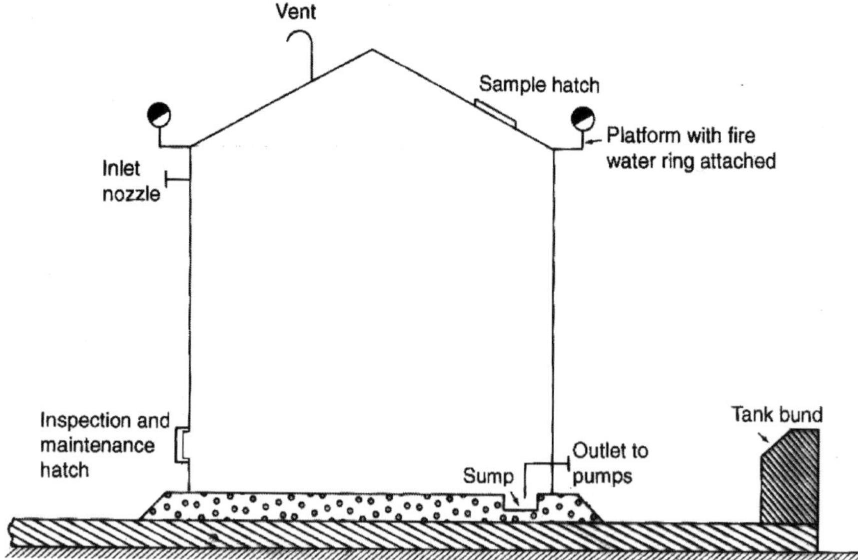

Figure 13.5.

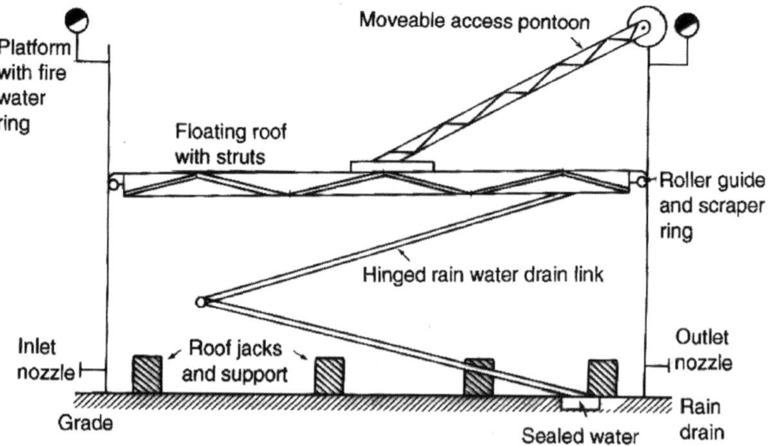

Figure 13.6.

Figure 19.O.1. Diagram of a cone roof tank and a flooting roof tank.

heater leaks. A calculation procedure for tank heater sizing is included in Chapter 13.

Product blending facilities

Blending is the combining of two or more components to produce a desired end product. The term in refinery practice usually refers to process streams being combined to make a saleable product leaving the refinery. Generally these include gasolines and middle distillates such as jet fuel, kerosene, diesel, and heating oil. Other blended finished products will include various grades of fuel oil and lube oil. The blending of the process streams is accomplished either by batch blending in blending tanks or in-line blending in the pipe line itself.

In-line versus batch blending
In batch blending the components are routed separately into a single receiver tank where they are mixed and tested to ensure the mixed product meets the finished product specification. In the case of in-line blending the component streams are routed through automatically operated flow valves to a finished product tank. With modern computerized control technology in-line blending is becoming the more common form of blending process.

The in-line blender operation
An in-line blender is essentially a multiple stream controller with feed-back. The controller itself is a computer into which the recipe for the blend is programmed. The controller automatically starts the pumps for the blend components and motivates the flow control valves on the component lines. These control valves are reset by computer inert-action to meet the required component quantities for finished product specification. A series of on line analyzers located in the blend run down lines, monitor the finished product properties and, in turn, reset the control valves.

Road and rail loading facilities

The extent of product shipping facilities required in a chemical or petroleum complex depends on the size of the complex, the local market, the number of different products to be shipped and the market to be supplied. Normally the shipping facilities installed in most plants are sufficient to cater for normal product handling and the flexibility required for seasonal demands. The capacity of these facilities will almost invariably exceed the plant's total production.

The most common method of shipping product is by road or rail in suitably designed tanker cars. In the case of large complexes located on coastal or river side sites shipping by barge or ships is feasible to carry the bulk of the plant products. Chapter 13 describes the loading of road and rail cars under the following subject headings:

- Loading rates
- Loading equipment
- Loading arrangements

Loading rates are described with details on the calculation procedures to arrive at these rates example calculations are provided. Drawings of loading equipment and their typical arrangements are also provided in this section of Chapter 13.

Jetty and dock facilities

Tankers and barges are loaded and unloaded at jetties or docks. In almost all circumstances these facilities for handling petroleum products are separate from those used for general cargo. Very often tankers, particularly the modern 'Super' tanker are loaded and unloaded by submarine pipelines at deep water anchorage. This section of Chapter 13 deals only with onshore docking facilities. The following items in this section of the chapter are:

- Jetty size and location
- Equipment
- Loading rates
- Ballast handling
- Slop and spill collection facilities

Jetty size and location covers the need to accommodate the various tanker or barge sizes. It continues with the parameters to be considered in the location of the jetty itself. This topic includes consideration of the access to the jetty and the depth of the water at the location. The jetty must be close to the refinery's tank farm and must cater for sufficient pipe-way. Just as important too is that the jetty has good approach road and parking facilities. These topics are discussed in some detail.

The equipment used for tanker loading and unloading is described in detail with supporting diagrams.

In the item on loading (and unloading) rates emphasis is placed on the relatively high pumping rates experienced for this function. The item covers the rates usually used for normal tanker size and those for the barge loading facilities. The other items cover the disposal of the ships' ballast and the collection of the spills and leaks that occur in the loading/unloading activities. The collection of these spills is achieved by temporary booms placed around the vessels. Other topics include communication with shore facilities and the mooring facilities. These topics are supported by diagrams where applicable.

Waste disposal and water effluent treatment

All process plants, including oil refineries produce large quantities of toxic and/or flammable material during periods of plant upset or emergencies. Properly designed Flare and Slop handling systems are therefore essential to the plant operation. This section of Chapter 13 describes and discusses typical disposal systems currently in use in the oil refining industry where the hydrocarbon is immiscible with water. Where the chemical is miscible in water special separation systems must be used. The item covering waste disposal systems is shown in a block flow diagram for a typical modern refinery. The system shown consists of three separate collection systems being integrated to a flare and a slops rerun system. A fourth system is included for the disposal of the oily water drainage with a connection to the flare and a separate connection for any oil laden skimming. Further description and discussion of these disposal systems is given in the following sections covering:

- Blow-down and slop disposal
- Flares

The description and discussion of these systems highlights the equipment used and, where applicable, the sizing parameters for some of these items. All the descriptions are further illustrated by diagrams.

This section of the offsite systems deals with the treating of waste water accumulated in a chemical process complex before it leaves the complex. Over the years requirements for safeguarding the environment has demanded close control on the quality of effluents discharged from chemical and oil refining plants. This includes effluents that contain contaminants that can affect the quality of the atmosphere and those that can be injurious to plant and other life in river waters and the surrounding seas. Effluent management in the oil industry has therefore acquired a position of importance and responsibility to meet these environmental control demands. This topic is further explored in Chapter 14, under the title of "Environmental Control and Engineering in Petroleum Refineries". The item in Chapter 13 deals with the broad principals and methodology of treating water effluents from paved area run off, ship ballast water, and effluent from the slop and spillage facilities. The major processing equipment to handle most effluent treating in the petroleum refinery is the API separator. A detailed description and design parameters of this equipment is given in Chapter 14. Other effluent treating processes such as flocculation and coagulation are described in Chapter 13.

Olefins

Although the major families or homologues of hydrocarbons found in all crude oils as described earlier are paraffins, cyclic paraffins and aromatics, there is a fourth group.

These are the unsaturated or olefin hydrocarbon. They are not naturally present in any great quantity in most crude oils, but are often produced in significant quantities during the processing of the crude oil to refined products. This occurs in those processes that subject the oil to high temperature for a relatively long period of time. Under these conditions the saturated hydrocarbon molecules break down permanently losing one or more of the four atoms attached to the quadravalent carbon. The resulting hydrocarbon molecule is unstable and readily combines with itself (forming double bond links) or with similar molecules to form polymers. An example of such an unsaturated compound is as follows:

```
    H   H
    |   |
H— C = C —H      Ethylene.
```

Note the double bond in this compound linking the two carbon atoms.

Overflash

Overflash is a term normally associated with the design of crude atmospheric or vacuum towers. Its objective is to provide additional heat (over and above that set by the product vaporization required) required by the process to generate the internal reflux required by the process. It also influences the flash zone conditions of temperature and partial pressure of the hydrocarbon vapor feed. Usually it is fixed at between 3 and 5 vol% on crude. This atmospheric flash temperature is now adjusted to the temperature at the previously calculated partial pressure existing in the flash zone. A further description and the purpose of overflash is given in Chapter 3.

P

Packed towers

Although the use of trays is generally the first choice for fractionation and absorption tower applications, there are two major instances where packed towers are preferable. These are:

- Small diameter towers (below 3 ft diameter)
- At the other end of the spectrum packing in the form of grids and large stacked packed beds have superceded trays in vacuum distillation towers whose diameter range up to 30 ft in some cases. This is because packing offers a much lower pressure drop than trays

The packing in the tower itself may be stacked in beds on a random basis or in a defined structured basis. For towers up to 10–15 ft the packing is usually dumped or randomly packed. Above this tower size and depending on its application the packing may be installed on a defined stacked or structured manner. For practical reasons and to avoid crushing the packing at the bottom of the bed the packing is installed in beds. As a rule of thumb packed beds should be around 15 ft in height. About 20 ft should be a maximum for most packed sections.

Properties of good packing are as follows:

- Should have high surface area per unit volume
- The shape of the packing should be such as to give a high percentage of area in active contact with the liquid and the gas or in the two liquid phases in the case of extractors
- The packing should have favorable liquid distribution qualities
- Should have low weight but high unit strength
- Should have low pressure drop, but high coefficients of mass transfer

Some data on the various common packings together with a typical packed tower layout are given in Chapter 18. This chapter also gives design procedures and data for packed towers in general. These procedures and characteristics of the packed tower are supported by diagrams and graphical correlations where applicable.

Petrochemical refineries

(See the previous item on 'Non-Energy Product Refineries').

Planning refinery operations

The basic organization of an oil company consists essentially of three main departments:

1.0 The marine department
2.0 The refining department
3.0 The marketing department

The functions of these three main departments are co-ordinated by a supply department. This department undertakes this role in accordance with the following sequence:

1.0 Marketing department informs the supply department with the quantities of products they can sell.
2.0 Supply department, after making adjustments for stock levels, advises the refinery of the quantity of each product required.

3.0 The refining department advises the supply department of the actual quantities they can produce and how much crude they will require to do so.

3.0 The supply department arranges with the marine department for the necessary shipping to:

- Provide the crude
- Lift the products
- Import the quantities that cannot be produced

  Each refinery organization contains a planning department whose function is to put together the Monthly Running Plan. The main uses of the running plan are to provide data for:
- Keeping the supply department and the company executives informed
- Provide information for long term chartering of ships
- Arrangement of import programs
- Pinpoint future product quality and equipment difficulties

The plans reflect inventory surplus or deficit in immediate stocks and those due in the short term and over a period of around 18 months to satisfy the objectives of the supply department. It is also the basis of the refinery's running program in the short term. By that is meant the establishment of the refinery's process plant conditions and product blending programs, storage schedule, etc., on a day by day basis. The program when extended to each quarter of the 18 months' duration, alerts the supply department of possible shortfalls, crude slate requirements, and the refinery's possible processing constraints. These constraints take into consideration scheduled shut downs, storage capabilities and the like. Full details of a refinery plan together with examples of its development are given in Chapter 17.

Plant commissioning

The sequence of events that ends with a plant or plants being fully operational follows the same lines whether it refers to a single unit or a complex of many units such as a refinery.

The commissioning activities fall into the following sequence of events:

- Pre-energizing activities
- Energizing the plant
- Conditioning equipment, calibrating instruments, and setting relief valves
- Final check out, and closing up all vessels
- Preparation for "start up"
- Start up
- Lining out
- Performance test runs and guarantee test run

Pre-energizing activities

These are the activities when the contractor or the refinery's own staff hands the plant over to the commissioning operators. At this stage the plant is not connected to power, steam, water, drains, etc and contains no hazardous material. The commissioning team's objective now is to check the hardware against specifications and the engineering drawings. This activity has been given the title 'Punch Listing'. Although the contractor has flushed the plant out before handover a further and more thorough flushout is usually done at this "safe plant" stage when utility lines and underground lines can also be flushed out.

Energizing the plant

When 'punch listing' is complete to everyone's satisfaction the plant can be energized. This includes the connection to power and the other utilities. The commissioning of the fuel systems indicates that the plant is now a 'Hazardous Area' and all regulations pertaining to this type of area come into effect.

Conditioning equipment

Reactors, certain pipelines carrying corrosive material, and other items such as the flare, and heater boxes which are refractory lined require some conditioning before being put into use. These conditioning procedures are usually detailed by the manufacturers. As most of these procedures call for an energized plant they are carried out by the commissioning team.

Final check out, and closing up of all vessels

This will be the last opportunity to check such items as the internals of towers, fractionation trays, condition of refractory and other linings, hold up grids, distributors, and the bottom of the tower baffling system (to and from the reboiler). A final check-out of the piping layout also needs to be carried out at this point. When satisfied that all is satisfactory the commissioning supervisor will authorize the following final pre-startup activities to be completed:

- Catalyst loading (where applicable)
- Loading the tower packing where applicable
- Closing up all vessels using the permanent gaskets

In cases where equipment has been subject to a caustic wash the temporary silica level gauges used during the wash are replaced by the specified operational ones.

With the completion of these final checks and vessel close up, the plant is now ready for start up.

Start up and lining out

The activities and their sequence for starting up the plant are carried out as described in the operating manual. In the case of an oil refining plant for example the first activity is to eliminate air from the plant systems. This is done by using water or steam or inert gas or a combination of all three.

After the purging comes the introduction of the cold feeds. In the case of units that contain reactors and use hydrogen under pressure a leak testing program is required. Where water has been used for purging, the water is replaced by the oil feed. This is termed the 'oil squeeze'. Start up may be defined as beginning when the purge program shows conditions to be safe to apply heat into the plant.

Pour points

The 'Pour Point' of an oil is the temperature at which the oil ceases to flow. It is usually a test applied to middle distillates, lube oils, and fuel oils. The test itself is quite simple and requires the sample oil to be carefully treated before the test and to reduce its temperature in a controlled and orderly way. Unlike most other petroleum properties pour points of two or more components cannot be blended directly to give a pour point of the blended stock. Blending indices are used with the volumetric composition of the blend components for this purpose. Details of the blending for 'Pour Point' are given in Chapter 1 while details of the test itself are given in Chapter 16.

Predicting product properties

Product properties are predicted from a composition of real components and of pseudo components that make up the product streams. Most crude oil assays will include the composition of the light ends in terms of volume or weight percent on crude. The light ends usually include C1 to C5's inclusive. The remainder of the crude oil may be divided into pseudo components based on the TBP curve. Details of this are given in Chapter 1 and further illustrated in Sections of Chapter 2 and Chapter 3.

The pseudo components are either expressed as wt% or vol% on whole crude, and are identified as either mid boiling point or mid volume on crude. This concept is further defined again in Chapter 1, Chapter 2, and Chapter 3 and under the title of Mid Boiling Points in this chapter. From the assay data and curves, properties such as specific gravity, sulfur content, pour points, etc. are given to these pseudo components either in terms of boiling point or their mid volume point on crude. In most cases the property of a cut is determined by the sum of these component properties times their vol% or wt% in the cut. In a few cases however the direct multiple of properties and percentages is not acceptable to define the overall property of the cut. The most notable are flash points of blended cuts, and pour points. Here blending indices are used as the multiple of volume composition. These indices are given in Chapter 1. The sulfur content of a product is always quoted as a percent by weight of the product. In this case therefore both the pseudo components and the sulfur content (from assay) are converted to weight.

The flash point of a product is related to its ASTM distillation by the expression:

$$\text{Flash Point} = 0.77\,(\text{ASTM } 5\% \text{ in } ^\circ\text{F}-150^\circ\text{F})$$

To determine the viscosity of a blend of two or more components, a blending index must be used. A graph of these indices is given in Maxwell's "Data Book on Hydrocarbons", and part of this graph is reproduced as Figure 1.08 in Chapter 1. Using the blending indices and having divided the TBP curve into components as before, the viscosity of the fraction can be predicted as shown in the following example:

Component	Vol%	Mid BPt °F	Viscosity Cs 100°F	Blending index	Viscosity factor
	(A)			(B)	(A × B)
1	13.0	410	1.49	63.5	825.5
2	16.5	460	2.0	58.0	957
3	21.0	489	2.4	55.0	1,155
4	18.0	520	2.9	52.5	945
5	18.5	550	3.7	49.0	906.5
6	13.0	592	4.8	46.0	598
Total	100.0				5,387.0

$$\text{Overall Viscosity index} = \frac{5{,}387}{100} = 53.87$$

From Figure 1.08 (in Chapter 1) an index of 53.87 = 2.65 Cs.
(Actual plant test data was 2.7 Cs).

The prediction of molecular weights of product streams are more often required for the design of the processes which are going to produce those products. There are other more rigorous calculations that can and are used for definitive design and in building up computer simulation packages. The method presented here is a simple method by which the mole weight of a product stream can be determined from a laboratory ASTM distillation test. The result is sufficiently accurate for use in refinery configuration studies and the like. A relationship exists between the mean average boiling point of a product (commonly designated as MEABP), the API gravity, and the molecular weight of petroleum fractions. This is described in detail in Chapter 1.

Preheat exchanger train

In most distillation processes heat recovery by heat exchange is of great importance. In the design of major processes such as the crude oil distillation unit, cracker recovery units and the like the optimization of this heat recovery concept is of paramount importance. The method best adopted to undertake this heat recovery concept is described in detail in Chapter 3. Briefly the method consists of examining several configurations of a heat transfer train, applying "pinch" analysis a cost data to the equipment, and to determine the terminal feed temperature of each configuration. This end temperature relates to the heater duty required and therefore to the fuel required by the heater. An economic balance may then be made to select the optimum heat exchanger configuration. Developing the various configurations is made by using the total enthalpy of the feed (in this case the crude oil feed to the atmospheric distillation unit) and the total enthalpies of the exchanged streams. This is shown by Figure 3.14 of Chapter 3, and is reproduced on page 1250.

Process guarantees

Among the items of major concern to the operating refinery staff and the engineering contractor in the design, procurement, and construction of a grass roots process or indeed a revamped process, is the final process guarantees that are developed and accepted. The process guarantees may begin to be developed as soon as a firm process has been established and manufacturers' guarantees obtained for the performance of the various manufactured items of equipment.

The process performance is tied also to a guarantee of its efficiency. This will be in terms of a guarantee of the utility consumption in the plant whilst operating on the design throughput and conditions. These guarantees as written will differ from process to process but will usually follow a pattern or format.

- Description of the feed in terms of throughput, composition, or source (in the case of crude oil)

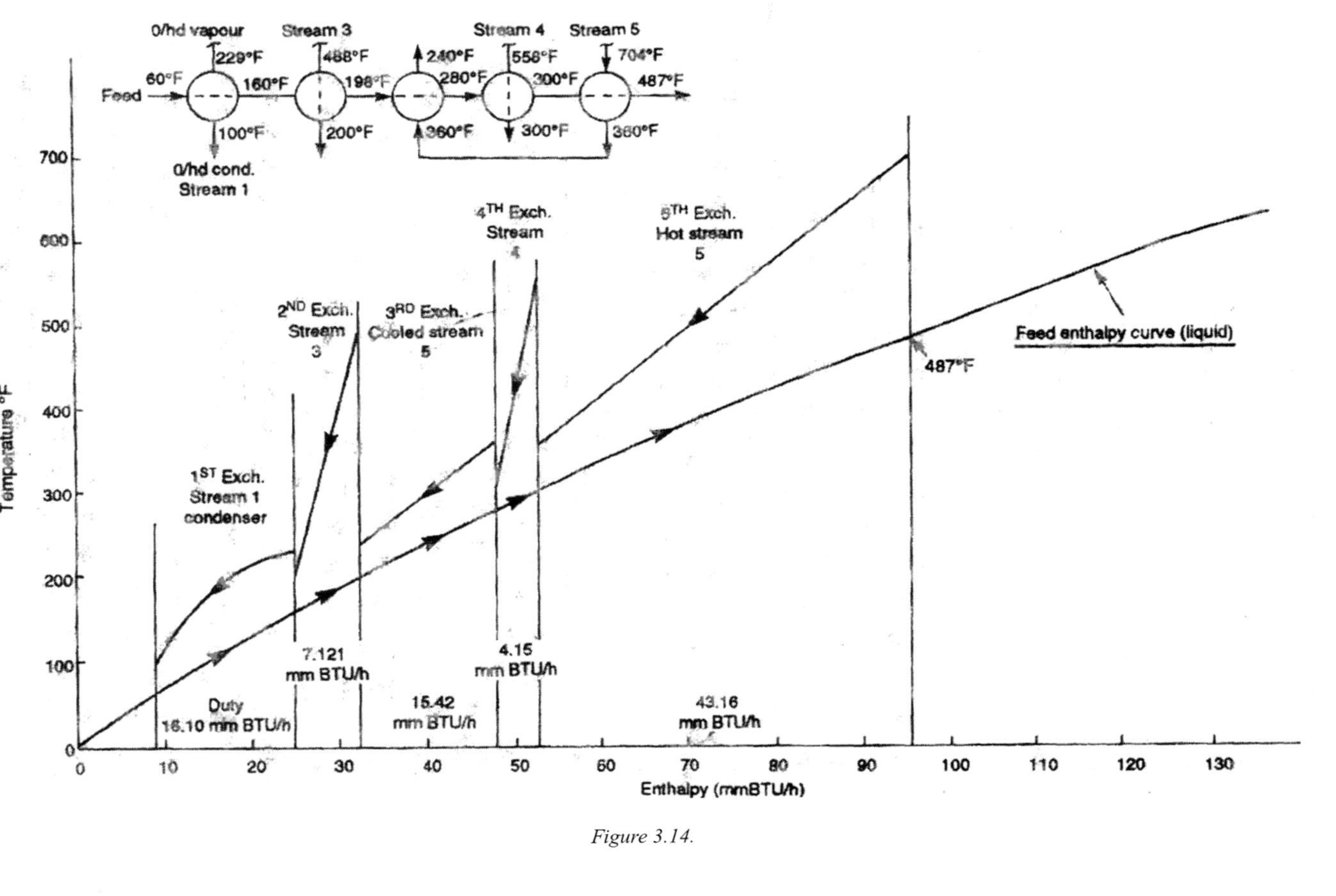

O/hd vapour
229°F
60°F
Feed
160°F
100°F
O/hd cond.
Stream 1
Stream 3
488°F
198°F
200°F
240°F
280°F
360°F
Stream 4
558°F
300°F
300°F
Stream 5
704°F
487°F
360°F
1ST Exch.
Stream 1
condenser
2ND Exch.
Stream
3
3RD Exch.
Cooled stream
5
4TH Exch.
Stream
4
5TH Exch.
Hot stream
5
Feed enthalpy curve (liquid)
487°F
Duty
16.10 mm BTU/h
7.121
mm BTU/h
15.42
mm BTU/h
4.15
mm BTU/h
43.16
mm BTU/h
Temperature °F
Enthalpy (mmBTU/h)

Figure 3.14.

- Design operating conditions and the guarantee of the product specification at these conditions
- A guarantee of the hydraulic capacity of the process system
- The utility consumption guarantees
- A list of the accepted test procedures that will be used
- The guarantee test run procedures which are written in some detail
- Description of the notices to be given in the event the guarantees are not met

These items are described and discussed in Chapter 17 and an example of both the performance guarantee and its associated utility guarantee is given in Appendix 17.9 of this Chapter.

Process configurations

Process configurations are represented in the form of block flow diagrams. They are prepared usually as the first step in deciding the type of process units that will make up a desired complex. These units are shown in sequence to each other by blocks which will be labeled with their throughput size (in the case of a petroleum refinery in barrels per stream day or in cubic meters per stream day). The diagram is further developed showing the product and feed lines from and to the unit blocks. The measure of flow is shown on each line. This will be in terms of barrels per calendar day or cubic meters per calendar day. Several block flow diagrams of differing configurations but meeting the end product objectives will be developed prior to the decision as to which route meets all the company's objectives. A simplified block flow diagram is shown as Figure 19.P.1.

Each of these process configurations will be cost estimated to provide a comparative capital or installed cost, their operating costs, and the refinery price for products

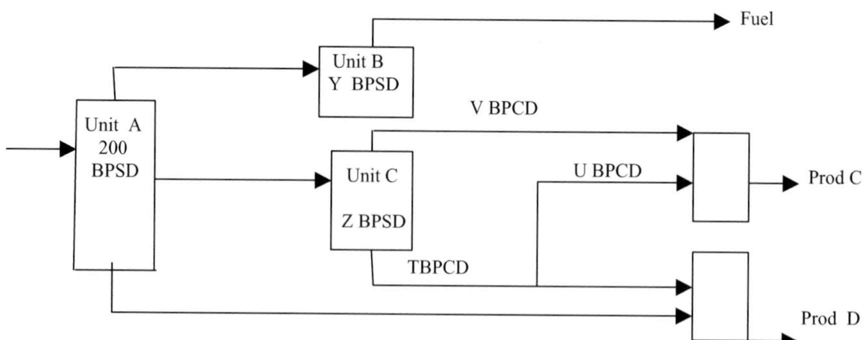

Figure 19.P.1. A typical block flow diagram.

and feedstock. With these parameters an economic evaluation to determine the best configuration can be made using the company's accepted evaluation procedures or by linear programming using a computer. Details of the development of a process configuration is given in Chapter 2. The use of process configuration in economic evaluation is described and discussed in Chapter 17. The sequence of events using process configurations that leads to a decision making item is given in Chapter 17 as Figure 17.3.

Project duty specification

Among the first activities to initiate a refinery project is the development of the "Project Duty Specification". This document must describe in detail the plant or complex of plants the company wishes to build. Among the major items of the Project Specification are:

The process specification
General design criteria
Any preliminary flow sheets (duly labeled "Preliminary")
Utilities specification
Basis for economic evaluations
Materials of construction
Equipment standards and codes
Instrument standards required

These items are described in detail in Chapter 17.

When completed and approved, the duty specification will form part of the enquiry document soliciting competitive bids from suitable contracting companies. More importantly it will form part of the contract awarded to the successful contractor. It will be accompanied in both the enquiry document and the contract with a detail of the contractor's scope of work, type of contract (i.e., Cost Plus, Lump Sum, etc.), guarantees required and other contractual items, including progress reports and meetings, approval requirements, etc.

Product blending

Finished products are blended with two or more components which are run down stream products from the refinery processes. The blending in most modern refineries is done "In Line". That is to say that measured amount of each component are mixed together in the line that finally enters the respective finished product storage tanks. The flow of each of the components is controlled 'on line' by analyzers which are programmed to open or close the component flow control valves to meet the specified product blend recipes. Although these analyzers are quite accurate the final contents of the finished tanks are always checked by laboratory tests before dispatch out of the refinery. Sufficient room is left in the finished tank to allow correction if this becomes necessary. Gasolines are always checked for octane number, Reid vapor pressure and volatility. Kerosenes are checked for flash point, and volatility. Aviation gasolines are subjected to much more stringent properties which would include water content, vapor pressure, and sulfur content. Gas oils (including diesel) are checked for diesel index, sulfur content, pour point, viscosity, and flash point. Fuel oils are tested for viscosity, sulfur content, flash point, and pour points. All finished products are tested for specific gravity. Chapter 2 describes and discusses the product properties. This chapter also illustrates the blending recipes in a typical process configuration.

Pseudo components

This subject has been described earlier under the item predicting product properties.

Pumparound

This is the term given to any reflux stream which is created inside the distillation tower by taking off a hot liquid stream, cooling it, and returning the stream back into the tower two or three trays above the draw off tray. The following Figure 19.P.2 shows a typical pumparound system.

Pumparound systems are usually associated with complex distillation processes such as the atmospheric and vacuum crude oil distillation, where a single reflux stream (i.e., overhead condensate) would result in too low a reflux in the lower section of the tower. A single overhead reflux also would require a large tower diameter to cater for

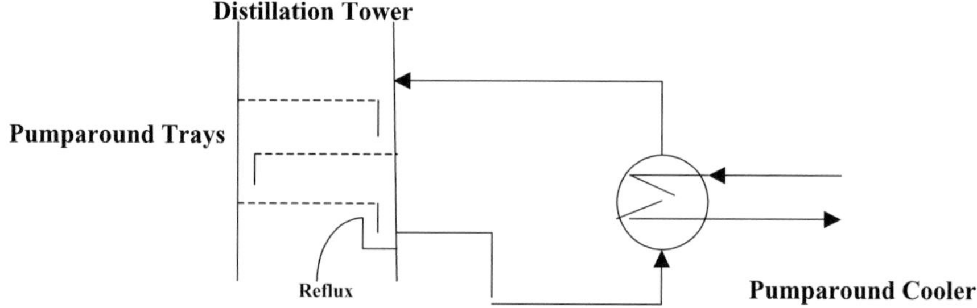

Figure 19.P.2. A typical pumparound arrangement.

the increased tower vapor and liquid traffic. The pumparound concept does smooth out the liquid/vapor traffic in the whole tower. The highest try loading in any distillation tower is below the bottom reflux stream. The addition of a pumparound therefore becomes a prime choice for a study to increase the distillation tower capacity on a revamp project.

Pumps

Types of pumps

Pumps in the petroleum and other process industries are divided into two general classifications which are

- Variable head-capacity
- Positive displacement

The variable head capacity types include centrifugal and turbine pumps whilst the positive displacement types cover reciprocating and rotary pumps. A summary of these types is as follows:

Centrifugal pumps. Centrifugal pumps comprise of a very wide class of pumps in which pumping of liquids or generation of pressure is effected by a rotary motion of one or several impellers. There are no valves in centrifugal type pumps (except of course, isolation valves). Flow is uniform and devoid of pulsation.

Turbine pumps. Turbine pumps are a type of centrifugal pumps designed to recover power in systems of high flow and high differential pressure. These pumps transmit some of the kinetic energy in the fluid into brake horsepower. The actual energy recovery is about 50% of the hydraulic horsepower available. This type of pump is expensive and is therefore not as widely used as the centrifugal pump.

Rotary pumps. Rotary pumps are positive displacement pumps. Unlike the centrifugal type pump these types do not throw the pumping fluid against the casing but push the fluid forward in a positive manner similar to the action of a piston. These pumps however do produce a fairly smooth discharge flow unlike that associated with a reciprocating pump. The types of rotary pumps commonly used in a process plant are:

Gear pumps
Screw pumps
Lobular pumps
Vane pumps

Reciprocating pumps. These are positive displacement pumps that use a piston within a fixed cylinder to pump a constant volume of fluid for each stroke of the piston. The discharge from reciprocating pumps is pulsating. Reciprocating pumps fall into two general categories. These are the simplex type and the duplex type. In the case of the simplex pump there is only one cylinder which draws in the fluid to be pumped on the back stroke and discharges it on the forward stroke. External valves open and close to enable the pumping action to proceed in the manner described. The duplex pump has a similar pumping action to the simplex pump. In this case however there are two parallel cylinders which operate on alternate strokes to one another. That is when the first cylinder is on the suction stroke the second is on the discharge stroke.

Other positive displacement pumps. There are other positive displacement pumps commonly used in the process industry for special services. Some of these are:

Metering or proportioning pumps—Which are small reciprocating plunger type pumps with an adjustable stroke.

Diaphragm pumps—These pumps are used for handling thick pulps, sludge, acid or alkaline solution, and fluids containing gritty solid suspensions. These are not usually used in petroleum refining.

Characteristic curves. Pump action and the performance of a pump are defined in terms of their *Characteristic Curves.* These curves correlate the capacity of the pump in unit volume per unit time versus discharge or differential pressures. Typical curves are shown in Figures 18.13–18.15 of Chapter 18.

Figure 18.13 is a characteristic curve for a reciprocating simplex pump which is direct driven. Included also is this reciprocating pump on a power drive.

Figure 18.14 gives a typical curve for a rotary pump. Here the capacity of the pump is plotted against discharge pressure for two levels of pump speed. The curves also show the plot of brake horsepower versus discharge pressure for the two pump speed levels.

Figure 18.15 is a typical characteristic curve for a centrifugal pump. This curve usually shows four pump relationships in four plots. These are:

- A plot of capacity versus differential head. The differential head is the difference in pressure between the suction and discharge.
- The pump efficiency as a percentage versus capacity.
- The brake horsepower of the pump versus capacity.
- The net positive suction head (NPSH) required by the pump versus capacity. The required NPSH for the pump is a characteristic determined by the manufacturer.

Detailed description and definition of terms associated with refinery pumps are given in Chapter 18. The chapter also describes the various pump drivers and supports these descriptions and discussions with worked examples. Finally the chapter gives the contents of a typical process specification for a centrifugal pump.

R

Reboilers

Reboilers are one of two heat energy input systems to a fractionation unit. The other source is the heat delivered by the feed or feeds to the unit. Reboilers are usually associated with the light ends distillation units and the product stabilizers and splitters on catalytic or thermal cracking units. In most cases, the reboiler is of a shell and tube type or a kettle type. In certain cases a fired heater may be used as a reboiler. This unit is fed either by the liquid phase from the bottom tray of the tower, or by vaporizing a portion of the bottom product. The first method uses thermosyphon as the driving force for flow through the heat exchanger. In the second case, the reboiler feed may be pumped or flow into the kettle section of the exchanger. In both cases, the flow from the reboiler is returned to the tower below the bottom fractionating tray. These two types are shown in Figure 19.R.1.

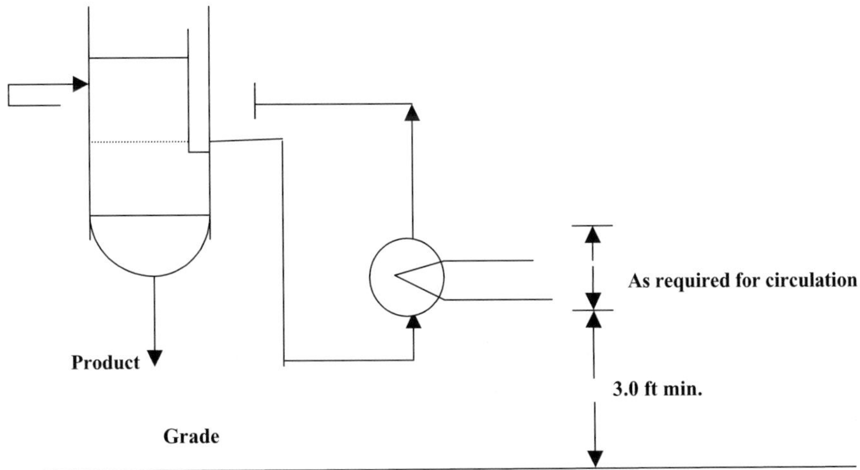

Figure 19.R.1. A once through thermosyphon reboiler.

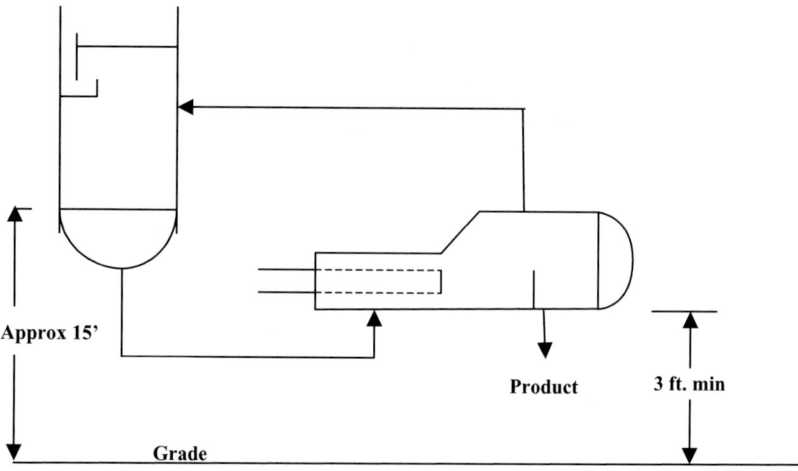

Figure 19.R.2. A kettle type reboiler.

There are two other types of reboilers used in the petroleum industry. These are the recirculating thermosyphon and the fired heater reboiler. Details of all types of reboilers are given in Chapter 18.

This chapter also discusses the operation of the reboilers with respect to their production of vapor and liquid flow objectives, which are:

1. To provide sufficient vaporising in the tower to strip the bottom product effectively.
2. To establish the loading on the bottom tray (and section) of the tower.
3. To establish the driving force for flow through the exchanger in the case of thermosyphon reboilers.

A typical bottom of the tower calculation is provided as an example. The heating medium for the heat exchange reboilers may be a hot process stream, or steam at an appropriate temperature and pressure level.

Reciprocating compressors

Reciprocating compressors are widely used in the petroleum and chemical industries. They consist of pistons moving in cylinders with inlet and exhaust valves. They are cheaper and more efficient than any other type in the fields in which they are used. Their main advantages are that they are insensitive to gas characteristics and they can handle intermittent loads efficiently. They are made in small capacities and are used in applications where the rates are too small for a centrifugal. Reciprocating compressors are used almost exclusively in services where the discharge pressures are above 5,000 PSIG. However, when compared with centrifugal compressors, the reciprocating compressors require frequent shutdowns for maintenance of valves and other wearing parts. For critical services, this requires either a spare compressor or a multiple compressor installation to maintain plant throughput. Chapter 18 describes and discusses reciprocating compressors in detail. The chapter includes discussion on the compressors ancillary equipment such as inter stage coolers, valve lifters, and compressor control in general. Table 18.22 in this chapter lists the services for which the various types of compressors are used. It also provides calculation procedures with example calculation for determining the compressor characteristics such as horse-power, suction, and discharge conditions, etc. Finally the chapter provides the input required in a typical process specification for a reciprocating compressor.

Refineries

Most of this work is devoted to the energy refinery. That is the refinery that converts the crude oil feed to energy products such as gasoline, diesel, aviation turbine gasoline, fuel oil, and the like. There are two major refinery complexes however that convert the crude oil into non-energy products. These are the lube oil refinery and the petrochemical refinery. Very often these complexes are located adjacent to the energy refinery and they are often integrated into one major refinery complex. Chapter 12 describe and discuss these two refinery complexes. They are also summarized in the introduction in Chapter 1.

The lube oil refinery

The process configuration for a typical lube oil refinery is given in Chapter 1 as Figure 1.13. Briefly, the light vacuum distillate from the crude vacuum distillation unit is routed to a secondary vacuum distillation unit where a light and heavy spindle oil cuts are removed as distillate and light motor grade oil as residue. Both the motor oil and heavy spindle oil are hydrotreated in blocked operation. The light and hydro treated heavy spindle oils, with a portion of the motor oil enter the spindle oil pool. The remainder of the motor oil is routed to the light engine lube pool. The medium vacuum gas oil distillate from the crude oil vacuum distillation unit is routed directly to a de-waxing unit (in this case a methyl ethyl ketone process). A portion of the de-waxed product is routed to the light engine oil pool while the remainder enters the heavy engine oil pool. The heavy vacuum gas oil cut from the vacuum unit is treated in the fufural extraction unit for the removal of the heavy olefin components before entering the de-waxing plant operating on a blocked operation. The de-waxed product from this operation is divided between the heavy engine oil pool and the turbine lube oil pool. The residue from the crude oil vacuum unit is de-asphalted in a propane de-asphalting unit. (Note: There are other de-asphalting processes but propane de-asphalting is the most common). The de-asphalted oil from this unit is called 'Bright Stock' and its processing follows that of the heavy vacuum gas oil, again on a blocked operation. After fufural extraction and dewaxing the bright stock is routed to the turbine lube oil pool.

Finally a portion of the vacuum residue is routed directly to the bitumen blending pool bypassing the PDA (propane de-asphalting) unit. A portion of the extracted asphalt stream from the PDA unit is routed to a bitumen blowing unit while the remainder joins the vacuum residue in the bitumen blending pool. The blown bitumen is the third stream that enters the bitumen blending pool. The various blended grades of bitumen are routed to storage and packaging. The base stocks of the turbine, engine oils, and spindle oils are sent to blending and packaging for further preparation of the respective commercial products ready for dispatch.

The petrochemical refinery

A configuration for a typical petrochemical refinery is given in Chapter 1 and shown as Figure 1.14 in that chapter. This configuration confines itself to the production of the olefin feedstock following steam cracking of a feedstock consisting of light atmospheric crude oil distillate, a naphtha cut, and the light distillate from a hydro-cracker. The refineries aromatic product stream is produced in a conventional catalytic reformer operating with a blend of straight run naphtha and hydro-cracker naphtha as feedstock. This reformate product enters an aromatic extraction unit where an aromatic rich stream leaves as the product. This configuration shows a coking unit taking the vacuum crude distillation residue as feed. The main purpose of this unit in the configuration is to provide increased feed to the hydrocracker and thus additional naphtha to the reformer and steam cracker. Other residue conversion units may be considered as an alternate to the coking unit, such as 'Deep Fluid Cracking' unit, or a residue hydrocracking unit.

Reformulated gasolines

The requirements of the clean air act of 1990 and additions to it since have changed refining requirements to meet this product's need quite significantly. Prior to this date much of the gasoline finished product recipe consisted of normal light naphtha, a reformate, usually some cracked naphtha and possibly an alkylate, with some butane added to meet volatility. The clean air requirement and its subsequent additions forced a reduction of both the reformate and the cracked stock and to replace them with oxygenates such as MTBE and TAME to meet octane number. Oxygenates were used originally simply as a additive to improve octane number. However, because of their oxygen content their addition was also required to reduce the carbon monoxide and hydrocarbons in the emission gases. There are a number of oxygenates used in gasoline manufactures; some of the more common are given in the following Table 19.R.1.

Table 19.R.1. Oxygenates commonly used in gasoline

Name	Formula	RON	RVP psig	Oxygen wt%	*Water solubility %
Methyl tertiary butyl ether (MTBE)	$(CH_3)_3COCH_3$	110–112	8	18	4.3
Ethyl tertiary butyl ether (ETBE)	$(CH_3)_3COC_2H_5$	110–112	4	16	1.2
Tertiary amyl methyl ether (TAME)	$(CH_3)_2(C_2H_5)COCH_3$	103–105	4	16	1.2
Ethanol	C_2H_5OH	112–115	18	35	100

*wt% soluble in water.

The EPA established limits for the use of each oxygenates in gasoline blends. For example, MTBE could be blended up to 15 vol% subject to an overall limit of 2.7 wt% oxygen content. The role of MTBE and other oxygenates was reviewed in the late 1990s after it was discovered that underground storage tanks had not been upgraded to retain reformulated gasoline. As a consequence tanks were leaking gasoline that contained MTBE into the ground and drinking water systems. Therefore, use of MTBE in the U.S. was essentially discontinued by the end of 2002. MTBE is still widely used outside the U.S. although the trend in western Europe is to use ETBE instead. More details on the manufacture of gasolines are given in Chapter 2.

Reid vapor pressure

This test is the standard test for low boiling point distillates. It is used for naphtha, gasoline, light cracked distillates, and aviation gasoline. For the heavier distillates with vapor pressures expected to be below 26 psig at 100°F the apparatus and procedures are different. Only the Reid vapor pressure for those distillates with vapor pressures above 26 psig at 100°F are described in Chapter 16. The apparatus used for this test is given as Figure 16.5.

Relief valves

Full details and discussion on relief systems which include the relief valves used in the petroleum industry are given in Chapter 13. The following is a list of those types of relief valves commonly used in industry. These have been approved according to ASME V111 “Boiler and Pressure Vessel” code:

Conventional safety relief valves

In a conventional safety relief valve the inlet pressure to the valve is directly opposed by a spring closing the valve, the back pressure on the outlet of the valve changes the inlet pressure at which the valve will open. A diagram of a conventional relief valve is shown in Chapter 13 as Figure 13.37.

Balanced safety relief valves

Balanced safety valves are those in which the back pressure has very little or no influence on the set pressure. The most widely used means of balancing a safety relief valve is through the use of a bellows. In the balanced bellows valve, the effective area of the bellows is the same as the nozzle seat area and back pressure is prevented from acting on the top side of the disk. Thus the valve opens at the same inlet pressure even though the back pressure may vary. A diagram of a balanced safety relief valve is shown in Chapter 13 as Figure 13.38.

Pilot operated safety relief valves

A pilot-operated safety relief valve is a device consisting of two principal parts, a main valve and a pilot. Inlet pressure is directed to the top of the main valve piston, and with more area exposed to pressure on the top of the piston than on the bottom; pressure, not a spring, holds the main valve closed. At the set pressure the pilot opens reducing the pressure on top of the piston and the main valve goes fully open.

Resilient seated safety relief valves

When metal-to-metal seated conventional or bellows type safety relief valves are used where the operating pressure is close to the set pressure, some leakage can be expected through the seats of the valve (Refer to API Standard 527, "Commercial Seat Tightness of Safety Relief Valves with Metal-to-Metal Seats").

Rupture disk

A rupture disk consists of a thin metal diaphragm held between flanges. The disk is designed to rupture and relieve pressure within tolerances established by ASME Code.

Residue conversion units

The conversion of residues to more commercially attractive products is described and discussed in Chapter 11. The processes addressed in this context are:

Thermal cracking
'Deep oil' fluid catalytic cracking
Residuum hydro cracking and desulfurization

Thermal cracking

The thermal cracking processes refer to those that convert the residuum feed (whether atmospheric or vacuum residues) into higher grade products such as naphtha and middle distillates, by heat at high temperature alone. That is, no catalyst or chemicals are used in the conversion. The processes themselves are:

Visbreaking
Thermal cracker
Cokers

Certain confusion exists in the definition of visbreaking and thermal cracking. Differentiation is based on the type of feedstock, severity of cracking or the final result. Strictly speaking, the term visbreaking should refer only to the viscosity reduction of heavy stock as the process main objective.

A residue feed stream (either from the atmospheric tower or the crude vacuum tower), is cracked in a specially designed heater. The effluent from the heater is quenched and routed to a fractionator, sometimes with a pre-flash. The products of cracking such as light gases, naphtha, gas oil, and residue are separated in the conventional manner. Some 20% of the residue feed can be converted into lighter products, mostly gas oil, by this process. Figure 11.1 of Chapter 11 shows a typical one stage thermal cracker. A visbreaker unit would have the same process configuration but the temperature severity and residence time in the heater would be less stringent than that required for product up grading.

There are two coker processes: the delayed coker, and the fluid coker. In the delayed coker the residuum feed is heated to above its dissociation temperature in a fired heater. Some coke is formed in the heater but the major portion of the coke is formed in the large drum to which the heater effluent is routed. The hot effluent is retained in this drum for a specified period of time (in some cases 8–12 hr). The liquid phase is drawn off slowly leaving the coke to remain in the drum. At the end of the specified period or when the drum is full, the coke is cooled and removed using high-pressure water jets.

Most present day fluid-coking units use a proprietary process licensed by Exxon called Flexi-Coking. Briefly, this process is an extension of the traditional fluid-coking process. The extension allows for the gasification of the major portion of the coke make to produce a low Btu gas.

Heavy residuum feed is introduced into the reactor vessel where it is thermally cracked. The heat for cracking is supplied by a fluidized bed of hot coke transferred to the reactor from the heater vessel. The vapor products of the reaction leave the reactor zone to enter the product recovery section consisting of a scrubber and a conventional fractionating facility. Steam is introduced to the bottom of the reactor to maintain a fluid bed of coke and to strip the excess coke leaving the reactor free from entrained oil.

The coke leaving the reactor enters the heater vessel, where some of the coke is converted into CO/CO_2 in the presence of air. This conversion of the coke provides the heat for cracking which is subsequently transmitted to the reactor by a hot coke stream. The net coke make leaves the heater and enters the gasifier vessel. Air and steam are introduced into the gasifier to react with the coke producing a low Btu gas consisting predominately of hydrogen, CO, CO_2, and nitrogen. This gas together with some excess air is transferred to the heater, and leaves this vessel to be suitably cleaned and cooled.

Both the delayed coker and the flexi-coker are described and discussed in Chapter 11 with process configuration drawings shown as Figures 11.4 and 11.5 in that chapter.

Deep oil cracking

This process is a form of the traditional Fluid Catalytic Cracking Process (FCCU). Originally FCC was restricted in the type of feedstock it could handle and one of the major constraints in this respect was the Conradson carbon contents of the feed. Even for the processing of distillate it had to be below 10 ppm. There were other constraints such as metal content that made the processing of residuum impossible. However in the late 1980s with a much improved catalyst and minor modification to the process configuration it became feasible to process most residues, including vacuum residue in the so-called "deep catalytic cracking" unit, or DC Full details of this process and discussion are given in Chapters 5 and 11. These chapters also contain process diagram of deep oil cracking process.

Residue hydrocracking

The most common hydro-cracking process for residue conversion is the fixed catalyst bed process. There is also a process that utilizes an ebullated catalyst bed but only the fixed bed process is described in Chapter 11. A typical process diagram for a complex utilizing the hydrocracker coupled with a thermal cracker or visbreaker is shown as Figure 11.15. Bitumen feed from the crude vacuum distillation unit enters the hydrocracker section of the plant to be preheated by hot flash vapors in shell and tube exchangers and finally in a fired heater. A recycle and make up hydrogen stream is similarly heated by exchange with hot flash vapors. The hydrogen stream is mixed with the hot bitumen stream before entering the hydrocracker heater. The feed streams are preheated to the reactor temperature in the fired heater and enter the top of the reactor vessel. The feed streams flow downwards through the catalyst beds contained in the reactor.

The reactor effluent leaves the reactor to enter a hot flash drum. Here the heavy bituminous portion of the effluent leaves from the bottom of the drum while the lighter oil and gas phase leaves as a vapor from the top of the drum. This vapor is subsequently cooled by heat exchange with the feed and further cooled and partially condensed by an air cooler. This cooled stream then enters a cold separator operating at a pressure only slightly lower than that of the reactor. A rich hydrogen gas stream is removed from this drum to be amine treated and returned as recycle gas to the process. The distillate liquid leaves from the bottom of the separator to join a vapor stream from the hot flash surge drum (thermal cracker feed surge drum). Both these streams enter the cold flash drum which operates at a much lower pressure than the upstream equipment. A gas stream is removed from the drum to be routed an absorber unit. The liquid distillate from the drum is routed to the de-butanizer in a light ends recovery unit and subsequently the product recovery process.

Residues

In petroleum refining the term 'Residue' refers to the un-vaporized portion of the heated crude oil entering either the atmospheric crude oil distillation tower or vacuum tower that leaves these towers as their bottom product. The stream from the atmospheric column is often referred to as the 'long' residue while that from the vacuum unit is often called the 'short' residue or bitumen. Both residues are black in color the atmospheric residue has a specific gravity usually between 0.93 and 0.96, whist the vacuum residue will be 0.99 and higher.

Road and rail loading facilities

Please refer to the item offsites in this chapter and Chapter 13.

S

Safety systems

The most hazardous occurrence in any oil refinery or really any establishment that handles hydrocarbons is that of a fire. Because of this, considerable emphasis in the safety policy of these plants is geared to fire prevention and fire fighting.

The fire safety measures are more important perhaps in oil refining than any other related facility because of the relative size of most refineries compared with petrochemical or chemical facilities. Refinery prevention and protection begins at the early stages of the refinery design and engineering. Chapter 6 details the development of the fire prevention and fire fighting through the early stages of the refinery design in describing the relevant passages that are usually contained in the project design specification. There are engineering and design standards that contractors must adhere to. They include the standards for mechanical equipment, electrical equipment (such as the 'Area Classification Code' which sets the parameters for equipment in terms of fire proofing (i.e., whether the item is to be spark proof, etc.) that will be located in the various areas of the refinery. The piping and layout specification will detail the piping codes to be used and the material break points. It will proceed to establish the criteria for equipment and tankage layout with respect to fire prevention (e.g., distance of fired heaters from other equipment). Other design specifications include amongst other requirements, fire prevention equipment such as sprays to be located on storage tanks, and vessels. Finally this Chapter 6 describes the location, size, and operation of refinery fire mains. On large installations these could amount to three or more separate fire main loops remotely and centrally controlled and operated.

Other safety systems that are part of petroleum refining concern the handling of hazardous chemicals. The item in Chapter 6 covers some of these compounds as to their storage and handling in the refinery. The chemicals selected as the most hazardous usually met with in the refining processes are: AHF (anhydrous hydrofluoric acid), amines, furfural, hydrogen sulfide, and MEK (methyl ethyl ketone). The composition of each of these chemicals is described, together with their injurious effect on humans and the remedial procedures to be adopted in the event of an accident. The chapter continues with a review of the materials of construction for these chemicals and the protective clothing that should be worn when handling them. The fire hazards of these chemicals are also highlighted.

Side stream stripping

Side streams from multi-component distillation towers are stripped free of entrained lighter products. Stripping may be accomplished either by injection of steam through the hot side-stream in a trayed column, or by injection of an inert gas instead of steam, or by reboiling the bottom product of the stripper tower. By far the most common method is that of steam stripping and the most common application is on the atmospheric crude distillation unit. Figure 19.S.1 shows the bottom distillate side stream stripper of a crude distillation unit.

Each side stream draw off from the main tower would be stripped free of entrained light ends in similar trayed columns. Normally these side stream columns would

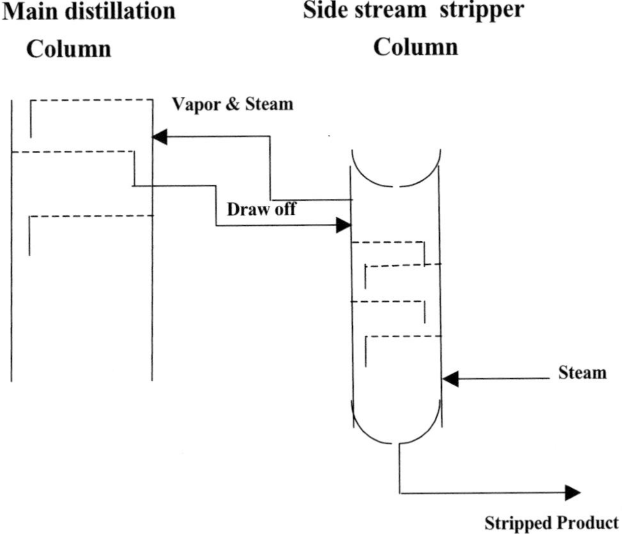

Figure 19.S.1. Side stream steam stripper.

be stacked to provide a single stripping tower. Such an arrangement is shown in Chapter 3 Figure 3.2. This chapter also describes the side stream stripper function with example calculations. It also provides recommended steam rates for the various stripping functions.

Soaking volume factor

The design of a thermal cracker is keyed to the configuration and temperature profile across the heater and soaking drum or soaking coil. The degree of cracking is dependant on this temperature profile and the residence time of the oil under these conditions. The soaking volume factor (SVF) is related to product yields and the degree of conversion. Definition of these items are given in Chapter 11. A design calculation using the SVF (soaking volume factor) is given as Appendix 1 of this chapter.

Specific gravity

The specific gravity of a liquid is the weight of a known volume of the liquid at a known temperature compared with water under the same conditions. The standard weight is taken as one gram and the standard temperature is usually 60°F or 15°C. The specific gravity of a petroleum compound is the basis for development of the material balance in design work, and most measurements within the refinery.

The basic specific gravities are given as an essential part of the crude assay. They are usually presented as a curve of specific gravities (usually quoted as °API) against mid point distillation temperatures. API gravities are related to specific gravities by the equation: specific gravity $= 141.5/(131.5 + {}^{\circ}\mathrm{API})$. The specific gravity of any petroleum compound may be calculated using the method provided in Chapter 1. The specific gravity of crude oil and its products are obtained in the refinery using the test method described in Chapter 16. The method uses a properly calibrated hydrometer under laboratory conditions.

Splitter, naphtha

In all hydro skimming refineries the key process, next to the crude distillation process is the catalytic reformer. The correct design and subsequent operation of this process produces the hydrogen stream that is required by many refinery operations. Important to the efficient operation of this process is the correct boiling point range of the naphtha feed. This is ensured by the fractionation of the full range naphtha stream from the crude unit overhead distillate. This is accomplished as part of the light ends unit complex. Typically the total overhead distillates plus in some cases other naphtha distillates (from thermal crackers) are first de-butanized in the light ends de-butanizer column. The bottom product from this column is the de-butanized full range naphtha. This stream is delivered hot to a naphtha splitter fractionator which produces a light naphtha overhead and a heavy naphtha the bottom product. The fractionation between these two products maximizes the naphthene content of the heavy naphtha. As this heavy naphtha is fed to the catalytic reformer the amount of naphthenes in its composition will, to large extent, determine the amount of hydrogen the unit will produce. Splitter towers contain between 25 and 35 actual distillation trays and operate at overhead reflux ratios of between 1.5 and 2.0. Further details and description is given in Chapter 4.

Stacks

Stacks are used to create an updraft of air from the firebox of a heater. The purpose of this is to cause a small negative pressure in the firebox and thus enable the introduction of air from the atmosphere. This negative pressure also allows for the removal of the products of combustion from the firebox. The stack therefore must have sufficient height to achieve these objectives and overcome the frictional pressure drop in the firebox and the stack itself.

The height required for a stack to achieve good draft can be estimated from the following equation:

$$\mathrm{D} = 0.187\,\mathrm{H}\,(\rho_{\mathrm{a}} - \rho_{\mathrm{g}})$$

where

D = draft in ins of water.
H = stack height in feet.
ρ_a = density of atmospheric air in lbs/cuft.
ρ_g = density of stack gasses in lbs/cuft at stack conditions.

For stack gas temperature use 100°F lower than gases leaving the convection section. Stack gasses are mostly nitrogen with some CO_2 with an average molecular weight of about 28. Details on stacks including velocity head, emissions, and environmental considerations are given in Chapter 18.

Steam and condensate systems

In most plants, steam condensate accumulated in the various processes is collected into a single header and returned to the boiler or steam generating plant. It is stored separate to the treated raw water because condensates may contain some oil contamination. A stream of treated water and condensate are taken from the respective storage tanks and pumped to the deaerator drum. The condensate stream passes through a simple filter on route to the deaerator to remove any oil contamination. Low-pressure steam is introduced immediately below the packing in the deaerator to flow upwards countercurrent to the liquid streams to remove any entrapped air in the liquid. The deaerated boiler feed water (BFW) is pumped by the boiler feed water pumps into the steam drum of the steam generator. There will normally be three 60% pumps for this service. Two will be operational and one will be on standby. These pumps are the most important in any chemical plant or a petroleum refinery. If they fail no steam can be generated and the whole complex is in danger of total shutdown or worse. Therefore, three separate pumps are used to cater for the normal high head and high capacity, and a separate pump driver operating on a completely different power source than electrical power or steam (usually a diesel engine) is mandatory for atleast one of the pumps to minimize the danger of complete shut down.

The steam drum is located above the generator firebox. The liquid in the drum flows through the generator coils located in the firebox by gravity and thermo-syphon. A mixture of steam and water is generated in the coils and flows back to the steam drum. Here, the steam and water are separated with the steam leaving the drum to enter the super-heater coil. This coil located in the lower section of the convection side of the heater super heats the saturated steam to the high-pressure refinery steam mains. Let down stations may be located at various points in the refinery to create lower pressure main systems. De-super heaters are used to establish the correct temperature levels in these lower pressure mains. Chapter 13 gives details of a steam generation systems and condensate recovery. Figure 13.27 provides a schematic diagram of a steam generation plant.

Storage facilities

Please refer to the item in Part 2 Offsites and Chapter 13.

Sulfur removal

Naphtha desulfurization

This uses catalytic ("cat") reformer hydrogen on a once through basis. Heavy naphtha feed to the cat reformer is fed to the naphtha hydrofiner from storage. The feed stream and the hydrogen gas stream are pre-heated by exchange with the hot reactor effluent stream. The feed then enters the fired heater which brings it up to the reactor temperatures (about 450°F) and leaves the heater to enter the reactor which operates at about 400–450 psig. Sulfur is removed from the hydrocarbon as hydrogen sulfide in this reactor and the reactor effluent is cooled to about 100°F by heat exchange with the feed. The cooled effluent is collected in a flash drum where the light hydrogen rich gas is flashed off. This gas enters the suction side of the booster compressor which delivers it to other hydrotreaters. The liquid phase from the drum is pumped to a reboiled stabilizer. The overhead vapor stream from the stabilizer is routed to fuel while the bottom product, cat reformer feed, is pumped to the cat reformer.

Gas oil desulfurization

This process uses a recycled hydrogen stream to desulphurise a gas oil feed. Figure 19.S.2 shows the gas oil feed entering the unit to be pre-heated with hot effluent stream before entering a fired heater. Where its temperature is increased to the reactor temperature of about 750°F. A hydrogen rich stream is introduced at the coil outlet prior to the mixed streams entering the reactor. The reactor contains a bed of cobalt molybdenum on alumina catalyst and desulfurization takes place over the catalyst with 70–75% of the total sulfur in the oil being converted to H_2S.

The reactor effluent is cooled by the cold feed stream, water or air. This cooled effluent enters a flash drum where the gas phase and liquid phase are separated. The gas phase rich in H_2S and hydrogen enters the recycle compressor. The gas stream then enters an amine contactor where the H_2S is absorbed into the amine and removed from the system. Although the diagram shows a purge stream before the amine absorber in most cases the purge is downstream after the amine cleanup. The purged gas is

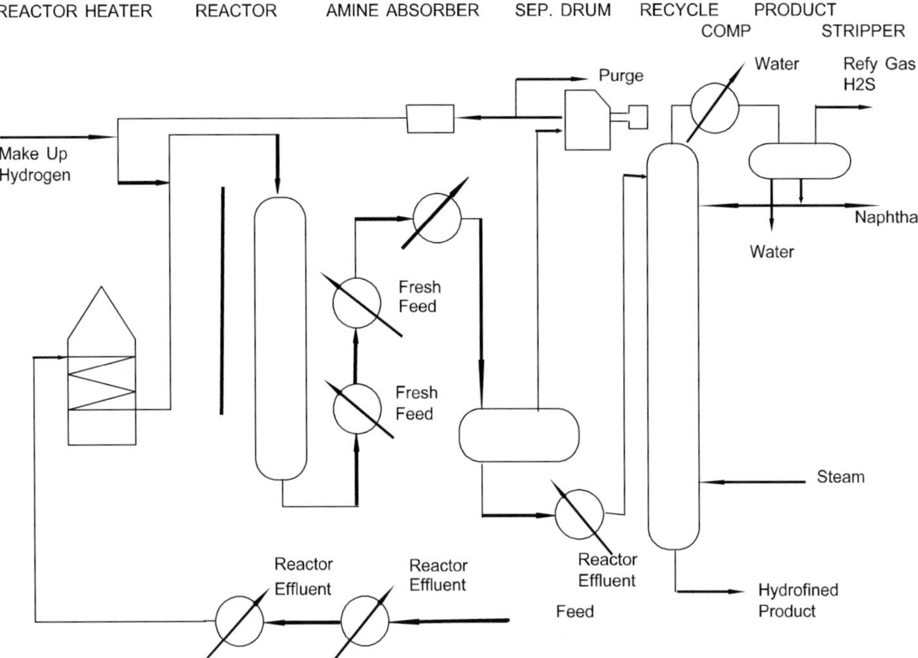

Figure 19.S.2. A typical middle distillate desulfurizer.

replaced by fresh hydrogen-rich make-up, thus maintaining the purity of the recycle gas.

The liquid phase leaving the flash drum is pre-heated before entering a stream stripping column where the light ends created in the process are removed as overhead products. The bottom product leaves the tower to be cooled and stored. Details of desulfurization of naphtha and heavier products are given in Chapter 8.

Other desulfurizing processes

Light naphtha and lighter hydrocarbons have sulfur content predominately in the form of mercaptans (see Chapter 1 for a definition of mercaptans.). These sulfur components are easily removed by a simple caustic wash or in licensed processes such as UOP's Merox process. Some refiners however elect to hydrotreat the debutanized full range naphtha before splitting the desulfurized product into the light and heavy product. Indeed several hydrotreat the whole of the crude unit overhead distillate, doing away with any further processing to eliminate the mercaptans. The problem here is however the loss of some LPG due to the high hydrogen content of the debutanizer overhead drum components.

At the other end of the crude oil product spectrum, in the area of the heavy vacuum gas oil and residue, desulfurization by hydrotreating requires a high severity operation. The reason for this is that the sulfur molecules in this case are in the form of thiophenes. These are sulfur molecules deeply encapsulated in the heavy hydrocarbon molecules. To expose the sulfur to the hydrogen stream requires some degree of cracking of the hydrocarbon molecule. Desulfurization in this case therefore becomes more of hydrocracking that hydrotreating.

Sulfur content

Sulfur content of a petroleum product or cut is always quoted as a percent by weight of the sample. The most common laboratory test for sulfur content is 'The Lamp Method ASTM D 1266' this is described in detail in Chapter 16. Briefly the test is as follows:

The sample is burned in a closed system using a suitable heat source and an artificial atmosphere of 70% carbon dioxide and 30% oxygen to prevent the formation of nitrogen oxides. The oxides of sulfur are absorbed and oxidized to sulfuric acid by means of a hydrogen peroxide solution. This solution is flushed with air to remove carbon dioxide. The sulfur as sulfate in the absorbent is determined by titration with standard sodium hydroxide. The calculation of the sulfur content from the titration is given by the following equation.

$$\text{Sulfur Content wt\%} = 16.03\,\text{M} \times (\text{A}/10\,\text{W})$$

where

A = milliliters of NaOH titrated
M = Molarity of the NaOH solution.
W = Grams of sample burned.

T

Tar

Tar is an ill-defined general term that describes heavy petroleum fractions that are solid or semi-solid at room temperature. An alternative term is *bitumen* although the latter is better used to denote naturally occurring tar deposits, as in tar pits or tar sands.

In addition to being very viscous or nonflowing materials (viscosity $>10{,}000$ cP), tars are also characterized by having relatively high densities lower than about 10 API degrees, which corresponds to a specific gravity (60/60) greater than 1.0.

$$\text{Degrees API} = \frac{141.5}{\text{sp. gr. at } 60^{\circ}\text{F}/60^{\circ}\text{F}} - 131.5$$

Depending on their physical properties, tars may be easily confused with asphalts. The main difference being that tars are usually either naturally occurring or unprocessed heavy fractions recovered as byproducts from other sources (e.g., petroleum residues or coal processing) while asphalts are typically processed or manufactured materials, whether air-blown or solvent extracted.

Sometimes, tars are further processed to recover the more volatile components. If so, the remaining very heavy residue is usually called *pitch* (1) (See also Chapter 12).

1. Gerd Collin, *"Tar and pitch." Ullmann's Handbook of Industrial Chemistry*, Wiley-VCH Verlag GmbH, 2002.

Tar sands

Tar sands or bituminous sands are several porous rock formations that contain highly viscous heavy hydrocarbon materials that cannot be recovered by conventional oil recovery methods, including enhanced oil recovery techniques. At present, the only practical method of recovering the hydrocarbons contained in tar sands is mining the tar sands followed by high-temperature retorting. Alternatively, in situ combustion or thermal processing techniques can be used to increase the fluidity of the hydrocarbons so as to enable their recovery out of the tar sands (1, 2).

Tar sands are characterized by having a relatively coarse porous structure. Some bituminous rock formations may, however, have a much tighter pore matrix that makes the recovery of the hydrocarbons even more difficult. An extreme case is evidenced in *oil shales* that are typically sedimentary clayish rock structures with a relatively high organic content (kerogen) that can be thermally decomposed to yield hydrocarbon-based oils. It is important to note that the hydrocarbons from oil shales, variously denoted as shale oils, cannot be recovered directly from the rock matrix, even using thermal means, but are only the products from the thermal decomposition of the kerogen. Often, shale oils have a relatively high nitrogen content that requires extensive hydrotreating before it can be used in standard petroleum refining applications.

Tar and shale deposits are extensively distributed worldwide and can potentially provide huge oil reserves for future generations. Some examples vary from the Brea tar pits in California, to the Athabasca tar sands in Canada, to the tar deposits in Trinidad or in the Orinoco river valley in Venezuela, to the tar deposits near the Caspian Sea, or to the shale oil deposits in Colorado and other parts in the western United States.

1. Frank J. Mink and Richard N. Houlihan, *"Tar sands." Ullmann's Encyclopedia of Industrial Chemistry*, Wiley-VCH Verlag GmbH, 2002.
2. R. W. Luhning, "Heavy oil, oil sands, and enhanced oil recovery: Where will the technology break-through come from?" Canadian Heavy Oil Association Conference on Heavy Oil—a New Direction, Calgary, Alberta, December 5, 1989.

Tetraethyl lead

A former additive used as an anti-knock additive to boost the octane number in gasoline. It was produced commercially from ethyl chloride. It was used at a dosage in 0–3 ml/gal range depending on the gasoline composition, sulfur content, and lead sensitivity. The use of tetraethyl lead has largely been discontinued worldwide, in particular as a result of the introduction of catalytic converters to clean up the exhaust from internal combustion engines. Use of lead irreversible poisons the oxidation catalysts used in such converters. Also, lead itself is a highly toxic substance that was present in automobile exhaust before it was banned with the advent of catalytic converters.

Other additives have been proposed or used to replace tetraethyl lead. Manganese derivatives (e.g., methyl cyclopentadienyl manganese tricarbonyl) were used for some time, principally in Canada. Methyl *tert*-butyl ether (MTBE) can be used both as an octane booster and as a gasoline pool extender, but MTBE has a foul smell and its use has fallen off in the United States as a result of ground water contamination from leaking gasoline tanks. It is, however, still used extensively in other countries along with ethyl *tert*-butyl ether (ETBE), and *tert*-amyl methyl ether (TAME). Passage of a legislatively mandated oxygenate requirement in the United States led to the increased usage of ethanol produced by the fermentation of grain, usually corn.

Thermal cracking

The earliest processing of crude oil involved its simple distillation (usually with batch stills) into various fractions that were variously called naphtha, kerosene, gas oil, etc. and with uses often limited to illumination and heating fuels. Neither the quantity nor the quality of the various fractions was particularly good. With the advent of the earlier internal combustion engines it became apparent that the more desirable naphtha fraction yielded only a poor performance and its availability was very limited.

Thermal cracking was the first commercial process used for the conversion of petroleum fractions into more useful products. Though largely superseded by other processes (in particular catalytic cracking), thermal cracking was used for many years for the decomposition (cracking) of heavy, high-molecular weight hydrocarbons into smaller molecules, and is still used commercially in the processing of very heavy fractions, as in visbreaking or coking.

In a different sense, thermal cracking is also the technology that is used universally for the production of light olefins (ethylene to butenes, in particular) from hydrocarbon feedstocks that range from ethane, to LPG, to naphthas, and all the way to gas oils. Cracking for the production of olefins is a far more specialized technology that is practically the basis for the entire petrochemical industry.

Within the refinery environment, thermal cracking started as a batch process in the early 1900s. Crude oil was heated in a still; the different hydrocarbons vaporized according to their respective boiling points and were then condensed in separate fractions. Gasoline was in increasing demand but the amount recovered by batch distillation was only a small fraction of the crude fed to the still. A much larger portion remained in the still as a heavy material. William M. Burton of the Standard Oil Company is credited with having discovered thermal cracking by applying heat and pressure to decompose part of the residue remaining after gasoline had been boiled off. The result was that the yields of gasoline virtually doubled relative to those obtained by straight distillation. The early years of thermal cracking were mired in litigation resulting from the timing of the filings for intellectual property protection. While Burton was the first to intentionally crack heavy oil by thermal means, his patent application filed in the summer of 1912 came much later than a 1909 filing by J. A. Dubbs for a pipe still for the demulsification of oil that incidentally also resulted in some thermal cracking. These early Burton cracking stills suffered from significant coking problems; similar coking problems were also encountered in Dubbs's pipe stills but less so when used with the demulsification water. Both technical and legal issues were eventually resolved by Carbon Petroleum (C.P.) Dubbs, son of J. A. Dubbs who, together with Gustav ("Gasoline Gus") Egloff, developed a "clean circulation" continuous thermal cracking process that largely avoided the formation of undesirable coke deposits. The success of this process led to the licensing in 1919 of a first unit for the processing of 250 barrels/day to Roxana Petroleum Corporation, a Shell subsidiary, in Wood River, Illinois. By 1923, there were 65 new Dubbs units with total processing capacity of 42,000 barrels/day and by 1926 there were 107 producing units and another 37 under construction. The process was later improved and used extensively for many years throughout the world.

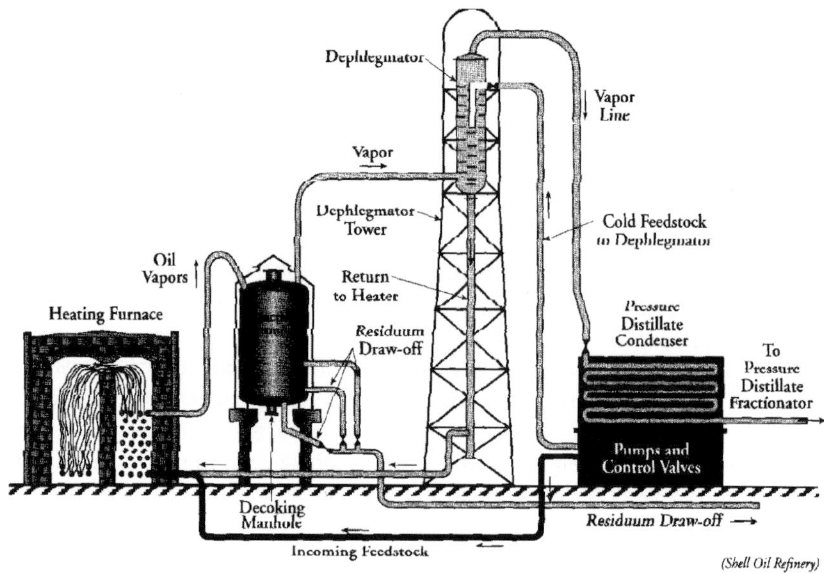

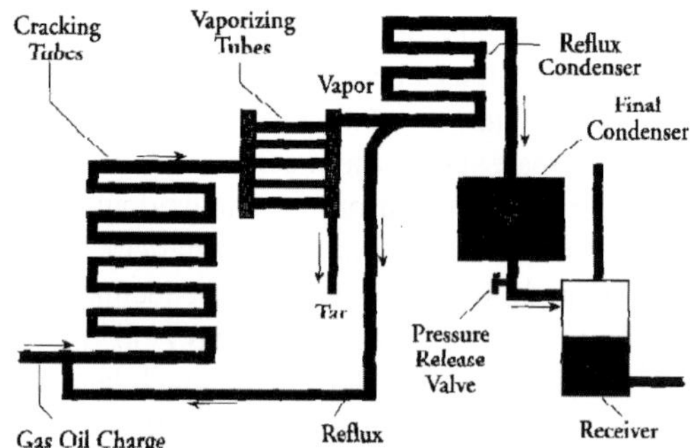

Figure 19.T.1. Flow schemes of the Dubbs circulation process.

Figure 19.T.1 illustrates a schematic of the Dubbs clean circulation process. The unit had four major components: a natural gas fired furnace where incoming oil was circulated through some 60 m of zig-zag tubes and heated to cracking temperature; an enlarged reaction chamber where the oil could remain or "stew" for a certain residence time and crack without providing additional heat; a dephlegmator where vapors from the cracked petroleum, including those of the gasoline product, could cool and partially condense before being separated; and a reflux line through which some of the heavier vapors that condensed in the dephlegmator could be rerouted back to the feed line and mixed with fresh incoming oil. The recirculation of the clean material relative to the fresh "dirty" oil was done at about a three-to-one ratio. The small amount of coke that settled in the reaction chamber could be drained off continuously into separate tanks (see also Chapter 11).

Thermal reforming

Thermal reforming is similar to thermal cracking applied to gasoline boiling range hydrocarbons. Because of consisting of smaller molecules, they are more difficult to crack and require higher severities, with furnace outlet temperatures of up to about 600°C. Good per-pass conversions and good octane improvements can be obtained while coke formation is limited because of the lighter nature of the feedstock.

An excellent review of thermal processes, both cracking and reforming, can be found in the Petroleum Processing Handbook, edited by John J. McKetta, Marcel Dekker (1992).

To a large extent thermal reforming has been superseded by catalytic reforming and thermal cracking by fluidized catalytic cracking (FCC) except as outlined above.

Thermofor

"Thermofor" is the commercial name of a continuous moving bed process used either for catalytic cracking or catalytic reforming. A distinctive feature of this process is that the catalyst, usually chromia/alumina, flows down through the reactor concurrently with the hydrocarbons. A mechanical conveying system is used to circulate the catalyst back to the top of the reactors. (1)

1. James G. Speight and Baki Özüm, *Petroleum Refining Processes*, Marcel Dekker, 2002.

Topping

"Topping" or "skimming" is the name used for the distillation of crude oil to remove the lighter fractions. The crude oil with such fractions removed is sometimes called "topped crude."

Crude oil is usually topped only down to some preselected intermediate temperature. The remaining heavy fraction may still contain substantial amounts of valuable intermediate hydrocarbons that can be recovered before the bottoms product is classed as a residuum or residue.

Most often the crude oil fractionation unit is a very large, complex column that separates the crude oil into multiple components: gas, light naphtha, heavy naphtha, kerosene, gas oil, and residue, and not just a topping operation.

Tower fractionation

A fractionation tower is a distillation column, typically with multiple trays or fractionation stages and with at least one feed and two product streams—top and bottom—but often having also provision for multiple feeds and multiple withdrawal points or side cuts.

The mechanical design of a fractionation tower is far more complex than just the specification of the operating conditions or the number of theoretical stages. Also needed will be the specification and design of its components, such as dimensions, plates or packing, risers, downcomers, internal supports, distributor nozzles, reboilers, condensers, etc.

The fractionation trays are normally divided into a minimum of three sections: one side is reserved for the reception of the liquid that descends from the plate immediately above, the middle part is reserved for liquid/vapor contacting, and the other side is reserved for the overflow of the liquid that will flow to the plate immediately below. These sections can vary in number and shape according to various criteria. Thus, for example, large columns may consist of multiple passes with downcomers and overflows arranged at various positions across the plates. The liquid may flow across the plate in the diagonal direction, or it may reversed by using suitable barriers, or it may flow radially. The downcomers and also the overflows may be straight along a chord of the cylindrical vessel or may consist of circular pipes, among other possibilities. The liquid overflow may flow over straight, curved, or cylindrical weirs. In exceptionally large columns, the trays may be stepped internally in tiers, with the liquid flowing over weirs from step to step. The intent is always to provide as much liquid/vapor contacting while avoiding the bypass of liquid across the plate or of vapor through the liquid.

One of the oldest types of contactors is the bubble-cap tray (Figure 19.T.2) in which the ascending vapors are introduced into the descending liquid through a number of slots or serrations cut along the periphery of cylindrical caps screwed onto the top of the plates and usually arranged according to a triangular pitch. Bubble-cap trays are fairly flexible, but are also fairly expensive to fabricate and install. Other types of trays have been used. The simplest and possibly also the most common are sieve trays, that consist simply of a stack of perforated plates, always with provisions for downcomers and overflows (Figure 19.T.3).

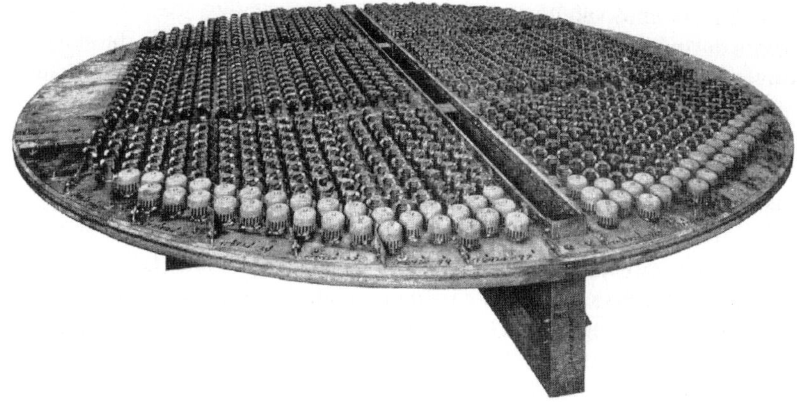

Figure 19.T.2. Conventional bubble-cap tray.

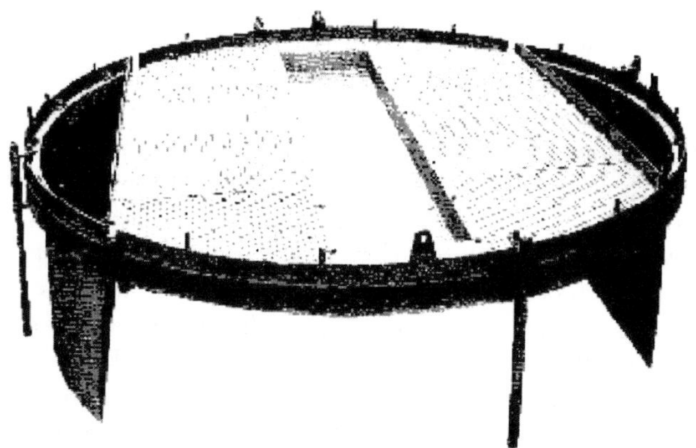

Figure 19.T.3. Conventional sieve tray.

The perforations may be in the form of round holes drilled in the plate, usually with a triangular pitch, or in the form of perforated slots cut or stamped into the trays using a variety of methods (Kittel plates, Turbogrid trays, etc.). Sieve trays are relatively inexpensive and are also fairly flexible but have more load limitations than bubble caps, both in terms of liquid weeping at low vapor loads or of vapor spouting in the case of low liquid loads. A suitable compromise is sometimes the use of valve trays in which metal valves, usually circular or rectangular, are allowed to ascend or descend within the liquid depending on the pressure differential exerted by the ascending vapors.

All these trays offer considerable mechanical complexity and are prone to fouling when used in "dirty" service. Other plates like disk-and-doughnut plates, slanted cascading plates, etc. can be used for dirty service since they may be more tolerant to fouling, avoid the accumulation of foulants, or may be easier to clean, but these types of plates usually have much lower contacting efficiencies than the traditional types.

It should evident that the correct operation of a fractionation column requires a delicate equilibrium among all the flows and the pressure gradients across the trays. If, for example, the liquid capacity of the column is increased, the quantity of liquid on the trays and in the downspouts will also increase, and so will the pressure differential across the plate. This may cause even more liquid to back up the downspouts and the system becomes unstable. The outcome usually is that the column fills up with liquid or, in common terms, it becomes *flooded*. It is, therefore, important to design the column for operation at a point sufficiently distant from *flooding*—usually about 70% or 80%—that this situation will not arise. *Flooding* can be related to a number of parameters: flow rate of the liquid or vapor is but one of them; another could be the separation between plates that affects the static pressure head, and also the frothing characteristics, if any, of the liquid. It is usually recommended to maintain a tray spacing equal at least to twice the height of the liquid in the downspout, but these rules are often violated in order to save in column height and cost. Other solutions that allow for the reduction of the spacing between trays may include the use of multiple downcomer trays, often denoted as MD trays (Figure 19.T.4).

MD trays may have different layouts with various downcomer arrangements available that can vary from parallel to cross flows at 90° angles. In the latter case, the trays have two-fold symmetry around two centerlines, which intersect at 90°. This symmetry arises from the rotation by 90° of the inlet downcomers with respect to the outlet downcomers on the next tray. This rotation also forces the liquid to flow across the tray in a 90° turn. MD trays allow for relatively short tray-to-tray distances. Several diameters up to 12–15 m have been commercialized. Multiple upcomer (MU) trays also exist and are often used in liquid–liquid extraction applications.

In all of the above the separations are effected by contacting the liquid and the vapor across mechanical (usually metallic) discrete units (trays or plates). Another possibility is to contact descending liquid with ascending vapor over an extended surface like that provided by the presence of a coarse or porous packing within the distillation vessel. Such packings usually have uniform metallic, ceramic, or polymeric structures like Raschig rings, or Berl saddles, or may even consist of rigid interlaced packing structures (usually metallic but sometimes also ceramic). The number of stages in this case is not discrete as in a tray column but, instead, is calculated on the basis of the *height equivalent of a theoretical plate* (HETP) or the *height of a transfer unit* (HTU). These columns usually have no upcomers or downcomers or other mechanical dividers (except for provisions to hold the packing in discrete segments for mechanical

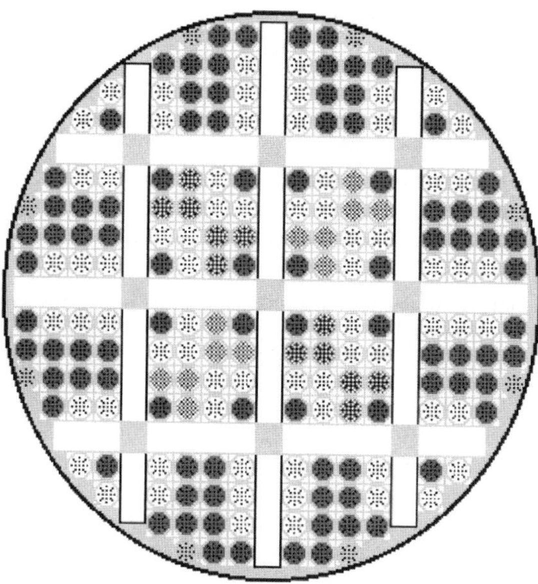

Figure 19.T.4. Typical layout of an MD tray.

purposes). Flooding is also a limitation that the designers have to contend with in packed columns (see also Chapter 18).

True boiling point (TBP)

The true boiling point distillation curve is obtained in a laboratory apparatus that is insulated from the surroundings. The outer jacket is maintained at substantially the same temperature as the temperature of the vapor within the column so that no heat is gained or lost across the fractionation stages and the column is as close to adiabatic as possible. This is often done by circulating heated air through the jacket but it can also be approached by carefully controlled zone heating. Depending on the material being distilled, true boiling point distillation may require operation under vacuum, so the apparatus is usually designed to accommodate such situations.

Distillation may be conducted either at a constant or at a variable rate. If operated at a constant rate, a smooth curve is obtained that reflects the variation in the still temperature as a function of the percentage distilled overhead. In the variable rate method the rate of distillation is adjusted depending on the amount of material that boils at each temperature. In all cases, sufficient reflux should be provided to keep the packing wet and at the maximum rate that the column will tolerate without flooding.

TBP distillation differs from an ASTM or *Engler* distillation in a number of ways. In an ASTM distillation, the hydrocarbons are distilled at a uniform rate of about 5 cc per min. The distillate is condensed and the temperature of the vapor when the first drop of condensate drips from the condenser is recorded as the Initial Boiling Point (IBP). The vapor temperature is then recorded at each successive 10% interval. When 95% had distilled the temperature of the still may have to be increased and the maximum temperature is recorded as the end point (EP). There is virtually no fractionation in an ASTM distillation. The hydrocarbons do not distill one by one according to their boiling points but only as successively higher and higher boiling mixtures.

Various correlations exist to relate TBP, ASTM, and other laboratory distillation procedures (1) (See also Chapter 1 and Chapter 3).

1. Nelson, W. L. *Petroleum Refining Engineering*, McGraw-Hill, 4th edition, 1958.

UOP K

The *UOP K*, or the *Watson K*, or the *Watson characterization factor*, is a parameter identified by Kenneth Watson (1) who defined it as follows:

$$K = [T_B]^{1/3}/(\text{sp.gr.})$$

where $[T_B]^{1/3}$ represents the cube root of the average molal boiling point of the hydrocarbon mixture $[T_B]$ in degrees Rankine ($^\circ R = {}^\circ F + 460$), divided by the specific gravity at 60°F, relative to water at 60°F.

This is a correlation parameter based on the observation that $K \sim 12.5$ corresponds to paraffinic materials while $K \sim 10.0$ indicates a highly aromatic material. It provides a means for roughly identifying the nature of a feedstock solely on the basis of two observable physical parameters.

The characterization factor has also been related to viscosity, aniline point, molecular weight, critical temperature, percentage of hydrocarbons, etc. so it can be estimated using a number of laboratory methods (2).

1. Characterization of petroleum fractions, *Ind. Eng. Chem.*, **27**, 1460, 1935.
2. Nelson, W. L., Petroleum Refinery Engineering, McGraw-Hill, 4th edition, 1958.

Urea dewaxing

A process for producing low pour point oils in which straight-chain paraffins are removed from the feedstock by complexing them with urea to form a crystalline adduct that can be separated by filtration.

It is interesting that, in the adducts, urea forms spiral structures connected by hydrogen bonds in a hexagonal crystalline structure with an internal channel diameter of about 5 angstrom into which only molecules with a smaller cross-sectional dimension can fit (1). In a sense, this behavior mimics the separation of n-paraffins from hydrocarbons with 5A molecular sieves that is extensively used in industry either for the production of heavy n-paraffins or, in combination with paraffin isomerization for the enhancement of the octane number of hydrocarbons in the gasoline boiling range. Heavy n-paraffins can be cracked to linear α-olefins or can be catalytically dehydrogenated to linear internal olefins; both are extensively used in the detergent industry as feedstocks for the production of detergent alcohols or of biodegradable linear alkylbenzenes. (See also Chapter 12.)

1. John J. McKetta (ed), *Petroleum Processing Handbook*, Marcel Dekker, 1992.

V

Vacuum distillation unit

Vacuum distillation unit refers to the further distillation of the residue portion of atmospheric distillation of the crude. The boiling curve range of this portion of the crude is too high to permit further vaporization at atmospheric pressure. Cracking of the residue would occur long before any temperature level for effective distillation would be reached. By reducing the pressure, however, the danger of cracking in further heating the residue oil for further distillation is reduced. Figure 19.V.1 is a process diagram of a typical crude oil vacuum distillation unit. These units operate at overhead pressures as low as 10 mmHg. Under these conditions, the hot residue feed from the atmospheric distillation unit is partially vaporized in a fired heater and enters the vacuum distillation tower at temperatures around 700°F. The hot vapors rise up the tower to be successively condensed by a cooled internal reflux stream moving down the tower as was the case in the atmospheric distillation unit. The condensed distillate streams are taken off as side stream distillates. There is no overhead distillate stream in this case. The high vacuum condition met with in these units is produced by a series of steam ejectors attached to the unit overhead system.

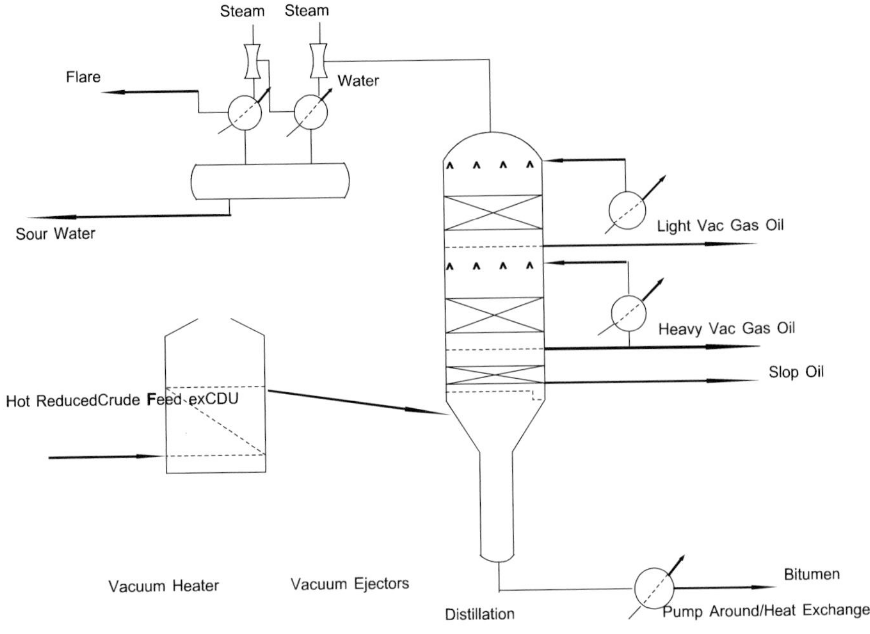

Figure 19.V.1. Vacuum distillation

A low-pressure drop within the tower allows for a flash zone pressure sufficiently low to accommodate the flash temperature of the feed below its cracking temperature (i.e., around the 700–750°F). Typical side stream products from this unit are as follows:

Top side stream	Light vacuum gas oil	690–750°F cut point
2nd side stream	Heavy vacuum gas oil	750–985°F cut point
Residue	Bitumen	+985°F cut point

The low-pressure drop tower internals are beds of proprietary grid packing. A portion of the respective side stream draw off is routed through coolers to be returned as a cool pumparound reflux stream to the packing section above each draw off tray. This unit is further described and discussed in Chapter 3.

Visbreaking process

The visbreaking process is a mild thermal cracking of crude oil residues. It is used to reduce the viscosity of vacuum residue to meet the fuel oil specification. The process configuration is very similar to the conventional once through thermal cracker except

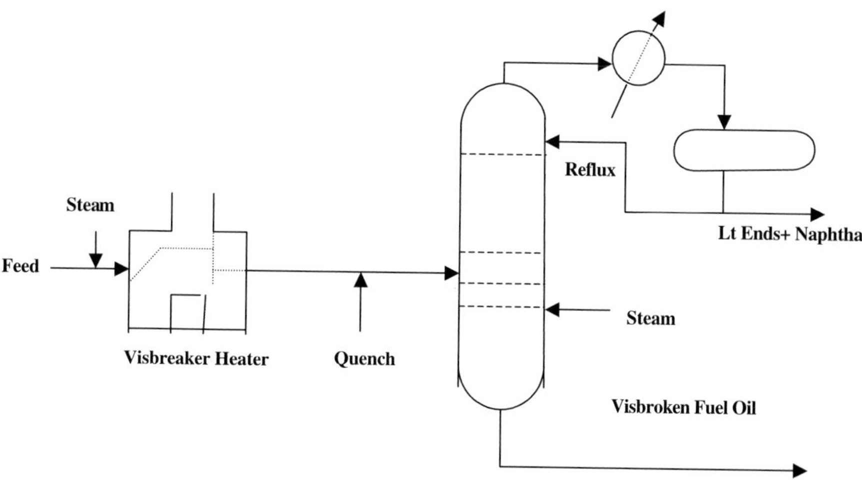

Figure 19.V.2. Typical visbreaker unit.

for the routing of the recovery products from the fractionator. Figure 19.V.2 shows the configuration of a typical visbreaker.

The cabin type heater has two sections: a heating section and a soaking section. A fire wall separates the two sections, a configuration similar to any thermal cracker. The residue feed enters the heater section together with high-pressure steam and is heated to a temperature of around 780–800°F before crossing over the separating fire wall to the soaking section. Here it is allowed to remain for a calculated period of time at a temperature slightly higher than the temperature it reached in the heater section. Some mild cracking occurs so as to produce some range of lower boiling material. The heater effluent leaves the heater to be quenched with cold fuel oil product before entering the fractionator. Unlike a thermal cracker, only the naphtha and lighter cracked distillate is taken off as an overhead product. There will be no middle distillate cut in this case. That material which has been formed by the cracking mechanism will remain in the visbroken bottom product as fuel oil. The design of the fired heater is the same as for any thermal cracker. Section of Chapter 11, describes 'The Soaking Volume Factor' concept.

Viscosity

The viscosity of any petroleum product is a measure of its resistance to flow. This measurement is important in many facets of process design and indeed is an essential quality of many finished products.

Chapter 16 describes the test method for kinematic viscosity. There are two basic viscosity parameters. They are:

Dynamic or absolute viscosity
Kinematic viscosity

Both are related since the kinematic viscosity may be obtained by dividing the dynamic viscosity by the mass density. The metric unit for viscosity is called the *poise* (P). The unit most often used in the petroleum industry for this measure is the *centipoise* (cP) which is the poise divided by 100. This dimensions of the poise are:

$$\frac{\text{gram}}{\text{cm} \times \text{sec}}.$$

The kinematic viscosity dimension in English units is the *square foot per second.* And in metric units is *square centimeter per second* called *the stoke.* In the petroleum refining industry, the stoke divided by 100, called *the centistoke*, is the unit most often used. When the terms centipoise and centistokes are used the mass density is numerically equal to the specific gravity.

Various types of instruments are available to determine viscosity in other terms. These terms are SSU which is Saybolt Seconds Universal and SSF which is Seconds Saybolt Furol. All viscosity terms can be converted by factors to one another. These factors may be found in most engineering data books such as Cameron Hydraulic Data and The GPSA (Gas Processors Suppliers Association) Engineering Hand Book. Viscosity is used extensively in product blending and quality control of petroleum products. In blending for viscosity, a blending index concept must be used. That is, one cannot blend directly using just volumetric proportions. The viscosity indices are given and discussed in Chapter 1 and again in Chapter 2.

Waste disposal facilities

All process plants including oil refineries produce large quantities of toxic and/or flammable materials during periods of plant upset or emergencies. Properly designed flare and slop handling systems are therefore essential to the plant operation. This section describes and discusses typical disposal systems currently in use in the oil refining industry where the hydrocarbon is immiscible with water. Where the chemical is miscible in water special separation systems must be used.

Figure 13.17 in Chapter 13 shows a completely integrated waste disposal system for the light end section of an oil refinery. Further description and discussion of these disposal systems is given in the following sections of that chapter:

- Blow-down and slop disposal
- Flares
- Water effluent treating

Blow-down and slop

This system generally consists of the following drums:

- Non condensable blow-down drum
- Condensable blow-down drum
- Water disengaging drum

Particular emphasis is put on the level control of liquid in these drums.

The flare

Vapors collected in a closed safety system are disposed of by burning at a safe location. The facilities used for burning are called flares. The most common flares used in industry today are:

- The elevated flare
- The multi jet ground flare

The elevated flare is used where some degree of smoke abatement is required. The flare itself operates from the top of a stack usually in excess of 150 ft high. Steam is injected into the gas stream to be burnt to complete combustion and thereby reduce the smoke emission.

The multi jet ground flare is selected where luminosity is a problem. For example at locations near housing sites. In this type of flare, the vapors are burned within the flare stack thus considerably reducing the luminosity. Steam is again used in this type of flare to reduce the smoke emission.

Effluent water treating facilities

This section of the offsite systems deals with the treating of waste water accumulated in a chemical process complex before it leaves the complex. Over the years, requirements for safeguarding the environment have demanded close control on the quality of effluents discharged from chemical and oil refining plants. This includes effluents that contain contaminants that can affect the quality of the atmosphere and those that can be injurious to plant and other life in river waters and the surrounding seas. Effluent management in the oil industry has therefore acquired a position of importance and responsibility to meet these environmental control demands.

Water effluents that are discharged from the process and other units are collected for treating and removal or conversion of the injurious contaminants. In most oil refineries imported water in the form of ship ballast water is also collected on shore for treatment before discharging back to the sea. Figure 13.26 is a schematic of the water effluent treating system for a major European oil refinery. More details and discussion on waste water disposal are given in Chapter 14.

Water systems

The major water systems generated in most chemical plants are:

- Cooling water
- Treated water for boiler feed water (BFW)

Potable water as raw water is usually drawn from municipal supply. Where water is required for cleaning or drinking this potable water is used without further processing.

Cooling water

The cooling water system is a circulating one. There is a cold supply line with an associated warmer return line. Chapter 13 describes and discusses a typical refinery cooling water supply and return. Figure 13.31 in that chapter shows such a system in some detail.

The water returned to the cooling tower by the return header enters the top of the cooling tower and flows down across the tower internals counter current to an air flow, either induced or forced by fans passing up through the tower. The water cooled by the air flow is collected in the cooling tower basin. Make up water (usually potable water) is added to the basin under level control. Vertical cooling water circulating water pumps take suction from the cooling tower basin sump to deliver the water into the distribution header. The supply header pressure is kept at around 30 psig and, very often in large plants covering long distances, booster stations are installed at predetermined locations to maintain the supply header pressure. The return flow is collected from each user into the return header and flows back to the cooling tower.

Boiler feed water treating

All water contains impurities no matter from what source the water comes from. Appendix 13.2 of Chapter 13, gives a listing of the common impurities found in water for industrial use. When it comes to generating steam and particularly high-pressure steam, these impurities become problematic. Appendix 13.2 also provides a description of the effect of these impurities on steam generators and gives the normal means of treating. In general there are three types of soluble impurities naturally present in water and which must be removed or converted in order to make the water suitable for boiler feed. These are:

- Scale forming impurities
- Compounds that cause foaming
- Dissolved gases

Solid build up in the boiler itself is removed or kept at a low level by blow down. Figure 13.32 of Chapter 13 gives an example of boiler blow down.

There are two types of external boiler feed water treatment in common use in the petroleum refining industry. These are.

- The "hot lime" process

This is a water softening process which uses a hot lime contact to induce a precipitate of the compounds contributing to the hardness.

- The ion exchange processes

As the name implies, this process exchanges undesirable ions contained in the raw water with more desirable ones that produce acceptable BFW. The ion exchange material needs to be regenerated after a period of operation. The operating period will differ from process to process and will depend to some extent on the amount of impurities in the water and the required purity of the treated water. Regeneration is accomplished in three steps:

- Back washing
- Regenerating the resin bed with regenerating chemicals
- Rinsing

These two systems are described in detail in Chapter 4.

X, Y, Z

Xylenes

These are aromatic compounds which are coupled with benzene, toluene, and ethyl benzene as the major products from a petrochemical petroleum refinery. These compounds as a whole are usually designated as BTX. The process begins in a conventional refinery with the catalytic reforming of a high naphthene content naphtha. This naphtha may be the product of a hydrocracker. The reformate is much richer in aromatics than that used for gasoline production. After stabilizing, the reformate enters an extraction unit where residual paraffins are removed, and the rich aromatic stream is routed to an aromatic splitter. Depending on the severity of the reformes the splitter may be located ahead of the extraction unit since the xylenes may not require extraction. The benzene and toluene components are taken off as overhead distillate and the bottom product enters a super fractionation unit, the xylene splitter. Here ethyl benzene, para-xylene, and meta-xylene are taken off as overhead distillates. Para-xylene may be separated from this overhead distillate by absorption or crystallization, or may continue with the remaining distillate to enter an isomerization unit, where they are isomerized into a rich ortho-xylene stream. Para-xylene product is by far the more valuable and important isomer of the C_8 aromatics.

Ortho-xylene stream leaves as the bottom product of the isomerizer splitter tower to be returned to the xylene splitter super fractionator tower. The bottom product of this tower is the ortho xylene product. Further details and description is given in Chapter 12. Figure 18 of this chapter, shows a process configuration of a typical BTX process. Ortho-xylene is used extensively in the production of phthalates, para-xylene is used in the production of terephtualic acid and polyesters, and ethylbenzene is used in the production of styrene.

'Z' Factor

"Z" is often the symbol used for the compressibility factor of a gas. This may be derived from the equation:

$$PV = ZnRT$$
$$PV = Z\frac{m}{M}RT$$
$$PM = Z\rho RT$$
$$Z = \frac{PM}{\rho RT}$$

Where:

V = volume of gas, m = mass of gas, R = gas constant
M = mole weight of the gas
P = gas pressure (absolute)
T = absolute temperature
ρ = density of the gas at gas temperature and pressure

Typical values for R are:
8.3143 J/(mol · °K)
0.08205341 atm · m^3/(kmol · °K)
1.98716759 cal/(mol · °K)
10.7313 psia · ft^3/(lb mol · °K)

Zeolite catalysts

Zeolite catalysts are used in catalytic cracking processes. This together with the technique of 'Riser Cracking' revolutionized these processes (distillate feed crackers and residuum feed). This is described and discussed in detail in Chapter 6.

APPENDICES

Appendix A

Examples of working flow sheets

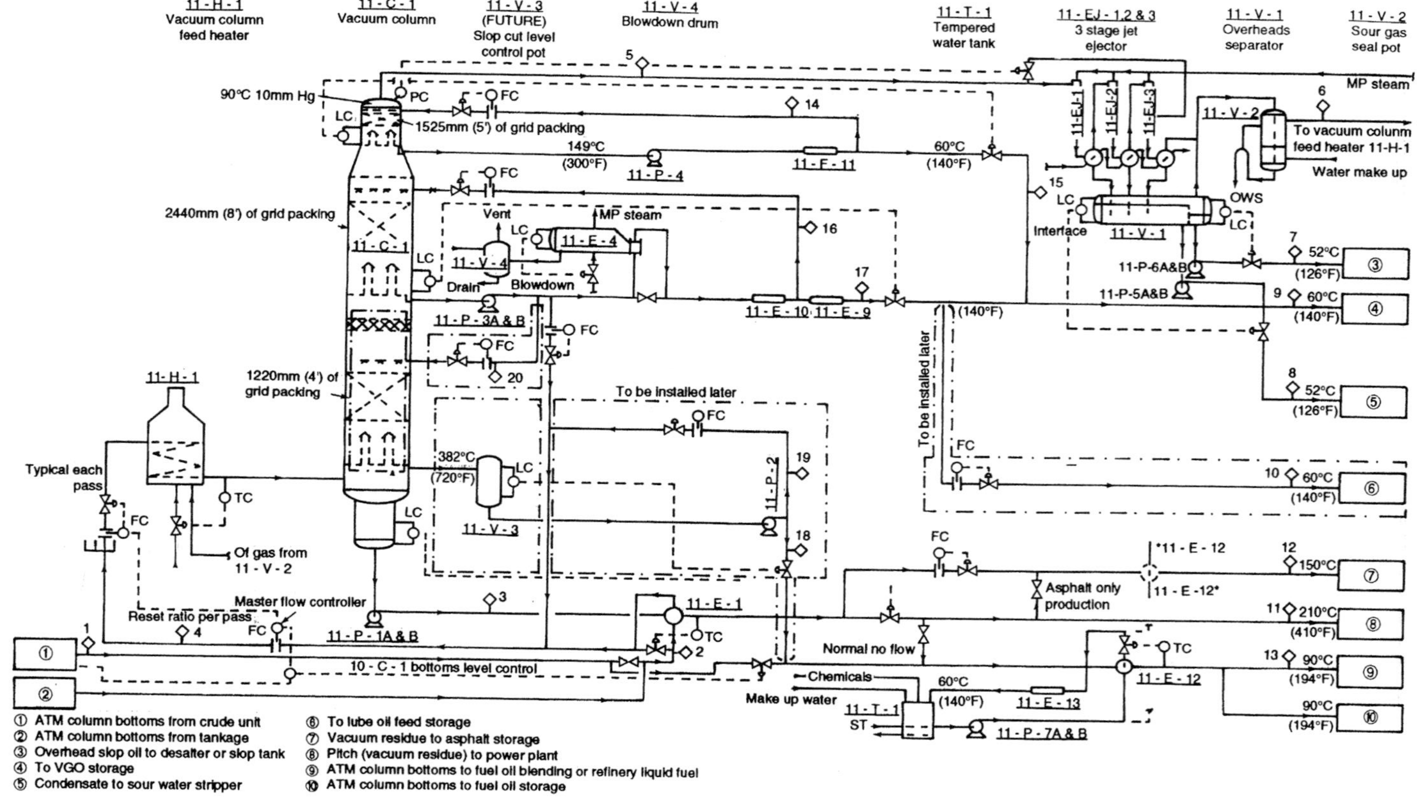

Figure A.1. A typical process flow sheet.

Stream up	1	2	3	4	5
Stream identification	ATM column bottoms from crude column	Feed to vacuum column	Vacuum column residue	Feed to vacuum column	Vacuum column overhead vapour
Normal Kg/ h	470,269	391,836	245,982	391,836	2971
BPSD based on standard conditions	71,800	60,000	36,000	60,000	
Cut	371°C+	371°C+	80 pen	371°C+	94.1 (Av. mol/wt)
Sg @ 15%	0.959	0.989	1.033	0.989	
API	11.5	11.5	5.5	11.5	
Max operating Kg/ h	470,300	457,070 (69,000 BPSD of 385°C+)	329,460 (48,500 BPSD @ 800 pen)	457,070 (385°C+ resid)	3000

Stream up	6	7	8	9	10
Stream identification	Vacuum column inerts exit stream	Vacuum column overhead slop	Condensate water from electors	Combined top side cut and heavy V.G.O.	Heavy vavuum gas oil
Normal Kg/ h	600	2371		137,900	
BPSD based on standard conditions		407		22,693	
Cut	29 (Av. mol/wt)	218 (Av. mol/wt)	18		
Sg @ 15%		0.883			
API		28.7		22.1	20.7
Max operating Kg/ h	600	2371	7710	137,901	

Stream up	11	12	13	14	15
Stream identification	Combined slop cut and vac. resid streams	Asphalt	ATM column botoms to fuel blending	Top circulating reflux	Top cut TBP 370 to 427°C (700 to 800°F)
Normal Kg/ h	251,565		78,433	174,603	51,279
BPSD based on standard conditions	36,900		11800		8593
Cut			371°C+		
Sg @ 15%			1.033	0.906	0.906
API	5.5	5.5	5.5	24.6	24.6
Max operating Kg/ h		11,900	457,000 (385°C+ resid)		51,280

Stream up	16	17	18	19	20
Stream identification	H.G.O. circulating reflux	Heavy vacu-um gas oil	Slop cut vacuum column product	Vac. column bottoms recycle	Metals wash out
Normal Kg/ h	149.206	86621	5583	0 to 16748	8481
BPSD based on standard conditions		14100	900	2700	
Cut					
Sg @ 15%	0.930	0.930	0.942	0.948	0.930
API	20.7	20.7	18.7	18.7	20.7
Max operating Kg/ h		86621	16,748	16,748	15,964

Figure A.1. The associated material balance.

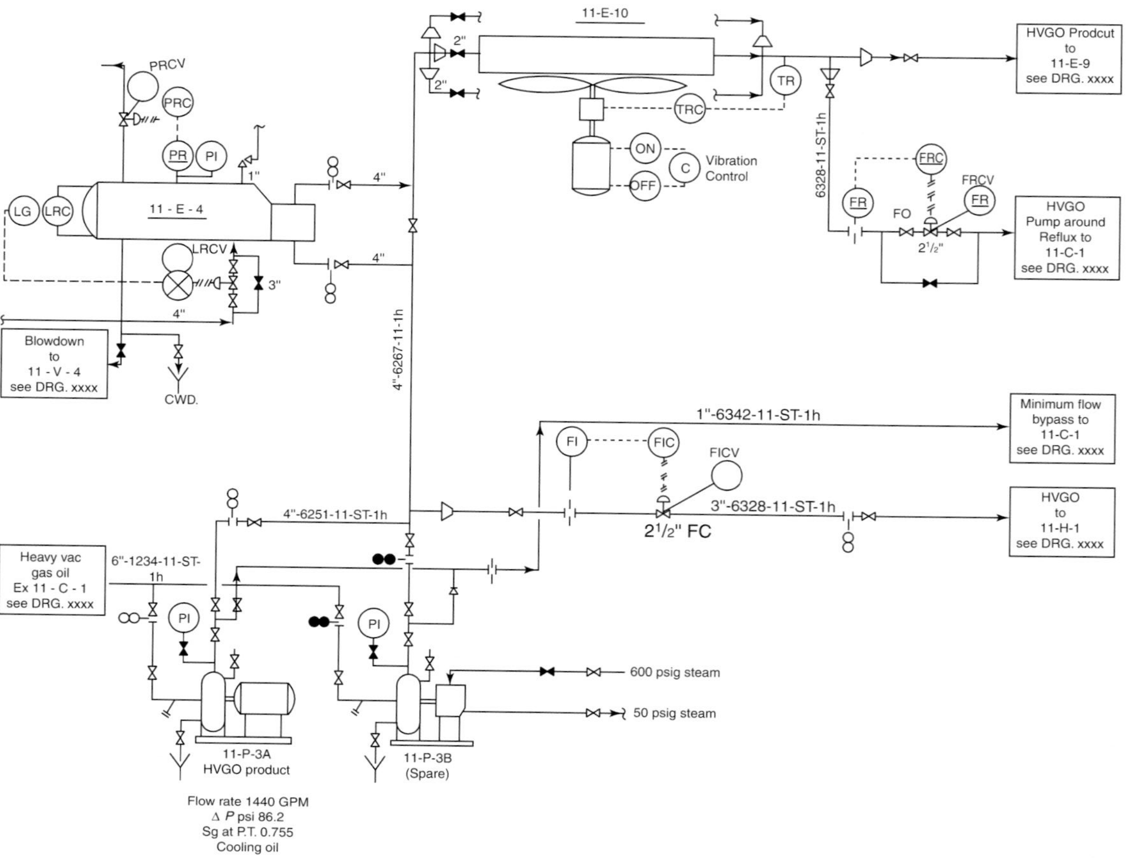

Figure A.2. A section of a typical mechanical flow sheet.

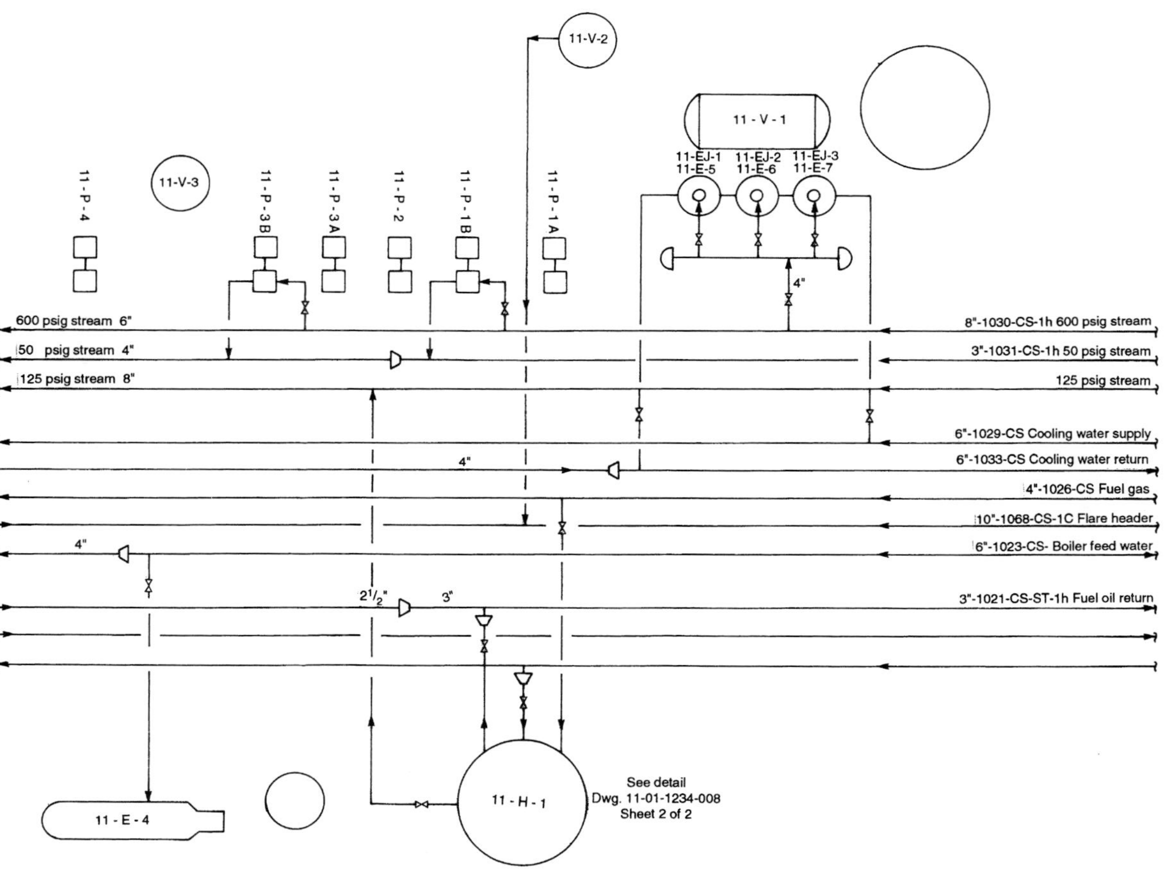

Figure A.3. A section of a typical utility flow sheet.

Appendix B

General data

Friction Loss for Viscous Liquids

(Based on Darcy's Formula)

Loss in Feet of Liquid per 1000 Feet of Pipe
1 Inch (1.049″ inside dia) Sch 40 New Steel Pipe

Flow: US gal per min	Flow: Bbl per hr (42 gal)	Kinematic viscosity—centistokes: 0.6	1.1	2.1	2.7	4.3	7.4	10.3	13.1	15.7	20.6
		Approx SSU viscosity	31.5	33	35	40	50	60	70	80	100
.5	.71	.29	.28	.55	.70	1.12	1.93	2.68	3.41	4.08	5.35
1	1.4	.96	1.13	1.09	1.41	2.24	3.86	5.36	6.82	8.16	10.7
2	2.9	3.23	3.72	4.41	4.80	4.48	7.72	10.7	13.6	16.3	21.4
3	4.3	6.84	7.63	9.04	9.48	10.8	11.6	16.1	20.5	24.5	32.1
4	5.7	11.4	12.2	14.9	15.9	17.6	15.4	21.5	27.3	32.6	42.8
5	7.1	17.2	19.2	22.1	23.4	26.3	19.3	26.8	34.1	40.8	53.5
6	8.6	24.2	26.8	30.5	32.1	36.8	41.3	32.2	40.9	49.0	64.2
7	10.0	32.3	35.4	40.8	42.6	47.4	53.8	37.5	47.7	57.2	74.9
8	11.4	41.6	45.5	51.1	54.3	59.9	68.2	75.3	54.5	65.2	85.6
9	12.9	51.8	56.2	63.5	66.3	73.4	84.9	91.7	61.4	73.4	96.3
10	14.3	62.7	68.1	76.2	80.1	88.2	101	111	115	81.6	107
12	17.1	89.3	95.3	106	111	122	142	151	162	167	129
14	20.0	120	129	140	147	160	185	198	212	221	150
16	22.8	155	164	181	188	205	234	257	268	279	295
18	25.7	194	202	224	233	254	286	305	326	341	365
20	28.6	237	250	272	281	308	342	372	394	405	437
25	35.7	368	383	410	429	464	501	551	583	599	651
30	42.9	523	545	382	600	640	712	759	803	842	904
35	50.0	708	735	780	795	852	933				

Flow: US gal per min	Flow: Bbl per hr (42 gal)	Kinematic viscosity—centistokes: 26.4	32.0	43.2	65.0	108.4	162.3	216.5	325	435	650
		Approx SSU viscosity: 125	150	200	300	500	750	1000	1500	2000	3000
.1	.14	1.37	1.66	2.25	3.38	5.65	8.45	11.3	16.9	22.6	33.8
.3	.43	4.12	4.98	6.75	10.2	17.0	25.3	33.8	50.7	67.8	102
.5	.71	6.86	8.32	11.3	16.9	28.3	42.3	56.4	85	113	169
1	1.4	13.7	16.6	22.5	33.8	56.5	84.5	113	169	226	338
2	2.9	27.5	33.2	45.0	67.6	113	169	226	338	452	676
3	4.3	41.2	49.8	67.5	102	170	253	338	507	678	
4	5.7	55.0	66.5	90.0	136	226	338	452	677	904	
5	7.1	68.7	83.2	113	169	283	423	564	846		
6	8.6	82.4	99.7	135	203	339	507	677			
7	10	96.2	117	158	237	395	591	790			
8	11.4	110	133	180	271	452	676	903			
9	12.9	124	150	203	303	508	760				
10	14.3	137	167	225	338	565	845				
12	17.1	165	200	270	406	678					
14	20.0	192	233	315	474	792					
16	22.8	220	266	360	541	904					
18	25.7	248	299	405	609						
20	28.6	470	332	450	677						

Friction Loss for Viscous Liquids *(Continued)*

(Based on Darcy's Formula)

Loss in Feet of Liquid per 1000 Feet of Pipe
1½ Inch (1.610″ inside dia) Sch 40 New Steel Pipe

Flow: US gal per min	Flow: Bbl per hr (42 gal)	Kinematic viscosity—centistokes: 0.6	1.1	2.1	2.7	4.3	7.4	10.3	13.1	15.7	20.6
		Approx SSU viscosity	31.5	33	35	40	50	60	70	80	100
1	1.4	.13	.10	.20	.25	.41	.69	.97	1.23	1.47	1.93
2	2.9	.42	.49	.39	.51	.83	1.39	1.93	2.46	2.94	3.86
3	4.3	.86	.98	1.17	1.25	1.24	2.08	2.89	3.68	4.41	5.79
4	5.7	1.43	1.63	1.92	2.07	1.65	2.78	3.86	4.91	5.88	7.72
5	7.1	2.11	2.42	2.83	3.05	2.06	3.47	4.82	6.14	7.35	9.65
6	8.6	2.90	3.36	3.89	4.18	4.69	4.17	5.79	7.37	8.82	11.6
8	11.4	4.97	5.60	6.44	6.87	7.77	9.02	7.72	9.83	11.8	15.5
10	14.3	7.51	8.34	9.58	10.1	11.6	13.2	9.65	12.3	14.7	19.3
12	17.1	10.4	11.6	13.4	14.0	15.6	17.9	19.8	14.7	17.6	23.2
15	21.4	16.0	17.4	19.6	20.7	23.2	26.5	29.1	31.1	21.0	29.0
20	28.6	27.2	29.5	32.9	34.6	38.2	43.9	47.6	51.1	53.8	38.6
25	35.7	41.4	44.8	49.5	51.8	57.5	64.6	70.1	75.2	78.9	84.7
30	42.9	58.8	63.0	69.1	72.0	79.0	89.3	97.1	103	109	116.5
40	57.1	102	107	117	122	132	150	160	170	178	191
50	71.4	157	164	178	183	198	222	237	251	263	281
60	85.7	224	233	249	259	279	306	330	347	362	388
70	100	300	312	333	343	369	402	436	457	477	508
80	114	389	403	427	440	470	516	551	580	602	643
90	129	498	508	536	550	585	634	681	715	746	792
100	143	601	624	656	670	714	774	820	863	898	949

Flow: US gal per min	Flow: Bbl per hr (42 gal)	Kinematic viscosity—centistokes: 26.4	32.0	43.2	65.0	108.4	162.3	216.5	325	435	650
		Approx SSU viscosity: 125	150	200	300	500	750	1000	1500	2000	3000
1	1.4	2.47	3.00	4.14	6.09	10.2	15.2	20.3	30.4	40.8	69.0
2	2.9	4.95	6.00	8.28	12.2	20.3	30.4	40.6	60.8	81.5	122
3	4.3	7.42	9.00	12.4	18.3	30.4	45.6	60.9	91.3	122	183
4	5.7	9.90	12.0	16.6	24.4	40.6	60.8	81.2	122	163	244
5	7.1	12.4	15.0	20.7	30.4	50.7	76.0	102	152	204	304
6	8.6	14.9	18.0	24.8	36.5	60.8	91.2	122	183	244	365
8	11.4	19.8	24.0	33.1	48.7	81.2	122	163	243	326	487
10	14.3	24.7	30.0	41.4	60.9	102	152	203	304	408	609
12	17.1	29.7	36.0	49.7	73.2	122	182	244	365	490	732
15	21.4	37.1	45.0	62.2	91.4	152	228	304	457	612	914
20	28.6	49.5	60.0	82.8	122	203	304	406	608	815	
25	35.7	61.9	75.0	103	152	254	380	507	760		
30	42.9	124	90.0	124	183	304	456	609	913		
40	57.1	204	216	166	244	406	608	812			
50	71.4	302	317	342	304	507	760				

Figure B.1. Line friction loss.

Friction Loss for Viscous Liquids *(Continued)*

(Based on Darcy's Formula)

Loss in Feet of Liquid per 1000 Feet of Pipe
2 Inch (2.067" inside dia) Sch 40 New Steel Pipe

Flow: US gal per min	Flow: Bbl per hr (42 gal)	Kinematic viscosity—centistokes .6 / Approx SSU viscosity —	1.1 / 31.5	2.1 / 33	2.7 / 35	4.3 / 40	7.4 / 50	10.3 / 60	13.1 / 70	15.7 / 80	20.6 / 100
1	1.4	.04	.04	.07	.09	.15	.26	.36	.45	.54	.71
2	2.9	.13	.15	.15	.19	.30	.51	.71	.90	1.08	1.42
4	5.7	.43	.50	.58	.63	.59	1.02	1.42	1.81	2.17	2.84
6	8.6	.87	1.00	1.20	1.27	1.46	1.53	2.13	2.71	3.25	4.26
8	11.4	1.47	1.68	1.97	2.13	2.38	2.04	2.84	3.61	4.33	5.68
10	14.3	2.20	2.55	2.90	3.09	3.52	2.56	3.56	4.52	5.42	7.11
12	17.1	3.06	3.46	3.97	4.23	4.78	5.57	4.27	5.43	6.51	8.53
14	20.0	4.07	4.51	5.22	5.51	6.26	7.28	4.98	6.33	7.59	9.96
16	22.8	5.17	5.79	6.65	7.01	7.92	9.16	9.95	7.23	8.67	11.4
18	25.7	6.44	7.16	8.18	8.63	9.67	11.2	12.6	13.1	9.76	12.8
20	28.6	7.82	8.64	9.77	10.4	11.6	13.5	14.8	15.5	10.8	14.2
25	35.7	11.9	13.0	14.7	15.4	17.2	19.9	21.6	22.8	24.1	17.8
30	42.9	17.0	18.2	20.4	21.5	23.8	27.2	29.9	31.2	33.0	35.6
35	50.0	22.3	24.1	27.0	28.2	31.2	35.6	38.6	40.4	43.4	47.3
40	57.1	28.8	31.0	34.2	36.0	39.6	44.6	48.8	52.9	54.1	60.7
50	71.4	44.1	47.2	52.0	54.0	59.2	66.4	72.0	76.9	78.4	86.2
60	85.7	62.7	66.5	72.2	74.2	82.3	91.8	98.6	105	111	119
70	100	84.1	83.5	95.8	99.4	108	120	130	137	145	156
80	114	109	114	123	127	138	154	166	174	182	195
90	129	137	143	154	158	171	192	204	215	225	239
100	143	167	176	188	193	208	230	244	260	269	289
110	157	202	211	225	231	246	275	290	307	319	335
120	171	238	249	265	273	290	321	341	358	375	396
130	186	277	290	307	316	335	372	392	411	432	459
140	200	320	347	352	364	383	424	449	472	491	521
150	214	366	382	403	415	437	479	510	529	553	586
160	228	414	431	457	469	494	536	572	595	621	659
170	243	467	485	513	522	553	601	639	665	694	729
180	257	524	543	572	583	619	665	714	743	767	804
190	271	584	602	634	649	688	733	792	825	846	884
200	286	643	666	699	716	756	808	851	901	927	977
210	300	709	731	768	786	826	880	935	975		
220	314	778	798	838	858	902	958				
230	328	851	873	912	934	982					
240	343	922	945	988							

Friction Loss for Viscous Liquids *(Continued)*

(Based on Darcy's Formula)

Loss in Feet of Liquid per 1000 Feet of Pipe
2 Inch (2.067" inside dia) Sch 40 New Steel Pipe

Flow: US gal per min	Flow: Bbl per hr (42 gal)	Kinematic viscosity—centistokes 26.4 / Approx SSU viscosity 125	32.0 / 150	43.2 / 200	65.0 / 300	108.4 / 500	162.3 / 750	216.5 / 1000	325 / 1500	435 / 2000	650 / 3000
1	1.4	.91	1.10	1.49	2.24	3.74	5.60	7.48	11.2	15.0	22.4
2	2.9	1.82	2.21	2.98	4.48	7.49	11.2	15.0	22.4	30.0	44.9
3	4.3	2.73	3.31	4.47	6.73	11.2	16.8	22.4	33.6	45.0	67.4
4	5.7	3.64	4.42	5.96	8.98	15.4	22.4	29.9	44.8	60.0	89.9
5	7.1	4.56	5.52	7.45	11.2	18.7	28.0	37.4	56.0	75.0	112
6	8.6	5.47	6.63	8.95	13.5	22.5	33.6	44.8	67.2	90.0	135
7	10.0	6.38	7.73	10.4	15.7	26.2	39.2	52.3	78.4	105	157
8	11.4	7.29	8.84	11.9	18.0	30.0	34.8	59.8	89.6	120	180
9	12.9	8.20	9.94	13.4	20.2	33.7	50.4	67.3	101	135	202
10	14.3	9.11	11.0	14.9	22.4	37.4	56.0	74.8	112	150	224
12	17.1	10.9	13.3	18.9	26.9	44.9	67.3	89.7	135	180	269
14	20.0	12.7	15.5	20.9	31.4	52.4	78.4	105	157	210	314
16	22.8	14.6	17.7	23.9	35.9	59.9	89.6	120	179	240	359
18	25.7	16.4	19.9	26.8	40.3	67.4	101	135	202	270	404
20	28.6	18.2	22.1	29.8	44.9	74.9	112	150	224	300	449
25	35.7	22.8	27.6	37.3	56.1	93.6	140	187	280	375	562
30	42.9	27.3	33.1	44.7	67.3	112	168	224	336	450	674
35	50.0	31.9	38.7	52.2	78.5	131	196	262	392	525	786
40	57.1	63.0	44.2	59.6	89.8	150	224	299	448	600	899
45	64.3	70.8	80.2	67.1	101	168	252	336	503	675	
50	71.4	92.8	97.1	74.5	112	187	280	374	560	750	
60	85.7	127	134	146	135	225	336	448	672	900	
70	100	162	176	189	157	262	392	523	784		
80	114	208	219	238	180	300	448	598	896		
90	129	257	270	293	327	337	504	673			
100	143	309	322	352	388	374	560	748			
110	157	364	379	412	465	412	617	823			
120	171	423	445	463	537	449	673	898			
130	186	487	510	549	618	487	728				
140	200	548	580	627	703	524	784				
150	214	622	655	705	792	909	840				
160	228	697	737	792	887		896				
170	243	775	808	882			952				
180	257	858	871	962							
190	271	947	986								

Figure B.1. Line friction loss.

Friction Loss for Viscous Liquids *(Continued)*

(Based on Dàrcy's Formula)

Loss in Feet of Liquid per 1000 Feet of Pipe
2½ Inch (2.469″ inside dia) Sch 40 New Steel Pipe

Flow		Kinematic viscosity—centistokes									
		.6	1.1	2.1	2.7	4.3	7.4	10.3	13.1	15.7	20.6
U S gal per min	Bbl per hr (42 gal)	Approx SSU viscosity									
			31.5	33	35	40	50	60	70	80	100
10	14.3	.92	1.05	1.23	1.31	1.48	1.26	1.75	2.22	2.66	3.50
12	17.1	1.28	1.47	1.70	1.80	2.04	1.51	2.10	2.67	3.19	4.19
14	20.0	1.68	1.92	2.23	2.37	2.65	3.09	2.45	3.11	3.73	4.89
16	22.8	2.15	2.43	2.81	2.99	3.34	3.86	2.80	3.56	4.26	5.59
18	25.7	2.68	2.99	3.46	3.68	4.21	4.77	5.26	4.00	4.80	6.28
20	28.6	3.23	3.61	4.27	4.42	4.94	5.73	6.26	4.44	5.33	6.98
25	35.7	4.88	5.39	6.17	6.55	7.31	8.42	9.20	9.79	6.66	8.73
30	42.9	6.87	7.57	8.55	9.07	10.1	11.5	12.6	13.5	14.1	10.5
35	50.0	9.18	9.97	11.3	11.8	13.3	15.1	16.4	17.7	18.5	20.0
40	57.1	11.8	12.8	14.3	15.0	16.8	19.0	20.7	22.1	23.2	25.0
45	64.3	14.8	15.9	17.7	18.7	20.8	23.5	25.6	27.1	28.4	30.6
50	71.4	18.1	19.3	21.4	22.4	24.9	28.3	30.6	32.5	34.1	36.8
60	85.7	25.6	27.2	29.9	31.1	34.2	39.1	42.1	44.7	46.7	50.4
70	100	34.2	36.2	39.6	41.4	45.2	51.4	55.4	58.5	61.5	66.0
80	114	44.1	46.7	50.6	53.0	57.3	64.5	69.7	74.2	77.5	82.7
90	129	55.2	58.8	63.2	65.4	70.9	80.0	86.6	91.2	95.3	103
100	143	67.2	72.3	76.7	79.4	85.8	96.3	104	109	115	123
110	157	80.9	86.2	92.4	94.8	103	113	121	130	135	146
120	171	95.7	102	108	112	121	133	143	151	158	169
130	186	112	118	125	130	139	155	166	176	181	194
140	200	129	136	144	150	160	176	188	198	209	221
150	214	147	155	165	170	181	198	212	223	234	250
160	228	167	175	187	191	203	224	238	253	262	281
170	243	188	196	210	214	226	248	267	283	290	312
180	257	210	219	234	239	253	277	297	311	322	345
190	271	233	243	260	265	280	306	328	339	356	378
200	286	258	269	286	292	308	334	357	373	391	417
220	314	310	322	343	351	369	400	427	448	461	486
240	343	367	381	404	416	436	469	494	522	539	573
260	371	429	445	470	482	505	543	575	599	621	660
280	400	497	513	540	556	580	630	657	686	710	758
300	429	568	586	617	632	659	705	748	775	803	849
320	457	643	663	695	716	747	799	837	875	903	952
340	486	725	745	776	800	839	894	933	980		
360	514	809	835	866	892	936	994				

Friction Loss for Viscous Liquids *(Continued)*

(Based on Darcy's Formula)

Loss in Feet of Liquid per 1000 Feet of Pipe
2½ Inch (2.469″ inside dia) Sch 40 New Steel Pipe

Flow		Kinematic viscosity—centistokes									
		26.4	32.0	43.2	65.0	108.4	162.3	216.5	325	435	650
U S gal per min	Bbl per hr (42 gal)	Approx SSU viscosity									
		125	150	200	300	500	750	1000	1500	2000	3000
1	1.4	.45	.54	.73	1.10	1.84	2.75	3.67	5.52	7.38	11.0
2	2.9	.90	1.09	1.47	2.20	3.68	5.50	7.35	11.0	14.8	22.0
4	5.7	1.79	2.17	2.93	4.41	7.36	11.0	14.7	22.1	29.5	44.1
6	8.6	2.69	3.26	4.40	6.62	11.0	16.5	22.0	33.1	44.3	66.2
8	11.4	3.58	4.34	5.87	8.82	14.7	22.0	29.4	44.1	59.1	88.2
10	14.3	4.48	5.43	7.33	11.0	18.4	27.5	36.7	55.2	73.8	110
12	17.1	5.38	6.51	8.80	13.2	22.1	33.0	44.1	66.2	88.6	132
14	20.0	6.27	7.60	10.3	15.4	25.7	38.5	51.4	77.2	103	154
16	22.8	7.16	8.68	11.7	17.6	29.4	44.0	58.8	88.2	118	176
18	25.7	8.06	9.77	13.2	19.8	33.1	49.5	66.1	99.3	133	198
20	28.6	8.96	10.9	14.7	22.0	36.8	55.0	73.4	110	148	220
25	35.7	11.2	13.6	18.3	27.6	46.0	68.8	91.8	138	185	276
30	42.9	13.4	16.3	22.0	33.1	55.2	82.5	110	165	222	331
35	50.0	15.7	19.0	25.6	38.6	64.4	96.3	129	193	258	386
40	57.1	17.9	21.7	29.3	44.2	73.6	110	147	221	295	441
45	64.3	33.0	24.4	33.0	49.6	82.8	124	165	248	332	496
50	71.4	39.2	27.2	36.6	55.2	92.0	138	184	276	369	551
60	85.7	54.0	56.5	44.0	66.2	110	165	220	331	443	662
70	100	70.0	73.5	51.3	77.2	129	193	257	386	517	772
80	114	87.7	93.4	101	88.3	147	220	294	441	591	882
90	129	110	115	125	99.3	166	248	330	497	665	993
100	143	130	137	148	110	184	275	367	552	738	
110	157	154	164	176	197	202	303	403	607	812	
120	171	180	188	205	226	221	330	441	662	886	
130	186	206	216	232	263	239	358	477	717	960	
140	200	234	247	267	299	257	385	514	772		
150	214	265	279	305	333	276	413	551	827		
160	228	296	312	338	374	294	440	588	882		
170	243	328	345	373	415	312	468	624	937		
180	257	364	384	412	461	530	495	661	993		
190	271	403	420	454	514	587	523	698			
200	286	438	457	493	550	628	550	734			
220	314	522	540	586	658	752	605	808			
240	343	612	633	682	760	866	660	881			
260	371	711	732	782	867		715	955			

Figure B.1. Line friction loss.

Friction Loss for Viscous Liquids *(Continued)*

(Based on Darcy's Formula)

Loss in Feet of Liquid per 1000 Feet of Pipe
3 Inch (3.068″ inside dia) Sch 40 New Steel Pipe

Flow		Kinematic viscosity—centistokes									
		.6	1.1	2.1	2.7	4.3	7.4	10.3	13.1	15.7	20.6
		Approx SSU viscosity									
U S gal per min	Bbl per hr (42 gal)		31.5	33	35	40	50	60	70	80	100
8	11.4	.22	.25	.29	.32	.24	.42	.59	.74	.89	1.18
10	14.3	.32	.37	.43	.47	.54	.53	.73	.93	1.11	1.47
15	21.4	.70	.76	.89	.94	1.07	.79	1.10	1.40	1.67	2.20
20	28.6	1.12	1.27	1.47	1.57	1.78	2.07	1.46	1.86	2.23	2.93
25	35.7	1.69	1.93	2.23	2.31	2.61	3.01	3.29	2.33	2.79	3.66
30	42.9	2.36	2.64	2.99	3.22	3.60	4.12	4.50	4.83	3.35	4.40
35	50.0	3.13	3.48	3.97	4.21	4.66	5.41	5.89	6.35	6.61	5.13
40	57.1	4.03	4.42	5.02	5.29	5.90	6.80	7.46	7.93	8.37	5.87
50	71.4	6.10	6.70	7.50	7.93	8.76	10.1	10.9	11.7	12.3	13.2
60	85.7	8.57	9.32	10.4	11.0	12.0	13.7	15.0	16.0	16.8	18.0
70	100	11.5	12.4	13.8	14.5	15.9	18.0	19.6	20.9	21.9	23.6
80	114	14.7	15.9	17.5	18.4	20.3	22.9	24.6	26.4	27.7	29.8
90	129	18.4	19.9	21.8	22.8	25.0	28.0	30.4	32.4	33.8	36.3
100	143	22.4	24.2	26.3	27.5	30.2	33.7	36.4	39.0	40.8	43.6
120	171	31.8	34.1	36.9	38.6	41.9	46.8	50.5	53.4	56.4	60.0
140	200	42.4	45.6	49.4	50.9	55.4	65.5	66.0	70.0	73.2	78.6
160	228	54.8	58.0	63.3	65.4	70.4	79.1	83.8	87.9	92.3	98.2
180	257	69.0	72.7	78.7	81.6	87.2	97.2	104	109	114	122
200	286	84.7	88.9	95.7	99.4	106	117	125	131	137	146
225	322	107	112	120	124	132	145	155	164	169	180
250	357	131	137	147	151	160	175	188	195	204	218
275	393	158	164	175	180	191	208	226	233	243	258
300	429	187	193	204	212	225	244	260	273	281	298
325	464	218	225	238	247	261	283	300	316	325	345
350	500	253	260	275	283	300	324	344	361	373	396
375	536	288	298	314	322	341	367	388	407	424	448
400	571	328	339	354	363	385	414	436	458	476	498
425	607	368	381	397	407	432	463	488	511	529	550
450	643	410	427	443	455	480	515	543	568	587	619
475	679	457	473	493	504	532	571	599	625	646	681
500	714	504	524	544	555	587	627	658	684	707	750
525	750	555	574	597	609	644	688	720	748	770	821
550	786	606	627	651	665	703	748	783	814	838	890
575	822	663	685	708	723	761	814	852	886	912	962
600	857	721	742	767	783	820	882	919	960	989	

Friction Loss for Viscous Liquids *(Continued)*

(Based on Darcy's Formula)

Loss in Feet of Liquid per 1000 Feet of Pipe
3 Inch (3.068″ inside dia) Sch 40 New Steel Pipe

Flow		Kinematic viscosity—centistokes									
		26.4	32.0	43.2	65.0	108.4	162.3	216.5	325	435	650
		Approx SSU viscosity									
U S gal per min	Bbl per hr (42 gal)	125	150	200	300	500	750	1000	1500	2000	3000
4	5.7	.75	.91	1.23	1.85	3.08	4.62	6.16	9.25	12.4	18.5
6	8.6	1.13	1.37	1.84	2.77	4.62	6.92	9.24	13.9	18.5	27.7
8	11.4	1.50	1.82	2.45	3.70	6.16	9.23	12.3	18.5	24.7	36.9
10	14.3	1.88	2.28	3.06	4.62	7.70	11.5	15.4	23.1	30.9	46.2
12	17.1	2.25	2.73	3.68	5.55	9.24	13.8	18.5	27.7	37.1	55.5
14	20.0	2.63	3.18	4.29	6.47	10.8	16.2	21.5	32.3	43.3	64.7
16	22.8	3.00	3.64	4.90	7.39	12.3	18.5	24.6	37.0	49.5	73.9
18	25.7	3.38	4.09	5.52	8.31	13.9	20.8	27.7	41.6	55.6	83.2
20	28.6	3.76	4.55	6.13	9.24	15.4	23.1	30.8	46.2	61.8	92.4
25	35.7	4.69	5.69	7.67	11.5	19.3	28.8	38.5	57.7	77.3	115
30	42.9	5.63	6.83	9.20	13.9	23.1	34.6	46.2	69.3	92.7	139
35	50.0	6.57	7.97	10.7	16.2	27.0	40.3	53.8	80.9	108	162
40	51.1	7.51	9.10	12.3	18.5	30.8	46.2	61.6	92.5	124	185
50	71.4	9.39	11.4	15.3	23.1	38.5	57.7	77.0	115	154	231
60	85.7	19.2	13.7	18.4	27.7	46.2	69.2	92.4	139	185	277
70	100	25.3	26.8	21.5	32.3	53.9	80.8	108	162	216	323
80	114	31.6	33.6	24.7	37.0	61.6	92.3	123	185	247	369
90	129	38.9	40.9	44.6	41.6	69.3	104	139	208	278	416
100	143	46.1	49.5	53.0	46.2	77.0	115	154	231	309	462
120	171	64.0	67.6	72.9	55.5	92.4	138	185	277	371	555
140	200	83.9	89.1	94.9	108	108	162	215	323	433	647
160	228	106	111	120	135	123	185	246	370	495	739
180	257	131	137	148	164	139	208	277	416	556	832
200	286	157	163	179	198	154	231	308	462	618	924
225	322	191	204	223	242	279	260	346	520	696	
250	357	229	242	261	291	332	288	385	577	773	
275	393	271	285	311	343	396	317	423	635	850	
300	429	316	331	361	398	456	346	462	693	927	
325	464	364	381	416	458	527	375	500	751		
350	500	415	436	467	523	593	672	538	809		
375	536	469	493	528	586	672	746	577	867		
400	571	526	550	592	656	751	843	616	925		
425	607	587	612	656	728	834	937	654	982		
450	643	652	675	728	802	928					
475	679	718	744	801	689						

Figure B.1. Line friction loss.

Friction Loss for Viscous Liquids *(Continued)*

(Based on Darcy's Formula)

Loss in Feet of Liquid per 1000 Feet of Pipe
3½ Inch (3.548″ inside dia) Sch 40 New Steel Pipe

Flow		Kinematic viscosity—centistokes									
		.6	1.1	2.1	2.7	4.3	7.4	10.3	13.1	15.7	20.6
		Approx SSU viscosity									
U S gal per min	Bbl per hr (42 gal)		31.5	33	35	40	50	60	70	80	100
20	28.6	.56	.63	.72	.78	.88	1.03	*.82*	*1.04*	*1.25*	*1.64*
25	35.7	.82	.93	1.08	1.14	1.29	1.52	1.67	*1.30*	*1.56*	*2.05*
30	42.9	1.15	1.29	1.50	1.57	1.78	2.09	2.30	*1.56*	*1.87*	*2.46*
35	50.0	1.53	1.70	1.94	2.08	2.31	2.68	2.97	3.18	*2.18*	*2.87*
40	57.1	1.95	2.17	2.49	2.68	2.91	3.35	3.70	4.00	4.21	*3.28*
45	64.3	2.46	2.68	3.03	3.22	3.59	4.11	4.57	4.91	5.19	*3.69*
50	71.4	2.95	3.25	3.68	3.90	4.32	4.98	5.42	5.87	6.20	6.64
60	85.7	4.17	4.54	5.12	5.38	6.00	6.82	7.41	7.97	8.41	9.21
70	100	6.78	6.02	6.78	7.11	7.84	8.87	9.76	10.4	10.9	11.9
80	114	7.19	7.72	8.57	8.79	9.81	11.2	12.3	13.0	13.7	14.9
90	129	8.92	9.59	10.6	11.2	12.5	13.9	15.0	16.1	16.8	18.3
100	143	10.8	11.7	12.9	13.6	14.8	16.8	18.0	19.3	20.2	21.7
120	171	15.4	16.4	18.1	18.7	20.5	22.9	25.0	26.3	27.7	29.7
140	200	20.5	22.0	23.9	24.9	27.3	30.5	33.1	34.6	36.0	39.1
160	228	26.4	28.3	30.9	31.9	35.0	38.4	41.8	43.9	45.6	49.0
180	257	33.0	35.0	38.0	39.6	42.5	47.7	50.8	54.0	56.3	60.1
200	286	40.3	43.0	46.3	48.0	51.4	57.1	61.8	65.5	68.0	72.2
225	322	50.7	53.2	57.7	59.7	64.7	71.2	75.9	80.1	84.3	88.8
250	357	62.6	65.0	70.9	72.7	78.9	86.5	92.7	96.9	101	107
275	393	75.4	77.9	84.6	87.1	93.3	102	110	115	120	127
300	429	89.2	92.2	99.6	103	109	119	128	135	140	148
325	464	104	108	116	120	127	138	147	154	161	171
350	500	121	124	133	138	146	159	169	178	183	194
375	536	138	142	152	157	167	181	191	200	207	220
400	571	156	161	171	178	188	203	213	225	233	248
425	607	176	181	192	200	211	225	238	252	259	279
450	643	196	203	213	221	234	250	265	280	287	309
475	679	219	225	235	244	257	276	292	310	319	336
500	714	241	249	259	268	282	304	321	340	350	368
550	786	290	300	311	323	340	365	385	407	415	440
600	857	343	355	367	377	399	426	452	466	480	510
650	929	400	414	428	440	461	498	522	540	557	587
700	1000	464	480	494	505	532	572	597	621	637	675
750	1070	532	548	567	576	604	651	682	704	725	769
800	1140	606	624	641	652	684	730	765	794	815	861

Friction Loss for Viscous Liquids *(Continued)*

(Based on Darcy's Formula)

Loss in Feet of Liquid per 1000 Feet of Pipe
3½ Inch (3.548″ inside dia) Sch 40 New Steel Pipe

Flow		Kinematic viscosity—centistokes									
		26.4	32.0	43.2	65.0	108.4	162.3	216.5	325	435	650
		Approx SSU viscosity									
U S gal per min	Bbl per hr (42 gal)	125	150	200	300	500	750	1000	1500	2000	3000
10	14.3	*1.05*	*1.27*	*1.72*	*2.58*	*4.32*	*6.46*	*8.62*	*12.9*	*16.9*	*25.8*
15	21.4	*1.57*	*1.91*	*2.58*	*3.88*	*6.47*	*9.68*	*12.9*	*19.4*	*25.4*	*38.8*
20	28.6	*2.10*	*2.54*	*3.44*	*5.17*	*8.63*	*12.9*	*17.3*	*25.8*	*33.8*	*51.7*
25	35.7	*2.62*	*3.18*	*4.29*	*6.47*	*10.8*	*16.1*	*21.6*	*32.3*	*42.3*	*64.7*
30	42.9	*3.15*	*3.82*	*5.15*	*7.76*	*13.0*	*19.4*	*25.9*	*38.8*	*50.7*	*77.6*
35	50.0	*3.67*	*4.45*	*6.01*	*9.05*	*15.1*	*22.6*	*30.2*	*45.3*	*59.2*	*90.6*
40	57.1	*4.20*	*5.09*	*6.87*	*10.3*	*17.3*	*25.8*	*34.5*	*51.7*	*67.6*	*103*
45	64.3	*4.72*	*5.73*	*7.73*	*11.6*	*19.4*	*29.0*	*38.8*	*58.2*	*76.1*	*116*
50	71.4	*5.25*	*6.36*	*8.59*	*12.9*	*21.6*	*32.3*	*43.1*	*64.7*	*84.5*	*129*
60	85.7	*6.30*	*7.64*	*10.3*	*15.5*	*25.9*	*38.8*	*51.8*	*77.6*	*101*	*155*
70	100	12.8	*8.91*	*12.0*	*18.1*	*30.2*	*45.2*	*60.4*	*90.6*	*118*	*181*
80	114	16.1	16.9	*13.7*	*20.7*	*34.5*	*51.6*	*69.0*	*103*	*135*	*207*
90	129	19.6	20.8	*15.5*	*23.3*	*38.8*	*58.1*	*77.6*	*116*	*152*	*233*
100	143	23.6	24.8	*17.2*	*25.9*	*43.2*	*64.6*	*86.2*	*129*	*169*	*258*
120	171	31.9	34.0	37.1	*31.0*	*51.8*	*77.5*	*104*	*155*	*203*	*310*
140	200	41.5	43.8	48.2	*36.2*	*60.4*	*90.4*	*121*	*181*	*236*	*362*
160	228	52.3	55.3	60.5	67.7	*69.1*	*103*	*138*	*207*	*270*	*413*
180	257	64.6	67.3	73.9	83.4	*77.7*	*116*	*155*	*233*	*304*	*466*
200	286	77.3	81.5	87.9	99.9	*86.3*	*129*	*173*	*258*	*338*	*517*
225	322	94.4	99.8	108	122	*97.1*	*145*	*194*	*291*	*380*	*528*
250	357	114	120	129	146	*108*	*161*	*216*	*323*	*422*	*647*
275	393	134	141	153	172	199	*177*	*237*	*356*	*465*	*711*
300	429	157	164	178	198	233	*194*	*259*	*388*	*507*	*776*
325	464	181	188	205	227	266	*210*	*280*	*420*	*549*	*840*
350	500	206	215	233	258	301	*226*	*302*	*452*	*592*	*906*
375	536	234	245	262	291	339	*242*	*323*	*485*	*634*	*970*
400	571	261	274	291	326	379	423	*345*	*517*	*676*	
425	607	291	303	325	363	419	473	*367*	*550*	*718*	
450	643	322	337	359	402	462	525	*388*	*582*	*761*	
475	679	353	368	395	440	505	572	*410*	*614*	*803*	
500	714	389	403	433	482	551	626	*431*	*647*	*845*	
550	786	459	481	514	566	648	733	795	*712*	*930*	
600	857	538	566	600	658	749	851	923	*776*		
650	929	620	648	690	755	865	978		*841*		
700	1000	708	735	789	863	987			*906*		

Figure B.1. Line friction loss.

Friction Loss for Viscous Liquids *(Continued)*

(Based on Darcy's Formula)

Loss in Feet of Liquid per 1000 Feet of Pipe
6 Inch (6.065″ inside dia) Sch 40 New Steel Pipe

Flow		Kinematic viscosity—centistokes									
U S gal per min	Bbl per hr (42 gal)	.6	1.1	2.1	2.7	4.3	7.4	10.3	13.1	15.7	20.6
		Approx SSU viscosity									
			31.5	33	35	40	50	60	70	80	100
75	107	.45	.49	.58	.61	.68	.80	.86	.93	.98	*.72*
100	143	.77	.85	.96	1.01	1.14	1.30	1.42	1.52	1.62	1.74
125	178	1.14	1.27	1.43	1.51	1.68	1.95	2.10	2.23	2.35	2.57
150	214	1.61	1.78	2.01	2.09	2.32	2.66	2.86	3.08	3.20	3.46
175	250	2.13	2.37	2.63	2.79	3.04	3.52	3.74	3.97	4.24	4.51
200	286	2.75	3.00	3.34	3.55	3.85	4.41	4.76	5.02	5.31	5.69
225	322	3.42	3.74	4.17	4.38	4.78	5.39	5.89	6.16	6.45	7.05
250	357	4.15	4.55	5.07	5.21	5.76	6.94	7.11	7.47	7.77	8.41
275	393	4.99	5.42	6.02	6.28	6.88	7.60	8.35	8.86	9.18	9.81
300	429	5.87	6.38	7.06	7.37	8.09	8.94	9.69	10.3	10.8	11.4
350	500	7.90	8.45	9.38	9.80	10.5	11.8	12.6	13.5	14.2	15.0
400	571	10.2	10.9	11.8	12.5	13.4	15.0	16.0	17.0	18.0	19.1
450	643	12.8	13.6	14.8	15.5	16.7	18.5	19.7	20.1	21.8	23.7
500	714	15.6	16.6	18.0	18.7	20.4	22.6	24.0	25.1	26.4	28.5
550	786	18.8	19.8	21.5	22.3	24.3	26.8	28.5	29.6	29.9	33.4
600	857	22.1	23.3	25.1	26.2	28.4	31.1	33.2	35.0	36.2	38.8
650	929	25.8	27.2	29.2	30.4	32.8	36.1	38.5	40.4	41.8	44.7
700	1000	29.7	31.2	33.5	34.9	37.5	41.1	44.2	46.3	47.9	51.3
750	1070	33.9	35.6	38.2	39.7	42.5	46.7	49.9	51.8	54.0	57.2
800	1140	38.3	40.5	43.2	44.4	47.8	52.7	56.1	58.5	60.9	64.3
900	1285	48.5	50.7	54.4	55.6	59.3	65.4	69.1	72.8	74.6	79.9
1000	1430	59.5	62.2	66.4	67.5	72.4	79.3	83.4	87.6	91.0	95.9
1100	1570	71.6	74.8	79.4	80.8	86.7	94.5	99.6	104	109	114
1200	1715	84.6	87.9	93.4	95.6	102	111	117	121	126	133
1400	2000	115	118	126	128	135	146	155	161	167	177
1600	2285	150	153	162	164	173	187	199	207	213	224
1800	2570	188	193	203	206	216	232	246	256	264	278
2000	2860	231	237	247	253	264	284	296	311	320	334
2200	3140	277	286	297	303	316	338	354	371	382	398
2400	3430	330	341	352	356	374	395	417	430	448	470
2600	3710	387	395	408	418	433	461	485	500	520	543
2800	4000	449	458	470	482	497	526	553	574	595	621
3000	4285	515	526	536	550	567	597	628	655	666	706
3250	4640	605	613	629	641	665	701	729	757	777	817
3500	5000	697	711	729	739	771	808	841	869	897	938

Friction Loss for Viscous Liquids *(Continued)*

(Based on Darcy's Formula)

Loss in Feet of Liquid per 1000 Feet of Pipe
6 Inch (6.065″ inside dia) Sch 40 New Steel Pipe

Flow		Kinematic viscosity—centistokes									
U S gal per min	Bbl per hr (42 gal)	26.4	32.0	43.2	65.0	108.4	162.3	216.5	325	435	650
		Approx SSU viscosity									
		125	150	200	300	500	750	1000	1500	2000	3000
50	71.4	*.62*	*.74*	*1.00*	*1.51*	*2.52*	*3.78*	*5.04*	*7.57*	*10.1*	*15.1*
75	107	*.92*	*1.12*	*1.51*	*2.27*	*3.78*	*5.66*	*7.56*	*11.4*	*15.2*	*22.7*
100	143	*1.23*	*1.49*	*2.01*	*3.03*	*5.05*	*7.55*	*10.1*	*15.1*	*20.3*	*30.2*
125	178	2.75	*1.86*	*2.51*	*3.79*	*6.31*	*9.45*	*12.6*	*18.9*	*25.3*	*37.8*
150	214	3.75	3.96	*3.01*	*4.54*	*7.58*	*11.3*	*15.1*	*22.7*	*30.4*	*45.4*
175	250	4.90	5.17	5.62	*5.30*	*8.84*	*13.2*	*17.6*	*26.5*	*35.5*	*53.0*
200	286	6.10	6.51	7.07	*6.06*	*10.1*	*15.1*	*20.2*	*30.3*	*40.5*	*60.6*
225	322	7.43	7.93	8.66	*6.82*	*11.4*	*17.0*	*22.7*	*34.1*	*45.6*	*68.1*
250	357	8.91	9.43	10.4	*7.57*	*12.6*	*18.9*	*25.2*	*37.8*	*50.7*	*75.7*
275	393	10.6	11.1	12.2	13.7	*13.9*	*20.8*	*27.7*	*41.7*	*55.8*	*83.2*
300	429	12.3	12.9	14.2	15.9	*15.1*	*22.6*	*30.2*	*45.4*	*60.9*	*90.8*
350	500	15.9	17.1	18.3	20.8	*17.7*	*26.4*	*35.3*	*63.0*	*71.0*	*106*
400	571	20.1	21.3	23.1	26.2	*20.2*	*30.2*	*40.3*	*60.6*	*81.1*	*121*
450	643	24.7	26.0	28.6	31.9	36.9	*34.0*	*45.3*	*68.2*	*91.3*	*136*
500	714	30.0	31.3	34.1	28.0	44.2	*37.8*	*50.4*	*75.7*	*101*	*151*
550	786	35.6	36.9	40.2	44.6	52.1	*41.6*	*55.4*	*83.3*	*112*	*166*
600	857	41.5	43.1	46.4	51.7	59.1	*45.3*	*60.5*	*90.9*	*122*	*182*
650	929	47.7	50.0	53.4	59.6	69.4	*49.1*	*65.5*	*98.5*	*132*	*197*
700	1000	54.1	57.0	60.8	68.6	78.8	88.3	*70.6*	*106*	*142*	*212*
750	1070	60.8	64.4	68.5	76.8	88.5	99.4	*75.6*	*114*	*152*	*227*
800	1140	68.0	72.1	76.9	85.7	97.8	111	*80.6*	*121*	*162*	*242*
900	1285	83.9	88.5	95.2	105	120	136	148	*136*	*183*	*272*
1000	1430	101	106	115	126	144	163	177	*151*	*203*	*302*
1100	1570	120	125	136	148	171	192	208	*167*	*223*	*333*
1200	1715	140	146	158	173	200	220	242	*182*	*243*	*363*
1400	2000	184	193	206	230	258	287	316	353	*284*	*424*
1600	2285	234	244	260	288	323	363	393	445	*324*	*484*
1800	2570	292	299	322	350	399	452	480	543	591	*545*
2000	2860	350	364	387	425	481	535	576	652	707	*605*
2200	3140	417	435	459	510	573	628	683	771	833	*666*
2400	3403	487	507	535	585	668	730	799	885	968	*726*
2600	3710	564	587	620	677	769	841	913			*787*
2800	4000	645	669	714	773	874	954				
3000	4285	734	751	805	867	993					
3200	4570	827	850	909	982						

Loss in lb per sq in = .433 (sp gr) (figures in table)

Figure B.1. Line friction loss.

Friction Loss for Viscous Liquids *(Continued)*

(Based on Darcy's Formula)

Loss in Feet of Liquid per 1000 Feet of Pipe
8 Inch (7.981″ inside dia) Sch 40 New Steel Pipe

Flow		Kinematic viscosity—centistokes									
		.6	1.1	2.1	2.7	4.3	7.4	10.3	13.1	15.7	20.6
U S gal per min	Bbl per hr (42 gal)	Approx SSU viscosity									
			31.5	33	35	40	50	60	70	80	100
150	214	.42	.47	.53	.56	.63	.72	.79	.84	.88	.96
200	286	.71	.78	.89	.94	1.05	1.15	1.30	1.38	1.45	1.57
250	357	1.07	1.18	1.33	1.40	1.56	1.77	1.92	2.04	2.14	2.30
300	429	1.50	1.65	1.85	1.94	2.15	2.43	2.65	2.80	2.93	3.15
350	500	2.01	2.19	2.45	2.57	2.81	3.20	3.46	3.69	3.85	4.13
400	571	2.58	2.78	3.12	3.26	3.58	4.04	4.37	4.64	4.83	5.21
450	643	3.21	3.48	3.85	4.05	4.42	5.08	5.39	5.70	5.96	6.38
500	714	3.94	4.23	4.69	4.90	5.33	5.98	6.49	6.82	7.18	7.91
600	857	5.54	5.95	6.61	6.82	7.44	8.27	8.89	9.48	9.83	10.6
700	1000	7.44	7.96	8.71	9.04	9.84	10.9	11.9	12.4	13.0	13.9
800	1140	9.66	10.2	11.1	11.7	12.6	13.9	14.8	15.8	16.4	17.4
900	1285	12.1	12.8	13.8	14.4	15.5	17.1	18.4	19.4	20.2	21.5
1000	1430	14.8	15.6	16.8	17.4	18.7	20.8	22.2	23.3	24.4	26.0
1200	1715	21.0	22.0	23.7	24.5	26.4	28.9	30.7	32.4	33.5	35.9
1400	2000	28.3	29.6	31.8	32.6	35.6	38.2	40.7	42.6	44.3	47.5
1600	2285	36.7	38.4	40.6	42.1	44.5	48.7	51.9	54.1	56.1	59.3
1800	2570	46.1	48.0	50.8	52.3	55.4	60.4	64.5	67.0	69.4	73.5
2000	2860	56.5	58.8	61.9	63.8	67.3	73.4	77.7	81.1	83.8	88.8
2200	3140	67.9	70.2	74.4	76.3	80.5	87.5	92.1	96.8	99.6	105
2400	3430	80.8	83.0	88.0	90.2	94.7	103	108	114	117	123
2600	3710	94.2	97.5	103	105	110	119	125	131	135	142
2800	4000	109	112	118	121	127	136	144	149	155	163
3000	4285	125	129	135	138	145	155	164	170	176	184
3200	4570	142	146	153	156	162	174	184	191	197	208
3400	4860	160	164	172	174	182	196	206	213	220	232
3600	5140	178	183	192	196	204	217	228	237	244	258
3800	5425	199	204	212	217	226	240	251	262	269	285
4000	5715	220	225	234	238	249	265	277	289	295	311
4500	6425	276	284	294	300	311	331	345	358	368	385
5000	7145	341	348	360	365	380	404	418	433	447	466
5500	7855	410	419	433	439	457	480	500	518	532	555
6000	8570	488	498	512	519	540	567	592	609	623	654
6500	9280	573	581	601	609	630	662	686	707	723	755
7000	10000	664	673	692	702	725	763	791	810	829	867
7500	10700	760	773	789	806	827	865	897	925	946	984

Friction Loss for Viscous Liquids *(Continued)*

(Based on Darcy's Formula)

Loss in Feet of Liquid per 1000 Feet of Pipe
8 Inch (7.981″ inside dia) Sch 40 New Steel Pipe

Flow		Kinematic viscosity—centistokes									
		26.4	32.0	43.2	65.0	108.4	162.3	216.5	325	435	650
U S gal per min	Bbl per hr (42 gal)	Approx SSU viscosity									
		125	150	200	300	500	750	1000	1500	2000	3000
50	71.4	.21	.25	.34	.50	.84	1.26	1.68	2.52	3.38	5.05
100	143	.41	.50	.67	1.01	1.68	2.52	3.36	5.04	6.76	10.1
150	214	1.03	.75	1.01	1.51	2.52	3.78	5.04	7.56	10.1	15.1
200	286	1.67	1.78	1.34	2.02	3.37	5.04	6.72	10.1	13.5	20.2
250	357	2.46	2.60	2.85	2.52	4.21	6.30	8.40	12.6	16.9	25.3
300	429	3.37	3.56	3.89	3.03	5.05	7.56	10.1	15.1	20.3	30.3
350	500	4.39	4.63	5.04	5.69	5.89	8.82	11.8	17.7	23.6	35.3
400	571	5.54	5.83	6.35	7.18	6.73	10.1	13.5	20.2	27.0	40.4
450	643	6.79	7.16	7.75	8.76	7.58	11.3	15.1	22.7	30.4	45.4
500	714	8.17	8.58	9.32	10.4	8.42	12.6	16.8	25.2	33.8	50.5
550	786	9.65	10.1	11.0	12.3	9.26	13.9	18.5	27.8	37.2	55.5
600	857	11.2	11.8	12.8	14.3	16.6	15.1	20.2	30.3	40.5	60.6
700	1000	14.7	15.5	16.7	18.6	21.7	17.6	23.5	35.3	47.3	70.6
800	1140	18.6	19.4	21.0	23.4	27.1	20.1	26.9	40.3	54.0	80.7
900	1285	22.9	24.0	25.9	28.7	33.1	37.4	30.2	45.4	60.8	90.8
1000	1430	27.4	28.8	31.0	34.4	39.7	44.9	33.6	50.4	67.6	101
1200	1715	37.8	39.4	42.8	47.3	54.3	60.9	66.4	60.5	81.0	121
1400	2000	49.7	52.0	56.1	62.0	70.8	79.3	86.7	70.6	94.5	141
1600	2285	63.2	65.7	70.6	78.0	89.3	99.4	108	80.7	108	161
1800	2570	77.9	81.3	86.9	96.2	110	122	132	150	122	182
2000	2860	93.4	98.0	105	116	132	147	159	180	135	202
2200	3140	111	116	124	137.	155	173	186	211	149	222
2400	3430	130	135	145	159	181	201	217	244	266	242
2600	3710	149	155	168	183	208	231	250	279	306	262
2800	4000	170	178	191	210	236	262	283	317	347	282
3000	4285	193	201	215	236	267	296	319	357	389	303
3200	4570	217	225	240	265	299	332	357	398	433	323
3400	4860	242	251	268	294	333	369	397	441	480	343
3600	5140	268	284	296	326	369	407	438	488	529	598
3800	5425	296	307	328	359	404	447	482	536	582	656
4000	5715	325	337	358	394	441	488	526	586	635	718
4500	6425	405	417	442	485	543	601	646	718	777	876
5000	7145	488	505	536	582	656	726	776	860	932	
5500	7855	579	605	634	689	773	855	913			
6000	8570	678	706	744	810	910	993				

Figure B.1. Line friction loss.

Friction Loss for Viscous Liquids *(Continued)*

(Based on Darcy's Formula)

Loss in Feet of Liquid per 1000 Feet of Pipe
10 Inch (10.02" inside dia) Sch 40 New Steel Pipe

Flow		Kinematic viscosity—centistokes									
		.6	1.1	2.1	2.7	4.3	7.4	10.3	13.1	15.7	20.6
US gal per min	Bbl per hr (42 gal)	Approx SSU viscosity									
			31.5	33	35	40	50	60	70	80	100
400	571	.83	.92	1.03	1.09	1.19	1.35	1.46	1.55	1.63	1.75
500	714	1.27	1.38	1.53	1.60	1.78	2.00	2.16	2.30	2.40	2.59
600	857	1.78	1.91	2.14	2.24	2.47	2.77	2.98	3.15	3.31	3.54
700	1000	2.39	2.55	2.84	2.97	3.26	3.62	3.93	4.14	4.32	4.67
800	1140	3.06	3.29	3.63	3.79	4.12	4.63	4.99	5.25	5.46	5.86
900	1285	3.84	4.12	4.49	4.72	5.09	5.72	6.14	6.46	6.74	7.19
1000	1430	4.68	4.99	5.42	5.70	6.13	6.90	7.36	7.83	8.10	8.63
1100	1570	5.63	5.97	6.49	6.82	7.34	8.20	8.76	9.25	9.62	10.3
1200	1715	6.61	7.05	7.63	7.85	8.65	9.58	10.3	10.8	11.3	11.9
1300	1855	7.71	8.18	8.85	9.16	9.95	11.0	11.8	12.4	13.0	13.7
1400	2000	8.88	9.42	10.2	10.6	11.4	12.6	13.5	14.2	14.7	15.7
1500	2140	10.1	10.8	11.7	12.0	12.9	14.3	15.3	16.1	16.6	17.8
1600	2285	11.5	12.2	13.2	13.6	14.6	16.0	17.2	18.1	18.7	20.0
1800	2570	14.3	15.1	16.2	16.7	17.9	19.7	20.9	22.1	22.9	24.3
2000	2860	17.8	18.6	19.8	20.6	21.8	24.0	25.5	27.0	28.0	29.5
2200	3140	21.3	22.2	23.7	24.6	26.1	28.6	30.3	32.1	31.8	35.0
2400	3430	25.2	26.3	28.0	28.9	30.7	33.4	35.5	37.3	38.9	41.0
2600	3710	29.6	30.6	32.5	33.5	35.6	38.7	41.0	42.9	44.8	47.3
2800	4000	34.1	35.3	37.4	38.4	40.8	44.5	47.1	49.0	51.0	54.1
3000	4285	39.1	40.2	42.7	43.5	46.6	50.7	53.2	55.7	57.7	61.3
3500	5000	52.5	54.4	57.4	58.9	62.3	66.4	70.6	73.6	76.2	80.8
4000	5715	68.0	70.5	73.9	75.9	79.9	85.8	90.2	94.2	97.1	102
4500	6430	86.1	88.6	92.3	94.8	99.2	107	112	117	120	127
5000	7145	106	109	113	116	122	130	136	142	146	153
5500	7855	128	131	136	139	145	156	162	169	173	182
6000	8570	152	154	161	164	172	183	191	197	204	213
6500	9280	177	180	187	191	201	212	221	228	236	246
7000	10000	205	208	217	220	231	243	255	263	369	282
7500	10700	236	239	248	251	262	277	291	298	303	321
8000	11400	266	272	282	286	296	314	329	337	345	360
8500	12100	301	307	318	321	334	352	367	378	387	403
9000	12900	337	341	354	359	372	392	407	422	429	447
10000	14300	416	422	434	441	453	478	492	511	524	542
11000	15700	503	511	522	533	544	574	593	611	626	649
12000	17150	599	603	617	630	643	679	701	719	737	763

Friction Loss for Viscous Liquids *(Continued)*

(Based on Darcy's Formula)

Loss in Feet of Liquid per 1000 Feet of Pipe
10 Inch (10.02" inside dia) Sch 40 New Steel Pipe

Flow		Kinematic viscosity—centistokes									
		26.4	32.0	43.2	65.0	108.4	162.3	216.5	325	435	650
US gal per min	Bbl per hr (42 gal)	Approx SSU viscosity									
		125	150	200	300	500	750	1000	1500	2000	3000
150	214	*.25*	*.30*	*.40*	*.61*	*1.02*	*1.52*	*2.03*	*3.04*	*4.08*	*6.09*
200	286	.58	*.40*	*.54*	*.81*	*1.35*	*2.03*	*2.71*	*4.06*	*5.43*	*8.12*
300	429	1.15	1.22	1.33	*1.22*	*2.03*	*3.04*	*4.06*	*6.09*	*8.15*	*12.2*
400	571	1.83	1.98	2.17	*1.62*	*2.71*	*4.06*	*5.41*	*8.12*	*10.9*	*16.2*
500	714	2.75	2.91	3.18	3.60	*3.39*	*5.07*	*6.77*	*10.1*	*13.6*	*20.3*
600	857	3.78	3.97	4.34	4.89	*4.06*	*6.08*	*8.12*	*12.2*	*16.3*	*24.4*
700	1000	4.94	5.19	5.66	6.37	*4.74*	*7.10*	*9.47*	*14.2*	*19.0*	*28.4*
800	1140	6.21	6.55	7.10	7.97	9.31	*8.12*	*10.8*	*16.2*	*21.7*	*32.5*
900	1285	7.66	8.04	8.71	9.76	11.4	*9.13*	*12.2*	*18.3*	*24.5*	*36.5*
1000	1430	9.21	9.61	10.5	11.7	13.6	*10.1*	*13.5*	*20.3*	*27.2*	*40.6*
1100	1570	10.9	11.4	12.3	13.8	16.0	18.0	*14.9*	*22.3*	*29.9*	*44.6*
1200	1715	12.6	13.4	14.4	16.0	18.6	21.0	*16.2*	*24.4*	*32.6*	*48.7*
1300	1855	14.5	15.4	16.4	18.4	21.2	24.0	*17.6*	*26.4*	*35.3*	*52.8*
1400	2000	16.6	17.5	18.7	20.9	24.1	27.2	*18.9*	*28.4*	*38.0*	*56.8*
1500	2140	18.7	19.6	21.2	23.6	27.2	30.6	33.3	*30.4*	*40.8*	*60.9*
1600	2285	21.0	21.9	23.7	26.3	30.4	34.1	37.2	*32.4*	*43.5*	*64.9*
1800	2570	26.0	27.3	29.2	32.2	37.1	41.7	45.5	*36.5*	*48.9*	*73.1*
2000	2860	31.3	32.7	35.0	38.6	44.4	49.8	54.4	*40.6*	*54.3*	*81.2*
2200	3140	37.0	38.6	41.3	45.8	52.3	58.8	64.0	71.9	*59.8*	*89.3*
2400	3430	43.1	45.1	48.3	53.4	60.9	68.3	74.2	83.8	*55.2*	*97.4*
2600	3710	49.8	52.1	55.4	61.4	70.0	78.3	85.0	96.2	*70.7*	*105*
2800	4000	56.8	59.2	63.5	70.1	79.7	88.9	96.1	109	*76.1*	*114*
3000	4285	64.3	66.8	71.8	78.8	89.8	99.8	108	122	133	*122*
3500	5000	84.9	88.3	94.4	103	117	131	142	159	180	*142*
4000	5715	108	112	119	131	149	165	178	199	217	*162*
4500	6430	133	139	147	162	182	202	218	244	266	300
5000	7145	160	168	179	195	218	241	262	293	318	360
5500	7855	191	199	212	231	258	286	309	345	373	423
6000	8570	223	232	247	268	302	334	359	399	435	489
6500	9280	258	267	286	310	348	384	411	460	499	560
7000	10000	296	305	326	355	396	436	468	522	566	637
7500	10700	335	347	369	402	447	492	529	589	638	766
8000	11400	377	389	414	452	505	550	594	659	710	797
9000	12900	469	482	512	557	624	679	729	809	869	976
10000	14300	567	582	619	666	743	817	872	964		

Figure B.1. Line friction loss.

Friction Loss for Viscous Liquids *(Continued)*
(Based on Darcy's Formula)
Loss in Feet of Liquid per 1000 Feet of Pipe
12 Inch (11.938″ inside dia) Sch 40 New Steel Pipe

Flow		Kinematic viscosity—centistokes									
		0.6	1.13	2.1	2.7	4.3	7.4	10.3	13.1	15.7	20.6
U S gal per min	Bbl per hr (42 gal)	Approx SSU viscosity									
			31.5	33	35	40	50	60	70	80	100
300	429	.21	.24	.27	.28	.31	.36	.38	.41	.43	.47
400	571	.36	.40	.45	.47	.51	.59	.64	.67	.70	.77
500	714	.54	.60	.67	.70	.75	.87	.95	.99	1.03	1.12
600	857	.76	.83	.93	.98	1.04	1.19	1.30	1.39	1.41	1.53
700	1000	.98	1.11	1.23	1.29	1.37	1.56	1.70	1.82	1.84	2.00
800	1140	1.26	1.42	1.57	1.64	1.74	1.98	2.15	2.30	2.36	2.51
900	1285	1.57	1.76	1.94	1.96	2.15	2.44	2.65	2.82	2.94	3.08
1000	1430	1.92	2.07	2.36	2.38	2.61	2.94	3.19	3.40	3.57	3.70
1200	1715	2.73	2.91	3.18	3.32	3.62	4.07	4.41	4.68	4.91	5.08
1400	2000	3.67	3.90	4.24	4.41	4.80	5.37	5.79	6.14	6.43	6.65
1600	2285	4.75	5.02	5.43	5.64	6.12	6.83	7.35	7.78	8.14	8.51
1800	2570	5.96	6.29	6.77	7.02	7.59	8.44	9.07	9.59	10.0	10.6
2000	2860	7.32	7.69	8.25	8.54	9.21	10.2	11.0	11.6	12.1	12.9
2500	3570	11.3	11.8	12.6	13.0	13.9	15.3	16.4	17.3	18.0	19.2
3000	4285	16.1	16.8	17.7	18.3	19.5	21.4	22.8	23.9	24.9	26.5
3500	5000	21.8	22.6	23.8	24.4	26.0	28.3	30.1	31.6	32.9	34.9
4000	5715	28.3	29.2	30.7	31.5	33.3	36.2	38.4	40.3	41.8	44.3
4500	6430	35.7	36.8	38.5	39.4	41.6	45.0	47.7	49.9	51.7	54.8
5000	7145	44.0	45.2	47.1	48.2	50.7	54.7	57.8	60.4	62.6	66.2
5500	7855	53.1	54.4	56.6	57.8	60.7	65.3	68.9	71.9	74.4	78.7
6000	8570	63.0	64.5	66.9	68.3	71.6	76.8	80.9	84.3	87.2	92.1
6500	9280	73.8	75.4	78.1	79.6	83.3	89.2	93.8	97.7	101	106
7000	10000	85.4	87.2	90.1	91.8	95.9	102	108	112	116	122
7500	10700	97.9	99.8	103	105	109	117	122	127	131	138
8000	11400	111	113	117	119	124	132	138	143	148	155
9000	12850	141	143	147	149	155	164	172	178	183	192
10000	14300	173	176	180	183	190	200	209	217	223	233
11000	15700	209	212	217	220	228	240	250	269	266	278
12000	17150	249	252	258	261	269	283	294	304	312	326
13000	18550	291	295	301	305	314	330	342	353	363	373
14000	20000	338	342	348	353	363	380	394	406	416	434
15000	21400	387	392	399	403	414	433	449	462	473	493
16000	22850	440	445	453	457	469	490	507	522	534	556
18000	25700	557	561	571	577	590	614	634	651	666	692
20000	28600	687	692	703	709	725	752	775	795	812	842

Friction Loss for Viscous Liquids *(Continued)*
(Based on Darcy's Formula)
Loss in Feet of Liquid per 1000 Feet of Pipe
12 Inch (11.938″ inside dia) Sch 40 New Steel Pipe

Flow		Kinematic viscosity—centistokes									
		26.4	32.0	43.2	65.0	108.4	162.3	216.5	325	435	650
U S gal per min	Bbl per hr (42 gal)	Approx SSU viscosity									
		125	150	200	300	500	750	1000	1500	2000	3000
100	143	.08	.10	.13	.20	.34	.51	.68	1.00	1.35	2.00
200	286	.16	.19	.27	.40	.67	1.00	1.37	2.05	2.74	4.01
300	429	.49	.53	.41	.62	1.01	1.51	2.00	3.08	4.11	6.16
400	571	.81	.86	.94	.82	1.34	2.02	2.68	3.98	5.46	8.21
500	714	1.22	1.25	1.37	1.02	1.71	2.50	3.37	5.01	6.84	10.3
600	857	1.66	1.71	1.87	2.12	1.97	3.02	4.05	6.03	7.99	12.3
700	1000	2.15	2.30	2.43	2.75	2.37	3.66	4.68	7.05	9.36	13.9
800	1140	2.70	2.88	3.05	3.45	2.63	4.04	5.36	8.09	10.7	15.9
900	1285	3.31	3.52	3.74	4.21	4.93	4.44	6.05	9.30	12.1	18.0
1000	1430	3.97	4.22	4.48	5.04	5.89	5.13	6.84	10.0	13.5	20.0
1200	1715	5.43	5.77	6.33	6.88	8.01	5.91	7.89	12.1	16.2	24.1
1400	2000	7.10	7.53	8.24	8.96	10.4	11.8	9.47	14.6	18.9	28.2
1600	2285	8.96	9.48	10.4	11.3	13.1	14.8	10.5	16.2	21.7	32.4
1800	2570	11.0	11.6	12.7	13.8	16.0	18.0	19.7	17.7	24.2	36.5
2000	2860	13.2	14.0	15.2	16.6	19.1	21.5	23.6	20.5	27.4	40.1
2500	3570	19.6	20.6	22.4	25.3	28.0	31.5	34.3	25.7	34.2	51.3
3000	4285	27.2	28.4	30.8	34.6	38.4	43.0	46.8	53.1	41.1	61.6
3500	5000	36.2	37.3	40.3	45.2	50.2	56.0	60.9	68.8	47.9	71.9
4000	5715	43.4	47.3	51.0	57.0	63.2	70.5	76.5	86.3	94.5	82.1
4500	6430	57.6	58.3	62.8	69.9	77.6	86.4	93.6	105	115	92.4
5000	7145	69.8	70.3	75.6	84.1	93.3	104	112	126	138	103
5500	7855	82.8	83.6	89.5	99.4	110	122	132	148	162	183
6000	8570	96.8	98.2	104	116	128	142	154	172	188	212
6500	9280	112	114	120	133	148	164	176	197	215	243
7000	10000	128	131	137	152	174	186	201	224	244	275
7500	10700	145	148	155	172	196	210	226	253	274	310
8000	11400	163	167	174	192	220	235	253	282	306	345
9000	12850	202	208	215	237	270	289	311	346	375	422
10000	14300	204	253	260	286	325	347	373	415	450	505
11000	15700	290	301	309	338	384	411	441	490	530	594
12000	17150	341	354	361	395	448	479	514	569	616	689
13000	18550	394	409	417	456	516	551	591	655	707	790
14000	20000	452	469	477	521	588	628	673	745	804	898
15000	21400	513	532	541	589	664	710	760	840	906	
16000	22850	578	598	608	662	745	796	851	940		

Figure B.1. Line friction loss.

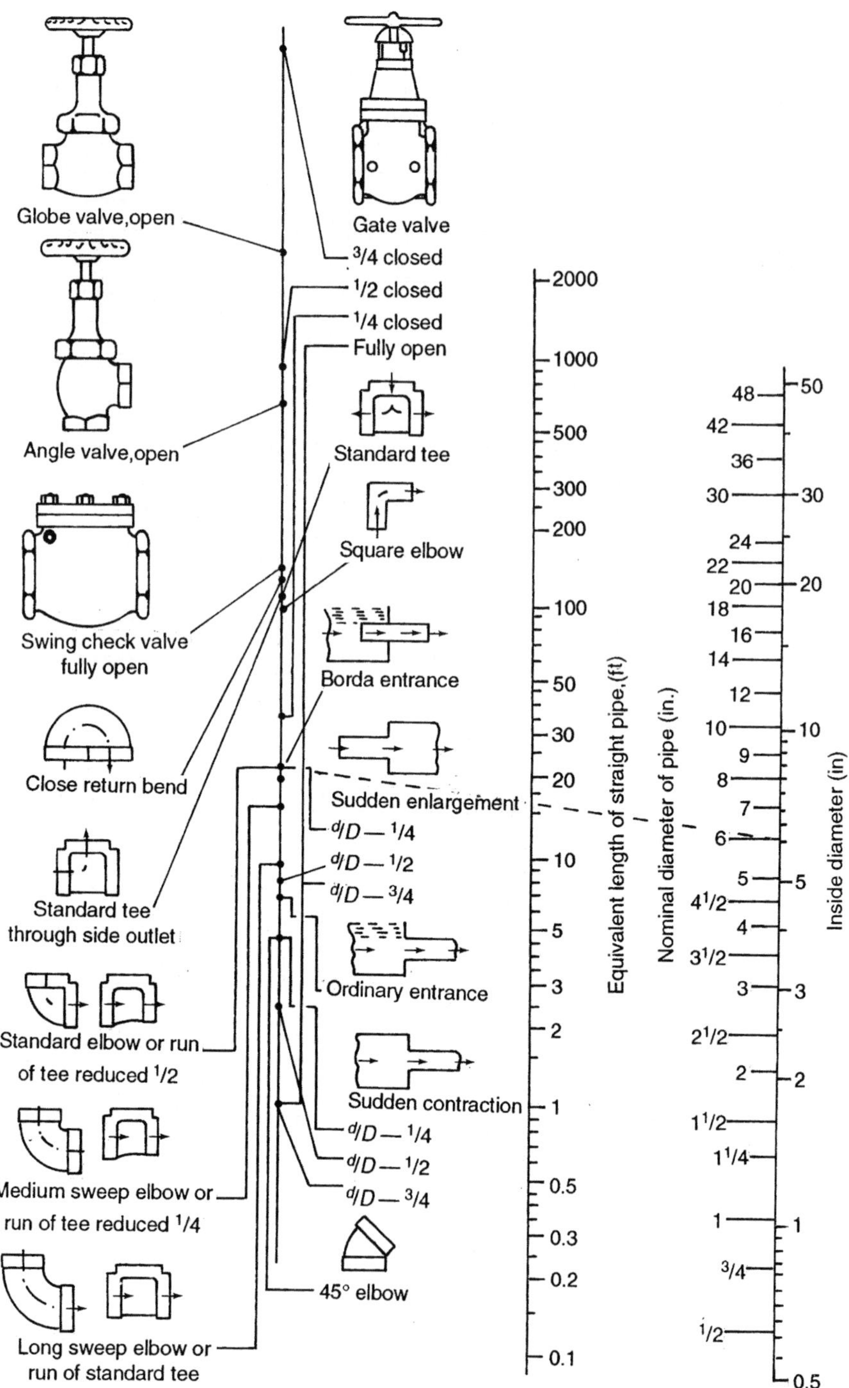

Figure B.2. Resistance of valves and fittings.

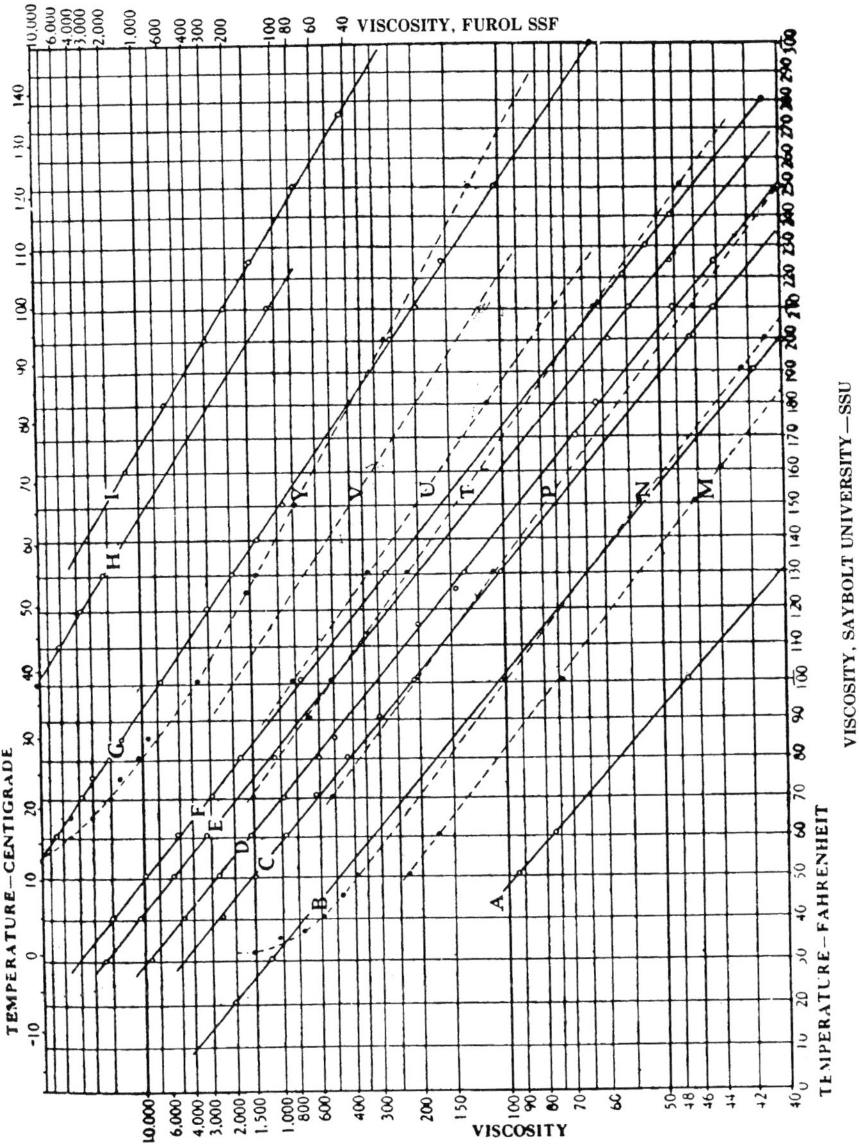

This chart may be used to determine the viscosity of an oil at any temperature provided its viscosity at two temperatures is known.

The lines on this chart show viscosities of representative oils.

Note: This chart is similar to ASTM tentative standard D341-32T which has a somewhat wider viscosity and temperature range.

Courtesy of Texaco, Inc.

Figure B.3. Viscosity versus temperature.

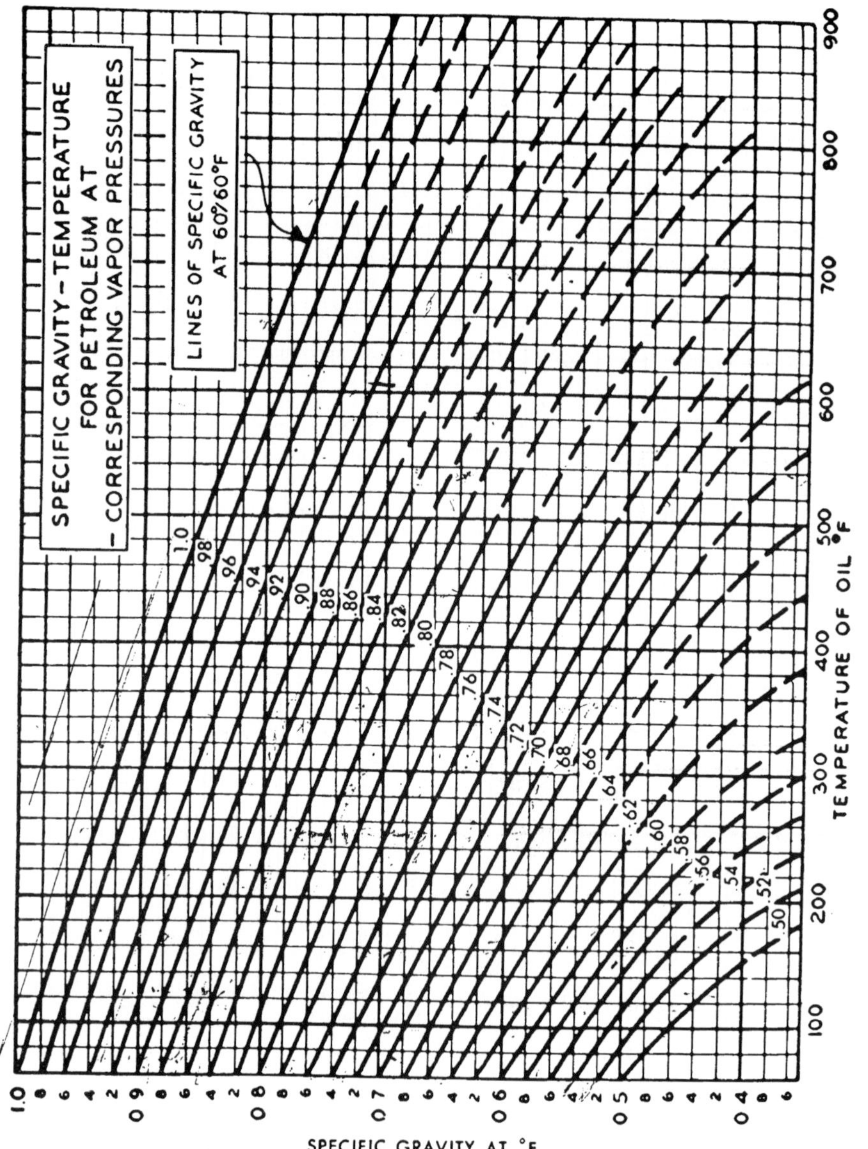

Figure B.4. Specific gravity versus temperature.

Specific Gravity—Referred to water at 60°F.
Example: oil with sp. gr. of 0.82 at 60°F will have sp. gr. of 0.64 at 500°F.

Courtesy of Hydraulic Institute.

Pounds per gallon and specific gravities corresponding to degrees API at 60°F

Formula—$\text{sp gr} = \dfrac{141.5}{131.5 + {}^\circ\text{API}}$

Deg API	Tenths of Degrees									
	0	1	2	3	4	5	6	7	8	9
10	8.328 1.0000	8.322 .9993	8.317 .9986	8.311 .9979	8.305 .9972	8.299 .9965	8.293 .9958	8.287 .951	8.282 .9944	8.276 .9937
11	8.270 .9930	8.264 .9923	8.258 .9916	8.252 .9909	8.246 .9902	8.241 .9895	8.235 .9888	8.229 .9881	8.223 .9874	8.218 .9868
12	8.212 .9861	8.206 .9854	8.201 .9847	8.195 .9840	8.189 .9833	8.183 .9826	8.178 .9820	8.172 .9813	8.166 .9806	8.161 .9799
13	8.155 .9792	8.150 .9786	8.144 .9779	8.138 .9772	8.132 .9765	8.127 .9759	8.122 .9752	8.116 .9745	8.110 .9738	8.105 .9732
14	8.099 .9725	8.093 .9718	8.088 .9712	8.082 .9705	8.076 .9698	8.071 .9692	8.066 .9685	8.061 .9679	8.055 .9672	8.049 .9665
15	8.044 .9659	8.038 .9652	8.033 .9646	8.027 .9639	8.021 .9632	8.016 .9626	8.011 .9619	8.006 .9613	8.000 .9606	7.995 .9600
16	7.989 .9593	7.984 .9587	7.978 .9580	7.973 .9574	7.967 .9567	7.962 .9561	7.956 .9554	7.951 .9548	7.946 .9541	7.940 .9535
17	7.935 .9529	7.930 .9522	7.925 .9516	7.919 .9509	7.914 .9503	7.909 .9497	7.903 .9490	7.898 .9484	7.893 .9478	7.887 .9471
18	7.882 .9465	7.877 .9459	7.871 .9452	7.866 .9446	7.861 .9440	7.856 .9433	7.851 .9427	7.846 .9421	7.841 .9415	7.835 .9408
19	7.830 .9402	7.825 .9396	7.820 .9390	7.814 .9383	7.809 .9377	7.804 .9371	7.799 .9365	7.793 .9358	7.788 .9352	7.783 .9346
20	7.778 .9340	7.773 .9334	7.768 .9328	7.762 .9321	7.757 .9315	7.752 .9309	7.747 .9303	7.742 .9297	7.737 .9291	7.732 .9285
21	7.727 .9279	7.722 .9273	7.717 .9267	7.711 .9260	7.706 .9254	7.701 .9248	7.696 .9242	7.691 .9236	7.686 .9230	7.681 .9224
22	7.676 .9218	7.671 .9212	7.666 .9206	7.661 .9200	7.656 .9194	7.651 .9188	7.646 .9182	7.641 .9176	7.636 .9170	7.632 .9165
23	7.627 .9159	7.622 .9153	7.617 .9147	7.612 .9141	7.607 .9135	7.602 .9129	7.597 .9123	7.592 .9117	7.587 .9111	7.583 .9106
24	7.578 .9100	7.573 .9094	7.568 .9088	7.563 .9082	7.558 .9076	7.554 .9071	7.549 .9065	7.544 .9059	7.539 .9053	7.534 .9047
25	7.529 .9042	7.524 .9036	7.519 .9030	7.514 .9024	7.509 .9018	7.505 .9013	7.500 .9007	7.495 .9001	7.491 .8996	7.486 .8990
26	7.481 .8984	7.476 .8978	7.427 .8973	7.467 .8967	7.462 .8961	7.458 .8956	7.453 .8950	7.448 .8944	7.443 .8939	7.438 .8933
27	7.434 .8927	7.429 .8922	7.424 .8916	7.420 .8911	7.415 .8905	7.410 .8899	7.406 .8894	7.401 .8888	7.397 .8883	7.392 .8877
28	7.387 .8871	7.383 .8866	7.378 .8860	7.373 .8855	7.368 .8849	7.364 .8844	7.360 .8838	7.355 .8833	7.350 .8827	7.346 .8822
29	7.341 .8816	7.337 .8811	7.332 .8805	7.328 .8800	7.323 .8794	7.318 .8789	7.314 .8783	7.309 .8778	7.305 .8772	7.300 .8767
30	7.296 .8762	7.291 .8756	7.287 .8751	7.282 .8745	7.278 .8740	7.273 .8735	7.268 .8729	7.264 .8724	7.259 .8718	7.255 .8713
31	7.251 .8708	7.246 .8702	7.242 .8679	7.238 .8692	7.233 .8686	7.228 .8681	7.224 .8676	7.219 .8670	7.215 .8665	7.211 .8660
32	7.206 .8654	7.202 .8649	7.198 .8644	7.193 .8639	7.188 .8633	7.184 .8628	7.180 .8623	7.176 .8618	7.171 .8612	7.167 .8607
33	7.163 .8602	7.158 .8597	7.153 .8591	7.149 .8536	7.145 .8581	7.141 .8576	7.137 .8571	7.132 .8565	7.128 .8560	7.123 .8555
34	7.119 .8550	7.115 .8545	7.111 .8540	7.106 .8534	7.102 .8529	7.098 .8524	7.093 .8519	7.089 .8514	7.085 .8509	7.081 .8504
35	7.076 .8498	7.072 .8493	7.067 .8488	7.063 .8483	7.059 .8478	7.055 .8473	7.051 .8468	7.047 .8463	7.042 .8458	7.038 .8453
36	7.034 .8448	7.030 .8443	7.026 .8438	7.022 .8433	7.018 .8428	7.013 .8423	7.009 .8418	7.005 .8412	7.001 .8408	6.997 .8403
37	6.993 .8398	6.989 .8393	6.985 .8388	6.980 .8383	6.976 .8378	6.972 .8373	6.968 .8368	6.964 .8363	6.960 .8358	6.955 .8353
38	6.951 .8348	6.947 .8343	6.943 .8338	6.939 .8333	6.935 .8328	6.930 .8324	6.926 .8319	6.922 .8314	6.918 .8309	6.914 .8304
39	6.910 .8299	6.906 .8294	6.902 .8289	6.898 .8285	6.894 .8280	6.890 .8275	6.886 .8270	6.882 .8265	6.878 .8260	6.874 .8256
40	6.870 .8251	6.866 .8246	6.862 .8241	6.858 .8236	6.854 .8232	6.850 .8227	6.846 .8222	6.842 .8217	6.838 .8212	6.834 .8208
41	6.830 .8203	6.826 .8198	6.822 .8193	6.818 .8189	6.814 .8184	6.810 .8178	6.806 .8174	6.802 .8170	6.798 .8165	6.794 .8160

Pounds per gallon and specific gravities corresponding to degrees API at 60°F (*Continued*)

Deg API	Tenths of Degrees									
	0	1	2	3	4	5	6	7	8	9
42	6.790 .8155	6.786 .8151	6.782 .8146	6.779 .8142	6.775 .8137	6.771 .8132	6.767 .8128	6.763 .8123	6.759 .8118	6.756 .8114
43	6.752 .8109	6.748 .8104	6.744 .8100	6.740 .8095	6.736 .8090	6.732 .8086	6.728 .8081	6.724 .8076	6.720 .8072	6.716 .8067
44	6.713 .8063	6.709 .8058	6.705 .8054	6.701 .8049	6.697 .8044	6.694 .8040	6.690 .8035	6.686 .8031	6.682 .8026	6.679 .8022
45	6.675 .8017	6.671 .8012	6.667 .8008	6.663 .8003	6.660 .7999	6.656 .7994	6.652 .7990	6.648 .7985	6.645 .7981	6.641 .7976
46	6.637 .7972	6.633 .7967	6.630 .7963	6.626 .7958	6.622 .7954	6.618 .7949	6.615 .7945	6.611 .7941	6.607 .7936	6.604 .7932
47	6.600 .7927	6.596 .7923	6.592 .7918	6.589 .7914	6.585 .7909	6.582 .7905	6.578 .7901	6.574 .7896	6.571 .7892	6.567 .7887
48	6.563 .7883	6.560 .7879	6.556 .7874	6.552 .7870	6.548 .7865	6.545 .7861	6.541 .7857	6.537 .7852	6.534 .7848	6.530 .7844
49	6.526 .7839	6.523 .7835	6.520 .7831	6.516 .7826	6.512 .7822	6.509 .7818	6.505 .7813	6.501 .7809	6.498 7805	6.494 .7800
50	6.490 .7796	6.487 .7792	6.484 .7788	6.480 .7783	6.476 .7779	6.473 .7775	6.469 .7770	6.466 .7766	6.462 .7762	6.459 .7758
51	6.455 .7753	6.451 .7749	6.448 .7745	6.445 .7741	6.441 .7736	6.437 .7732	6.434 .7728	6.430 .7724	6.427 .7720	6.423 .7715
52	6.420 .7711	6.416 .7707	6.413 .7703	6.410 .7699	6.406 .7694	6.402 .7690	6.399 .7686	6.396 .7682	6.392 .7678	6.389 .7674
53	6.385 .7669	6.381 .7665	6.378 .7661	6.375 .7657	6.371 .7653	6.368 .7649	6.365 .7645	6.360 .7640	6.357 .7636	6.354 .7632
54	6.350 .7628	6.347 .7624	6.344 .7620	6.340 .7616	6.337 .7612	6.334 .7608	6.330 .7603	6.326 .7599	6.323 .7595	6.320 .7591
55	6.316 .7587	6.313 .7583	6.310 .7579	6.306 .7575	6.303 .7571	6.300 .7567	6.296 .7563	6.293 .7559	6.290 .7555	6.287 .7551
56	6.283 .7547	6.280 .7543	6.276 .7539	6.273 .7535	6.270 .7531	6.266 .7527	6.263 .7523	6.259 .7519	6.256 .7515	6.253 .7511
57	6.249 .7507	6.246 .7503	6.243 .7499	6.240 .7495	6.236 .7491	6.233 .7487	6.229 .7483	6.226 .7479	6.223 .7475	6.219 .7471
58	6.216 .7467	6.213 .7463	6.209 .7459	6.206 .7455	6.203 .7451	6.199 .7447	6.196 .7443	6.193 .7440	6.190 .7436	6.187 .7432
59	6.184 .7428	6.180 .7424	6.177 .7420	6.174 .7416	6.170 .7412	6.167 .7408	6.164 .7405	6.161 .7401	6.158 .7397	6.154 .7393
60	6.151 .7389	6.148 .7385	6.144 .7381	6.141 .7377	6.138 .7374	6.135 .7370	6.132 .7366	6.129 .7362	6.125 .7358	6.122 .7354
61	6.119 .7351	6.116 .7347	6.113 .7343	6.109 .7339	6.106 .7335	6.103 .7332	6.100 .7328	6.097 .7324	6.094 .7320	6.090 .7316
62	6.087 .7313	6.084 .7309	6.081 .7305	6.078 .7301	6.075 .7298	6.072 .7294	6.068 .7290	6.065 .7286	6.062 .7283	6.059 .7279
63	6.056 .7275	6.053 .7271	6.050 .7268	6.047 .7264	6.044 .7260	6.040 .7256	6.037 .7253	6.034 .7249	6.031 .7245	6.028 .7242
64	6.025 .7238	6.022 .7234	6.019 .7230	6.016 .7227	6.013 .7223	6.010 .7219	6.007 .7216	6.004 .7212	6.000 .7208	5.997 .7205
65	5.994 .7201	5.991 .7197	5.988 .7194	5.985 .7190	5.982 .7186	5.979 .7183	5.976 .7179	5.973 .7175	5.970 .7172	5.967 .7168
66	5.964 .7165	5.961 .7161	5.958 .7157	5.955 .7154	5.952 .7150	5.949 .7146	5.946 .7143	5.943 .7139	5.940 .7136	5.937 .7132
67	5.934 .7128	5.931 .7125	5.928 .7121	5.925 .7118	5.922 .7114	5.919 .7111	5.916 .7107	5.913 .7013	5.910 .7100	5.907 .7096
68	5.904 .7093	5.901 .7089	5.898 .7086	5.895 .7082	5.892 .7079	5.889 .7075	5.886 .7071	5.883 .7068	5.880 .7064	5.877 .7061
69	5.874 .7057	5.871 .7054	5.868 .7050	5.866 .7047	5.863 .7043	5.860 .7040	5.857 .7036	5.854 .7033	5.851 .7029	5.848 .7026
70	5.845 .7022	5.842 .7019	5.839 .7015	5.836 .7012	5.833 .7008	5.831 .7005	5.828 .7001	5.825 .6998	5.823 .6995	5.820 .6991
71	5.817 .6988	5.814 .6984	5.811 .6981	5.808 .6977	5.805 .6974	5.802 .6970	5.799 .6967	5.796 .6964	5.793 .6960	5.791 .6957
72	5.788 .6952	5.785 .6950	5.782 .6946	5.779 .6943	5.776 .6940	5.773 .6936	5.771 .6933	5.768 .6929	5.765 .6926	5.762 .6923
73	5.759 .6919	5.757 .6916	5.754 .6913	5.751 .6909	5.748 .6906	5.745 .6902	5.743 .6899	5.740 .6896	5.737 .6892	5.734 .6889
74	5.731 .6886	5.728 .6882	5.726 .6879	5.723 .6876	5.720 .6872	5.718 .6869	5.715 .6866	5.712 .6862	5.709 .6859	5.706 .6856

Figure B.5. Relationship of SG, °API, Lbs/Gal.

Pounds per gallon and specific gravities corresponding to degrees API at 60°F (*Continued*)

Deg API	Tenths of Degrees									
	0	.1	2	3	4	5	6	7	8	9
75	5.703 .6852	5.701 .6849	5.698 .6846	5.695 .6842	5.693 .6839	5.690 .6836	5.687 .6832	5.685 .6829	5.682 .6826	5.679 .6823
76	5.676 .6819	5.673 .6816	5.671 .6813	5.668 .6809	5.665 .6806	5.662 .6803	5.660 .6800	5.657 .6796	5.654 .6793	5.652 .6790
77	5.649 .6787	5.646 .6783	5.643 .6780	5.641 .6777	5.638 .6774	5.635 .6770	5.632 .6767	5.630 .6764	5.627 .6761	5.624 .6757
78	5.622 .6754	5.619 .6751	5.617 .6748	5.614 .6745	5.611 .6741	5.608 .6738	5.606 .6735	5.603 .6732	5.600 .6728	5.598 .6725
79	5.595 .6722	5.592 .6719	5.590 .6716	5.587 .6713	5.584 .6709	5.582 .6706	5.579 .6703	5.577 .6700	5.574 .6697	5.571 .6693
80	5.568 .6690	5.566 .6687	5.563 .6684	5.561 .6681	5.558 .6678	5.556 .6675	5.553 .6671	5.550 .6668	5.548 .6665	5.545 .6662
81	5.542 .6659	5.540 .6656	5.537 .6653	5.534 .6649	5.532 .6646	5.529 .6643	5.526 .6640	5.524 .6637	5.522 .6634	5.519 .6631
82	5.516 .6628	5.514 .6625	5.511 .6621	5.508 .6618	5.506 .6615	5.503 .6612	5.501 .6609	5.498 .6606	5.496 .6603	5.493 .6600
83	5.491 .6597	5.489 .6594	5.486 .6591	5.483 .6588	5.480 .6584	5.477 .6581	5.475 .6578	5.472 .6575	5.470 .6572	5.467 .6569
84	5.465 .6566	5.462 .6563	5.460 .6560	5.458 .6557	5.455 .6554	5.453 .6551	5.450 .6548	5.448 .6545	5.445 .6542	5.443 .6539
85	5.440 .6536	5.437 .6533	5.435 .6530	5.432 .6527	5.430 .6524	5.427 .6521	5.425 .6518	5.422 .6515	5.420 .6512	5.417 .6509
86	5.415 .6506	5.412 .6503	5.410 .6500	5.407 .6497	5.405 .6494	5.402 .6491	5.400 .6488	5.397 .6485	5.395 .6482	5.392 .6479
87	5.390 .6476	5.387 .6473	5.385 .6470	5.382 .6467	5.380 .6464	5.377 .6461	5.375 .6458	5.372 .6455	5.370 .6452	5.367 .6449
88	5.365 .6446	5.363 .6444	5.361 .6441	5.358 .6438	5.356 .6435	5.353 .6432	5.351 .6429	5.348 .6426	5.346 .6423	5.343 .6420
89	5.341 .6417	5.338 .6414	5.336 .6411	5.334 .6409	5.331 .6406	5.329 .6403	5.326 .6400	5.324 .6397	5.321 .6394	5.319 .6391
90	5.316 .6388	5.314 .6385	5.312 .6382	5.310 .6380	5.307 .6377	5.305 .6374	5.302 .6371	5.300 .6368	5.297 .6365	5.295 .6362
91	5.293 .6360	5.291 .6357	5.288 .6354	5.286 .6351	5.283 .6348	5.281 .6345	5.278 .6342	5.276 .6340	5.274 .6337	5.271 .6334
92	5.269 .6331	5.266 .6328	5.264 .6325	5.262 .6323	5.260 .6320	5.257 .6317	5.254 .6314	5.252 .6311	5.250 .6309	5.248 .6306
93	5.245 .6303	5.243 .6300	5.241 .6297	5.238 .6294	5.236 .6292	5.234 .9289	5.232 .6286	5.229 .6283	5.227 .6281	5.225 .6278
94	5.222 .6275	5.220 .6272	5.217 .6269	5.215 .6267	5.213 .6264	5.211 .6261	5.208 .6258	5.206 .6256	5.204 .6253	5.201 .6250
95	5.199 .6247	5.196 .6244	5.194 .6242	5.192 .6239	5.190 .6236	5.187 .6233	5.185 .6231	5.183 .6228	5.180 .6225	5.179 .6223
96	5.176 .6220	5.174 .6217	5.172 .6214	5.170 .6212	5.167 .6209	5.164 .6206	5.162 .6203	5.160 .6201	5.158 .6198	5.156 .6195
97	5.154 .6193	5.151 .6190	5.149 .6187	5.146 .6184	5.144 .6182	5.142 .6179	5.140 .6176	5.138 .6174	5.136 .6171	5.133 .6168
98	5.131 .6166	5.129 .6163	5.126 .6160	5.124 .6158	5.122 .6155	5.120 .6152	5.118 .6150	5.116 .6147	5.113 .6144	5.111 .6141
99	5.109 .6139	5:107 .6136	5.104 .6134	5.102 .6131	5.100 .6128	5.098 .6126	5.096 .6123	5.093 .6120	5.091 .6118	5.089 .6115
100	5.086 .6112	5.09 .6110	5.09 .6107	5.08 .6104	5.08 .6102	5.08 .6099	5.08 .6097	5.07 .6094	5.07 .6091	5.07 6089
101	5.07 .6086	5.07 .6083	5.06 .6081	5.06 .6078	5.06 .6076	5.06 .6073	5.05 .6070	5.05 .6068	5.05 .6065	5.05 .6063
102	5.05 .6060	5.04 .6058	5.04 .6055	5.04 .6052	5.04 .6050	5.04 .6047	5.03 .6044	5.03 .6042	5.03 .6039	5.03 .6037
103	5.02 .6034	5.02 .6032	5.02 .6029	5.02 .6026	5.02 .6024	5.01 .6021	5.01 .6019	5.01 .6016	5.01 .6014	5.01 .6011
104	5.00 .6008	5.00 .6006	5.00 .6003	5.00 .6001	4.99 .5998	4.99 .5996	4.99 .5993	4.99 .5991	4.99 .5988	4.98 .5986
105	4.98 .5983	4.98 .5981	4.98 .5978	4.98 .5976	4.97 .5973	4.97 .5970	4.97 .5968	4.97 .5965	4.97 .5963	4.96 .5960
106	4.96 .5958	4.96 .5955	4.96 .5953	4.95 .5950	4.95 .5948	4.95 .5945	4.95 .5943	4.95 .5940	4.94 .5938	4.94 .5935
107	4.94 .5933	4.94 .5930	4.94 .5928	4.93 .5925	4.93 .5923	4.93 .5921	4.93 .5918	4.93 .5916	4.92 .5913	4.92 .5911

Pounds per gallon and specific gravities corresponding to degrees API at 60°F (*Continued*)

Deg API	Tenths of Degrees									
	0	1	2	3	4	5	6	7	8	9
108	4.92 .5908	4.92 .5906	4.92 .5903	4.91 .5901	4.91 .5898	4.91 .5896	4.91 .5893	4.91 .5891	4.90 .5888	4.90 .5886
109	4.90 .5884	4.90 .5881	4.90 .5879	4.89 .5876	4.89 .5874	4.89 .5871	4.89 .5869	4.89 .5867	4.88 .5864	4.88 .5862
110	4.88 .5859	4.88 .5857	4.87 .5854	4.87 .5852	4.87 .5850	4.87 .5847	4.87 .5845	4.86 .5842	4.86 .5840	4.86 .5837
111	4.86 .5835	4.86 .5833	4.85 .5830	4.85 .5828	4.85 .5825	4.85 .5823	4.85 .5821	4.84 .5818	4.84 .5816	4.84 .5813
112	4.84 .5811	4.84 .5909	4.83 .5806	4.83 .5804	4.83 .5802	4.83 .5799	4.83 .5797	4.82 .5794	4.82 .5792	4.82 .5790
113	4.82 .5787	4.82 .5785	4.82 .5783	4.81 .5780	4.81 .5778	4.81 .5776	4.81 .5773	4.81 .5771	4.80 .5768	4.80 .5766
114	4.80 .5764	4.80 .5761	4.80 .5759	4.79 .5757	4.79 .5754	4.79 .5752	4.79 .5750	4.79 .5747	4.78 .5745	4.78 .5743
115	4.78 .5740	4.78 .5738	4.78 .5736	4.77 .5733	4.77 .5731	4.77 .5729	4.77 .5726	4.77 .5724	4.76 .5722	4.76 .5719
116	4.76 .5717	4.76 .5715	4.76 .5713	4.76 .5710	4.75 .5708	4.75 .5706	4.75 .5703	4:75 .5701	4.75 .5699	4.74 .5696
117	4.74 .5694	4.74 .5692	4.74 .5690	4.74 .5687	4.73 .5685	4.73 .5683	4.73 .5680	4.73 .5678	4.73 .5676	4.73 .5674
118	4.72 .5671	4.72 .5669	4.72 .5667	4.72 .5665	4.72 .5662	4.71 .5660	4.71 .5658	4.71 .5655	4.71 .5653	4.71 .5651
119	4.70 .5649	4.70 .5646	4.70 .5644	4.70 .5642	4.70 .5640	4.69 .5637	4.69 .5635	4.69 .5633	4.69 .5631	4.69 .5628
120	4.69 .5626	4.68 .5624	4.68 .5622	4.68 .5620	4.68 .5617	4.68 .5615	4.67 .5613	4.67 .5611	4.67 .5608	4.67 .5606
121	4.67 .5604	4.67 .5602	4.66 .5600	4.66 .5597	4.66 .5595	4.66 .5593	4.66 .5591	4.65 .5588	4.65 .5586	4.65 .5584
122	4.65 .5582	4.65 .5580	4.64 .5577	4.64 .5575	4.64 .5573	4.64 .5571	4.64 .5569	4.64 .5566	4.63 .5564	4.63 .5562
123	4.63 .5560	4.63 .5558	4.63 .5556	4.62 .5553	4.62 .5551	4.62 .5549	4.62 .5547	4.62 .5545	4.62 .5542	4.61 .5540
124	4.61 .5538	4.61 .5536	4.61 .5534	4.61 .5532	4.61 .5530	4.60 .5527	4.60 .5525	4.60 .5523	4.60 .5521	4.60 .5519
125	4.59 .5517	4.59 .5514	4.59 .5512	4.59 .5510	4.59 .5508	4.59 .5506	4.58 .5504	4.58 .5502	4.58 .5499	4.58 .5497
126	4.58 .5495	4.57 .5493	4.57 .5491	4.57 .5489	4.57 .5487	4.57 .5484	4.57 .5482	4.56 .5480	4.56 .5478	4.56 .5476
127	4.56 .5474	4.56 .5472	4.56 .5470	4.55 .5468	4.55 .5465	4.55 .5463	4.55 .5461	4.55 .5459	4.54 .5457	4.54 .5455
128	4.54 .5453	4.54 .5451	4.54 .5449	4.54 .5446	4.53 .5444	4.53 .5442	4.53 .5440	4.53 .5438	4.53 .5436	4.53 .5434
129	4.52 .5432	4.52 .5430	4.52 .5428	4.52 .5426	4.52 .5424	4.51 .5421	4.51 .5419	4.51 .5417	4.51 .5415	4.51 .5413
130	4.51 .5411	4.50 .5409	4.50 .5407	4.50 .5405	4.50 .5403	4.50 .5401	4.50 .5399	4.49 .5397	4.49 .5395	4.49 .5393
131	4.49 .5390	4.49 .5388	4.49 .5386	4.48 .5384	4.48 .5382	4.48 .5380	4.48 .5378	4.48 .5376	4.48 .5374	4.47 .5372
132	4.47 .5370	4.47 .5368	4.47 .5366	4.47 .5364	4.47 .5362	4.46 .5360	4.46 .5358	4.46 .5356	4.46 .5354	4.46 .5352
133	4.46 .5350	4.45 .5348	4.45 .5346	4.45 .5344	4.45 .5342	4.45 .5340	4.45 .5338	4.44 .5336	4.44 .5334	4.44 .5332
134	4.44 .5330	4.44 .5328	4.44 .5326	4.43 .5324	4.43 .5322	4.43 .5320	4.43 .5318	4.43 .5316	4.43 .5314	4.42 .5312
135	4.42 .5310	4.42 .5308	4.42 .5306	4.42 .5304	4.42 .5302	4.41 .5300	4.41 .5298	4.41 .5296	4.41 .5294	4.41 .5292
136	4.41 .5290	4.40 .5288	4.40 .5286	4.40 .5284	4.40 .5282	4.40 .5280	4.40 .5278	4.39 .5276	4.39 .5274	4.39 .5272
137	4.39 .5270	4.39 .5268	4.39 .5266	4.38 .5264	4.38 .5262	4.38 .5260	4.38 .5258	4.38 .5256	4.38 .5254	4.37 .5252
138	4.37 .5250	4.37 .5249	4.37 .5247	4.37 .5245	4.37 .5243	4.36 .5241	4.36 .5239	4.36 .5237	4.36 .5235	4.36 .5233
139	4.36 .5231	4.35 .5229	4.35 .5227	4.35 .5225	4.35 .5223	4.35 .5221	4.35 .5219	4.35 .5218	4.34 .5216	4.34 .5214

Figure B.5. Relationship of SG, °API, Lbs/Gal.

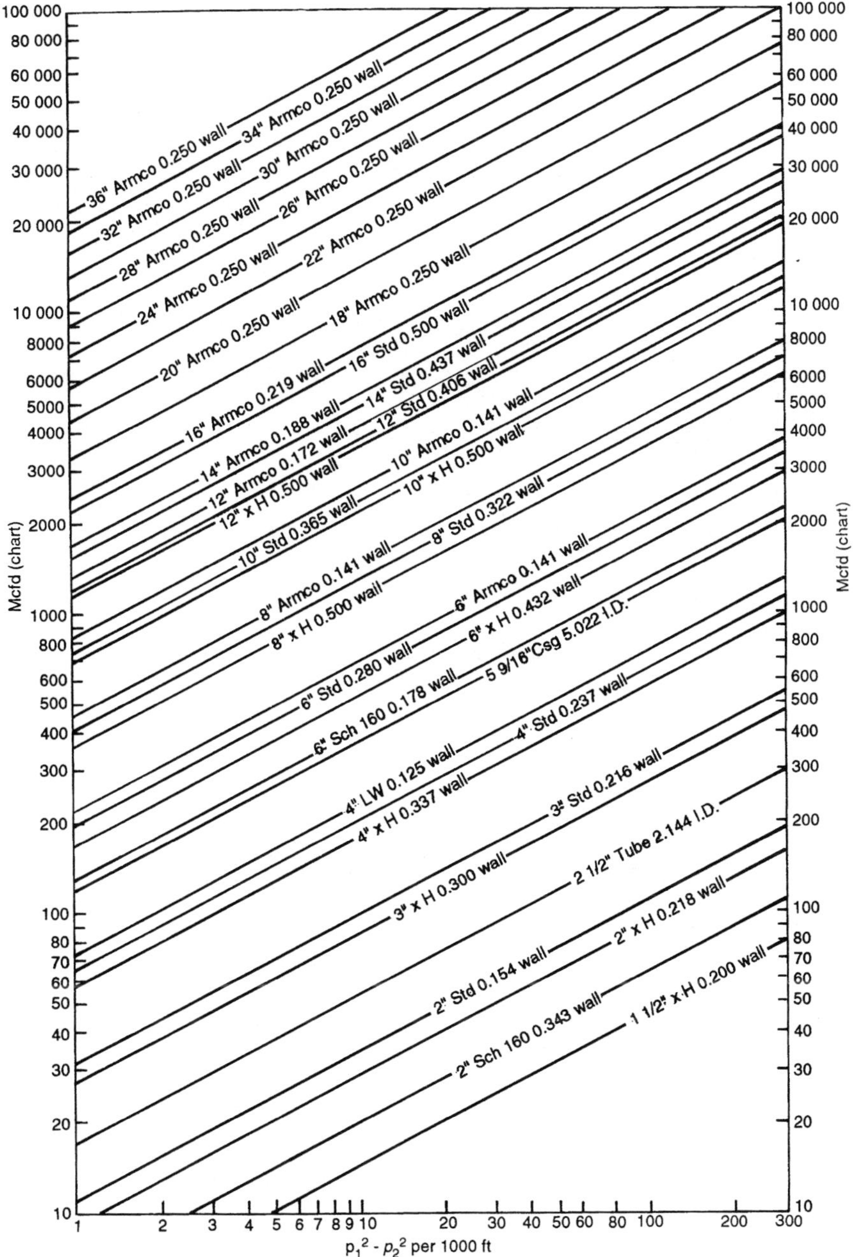

1. Chart— Mcfd—based on 'Weymouth Formule' where specific gravity = 0.9 (air = 1), flowing temperature = 90°F and pressure base = 14.65 psia
2. Simplified Mcfd = $1.59\ d^{2/3}\ (P_1^2 — P_2^2/1000\ \text{ft})^{1/2}$

Figure B.6. Flow pressure drop for gas streams.

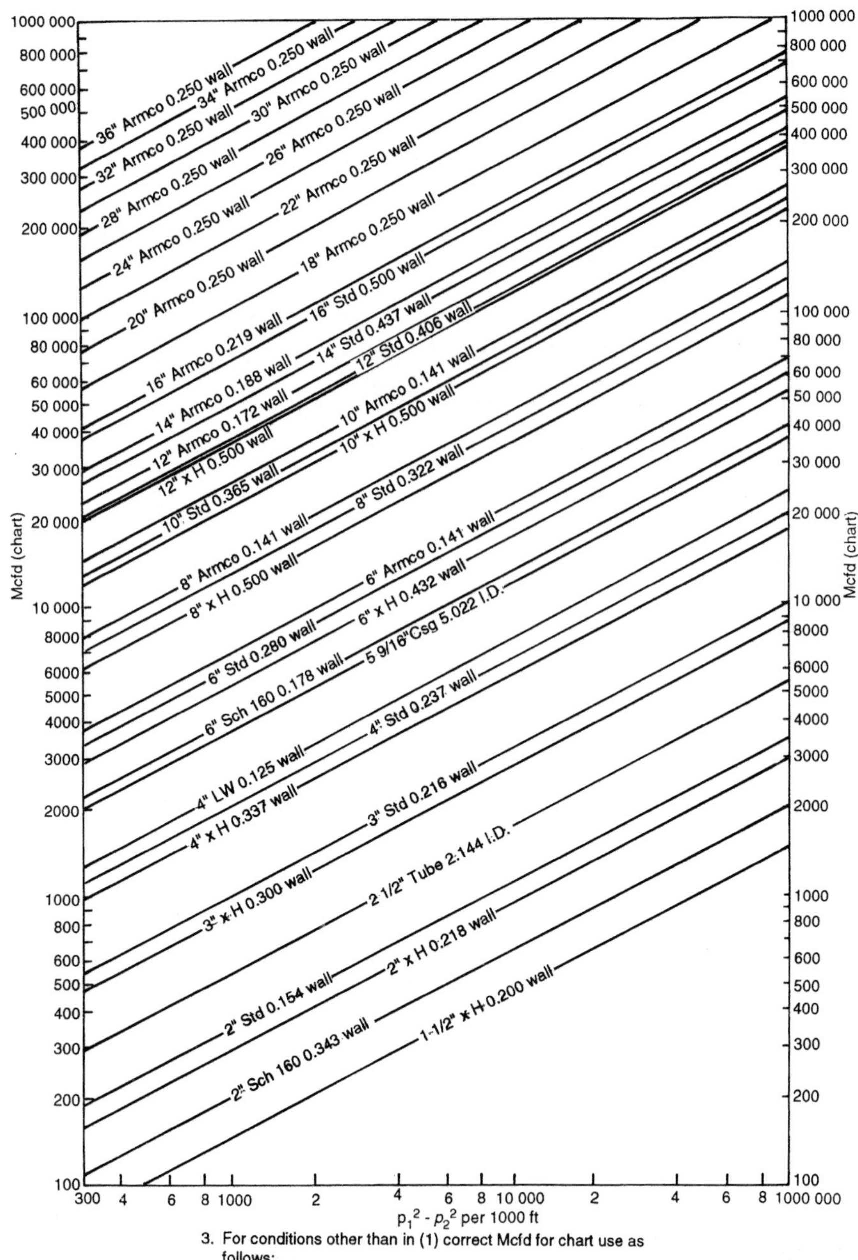

3. For conditions other than in (1) correct Mcfd for chart use as follows:

$$\text{Mcfd(chart)} = \frac{P}{14.65}\left(\frac{SG}{0.9} \times \frac{T}{550} \times \frac{Z}{1}\right)^{1/2} \times \text{Mcfd}$$

Z = Compressibility factor as determined from 'Natural Gas Under Pressure' in the GPSA Engineering Data Book

P = Pressure base other than 14.65 psia

T = Flowing temperature in degrees Rankine (460 + °F)

Figure B.6. Flow pressure drop for gas streams.

R*	L*	A*	R*	L*	A*	R*	L*	A*	R*	L*	A*	R*	L*	A*	R*	L*	A*
0.070	0.511	0.0308	0.120	0.650	0.0680	0.170	0.751	0.113	0.220	0.828	0.163	0.280	0.898	0.230	0.390	0.977	0.361
1	0.514	0.0315	1	0.652	0.0688	1	0.753	0.114	1	0.829	0.164	5	0.903	0.236	5	0.979	0.367
2	0.517	0.0321	2	0.654	0.0697	2	0.755	0.115	2	0.831	0.165						
3	0.521	0.0328	3	0.657	0.0705	3	0.756	0.116	3	0.832	0.166	0.290	0.908	0.241	0.400	0.980	0.374
4	0.524	0.0335	4	0.659	0.0714	4	0.758	0.117	4	0.834	0.167	5	0.913	0.247	5	0.982	0.380
5	0.527	0.0342	5	0.661	0.0722	5	0.760	0.117	5	0.835	0.169	0.300	0.917	0.252	0.410	0.984	0.386
6	0.530	0.0348	6	0.663	0.0731	6	0.762	0.118	6	0.836	0.170	5	0.921	0.258	5	0.986	0.392
7	0.533	0.0355	7	0.665	0.0739	7	0.763	0.119	7	0.838	0.171						
8	0.536	0.0362	8	0.668	0.0748	8	0.765	0.120	8	0.839	0.172	0.310	0.925	0.264	0.420	0.987	0.398
9	0.539	0.0368	9	0.670	0.0756	9	0.786	0.121	9	0.841	0.173	5	0.930	0.270	5	0.989	0.405
0.080	0.542	0.0375	0.130	0.672	0.0765	0.180	0.768	0.122	0.230	0.842	0.174	0.320	0.933	0.276	0.430	0.991	0.412
1	0.545	0.0382	1	0.674	0.0774	1	0.770	0.123	1	0.843	0.175	5	0.937	0.282	5	0.993	0.418
2	0.548	0.0389	2	0.677	0.0782	2	0.772	0.124	2	0.845	0.176						
3	0.552	0.0396	3	0.679	0.0791	3	0.773	0.125	3	0.846	0.177	0.330	0.941	0.288	0.440	0.994	0.424
4	0.555	0.0403	4	0.682	0.0799	4	0.775	0.126	4	0.848	0.178	5	0.945	0.294	5	0.995	0.430
5	0.558	0.0410	5	0.684	0.0808	5	0.777	0.127	5	0.849	0.179	0.340	0.948	0.300	0.450	0.996	0.437
6	0.561	0.0418	6	0.686	0.0817	6	0.778	0.128	6	0.850	0.180	5	0.951	0.306			
7	0.564	0.0425	7	0.688	0.825	7	0.780	0.129	7	0.851	0.181				0.460	0.997	0.450
8	0.567	0.0432	8	0.691	0.0834	8	0.781	0.130	8	0.853	0.182	0.350	0.955	0.312			
9	0.570	0.0439	9	0.693	0.0842	9	0.783	0.131	9	0.854	0.183	5	0.958	0.318	0.470	0.998	0.462
0.090	0.573	0.0446	0.140	0.695	0.0851	0.190	0.784	0.132	0.240	0.855	0.184	0.360	0.961	0.324	0.480	0.998	0.475
1	0.576	0.0454	1	0.697	0.0860	1	0.786	0.133	5	0.860	0.190	5	0.964	0.330			
2	0.578	0.0461	2	0.699	0.0869	2	0.787	0.134							0.490	0.999	0.488
3	0.581	0.0469	3	0.700	0.0878	3	0.789	0.135	0.250	0.866	0.196	0.370	0.967	0.337			
4	0.583	0.0476	4	0.702	0.0887	4	0.790	0.136	5	0.872	0.202	5	0.969	0.343	0.500	1.0	0.50
5	0.586	0.0484	5	0.704	0.0896	5	0.792	0.137	0.260	0.878	0.207	0.380	0.971	0.348			
6	0.589	0.0491	6	0.706	0.0905	6	0.794	0.138	5	0.883	0.213	5	0.977	0.354			
7	0.592	0.0499	7	0.708	0.0914	7	0.795	0.139									
8	0.594	0.0506	8	0.710	0.0923	8	0.797	0.140	0.270	0.888	0.218						
9	0.597	0.0514	9	0.712	0.0932	9	0.798	0.141	5	0.893	0.224						
0.100	0.600	0.0521	0.150	0.714	0.0941	0.200	0.800	0.142									
1	0.603	0.0529	1	0.716	0.0950	1	0.802	0.143									
2	0.605	0.0537	2	0.718	0.0959	2	0.803	0.144									
3	0.608	0.0545	3	0.720	0.0969	3	0.805	0.145									
4	0.610	0.0555	4	0.722	0.0978	4	0.806	0.146									
5	0.613	0.0561	5	0.724	0.0987	5	0.808	0.148									
6	0.615	0.0568	6	0.726	0.0996	6	0.809	0.149									
7	0.618	0.0576	7	0.728	0.1005	7	0.810	0.150									
8	0.620	0.0584	8	0.729	0.1015	8	0.812	0.151									
9	0.623	0.0592	9	0.731	0.102	9	0.813	0.152									
0.110	0.625	0.0600	0.160	0.733	0.103	0.210	0.814	0.153									
1	0.628	0.0608	1	0.735	0.104	1	0.816	0.154									
2	0.630	0.0616	2	0.737	0.105	2	0.817	0.155									
3	0.633	0.0624	3	0.738	0.106	3	0.819	0.156									
4	0.635	0.0632	4	0.740	0.107	4	0.820	0.157									
5	0.638	0.0640	5	0.742	0.108	5	0.822	0.158									
6	0.640	0.0648	6	0.744	0.109	6	0.823	0.159									
7	0.643	0.0656	7	0.746	0.110	7	0.824	0.160									
8	0.645	0.0664	8	0.747	0.111	8	0.826	0.161									
9	0.648	0.0672	9	0.749	0.112	9	0.827	0.162									

* This table relates the downcomer area the weir length, and the height of the circular segment formed by the weir

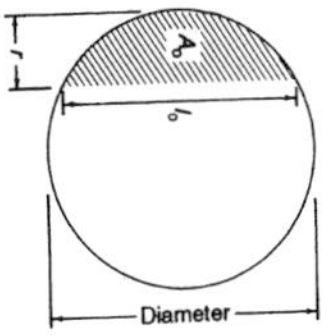

Diameter

$$R = \frac{\text{*Downcomer rise}}{\text{Diameter}} = \frac{r}{\text{Dia.}} \qquad L = \frac{\text{* Weir length}}{\text{Diameter}} = \frac{l_O}{\text{Dia.}} \qquad A = \frac{\text{* Downcomer area}}{\text{Tower area}} = \frac{A_D}{A_S}$$

Figure B.7. Relationship of chords, diameters, and areas.

Appendix C

A selection of crude oil assays

Note: For final studies and definitive engineering the up to date assay from the crude oil supplier should be used.

Amna (high pour), Libya

Crude
Gravity, °API: 36.1
Sulfur, wt %: 0.15
Pour point, °F.: 75
Vis kin. 37.8° C., cSt: 13.70
Car. residue (CCR), wt %: 3.7
RVP, psi: 3.9
Salt, lb/1,000 bbl: 8.2
V, ppm: 0.6
Ni, ppm: 5.0
IBP-60° F., yield, vol %: 2.11

Light naphtha
Range, °F.: 60-120
Yield, vol %: 2.4
Gravity, °API: 93.6
Sulfur, wt %: 0.01
Paraffins, vol %: 99.0
Naphthenes, vol %: 1.0
RON, clear: 78.0
RON + 3 ml TEL: 94.0

Light naphtha
Range, °F.: 120-250
Yield, vol %: 7.7
Gravity, °API: 69.5
Sulfur, wt %: 0.01
Paraffin, vol %: 69.9
Naphthenes, vol %: 26.5
Aromatics, vol %: 3.7

Heavy naphtha
Range, °F.: 250-330
Yield, vol %: 6.9
Gravity, °API: 57.4
Sulfur wt %: 0.02
Paraffin, vol %: 60.5
Naphthenes, vol %: 33.6
Aromatics, vol %: 5.9

Kerosine
Range, °F.: 330-443
Yield, vol %: 9.4
Gravity, °API: 49.9
Sulfur, wt %: 0.05
Paraffin, vol %: 60.9
Naphthenes, vol %: 29.9
Aromatics, vol %: 9.2
Pour point, °F.: —75
Freezing point, °F.: —58
Aniline point, °F.: 150.6
Diesel index: 75
Smoke point, mm: 31

Amna (high pour)

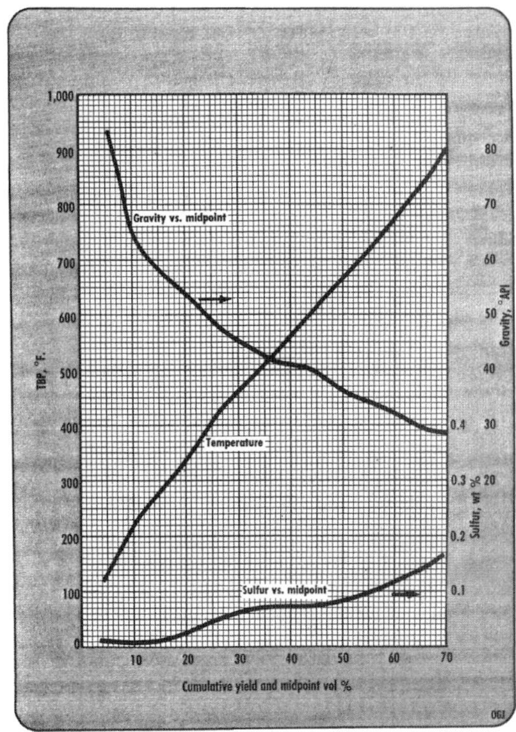

Gas oil
Range, °F.: 443-600
Yield, vol %: 15.0
Gravity, °API: 42.3
Sulfur, wt %: 0.07
Pour point, °F.: 0
Aniline point, °F.: 169.1
Diesel index: 71

Gas oil
Range, °F.: 600-850
Yield, vol %: 23.3
Sulfur, wt %: 0.13
Pour point, °F: 80
Aniline point, °F.: 202.0
Diesel index: 69

Residual oil
Range, °F.: 655+
Yield, vol %: 55.4
Gravity, °API: 25.4
Sulfur, wt %: 0.22
Pour point, °F. 100

Carbon residue CCR, wt %: 6.22
V, ppm: 1.1
Ni, ppm: 8.5

Residual oil
Range, °F.: 1,000+
Yield, vol %: 21.8
Gravity, °API: 16.7
Sulfur, wt %: 0.31
Pour point, °F.: 130
Car. residue CCR, wt %: 14.50
V, ppm: 2.5
Ni, ppm: 20.0

Arabian heavy, Saudi Arabia

Crude
Gravity, °API: 28.2
Sulfur, wt %: 2.84
Pour point, °F.: −30
RVP, psi: 8.5
Viscosity
Kin. cSt @ 70° F.: 35.8
Kin. cSt @ 100° F.: 18.9

Light naphtha
Range, °FVT: 68-212
Yield, vol %: 7.9
Gravity, °API: 80.1
Sulfur, wt %: 0.0028
RVP, psi: 10.2
Paraffins, vol %: 89.6
Naphthenes, vol %: 9.5
Aromatics, vol %: 0.9
RON, clear: 59.7

Heavy naphtha
Range, °FVT: 212-302
Yield, vol %: 6.8
Gravity, °API: 60.6
Sulfur, wt %: 0.018
Paraffins, vol %: 70.3
Naphthenes, vol %: 21.4
Aromatics, vol %: 8.3

Kerosine
Range, °FVT: 302-455
Yield, vol %: 12.5
Gravity, °API: 48.3
Sulfur, wt %: 0.19
Paraffins, vol %: 58.0
Naphthenes, vol %: 23.7
Aromatics, vol %: 18.3
Freeze point, °F.: −64
Smoke point, mm: 26
Luminometer no.: 60
Aniline point, °F: 138
Viscosity
Kin., cSt @ −30° F.: 4.74
Kin., cSt @ 100° F.: 1.12

Light gas oil
Range, °FVT: 455-650
Yield, vol %: 16.4
Gravity, °API: 35.8
Sulfur, wt %: 1.38
Pour point, °F.: 5
Aniline point, °F.: 156
Viscosity
Kin., cSt @ 100° F.: 3.65
Kin., cSt @ 210° F.: 1.40

Arabian heavy

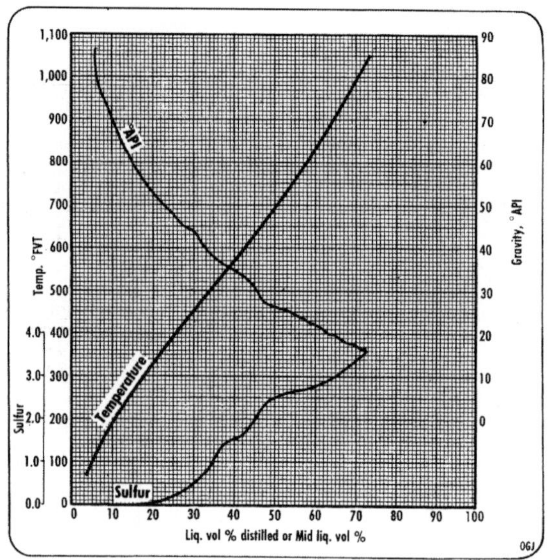

Heavy gas oil
Range, °FVT: 650-1,049
Yield, vol %: 26.3
Gravity, °API: 21.8
Sulfur, wt %: 2.88
Pour point, °F.: 90
Aniline point, °F.: 172
Viscosity
Kin., cSt @ 100° F.: 62.5
Kin., cSt @ 210° F.: 7.05

Residual oil
Range, °FVT: 650+
Yield, vol %: 53.1
Gravity, °API: 12.3
Sulfur, wt %: 4.35
Pour point, °F.: 55
Con carbon, wt %: 13.2
Viscosity
Kin., cSt @ 100° F.: 5,400
Kin., cSt @ 210° F.: 106

Residual oil
Range, °FVT: 1,049+
Yield, vol %: 26.8
Gravity, °API: 4.0
Sulfur, wt %: 5.60
Pour point, °F.: 120
Con carbon, wt %: 24.4
Viscosity
Kin., cSt @ 210° F.: 13,400
Furol, sec @ 275° F.: 490
Metals
Vanadium, ppm: 171
Nickel, ppm: 53
Iron, ppm: 28

Arabian light, Saudi Arabia

Crude
Gravity, °API: 33.4
Sulfur, wt %: 1.80
Pour point, °F.: −30
RVP, psi: 4.2
Viscosity
Kin., cSt @ 70° F.: 10.4
Kin., cSt @ 100° F.: 6.14

Light naphtha
Range, °FVT: 68-212
Yield, vol %: 9.0
Gravity, °API: 78.5
Sulfur, wt %: 0.024
RVP, psi: 8.3
Paraffins, vol %: 87.2
Naphthenes, vol %: 10.4
Aromatics, vol %: 2.4
RON, clear: 54.7

Heavy naphtha
Range, °FVT: 212-302
Yield, vol %: 8.4
Gravity, °API: 59.6
Sulfur, wt %: 0.027
Paraffines, vol %: 69.5
Naphthenes, vol %: 18.2
Aromatics, vol %: 12.3

Kerosine
Range, °FVT: 302-455
Yield, vol %: 15.0
Gravity, °API: 38.5
Sulfur, wt %: 0.094
Paraffins, vol %:
Naphthenes, vol %:
Aromatics, vol %: 20.4
Freeze point, °F.: −69
Smoke point, mm: 23
Luminometer No.: 52
Aniline point, °F.: 135

Viscosity
Kin., cSt @ −30° F.: 5.13
Kin., cSt @ 100° F.: 1.11

Light gas oil
Range, °FVT: 455-650
Yield, vol %: 19.8
Gravity, °API: 37.1
Sulfur, wt %: 1.05
Pour point, °F.: 0
Aniline point, °F.: 156
Viscosity
Kin., cSt @ 100° F.: 3.28
Kin., cSt @ 210° F.: 1.24

Heavy gas oil
Range, °FVT: 650-1,049
Yield, vol %: 32.5
Gravity, °API: 22.8
Sulfur, wt %: 2.46
Pour point, °F.: 90
Aniline point, °F.: 179
Viscosity
Kin., cSt @ 100° F.: 52.5
Kin., cSt @ 210° F.: 6.85

Arabian light

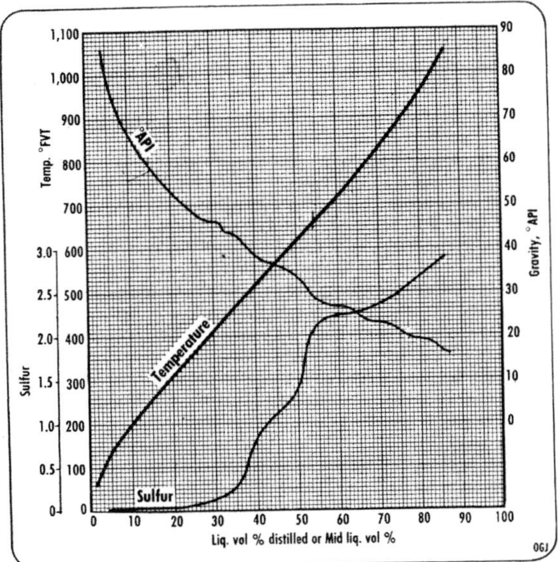

Residual oil
Range, °FVT: 650+
Yield, vol %: 46.1
Gravity, °API: 17.6
Sulfur wt %: 3.08
Pour point, °F.: 40
Con Carbon, wt %: 7.6
Viscosity
Kin., cSt @ 100° F.: 367
Kin., cSt @ 210° F.: 21.0

Residual oil
Range, °FVT.: 1,049+
Yield, vol %: 13.6
Gravity, °API: 6.5
Sulfur, wt %: 4.40
Pour point, °F.: 115
Con Carbon, wt %: 22.4
Viscosity
Kin., cSt @ 210° F.: 2,017
Furol, sec @ 275° F.: 132
Metals
Vanadium, ppm: 94
Nickel, ppm: 22
Iron, ppm: 25

Arabian medium, Saudi Arabia

Crude
Gravity, °API: 30.8
Sulfur, wt %: 2.40
Pour point, °F.: +5
RVP, psi: 3.2
Viscosity
Kin. cSt @ 70° F.: 16.2
Kin. cSt @ 100° F.: 9.41

Light naphtha
Range, °FVT: 68-212
Yield, vol %: 8.9
Gravity, °API: 77.7
Sulfur, wt %: 0.043

RVP, psi: 7.9
Paraffins, vol %: 85.3
Naphthenes, vol %: 12.3
Aromatics, vol %: 2.4
RON, clear 54.5

Heavy naphtha
Range, °FVT: 212-302
Yield, vol %: 7.7
Gravity, °API: 59.1
Sulfur, wt %: 0.050
Paraffins, vol %: 68.5
Naphthenes, vol %: 18.7
Aromatics, vol %: 12.7

Kerosine
Range, °FVT: 302-455
Yield, vol %: 14.5
Gravity, °API: 48.0
Sulfur, wt %: 0.14
Paraffins, vol %:
Naphthenes, vol %:
Aromatics, vol %: 20.6
Freeze point, °F: −62
Smoke point, mm: 23
Luminometer No.: 52
Aniline point, °F.: 136
Viscosity
Kin., cSt @ −30° F.: 5.10
Kin., cSt @ 100° F.: 1.13

Light gas oil
Range, °FVT: 455-650
Yield, vol %: 18.1
Gravity, °API: 36.0
Sulfur, wt %: 1.24
Pour point, F.: 15
Aniline point, °F.: 157
Viscosity
Kin., cSt @ 100° F.: 3.53
Kin., cSt @ 210° F.: 1.32

Heavy gas oil
Range, °FVT: 650-1,049
Yield, vol %: 30.9
Gravity, °API: 21.9
Sulfur, wt %: 2.91
Pour point, °F.: 100
Aniline point, °F.: 176
Viscosity
Kin., cSt @ 100° F.: 49.2
Kin., cSt @ 210° F.: 6.40

Arabian medium

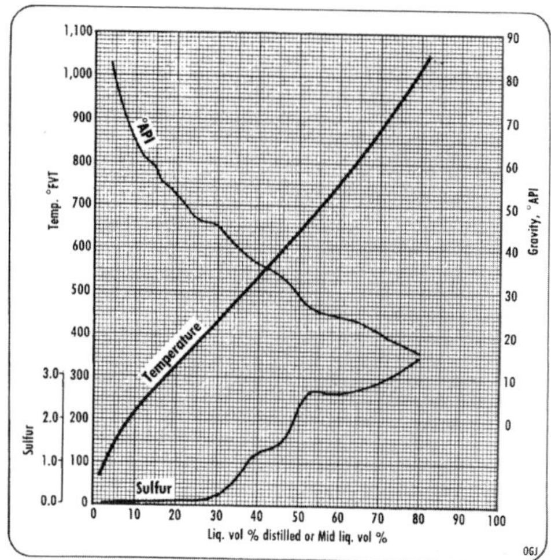

Residual oil
Range, °FVT: 650+
Yield, vol %: 49.6
Gravity, °API: 15.0
Sulfur, wt %: 3.90
Pour point, °F.: 55
Con carbon, wt %: 9.9
Viscosity
Kin., cSt @ 100° F.: 850
Kin., cSt @ 210° F.: 36.0

Residual oil
Range, °FVT: 1,049+
Yield, vol %: 18.7
Gravity, °API: 4.9
Sulfur, wt %: 5.35
Pour point, °F.: 115

Con carbon, wt %: 23.3
Viscosity
Kin., cSt @ 210° F.: 3,847
Furor, sec @ 275° F.: 226
Metals
Vanadium, ppm: 96
Nickel, ppm: 32
Iron, ppm: 30

Arjuna, Java, Indonesia

Crude
Gravity, °API: 37.7
Sulfur, wt %: 0.12
Pour point, °F.: 80
RVP, lb: 7.0
Vis. SUS @
90° F.: ?9.2
100° F.: 37.7
Salt, lb/1,000 bbl. ∠10

Naphtha
Range TBP, °F.: C_5−230° F.
Yield, vol %: 13.8
Gravity, °API: 68.2
Sulfur, wt %: ∠0.01
RVP, lb: 6.9
RON, clear: 72.2
RON + 3.17g TEL: 89.4
Paraffin C_6 +, vol %: 41.0
Naphthenes, vol %: 34.9
Aromatics, vol %: 5.7

Naphtha
Range TBP, °F.: 230-380
Yield, vol %: 17.9
Gravity, °API: 48.5
Sulfur, wt %: ∠0.01
RON, clear: 61.6
Paraffin C_6+, vol %: 43.8
Naphthenes, vol %: 29.1
Aromatics, vol %: 27.1

Distillate
Range TBP, °F.: 380-640
Yield, vol %: 28.7
Gravity, °API: 33.0
Sulfur, wt %: 0.09
Pour point, °F.: 15
Cetane No.: 41

Gas oil
Range TBP, °F.: 640-780
Yield, vol % 10.7
Gravity, °API: 31.3
Sulfur, wt %: 0.12
Aniline point, ° C.: 90.2
Pour point, °F.: 90
Car. residue, wt %: 0.08
V/Ni, ppm: 0.0/0.3

Gas oil
Range TBP, °F.: 780-1,049
Yield, vol %: 20.2
Gravity, °API: 25.8
Sulfur wt %: 0.15
Aniline point, °C.: 98.4
Pour point, °F.: 120
Car. residue, wt %: 0.42
V/Ni, ppm: 0.3/0.9

Residual oil
Range, °F.: 1,049 +
Yield, vol %: 5.4
Gravity, °API: 5.6
Sulfur, wt %: 0.30
Kin. vis. @ 275° F.: 4,100

Arjuna

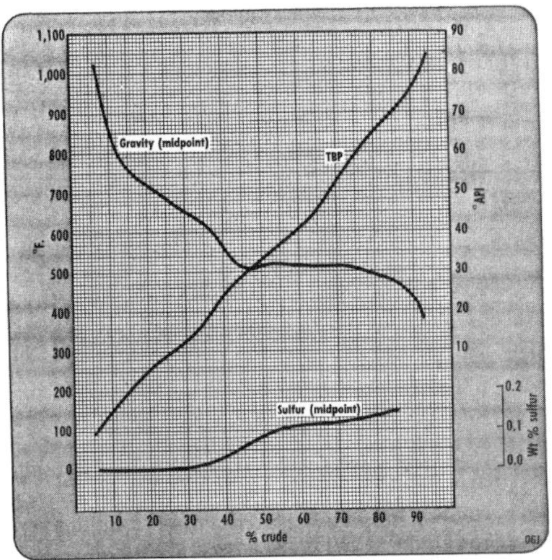

Bachequero, 16.8° API (Bachequero heavy), Venezuela

Crude
Gravity, °API: 16.8
Salt, lb/1,000 bbl: 6.0
Sulfur wt %: 2.40
Pour pt., °F.: −10
Vis. SUS @ 100° F.: 1,362

Light naphtha
Range, °F.: 90-200
Yield, vol %: 2.5
Gravity, °API: 65.0
RVP, psi: 2.8
Paraffin/naphthenes, vol %: 43.4/51.9
Aromatics, vol %: 4.7

Heavy naphtha
Range, °F.: 200-350
Yield, vol %: 6.0

Light distillate
Range, °F.: 350-475
Yield, vol %: 5.0
Gravity, °API: 36.4
Sulfur, wt %: 0.48
Paraffins /naphthenes, vol %: 19.2/54.8
Aromatics, vol %: 26.0
Smoke point, mm: 16
Aniline point, °F.: 125
Diesel index: 46
Pour point, °F.: −80

Heavy distillate
Range, °F.: 475-650
Yield, vol %: 15.5
Sulfur, wt %: 0.99
Aniline pt., °F.: 134
Diesel index: 40

Residual oil
Range, °F.: 9.7
CCR, wt %: 14.2
Sulfur, wt %: 3.0
V/Ni., ppm: 437/75
Fe, ppm: 36
Pour point, °F.: +60

Gravity, °API: 49.0
Paraffins/naphthenes, vol %: 27.6/58.5
Aromatics, vol %: 13.9

Bachequero, 16.8° API

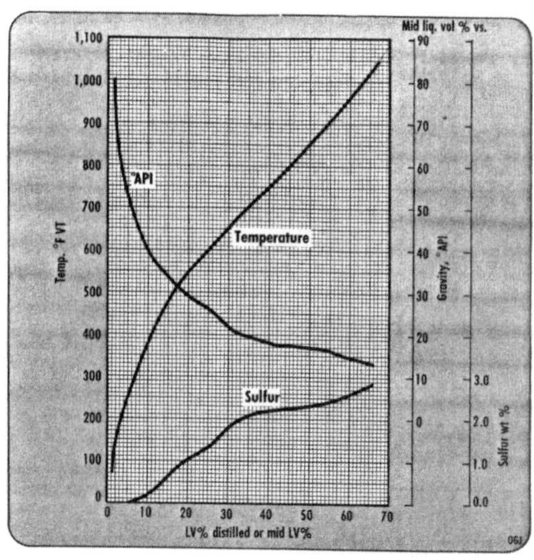

Bonny light, Nigeria

Crude
Gravity, °API: 37.6
Sulfur, wt %: 0.13
Vis. at 100° F., SUS: 36.0
Pour point, °F.: 5
C_4 and lighter, vol %: 2.2
Vanadium, ppm: <0.5
Nickel, ppm: 4
Car. residue (Conradson), wt %: 1.1

Light naphtha (debutanized)
Yield, vol %: 6.4
TBP range, °F.: 60-167
Gravity, °API: 79.9
Sulfur, wt %: 0.0002
Paraffins, vol %: 77
Naphthenes, vol %: 21.5
Aromatics, vol %: 1.5
Research ON, clear: 78
Research ON, + 3 ml TEL/U.S. gal: 93.5

Heavy naphtha
Yield, vol %: 22.0
TBP range, °F.: 167-347
Gravity, °API: 53.6
Sulfur, wt %: 0.003
Paraffins, vol %: 34
Naphthenes, vol %: 55
Aromatics, vol %: 11
Characterization factor: 11.7
Aniline point, °F.: 120

Kerosine
Yield, vol %: 15.4
TBP range, °F.: 347-482
Gravity, °API: 40.2
Sulfur, wt %: 0.030
Smoke point (IP), mm: 19
Freezing point, °C.: −47
Aniline point, °F.: 136
Diesel index: 55
Pour point, °F.: −70

Light gas oil
Yield, vol %: 23.2
TBP range, °F.: 482-662
Gravity, °API: 33.2
Sulfur, wt %: 0.13
Cetane index: 51
Diesel index: 53
Pour point, °F.: 20
Vis. at 100° F., SUS: 40.3

Heavy gas oil
Yield, vol %: 23.1
TBP range, °F.: 662-977
Gravity, °API: 25.4
Sulfur, wt %: 0.21
Nitrogen (total), ppm: 1,150
Vanadium, ppm: <0.1
Nickel, ppm: <0.1
Aniline point, °F.: 190
Pour point, °F.: 105
Vis. at 210° F., SUS: 48.1

Residual oil
Range, °F.: +
Yield, vol %: 7.7
Gravity, °API: 11.8
Sulfur, wt %: 0.39
Vis. at 210° F., SUS: 2,030
Carbon residue, wt %: 12.0
Vanadium, ppm: 3
Nickel, ppm: 40

Bonny light

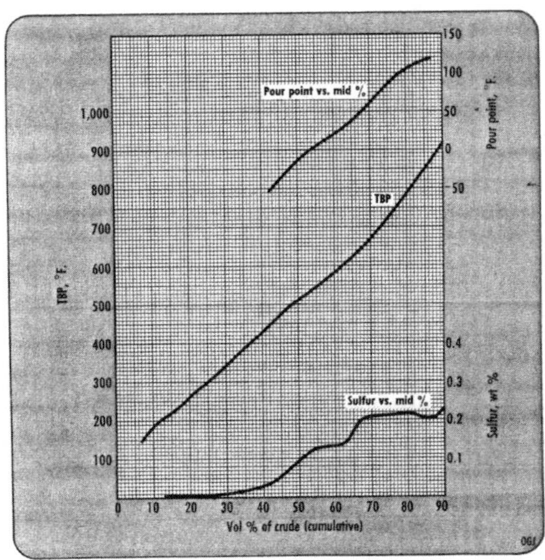

Bonny medium, Nigeria

Crude
Gravity, °API: 26.0
Sulfur, wt %: 0.23
Vis. at 100° F., SUS: 60.7
Pour pt, °F.: <−5
C_4 and lighter, vol %: 0.7
Vanadium, ppm: 1
Nickel, ppm: 7
Carbon residue (Conradson), wt %: 1.8

Light naphtha (debutanized)
Yield, vol %: 2.1
TBP range, °F.: 60-167
Gravity, °API: 79.2
Sulfur, wt %: 0.001
Paraffins, vol %: 73
Naphthenes, vol %: 24
Aromatics, vol %: 3
Research ON, clear: 80
Research ON, + 3 ml TEL/U.S. gal: 94.5

Heavy naphtha
Yield, vol %: 8.7
TBP range, °F.: 167-347
Gravity, °API: 50.1
Sulfur, wt %: 0.010
Paraffins, vol %: 27.5
Naphthenes, vol %: 58.5
Aromatics, vol %: 14
Characterization factor: 11.5
Aniline point, °F.: 110

Kerosine
Yield, vol %: 14.7
TBP range, °F.: 347-482
Gravity, °API: 34.4
Sulfur, wt %: 0.063
Smoke pt. (IP), mm: 17
Freezing point, °C.: −65
Aniline point, °F.: 123
Diesel index: 42
Pour point, °F.: <−70

Light gas oil
Yield, vol %: 29.7
TBP range, °F.: 482-662
Gravity, °API: 27.5
Sulfur, wt %: 0.18
Cetane index: 40
Diesel index: 37
Pour point, °F.: −15
Vis. at 100° F., SUS: 44.6

Heavy gas oil
Yield, vol %: 31.3
TBP range, °F.: 662-977
Gravity, °API: 19.7
Sulfur, wt %: 0.31
Vanadium, ppm: <0.1
Nickel, ppm: <0.1
Aniline point, °F.: 163
Pour point, °F.: 80
Vis. at 210° F., SUS: 53.1

Residual oil (977° F. +):
Yield, vol %: 12.8
Gravity, °API: 10.1
Sulfur, wt %: 0.48
Vis. at 210° F., SUS: 3,690
Carbon residue, wt %: 12.7
Vanadium, ppm: 7
Nickel, ppm: 52

Bonny medium

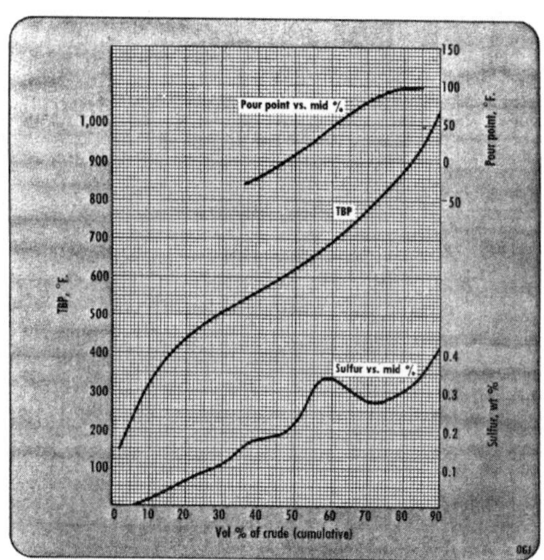

Brega, Libya

Crude
Gravity, °API: 40.4
Sulfur, wt %: 0.21
Pour point, F: +30
RVP, psi: 6.4
Kin., cS @ t70° F.: 5.58
100° F: 3.56

Light naphtha
Range, °F.: 68-212
Yield, vol %: 12.4
Gravity, °API: 73.8
Sulfur, wt %: 0.014
RVP, psi: 7.4
Paraffins, vol %: 72.5
Naphthenes, vol %: 25.3
Aromatics, vol %: 2.2
RON, clear: 59.9

Heavy naphtha
Range, °F.: 212-302
Yield, vol %: 10.7
Gravity, °API: 56.8
Sulfur, wt %: 0.020
Paraffins, vol %: 53.0
Naphthenes, vol %: 39.3
Aromatics, vol %: 7.7

Kerosine
Range, °F.: 302-455
Yield, vol %: 17.4
Gravity, °API: 46.7
Sulfur, wt %: 0.035
Paraffins, vol %: 51.2
Naphthenes, vol %: 34.7
Aromatics, vol %: 14.1
Freezing Point, °F: −68
Smoke Point, mm: 27
Luminometer No.: 61
Aniline Point, °F. 144
Kin. cSt @ 30° F.: 5.50
Kin. cSt @ 100° F.: 1.17

Light gas oil
Range, °F.: 455-650
Yield, vol %: 20.3
Gravity, °API: 38.1
Sulfur, wt %: 0.10
Pour point, °F.: +15
Aniline point, °F.: 172
Kin. cSt @ 100° F.: 3.61
Kin. cSt @ 210° F.: 1.37

Heavy gas oil
Range, °F.: 650-1049
Yield, vol %: 27.8
Gravity, °API: 27.7
Sulfur, wt %: 0.30
Pour point, °F.: 105
Aniline point, °F.: 206
Kin. cSt @ 150° F.: 14.9
Kin. cSt @ 210° F.: 6.10

Brega

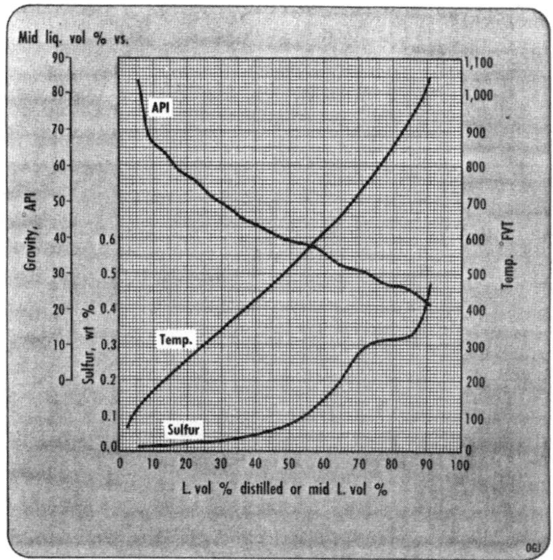

Residual oil:
Range, °F.: 650+
Yield, vol %: 36.8
Gravity, °API: 23.6
Sulfur, wt %: 0.41
Pour point, °F.: 95
Con. carbon wt %: 4.1
Kin. cSt @ 100° F.: 184
Kin. cSt @ 122° F.: 92.0
Kin. cSt @ 210° F.: 14.1

Residual oil
Range, °F.: 1,049+
Yield, vol %: 9.0
Gravity, °API: 12.3
Sulfur, wt %: 0.69
Pour point, °F.: >120
Con. Carbon, wt %: 14.6
Kin. cSt @ 210° F.: 620
Vanadium, ppm: 24
Nickel, ppm: 32
Iron, ppm: 124

Darius, Iran

Crude
Gravity, °API: 33.9
Sulfur, wt %: 2.45
Pour point, °F.: 0
Vis. SUS @ 68° F.: 58
130° F.: 40
C_4 & lighter, vol %: 3.05

Lt. straight run
Range, TBP, °F.: C_5 — 200
Yield, vol %: 9.95
Gravity, °API: 83.3
Sulfur, wt %: 0.07
Motor octane, clear: 57.2

Cat. reformer feed
Range, TBP, °F.: 200-350
Yield, vol %: 15.30
Gravity, °API: 56.9
Sulfur, wt %: 0.13
Arom. + naph., vol %: 30.2

Heater oil dist.
Range, TBP, °F.: 350-540
Yield, vol %: 17.60
Gravity, °API: 42.7
Sulfur, wt %: 0.70
Blend pour, °F.: −23
Cetane index: 51.5

Furnace oil dist.
Range, TBP, °F.: 540-620
Yield, vol %: 6.99
Gravity, °API: 33.4
Sulfur, wt %: 1.84
Blend pour, °F.: 57
Cetane index: 54.0

Lt. FCU feed
Range, TBP, °F.: 620-690
Yield, vol %: 6.36
Gravity, °API: 26.4
Sulfur, wt %: 2.67
CA*, wt %: 15.1
Nitrogen, wt %: 0.055

Heavy FCU feed
Range, TBP, °F.: 690-1,010
Yield, vol %: 22.18
Gravity, °API: 23.4
Sulfur, wt %: 2.97
CA*, wt %: 14.7
Nitrogen, wt %: 0.093
Equiv. nickel, ppm: 1.4

Reduced crude:
Range, TBP, °F.: 1,010+
Yield, vol %: 18.85
Gravity, °API: 5.3
Sulfur, wt %: 5.57
Rams car., wt %: 18.9
V + Ni, ppm: 195

* Carbon atoms (aromatics form)

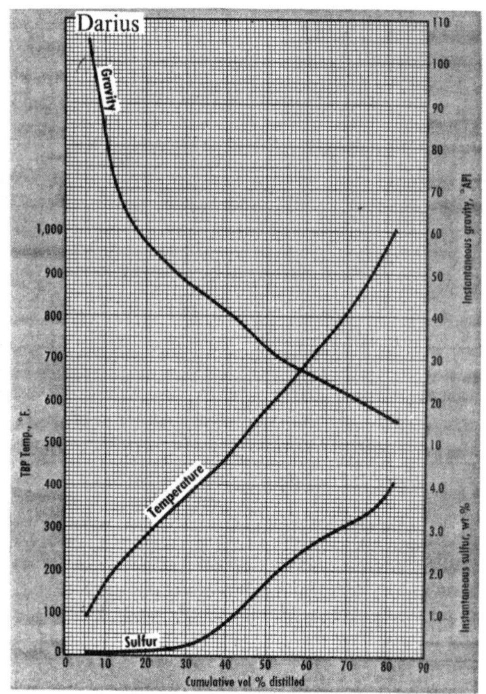

Ecuador (Oriente), Ecuador

Crude
Gravity, °API: 30.4
Sulfur, wt %: 0.87
Viscosity, SUS at 100° F.: 61.8
Pour point, °F.: +20
C_4 and lighter, vol %: 1.53
Reid vapor pressure, lb: 4.8
Vanadium, ppm: 61
Nickel, ppm: 28
Salt, lb/1,000 bbl: 5

Light naphtha
TBP range, °F.: IBP-140*
Yield, vol %: 3.20
Gravity, °API 90.6
Sulfur, wt %: 0.019
Paraffins, vol %: —
Naphthenes, vol %: —
Aromatics, vol %: —
RON, cear: 80.1
RON, 3+ ml. TEL 94.8

Light naphtha
TBP range, °F.: 140-170
Yield, vol %: 1.64
Gravity, °API: 74.7
Sulfur, wt %: 0.017
Paraffins, vol %: 73.1
Naphthenes, vol %: 25.3
Aromatics, vol %: 1.6
RON, clear: 72.0
RON, +3 ml TEL: 90.0

Heavy naphtha
TBP range, °F.: 170-310
Yield, vol %: 11.91
Gravity, °API: 58.2
Sulfur, wt %: 0.015
Paraffins, vol %: 49.4
Naphthenes, vol %: 43.9
Aromatics, vol %: 6.7
Aniline point, °F.: 131
ASTM, 50% point, °F.: 244

Kerosine
TBP range, °F.: 310-520
Yield, vol %: 18.44
Gravity, °API: 41.9
Sulfur, wt %:.0.144
Aromatics, vol %: 18
Freezing point, °F.: −52
Aniline point, °F.: 137
Cetane index: 46
Smoke point (ASTM), mm: 20

Ecuador (Oriente)

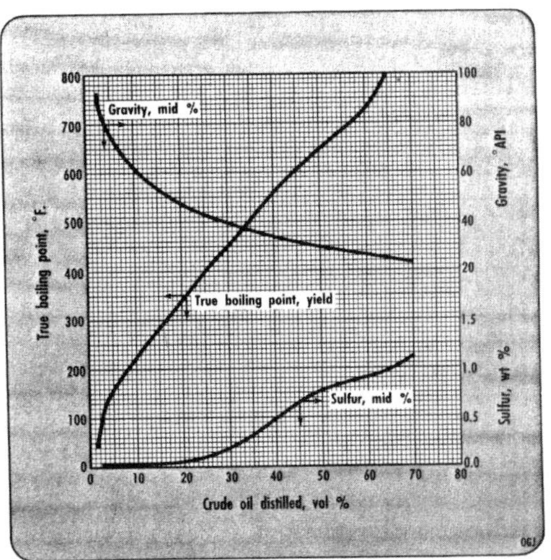

Light gas oil
TBP range, °F.: 520-680
Yield, vol %: 18.94
Gravity, °API: 32.0
Viscosity, SUS at 100° F.: 44.3
Sulfur, wt %: 0.65
Pour point, °F.: +25
Diesel index: 51.2
Cetane index: 52.4
ASTM, 50% point, °F.: 580

Heavy gas oil
TBP to EFV range, °F.: 680-1,000
Yield, vol %: 27.66
Gravity, °API: 23.3
Viscosity, SUS at 210° F.: 55.5
Sulfur, wt %: 1.11
Nitrogen (total), ppm: 1,300
Pour point, °F.: +105
Aniline point, °F.: 198.0
Vanadium, ppm: <0.1
Nickel, ppm: <0.1
ASTM, 50% point, °F.: 845

Residual oil
TBP or EFV, °F.: 680+
Yield, vol %: 44.94
Gravity, °API: 15.2
Viscosity, SUS at 210° F.: 325
Sufur, wt %: 1.43
Con. carbon residue, wt %: 11.3
Pour point, °F.: +95
Vanadium, ppm: 123
Nickel, ppm: 56

*Does not include uncondensed gases of 0.93 vol % to crude.

Ekofisk, Norway

Crude
Gravity, °API: 36.3
BS&W, vol %: 1.0
Sulfur, wt %: 0.21
Pour test, °C.: + 20
Vis. SUS @ 100° F.: 42.48
RVP, psi @ 100° F.: 5.1
Salt, lb/1,000 bbl: 14.5
NC_4 & lighter, vol %: 1.0

Gasoline
Range, °F.: 60-200
Yield, vol %: 10.7
Gravity, °API: 77.2
Sulfur, wt %: 0.003
RON, clear: 74.4
RON + 3 ml TEL: 90.0

Gasoline
Range, °F.: 60-400
Yield, vol %: 31.0
Gravity, °API: 60.1
Paraffins, vol %: 56.52
Naphthenes, vol %: 29.52
Aromatics, vol %: (0 +A): 13.96
Sulfur, wt %: 0.0024
RON, clear: 52.0
RON + 3 ml TEL/gal: 76.0

Kerosine
Range, °F.: 400-500
Yield, vol %: 13.5
Gravity, °API: 40.2
Vis., SUS @ 100° F.: 32.33
Freezing point, °F.: −38
Aromatics, vol % (0 + A): 13.1
Sulfur, wt %: <0.05
Aniline point, °F.: 146.2
Smoke point, mm: 21

Light gas oil
Range, °F.: 500-650
Yield, vol %: 15.7
Gravity, °API: 33.7
Vis. SUS @ 100° F.: 43.83
Pour point, °F.: +25
Sulfur, wt %: 0.11
Aniline point, °F.: 164.3
Car. res., Rams., wt %: 0.08
Cetane index: 56.5

Topped crude
Range, °F.: 650 +
Yield, vol %: 38.8
Gravity, °API: 21.5
Vis. SFS @ 122° F.: 80.25
Pour point, °F.: +85
Sulfur, wt %: 0.39
Car. res. Rams., wt %: 4.0
Ni/V, ppm: 5.04/1.95

Ekofisk

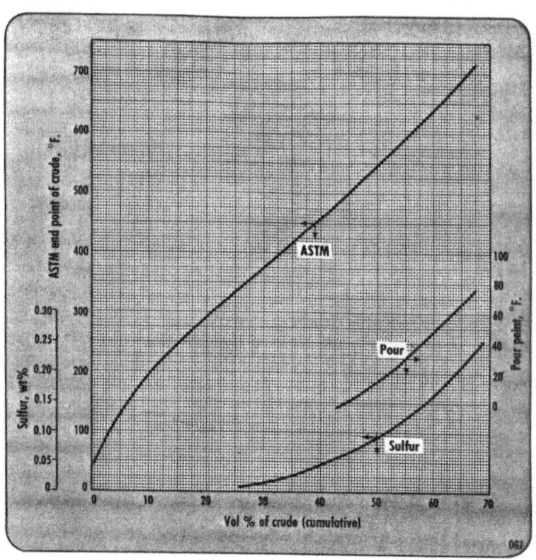

Escravos, Nigeria

Crude
Gravity, °API: 36.2
Sulfur, wt %: 0.16
Vis. SUS at 100° F.: 38.0
Pour point, °F.: +50
C_4 and lighter, vol %: 2.42
RVP, lb: 4.5
Vanadium, ppm: 0.2
Nickel, ppm: 5.1
Con. carbon residue, wt %: 1.3
Salt, lb/1,000 bbl: 10

Light Naphtha
TBP range, °F.: IBP-140*
Yield, vol %: 3.35
Gravity, °API: 88.1
Sulfur, wt %: 0.008
Paraffins, vol %: 89
Naphthenes, vol %: 10
Aromatics, vol %: 1
RON, clear: 81.2
RON, +3 ml TEL: 96.1

Light naphtha
TBP range, °F.: 140-170
Yield, vol %: 2.12
Gravity, °API: 69.6
Sulfur, wt %: 0.005
Paraffins, vol %: 64.9
Naphthenes, vol %: 29.7
Aromatics, vol %: 5.4

RON, clear: 72.0
RON, +3 ml TEL: 89.3

Heavy Naphtha
TBP range, °F.: 170-310
Yield, vol %: 17.03
Gravity, °API: 54.0
Sulfur, wt %: 0.006
Paraffins, vol %: 38.0
Naphthenes, vol %: 47.1
Aromatics, vol %: 14.9
Aniline point, °F.: 117.0

Kerosine
TBP range, °F.: 310-520
Yield, vol %: 26.15
Gravity, °API: 39.6
Sulfur, wt %: 0.057
Aromatics, vol %: 21.5
Freezing point, °F.: −48
Aniline point, °F.: 133.0
Cetane index: 42
Smoke point (ASTM), mm: 19

* Does not include uncondensed gases of 2.26 vol % to crude.

Escravos

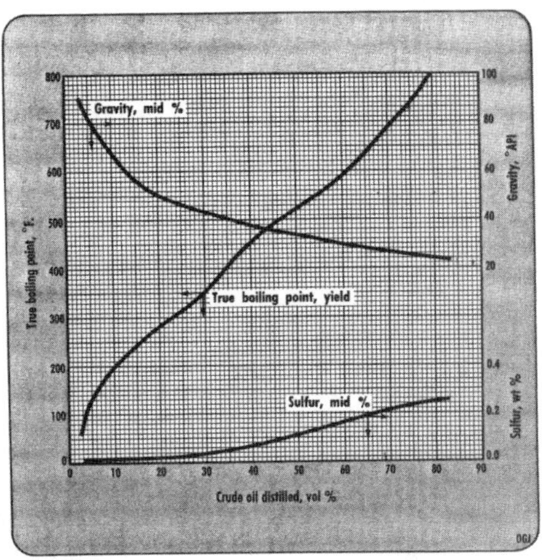

Light gas oil
TBP range, °F.: 520-680
Yield, vol %: 18.58
Gravity, °API: 31.4
Vis. SUS at 100° F.: 44.1
Sulfur, wt %: 0.17
Pour point, °F.: +30
Diesel index: 49.9
Cetane index: 50.5

Heavy gas oil
TBP to EFV range, °F.: 680-1,000
Yield, vol %: 24.87
Gravity, °API: 23.2
Vis. SUS at 210° F.: 56.4
Sulfur, wt %: 0.25
Nitrogen (total), ppm: 1,200
Pour point, °F.: +110
Aniline point, °F.: 194
Vanadium, ppm: <0.1
Nickel, ppm: 1.0

Residual oils
TBP or EFV, °F.: 680+
Yield, vol %: 30.51
Gravity, °API: 20.4
Vis. SUS at 210° F.: 93.2
Sulfur, wt %: 0.30
Con. carbon residue, wt %: 3.62
Pour point, °F.: +105
Vanadium, ppm: 0.7
Nickel, ppm: 15

Iranian heavy, Iran

Crude
Gravity, °API: 30.8
Sulfur, wt %: 1.6
Pour Point, °F.: −5
Viscosity, cSt at 100° F.: 9.81
Viscosity, cSt at 130° F.: 7.55
RVP, psi: 6.6
Ramsbottom carbon, wt %: 5.0
Vanadium, ppm: 88
Nickel, ppm: 30
Ethane, vol %: 0.1
Propane: 0.5
Isobutane: 0.3
Normal butane: 1.2

Light gasoline
Range, °F.: C_5−200°
Yield, vol %: 7.9
Gravity, °API: 78.8
Sulfur, wt %: 0.10
Mercaptan sulfur, ppm: 340
RON clear: 66.1
RON + 3 ml TEL/gal: 81.0

Light naphtha
Range, °F.: 200-300
Yield, vol %: 9.6
Gravity, °API: 60.0
Sulfur, wt %: 0.13
Mercaptan sulfur, ppm: 340
Paraffins, vol %: 53
Naphthenes: 34
Aromatics: 13
RON clear: 49.6
RON + 3 ml TEL/gal.: 70.0

Heavy naphtha
Range °F.: 300-400
Yield, vol %: 9.4
Gravity, °API: 48.2
Sulfur, wt %: 0.22
Mercaptan sulfur, ppm: 100
Paraffins, vol %: 50
Naphthenes: 35
Aromatics: 15
Aniline point, °F.: 130
Smoke point, ASTM, mm: 23

Kerosine
Range, °F.: 400-500
Yield, vol %: 9.2
Gravity, °API: 40.1
Sulfur, wt %: 0.44
Mercaptan sulfur, ppm: 14
Paraffins, vol %: 27
Naphthenes: 43
Aromatics: 30
Freezing point, °F.: −35
Pour point, °F.: −40
Aniline point, °F.: 140
Smoke point, ASTM, mm: 19

Light gas oil
Range, °F.: 500-650
Yield, vol %: 14.0
Gravity, °API: 34.0
Sulfur, wt %: 1.1
Pour point, °F.: +15
Aniline point, °F.: 157

Iranian heavy

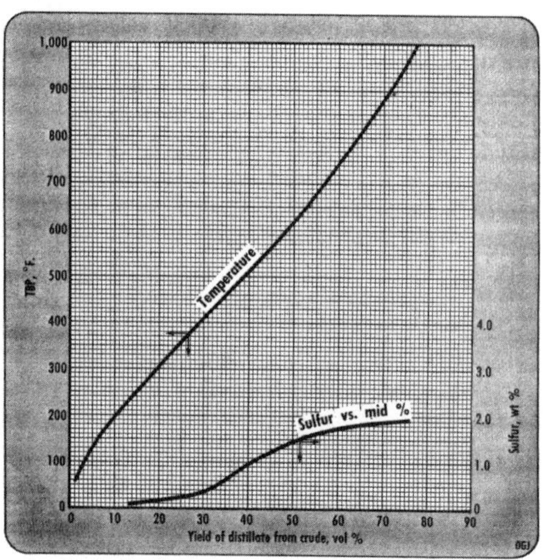

Heavy gas oil
Range, °F.: 650-1,000
Yield, vol %: 26.0
Gravity, °API: 23.0
Sulfur, wt %: 1.8
Nitrogen, ppm: 1,600
Pour point, °F.: 90
Aniline point, °F.: 176
Viscosity, cSt at 130° F.: 30.0

Long residuum
Range, °F.: 650+
Yield, vol %: 47.8
Gravity, °API: 14.4
Sulfur, wt %: 2.5
Nitrogen, wt %: 4,700
Pour point, °F.: +75
Viscosity, cSt at 130° F.: 500
Ramsbottom carbon, wt %: 9.5

Short residuum
Range, °F.: 1,000+
Yield, vol %: 21.8
Gravity, °API: 6.3
Sulfur, wt %: 3.2
Nitrogen, wt %: 8,300
Pour point, °F.: +135
Viscosity, cSt at 130° F.: 1.1×10^6
Viscosity, cSt at 210° F.: 4,250
Penetration at 77° F.: 47

Iranian light, Iran

Crude
Gravity, °API: 33.5
Sulfur, wt %: 1.4
Pour point, °F.: −20
Viscosity, cSt at 100° F.: 6.41
Viscosity, cSt at 130° F.: 4.83
RVP, psi: 6.5
Ramsbottom carbon, wt %: 3.4
Vanadium, ppm: 35
Nickel, ppm: 13
Ethane, vol %: 0.1
Propane: 0.4
Isobutane: 0.3
Normal butane: 1.1

Light gasoline
Range, °F.: C_5−200
Yield, vol %: 8.1
Gravity, °API: 75.0
Sulfur, wt %: 0.076
Mercaptan sulfur, ppm: 210
RON clear: 64.7
RON + 3 ml TEL/gal: 79.5

Light naphtha
Range, °F.: 200-300
Yield, vol. %: 10.0
Gravity, °API: 57.8
Sulfur, wt %: 0.09
Mercaptan sulfur, ppm: 210
Paraffins, vol %: 50
Naphthenes: 33
Aromatics: 17
RON clear: 47.4
RON + 3 ml TEL/gal: 67.0

Heavy naphtha
Range, °F.: 300-400
Yield, vol %: 10.1
Gravity, °API: 49.6
Sulfur, wt %: 0.11
Mercaptan sulfur, ppm: 20
Paraffins, vol %: 54
Naphthenes: 30
Aromatics: 16
Aniline point, °F.: 135
Smoke point, ASTM, mm: 25

Kerosine
Range, °F.: 400-500
Yield, vol %: 10.5
Gravity, °API: 40.9
Sulfur, wt %: 0.34
Mercaptan sulfur, ppm: 23
Paraffins, vol %: 30
Naphthenes: 41
Aromatics: 29
Freezing point, °F.: −30
Pour point, °F.: −35
Aniline point, °F.: 146
Smoke point, ASTM, mm: 21

Light gas oil
Range, °F.: 500-650
Yield, vol %: 14.0
Gravity, °API: 34.4
Sulfur, wt %: 1.0
Pour point, °F.: +20
Aniline point, °F.: 162

Iranian light

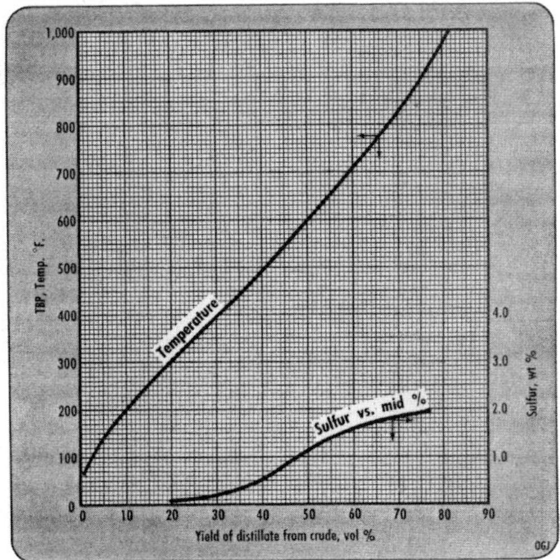

Heavy gas oil,
Range, °F.: 650-1,000
Yield, vol %: 26.8
Gravity, °API: 23.7
Sulfur, wt %: 1.8
Nitrogen, ppm: 1,300
Pour point, °F.: 90
Aniline point, °F.: 183
Viscosity, cSt at 130° F.: 20.0

Long residuum, 650° F.+
Range, °F.: 650° +
Yield, vol %: 45.4
Gravity, °API: 17.0
Sulfur, wt %: 2.4
Nitrogen, wt %: 2,900
Pour point, °F.: +75
Viscosity, cSt at 130° F.: 190
Ramsbottom carbon, wt %: 6.8

Short residuum, 1,000° F.+
Range, °F.: 1,000+
Yield, vol %: 18.6
Gravity, °API: 11.0
Sulfur, wt %: 3.3
Nitrogen, wt %: 5,100
Pour point, °F.: +100
Viscosity, cSt at 130° F.: 45,300
Viscosity, cSt at 210° F.: 930
Penetration at 77° F.: 285

Kirkuk, Iraq

Crude
Gravity, °API: 35.9
Sulfur, wt %: 1.95
Pour point, °C.: −36
Kin. vis. cSt @ 100° F.: 4.61
Wax content, wt %: 3.9
Asphaltenes, wt %: 1.5
Con. car. residue, wt %: 3.8

Light straight run
Range, °C.: C_5 −65
Yield, vol %: 6.1
RON, clear: 73
Paraffins, vol %: 97

Light naphtha
Range, °C.: 65-100
Yield, vol %: 6.4
S.G. 15/4° C.: 0.701
Sulfur, ppm: 333
RON clear: 52
Paraffins, vol %: 80
Naphthenes, vol %: 18
Aromatics, vol %: 2

Heavy naphtha
Range, °C.: 100-150
Yield, vol %: 10.0
S.G. 15/4° C.: 0.746
Sulfur, ppm: 903
RON clear: 38
Paraffins, vol %: 69
Naphthenes, vol %: 21
Aromatics, vol %: 10

Kerosine
Range, °C.: 150-200
Yield, vol %: 10.5
S.G., 15/4° C.: 0.782
Sulfur, ppm: 1,810
Smoke point, mm: 24
Aniline point, °C.: 54.8

Light gas oil
Range, °C.: 200-250
Yield, vol %: 9.9
Kin. vis. cSt @ 100° C.: 1.68
Sulfur, wt %: 0.30
Cetane No.: 53

Medium gas oil
Range, °C.: 250-300
Yield, vol %: 9.4
S.G., 15/4° C.: 0.835
Kin. vis. cSt @ 100° F.: 3.06
Sulfur, wt %: 0.85
Aniline point, °C.: 71.8

Heavy gas oil
Range, °C.: 300-350
Yield, vol %: 8.8
Sulfur, wt %: 1.60

Kirkuk

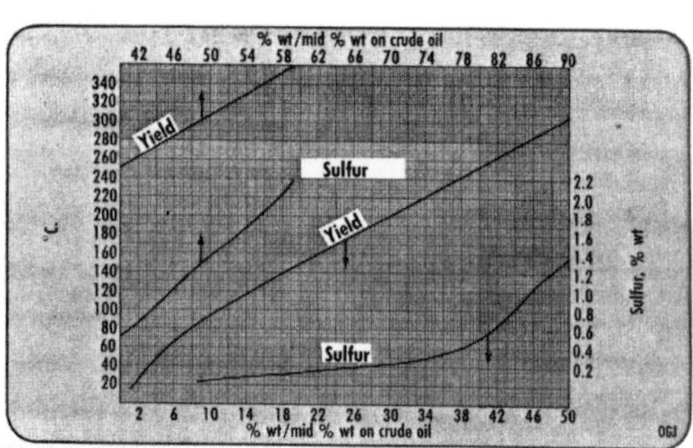

Residual oil
Range, °C.: 350-370
Yield, vol %: 3.1
S.G. 15/4 °C.: 0.884
Sulfur, wt %: 2.17

Residual oil
Range, °C.: 370+
Yield, vol %: 34.4
S.G., 15/4 °C.: 0.963
Kin vis. cSt @ 140° F.: 280
Kin vis. cSt @ 210° F.: 43.9
Sulfur, wt %: 4.0
Pour point, °F.: 86
Con. car. Residue, wt %: 10.0
Vanadium, ppm: 58
Nickel, ppm: <3

Kuwait crude, Kuwait

Crude
Gravity, °API: 31.2
Sulfur, wt %: 2.50
Viscosity, SUS at 100° F.: 58.7
Pour point, °F.; 0
C_4 and lighter, vol %: 2.46
Reid vapor pressure, lb.: 5.4
Vanadium, ppm: 31
Nickel, ppm: 9.6
Con. carbon residue, wt %: 5.3
Salt, lb/1,000 bbl: 3
ASTM, 50% point, °F.: >590

Light naphthas
TBP range, °F.: IBP-140*
Yield, vol %: 5.49
Gravity, °API: 94.6
Sulfur, wt %: 0.01
Paraffins, vol %: 98.5
Naphthenes, vol %: 1.4
Aromatics, vol % 0.1
Res. octane, clear: 78.5
Res. octane, +3 ml TEL: 92.0
ASTM, 50% point, °F.: 96

Light naphthas
TBP range, °F.: 140-170
Yield, vol %: 1.85
Gravity, °API: 78.2
Sulfur, wt %: 0.02
Paraffins, vol %: 86.2
Naphthenes, vol %: 12.5
Aromatics, vol %: 1.3
Res. octane, clear: 58.8
Res. octane, +3 ml TEL: 76.3
ASTM, 50% point, °F.: 156

Heavy naphtha
TBP range, °F.: 170-310
Yield, vol %: 12.03
Gravity, °API: 62.2
Sulfur, wt %: 0.02
Paraffins, vol %: 67.9
Naphthenes, vol %: 22.1
Aromatics, vol %: 10.0
Aniline point, °F.: 132.8
ASTM, 50% point, °F.: 244

Kerosine
TBP range, °F.: 310-520
Yield, vol %: 18.20
Gravity, °API: 45.9
Sulfur, wt %: 0.28
Aromatics, vol %: 19.5
Freezing point, °F.: −46
Aniline point, °F.: 143
Cetane Index: 52.3
Smoke point (ASTM), mm: 24
ASTM, 50% point, °F.: 403

Kuwait crude

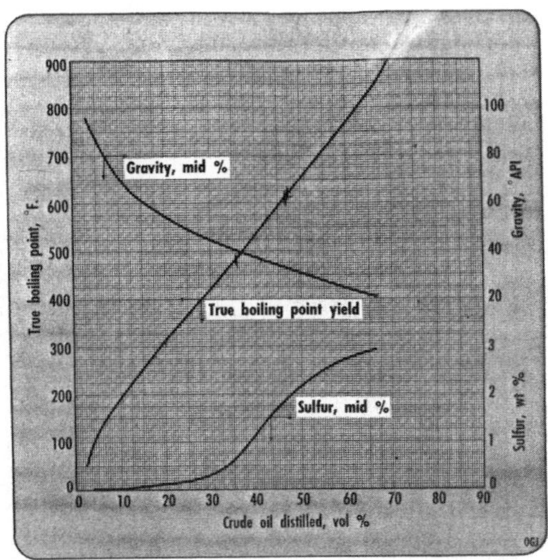

Light gas oil
TBP range, °F.: 520-680
Yield, vol %: 14.11
Gravity, °API: 33.7
Viscosity, SUS at 100° F.: 41.3
Sulfur, wt %: 1.66
Pour point, °F.: +20
Diesel index: 53.6
Cetane index: 55
ASTM, 50% point, °F.: 573

Heavy gas oil
TBP to EFV range, °F.: 680-1,000
Yield, vol %: 26.59
Gravity, °API: 21.7
Viscosity, SUS at 210° F.: 53.9
Sulfur, wt %: 2.91
Nitrogen (total), ppm: 950
Pour point, °F.: +100
Aniline point, °F.: 178
Vanadium, ppm: 0.4
Nickel, ppm: 0.1
ASTM, 50% point, °F.: 850

Residual oils
TBP or EFV, °F.: 680+
Yield, vol %: 47.53
Gravity, °API: 14.0
Viscosity, SUS at 210° F.: 267
Sulfur, wt %: 4.14
Con. carbon residue, wt %: 9.37
Pour point, °F.: +70
Vanadium, ppm: 59
Nickel, ppm: 18

*Does not include uncondensed gases of 0.79 vol % to crude.

Murban, Abu Dhabi

Crude
Gravity, °API: 39.4
Sulfur, wt %: 0.74

Rvp, psi: 5
Pour point, °C.: −15
Wax content, wt %: 8.0
Vis., cSt @ 21° C.: 5.0
Salt content, lb/1,000 bbl: <5
V/Ni, ppm: 0.8/0.58

Straight-run gasoline
Range, °C.: C_5-75
Yield, vol %: 6.78
Gravity, °API: 82.2
Sulfur, wt %: 0.012
Rvp, psi: 10.1
RON, clear: 69
RON + 0.5 ml TEL/liter: 84

Naphtha
Range, °C.: 75-175
Yield, vol %: 21.22
Gravity, °API: 56.9
Sulfur wt %: 0.013
Paraffins, wt %: 63
Naphthenes, wt %: 20
Aromatics, wt %: 17
Aniline point, °C.: 51.3

Kerosine

Range, °C.: 175-250
Yield, vol %: 16.14
Gravity, °API: 45.4
Sulfur, wt %: 0.058
Aniline point, °C.: 61.8
Diesel index: 65
Smoke point, mm: 24
Freezing point, °C.: −43
Vis., cSt @ 20° C.: 1.8

Gas oil
Range, °C.: 250-300
Yield, vol %: 10.40
Gravity, °API: 37.8
Sulfur, wt %: 0.47
Diesel index: 59
Cetane index: 54
Pour point, °C.: −18
Vis., cSt @ 20° C.: 4.2

Gas oil
Range, °C.: 300-350
Yield, vol %: 9.24
Gravity, °API: 33.6
Sulfur, wt %: 1.06
Diesel index: 58
Wax content, wt %: 17.5
Pour point, °C.: +4
Vis., cSt @ 20° C.: 9.5

Murban

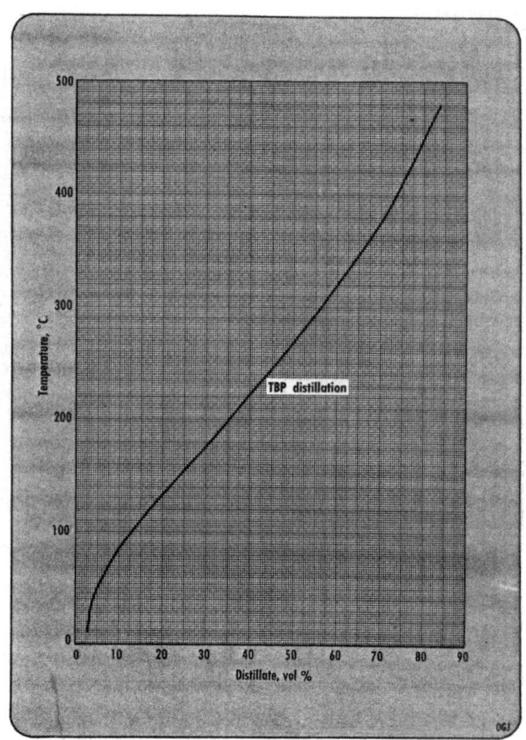

Residual oil
Range, °C.: 350+
Yield, vol %: 34.51
Gravity, °API: 22.6
Sulfur, wt %: 1.49
Pour point, °C.: +35
Wax content, wt %: 19.5
Conradson carbon, wt %: 3.6
Vis., cSt @ 37.8° C.: 104
V/Ni, ppm 2/2

Pennington, Nigeria

Crude
Gravity, °API: 37.7
Sulfur, wt %: 0.076
Vis., at 100° F.: SUS: 36
Pour point, °F.: 37
C_4 and lighter, vol %: 1.1
Salt, lb/1,000 bbl: 2.9

Light naphtha
TBP range, °F.: C_5-200
Yield, vol %: 6.0
Gravity, °API: 70.1
Sulfur, wt %: 0.001
Aromatics, vol %: 1.5
Naphthenes, vol %: 24.0
Paraffins, vol %: 74.0

Heavy naphtha
TBP range, °F.: 200-340
Yield, vol %: 16.3
Gravity, °API: 51.2
Sulfur, wt %: 0.004
Aromatics, vol %: 11.0
Naphthenes, vol %: 56.5
Paraffins, vol %: 32.0

Kerosine
TBP range, °F.: 340-470
Yield, vol %: 18.0
Gravity, °API: 40.1
Sulfur, wt %: 0.018
Smoke point, mm: 21.0
Freezing point, °F.: −74

Gas oil
TBP range, °F.: 470-650
Yield, vol %: 33.5
Gravity, °API: 34.5
Sulfur, wt %: 0.058
Pour point, °F.: 10

Reduced crude
TBP range, °F.: 650+
Yield, vol %: 25.1
Gravity, °API: 23.6
Sulfur, wt %: 0.19
Pour point, °F.: 88
Vis., at 210° F.: SUS: 6.9
Conradson Carbon, wt %: 1.3

Pennington

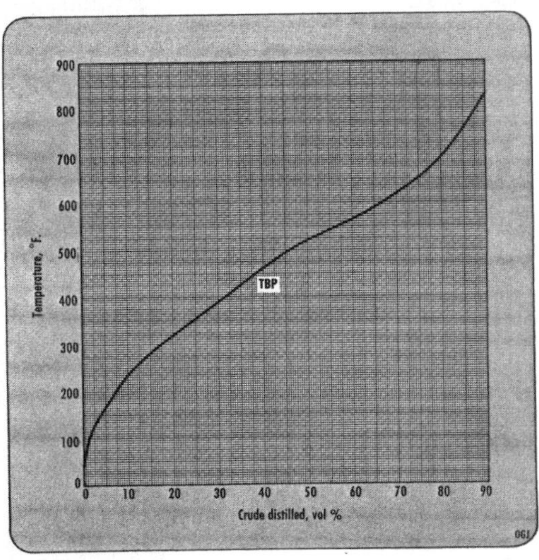

Qatar Marine, Qatar

Crude
Gravity, °API: 37.0
Sulfur, wt %: 1.50

Rvp, psi: 5.7
Salt, lb/1,000 bbl: 9.0
Pour point, °F.: +25
Vis. SUS @ 40° F.: 70.6
Vis. SUS @ 80° F.: 42.8
Yield, vol %: IBP-113° F.: 5.41

Straight-run gasoline
Range, °F.: 113-220
Yield, vol %: 9.11
Gravity, °API: 69.0
Rvp, psi: 3.9

RON + 3 cc Tel/gal: 74.4
Sulfur, wt %: 0.04
Paraffins, vol %: 71.3
Naphthenes, vol %: 22.4
Aromatics, vol %: 6.3

Naphtha
Range, °F.: 220-390
Yield, vol %: 19.57
Gravity, °API: 52.0
Sulfur, wt %: 0.07
Paraffins, vol %: 59.2

ASTM, 50% point, °F.: 578

Heavy gas oil
TBP to EFV range, °F.: 680-1,000
Yield, vol %: 26.45
Gravity, °API: 21.5
Vis., SUS at 210° F.: 58.8
Sulfur, wt %: 2.24
Nitrogen (total), ppm: 1,400
Pour point, °F.: +100
Aniline point, °F.: 176.0
Vanadium, ppm: 0.6
Nickel, ppm: 0.5
ASTM, 50% point, °F.: 850

Residual oil
TBP or EFV, °F.: 680+
Yield, vol %: 43.57
Gravity, °API: 15.3
Vis., SUS at 210° F.: 236
Sulfur, wt %: 2.69
Con. carbon residue, wt %: 9.27
Pour point, °F.: +85
Vanadium, ppm: 110
Nickel, ppm: 44

Residual oil
TBP or EFV, °F.: Asphalt 1,000+
Yield, vol %: 17.12
Gravity, °API: 6.7
Vis., SUS at 210° F.: 36 PEN
Sulfur, wt %: 3.32
Con. carbon residue, wt %: 21.6
Pour point, °F.: 127 SP
Vanadium, ppm: 263
Nickel, ppm: 107
*Does not include uncondensed gases of 1.53 vol % to crude.

Qatar marine

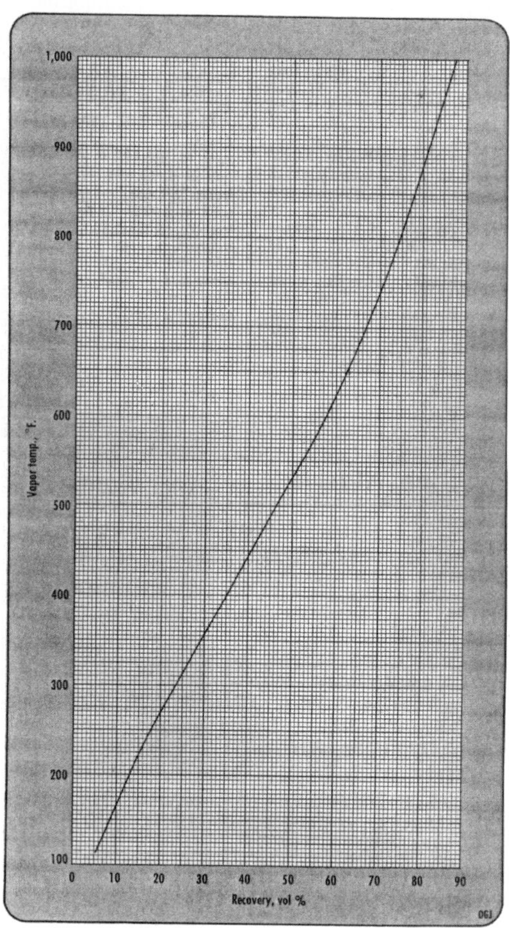

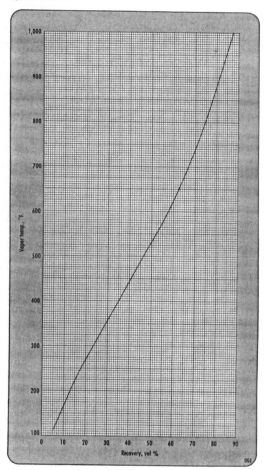

Sassan, Iran

Crude
Gravity, °F.: 33.9
Paraffin wax, wt %.: 1.9
Sulfur, wt %.: 1.91
Pour point, °F.: −5
Viscosity, SUS @ 77° F.: 52.1
100° F.: 44.2
Bs&w: 17,550

Straight-run gasoline
Range, vapor-cut points, °F.: 50-200
Yield, vol %: 10.5
Gravity, °API: 77.6
Sulfur, wt %: 0.081
Mercaptan sulfur: 0.067
Octane no., F-1 clear: 61.0

Naphtha
Range, VCP, °F.: 200-375
Yield, vol %: 16.4
Gravity, °API: 53.3
Sulfur, wt %: 0.0820
Mercaptan S, wt %: 0.002
Aniline point, °F.: 121.5
Octane no., F-1 clear: 35.4
Arom. + naphthenes, vol %: 41

Kerosine
Range VCP, °F.: 375-450
Yield, vol %: 7.5
Gravity, °API: 45.1
Sulfur, wt %: 0.19
Freezing point, °F.: −50
Aniline point, °F.: 137.9
Smoke point, mm: 20
Vis, kin, cSt @ 60° F.: 2.05
100° F.: 30

Gas oil
Range VCP, °F.: 450-700
Yield, vol %: 23.8
Gravity, °API: 34.1
Sulfur, wt %: 1.2
Pour point, °F.: +15
Aniline point, °F.: 153.3
Vis. kin cSt @ 60° F.: 7.52
100° F.: 4.03
Cetane No.: 56.3

Residual oil
Range VCP, °F.: 700+
Yield, vol %: 39.7
Gravity, °API: 15.3
Sulfur, wt %: 3.5
Pour point, °F.: + 75
Vis. SUS @ 210° F.: 154
275° F.: 67.8
V + Ni, ppm: 44

Residual oil
Range, VCP, °F.: 1,070+
Yield, vol %: 12.0
Gravity, °API: 1.3
Sulfur, wt %: 5.0
Vis, SUS @ 210° F.: 8,000
275° F.: 1,000
V + Ni, ppm: 127

Sassan

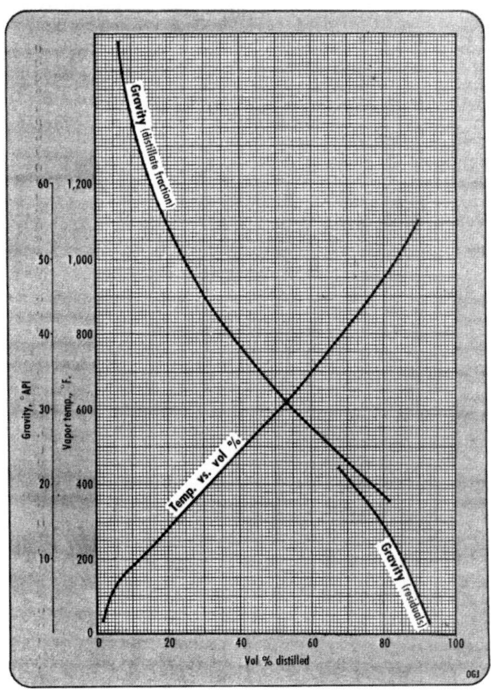

Appendix D

Conversion factors

To Convert From	To		Multiply By
Length			
Feet	Meter	m	0.304
Inch	Millimeter	m m	25.4
Statute Mile	Kilometer		1.609
Area			
Square inches	Square millimeters		645.2
Square inches	Square centimeters		6.452
Square inches	Square meters		0.000645
Square feet	Square meters		0.0924
Acres	hectare		0.4047
Volume			
Cubic inches	Cubic millimeter		16387
Cubic inches	Cubic centimeter		16.387
Cubic inches	Cubic meter		0.00001639
Cubic feet	Cubic meter		0.0281
Fluid ounces	Milliliter		29.57
Gallons (US)	Liter		3.785
Mass			
Pounds	Kilogram	kg	0.4536
Ton (short)	Metric ton	tonne	0.9072
Ton (long))	Metric ton	tonne	1.016
Pressure			
Pounds per square inch	Pascal		6895
Pounds per square inch	Kilopascal		6.895
Kilograms per sq meter	Pascal		9.807
Bar	Kilopascal		100
Force			
Pounds force	Newton		4.448
Kilogram force	Newton		9.807
Work			
British thermal unit	Joule		1055
Foot pound	Joule		1.356
Calorie	Joule		4.186
Power			
Btu / hour	Watt		0.293
Btu / sec	Watt		1055
Horsepower	Kilowatt		0.746
Flow Rate			
Cubic feet per minute	Cubic meters / minute		0.0283
Gallons (US) / minute	Liter / minute		3.785
Barrels (US) oil	Gallons Oil (US)		42
Specific Energy. Latent Heat			
BTU / Pound	Joule / Kilogram		2326
Specific Heat, Specific Entropy			
BTU / Pound – deg F	Joule / Kilogram - Kelvin		4184
Miscellaneous			
(US) Barrels per day	(US) Gallons / hour		1.75
(US) Gallons	Imperial gallons		0.8326

Appendix E

An example of an exercise using linear programming

Linear programming aids decisions on refinery configurations

D. S. J. Jones
Fluor (England) Ltd.
and
J. N. Fisher
Bonner & Moore Associates, Inc., Houston, Texas

Mathematical modeling using linear programming can solve many problems associated with refinery operation and planning. The technique, compared with the cost of an error of judgment, represents only an insignificant financial outlay.

It is becoming more and more evident that there is a definite economic incentive to studying problems associated with refinery planning by mathematical modeling using linear programming.

Ever-increasing investment costs associated with the tightening of product specifications, changing crude slates, alterations in the energy pattern in marketing countries, and expanding petrochemical requirements, all make an error in decision judgment of refinery processing increasingly costly. Whereever this basic risk can be reduced by modern mathematical techniques, the potential saving in capital or investment could make the financial outlay on such a study insignificant.

What follows is a description of a typical refinery simulation study but this is only one of an increasing number of problems that can be solved by mathematical modeling using linear programming.

Wider impetus. The application of linear programming techniques using computers has long been used in the oil refining industry for the development of planning and operating policies. With the introduction of modern high-speed, large-capacity computers, this technique grew considerably within the industry.

Briefly, linear programming is the developing of linear sub models which mathematically describe many of the various operations within the industry, such as refinery processing, crude and product flows, marketing demands. etc. These linear sub models are inter-related to build up a complete mathematical model of the specific operation. By the use of the computer, the equations within the model can be solved to optimize, on a selective basis, the operation under study.

The growth of mathematical models using this technique provides management with a means of making an increasing number of decisions which do have a calculable basis. Thus, in many cases, the need for decisions based only on individual experience or 'feel', with its obvious inherent dangers, is being eliminated.

One such study illustrates a relatively simple refinery problem application of mathematical modeling and linear programming techniques.

Definition of problem

The client, wished to build a new refinery. It had already executed a marketing survey in the area and could specify quantity and quality, together with prices, of the products which would meet its market requirements. Its management now had to decide the economic optimum refinery configuration that would satisfy its crude and product slate. At this stage only one type of crude was intended for the refinery, and to some extent this simplified the problem. However, to satisfy other considerations, management required the solutions to the following premises:

- The refinery configurations, which would satisfy a MINIMUM investment, when producing a high volume of gasoline with and without a low sulfur content limitation of the fuel oil. All other products were to meet quality and quantity requirements.
- The refinery configurations which would give the MAXIMUM return on investment to satisfy a fixed crude throughput with no quantity restriction on the product slate, and then to satisfy a limited restriction on the product slate with no limit to the crude throughput.

Such a problem lends itself readily to linear programming and thus a refinery simulation model was developed to solve these two premises.

Process consideration

The first step in constructing the model was to establish as many processing units as could conceivably contribute to the solution of the problem. For instance, with such large requirements of gasoline there would obviously be required a cracking unit of some kind. Thus, the model included a reformer, fluid cracker, hydrocracker, coker and visbreaker. Some combination of these processes must satisfy the premises of the problem.

Similarly, the lower sulfur content of the fuel oil would probably require some form of residue treating. Thus two severity desulfurizers for both short and long residue respectively were included together with a process for hydrocracking these residues.

The many process units now included were then defined in terms of feed streams, product yields and quality, and operating costs, all based as a percentage on the feed streams. This part of the study was the first important step which required the expertise of specialists. This data forms the basis for the rational solution to the problem, and therefore it was necessary to be accurate and to augment prediction and theory with realism and technical experience.

For instance, in arriving at the yields from the crude and vacuum units the effect of fractionation on the product yield was considered. Realistic ASTM distillation gaps were used that could be met by a commercial distillation unit.

In the fluid catalytic cracking units a more sophisticated approach was needed to correlate the yields from the many feedstocks which would be independent of thermodynamic considerations. Here a base case feed yield data (in this case a straight run waxy distillate) at a conversion of 75% using zeolite catalyst was used.

Yields from all other feedstocks (including those which had been hydrotreated) were related by first principle kenetic and thermodynamic considerations to the base case. A short and simple computer program was used for this purpose, and it was also possible to simulate the effect of changing the quantity of zeolite catalyst by this means. The results of these computerized calculations were checked against existing plant data before being incorporated into the study.

In other processes such as hydrocracking, hydrotreating, visbreaking, etc., care was taken that only proven yield data or correlations were used.

Catalytic reforming yield data was obtained from a correlation which related yield to severity for a basic naphthene and aromatic content of the feed-stock. A whole range of severity operations from 95 to 105 O.N. (Research) clear was encompassed in the study. Spot checks of the predicted yield by this method against actual yield from an operating unit showed that the method was viable and acceptably accurate.

Basic economic data

Having developed the physical yield structure of the 'model', the next step was to complete the basic data by providing investment and maintenance costs.

There is, of course, a considerable wealth of plant cost data available to a contractor from the projects he has completed over the years. However, there is always the need to analyze these costs, and to review them in terms of up to date material and labor cost changes.

For this study, a large amount of cost data was statistically analyzed for each type of plant. From this analysis, a base cost and an empirical expotential factor was developed in order to relate a total investment cost to capacity in as realistic a way as possible. This realtionship can be expressed mathematically in a non-linear form.

$$C = C_o(T/T_o)^k$$

where

C = Investment cost
C_o = Base cost at a base throughput T_o
T = New throughput
K = An empirical constant

The inclusion of a non-linear form for investment costs in a linear program required special consideration, and we shall see later how this was utilized.

In the models, many of the units considered were licensed processes, for which a royalty would be paid. A value in terms of a paid up royalty in dollars per barrel of throughput was included in the investment. Where chemicals and catalysts were used, the first inventory of these was also included as part of investment.

Chemicals and loss of catalyst was considered as an operating cost based on usage as were utilities. Labor, a fixed operating cost, was included with the return on investment. Maintenance cost was fixed as a percentage of the total maintenance cost.

Model Development

At the same time as the process and cost data were being generated, the basic form that the model would finally take was also being developed. This consisted of defining the various optional routes of each stream within the simulated refinery model.

The optional routing of the streams was carefully selected. This selection had to satisfy at least one of two requirements. Firstly, would such a routing actually contribute to satisfying the product slate and the premises of the problem? Secondly, would such a routing be feasible under actual operating conditions?

Just as a refinery is described by the units of which it is comprised, so also was the refinery linear model described. Here, each processing unit was considered a submodel in itself and these submodels were defined by their process and economic data.

These data were arranged in tabular from which were easily accessed and listed in recognizable terminology. An example of a submodel tabulation as used in this study

Table E.1. An extract from a typical base data file. TABLE TEF—H2 TREAT OF CRACKED GAS OILS.

	KGO COKER GAS OIL	VBO VIS BR GAS OIL	CYO CAT CY LE OIL	
REFORMER H2 FOEB	0.0244	0.0209	0.0096	XX+H2F
GAS, FOEB	–	–	–	XX+GS1
PROPANE	–	–	–	XX+C3S
ISOBUTANE	–	–	–	XX+IC4
N-BUTANE	–	–	–	XX+NC4
C5-380 HYDRO GASL	−0.0916	−0.0550	−0.0527	XX+TNP
LOSS OR GAIN	−0.0169	−0.0109	−0.0034	XX+LOS
DESULF COKER GAS O	−0.9159	–	–	XX+SKO
LT COKER GAS OIL	1.0000	–	–	XX+KGO
LT VIS BR GAS OIL	–	1.0000	–	XX+VBO
H2 TREAT LT VIS GO	–	−0.9550	–	XX+SVO
H2 TRTED CYCLE OI	–	–	−0.9603	XX+SCY
CYCLE OIL	–	–	1.0000	XX+CYO
FUEL MMBTU/UNIT	0.1113	0.1113	0.1113	XX+FUL
ELEC KWH/UNIT	3.0764	3.0764	3.0764	XX+KWH
STEAM MLB/UNIT	0.0029	0.0029	0.0029	XX+STM
CHEM ROYALTY CATAL	0.0023	0.0023	0.0023	XX+CRC
REPT FEED COLLECTO	1.0000	1.0000	1.0000	XX+FOR
******	–	–	–	STOPIT
REPORT WRITER AID	1.0000	1.0000	1.0000	UNPACK
SULFUR TO RECOVER	−8.8500	−5.1400	−5.9800	XX+SUA
H2 TRT LT GO CAP	1.0000	1.0000	1.0000	HTXCPE
GAS PLANT MAX CAP	–	–	–	GPXCAP

is shown in Table E.1. From this tabulation, a matrix generator, called GAMMA[1,2] was used to assemble the many submodels into a complete LP matrix. The matrix was solvable by a linear programming system called OMEGA[3]. These tabular input arrays were also used by the solution report writers as we shall see later.

This complete mathematical model of the refinery was displayed by an equation listing of the entire contents of the data. The equations showed the inter-relationships of the many variables, including the refinery streams, the blending constraints, the unit to investment ties, etc. These were also in a form which was ecognizable to the engineers working on the project.

Having now assembled all the data in a manner usable for linear programming, it was necessary to check it for errors. Various computer techniques had been developed for this purpose, and these, together with a secondary check by the process engineers, substantially eliminated the possibilities of obvious error and invalid data.

However, as an added safeguard, a final checkout was carried out by actually solving a test case. These results were scrutinized to ensure that the output gave a realistic refinery configuration and that all was in balance.

Optimizing and other techniques

A major value of linear programming is that once environment is reflected within a LP framework, this environment can be optimized. In this study, optimization could be accomplished either by maximizing profit or minimizing expense, and in this specific case, the former was selected. It should be emphasized that optimization can only be achieved for the environment reflected in the model. Great care was taken therefore to reflect all the meaningful, worthwhile options known to be available.

There are many refinery variables that are of a non-linear nature. Among these non-linearities are the effects of blending on motor gasoline, the capital expenditures in relation to size of the units, and many severity effects within the various processes. When these effects could be described on a cost basis by a *CONVEX* curve, they were generally included in the model as linear segments of a curve (see Fig. E.1).

If the severity to value of product relationship was a concave curve, only one variable could be used to reflect changing severity. This by nature has to be an estimate with a review of the estimate upon solution.

The blend to octane relationship is highly non-linear and a new approach was used to reflect this in the model. This approach had considerable advantage over the older techniques in that it was relatively easy to understand and use.

It was capable of reflecting the value of octane susceptibility to the individual components available for the blends. It also had the capability to represent accurately more than one type of octane (i.e., research, motor, road, etc.) with these effects also reflected back to the various components.

This technique required the aid of a recursive routine to update various model coefficients that reflected the actual susceptibilities at the solution point. The model then re-optimized with the recalculated octane for various blendstocks and again they were checked. This was repeated until no further change was required.

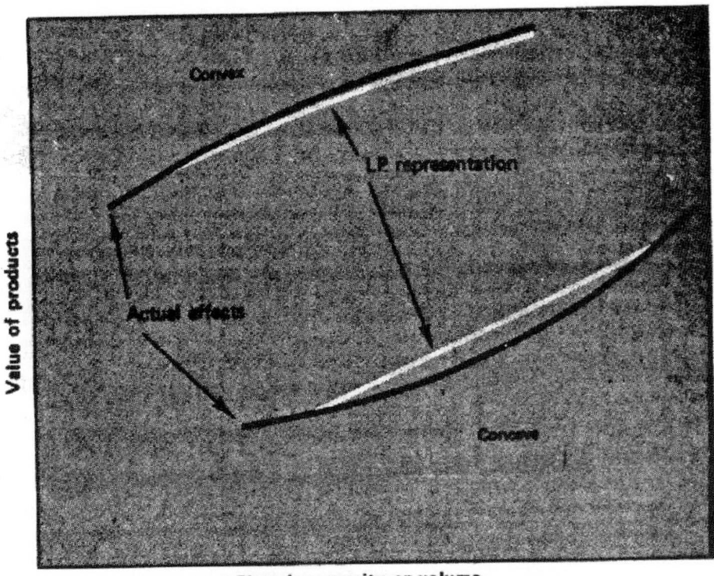

Figure E.1. LP representation of the convex and concave.

The non-linearities of the capital expenditures are concave in nature. Thus, the initial investment cost estimate was updated by means of a recursive routine which calculated the investment cost per unit of activity at the solution level. This recursive routine is described in detail next. Among the various factors that could be considered in investment costs are total investment cost to size relationships, offsites, insurances, taxes (both income and property), overhead, maintenance, labor, royalties, escalation, plant or service factors, depreciation, and the expected economic life of the various facilities.

Solution approach

In a study of capital expenditure (commonly called facilities planning) such as this, there are many possible mathematical techniques which can be used to obtain a solution. Experience has shown, however, that linear programming on a computer is by far the most economic approach to solving this type of problem. By this method, most of the many possible solutions can be examined quickly and effectively. Before discussing the solution approach used in this study, let us quickly review the investment environment for any typical refinery unit. This is illustrated in a simplified form in Fig. E.2.

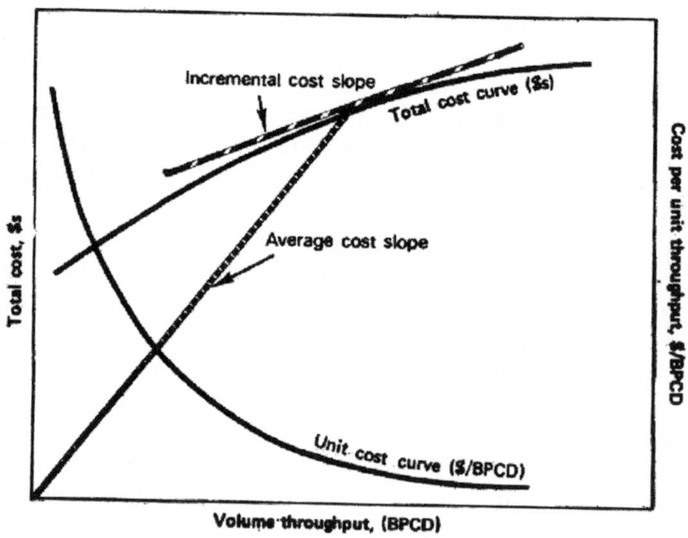

Figure E.2. Cost parameters for any typical refinery unit.

The total cost curve, on linear graph paper, shows that as a volume throughput is increased, the total cost of a unit will increased, However, the plot of unit volume cost ($/bpcd) against throughput shows the reverse; that is, the cost per unit of throughput will decrease as the volume of throughput increases. It is this type of cost that is reflected in an LP model.

Figure E.2 also shows a plot of the average cost slope and the incremental cost slope at a given throughput level. This *average cost* slope includes all the costs associated with a particular unit. The slope is linear and must pass through the origin. The *incremental cost* slope reflects the change in cost per unit throughput over a short range. This slope is always tangential to the total cost curve at any throughput under consideration.

It can be seen that the *average cost* gives the model 'greater than to be expected' incentive for changing the size of the unit, while the incremental cost curve on the other hand gives the expected incentive. The incremental cost ignores all fixed costs at the solution level, and consists mainly of the expected return on investment.

The technique used in this study was to begin the solution by establishing a very nominal cost on all units. This allowed any unit to be chosen in the solution. Once the LP had selected an optimal unit configuration with those nominal costs, a recursive program for investment cost estimating was used to determine *average cost* of the units at the solution throughputs. These average costs were based on a minimum return of investment, and an expected economic life calculated on a discounted cash flow basis.

These new calculated costs were then substituted for the original arbitrary cost in the model. This average costing tended to delete the very unrealistically small units that may have been chosen in the unrestricted configuration. Again, the model was optimized with these new costs.

The recursions and solutions were repeated until no further cost, configuration or size changes were required. All the items described above were carried out in a single computer run, and this solution was saved on tape and reported. Close scrutiny of the results then followed to make sure that they were reasonable and that there was no automatically restricted unit that might have been selected had a different solution path been used.

A second step was then commenced which restored the solution of the first step based on the *average cost* to the computer. The investment costs for the configuration were then recalculated using the *incremental cost* concept. This incremental cost required no return on investment and an infinite economic life was assumed for the units. The only costs that were recognized in this step were the incremental maintenance for each unit. Therefore, the building costs did not economically suppress the size of units or the total investment. The unit sizes and total investment were optimized on the premised product slate and available raw materials. A similar recursive step was again used, but this time it included only the incremental maintenance cost.

The model contains a variable that carries the sum total of the investment. This variable is updated at the same time as the unit costs are updated by the recursive operations described. Using this total investment variable it was possible to step down (paramatrize) on the total investment, until a feasible solution was no longer possible. An infeasible solution in this context is that in which the cash flow for the configuration becomes less than zero.

As the total investment are reduced, the unit sizes, the product slates and raw materials are changed within the framework of the overall premises. Therefore, by this parametric sweep, the refinery configurations (both in size and form) could now be found which would satisfy the following requirements:

- Maximum investment
- Maximum return on investment
- Maximum expected cash flow
- Minimum investment

The results of each parametric step were reported and documented.

Table E.2. An extract of a typical BI/DJ output

Status	Label	Cost	BI/DJ	Status	Label	Cost	BI/DJ
BI	FU+VIS	.000	12.9443	BI	FU/VBR	.000	.0000
BI	FU/CYO	.000	1.4567		GO−DIN	.000	.0574
BI	FU/HBA	.000	.0000	BI	GO−SPG	.000	.0029
	FU/HBB	.000	.0591		GO−XXX	−10.000	14.1623
BI	FU/HBE	.000	.0000	BI	GO−446	.000	8.7949
	FU/HBF	.000	.0000	BI	GO−675	.000	12.5429
	FU/HBG	.000	.1084	BI	GO+POR	.000	1.9087
	FU/HBH	.000	1.3164	BI	GO+SPG	.000	.0097
	FU/HBM	.000	.6095		GO+SUL	.000	.0754

Solution analysis

Besides the specially designed report writing technique already discussed, the LP also has a standard number of solution reports and these are generally of a more technical nature. Although of great value, they do not readily lend themselve to immediate and apparent interpretation. It is imperative, however, that some members of a team using an LP can read and interpret these outputs, particularly as one major use of these reports is to highlight any obvious errors that may have been overlooked. Some of these reports used in this study were called the BI/DJ and Range output. They warrant a brief explanation.

The BI/DJ output (see Table E.2) gives the solution level activities (BI) for all the variables that were selected in the optimum solution. For those variables that were not selected, the cost or decrease in profit that would occur were they forcibly included in the configuration (or basis) is given by the DJ value.

The Range information is a complementary report to the BI/DJ. The ranges give the incremental volume associated with the DJ's and the cost ranges associated with the basis variables. The valuable use of the BI/DJ and the range files can best be expressed by an example, using the extracts shown in Tables E.2, E.3, and E.4.

Consider the component CYO that has been selected in the configuration. (The code FU/CYO in this case indicates that the refinery stream CYO is routed to fuel.) The quantity of the stream CYO that enters the fuel blend is 1,457 bpcd. Now consider the component 'HBM'. This has not been selected in the basis, and this stream has a DJ value (i.e., no prefix). To route the 'HBM' stream to fuel forcibly would cost $0.6095/bbl. This information is interesting, but has no real value unless the ranges for their streams are known.

Table E.3. An extract of the corresponding primal range output

		Limits of range			
		Negative		Positive	
LP label	LP cost	Variable affected	Cost increment	Variable affected	Cost increment
MD/KER	.000000	MD/HDF	−.005291	MD+SUL	.131973
FU/CYO	.000000	HF/CYO	−.000000	XX+VBR	.028074
FU/HBA	.000000	HF/HBA	−.000000	GO/HDA	1.293871
FU/HBE	.000000	HF/HBE	−.000000	H3FHCE	1.085441
FU/HRB	.000000	HF+POR	−.000000	HF+VIS	.000000
FU/HRE	.000000	HF/HRE	−.000000	HVFHBE	.164348
FU/HRG	.000000	HF/HRG	−.000000	GO/HDG	1.243582
FU/HSB	.000000	HF/HSB	−.000003	GO/HDB	1.510931
FU/KEX	.000000	XX+KEX	−2.541661	SLYKER	19.910502

Because the FU/CYO item is selected in the basis, the range data for this is found in the primal range output (Table E.4). Interpreting the statement for this stream in this output means that optimum volume levels of the solution would not have changed even if the variable had a very small *negative* incentive (less than $0.000001). Further, the situation would not have changed even if this variable had a *positive* cost incentive of up to $0.02807/bbl.

Let us now look at the 'HBM' variable. this was not chosen in the basis and it appears in the dual range output (Table E.5). Interpreting the data for this item shows that to route 'HBM' to fuel would cost 60 cents/bbl for the first 549 bpcd. All that is known thereafter is that the cost per barrel over 549 bpcd would increase.

Table E.4. An extract of the corresponding dual range output

		Limits of range	
LP label	Original activity	Variable affected	Positive volume increment
FU/HBB	.000000	FU/HRB	.000000
FU/HBF	.000000	HF/HBF	.000000
FU/HBG	.000000	CCFHSG	.000000
FU/HBH	.000000	TEFCYO	.109287
FU/HBM	.000000	FU/CYO	.549070
FU/HRA	.000000	FU/VBR	.000000
FU/HRF	.000000	FU/VBR	.000000
FU/HRH	.000000	FU/VBR	.000000
FU/HSA	.000000	CCFHSA	.000000

Table E.5. Material and economic balance

Product or feed	price	M B/CD	M $/CD	MM $/YR
Premium gasoline	–	–	–	–
Inter. gasoline	–	–	–	–
Regular gasoline	–	–	–	–
Hi. vis. hvy fuel	–	–	–	–
Kerosene, regular	–	–	–	–
Propane LPG	–	–	–	–
Marine diesel	–	–	–	–
Sulfur MM LBS	*(Actual data has been deleted)*			
Shortage and fuel	–	–	–	–
Total production	–	–	–	–
Crude	–	–	–	–
Total feedstocks	–	–	–	–
Tel. in liters	–	–	–	–
Production margin	–	–	–	–
Expenses				
Utilities chem. and royalties	–	–	−6.064	–
Operating labor, super, and lab.	–	–	−2.930	–
Maint., ins., tax, and overhead	–	–	−3.061	–
Capital recovery	–	–	−12.817	4.678
Total expenses	–	–	−24.872	–
Earnings (loss)	–	–	−4.202	−1.534
Cash flow, earning plus capital recovery	–	–	–	3.144
Investment				*MM* $
Plant				27.929
Offsite				–
Catalyst and royalties				–
Added offsites, wharfs, etc.				–
Total				27.929
Year to payout, inv./cash flow*				8.882
ROI 7 years, percent 3.0				
ROI 10 years, percent 10.5				
ROI 16 years, percent 15.1				

*Figures reported before income tax withdrawn.

The example chosen here describes the economic analysis of two optional streams which can be logically routed ao a fuel blend. It is emphasized that the BI/DJ and range outputs, however, contain similar information for *all* variables whether process units, refinery streams, product specifications, etc. contained in the model.

Computer report writer

The data generated by the computer contains all facts relevant to the solution. However, to all but a few highly trained people the data in this form would be meaningless and of no practical use. The LP system used in this study contained a specially designed report writer, coded in a language called DART[4,5]. This converted and assembled the computer LP output into management orientated reports that could easily be read and understood without sacrificing the relevant technical content.

In this particular case, too, much of data, as produced by the report writing sequence, was in such a form as to be reproducible and able to be included in the final documentation. Table E.5 shows an example of such a report. (Note: the actual calculated data in this example has been deleted).

For the parametric series discussed above, a special report writing technique was developed which allowed each succeeding parametric step to be repeated in a case stacking fashion. This type of report was considerably condensed from the reports described earlier.

Final documentation

When the solutions to the four premises of the problem had been determined, using the techniques described, they existed, hidden among the mass of tabulated data that formed the computer output. It remained now to extract the pertinent section of the output and to present it so that the objective of the study, which was to provide management with information to make a good decision, could be achieved. The most common means of doing this—and the one chosen on this occasion—is by a written report in which the data is summarized, discussed and the conclusion stated.

Although it is not proposed to discuss the general techniques of technical report writing here, some fundamental requirements of a complex presentation such as this are worth highlighting. This report had to satisfy two principal functions. The first, to present as succinctly as possible the conclusions, and the interpretation of those conclusions, for the convenience of the client's management. Secondly, it had to present all the back-up data in as short a form as possible that would be necessary to enable the client's own staff to check and confirm the conclusions reached.

This second function was satisfied in this report in the form of an appendix. This included copies of the actual pertinent computer printouts complete with tabular listing of the submodels, economic balances, etc. These data were further augmented by the summary of the economic and yield output for the respective parametric runs.

The main body of the report consisted of a short description of the study, together with discussion of the result. The results were however succinctly described by two illustrations for each of the four cases of the problem. The first illustration showed the ultimate refinery complex which satisfied the premise of the case studied, and a typical example is shown in Fig. E.3. The second illustration, typified by Fig. E.4,

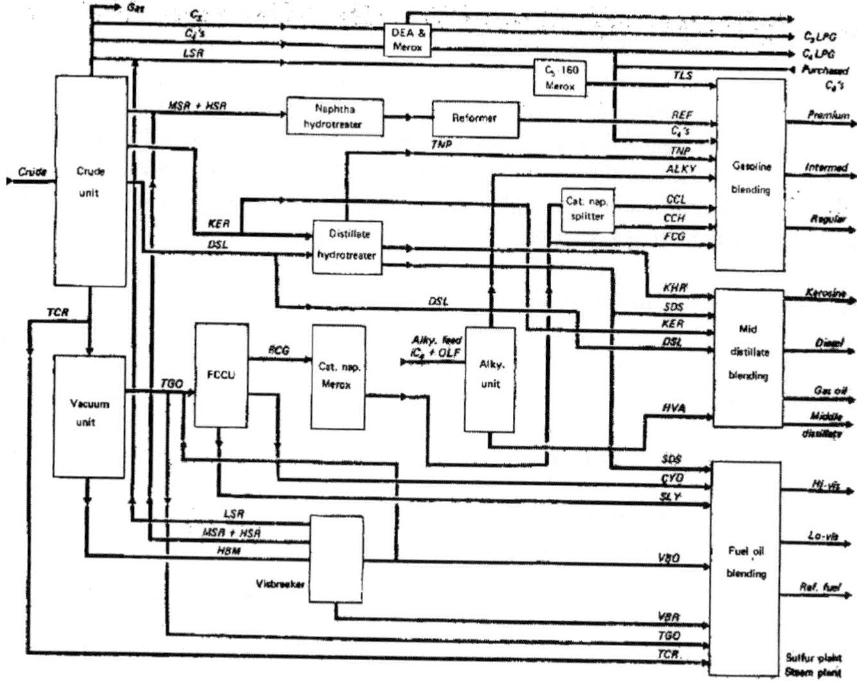

Figure E.3. Typical final optimum configuration.

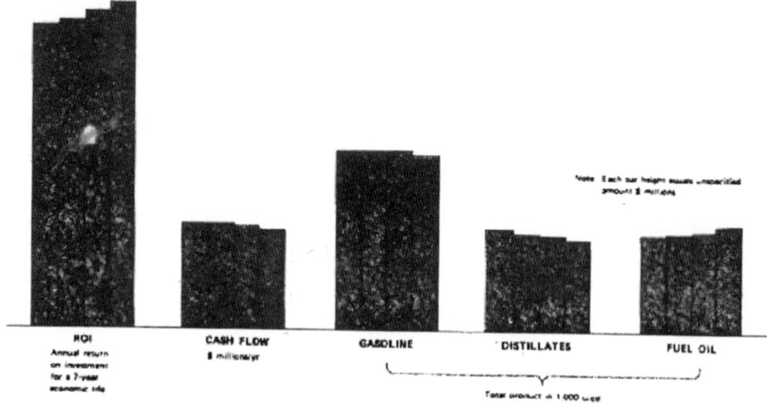

Figure E.4. Bar chart summarising economic studies for configuration in Fig E.3.

gave the basic economic trend for this configuration and also described the yields of major products for each parametric case.

The charts shown are meant only as an example and the figures are fictitious or have been purposely deleted. However, it can be seen that Fig. E.3 describes the result in terms of the processes that must be built to satisfy the premise. Figure E.4 shows why such a configuration is the optimum, and what the ultimate product slate would look like. Such a chart also gives the client's management an opportunity to assess quickly the effect of changing a basic premise such as maximum return on investment or minimum capital investment.

Bibliography

1. Hearn. J.: "GAMMA—A General Description", Bonner and Moore Publication. Document C 139–1.
2. Hearn. J.: "GAMMA Primer", Bonner and Moore Publication. Document M G1-1.
3. Romberg, F. A.: "OMEGA—A General Description", Bonner and Moore Publication, Document C 74–2.
4. Romberg, F. A.: "Dart Systems for Larger Computers", Bonner and Moore Publication. Document C 55–2.
5. Romberg, F. A.: "Dart 2—A General Description", Bonner and Moore Publication. Document C 53–2.

Index